Beyer/Walter
ORGANIC CHEMISTRY
A comprehensive degree text and source book

Wolfgang Walter
Institut für Organische Chemie
Universität Hamburg, Germany

Translated and Edited by **Douglas Lloyd,**
Department of Chemistry, University of St. Andrews, Scotland

Albion Publishing
Chichester

Originally published in German as Beyer/Walter *Lehrbuch der Organischen Chemie* 22nd edition, © (1991) by S. Hirzel Verlag, Birkenwaldstr. 44, 70191 Stuttgart, Germany. **All rights reserved.**
First published in English by Prentice-Hall Europe (1996)

This edition published by **Albion Publishing Limited, Chichester, West Sussex PO20 6QL England** © **(1997)**

The publication of this work has been sponsored by means of Inter Nationes, Bonn.

British Library Cataloguing in Publication Data
A catalogue record of this book is available from the British Library

ISBN 1-898563-37-3

Printed and bound in Great Britain by Hartnolls, Bodmin, Cornwall

Table of Contents

2 Aliphatic Compounds   51

3 Alicyclic Compounds   385

4 Carbohydrates   423

5 Aromatic Compounds. Derivatives of Benzene   467

6 Isoprenoids (Terpenes and Steroids)   661

7 Heterocyclic Compounds   701

8 Amino Acids, Peptides and Proteins   819

9 The Chemistry and Function of Nucleic Acids   849

10 Enzymes   883

11 Metabolic Processes   899

Appendix: Hazardous Substances and Carcinogenic Materials. Named Reactions and Concepts   909

Index of Names   953

Subject Index   962

Author's Foreword to the English Translation of the 22nd Edition of Beyer •Walter: Lehrbuch der Organischen Chemie

In 1953 the publisher Hirzel Verlag of Leipzig in the former DDR (Eastern Germany) produced the LEHRBUCH DER ORGANISCHEN CHEMIE by *Hans Beyer*, Professor of Chemistry at the University of Greifswald. Its use spread rapidly through German-speaking countries because it provided, in a readily understandable way, an ideal combination of basic fundamental material, factual information and theoretical considerations; it also paid appropriate attention to the relationship of chemistry to biochemistry.

Dr Hans Beyer died on February 1, 1971, and in the spring of 1973 I took over responsibility for the further developments of his book and edited the newly revised manuscript of its 17th edition, but by this time for Hirzel Verlag Stuttgart in the Federal German Republic (West Germany). Since then five further editions have appeared; these have repeatedly been made necessary by the vigorous development of organic chemistry and of the areas of biochemistry which are referred to in its text. In addition, attention is drawn to important synthetic methods which should be known in view of their regular use in recent times.

During this period Editorial Reverté in Barcelona has produced a Spanish translation of the 19th edition. This was undertaken by Dr José Barlengua Mur, Professor at the University of Oviedo, in Asturias. Its title is 'Manual de Quimica Organica'. In 1989 an unauthorised translation of the 18th edition into Chinese appeared; this was published by the High School Publishing House in Beijing under the title of 'Student Course in Organic Chemistry'.

Because of my insufficient knowledge of the language I was unable to contribute to the Spanish edition. I only became aware of the Chinese edition by chance a year after it had appeared. In contrast, in the present English edition I have collaborated closely with Douglas Lloyd throughout the enterprise.

Initially the 21st edition formed the basis of this translation but with the appearance of the 22nd edition the text was modified to bring it into line with the new edition. In addition, corrections and extra material which have already been gathered together in readiness for the forthcoming 23rd edition have been included in this English edition. Older nomenclature, as well as IUPAC nomenclature, is mentioned in order to improve the ease of access to older literature. An observant reader will not fail to notice the associations which are made between chemistry and general culture at appropriate places. Thus, because of chemical connections the names of *Plato* and *Hannibal* may be found in the name index of the 22nd German edition, and in the present translation the name of *Alexander Borodin* also appears.

In the course of working together the author and the translator have gained much first-hand experience of the interweave of language and culture and have derived great enjoyment from it. Stimulating discussions of a variety of chemical matters have led to numerous improvements of the text, resulting in a translation that in many places is also a revision. Dr John Mellor also played an important role thanks to his collaboration in the project. Special thanks must go to my collaborator of many years, Anneliese Kuhlmann, who has dealt so efficiently with the large amount of correspondence associated with the translation.

I hope that this result of our work, which has occupied nearly five years will lead to its reaching many new readers. We will be very grateful to them for any criticisms or suggestions which may lead to further improvement of the book.

Hamburg, Easter 1995

Wolfgang Walter

Translator's Foreword

Translating another author's work is a difficult task which provides many pitfalls. Apart from inadequacies in the knowledge of the original language, the major challenge is to maintain the style and character of the author's language and approach so that it is still *his* book. This I have tried my utmost to do. Linguistic differences rule out the possibility of a completely literal translation, but I have endeavoured throughout to stay as closely as possible to the language of the original text, so that it retains the style and flavour of *Beyer–Walter*; the publishers have collaborated in this attempt by making the text appear as similar as possible to the German text. The blocks used for formulae and illustrations are those used for the 22nd, or in some cases the forthcoming 23rd, German editions.

When I was approached in 1990 about taking on this task, a first sight of the size of the book made me wonder. Looking at the cover, I read a review from *J. Am. Chem. Soc.* which was reprinted there and said 'This text is a prime example of the European "Lehrbuch" at its best. It is up to date, thorough, comprehensive, and written lucidly. ... To my knowledge, there is no comparable book in English. (It) seems to be meant for students with some basic knowledge of organic chemistry, but it is also an excellent reference book for chemists at any stage of their career. ....'. This suggested that time spent on a translation could be time well spent and, as soon as I opened the book and started to delve into its contents, I realised that someone had to translate it into English, because indeed no comparable English-language text was available, and that it would be a privilege and a pleasure to be associated with the project.

Ellis Horwood, with great wisdom, invited Professor Walter and me to meet as his guests in Chichester and to spend some time together by ourselves to see how we got on with one another. This was highly successful and our friendship has continued to grow throughout the project and, as mentioned in the author's preface, both of us have derived much stimulation and pleasure from the enterprise. I must warmly thank Wolf Walter for his unstinting support and critical assistance throughout the enterprise, not least for sometimes correcting and/or improving my English.

The publishers, author and translator had some difficulty in deciding what was the best translation of the German title 'Lehrbuch der Organischen Chemie' to use. It is so much more than a textbook. Rather it is both a background book for lecture courses in the European tradition, and also a companion, containing a wealth of material gathered together over a very wide range and not otherwise assembled in this way in an English text. It should be of value from one's first day as an organic chemist to one's last.

Furthermore it contains much material which is of interest not only to chemists but to others with scientific interests, perhaps especially biochemists and biologists (botanists and zoologists), but also to other physical scientists and industrial scientists. A unique feature for an English text is the manner in which industrial chemistry and processes are woven into the text and given the attention which they deserve. Indeed the text should serve as an invaluable reference source to the role of chemistry in the widest sense, in everyday life and in its applications, as well as in the chemical laboratory. Bearing all these points in mind, the title 'Comprehensive Organic Chemistry' was chosen.

A characteristic feature of the book is that it has its roots firmly in facts and that these facts form the basis for mechanistic discussion and theory rather than vice versa. It is intended to be a European textbook, not just a German or a British one. Reflecting this wider outlook German language references have for the most part been retained but are supplemented by British and American language references, most of which, moreover, are present in the original German text; it is essential for modern chemists to be in touch with German as well as British and American approaches to the subject. In particular 'Houben–Weyl' is frequently referred to. We trust that our American friends will also welcome this breadth. References are not limited to recent publications, although every attempt has been made to keep them up to date. In so doing, however, older references have often been retained, especially those to older review articles on a variety of topics of importance and some to the first reports on topics; such articles often provide an invaluable and irreplaceable introduction to these subjects.

Spelling is British; on occasions when alternatives appeared to exist, I commonly referred to the Penguin Dictionary of Chemistry (ed. *D.W.A. Sharp*) as an arbiter. Translation had passed the halfway mark when 'sulfur' was authorised as a permitted alternative to 'sulphur' in British publications. Since 'sulfur' is only an approved alternative, and less remained to be translated than had already been accomplished, 'sulphur' has been retained throughout. In any case it seemed somewhat more consistent to match 'phosphorus' with 'sulphur'.

Nomenclature is always controversial; should trivial names be used or systematic nomenclature be the rule? Both the German and English texts meet this problem by using both, indeed a variety of names, ancient and modern, are often given for a compound. A knowledge of trivial names makes access to older literature and to technical literature much simpler. It should be remembered that the objective of the Geneva rules was not to supersede trivial names but to provide 'official names' for purposes of indexes and dictionaries. Furthermore, in much official legislation it is the trivial names which are commonly cited. It is a feature of everyday life that nicknames frequently take the place of 'real' names and need to be known. Although both IUPAC and other names appear for most compounds, where applicable, a choice often had to be made as to which name actually to use in the text. The answer here has always been the one which is most commonly used, and, if I have needed external guidance I have always sought it from the various catalogue handbooks issued by the larger suppliers of fine chemicals, and selected the ones of their choice; there is some advantage in knowing what to ask for when one goes to a shop[1].

[1] Extract from Guide to IUPAC Nomenclature of Organic Compounds, Recommendations 1993 (IUPAC 1993):
In contrast to such systematic names, there are traditional names, semisystematic or trivial, which are widely used for a core group of common compounds. Examples are 'acetic acid', 'benzene', 'cholesterol', 'styrene', 'formaldehyde', 'water', 'iron'. Many of these names are also part of general non-scientific language and are thus not confined to use within the science of chemistry. They are useful and in many cases indispensable (consider the alternative systematic name for cholesterol, for example). Little is to be gained, and certainly much to be lost, by replacing such names. Therefore where they meet the requirements of utility and precision, and can be expected to continue to be widely used by chemists and others, they are retained and, for the most part, preferred in this Guide. Semisystematic names also exist such as 'methane', 'propanol', and 'benzoic acid', which are so familiar that few chemists realize that they are not fully systematic. They are retained, and indeed, in some cases there are no better systematic alternatives.

Two terms which I have decided to use perhaps need comment. I have followed the German text in using *mesomerism* rather than *resonance*. More unusually I have used the literal translation of German *grenzformeln*, namely *limiting structures*, or sometimes, when apposite, *contributing structures*, rather than *canonical forms*, because it seems to me that the German term provides a more evident description.

I have already mentioned my indebtedness and gratitude to Professor Walter for his unstinting help with the translation. A number of other people have also given me great help, first and foremost my wife Lydia, and my colleague and friend Dr Raymond K. Mackie. My wife is a zoologist not a chemist but was gallant enough to read portions and advise me on their improvement; in particular she guided me away from misleading translations of biological material. In addition I am grateful to her for her general support throughout the exercise. Dr Mackie, in addition to being a friend, is also a neighbour in the department, and was therefore all too handily placed for me to pester him with queries and requests for advice and information. All this he gave with great good grace and he has contributed a lot to the translation. Various other colleagues also helped with specialist knowledge; particular thanks go to Dr Alan Aitken, Dr Nigel Botting and Dr John Walton in the Department of Chemistry and Professor H.G. Callan and Dr C. Muir, both formerly of the Department of Zoology in the University of St Andrews, and Dr Heather McOmie (Bristol), who helped me over some medical queries. Dr John Mellor (University of Southampton), the publisher's consultant, also very kindly read the whole text, and the proofs, and offered many valuable suggestions. To all of these people I, and the book, owe a big debt of gratitude. Last but not least an enormous expression of thanks must go to Mrs. Lynn Marouf (Department of Chemistry, The University of Edinburgh) who typed the whole work and prepared the disks which were used in its production. Her very able and willing collaboration made life very much easier and happier for me.

I am also grateful to the staff of Ellis Horwood, and to Ellis Horwood himself for all the practical help in the preparation of the first English version, published in hardback binding in 1996 by Prentice-Hall Europe as *Handbook of Organic Chemistry*, and in the preparation of the present edition. I also thank Rosemary Harris for her skills in implementing the emendations to this first paperback edition published by Albion Publishing limited and retitled *Organic Chemistry: a comprehensive degree text and source book*. The valuable help and co-operation of Sabine Körner of S. Hirzel Verlag, is also gratefully acknowledged.

It would be unnatural if I had perpetrated no errors of omission or commission: I apologise for them to Professor Walter and trust that kind friends will point them out to me.

Finally, I would like to express the pleasure which I have received from carrying out this task; I am grateful to the author and to the publisher for trusting me with it. I hope that readers may gain as much pleasure and learn as much of interest as I have.

Douglas Lloyd
University of St Andrews
Martinmas term 1996

Periodic System of the Elements

	Ia (1)	IIa (2)	IIIa (3)	IVa (4)	Va (5)	VIa (6)	VIIa (7)	VIII (8)	VIII (9)	VIII (10)	Ib (11)	IIb (12)	IIIb (13)	IVb (14)	Vb (15)	VIb (16)	VIIb (17)	O (18)
1	1 **H** 1.008																	2 **He** 4.003
2	3 **Li** 6.94	4 **Be** 9.01											5 **B** 10.81	6 **C** 12.011	7 **N** 14.01	8 **O** 16.00	9 **F** 19.00	10 **Ne** 20.18
3	11 **Na** 22.99	12 **Mg** 24.31											13 **Al** 26.98	14 **Si** 28.09	15 **P** 30.97	16 **S** 32.06	17 **Cl** 35.45	18 **Ar** 39.95
4	19 **K** 39.10	20 **Ca** 40.08	21 **Sc** 44.96	22 **Ti** 47.90	23 **V** 50.94	24 **Cr** 52.00	25 **Mn** 54.94	26 **Fe** 55.85	27 **Co** 58.93	28 **Ni** 58.71	29 **Cu** 63.55	30 **Zn** 65.37	31 **Ga** 69.72	32 **Ge** 72.59	33 **As** 74.92	34 **Se** 78.96	35 **Br** 79.90	36 **Kr** 83.80
5	37 **Rb** 85.47	38 **Sr** 87.62	39 **Y** 88.91	40 **Zr** 91.22	41 **Nb** 92.91	42 **Mo** 95.94	43 **Tc** 98.91	44 **Ru** 101.07	45 **Rh** 102.91	46 **Pd** 106.4	47 **Ag** 107.87	48 **Cd** 112.40	49 **In** 114.82	50 **Sn** 118.69	51 **Sb** 121.75	52 **Te** 127.60	53 **I** 126.90	54 **Xe** 131.30
6	55 **Cs** 132.91	56 **Ba** 137.34	57 **La** 138.91	72 **Hf** 178.49	73 **Ta** 180.95	74 **W** 183.85	75 **Re** 186.2	76 **Os** 190.2	77 **Ir** 192.22	78 **Pt** 195.09	79 **Au** 196.97	80 **Hg** 200.59	81 **Tl** 204.37	82 **Pb** 207.2	83 **Bi** 208.98	84 **Po** 208.98	85 **At** 209.99	86 **Rn** 222.02
7	87 **Fr** 223.02	88 **Ra** 226.03	89 **Ac** 227.03	104 **Rf/Ku** 261.11	105 **Ha** 262.11	106 **Sg¹⁾** 263.12	107 **Ns** 262.12	108 **Hs**	109 **Mt**									

Lanthanides	58 **Ce** 140.12	59 **Pr** 140.91	60 **Nd** 144.24	61 **Pm** (145)	62 **Sm** 150.4	63 **Eu** 151.96	64 **Gd** 157.25	65 **Tb** 158.93	66 **Dy** 162.50	67 **Ho** 164.93	68 **Er** 167.26	69 **Tm** 168.93	70 **Yb** 173.04	71 **Lu** 174.94
Actinides	90 **Th** 232.04	91 **Pa** 231.04	92 **U** 238.03	93 **Np** 237.05	94 **Pu** 244.06	95 **Am** 243.06	96 **Cm** 247.07	97 **Bk** 247.07	98 **Cf** 251.08	99 **Es** 252.08	100 **Fm** 257.10	101 **Md** 258.10	102 **No** 259.10	103 **Lr** 262.11

¹⁾The name Seaborgium (Sg), which was proposed by the American Chemical Society, has not yet been accepted by IUPAC

1 General

1.1 Introduction

The beginnings of organic chemistry date from the seventeenth century when, in addition to minerals, materials derived from the plant and animal kingdoms became the objects of chemical investigation. A clearer insight into their elemental composition first became possible as a result of *Lavoisier's* demonstration in 1774, following his studies of the combustion of metals (mercury, lead), that combustion was associated with the uptake of oxygen. The combustion products which were obtained from materials of plant or animal origin always included carbon dioxide and water. Therefore the elements carbon and hydrogen must be building blocks in their structure. Sometimes nitrogen or oxides of nitrogen were also obtained, showing that nitrogen also was present in some natural products. Combustion of some substances in the absence of air also resulted in the formation of carbon dioxide and water, which showed that oxygen must be present in these substances.

The proper scientific investigation of chemical problems began with the discovery of the qualitative composition of chemical compounds. This started in the middle of the eighteenth century, above all with *Scheele's* analysis of various products isolated from plant or animal sources, including oxalic acid, malic acid, tartaric acid, citric acid, lactic acid, uric acid, and the fatty acids obtained from fats and oils. This led to the first appreciation of the broad general similarity of the organic materials obtained from plants and animals and, conversely, to the very different chemical nature of inorganic substances. This discovery led in the course of time to a greater differentiation between organic and inorganic compounds.

The term 'organic chemistry' was first used as a definitive term by *T. Bergman* (1784). *Berzelius* (1808) compared plant and animal organisms to chemical workshops in which the different parts of the living organisms were manufacturing substances essential to them. Since at that time it was only possible to carry out decomposition reactions of these substances, it was thought that reactions leading to their formation were only possible in living material. This so-called 'vital force' theory was shaken by *Wöhler's* conversion, in 1828, of ammonium cyanate into urea. Here, for the first time, a substance

$$[NH_4]^{\oplus} [OCN]^{\ominus} \longrightarrow OC(NH_2)_2$$

Ammonium cyanate   Urea

from a living organism, first isolated by *Rouelle* in 1773 as a metabolic product from urine, had been prepared from inorganic material. The consequences of this finding on the theory of vital force were only realised slowly. In the first place, in the following years various new organic compounds which did not occur in nature were prepared, as were compounds earlier isolated from natural sources. From this the idea spread that a large-scale synthetic manufacture of organic compounds was possible. As a result of further studies in the middle of the nineteenth century the presence of *carbon* was taken to be the basic feature of an organic substance. This, in turn, led to the definition of organic chemistry as the 'study of the chemistry of carbon compounds' (*Gmelin, Kolbe, Kekulé*).

Most organic compounds are built up from relatively few elements. Besides carbon, the commonest are hydrogen, oxygen and nitrogen. Less common are sulphur, phosphorus, halogens, calcium, magnesium, iron, cobalt and others. In principle, all elements of the Periodic Table can be incorporated into organic compounds.

The chemical properties and reactivities of organic compounds are rather different from those of inorganic compounds. While the latter are often stable at high temperatures, organic compounds are usually sensitive to heat and to chemical reagents, and readily undergo chemical change and decomposition. Hence the study of organic chemistry often demands different methodology and techniques. Also, in contrast to the rapid ionic reactions common in inorganic chemistry, organic reactions are often much slower. According to *van't Hoff's rule*, a temperature rise of 10 K results, on average, in a two- to three-fold increase in the rate of reaction. In consequence, organic reactions are often carried out in selected solvents heated to their boiling points. Another method which is often used to speed up organic reactions, or indeed to enable them to take place at all, involves the use of *catalysts*, which are frequently inorganic materials.

Thus the separate treatment of organic chemistry is not only a matter of convenience, but is also based on the structure and chemical behaviour of carbon compounds. Furthermore, even when the sheer number of organic compounds is considered—it exceeds twelve million—none the less through all organic chemistry there run, like a continuous thread, common structural principles, which serve to clarify the understanding and the unity of organic chemistry.

In ending these introductory remarks, the great importance of organic chemistry to biochemists, biologists, medical scientists and pharmacologists must be emphasised. While the experimental skill of the organic chemist has contributed in a variety of ways to the study of living organisms, nature is capable of much finer control in the biochemical interactions of organic materials. Advances in this direction can only come about through interdisciplinary scientific collaboration. If antibiotics, metabolites and other natural products are to be synthesised, then collaboration between chemistry and biology must be encouraged. With the help of *isotopic labelling* it is possible to gain a deeper insight into the mechanisms of the life processes of animals and of plants, and into the biogenesis of natural products. The recent developments in biochemistry and in *molecular biology*, which contribute so much to the study of biology and of medicine, are closely tied up with this. Likewise, *chemotherapy* has been greatly enriched by the collaboration between different disciplines; this is especially evident in the fields of antibiotics, tuberculostatics and cytostatics. The research results obtained in this way and their practical applications lead in an increasing number of ways to the control of nature by mankind, and offer the possibility of their responsible applications to the development of mankind.

1.2 Pure Substance

The goal of preparative organic chemistry is the isolation of pure materials, and intensive work is directed towards this end. The products are usually either crystalline or liquid, less often gaseous. A pure substance is made up, as far as is attainable, of only one kind of molecule; it is not possible for it to be significantly purified any further by other operations. For this reason, the common conception that 'natural' caffeine or 'natural' vitamins have different actions from the corresponding synthetic compounds, with which they are identical, is clearly untenable. It is possible, however, that impurities would still remain in a sample, despite its most efficient purification, and that these in turn might provide some difference in its character.

In contrast to the ionic reactions of inorganic compounds, the outcome of organic reactions is seldom quantitative. Because there are many possibilities for different reactions to take place, and because organic reactions take longer, in almost all instances, and especially when reactions are heated, alternative competing reactions take place as well and in turn lead to alternative products, thus lowering the yield of the intended main product. Increasing the yield of a single product is always one of the main problems in organic chemistry.

In order to obtain a proper characterisation of a substance, and to determine its stoichiometric make-up, it is essential that the main product from a reaction should be isolated from the complex mixture which is obtained. It must be separated from any amorphous or resinous impurities, which are frequently dark coloured, and finally obtained in a pure state.

An account of fundamental methods of purification, which an organic chemist must know and be able to apply, follows.[1]

1.2.1 Recrystallisation

This method of purification depends upon the different solubilities of solid organic compounds in the appropriate solvents. If the by-products are more soluble in the selected solvent than the main product is, then isolation of the main product presents little difficulty. It crystallises out from a hot saturated solution, whilst the impurities remain dissolved in the solution as it cools and in the resultant 'mother liquors'. Because of the different rates of crystallisation of different compounds, an almost complete separation of relatively pure product is obtained within hours, or sometimes days. It is often possible to induce crystallisation of the required product from a saturated solution by *seeding* it with a crystal of this material. By evaporation of the mother liquors it is often possible not only to increase the yield of the main product, but also to isolate by-products. Not infrequently a further recrystallisation of the main product from a different solvent, often with active *charcoal* present to absorb impurities, provides a pure product. This method of purification may fail if the required product forms *mixed crystals* with the by-products.

---

[1] See *B. L. Karger* et al., An Introduction to Separation Science, 2nd edn (Wiley, New York 1981); *J. T. Sharp, I. Gosney* and *A. G. Rowley*, Practical Organic Chemistry (Chapman & Hall, London 1989).

The production of single crystals is becoming an increasingly important process and significant advances have been achieved in this field.[1]

Meanwhile it has been shown that it is possible to obtain better protein crystals in space laboratories under conditions of *microgravity* than can be produced on the earth. This is because under these conditions movements induced by concentration gradients have been almost entirely eliminated.

1.2.2 Distillation

In a simple *distillation*, a liquid is evaporated, and the resultant vapour is condensed to provide a *distillate*. In the laboratory this technique provides a method for separating a liquid from impurities and from other substances with sufficiently higher boiling points.

The method is dependent upon the fact that different liquids have different boiling points. The larger the difference in boiling points the easier it is to achieve separation. By means of *fractional distillation* a mixture may be split into its component *fractions*, which can each be purified by repeated distillation. The method fails if *azeotropic mixtures* result, in which two or more compounds co-distil at a constant boiling point; this may be higher or lower than the boiling points of the individual components of the mixture.

In simple cases, straightforward distillation is used, but the presence of compounds of very similar boiling points requires the use of a *fractionating column*, which is included between the distillation flask and the condenser. In this long column the hot vapours move upwards, while at the same time some cooler condensed liquid flows downwards, resulting in an exchange of heat and material between the two phases. In the process the vapour is cooled and its higher boiling point components are condensed. The nascent heat of condensation which is evolved volatilises the lower boiling point material. Gradually an enrichment of the lower boiling point material builds up in the ascending vapour phase, and the higher boiling material is concentrated in the descending liquid phase. In industry, columns filled with plates are preferred. In laboratories, columns with various sorts of fillers, which always have a large surface area, such as glass rings or helices, are used. In each case the objective is to achieve the most intimate contact that is possible between the ascending vapour and the descending condensate. The efficiency of a column can be derived from phase diagrams and is defined in terms of 'theoretical plates'. Columns of suitable efficiency to obtain the required degree of separation can then be used. If the liquids to be distilled have very high boiling points, or decompose easily, then distillation is carried out *in vacuo* at about 16 mbar (16 hPa) pressure.

1.2.3 Distillation and Sublimation in Low or High Vacuum

If substances which are sensitive to heat, or have very high boiling points, need to be distilled, it is better to do it either in low vacuum ($1-10^{-3}$ mbar/$1-10^{-3}$ hPa) or in high vacuum ($10^{-3}-10^{-6}$ mbar/$10^{-3}-10^{-6}$ hPa).

Another technique makes use of a *falling-film* still. A thin film of the material to be distilled flows down the inside of a column, which is heated by a surrounding jacket and the liquid is thus evaporated. Another concentric tube occupies the centre of the column; this is cooled and the vapour condenses on it and is run off at the bottom. This method

---

[1] See *J. Hülliger*, Chemistry and Crystal Growth, Angew. Chem. Int. Ed. Engl., *33*, 143 (1994).

can be used at different pressures from normal to about 10^{-3} mbar (10^{-3} hPa), depending upon the boiling point and thermal stability of the liquid.

When the free-falling technique is used in high vacuum conditions ($<10^{-3}$ mbar/ 10^{-3} hPa), it is known as falling-film *molecular distillation*. In these conditions, the distance which a molecule evaporated from the outer jacket has to travel to reach the cooled inner tube is less than the mean free path of the molecule, *i.e.* the mean distance it must travel before meeting another molecule. In this way fats and other high molecular weight materials such as vitamins and hormones have been distilled. A further advantage of this method is that compounds having the same vapour pressure may be separated as a consequence of their having different molecular weights. The normal concept of boiling point does not apply since there is no longer an equilibrium between liquid and vapour.

In *sublimation* a solid is evaporated, and its separation may be achieved without involving a liquid phase, by means of the general scheme: solid → vapour → solid. This method of purification is particularly useful for materials which, because of their low solubility or for some other reasons, are difficult to recrystallise. Sublimation is conveniently carried out in either low or high vacuum. It usually provides clean sublimates, and can be carried out with very small amounts of material without serious loss.

1.2.4 Zone Melting[1]

The process of zone melting provides a method for loss-free separation and purification of substances, using crystallisation from a melt. It makes use of differences in the readiness to crystallise of a main product and the accompanying impurities.

The impure material is made into a bar or ingot and is placed in a tube in which there are alternate hot and cold zones. Thus small zones of the bar are melted which are separated from one another by set distances. The melted zones are then moved along the bar. They carry with them impurities which lower the melting point of the main product, often with the formation of eutectic mixture, which sets when it reaches the end of the bar. Impurities which raise the melting point are concentrated at the other end of the bar, often with formation of mixed crystals. The purified main product is found in the middle of the bar. In practice, this method provides in effect repeated recrystallisations, leading to the formation of an increasingly pure product. This process can be carried out as a continuous process. By means of this elegant method it is possible to purify very small samples consisting of only a few milligrams.

1.2.5 Steam Distillation

Steam distillation is widely used both in the laboratory and in industry. It is based upon the fact that many substances whose boiling points are appreciably higher than that of water are, because of their vapour pressure, evaporated when steam is passed through them, and condense again with the water in a cooled receiver. If the material to be purified is in effect insoluble in water, then their individual vapour pressures are scarcely affected by each other. Hence, so long as the sum of their partial pressures is equal to the atmospheric pressure at some temperature, boiling occurs at this temperature. Substances

---

[1] See H. *Schildknecht*, Zonenschmelzen (Hüthig-Verlag, Heidelberg, 1987).

which are difficult to vaporise may consequently be treated with superheated steam, and thus distilled and purified.

1.2.6 Extraction

Another method used in the isolation and purification of organic compounds is *extraction*. In this process, a mixture of solid or liquid substances is treated with a selected suitable solvent, which extracts one or other of the substances by dissolving it. By using an easily evaporated solvent such as ether or chloroform which is immiscible with water, organic compounds can be extracted from an aqueous solution. The organic solvent is separated and distilled off to leave the extracted material. Highly compressed supercritical gases have proved to be highly effective and inert extraction solvents.[1]

1.2.7 Adsorption Chromatography[2]

This procedure, which was developed by *Day, Engler* and *Tswett*, has been especially valuable for the separation and isolation of coloured natural products such as carotenoids and chlorophylls, which are not easily separated by fractional crystallisation. Such materials, which commonly have different colours, are very similar chemically, but in solution they show differing, but reversible, affinity towards appropriate adsorbent materials. Their separation depends upon the differing extent of transfer of dissolved materials between a stationary phase consisting of an adsorbent material, and a mobile phase, which is the solution of the materials to be separated. Between two such phases there is a characteristic equilibrium for each substance, between the amount of it which is adsorbed and the amount that remains in solution. Commonly used adsorbents include alumina (which is standardised on a scale introduced by *Brockmann*), silica gel, calcium carbonate, calcium oxide, powdered sugar, cellulose, etc. The degree of separation achieved on a stationary phase can be improved significantly by adding to the column either a substance which reacts specifically with *one* constituent of the mixture to be separated or which forms a complex with it. This technique, which is called *affinity chromatography*, has proved valuable in the isolation of enzymes.[3]

In carrying out this process, a solution of a mixture in an organic solvent is either poured or sucked slowly through a column filled with one or more adsorbent materials. Hence it is called *column chromatography*. Then the *chromatogram* is developed. To do this, the initially mixed-up colour zones are separated from one another, by washing them through the column, using either the original solvent, or, more commonly, a different solvent. After separation of the individual zones or *fractions*, each of them can be extracted with a suitable solvent to dissolve the separated organic material, and hence each constituent may be obtained pure. If the column is shortened and pressure is applied to it, separation can be achieved in about 5–10 minutes (*flash chromatography*[4]).

---

[1] See *G. Wilke*, Extraction with Supercritical Gases, Angew. Chem. Int. Ed. Engl., *17*, 701 (1978), and some following papers.
[2] See *E. Heftmann*, Chromatography, 2nd edn (Reinhold, New York 1967); *L. R. Snyder*, Principles of Adsorption Chromatography (Marcel Dekker, New York 1968).
[3] See *W. H. Scouten*, Affinity Chromatography (Wiley, New York 1981).
[4] See *W. C. Still* et al., Rapid Chomatographic Technique for Preparative Separation with Moderate Solution, J. Org. Chem. *43*, 2923 (1978).

In an alternative procedure, an *eluting* solvent is used which has a greater affinity for the adsorbed material, and which washes the separated constituents of the mixture in turn from the column.

Colourless substances can also be separated by means of column chromatography. The substances on the column are eluted and the material leaving the column is collected in separate fractions. This is best achieved by the use of an automatic *fraction cutter*. Each fraction is investigated separately and the various components can be identified and separated, providing thereby pure products.

1.2.8 Gel Chromatography

This method allows larger molecules to be separated from smaller molecules. As the stationary phase a gel (Sephadex) is used, which has pores into which the smaller molecules can pass. Larger molecules, which cannot enter the pores, pass through the column with the solvent and appear in the eluate. Because the stationary phase in effect sieves out the smaller molecules, it is described as a *molecular sieve*. Using this method, it is possible to isolate compounds with relative molecular masses of between 700 and 200 000.

The sharpness of the separation can be substantially improved by applying an electric current.[1] Such *two-dimensional* (2D) *gel electrophoresis* has been used very successfully in the study of proteins and nucleic acids.

1.2.9 Partition Chromatography

This is a method of chromatography which makes use of *Craig's* principle of counter-current separation, which involves the fractional separation of substances between two liquid phases which are only partially miscible with one another. The liquid which acts as the stationary phase is adsorbed onto a support (silica gel, kieselguhr, etc.). This is described as *Counter-current Chromatography* (CCC). The mobile phase is pumped through the column under pressure (40–200 bar/4–20 MPa), at a controlled speed and constant temperature, and the compounds leaving the column are detected by, for example, UV light. This technique is known as *High Performance* (or *Pressure*) *Liquid Chromatography* or HPLC[2]. The method is also applicable to adsorption, ion-exchange and gel chromatography.

If the stationary phase is aqueous and the mobile phase is an organic solvent, the rates of movement through the column of the different constituents of the mixture increase, the lower their solubility in water. The method can also be used with compressed supercritical gases as the mobile phase (*Supercritical Fluid Chromatography*, SFC)[3].

1.2.10 Paper Chromatography

Paper chromatography is a special form of partition chromatography. It was developed by *Consden*, *Gordon* and *Martin* (1944), and has been of special importance in the separation of amino acids. The carrier of the aqueous phase is cellulose in the form of

---

[1] See *P. D. Grossmann*, Capillary Electrophoresis: Theory and Practice (Academic Press, San Diego, Cal. 1992).

[2] See *J. H. Knox*, High Performance Liquid Chromatography (Edinburgh University Press, Edinburgh 1978); *P. R. Brown* et al. (Eds) High Performance Chromatography (Wiley, New York 1989).

[3] See *U. van Wansen* et al., Physicochemical Principles and Applications of Supercritical Fluid Chromatography, Angew. Chem. Int. Ed. Engl., *19*, 575 (1980).

specially prepared filter paper, which can take up about 6–7% moisture. A drop of solution of a mixture is spotted onto the top end of a piece of this paper and allowed to dry off. The paper is then put in a closed vessel where it is exposed to the vapour of a solvent which is only slightly miscible with water, but is saturated with it. Typical solvents are butanol, pentanol, phenol and collidine. The top of the paper is folded over and immersed in a trough containing the solvent. Because of the capillary action of the filter paper, the solvent is adsorbed onto the paper and the impure mixture is separated out as a one-dimensional chromatogram as the solvent descends the paper. Alternatively the impure mixture can be spotted near the bottom of the piece of filter paper, which just dips into a shallow layer of solvent in the bottom of the vessel, and separation takes place up the paper instead of downwards.

If this procedure does not bring about a satisfactory separation, the process can be repeated using a different solvent, and allowing it to run in a direction at right angles to the original separation, thus providing a two-dimensional chromatogram.

The ratio of the distance moved by a particular substance to the distance moved by the solvent front is described as the R_f-value (Retention factor). This depends on a number of variables such as the temperature, composition of the two phases, the nature of the paper and so on. The method can also be modified in other ways, for example by the use of round filter paper. The separated constituents from a mixture, in the form of spots or rings, are usually made visible by adding suitable reagents called developers, which colour them. The sizes of the bands or rings give an indication of the amounts of each component present.

Paper chromatography can be used to analyse very small amounts (5–50 µg) of material. It has been of particular value in the study of proteins, hormones and antibiotics, and has provided valuable insights into the structures of such natural products.

A related method is *paper electrophoresis*, in which the migration on filter paper of amphoteric electrolytes in a buffered solution is followed. It has been used especially in the study of albumins, and allows the colloidal particles of proteins to be separated because of their different mobilities in the presence of an electric field. When applied to smaller ions it is known as *ionophoresis*.

1.2.11 Thin-layer Chromatography (TLC)[1]

The basic principles of *thin-layer chromatography* were recognised by *Ismailow* and *Schraiber* (1938). It first received a wide practical application following its development as a reliable laboratory method by *Stahl*. In this process a layer, about 250 µm thick, of an adsorbent such as kieselgel, alumina, etc. is coated onto a glass plate or onto metal foil. Then a solution of the material to be separated is spotted onto the lower end of the plate. This is put into a vessel, which has a small amount (ca 0.5 cm) of the developing solvent in the bottom. The solvent moves upwards through the layer, carrying the constituents of the mixture with it. They separate into individual lines or spots, which may be detected by using appropriate methods. By suitable choice of material for the layer, and by preparing the layer in a suitable way, this can be applied as an analytical technique to *hydrophilic* as well as to *lipophilic* substances. The great advantages of this technique are the clean separation achieved and the speed of separation; it can usually be carried out in 10–30 min. A further development is known as *High Pressure Planar Liquid Chromatography* or HPPLC. In this process the thin-layer plate is replaced by a thin-layer filling between two glass plates, which are forced together under a pressure of

---

[1] See *F. Geiss*, Fundamentals of Thin Layer Chromatography (Hüthig-Verlag, Heidelberg, 1987).

3000 bar (300 MPa). The technique, which makes use of a controlled stepwise injection system, can also be used for two-dimensional chromatography.

Thin-layer chromatography has been applied to the separation of vitamins, terpenes, steroids and natural pigments, and also to amino acids, sugars, nucleotides, nucleic acids, alkaloids and pharmaceutical materials. It has also been used for the analysis of inorganic anions and cations.

Separation can also be assisted by the use of electrophoresis, and a combination of electrophoresis and chromatography has been used in the identification of materials of high molecular weights (*blotting*).[1]

1.2.12 Gas Chromatography (GC)[2]

In principle, this process is analogous to the chromatographic methods described previously, but in this case exchange takes place between a gas phase and a solid or liquid phase. There is a further sub-division between *gas adsorption chromatography* and *gas partition chromatography* (*James* and *Martin*, 1952). These methods can be used for the separation of substances which can be volatilised without decomposition up to a temperature of 500°C (775 K). The separating column is made up with either a solid *adsorbent* (active charcoal, alumina, kieselguhr, molecular sieve) or with a carrier impregnated with an involatile liquid. Glass, quartz or metal capillaries, whose inner surfaces have been impregnated with the liquid stationary phase, can also be used. By changing the polarity of the stationary phase different degrees of separation can be achieved. The mixture to be analysed is volatilised in the presence of a carrier gas, often helium, nitrogen or argon, which carries it through the column. Depending upon the affinity of the different substances for the stationary phase, they are to a greater or lesser extent retained by it and released again, and finally leave the column. The separating efficiency of the method is equivalent to that of a column having up to 5000 theoretical plates per metre, and suitable columns have even made it possible to achieve the complete separation of enantiomers (see p. 348 *ff.*). The separated components are detected as they leave the column; the *detectors* include thermal conductivity and ionisation devices and mass spectrometers.

Gas chromatography is useful for qualitative and quantitative analysis and in preparative work.

It has particular value in monitoring continuous production processes. It is also used in quantitative carbon, hydrogen and nitrogen analyses. In addition, gas chromatography is able, depending upon the efficiency of the stationary phase, to provide ideas about the constitution of unknown volatile substances.

The following abbreviated descriptions are in common use: GLC (gas–liquid chromatography), VPC (vapour phase chromatography), GC–MS (a combination of GLC with mass spectrometry as the detector).

1.2.13 Criteria for Purity of Substances

The correct choice of the methods described above for the separation and purification of organic compounds, and the manual dexterity in handling them, make high demands on the art and skill of chemists, especially when working with small amounts of material.

---

[1] See *U. Beisiegel*, Protein-blotting, Electrophoresis, *8*, 1 (1986).
[2] See *D. W. Grant*, Gas–Liquid Chromatography (Van Nostrand Reinhold, New York 1971). *H. Purnell*, New Developments in Gas Chromatography (Wiley–Interscience, New York 1973); *R. M. Smith*, Gas and Liquid Chromatography in Analytical Chemistry (Wiley, Chichester 1988).

When a chemically pure compound has been isolated, it must then be characterised by suitable physical properties.

Usually chemically pure substances have sharp and constant melting points and boiling points so long as they do not decompose or form *liquid crystals*. The latter have two melting points. At the lower melting point they form a turbid liquid, which suddenly becomes clear at the higher melting point (*clarification point*).[1]

To check the purity of a crystalline organic substance, the *melting point* is first determined. This is an important constant, which needs to be reported for novel compounds, and is also important for the identification of compounds which have already been described. If two samples have the same melting point and mixed melting point then they are considered to be identical compounds. If two different components are mixed, then the melting point is lowered because each component is an impurity added to the other.

To determine *mixed melting points*, small amounts of three samples (A; B; and a mixture of A and B) are placed in separate melting-point tubes, attached to one and the same thermometer, and the three melting points are thus all recorded under identical conditions. This test may fail for *isomorphous* materials.

The purity of an organic liquid can be judged from its boiling point, which is constant under identical conditions, and so long as no decomposition takes place during the distillation. This criterion fails for azeotropic mixtures, since such mixtures also have constant boiling points.

For all organic compounds their physical constants[2] and spectroscopic data (see p. 29 *ff.*) are of the greatest value for confirming their purity.

1.3 Qualitative Organic Elemental Analysis

Whenever a new organic compound has been purified, one of the first things which is done with it is to find out what elements it contains. The following tests have been used for this purpose.

1.3.1 Carbon

When a carbon compound is heated it is commonly the case that it either chars or burns with a sooty flame. However, volatile materials, high-melting compounds, or compounds with a low carbon content, often do not behave in this way, and so the test fails for them.

A more reliable test for carbon is to burn an intimate mixture of the compound with fine copper(II) oxide. In these circumstances, *carbon dioxide* is formed, which, if passed through an aqueous solution of barium hydroxide, causes the precipitation of barium carbonate:

$$2\,CuO + C = 2\,Cu + CO_2 \quad\quad Ba(OH)_2 + CO_2 = BaCO_3 + H_2O$$

[1] See *G. W. Gray*, Molecular Structure and the Properties of Liquid Crystals (Academic Press, New York 1962); Thermotropic Liquid Crystals (Wiley, Chichester 1987); *G. A. Jeffrey*, Carbohydrate Liquid Crystals, Accounts Chem. Res. *19*, 168 (1986).

[2] *A. C. Weast*, Handbook of Chemistry and Physics (CRC Press, Cleveland, Ohio) published annually.

1.3.2 Hydrogen

The same test, namely burning the compound with copper(II) oxide, can also be used to detect hydrogen. This is oxidised to form water which condenses on the cooler surfaces of the vessel.

$$CuO + H_2 = Cu + H_2O$$

1.3.3 Nitrogen

A reliable test for nitrogen is the *Lassaigne test* (1843). A sample of the material to be tested is heated with a small piece of sodium (about the size of a lentil) in an ignition tube. The carbon and nitrogen in the material are thereby converted into sodium cyanide. The residue is extracted with water and filtered, and the alkaline solution is boiled with a small amount of iron(II) sulphate. Sodium hexacyanoferrate(II) is formed and, when this solution is acidified with dilute hydrochloric acid or dilute sulphuric acid, this reacts with iron(III) ions which have been formed by atmospheric oxidation of some of the iron(II) ions and gives the characteristic colour of Prussian Blue. If the test compound contains only a small amount of nitrogen, then a blue-green solution results, from which a blue precipitate settles out when it is kept.

$$Fe^{2\oplus} + 6\ CN^{\ominus} \rightarrow [Fe(CN)_6]^{4\ominus} \qquad 4\ Fe^{3\oplus} + 3\ [Fe(CN)_6]^{4\ominus} \rightarrow Fe_4^{III}[Fe^{II}(CN)_6]_3$$

$$\text{Prussian Blue}$$

Not only the structure of Prussian Blue, but also the reason for its colour, was a matter of debate. It has been shown, with the help of *Mössbauer* effects,[1] that insoluble Prussian Blue has the structure shown above, and that the iron atoms can be assigned definite oxidation states of $+2$ and $+3$ respectively. The deep colour can be attributed to a metal–metal interaction, which has been called intervalence absorption.

1.3.4 Sulphur

The qualitative test for sulphur is carried out similarly to that for nitrogen. When a compound containing sulphur is heated with sodium, sodium sulphide is formed. If the aqueous filtrate is treated with lead acetate and acetic acid, a black precipitate of lead sulphide separates out. If only a small amount of sulphur is present the solution turns dark brown. In addition, another portion of the filtrate can be tested with a freshly prepared solution of sodium pentacyanonitrosylferrate(II) $Na_2[Fe^{II}(CN)_5NO].2H_2O$ (sodium nitroprusside). If sulphur is present, a red-violet colour appears, which is due to the formation of the complex salt $Na_4[Fe(CN)_5NOS]$.

1.3.5 Halogens

The *Beilstein* test has been used to detect the presence of halogens in organic compounds. A small grain of material is put onto a hot copper wire which is then heated further in

[1] See E. *Fluck* et al., The Mössbauer Effect and its Significance in Chemistry. Angew. Chem. *75*, 461 (1963); R. L. *Mössbauer*, Gamma-ray Resonance Spectroscopy and Chemical Bonding, Angew. Chem. Int. Ed. Engl., *10*, 462 (1971); N. N. *Greenwood* and T. C. *Gibb*, Mössbauer Spectroscopy (Chapman & Hall, London 1971).

a non-luminous Bunsen flame. If the sample contains halogens the flame takes on a characteristic green to blue-green colour, associated with the volatilisation of copper halides.

1.3.6 Other Elements

The presence of other elements, such as phosphorus, arsenic or metals, in organic compounds has been detected by destructive oxidation of a sample, either by heating it with fuming nitric acid in a sealed tube or by melting it with a mixture (1:2) of potassium nitrate and sodium carbonate. The residue is diluted with water and the standard tests for inorganic groups are used to identify the elements present.

Mass spectrometry is also of great use in identifying the presence of some elements in organic compounds, especially those elements, such as chlorine and bromine, which have characteristic and readily observable isotope ratios.

1.4 Quantitative Organic Elemental Analysis

The quantitative determination of single elements in an organic compound is now carried out on a micro-scale, often using about 1 mg of substance, and is usually done in appropriate analytical laboratories.

1.4.1 Carbon, Hydrogen and Nitrogen

Quantitative analysis for these elements was first developed by *J. v. Liebig* (1831) and later refined by *F. Pregl* (1912) and other analytical chemists. A weighed amount of the material to be tested is burned in the presence of copper(II) oxide in a special apparatus. The carbon and hydrogen are quantitatively oxidised to carbon dioxide and water.

Nitrogen analysis is carried out similarly. In the process nitrogen is the main product but accompanied by varying small amounts of nitrogen oxides; the latter are reduced to nitrogen by a layer of red-hot copper.

The carbon dioxide, water and nitrogen which are formed are then separated by gas chromatography. An accurately weighed sample of about 1–3 mg is used for the analysis. In an analogous way it is possible to obtain a direct measure of the oxygen content.[1]

The method of *Dumas* (1830) for the quantitative estimation of nitrogen has been completely superseded by gas-chromatographic methods. In contrast, the other traditional method, that of *Kjeldahl* (1883) is still used. In this method a weighed amount of substance is heated with concentrated sulphuric acid in the presence of a catalyst (powdered selenium, copper(II) sulphate or mercury(II) oxide) in a *Kjeldahl* flask. The organic material is oxidatively destroyed and any nitrogen is converted into ammonium sulphate. The resultant solution is basified with sodium hydroxide, ammonia is distilled out into a known amount of 0.1N HCl, and the excess of the latter is back-titrated with 0.1N sodium hydroxide.

---

[1] See *F. Ehrenberger* and *S. Gorbach*, Methoden der Organischen Elementar- und Spuren-analyse (Methods of Organic Elemental and Trace Analysis) (Verlag Chemie, Weinheim 1973).

The *Kjeldahl* method is not universally applicable. It does not work properly for nitroso-, nitro- or azo-compounds. Its main use is in routine analysis in industrial laboratories associated with nitrogen-containing compounds, especially in food and agricultural chemistry.

1.4.2 Sulphur

Schöniger **Method**[1] (1961). This method is very valuable because it involves no costly apparatus. The substance is weighed onto an ash-free filter paper, which is placed on platinum gauze and burned in a flask full of oxygen. The gases produced are absorbed into 3% hydrogen peroxide solution. Sulphuric acid is formed and is measured volumetrically.

1.4.3 Halogens (Chlorine, Bromine, Iodine)

1.4.3.1. *Wurzschmitt* **Method**[2] (1950). A weighed analysis sample, ethylene glycol and sodium peroxide are burned in a nickel bomb. The reaction mixture is washed out and chloride or bromide which have been formed are estimated either as their silver salts or potentiometrically. Iodine gets oxidised to iodate, which is estimated iodometrically in the standard way.

1.4.3.2. *Schöniger* **Method**. Chlorine and bromine can be analysed very satisfactorily using the method given for the estimation of sulphur. In this case the combustion gases are absorbed in a weakly alkaline solution of hydrogen peroxide. The resultant chloride and bromide can again be estimated either as their silver salts or potentiometrically.

1.5 The Determination of Chemical Formulae

1.5.1 Empirical Formulae

The determination of the chemical formula of an unknown organic substance starts with a quantitative elemental analysis, which shows the percentages of the different elements present in the substance. If each of these percentages is divided by the atomic weight of the particular element, this provides the ratios of the numbers of atoms of each of these elements. The following calculation shows the way in which this is done, starting from a typical quantitative elemental analysis.

---

[1] See *W. Schöniger*, Die Kolbenmethode in der organischen Mikroelementaranalyse (Flask Methods in Organic Elemental Microanalysis) Z. Analyt. Chem. *181*, 28 (1961); *J. Bassett, R. C. Denney, G. H. Jeffrey* and *J. Mendham*, Quantitative Inorganic Analysis, 4th edn, p. 115 (Longman Group Ltd., London 1978).

[2] See *B. Wurzschmitt* et al., Die Metallbombe als Hilfsmittel in der Elementaranalyse (Metal Bombs as a Help in Elemental Analysis) Fortschr. Chem. Forsch. *1*, 485 (1950).

	Percentage composition	Atomic weight	Percentage/ Atomic weight
	C = 40.82%	C = 12	40.82/12 = 3.40
	H = 8.63%	H = 1	8.63/1 = 8.63
	N = 23.75%	N = 14	23.75/14 = 1.69
Total	= 73.20%		
Remainder = O =	26.80%	O = 16	26.80/16 = 1.67
	100.00%		

This provides atomic ratios of C:H:N:O = 3.40:8.63:1.69:1.67. If each of these figures is divided by the one with the lowest numerical value, 1.67, the ratio C:H:N:O = 2:5:1:1 is obtained. From this the simplest formula is C_2H_5NO, but multiples of this simple formula, $C_4H_{10}N_2O_2$, $C_6H_{15}N_3O_3$ and, in general terms, $C_{2n}H_{5n}N_nO_n$ (n = 1,2,3,4, etc.) are also consistent with the ratio. In order to distinguish between these possible formulae it is necessary to know the molecular weight of the compound.

The unit involved is known as the *mole* (unit abbreviation: mol). One mol is the amount of a substance which contains as many individual units of that substance as there are atoms in 12 g of pure ^{12}C. The nature of the unit must be specified, and may be an atom, a molecule, an ion, a radical, an electron, a photon, etc. or a specified group of these entities.[1] The *molecular weight* or the *relative molecular mass* of a compound is equal to the sum of its atomic weights (or relative atomic masses).[1,2] It is equal to the ratio of the mass of 1 mol of the molecule to the mass of 1 mol of ^{12}C.

Thus a mol of a compound is the amount of material whose weight in grams is numerically the same as its relative molecular mass (g/relative molecular mass = 1). For water this is 18.12 g. Fractions or multiples of a mol are expressed in g/mol. 1 mol contains 6.022×10^{23} particles; this number is known as the *Avogadro* constant N_A.

1.5.2 Determination of the Relative Molecular Mass and the Molecular Formula

In continuing this discussion, it is desirable to mention briefly the most important physical methods which are used for the determination of relative molecular masses. The classical methods are based on the properties of dilute solutions, but in recent times, the technique of *mass spectrometry* (p. 30 *ff*) has become of major importance.

1.5.2.1 Depression of the Freezing Point

(a) *Beckmann* Method. If an organic substance is dissolved in a solvent and the solution is cooled to its freezing point, then the lowering of the freezing point of the solvent which results is proportional to the concentration of the dissolved material. This lowering is dependent only upon the number of dissolved molecules. For example, if quantities of two different substances, which are in the ratio of their relative molecular masses, are each dissolved in 1000 g of solvent, then each of these solutions will have the same freezing point, since in each case the solution will contain the same number of dissolved molecules.

---

[1] See *M. L. McGlashan*, Physico-chemical Quantities and Units, R. S. C. Monographs for Teachers, No. 15.
[2] See Manual of Symbols and Terminology for Physicochemical Quantities and Units (Butterworth, London 1970).

This fact provides the basis for *Raoult's* law: equimolar solutions in the same solvent have the same freezing point.

It follows, therefore, that the *depression of the freezing point* brought about when 1 mol of some substance is dissolved in 1000 g of a certain solvent is a constant, K_f. The values of the cryoscopic constants for some solvents are as follows: water, 1.863 K; acetic acid, 3.90 K; benzene, 5.10 K.

These cryoscopic constants thus provide a means for the determination of the relative molecular masses of organic compounds. If *a* grams of a substance are dissolved in *b* grams of a solvent having a cryoscopic constant K_f, and the depression of freezing point Δt is measured, then the relative molecular mass is derived from the equation

$$M_r = \frac{K_f.\ a.\ 1000}{b.\ \Delta t}$$

(b) *Rast* **Method.** This method is also based on the depression of the freezing point. The procedure is simplified in that only a test tube and a thermometer graduated in tenths of a degree are required. Camphor is used as solvent. This has a very high cryoscopic constant, $K_f = 40$ K. The melting point or freezing point of pure camphor is 177°C (450 K). Hence, if 1 mol of the substance to be examined is dissolved in 1000 g of camphor, then the freezing point is lowered to 137°C (410 K). In practice the method is the same as in (a). It is practicable only if the test sample is sufficiently soluble in camphor, does not react with it, and is stable at the melting point. Other solvents with high cryoscopic constants can be used instead of camphor, *e.g.* camphene (m.p. 49°C [322 K], $K_f = 31$ K).

1.5.2.2 Raising of the Boiling Point

According to *Raoult's* law, equimolar solutions in the same solvent not only have the same freezing point, but also have the same boiling point. It follows that the relative molecular mass of an unknown substance can be measured in a similar way, and indeed more simply, by its effect on the boiling point. The boiling point constant K_b (1 mol substance in 1000 g solvent) is generally smaller than the cryoscopic constant K_f; examples are as follows: water, 0.515 K; benzene, 2.57 K; acetic acid, 3.07 K. The methodology resembles that in section 1.5.2.1(a), with the measured raising of boiling point Δt used in conjunction with K_b instead of K_f in order to determine the relative molecular mass.

1.5.2.3 Measurement of the Temperature Difference between a Solution and its Solvent in a Vapour-pressure Osmometer

A solution has a lower vapour pressure than the corresponding pure solvent has. Consequently, if a small amount of solvent is placed immediately adjacent to a small amount of a solution in a sealed, completely isolated system, then some of the solvent sample evaporates and condenses into the sample of solution. Because of this, a small difference in temperature is set up between the two samples. From this difference, measured at a series of different concentrations of solute, and with the aid of a calibration curve, the relative molecular mass of the solute can be found.

This measurement of the relative molecular mass often provides results which are not very accurate, but this is not vital in its use for the elucidation of a *molecular formula*, since the mass must in fact either be identical with the empirical formula or be a simple multiple of it. It is thus easy to see which of these possibilities is most in accord with the measured relative molecular mass.

Very accurate molecular weights or relative molecular masses of unidentified organic compounds can be obtained by the use of *mass spectrometry*.

1.5.2.4 Special Methods for Macromolecular Compounds

The methods described above are not suitable for the determination of the relative molecular masses of macromolecular compounds. Compounds are described as *macromolecular compounds* or as *high polymers* if their relative molecular masses are of the order of 10^4 or greater. To obtain results for them, techniques involving measurements of viscosity, osmotic pressure, sedimentation rate in an ultra-centrifuge and the scattering of light (*Tyndall* effect) are used.[1] These methods provide only an *average molecular weight* since a synthetic high polymer is made up of individual molecules with a range of sizes, while natural products occur in different states of aggregation, depending upon pH, and in consequence molecular sizes also vary. So, up till now it is only possible to find approximate values for the relative molecular masses of natural products such as polysaccharides, proteins and nucleic acids.

1.5.3 Structural Formulae

Even when the molecular formula has been found, by using the above methods, the actual arrangement and bonding of the different atoms is still not known. Organic compounds which have identical molecular formulae may have very different physical and chemical properties. The reasons for this may be found in the variety of ways in which the constituent atoms in organic molecules can be arranged. The first insight into the interrelationships between atoms came from *Kekulé* and *Couper* (1857). Quite independently, they established the concept that atoms of any one element are not bound to other atoms in a random fashion, but are bonded only to certain defined numbers of other atoms. They assigned to each atom a definite number of *units of affinity*, which was described as its 'atomicity'. Depending upon the numbers of units of affinity of a particular atom, in particular whether it had one or more units, it was possible to differentiate between monoatomic and polyatomic elements. At about the same time, *Kekulé* deduced, quite empirically, from the stoichiometric composition of organic compounds, that carbon had an 'atomicity' of four. He also recognised the a number of carbon atoms could be joined together to satisfy this atomicity. Later the term 'atomicity' was replaced by the word *valence*. *A. Crum Brown* (1865) introduced the use of a dash as a formal representation of valence. *Butlerov*, in 1861, first introduced the concept of chemical structure and the binding together of atoms to form molecules. At the same time he put forward the idea that the chemical properties of substances were determined by their structures.[2] A further important advance in the understanding of organic chemical structure came from *van't Hoff* (1874), who realised that, because of the optical activity of certain carbon compounds, the valences of the carbon atoms did not lie all in one plane, but that rather they must be arranged in three-dimensional structures.

These significant advances in the understanding of molecular structure, together with the systematic expansion in experimental organic chemical research,[3] led to the currently accepted structural theory of organic compounds. Since then, one of the most important functions of organic chemists has been to find the correct structural formulae of any new organic compounds. First of all, the structural formula, which indicates the atomic make-up and which atoms are linked to one another, tells us what kind of compound it is, such as an alcohol, aldehyde, ketone, carboxylic acid, ester, amide, amine, etc. A compound

---

[1] See *Houben-Weyl*, Methoden der organischen Chemie, 4th edn, Vol. 3/1, pp. 371 *ff*. (Thieme-Verlag, Stuttgart 1955).
[2] See *C. A. Russell*, The History of Valency (Leicester University Press, 1971).
[3] See *N. J. Turro*, Geometric and Topological Thinking in Organic Chemistry. Angew. Chem. Int. Ed. Engl. *25*, 882 (1986).

may be further characterised by conversion into *derivatives*, or by an independent synthesis. The latter is always one of the safest proofs of the constitution. Nowadays, physical methods such as mass spectroscopy and infra-red and nuclear magnetic resonance spectroscopy are of vital importance in proving molecular structure.

Structural formulae include information about the spatial relationships of the constituent atoms to one another. X-ray structure analysis is increasingly in use for the determination of structural formulae.

1.6 The Nature of Chemical Bonding

To provide a full understanding of structural formulae, it is necessary next to clarify the structure of atoms and their modes of bonding in molecules, and also the special nature of carbon atoms.

In older ideas about chemical bonding forces, the way in which atoms are held together in molecules was attributed to mutual attractions between them, which some investigators (*Davy*, 1807; *Berzelius*, 1812) thought to be electrical forces and later workers (*Kekulé*, *Kolbe*) thought to be mechanical forces. Although chemical bonding between atoms was represented in structural formulae by a dash, the explanation of this bonding remained an unsolved problem for many years. Later discoveries concerning atomic structure confirmed ideas derived from quantum mechanics that *chemical bonding* came about because of the interaction of electrical forces.

1.6.1 Atomic Structure[1]

The chemical properties of atoms are of prime interest for chemists, and of decisive importance for these properties are the shells of electrons. The first insights into their very complicated structure were made possible by the combination of spectroscopic data and the quantum theory (*Planck, Bohr, Sommerfeld*). By this means the line spectra of the simplest elements were interpreted in terms of groups of electrons occupying different energy levels (K-, L-, M-, N-shells, etc.). According to the older concepts, in the *Rutherford–Bohr–Sommerfeld* model of the atom, within these shells the electrons occupied various circular or elliptical orbits, which were described as the *s, p, d* and *f* orbits. This model does indeed provide a clear picture of the movements of the electrons with respect to the nucleus, but it does not give a true picture of the overall situation. A satisfactory theory of chemical bonding was first made possible by the newer quantum or wave mechanics (*de Broglie, Schrödinger, Heisenberg, Dirac*). In this, the different states (*wave functions*) of the electrons moving around the nucleus were decribed mathematically by means of approximate solutions of the *Schrödinger* equation. Since the quantum mechanics cannot be expounded in a short easily comprehensible form, in the present text only a few results,

---

[1] See *R. McWeeney, Coulson's* Valence (Oxford University Press, Oxford 1980); *W. J. Jorgensen* and *L. Salem*, The Organic Chemist's Book of Orbitals (Academic Press, New York 1973); *M. J. S. Dewar*, The Molecular Orbital Theory of Organic Chemistry (McGraw–Hill, New York 1969); *P. W. Atkins*, Quanta: A Handbook of Concepts (Clarendon Press, Oxford 1974); *J. M. Tedder* and *A. Nechvatal*, Pictorial Orbital Theory (Pitman, London 1985).

which are of relevance to the structure of the electron shells and molecules, will be mentioned. According to this theory it is not possible, in principle, to attribute electrons to specific circular or elliptical orbits. Instead it is only possible to indicate the locations or the regions in space in which there is a greater probability of the electron being found. These regions in space are known as *orbitals*. The electron occupies an orbital rather than moving in a specific orbit. Of the various types of orbital which are possible, the ones which are important for the organic chemist are *s*, *p* and *d* orbitals, and these will be the only ones to be mentioned further.

An understanding of the *atomic orbitals* can be gained as follows. If one were able to determine the position of an electron over a longer period of time—although on the basis of the *Heisenberg* uncertainty principle this is intrinsically impossible—and record this graphically, then, for each atomic orbital, a certain 'electron distribution' would appear, as shown in Figs. 1 and 3. The surfaces shown in these diagrams enclose a space within which there is a 90% probability that the electron would be found. Often the three-dimensional representation is not used, and instead the orbital is drawn only in section (Figs. 2 and 4). As can be seen in Figs. 1 and 2, an *s* orbital has the shape of a sphere, while a *p* orbital (Figs. 3 and 4) consists of two lobes which are separated from one another by a nodal plane, in which the probability of finding an electron is nil, although more sophisticated treatments suggest that nodes do have a very small electron density.[1] Because of its spherical symmetry only one *s* orbital is possible, but there are three possible *p* orbitals which are orientated, respectively, along the *x*, *y* and *z*-axes in a right-angled co-ordinate system.

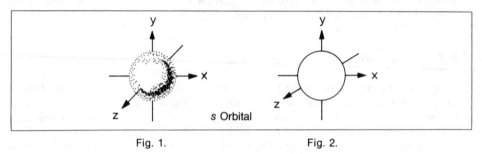

Fig. 1.   Fig. 2.

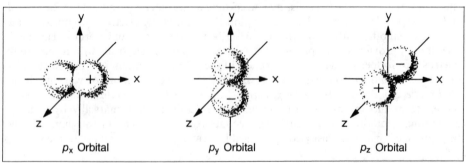

Fig. 3.

---

[1] See *F. O. Ellison* and *C. A. Hollingsworth*, J. Chem. Educ. *53*, 767 (1976).

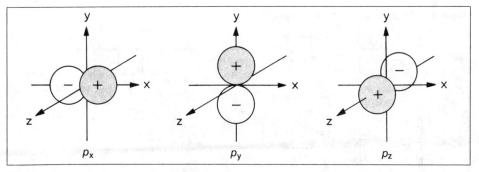

Fig. 4.

The signs of the *wave functions* of the p orbitals are positive in the one lobe and negative in the other (see Figs 3 and 4), i.e. the wave function changes its sign at the *nodal plane*. In the cases of elements in the lower rows of the Periodic Table, such as phosphorus and sulphur, d orbitals can also participate in chemical bonding. In transition metal complexes, d orbitals are also involved. Fig. 5 illustrates one of the five d orbitals, namely the d_{xy} orbital.

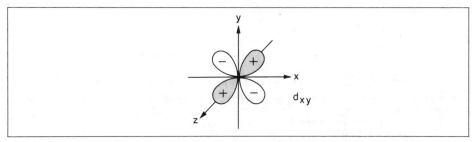

Fig. 5.

Single electrons, which are unpaired and remain isolated, are also described as *lone electrons* or *free electrons*. Two electrons, which have *opposite* or *anti-parallel spins*, can form an *electron pair*. Such pairing has been described as the *spin-compensation principle*. Electron pairs exhibit *diamagnetism*, while the uncompensated spin of an unpaired electron in an atom results in *paramagnetic* properties. A paramagnetic substance is attracted into a magnetic field, whereas a diamagnetic substance tends to be repelled by it.

The various orbitals are filled with electrons in turn, depending upon their energies (Table 1, *Aufbau principle*). Another important factor is the *Pauli* principle, which stipulates that each orbital cannot contain more than two electrons, and that these must be of opposite spin. If, within the electron shell, orbitals of equal energy, described as *degenerate orbitals*, are available for occupation by electrons, then two electrons will go into separate orbitals, each of which thus becomes singly occupied. Both electrons will have the same, or *parallel* spins, because this results in the lowest energy state (Table 2). This arrangement is said to possess the highest *multiplicity*.[1] (*Hund's* first rule; spin correlation).

The spins of the electrons are indicated (Table 2) by the directions of the arrows. The ground state of the electrons is shown, using the usual shorthand form, in the third

---

[1] For two unpaired electrons the multiplicity is $n + 1$ (singlet, $n = 0$; doublet, $n = 1$; triplet, $n = 2$).

Table 1. Energy levels of atomic orbitals

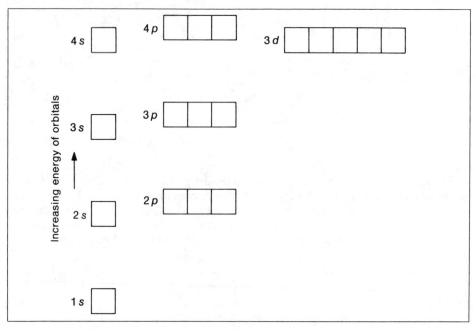

Table 2. Electronic configuration of elements in the first rows of the Periodic Table

Element	Orbitals					Ground state	Excited state	Stoichio-metric valence towards H
	K-Shell	L-Shell						
	$1s$	$2s$	$2p_x$	$2p_y$	$2p_z$			
H	↑					$1s^1$		1
He	↑↓					$1s^2$		0
Li	↑↓	↑				$1s^2\,2s^1$		1
Be	↑↓	↑↓				$1s^2\,2s^2$	$1s^2\,2s^1\,2p$	2
B	↑↓	↑↓	↑			$1s^2\,2s^2\,2p^1$	$1s^2\,2s^1\,2p^2$	3
C	↑↓	↑↓	↑	↑		$1s^2\,2s^2\,2p^2$	$1s^2\,2s^1\,2p^3$	4
N	↑↓	↑↓	↑	↑	↑	$1s^2\,2s^2\,2p^3$		3
O	↑↓	↑↓	↑↓	↑	↑	$1s^2\,2s^2\,2p^4$		2
F	↑↓	↑↓	↑↓	↑↓	↑	$1s^2\,2s^2\,2p^5$		1
Ne	↑↓	↑↓	↑↓	↑↓	↑↓	$1s^2\,2s^2\,2p^6$		0

column. In this, the number of electrons associated with each energy level is shown as an exponent term.

In the case of hydrogen there is one electron, which is in the one possible orbital, *i.e.* the 1s orbital, of the K-shell. In helium, the second electron which is present differs from the first only in spin. In this case, in accordance with the *Pauli* principle, the K-shell is filled with two electrons. In a lithium atom the third electron goes into a 2s orbital of the L-shell, and in beryllium a second electron is also present in this shell. The two electrons have opposite spins and form an electron pair. From boron onwards, the extra electrons are found in the 2p orbitals. Since orbitals of equal energy are available, first of all only a single electron will enter each orbital. Thus the extra electron associated with a carbon atom, as compared to a boron atom, does not join the single electron which, in the case of boron, is in a $2p_x$ orbital, but instead occupies the $2p_y$ orbital, in accordance with *Hund's* rule. The same situation obtains in the case of nitrogen. Only with oxygen can double occupancy of a 2p orbital begin. In the ground state, oxygen exists as a diradical, as does the oxygen molecule O_2, which is a diradical with two unpaired electrons (triplet oxygen). If the experimentally determined stoichiometric valences towards hydrogen, as shown in the last column, are compared with the electronic configurations, it may be seen that in the cases of hydrogen, lithium, nitrogen, oxygen and fluorine these valences correspond to the numbers of unpaired electrons. Beryllium, boron and carbon are exceptions to this generalisation; in each case the numbers of unpaired electrons are smaller by two than the stoichiometric valences. This difference is explained by the following hypothesis, which results from calculations based on quantum chemical theories. Boron, beryllium and carbon react with other atoms through an excited state, in which an increase in energy results in one electron of the doubly occupied 2s orbital being raised into an unoccupied 2p orbital. This can be illustrated as follows, taking carbon as an example.

Table 3. Excited state of carbon

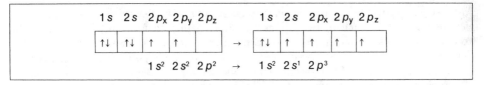

$$1s \quad 2s \quad 2p_x \quad 2p_y \quad 2p_z \qquad 1s \quad 2s \quad 2p_x \quad 2p_y \quad 2p_z$$

$$1s^2 \; 2s^2 \; 2p^2 \quad \rightarrow \quad 1s^2 \; 2s^1 \; 2p^3$$

The promotion of an electron requires an input of energy of about 377–418 kJ/mol, but this is more than compensated for by the energy which is released when a chemical bond is formed. In their excited states, beryllium, boron and carbon thus have, respectively, two, three and four single electrons in separate orbitals; that is, they are unpaired, as shown in the last but one column of Table 2.

It may be seen from Table 2 that with the first two elements, hydrogen and helium, which have respectively one and two electrons, the K-shell is filled, since it only requires two electrons to fill it. In the following sequence of elements the number of electrons rises regularly from one to eight. It can also be seen that the noble gases (He, Ne) have no unpaired electrons, but only pairs of electrons. Hence the stoichiometric valence, and with it the chemical properties of the elements, must depend very much upon the electronic configuration of the outer shell of electrons, or upon the presence of an unpaired electron. This consideration led *Kossel* and *Lewis* to the formulation of the *octet* principle. This says that the chemical reactivities of atoms depend upon their striving to achieve the octet configuration of a noble gas. In keeping with the classical concept of valence, the unpaired electrons in the L-shell (Table 2) are called *valence electrons*. The possibilities which the various atoms have for taking up stable noble-gas type octet configurations must now be examined.

In the case of the transition metals the noble-gas type of configuration is achieved by building up configurations of 18 electrons (see the Periodic Table on page XX). Many complexes of transition metals are formed in keeping with the 18-electron rule.

1.6.2 Ionic Bonding

One way of setting up a noble-gas configuration is by certain atoms either gaining or losing electrons. For example, if sodium and chlorine interact with one another, then the neutral sodium atom can give up one of its electrons and attain the octet configuration of the next lowest noble gas, neon, whilst the neutral chlorine atom takes the valence electron provided by sodium into its electronic shell and thus attains the outer configuration of the next highest noble gas, argon. The resultant salt, sodium chloride, consists of electrically charged ions, the cation Na^+ and the anion Cl^-.

$$Na^{\cdot} + \cdot \ddot{\underset{..}{Cl}} : \longrightarrow Na^{\oplus} + : \ddot{\underset{..}{Cl}} : ^{\ominus}$$

In the crystal these ions are arranged in a cubic face-centred lattice, which is held together by electrostatic forces. This *ionic bonding* is sometimes also described as *heteropolar bonding* or *electrovalent bonding*. In dilute solution, the ions move about freely, and no ion is bound to any other particular *counter ion*.

1.6.3 Covalent Bonding

Another way in which a noble-gas electronic structure is achieved is by two unpaired electrons, one from each of two atoms, getting together to form an electron pair. In accomplishing this, neither electron loses contact with the atom from which it was derived. On the basis of the octet principle, *Lewis* put forward the hypothesis that, for example in the cases of the H_2 or Cl_2 molecules, the noble-gas structure is achieved by the pairing of two electrons with opposing spins, in this way providing a bonding pair of electrons. The bonding electrons are shared between the two atoms they link, which remain electrically neutral.

$$H \cdot + \cdot H \longrightarrow H : H \qquad \qquad : \ddot{\underset{..}{Cl}} \cdot + \cdot \ddot{\underset{..}{Cl}} : \longrightarrow : \ddot{\underset{..}{Cl}} : \ddot{\underset{..}{Cl}} :$$

In the hydrogen molecule two helium-type configurations result; in the chlorine molecule, two argon-type configurations. The shared electron pair thus provides an octet configuration for both of the atoms it links. In contrast to ionic bonding there is in this case true chemical bonding of atom to atom. This has been called *atomic bonding* (*C. A. Knorr*) or, more commonly, *homopolar* or *covalent bonding* (*Langmuir*).

Covalent bonding is associated with the quantum mechanical concept that the pair of electrons lies within the attractive forces of both atoms.[1] The probability of the locations

---

[1] This is a greatly simplified statement of the position. For a deeper but easily understandable account of the problem of bonding see *W. Kutzelnigg*, The Physical Mechanism of the Chemical Bond. Angew. Chem. Int. Ed. Engl. *12*, 546 (1973).

of these electrons spreads around both of the nuclei, in a so-called *molecular orbital*, rather than in overlapping but separate atomic orbitals. The *Pauli* principle also applies; bonding is only possible if the electrons which occupy the new molecular orbital have anti-parallel spins. *Chemical bonding* can thus be defined as the combination, by mutual overlapping, of atomic orbitals to provide a molecular orbital. In the case of the hydrogen molecule, wherein two 1s orbitals overlap, the molecular orbital (MO) (Fig. 6a, b) can be illustrated as follows.

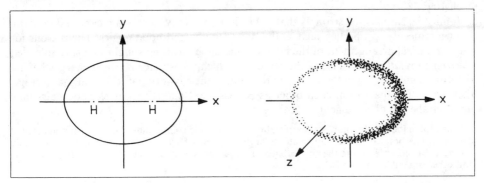

Fig. 6a. Molecular orbital of hydrogen.   Fig. 6b. Molecular orbital of hydrogen.

In the case of the chlorine molecule, the overlap is between two occupied $3p_x$ orbitals arranged spatially so as to provide maximum overlap. Hence the axis of the molecule lies in the direction of the x co-ordinate. This type of bonding, in which the molecular orbital has rotational symmetry about a line joining the two nuclei is known as *σ-bonding*.

From spectroscopic studies and from considerations of the retention of orbital symmetry (see p. 104 *ff.*), it is seen that the combination of two equivalent atomic orbitals leads to the formation of two molecular orbitals, one a *bonding* and the other an *anti-bonding* orbital. The bonding combination is characterised by a positive overlap and an increased electron density between the atomic nuclei. Such overlap only takes place between two orbital lobes with the same sign (in the wave function). In an anti-bonding combination electrons with the same spin are involved; in this case the node of the molecular wave function comes between the two atoms and interaction is between orbitals of opposite sign. Anti-bonding orbitals are indicated by a star, for example $σ^*$.

1.6.4 C—H and C—C Bonding

When these considerations are applied to the carbon atom, and to C—H and C—C bonding, the following picture arises. In the excited state of a carbon atom (see p. 21) there is one electron in the spherical 2s orbital and one in each of the $2p_x$, $2p_y$ and $2p_z$ orbitals. A methane molecule derived from the overlap of these orbitals with the spherically symmetrical 1s orbitals of four hydrogen atoms would have three C—H bonds stronger than the fourth, since the three 2p orbitals have more directional character than the spherically symmetric 2s orbital. In practice all the four C—H bonds are equivalent. This was explained by *Pauling* by introducing the mathematical concept of *hybridisation*. He showed that four fully equivalent sp^3 hybrid orbitals could be constructed out of the spherical 2s orbital and the three 2p orbitals (p_x, p_y, p_z, all at right angles to each other) of carbon.

A single sp^3 hybridised orbital is shown schematically in Fig. 7. It has more directional character, like a p orbital, and can therefore overlap effectively with other orbitals.

The four sp^3 hybrid orbitals point in the directions of the corners of a tetrahedron, as illustrated in Figs. 8 and 9. This arrangement also results in the formation of bonds, e.g. by the overlap of the sp^3 orbitals of carbon with the spherically symmetrical $1s$ orbitals of four hydrogen atoms in a molecule of methane. The hydrogen atoms lie at the corners of a tetrahedron; at its centre is the carbon atom, and the angles between the bonds are each 109° 28'.

It must be emphasised strongly that hybridisation is only used as a means of describing known molecular properties such as the bond lengths and bond angles. The reasons for the tetrahedral arrangement of the bonds in methane has been sought by using so-called *electron correlation*;[1] because of the repulsion which takes place between charges having the same sign, they have the tendency to arrange their co-ordinates so that, on average, they are as far apart from each other as is possible. Hence the four electron pairs take up a tetrahedral arrangement.

This explanation is not, however, applicable to all organic compounds. An increasing number of carbon atoms bonded to four other atoms are known in which the bonding is not tetrahedrally arranged.[2] There are limits to the applicability of simple ideas of hybridisation to both organometallic and other organic compounds.[3]

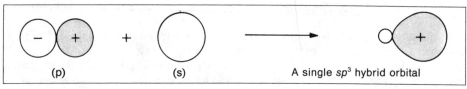

(p)	(s)	A single sp^3 hybrid orbital

Fig. 7.

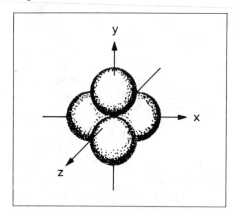

Fig. 8. sp^3 hybrid orbitals of carbon.

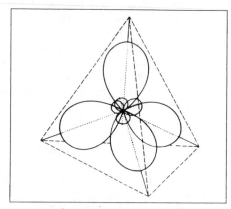

Fig. 9. sp^3 hybrid orbitals of carbon.

[1] See *R. J. Gillespie*, The Valence-Shell Electron Pair Repulsion (VSEPR) Theory of Directed Valency, J. Chem. Educ. **40**, 295 (1963); *W. Kutzelnigg* in *H. F. Schaefer*, Methods of Electronic Structure Theory (Plenum Press, New York, 1977).

[2] See Propellane and Fenestrane, p. 419 *ff*.

[3] See *P. von R. Schleyer* et al., Stabilization of Planar Tetracoordinate Carbon, J. Am. Chem. Soc. **98**, 5419 (1976).

The structure of the higher alkanes, such as ethane, H_3C—CH_3, are similar. In the case of ethane, only three of the sp^3 hybridised orbitals of each carbon atom are involved in overlap with $1s$ orbitals of the six hydrogen atoms; the remaining sp^3 hybridised orbitals interact to form a C—C bond (Fig. 10). There are seven σ-bonds in the molecule as a whole.

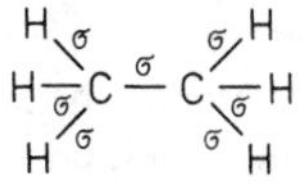

Ethane

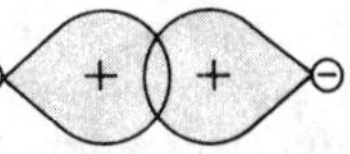

Fig. 10. C—C bonding.

The concept of hybridisation can also be used to explain multiple bonding between carbon atoms. This involves the assumption that sp^2 and sp hybridisation is also possible.

As an alternative to the use of a colon as a symbol to represent a pair of electrons (*Lewis*), *R. Robinson* suggested the use of a dash, as used in structural formulae (*Couper*) for this purpose. In the present text, lone pairs of electrons are represented by a dash or dashes adjacent to the symbol for the atom (*Eistert, P. Baumgarten*). A dot is used as the symbol for an unpaired electron.

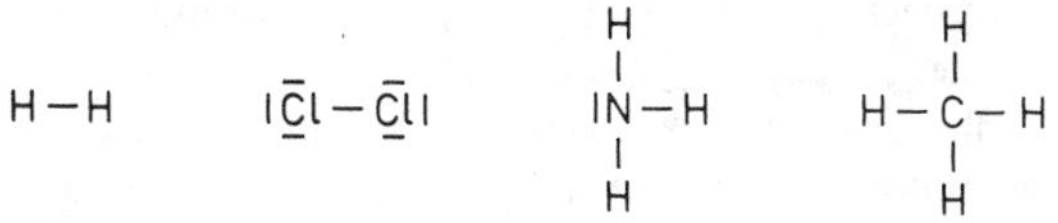

$$H-H \qquad I\bar{C}l-\bar{C}lI \qquad \begin{array}{c} H \\ | \\ IN-H \\ | \\ H \end{array} \qquad \begin{array}{c} H \\ | \\ H-C-H \\ | \\ H \end{array}$$

The number of bonds (or bonding electron pairs) which an atom makes with other atoms is known as its *co-ordination number* or *covalence*. In simple binary compounds this corresponds to the stoichiometric valence, i.e. the hydrogen atoms in a molecule of hydrogen and the chlorine atoms in a molecule of chlorine are monovalent, the nitrogen atom in ammonia is tervalent, and the carbon atom in methane is tetravalent. As will be seen in the following section, atoms which have lone pairs of electrons, such as nitrogen, oxygen and sulphur, can form additional bonds, *i.e.* they can increase their co-ordination number.

It is worth noting that, because of their strong adherence to the octet principle, the atoms of the first periods of the Periodic Table show a maximum co-ordination number of four.

1.6.5 Onium Complexes

Onium complexes are of particular importance as examples of species which have increased valence or co-ordination numbers. For instance, the ammonia molecule, which has three covalent bonds, involving three electrons from the nitrogen atom, has in addition a lone, unbonded pair of electrons. Although the nitrogen atom has an octet of electrons, it is only tervalent. However its lone pair of electrons is capable of forming a bond. In consequence of this, ammonia behaves as an *electron donor* and has *nucleophilic* properties. When ammonia is treated with hydrogen chloride, the proton of the latter acts as an *electron acceptor* and ammonium chloride is formed; this is dissociated into a positively charged ammonium ion and a negatively charged chloride ion.

$$H-\overset{\overset{\displaystyle H}{|}}{\underset{\underset{\displaystyle H}{|}}{N}}|+H^{\oplus}Cl^{\ominus} \longrightarrow \left[H-\overset{\overset{\displaystyle H}{|}}{\underset{\underset{\displaystyle H}{|}}{\overset{\oplus}{N}}}-H \right] Cl^{\ominus}$$

Thus a covalent bond may be formed, not only by the coupling of two single electrons, each provided by one atom, but also as a result of one atom donating two electrons to another. In the ammonium ion all the hydrogen atoms are identically bound to the nitrogen atom. This kind of simple positively charged complex ion, in which the central atom (in this case N) can, because of the number of valence electrons associated with it, form an extra bond, is known as an *onium complex*. Further examples are the sulphonium and oxonium complexes, which involve three-coordinate sulphur or oxygen.

1.6.6 Polar Covalent Bonds

The bonding electrons which link two atoms are only shared equally between these atoms if they both have the same *electron affinity*, for example when they are identical, as in hydrogen or chlorine molecules. The factor which determines the reactions of most organic compounds is the presence of bonds linking carbon to other atoms, such as nitrogen, oxygen or sulphur, which link atoms having different electron affinity. In such bonds the electrons are not symmetrically disposed with respect to the two atoms, but are associated more with the atom which has the higher electron affinity. The tendency for atoms to draw bonding electrons towards themselves is described as their *electronegativity*[1]. It is dependent upon the magnitude of the positive charge on the nuclei of the atoms concerned; this, despite shielding due to electrons in the inner shells, has an effect on the valence electrons. The positive charge on the nuclei of the elements in the upper periods of the Periodic Table increases from the left to the right of the Table and, despite the fact that the number of electrons in the inner shells between the nucleus and the valence electrons increases, the electronegativity also increases from left to right in the Table. This effect is less as one goes down the Table, because the electrostatic charge on the nucleus has less influence on the valence electrons, due to the increased shielding by the inner electrons and the increased atomic radius.

The more electronegative atom attracts the bonding electrons nearer to itself and thus acquires a negative charge, whilst the partnering atom becomes positively charged. The resultant bonding is called *polar covalent bonding*. This results in a *polarity* within a molecule, that is, an electrical dipole moment, whose magnitude can be measured.

The dipole moment μ is the product of the charge e and the distance apart, a, of the centres of charge:

$$\mu = e \cdot a$$

The SI unit for dipole moments is the coulomb.metre (C m). It is related to the Debye unit (D), which was used previously, by the following equation:

$$\mu = 1(D) = 3.336 \times 10^{-30} \text{ C m}$$

If the centres of positive and of negative charge of a molecule are coincident, and no external field is present, then $a = 0$, and the dipole moment $= 0$.

---

[1] See *G. Simons* et al., Nonempirical Electronegativity Scale. J. Am. Chem. Soc. *98*, 7869 (1976).

A way of symbolising a polar covalent bond is by representing the charges by the symbols $\delta+$ and $\delta-$, which signify the presence of a significant but unquantified charge on the atoms concerned. The following series represents the covalent bonding in the first period of the Periodic Table.

$$\overset{\delta+\ \ \delta-}{Li-C} \quad \overset{\delta+\ \ \delta-}{Be-C} \quad \overset{\delta+\ \ \delta-}{B-C} \quad C-C \quad \overset{\delta+\ \ \delta-}{C-N} \quad \overset{\delta+\ \ \delta-}{C-O} \quad \overset{\delta+\ \ \delta-}{C-F}$$

There is an equal sharing of electrons in the carbon—carbon bond in the middle of the series. It is therefore *non-polar* and, in consequence, in the appropriate reaction conditions can be broken homolytically, especially in non-polar solvents or in the gas phase, *e.g.*

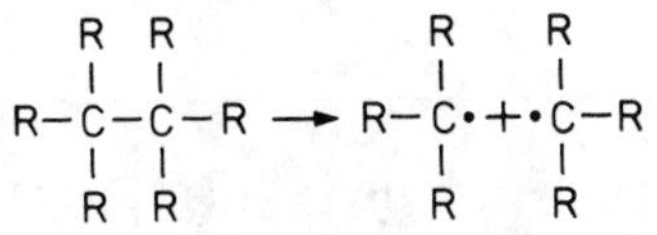

$$\begin{array}{c} R\ R \\ |\ \ | \\ R-C-C-R \\ |\ \ | \\ R\ R \end{array} \longrightarrow \begin{array}{c} R\ \ \ \ R \\ |\ \ \ \ | \\ R-C\bullet+\bullet C-R \\ |\ \ \ \ | \\ R\ \ \ \ R \end{array} \quad (R = H \text{ or an organic residue})$$

The two fragments which are formed each have an unpaired electron and are examples of carbon *free radicals*.

All the other examples in the above series are *polar*, and the polarity increases as one goes away from the middle. This polarity can be increased by external factors, such as the presence of polar solvents or the reaction partners. The bond thereby undergoes further *polarisation*. In the extreme case, a complete dissociation into ions can ensue; this is a result of *heterolytic cleavage*. Depending upon the direction of the polarisation, two possible routes may be followed:

1.6.6.1 If the carbon atom is bound to a more electronegative element, the latter can separate together with the bonding electrons, and the carbon atom is left as a three-coordinate *carbenium* ion[1] with six valence electrons, *e.g.*

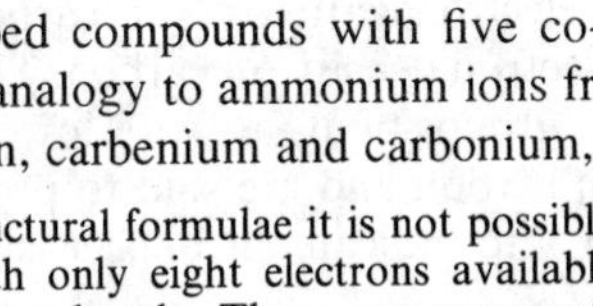

$$\begin{array}{c} R \\ |\ \overset{\delta+\ \delta-}{} \\ R-C-\underline{\overline{X}}| \\ | \\ R \end{array} \longrightarrow \begin{array}{c} R \\ |\oplus \\ R-C + |\underline{\overline{X}}|^{\ominus} \\ | \\ R \end{array} \qquad \left[\begin{array}{c} H \\ | \\ H-C\cdots \overset{H}{\underset{H}{<}} \\ | \\ H \end{array}\right]^{\oplus}$$

(R = H or an organic residue, X = halogen)

Carbenium ion   Carbonium ion

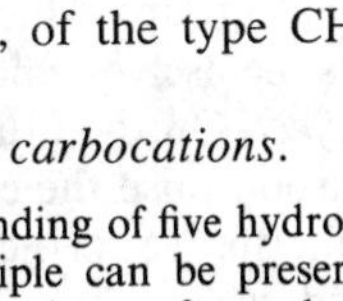

Olah[2] has described compounds with five co-ordinate carbons, of the type $CH_5^{\oplus}$, as *carbonium ions*, by analogy to ammonium ions from ammonia.

Both classes of ion, carbenium and carbonium, have been called *carbocations*.

Using standard structural formulae it is not possible to denote the bonding of five hydrogens to the carbon atom, with only eight electrons available. The octet principle can be preserved by formulating three-centre bonds. These are represented by dotted lines drawn from the atoms concerned and meeting at a point. The point where the lines meet does not represent any extra atoms.[3]

---

[1] See *G. A. Olah*, Carbocations and Electrophilic Reactions, Angew. Chem. Int. Ed. Engl. *12*, 173 (1973).

[2] See *G. A. Olah*, Chem. Brit., 281 (1972).

[3] For other means of formulation see *P. Vogel*, Carbocation Chemistry, pp. 62 *ff.* (Elsevier, Amsterdam 1985).

1.6.6.2 Conversely, if the bond to be split is between a carbon atom and an atom of lower electronegativity, then the electrons remain with the carbon atom and a *carbanion*[1] is formed, *e.g.*

$$
\begin{array}{ccc}
\text{R} & & \text{R} \\
\overset{|\delta^-\ \ \delta^+}{\text{R}-\text{C}-\text{Na}} & \longrightarrow & \overset{|}{\text{R}-\overset{\ominus}{\text{C}}}+\text{Na}^{\oplus} \quad (\text{R}=\text{C}_6\text{H}_5) \\
| & & | \\
\text{R} & & \text{R} \\
& \text{Carbanion} &
\end{array}
$$

Polar covalent bonds are thus intermediate between non-polar covalent bonding and ionic bonding, and are of great significance in the chemical properties of organic compounds.

1.7 Functional Groups. The Inductive Effect

Since the chemical reactions of organic compounds take place preferentially at polar bonds, the reactive sites in the molecules are usually those bearing *functional groups*. Some of the commonest of these groups are as follows:

$$-\text{OH} \qquad -\text{NH}_2 \qquad -\text{NO}_2 \qquad -\text{SO}_3\text{H}$$

Hydroxy group   Amino group   Nitro group   Sulphonyl group

$$\overset{\diagdown}{\underset{\diagup}{}}\text{CO} \qquad -\text{COOH} \qquad -\text{CN}$$

Carbonyl group   Carboxyl group   Nitrile group (carbonitrile)

Halogen atoms are also classified as functional groups.

A functional group affects the carbon atom to which it is attached; this effect is usually described as an *inductive effect* (I-effect) and will now be discussed.

I-Effect. This effect results from the different electronegativities of two atoms linked by a polar covalent bond. The partial charges associated with it set up an electrostatic field in the molecule. There is therefore some change in the electron density on neighbouring atoms and also, although to a lesser degree, on atoms which are further away. Atoms or functional groups which have a greater electronegativity than carbon attract electrons towards themselves, and thus lower the electron density on carbon. They are said to exert a −*I-effect*. On the other hand, atoms or groups of atoms with lower electronegativities than carbon raise the electron density on carbon and are said to produce a +*I-effect*.

An alternative to the use of $\delta+$ and $\delta-$ to indicate an I-effect is to add an arrow to the bond, viz.

$$\text{H}_3\text{C}\longrightarrow\text{OH} \qquad\qquad \text{H}_3\text{C}\longleftarrow\text{Li}$$

−I-Effect   +I-Effect

The I-effect provides an explanation for the occurrence of many organic reactions, which will be dealt with later, when considering the different classes of compounds involved. At

---

[1] See *R. B. Bates*, Carbanion Chemistry (Springer-Verlag, Berlin 1983).

this point only one consequence of the action of an I-effect will be mentioned. This concerns the influence which any groups having either a large or small *electron-withdrawing* group have on the adjacent carbon atom (*−I-effect*). This effect in turn affects the reactivity of hydrogen atoms attached to this carbon atom. The carbon atom tries to compensate for the lowering of its electron density by attracting the electrons of the C—H bond nearer to itself. This results in a weakening of the C—H bond and increased reactivity of the hydrogen atom. This can be demonstrated as follows, using the nitro group as an example of an electron-withdrawing group:

$$R-C \longrightarrow NO_2$$

The commonest examples of atoms or groups which are *electron-donating* and cause a *+I-effect* are the metal atoms of organometallic compounds. In certain circumstances, for example in carbenium ions, in carboxylic acids and in aryl derivatives, *alkyl groups* also cause +I-effects.

The inductive effect is not the only effect which influences the reactivity of organic molecules; other effects will be discussed in later chapters.

1.8 Physical Methods in the Determination of Molecular Structure

Spectroscopic methods are invaluable in the determination of the structure of organic molecules. Various molecular processes, such as electronic excitation, or vibrational and rotational excitation, which result from interactions between the molecules and electromagnetic radiation, are investigated and the results recorded. Because of their structures, different molecules absorb electromagnetic radiation at quite specific wavelengths (λ). The spectra obtained provide evidence about the presence of certain specific atoms or groups of atoms, about the nature of the bonding between atoms, and about the intermolecular interactions between molecules. The schematic diagram in Table 4 shows the ranges of the electromagnetic spectra which are of especial interest for the study of the molecular structure and properties of organic compounds.

The choice of the appropriate spectroscopic techniques as tools for the organic chemist, and the detailed interpretation of the spectra obtained, require more specialised discussion than is appropriate for this text. A detailed account of the theoretical basis for such work is not given here. This is justified since a number of excellent monographs on spectroscopy are available.[1] However, some mention is made of the more important methods, and interpretations of a few characteristic spectra are given. In the cases of UV, IR and NMR spectra, the most important features are shown in tabular form.

---

[1] e.g. *D. H. Williams* and *I. Fleming*, Spectroscopic Methods in Organic Chemistry, Revised 4th edn (McGraw-Hill (UK), London, 1989); *W. Kemp*, Organic Spectroscopy, 2nd edn (Macmillan, London 1987).

Table. 4. Regions of electromagnetic spectra which are of interest for the investigation of molecular properties

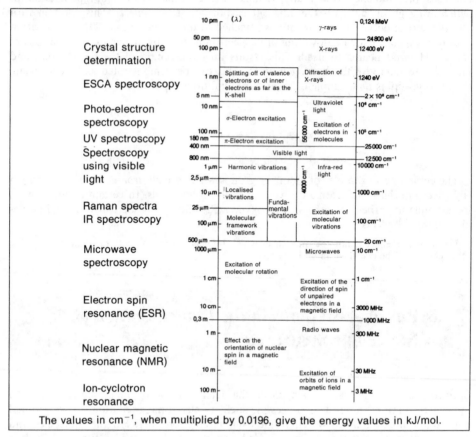

The values in cm^{-1}, when multiplied by 0.0196, give the energy values in kJ/mol.

Another technique which must be added to those listed in Table 4 is mass spectrometry. Mass spectrometry was later in being applied to organic chemistry than many of these techniques, but nowadays it is commonplace for the investigation of the structure of an unknown compound to commence with the recording of its mass spectrum. This is because it provides in one go the relative molecular mass, and also further information derived from the fragments which are formed in the breakdown of the molecule. The different spectroscopic methods are now considered in turn.

1.8.1 Mass Spectrometry[1]

In this technique, organic compounds are volatilised in a high vacuum and bombarded with electrons. The molecules are thereby broken down into fragments, which are for the

[1] See *D. H. Williams* and *I. Howe*, Principles of Organic Mass Spectrometry (McGraw–Hill (UK), London 1972); *M. J. McLafferty*, Interpretation of Mass Spectra, 2nd edn (Benjamin, Reading, MA 1973).

most part *positively charged ions*, together with some uncharged radicals and small neutral molecules. The different ions which are formed are then separated out, by means of an applied magnetic field, according to their mass and charge (m/e), and pass into a recording system. The latter provides a spectrum which resembles an optical spectrum.

The energy of the beam of electrons which is required to remove an electron from an organic molecule is known as the ionisation potential, and is of the order of 10 eV. π-Electrons and lone pairs of electrons are more susceptible to removal than are σ-electrons. When an electron is removed from a molecule in this way, a molecular *radical ion* is formed.

$$M + e^{\ominus} \rightarrow M^{\oplus} + 2\,e^{\ominus}$$

Molecular radical ion

The mass number $M^{\oplus}$ represents the *relative molecular mass* of the compound. Since the energy of the applied electrons amounts for the most part to 70 eV, the excess energy in the radical ion $M^{\oplus}$ causes some breaking of bonds, with the formation of smaller, charged, *fragment ions*. This fragmentation is comparable to a chemical decomposition and follows established patterns. A knowledge of these patterns makes it possible to gain information, from fragments which are formed, about the presence of functional groups or structural units in the original molecule. It sometimes happens that no molecular ion signal appears in the spectrum, because the molecule fragments whenever it is ionised. Each fragment ion has a specific value for m/e, the ratio of its mass to its charge. Most ions carry a charge of 1, so that the m/e value is usually the relative mass of the ion. The intensities of the signals reflect the relative abundance in which the ions are present. Breakdown of the molecule takes place for the most part in such a way that

(1) fragments are formed in which the charge is stabilised by mesomerism or by inductive effects; or
(2) fragmentation involves the loss of stable neutral fragments (H_2O, H_2S, NH_3, C_2H_2, HCN, CO, CO_2, N_2, SO_2, etc.).

From the mass spectra of alkanes it is evident that, in accord with the relative stabilities of *tertiary*, *secondary* and *primary* ions, bond-breaking occurs at the carbon atom, which is associated with the most branching in the molecule (see Figs. 11 and 12).

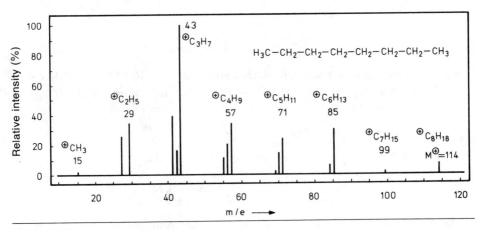

Fig. 11. Mass spectrum of n-octane.

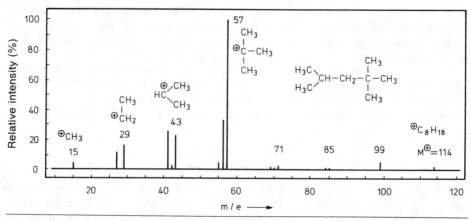

Fig. 12. Mass spectrum of iso-octane.

In an alkene, the presence of the double bond makes bond-breaking easier at the *allyl position*, since a stable allyl cation is formed thereby.

$$H_2C = CH-CH_2-R + e^\ominus \rightarrow 2\,e^\ominus + H_2C \cdots \overset{\oplus}{CH} \cdots CH_2 + R\cdot$$

Hetero-atoms which possess a lone pair of electrons promote the breaking of the nearest carbon–carbon bond in an aliphatic compound, since the resultant cation is stabilised by the *nucleophilic* interaction from the hetero-atom.

$$R-CH_2-CH_2-X + e^\ominus \rightarrow 2\,e^\ominus + R-\dot{C}H_2 + \left[\overset{\oplus}{C}H_2-X \longleftrightarrow CH_2 = \overset{\oplus}{X}\right] X = \underline{O}H, \underline{N}H_2, \underline{S}H$$

Breakage of bonds can be followed by *rearrangement* of the fragment so as to provide a *more stable* ion; sometimes this is accompanied by the splitting off of a neutral molecule. For example, in the mass spectrometer benzyl compounds are cleaved to provide the stable tropylium ion:

$$\text{benzyl}-CH_2-R + e^\ominus \longrightarrow 2e^\ominus + R\cdot + \text{benzyl}-\overset{\oplus}{CH}_2 \longrightarrow \text{tropylium}$$

Benzyl cation   Tropylium ion

If a compound contains a multiply bonded group such as $>C=O$, $>C=NR$, etc. in a *γ*-position with respect to a CH group, it is usual for the so-called *McLafferty rearrangement* to take place.

= Homolysis

X = CH$_2$, O, S, NR
Y = OH, SH, NH$_2$ . . .

This proceeds *via* a six-membered ring transition state. A hydrogen atom is transferred to the charged unsaturated centre, and an alkene is eliminated.

Mass spectrometry is used in a number of ways in organic chemistry. Thus, *high resolution* mass spectrometry provides the molecular formula of an organic compound as well as information about its structure. Another application is to show the identity of two compounds. In addition, mass spectrometry can be used for both the qualitative and quantitative analysis of mixtures, as well as for the detection of trace amounts, and for isotopic analysis. Finally, with its help, isotope effects[1] can be determined and hence information may be obtained about reaction mechanisms.

The most intense peak in a mass spectrum is called the *base peak*. Its intensity is taken to be 100% and all other peaks are related to it. Only in the case of stable organic compounds is the base peak also the molecular ion peak.

Since chemical elements exist as mixtures of various isotopes, of which the lightest isotope is predominant, then as well as the main molecular ion peak, one or more *isotope peaks* with higher *m/e* are seen. The relative intensities of such a group of peaks indicate the natural abundance of the different isotopes in isotopomeric molecules. Determination of the isotopic make-up from a mass spectrum provides another way of ascertaining the molecular formula of an organic compound. On account of the relatively large proportions of the heavier isotopes of sulphur, chlorine and bromine, compounds containing these elements are relatively easily identified.

When the ionisation of sensitive compounds by electron bombardment results in too much fragmentation, more gentle methods for bringing about ionisation are used. The molecule to be investigated can be ionised by *chemical ionisation* induced by proton transfer from a reactant such as $[CH_5]^+$. This is known as CID (*Collision-Induced Decomposition*). The reactant ion is generated from a suitable gas by electron bombardment; thus $[CH_5]^+$ is obtained from methane. In contrast to normal electron bombardment mass spectrometry, the most intensive peak observed in this method is from the $[M + H]^+$ ion. In this method fragmentation is suppressed.[2]

1.8.2 IR Spectroscopy[3]

In the region covered by IR absorption spectroscopy, the transitions between molecular states which are recorded are, above all, vibrational and rotational processes. IR spectra are very important for organic chemists because they provide information about the type of bonding which is involved and its location in the molecule, as well as about the presence of particular functional groups. Infra-red radiation (see Table 4, p. 30) is passed through a sample to be investigated, either in its solid state or in solution, and the absorption which takes place at different frequencies is displayed by the IR spectrometer.

Both single atoms and groups of atoms in a molecule can undergo different vibrations with respect to one another. There are molecular vibrations, to which the whole molecule contributes, and also localised vibrations. Vibrations occurring in the axes of the bonds are described as *stretching* vibrations, while vibrations out of the axes are called *bending* or *deformation* vibrations. Both types are illustrated in the case of a methylene group in Fig. 13. The absorption maximum associated with the stretching frequency of a group is always at a higher wavenumber than the absorption maximum associated with the bending vibration.

[1] See *A. V. Willi*, Isotopeneffekte bei chemischen Reaktionen (Thieme Verlag, Stuttgart 1983).
[2] See *W. J. Richter* and *H. Schwartz*, Chemical Ionisation—A Mass Spectrometric Analytical Procedure of Rapidly Increasing Importance, Angew. Chem. Int. Ed. Engl., 17, 424 (1978).
[3] See *L. J. Bellamy*, The Infrared Spectra of Complex Molecules (Chapman & Hall, London 1975).

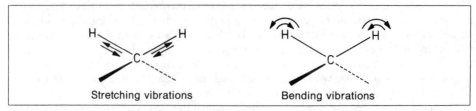

Fig. 13. Modes of vibration of a methylene group.

Optical stimulation of vibration is only possible when, because of the vibration process, there is a change in the charge distribution in the molecule, *i.e.* an electric field results which can interact reciprocally with the radiation. Such a vibrating dipole absorbs electromagnetic radiation only if it has the same frequency as that of the dipole. Molecules of high symmetry thus only show a few bands in their IR spectra, since some of the vibrations do not interact, and are *IR-inactive*. This may be illustrated in the case of acetylene:

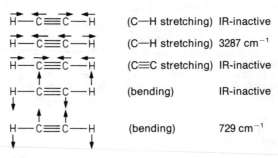

(C—H stretching)	IR-inactive
(C—H stretching)	3287 cm^{-1}
(C≡C stretching)	IR-inactive
(bending)	IR-inactive
(bending)	729 cm^{-1}

The numerical values for absorption bands in IR spectra are generally expressed in *wavenumbers* (vibrations per cm), and their intensities as either strong (s), medium (m), weak (w), variable (v) or shoulder (sh).

The most common fundamental vibrations of organic molecules lie in the range 4000–400 cm^{-1}. Within this region the characteristic absorption bands of simple functional groups occur between 4000 and 1450 cm^{-1}. They appear at sharply defined positions on the scale, which vary only slightly depending on the nature of the remainder of the molecule to which the functional group is attached. However, because of this variation it may be possible not only to detect functional groups, but also to gain some information about their structural environment (see Table 5).

It is also possible to show that two compounds are identical by means of IR spectroscopy. For this purpose the so-called 'fingerprint region' of the spectrum between 1450 and 650 cm^{-1} is used. Signals in this region are characteristic of the molecule as a whole. In some instances, IR spectroscopy can be used as a tool in quantitative analysis. *Fourier transform IR spectroscopy* (FTIR)[1] is especially suitable for this purpose.

The IR spectra of iso-octane, isopropanol and acetophenone (acetylbenzene) are shown as examples in Figs. 14–16 (p. 36).

Raman spectroscopy is closely similar to IR spectroscopy. It also indicates the presence of particular groups of atoms by their absorptions at characteristic frequencies. It has grown in importance

---

[1] See *P. R. Griffiths*, Fourier Transform Infrared Spectroscopy (Wiley, New York 1986).

Table 5. Some important absorption bands in IR spectra (stretching bands: ν)

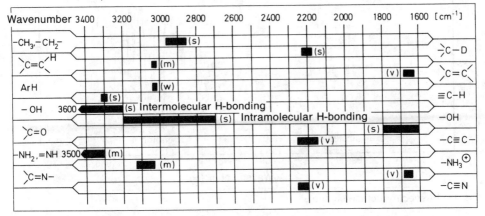

following the introduction of lasers as light sources, and is a valuable alternative to IR spectroscopy, since the IR-inactive vibrations can be detected in *Raman* spectroscopy.[1] On the other hand, it does not detect a number of IR-active vibrations.

1.8.3 Electronic Spectra (UV and Visible Spectra)

In both the visible and UV regions (see Table 4, p. 30) absorption of light by molecules leads to promotion of electrons from occupied bonding or non-bonding orbitals into empty non-bonding or anti-bonding orbitals. Hence they are both commonly described together as electronic spectra. The move from one orbital into another involves transfer between discrete energy levels. This absorption is also accompanied by vibrational and rotational transitions, so that a more widespread change of state is involved. The wavelength of the absorption provides a measure of the difference in energy between the respective energy levels. The shorter the wavelength of the absorbed light, the greater the energy which is absorbed in bringing about excitation of the electron, signifying that the orbitals are further apart from one another in energy. It requires radiation of a higher energy to bring about excitation of a σ-electron than it does to excite a π-electron or an electron from a lone pair of electrons (see Table 4, p. 30).

The groups which absorb radiation, and thus cause electronic excitation, are called *chromophores*. In normal practice, chromophores are only of use if the wavelengths of their absorption maxima are greater than 200 nm. The π–π* transition of an isolated C=C double bond produces an absorption maximum at about 190 nm and thus lies outwith the normal range. In general, only chromophores which are associated with conjugated systems are significant. Because of the conjugation, the energy levels of the highest occupied π-orbitals (HOMO) are raised in energy while the energy levels of the lowest anti-bonding (unoccupied) π*-orbitals (LUMO) are lowered. The more conjugated double bonds there are in a polyene, the lower the difference in energy between the highest bonding and the lowest anti-bonding orbitals. This means that light of longer wavelength is required to excite a π-electron. Table 6 gives the data for some unsaturated

---

[1] See *S. K. Freeman*, Applications of Laser Raman Spectroscopy (Wiley, New York 1974).

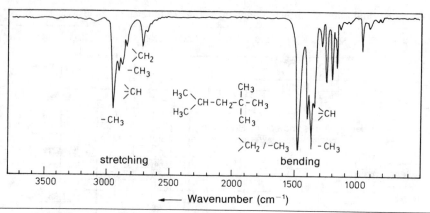

Fig. 14. IR spectrum of iso-octane.

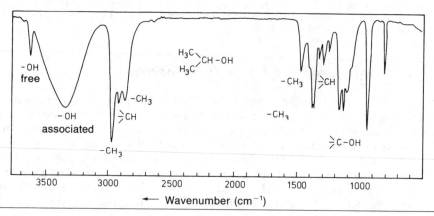

Fig. 15. IR spectrum of isopropanol.

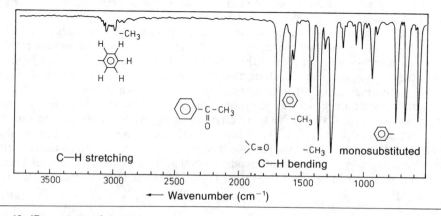

Fig. 16. IR spectrum of acetophenone.

organic compounds; the differences in the absorption maxima, depending upon the number of conjugated bonds, is evident.

Thus it may be seen that electronic spectra are used essentially for the characterisation of unsaturated aliphatic and aromatic compounds.

Table 6

Compound	Absorption maximum [nm]	Transition
Ethylene	190	$\pi \to \pi^*$
Buta-1,3-diene	217	$\pi \to \pi^*$
trans-Deca-2,4,6,8-tetraene	273, 283, 296, 310	$\pi \to \pi^*$
trans-α-Carotene	445, 474	$\pi \to \pi^*$
Benzene	184, 204, 255	$\pi \to \pi^*$
Naphthalene	220, 276, 311	$\pi \to \pi^*$
Anthracene	253, 356	$\pi \to \pi^*$
Pyridine	175, 194, 251	$\pi \to \pi^*$
Formaldehyde	270	$n \to \pi^*$

From Table 6 it may be seen that benzene (Fig. 17) and pyridine, both of which have six delocalised π-electrons, and whose molecular orbitals are very similar in energy, have very similar absorption maxima in their UV spectra.

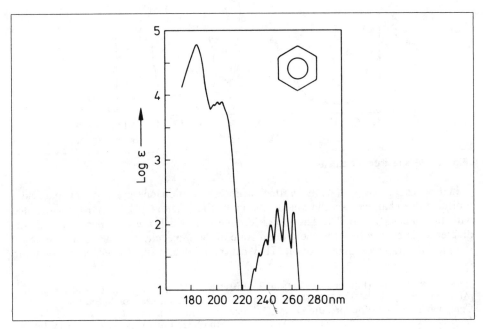

Fig. 17. UV spectrum of benzene.

Less information about structure can be gained from an electronic spectrum than from an IR spectrum. Electronic spectra consist of a small number of broad bands.

On the other hand, electronic spectra are very valuable for accurate quantitative analysis, making use of the *Beer–Lambert* law.

1.8.4 Photoelectron Spectroscopy (PE Spectroscopy)[1]

When radiation of higher energy, and thus of shorter wavelength, than UV light interacts with molecules, it brings about emission of electrons. When they leave the molecules, these electrons possess kinetic energy; the energy depends upon the way in which the electrons were bound in the molecule. The following relationship applies:

$$T_e = h\nu - I$$

This means that the higher the kinetic energy, T_e, of the emitted electron, the smaller is the ionisation energy, I, of that particular electron ($h = Planck's$ constant).

The spectrometer records how many electrons of a particular energy are emitted from the probe (the count rate). This provides a *photoelectron spectrum* (*PE spectrum*). The spectrum of ethylene is shown in Fig. 18.

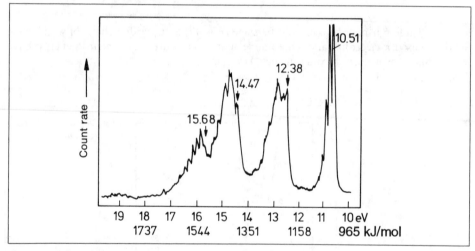

Fig. 18. PE spectrum of ethylene.

In Fig. 18 the count rate is plotted against ionisation energy. For any particular molecule this characterises the kinetic energy of the different photoelectrons associated with the values $T_e = h\nu - I$. The ionisation energies which are obtained represent the differences in energy between the ground state of the molecule, and the various states of the radical cations obtained from it. They can be related, using *Koopmans'* theorem, to

[1] See *C. R. Brundle* (Ed.) et al., Electron Spectroscopy : Theory, Techniques and Application (Academic Press, New York 1977); *H. Bock* and *B. G. Ramsey*, Photoelectron Spectra of Non-Metal Compounds and their Interpretation by M.O. Models, Angew. Chem. Int. Ed. Engl., *12*, 734 (1973); *H. Bock* and *B. Solouki*, Photoelectron Spectra and Molecular Properties, Angew. Chem. Int. Ed. Engl., *20*, 427 (1981).

the orbital energies, which can be obtained from SCF quantum chemical calculations.

The spectrum in Fig. 18 has four bands, which arise from electrons that are associated with the four highest occupied orbitals. Radiation from a helium discharge tube, which has an energy of 21.22 eV (= 2050 kJ/mol) is suitable for bringing about the emission of electrons. It is not possible to remove the more strongly bound electrons from the lower-lying energy states of ethylene. PE spectroscopy carried out using this type of excitation is known as UPS (*Ultraviolet Photoelectron Spectroscopy*) (*Turner*).

The π-electrons in ethylene occupy the orbitals with the lowest ionisation energy. The band at 10.51 eV (1014 kJ/mol) is attributed to them. The other bands are associated with σ-electrons of higher ionisation energy. To a first approximation, the method provides a direct measure of the energies of the occupied molecular orbitals, whilst electronic spectroscopy provides the differences in energy between orbitals. Substitution, conjugation and steric effects can all have a considerable influence on ionisation energies.

When it is photoionised a molecule is turned into a radical cation.

$$M \overset{h\nu}{\to} M^{\oplus} + e$$

The fine structure in the bands in Fig. 18 can be attributed to, among other things, the fact that in the ionisation, the excited vibrational states of the radical cations are generated. Information about the bonding in the radical cation can be derived from this vibrational fine structure.

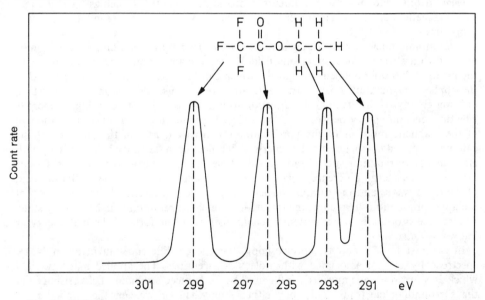

Fig. 19. ESCA spectrum of the 1s electrons of the carbon atoms in ethyl trifluoroacetate.

By using X-rays as the radiation, sufficient energy (more than 1000 eV) is available to extrude electrons from all of the orbitals. This method, introduced by *Siegbahn*, is called ESCA (*Electron Spectroscopy for Chemical Analysis*)[1] or XPS (*X-ray Photoelectron*

---

[1] See C. *Nordling*, ESCA : Electron Spectroscopy for Chemical Analysis, Angew. Chem. Int. Ed. Engl., *11*, 83 (1972).

Spectroscopy). The Mg K^α line (1245 eV), for example, may be used as a source of excitation. In this way, even $1s$ electrons in the innermost shells of heavy elements can be detected. These electrons are for the most part neglected by chemists since they play practically no part in chemical bonding. Their ionisation energy is dependent on the environment in which the particular atom exists, as is shown by the ESCA spectrum of ethyl trifluoroacetate (Fig. 19). Whereas the valence electrons are affected by other electron-attracting groups, the $1s$ electrons are so strongly bound as to be almost unaffected. However the effects of inductive groups on carbon atoms, discussed previously in this chapter, is evident in Fig. 19.

1.8.5 Nuclear Magnetic Resonance Spectroscopy (NMR Spectroscopy)[1]

Nuclear magnetic resonance (NMR) was discovered in 1946 by *Purcell* and *Bloch*. In this technique, the interaction which is induced under certain conditions between atoms in a powerful magnetic field and electromagnetic radiation in the form of radio waves is recorded. *Proton nuclear magnetic resonance spectroscopy* (^{1}H-NMR spectroscopy) and ^{13}C-NMR spectroscopy are widely used in organic chemistry. Atoms which have *odd* atomic numbers, such as $^1_1$H, $^{19}_1$F, $^{31}_{15}$P, and also the isotopes $^{13}_6$C, $^{15}_7$N have a *magnetic moment*. Atoms made up of *even* numbers of protons and neutrons, such as $^{12}_6$C, $^{16}_8$O, $^{32}_{16}$S, have magnetic moments equal to zero. Hence they have no perturbing influences on the properties of $^1_1$H and $^{13}_6$C.

If an atomic nucleus with a magnetic field other than zero, for example a proton, is exposed to a strong applied magnetic field, then the magnetic moment associated with the proton takes up a defined orientation, which can be either *parallel* or *anti-parallel* to the applied magnetic field, i.e. in the same direction as this field (the more stable state, of lower energy), or in the opposite direction to it (the less stable state, of higher energy). The difference in energy between the two different orientations is a function of the size of the magnetic moment of the particular atom involved and of the strength of the externally applied magnetic field. The energy differences lie in the range of long-wavelength electromagnetic radiation such as radio waves. If the energy of the radio waves corresponds exactly to this energy difference, *magnetic resonance* occurs between the oscillating field of the radio waves and the rotating nucleus and, as a consequence of the absorption of the appropriate radio frequency, the magnetic moment 'flips' into its higher energy state. The reverse process, in which potential energy is lost in the form of thermal energy, is known as *relaxation*.

In contrast to UV and IR spectroscopy, the energy differences involved in NMR spectroscopy are not associated with molecules as entities. The applied magnetic field which acts on an atomic nucleus is influenced by an induced secondary field arising from the surrounding electrons. This may either strengthen or weaken the applied field; hence protons in different electronic environments provide separate signals at different resonances when the applied field ($H_0 = 1.4$ or 2.35 tesla) changes whilst the radio frequency (say 60 or 100 MHz) is kept constant.

[1] *R. J. Abraham, J. Fisher* and *P. Loftus*, Introduction to NMR Spectroscopy (Wiley, Chichester 1988); *R. R. Ernst* et al., Principles of Nuclear Magnetic Resonance in One and Two Dimensions (Oxford University Press, Oxford 1987); *A. E. Derome*, Modern NMR Techniques for Chemistry Research 2nd ed. (Pergamon, Oxford 1990); *J. K. M. Sanders* and *B. K. Hunter*, Modern NMR Spectroscopy: A Guide for Chemists (Oxford University Press, Oxford 1987).

1.8.5.1 ^{1}H-NMR Spectroscopy

NMR spectra of organic compounds are usually determined using solutions, for example in CCl_4 or $CDCl_3$. In order to give numerical values to the signals in spectra, a reference signal is used. For this purpose tetramethylsilane (TMS), $(CH_3)_4Si$, is added to the solution to provide an internal standard. The twelve chemically equivalent protons of this molecule provide a single sharp signal in an NMR spectrum, whose position is used, arbitrarily, as the zero point of the scale. TMS is very well suited for this purpose because it does not participate in intermolecular interactions with other molecules. It also happens that, for the most part, protons in organic molecules provide signals at lower field than that associated with TMS. In order to compare spectra recorded at different frequencies, say 60 MHz or 100 MHz, with one another, the scale must also take into account the radio frequency. To this end, the parameter δ (or, in older work, $\tau = 10 - \delta$), which indicates the *chemical shift*, is used. It is defined as follows:

$$\delta = \frac{\text{Resonance frequency of sample relative to TMS, in Hz}}{\text{Operating frequency of instrument, in MHz}}$$

δ is expressed in parts per million (ppm).

The positions of the proton resonance signals in a ^{1}H-NMR spectrum are influenced by a number of factors. One such factor involves the σ-electrons of the C—H bond. In the magnetic field the electrons associated with the proton generate a small secondary magnetic field which opposes the applied field. This is described as *shielding* of the hydrogen nucleus. In alkenes, alkynes and aryl derivatives the π-electrons may either strengthen or weaken the magnetic field around a proton depending upon their steric relationship to the proton (*anisotropic effect*). The type of influence, shielding or deshielding, of such secondary fields on a proton, is dependent upon the relative siting of the proton, as illustrated in Figs. 20 and 21.

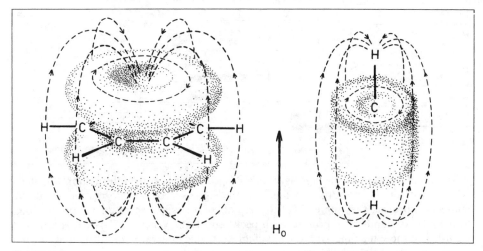

Fig. 20. Benzene.   Fig. 21. Acetylene

Inductive effects, intermolecular interactions such as *van der Waals* forces or hydrogen bonding, and various other influences change the electron density and hence the strength of the induced secondary field, at different sites in a molecule, and this is shown by the shifts of the resonance signals in an NMR spectrum. A distinction is made between *paramagnetic* shifts (to lower field = higher δ value) and *diamagnetic* shifts (to higher field = lower δ value) of the resonance signals. Paramagnetic shifts result from the deshielding of protons by secondary fields due to electrons; diamagnetic shifts result from increased shielding. Table 7 shows some characteristic values for the chemical shifts associated with protons in the commoner types of organic compounds. In contrast to the shifts associated with some other nuclei, they are spread over only a small range of about 15 ppm. This is because protons have no p-electrons, which can exert a paramagnetic effect, causing large chemical shifts to lower field. (Compare Table 8).

In the case of aliphatic hydrocarbons the increasing paramagnetic shift of protons on going from primary to secondary to tertiary carbon atoms is notable. It is also evident that protons attached to alkene, alkyne and aryl groups appear at lower field than do protons which are part of alkyl groups. Aryl protons are shifted even further to low field than are alkene or alkyne protons. This is due to the effect of the strong applied magnetic field on the π-electrons, which induces a ring current that in turn produces a secondary magnetic field. This reinforces the applied field in the vicinity of the protons, causing deshielding of them. This effect is illustrated in Fig. 20 for the case of benzene, which has a sextet of π-electrons and is stabilised by mesomerism. (See also the structure of benzene, pp. 467 *ff.*) Fig. 21 similarly illustrates the shielding of the protons of acetylene as a consequence of the secondary field set up by the π-electrons.

Table 7. Some typical δ-values in ^{1}H-NMR spectroscopy

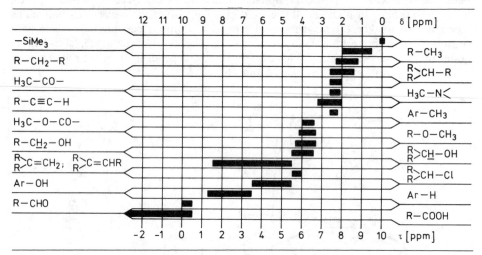

While the position of the signals in NMR spectra gives some information about the environment of the protons, the areas of the signals provide information about the numbers of protons they represent. The areas of the peaks in a spectrum are proportional to the relative numbers of protons each peak represents. When a ^{1}H-NMR spectrum is recorded, at the same time, with the aid of an electronic integrator, another stepped curve is printed in which the heights of the steps are proportional to the area of each NMR signal. From the information thus derived, together with the molecular formula, much can be learned about the types and the numbers of protons in the various structural units which are present in organic molecules. This is demonstrated by the NMR spectrum of 4-t-butyltoluene shown in Fig. 22.

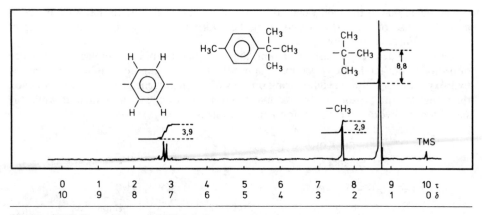

Fig. 22. ¹H-NMR spectrum of 4-t-butyltoluene.

If protons are chemically equivalent they do not show any magnetic interaction with one another, but interaction is observed between the magnetic moments of chemically non-equivalent protons on adjacent carbon atoms or indeed on one and the same carbon atom. This is known as *spin–spin* coupling. (The magnetic moment of each proton has two possible orientations in the magnetic field, and hence this orientation may be either aligned with or opposed to the moment of a proton on a neighbouring group.) The resultant interactions lead to a number of different modes of coupling, which results in a splitting of the NMR signal (a so-called *spin-multiplicity*), which may appear as a doublet, triplet, quartet, etc. The spin-multiplicity of a particular proton depends on the product $(n_1 + 1)(n_2 + 1)\ldots$, where n_1, n_2, etc. are the numbers of chemically equivalent protons attached to the atoms adjacent to the site of the first proton. This is exemplified below using the ¹H-NMR spectra of ethanol (Fig. 23), propan-1-ol, propan-2-ol and t-butanol.

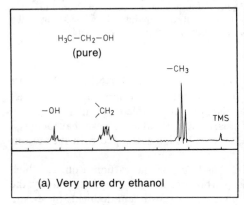

(a) Very pure dry ethanol

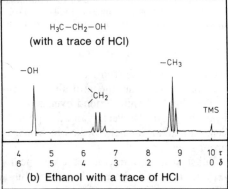

(b) Ethanol with a trace of HCl

Fig. 23. NMR spectra of ethanol.

In ethanol the methylene group has as neighbours both the methyl group and the hydroxy group. As a result the signals for both of the latter groups appear as triplets ($n_1 = 2$). The methylene protons couple on one side with the three chemically equivalent protons of the methyl group ($n_1 = 3$) and on the other side with the single proton of the hydroxy group ($n_2 = 1$). Hence the methylene signal appears as an octet.

The signal due to the methylene group appears at lower field than that for the methyl group because of the inductive effect of the hydroxy group, and the resultant lower shielding (Fig. 23).

If a trace of acid, or of base, is present, this catalyses a rapid intermolecular exchange of protons from the hydroxy group. This results in there being no coupling between the hydroxy protons and the neighbouring protons of the methylene groups. In consequence, the signal for the protons of the methylene group now appears as a quartet while the signal for the hydroxy group proton is not split (Fig. 23b).

$$H_3C-CH_2-OH$$

Ethanol

$$H_3C-CH_2-CH_2-OH$$

Propan-1-ol

$$\begin{array}{c} H_3C \\ {}^{\diagdown}CH-OH \\ H_3C \diagup \end{array}$$

Propan-2-ol

$$\begin{array}{c} H_3C \\ {}^{\diagdown} \\ H_3C-C-OH \\ {}^{\diagup} \\ H_3C \end{array}$$

t-Butanol

The three methyl protons of propan-1-ol couple only with the two methylene protons of the next carbon atom, and hence their signal is a triplet. The methylene group next to the methyl group has as neighbours the methyl group on one side and a *different* methylene group on the other side ($n_1 = 3$, $n_2 = 2$), and its signal is consequently split 12-fold. The methylene group next to the hydroxy group ($n_1 = 2$, $n_2 = 1$) appears as a sextet and the hydroxy group ($n_1 = 2$) as a triplet.

In the case of propan-2-ol there are six equivalent protons in the two methyl groups; their signal appears as a doublet since their only neighbour is a methine (CH) proton. The signal for the latter is split into 14, since it has six equivalent protons of the methyl groups ($n_1 = 6$) and the proton of the hydroxy group ($n_2 = 1$) next to it. The hydroxy signal appears as a doublet. t-Butanol provides a sharp unsplit signal for the nine equivalent protons of its three methyl groups and also a singlet for the hydroxy-group proton.

As in the case of ethanol, coupling with the hydroxy protons of the propanols is not possible in the presence of acids or bases, and, in consequence, splitting of the relevant signals is either of lower multiplicity or is absent.

The multiplicity of the spin coupling is also characterised by a coupling constant J, expressed in hertz (unit abbreviation: Hz). The numerical value of J defines the separation of the splitting and is a measure of the magnitude of the coupling. The coupling which has been considered above is known as ^3J-coupling, since it involves coupling between atoms separated from one another by three bonds. In the case of open-chain compounds it usually falls in the range 2–10 Hz.

1.8.5.2 ^{13}C-NMR Spectroscopy

As a result of the introduction of *Fourier transform spectroscopy* (FT-NMR), ^{13}C-NMR spectra, derived from the naturally occurring 1.11% of this isotope, are readily available.[1]

The chemical shifts extend over a range of 350 ppm, and Table 8 provides a selection from the region between 0 and 220 ppm. Comparison with Table 7 shows that there is a general similarity in the orders in which the chemical shifts associated with the various groups appear.

Substituent groups affect the chemical shifts not only of the carbon atom to which they are attached (the α-carbon atom) but also carbon atoms which are further away; the effect is transmitted through several bonds. In general, a substituent deshields the α- and β-carbon atoms, and shields the γ-position. This γ-effect is evident in cyclic as well as acyclic compounds.

[1] G. C. Levy, R. L. Lichter and G. L. Nelson, Carbon-13 Nuclear Magnetic Resonance for Organic Chemists, 2nd ed. (Wiley, New York 1980); F. W. Wehrli, A. P. Marchand and S. Wehrli, Interpretation of Carbon 13 NMR Spectra (Wiley, Chichester 1988); E. Breitmaier, G. Haas and W. Voelter, Atlas of Carbon-13 NMR Data Vols 1 and 2 (Heyden, London 1979).

The $^{1}J_{(C-H)}$ coupling constants lie in the range 150–250 Hz; the values are associated with the hybridisation of the carbon atoms (see p. 93).

Table 8. Some typical δ-values in ^{13}C-NMR spectroscopy. Internal standard $(CH_3)_4Si$
▯▯▯▯ conjugated

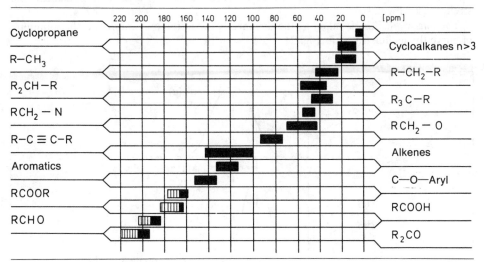

In order to help clarify complicated NMR spectra, use is made of paramagnetic rare earth (lanthanoid) compounds.[1] In the presence of small amounts of these compounds, large chemical shifts are induced for the signals due to groups which can form complexes with them. They are called *lanthanide induced shift reagents*. Choice of a suitable solvent can also bring about increased separation of signals; benzene can often be very effective for this.

The use of two-dimensional NMR is of especial importance in the interpretation of spectra.[2]

1.8.6 Electron Spin Resonance (ESR)[3]

For the study of *free radicals* the most valuable technique is *electron spin resonance* (ESR) (or *electron paramagnetic resonance* (EPR)) spectroscopy. This is based on the same principle as NMR spectroscopy. ESR spectra provide information about the radical character and the possible site of single unpaired electrons in a molecule. The measurement of ESR makes use of the magnetic moment of electrons, and a similar procedure to NMR is employed. According to the *Pauli* principle, in most molecules all the electrons are spin-paired (anti-parallel). Consequently they have no magnetic moment and do not provide ESR spectra. This is not so in the case of free *radicals* (see pp. 612 *ff.*), *radical cations* or *radical anions*, all of which have unpaired electrons. They are therefore suitable

[1] See *R. E. Sievers* (Ed.) NMR Shift Reagents (Academic Press, New York 1973); *T. J. Wenzel*, Nuclear Magnetic Resonance Shift Reagents (CRC Press, Cleveland, Ohio 1987).

[2] See *H. Kessler, M. Gehrke* and *C. Griesinger*, Two-Dimensional NMR Spectroscopy: Background and Overview of the Experiments, Angew. Chem. Int. Ed. Engl., *27*, 490 (1988).

[3] See *J. E. Wertz* et al., Electron Spin Resonance. Elementary Theory and Practical Applications (Chapman & Hall, London 1986).

species for study by ESR spectroscopy. The magnetic moment of an electron is about 200 times greater than that of a proton, so that the resonance frequency in a field of about 0.3 T is about 10 GHz (10×10^9), in the microwave region. For technical reasons, whereas NMR signals are a plot of absorption against field strength, ESR are plotted as first derivative spectra, i.e. they are a plot of the *rate of change of absorption* against field strength (see Fig. 24).

The *line-width*, ΔH, is taken as the distance between the maximum and the minimum of curve (b), as illustrated in Fig. 24.

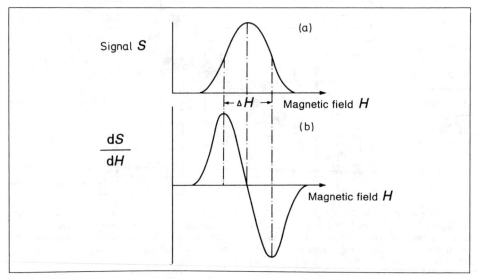

Fig. 24. ESR signals: (a) absorption signal, (b) first derivative.

Akin to the use of the chemical shift (δ) in NMR spectroscopy, the position of the resonance signal in ESR is expressed as the g-value of the unpaired electron. The g-value for a free electron, g_e, is 2.0023. In an organic substrate, magnetic interactions shift g from this value. This shift is small; for example, for the allyl radical, ($CH_2=CH-CH_2\cdot$), the g-factor is 2.0026.

The most important feature of an ESR spectrum is its *hyperfine structure*, which arises from the magnetic interaction, or coupling, of the electron spin with the nuclear spins of ^1H, ^{14}N, ^{13}C, or other nuclei with $I \neq 0$. An example of an ESR spectrum, that of the radical anion of benzene, $C_6H_6^{-}$, is shown in Fig. 25.

The six equivalent protons in this radical give rise to ($n + 1$) = 7 lines, with relative intensities of 1:6:15:20:15:6:1. These intensities may, in general, be derived from *Pascal's* triangle, as illustrated in Fig. 26. As in NMR, the distance between the lines is characterised by the coupling constant a. For the radical anion of benzene, $a = 0.375$ mT (millitesla) = 0.375 G (gauss).

In the ethyl radical $CH_3-CH_2\cdot$ there are two equivalent protons in the CH_2 group and three in the CH_3 group. The total number of lines calculated from the formula ($n_1 + 1$)($n_2 + 1$) is thus (2 + 1)(3 + 1) = 12. This may be seen in the spectrum shown in Fig. 27.

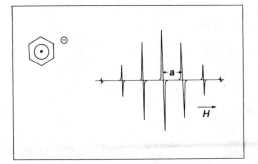

Fig. 25. ESR spectrum of the radical anion of benzene.

														n
						1								0
					1		1							1
				1		2		1						2
			1		3		3		1					3
		1		4		6		4		1				4
	1		5		10		10		5		1			5
1		6		15		20		15		6		1		6
1	7		21		35		35		21		7		1	7

Fig. 26. *Pascal's* triangle of the binomial coefficients for $(1 + x)^n$.

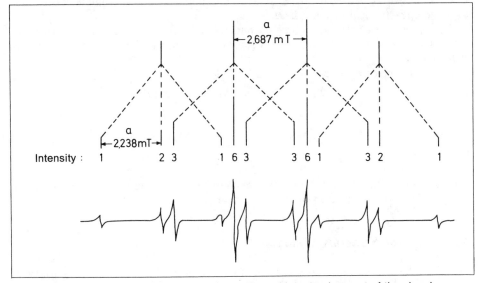

Fig. 27. ESR spectrum of the ethyl radical, together with an assignment of the signals.

The binomial distribution of intensity of the signals is also in agreement with the numbers in Fig. 26. Each of the two overlapping signals show intensities which are in accord with binomial coefficients. The coupling constants are obtained from the relevant line separation.

The large difference between the coupling constants of the two radicals may be understood by applying the *McConnell* equation $a = Q \cdot \rho$; Q ($\approx 2.3 - 2.7$) is an empirical parameter, and ρ is the so-called *spin density* or *spin population*, which, from quantum chemical calculations, gives the probability of finding the unpaired electron on the carbon atoms of the π-radical. For the ethyl radical, the *McConnell* equation indicates that $\rho \approx 1$ for the α-carbon atom; that is, the unpaired electron is localised almost completely on the carbon atom of the methylene group, as the formula $CH_3 - CH_2 \cdot$ suggests. The fact that the methyl protons also produce a hyperfine structure is surprising, and is attributed to hyperconjugation. In the case of the benzene radical ion, with $Q = 2.5$, the results indicate that each carbon atom has a spin density of $\frac{1}{6}$. In this case the unpaired electron is delocalised over the carbon atoms (see page 469 *ff.*).

This method has been still further developed and, using ENDOR spectroscopy (*electron nuclear double resonance*), it is possible to interpret the coupling constants with more certainty, and to investigate polyradicals.[1]

1.8.7 Ion Cyclotron Resonance (ICR)[2]

Ion cyclotron resonance is a combination of mass spectrometry and NMR spectroscopy. In this technique, ions, which are made to travel in circular orbits in a magnetic field that are defined by the ratio e/m, are exposed to irradiation by radio waves. If the radio wave frequency corresponds to the frequency of rotation of the ions in the cyclotron, then the ions absorb energy and their rate of rotation is raised. As a result the orbit increases in size so that the frequency remains constant. Using this technique it is possible to observe reactions in the gas phase.

If mixtures of alcohols and water, *e.g.* t-butanol and water, are bombarded by electrons, hydroxide ions, HO^-, are formed, and it is possible to observe these ions because of their characteristic cyclotron frequency $e/m = \frac{1}{17}$.

Within the mixture an acid–base reaction follows:

$$(H_3C)_3COH + OH^\ominus \longrightarrow (H_3C)_3CO^\ominus + H_2O$$

t-Butanol   t-Butoxide

A new frequency of $e/m = \frac{1}{16}$, characteristic of the t-butoxide ion, now appears. The reverse reaction:

$$(H_3C)_3CO^\ominus + H_2O \longrightarrow (H_3C)_3COH + OH^\ominus$$

is not observed. It follows, therefore, that in contrast to the situation in aqueous solution (see p. 111), in the gas phase t-butanol is a stronger acid than water.

1.8.8 Crystal Structure Analysis

Valuable information about the structure of molecules can be obtained by the use of X-ray diffraction[3] or neutron diffraction[4] techniques. This provides information about the spatial arrangement of atoms in molecules, about atomic distances (bond lengths), bond angles, etc. By this means the structures of highly complex compounds such as vitamin B_{12}, proteins and nucleic acids have been determined. In particular, the use of high-powered computers has greatly assisted and extended the use of X-ray analysis. A large amount of X-ray structural data is readily available from a central data-bank.[5]

---

[1] See *H. Kurreck*, ENDOR Spectroscopy—A Promising Technique for Investigating the Structure of Organic Radicals, Angew. Chem. Int. Ed. Engl., *23*, 173 (1984).
[2] See *T. A. Lehman* et al., Ion Cyclotron Resonance Spectroscopy (Wiley, New York 1976).
[3] *C. Krüger*, High Resolution X-ray Crystallography—An Experimental Method for the Description of Chemical Bonds, Angew. Chem. Int. Ed. Engl., *24*, 237 (1985); *J. D. Dunitz*, X-ray Analysis and the Structure of Organic Molecules (Cornell University Press, 1979).
[4] *G. Will*, Crystal Structure Analysis by Neutron Diffraction, Angew. Chem. Int. Ed. Engl., *8*, 356, 950 (1969).
[5] 'Cambridge File', see *O. Kennard* et al., Systematic Analysis of Structural Data as a Research Technique in Organic Chemistry, Accounts Chem. Res., *16*, 146 (1983).

Other methods used in the determination of molecular structure include the measurement of dipole moments,[1] optical rotation (polarimetry), optical rotatory dispersion (ORD),[2] and polarography.[3] For the direct determination of the structures of molecules in the gas phase, electron diffraction[4] and microwave spectroscopy[5] are used. The latter not only provides information about the spatial arrangements of atoms, but also about the vibration and rotation in the free molecule. The surfaces of molecules can be made visible at the atomic level, even for non-crystalline samples, by means of the scanning tunnelling microscope.[6]

Satisfactory work in the field of organic chemistry depends upon the thorough application of all the methods which have been discussed. Hence it seemed apposite to present some account of the practical application of these methods to the solution of the problems with which one is confronted in organic chemistry.

1.9 Classification of Organic Chemistry

There are two possible ways of systematising the treatment of organic compounds. On the one hand, they can be classed according to their functional groups, and the chemical reactivity of these groups is emphasised while the reactive influence of the carbon residue is only considered secondarily. This type of classification can lead to difficulties, above all with heterocyclic compounds. On the other hand, since there are usually some interactions between functional groups and the carbon skeleton, which are of some significance for the distinctive overall chemical properties of compounds, it is possible to keep the consideration of this interaction systematically in the foreground. This latter approach is used as the basis for this textbook, and organic chemistry is classified herein into the following large sections.

Aliphatic Compounds

These compounds are based on chains of carbon atoms. The chains may be unbranched or branched. The name is derived originally from the fatty acids which were studied at an earlier stage of the development of organic chemistry (Grk. *aliphos* = fat).

Alicyclic Compounds

These compounds are based on rings of carbon atoms. The rings are of different sizes, and can be regarded as arising from the cyclisation of aliphatic compounds.

---

[1] See *O. Exner*, Dipole Moments in Organic Chemistry (Thieme Verlag, Stuttgart 1975).
[2] See *P. Crabbé*, ORD and CD in Chemistry (Academic Press, New York 1972); *G. Snatzke*, Circular Dichroism and Optical Rotatory Dispersion—Principles and Application to the Investigation of the Stereochemistry of Natural Products, Angew. Chem. Int. Ed. Engl., *7*, 14 (1968).
[3] See *P. Zuman*, Polarography in Organic Chemistry, Fortsch. Chem. Forsch., *12*, 1 (1969).
[4] See *O. Bastiansen*, Structure Determination of Free Molecules by Electron Diffraction, Angew. Chem. Int. Ed. Engl., *4*, 819 (1965); *I. Hargittai*, Stereochemical Applications of Gas Phase Electron Diffraction, 2 vols. (VCH Verlag, Weinheim 1988).
[5] See *T. M. Sugden* and *C. N. Kenney*, Microwave Spectrocopy of Gases (van Nostrand, London 1965).
[6] See *G. Binnig* and *H. Rohrer*, Scientific American, *253*, August issue, p. 50 (1985).

Aromatic Compounds

This class is used herein to typify compounds which contain a benzene ring and show 'aromatic character'. 'Non-benzenoid aromatic' compounds are also dealt with in this section.

Heterocyclic Compounds

This is a very large class of compounds. They are based on rings in which one or more of the carbon atoms has been replaced by a so-called hetero-atom, such as nitrogen or oxygen or sulphur.

Natural Products

The commonest types of natural products are carbohydrates, terpenes and steroids, amino acids, peptides and proteins, nucleic acid, enzymes and metabolites. All of these types are dealt with in separate sections.

In the following descriptions of organic compounds, for pedagogic reasons their synthesis,[1,2], properties and characteristic reactions are first of all considered on the basis of their structural formulae. A complete understanding of their electronic structure is required for a proper theoretical explanation of their molecular consititution and the mechanisms of the reactions.[2] An appreciation of their stereochemistry[3] is also essential for the full understanding of the properties and reactivities of organic compounds.

[1] See *L. F. Fieser* and *M. Fieser* et al., Reagents for Organic Synthesis, Vols 1–16 (Wiley, New York 1967–1992).

[2] See *J. March*, Advanced Organic Chemistry, 4th edn (Wiley, New York 1992).

[3] See *E. L. Eliel* and *S. H. Wilen*, Stereochemistry of Organic Compounds (Wiley, Chichester 1994); *K. Mislow*, Introduction to Stereochemistry (Benjamin, New York 1965).

2 Aliphatic Compounds

The simplest aliphatic compounds are the *hydrocarbons*. On the basis of their structure and of their chemical reactivity they are divided into *alkanes, alkenes* and *alkynes*.

2.1 Alkanes (Paraffins), C_nH_{2n+2}

The simplest aliphatic compound is methane, CH_4

$$H - \overset{\displaystyle H}{\underset{\displaystyle H}{\overset{|}{\underset{|}{C}}}} - H$$

Its three-dimensional structure can be illustrated in a number of ways, for example, as a tetrahedron (Fig. 28a), using ball-and-stick models (Fig. 28b) or with the use of space-filling models (Fig. 28c).

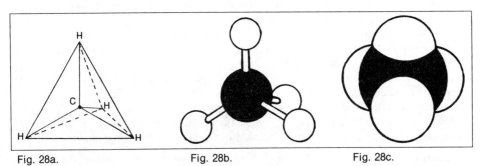

Fig. 28a.   Fig. 28b.   Fig. 28c.

Three-dimensional models of methane.

When a number of carbon atoms are linked together to form a chain and their remaining valences are used to form C—H bonds, then the resultant hydrocarbons have their terminal carbon atoms each linked to three hydrogen atoms while all the other carbon atoms have two hydrogen atoms bound to them. Thus n carbon atoms are linked to $(2n + 2)$ hydrogen atoms. These hydrocarbons thus have the general molecular formula C_nH_{2n+2}. They are called *saturated hydrocarbons*, because all their valences are used up by attached hydrogen atoms. In the IUPAC system of nomenclature[1] they are called *alkanes*. Sometimes, in technical literature, the older name, *paraffins*, is found (from Lat. *parum affinis* = little reactivity).

This description has been overtaken by events, because it is now realised that, under the appropriate reaction conditions, alkanes can undergo such reactions as protonation, halogenation, nitration, conversion into sulphonyl chlorides and sulphonic acids, and oxidation to carboxylic acids.

Each compound in this series differs from its immediate neighbours by having one more or one less CH_2 group. They are said to form a *homologous series*. The simpler members of this series of alkanes, which have unbranched chains of carbon atoms and are called *normal* alkanes, are listed in the following table. (In the *skeletal* or *line formulae* given for propane and higher members, the junction point of two lines represents a CH_2 group, the ends of the lines represent CH_3 groups).

		Boiling point
Methane	CH_4	$-161,5\ °C$ (111,7 K)
Ethane	H_3C-CH_3	$-88,6\ °C$ (184,6 K)
Propane	$H_3C-CH_2-CH_3$;	$-42,2\ °C$ (231,0 K)
n-Butane	$H_3C-CH_2-CH_2-CH_3$;	$-0,5\ °C$ (272,7 K)
n-Pentane	$H_3C-CH_2-CH_2-CH_2-CH_3$;	$36,1\ °C$ (309,3 K)
n-Hexane	$H_3C-CH_2-CH_2-CH_2-CH_2-CH_3$;	$68,8\ °C$ (342,0 K)

Then follow n-heptane, C_7H_{16}; n-octane, C_8H_{18}; n-nonane, C_9H_{20}; n-decane, $C_{10}H_{22}$; n-undecane, $C_{11}H_{24}$; n-dodecane, $C_{12}H_{26}$; n-eicosane, $C_{20}H_{42}$; n-heneicosane, $C_{21}H_{44}$; n-docosane, $C_{22}H_{46}$; n-tricosane, $C_{23}H_{48}$, n-hentricontane, $C_{31}H_{64}$, etc. In 1954, an alkane with 100 carbon atoms, n-hectane, was synthesised. It is common practice to omit the prefix n-; absence of any prefix implies that the chain of carbon atoms is unbranched. X-ray structure analysis of solid alkanes shows that the chain is kinked and takes up a zigzag conformation as indicated by the line formulae.

If one hydrogen atom is removed from an alkane, the remainder is known as an *alkyl group*, e.g. $-CH_3$, methyl; $-CH_2-CH_3$, ethyl. It is frequent practice to represent alkyl groups in general by the symbol R.

For the first three members of the alkane series it is only possible to arrange the carbon atoms in one way. But the formula C_4H_{10} can represent two different hydrocarbons, n-butane (or butane) and isobutane or 2-methylpropane.

These *isomers* differ in the way in which their atoms are arranged. Butane has an unbranched chain of carbon atoms whereas isobutane has a branched chain.

Isomerism (*Berzelius*, 1830). Isomeric compounds have the same molecular formulae but different physical and chemical properties. This results from the different arrangements

---

[1] The IUPAC, or Geneva, system of nomenclature was drawn up by the International Union of Pure and Applied Chemistry. See Nomenclature of Organic Chemistry, 1979 edn (Pergamon Press, Oxford); *P. Fresenius*, Organic Chemical Nomenclature, Introduction to the Basic Principles (Ellis Horwood, Chichester 1989). Alongside IUPAC nomenclature, so-called Trivial Names are widely and indispensably used, and are officially accepted. See, for example, *A. Nickon* et al., Organic Chemistry: The Name Game (Pergamon Press, New York 1987).

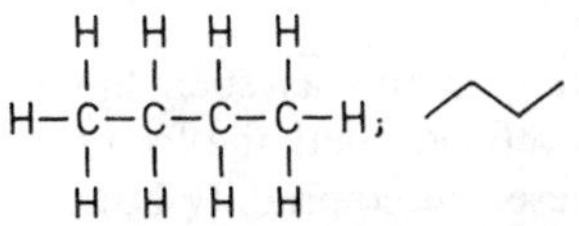

H H H H
| | | |
H−C−C−C−C−H;   ∿
| | | |
H H H H

H H H
| | |
H−C−C−C−H;   ⌉⌐
| | |
H | H
|
H−C−H
|
H

n-Butane
(Butane)

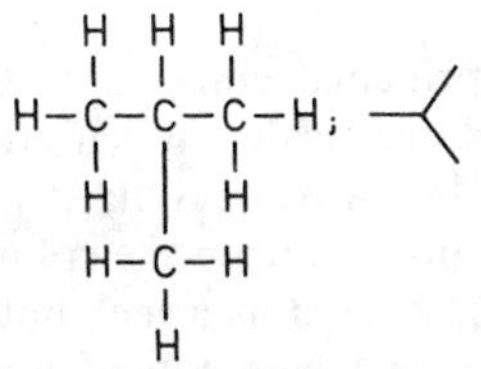

Isobutane or 2-Methylpropane
(b.p. −11.7°C [261.5 K])

of the atoms in the molecules. Isomers have the same molecular formula but different structural formulae.

The Most Important Alkanes

2.1.1 Methane

Above all, methane is the major constituent of natural gas, found in the oil and gas fields of the world. Mixed with air and oxygen it forms explosive mixtures which detonate violently when ignited, and such mixtures have been responsible for disastrous explosions in coal mines, where the presence of methane is frequently a hazard. In mines it is known as *fire-damp*. Mixed with carbon dioxide it is also found in the sludge at the bottom of lakes and marshes, from which it bubbles to the surface, and is described as *marsh gas*. Here it results from the anaerobic decomposition of plants. Methane can be obtained from sewage and sludge by fermentation processes, and this has been done in some places to provide gas for heating purposes. Methane occurs to a small extent (1.3 ppm) in the atmosphere; the amount has doubled since 1750 and is presently increasing by about 1–2% per annum.

Preparation

2.1.1.1 Methane can be made directly from its elemental constituents by passing an electric arc between carbon electrodes in an atmosphere of hydrogen at about 1200°C (1475 K) (*Bone* and *Jerdan*).

$$C + 2H_2 \rightleftharpoons CH_4 \quad \Delta H = -75 \text{ kJ/mol}$$

The yield is very small, however, since it tends to decompose above 1000°C (1275 K).

2.1.1.2 Nowadays the principal source of methane is from natural gas. It is also obtained from the distillation of coal and from the volatile gases produced in the refining of oil.

Properties

Methane is a colourless and odourless gas. It burns with a non-luminous flame. When it is mixed with oxygen in a molar ratio of 1:2 it forms a readily ignited explosive mixture.

$$CH_4 + 2O_2 \rightarrow CO_2 + 2H_2O \quad \Delta H = -803 \text{ kJ/mol}$$

Uses

Methane is the principal constituent of natural gas, and as such has enormous use industrially and domestically. When crude oil is refined, methane forms a large part of the volatile materials which result; at present these are commonly burned off in flares. Methane is also obtained from the coking of coal, lignite, peat and wood, and, prepared in this way, has been used as a fuel, both for industry and for domestic purposes.

On the technical scale, a mixture of methane and steam is heated at about 800°C (1075 K) in the presence of nickel to provide a mixture of hydrogen and carbon monoxide (synthesis gas).

$$CH_4 + H_2O \overset{(Ni)}{\rightleftharpoons} CO + 3H_2 \quad \Delta H = +205\,kJ/mol$$

Unchanged methane can be converted by the action of air in a second reactor to give more *synthesis gas*.

$$CH_4 + 1/2\,O_2 \rightarrow CO + 2H_2$$

Methane serves as the starting point for a number of important industrial synthetic processes, for example to prepare chloromethane, dichloromethane, methanol, formaldehyde, acetylene, hydrogen cyanide and carbon disulphide.

The preparation of carbon black is based on the incomplete combustion in air of methane (obtained from natural gas) or of anthracene oil (a high-boiling fraction from the distillation of coal tar; see p. 646), or by the thermal decomposition of methane to give carbon and hydrogen:

$$CH_4 + O_2 \rightarrow C + 2H_2O \qquad CH_4 \rightarrow C + 2H_2$$

The very finely divided amorphous carbon (soot) is used in the rubber industry as a filler for motorcar tyres, in the dyestuffs industry as a pigment for use in printers' ink, in cosmetics, and in the preparation of dry batteries.

2.1.2 Ethane

Ethane, H_3C-CH_3, occurs in natural gas and in gases produced in the refining of oil, and these are the sources from which it is obtained industrially. It is a colourless, flammable gas which is used for heating.

Preparation[1]

2.1.2.1 By reaction of diethylzinc with water (*Frankland*, 1850):

$$Zn(C_2H_5)_2 + 2H_2O \rightarrow Zn(OH)_2 + 2C_2H_6$$

2.1.2.2 *Kolbe* **Electrolysis Method** (1849). Electrolysis of a concentrated solution of sodium acetate results in the evolution of hydrogen at the cathode and of ethane and carbon dioxide at the anode:

$$\begin{matrix} H_3CCOO^{\ominus} \\ \\ H_3CCOO^{\ominus} \end{matrix} \xrightarrow{-2e} \begin{matrix} CH_3 \\ | \\ CH_3 \end{matrix} + 2\,CO_2$$

---

[1] The methods listed are not limited to the preparation of ethane. See also *Houben-Weyl*, Methoden der Organischen Chemie, 4th edn, Vol. *5/1a* (Thieme Verlag, Stuttgart 1970).

Reaction proceeds *via* an anodic decarboxylation which gives rise to methyl radicals, ·CH$_3$, which dimerise to give ethane.

Kolbe's method was the first example of an *organic electrosynthesis*. This technique is now of ever-increasing importance in industrial chemistry.

2.1.2.3 *Wurtz* Synthesis (1855). Reaction between methyl halides and sodium leads to the formation of ethane. Methylsodium is formed as an intermediate:

$$CH_3I + 2\,Na \longrightarrow CH_3Na + NaI$$
$$CH_3Na + ICH_3 \longrightarrow H_3C{-}CH_3 + NaI$$

Ethane

The *Wurtz* reaction is of particular importance for the preparation of higher alkanes from the appropriate alkyl halides. If a mixture of two different alkyl halides is used, a mixture of hydrocarbons results. The most suitable reagents are alkyl iodides, which are better for this purpose than bromides and chlorides.

2.1.2.4 *Corey–House* Synthesis. Lithium dialkyl cuprates react with alkyl halides to give alkanes.[1] By this method it is possible to get alkanes with an odd number of carbon atoms, e.g.

$$(H_3CCH_2CH_2CH_2)_2CuLi + H_3C(CH_2)_6Cl \longrightarrow H_3CCH_2CH_2CH_2(CH_2)_6CH_3$$

Undecane

2.1.3 Propane and Butanes

Propane, H$_3$C—CH$_2$—CH$_3$, n-butane (butane), H$_3$C—CH$_2$—CH$_2$—CH$_3$, and isobutane (2-methylpropane), H$_3$C—CH(CH$_3$)—CH$_3$, occur in natural gas and among the products which arise from the cracking of oil. They are obtained in large quantities from the latter process and are important raw materials in the petrochemical industry. They are compressed in steel vessels for use as 'bottled gas' for heating, camping, etc. They disperse rapidly in the atmosphere and hence are used increasingly as propellants in aerosols, notwithstanding their high flammability.

By removal of a hydrogen atom from different carbon atoms in either of these alkanes, more than one isomeric alkyl group may be obtained, as follows:

—CH$_2$—CH$_2$—CH$_3$	—CH$_2$—CH$_2$—CH$_2$—CH$_3$	—CH$_2$—CH(CH$_3$)$_2$
n-Propyl	n-Butyl	Isobutyl
—CH(CH$_3$)$_2$	—CH(CH$_3$)—CH$_2$—CH$_3$	—C(CH$_3$)$_3$
Isopropyl	*sec*. Butyl	t-Butyl

The prefix 'iso' indicates that a hydrocarbon chain has two methyl groups at the end of it.

[1] See *G. H. Posner*, Substitution Reactions using Organo Copper Reagents, Org. Reactions *22*, 253 (1975).

2.1.4 Pentanes and Higher Homologues

There are three isomers of pentane, C_5H_{12}. According to IUPAC rules they are named as follows. The longest carbon chain present is numbered and the site of each attached side chain is designated by the number of the carbon atom to which it is attached. Numbering of the main chain starts at the end nearest to a side chain.

			b.p.
$H_3C-CH_2-CH_2-CH_2-CH_3$;		Pentane	36 °C (309 K)
(1) (2) (3) (4) $H_3C-CH-CH_2-CH_3$; $\quad\quad\ \ |$ $\quad\quad CH_3$		2-Methylbutane (isopentane)	28 °C (301 K)
$\quad\quad CH_3$ $\quad\quad |(2)$ $H_3C-C-CH_3$ $\quad\quad |$ $\quad\quad CH_3$		2,2-Dimethylpropane (neopentane)	9,4 °C (282,6 K)

A carbon atom which is attached to only one other carbon atom is called a *primary carbon atom*. Similarly, carbon atoms which are attached to two, three or four other carbon atoms are known, respectively, as *secondary, tertiary* or *quaternary carbon atoms*. The carbon atom in methane comes into neither of these categories since it has only hydrogen atoms joined to it.

In the cases of the higher alkanes, the number of possible isomers increases dramatically. Thus there are nine isomers of heptane. In the cases of octane and of dodecane there are, respectively, 18 and 355 possible isomers.[1]

Properties

Alkanes with up to four carbon atoms are all gases at room temperature. In the case of the normal (unbranched) alkanes, those from n-pentane to n-heptadecane are liquids at room temperature; the higher homologues are solids. While the gaseous and solid alkanes are odourless, the liquid alkanes have a characteristic petrol-like smell. They are all insoluble in water. All of them have very similar chemical properties.

The boiling points of the n-alkanes are higher as one goes up the homologues series, but the differences between the boiling points of successive neighbours in the series decrease as their molecular masses get higher. In consequence, the lower members of the series are easier to separate by fractional distillation than the higher homologues are.

The boiling points of the n-alkanes are always higher than the boiling points of their branched isomers. Thus n-butane has b.p. -0.5°C (272.7 K), while isobutane boils at -11.7°C (261.5 K).

There is also a regular pattern for the melting points of n-alkanes with up to 24 carbon atoms. From methane upwards melting points alternate, n-alkanes with even numbers of carbon atoms melting at relatively higher temperatures than n-alkanes having odd numbers of carbon atoms. This is because of the different lattice structure associated with n-alkanes with even or odd numbers of carbon atoms.

---

[1] See *H. R. Henze* et al., The Number of Isomeric Hydrocarbons of the Methane Series, J. Am. Chem. Soc. *53*, 3077 (1931).

2.1.5 Conformations of Ethane

The concept of *conformation* (*Haworth*, 1929) is now applied to the three-dimensional structure of alkanes. This permits an understanding of the relative arrangements of the atoms in a molecule in the gaseous or liquid states.

The *conformations of a molecule* are defined as the different steric structures which it can take up solely by rotation about its single bonds. *Conformers* are not, in general, isolable.

In the case of ethane, the two methyl groups should be able to rotate freely, with respect to one another, about the C—C axis. It appears, however, that the rotation about the single bond has certain restrictions.

Closer investigations reveal that of all the possible geometrical arrangements, one is particularly favoured. The *energy barrier* which tends to inhibit rotation is about 13 kJ/mol (*Pitzer*) and results in most of the molecules taking up the most stable conformation, that of lowest energy. In the case of ethane this is achieved when the six hydrogen atoms are arranged in such a way that they are as far away from one another as possible; when looked at along the C—C axis they are in a *staggered* arrangement. When the hydrogen atoms are arranged in such a way that, viewed along the C—C axis, they are as near to each other as possible, this is the least stable arrangement, and is described as the *eclipsed* form. If the molecule is drawn with the C—C bond perpendicular to the plane of the paper, the arrangements of the hydrogens in the *staggered* and *eclipsed* conformations can be represented as shown, respectively, in Figs. 29 and 30. In these projection formulae, which are called *Newman* projections, the intersection point of the three lines represents the nearest carbon atom, the circle the other carbon atom. The two conformations can be illustrated by the use of space-filling models (Figs. 31, 32).

In the case of butane, six conformations are possible, three staggered and three eclipsed, which are interchangeable by rotation, in a series of 60° steps, about the central C—C bond (Figs. 33–38). In these *Newman* projections the front C-atom of this bond is held in one position, while the rear one is twisted in steps of 60° to give different torsion angles φ. Other terms for these conformations which are used in the literature are given beneath each conformation.[1]

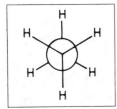

Fig. 29.

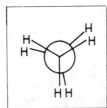

Fig. 30.

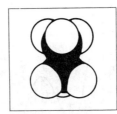

Fig. 31.

Fig. 32.

From molecular models it may be seen that the forms φ_1 (Fig. 34) and φ_5 (Fig. 38) are mirror images of each other and are thus of equal energy, as are forms φ_2 (Fig. 35) and φ_4 (Fig. 37).

---

[1] For an account of stereochemical nomenclature see IUPAC, Nomenclature of Organic Chemistry, Section E: Stereochemistry (Recommendations 1974) (Pergamon Press, Oxford 1979).

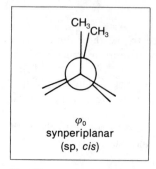

φ_0
synperiplanar
(sp, *cis*)

Fig. 33.

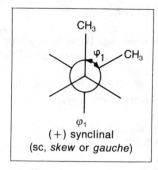

φ_1
(+) synclinal
(sc, *skew* or *gauche*)

Fig. 34.

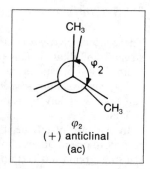

φ_2
(+) anticlinal
(ac)

Fig. 35.

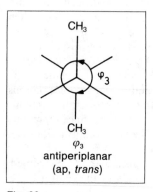

φ_3
antiperiplanar
(ap, *trans*)

Fig. 36.

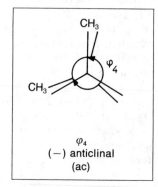

φ_4
(−) anticlinal
(ac)

Fig. 37.

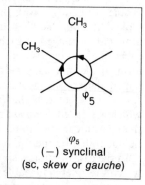

φ_5
(−) synclinal
(sc, *skew* or *gauche*)

Fig. 38.

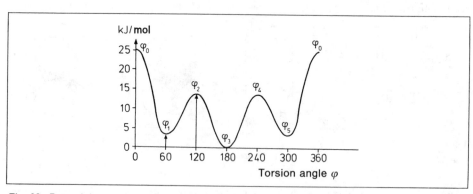

Fig. 39. Potential energies of the conformations of butane.

Electron diffraction studies show that in the gas phase, butane has about 75% of the molecules in conformation φ_3; the remaining 25% are in conformations φ_1 and φ_5. Thus the φ_3 form, in which the two methyl groups are as far away from one another as is possible, is the most stable (Fig. 39). In conformations φ_1 and φ_5, in which the methyl groups are in a skew conformation to each other, there is more interaction between these groups. There is no evidence for any molecules being in the φ_0, φ_2 or φ_4 conformations. These are eclipsed conformations and represent the peaks of the energy barriers.

Conformational analysis[1] is important not only in the case of alkanes and their substitution products, but especially in the study of cyclohexane derivatives.

2.2 Alkenes (Olefins), C_nH_{2n}

The alkenes are a homologous series of compounds with the overall formula C_nH_{2n}; they thus contain two fewer hydrogen atoms than the corresponding alkanes. Since not all of the valences of the carbon atoms are taken up by hydrogen atoms, they are described as *unsaturated hydrocarbons*.

For a long time, the constitution of these hydrocarbons presented great difficulties if the tetravalence of carbon was to be maintained. Eventually, on the basis of a large amount of chemical information which had accrued, the idea developed that the properties and reactions of alkenes were best explained by postulating the presence of a $C{=}C$ *double bond*.

On this basis the lower members of this series were formulated as follows:

		b.p.
$H_2C{=}CH_2$	Ethylene, Ethene	$-103{,}7\ ^\circ C$ (169,5 K)
$H_3C{-}CH{=}CH_2$	Propene (Propylene)[2]	$-47{,}7\ ^\circ C$ (225,5 K)
$H_3C{-}CH_2{-}CH{=}CH_2$	But-1-ene (Butylene)	$-6{,}5\ ^\circ C$ (266,7 K)
$H_3C{-}CH{=}CH{-}CH_3$	But-2-ene *cis* form,[3] 3,7 $^\circ C$ (276,9 K)	
$H_3C{-}\underset{\underset{CH_3}{|}}{C}{=}CH_2$	Isobutene (Isobutylene), Methylpropene $6{,}6\ ^\circ C$ (279,8 K)	
$H_3C{-}CH_2{-}CH_2{-}CH{=}CH_2$	Pent-1-ene	$30{,}0\ ^\circ C$ (303,2 K)
$H_3C{-}CH_2{-}CH{=}CH{-}CH_3$	Pent-2-ene *cis* form,[3] 36,5 $^\circ C$ (309,7 K)	
$H_2C{=}\underset{\underset{CH_3}{|}}{C}{-}CH_2{-}CH_3$	2-Methylbut-1-ene	$31{,}0\ ^\circ C$ (304,2 K)
$H_3C{-}\underset{\underset{CH_3}{|}}{C}{=}CH{-}CH_3$	2-Methylbut-2-ene	$38{,}5\ ^\circ C$ (311,7 K)
$H_2C{=}CH{-}\underset{\underset{CH_3}{|}}{CH}{-}CH_3$	3-Methylbut-1-ene	$20{,}1\ ^\circ C$ (293,3 K)

---

[1] See *E. L. Eliel* et al., Conformational Analysis (Interscience Publishers, New York 1981).
[2] *Propylene* is also the IUPAC name for the diradical $CH_3{-}\dot{C}H{-}\dot{C}H_2$.
[3] *cis*-Form, see p. 64.

Ethylene, C_2H_4, was formerly described as an 'olefin' (from *olefiant* = oil forming), since it reacted with halogens to give oily liquids which were insoluble in water, e.g. 1,2-dibromoethane. The members of the homologous series were described as *ethylenes*, after the first member, as alkylenes, or as olefins. Following the introduction of the IUPAC rules of nomenclature, they are now described as *alkenes*. The ending '-ene' in organic chemistry is indicative of the presence of unsaturation in the molecule.

In the case of branched-chain alkenes, the carbon chain which includes the double bond is taken as the main chain and is numbered starting from the end nearest to the double bond. The site of the double bond is indicated by inserting the number of the carbon atom at which it begins; see but-1-ene and but-2-ene above.

But-1-ene and but-2-ene differ only in the position of the double bond; there is a third isomer, isobutene or methylpropene, which has a branched chain. With higher alkenes, C_5H_{10}, C_6H_{12}, etc. even more isomers are possible than is the case for the corresponding alkanes, since extra isomers exist which differ in the position of the double bond. *Alkenyl* groups are derived by removal of a hydrogen atom from an alkene, e.g. vinyl, $-CH=CH_2$; prop-1-enyl, $-CH=CH-CH_3$; prop-2-enyl or allyl, $-CH_2-CH=CH_2$; but-1-enyl, $-CH=CH-CH_2-CH_3$, etc. In each case the carbon with the free valency is numbered as 1.

A group which ends with a double bond, $RCH=$, is known as an *alkylidene* group. Examples are ethylidene $CH_3-CH=$, isopropylidene, $(CH_3)_2C=$ (given a grey background in the formula of 2-methylbut-2-ene, p. 59).

The Most Important Alkenes[1]

2.2.1 Ethylene (Ethene) and Propene

Ethylene and its homologues used to be prepared industrially by removal of water from alcohols, but nowadays they are largely produced by cracking of hydrocarbons obtained in the distillation of crude oil. Ethylene is the organic compound which is produced in the greatest quantity; world production is more than 25 million t per year.[2]

Ethylene is a plant hormone (phytohormone), which plays a number of roles in developing plants. In plants it is formed from 1-aminocyclopropane-1-carboxylic acid.

Preparation

2.2.1.1 From Ethanol. In the laboratory ethylene is obtained by heating ethanol with concentrated sulphuric acid (or phosphoric acid or zinc chloride). Ethyl hydrogen sulphate (ethylsulphuric acid) is formed first and decomposes at 150°C (425 K) to give ethylene and sulphuric acid.

$$H_3C-CH_2-O-H+H_2SO_4 \xrightleftharpoons[-H_2O]{} H_3C-CH_2-O-SO_3H \xrightleftharpoons[]{425\ K} H_2C=CH_2+H_2SO_4$$

Ethanol   Ethyl hydrogen sulphate   Ethylene

[1] See *S. Patai, J. Zabicky*, The Chemistry of Alkenes, 2 vols (Interscience Publishers, New York 1964, 1970); *Houben-Weyl*, Methoden der Organischen Chemie, 4th edn, Vol. 5/1b (Thieme Verlag, Stuttgart 1972).
[2] See *S. A. Miller*, Ethylene and its Industrial Derivatives (Benn, London 1969).

The acid-catalysed elimination of water from alcohols begins with protonation of a lone pair of electrons on the oxygen atom, leading to formation of an *oxonium ion*. Loss of water from this gives a carbenium ion, which then loses a proton to form the alkene.

$$R-CH_2-CH_2-\overline{O}H + H_2SO_4 \;\rightleftharpoons\; R-CH_2-CH_2-\overset{\oplus}{\underset{H}{O}}-H + HSO_4^{\ominus}$$

Oxonium ion

$$R-\overset{H}{\underset{H}{C}}\overset{\oplus}{-}CH_2 \xrightarrow{-H^\oplus} R-\overset{H}{C}=CH_2$$

Carbenium ion   Alkene

This proton is transferred to another alcohol molecule to form another oxonium ion. Loss of a proton to provide an alkene takes place most readily when a *tertiary* carbenium ion is involved, and least readily when a *primary* carbenium ion is involved. In other words *tert.* alcohols more readily lose a molecule of water, and the alkene which is formed is that with the greatest number of alkyl groups attached to the resultant double bond (*Saytzeff* rule, 1875) (see also p. 157).

2.2.1.2 Thermal Cracking. When a mixture of hydrocarbons (crude oil) is pyrolysed at about 1000°C (1275 K) and at normal pressure, in the presence of steam—to limit the formation of coke—ethylene is formed, together with higher alkenes and dienes. Thermal dehydrogenation of ethane gives ethylene.

$$H_3C-CH_3 \rightarrow H_2C=CH_2 + H_2$$

As the chain length of alkanes increases, so pyrolysis, which involves the formation of radicals, results in the splitting of C—C bonds rather than C—H bonds. For example, pyrolysis of propane provides propene, ethylene, methane and hydrogen:

$$2H_3C-CH_2-CH_3 \quad\Big\langle\begin{array}{l} H_3C-CH=CH_2 + H_2 \\ H_2C=CH_2 + CH_4 \end{array}$$

Propane

By such *steam cracking* of crude oil, ethylene, propene and, in the C_4 fraction, the isomeric butenes and butadiene, can be obtained in very pure form. They are important starting materials for a variety of processes in the petrochemical industry.

2.2.2 *Butenes and Higher Homologues*

But-1-ene and but-2-ene are present in the C_4 fraction obtained from cracking, and butadiene is made from them.

But-1-ene and other unbranched alk-1-enes are obtained from ethylene by interaction with trialkylaluminiums by the following processes.

Mülheimer Alkene Synthesis (*K. Ziegler*).[1] Ethylene interacts with triethylaluminium at 100–120°C (375–395 K) and under a pressure of about 100 bar (10 MPa), resulting in the

[1] See Metallorganische Synthese höherer aliphatischer Verbindungen aus niederen Olefinen in Praxis und Theorie (The Synthesis of Higher Molecular Weight Aliphatic Compounds from Lower Alkenes in Practice and Theory), Angew. Chem. **72**, 829 (1960).

insertion of ethylene molecules between the aluminium and the ethyl groups and the formation of higher molecular weight trialkylaluminium compounds of the structure shown:

$$Al \begin{array}{l} / \ \\ \ \\ \backslash \end{array} \begin{array}{l} (CH_2-CH_2)_x-C_2H_5 \\ (CH_2-CH_2)_y-C_2H_5 \\ (CH_2-CH_2)_z-C_2H_5 \end{array}$$

Hydrolysis of these solid trialkylaluminium compounds results in the separation of the hydrocarbon part as a mixture of alkanes from which can be obtained both hard and flexible hydrocarbons resembling the products from the polymerisation of ethylene (see p. 76).

In competition with this reaction, and especially at higher temperatures and in the presence of metallic nickel or cobalt or platinum, a breakdown reaction takes place, in which ethylene removes the alkyl groups from the aluminium to provide *alk-1-enes* with even numbers of carbon atoms, *e.g.* but-1-ene, hex-1-ene, oct-1-ene, dec-1-ene, etc.:

$$al-(CH_2-CH_2)_n-CH_2-CH_3+H_2C=CH_2 \xrightarrow[\text{(heat)}]{\text{(Ni)}}$$

$$al-CH_2-CH_3+H_2C=CH-(CH_2-CH_2)_{n-1}-CH_2-CH_3$$

al = Al-equivalent α-Alkenes (alfenes)

The alk-1-enes, sometimes called α-alkenes or *alfenes*, with 12–18 carbon atoms which are obtained in this way, are used, among other things, as the starting materials for making *poly-α-olefins* such as poly-α-decene which are used as components of high-grade lubricating oils. In a similar manner, from the reaction of propene with tripropylaluminium at about 200°C (475 K) and 100 bar (10 MPa) the dimer 2-methylpent-1-ene, $H_3C-CH_2-CH_2-\underset{\underset{CH_3}{|}}{C}=CH_2$ is obtained. This serves as a starting material for the

production of isoprene (see p. 100 *ff*).

Alkene-Metathesis or *Alkene-Disproportionation*. When propene is heated at 120–210°C (395–485 K) and at 25–30 bar (2.6–3 MPa) in the presence of catalysts containing tungsten or molybdenum, the following equilibrium is set up:

$$2 H_3C-CH=CH_2 \rightleftharpoons H_3C-CH=CH-CH_3 + H_2C=CH_2$$

Propene   But-2-ene   Ethylene

This reaction, which was discovered by *Calderon* in 1967, is in principle applicable to all hydrocarbons which have double bonds. Formally it can be considered to be an *exchange of alkylidene groups,*$=CHR$:

$$\begin{array}{c} R^1-CH=CH-R^2 \\ + \\ R^3-CH=CH-R^4 \end{array} \rightleftharpoons \begin{array}{c} R^1-CH \quad CH-R^2 \\ \| \; + \; \| \\ R^3-CH \quad CH-R^4 \end{array}$$

The reaction involves an oxygen-containing metal–carbene complex,[1] $RCH=M\overset{\nearrow O}{\underset{\searrow}{\rule{0pt}{0pt}}}$ (see p. 376 *ff*).

---

[1] See R. R. Schrock et al., J. Am. Chem. Soc. *102*, 4515 (1980).

Isobutene, methylpropene (isobutylene) is used both as the starting material for the technical synthesis of isoctane and for polymerisation to polyisobutene. It is made either by removal of water from isobutanol or t-butanol or by catalytic dehydrogenation of isobutane:

$$H_3C-CH-CH_2OH \qquad\qquad H_3C-C=CH_2 \qquad\qquad H_3C-CH-CH_3$$

Isobutanol   Isobutene   Isobutane

$-H_2O$ $-H_2$

Physical Properties

The alkenes closely resemble the alkanes. Lower members of the series, with 2–4 carbon atoms, are gases at room temperature, those with 5–15 carbon atoms are liquids, and the higher members are solids. They are immiscible with water and burn with a luminous flame.

2.2.3 The C=C Double Bond

When, in the 1860s, the C=C symbolism for the double bond was introduced, it indicated that the two carbon atoms were held together by two bonds. It seemed, therefore, that they should be more strongly bound together than the carbon atoms joined by a single C—C bond. However, in practice it was found that a variety of *addition* and *oxidation* reactions take place at the C=C double bond and that it is a reactive site in a molecule. The tendency of the C=C double bond in ethylene to add on other atoms, and attain thereby a saturated state as a C—C bond, justifies its description as an unsaturated hydrocarbon.

The C=C double bond can be described, with the help of hybridisation, by the *Hückel* model. As was seen on p. 23 *f*, in the case of sp^3 hybridisation four entirely equivalent sp^3 orbitals are formulated which can overlap the orbitals of other atoms to form four tetrahedrally arranged σ-bonds. If instead only two $2p$ orbitals, say the p_x and p_y orbitals, are combined with the $2s$ orbital (Fig. 40), then the result is sp^2 hybridisation. In this case there are three sp^2 hybrid orbitals, which are all in one plane with an angle of 120° between them, a so-called *trigonal* arrangement (Fig. 41).

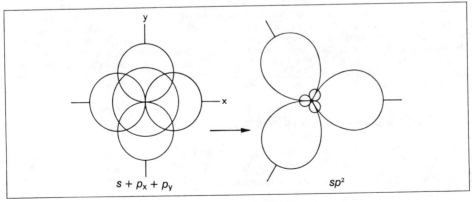

Fig. 40.   Fig. 41.

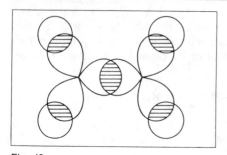

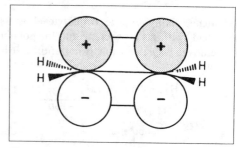

Fig. 42.   Fig. 43.

Overlap between two sp^2 orbitals leads to the formation of a σ-bond between the two carbons. The remaining four sp^2 orbitals can each overlap an s orbital of a hydrogen atom to give four C—H σ-bonds (Fig. 42). The two p_z orbitals, which take no part in the hybridisation, can, however, overlap one another, although to a lesser extent and in a direction perpendicular to their main orientation, and then only if they both lie in the same plane. This overlap is indicated schematically in Fig. 43 as a horizontal line between the p orbitals. This is described as π-*bonding*. The resultant distribution of the electrons is as shown in Fig. 44. From Figs. 43 and 44 it may be seen that the π-bonding has a nodal plane in the plane of the molecule, which is in fact a mirror plane. This contrasts with σ-bonding, which is rotationally symmetric about a line joining the carbon nuclei.

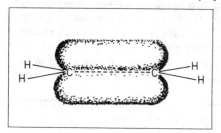

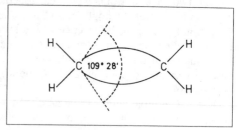

Fig. 44.   Fig. 45.

In analogous fashion to the wave mechanical treatment of σ-bonding, so for π-bonding there are bonding and anti-bonding orbitals. The anti-bonding orbitals are designated π^* (cf. p. 23).

In contrast to the $\sigma\pi$ model which has just been discussed, the C=C double bond has also been described quantum mechanically with the help of *equivalent orbitals*.[1] A very similar suggestion had been made in 1885 by *A.v. Baeyer*. In this treatment, as in ethane, both of the carbon atoms are regarded as sp^3 hybridised, and *two* sp^3 orbitals overlap one another between the carbon atoms. These overlaps cannot be colinear but must be at an angle, and hence the molecular orbitals are bent. They have been called *bent bonds* or *banana bonds*. The other two sp^3 orbitals consequently lie in the same plane as one another and can overlap the s orbitals of four hydrogen atoms to form a molecule of ethylene (Fig. 45).

2.2.4 Cis–trans *Isomerism in Alkenes*

From his studies concerning the tetrahedral model for carbon, *van't Hoff* suggested in 1874 that there should be *stereoisomeric* ethylene derivatives. These were first recorded in the case of the isomeric maleic and fumaric acids (*Wislicenus*, 1887).

---

[1] See *L. Pauling*, The Nature of the Chemical Bond, 3rd edn, p. 175 *ff.* (Oxford University Press, Oxford 1950).

According to either the $\sigma\pi$ model or the bent-bond model for ethylene all the six constituent atoms lie in one plane. Both models suggest the existence of two isomers of 1,2-dichloroethylene.

In fact there are two such *stereoisomers*. The two separate molecules are known as *diastereoisomers* or *diastereomers* (see p. 348). In this case they are *cis* and *trans* isomers. If the two substituents are on the same side of the double bond and are neighbours, it is called the *cis* form. In the *trans* isomer they are on opposite sides and diametrically opposed to each other.[1] In *stereoisomerism* molecules have the same atoms (or groups of atoms) linked to one another but their spatial configurations are different.

cis-1,2-Dichloroethylene
(b.p. 60.3 °C [333.5 K])
$\mu = 6.17 \cdot 10^{-30}$ C m

trans-1,2-Dichloroethylene
(b.p. 48.4 °C [321.6 K])
$\mu = 0$

The distance apart of the two chlorine atoms in both isomers was determined by X-ray crystallography. It was found that the distance for the *cis* isomer was 370 pm and for the *trans* isomer 470 pm, in accord with expectations.

The specific arrangement of the atoms or groups of atoms about the double bond in *cis–trans* isomers is called the *configuration*. *Cis* and *trans* isomers differ from one another in both their physical and chemical properties. One way which was used to distinguish between them involved the measurement of their *dipole moments*. If both substituents are identical, in the case of the *trans* form there is a centre of symmetry and moments due to the two groups are equal and opposite. The dipole moment μ of the *trans* form is thus equal to zero. In contrast, the *cis* form does have a dipole moment; for example, that of *cis*-1,2-dichloroethylene is equal to 6.17×10^{-30} C m. It is possible to distinguish between *cis* and *trans* isomers by means of their IR and NMR spectra.

It is usually the case that the *trans* form is of lower energy than the *cis* form, although the differences in energy are for the most part small. For example, *trans*-but-2-ene is lower in energy by 5.4 kJ/mol, and hence more stable, than *cis*-but-2-ene. They can be separated satisfactorily by gas chromatography.

cis-but-2-ene
(b.p. 3.73 °C[276.88 K])

trans-but-2-ene
(b.p. 0.96 °C[274.11 K])

tetraisopropylethylene

The 1,2-dichloroethylenes are exceptional in that the *cis* form is more stable than the *trans* isomer by about 5 kJ/mol (*Stuart*).

To bring about interchange between *cis* and *trans* isomers an *isomerisation energy* of about 250 kJ/mol is needed, which can be provided either by heating or by UV light.

---

[1] This type of stereoisomerism used to be known as *geometric* isomerism.

Thus, depending upon the reaction conditions, an equilibrium may be set up between *cis* and *trans* isomers.[1]

For interchange between *cis* and *trans* isomers to take place, the transition involves *homolysis* of the electron pairs of the double bonds, leaving two unpaired electrons, one on each of the carbon atoms; rotation by 180° about the residual σ-bond is then possible. If this is followed by re-formation of the π-bond then the alternative isomer may be formed.

In the case of tetraisopropylethylene, at low temperatures the protons H_a, H_b (see p. 65) are arranged so that H_a are directed towards the double bond, whilst H_b are directed away from the double bond. This can only be changed by the simultaneous rotation of all four isopropyl groups with respect to each other. This is known as the *cogwheel effect* (*Kwart*, *Roussel*), because the methyl groups must move past one another rather than as a cogwheel moves.[2] The isomerisation energy amounts to 70 kJ/mol; this is available at room temperature.

The *cis–trans* terminology is not satisfactory when different substituents are present on a double bond. For this reason IUPAC has introduced an alternative more general system of nomenclature,[3] which is applicable to C=N double bonds (e.g. *syn, anti* isomerism in oximes) and N=N double bonds (see p. 565) as well as to C=C double bonds. If one looks along a double bond, the following arrangements of 4 substituents a–d about the double bond A=B are possible.

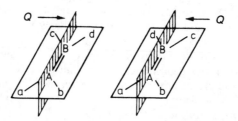

This can be illustrated by reference to 1-bromo-1,2-dichloroethylene. The substituents are assigned a priority based on their atomic numbers, according to the *Cahn–Ingold–Prelog* system, namely Br > Cl > H (see p. 274 *ff*). The two isomers are then described as follows:

$$\underset{\text{(Z)-Form}}{\overset{Cl}{\underset{H}{\diagdown}}C=C\overset{Br}{\underset{Cl}{\diagup}}} \qquad \underset{\text{(E)-Form}}{\overset{Cl}{\underset{H}{\diagdown}}C=C\overset{Cl}{\underset{Br}{\diagup}}}$$

The *Z–E* convention says that if the elements of higher priority on each carbon atom lie on the same side of the double bond it is the *Z* isomer (*Z* from *zusammen* [Ger.], together); if they are on opposite sides of the double bond it is the *E* isomer (*E* from *entgegen* [Ger.], opposite).

As another example, in the case of 1-bromo-2-chloro-1-iodoethylene the priorities of the substituents are I > Br > Cl > H. Hence the *Z* and *E* isomers are as follows:

---

[1] There are some problems about the distinction between the concepts of *configuration* and *conformation*. See IUPAC Rules, J. Org. Chem. *35*, 2849 (1970).

[2] See K. *Mislow*, Molecular Machinery in Organic Chemistry, Chemtracts-Organic Chemistry, *2*, 151 (1989).

[3] See Nomenclature of Organic Chemistry, 1979 edn (Pergamon Press, Oxford); A Guide to IUPAC Nomenclature of Organic Compounds. Recommendations 1993 (Blackwell Scientific Publications, Oxford 1993).

$$\begin{array}{cc} \text{Cl} & \text{I} \\ & \diagup \\ \text{C}=\text{C} \\ \diagup & \diagdown \\ \text{H} & \text{Br} \end{array} \qquad \begin{array}{cc} \text{Cl} & \text{Br} \\ & \diagup \\ \text{C}=\text{C} \\ \diagup & \diagdown \\ \text{H} & \text{I} \end{array}$$

(*Z*)-Form   (*E*)-Form

It must be recognised that the *Z–E* names do not always coincide with the older *cis–trans* designations.

Chemical Properties

The typical reactions of alkenes are addition reactions, which also include polymerisation reactions. The former have also been used for the detection of C=C double bonds.

2.2.5 Addition Reactions

In general the addition reactions of alkenes involve *electrophilic* attack on the double bond. In the first step of the reaction, an electrophile $X^{\oplus}$, often a cation with a deficiency of electrons, reacts with the π-electrons of the double bond, to form either a π-complex or a *carbonium ion*, which can in turn provide a *carbenium ion*:

π-Complex   Carbonium ion   Carbenium ion

In the next step a nucleophilic species $Y^{\ominus}$ adds to the intermediate, thereby redressing the electron deficiency of the carbenium ion. In this step it is possible for $Y^{\ominus}$ to approach the carbon atom from either side, to produce either *cis*- or *trans*-addition to the double bond. If the alkenes are 1,2-disubstituted or are cyclic alkenes then two distinct adducts may be obtained (see p. 71).

The way in which 'cis' and 'trans' were used in the previous paragraph to denote different modes of addition has no connection with their other usage to denote different stereoisomers of alkenes, and the two usages should not be mixed up. The one serves to describe a direction of attack on a molecule, the other describes a static structure of a molecule.

The most important addition reactions of alkenes will now be considered.

2.2.5.1 Addition of bromine or chlorine to an alkene gives vicinal dibromo- or dichloro-alkanes:

$$\begin{array}{c} \text{Br} \\ | \\ \text{R}-\text{CH}=\text{CH}-\text{R} + \text{Br}_2 \longrightarrow \text{R}-\text{CH}-\text{CH}-\text{R} \\ | \\ \text{Br} \end{array}$$

(Vicinal or *vic.* indicates that two substituents are attached to two neighbouring carbon atoms.)
The first step involves addition of a bromine cation to the C=C double bond and formation of a π-complex which is transformed into a *bromonium ion*. This then undergoes *trans*-addition by a bromide ion:

$$\text{(Bromonium ion reaction scheme)}$$

Bromonium ion

Trans-attack is favoured because the bromine atom is large and gets in the way of any attack at the *cis*-side. Both bromonium and bromide ions may be formed from a bromine molecule in the first step:

$$\text{(reaction scheme: C=C + Br—Br} \longrightarrow \text{Br}^\oplus + \text{Br}^-)$$

In this reaction the colour of the bromine solution disappears and this has been used as a qualitative test for the presence of double bonds.

If chlorine or bromine react with an alkene R—CH=CH—CH$_3$ at higher temperature and in the gas phase, then the only reaction which occurs is a substitution reaction in the methyl group instead of addition to the double bond. For example, at 500°C (775 K) propene and chlorine give allyl chloride:

$$H_2C = CH - CH_3 + Cl_2 \rightarrow H_2C = CH - CH_2Cl + HCl$$

Propene   Allyl chloride (3-Chloropropene)

2.2.5.2 Addition of hydrogen halides to alkenes provides a method for the preparation of alkyl halides. Hydrogen iodide reacts most readily, followed by hydrogen bromide; hydrogen chloride is the least reactive.

$$H_2C = CH_2 + HI \rightarrow H_3C - CH_2I$$

Ethylene   Iodoethane (Ethyl iodide)

Here again an electrophilic attack on the alkene is the first step. The polarity of the H–X bond means that it can readily be broken to give a proton and a halide ion.

Regioselective Reactions (Markovnikov's rule). If the two carbon atoms linked by a double bond have different numbers of hydrogen atoms attached directly to them, then, on addition of a hydrogen halide, the halide atom becomes bound to the carbon atom which had the fewest hydrogens attached to it. Thus addition of hydrogen iodide to propene gives 2-iodopropane:

$$H_3C - CH = CH_2 + HI \rightarrow H_3C - CH - CH_3$$
$$| \quad\quad\;\; I$$

Propene   2-Iodopropane (Isopropyl iodide)

The *Markovnikov* rule (1870) complements the *Saytzeff* rule (see p. 157).

Reactions in which more than one site of attack in a molecule is possible, but in which reaction in fact takes place preferentially at one of these sites, are described as *regioselective* reactions.[1] In the above reaction electrophilic attack on the double bond leads to the formation of a *carbenium ion*, which then reacts with a nucleophilic iodide anion to give 2-iodopropane.

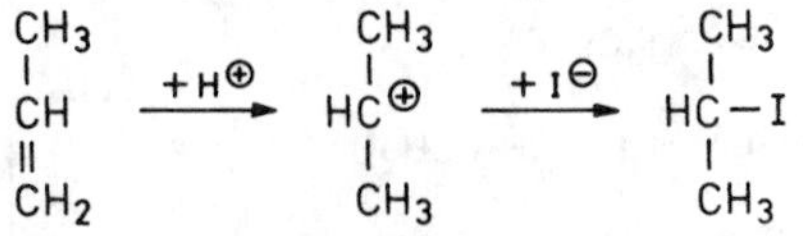

$$
\begin{array}{ccccc}
CH_3 & & CH_3 & & CH_3 \\
| & & | & & | \\
CH & \xrightarrow{+H^{\oplus}} & HC^{\oplus} & \xrightarrow{+I^{\ominus}} & HC-I \\
\| & & | & & | \\
CH_2 & & CH_3 & & CH_3
\end{array}
$$

Carbenium ion

It is possible to form two different carbenium ions, with different structures, either that shown, with a positive charge on the middle carbon atom or, by attack of the proton on the middle carbon atom, with a charge on the terminal carbon atom. Since alkyl groups are electron-donating in character, they stabilise an adjacent carbenium ion by partially delocalising the positive charge. Because of this the stability of carbenium ions rises in the order of primary < secondary < tertiary. Hence, in the reaction above, formation of the secondary carbenium ion is the preferred reaction (see p. 101 *ff*). This isopropyl cation can be detected in solution by means of *NMR spectroscopy*.

Reactions of *allyl derivatives* may go against *Markovnikov*'s rule.[2] For example, addition of hydrogen bromide to allyl bromide provides a mixture of 1,2- and 1,3- dibromopropanes:

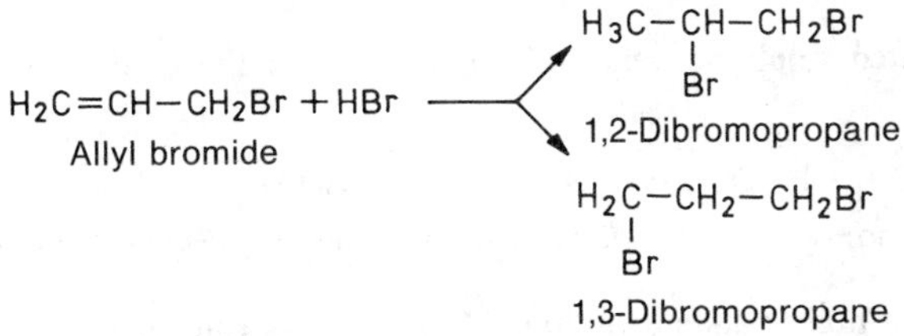

$$
H_2C=CH-CH_2Br + HBr \longrightarrow
\begin{cases}
H_3C-CH-CH_2Br \\
\quad\quad | \\
\quad\quad Br \\
\text{1,2-Dibromopropane} \\
\\
H_2C-CH_2-CH_2Br \\
\quad | \\
\quad Br \\
\text{1,3-Dibromopropane}
\end{cases}
$$

Allyl bromide

The formation of 1,2-dibromopropane is not only in agreement with *Markovnikov*'s rule; it is also in accord with the ionic reaction mechanism. For a complete understanding of reaction mechanisms, it is necessary to consider radical mechanisms as well as ionic mechanisms.

2.2.5.3 Peroxide Effect. In 1933, *Kharasch* and *Mayo* showed that addition of hydrogen bromide to an alkene in the presence of a small amount of a diacyl peroxide led to the formation of the alternative product from that expected according to *Markovnikov*'s rule, namely 1,3-dibromopropane.

This was described as an '*anti-Markovnikov*' product, and the action of the peroxide was called a *peroxide effect*.

The reaction mechanism involves the dissociation of the peroxide into free radicals, which then bring about chain reactions, as follows:

---

[1] This type of definition is more general and does not only apply to *Markovnikov*'s rule; see *A. Hassner*, J. Org. Chem. *33*, 2684 (1968).
[2] The deviation from *Markovnikov*'s rule which is seen in reactions of hydrogen halides with 3,3,3-trifluoropropene, which only proceed in the presence of AlCl$_3$, is probably associated with the formation of an intermediate allyl cation, $F_2\overset{\oplus}{C}-CH=CH_2$ (see p. 103); see *P. C. Myrhe* et al., J. Am. Chem. Soc. *92*, 7596 (1970).

$$H_3C-C \overset{\bar{O}l}{\underset{\bar{O}-\bar{O}}{}} \overset{l\bar{O}}{\underset{}{}} C \cdot CH_3 \longrightarrow 2\,H_3C-C \overset{\bar{O}l}{\underset{\dot{O}}{}} \longrightarrow H_3C-C \overset{\bar{O}l}{\underset{\dot{O}}{}} + \cdot CH_3 + CO_2$$

Diacetyl peroxide

$$H_3CCOO \cdot + HBr \longrightarrow H_3CCOOH + \cdot Br \qquad\qquad \text{Initiation}$$

$$H_3C-CH=CH_2 + \cdot Br \longrightarrow H_3C-\underset{\cdot}{C}H-CH_2Br \qquad (1)$$

Propene

$$H_3C-\underset{\cdot}{C}H-CH_2Br + HBr \longrightarrow H_3C-CH_2-CH_2Br + \cdot Br \qquad (2)$$

Propagation of chain reactions

1-Bromopropane

In the initiation reaction a bromine atom is generated, which attacks the alkene and produces a carbon radical. The bromine atom is highly reactive and attacks the most readily available site, which is the terminal carbon atom, providing a secondary radical (reaction 1, above). This removes a hydrogen atom from a molecule of hydrogen bromide, generating thereby another bromine atom, which can continue the chain (reaction 2). Termination of the chain involves combination of two radicals.

This type of chain reaction, promoted by the presence of peroxide, is only applicable to reactions involving hydrogen bromide, but not those involving HCl or HI. This is because reactions of type (1) are highly endothermic in the case of HI, whilst reactions of type (2) are highly endothermic in the case of HCl.

2.2.5.4 Concentrated sulphuric acid adds to alkenes in the cold, giving *alkyl hydrogen sulphates*:

$$H_2C=CH_2 + H-OSO_3H \rightarrow H_3C-CH_2-O-SO_3H$$

Ethylene   Ethyl hydrogen sulphate (Ethylsulphuric acid)

The reactions of higher alkenes follow *Markovnikov*'s rule, e.g.

$$H_3C-CH=CH_2 + H-OSO_3H \longrightarrow H_3C-\underset{\underset{OSO_3H}{|}}{C}H-CH_3$$

Propene

Since alkylsulphuric acids are readily hydrolysed to alcohols, this reaction provides an indirect, but useful, method for the addition of water to double bonds (see pp. 181, 188 *ff*). Industrially this is the so-called *sulphuric acid process* for the preparation of alcohols from alkenes.

2.2.5.5 Addition of hypochlorous acid to alkenes leads to the formation of 2-chloroalcohols, e.g.

$$H_2C=CH_2 + HO-Cl \longrightarrow H_2CCl-CH_2-OH$$

Ethylene   2-Chloroethanol

For higher alkenes, the reaction is regioselective, with the chlorine atom going to the carbon atom which has the most hydrogen atoms attached to it.

2.2.5.6 Nitrosyl chloride, NOCl, and nitrosyl bromide, NOBr, both add to alkenes, *e.g.*

$$H_2C=CH_2 + ON-Cl \longrightarrow H_2C-CH_2$$
$$\qquad\qquad\qquad\qquad\qquad \underset{NO}{|}\ \underset{Cl}{|}$$

1-Chloro-2-nitrosoethane

The nitrosoalkyl halides normally exist as dimers.

2.2.5.7 Oxidising agents react with alkenes to give a variety of products.

(a) Epoxidation of ethylene takes place with air or oxygen, in the presence of silver as a catalyst, at 220–280°C (495–555 K) and under pressure. *Ethylene oxide*, or *oxirane* is formed; it is made industrially in large quantities.

$$H_2C=CH_2 \xrightarrow[\text{(Ag)}]{+1/2\ O_2} H_2C\underset{O}{-}CH_2$$

Ethylene oxide (oxirane)

This method has not been used hitherto for higher alkenes.

(b) Prilezhaev Reaction (1909). Almost quantitative yields of *epoxides* (*oxiranes*) are obtained by the action of peracids, and especially *m*-chloroperbenzoic acid, on alkenes in inert solvents. Care is needed in handling peracids, see p. 547.

$$R-CH=CH-R' + \text{(m-ClC}_6\text{H}_4)\text{C}-\text{OOH} \longrightarrow R-CH-CH-R' + \text{(m-ClC}_6\text{H}_4)\text{C}-\text{OH}$$

m-Chloroperbenzoic acid   Epoxide

The reaction is stereospecific. Thus (*Z*)- and (*E*)-cyclooctene provide, respectively, only the *cis-* and *trans*-epoxides:

(*Z*)-Cyclooctene   *cis*-Epoxide   (*E*)-Cyclooctene   *trans*-Epoxide

Alkenes can be *hydroxylated* by *hydrogen peroxide* in the presence of acetic acid. Reaction proceeds by the following mechanism; an intermediate epoxide is converted into a *glycol* or *1,2-diol*:

Epoxide   A glycol (1,2-Diol)

Acetic acid and hydrogen peroxide form, and are in equilibrium with, peracetic acid. This makes an electrophilic attack on the C=C bond, forming an epoxide. In acid the peroxide is protonated. This increases the polarity of the C—O bonds and makes possible nucleophilic attack by water. For steric reasons water can only approach the epoxide from the rear. Loss of a proton provides the glycol.

(c) Oxidation of alkenes by osmium(VIII) oxide or by weakly alkaline potassium permanganate proceeds in the first place by *cis-addition*, giving a cyclic ester, which in the former case can be isolated. This is hydrolysed to a glycol.

At higher temperatures the oxidation with potassium permanganate proceeds further and results in splitting of the carbon chain to give carboxylic acids. The loss of the purple colour of a solution of potassium permanganate can be used as a qualitative test (*Baeyer* test) for the presence of C=C double bonds, but it is not a very specific test.

2.2.5.8 Ozonolysis of the C=C double bond was first carried out by *Harries* (1904) by passing ozone through a dry solution of the unsaturated material.

Petrol, chloroform, tetrachloromethane and acetic acid have been used as solvents. Ozonides are formed, which are explosive when dry. Their structure was suggested by *Staudinger* (1925) and later proved by *Pummerer* (1938) and *Rieche* (1942).

In this reaction the two carbon atoms formerly joined to one another by the double bond become separated by oxygen bridges. Hydrolysis of an ozonide breaks the ring and aldehydes and/or ketones are formed. In this reaction hydrogen peroxide is formed. This oxidises some of any aldehyde present, so that it is preferable to decompose an ozonide by catalytic hydrogenation.

Ozonolysis has been used to determine the constitution of alkenes, since the cleavage products provide information about the environment of the original double bond. Ozone is generated by passing an electric discharge through a stream of oxygen.

The method of formation of ozonides was first put forward by *Criegee*, and not all the details of the mechanism are yet known. *Criegee's* mechanism was modified by *Bailey* to account for some limited, but real, stereo-specificity which has been observed.[1] The alkene and ozone first of all undergo a 1,3-*dipolar cycloaddition reaction* to give an unstable 'initial' or 'primary' ozonide, which is a 1,2,3-trioxolan. This decomposes spontaneously to a carbonyl oxide, which may be dipolar or a diradical. These products recombine to give the final ozonide, which is a 1,2,4-trioxolan:

[1] P. S. *Bailey*, Ozonation in Organic Chemistry, 2 vols (Academic Press, New York 1982).

2.2.5.9 In the chemistry taking place in the atmosphere, peroxyradicals $RCH_2COO\cdot$ are of importance not only in the photochemical cleavage of alkanes but also for the generation of ozone in the troposphere.

2.2.5.10 Catalytic hydrogenation of C$=$C double bonds, to convert alkenes into alkanes, can be carried out at room temperature using hydrogen at a little above atmospheric pressure, in the presence of platinum black (*Willstätter*) or palladium (*Paal* and *Skita*) as heterogeneous catalysts.

$$R-CH=CH-R \xrightarrow{+2H} R-CH_2-CH_2-R$$

In the laboratory, *Adams'* PtO_2-catalyst is often used. This is made by the reaction of molten hexachloroplatinic(IV) acid with sodium nitrate. At the beginning of the hydrogenation the platinum(IV) oxide is reduced to give a highly active fine suspension of platinum. Raney nickel is another catalyst which has been used successfully. It is made by treating a nickel–aluminium alloy with a solution of sodium hydroxide.

Catalytic hydrogenation can also be carried out using *Wilkinson*'s catalyst at room temperature and with the hydrogen at normal pressure (see p. 381 *ff*).

In industry, as well as nickel (*Sabatier* and *Senderens*) other metals such as iron, chromium, cobalt or copper are also used as hydrogenation catalysts. Since these metals are less active, high pressure and temperatures of about 200–300°C (475–575 K) are required. These hydrogenations are normally carried out in the gas phase.

2.2.5.11 For **the addition of diborane** to alkenes see p. 180 *ff*).

2.3 Polymerisation of Alkenes and Vinyl Derivatives

Polymerisation is one of the most important of the reactions which are characteristic of alkenes and their derivatives, and may be formulated in general terms as follows:

$$n\,H_2C=CH \longrightarrow \left[\!\!\begin{array}{c} -CH_2-CH- \\ | \\ R \end{array}\!\!\right]_n$$

Monomer   Polymer

The vital step is the activation of the C$=$C double bond, which may be brought about in a number of ways. Once this is achieved, the polymer chain grows, normally providing a straight chain or *linear polymer*. Under certain conditions the chain may at some point branch into two separate strands, resulting in *chain branching.*

Substances which bring about polymerisation, with the formation of *macromolecules*, are called *initiators*, or sometimes *activating agents*, or just *catalysts*. They are, for the most part, consumed in the course of the polymerisation, but only catalytic amounts are required. They can be classified generally into three types, depending on whether they bring about radical polymerisation, ionic polymerisation or co-ordinative polymerisation.

In the case of radical processes, some monomers can be activated by heat or by short-wavelength irradiation.[1]

2.3.1 Radical-induced Polymerisation

This type of polymerisation is the most economic, because it can be carried out in aqueous emulsions or suspensions. The macromolecules produced in this way have the widest range of chain lengths. Polymerisation requires an *initiator* (a *radical generator*); common examples are oxygen, hydrogen peroxide, organic peroxides such as a diacyl peroxide or cumyl hydroperoxide or aliphatic azo-compounds such as α, α'-azoisobutyrylnitrile (1,1'-dimethyl-1,1'-dicyanoazoethane, $(H_3C)_2(CN)C-N=N-C(CN)(CH_3)_2$.

The mechanism of radical chain polymerisation is as follows:

$$\bullet R' + H_2C{=}CH{\underset{R}{|}} \longrightarrow R'-CH_2-\overset{\bullet}{C}H{\underset{R}{|}} \qquad \text{Initiation reaction}$$

$$R'-CH_2-\overset{\bullet}{C}H{\underset{R}{|}} + nH_2C{=}CH{\underset{R}{|}} \longrightarrow R'-CH_2-CH{\underset{R}{|}}{\left[CH_2-CH{\underset{R}{|}}\right]}_{n-1}CH_2-\overset{\bullet}{C}H{\underset{R}{|}} \qquad \text{Chain-lengthening (Propagation)}$$

Radical   Macroradical

The initiator provides a radical $\cdot R'$, which activates the double bond by adding to it to form a new radical, which then initiates the propagation reactions. This proceeds by successive addition of monomer molecules to the growing chain, producing a macroradical. When this chain reaction comes to an end, the stable polymer results. The *termination reaction* can be brought about in a number of ways. The regioselectivity of the chain-lengthening process depends largely upon steric effects.[2]

Most radical chain polymerisations can be slowed down or totally stopped by the presence of *inhibitors* (*antioxidants*) such as hydroquinone, pyrogallol, thiophenol, etc. These are used as *regulators* to control the speed of polymerisations which are inherently too rapid, and as *stabilisers* for monomeric vinyl derivatives.

An important function of regulators is to bring about *chain* transfer, in which the growth of one chain is stopped whilst a new chain is started:

$$R'{\sim}CH_2-CH\bullet{\underset{R}{|}} \ + \ HSC_6H_5 \longrightarrow R'{\sim}CH_2-CH_2{\underset{R}{|}} \ + \ C_6H_5S\bullet$$

Macroradical   Thiophenol

$$C_6H_5S\bullet \ + \ CH_2{=}CH{\underset{R}{|}} \longrightarrow C_6H_5S-CH_2-CH\bullet{\underset{R}{|}}$$

In this way polymers with fairly uniform chain lengths can be obtained.

---

[1] See *K. J. Saunders*, Organic Polymer Chemistry, 2nd edn. (Chapman & Hall, London 1988); *Houben-Weyl*, Methoden der Organischen Chemie, Makromolekulare Stoffe, 4th edn, *E20* (Thieme Verlag, Stuttgart 1987).
[2] See *B. Giese*, Formation of C—C Bonds by Addition of Free Radicals to Alkenes, Angew. Chem. Int. Ed. Engl. *22*, 753 (1983); Radicals in Organic Synthesis: Formation of Carbon Bonds (Pergamon Press, Oxford 1987).

2.3.2 Ionic Polymerisation

The second type of polymerisation is ionic chain polymerisation. This is started by using either cationic (electrophilic/acidic) or anionic (nucleophilic/basic) initiators. Water tends to destroy these initiators; hence ionic polymerisation is carried out only in organic solvents with which the initiators do not react.

2.3.2.1 Cationic Polymerisation. Typical catalysts for these reactions are *Lewis* acids, such as boron trifluoride, aluminium trichloride, titanium(IV) chloride and tin(IV) chloride; traces of water or of acid are needed to initiate reaction. A suitable alkene, e.g. a vinyl ether, reacts with the catalyst to produce a *carbenium* ion, which then takes part in an ionic polymerisation chain reaction:

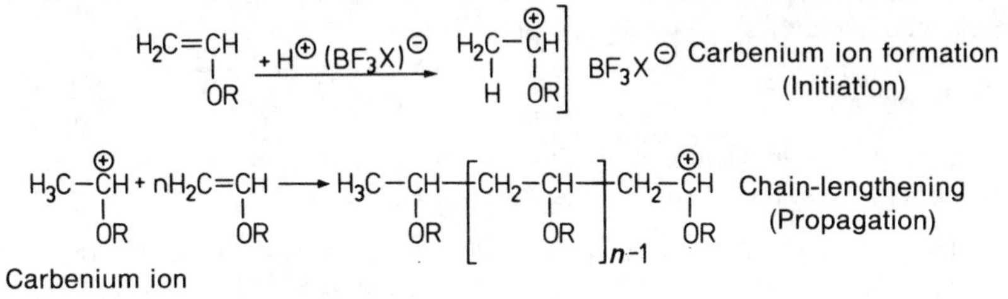

Chain growth is stopped by anions.

2.3.2.2 Anionic Polymerisation. The catalysts for this process are bases, for example, metal alkyls, alkoxides, amides and hydroxides. When sodamide reacts with a suitable alkene ($R''=COOH, CN$) a *carbanion* is formed. This in turn initiates the anionic chain polymerisation:

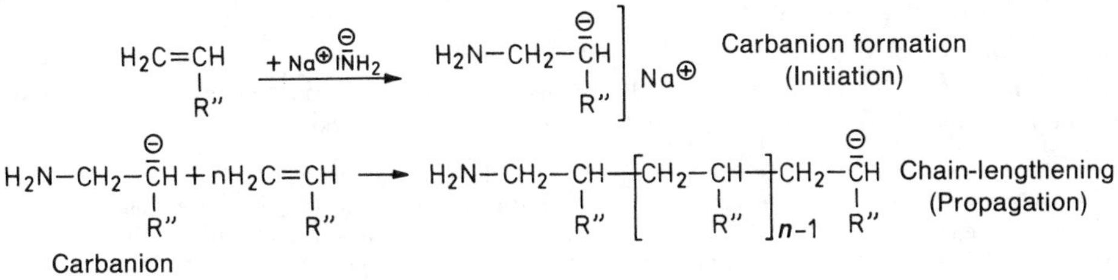

Chain growth is stopped by cations.

2.3.3 Co-ordinative Polymerisation

Ziegler–Natta catalysts are a group of mixed catalysts which promote chain polymerisation by forming metal complexes. In the following reaction scheme it is shortened to Al/Ti which represents the composition of the first *Ziegler* catalyst. The monomers are represented by the letter E, and are numbered in sequence in the order in which they take part in the reaction.

$$\text{Al/Ti} \xrightarrow{E_1} \text{Al/Ti-E}_1 \xrightarrow{E_2} \text{Al/Ti-E}_2\text{-E}_1 \rightarrow \rightarrow \rightarrow \text{Al/Ti-E}_n \text{ - - - } E_2\text{-E}_1$$

The underlying difference between co-ordinative polymerisation and the other methods of polymerisation discussed previously lies in the fact that the polymer chain is not built up at its *free* end, but rather from the end attached to the catalyst. Each new monomer unit is first co-ordinated to the metal complex, activated in the process, and inserted between the existing chain and the metal complex. In doing this it can be so orientated that a *stereoregular polymerisation* ensues, with all the constituent units arranged in the same direction. Chain lengthening can be halted by thermal cleavage of the bond linking the chain to the catalyst.

2.3.4 *Polymerisation of Alkenes*

The extent to which polymerisation of alkenes proceeds depends upon the catalyst which is present, and the degree of polymerisation determines the nature of the resultant polymers, whether they are viscous liquids or solids.

2.3.4.1 Polyethylene

The *polymerisation of ethylene* can proceed *via* ionic, radical or co-ordinative mechanisms. Usually, aliphatic, alicyclic or benzenoid hydrocarbons are used as solvents or as diluents.

(a) **In the presence of** *Lewis acids* ($AlCl_3$, BF_3), solutions of ethylene undergo cationic polymerisation at 30–50 bar (3–5 MPa) to provide low molecular weight polymers (mol. mass ~ 400), which, because of their high viscosity and low freezing point, are used as lubricating oils. Aluminium chloride also brings about isomerisation, giving highly branched paraffins.

(b) **High Pressure Process of ICI** (Imperial Chemical Industries). Radical chain polymerisation of ethylene, giving high molecular weight products, is brought about by the use of very high pressures (1000–2000 bar, 100–200 MPa) at 200°C (475 K), using a precise amount of oxygen (0.05–0.10%) as initiator. The process is a continuous one, providing a 10–15% yield for each time of passage. Liquid polyethylene is separated at normal pressure and, when the product has set, it is granulated.

High pressure polyethylene is made up of long branched alkane chains with a relative molecular mass above 10000. It is characterised by high elasticity and flexibility. Its trade names include *Alkathene* (ICI). *Baylon* (Bayer), *Mirathen* (Leuna), *Lupolen* (BASF), *Hostalen* (Hoechst) and *Polythene* (Wolff Walsrode AG). It melts at about 115°C (390K) and can be moulded when heated. It finds a variety of uses as foil, films, household implements, protective packaging, armatures, gears and cables and other machine parts, and as insulating material in electrical apparatus and cables. Because of its lower density it is generally known as LDPE (*low density polyethylene*).

(c) If the polymerisation of ethylene is carried out **in solution in methanol**, using dibenzoyl peroxide as initiator, at 110–120°C (385–395 K) and under a pressure of 200–300 bar (20–30 MPa), a waxy polymer (Lupolen N) with a relative molecular mass of 2000–3000 is formed.

(d) **Medium Pressure Process** (Phillips Petroleum Company). In this process, ethylene is polymerised at 150–180°C (425–455 K) and 35 bar (3.5 MPa) in a solvent such as a xylene fraction, in the presence of a chromium(VI) oxide/aluminium silicate catalyst. The product is a completely linear polymer (Marlex), which is crystalline and has a relative molecular mass of 5000–20 000.

(e) **Normal Pressure Process** (*Ziegler* Process). In 1955 *K. Ziegler*[1] described the polymerisation of ethylene at normal pressure and lower temperature (70°C [345K]), in

[1] See *K. Ziegler* et al., Das Mülheimer Normaldruck-Polyethylen-Verfahren (The Mülheim Normal Pressure Polyethylene Process), Angew. Chem. 67, 541 (1955) and 71, 623 (1959).

the presence of a mixed organometallic catalyst, for example a mixture of triethylaluminium and titanium(IV) chloride, in a solvent such as isododecane. In contrast to high pressure polyethylene, the product is made up from unbranched paraffin chains and is 80% crystalline. The relative molecular mass varies between 10 000 and 6 000 000. Of special industrial importance are the polyethylenes with relative molecular masses of 50 000 to 100 000.

$$n\,H_2C = CH_2 \rightarrow \;+\!\!\!-CH_2-CH_2\!\!-\!\!\!\!\;]_n$$

Normal pressure (low pressure) polyethylene

This thermoplastic has a higher density than LDPE and is consequently known as HDPE or *high density polyethylene*. Because of its higher rigidity it is useful for making products which are produced by means of injection moulding processes, *e.g.* household products, and fibrous material used for making filterpads. In the pharmaceutical industry it is mixed with paraffin oil in the preparation of 'Vaseline', which is characterised by its constant yield point.

2.3.4.2 Polypropylene; Polypropene

Propene is obtained from the distillation, under pressure, of gases obtained by the cracking of oil. It can be converted into polymers *only* by co-ordinative chain polymerisation.

(a) **Stereoregular Polymerisation**[1] (*Natta* Process). In 1956, *Natta* achieved the *stereoregular* polymerisation of unsymmetric monoalkenes, such as propene, vinyl ether and styrene, using modified *Ziegler* catalysts such as triethylaluminium and violet α-titanium(III) chloride to bring about a co-ordinative chain polymerisation. A heterogenous initiation system was used. From propene, up to 95% *isotactic* polypropene of m.p. 176°C (449 K) was obtained.

The chains of the isotactic polymer have a regular ordering of the monomer units and all the tertiary carbon atoms take up the same conformation. If one thinks of the main chain as taking up a zigzag form in one plane, then the substituents (R = methyl) all lie on the same side of this plane (Fig. 46).

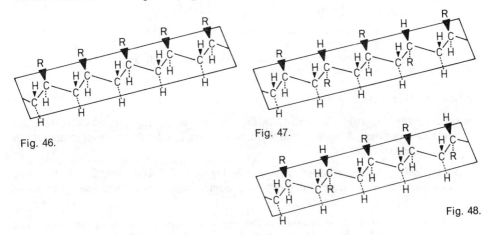

Fig. 46.

Fig. 47.

Fig. 48.

---

[1] *G. Natta*, Von der stereospezifischen Polymerisation zur asymmetrischen autokatalytischen Synthese von Makromolecülen, Angew. Chem. *76*, 553 (1964).

The regular arrangement of the side chains must be a consequence of the monomer units all co-ordinating to the metal complex in the same steric fashion.[1]

Instead of being arranged in this isotactic manner, it could be possible for the substituent groups R to be alternatively on either side of the plane (Fig. 47), or to be arranged quite randomly (Fig. 48). These two structures are described, respectively, as *syndiotactic* and *atactic*. The general term for the arrangement of the units in the polymers is known as their *tacticity*. Rigorously isotactic and syndiotactic polymers are achiral.[2]

By appropriate choice of the polymerisation process it is possible to obtain from propene polymers having either isotactic, syndiotactic or atactic structures. The first two are crystalline; but atactic polymers, because of their disordered structure, are amorphous.

Isotactic polypropylene is harder, stiffer and more transparent than polyethylene and in consequence is used for the production of fibres and of films. Trade names include *Hostalen PP* (Hoechst), *Vestolen P* (Hüls), *Moplen* (Montecatini) and *Profax* (Hercules).

(b) When propene and ethylene are *copolymerised* in the presence of a special *Ziegler–Natta* catalyst, for example diethylaluminium chloride and triethyl orthovanadate, $OV(OC_2H_5)_3$, a rubber-like *elastomer* (EPM) is obtained, which can be cross-linked by the action of peroxides (*Natta*, 1957).

If a diene, such as 5-ethylidenenorbornene (p. 397), is added as a third component in the polymerisation (*terpolymerisation*), a product is formed which has side chains with double bonds in them. In consequence, the polymer can be vulcanised; the processes involving the double bonds in the side chains do not affect the main chains of the polymers.

2.3.4.3 Polybut-1-ene

Like propene, but-1-ene can be polymerised in the presence of a *Natta* catalyst to give a stereoregular isotactic polybut-1-ene:

$$n\ \underset{\substack{|\\CH_2\\|\\CH_3}}{CH}=CH_2 \longrightarrow \left[\underset{\substack{|\\CH_2\\|\\CH_3}}{CH}-CH_2\right]_n$$

Isotactic polybut-1-ene, m.p. 125°C (398 K)[3]

This crystalline polymer is characterised by its high tensile strength and resistance to breakage. Therefore it is used for making foil and tubing.

2.3.4.4 Poly(isobutene)

Isobutene, which is available both from the cracking of oil and synthetically, does not polymerise under the influence of heat, light, or radicals, but does so at lower temperatures in the presence of *Lewis* acids.

(a) Cationic polymerisation of isobutene is brought about at $-80°C$ (195 K) using boron trifluoride as catalyst and provides products, which, with increasing degrees of

---

[1] See *W. Kaminsky* et al., Polymerisation of Propene and Butene with a Chiral Zirconocene and Methylalumoxane as Cocatalyst, Angew. Chem. Int. Ed. Engl. *24*, 507 (1985).
[2] *G. Wulff*, Main-Chain Chirality and Optical Activity in Polymers Consisting of C—C Chains, Angew. Chem. Int. Ed. Engl. *28*, 21 (1989).
[3] According to IUPAC rules, the repeating unit enclosed within the square brackets is drawn so that the more substituted carbon atom is on the left.

polymerisation, range from lubricating oils, through sticky solids, to rubber-like products (Oppanol B, Vistanex).

$$n \; \underset{\underset{CH_3}{|}}{\overset{\overset{CH_3}{|}}{C}}{=}CH_2 \quad \xrightarrow{(BF_3)} \quad \left[\underset{\underset{CH_3}{|}}{\overset{\overset{CH_3}{|}}{C}}{-}CH_2 \right]_n$$

Isobutene   Poly(isobutene)

In contrast to rubber, these polymers have no double bonds in the molecules. In consequence they are unaffected by air, and they cannot be vulcanised with sulphur. They are used as linings and as insulators for cables.

(b) When isobutene is copolymerised with a small amount of either isoprene or butadiene (1–5%) at $-80°C$ (195 K), with aluminium chloride as catalyst, a vulcanisable rubber called *butylrubber*, is formed, which has a small number of double bonds in the molecules. Because of its very low permeability to gases it is used for the manufacture of tyres for motorcars.

2.3.5 Polymerisation of Vinyl Derivatives

Among readily polymerisable vinyl compounds are vinyl chloride, $H_2C{=}CHCl$, vinylidene chloride (1,1-dichloroethylene), $H_2C{=}CCl_2$, vinyl ethers, $H_2C{=}CH{-}OR$, and vinyl acetate, $H_2C{=}CH{-}O{-}COCH_3$. When either one hydrogen atom, or two hydrogen atoms at the same end, of an ethylene molecule are replaced by atoms or groups of atoms which are more electronegative than carbon, then the $C{=}C$ double bond is polar in character and ionic polymerisation is hence favoured. It would be expected, therefore, that the preferred method of polymerisation of these compounds would be ionic. However this is only so in the case of vinyl ethers. For vinyl halides and vinyl acetate, radical polymerisation is the method of choice.

Other ethylene derivatives which are readily polymerised include derivatives of acrylic acid, styrene (vinylbenzene) and tetrafluoroethylene. They will be dealt with in later chapters.

2.3.5.1 Poly(vinyl chloride) (PVC)

The polymers derived from vinyl chloride are among the most important of *thermoplastics*. Vinyl chloride is prepared from either ethylene or acetylene; it is a *carcinogen*.

Preparation of Vinyl Chloride

(a) By *addition of hydrogen chloride to acetylene* at 200°C (475 K) under pressure, in the presence of mercury(II) chloride on active charcoal:

$$HC{\equiv}CH + HCl \xrightarrow{(Hg^{2\oplus})} H_2C{=}CHCl$$

Acetylene   Vinyl chloride (b.p. -13.8 °C [259.4])

(b) *Addition of chlorine to ethylene*, obtained from the cracking of oil, gives *1,2-dichloroethane*; hydrogen chloride is removed from this in the presence of aluminium oxide at 300–350°C (575–625 K):

$$H_2C = CH_2 \xrightarrow[\text{+ Cl}_2]{} ClH_2C-CH_2Cl \xrightarrow[-HCl]{(Al_2O_3)} H_2C = CHCl$$

Ethylene   1,2-Dichloroethane   Vinyl chloride

The principal method for converting ethylene into 1,2-dichloroethane is by *oxidative chlorination* with hydrogen chloride and oxygen in the gas phase, using copper(II) chloride as catalyst:

$$H_2C = CH_2 + 2\,HCl + 1/2\,O_2 \rightarrow ClH_2C-CH_2Cl + H_2O \quad \Delta H = -239\ \text{kJ/mol}$$

Radical polymerisation of vinyl chloride is usually carried out in aqueous emulsion or suspension and in the presence of a peroxide. Solid poly(vinyl chloride) (PVC) is formed:

$$n\ \underset{\underset{Cl}{|}}{CH{=}CH_2} \longrightarrow \left[\underset{\underset{Cl}{|}}{CH{-}CH_2} \right]_n$$

Vinyl chloride   Poly(vinyl chloride)

In the working-up process, the *PVC powder*, which contains 56.6% chlorine and begins to soften at about 80°C (355 K), is sintered together at 150–170°C (420–440 K) and pressed in a mould. In the thermal treatment of PVC some hydrogen chloride is formed, which acts as a catalyst for further decomposition. Consequently, salts of weak acids, such as sodium carbonate, or organotin compounds, are added to act as *stabilisers*.

Since as long ago as 1913, PVC has found many technical applications, for example for domestic articles, discs, tubing, valves and stopcocks, as well as for coatings for apparatus. When it is mixed with plasticisers, such as esters of phthalic acid or of adipic acid, a malleable PVC results, which has some elasticity, and is suitable for use as artificial leather, raincoats, floor coverings and piping.

For special purposes, which require a lower moulding temperature and higher solubility of the material, vinyl chloride is *copolymerised* together with vinyl acetate, vinyl ethers, esters of acrylic acid or acrylonitrile.

2.3.5.2 Poly(vinylidene chloride)

Vinylidene chloride, $H_2C{=}CCl_2$, b.p. 37°C (310 K), is obtained by removal, by means of sodium hydroxide, of hydrogen chloride from *1,1,2-trichloroethane*, which is itself prepared by the action of chlorine on either vinyl chloride or 1,2-dichloroethane:

$$
\begin{array}{c}
H_2C{=}CHCl \\
\text{Vinyl} \\
\text{chloride} \\
ClH_2C-CH_2Cl \\
\text{1,2-Dichloroethane}
\end{array}
\begin{array}{c}
\xrightarrow{+Cl_2} \\
\xrightarrow[-HCl]{+Cl_2}
\end{array}
ClH_2C-CHCl_2 \xrightarrow[-HCl]{(NaOH)} H_2C{=}CCl_2
$$

1,1,2-Trichloroethane   Vinylidene chloride

The polymerisation of vinylidene chloride proceeds similarly to that of vinyl chloride:

$$n\ \underset{\underset{Cl}{|}}{\overset{\overset{Cl}{|}}{C}{=}CH_2} \longrightarrow \left[\underset{\underset{Cl}{|}}{\overset{\overset{Cl}{|}}{C}{-}CH_2} \right]_n$$

Poly(vinylidene chloride)

For various technical purposes vinylidene chloride and vinyl chloride are *copolymerised*. The polymer so obtained (Diophan, Vertan) is used for films, as fibres for filter cloths, conveyer belts and acid-proof ropes. The copolymer of vinylidene chloride and acrylonitrile (Diorit) is used, for example, for making bristles.

2.3.5.3 Poly(vinyl ethers)

Vinyl ethers, $H_2C{=}CH{-}OR$, are examples of enol ethers (see p. 213), which are prepared industrially by the addition of alcohols to acetylene. Boron trifluoride is used as a catalyst for the polymerisation of vinyl ethers,[1] which proceeds by a cationic mechanism:

$$n\ \underset{\underset{OR}{|}}{CH{=}CH_2} \longrightarrow \left[\underset{\underset{OR}{|}}{CH{-}CH_2}\right]_n$$

Poly(vinyl ether)

When vinyl ethers are copolymerised with other vinylic monomers, radical polymerisation is employed.

Poly(vinyl ethers) are used as finishings for textiles, in varnishes, and as sizing materials and adhesives; poly(isobutylvinyl ether) is used as dubbing for leather.

2.3.5.4 Poly(vinyl acetate)

Vinyl acetate, $H_2C{=}CH{-}O{-}COCH_3$, b.p. 72°C (345 K)/970 mbar (970 hPa) is made industrially by the *addition of acetic acid to ethylene* in the presence of palladium salts, copper(II) chloride and oxygen (oxidative acetylation):

$$H_2C{=}CH_2 + H_3CCOOH + \tfrac{1}{2}O_2 \xrightarrow{\text{(Cat.)}} H_2C{=}CH{-}O{-}COCH_3 + H_2O \quad \Delta H = -176\ \text{kJ/mol}$$

Ethylene   Acetic acid   Vinyl acetate

Vinyl acetate polymerises readily under the influence of heat, light or peroxides to give poly(vinyl acetate):

$$n\ \underset{\underset{O{-}COCH_3}{|}}{CH{=}CH_2} \longrightarrow \left[\underset{\underset{O{-}COCH_3}{|}}{CH{-}CH_2}\right]_n$$

Poly(vinyl acetate)

These polymers are readily soluble in organic solvents. They are used as binders for paints and as adhesives.

It is interesting that poly(vinyl esters) undergo ester exchange reactions with methanol, leading to the formation of poly(vinyl alcohol). The framework of the macromolecular chains remains unaltered.

$$\left[\underset{\underset{O{-}COCH_3}{|}}{CH{-}CH_2}\right]_n \xrightarrow[-\ n\ H_3CCOOCH_3]{+\ n\ H_3COH} \left[\underset{\underset{OH}{|}}{CH{-}CH_2}\right]_n$$

Poly(vinyl acetate)   Poly(vinyl alcohol)

---

[1] See *G. Natta* et al., Stereospezifische Polymerisation von Vinyläthern (Stereospecific Polymerisation of Vinyl Ethers), Angew. Chem. *71*, 205 (1959).

Industrially, about half of the production of poly(vinyl acetate) is converted into poly(vinyl alcohol). Because it has very similar properties to starch, the latter is used in the textile industry, as an adhesive for the light-sensitive layer in television tubes, and as a protective colloid. Poly(vinyl alcohol) is not prepared directly from vinyl alcohol, $H_2C{=}CHOH$, since the monomer is very unstable and very readily isomerises to acetaldehyde (see p. 95).

2.4 The Chemistry of Mineral Oil (Petroleum)

The oil industry is responsible for extracting mineral oil together with natural gas from its natural sources, and obtaining from it petrol, diesel oil, heating oil, lubricants, paraffin and bitumen.

Closely connected with this, and of especial importance in countries which lack natural oil resources, is the production of fuels from coal and brown coal. This will also be discussed, therefore, in this section.

2.4.1 Mineral Oil, Petroleum; Natural Gases; Oil Shale

Mineral Oil. This is derived largely from the bacterial decomposition of organic material which is of maritime origin. It is a natural fossil raw material.[1] Its economic importance is due to the very many products which can be obtained when it is refined.

The main sources of oil are in the Near East (Saudi Arabia, Kuwait, Iran, Iraq, Abu Dhabi), in North America (Texas, Pennsylvania, California, Alaska, Canada, Mexico), in the former Soviet Union (the Ural–Volga region, Caucasus, Siberia), in South America (Venezuela, Argentina, Brazil, Chile, Peru), in the Far East (China, India, Malaysia), in Africa (Libya, Nigeria, Algeria, Egypt) and in Western Europe (Great Britain, Norway, the North Sea, Italy, and a small amount in Germany).[2] The known reserves at present are estimated to be more than 10^{11} tonnes.[3] About 50% of this is in the Near East. The known reserves of natural gas are at least $10^{12} \, m^3$; about 40% of this is in the former Soviet Union.

The constitutions of mineral oils differ according to their sources. They consist of complex mixtures of hydrocarbons, including alkanes, cycloalkanes and arenes. For example, oil from Pennsylvania consists largely of alkanes, while that from the former Soviet Union and Romania contains up to 80% cycloalkanes. Because of the variations, a distinction is made between *alkane-based* and *cycloalkane-based* mineral oils. The former contain only very small amounts of arenes, the latter contain rather more.

The commonest cycloalkanes are cyclopentane and cyclohexane and also alkyl derivatives thereof.

---

[1] See *J. M. Hunt*, Petroleum Geochemistry and Geology (Freeman, San Francisco 1979).
[2] See Weltatlas Erdöl und Erdgas, 3 Aufl. (World Atlas of Oil and Natural Gas, 3rd edn) (Westermann Verlag, Brunswick 1982).
[3] It is estimated that of the known deposits, only about 30% will be recoverable.

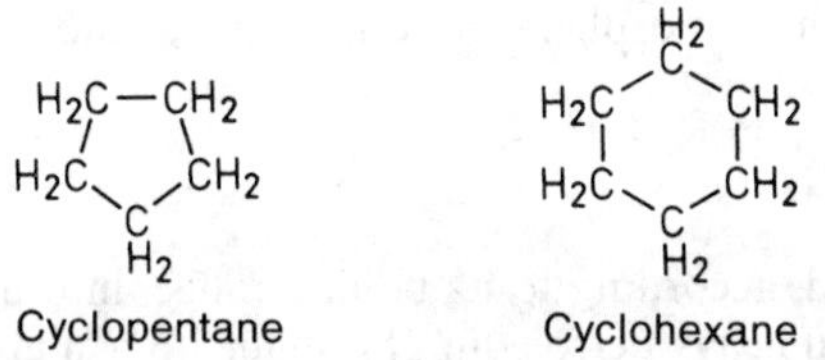

H2C—CH2
H2C   CH2
        C
        H2

Cyclopentane

       H2
       C
H2C   CH2
H2C   CH2
       C
       H2

Cyclohexane

Oil from Indonesia contains up to 40% arenes, such as *benzene, toluene, xylene* (dimethylbenzene) and other higher homologues.

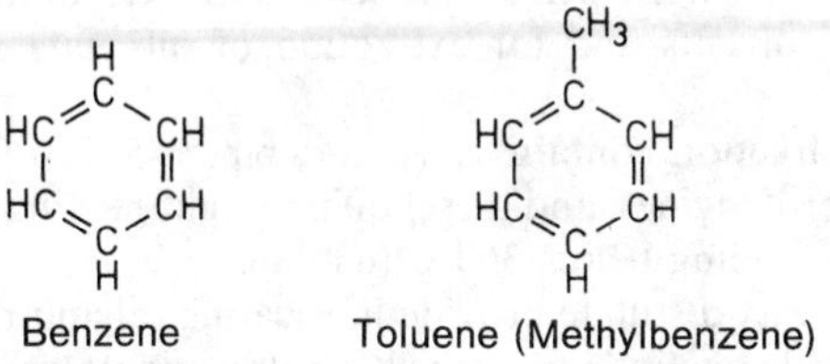

        H
        C
HC   CH
HC   CH
        C
        H

Benzene

        CH3
        C
HC   CH
HC   CH
        C
        H

Toluene (Methylbenzene)

Mineral oil also contains between 0.1 and 10% of organic sulphur compounds, which may be mercaptans (thiols), sulphides, disulphides, thiophenes and other thiacycles. In addition there may be small amounts of nitrogen-containing compounds such as pyridine and quinoline derivatives, and of oxygen-containing compounds such as cycloalkylcarboxylic acids. The latter have been identified as cyclopentane derivatives (*Markovnikov, Nenitzescu*).

Extraction of Mineral Oil from the Earth. Since the oil-bearing layers in the earth are under high pressure, it may only be necessary to bore a hole to bring the dark brown *crude oil*, which is contaminated with sand and water, to the surface. If the pressure is insufficient or if it drops, the oil is pumped to the surface. The first rough separation of oil from gas then takes place in a separator.

Natural Gas. The gas obtained from crude oil contains alkanes which are liquids at normal pressure and temperature. They are removed either by compressing the gas, when they condense, or by absorption in oil. *Natural gas* consists largely of methane, together with small amounts of ethane, propane, butane and isobutane, and also, possibly, some nitrogen, hydrogen sulphide, carbon dioxide and helium.

In recent years there has been a series of new discoveries of large reserves of natural gas, for example in the North Sea and in the Netherlands. The gas from these fields is taken by pipelines directly to industrial centres, or for use as mains gas, which is distributed to domestic and industrial users. It has about double the calorific value of coal gas. The hydrocarbons in natural gas can be liquified under pressure and separated by low-temperature distillation.

Industrially natural gas is used not only as a fuel and for heating but also as a starting material for the preparation of acetylene, hydrogen cyanide, carbon black, and of hydrogen for *synthesis gas* (see p. 54). Synthesis gas is also produced by the gasification of coal with steam and oxygen, despite the high energy input required in the process (see p. 373).

Industrial Work-up of Crude Oil. The crude oil is either worked up in the vicinity of the oil field or transported by pipelines or in ships (tankers) to special oil refineries. Fractional distillation of crude oil at atmospheric pressure provides gas (CH_4, C_2H_6), *liquified gases* (C_3H_8, C_4H_{10}) and also the following fractions:

Up to 200°C (475 K): gasoline (straight-run gasoline, naphtha)
175–350°C (445–625 K) (middle distillate): kerosene (paraffin) and diesel oil (gas oil)
Above 350°C (625 K): atmospheric residue (heavy heating oil).

Vacuum distillation of the atmospheric residue provides the following fractions:

Vacuum gas oil
Wax distillate
Vacuum residue (bitumen).

Gasoline is subdivided, according to its boiling range, into light petroleum, b.p. up to about 100°C (375 K), and heavy petroleum. It is made up of a mixture of different alkanes, cycloalkanes and arenes, containing from five to about twelve carbon atoms, the nature of the mixture depending very much upon the source of origin. It is further fractionated to give different grades of light petroleum with narrower boiling-point ranges. These fractions are used as solvents and for extraction of oils, fats and resins, and in dry-cleaning.

The middle-boiling fraction contains kerosene, b.p. 175–280°C (450–555 K), which is used as a fuel for jet aero-engines, and diesel oil and light heating oil, which must contain up to 90% of material boiling below 360°C (630 K).

Vacuum gas oil and wax distillate provide lubricating oil and paraffin oil (white spirit). In the process, unwanted constituents, such as paraffin and polycyclic arenes, are removed by further refining. Heavy lubricating oils are obtained by extraction of the vacuum residue with various solvents.

The paraffins obtained by the refining of lubricating oil are converted, by means of oxidation and reaction with sulphur dioxide and chlorine (chlorosulphonation), into alkyl sulphonyl chlorides, which are used in the manufacture of detergents, wetting agents, etc. The paraffins are also used in the manufacture of candles and for impregnating materials.

Bitumen is used, among other things, for making road surfaces (tar, asphalt), for making brickettes for fuel, and in the preparation of roofing material.

Oil Shale is a sedimentary rock which contains a complicated mixture of organic compounds known as *kerogen*. *Shale oil* is obtained from it by heating. It consists largely of condensed hydrocarbons, which can be converted by hydrogenation into fuel oil.

2.4.2 Fuels from Mineral Oil

The great increase in the number of motorcars and the associated growth of the motorcar and aircraft industries has led to a constantly increasing demand for fuel oils, above all petrol. This has provided the petroleum industry with new problems concerning the quality and the antiknock properties of petrol.

When 'knocking' occurs, there is a premature ignition of the mixture of petrol and air when they are compressed, resulting in a considerable loss of energy. It is necessary to produce 'antiknock' petrol in order to increase the efficiency of the engine by higher compression, while at the same time lowering the fuel consumption. The term *octane number* was introduced in 1927 as an indication of the quality of a petrol. Quite arbitrarily, n-heptane, which has a great tendency to cause knocking, was given an octane number = 0, and isooctane, which only ignites at much higher compression, was given an octane number = 100.

$$\underset{\underset{CH_3}{|}}{\overset{\overset{CH_3}{|}}{H_3C-C-CH_2-CH-CH_3}}\underset{CH_3}{|} \qquad \text{Isooctane (2,2,4-Trimethylpentane)}$$

The octane rating of any particular petrol is then derived from the percentage of isooctane in an isooctane–heptane mixture which would have the same tendency to knock as that petrol.

Knocking is also decreased by *additives* which are included in the petrol in small amounts, and which are described as antiknock agents.

The commonest antiknock agent hitherto has been tetraethyl-lead, $Pb(C_2H_5)_4$. Halogenoalkanes such as 1,2-dichloro- or 1,2-dibromo-ethane are also added to trap lead which is formed, and to prevent its deposition in the cylinders of the engine. Lead in the exhaust gases, together with nitrogen oxides (NO_x) and incompletely burned hydrocarbons, pollute the environment. The production of lead-free petrol with comparable resistance to knocking results in higher cost, since the processes outlined below have to be used to an increasing extent.

The tendency of a diesel fuel to ignite is characterised by its *cetane number*. Hexadecane (cetane), $CH_3(CH_2)_{14}CH_3$ is assigned a value = 100, and 1-methylnaphthalene, which ignites very sluggishly in a mixture, is rated = 0.

The increased demand for fuels and for light heating oils cannot be met solely by the distillation of crude oil, and, as a result, the cracking of heavy oils to provide fuels and light oils has been developed (conversion processes).

2.4.2.1 Thermal Cracking (1913). This process involves the thermal breakdown of the middle distillate and the distillation residues to provide a mixture of lower-boiling hydrocarbons. It was first in general use in the USA. Cracking takes place at about 400–500°C (675–775 K), and under various pressures ranging from 1 bar (0.1 MPa) (gas phase) up to about 85 bar (8.5 MPa) (liquid phase).

This cracking is a *radical* process and brings about not only the splitting of higher molecular weight alkanes into alkanes of lower molecular weight, but also their dehydrogenation to form alkenes. It is especially useful that distillates with smaller ranges of boiling point can be obtained by 'selective' cracking. The petrol which is produced is characterised by having relatively good antiknock properties. In addition, alkenes with two to four carbon atoms are formed, which are useful for the petrochemical industry.

2.4.2.2 Catalytic Cracking (*Houdry* Process, 1934). In the presence of a catalyst (Al_2O_3/SiO_2), the middle distillate and the vacuum distillate can be cracked in the gas phase at about 500°C (775 K) and at relatively low pressure (2 bar; 0.2 MPa) to give a petrol, which contains considerable amounts of alkenes, isoalkanes and arenes, and has an octane rating greater than 90.

In all catalytic cracking, after some time the catalyst becomes deactivated because of the unavoidable deposition of coke on it. Hence it must be continually renewed. For this reason it is better to use a continuously moving bed or a fluidised bed of catalyst rather than a stationary catalyst.

2.4.2.3 Hydrogen Cracking (Hydrocracking). In this process catalytic cracking is accompanied by hydrogenation. This results in an increase in the yield of saturated hydrocarbons. Sulphur-containing compounds are thereby converted into saturated hydrocarbons and hydrogen sulphide.

2.4.2.4 Catalytic Reforming. In this process heavy petroleum is exposed briefly to a catalyst at temperatures above 500°C (775 K) and at high pressure (15–70 bar; 1.5–7.0 MPa) in the presence of hydrogen. In these conditions the main reactions which ensue are as follows: dehydrocyclisation of alkanes to arenes, conversion of alkylcyclopentanes into cyclohexane derivatives, dehydrogenation of the latter to arenes. By this process octane ratings greater than 100 can be achieved.

In the so-called **Platforming Process (1949)** a Pt/Al_2O_3 catalyst is used. Better yields of liquid hydrocarbons result if a bimetallic catalyst such as $Re/Pt/Al_2O_3$ is used. Petrols produced in this way have an octane number of 80–90, and are a useful source for the preparation of benzene, toluene and xylenes for the petrochemical industry.

2.4.2.5 Other Ways of Obtaining High Antiknock Fuels. Simple distillation of crude oil leads to products with very different compositions, and not all fractions can be converted by cracking or reforming into petrol with high octane rating. On the other hand, because of the higher compression and increased rotation rate of the petrol engine, even greater demands are made on the quality of the fuel it uses. This has led to the development of further refining processes.

(a) **Polymerisation**. The gases obtained from cracking processes, which contain a considerable amount of alkenes, are used as intermediates for the preparation of liquid fuels. The *alkenes* are *polymerised* in the presence of sulphuric acid (or phosphoric acid) as catalyst to provide *polymer-petrol*. For example, one industrial process involves passing isobutene, obtained from the C_4 fraction of the cracking product, into 70% sulphuric acid at about 30°C (300 K) and the t-butyl sulphate which results is heated for a short time under pressure at 100°C (375 K). This provides a mixture of two isomeric isooctenes [di(isobutenes)], together with some tri(isobutenes). The mixture of isooctenes is separated by distillation, in a yield of about 75%. This is hydrogenated quantitatively to a technical iso-octane with an octane number between 92 and 97 which is especially used as a component of aviation fuel.

Di- and tri-(isobutenes) can be obtained in a single-stage *oligomerisation*, in the presence of a cationic ion-exchange resin, of the isobutenes from the C_4 fraction.

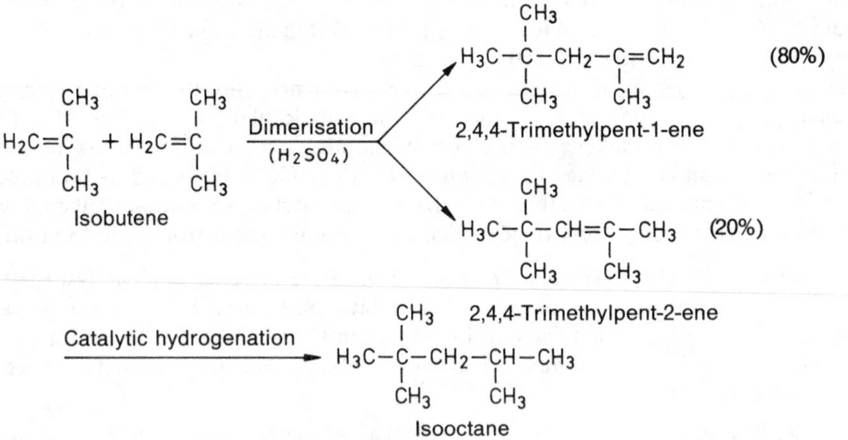

(b) **Alkylation**. In this process, isoalkanes (principally isobutane) react with alkenes in the presence of boron trifluoride to give higher isoalkanes (*Ipatiev*, 1935). Industrially, alkylation is carried out between −10°C (265 K) and +35°C (310 K), using sulphuric acid or dry hydrogen fluoride as catalyst (*Morrell*, 1939). For example, three alkanes are obtained from isobutane and propene, as shown below:

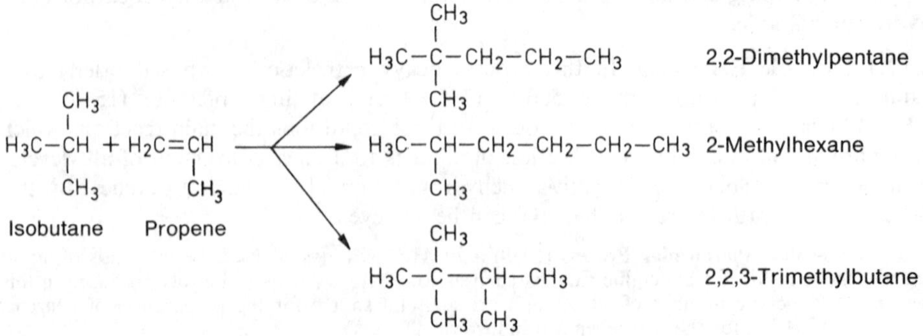

The resultant product has a high octane number. Alkylation is not limited to isoalkanes. In the presence of *aluminium chloride* as catalyst, n-alkanes, except methane and ethane, can be 'alkylated' by alkenes.

(c) Isomerisation. By this method n-alkanes can be converted into isoalkanes, using a *Lewis* acid as catalyst. A *molecular rearrangement* takes place, and the n-alkane is isomerised. For example, in the presence of *aluminium chloride* at 25°C (300 K), n-butane is involved in an equilibrium with isobutane:

$$H_3C-CH_2-CH_2-CH_3 \xrightarrow{\text{(AlCl}_3)} \begin{array}{c} CH_3 \\ | \\ H_3C-CH \\ | \\ CH_3 \end{array}$$

n-Butane (20%)   Isobutane (80%)

Since the octane number is appreciably raised by chain-branching in an alkane, isomerisation of the initial fractional distillate leads to a marked improvement in its octane rating (see reforming processes, p. 85).

2.4.3 Fuel-oils from Coal

In many countries which have relatively little or no mineral oil, the demand for fuel was, until about 40 years ago, met (apart from imported oil) largely from coal, which is also a fossil fuel. The conversion of coal, or of brown coal, into petrol and diesel oil is an industrially important problem, which has been achieved either by hydrogenation of coal (*Bergius*) or by *Fischer–Tropsch* synthesis, for example in the Sasol process developed in South Africa, which is short of deposits of mineral oil. The industrial value of both processes is dependent upon the relative prices of coal and oil. Because of the possible imbalance between demand for and supply of oil, chemical refining of coal and brown coal has great possibilities for further development.

2.4.3.1. Low Temperature Carbonisation of Coal. Low temperature carbonisation of brown coal at about 600°C (875 K) results in the formation of coal tar, a light oil (petrol), middle oil, producer gas (carbon monoxide and nitrogen) and an aqueous fraction. Coke remains as a residue.

If suitable bitumen-rich brown coals are extracted with organic solvents, for example ethanol–benzene mixtures, a mixture of wax and resin, with m.p. 80–90°C (355–365 K), which is known as *Montanwax*, is obtained. This has many uses.

Low temperature carbonisation of coal at 450–600°C (725–875 K), produces coke, coal tar, lubricating oil and a petrol fraction. Refining and hydrogenation provides high quality diesel oil and petrol.

2.4.3.2 Low Temperature Hydrogenation. Coal tar obtained from brown coal is hydrogenated in the liquid phase in the presence of a WS_2/NiS catalyst, and in a stepwise procedure, at 350°C (625 K) and 300 bar (300 MPa), providing diesel oil, lubricating oil, petrol and liquid and solid paraffin.

2.4.3.3 High Pressure Dehydrogenation. The hydrogen-rich light oil obtained from brown coal is dehydrogenated over a MoO_3/Al_2O_3 catalyst, or in more recent times a Pt/Al_2O_3 catalyst, at about 500°C (775 K) under a partial pressure of hydrogen of 25–35 bar (2.5–3.5 MPa). Not only are the cycloalkanes dehydrogenated, but also the n-alkanes undergo cyclisation and dehydrogenation to give arenes. The resultant petrol can achieve an octane rating of $\sim$ 90 if tetraethyl lead is added to it.

2.5 The Chemistry of Petroleum

In considering the chemistry of mineral oil, the main concern has been with its work-up and refinement. The chemistry of petroleum is especially concerned with the preparation of starting materials and products, aliphatic and aromatic, from natural gas and mineral oil. The most important starting materials are *methane* from natural gas, and *alkenes* such as ethylene, propene, but-1-ene, but-2-ene and isobutene obtained from cracking and refining of mineral oil, as well as *arenes* such as benzene and its alkyl derivatives obtained by dehydrogenation of cycloalkanes and cyclodehydrogenation of n-alkanes.

The first petrochemical to be made industrially on a large scale was *propan-2-ol* (*isopropanol*). This was made in 1920 in the USA from *propene*, *via* isopropyl hydrogen sulphate as an intermediate (see p. 116). This was followed by the preparation of *butan-2-ol* from *butene*. These two alcohols were also dehydrogenated to the corresponding ketones (*acetone, butan-2-one*). In 1930 came the production of *ethylene oxide* from ethylene, obtained from the refinery gases. The *chlorination of propene* at > 500°C (775 K) led to a technical synthesis of *glycerol*. *Butadiene*, important for the preparation of artificial rubber, was produced by dehydrogenation of the C_4-fraction of refinery gases. The hydrocarbons obtained from oil were also used for the manufacture of plastics, synthetic fibres and materials used in the textile industry. Nowadays the ever increasing production of *polyethylene* and *polypropene* from alkenes obtained from mineral oil is of the greatest industrial importance.

Refining capacity and the petrochemical industry have expanded enormously since the 1939–45 *war*.[1] Whilst Table 9 presents a picture of the individual reactions involving the different compounds, it should also provide an overall picture of the importance of the synthetic products derived from oil, using ethylene and propene as starting materials. The arenes obtained by the dehydrogenation and cyclisation of alkanes and cycloalkanes are considered further in the section on 'Aromatic Compounds'.

2.6 Alkynes (Acetylenes), C_nH_{2n-2}

As well as the alkenes, there is another *homologous series* of unsaturated hydrocarbons, the alkynes (acetylenes), with general formulae C_nH_{2n-2}. The characteristic feature of these compounds is a $C{\equiv}C$ *triple bond*.

The most important members of this series are:

		b.p.
H–C≡C–H	Acetylene (Ethyne)	– 83.6 °C (189.6 K)
H₃C–C≡C–H	Propyne (Methylacetylene)	– 27.5 °C (245.7 K)
H₃C–CH₂–C≡C–H	But-1-yne (Ethylacetylene)	+ 18 °C (291 K)
H₃C–C≡C–CH₃	But-2-yne (Dimethylacetylene)	+ 27.2 °C (300.4 K)

---

[1] See *J. M. Tedder, A. Nechvatal* and *A. H. Jubb*, Basic Organic Chemistry, Part 5. Industrial Products (Wiley, Chichester 1975); *H. A. Whitcoff* and *B. G. Reuben*, Industrial Organic Chemicals in Perspective (Wiley, New York 1980); *K. Weissermel* and *H.-J. Arpe*, Industrial Organic Chemistry, 2nd edn. (V.C.H., Weinheim 1993).

Table 9. Petrochemical products from ethylene and propene

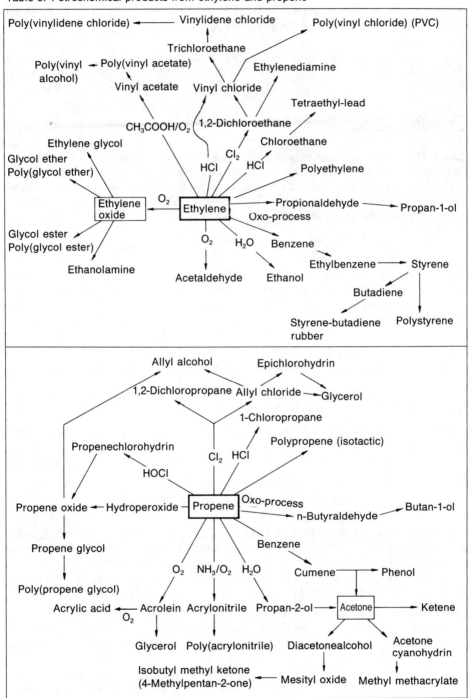

This series of compounds is described in IUPAC rules as *alkynes*, and the groups —C≡CR as *alkynyl* groups, e.g. —C≡CH, ethynyl; —C≡C—CH₃, prop-1-ynyl; —CH₂—C≡CH, prop-2-ynyl (formerly propargyl).

Acetylene[1]

2.6.1 Preparation

2.6.1.1 The formation of acetylene from its elements is a strongly endothermic reaction and requires extremely high temperatures; for example, dry distillation of coal produces acetylene in up to 0.06% yield. If an electric arc passes between two graphite (or other carbon) electrodes in an atmosphere of hydrogen (*Berthelot*), even at about 2500°C (2800 K) only about 4% of acetylene is produced.

$$2\,C + H_2 \rightleftharpoons HC\equiv CH \qquad \Delta H = +\,226\,kJ/mol$$

2.6.1.2 Formerly acetylene was made on the industrial scale from *calcium acetylide* (*calcium carbide*), which was itself made by heating calcium oxide (lime) and coke together in an electric furnace at about 2200°C (2475 K). Treatment of the salt with water gave acetylene (*Wöhler*, 1862).

$$CaO + 3\,C \rightleftharpoons CaC_2 + CO \qquad \Delta H = +\,461\,kJ/mol$$
$$CaC_2 + 2\,H_2O \rightarrow HC\equiv CH + Ca(OH)_2 \qquad \Delta H = -\,130\,kJ/mol$$

Because of the availability of cheap ethylene from petrochemical processes, this preparation of acetylene has greatly diminished in importance.

2.6.1.3 Preparation from Methane in an Electric Arc. In this process methane is pyrolysed for a very short time at above 1400°C (1675 K). Since acetylene quickly decomposes into its constituent elements above 1000°C (1275 K), the reaction gases must be quickly cooled with water.

$$2\,CH_4 \rightarrow HC\equiv CH + 3\,H_2 \qquad \Delta H = +\,398\,kJ/mol$$

If the gases leaving the furnace at 1500°C (1775 K) are initially cooled with liquefied gas, instead of with water, the yield of acetylene is raised from 15% to 25–30%. In addition, ethylene and hydrogen are formed, together with carbon black.

The acetylene is extracted from the mixture of gases by a suitable selective solvent such as *N*-methyl-2-pyrrolidone (NMP) or dimethylformamide (DMF).

2.6.1.4 *Sachsse–Bartholomé* Process. If methane is incompletely burned in oxygen at about 1500°C (1775 K), and the products are cooled with water, acetylene (8–9%) is formed together with a residual mixture of gas (56% H_2 + 25% CO) which has great use

---

[1] See *H. G. Viehe*, Chemistry of Acetylenes (Dekker, New York 1969); *V. Jäger* and *H. G. Viehe*, *Houben-Weyl*, Methoden der Organischen Chemie, 4th edn, Vol. *5/2a* (Thieme Verlag, Stuttgart 1977).

and is known as *synthesis gas*. This process involves an energetic yield of about 75%, while the formation of acetylene from calcium carbide or by the electric furnace processes involve energetic yields of 50% and 66% respectively.

2.6.1.5 Plasma Process. There is a process, which is in the pilot-plant stage of development, to make acetylene by thermal cracking of hydrocarbons with hydrogen at high temperature. To this end, hydrogen is heated in an electric arc to about 3500–4000°C (3775–4275 K). When it leaves the arc this hot gas (plasma) reacts in an adjacent zone with an excess ($\sim$ 15%) of gaseous or volatilised hydrocarbons (petrol). The end products are a mixture of acetylene and ethylene (together making up more than 70%), with some methane and hydrogen.

2.6.1.6 Higher members of the alkyne series are made by elimination reactions from 1,2-dibromoalkanes, which are heated with an alcoholic solution of potassium hydroxide or, better, with sodamide, e.g.:

$$H_3C-CHBr-CH_2Br + 2\,NaNH_2 \rightarrow H_3C-C\equiv C-H + 2\,NaBr + 2\,NH_3$$

1,2-Dibromopropane   Propyne (Methylacetylene)

A similar method is used to produce an *octadecynoic acid* from an ester of oleic acid:

$$H_3C(CH_2)_7-CH=CH-(CH_2)_7-COOCH_3 \xrightarrow{\;Br_2\;} H_3C(CH_2)_7-\underset{|}{C}H-\underset{|}{C}H-(CH_2)_7-COOCH_3$$
$$Br \quad Br$$

$$\xrightarrow[\text{2. HCl}]{\text{1. KOH, n-C}_5\text{H}_{11}\text{OH}} H_3C(CH_2)_7-C\equiv C-(CH_2)_7\,COOH$$

Octadec-9-ynoic acid

Physical Properties

Pure acetylene is a colourless gas and, when pure, has an ethereal odour. It has narcotic properties. It burns with a luminous flame and, in contrast to ethane and ethylene, is somewhat soluble in water. It is readily soluble in acetone. It is thermodynamically unstable with reference to its constituent atoms carbon (solid) and hydrogen:

$$HC\equiv CH \rightarrow 2[C] + H_2 \qquad \Delta H = -226 \text{ kJ/mol}$$

As a result of this, liquefied acetylene explodes violently, either on impact or when heated. Mixtures of acetylene with air, over the range 3–70% acetylene, are extremely explosive. The unpleasant smell of industrial acetylene arises from impurities such as hydrogen sulphide, phosphine and organic sulphur and phosphorus compounds.

2.6.2 The C≡C Triple Bond

The C≡C triple bond can be described in an analogous way to the C=C double bond, using a $\sigma\pi$ model. In this case, only one p orbital, say the p_x orbital, is hybridised with an s orbital (Figs. 49 and 50). This leads to a linear arrangement of the two sp-hybridised orbitals. It results in sp *hybridisation* of the carbon atom. In acetylene an sp orbital from each carbon atom overlaps to form a C—C -bond and the remaining sp orbital of each of these atoms overlaps an s orbital of a hydrogen atom to give two C—H σ-bonds (Fig. 51). The two remaining p orbitals on each carbon atom, that is the p_y and p_z orbitals, overlap the corresponding orbitals on the other carbon atom, in this way giving rise to two π-bonds, which lie in two planes which are at 90° to each other, as shown schematically in Fig. 52. Since these two pairs of π-electrons cannot in principle be distinguished from one another, the resultant distribution of these electrons has a *rotational symmetry* about the σ-bond (the line joining the two atomic nuclei) (Fig. 53, see also Fig. 21, p. 41).

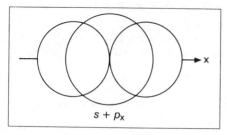

$s + p_x$

Fig. 49.

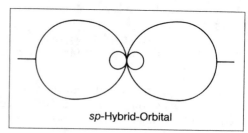

sp-Hybrid-Orbital

Fig. 50.

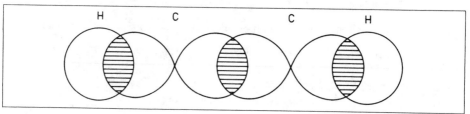

Fig. 51.

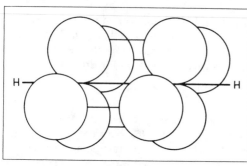

Fig. 52.

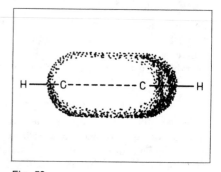

Fig. 53.

Alternatively, like the C=C double bond, the C≡C triple bond can be described in terms of three *equivalent orbitals*; in this case each carbon atom has three sp^3-hybridised orbitals which overlap in the form of *banana-bonds*. The two remaining sp^3 orbitals, one on each carbon atom, are collinear (Fig. 54).

Fig. 54.

Bond Lengths and Hybridisation.[1] Bond lengths in organic molecules can be determined with the aid of electron diffraction and from microwave, IR and *Raman* spectroscopy. The *bond distances* and the *bonding enthalpies* between the two carbon atoms in ethane, ethylene and acetylene are as follows:

C–C-		153.4 pm		346 kJ/Mol
C=C-	Bond distances	133.7 pm	Bonding enthalpies	602 kJ/Mol
C≡C-		120.7 pm		836 kJ/Mol

$$
\begin{array}{ccc}
153.4 & 133.7 & 120.7 \\[4pt]
\underset{110.2}{\overset{\text{H}\diagdown\,\big|\,\diagup\text{H}}{\underset{sp^3}{\text{H–C–C–H}}}} &
\underset{108.6}{\overset{\text{H}\diagdown\quad\diagup\text{H}}{\underset{sp^2}{\text{C=C}}}} &
\underset{105.9}{\underset{sp}{\text{H–C≡C–H}}}
\end{array}
$$

Whilst the shortening of the bond distances associated with multiple bonding is understandable, it is rather surprising that the lengths of the C—H bonds differ. It has been shown that the lengths of chemical bonds are closely connected with the *hybridisation* of the atoms they connect. Increasing *s*-contribution to the hybridisation of carbon atoms leads to a shortening of C—H bonds to these atoms. The shortening is directly proportional to the amount of *s*-character, and consequently the change in going from ethylene (33% *s*-character) to acetylene (50% *s*-character) is about twice as great as the change in going from ethane (25% *s*-character) to ethylene.

There is also a relationship between the percentage *s*-character and the coupling constants between ^{13}C and ^1H in NMR spectra, as follows: $^1J_{(CH)} = 500\%$ *s*-character. This is illustrated by the following values: $^1J_{(CH)}$, ethane = 125 Hz; ethylene = 156 Hz; acetylene = 248 Hz.

The C—C single bond is found to be shorter on going from propane to propene to propyne, and this change is also associated with the change in hybridisation of the central carbon atom:

$$
\begin{array}{ccc}
152.6 & 150.1 & 145.9 \\[4pt]
\underset{sp^3}{\text{H}_3\text{C–CH}_2\text{–CH}_3} &
\underset{sp^2}{\text{H}_3\text{C–CH} = \text{CH}_2} &
\underset{sp}{\text{H}_3\text{C–C} \equiv \text{C–H}}
\end{array}
$$

A similar situation obtains in the case of cumulated C=C double bonds, as well as for carbon atoms linked by single or multiple bonds to hetero-atoms such as halogens, oxygen or nitrogen.

Associated with the shortening of bond distances is an increase in the *bonding enthalpies*. In addition, the *electronegativity* of a carbon atom increases with the type of hybridisation in the order $sp^3 < sp^2 < sp$; this is a factor contributing to the acidity of the hydrogen atoms in acetylene.

Another result of the increased electronegativity associated with increased *s*-character is that an *sp*- or *sp²*-hybridised carbon atom has an electron-withdrawing effect on an adjacent *sp³*-hybridised carbon atom. Hence the groups

$$-\text{C} \equiv \text{CH} \quad \text{or} \quad -\text{CH} = \text{CH}_2$$

---

[1] See *M. J. S. Dewar*, Hyperconjugation (Ronald Press Co., New York 1962); *W. Kurzelnigg* et al. σ and π Electrons in Theoretical Organic Chemistry, Fortschr. Chem. Forsch. *22* (1971).

cause a marked $-I$-effect on a carbon atom. For example, vinylacetic acid, $H_2C=CH-CH_2-COOH$, in which the $H_2C=CH-$ group replaces a hydrogen atom of acetic acid, is twice as strong an acid as the latter, because of this $-I$-effect. (Compare the $-I$-effect in chlorinated acetic acids.)

Chemical Properties

The acidity of acetylene itself is so weak that it cannot be detected in aqueous solution. However, in contrast to alkenes, the hydrogen atom attached to a $C\equiv C$ triple bond in alkynes can be replaced by a metal atom (metallation) in liquid ammonia, e.g.

$$H-C\equiv C-H \xrightarrow[-NH_3]{+NaNH_2} H-C\equiv CNa$$

A *Grignard* compound (see p. 185) can be prepared by the reaction of acetylene with ethyl magnesium bromide in tetrahydrofuran.

$$H-C\equiv C-H + C_2H_5MgBr \rightarrow H-C\equiv CMgBr + C_2H_6$$

The resultant metal derivative is called an *acetylide*, or, if it contains no hydrogen, a *carbide*. If a solution of acetylene in ammonia is treated with a silver or a copper(II) salt, a colourless or a reddish brown precipitate of, respectively, Ag_2C_2 or Cu_2C_2 is formed. These exist as polymers. When dry they explode violently.

$$2\,[Ag(NH_3)_2]^{\oplus}NO_3^{\ominus} + HC\equiv CH \rightarrow AgC\equiv CAg + 2\,NH_4NO_3 + 2\,NH_3$$

The precipitation of silver and copper acetylides has been used as a *qualitative* test for acetylene and its derivatives which have the general formula $R-C\equiv C-H$.

Alkali and alkali-earth metal carbides react with water to form acetylene, but heavy metal acetylides require dilute hydrochloric acid for conversion into acetylene, or, in the case of higher alkynes, into mixtures of hydrocarbons.

2.6.3 Addition Reactions

Like alkenes, alkynes are *unsaturated* and undergo a wide range of addition reactions, which proceed smoothly. But, because of the rotationally symmetrical distribution of charge, the two pairs of π-electrons of the $C\equiv C$ triple bond have a reduced reactivity towards electrophilic reagents. Further, the reactivity of acetylenes towards nucleophilic reagents, such as alkoxide anions or amines, becomes more important.[1]

2.6.3.1 Catalytic hydrogenation of alkynes, using hydrogen in the presence of platinum, palladium or nickel, results in the formation of alkanes, with alkenes as intermediates, e.g.

$$HC\equiv CH \xrightarrow{+2H} H_2C=CH_2 \xrightarrow{+2H} H_3C-CH_3$$

Acetylene   Ethylene   Ethane

The alkene can itself be obtained under special conditions.

---

[1] See *F. Bohlmann*, Struktur und Reaktionsfähigkeit der Acetylenbindung (Structure and Reactivity of the Acetylene Bond), Angew. Chem. *69*, 82 (1957); *R. A. Raphael*, Acetylenic Compounds in Organic Synthesis (Butterworths, London 1954).

2.6.3.2 Addition of chlorine (or bromine) provides, in turn, (E)-1,2-dichloroethylene and 1,1,2,2-tetrachloroethane:

$$HC \equiv CH \xrightarrow{+Cl_2} \underset{Cl}{\overset{H}{\diagdown}} C = C \underset{H}{\overset{Cl}{\diagup}} \xrightarrow{+Cl_2} Cl_2HC-CHCl_2 \xrightarrow[- CaCl_2, - H_2O]{Ca(OH)_2} ClHC = CCl_2$$

Acetylene *trans*-1,2-Dichloroethylene 1,1,2,2-Tetrachloroethane Trichloroethylene
(*E*)-1,2-Dichloroethylene

This reaction is catalysed by light or by metal halides, e.g. $FeCl_3$, $SbCl_5$.

Tetrachloroethane, b.p. 146°C (419 K) is rather poisonous, and is sensitive to air. When treated with calcium hydroxide it loses HCl and provides the less toxic and more stable *trichloroethylene* (trichloroethene, TRI), b.p. 87°C (360 K), which is used industrially as a solvent for fats, oils, resins and varnishes, as is *tetrachloroethylene* (tetrachloroethene, perchloroethylene, PER, Perc.), b.p. 121°C (394 K).

In more recent times, when ethylene has largely replaced acetylene as a starting material, both of these chloroethylenes have been made industrially from the reaction of 1,2-dichloroethane with chlorine in the presence of a catalyst at 350–450°C (620–720 K).

$$ClH_2C-CH_2Cl + Cl_2 \xrightarrow[- HCl]{Catalyst} ClHC = CCl_2 + Cl_2C = CCl_2 + HCl$$

Trichloroethylene Tetrachloroethylene

Tetrachloroethylene is the most stable of all the chlorinated ethanes and ethylenes. It is widely used in dry-cleaning and has led to environmental problems.

2.6.3.3 The addition of hydrogen halides to acetylene also proceeds stepwise and above all is useful for the preparation of vinyl halides, e.g.

$$HC \equiv CH \xrightarrow{+HI} H_2C = CHI \xrightarrow{+HI} H_3C-CHI_2$$

Acetylene Vinyl iodide 1,1-Diiodoethane

The regioselective addition in the second step is in accord with *Markovnikov*'s rule; the product is a *geminal* (gem.) diiodo compound.

2.6.3.4 Water adds to acetylene in the presence of mercury(II) sulphate in sulphuric acid as a catalyst. The initial product is vinyl alcohol, which is unstable under these conditions,[1] and rearranges, by transfer of a proton, to give *acetaldehyde*, which is more stable.

$$HC \equiv CH \xrightarrow[(H_2SO_4)]{+H_2O(Hg^{2\oplus})} \left[\begin{array}{c} H_2C = CH \\ | \\ OH \end{array} \right] \longrightarrow H_3C-C \underset{O}{\overset{H}{\diagdown}}$$

Acetylene Vinyl alcohol Acetaldehyde

2.6.3.5 In a solution of copper(I) chloride and ammonium chloride in hydrochloric acid acetylene dimerises to *vinylacetylene* (but-1-en-3-yne) (*Nieuwland*, 1931):

$$H-C \equiv C-H + H-C \equiv C-H \xrightarrow{(CuCl, NH_4Cl)} H_2C = CH-C \equiv C-H \qquad \Delta H = -264 \text{ kJ/mol}$$

Vinylacetylene

2.6.3.6 In the presence of a *Ziegler–Natta* catalyst acetylene polymerises to give a polyene, with a continuous series of conjugated double bonds (see p. 103 *ff*).

$$\sim CH=CH-CH=CH-CH=CH \sim \equiv \left[CH=CH \right]_n \qquad \left[\diagup \diagdown \diagup \right]_n$$

Polyene   *trans* Form

---

[1] Conditions have been found under which vinyl alcohol is sufficiently stable to enable its ^{1}H-NMR spectrum to be recorded. See B. *Capon* et al., Acc. Chem. Res. *21*, 135 (1988).

This polymer, which is called *polyacetylene* exists in the *trans* form at room temperature. It shows an electrical conductivity of about 10^{-9} S/cm, which can be raised by doping, either oxidatively with AsF_5 or reductively with sodium, to about 10^4 S/cm, which is comparable to that of mercury. The polymer behaves like an organic metal.

Uses

For use in industry, acetylene is made more safe to handle by keeping it in solution in steel cylinders (12–20 bar, 1.2–2 MPa) which are filled with either kieselguhr, pumice or charcoal, and acetone. At 15°C (290 K) and normal pressure acetone dissolves about 25 times its own volume of acetylene, and at 12 bar (1.2 MPa) 300 times its own volume. A mixture of acetylene and pure oxygen generates a temperature of 2800°C (3075 K) when it burns and is used for the cutting and welding of steel.

In the chemical industry, acetylene has been an important starting material in synthetic processes, but in large part its role has been usurped by ethylene, which is readily available from oil. The reason for this can be found in the fact that the formation of C≡C triple bonds always requires severe conditions, and the cost in energy is greater than in the case when ethylene is used.

2.6.4 Reppe Syntheses[1]

There are four main reactions found by *Reppe* which are of importance in the chemistry of acetylene.

2.6.4.1 Vinylation. This reaction is based on the conversion of acetylene into monosubstituted alkenes by allowing it to react with organic compounds which have functional groups with reactive hydrogen atoms: —OH, —SH, —NH_2, =NH, —$CONH_2$ or —COOH. In the process the C≡C triple bond is converted into a C=C double bond, thereby becoming a vinyl group. For example, addition of ethanol to acetylene at 130–180°C (400–450 K) under pressure and in the presence of ethoxide as a catalyst gives *ethyl vinyl ether* (methoxyethylene):

$$HC \equiv CH + H-OC_2H_5 \rightarrow H_2C = CH-O-C_2H_5$$

Acetylene   Ethyl vinyl ether
                             (Methoxyethylene)

Addition of carboxylic acids gives rise to vinyl esters, e.g. from diluted acetylene at 10–15 bar (1–1.5 MPa) and stearic acid, in the presence of zinc stearate, at 155°C (428 K), vinyl stearate is obtained:

$$C_{18}H_{37}COOH + HC \equiv CH \rightarrow C_{18}H_{37}COO-CH = CH_2$$

Stearic acid   Vinyl stearate

The mechanism of vinylation is as follows, using addition of an alcohol as an example. First of all an alkoxide anion undergoes *nucleophilic* addition to the triple bond. The carbanion which is formed then reacts with an alcohol molecule to form the vinyl ether, at the same time regenerating the alkoxide anion.

$$HC \equiv CH \xrightarrow{+ \overline{|\underline{O}R}} H\underline{C} = \overset{H}{\underset{}{C}}-\overline{O}R \xrightarrow[-\Theta|\underline{O}R]{+ H-\overline{O}R} H_2C = CH-\overline{O}R$$

Carbanion   Vinyl ether

---

[1] See *W. Reppe*, Neue Entwicklungen auf dem Gebiet der Chemie des Acetylenes und Kohlenoxyds (New Developments in the Chemistry of Acetylene and of Carbon Monoxide) (Springer-Verlag, Berlin 1949).

2.6.4.2 Addition to Aldehydes and Ketones. Acetylene and its monosubstituted derivatives add to aldehydes or ketones in the presence of copper acetylide as a catalyst. The $C \equiv C$ triple bond is retained in the product. In the case of acetylene itself a twofold addition is possible.

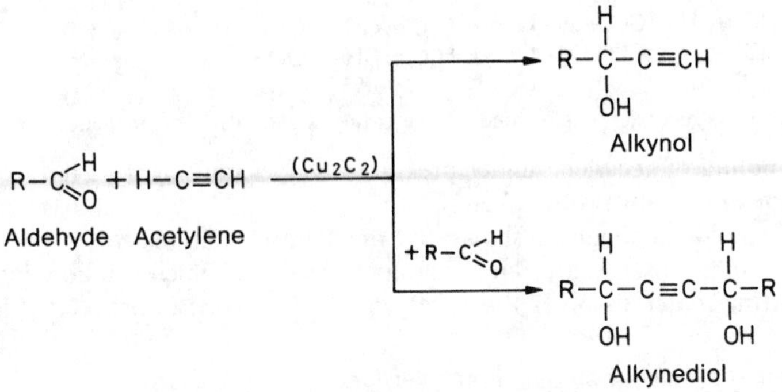

The product is either an alkynol or an alkynediol. The tendency for diol formation decreases for higher aldehydes. Ketones react similarly.

2.6.4.3 Cyclisation. Cyclic polymerisation of acetylene takes place under the influence of certain selective catalysts and leads to the formation of cyclic polyalkenes with the general formula $C_{2n}H_{2n}$, where $n \geqslant 3$.

As long ago as 1866 *Berthelot* showed that at high temperatures (400–500°C [675–775K]) acetylene trimerised to give benzene (see Structure of Benzene):

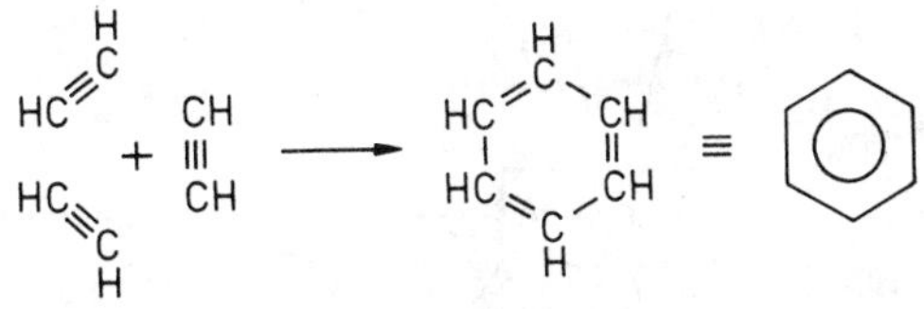

Benzene

The yield is however small; as by-products biphenyl, naphthalene and other arenes are formed.

The yield of benzene can be increased markedly if *triphenylphosphine-nickel carbonyls* such as $[(C_6H_5)_3P]_2Ni(CO)_2$, preferably in solution in benzene, are used as catalysts, reaction being carried out at 60–70°C (335–345 K) and 15 bar (1.5 MPa). In this way 88% benzene and 12% styrene (vinylbenzene) are produced, but it is not used industrially.

In addition to this trimerisation, Reppe[1] showed that tetramerisation of acetylene to give *cyclooctatetraene*, in a yield of 80–90%, took place in tetrahydrofuran with nickel cyanide as catalyst.

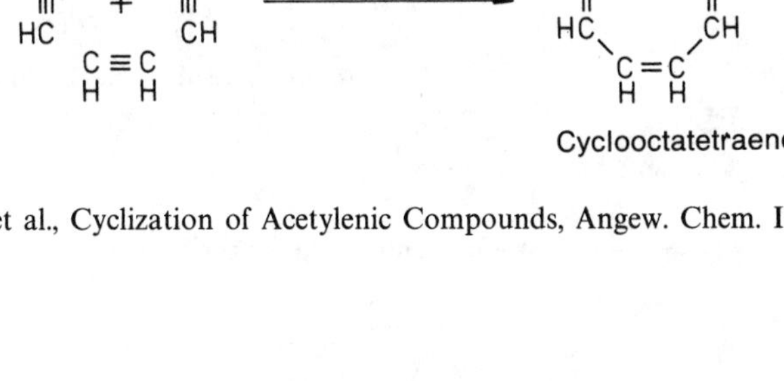

Cyclooctatetraene

---

[1] See *W. Reppe* et al., Cyclization of Acetylenic Compounds, Angew. Chem. Int. Ed. Engl. *8*, 727 (1969).

2.6.4.4 Hydrocarboxylation.[1] Acetylene reacts with carbon monoxide in the presence of compounds having an acidic hydrogen atom, such as water or alcohols, under pressure, and in the presence of a metal carbonyl, most commonly $Ni(CO)_4$, as a catalyst, to give an unsaturated carboxylic acid or ester, e.g.

$$HC \equiv CH + CO + H-OH \rightarrow H_2C = CH-COOH \quad \text{Acrylic acid}$$
$$HC \equiv CH + CO + H-OR \rightarrow H_2C = CH-COOR \quad \text{Acrylic esters}$$

At present this process has been superseded by the catalytic oxidation of propene.

Since the handling of acetylene under pressure involves considerable hazard, suitable safety precautions must always be taken.

In Table 10, below, a diagram shows the most important processes starting from acetylene; they will be discussed in later chapters. Products which are now increasingly made from starting materials other than acetylene are enclosed in brackets [ ].

Table 10. Industrial syntheses starting from acetylene

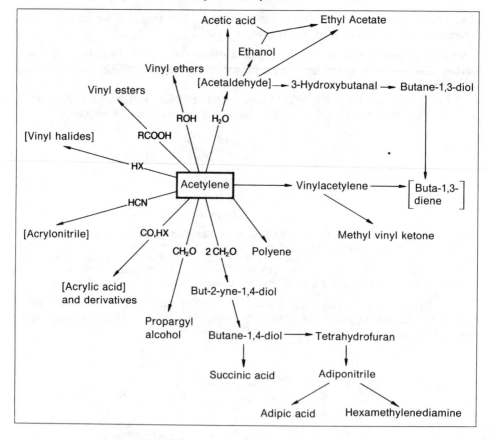

---

[1] See *J. Falbe*, New Syntheses with Carbon Monoxide (Springer-Verlag, New York 1980).

2.7 Hydrocarbons with Two or More C=C Double Bonds (Polyenes)

Hydrocarbons with two C=C double bonds can be classified into three types, depending on the relative disposition of the double bonds:

(a) with *cumulated* (adjacent) double bonds: $H_2C{=}C{=}CH_2$, *allenes*
(b) with *conjugated* double bonds (separated from one another by one C—C single bond): $H_2C{=}CH{-}CH{=}CH_2$, *conjugated dienes*
(c) with *isolated* double bonds (separated from one another by more than one single bond): $H_2C{=}CH{-}(CH_2)_n{-}CH{=}CH_2$.

2.7.1 Allenes

Allene (propadiene), $H_2C{=}C{=}CH_2$ (skeletal formula: $={=}$) is the simplest hydrocarbon with cumulated double bonds. It can be prepared from *2,3-dichloroprop-1-ene* by dechlorination with *zinc* in aqueous ethanol (yield 80%).

$$H_2C{=}\underset{\underset{Cl}{|}}{C}{-}\underset{\underset{Cl}{|}}{CH_2} + Zn \longrightarrow H_2C{=}C{=}CH_2 + ZnCl_2$$
Allene (b.p. $-34°C$ [239 K])

Allene[1], is made up of two sp^2-hybridised carbon atoms and one sp-hybridised carbon atom. It has both alkene- and alkyne-like character; this is particularly evident in its addition and polymerisation reactions.

Allene can be obtained by the cracking of mineral oil fractions and especially from propene and isobutene; however it is obtained as a mixture, together with its isomer, propyne. At 300°C (575 K) in the gas phase an equilibrium is set up between allene (20%) and propyne (80%).

$$H_2C = C = CH_2 \rightleftarrows H_3C{-}C{\equiv}CH$$
Allene   Propyne

This isomerisation takes place only in the presence of a catalyst; the most effective is activated charcoal. Pure allene is stable up to 400°C (675 K).

Compounds with an allene-type structure involving more than three carbon atoms are known as *cumulenes*. The simplest aliphatic cumulene is *butatriene*, which can be prepared by the debromination, using zinc, of 1,4-dibromobut-2-yne (*Schubert*, 1954):

$$H_2\underset{\underset{Br}{|}}{C}{-}C{\equiv}C{-}\underset{\underset{Br}{|}}{CH_2} + Zn \longrightarrow H_2C{=}C{=}C{=}CH_2 + ZnBr_2$$
Butatriene

It is a low-boiling light-sensitive liquid, which polymerises very readily.

The stereochemistry of cumulenes is discussed later, in the chapter on aromatic compounds.

---

[1] See *S. R. Landor* (Ed.), The Chemistry of the Allenes (Academic Press, New York 1982); *L. Brandsma* et al., Synthesis of Acetylenes, Allenes and Cumulenes (Elsevier, Amsterdam 1981).

2.7.2 Conjugated Dienes

The three most important conjugated dienes are:

$$(1)\ (2)\ (3)\ (4)$$
$$H_2C=CH-CH=CH_2$$

$$H_2C=C-CH=CH_2$$
$$\quad\quad |$$
$$\quad\quad CH_3$$

$$H_2C=C-C=CH_2$$
$$\quad\quad | \quad |$$
$$\quad\quad H_3C \quad CH_3$$

Buta-1,3-diene
(b.p. −4.5°C [268.7 K])

2-Methylbuta-1,3-diene
(Isoprene, b.p. 34°C [307 K])

2,3-Dimethylbuta-1,3-diene
(b.p. 70°C [343 K])

2.7.2.1 Butadiene

Butadiene has importance industrially as the starting material from which the synthetic rubber Buna is made, and theoretically as a model compound for hydrocarbons with conjugated systems.

Preparation. (a) The earlier methods for the industrial preparation of butadiene almost always involved forming the C_4 chain by joining together two C_2 units. The following method serves as an example. Acetaldehyde was dimerised in a weakly alkaline solution. The resultant product was catalytically reduced to butane-1,3-diol, which, by loss of two molecules of water, gave butadiene.

$$H_3C-C\!\!\nwarrow_O^H + H_2C-C\!\!\nwarrow_O^H \xrightarrow{(OH^\ominus)} H_3C-CH(OH)-CH_2-C\!\!\nwarrow_O^H \xrightarrow{+2H(Ni)}$$

Acetaldehyde   3-Hydroxybutanal

$$H_3C-CH(OH)-CH_2-CH_2OH \xrightarrow{-2H_2O} H_2C=CH-CH=CH_2$$

Butane-1,3-diol   Butadiene

(b) Industrially butadiene is obtained from the C_4 fraction of gases obtained by the cracking of oil. They are formed during the production of ethylene by steam cracking and contain about 13% butadiene. They are isolated by extractive distillation with acetonitrile or N-methylpyrrolid-2-one. Intensive research has shown butadiene to be a carcinogen in some animals.

2.7.2.2 Isoprene

Isoprene is the building block (monomer) from which natural rubber is derived, and can be obtained from rubber by dry distillation (*Williams*, 1861).

Preparation. Isoprene is obtained industrially in various ways, *e.g.*

(a) *From the C_5 fraction of mineral oil* by extractive distillation.
(b) *By the dimerisation of propene* with a *Ziegler* catalyst to give 2-methylpent-1-ene, which is in turn split by cracking into isoprene and methane.

$$2\,H_2C=CH \longrightarrow H_3C-CH_2-CH_2-C=CH_2 \xrightarrow[-CH_4]{\Delta} H_2C=CH-C=CH_2$$
$$\quad\quad | \quad\quad\quad\quad\quad\quad\quad\quad\quad\quad\quad | \quad\quad\quad\quad\quad\quad\quad\quad\quad |$$
$$\quad\quad CH_3 \quad\quad\quad\quad\quad\quad\quad\quad\quad\quad CH_3 \quad\quad\quad\quad\quad\quad\quad\quad CH_3$$

Propene   2-Methylpent-1-ene   Isoprene

(c) *By loss of water from 3-methylbutane-1,3-diol*, which is obtained from isobutene and formaldehyde by means of the *Prins* reaction:[1]

$$
\begin{array}{c}
\underset{\underset{CH_3}{|}}{\overset{\overset{CH_3}{|}}{C}}=CH_2 + H_2CO \xrightarrow{H_2O,H^{\oplus}} HO-\underset{\underset{CH_3}{|}}{\overset{\overset{CH_3}{|}}{C}}-CH_2-CH_2-OH \xrightarrow[-2\,H_2O]{} \underset{\underset{CH_3}{|}}{H_2C}=C-CH=CH_2
\end{array}
$$

Isobutene   3-Methylbutane-1,3-diol

2.7.3 1,2- and 1,4-Addition; Mesomerism

For dienes with two isolated double bonds, such as penta-1,4-diene, $H_2C=CH-CH_2-CH=CH_2$, addition reactions with bromine, hydrogen or hydrogen bromide or chloride proceed just as they do with monoenes. On the other hand, dienes with conjugated double bonds behave differently.

When bromine reacts with butadiene, addition of the first molecule of bromine takes place rather quicker than addition of the second molecule. In consequence the dibromo compound can be isolated. Bromine is found to have added not only to two carbon atoms of one of the double bonds (*1,2-addition*), but in other cases to the two ends of the conjugated system (*1,4-addition*). *In the latter case a new* $C=C$ *double bond results*, between the carbon atoms 2 and 3.

$$
H_2C=CH-CH=CH_2 \xrightarrow{+\,Br_2} \underset{\underset{Br}{|}\;\underset{Br}{|}}{H_2C-CH-CH=CH_2} \text{ and } \underset{\underset{Br}{|}\qquad\underset{Br}{|}}{H_2C-CH=CH-CH_2}
$$

3,4-Dibromobut-1-ene   1,4-Dibromobut-2-ene

The mechanism of 1,2- and 1,4-addition to butadiene is as follows. As in the case of ethylene a bromine cation makes an *electrophilic* attack on a π-bond, giving rise, via a π-complex, to a carbenium ion. In the present case this is a substituted *allyl cation* (see p. 107) which cannot be represented satisfactorily by a single classical formula. They can be shown as two limiting structures (contributing structures, canonical forms); the real structure lies somewhere between these alternative forms. The double-headed arrow ↔ between them indicates *not* that there is an *equilibrium* between two different kinds of molecules, but an *intermediate state* (*Arndt*, 1924). This phenomenon is described as *mesomerism* (*Ingold*, 1934) or as *resonance* (*Heisenberg*, 1926; *Pauling*). An alternative representation uses dotted lines to portray the intermediate structure:

$$
\underset{\text{Butadiene}}{H_2C=CH-CH=CH_2} \xrightarrow{+\,Br^{\oplus}}
\left[
\begin{array}{c}
\overset{\oplus}{H_2C}-\underset{\underset{Br}{|}}{CH}-CH=CH_2 \\
\updownarrow \\
\underset{\underset{Br}{|}}{H_2C}-CH=CH-\overset{\oplus}{CH_2}
\end{array}
\right]
\equiv
\left[
\underset{\underset{Br}{|}}{H_2C}-CH\!\!\cdots\!\!CH\!\!\cdots\!\!CH_2
\right]^{\oplus}
$$

(Bromomethyl)allyl cation

---

[1] See *D. R. Adams* et al., The *Prins* Reaction, Synthesis, *1977*, 661.

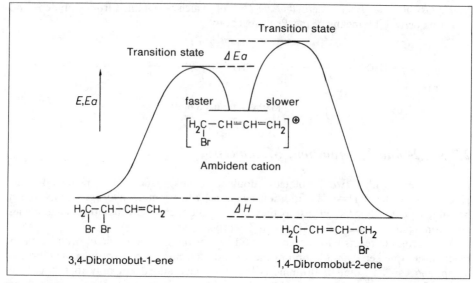

Fig. 55. Kinetic (ΔE_a) and thermodynamic (ΔH) effects in the addition of bromine to butadiene.

The mesomeric (bromomethyl)allyl cation can undergo nucleophilic attack by a bromide anion at either the α- or γ-position, to form either *3,4-dibromobut-1-ene* or *1,4-dibromobut-2-ene*. Because the mesomeric cation has two reactive sites, it is called *ambident* (see Fig. 55).

Since 1,2-addition has a lower activation energy E_a than 1,4-addition it takes place more rapidly, and is said to be kinetically favoured. If reaction takes place at a low temperature in a non-polar solvent, formation of the 1,2-adduct is favoured (see Fig. 55). Higher temperatures or the presence of an acid catalyst lead to the formation of the thermodynamically more stable 1,4-adduct, in a more exothermic reaction (note ΔH).

This relationship between *kinetic* and *thermodynamic* control of reactions can be important in industrial practice, as illustrated by the following industrial preparation of *chloroprene* from *butadiene*. If butadiene is treated with chlorine in the gas phase at about 300°C (570 K), a mixture of 1,2- and 1,4-dichlorobutenes is obtained. But chloroprene can only be prepared from the 1,2-adduct:

$$H_2C=CH-\underset{\underset{Cl}{|}}{CH}-CH_2Cl \xrightarrow{-HCl} H_2C=CH-\underset{\underset{Cl}{|}}{C}=CH_2$$

3,4-Dichlorobut-1-ene   2-Chlorobutadiene
(Chloroprene)

The 1,4-adduct, which is of no use in this reaction, can be isomerised to the 1,2-adduct by heating it with catalytic amounts of copper(I) chloride. The equilibrium is then displaced in the desired direction by distilling off the more volatile 1,2-adduct, thus favouring formation of the thermodynamically less favoured isomer.

$$H_2C-CH=CH-CH_2 \underset{CuCl}{\rightleftharpoons} H_2C=CH-CH-CH_2Cl$$
$$\;\;\;\;|\;\;\;\;\;\;\;\;\;\;\;\;\;\;|\;|$$
$$\;\;\;\;Cl\;\;\;\;\;\;\;\;\;\;\;Cl\;Cl$$

1,4-Dichlorobut-2-ene   3,4-Dichlorobut-1-ene
1,4-adduct, b.p. 155°C (428 K)   1,2-adduct, b.p. 123°C (396 K)

The concept of mesomerism is applicable not only to cations, but also to anions (e.g. the carboxylate anion) and to neutral molecules (e.g. benzene), in which there are conjugated multiple bonds or double bonds adjacent to atoms which have free pairs of electrons or which are deficient in electrons.

In order to ensure that the precise meaning of mesomerism is not lost or used arbitrarily, the following rules must be adhered to.

1. In the contributing (limiting) structures (canonical forms) the atomic nuclei in the molecule must always occupy the same relative positions to each other. The respective bond distances and valence angles must not vary too greatly from structure to structure.
2. All the parts of a molecule which are concerned in a mesomeric system must be nearly coplanar, since only in this way can all the p orbitals, which are perpendicular to this plane, overlap one another satisfactorily.
3. The contributing structures must all have the same number, respectively, of paired and unpaired electrons.
4. The structures contributing to a mesomeric system must have relatively similar energies, and also essentially the same total number of atomic bonds.

The energy of a mesomeric compound is in fact lower than would be that of the various contributing structures, so a molecule is *mesomerically stabilised.*

The *addition of hydrogen halides*, e.g. hydrogen chloride, to butadiene is similar to the addition of halogens, and a mixture of *3-chlorobut-1-ene* and *1-chlorobut-2-ene* results. In this case also, lower temperatures and non-polar solvents favour 1,2-addition, whereas higher temperatures and polar solvents favour 1,4-addition. The first step is the *protonation* of butadiene at the 1-position to give a stabilised carbenium ion. This mesomeric methylallyl cation then undergoes *nucleophilic* attack by a chloride ion at either the α- or the γ-position:

$$H_2C=CH-CH=CH_2 \quad \xrightarrow{+H^\oplus} \quad \begin{bmatrix} H_3C-\overset{\oplus}{C}H-CH=CH_2 \\ \updownarrow \\ H_3C-CH=CH-\overset{\oplus}{C}H_2 \end{bmatrix} \quad \xrightarrow{+Cl^\ominus}$$

Butadiene

Methylallyl cation

$$H_3C-CH-CH=CH_2$$
$$|$$
$$Cl$$
3-Chlorobut-1-ene
(α-Methyl-allylchloride)

$$H_3C-CH=CH-CH_2-Cl$$
1-Chlorobut-2-ene
(γ-Methyl-allylchloride, crotyl chloride)

Structure of Butadiene. The two conjugated C=C double bonds, which in butadiene are separated from one another by a C—C single bond, result, as in the case of ethylene, from overlap of p orbitals, in this case between orbitals on C(1) and C(2) and on C(3) and C(4). The lengths of these bonds, 133.7 pm, is the same as that of the C=C bond in ethylene.

In contrast, the length of the central single bond in butadiene, 147.6 pm, is shorter than that in butane (153.3 pm). Formerly this was thought to be due to there being some overlap between the p orbitals on C(2) and C(3), resulting in the central bond assuming a certain amount of double-bond character, which could be expressed by means of a suitable contributing structure. According to Dewar, there was no call for this assumption, since the length of a C—C single bond depends upon the hybridisation of the carbon atoms it links. The central bond in butadiene is between two sp^2-hybridised carbon atoms and will therefore be markedly shorter than the corresponding bond in butane, which links two sp^3-hybridised carbon atoms.

Similarly, the central bond in buta-1,3-diyne, which is between two sp-hybridised carbon atoms, is even shorter, 137.6 pm; again the decrease in bond length is directly proportional to the change in the s-contribution to the hybridised orbitals (cf. p. 93).

	153.3		147.6		137.6

$$H_3C-CH_2-CH_2-CH_3 \qquad H_2C=CH-CH=CH_2 \qquad HC\equiv C-C\equiv CH$$

	sp^3 sp^3		sp^2 sp^2		sp sp
	Butane		Butadiene		Butadiyne

Rotation about the central *single* bond of butadiene is possible, and two *planar conformations* are favoured, which are known as the *s-trans (transoid)* and *s-cis (cisoid)* conformations. The energy barrier to rotation amounts to about 20.5 kJ/mol; at room temperature most of the molecules exist in the *transoid* form, which is about 9.6 kJ/mol lower in energy.

s-trans   s-cis

Conformations of butadiene

Also in the case of longer chain *polyenes*, $H_2C=CH-(CH=CH)_n-CH=CH_2$, and again contrary to older ideas, delocalisation of the π-electrons and mesomeric character plays no part. The bond lengths alternate and do not tend to average out with increase in length of the conjugated chain. An exception is provided by cyclic conjugated molecules of the benzene type ('cyclohexatrienes'), which will be discussed later in connection with their aromatic character.

2.7.4 Woodward–Hoffmann–Rules[1]

The recognition of the concepts of the *Principle of the Conservation of Orbital Symmetry* arose out of studies of the stereochemical paths of intramolecular cycloaddition and cycloreversion reactions, and of electrocyclic reactions, and of attempts to explain them. Electrocyclic reactions are reactions which take place between the ends of linear conjugated systems, resulting in their being linked together by a single bond, and also the converse ring-opening reactions.

Examples are the conversion of cyclobutene derivatives into butadiene derivatives and the reverse reactions:

---

[1] See *R. Hoffmann* and *R. B. Woodward*, The Conservation of Orbital Symmetry, Accounts Chem. Res. *1*, 17 (1968); The Conservation of Orbital Symmetry, Angew. Chem. Int. Ed. Engl., *8*, 781 (1969); *M. J. S. Dewar*, Aromaticity and Pericyclic Reactions, Angew. Chem. Int. Ed. Engl., *10*, 761 (1971); *T. L. Gilchrist* and *R. C. Storr*, Organic Reactions and Orbital Symmetry (Cambridge University Press, Cambridge 1972 and later editions).

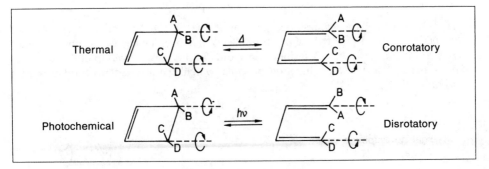

Fig. 56.

In these reactions the substituents (A–D) may all be rotated in the same direction (*conrotatory*) or A,B may be rotated in the opposite direction to C,D (*disrotatory*). By application of the principle of conservation of orbital symmetry it can be deduced which of these processes will occur.

The symmetries of the molecular orbitals (MO) of *butadiene* are shown in Table 11.

Table 11. Diagram of the wave functions of butadiene

Bonding states of the energy levels	Schematic representation of MO wave functions	Designation of wave function	MO picture	Nodes	Symmetry
Antibonding { Lowest unoccupied MO (LUMO)		ψ_4		3	A
		ψ_3		2	S
Bonding { Highest occupied MO (HOMO)		ψ_2		1	A
		ψ_1		0	S

S = symmetric,   A = anti-symmetric

a) LUMO: lowest unoccupied MO; b) HOMO: highest occupied MO

In the MO pictures, the positions of the nodes are represented by dots. The planes of symmetry are marked by dotted lines. HOMO and LUMO are known as frontier orbitals, because between them lies the boundary line between bonding and non-bonding orbitals, shown as —·—·—·— in Table 11.[1]

---

[1] See *K. Fukui*, Recognition of Stereochemical Paths by Orbital Interaction, Accounts Chem. Res. *4*, 57 (1971).

For *electrocyclic reactions* the following rules apply:

1. The symmetry of the highest occupied orbital (HOMO) determines the course of *thermal* reactions. In the case of the example given above, the highest occupied orbitals must rotate in a *conrotatory* fashion, since only in this way is overlap of orbitals with the same sign in the wave function possible, which in turn leads to formation of a σ-bond.

Highest occupied orbital
of butadiene (HOMO)

2. For *photochemical* electrocyclic reactions the decisive feature is the symmetry of the lowest unoccupied orbitals (LUMO). In the example given above, this results in a *disrotatory* course of reaction.

Lowest unoccupied
orbital (LUMO)

The preferred alignment of the frontier orbitals forms the basis of the *Woodward–Hoffmann* rules. Very reasonably they have been criticised because they involve an oversimplification of the situation. For a more accurate approach it is necessary to consider the full correlation diagrams. Appropriate monographs or the original literature should be consulted for more details (see p. 104).

The migration of a σ-bond within a system of π-bonds is also an example of a *sigmatropic* reaction.[1]

$$H_2C-(CH=CH)_n-CH=CH_2 \longrightarrow H_2C=CH-(CH=CH)_n-CH_2$$

(with R substituent on the first carbon in both structures)

In this reaction the π-bonds are shifted. The locations of the group R before and after the shift are indicated by numerals in square brackets [ ], as illustrated in the following example, which involves a [1,5] hydrogen shift in 1,3-pentadiene:

455–475 K

The stereochemistry of this reaction, in which the migrating group stays on the same side of the conjugated system, can be explained in terms of the *Woodward–Hoffmann* rules, if one makes the following assumptions about the transition state (see p. 126 *ff*):

[1] See *J. J. Gajewski*, Energy Surfaces of Sigmatropic Shifts, Accounts Chem. Res., *13*, 142 (1980).

1. The migrating hydrogen atom is attached at this time to both the site it moves from and that to which it moves.
2. The bonding then involves both the hydrogen atom and a radical derived from the π-system.

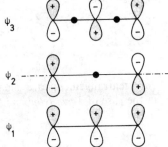

Then the frontier orbitals of the π-electron system ψ_3 which includes all the carbon atoms in the part of the molecule involved are considered. For uncharged compounds this always involves an odd number of carbon atoms. ψ_2

The simplest example of such a system is the allyl group, whose molecular orbitals, displayed as in Table 11, are as shown alongside: ψ_1

In the case of the allyl cation $[H_2C=CH-\overset{\oplus}{C}H_2 \leftrightarrow H_2\overset{\oplus}{C}-CH=CH_2]$, which has only two π-electrons, ψ_1 is the HOMO; ψ_2 is the HOMO for the allyl anion $[H_2C=CH-\overset{\ominus}{C}H_2 \leftrightarrow H_2\overset{\ominus}{C}-CH=CH_2]$, which has four π-electrons. This orbital lies on the boundary between bonding and anti-bonding in the energy diagram; it is called a non-bonding molecular orbital, NBMO.[1] In the case of an allyl radical the NBMO is occupied by a single electron.

In ψ_2 a node coincides with the central carbon atom. The NBMO of the *pentadienyl* radical is ψ_3. From Table 11 it can be seen that this orbital must have two nodes. It is hence written as follows:

NBMO of the pentadienyl radical

It can be seen that, in order to achieve appropriate orbital overlap at the two ends of this system, the hydrogen atom must be on the same side of the pentadienyl radical before and after migration (*suprafacial*) for this shift to take place.

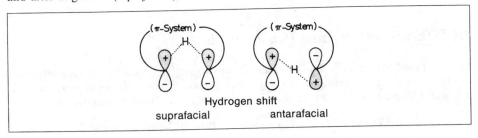

Fig. 57.

This reaction is energetically favoured. A different situation arises in the case of the NBMO of the allyl radical. Here the hydrogen atom must move to the opposite side of

---

[1] See M. J. S. *Dewar* et al., The PMO theory of Organic Chemistry (Plenum Press, New York 1975).

the allyl system (*antarafacial*). Because of the unfavourable geometry in the transition state this is difficult, and thermal [1,3] sigmatropic shifts are subject to steric hindrance.

Non-conjugated 1,5-dienes also undergo isomerisation when they are heated (*Cope rearrangement*):

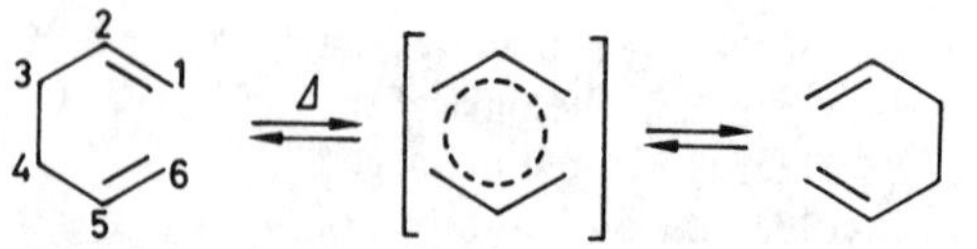

Cope rearrangements are usually only detectable when the 1,5-diene is unsymmetric, as in the case of *3-methylhexa-1,5-diene*, which at 300°C (575 K) is converted into *hepta-1,5-diene*:

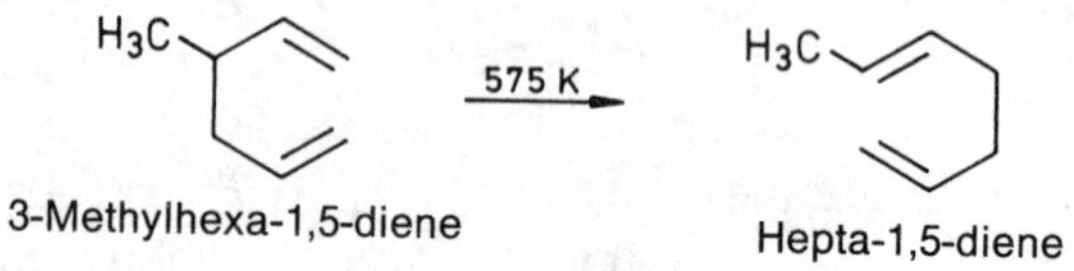

3-Methylhexa-1,5-diene   Hepta-1,5-diene

Most *degenerate rearrangements*, such as that of hexa-1,5-diene, in which the product is identical to the starting material, are not in fact observable.

This is a sigmatropic [3,3] shift, which proceeds through a cyclic transition state as shown. In this transition state the NBMOs of two allyl radicals interact with one another.

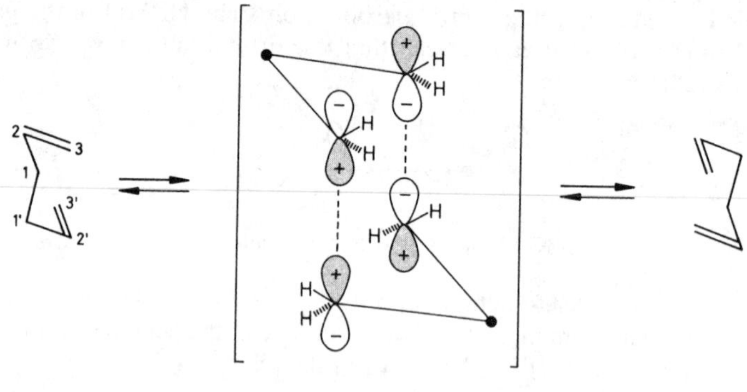

2.7.5 Dienes with Isolated Double Bonds

The diene used to exemplify the *Cope* rearrangement, *hexa-1,5-diene* (*diallyl*) is a molecule having isolated double bonds. It can be prepared from *allyl chloride* (*3-chloropropene*) by reaction with magnesium (this reaction is analogous to the *Wurtz* synthesis):

$$2\,H_2C = CH-CH_2Cl + Mg \xrightarrow{\text{Ether}} H_2C = CH-CH_2-CH_2-CH = CH_2$$

Allyl chloride   Hexa-1,5-diene
                                 b.p. 59.5°C [332.7 K]

In the normal way, isolated double bonds react quite independently of one another. The *Cope* rearrangement is an instructive example of a concerted reaction and provides a warning that the word 'isolated' should not be taken too literally.

2.8 Monohydric Alcohols (Alkanols)

Monohydric alcohols are compounds in which a *hydroxy* group (OH) is directly attached to an sp^3-hybridised carbon atom which is itself attached to either hydrogen atoms or other carbon atoms. They are derived from alkanes by replacing a hydrogen atom with a hydroxy group. The saturated alcohols, or alkanols, form a *homologous series* with the general formula $C_nH_{2n+1}OH$.

There are two ways of naming alcohols. Either the suffix 'ol' is attached to the name of the corresponding alkane as in methanol, ethanol, propanol, butanol, *etc.*, or alternatively the name of the alkyl group precedes the word alcohol as in methyl alcohol, ethyl alcohol, propyl alcohol, butyl alcohol, *etc.* The first is the common form in modern literature and is the recommended usage. An older name for methanol was 'carbinol'.

The first members of the series of monohydric alcohols are:

H_3C-OH Methanol, Methyl alcohol, Wood spirit
H_3C-CH_2-OH Ethanol, Ethyl alcohol, (Alcohol), Wine spirit

In the case of propane the hydroxy group can replace a hydrogen atom either in a methyl group or in the methylene group. Thus there are *two* isomeric propanols with different structural formulae:

$H_3C-CH_2-CH_2-OH$ Propan-1-ol, n-Propanol, n-Propyl alcohol

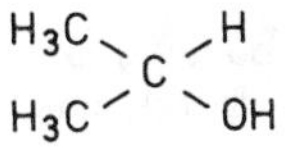

Propan-2-ol, Isopropyl alcohol

For alcohols with 4 or 5 carbon atoms derived from the isomeric butanes and pentanes there are, respectively, 4 and 8 structural isomers.

Alcohols are defined as *primary*, *secondary* or *tertiary* alcohols, depending upon whether the hydroxy group is attached to a primary, secondary (*sec.* or s-) or tertiary (*tert.* or t-) C-atom.

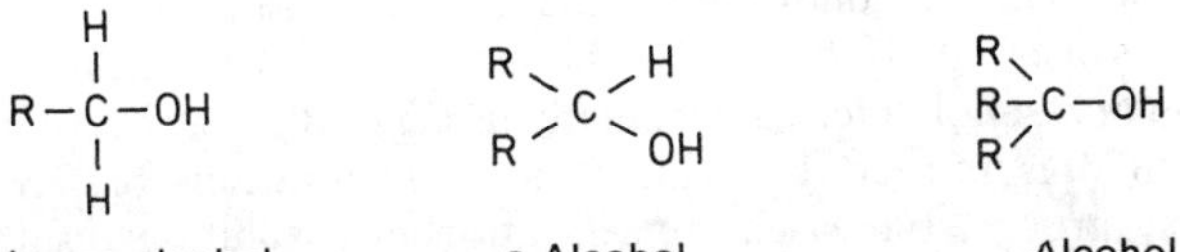

Primary alcohol   s-Alcohol   .-Alcohol

Methanol is also a primary alcohol, but it is a special case since the CH_2OH group is attached only to another hydrogen atom.

Physical Properties

The lower monohydric alcohols are colourless, neutral liquids with a characteristic odour and a burning taste. Methanol, ethanol and the propanols are miscible with water in all proportions. Alcohols with 4 to 11 C atoms are oily liquids only slightly miscible with water; higher alcohols are tasteless and odourless solids which are insoluble in water. As the relative molecular masses of the alcohols increase, so the influence of the carbon chain on properties overrides that of the hydroxy group. The physical properties of the higher alcohols resemble those of alkanes.

The *boiling points* of the first five normal primary alcohols are as follows:

Methanol	b.p. 64.7 °C (337.9 K)	n-Butanol	117.5 °C (391.0 K)
Ethanol	78.4 °C (351.6 K)	n-Pentanol	138.0 °C (411.2 K)
n-Propanol	97.2 °C (370.4 K)		

The difference in boiling point between neighbouring members of this homologous series up to C_{10} is generally about 18–20 K; thereafter it decreases. This is a result of the similar steric structure of all the normal primary alcohols. The normal alcohol always has the higher boiling point of any pair of structurally isomeric alcohols.

2.8.1 Hydrogen Bonding (Hydrogen Bridging)[1]

The relatively high boiling points of the monohydric alcohols, like that of water, are attributable to *molecular association*. Detailed studies showed that this involves the hydrogen atom of the hydroxy group. This results in *hydrogen bonding* (*Latimer* and *Rodebush*, 1920) or *hydrogen bridging* (*Huggins*, 1937) between the hydrogen atom of a hydroxy group and a lone pair of electrons on the oxygen atom of another molecule. In consequence the alcohol molecules are linked together in chains, as illustrated in the following diagram:

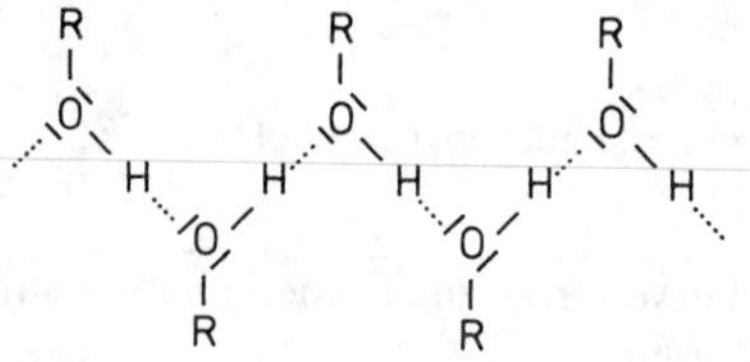

(The hydrogen bonding is represented by dotted lines)

There is in fact a dynamic equilibrium in which hydrogen bonds are broken and new ones are formed. In the vapour phase and in very dilute solutions in non-polar solvents, however, the molecules of alcohol are for the most part isolated from one another and not linked by hydrogen bonding. This can be seen, for example, from the *IR spectrum* of a 0.03% solution of ethanol in tetrachloromethane (Fig. 58), in which the sharp absorption band at 3640 cm^{-1} is produced by the stretching vibrations of *free* (non-hydrogen-bonded) hydroxy groups. Increased concentration of a solution results in there being hydrogen bonding between alcohol molecules, and this leads to a shift in the position of the OH absorption by about 300 cm^{-1}. Thus the IR spectrum of a 0.75% solution of ethanol in tetrachloromethane shows, as well as a weaker band associated with free hydroxy groups, a strong broad absorption band at 3600–3200 cm^{-1}.

The relative intensities of the two bands (see Fig. 58) gives an indication of the *degree of association*.

---

[1] See *P. Schuster* et al., The Hydrogen Bond, 3 vols (North Holland Publishing Co., Amsterdam, 1976); *C. Reichardt*, Solvents and Solvent Effects in Organic Chemistry, 2nd edn (VCH, Weinheim 1988).

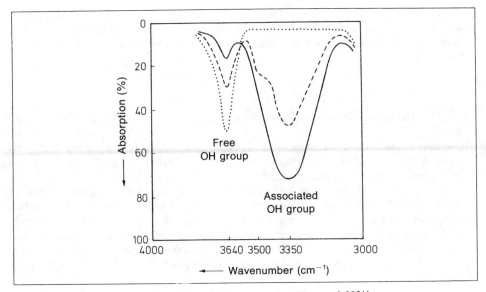

Fig. 58. Ethanol in tetrachloromethane (— 3%; ---0.75%; ... 0.03%)

This provides a way of estimating the energy of the hydrogen bond, which corresponds to its bonding enthalpy. The experimentally determined value, about 21–29 kJ/mol, is considerably less than that of simple atomic bonding, which involves bonding enthalpies in the range 150–500 kJ/mol. Hydrogen bonding is also found in, for example, carboxylic acids, amines, chelates, peptides and nucleic acids.

Chemical Properties

Because of the greater electronegativity of the oxygen atom some of the bonds in alcohols are polarised, as follows:

$$\begin{array}{c} H \\ | \; {\scriptstyle \delta^+ \; \delta^- \; \delta^+} \\ R-C-\overline{O}-H \quad (R = \text{Alkyl group}) \\ | \\ H \end{array}$$

However, the polarity of the C—O bond is too small for there to be dissociation in aqueous solution into a carbenium ion and a hydroxide anion.

Rather, the alcohols may be regarded as extremely weak acids, in which the hydrogen of the hydroxy group can be replaced by metals such as sodium; in this respect they behave analogously to water. The salt-like products are known as *alcoholates* or *alkoxides*. They are only stable in the absence of moisture:

$$R-\overline{O}-H + Na \rightarrow R-\overline{O}|^{\ominus} Na^{\oplus} + 1/2\, H_2$$

Alkoxides react with water to give the corresponding alcohols and a metal hydroxide:

$$R-\overline{O}|^{\ominus} Na^{\oplus} + H_2O \rightleftharpoons R-\overline{O}-H + Na^{\oplus}\, OH^{\ominus}$$

This is in clear contrast to their properties in the gas phase (see p. 48).

Drying agents, such as calcium chloride which reacts with water to form a hexahydrate, $CaCl_2.6H_2O$, similarly associate with methanol or ethanol to form crystals containing alcohol of crystallisation. Hence alcoholic solutions *cannot* be dried by the use of calcium chloride.

2.8.2 Oxidation Products from Alcohols

Especially characteristic reactions of alcohols are those which take place when they are treated with oxidising (or dehydrogenating) reagents.[1] These differ according to whether the alcohol is primary, secondary or tertiary.

The *mechanisms* of these oxidation reactions are complex, but two general types are possible. In the one, a hydrogen atom, which is attached to the same carbon atom as is the hydroxy group, is converted into a hydroxy group.[2] According to *Erlenmeyer's* rule, two hydroxy groups attached to the same carbon atom are in general unstable and lose a molecule of water, resulting in the formation of a carbonyl group, C=O. In this way *primary* alcohols are converted first of all into *aldehydes* and then into *carboxylic acids*, *e.g.*

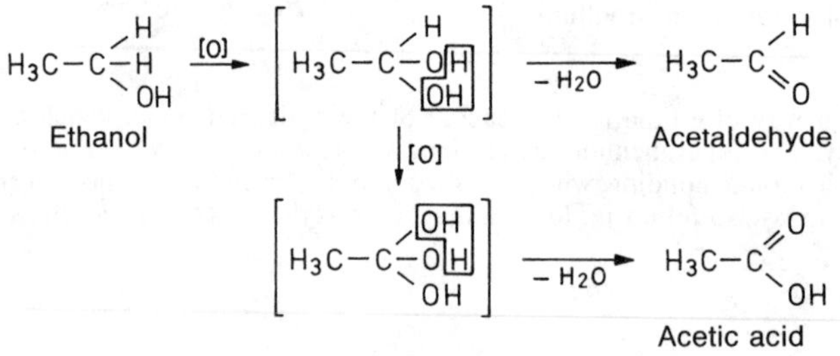

On the other hand, *H. Wieland* noted that *primary* alcohols can often be converted into aldehydes in the absence of any oxygen. Reagents such as palladium act instead as *hydrogen* acceptors:

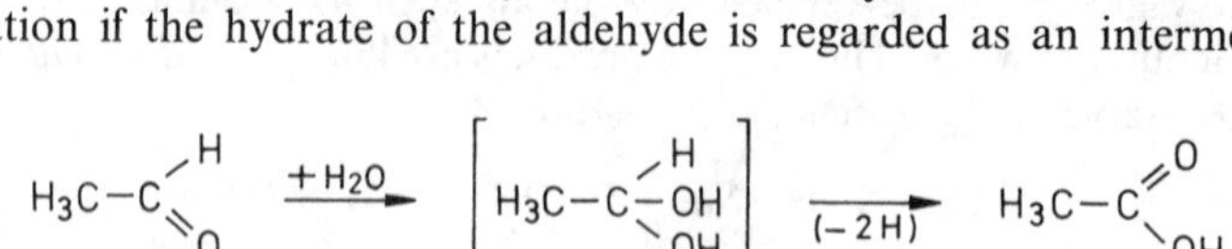

Since in this case the alcohol loses two hydrogen atoms, it is described as *dehydrogenation* rather than as oxidation.

The further conversion of an aldehyde into a carboxylic acid can also be seen as a dehydrogenation if the hydrate of the aldehyde is regarded as an intermediate in the reaction:

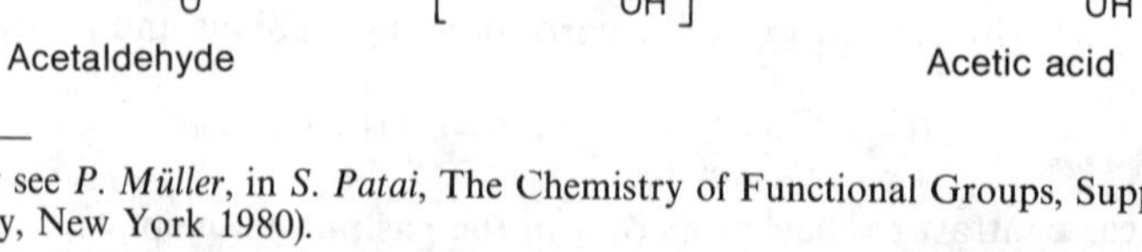

[1] For a review see P. Müller, in S. Patai, The Chemistry of Functional Groups, Supplement E, *pp.* 469–538 (Wiley, New York 1980).

[2] See Houben-Weyl, Methoden der organischen Chemie, 4th edn, Vol. 4/la (Thieme Verlag, Stuttgart 1981).

Dehydrogenation of *secondary* alcohols leads to the formation of *ketones*, *e.g.* acetone is obtained from isopropanol:

$$(H_3C)_2C\overset{H}{\underset{OH}{\diagdown}} \xrightarrow{-2H} (H_3C)_2C{=}O$$

Propan-2-ol   Acetone

A common reagent which is used for converting primary and secondary alcohols into, respectively, aldehydes and ketones, is chromic acid. Reaction proceeds via formation of a chromate ester:

$$\underset{OH}{\overset{R_2C-H}{|}} + H_2CrO_4 \longrightarrow \underset{O-CrO_3H}{\overset{R_2C-H}{|}} \longrightarrow R_2C{=}O + H_2CrO_3$$

Tertiary alcohols, $R_3C{-}OH$, *cannot* be oxidised without further cleavage of the molecules.

The Most Important Monohydric Alcohols

2.8.3 Methanol (Methyl Alcohol, Carbinol)

Preparation

2.8.3.1 Methanol was first obtained by the *dry distillation of wood* (*Boyle*, 1661). A flammable gas, containing hydrogen, carbon monoxide and ethylene, was evolved, a mixture of watery and tarry liquids distilled, and a residue of wood charcoal remained. The distillate contained 1.5–3% methanol, 10% acetic acid and 0.5% acetone, together with acetaldehyde, methyl acetate, organic bases and wood-tar. (The methanol is derived from the methoxy groups of hemicelluloses and lignins.)

The acetic acid was precipitated out as calcium acetate by addition of calcium hydroxide.

2.8.3.2 High-pressure Process. Industrially methanol is prepared from carbon monoxide and hydrogen (synthesis gas) in the presence of a ZnO/Cr_2O_3 catalyst at about 400°C (675 K) and 200 bar (20 MPa):

$$CO + 2H_2 \xrightarrow{(ZnO/Cr_2O_3)} H_3COH \quad \Delta H = -119\,kJ/mol$$

Whilst the above conditions provide fairly pure methanol, at the same pressure but at a 30–40 K higher temperature and in the presence of an alkaline ZnO/Cr_2O_3 catalyst not only methanol but also higher alcohols such as n-propanol (2%) and especially isobutanol [$(CH_3)_2CH{-}CH_2OH$] (11%) are formed. This is used as a way of synthesising isobutanol.

2.8.3.3 Low-pressure Method of ICI. In this method a Cu/Zn/Al catalyst is used to convert *synthesis gas* at about 250°C (525 K) and 50 bar (5 MPa) into methanol. It requires less energy than the high-pressure process.

Properties

Methanol is a liquid which mixes completely with other organic solvents. It burns with a non-luminous pale blue flame. Unlike ethanol, methanol can be obtained with 99% purity by fractional distillation from aqueous solutions. To obtain 'absolute' (very dry) methanol, magnesium is added, which dissolves, forming magnesium methoxide (methanolate). The latter is converted by water present into insoluble magnesium oxide and methanol, in the process also removing the water:

$$2\,H_3COH \xrightarrow[-H_2]{+\,Mg} (H_3CO)_2Mg \xrightarrow{+\,H_2O} MgO + 2\,H_3COH$$

In living organisms, methanol is oxidised to *formaldehyde* and *formic acid*. This is the reason for its being a *poison*, which can cause severe and permanent damage. (It causes blindness; the fatal dose for men is about 25 g.)

Uses

Methanol has many uses. For example, it is added to petroleum, and is used as a solvent and for the preparation of formaldehyde, polyesters, adhesives and dyestuffs and, in addition, as a *methylating agent*.[1]

2.8.4 Ethanol (Ethyl Alcohol)

Preparation

2.8.4.1 From ethylene. (a) Industrially this is carried out by *absorption of ethylene* (from gases from cracking of oil) in *concentrated sulphuric acid* at 55–80°C (330–355 K). Ethyl hydrogen sulphate and diethyl sulphate are formed in an exothermic reaction (sulphuric acid process).

$$H_2C = CH_2 \xrightarrow{+\,HO-SO_3H} C_2H_5O-SO_3H \xrightarrow{+\,C_2H_4} (C_2H_5O)_2SO_2$$

Ethyl hydrogen sulphate   Diethyl sulphate

The reaction mixture is hydrolysed with an equivalent amount of water at 70–100°C (345–375 K) and ethanol is distilled out. At the same time some diethyl ether (about 2%) is also formed.

$$C_2H_5OSO_3H + H_2O \rightarrow C_2H_5OH + H_2SO_4$$
$$(C_2H_5)_2SO_2 + 2\,H_2O \rightarrow 2\,C_2H_5OH + H_2SO_4$$
$$(C_2H_5O)_2SO_2 + C_2H_5OH \rightarrow (C_2H_5)_2O + C_2H_5OSO_3H$$

Diethyl ether

(b) The conversion of ethylene into ethanol is also carried out on a large scale industrially by its reaction with excess steam under pressure at 300°C (575 K) in the presence of phosphoric acid on kieselgel:

$$H_2C = CH_2 + H_2O \xrightarrow{(H_3PO_4)} H_3C-CH_2-OH \qquad \Delta H = -48\,kJ/mol$$

2.8.4.2 By hydrogenation of acetaldehyde at 100–130°C (375–405 K) over a nickel catalyst:

$$H_3C-C\overset{H}{\underset{O}{\diagup}} \xrightarrow[(Ni)]{+2\,H} H_3C-CH_2-OH$$

2.8.4.3 By alcoholic fermentation. The term *fermentation* means a biochemical process brought about by micro-organisms. In 'alcoholic fermentation' sugars (hexoses), $C_6H_{12}O_6$,

---

[1] See *F. Asinger*, Methanol—Chemie und Energierohstoff (Methanol — Source of Chemicals and Energy) (Springer-Verlag, Berlin 1986).

especially glucose (dextrose, grape sugar) and fructose (fruit sugar) are converted by yeast, which contains the enzyme complex *zymase*, into ethanol and carbon dioxide:

$$C_6H_{12}O_6 \xrightarrow{\text{(Zymase)}} 2\,C_2H_5OH + 2\,CO_2 \qquad \Delta H = -105\,kJ/mol$$

An *enzyme* is an organic catalyst, which is specific and only present in very small quantity. The majority of enzymes are proteins (but see p. 869).

Properties

It is not possible to remove all water from ethanol by distillation, because 4.4% water and 95.6% ethanol form a constant boiling point (*azeotropic*) mixture, which boils at 78.2°C (351.4 K). The water can be removed by distillation over calcium oxide (quicklime). To prepare 'absolute alcohol' industrially, *benzene* (b.p. 80.1°C [353.3 K]) is added to the alcohol–water mixture, and it is distilled. At 64.9°C (338.1 K) a ternary mixture of benzene, alcohol and water comes over, which contains all the water from the wet alcohol; this is followed at 68.2°C (341.4 K) by a binary mixture (benzene–alcohol). When this fraction has distilled, absolute alcohol remains, which boils at 78.4°C (351.6 K). This is known as the *Young* process. Trichloroethylene can be used instead of benzene (Drawinol process). To remove the last traces of water (up to 1%) from ethanol, magnesium is used, as described for methanol.

Ethanol is a liquid with a characteristic odour. It burns with a pale blue flame and will do so even if it contains up to 50% water. The water content of a sample of alcohol can be determined by measuring its specific gravity.

Uses

Ethanol is widely used as a solvent. In countries which are rich in sources of carbohydrates or of wood, it is used as a starting material in chemical processes; for example, it is oxidised to obtain acetaldehyde, acetic acid and acetic anhydride, and is used in the preparation of dyestuffs, perfumes and pharmaceuticals, and as an additive to petrol. In medicine ethanol is used for the preservation of anatomical specimens.

Alcohols are also used for the preparation of metal alkoxides,[1] which are valuable reagents in organic synthesis. Sodium methoxide, H_3CONa, or sodium ethoxide, C_2H_5ONa, are widely used as *condensing agents*. They are made by dissolving sodium in excess alcohol; hydrogen is evolved. Since such a solution of sodium ethoxide readily turns brown owing to the formation of acetaldehyde resins, sodium methoxide is commonly the preferred reagent. Sodium t-butoxide, $(CH_3)_3CONa$, is also used similarly. As well as sodium alkoxides, magnesium ethoxide, $Mg(OC_2H_5)_2$, and aluminium ethoxide, $Al(OC_2H_5)_3$, have also proved useful in synthesis, as have alkoxides derived from higher, usually branched-chain alcohols, *e.g.* potassium and aluminium t-butoxides.

A large amount of alcohol is made for alcoholic drinks. These may be divided into *distilled* and *undistilled* drinks. In the first group are the various spirits, in the second beer, cider and wines.

Spirits are made by fermenting materials which contain starch or sugar, and distilling the resultant product. In this way *whisky* is made from barley or maize, *rum* from molasses, *arrack* from rice, *corn spirits* from cereals (wheat, rye). Spirits derived from fruit include *brandy* from grapes and other spirits from, for example, cherries (*kirsch*), apples (*calvados*) and plums. The alcohol content is usually between 30 and 50%.

In the *brewing* of *beer*, the starch from barley is converted into sugars by enzymes, and the sugar is then fermented by yeast. The alcohol content of beers is about 3–6%.

Wine arises from the alcoholic fermentation by yeasts of the sugar in grapes. The ripe grapes are pressed to obtain the grape juice, which is then allowed to ferment. The resultant wine contains

---

[1] See *D. C. Bradley* et al., Metal Alkoxides (Academic Press, New York 1978).

8–14% alcohol. Making use of the natural deuterium content it is possible to distinguish between natural wine and wine which has been 'improved' by addition of sugar (chaptalisation) or sugar solution (SNIF = site-specific natural isotope fractionation).[1]

Cider and perry are similarly produced from, respectively, apples and pears.

In small quantities ethanol acts as a stimulant and then as a narcotic, but in larger amounts it is a poison and can cause death. Industrial alcohol is made unpleasant and unpalatable by addition of butanone or pyridine.

2.8.5 Propanols (Propyl Alcohols)

Propan-1-ol, 1-propanol, n-propanol, is a colourless liquid with a pleasant smell. It boils at 97.2°C (370.4 K). Industrially, it is made by the oxo-process from ethylene (see p. 204).

Propan-2-ol, 2-propanol, isopropanol, is a colourless liquid with b.p. 82.4°C (355.6 K). Industrially it is made from propene obtained from the cracking of oil. Propene is absorbed in sulphuric acid and the resultant esters are then hydrolysed:

$$H_3C-CH=CH_2 \xrightarrow{+ H_2SO_4} (H_3C)_2CH-O-SO_3H \xrightarrow[- H_2SO_4]{+ H_2O} (H_3C)_2CH-OH$$

Propene   Isopropyl hydrogen sulphate   Isopropanol

Di-isopropyl ether, $(H_3C)_2CH-O-CH(CH_3)_2$, is formed as a by-product.

Both propanols are widely used as solvents. Propan-2-ol is added to petrol to act as an antifreeze. Both are more poisonous than ethanol. Propan-2-ol is important industrially for conversion into acetone.

2.8.6 Butanols (Butyl Alcohols)

The four isomeric butanols are as follows:

		b.p.
$H_3C-CH_2-CH_2-CH_2-OH$	Butan-1-ol, 1-Butanol, n-Butanol, n-Butyl alcohol	117,5 °C (390.7 K)
$H_3C-CH_2-CH-CH_3$ $\quad\quad\quad OH$	Butan-2-ol, 2-Butanol, sec. Butyl alcohol	99,5 °C (372.7 K)
$H_3C-CH-CH_2OH$ $\quad\; CH_3$	2-Methylpropan-1-ol, 2-Methyl-1-propanol, Isobutanol, Isobutyl alcohol	108,4 °C (381.6 K)
CH_3 $H_3C-C-OH$ $\quad\; CH_3$	2-Methylpropan-2-ol, 2-Methyl-2-propanol, t-Butanol, tert.Butyl alcohol	82,5 °C (355.7 K)

Butan-1-ol is often called just 'butanol'.

---

[1] See H. O. Kalinowski, Der Schrecken der Weinpanscher (The Terror of the Wine Adulterer), Quantitative Deuterium-NMR- Spektroskopie, Chemie in unserer Zeit 22, 162 (1988).

Preparation

2.8.6.1 Industrially butan-1-ol is made by *hydrogenation, under pressure, of crotonaldehyde* in the presence of a copper catalyst:

$$H_3C-CH=CH-C\overset{H}{\underset{O}{\diagdown}} \xrightarrow[\text{(Cu)}]{+2H} H_3C-CH_2-CH_2-C\overset{H}{\underset{O}{\diagdown}} \xrightarrow[\text{(Cu)}]{+2H} H_3C-CH_2-CH_2-CH_2-OH$$

Crotonaldehyde   n-Butyraldehyde   Butan-1-ol

2.8.6.2 In the large industrial countries it is now made by the *oxo-process* from propene.

Properties and Uses

Butan-1-ol is only partially soluble in water. Like methanol and ethanol it is of industrial importance. It is used as a solvent for resins, as a dispersant for lacquers and for making esters such as butyl acetate and butyl phthalate.

Butan-2-ol is prepared industrially from butene, itself obtained from the cracking of oil:

$$H_3C-CH_2-CH=CH_2 \xrightarrow{+H_2SO_4} H_3C-CH_2-\underset{\underset{OSO_3H}{|}}{CH}-CH_3 \xrightarrow[-H_2SO_4]{+H_2O} H_3C-CH_2-\underset{\underset{OH}{|}}{\overset{*}{CH}}-CH_3$$

But-1-ene   Butan-2-ol

In the laboratory it is made from acetaldehyde and ethyl magnesium bromide (see p. 185).

$$H_3C-C\overset{H}{\underset{O}{\diagdown}} \xrightarrow{+H_3C-CH_2-MgBr} H_3C-CH_2-\underset{\underset{OMgBr}{|}}{CH}-CH_3 \xrightarrow{+H_2O(H)^{\oplus}} H_3C-CH_2-\underset{\underset{OH}{|}}{\overset{*}{CH}}-CH_3$$

The general use of this method for the preparation of alcohols will be discussed at that point. Butan-2-ol contains an *asymmetric carbon atom* (*); its racemic form can be resolved into two *enantiomers*.[1]

Isobutanol is found in fusel oil, the higher boiling mixture of other alcohols which is formed as a by-product in the preparation of ethanol by fermentation, and which is poisonous. Industrially it is made by the *oxo process*.

t-Butanol is the only member of this group which can be a solid at ambient temperature. Its m.p. is 25.5°C (298.7 K). It is made from *isobutene*, which is formed in the cracking of oil, by treatment with sulphuric acid and hydrolysis of the resultant ester. If it is treated with sulphuric acid and hydrogen peroxide it is converted into t-butyl hydroperoxide (TBHP), $(H_3C)_3COOH$, b.p.$_{15}$ 33°C (306 K).

2.8.7 Pentanols (Pentyl or Amyl Alcohols)

Pentanols are made from alkenes by a hydroxymercuration process (see p. 188).

---

[1] In older literature enantiomers were described as optical isomers or optical antipodes, but these terms have now passed out of use.

The most important member of this group is **pentan-1-ol** (1-pentanol, n-pentyl or n-amyl alcohol), H_3C—$(CH_2)_3CH_2OH$, b.p. 138°C (411 K). It is a stronger intoxicant and poison than ethanol, as are the two isomeric pentanols found in *fermentation amyl alcohol*. These are found in fusel oil, a by-product in the alcoholic fermentation process, and are very injurious to humans. They do not arise from the fermentable sugars, but from the *albumens*, which occur in the starchy materials present in the starting matter (potatoes, grain).

Fermentation amyl alcohol consists of the two following isomers:

$$H_3C-CH-CH_2-CH_2-OH \quad und \quad H_3C-CH_2-CH-CH_2-OH$$
$$| \qquad\qquad\qquad\qquad\qquad\qquad | $$
$$CH_3 \qquad\qquad\qquad\qquad\qquad\qquad CH_3$$

3-Methylbutan-1-ol (85%)   2-Methylbutan-1-ol (15%)
(Isopentyl alcohol)

2.8.8 Optical Isomerism, Chirality

2-Methylbutan-1-ol, and also *sec.*butanol are examples of substances which rotate the plane of linearly polarised light. They are said to be *optically active*. This property was investigated in detail by *Pasteur* (1848) for the optically active tartaric acids. He showed that solutions, of the same concentration, of the two acids rotated the plane of polarisation by the *same amounts*, but in *opposite directions*. The one rotated it to the right ($+$), the other to the left ($-$). Furthermore, it was evident that the crystalline forms of these two tartaric acids were *enantiomorphs*; that is, they had identical faces and angles but were not superimposable. They were *mirror images* of one another, and were described as *enantiomers*. The reason for the optical activity of compounds may be found in their three-dimensional structures. *Pasteur* (1860) postulated that all optically active molecules must have *asymmetric*[1] structures; *no plane of symmetry* can be found in them.

In 1874 *van't Hoff* and *le Bel*, quite independently of each other, noted that all the optically active compounds known at that time had at least one asymmetric carbon atom, that is, a carbon atom with four different atoms or groups of atoms attached to it. From this observation *van't Hoff* concluded that the four substituents attached to a carbon atom did not all lie in one plane, but were situated at the corners of a tetrahedron at the centre of which the carbon atom itself was located.

If one considers 2-methylbutan-1-ol in this way, it is seen that C-atom (2) has four different ligands (H, CH_3, C_2H_5, CH_2OH) attached to it and is thus *asymmetric*.

$$\begin{array}{c} H \\ |(2) \\ H_3C-CH_2-C-CH_2OH \\ | \\ CH_3 \end{array}$$

2-Methylbutan-1-ol

The appearance of the two *optically active* forms of 2-methylbutan-1-ol is made clear by tetrahedral models (Figs. 59 and 60).

The two structures are mirror images of one another and are not superimposable. They are called *enantiomers*. Such compounds rotate the plane of linearly polarised light by the same extent but in opposite directions.

---

[1] *Pasteur* actually used the term *dissymétrique*, but it was incorrectly translated as *asymmetric*.

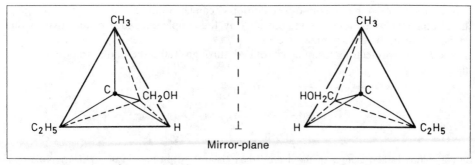

Fig. 59.   Fig. 60.

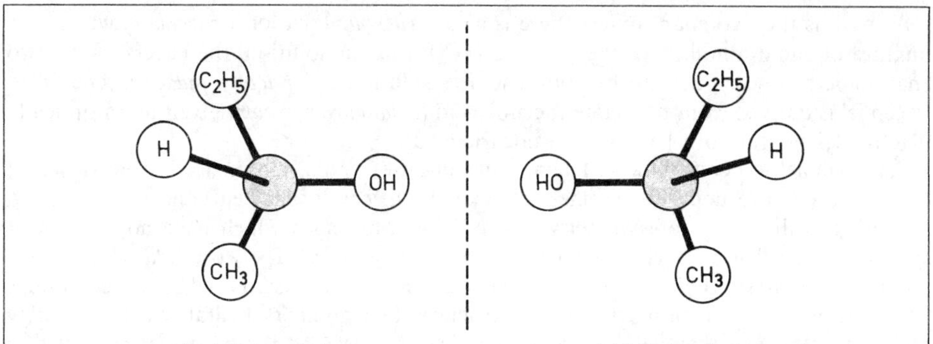

(−)-Butan-2-ol   (+)-Butan-2-ol

Fig. 61.

(-)-2-Methylbutan-1-ol is found in fermentation amyl alcohol; there is also a (+)-2-methylbutan-1-ol.

The *steric structure of enantiomers* is demonstrated well by the use of molecular models; see Fig. 61.

In order to depict the shape of enantiomers on paper, use is made of *Fischer* projection formulae. In this convention, groups attached to the asymmetric centre by horizontal lines—H and HO in the example below—lie *above* the plane of the paper, whilst groups attached by vertical lines are *beneath* the plane of the paper. Another common convention is to depict bonds to groups above the plane of the paper by thick lines, and bonds to groups below the plane of the paper by dotted lines. As examples, the enantiomeric butan-2-ols may be represented on paper as follows:

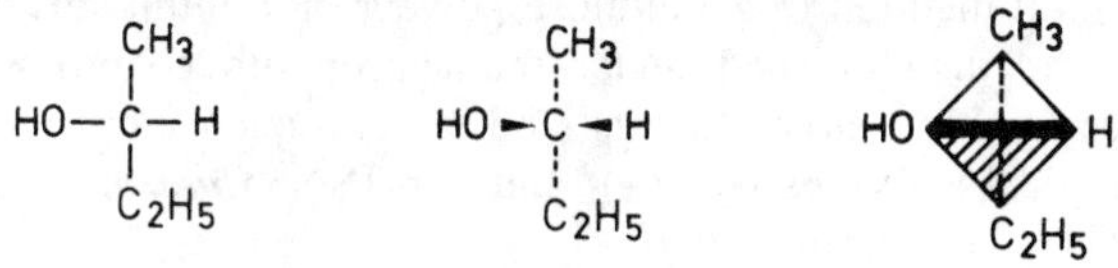

(−)-Butan-2-ol

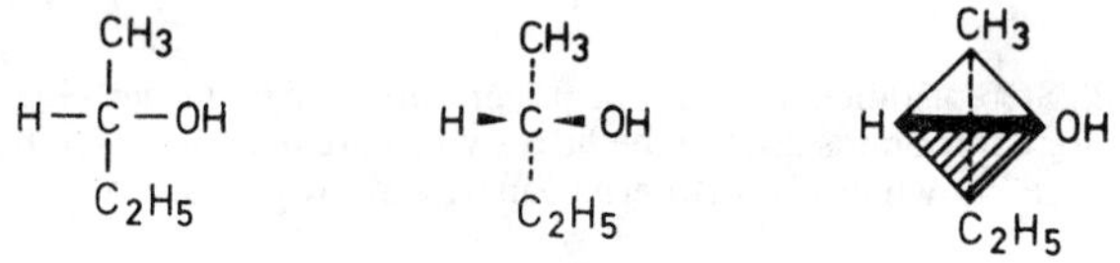

(+)-Butan-2-ol

It should be noted that the opposite enantiomer can be obtained, not only by exchanging the H and HO as is done above, but also by the simple exchange of any two of the ligands attached to the asymmetric centre.

If a *Fischer*-projection formula is rotated through 180°, the configuration remains unaltered, e.g.

$$
\begin{array}{ccc}
\text{CH}_3 & & \text{C}_2\text{H}_5 \\
| & & | \\
\text{HO}-\text{C}-\text{H} & \equiv & \text{H}-\text{C}-\text{OH} \\
| & & | \\
\text{C}_2\text{H}_5 & & \text{CH}_3
\end{array}
$$

By convention carbon atom 1 of the main carbon chain is put at the top of a *Fischer*-projection formula: this is often the carbon atom with the highest oxidation state (see p. 284).

As well as the two enantiomers there is also a *racemic* form or a *racemate*, which is a mixture of equal amounts of the $(+)$ and $(-)$ forms. In solution, the effects of the two enantiomers cancel one another out and the solution is *optically inactive*. The word 'racemic' is derived from the name racemic acid (*acidum racemicum*) given to an optically inactive 1:1 mixture of $(+)$- and $(-)$-tartaric acids.

An asymmetric centre is a sufficient, but not an essential, reason for the existence of enantiomers. For example enantiomers can exist in the case of molecules having C_2 and D_2 symmetry, with an axis of symmetry (see p. 314). Molecules which have no element of symmetry other than an axis of symmetry are said to be *chiral*. The word '*chirality*', which means 'handedness; (Grk. *cheir*, hand), provides a precise description of the conditions which are essential and sufficient for the existence of enantiomers. Chirality is the property of a molecule, or of a model; optical activity is only observed if a sufficient amount of a homochiral molecule is present. The so-called *asymmetric carbon atom* is either a *chiral centre (Prelog)* or a *stereogenic element* (in the purely geometric sense, *Mislow*).[1]

Mislow has pointed out the conceptual difference between a stereogenic element and its environment. He designated butan-2-ol, for example, as chirotopic, because the stereogenic central element, in this case the carbon atom, is located in a chiral environment, in which exchange of two of the ligands attached to it results in formation of the enantiomer. Its *local symmetry* is chiral. The ligands bound to this carbon atom are likewise chirotopic, but are not stereogenic. There are also stereogenic elements with achiral local symmetry, for example a carbon atom of 1-bromo-2-chloro-1-iodoethylene. Exchange of two ligands on such a carbon atom results in the formation of a *diastereomer* and not an *enantiomer*. The environment of such a stereogenic element is achirotopic, because it is not itself in a chiral environment.

Enantiomers differ from each other solely in their effect on polarised light. So long as they are in an achiral environment they are identical in all their physical and chemical properties, and cannot be distinguished then by either chemical or physical methods. In the presence of a chiral medium (*e.g.* cellulose powder or a chiral solvent) this is not the case and they can be distinguished under the appropriate conditions. All biological processes take place in chiral environments, and in consequence enantiomers often show quite different biological activities (see $(+)$- and $(-)$-(INN[2] *Ibuprofen*, p. 236 and L- and D-asparagine, p. 823).

---

[1] See *K. Mislow* et al., Stereoisomerism and Local Chirality, J. Am. Chem. Soc. *106*, 3319 (1984).

[2] Abbreviated names for compounds used in medicine which are recommended by the World Health Organisation (WHO) are known as INN (International Non-proprietary Names) and are indicated by heavy italic type.

Measurement of Specific Rotations (*Biot*, 1836), Polarimetry When linearly polarised light is passed through a solution of an optically active compound, the rotation of the plane of polarisation is proportional to the percentage present p (g active substance in 100 g solution) at a density d, or the concentration c (g active substance in 100 ml solution) of the solution, and the length l in decimetres of the sample through which the light passes. The angle of rotation α produced by 1 g active substance in 1 ml of solution in a tube of length 1 dm is known as the *specific rotation*:

$$[\alpha]_\lambda^t = \frac{\alpha \cdot 100}{l \cdot p \cdot d} = \frac{\alpha \cdot 100}{l \cdot c}$$

The specific rotation is not a constant but depends upon the temperature t and the wavelength of the light used. It is common practice to use monochromatic sodium light (D line, 589.3 nm) and a temperature of 20°C (293 K); this is indicated as follows: $[\alpha]_D^{20}$. The results can be used to help in the characterisation of an optically active substance. The values are also dependent upon the concentration of the solution and on the solvent which is used.

If in a mixture of enantiomers, one is preponderant, then the *optical purity* is defined as a percentage, $100 \cdot [\alpha]/[\alpha]_0$. $[\alpha]$ is the specific rotation of the mixture of enantiomers and $[\alpha]_0$ is the value for the pure enantiomer.

The *enantiomeric excess*, (symbol: = ee), is given where, for example, of the two enantiomers $R > S$, by the equation:

$$ee\ [\%] = \frac{R-S}{R+S} \cdot 100$$

R and S are used to define the two enantiomers, as described on p. 275. Enantiomerically pure compounds (ee = 100%) are known as *homochiral compounds*.

If the specific rotation of a chiral compound is not known, then ee can be determined by NMR spectroscopy, with the aid of *chiral shift reagents*, because in their presence the chemical shifts of the NMR signals are not the same for the two enantiomers.

2.8.9 Higher Alcohols, $C_nH_{2n+1}OH$

Esters, which occur in natural products and are derived from monohydric alcohols, are in general derivatives of higher *primary* alcohols. For example in plant oils may be found *hexan-1-ol*, $C_6H_{13}OH$, b.p. 157.5°C (430.7 K), *octan-1-ol*, $C_8H_{17}OH$, b.p. 194.5°C (467.7 K), and *dodecan-1-ol*, or *lauryl alcohol*, $C_{12}H_{25}OH$, m.p. 26°C (299 K). In addition, there are the so-called 'wax alcohols', which occur in natural waxes, such as *hexadecan-1-ol* or *cetyl alcohol*, $C_{16}H_{33}OH$, m.p. 50°C (323 K) (in spermaceti), *hexacosan-1-ol* or *ceryl alcohol*, $C_{26}H_{53}OH$, m.p. 79.5°C (352.7 K) (in Chinese wax) and *myricyl alcohol*, a mixture of $C_{30}H_{61}OH$ and $C_{32}H_{65}OH$ (in beeswax).

The preparation of 'fatty alcohols' from the higher 'fatty acids' and their use as starting materials for the preparation of synthetic detergents are mentioned later.

2.8.10 Unsaturated Alcohols (Alkenols and Alkynols)

The monohydric unsaturated alcohols are structurally derivatives of alkenes and alkynes in which a hydrogen atom has been replaced by a hydroxy group. If the HO group is attached to a doubly bonded carbon atom, the compound is called an *enol*. Enols are for the most part unstable and rearrange to give carbonyl compounds. For example, *vinyl alcohol* (ethenol), which is unstable under normal conditions isomerises to give acetaldehyde (see p. 95). In contrast, *allyl alcohol*, prop-2-enol, $H_2C=CH-CH_2-OH$, and *propargyl alcohol*, prop-2-ynol, $HC\equiv C-CH_2OH$, in both of which the hydroxy group is attached to a saturated carbon atom, are relatively stable.

Allyl alcohol is present in garlic (*allium sativum*). It is a liquid with a pungent smell. It boils at 97°C (370 K) and is miscible with water. It is used as an intermediate in organic synthesis; its esters, such as *allyl phthalate*, are polymerised to provide synthetic resins.

Preparation

2.8.10.1 In the laboratory allyl alcohol is made *by heating glycerol with formic acid*; a formate ester is formed as an intermediate:

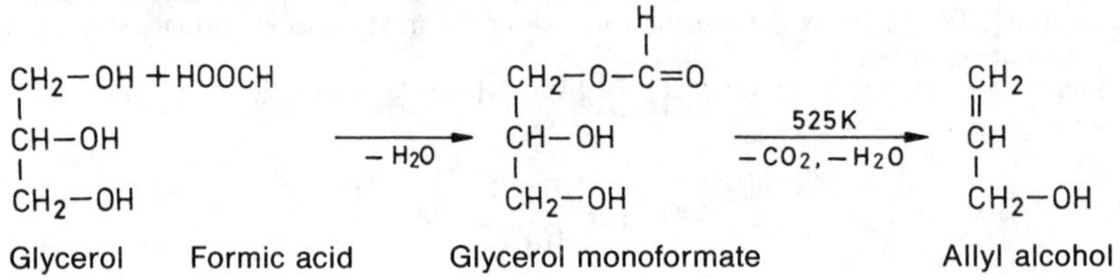

Glycerol   Formic acid   Glycerol monoformate   Allyl alcohol

2.8.10.2 Industrially it is made by *alkaline hydrolysis of allyl chloride (3-chloropropene)*:

$$H_2C=CH-CH_2-Cl + NaOH \rightarrow H_2C=CH-CH_2-OH + NaCl$$

2.8.10.3 or by *catalytic isomerisation of propene oxide*:

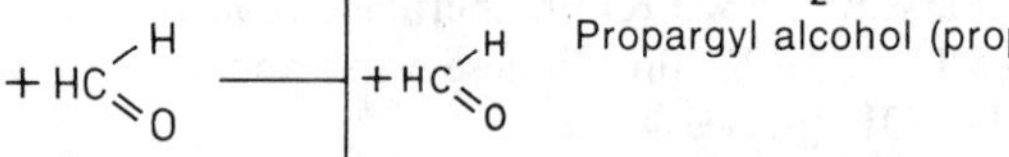

Propargyl alcohol (prop-2-ynol) is a liquid, b.p. 114°C (387 K) with an agreeable odour. Industrially it is obtained, together with but-2-yn-1,4-diol, by the *Reppe* process, from acetylene and formaldehyde in the presence of copper acetylide:

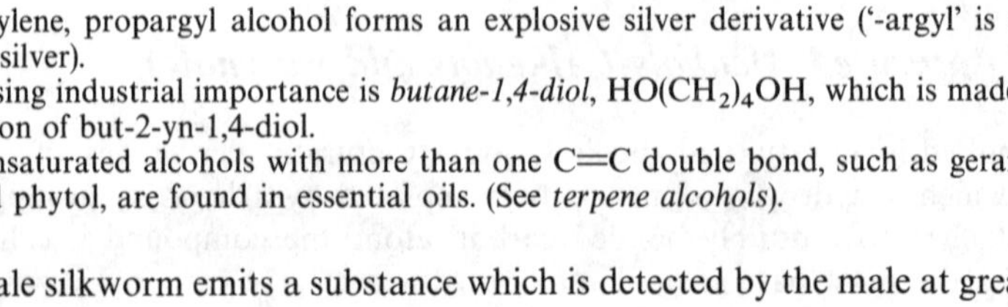

HC≡C-CH₂OH
Propargyl alcohol (prop-2-ynol)

HOCH₂-C≡C-CH₂OH
But-2-yn-1,4-diol

Like acetylene, propargyl alcohol forms an explosive silver derivative ('-argyl' is derived from *argentum* = silver).

Of increasing industrial importance is *butane-1,4-diol*, $HO(CH_2)_4OH$, which is made by catalytic hydrogenation of but-2-yn-1,4-diol.

Higher unsaturated alcohols with more than one C=C double bond, such as geraniol, linalool, farnesol and phytol, are found in essential oils. (See *terpene alcohols*).

The female silkworm emits a substance which is detected by the male at great distances.

The compound is (*E,Z*)hexadeca-10,12-dienol, also sometimes called *bombykol* (*Butenandt*). It is only effective between two individuals of the same species.

$$H_3C \diagup\diagdown\diagup\diagdown = \diagdown\diagup\diagdown\diagup\diagdown\diagup\diagdown\diagup\diagdown CH_2-OH$$

Such information-carrying substances are in general termed *semiochemicals*. *Pheromones* (Grk. *pherein*, to carry; *hormon*, an active substance) *are intraspecific messenger substances*. Bombykol induces a *specific behaviour* and is classified as a *releaser*; a *physiologically active* pheromone is called a primer. *Interspecific messenger substances* are known as *allelochemicals*. In this field, distinction is made between *kairomones* (a signal of value to a receptor belonging to another species) and *allomones* (signals important for the sender). All these substances are carried, for the most part through the air, to the sites where they act. They are among the most active of all biologically active substances and are widespread, especially in the insect kingdom.[1]

2.9 Halogen Derivatives of Alkanes

Whilst direct introduction of an HO group into an alkane is not successful, alkanes can be *halogenated* under the appropriate conditions, whereby one or more hydrogen atoms of the alkane are replaced by halogen atoms (chlorine, bromine).

The term *substitution* means in general the exchange of an atom or group of atoms by an equivalent atom or group of atoms.

It is worthwhile to consider *monohalogenated* and *polyhalogenated* alkanes separately. The former, known as *alkyl halides*, will be considered first. Fluorinated hydrocarbons are dealt with in a separate section.

2.9.1 Alkyl Halides (Halogenoalkanes)

The first members of the homologous series $C_nH_{2n+1}X$ (X = halogen) are the methyl and ethyl halides, which are derived, respectively, from methane and ethane by replacement of a hydrogen atom with a halogen atom. From propane and higher alkanes, isomeric alkyl halides are possible, as in the case of alcohols. Examples are:

$H_3C-CH_2-CH_2-Cl$ 1-Chloropropane, n-propyl chloride, b.p. 46.6°C (319.8 K)

$H_3C-CH-CH_3$ 2-Chloropropane, isopropyl chloride, b.p. 34.8°C (308.0 K)
    |
    Cl

As with alcohols, the numbers of isomers increase with increasing chain length.

---

[1] See *W. Francke*, Chemische Ökologie/Ökologische Chemie (Chemical Ecology; Ecological Chemistry) Nachr. Chem. Tech. Lab. *37*, 478 (1989).

Preparation

2.9.1.1 A general method for the preparation of alkyl halides, used both industrially and in the laboratory, involves the treatment of alcohols with hydrogen halides. Hydrogen iodide reacts most readily; when hydrogen bromide or hydrogen chloride are used it is generally necessary to use higher temperatures or a catalyst such as concentrated sulphuric acid or dry zinc chloride. The latter speed up the setting up of the following equilibrium:

$$R-OH + H-X \rightleftharpoons R-X + H_2O$$

Alcohol + Acid $\rightleftharpoons$ Ester + Water

This general reaction is known as *esterification*. Alkyl halides may be regarded as esters of the halogen acids HX (X = I, Br, Cl).

The reverse of esterification is *hydrolysis*. Because of its use in the making of soap from fats, it has also been described as *saponification*; this latter process involves the use of an alkali to bring about the reaction.

The mechanism of the conversion of an alcohol into an alkyl halide is as follows:

$$C_2H_5OH + HI \longrightarrow C_2H_5\overset{\oplus}{O}H_2 \quad I^{\ominus}$$

$$I^{\ominus} \searrow$$

$$\underset{CH_3}{\overset{}{CH_2}} - \overset{\oplus}{O}H_2 \longrightarrow CH_3CH_2I + H_2O$$

2.9.1.2 In the laboratory alkyl halides have often been prepared *from alcohols by treating them with phosphorus halides* (PCl_3, PCl_5, PBr_3, PI_3, etc.); for example:

$$3\,C_2H_5-OH + PBr_3 \longrightarrow 3\,C_2H_5-Br + H_3PO_3$$

Bromoethane (Ethyl Bromide) b.p. 38.4°C [311.6 K])

In this case the phosphorus(III) bromide is made from phosphorus and bromine. Bromine is dripped slowly into a solution of phosphorus in ethanol; the bromoethane is separated off from the reaction mixture by distillation.

Thionyl chloride is often used instead of a phosphorus chloride. This first reacts with the alcohol to form an unstable alkyl chlorosulphite, which decomposes when gently heated to give the alkyl halide and sulphur dioxide:

$$R-OH + SOCl_2 \xrightarrow[-HCl]{} \left[\begin{matrix} RO-S-Cl \\ \underset{O}{\overset{\|}{}} \end{matrix} \right] \longrightarrow R-Cl + SO_2$$

2.9.1.3 *Hunsdiecker* **Reaction,** *Borodin* **Reaction.**[1] In this reaction the silver salt of a carboxylic acid reacts with bromine (or iodine), preferably in tetrachloromethane as solvent. Carbon dioxide is evolved and an alkyl halide is formed:

$$R-COOAg + Br_2 \longrightarrow R-Br + AgBr + CO_2$$

This method is especially useful for the selective preparation of specific open-chain or cyclic monohalogeno-compounds. It proceeds via radicals in a chain reaction (see p. 125).

[1] See *C. V. Wilson*, The Reactions of Halogens with Silver Salts of Carboxylic Acids, Org. Reactions **9**, 322 ff. (1957).

2.9.1.4 The formation of alkyl halides by the *addition of hydrogen halides to alkenes* has been dealt with previously.

2.9.1.5 Halogenation of Alkanes. The *chlorination of methane* in the gas phase is important in industry. It leads in turn to the following products:

$$CH_4 \xrightarrow[-HCl]{+Cl_2} CH_3Cl \xrightarrow[-HCl]{+Cl_2} CH_2Cl_2 \xrightarrow[-HCl]{+Cl_2} CHCl_3 \xrightarrow[-HCl]{+Cl_2} CCl_4$$

Chloromethane (37)	Dichloromethane (41)	Trichloromethane (19)	Tetrachloromethane (3)
Methyl chloride	Methylene chloride	Chloroform	Carbon tetrachloride

The numbers next to the names indicate the yields, in mol-%, which are obtained when equimolar amounts of chlorine and of methane are used.

By variation of the reaction conditions (photohalogenation, thermal or catalytic halogenation, or the presence of inhibitors, e.g. hydroquinone) it is possible to make one or other of these products the main product.

The higher alkanes can similarly be chlorinated or brominated. *Chlorination of propane* provides the two monochloro-derivatives in about equal amounts:

$$H_3C-CH_2-CH_3 \xrightarrow[575\,K]{(Cl_2)}$$

$$H_3C-CH_2-CH_2-Cl \quad (48\%)$$
1-Chloropropane (Propyl chloride)

$$\underset{\underset{Cl}{|}}{H_3C-CH-CH_3} \quad (52\%)$$
2-Chloropropane (Isopropyl chloride)

It is not possible to iodinate alkanes in this way.

Substitution reactions which take place under the influence of light or in the presence of a peroxide as catalyst, proceed by means of a mechanism involving radicals.

In the *photochlorination of alkanes*, the initial reaction is the homolytic cleavage of a chlorine molecule into two chlorine atoms, which can be regarded as radicals. A chlorine atom then removes a hydrogen atom from a molecule of hydrocarbon to form hydrogen chloride. This leaves an alkyl radical, which in turn reacts with another molecule of chlorine to form an alkyl chloride and another chlorine atom; this can initiate a similar sequence of reactions, leading to the setting up of a chain reaction. If a chlorine atom should happen to encounter, for example, an alkyl radical, then the chain reaction is terminated.

$$|\overline{Cl}-\overline{Cl}| \longrightarrow |\overline{Cl}\bullet +\bullet\overline{Cl}| \qquad \text{Initiation } \varDelta H = +242\,kJ/mol$$

$$RH +\bullet\overline{Cl}| \longrightarrow H-\overline{Cl}| +R\bullet$$
$$R\bullet + |\overline{Cl}-\overline{Cl}| \longrightarrow R-\overline{Cl}| +\bullet\overline{Cl}| \qquad \text{Propagation of chain}$$

$$R\bullet +\bullet\overline{Cl}| \longrightarrow R-\overline{Cl}| \qquad \text{Termination}$$

Although the C—C bond enthalpies (or their bond dissociation energies) are lower than those of the C—H bonds in alkanes (346 kJ/mol against 414 kJ/mol), it is the latter which are exclusively attacked by radicals. The carbon chains of alkanes are surrounded with C—H bonds, and attack by radicals on the C—C bonds is sterically hindered.

Properties

Chloromethane (b.p. $-23.7°C$ [249.5 K]) and chloroethane (b.p. $+12.3°C$ [285.5 K]), as well as bromomethane (b.p. $+3.5°C$ [276.7 K]), are gases at room temperature. Most of the higher members of this homologous series are liquids and only a few are solids. Lower members of the series have a sweetish odour, and burn with a green-edged flame. They are almost insoluble in water, but are miscible with ether or alcohols.

Uses

The overriding importance of alkyl halides is for their use in synthesis. They serve as the starting materials in the preparation of a wide range of organic compounds: alkanes via the *Wurtz* reaction, alkenes, alcohols, thiols (mercaptans), ethers, sulphides (thioethers), amines, nitriles, as well as for *Grignard* and *Friedel–Crafts* reagents.

Chloromethane and iodomethane are also used as *methylating agents*. It should be noted that tests on animals have shown iodomethane to be carcinogenic.

Chloroethane (*Chloroethyl*[1]) can be used as a local anaesthetic (by freezing) for sporting injuries and small operations. Industrially it has been used in the preparation of tetraethyl lead, $Pb(C_2H_5)_4$, used as an antiknock agent.

2.9.2 *Mechanism of Nucleophilic Substitution at Saturated Carbon Atoms (C. K. Ingold, 1930)*

The preparation of alkyl halides from alcohols and their hydrolysis to give alcohols have been described above. Kinetic studies of these reactions have provided firm evidence concerning their *mechanisms*.

$$|Y^\ominus + R{-}X \rightleftharpoons R{-}Y + |X^\ominus$$

In the breaking of the polar bond between a carbon atom and the ligand X, the bonding pair of electrons remains with X and a *carbenium ion* is generated. The anion $|Y^\ominus$, with its lone pair of electrons, becomes attached to this carbon atom. The incoming substituent is described as a *nucleophile* (i.e. nucleus-seeking) and this type of reaction is known as *nucleophilic substitution* and is represented by the symbol S_N.

Among the commonest anionic nucleophiles are the following:

$$|Y^\ominus = {}^\ominus|\overline{O}H, {}^\ominus|\overline{O}R, {}^\ominus|\overline{N}H_2, {}^\ominus|\overline{S}R \text{ und } {}^\ominus|C\equiv N|$$

The splitting off of $X^\ominus$ and the combination with $Y^\ominus$ can take place either at the same time (synchronous) or separately (stepwise).

2.9.2.1 S_N2 Reaction. In the synchronous process, both reactants are involved in the initial reaction. It is a bimolecular reaction, designated by the symbol S_N2. The exchange of substituents has to be postulated in such a way that the C—Y bond is formed as the C—X bond is broken. This involves the formation of an intermediate *transition state*, in which the three substituents R^1, R^2 and R^3 attached to the central carbon atom lie in a plane perpendicular to the plane of the paper. The ligands X and Y and the central carbon atom lie on a straight line in the plane of the paper:

Reactants   Transition state   Products

---

[1] Medical name outwith the INN list (such names are always printed in italics), see p. 120.

The course of the reaction can be presented graphically by plotting the potential energy E of the participants in the reaction against the progress of the reaction, described as the reaction co-ordinate, in an *energy diagram* or *energy profile* (Fig. 62).[1]

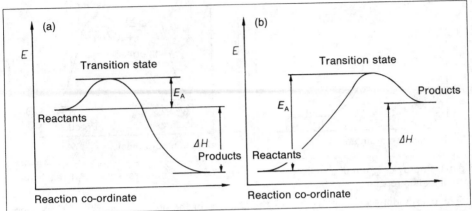

Fig. 62. Energy diagrams of S_N2 reactions. (a) An energetically favoured reaction: exothermic (ΔH negative); (b) an energetically unfavourable reaction: endothermic (ΔH positive).

It may be seen from Fig. 62 that in an S_N2 reaction there is an energy barrier between the reactants and the products, irrespective of whether the reaction is exothermic or endothermic. The height of the barrier is called the *activation energy* E_A. This is larger, the greater the energy required to achieve the necessary structure and geometry of the transition state; this is assumed to be reached at the peak of the curve. Another point which is brought out by Fig. 62 is that the transition state more closely resembles the reactants, the less stable they are. This is generalised in *Hammond's* postulate that, if a reaction step is strongly endothermic, the transition state more resembles the products (a 'late' transition state), whereas for a strongly exothermic reaction step, the transition state more resembles the reactants (an 'early' transition state).

The decisive factor in the value of E_A is the energy which is required to stretch the C—X bond. Therefore, reactions which proceed without involving the breaking of bonds often involve no activation energy. An example is provided by the *recombination of radicals* (Fig. 63, p. 128), which leads to the termination of radical chain reactions.

The one-dimensional energy diagrams which are used in this book are greatly simplified pictures. If a three-dimensional picture is used the transition state lies at the top of a pass which goes over a saddle between two lines of hills.[2]

2.9.2.2. S_N1 Reaction. The second possible mechanism for an S_N reaction involves the initial complete breaking of the C—X bond, before the new substituent $|Y^\ominus$ takes any part in the reaction. Here the rate-determining first step is a *monomolecular* reaction, and it is given the symbol S_N1. The overall process takes place in two steps:

$$R_3C\overset{\frown}{-}X \rightleftharpoons R_3C^\oplus + |X^\ominus \quad \text{slow}$$
$$R_3C^\oplus + |Y^\ominus \longrightarrow R_3C-Y \quad \text{fast}$$

[1] Such representations have been used earlier, in Fig. 39 (p. 58) and Fig. 55 (p. 102). In Fig. 39 it was used to demonstrate movements within a molecule.
[2] See *K. Müller*, Reaction Paths on Multidimensional Energy Hypersurfaces, Angew. Chem. Int. Ed. Engl., *19*, 1 (1980).

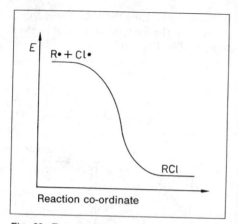

Fig. 63. Energy diagram of a radical recombination reaction.

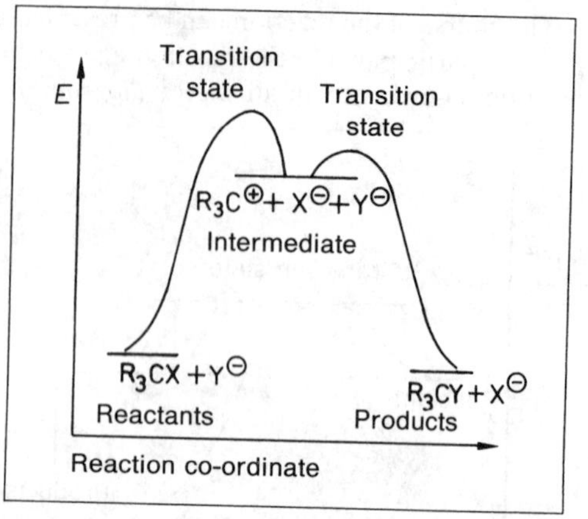

Fig. 64. Energy diagram of an S_N1 reaction.

The energy diagram, Fig. 64, shows the changes in the potential energy of the system as reaction proceeds. There is a higher energy barrier between the initial reactants and the reactive intermediate which is formed than there is between this intermediate and the final products. In consequence, the rate-determining step of an S_N1 reaction is the first one, involving conversion of a reactant into a reactive intermediate.

It is not usually possible to observe a reactive intermediate, although it persists long enough for at least five molecular vibrations before it reacts further. The transition state, however, has a much shorter life-time.

Which of the two mechanisms operates depends on a variety of factors such as the steric structure of the alkyl group, the nucleophilicity of substituents X and Y, the polarity of the solvent and the stability of the intermediate carbenium ion. The commonest solvents are aqueous acetone, ethanol, acetonitrile, aqueous dioxan, dimethylformamide or dimethyl sulphoxide. The S_N1 and S_N1 mechanisms are limiting cases, and only operate in exceptional cases. In fact, many reactions involve ion pairs or *preassociation*, and are only incompletely described by either of the two limiting cases.[1] Especially in the case of alkyl iodides, radical reactions can also be involved, initiated by electron transfer (SET = single electron transfer).[2]

2.9.2.3 Examples of Nucleophilic Substitution Reactions

(a) **Alkaline Hydrolysis of Methyl Halides.** Kinetic measurements show that these reactions are bimolecular (S_N2 mechanism).

$$H_3C-\overline{\underline{X}}| + {}^{\ominus}|\overline{\underline{O}}H \rightarrow H_3C-\overline{\underline{O}}H + |\overline{\underline{X}}|^{\ominus}$$

The rate of the reaction depends on the concentration of both reactants. The rate of hydrolysis increases in the order F < Cl < Br < I. This reflects the efficiency of the different halide ions as *leaving groups* (nucleofuges), which influences the progress of the reaction. This can be related to the strength of the acid H—X. The stronger the acid H—X, the weaker is the basicity of the anion $X^{\ominus}$, and the better leaving group is $X^{\ominus}$. This relationship is generally applicable, for example, the trifluoromethanesulphonyl group (triflate group) $CF_3SO_3^{\ominus}$ is a much more effective leaving group than the methanesulphonyl group, and methyl fluoro-sulphonate (FSO_3OCH_3) is the most efficient methylating agent available ('magic methyl').

---

[1] See *W. P. Jencks*, Accounts Chem. Res., *13*, 161 (1980).
[2] See *E. C. Ashby*, Accounts Chem. Res., *21*, 414 (1988).

(b) Esterification of Primary Alcohols with Halogen Acids. Whilst the basicity of the halide ions decreases in the order $F^\ominus > Cl^\ominus > Br^\ominus > I^\ominus$, their nucleophilicity increases in this order. This is because the larger ions are less solvated and are more easily polarised than the smaller anions are. As a consequence, primary alcohols only react readily with HI, whilst acid catalysts are required to bring about esterification satisfactorily when HBr or HCl are used. It is not possible to make alkyl fluorides in this way. It is necessary to distinguish between basicity and nucleophilicity. The former is a thermodynamic factor, concerned with equilibrium reactions; the latter is a kinetic factor.

The catalytic action of acids is a result of their protonation of the alcohol molecules. This makes H_2O available as a leaving group, rather than $HO^\ominus$, which is a much poorer leaving group.

$$R-CH_2-\overline{O}H \xrightarrow{+H^\oplus} R-CH_2-\overset{\oplus}{O}H_2 \xrightarrow{+Br^\ominus} R-CH_2Br + H_2O$$

It is characteristic of the iodide ion that, not only is it a good nucleophile, but it is also readily displaced again. These properties lead to the use of hydriodic acid to catalyse other nucleophilic substitution reactions.

(c) Alkaline Hydrolysis of Higher 2-Bromo-alkanes, $R-CH_2-\overset{*}{C}HBr-CH_3$. It should be noted that in these reactions, exchange of substituents takes place at a *chiral centre* *. Since the approach of the newly entering substituent takes place from the rear of the tetrahedron, it follows (see Fig. 65) that introduction of the new substituent results in an *inversion* of configuration (*epimerisation*). In the transition state groups R^1, R^2, R^3 lie in one plane, as part of a trigonal bipyramid (see Fig. 72a, p. 175), and then invert from their original shape. In general, reactions of this sort are accompanied by this inversion, known as the *Walden* inversion, which was discovered in 1895. Conversely, if such a reaction is seen to involve inversion, then it may be concluded that it goes by an S_N2 mechanism.

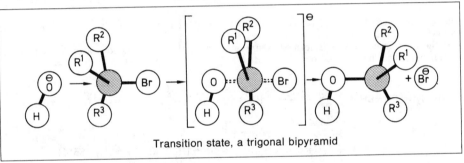

Transition state, a trigonal bipyramid

Fig. 65. Stereochemistry of the S_N2 reaction.

(d) Alkaline Hydrolysis of t-Butyl Bromide, $(H_3C)_3C-Br$. The dissociation of t-butyl bromide is favoured by the fact that the resultant carbenium ion is stabilised by the $+I$-effect of the alkyl groups. As a consequence, its hydrolysis proceeds by the S_N1 mechanism, and its rate increases with increasing polarity of the solvent, which leads to the possibility of increased solvation of the ion. This is illustrated by the fact that t-butyl bromide is hydrolysed much faster in 50% aqueous ethanol than in pure ethanol. The incoming substituent exerts no influence on the rate of reaction, since it plays no part in the rate-determining step (see Fig. 64).

$$(H_3C)_3C-Br \rightleftharpoons (H_3C)_3C^\oplus + Br^\ominus$$

$$(H_3C)_3C^\oplus + H_2O \longrightarrow (H_3C)_3C-OH_2^\oplus \longrightarrow (H_3C)_3C-OH + H^\oplus$$

No general statements can be made about the steric course of S_N1 reactions of alkyl halides which have chiral centres at the halogen-bearing carbon atoms. If the assumption is made that the three substituents R^1, R^2 and R^3 take up a planar shape in the intermediate *carbenium ion*, then the possibility of attack by the water molecule should be identical for both sides of this plane. This should lead to complete *racemisation*.

In fact, one or other of the two optically active forms usually predominates, depending on the constitution of the alkyl halide and of the solvent.

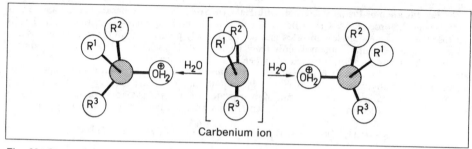

Fig. 66. Stereochemistry of the S_N1 reaction.

(e) **Alkaline Hydrolysis of Allyl Chloride.** This reaction also proceeds by an S_N1 mechanism. The decisive factor here is the *mesomeric stabilisation* of the allyl cation which is formed as an intermediate:

$$H_2C=CH-CH_2-Cl \xrightarrow{-Cl\ominus} \left[H_2C=CH-\overset{\oplus}{C}H_2 \longleftrightarrow H_2\overset{\oplus}{C}-CH=CH_2 \right]$$

Allyl chloride   Allyl cation

$$\xrightarrow{+OH\ominus} H_2C=CH-CH_2-OH$$

Allyl alcohol

The greater reactivity of allyl halides can also be attributed to the formation of the mesomeric cation.

If the allyl cation which is formed is unsymmetrical, then the reaction often leads to some rearrangement in the product, e.g.

$$(H_3C)_2C=CH-CH_2Cl \xrightarrow{Ag_2O,H_2O} \left[(H_3C)_2C=CH-\overset{\oplus}{C}H_2 \longleftrightarrow (H_3C)_2\overset{\oplus}{C}-CH=CH_2 \right]$$

$$\longrightarrow (H_3C)_2C=CH-CH_2OH \ (15\%) + (H_3C)_2\overset{OH}{\underset{|}{C}}-CH=CH_2 \ (85\%)$$

The formation of 2-methylbut-3-en-2-ol from 4-chloro-2-methylbut-2-ene results from an *allyl rearrangement*.

(f) **Behaviour of Vinyl Chloride,** $H_2C=CH-Cl$. In contrast to allyl chloride, vinyl chloride reacts only very slowly. This is a result of the different hybridisation of the carbon atom to which the chlorine atom is attached. Since an sp^2-hybridised carbon atom has a higher electronegativity than an sp^3-hybridised carbon, the difference in electronegativity between the carbon and chlorine atoms is lower in vinyl chloride than in other alkyl halides.

Because of the lower polarity of this bond, the reactivity is lowered; there is less tendency for nucleophilic attack. Hence vinyl chloride is not hydrolysed, nor does addition of a solution of silver nitrate in ethanol lead to any precipitation of silver chloride. Alkynyl halides, $R-C\equiv C-X$, and aryl halides, C_6H_5-X, behave like vinyl halides and are resistant to hydrolysis.

2.9.3 Elimination Reactions

These reactions involve the loss of two atoms or groups of atoms from a molecule, without their being replaced by other atoms. In most cases the loss is from two neighbouring carbon atoms.

Since a nucleophile is also a base it can attack not only a positively charged carbon atom, but can also remove a proton. If an anion $X^\ominus$ is removed at the same time, then an alkene is formed. Since the proton is removed from a carbon atom which is β to the carbon atom to which the substituent X is attached, this reaction is described as β-elimination. The first step is bimolecular, and is said to follow an E2 mechanism.

$$R-CH_2-CH_2-X+IY^\ominus \quad \begin{array}{l} R-CH_2-CH_2-Y+IX^\ominus \qquad S_N2 \text{ Substitution} \\[2mm] HY+R-CH=CH_2+IX^\ominus \qquad \beta\text{-Elimination } (E2) \end{array}$$

$$\underset{\beta \quad \alpha}{R-CH_2-CH_2-X}$$

The removal of hydrogen halides from alkyl halides by bases, and the acid-catalysed loss of water from alcohols, are reactions of this sort.

The elimination reactions are dependent upon the structure of the alkyl group R, and take place more readily the more branched R is. Hence the rate of reaction increases as one goes from primary to secondary to tertiary alkyl halides. In contrast to S_N2 reactions, the effect of steric hindrance on β-elimination reactions is small. The action of a base on t-butyl chloride leads to removal of a proton and formation of isobutene by an E2 elimination reaction.

$$\underset{\overset{|}{CH_3}}{\overset{\overset{CH_3}{|}}{H_3C-C-Cl}} + OH^\ominus \longrightarrow H_2C=C\overset{CH_3}{\underset{CH_3}{\diagup}} + H_2O + Cl^\ominus$$

t-Butyl chloride   Isobutene

The mechanism may be depicted as follows:

$$HO^\ominus \qquad \underset{\overset{|}{H} \quad \overset{|}{CH_3}}{\overset{\overset{H}{|} \quad \overset{CH_3}{|}}{H-C-C-Cl}} \longrightarrow H_2C=C\overset{CH_3}{\underset{CH_3}{\diagup}}$$

Analogous to an S_N1 reaction, there is also a *monomolecular* elimination reaction (symbol = E1), in which the first step is the dissociation of the anion $X^\ominus$. An example is provided by the acid-catalysed hydrolysis of t-butyl bromide. After formation of the carbenium ion, both substitution and β-elimination (E1 mechanism) ensue:

$$(H_3C)_3C-Br \longrightarrow (H_3C)_3C^\oplus + Br^\ominus$$

$$\underset{\overset{|}{CH_3}}{\overset{\overset{CH_3}{|}}{H_3C-C^\oplus}} \xrightarrow{(H_2O)} \begin{array}{l} (H_3C)_3C-OH+H_3O^\oplus \qquad S_N1 \text{ Substitution} \\ \text{t-Butanol} \\[3mm] H_2C=C(CH_3)_2+H_3O^\oplus \qquad \beta\text{-Elimination } (E1) \\ \text{Isobutene} \end{array}$$

Nucleophilic substitution and elimination are closely related to one another and frequently take place side by side.

The influence of structure on $E1$ reactions resembles its influence on S_N1 reactions, that is, the reactivity increases $F < Cl < Br < I$ (increasingly better leaving groups) and also from primary to secondary to tertiary alkyl groups (giving increasingly more stable carbenium ions).

As well as β-elimination there are also a few examples of α-elimination reactions, in which both of the atoms or groups of atoms which are eliminated come from the same carbon atom.[1]

The best known example is the hydrolysis of chloroform by strong bases:

$$Cl_3C{-}H + \ominus OH \rightleftharpoons H_2O + \ominus CCl_3 \quad \text{(fast)}$$

$$\ominus CCl_3 \longrightarrow :CCl_2 + \ominus Cl \quad \text{(slow)}$$
$$\text{Dichlorocarbene}$$

The elimination of hydrogen chloride from chloroform is a *two-step process* leading to the formation of *dichlorocarbene*. The increased acidity of the proton of chloroform is a consequence of the electron-attracting character of the three chlorine atoms. Dichlorocarbene, whose carbon atom has only a sextet of electrons, is not isolable, but its existence can be demonstrated by the various reactions which it undergoes.

2.9.4 Phase-Transfer Catalysts (PTC)[2]

Under the conditions in which dichlorocarbene is prepared it is also very easily hydrolysed:

$$:CCl_2 \xrightarrow{H_2O, \ominus OH} HCOO^\ominus + CO + Cl^\ominus$$
$$\text{Formate anion}$$

This problem can be overcome by the use of *phase-transfer catalysts*. A mixture of *50% aqueous sodium hydroxide* and *chloroform* exists in two *phases*. If a small amount of tetraalkylammonium salt ($TAA^\oplus Cl^\ominus$, see p. 162) is added, the processes, shown schematically below, take place at the boundary between the phases:

(H$_2$O)	Na$^\oplus$OH$^\ominus$		Na$^\oplus$ + H$_2$O		Na$^\oplus$Cl$^\ominus$	
(CHCl$_3$)	HCCl$_3$		$\ominus$CCl$_3$	TAA$^\oplus$Cl$^\ominus$ $\ominus$CCl$_3$TAA$^\oplus$	:CCl$_2$ + TAA$^\oplus$Cl$^\ominus$	

The base which is necessary to bring about α-elimination remains in the aqueous phase. The only other substance in the organic phase, in which the dichlorocarbene is formed, is the innocuous $TAA^\oplus Cl^\ominus$; the latter can also react without trouble with a suitable reagent in the organic phase.

The method was developed by *Makosza* in 1965, and has proved to be outstandingly useful for reactions involving organic anions. If *chiral ammonium salts* are used as catalysts (see p. 810), asymmetric alkylation can be achieved with a high degree of selectivity (ee $= 84-95\%$).

[1] W. Kirmse, Intermediates of α-Eliminations, Angew. Chem. Int. Ed. Engl., 4, 1 (1965).
[2] See E. V. Dehmlow, Advances in Phase-Transfer Catalysis, Angew. Chem. Int. Ed. Engl., 16, 493 (1977); Phase Transfer Catalysis (Verlag Chemie, Weinheim 1980); W. E. Keller, Phase Transfer Reactions, Fluka-Compendium, 2 vols (Thieme Verlag, Stuttgart 1986, 1987).

2.9.5 Fragmentation Reactions[1]

In addition to nucleophilic substitution and elimination, another type of reaction is possible in which a substituent is removed from a molecule together with its bonding electron pair.

This was described by *Grob* as a *fragmentation* reaction. An example is the cleavage of a 1,3-disubstituted carbon chain to give breakdown products, as follows:

$$\overset{\ominus}{Z}-\overset{|}{\underset{|}{C}}-\overset{|}{\underset{|}{C}}-\overset{|}{\underset{\underset{\alpha}{\overset{\beta}{|}}}{\underset{\gamma}{C}}}-X \longrightarrow \overset{\oplus}{Z}=\overset{|}{C}+\overset{|}{\underset{|}{C}}=\overset{|}{C}+X^{\ominus} \qquad \begin{array}{l} Z = OH, NH_2, \text{etc.} \\ X = \text{halogen, etc.} \end{array}$$

This resembles β-elimination in that a bond between the β-carbon atom and a γ-atom (in this case carbon) is broken, while at the same time a double bond is formed between the α- and β-carbon atoms. The remaining fragment is stabilised by the electron-donating character of the γ-substituent. The fragmentation process and the loss of the substituent X may take place synchronously or the reaction may take place in two steps.

In the alkaline hydrolysis of 1-chloro-1,1-dimethyl-3-dimethylaminopropane, even at 0°C (273 K), the first-formed carbenium ion reacts further in three different ways: by substitution (38%), elimination (9%) and fragmentation (44%).

The iminium ion which is formed along with isobutene is hydrolysed in the alkaline medium to give dimethylamine and formaldehyde.

2.9.6 Polyhalogenoalkanes

The most important members of this class of compounds are the *dihalogenomethanes* (*methylene halides*), the *trichloromethanes* (*haloforms*, chloroform, bromoform and iodo-form) and *tetrachloromethane*. Fluorinated alkanes and alkenes are dealt with in the next section.

Preparation

2.9.6.1 Dihalogenomethanes are prepared by the *partial reduction of haloforms*, for example, diiodomethane is obtained in good yield by the reaction of iodoform with a solution of sodium arsenite.

---

[1] See *C. A. Grob* et al., Angew. Chem. Int. Ed. Engl., **6**, 1 (1967); **8**, 535 (1969).

$$CHI_3 + Na_3AsO_3 + NaOH \longrightarrow CH_2I_2 + NaI + Na_3AsO_4$$

Dichloromethane is made industrially by the chlorination of methane (see p. 125).

2.9.6.2 Chloroform (trichloromethane) is made industrially by the *chlorination of methane* or by the following methods:

(a) By the action of bleaching powder (*calcium hypochlorite*) *on ethanol* (*or acetone*) in aqueous solution. The hypochlorite first converts the ethanol into *acetaldehyde*, and then into trichloroacetaldehyde (*chloral*). This is then split by hydroxide ions into chloroform and formate ions (*Liebig and Soubeiran*, 1831).

This may be represented by the following sequence of reactions:

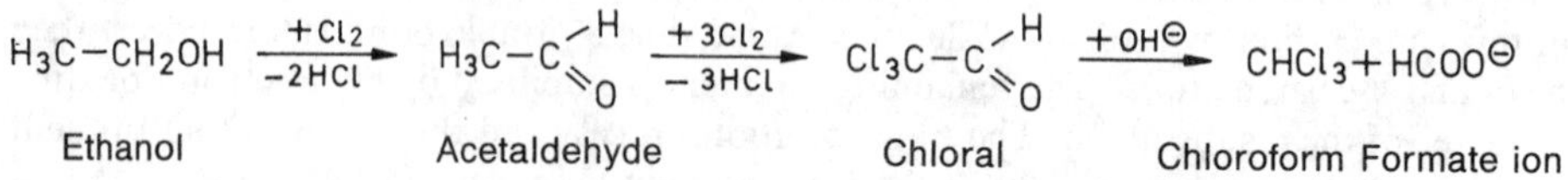

$$H_3C-CH_2OH \xrightarrow[-2HCl]{+Cl_2} H_3C-C\!\!\begin{array}{c}^H\\_O\end{array} \xrightarrow[-3HCl]{+3Cl_2} Cl_3C-C\!\!\begin{array}{c}^H\\_O\end{array} \xrightarrow{+OH^{\ominus}} CHCl_3 + HCOO^{\ominus}$$

Ethanol   Acetaldehyde   Chloral   Chloroform Formate ion

Starting from acetone, this is first chlorinated to give trichloroacetone, which is then split to give chloroform and acetate ions:

$$H_3C-CO-CH_3 \xrightarrow[-3HCl]{+3Cl_2} H_3C-CO-CCl_3 \xrightarrow{+OH^{\ominus}} CHCl_3 + CH_3-COO^{\ominus}$$

Acetone   Trichloroacetone   Choroform Acetate ion

Bromoform, $CHBr_3$, (b.p. 149.5° [422.7K]) and *iodoform*, CHI_3, can be prepared in an analogous way.

(b) Very pure chloroform can be obtained by heating either *chloral hydrate* or *trichloroacetic acid with alkali*:

$$Cl_3C-CH(OH)_2 + NaOH \longrightarrow CHCl_3 + HCOONa + H_2O$$

Chloral hydrate   Chloroform

$$Cl_3C-COOH + NaOH \longrightarrow CHCl_3 + NaHCO_3$$

Trichloroacetic acid

If the splitting of chloral hydrate by alkali is carried out in deuterium oxide (D_2O), then deuteriochloroform (chloroform-*d*), $CDCl_3$, is formed; this has wide use as a solvent in NMR spectroscopy.

2.9.6.3 Tetrachloromethane (carbon tetrachloride) is obtained industrially by chlorination of heated chlorine-containing organic residues. This chlorination/pyrolysis has been called *chlorolysis*. With a mixture of propane and propene chlorolysis proceeds as follows (for propene):

$$C_3H_6 + 7Cl_2 \xrightarrow[\text{200 mbar (20 MPa)}]{\text{870 K,}} CCl_4 + Cl_2C=CCl_2 + 6HCl$$

2.9.6.4 Tetrachloroethylene is formed as a by-product in such chlorolyses; the proportion of the products is controlled by a temperature- dependent equilibrium:

$$2CCl_4 \rightleftharpoons Cl_2C=CCl_2 + 2Cl_2$$

Properties and Uses

Polyhalogenated alkanes have chemical properties similar to those of alkyl halides. Because they are poisonous, care is needed in handling them.

Dichloromethane (*methylene chloride*), CH_2Cl_2, is a liquid with b.p. 40.6°C (313.8 K), which is used as a solvent.

Dibromomethane (*methylene bromide*), CH_2Br_2, (b.p. 96.5°C [369.7 K]) and **diiodomethane** (b.p. 181°C [454 K]) are colourless liquids which do not burn. Together with dichloromethane they are often used as starting materials in organic syntheses.

Chloroform, $CHCl_3$, is a colourless sweet-smelling liquid of b.p. 61.2°C (334.4 K), which is only slightly soluble in water but is totally miscible with ethanol and ether.

When inhaled, chloroform induces unconsciousness (*Simpson*, 1848), and it was formerly used in medicine as an anaesthetic. On long standing in moist air, and especially in light, chloroform is decomposed, forming phosgene, $COCl_2$, which is extremely poisonous. Because of this, chloroform is kept in brown bottles. If it was to be used as an anaesthetic about 1% of ethanol was added to it. This reacts with any phosgene which is formed, converting it into diethyl carbonate, $OC(OC_2H_5)_2$, which is harmless.

Chloropicrin (*Nitrochloroform*), a liquid which boils at 112°C (385 K), is formed by the action of conc. nitric acid on chloroform, and is used as an insecticide.

$$CHCl_3 + HNO_3 \rightarrow Cl_3C-NO_2 + H_2O$$

Bromoform, $CHBr_3$, boils at 149.5°C (422.7 K), and closely resembles chloroform in its chemical properties. It is found in sea water at a concentration of ng/l (10^{-9} g/l), and, as a metabolic product from algae, passes into the atmosphere.

Iodoform, CHI_3, forms leaflets of m.p. 119°C (392 K). It has a characteristic smell, and is insoluble in water but soluble in ethanol and ether. Its formation serves as a *test for ethanol*, *acetone* or other compounds containing an acetyl, H_3C-CO-, group.

Iodoform test. The liquid to be tested is treated with iodine and aqueous potassium hydroxide. If the test is positive a yellow crystalline precipitate of iodoform settles out:

$$C_2H_5OH + 4 I_2 + 6 KOH \rightarrow CHI_3 + HCOOK + 5 KI + 5 H_2O$$
$$R-COCH_3 + 3 I_2 + 4 KOH \rightarrow CHI_3 + R-COOK + 3 KI + 3 H_2O$$

Tetrachloromethane (carbon tetrachloride), CCl_4, is a colourless liquid, b.p. 76.7°C (349 K), with a sweetish smell. It is non-flammable and is used as a solvent, and for the extraction of fats, oils and resins. It is also used as an ingredient in cleaning-fluids and was used as the filling for fire-extinguishers.

2.9.7 Fluorinated Hydrocarbons[1]

Some chlorofluorocarbons derived from methane and ethane (CFCs), because of their low boiling points, non-toxicity and chemical inertness, have been used as *coolants* (refrigerants) in refrigerators and cooling plants, as propellants for aerosols and foams, and as dry-cleaning fluids. They have been described as freons or as CFCs. The commonest

---

[1] See *G. A. Olah, R. D. Chambers* and *G. K. Prakash*, Synthetic Fluorine Chemistry (Wiley, Chichester 1992).

coolants include *trichlorofluoromethane*, CCl_3F, (CFC 11, R11), b.p. 24.9°C (298.1 K), *dichloro-difluoromethane*, CCl_2F_2 (*Halon 122, R12*), b.p. −30°C (243 K), 1,2,2-trichloro-1,1,1-trifluoro-ethane, $CClF_2$—CCl_2F, b.p. 48°C (321 K) and 1,2-dichloro-1,1,2,2-tetrafluoroethane, $CClF_2$—$CClF_2$ (INN: *Cryofluoran*), b.p. 3.5°C (276.7 K).

The numbering system used for CFCs is based on a three-figured number, which is derived in the following way. The final figure signifies the number of fluorine atoms. The last but one figure represents the number of hydrogen atoms plus one. The first figure represents the number of carbon atoms in the molecule minus one; in the case of methane derivatives it is thus zero and is omitted. All the remaining atoms in the molecule are assumed to be chlorine atoms. Bromine-containing derivatives are used in fire-extinguishers. An extension of the system is needed to represent other polyhalogeno-compounds, which are designated as *halons*; five numbers are used to represent respectively and in turn the numbers of C, F, Cl, Br and I atoms. Thus CF_2ClBr is Halon 1211.

Because of their great volatility and chemical stability,[1] CFCs are transported to the higher layers of the earth's atmosphere, the stratosphere, where they react with ozone in the following way:

$$CF_2Cl_2 \xrightarrow{h\nu} CF_2Cl + Cl \qquad Cl + O_3 \rightarrow ClO + O_2 \qquad ClO + O_3 \rightarrow Cl + 2O_2$$

There is therefore a danger that the shielding action of the ozone layer, which protects the earth from the influx of UV radiation from outer space, will be impaired.[2] In addition, because CFCs absorb IR radiation much more strongly than CO_2 does, their increasing presence in the atmosphere may lead to its heating up in the same way that CO_2 does (see p. 351). If a compound contains hydrogen, as in the case of chlorodifluoro-methane, $CHClF_2$ (CFC 22, b.p. −40°C [233 K]), its decomposition occurs in lower layers of the atmosphere, in the troposphere. Its action on the ozone layer of the stratosphere is thus less than that of other chlorofluorocarbons.

Preparation

2.9.7.1 The direct fluorination of alkanes is only accomplished with difficulty, since the reaction is highly exothermic and usually explosive; it almost always leads to a mixture of perfluorinated compounds.

2.9.7.2 *Swarts* Reaction. In this, instead of fluorine, the less reactive *metal fluorides*, such as silver, mercury(I) or potassium fluorides are used as fluorinating agents. These react with alkyl halides, with exchange of the halogen atom by fluorine; mercury(I) fluoride is best for converting them into alkyl fluorides:

$$2\,R\text{--}X + Hg_2F_2 \rightarrow 2\,R\text{--}F + Hg_2X_2 \qquad X = Cl, Br, I$$

2.9.7.3 In industry, chlorofluoroalkanes are obtained by fluorination of suitable *chloroalkanes* with dry hydrogen fluoride over a *catalyst bed* consisting of *aluminium* or *chromium fluorides*:

$$CCl_4 \xrightarrow[\text{− HCl}]{\text{Catalyst}} CCl_2F_2 + CCl_3F$$

Dichloro-   Trichloro-
difluoromethane   fluoromethane

---

[1] The life-time of $CClF_2$—$CClF_2$ in the atmosphere is up to about 200 years.
[2] See *S. C. Wafsy* et al., Freon Consumption, Implications for Atmospheric Ozone, Science *187*, 535 (1975).

2.9.7.4 Tetrafluoroethylene is prepared industrially by *pyrolysis of chlorodifluoro-methane* (CFC 22), which is itself obtained by reaction of chloroform with dry hydrogen fluoride in the presence of antimony(V) chloride:

$$2\ CHCl_3 \xrightarrow[-2\ HCl]{+\ 2\ HF(SbCl_5)} 2\ CHClF_2 \xrightarrow[-2\ HCl]{(975\ K)} F_2C = CF_2$$

Chloroform   Chloro-difluoromethane   Tetrafluoroethylene

2.9.7.5 A generally applicable method is that of **electrofluorination** (*Simons*). *Anodic fluorination* is carried out by applying a voltage across dry hydrogen fluoride which is not great enough to liberate elementary fluorine. It leads to perfluorinated compounds, e.g.

$$H_3C-COOH \xrightarrow[-8^{\ominus},\ -\frac{1}{2}\ O_2,\ -4\ H^{\oplus}]{4\ F^{\ominus}} CF_3COF$$

Trifluoroacetyl fluoride

The especial advantage of this method is that functional groups remain unchanged in the fluorination.

2.9.7.6 A valuable fluorinating agent is **sulphur tetrafluoride**, SF_4, which replaces oxygen atoms in ketones or in carboxylic acids by fluorine atoms:

$$\underset{O}{R-\overset{\|}{C}-R'} + SF_4 \longrightarrow R-CF_2-R' + SOF_2$$

Ketone

$$R-C\underset{OH}{\overset{O}{\diagup}} \xrightarrow[-SOF_2,\ -HF]{+SF_4} R-C\underset{F}{\overset{O}{\diagup}} \xrightarrow[-SOF_2]{+SF_4} R-CF_3$$

Carboxylic acid   **Acyl fluoride**

Unlike phosphorus(V) halides, sulphur tetrafluoride reacts further with the first-formed acyl fluoride to give fluorohydrocarbons. C=C double or C≡C triple bonds in the starting materials retain their identity and are not attacked by SF_4.[1]

Care: SF_4 is volatile (b.p. $-38°C$, 235 K) and *very poisonous.*

Properties

Tetrafluoroethylene is a colourless gas (b.p. $-78.4°C$ [194.8 K]) and, in the presence of peroxide as an initiator and under pressure, is polymerised to poly- (tetrafluoroethylene):

$$nF_2C = CF_2 \rightarrow [-CF_2-CF_2-CF_2-CF_2]_{n/2} \quad \text{Poly(tetrafluoroethylene) (PTFE)}$$

This polymer, which is known by its trade names *fluon* (UK), *teflon* (USA), *hydeflon* and *hostaflon TF* (Germany) is notable because of its relatively high resistance to heat and its chemical inertia. Because of its high softening point (> 250°C [525 K]), its processing requires special sintering methods. It is used for the manufacture of seals and of tubes, as well as for water-repellant films, protective coatings and elastomers. Fully fluorinated polymers, which can be processed by injection moulding, are obtained, for example, by the copolymerisation of tetrafluoroethylene with hexafluoropropene.

2-Bromo-2-chloro-1,1,1-trifluoroethane, F_3C—CHBrCl, is used as a narcotic (INN: *Halothane*) (see p. 120) in anaesthesia. It is three to four times more effective than ether.

---

[1] See *C-L. J. Wang*, Fluorination by Sulphur Tetrafluoride, Org. Reactions, *34*, 319 (1985).

2.10 Esters of Inorganic Acids

Alkyl halides have already been described as esters of halogen acids.

Whilst, in the reaction of alcohols with halogen acids to give alkyl halides, the HO groups of the alcohol are split off and react with protons from the acid to give water, the formation of water from alcohols and oxyacids can involve the HO groups of the acid reacting with protons from the alcohol:

$$R-OH + H-X \rightleftharpoons R-X + H_2O$$

$$RO-H + HO-SO_3H \rightleftharpoons RO-SO_3H + H_2O$$

This *mechanism* has been established for esterification reactions of carboxylic acids.

2.10.1 Esters of Sulphuric Acid. Alkyl Sulphates

Sulphuric acid can form both acid esters and neutral esters.

The *acid esters*, such as *ethylsulphuric acid* (ethyl hydrogen sulphate) can be obtained from *ethanol and conc. sulphuric acid*:

$$C_2H_5OH + H_2SO_4 \rightleftharpoons C_2H_5OSO_3H + H_2O$$

Ethylsulphuric acid

The mechanism of the reaction has not been completely cleared up.

Ethylsulphuric acid is an intermediate in the preparation of ethylene from ethanol, and is converted into ethylene by the addition of conc. sulphuric acid.

Pure ethylsulphuric acid is obtained by treatment of ethanol with sulphur trioxide at 0°C (273 K). SO$_3$ first of all adds to the oxygen atom of the alcohol:

$$C_2H_5\bar{O}-H \xrightarrow{+ SO_3} C_2H_5\overset{\oplus}{\bar{O}}\underset{H}{-}SO_3^{\ominus} \longrightarrow C_2H_5\bar{O}-SO_3H$$

In contrast to barium sulphate, its barium salt is soluble in water.

More important are the neutral *esters*, especially *dimethyl sulphate*, $(H_3C)_2SO_4$, which is made industrially from dimethyl ether and sulphur trioxide:

$$(H_3C)_2O + SO_3 \rightleftharpoons (H_3CO)_2SO_2$$

Dimethyl sulphate, $(H_3C)_2SO_4$, is a liquid with b.p. 188°C (461 K). Its vapours are highly toxic. It is used above all as a *methylating agent*. In the presence of alkali it replaces the *acidic* H-atoms of hydroxy-, mercapto-, amino- or imino-groups by methyl groups, e.g.

$$R-C\overset{\diagup O}{\diagdown OH} + \overset{H_3CO}{\underset{H_3CO}{\diagdown\diagup}}SO_2 + NaOH \longrightarrow R-C\overset{\diagup O}{\diagdown OCH_3} + \overset{H_3CO}{\underset{Na^\oplus O^\ominus}{\diagdown\diagup}}SO_2 + H_2O$$

Methyl carboxylate

Usually only one of the methyl groups of dimethyl sulphate is transferred in this reaction.

It has recently been found that dimethyl sulphate can act as a *carcinogen*.

Diethyl sulphate, $(C_2H_5)_2SO_4$, is formed when *ethylene* (in excess) is passed into cold conc. sulphuric acid:

$$2\,C_2H_4 + H_2SO_4 \rightleftharpoons (C_2H_5O)_2SO_2$$

It is a liquid with b.p. 208°C (481 K) which can be used as an ethylating agent. It is poisonous, and has been shown to act as a carcinogen in animals.

2.10.2 Esters of Nitric Acid

Esters of nitric acid can be made from alcohols and conc. nitric acid, e.g.

$$C_2H_5O-H + HO-NO_2 \rightleftharpoons C_2H_5O-NO_2 + H_2O$$
Ethyl nitrate (b.p. 88.7°C [361.9 K])

Oxidation of the alcohol also takes place as a side reaction. This results in the formation of some nitrous acid and this can hasten further oxidation, which may be explosive. To remove any nitrous acid which is formed, urea is added to the reaction mixture:

$$O=C(NH_2)_2 + 2\,O=N-OH \rightarrow 3\,H_2O + 2\,N_2 + CO_2$$

Alkyl nitrates are liquids with pleasant smells. When heated above their boiling points they very frequently explode, so that the greatest care must be taken in any attempts to distil them.

Among the industrially important nitrate esters are *glyceryl trinitrate* (nitroglycerol) and *cellulose nitrate* (nitrocellulose).

2.10.3 Esters of Nitrous Acid

The esters of nitrous acid, such as **ethyl nitrite** and **isopentyl nitrite** are obtained by the action of *dinitrogen trioxide on alcohols* or by adding one equivalent of conc. hydrochloric acid or sulphuric acid dropwise into a mixture of the alcohol and aqueous sodium nitrite, e.g.

$$2\,C_2H_5O-H + N_2O_3 \rightleftharpoons 2\,C_2H_5O-NO + H_2O \qquad C_5H_{11}O-H + HO-NO \rightleftharpoons C_5H_{11}O-NO + H_2O$$

Ethanol   Ethyl nitrite   Isopentyl alcohol   Isopentyl nitrite
(b.p. 17 °C [290 K])   (b.p. 97 °C [370 K])

Since alkyl nitrites readily react with mineral acids, in preparative chemistry they are used rather than inorganic nitrites to provide nitrous acid when it is required for use in reactions in organic solvents. (See *diazotisation*.)

Isopentyl nitrite is used in medicine to treat spasms, *e.g.* in *angina pectoris*.

***Barton* reaction**: When alkyl nitrites, which have hydrogen atoms available at the 4-position, are irradiated, then the following radical reaction ensues:

As a result of this *photolysis* of the *nitrite*, an oxime group is introduced at the 4-position, which can easily be hydrolysed to a carbonyl group. Much use has been made in steroid chemistry of this means of introducing a functional group at a non-activated carbon atom.

2.10.4 Esters of Phosphoric Acid[1]

There are three series of esters derived from *ortho*-phosphoric acid:

$$O=P\begin{subarray}{l} \diagup OR \\ -OH \\ \diagdown OH \end{subarray} \qquad O=P\begin{subarray}{l} \diagup OR \\ -OR \\ \diagdown OH \end{subarray} \qquad O=P\begin{subarray}{l} \diagup OR \\ -OR \\ \diagdown OR \end{subarray}$$

Monoalkyl phosphate   Dialkyl phosphate   Trialkyl phosphate
(Monoalkyl dihydrogen phosphate) (Dialkyl hydrogen phosphate)

Of particular preparative importance are the *neutral phosphate esters*, which are made from the appropriate alcohol and phosphorus oxychloride.

$$O=PCl_3 + 3\,H{-}OR \rightarrow O=P(OR)_3 + 3\,HCl$$

The neutral esters are fairly high-boiling liquids; for example, **trimethyl phosphate** boils at 197°C (470 K) and **triethyl phosphate** at 215°C (488 K).

Tricresyl phosphate, m.p. 78°C (351 K), is used as a plasticiser for PVC and nitrocellulose; its high toxicity must be borne in mind.

Some esters of phosphoric and thiophosphoric acids are used as *insecticides* for the control of pests; among these are *O,O*-diethyl-*O*-(4-nitrophenyl)-thiophosphate (Parathion, E605) and *O,O*-dimethyl-*S*-(2-ethyl-sulphinylethyl)-thiophosphate (Oxydemethon-methyl, Metasystox R):

Parathion   Metasystox R   Tabun   Methylfluorophosphonate ester

The chemical nerve agent Tabun is cyano-diethylamino-ethyl phosphate. The phosphonic acid derivatives *Sarin* (isopropyl methylfluorophosphonate $[R = (CH_3)_2CH{-}]$) and *Soman* (1,2,2-trimethylpropyl methylfluorophosphonate $[R = (CH_3)_3C{-}CH(CH_3){-}]$) show similar activity.

Certain monoesters of phosphoric acid and of polyphosphoric acid, and especially **adenosine triphosphate** (ATP), discovered by Lohmann in 1929, are essential phosphorylating agents in living organisms.

Phospholipids are diesters of phosphoric acid.

2.10.5 Esters of Boric Acid

When boron oxide and methanol are heated together, volatile trimethyl borate is formed. It burns with a green flame, and this is used for the detection of boric acid:

$$B_2O_3 + 6\,H_3COH \underset{}{\overset{H^\oplus}{\rightleftharpoons}} 2\,B(OCH_3)_3 + 3\,H_2O$$

A trace of conc. sulphuric acid catalyses the esterification of boric acid.

---

[1] See *Houben-Weyl*, Methoden der Organischen Chemie, 4th edn, Vol. 12/2, p. 143 (1964).

2.11 Ethers

This class of compounds can be regarded either as anhydrides of alcohols, or as derivatives of water in which both hydrogen atoms are replaced by alkyl groups.

$$H \diagdown_O\diagup H \qquad R \diagdown_O\diagup H \qquad R \diagdown_O\diagup R$$

| Water | Alcohol | Ether |

Depending upon whether the two alkyl groups are the same or different, ethers are described as *simple* (R—O—R) or *mixed* (R—O—R') ethers.

The names of individual ethers are formed by adding the names of the substituent alkyl groups to the word 'ether', e.g. dimethyl ether, H_3C—O—CH_3; diethyl ether, H_5C_2—O—C_2H_5; ethyl methyl ether, H_3C—O—C_2H_5. According to IUPAC rules, ethers may also be named as alkoxyalkanes. The larger group is taken as the main chain, e.g. ethyl methyl ether = methoxyethane.

The most important member of the series is **diethyl ether**, generally known just as 'ether'.

2.11.1 Preparation

2.11.1.1 Williamson synthesis (1850). This method involves the reaction of a metal alcoholate with an alkyl halide:

$$R-\overline{\underline{O}}|^{\ominus}Na^{\oplus} + I-R' \longrightarrow R-O-R' + NaI$$

It can be used for simple or mixed ethers, and at the same time proves the structure of the ether.

Reaction proceeds by an S_N2 mechanism, and, depending upon the nature of the alkyl group of the alkyl halide, may also be accompanied by an $E2$ reaction, e.g.

$$H_3C-I + (H_3C)_2\overset{H}{\underset{|}{C}}-O^{\ominus} \longrightarrow (H_3C)_2\overset{H}{\underset{|}{C}}-OCH_3 + I^{\ominus} \quad (S_N2)$$

| Methyl iodide | Isopropoxide ion | | Isopropyl methyl ether | |

$$(H_3C)_2CH-I + H_3CO^{\ominus} \longrightarrow H_3C-CH=CH_2 + H_3COH + I^{\ominus} \quad (E2)$$

Isopropyl iodide   Methoxide ion   Propene

2.11.1.2 Diethyl ether is formed when an excess of ethanol is heated with conc. sulphuric acid at about 130°C (400 K). Ethylene is formed as a by-product:

$$2\,C_2H_5OH \xrightarrow[-H_2O]{(H_2SO_4)} C_2H_5-O-C_2H_5$$

According to *Meerwein* this reaction is catalysed by $H^\oplus$ ions. First of all the alcohol reacts with a proton to form an *oxonium ion*. In this the C—O bonding is strongly polarised, and another alcohol molecule can attack the alkyl group, as shown in the mechanism given below. The resultant dialkyloxonium ion loses a proton to give the ether:

$$
\underset{\text{Alcohol}}{R-\underset{|}{\underset{H}{O}}|} \xrightarrow{+H^\oplus} \underset{\text{Oxonium ion}}{R-\underset{|}{\underset{H}{O}}{}^\oplus-H} \xrightarrow{+R-\bar{O}-H(-H_2O)} \underset{\text{Dialkyloxonium ion}}{R-\underset{|}{\underset{R}{O}}{}^\oplus-H} \xrightarrow{-H^\oplus} \underset{\text{Ether}}{R-\underset{|}{\underset{R}{O}}|}
$$

Mechanistically the formation of the dialkyloxonium ion can be represented as:

$$
\underset{H}{\overset{R}{>}}O \longrightarrow R-\underset{|}{\underset{H}{O}}{}^\oplus-H \longrightarrow R-\underset{|}{\underset{H}{\overset{\oplus}{O}}}-R + H_2O
$$

2.11.1.3 Industrially the conversion of ethanol into diethyl ether is carried out in the presence of an alumina catalyst at about 300°C (575 K):

$$
2\ C_2H_5OH \xrightarrow[-H_2O]{(Al_2O_3)} H_5C_2-O-C_2H_5
$$

In countries where ethanol is prepared synthetically, enough ether arises as a by-product to make its separate synthesis unnecessary.

2.11.2 Properties

For the most part, ethers are very mobile liquids of high volatility, which are highly flammable. Whilst the properties of alcohols show considerable similarities to those of water, this is not so in the case of ethers. There is no hydrogen atom attached to the oxygen atom and so they cannot associate by means of hydrogen bonding. This is the cause of the great differences in boiling point between alcohols and ethers with the same molecular formulae:

Ethanol	+ 78.4 °C (351.6 K)	Dimethyl ether	−24.9 °C (248.3 K)
n-Butanol	+117.5 °C (390.7 K)	Diethyl ether	+34.6 °C (307.8 K)

The boiling points of the ethers are markedly lower than those of the isomeric alcohols.

Diethyl ether is soluble in alcohols, but only slightly soluble in water. Since salts or other salt-like compounds do not dissolve in ether, they can be precipitated from solutions in alcohol by addition of ether. In general, ethers do not react with *sodium* at room temperature. Hence sodium wire is used as a drying agent in the preparation of 'absolute' ether. Diethyl ether vapour is highly flammable and forms an explosive mixture with air.

On long exposure to air in the presence of light, ethers form rather involatile *peroxides* (peroxo compounds) which can lead to dangerous explosions when the ether is evaporated. The presence of peroxide in ether can be detected by shaking a test portion with a solution of potassium iodide in acetic acid; iodine is formed if peroxide is present. Peroxide can be destroyed by reduction, by addition of a solution of titanium(III) sulphate in 50% sulphuric acid or of iron(II) salts.

A *radical mechanism* is responsible for the autoxidation of diethyl ether to the highly explosive peroxide:

$$H_3C-CH_2-O-C_2H_5 \xrightarrow{h\nu} H_3C-\overset{\bullet}{C}H-O-C_2H_5 \xrightarrow{+O_2} H_3C-CH-O-C_2H_5$$

Diethyl ether

$$\underset{OO\bullet}{|}$$

Alkylperoxyl radical

$$\frac{+H_3C-CH_2-O-C_2H_5}{-H_3C-\overset{\bullet}{C}H-O-C_2H_5} \quad H_3C-CH-O-C_2H_5 \quad \text{1-Ethoxyethylhydroperoxide}$$

$$\underset{OOH}{|}$$

The hydroperoxide, which is not isolable in this case, can, however, be prepared from hydroxyethylhydroperoxide, $H_3C-CH(OH)-OOH$ and ethanol in the presence of phosphorus(V) oxide (*Rieche*).

1-Ethoxyethylhydroperoxide self-condenses, with extrusion of ethanol, to give the highly explosive polymer 'ether peroxide':

$$nH_3C-\underset{OOH}{\underset{|}{CH}}-O-C_2H_5 \xrightarrow{-nC_2H_5OH} \left[\begin{array}{c} CH_3 \quad\quad CH_3 \\ | \quad\quad\quad | \\ -CH-O-O-CH-OO- \end{array} \right]_{n/2}$$

Hydroperoxide

Similar *autoxidation processes* also take place with hydrocarbons.

In *conc. hydrochloric acid* (or conc. sulphuric acid) ether takes up a proton from the acid which attaches itself to a lone pair of electrons of the oxygen atom to form a secondary oxonium salt:

$$\begin{array}{c} R \\ \diagdown \\ O \\ \diagup \\ R' \end{array} + H^{\oplus}Cl^{\ominus} \longrightarrow \left[\begin{array}{c} R \\ \diagdown \quad\oplus \\ O-H \\ \diagup \\ R' \end{array} \right] Cl^{\ominus}$$

sec-Oxonium chloride

The prefix 'sec' in this case does not refer to a carbon atom, but to the oxygen atom. In an oxonium ion the oxygen atom is linked to three atoms. The link between this oxonium complex and the chloride ion is ionic (see onium complexes).

Tertiary oxonium salts have also been prepared by *Meerwein*, for example by treating boron trifluoride–ether complexes with alkyl fluorides:

$$R_2\overset{\oplus}{\underset{=}{O}}-\overset{\ominus}{B}F_3 + R'-F \longrightarrow \left[\begin{array}{c} R \\ \diagdown\quad\oplus \\ O-R' \\ \diagup \\ R' \end{array} \right] BF_4^{\ominus}$$

tert-Oxonium tetrafluoroborate

tert-Oxonium salts react readily with nucleophilic compounds; for example, they react with a lone pair of electrons in dialkyl sulphides to give sulphonium salts:

$$\left[R_3\overset{\oplus}{\underset{=}{O}} \right] BF_4^{\ominus} + \begin{array}{c} R' \\ \diagdown \\ S \\ \diagup \\ R'' \end{array} \longrightarrow \left[\begin{array}{c} R' \\ \diagdown\quad\oplus \\ S-R \\ \diagup \\ R'' \end{array} \right] BF_4^{\ominus} + R_2O$$

Trialkylsulphonium tetrafluoroborate

tert-Oxonium salts are therefore used as *alkylating agents*.

When ethers are heated with hydriodic acid (density = 1.7) a carbon–oxygen bond is broken and an alkyl iodide is formed. Formation of an oxonium salt is followed by nucleophilic attack by iodide:

$$CH_3-O-R \xrightarrow{\ HI\ } CH_3-\overset{\overset{\displaystyle H}{|}}{\underset{\oplus}{O}}-R \longrightarrow ICH_3 + HO-R$$

$$I^{\ominus}$$

This reaction forms the basis of the *quantitative determination* of alkoxy groups such as OCH_3, OC_2H_5, by the *Zeisel* method. The alkoxy-containing material is heated with hydriodic acid, and the methyl or ethyl iodide which is evolved is trapped in alcoholic silver nitrate; the silver iodide which separates out is estimated either gravimetrically or volumetrically.

Uses

Diethyl ether has great use in the laboratory for the extraction of substances out of aqueous solution. Its vapour is heavier than air and can form explosive mixtures with it. In industry other less volatile and non-flammable solvents are used instead. Because of the large amount of cooling it provides when it evaporates, it is used to produce low temperatures. A mixture of ether and solid carbon dioxide provides a temperature of $-80°C$ (190 K).

Diethyl ether has a pleasant sweetish smell. Prolonged inhalation of it induces unconsciousness (*Jackson* and *Morton*, 1846). It was therefore used in medicine as an *anaesthetic* (*aether pro narcosi*). *Hoffmann*'s drops (1 part ether and 3 parts ethanol) act as a resuscitating agent.

t-Butyl methyl ether (TBME) is prepared in large quantity from *isobutene* and *methanol*:

$$(H_3C)_2C=CH_2 + H_3COH \xrightarrow{H_2SO_4} (H_3C)_3COCH_3$$
Isobutene   t-Butyl methyl ether (b.p. 55°C [328 K])

This compound is rather more stable than diethyl ether is towards oxygen. It is used as a solvent in chromatography and also as an antiknock agent in petrol, following concern over the use of lead; it plays a role in the formation of organic peroxide radicals in the atmosphere.

2.12 Alkanethiols (Mercaptans)

Alkanethiols (mercaptans) are the sulphur analogues of alcohols and can be regarded as monoalkyl derivatives of hydrogen sulphide, R—S—H. The SH group is known as the *thiol* group, or as a mercapto group. The chemistry of thiols in many ways resembles that of alcohols.

Thiols and other sulphur-containing compounds can be found in crude oil, in quantities amounting to several per cent, and may cause difficulties in the refining process, especially by poisoning catalysts.

2.12.1 Preparation

2.12.1.1 By heating alkyl halides with alkali-metal hydrogen sulphides in solution in ethanol in the presence of an excess of hydrogen sulphide, e.g.

$$H_3C-Br + Na-SH \xrightarrow{(H_2S)} H_3C-SH + NaBr$$

Methanethiol (Methyl mercaptan)
(b.p. 5.9°C [279.1 K])

Industrially methanethiol is made from methanol and hydrogen sulphide at about 400°C (675 K) in the presence of a tungsten activated alumina catalyst:

$$H_3COH + H_2S \xrightarrow{Al_2O_3/K_2WO_4} H_3C-SH + H_2O$$

2.12.1.2 By distillation of an aqueous solution of a potassium alkyl sulphate and potassium hydrogen sulphide, e.g.

$$C_2H_5-O-SO_3K + K-SH \rightarrow C_2H_5-SH + K_2SO_4$$

Ethanethiol (Ethyl mercaptan) (b.p. 35°C [308 K])

Both reactions proceed in the same way as the analogous syntheses of alcohols, by an S_N2 mechanism.

2.12.1.3 By heating S-alkylisothiouronium salts, which are made from thiourea and alkyl halides, with aqueous sodium hydroxide:

$$2S=C\begin{smallmatrix}NH_2\\NH_2\end{smallmatrix} \xrightarrow{+2R-Br} \left[2R-S-C\begin{smallmatrix}\oplus\\NH_2\\NH_2\end{smallmatrix}\right]Br^\ominus \xrightarrow{+2NaOH}$$

Thiourea

$$2R-SH + 2NaBr + HN=C\begin{smallmatrix}NH_2\\NH-CN\end{smallmatrix} + 2H_2O$$

Alkanethiol   Dicyanodiamide
                     (cyanoguanidine)

2.12.2 Properties and Uses

With the exception of methanethiol, which is a gas, alkanethiols are liquids with extremely unpleasant odours. Because of their inability to form intermolecular hydrogen bonds, they boil at significantly lower temperatures than the corresponding alcohols, e.g. ethanethiol, b.p. 35°C (308 K) (cf. ethanol, b.p. 78.4°C [351.6K]). Like hydrogen sulphide, the thiols are weakly acidic and dissolve in dilute solutions of alkali:

$$R-S-H + NaOH \rightleftharpoons R-\overline{\underline{S}}|^\ominus Na^\oplus + H_2O$$

Their salts, the *mercaptides*, are less easily hydrolysed than are alkoxides. They form characteristic mercury salts, which are only very slightly soluble in water:

$$2R-S-H + HgO \rightarrow (R-S)_2Hg + H_2O$$

It is from the latter reaction that the name mercaptan was derived (Lat. *corpus mercurium captans* = mercury-precipitating body, *Zeise*, 1834).

Also like hydrogen sulphide, thiols are slowly oxidised in air to disulphides, e.g.

$$C_2H_5-S-H + H-S-C_2H_5 + 1/2O_2 \rightarrow C_2H_5-S-S-C_2H_5 + H_2O$$

Diethyl disulphide

If stronger oxidising agents, such as nitric acid, are used, thiols are oxidised to alkanesulphonic acids.

Methanethiol is an important starting material in various syntheses, for example, of the amino-acid methionine, and of dimethyl sulphoxide. It is also used in the formation of polymers and resins.

2.13 Dialkyl Sulphides (Thioethers)

The analogues of the ethers are *thioethers* or *dialkyl sulphides*, which are derivatives of hydrogen sulphide in which both hydrogen atoms are replaced by alkyl groups. According to whether the two alkyl groups are the same or not, they are described, respectively, as *simple* (R—S—R) or *mixed* (R—S—R') thioethers or dialkyl sulphides. They are obtained in the following ways, which both involve S_N2 reactions.

2.13.1 Preparation

2.13.1.1 By heating alkyl halides with alkali metal mercaptides, e.g.

$$H_3C-Br + Na^{\oplus}|\overset{\ominus}{\underline{S}}-CH_3 \longrightarrow H_3C-S-CH_3 + NaBr$$

Dimethyl sulphide (Methylthiomethane)
(b.p. 37.3°C [310.5 K])

Mixed, as well as simple, thioethers are obtainable in this way.

2.13.1.2 By heating alkyl halides with potassium sulphide, *e.g.*

$$C_2H_5-Cl + K-S-K + Cl-C_2H_5 \longrightarrow C_2H_5-S-C_2H_5 + 2 KCl$$

Diethyl sulphide (Ethylthioethane)
(b.p. 92°C [365 K])

Bis(2-chloroethyl)sulphide (β, β'-dichloro-diethyl sulphide), Cl—CH$_2$—CH$_2$—S—CH$_2$—CH$_2$—Cl, was used in the First World War with the name *mustard gas*; it is still used as a chemical warfare agent. The vapours of the liquid, which boils at 217°C (490 K), have a destructive effect on the bronchia and on the skin, forming blisters.

2.13.2 Properties

Thioethers are insoluble in water. They are liquids with penetrating unpleasant smells. With halogens, alkyl halides or metal salts they form salt-like addition products. By analogy to oxonium salts they are known as sulphonium salts:

$$\begin{matrix} R \\ \diagdown \\ S \\ \diagup \\ R' \end{matrix} + R'-I \longrightarrow \left[\begin{matrix} R & \oplus & R' \\ \diagdown & S & \diagup \\ R' \end{matrix} \right] I^{\ominus}$$

Trialkylsulphonium iodide

The bonding is like that in oxonium salts, with a three-coordinate S atom.

When sulphonium iodides are treated with moist silver oxide, sulphonium hydroxides are formed. These are strong bases. When heated they decompose to give a thioether, an alkene and water, e.g.

$$(C_2H_5)_3\overset{\oplus}{S}]OH^{\ominus} \longrightarrow (C_2H_5)_2S + C_2H_4 + H_2O$$

Sulphonium salts in which the S atom is attached to three different substituents, can be resolved into *enantiomers*, (*Smiles*, 1900), e.g.

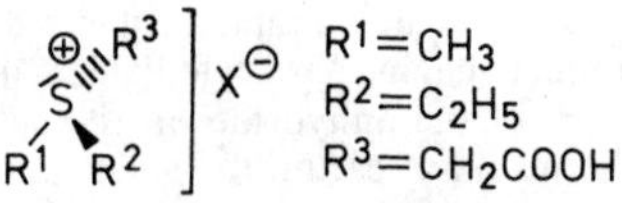

$$\left[\overset{\oplus}{\underset{R^1 \quad R^2}{S}}{\cdots}^{R^3}\right] X^{\ominus} \qquad \begin{array}{l} R^1 = CH_3 \\ R^2 = C_2H_5 \\ R^3 = CH_2COOH \end{array}$$

It follows, therefore, that the substituents on the central S atom take up a tetrahedral configuration, with the single lone pair of electrons on the sulphur assuming the function of the fourth substituent (see p. 164).

2.13.3 Sulphoxides and Sulphones

When *thioethers* are *oxidised*, using the correct calculated amount of hydrogen peroxide or dilute nitric acid, *sulphoxides* are formed, which are easily oxidised further by excess hydrogen peroxide, conc. nitric acid or potassium permanganate, giving sulphones:

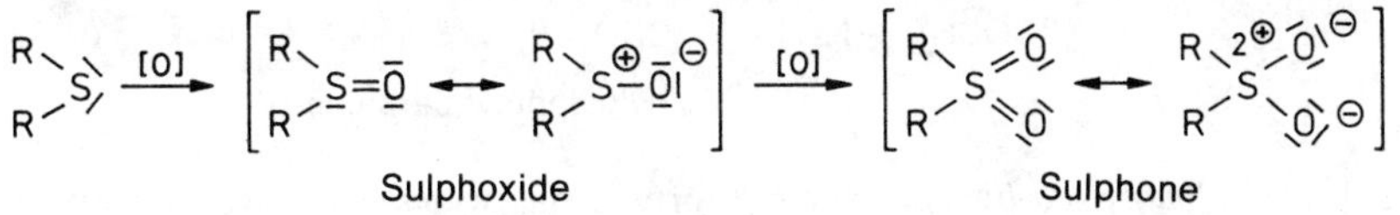

$$\underset{R'}{\overset{R}{>}}S \xrightarrow{[O]} \left[\underset{R'}{\overset{R}{>}}S=\bar{O} \longleftrightarrow \underset{R'}{\overset{R}{>}}\overset{\oplus}{S}-\bar{\underline{O}}^{\ominus}\right] \xrightarrow{[O]} \left[\underset{R'}{\overset{R}{>}}\underset{\underset{O}{\diagdown}}{\overset{=O}{S}} \longleftrightarrow \underset{R'}{\overset{R}{>}}\overset{2\oplus}{\underset{\underset{O}{\diagdown}\ominus}{S}}{\overset{\bar{O}|\ominus}{}}\right]$$

Sulphoxide   Sulphone

Industrially sulphoxides and sulphones are obtained from thioethers by catalytic oxidation, making use of atmospheric oxygen.

Both classes of compounds are often drawn with S=O double bonds, since sulphur, as an element of the third period of the Periodic Table, has available energetically favourable vacant orbitals above the $3p$ orbitals, and can hence expand its electron shells to decets or dodecets. There are, however, strong arguments for representing sulphoxides and sulphones by means of semipolar $S^{\oplus}$—$O^{\ominus}$ bonds (see p. 151).[1]

Pummerer **Rearrangement**. When treated with acid anhydrides, sulphoxides are converted into dialkyl sulphides or, more precisely, acyloxysulphides:

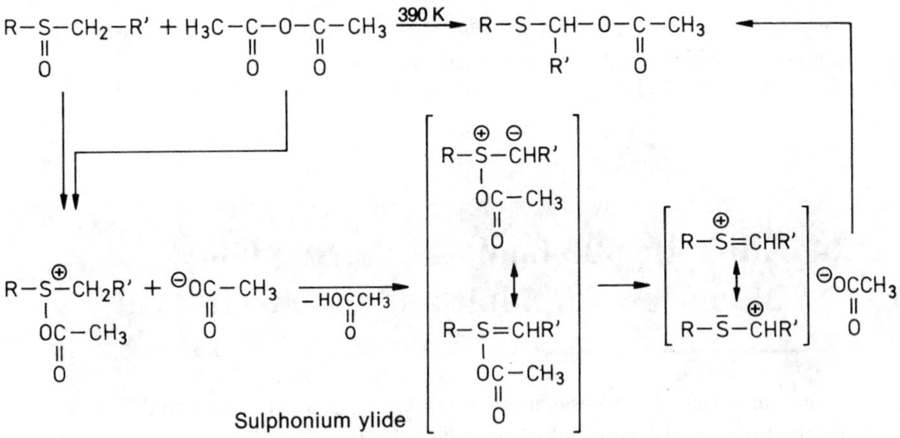

Sulphonium ylide

---

[1] See *H. Kwart* et al., d-Orbitals in the Chemistry of Silicon, Phosphorus and Sulphur (Springer-Verlag, Berlin 1977).

The overall process can be regarded as an intramolecular redox reaction. It is of preparative importance since it gives rise to products having 'protected aldehyde groups' (see p. 196).

Reaction is initiated by the attack of the anhydride on the sulphur atom of the sulphoxide. A proton is then transferred from the resultant cation to the acetate anion, leading to the formation of a sulphonium ylide. This can lose an acetate ion to give a mesomerically stabilised carbenium ion, which adds acetate again, but at the carbon atom, to form the final product. The overall result is an α-shift of the oxygen function.

Dimethyl sulphoxide (DMSO), $(H_3C)_2SO$, is a colourless hygroscopic liquid with b.p. 189°C (462 K). It is miscible with water, alcohols, acetone, chloroform and benzene, but not with alkanes. It can be used as a solvent for PVC, poly(acrylonitrile) and their copolymers, polyurethanes, terylene, perlon and cellulose derivatives. Dimethyl sulphoxide, like dimethylformamide, is an example of a *nucleophilic aprotic* solvent. Because of its nucleophilicity it solvates cations such as $H^\oplus$, $Na^\oplus$ or $K^\oplus$ rather well, but not anions, wherefore in this solvent they are very reactive, and are described as 'naked anions'.

Dimethyl sulphoxide is easily metallated by means of sodium hydride:

$$H_3CSOCH_3 + NaH \rightarrow H_3CSOCH_2^\ominus Na^\oplus + \tfrac{1}{2}H_2$$

Methylsulphinyl carbanion

The *methylsulphinyl carbanion* which is formed has widespread use as a nucleophilic reagent. For example, it is used to convert an ester into an α-ketosulphoxide by replacement of the alkoxy group.[1]

$$RCOOC_2H_5 + H_3CSOCH_2^\ominus Na^\oplus \rightarrow RCOCH_2SOCH_3 + C_2H_5O^\ominus Na^\oplus$$

α-Ketosulphoxide

The α-ketosulphoxide, or better *methylsulphinyl ketone*, which is formed, can be reduced by *aluminium amalgam* in 90% tetrahydrofuran (THF) to the related ketone:

$$RCOCH_2SOCH_3 \xrightarrow{\text{Al/Hg, 90\% THF}} RCOCH_3$$

The similarity of this reaction to the *Pummerer* rearrangement should be noted. Dimethyl sulphoxide is of considerable preparative importance (see p. 191) as a consequence of the many ways in which it can be used as a selective oxidising agent (*Kornblum*, 1957).

2.14 Aliphatic Sulphonic Acids, Sulphonyl Chlorides, Sulphinic and Sulphenic Acids

The homologous series of *alkanesulphonic acids*, $C_nH_{2n+1}SO_3H$, is derived from alkanes by replacement of a hydrogen atom by a *sulphonyl* group, $-SO_3H$. The *alkanesulphonyl chlorides*, $R-SO_2Cl$, can be prepared from the alkanesulphonic acids, and can be reduced to alkanesulphinic acids, $R-SO_2H$.

---

[1] See *T. Durst*, Dimethyl sulphoxide (DMSO) in Organic Synthesis, Adv. Org. Chem. *6*, 285 (1969).

2.14.1 Alkanesulphonic Acids

Alkanesulphonic acids are very hygroscopic materials. Their acidity approaches that of mineral acids. *Trifluoromethanesulphonic acid* is even stronger than perchloric acid; it belongs to the group of acids known as superacids.[1] The alkali-metal and alkaline earth metal salts are soluble in water. In contrast to arenesulphonic acids, alkanesulphonic acids cannot be made by direct sulphonation of alkanes. The monosulphonic acids derived from higher alkanes serve as starting materials for the preparation of synthetic detergents.

Preparation

2.14.1.1 By the oxidation of alkanethiols with nitric acid, e.g.

$$H_3C-SH \xrightarrow{+\,\frac{3}{2}\,O_2} H_3C-SO_3H$$

Methanethiol   Methanesulphonic acid

2.14.1.2 By nucleophilic attack of alkali-metal sulphites on alkyl iodides, e.g.

$$C_2H_5-I + K_2SO_3 \rightarrow C_2H_5-SO_3^{\ominus}K^{\oplus} + KI$$

Potassium ethanesulphonate

2.14.1.3 By sulphoxidation of alkanes. When higher alkanes are treated with sulphur dioxide and oxygen in the presence of a radical source (UV light, ozone and peracids) a mixture of alkanesulphonic acids is formed, since the site of attack is statistically controlled:

$$R-H + SO_2 + 1/2\,O_2 \rightarrow R-SO_3H$$

Trifluoromethanesulphonic acid is made from *methanesulphonyl chloride* by *electrofluorination*, followed by hydrolysis of the resultant acid chloride:

$$H_3CSO_2Cl \xrightarrow[-6^{\ominus},\,-3\,H^{\oplus}]{3\,F^{\ominus}} F_3CSO_2Cl \xrightarrow[-HCl]{H_2O} F_3CSO_3H$$

Methane sulphonyl chloride   Trifluoromethane   Trifluoromethane
                                   sulphonyl chloride   sulphonic acid

2.14.2 Alkanesulphonyl Chlorides, Alkanesulphinic and Alkanesulphenic Acids

Treatment of alkanesulphonic acids with phosphorus(V) chloride gives rise to alkanesulphonyl chlorides. The latter can be reduced to either sulphinic acids or alkanethiols, e.g.

$$R-SO_2-OH + PCl_5 \longrightarrow R-SO_2Cl + POCl_3 + HCl$$

Alkanesulphonyl chloride

$$R-SO_2Cl \xrightarrow{(2\,H)} R-SO_2H + HCl$$

Alkanesulphinic acid

$$R-SO_2Cl \xrightarrow{(6\,H)} R-SH + HCl + 2\,H_2O$$

Alkanethiol

[1] See *R. J. Gillespie*, Fluorosulphuric Acid and Related Superacid Media, Accounts Chem. Res. *1*, 202 (1968).

Reaction of alkali-metal salts of sulphinic acids with alkyl iodides gives rise not to the corresponding sulphinic acid ester, but to a sulphone.

$$R-\underset{.}{S}O_2{}^{\ominus}Na^{\oplus} + R'-I \longrightarrow \begin{array}{c} R \\ \diagdown \\ R' \diagup \end{array} SO_2 + NaI$$

The sulphur atom in the sulphinate anion is more nucleophilic than the oxygen atoms.

2.14.2.1 Sulphochlorination. Sulphur dioxide and chlorine (1:1) react with alkanes in the presence of UV light or energy-rich radiation (^{60}Co) to give alkanesulphonyl chlorides (*Reed*, 1936):

$$R_2CH-CH_3 + SO_2 + Cl_2 \longrightarrow R_2CH-CH_2-SO_2Cl + HCl$$

As in the case of the chlorination of alkanes, a radical chain reaction is involved. It is initiated as follows: Initiation reaction

$$|\underline{\overline{C}}l-\underline{\overline{C}}l| \overset{h\nu}{\longrightarrow} |\underline{\overline{C}}l\cdot + \cdot\underline{\overline{C}}l| \quad \text{Initiation reaction}$$

$$R-H + \cdot\underline{\overline{C}}l| \longrightarrow R\cdot + H-\underline{\overline{C}}l|$$

Before the alkyl radical can react with a chlorine atom it is captured by an SO_2 molecule, producing a sulphonyl radical which reacts with a molecule of chlorine to form the alkanesulphonyl chloride:

$$R\cdot + SO_2 \longrightarrow R-\underset{.}{S}O_2$$

$$R-\underset{.}{S}O_2 + |\underline{\overline{C}}l-\underline{\overline{C}}l| \longrightarrow R-SO_2-\underline{\overline{C}}l| + |\underline{\overline{C}}l\cdot$$

The following reaction is important as a chain-breaking reaction:

$$R-\underset{.}{S}O_2 + R\cdot \longrightarrow R-SO_2-R$$

$$\text{Sulphone}$$

In the sulphochlorination of higher alkanes a mixture of isomers always results, since the site at which reaction takes place is statistically governed. In addition, small amounts of chlorinated alkanes are present.

On the industrial scale long-chain alkanes are converted in this way into sulphonyl chlorides (Mersolat, Levapon), which serve as starting materials for the preparation of detergents, wetting agents and plasticisers (Mesamoll) and for use in tanning.

Alkanesulphenic acids, R—S—OH, which can be regarded as the initial oxidation products from alkanethiols, are short-lived owing to their being readily further oxidised; they can only be isolated in a few exceptional cases. The sulphenyl chlorides, R—S—Cl, sulphenyl esters, R—S—OR', and sulphenyl amides, R—S—NR$_2'$, are relatively stable derivatives.

2.15   Nitroalkanes

Nitroalkanes are derived from the alkanes by replacement of a hydrogen atom by a nitro group —NO$_2$, which is attached to the carbon atom through the nitrogen atom. The simplest member of the series is *nitromethane*.

Nitroalkanes are isomers of the alkyl esters of nitrous acid, the alkyl nitrites. In the latter, the NO_2 group is bound to a carbon through an oxygen atom, e.g.

$$H_3C-NO_2 \qquad\qquad H_3C-O-NO$$

Nitromethane (b.p. 101°C [374 K])   Methyl nitrite (b.p. −12°C [261 K])

Besides the great difference in their boiling points, the two compounds show completely different chemical properties. For example, a nitroalkane is reduced to a primary amine, whereas with nascent hydrogen an alkyl nitrite is converted into the alcohol from which it is derived and ammonia (or hydroxylamine):

$$R-NO_2 \xrightarrow{+6H} R-NH_2 + 2\,H_2O$$
$$R-O-NO \xrightarrow{+6H} R-OH + NH_3 + H_2O$$

Since a nitrogen atom has a maximum valence of four, the nitro group is described as a hybrid of two canonical forms (limiting structures):

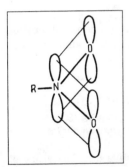

The above formulae include a covalent N—O single bond, which in addition has a dipolar character. This is called a *semipolar* bond.

The molecular orbital description has a p orbital of the N atom which partially overlaps p orbitals on both of the oxygen atoms, resulting in the formation of a molecular π-orbital which extends over all three atoms (Fig. 67).

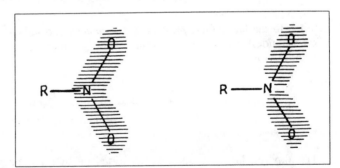

Fig. 67.   Fig. 68.

The four π-electrons (two from an N=O double bond and two from the other oxygen atom) occupy the two molecular orbitals (MO) of lowest energy, according to the Pauli principle; these are represented pictorially in Fig. 68. The second MO differs from the first by the presence of a node at the N atom. In accord with the orbital schemes shown on p. 107 it may be seen that these orbitals are analogous to those in an allyl anion.

As well as mononitroalkanes there are polynitroalkanes, such as trinitromethane (Nitroform), $HC(NO_2)_3$, and tetranitromethane $C(NO_2)_4$.

2.15.1 Preparation

2.15.1.1 Direct nitration of alkanes with nitric acid is only possible in the gas phase at 400–500°C (675–725 K), with a short contact time (*Hass*, 1936), e.g.

$$CH_4 + HO-NO_2 \rightarrow H_3C-NO_2 + H_2O$$
Nitromethane

When higher alkanes are nitrated, a mixture of isomeric nitro compounds results; for example, with propane:

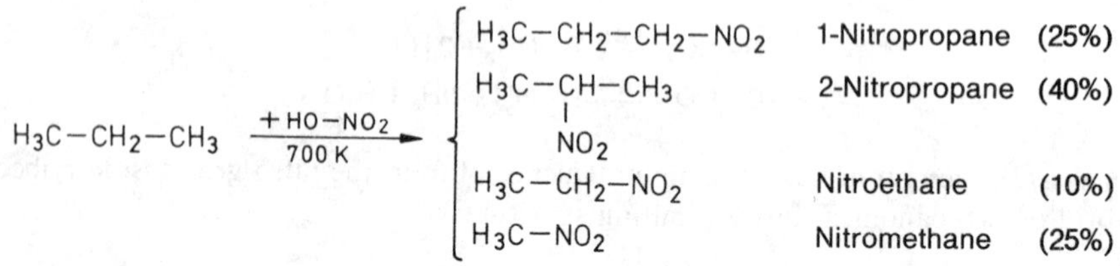

$$H_3C-CH_2-CH_3 \xrightarrow[700\ K]{+ HO-NO_2}$$

$$\begin{cases} H_3C-CH_2-CH_2-NO_2 & \text{1-Nitropropane (25\%)} \\ H_3C-\underset{\underset{NO_2}{|}}{CH}-CH_3 & \text{2-Nitropropane (40\%)} \\ H_3C-CH_2-NO_2 & \text{Nitroethane (10\%)} \\ H_3C-NO_2 & \text{Nitromethane (25\%)} \end{cases}$$

The yields are generally about 35–45%. The mixture of nitroalkanes usually has a percentage composition on the lines of that shown above. By means of this procedure it is possible to prepare lower *primary* nitroalkanes industrially.

It is evident from the above range of products that in gas-phase nitration some cracking of alkanes also takes place. The various fragments which result are then nitrated in a radical chain reaction.

$$R-H \rightarrow R\cdot + \cdot H \quad \text{or} \quad R-R' \rightarrow R\cdot + \cdot R'$$
$$R\cdot + H\underline{O}-NO_2 \rightarrow R-NO_2 + \cdot \underline{O}H$$
$$R-H + \cdot \underline{O}H \rightarrow R\cdot + H_2O$$

2.15.1.2 In the laboratory, *primary* **and** *secondary* **nitroalkanes** are prepared by heating alkyl bromides or iodides with sodium nitrite in dimethylformamide, with some urea also present (*Kornblum*).

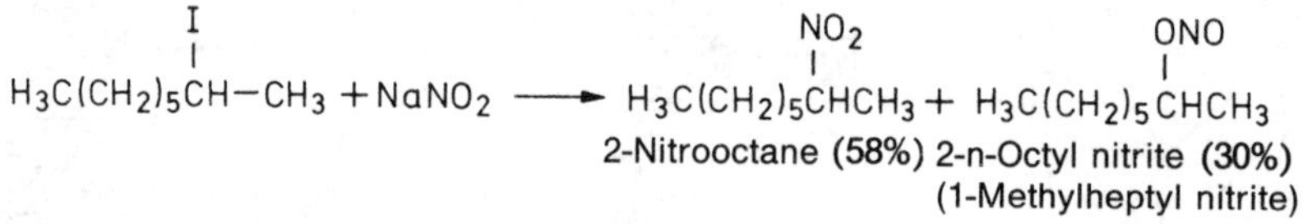

$$H_3C(CH_2)_5\underset{\underset{I}{|}}{CH}-CH_3 + NaNO_2 \longrightarrow H_3C(CH_2)_5\underset{\underset{NO_2}{|}}{CH}CH_3 + H_3C(CH_2)_5\underset{\underset{ONO}{|}}{CH}CH_3$$
2-Nitrooctane (58%)   2-n-Octyl nitrite (30%)
(1-Methylheptyl nitrite)

The alkyl nitrites may easily be separated off by fractional distillation. The yields of primary nitroalkanes are improved if silver nitrite is used in place of sodium nitrite (*V. Meyer*), e.g.

$$H_3C(CH_2)_6CH_2I + AgNO_2 \rightarrow H_3C(CH_2)_6CH_2NO_2 + H_3C(CH_2)_6CH_2ONO$$
(83%)   (11%)

The nitrite ion is a good example of an *ambident anion*, in which the two oxygen atoms, as well as the nitrogen atom, can function as nucleophilic reactant sites:

$$[^{\ominus}|\underline{O}-\overline{N}=\underline{O} \leftrightarrow \underline{O}=\overline{N}-\underline{O}|^{\ominus}]$$

The ratios of the products depend upon many things and cannot be readily clarified. It appears from the work of *Kornblum*[1] and *Pearson*[2] that it is dependent upon a variety of different factors.

2.15.1.3 *Tertiary* **nitroalkanes** are best obtained by oxidation of t-alkylamines with potassium permanganate (*Kornblum*):

$$R_3C-NH_2 \xrightarrow[-H_2O]{+3O} R_3C-NO_2$$

2.15.1.4 Nitromethane can be made by heating chloroacetic acid with aqueous sodium nitrite; the nitroacetic acid which is formed as an intermediate at once loses carbon dioxide (*Kolbe*):

$$Cl-CH_2-COOH \xrightarrow[-NaCl]{+NaNO_2} [O_2N-CH_2-COOH] \xrightarrow[-CO_2]{} H_3C-NO_2$$

| Chloroacetic acid | Nitroacetic acid | Nitromethane |

2.15.1.5 Tetranitromethane (b.p. 126°C [399 K]) is formed by the action of fuming nitric acid on acetic anhydride in the cold.

$$(H_3CCO)_2O + 4 HNO_3 \rightarrow C(NO_2)_4 + H_3CCOOH + CO_2 + 3 H_2O$$

2.15.2 Properties

Nitroalkanes are colourless liquids with agreeable smells. They are only slightly soluble in water. The *primary* and *secondary* nitroalkanes, but not the *tertiary*, dissolve slowly in aqueous alkali with the formation of salts. The sodium salts precipitate out from solutions in ethanolic sodium hydroxide. In keeping with this, *primary* and *secondary* nitroalkanes show acidic properties. The nitro group exerts a $-$I-effect, which makes loss of a proton from the α-carbon atom easier. The resultant anion is stabilised by mesomerism, and the negative charge is delocalised over the carbon atom and the oxygen atoms of the nitro group:

primary Nitroalkane   Anion of the nitro compound

If an alkaline solution of a simple aliphatic nitro compound such as nitromethane is acidified, then the nitroalkane itself is reformed.

In the case of *phenylnitromethane* an *isomeric* compound is isolated first, which is strongly acidic and conducts electricity in aqueous solution. This is called aci-phenylnitromethane and described as a *nitronic acid*. It is unstable, and in a short time changes *irreversibly* into the normal form, *i.e.* $C_6H_5CH_2NO_2$.

---

[1] See *N. Kornblum* et al., J. Am. Chem. Soc. *77*, 6269 (1955).
[2] See *R. G. Pearson* et al., J. Am. Chem. Soc. *89*, 1827 (1967).

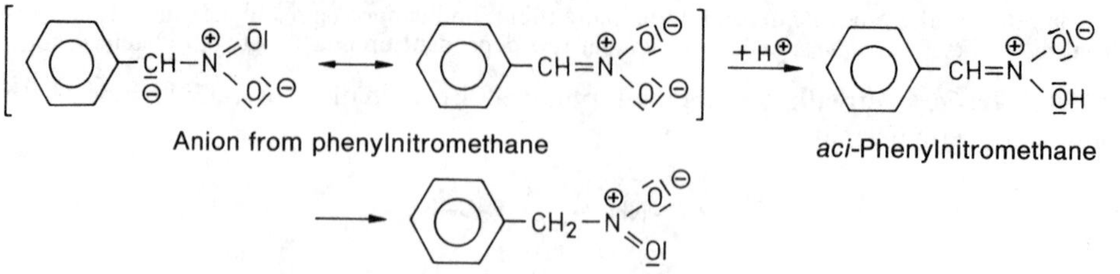

Anion from phenylnitromethane   aci-Phenylnitromethane

Phenylnitromethane

In some cases an *equilibrium* between normal and aci-forms has been detected. In these cases there is nitro–aci-nitro tautomerism, in which there is a *tautomeric equilibrium* between two isomeric compounds. A proton moves from one site to another, and the shift is accompanied by a reorganisation of the bonding between the two isomers. Thus in the case of *p*-nitrophenylnitro-methane, in aqueous methanol there is 0.79%, and in pyridine up to 16%, of the *aci*-nitro form, which is *stabilised* by mesomerism:

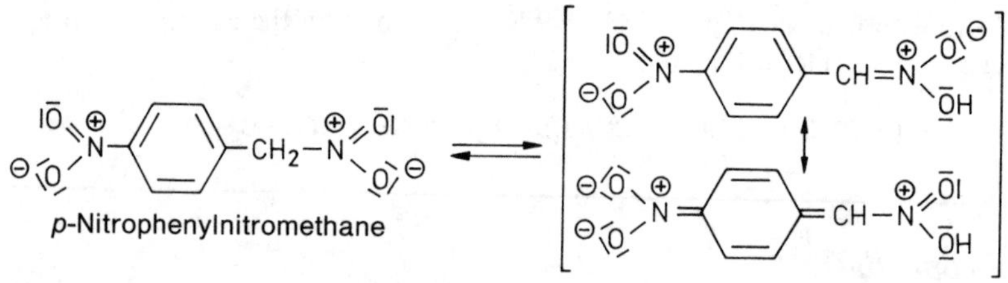

p-Nitrophenylnitromethane

Tautomerism   Mesomerism

Nitroalkanes react with carbonyl compounds in the presence of base to give β-nitroalcohols, *e.g.*

$$H_3C-CHO + H_3CNO_2 \xrightarrow{OH^\ominus} H_3C-CH(OH)-CH_2-NO_2$$

Acetaldehyde Nitromethane   1-Nitropropan-2-ol

2.15.2.1 Cleavage of Nitroalkanes (*V. Meyer*).

The C—N bond in *primary* and *secondary* nitroalkanes can be split by strong mineral acids, in which an intramolecular redox process is involved. *Primary* nitroalkanes are converted into *hydroxamic acids*, which are hydrolysed to hydroxylamine and a carboxylic acid. Hydroxylamine is prepared industrially from 1-nitropropane in this way.

$$H_3C-CH_2-CH_2-NO_2 \xrightarrow{(H^\oplus)} H_3C-CH_2-C\overset{\displaystyle O}{\underset{\displaystyle NHOH}{<}} \xrightarrow{+H_2O} H_3C-CH_2-COOH + NH_2OH$$

1-Nitropropane   Hydroxamic acid   Propionic acid   Hydroxylamine

2.15.2.2 *aci*-Nitroalkane Cleavage (*Nef* Reaction).

Ketones or aldehydes are obtained in yields of up to 85% by treating the salts of the appropriate nitro compounds with 50% sulphuric acid. Dinitrogen monoxide is also formed:

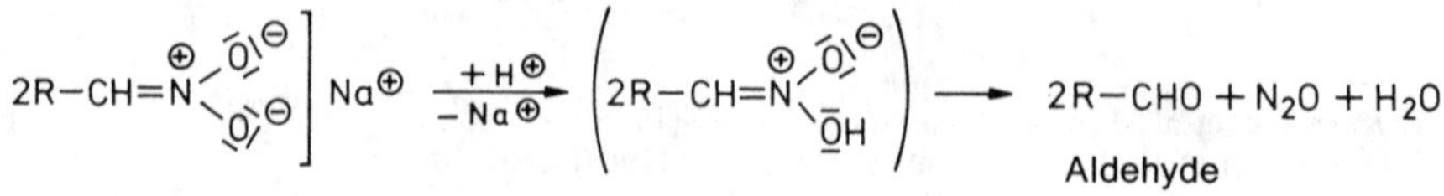

Aldehyde

Uses

The lower primary nitroalkanes are excellent solvents for vinyl resins, polystyrene, nitrocelluloses, acetylcelluloses and polyacrylonitrile. They are also increasingly used in synthetic processes; for example the *insecticide* chloropicrin is made by chlorinating nitromethane:

$$H_3C-NO_2 + 3\,Cl_2 \rightarrow Cl_3C-NO_2 + 3\,HCl$$

Chloropicrin

2.16 Aliphatic Amines

2.16.1 Monoamines

The homologous series of amines is derived from ammonia. As the three hydrogens of ammonia are replaced in turn by alkyl groups, so *primary*, *secondary* and *tertiary* amines are formed.

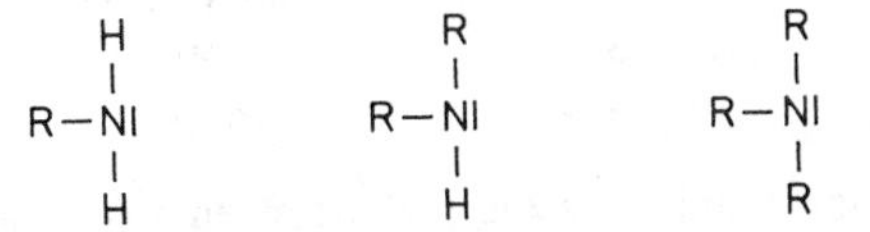

primary Amine   secondary Amine   tertiary Amine

In contrast to the carbon compounds discussed previously, the names *primary*, *secondary* and *tertiary* do not refer to the number of groups attached to a carbon atom, but to the number attached to a nitrogen atom. Thus t-butylamine is in fact a *primary amine*. The N-alkyl groups in secondary or tertiary amines may or may not be all the same.

The $-NH_2$ group is an *amino* group, $=NH$ is an *imino* group.

The lone pair of electrons on the nitrogen atom results in amines being both bases and nucleophiles. They readily add protons from acids to form salts:

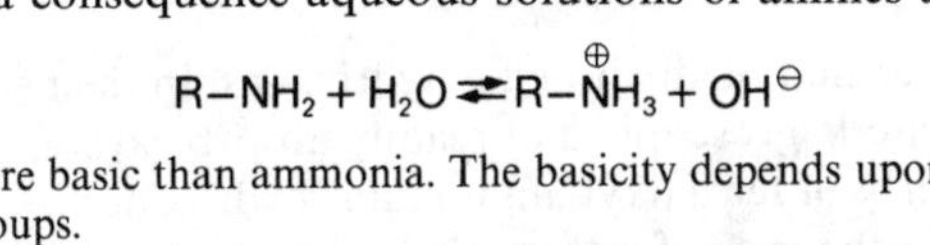

primary Alkylamine   Alkylammonium chloride

The protons can also come from water, which is involved in the following *equilibrium* with the amine. As a consequence aqueous solutions of amines are alkaline:

$$R-NH_2 + H_2O \rightleftharpoons R-\overset{\oplus}{N}H_3 + OH^{\ominus}$$

Alkylamines are more basic than ammonia. The basicity depends upon both the number and the nature of the alkyl groups.

Secondary amines are more basic than *primary* amines, but tertiary amines are markedly less basic. This can be related to the fact that steric hindrance to *solvation* in tertiary ammonium ions is stronger than it is for secondary or primary ammonium ions. Support for the influence of *solvation effects* comes from a study of the basicity of tertiary amines in chloroform. In this solvent they are stronger bases than are secondary or primary amines. The same applies to their relative basicities in the gas phase.[1]

Since in the presence of a large excess of water the concentration of water in the equilibrium can be regarded as constant, the following equation gives the measure of the basicity constant K_b for a primary amine:

$$K_b \approx \frac{[\overset{\oplus}{R}NH_3][OH^{\ominus}]}{[RNH_2]}$$

It is more convenient to express the basicity in terms of the pK_b, which is derived by the following relationship:

$$pK_b = -\log_{10} K_b$$

Alternatively the strength of a base can also be specified by the pK_a value, which is related to the pK_b value as shown:

$$pK_a + pK_b = 14.00 \ (25 \ °C \ [298 \ K])$$

Some examples are as follows:

	Ammonia NH_3	Methylamine H_3C-NH_2	Dimethylamine $(H_3C)_2NH$	Trimethylamine $(H_3C)_3N$
pK_b	4,75	3,34	3,27	4,19
pK_a	9,25	10,66	10,73	9,81

The pK_a value provides a unified scale for the strengths of acids and bases. It is a measure of how readily in the equilbrium

$$R-\overset{\oplus}{N}H_3 + H_2O \rightleftharpoons R-NH_2 + H_3O^{\oplus}$$

the conjugate acid $R-NH_3^{\oplus}$ of the base RNH_2 gives up a proton, i.e. it is a measure of the acidity of the conjugate acid. To a first approximation, the pK_a value equals the pH value of a half neutralised solution of the base.

Primary and secondary amines can also act as very weak acids, for example they form lithium salts on the treatment with phenyl lithium in ethereal solution:

$$(C_2H_5)_2\overline{N}H + C_6H_5Li \longrightarrow (C_2H_5)_2\overset{\ominus}{N}Li^{\oplus} + C_6H_6$$

If a tertiary amine is heated with an alkyl halide, a tetraalkylammonium salt is formed:

$$R_3NI + R-I \longrightarrow R_4N]^{\oplus}I^{\ominus}$$

When such a salt is treated with moist silver oxide, a *quaternary* ammonium hydroxide is generated; this is completely dissociated in aqueous solution:

$$R_4N]^{\oplus}I^{\ominus} + AgOH \longrightarrow R_4N]^{\oplus}OH^{\ominus} + AgI$$

The basicity is similar to that of alkali-metal hydroxides. In their solid state quaternary ammonium hydroxides are hygroscopic and readily absorb carbon dioxide from the air.

A characteristic property of tetraalkyl-ammonium hydroxides is that they decompose when they are heated. In the case of tetramethyl-ammonium hydroxide trimethylamine and methanol are formed:

---

[1] See E. M. Arnett, Gas-Phase Proton Transfer—a Breakthrough for Solution Chemistry, Accounts Chem. Res. 6, 404 (1973).

$$(H_3C)_4N]^{\oplus}OH^{\ominus} \rightarrow (H_3C)_3NI + H_3COH$$

2.16.1.1 *Hofmann* Elimination Reaction.[1] This reaction concerns higher tetraalkylammonium salts, which, when they are heated, are split into a tertiary amine, an alkene and water. For example, loss of an ethyl group gives rise to ethylene as the cleavage product, *e.g.*

$$\left[\begin{array}{c} CH_3 \\ | \oplus \\ H_3C-N-CH_2-CH_3 \\ | \\ C_3H_7 \end{array} \right] OH^{\ominus} \longrightarrow \begin{array}{c} CH_3 \\ | \\ H_3C-NI + H_2C=CH_2 + H_2O \\ | \\ C_3H_7 \end{array}$$

Reaction usually involves removal of a hydrogen atom as a proton from the β-carbon atom and it is an example of β-elimination:

$$\begin{array}{ccc} CH_3 & H\ OH & CH_3 \\ | \oplus & | & | \\ H_3C-N-CH_2-C-H & \longrightarrow & H_3C-N + CH_2=CH_2 + H_2O \\ | & | & | \\ C_3H_7 & H & C_3H_7 \end{array}$$

This thus requires the presence of at least one hydrogen atom on a β-carbon in the cation; such a reaction is impossible for tetramethyl-ammonium hydroxide.

If the tetralkylammonium cation is made up from alkyl groups which could provide different alkenes, the alkene which is formed predominantly is that with the least number of alkyl substituents (*Hofmann* rule):

$$\left[\begin{array}{c} R \\ | \\ C-C-C-C \\ \oplus | \\ NR_3 \end{array} \right] OH^{\ominus} \xrightarrow[-NR_3,\,-H_2O]{} \begin{array}{c} R \\ | \\ C=C-C-C \end{array}$$

In contrast, in the case of β-eliminations involving alkyl halides, elimination follows the *Saytzeff* rule and provides the most substituted alkene:

$$\begin{array}{c} R \\ | \\ C-C-C-C \\ | \\ X \end{array} \xrightarrow[-HX]{} \begin{array}{c} R \\ | \\ C-C=C-C \end{array}$$

Particularly selective reagents for use as proton acceptors in dehydrohalogenation reactions are the bicyclic amidines 1,5-diaza-bicyclo[4.3.0]non-5-ene (DBN) and 1,8-diaza-bicyclo[5.4.0]undec-7-ene (DBU).

Preparation

2.16.1.2 By reduction of nitroalkanes, using tin and hydrochloric acid, tin(II) chloride, or catalytic hydrogenation, e.g.

$$H_3C-CH_2-NO_2 \xrightarrow[-H_2O]{+4H} [H_3C-CH_2-NHOH] \xrightarrow[-H_2O]{+2H} H_3C-CH_2-NH_2$$

Nitroethane   Ethylamine

Alkyl derivatives of hydroxylamine are intermediates in the process.

---

[1] See *A. C. Cope*, Org. Reactions 11, 317 (1960).

2.16.1.3 Industrially by the catalytic reduction, under pressure, of nitriles. In this way primary (but also secondary) amines are formed. In order to make sure that primary amines are the predominant product, the hydrogenation is carried out in the presence of a large amount of ammonia.

$$R-C \equiv N \xrightarrow[\substack{(475\ K)}]{+\ 4\ H\ (Ni\ or\ Co)} R-CH_2-NH_2$$

Nitrile   Primary amine

2.16.1.4 By the action of ammonia on alkyl halides in aqueous or alcoholic solution (*A. W. v. Hofmann*, 1850).

Alkylation of ammonia in this way leads to a mixture of primary, secondary and tertiary amines and quaternary ammonium salts, since the alkyl halide can react not only with the ammonia but also with the amines which are formed, e.g.

$$NH_3 \xrightarrow{+ CH_3I} H_3C-\overset{\oplus}{N}H_3]\ I^{\ominus} \xrightarrow[-NH_4I]{+ NH_3} H_3C-NH_2 \quad \text{Methylamine}$$

$$H_3C-NH_2 \xrightarrow{+ CH_3I} (H_3C)_2\overset{\oplus}{N}H_2]\ I^{\ominus} \xrightarrow[-NH_4I]{+ NH_3} (H_3C)_2NH \quad \text{Dimethylamine}$$

$$(H_3C)_2NH \xrightarrow{+ CH_3I} (H_3C)_3\overset{\oplus}{N}H]\ I^{\ominus} \xrightarrow[-NH_4I]{+ NH_3} (H_3C)_3N \quad \text{Trimethylamine}$$

$$(H_3C)_3N \xrightarrow{+ CH_3I} (H_3C)_4\overset{\oplus}{N}]\ I^{\ominus} \qquad\qquad \text{Tetramethylammonium chloride}$$

The resultant mixture is made strongly alkaline and distilled; the amines boil out and the quaternary salt remains behind.

Separation of the amines can be achieved by fractional distillation or by treating the mixture with benzenesulphonyl chloride, $C_6H_5-SO_2Cl$. This reacts with primary and secondary amines giving crystalline *benzenesulphonamides*; only those prepared from primary amines are soluble in alkali:

$$C_6H_5-SO_2-NHR \qquad\qquad C_6H_5-SO_2-NR_2$$

Soluble in alkali   Insoluble in alkali

The solubility in alkali is a consequence of the fact that the hydrogen atom attached to the nitrogen atom is acidic, because of the $-$I-effect of the neighbouring SO_2 group. The '*benzenesulphonyl chloride method*' is also used as a general method for the separation of primary and secondary amines (*Hinsberg*, 1890). If the sulphonamides are heated with strong mineral acid, the benzenesulphonyl groups are removed and the amines are reformed and may be isolated.

The preparation of tertiary amines by this method is rendered difficult because the more strongly basic secondary amine, instead of undergoing further alkylation, reacts preferentially to give a dialkylammonium salt:

$$\left[\begin{array}{c} R \overset{\oplus}{\underset{R}{\diagdown}}\overset{R}{\underset{N}{\diagup}} \\ R \diagdown N \diagup H \end{array}\right] X^{\ominus} + \begin{array}{c} R \\ R \diagdown N H \end{array} \rightleftharpoons \begin{array}{c} R \\ R \diagup N \diagdown R \end{array} + \left[\begin{array}{c} R \overset{\oplus}{\underset{R}{\diagdown}}\overset{H}{\underset{N}{\diagup}} \\ R \diagdown N \diagup H \end{array}\right] X^{\ominus}$$

If another base, which is stronger than the secondary amine, is added, then the latter is more rapidly alkylated. This added base must, despite its strength as a base, not be a good enough nucleophile to compete for the alkyl halide. Ethyldiisopropylamine (*Hünig's* base), $C_2H_5-N[CH(CH_3)_2]_2$ fulfils this role and has therefore become of great importance in preparative chemistry. Its nucleophilicity is lowered because of steric hindrance.

2.16.1.5 In industry methylamines are made from *methanol* and *ammonia* in the presence of an Al_2O_3 catalyst at 450°C (725 K) and under a pressure of 15 bar (1.5 MPa). The resultant mixture of mono-, di- and tri-methylamines is separated by distillation under pressure.

2.16.1.6 Four named reactions[1] are now listed, in which one starts with a carboxylic acid or a derivative thereof, and ends up exclusively with the *primary amine* which has one less carbon atom than the original carboxylic acid had.

(a) *Hofmann* **Reaction[2] (1881).** When the amide of a carboxylic acid is dissolved in a small excess of an aqueous solution of sodium hypobromite (bromine and sodium hydroxide) and heated quickly to about 70°C (345 K), loss of carbon dioxide, accompanied by a molecular rearrangement reaction, leads to the formation of a primary amine:

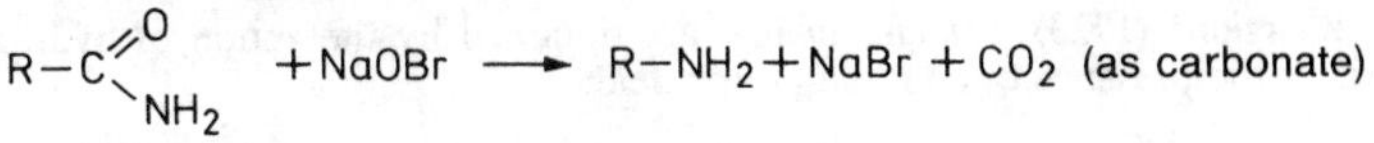

$$R-C\!\!\begin{array}{c}{}^{\displaystyle O}\\{}_{\displaystyle NH_2}\end{array} \;+NaOBr \longrightarrow R-NH_2+NaBr+CO_2 \text{ (as carbonate)}$$

The *reaction mechanism* can be formulated as follows:

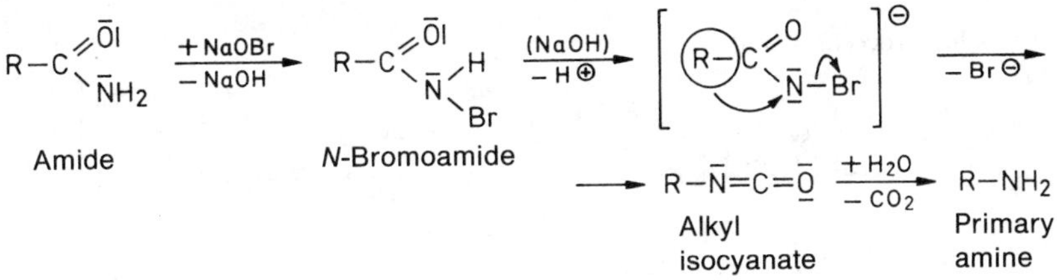

The first step is the replacement of a hydrogen atom of the amide group by bromine to give an N-bromoamide, which is isolable. When this is heated with excess alkali it loses a proton, forming an unstable anion. In the decomposition of this anion, loss of a bromide ion is accompanied by migration of the alkyl group R, together with its bonding electrons, to the nitrogen atom. It is evident that these steps are concerted, i.e. take place simultaneously, since, were the bromide ion to separate completely before the alkyl group migrated, the transitorily formed acyl nitrene would be expected to react with the water present to give a hydroxamic acid:

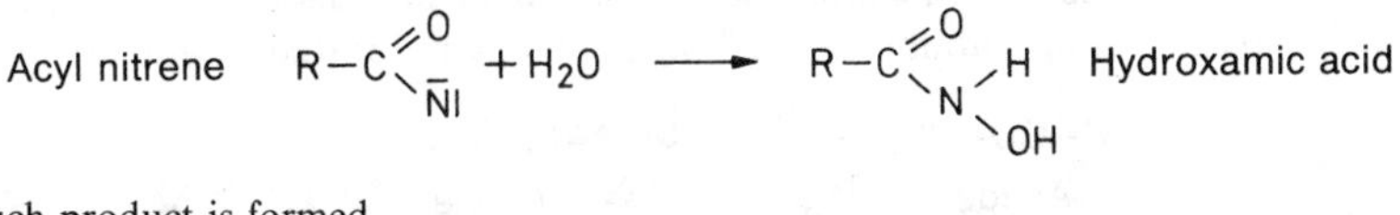

$$\text{Acyl nitrene} \quad R-C\!\!\begin{array}{c}{}^{\displaystyle O}\\{}_{\displaystyle \underline{N}I}\end{array} +H_2O \longrightarrow R-C\!\!\begin{array}{c}{}^{\displaystyle O}\\{}_{\displaystyle N-OH}^{\displaystyle\;\;H}\end{array} \quad \text{Hydroxamic acid}$$

No such product is formed.

The alkyl isocyanate is hydrolysed, with loss of carbon dioxide, to provide the primary amine. This reaction was responsible for the discovery of amines by *Wurtz* (1849).

(b) *Curtius* **Reaction[3] (1894).** When an acid azide is warmed in alcoholic solution, nitrogen is lost, and, analogously to (a) above, an alkyl group migrates from the carbonyl group to the remaining nitrogen atom, thereby forming an alkyl isocyanate. Addition of alcohol to this gives a urethane, which can be hydrolysed in the presence of acid or alkali to provide a *primary* amine:

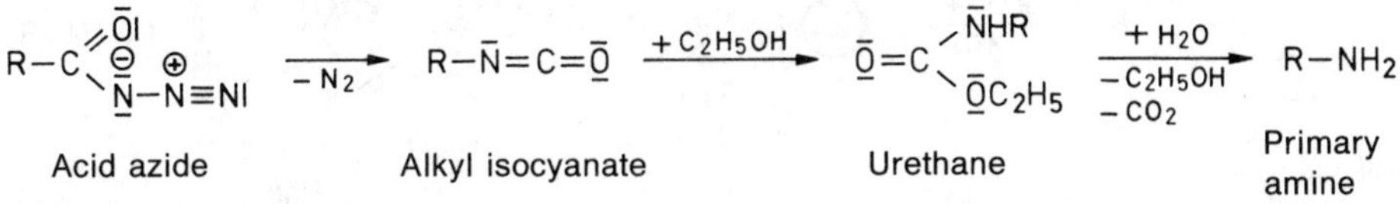

---

[1] See *L. Birladeanu*, Organic Name Reactions, in the Merck Index, 10th edn (Merck & Co., Rahway, N.Y., 1983).

[2] See *E. S. Wallis* and *J. F. Lane*, The Hofmann Reaction, Org. Reactions *3*, 267 ff. (1946).

[3] See *P. A. S. Smith*, The Curtius Reaction, Org. Reactions *3*, 337 ff. (1946).

If decomposition of the acid azide is brought about photochemically, then products are formed which indicate that, at least to some extent, reaction involves an acylnitrene as intermediate.

(c) *Lossen* **Reaction (1875).** When hydroxamic acids are heated in an inert solvent, especially in the presence of phosphorus(V) oxide or acetic anhydride, they lose H_2O and form, for example in the presence of acetic anhydride, an acetyl derivative which undergoes rearrangement to give an alkyl isocyanate. This is then hydrolysed to a *primary* amine:

Hydroxamic acid   Alkyl isocyanate   Primary amine

(d) *Schmidt* **Reaction[1] (1923).** *Primary amines* are obtained by the action of hydrazoic acid on carboxylic acids in the presence of conc. sulphuric acid:

$$R-C{\overset{O}{\underset{OH}{}}} + HN_3 \xrightarrow{\text{(conc. } H_2SO_4\text{)}} R-NH_2 + CO_2 + N_2$$

Reaction probably proceeds as follows:

Carboxylic acid

$$\xrightarrow[-H^{\oplus}]{-N_2} R-\bar{N}=C=\bar{O} \xrightarrow[-CO_2]{+H_2O} R-NH_2$$

Alkyl isocyanate   Primary amine

First of all there is nucleophilic attack by HN_3 on the carbon atom of the carboxyl group. Loss of water and nitrogen from the adduct, accompanied by migration of the alkyl group, leads to the formation of an alkyl isocyanate, which in turn is hydrolysed to the amine.

2.16.1.7 Reduction of amides with lithium aluminium hydride[2] leads to the formation of amines which have the same number of carbon atoms as the initial amide (*Schlittler*):

$$R-CO-NH_2 \xrightarrow[-H_2O]{+4H(LiAlH_4)} R-CH_2-NH_2$$

Amide   Amine

2.16.1.8 *Gabriel* **Synthesis (1887).** When alkyl halides are heated with the potassium salt of phthalimide, N-alkylphthalimides are formed. On treatment with hydrazine the latter are split into *a primary* amine and phthalhydrazide:

Potassium salt of phthalimide   N-Alkyl-phthalimide   Phthal-hydrazide   Primary amine

---

[1] See H. *Wolff*, The Schmidt Reaction, Org. Reactions *3*, 307 *ff* (1946).
[2] See N. G. *Gaylord*, Reduction with Complex Metal Hydrides (Wiley, New York 1965); A. *Hajós*, Complex Hydrides (Elsevier, New York 1979).

Properties

Methylamine, dimethylamine and trimethylamine are gases at room temperature; they are readily soluble in water. Ethylamine boils at 16.6°C (289.8 K). Most other common alkylamines are liquids which smell very like ammonia but are less pungent. Only a few higher homologues are solids. Although in most ways amines closely resemble ammonia, they differ in that they will burn; this indeed led to their original discovery. Also, unlike ammonium salts, salts of amines dissolve in absolute ethanol. The methylamines are found in *pyroligneous acid*, the aqueous fraction obtained from the distillation of wood, and in herrings; the typical smell of seafish and of lobsters comes from trimethylamine.

Isonitrile Reaction: A test for primary amines derives from the fact that, when treated with chloroform and alkali, repulsively smelling isonitriles (isocyanides) are formed (for the *reaction mechanism*, see p. 375).

$$R-NH_2 + HCCl_3 + 3\,NaOH \rightarrow R-\overset{\oplus}{N} \equiv \overset{\ominus}{C}l + 3\,NaCl + 3\,H_2O$$

Isonitrile

2.16.1.9 Behaviour of Amines towards Nitrous Acid. When *primary*, *secondary* or *tertiary* amines are treated with nitrous acid they give rise to different products. In all three cases an N-nitrosation reaction is involved.[1] The actual nitrosating agent is the anhydride of nitrous acid, N_2O_3 (dinitrogen trioxide, with a *nitroso–nitro* constitution $ON-NO_2$), or the nitrosyl cation, $^{\oplus}NO$.

$$2\,HNO_2 \rightleftharpoons ON-NO_2 + H_2O$$

$$\begin{matrix} R^1 \searrow \\ R^2-N| \\ R^3 \nearrow \end{matrix} + O{=}N-NO_2 \longrightarrow \begin{matrix} R^1 \searrow \overset{\oplus}{} \\ R^2-N-\bar{N}{=}\underline{\bar{O}} \\ R^3 \nearrow \end{matrix} + N\bar{O}_2^{\ominus}$$

(a) In the case of a primary amine ($R^2=R^3=H$), the N-nitroso compound undergoes a prototropic shift to form an unstable diazohydroxide, which is protonated and then loses water leaving a diazonium ion. This at once loses nitrogen with concomitant formation of a carbenium ion, which reacts further, by S_N1 or $E1$-mechanisms, giving primary alcohols and/or alkenes, *viz.*

$$R-CH_2-CH_2-\overset{H}{\underset{H}{N|}} \xrightarrow{+ON-NO_2} \left[R-CH_2-CH_2-\overset{H}{\underset{H}{\overset{\oplus}{N}}}-NO \right] NO_2^{\ominus} \xrightarrow{-HNO_2}$$

Primary amine

$$R-CH_2-CH_2-\underset{H}{\overset{}{N}}-NO \longrightarrow R-CH_2-CH_2-N{=}N-OH \xrightarrow[-H_2O]{+H^{\oplus}}$$

Diazohydroxide

$$R-CH_2-CH_2-\overset{\oplus}{N}{\equiv}N| \xrightarrow{-N_2} R-CH_2-CH_2^{\oplus} \underset{\substack{+H_2O \\ -H^{\oplus}}}{\overset{}{\underset{-H^{\oplus}}{}}} \begin{matrix} \xrightarrow{+H_2O,\,-H^{\oplus}} R-CH_2-CH_2-OH \quad \text{Primary alcohol} \\ \xrightarrow{-H^{\oplus}} R-CH{=}CH_2 \quad \text{Alkene} \end{matrix}$$

Diazonium ion   Carbenium ion   Alkene

---

[1] See *J. H. Ridd*, Nitrosation, Diazotisation and Deamination, Quart. Rev. *15*, 418 (1961).

Higher carbenium ions can also undergo rearrangement reactions leading to the formation of an increased number of products; for example, from n-propylamine propan-1-ol (7%), propan-2-ol (32-40%) and propene (28%) are obtained.

(b) The N-nitroso compound formed from a *secondary* amine (R^3=H) and nitrous acid loses a proton to give a stable nitrosamine:

Secondary amine   Nitrosamine

Nitrosamines are orange-yellow coloured oils which are slightly soluble in water. Conc. hydrochloric acid converts them back into the original amine. They can be used both in the identification and the purification of secondary amines. As nitroso compounds they give the *Liebermann* reaction (see p. 505).

Many nitrosamines, and especially dimethylnitrosamine, have been shown to be strongly *carcinogenic* in many different kinds of animals (*Druckrey*, 1966). The same applies to nitrosamides, for example *N*-methylnitrosourea (see p. 167).

(c) *Tertiary* amines[1] are not affected by nitrous acid in the cold in fairly acid solution (pH < 3). In weaker acid (pH 3-6) complex products are formed, among them nitrosamines, whose formation must involve loss of an alkyl group from the tertiary amine.

Alkyl-bis(β-chloroethyl)amines, $R-N(CH_2-CH_2-Cl)_2$, like their sulphur analogue *mustard gas*, are vesicants (*Nitrogen mustards*). Compounds of this kind are used in the chemotherapy of cancers;[2] thus 2-[*bis*(2-chloroethyl)amino]-*1,3,2-oxaazaphosphinan-2-oxide* (N-mustard-phosphamideester, *INN*:*Cyclophosphamide, Endoxan*) is used commercially as a *cytostatic*, and *immune depressive. In vitro* this compound behaves as the almost inactive transport form for the cytostatically active form *N,N*-bis(2-chloroethyl)phosphoric acid diamide, which is liberated biologically in the organism.

Nitrogen   INN: N,N-Bis(2-chloroethyl)-
mustard   **Cyclophosphamide**   phosphoric acid diamide

Tetraalkylammonium salts ($TAA^{\oplus}Cl^{\ominus}$) have become important as phase transfer catalysts.[3] *Benzyltriethylammonium chloride*, $C_6H_5-CH_2N^{\oplus}(C_2H_5)_3]Cl^{\ominus}$ (TEBA) has proved useful in the generation of dichlorocarbene described on p. 132, since in this process a concentrated solution of alkali-metal hydroxide is used as the aqueous phase. For dilute aqueous solutions TEBA is too hydrophilic, and, in this case, to be of use the catalyst must have 15 or more carbon atoms. *Methyltrioctylammonium chloride*, $H_3CN^{\oplus}(C_8H_{17})_3]Cl^{\ominus}$, which has 25 carbon atoms, is highly effective. The trade names for the technical and not completely pure products are *Aliquat 336* and *Adogen 464*.

[1] See *G. E. Hein*, The Reaction of Tertiary Amines with Nitrous Acid, J. Chem. Educ. *40*, 181 (1963).
[2] See *H. Arnold*, Synthese und Abbau cytostatisch wirksamer cyclischer *N*-Phosphamidester des Bis-(β-chloräthyl)-amins, Angew. Chem. *70*, 539 (1958).
[3] See *E. V. Dehmlow* and *S. Dehmlow*, Phase Transfer Catalysis, 2nd edn. (Verlag Chemie, Deerfield Beach, Florida, USA 1983).

2.16.1.10 Semipolar Bonding

Amine oxides. Tertiary amines react with oxygen-providing reagents such as hydrogen peroxide and other peroxides to form *amine oxides*. Their formation results from the interaction of the lone pair of electrons on the nitrogen atom with the oxygen atom, thereby providing the latter with an octet of electrons:

$$R-\overset{\underset{|}{R}}{\underset{|}{N}}| + \bar{O}| \longrightarrow R-\overset{\underset{|}{R}}{\underset{|}{\overset{\oplus}{N}}}-\bar{O}|^{\ominus}$$

Tertiary amine   Amine oxide

As a result, the nitrogen atom is linked to four atoms and bears a positive charge, whilst the oxygen atom is negatively charged. In addition to the single N—O bonding there is also ionic interaction between the two atoms, *i.e.* there is *semipolar* bonding (*cf.* the nitro group). The polarity is shown by the symbols $\oplus$ and $\ominus$, as in the formula above.

Dipole moment measurements on amine oxides show the overriding influence of the Coulombic forces. The bond distance between the nitrogen and oxygen atoms in these compounds is 136 pm.

Because of the electronegativity of the oxygen atom, amine oxides are weak bases and dissolve in, for example, dilute hydrochloric acid giving a salt:

$$R-\overset{\underset{|}{R}}{\underset{|}{\overset{\oplus}{N}}}-\bar{O}|^{\ominus} + HCl \longrightarrow \left[R-\overset{\underset{|}{R}}{\underset{|}{\overset{\oplus}{N}}}-\bar{O}-H \right] Cl^{\ominus}$$

There are appreciable amounts of trimethylamine oxide in sea-fish. When they are caught it decomposes to produce trimethylamine, which is responsible for the characteristic smell of fish.

2.16.1.11 *Cope* Elimination

The presence of the positive charge on nitrogen in amine oxides results in the C—N bond being more easily broken, with the formation of an alkene and a hydroxylamine derivative, e.g.

$$\underset{\underset{\underset{|}{\overset{|}{\underset{\bar{O}}{}}}}{H}}{\overset{|}{-}\underset{|}{C}-\underset{\overset{|}{\overset{\oplus}{NR_2}}}{C}-} \xrightarrow{360-420\ K} \quad \overset{}{\underset{}{>}}C=C\overset{}{\underset{}{<}} + R_2NOH$$

To bring about this reaction it is only necessary to heat the tertiary amine with hydrogen peroxide; there is no need to isolate the intermediate amine oxide.

Related to these dipolar compounds are the *ylides* prepared by *Wittig*.[1] In these compounds there is semipolar bonding between a nitrogen atom and a carbon atom. They are prepared by the action of phenyl lithium on suitable quarternary ammonium salts, e.g.

$$(H_3C)_3\overset{\oplus}{N}-CH_3] \ Br^{\ominus} \xrightarrow[-C_6H_6]{+\ LiC_6H_5} (H_3C)_3\overset{\oplus}{N}-\overset{\ominus}{C}H_2 \cdot LiBr$$

In this reaction a proton is removed from a methyl group and the products are benzene and trimethylammonium methylide. The ylide may be regarded as an intramolecular (or inner)

---

[1] See *A. W. Johnson*, Organic Chemistry, A Series of Monographs, Vol. 7. Ylid Chemistry (Academic Press, New York 1966); *W. K. Musker*, Nitrogen Ylids, Fortschr. Chem. Forsch. *14*, 295 (1970).

ammonium salt (*betaine; zwitterion*). It can only be isolated in the form of a lithium bromide adduct.

The name 'ylide' signifies that the carbon atom of the methylene group is linked to the nitrogen atom not only by covalent bonding (yl) but also by ionic bonding (ide). Ylides are not confined to nitrogen compounds. For instance, the intermediate in the *Pummerer* rearrangement (p. 147) is an example of a sulphonium ylide. Phosphoranes which are of importance in the *Wittig* reaction (p. 173), have a phosphonium ylide structure as one of their canonical forms (limiting structures).

If the ylide shown above is treated with methyl iodide, ethyltrimethylammonium iodide is formed:

$$(H_3C)_3\overset{\oplus}{N}-\overset{\ominus}{C}H_2 + H_3C-I \rightarrow (H_3C)_3\overset{\oplus}{N}-CH_2-CH_3]\ I^{\ominus}$$

2.16.2 Optical Activity in Compounds with 3- and 4-Coordinate Nitrogen Atoms

In the case of optically resolvable sulphonium salts with tervalent sulphur, a *lone* pair of *electrons* of the central sulphur atom acts as the fourth substituent in a tetrahedral structure (Fig. 69a). A similar configuration is found in tertiary amines, in which the nitrogen atom is bound to three different substituents. In this case also the nitrogen atom has a lone pair of electrons. Like ammonia, an amine takes up the shape of a regular three-sided pyramid (Fig. 69b). The dipole moment of ammonia is 4.87×10^{-30} Cm, that of a tertiary amine about 2.17×10^{-30} Cm.

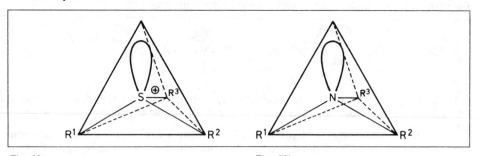

Fig. 69a.   Fig. 69b.

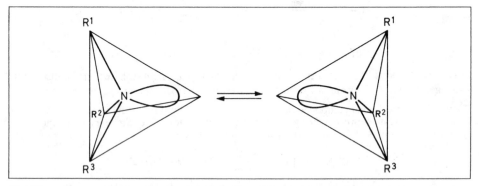

Fig. 70.

From this it would appear that tertiary amines with three different substituents should be resolvable into enantiomers, but all attempts to achieve this so far have been unsuccessful. This lack of success has been attributed to the central nitrogen atom of ammonia or amines being able to oscillate between the two sides of the plane formed by the three ligands, thereby bringing about racemisation (Fig. 70).

If this hypothesis is correct, then inclusion of nitrogen atoms in a suitable ring system could render such oscillation impossible. This has been confirmed by *Prelog*, who prepared the *Tröger's Base* shown below, which has two tertiary amine groups, by a condensation reaction between p-toluidine and formaldehyde, and resolved it into two enantiomeric forms (Fig. 71).

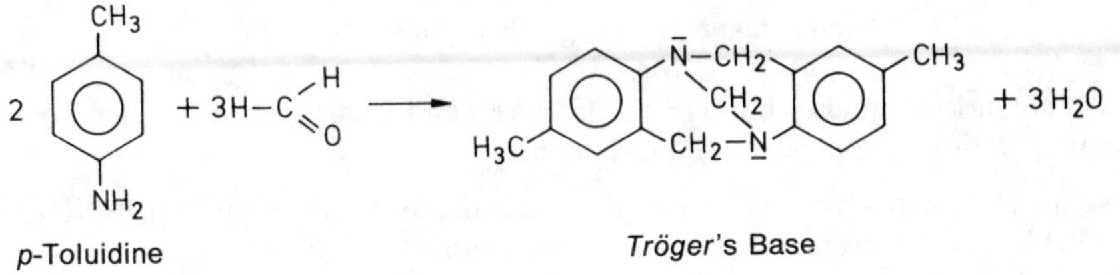

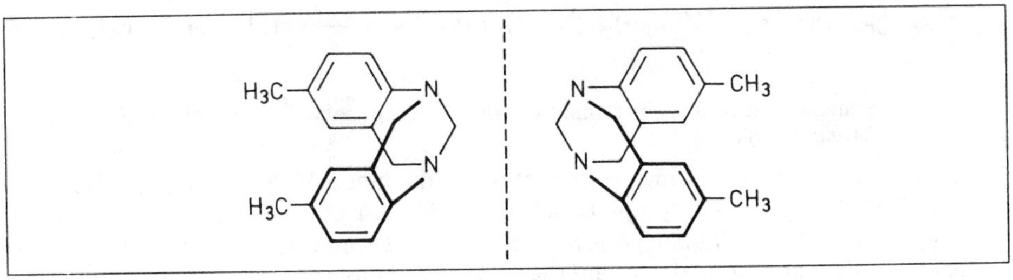

p-Toluidine   Tröger's Base

Fig. 71. Mirror image forms of *Tröger's* Base.

Optically Active Ammonium Salts

If the fourth corner of the tetrahedron in Fig. 69b is occupied by another substituent R^4 different from the other three present, then, by analogy with an asymmetric carbon atom, optical activity must arise. This was confirmed by *Pope* and *Peachey* (1899) who resolved allyl-benzyl-methyl-phenylammonium iodide into enantiomeric forms by the use of (+)-3-bromocamphor-3-sulphonic acid as resolving agent.

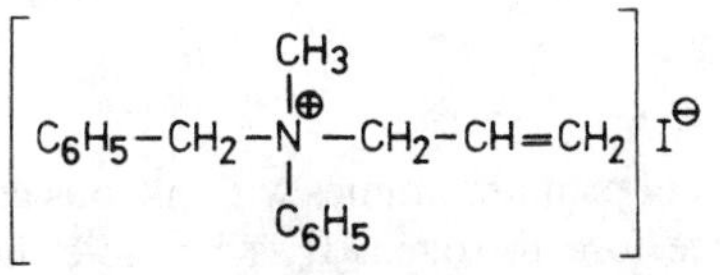

This was the first time that a pure optically active compound had been obtained which, instead of carbon, had an element of the fifth main group of the Periodic Table (in this case nitrogen) as the *central atom* in a tetrahedron.

2.16.3 Unsaturated Amines (Enamines)

As has been done in the case of unsaturated alcohols or *enols*, *enamines* will be dealt with in a separate place. Because of their close connection with ketones they are considered in that context.

2.16.4 Diamines

Diamines are made from dihalogenoalkanes or from dinitriles in similar ways to those used for monoamines.

Ethylenediamine (1,2-diaminoethane), H_2N—CH_2—CH_2—NH_2 is formed from the reaction of *1,2-dibromoethane with an excess of ammonia* at 180°C (455 K) under pressure:

$$Br-CH_2-CH_2-Br + 2NH_3 \rightarrow H_2N-CH_2-CH_2-NH_2 + 2HBr$$

1,2-Dibromoethane   Ethylenediamine

It is a colourless liquid of b.p. 116.5°C (389.7 K) and is miscible with water. It is used in the preparation of complexes and of piperazine

Putrescine (1,4-Diaminobutane, Tetramethylene-diamine), H_2N—$(CH_2)_4$—NH_2, m.p. 27°C (300 K) is made from 1,2-dibromoethane as follows:

$$Br-CH_2-CH_2-Br \xrightarrow[-2KBr]{+2KCN} N\equiv C-CH_2-CH_2-C\equiv N \xrightarrow{+8H} H_2N-(CH_2)_4-NH_2$$

Putrescine is a component of the perfume from the flowers of some varieties of the family of the cuckoo-pints (*Arum maculatum*).

Cadaverine (1,5-diaminopentane, pentamethylene-diamine), H_2N—$(CH_2)_5$—NH_2, b.p. 178–180°C (451–453 K), is a syrupy liquid. It is made in a similar way to putrescine but starting instead with *1,3-dibromopropane*. Both have extremely unpleasant smells and are poisonous. They are formed during the putrefaction of proteins, arising from the decarboxylation of the amino-acids ornithine and lysine.

2.17 Aliphatic Diazo-Compounds, Diazirines and Diaziridines

2.17.1 Diazo-compounds[1]

As already mentioned, treatment of primary amines with nitrous acid gives rise to unstable diazo-compounds which at once lose nitrogen. It is possible, however, to use indirect routes to obtain diazo-compounds, for example, among others, diazoalkanes, diazoketones and diazoacetic ester.

Diazomethane, H_2CN_2, is the simplest aliphatic diazo-compound. Physical measurements show that the C and N atoms are arranged *linearly*. The ground state of diazomethane can be described in terms of the following *canonical forms* or *limiting structures*; the left-hand structure indicates that this compound should have some ylidic properties.

---

[1] See *R. Huisgen*, Altes und Neues über aliphatische Diazoverbindungen (Old and New about Aliphatic Diazo-compounds), Angew. Chem. *67*, 439 (1955); *M. Regitz*, Transfer of Diazo Groups, Angew. Chem. Int. Ed. Engl *6*, 733 (1967).

$$\left[\begin{array}{cc} H \diagdown \overset{\ominus}{\underset{H \diagup}{C}} - \overset{\oplus}{N} \equiv N| & \longleftrightarrow & \overset{H}{\underset{H}{\diagdown}} C = \overset{\oplus}{N} = \overset{\ominus}{N} \end{array}\right]$$

Preparation

Treatment of an ethereal solution of *N-methyl-N-nitrosotoluene-p-sulphonamide* with ethanolic potassium hydroxide at 60°C (330 K) leads to the formation of diazomethane, which can be codistilled with ether out of the reaction mixture:

$$H_3C - \bigcirc - SO_2 - N \overset{CH_3}{\underset{NO}{\diagdown}} + KOH \longrightarrow H_2CN_2 + H_3C - \bigcirc - \overset{\ominus}{S}\overset{\oplus}{O_3K} + H_2O$$

N-Methyl-*N*-nitrosotoluene-*p*-sulphonamide m.p. 62°C [335 K]	Potassium Salt of toluene-*p*-sulphonic acid

Formerly it was usually prepared from *nitroso-N-methylurethane* or *nitrosomethylurea*, but they are no longer used since both compounds have been shown to have carcinogenic behaviour in rats. *Diazomethane* itself has been shown to be a carcinogen towards some animals. When it is used, special care should therefore be taken.[1]

Properties and Uses

Diazomethane is an extremely poisonous yellow compound with b.p. $-24°C$ (249 K). It is highly explosive in the gaseous or liquid state, but in solution (in ether, ethanol or benzene) it is relatively safe, and can be kept for several days in the cold. It is used as a methylating agent for compounds which have sufficiently acidic hydrogen atoms, such as carboxylic acids, phenols and enols:

$$R - C \overset{\diagup O}{\underset{OH}{\diagdown}} + H_2CN_2 \longrightarrow R - C \overset{\diagup O}{\underset{OCH_3}{\diagdown}} + N_2$$

The mechanism of this methylation reaction is as follows:

$$\overset{H}{\underset{H}{\diagdown}} \overset{\ominus}{C} - \overset{\oplus}{N} \equiv N| \xrightarrow{+H^{\oplus}} \overset{H}{\underset{H}{\diagdown}} \underset{H}{\overset{\oplus}{C}} - N \equiv N| \xrightarrow[-N_2]{+A^{\ominus}} \overset{H}{\underset{H}{\diagdown}} \underset{H}{C} - A \qquad \overset{\oplus}{H_2C} = N = NH$$

Methyldiazonium ion		Methylenediazenium ion

A proton from the compound being methylated transfers to the lone pair of electrons on the carbon atom of diazomethane, forming a methyldiazonium ion, which very readily loses nitrogen. This leaves a carbon atom deficient in electrons, to which the anion $A^{\ominus}$, derived by loss of a proton from the substrate, or the *nucleophile* HA, can donate a pair of electrons to complete its octet.

The methyldiazonium ion is detectable, using NMR spectroscopy, in solutions in superacids such as fluorosulphonic acid ($HOSO_2F$) at $-120°C$ (153 K). In the still stronger acid system $HOSOF_2/SbF_5$ diazomethane is protonated on nitrogen to give a methylenediazenium ion.[2]

Because of their lower acidity, alcohols cannot be methylated directly with diazomethane, except in the presence of silica gel. Alcohols which are rather more acidic due to the inductive (-I) effect

---

[1] See Org. Syntheses Coll. Vol. *4*, 250 (1963).
[2] *J. F. McGarrity* and *D. P. Cox*, J. Am. Chem. Soc. *105*, 3961 (1983).

of suitable substituent groups, e.g. trichloroethanol, Cl_3C-CH_2OH, can be methylated by diazomethane alone.

A general method for the preparation of *diazoalkanes* is by *diazo-group* transfer reactions[1] (see p. 648).

2.17.2 Diazirines and Diaziridines[2]

Diazirines are cyclic isomers of aliphatic diazo-compounds and are therefore discussed here. They, and their dihydro derivatives, the diaziridines, are *three-membered ring heterocyclic compounds* with two nitrogen atoms.

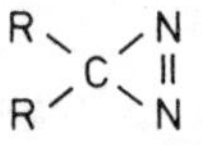

R   N   R   NH
 \   ‖   \ |
  C   C
 /   ‖   / |
R   N   R   NH

Diazirine 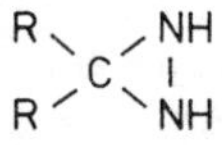   Diaziridine

Preparation

2.17.2.1 The synthesis of diaziridines was achieved by *J. H. Abendroth* and *S. R. Paulsen*— quite independently—by the reaction of ammonia and chlorine with ketones (or aldehydes) in the gas phase, and by *E. Schmitz* by the reaction of *Schiff's* bases with chloramine (or *N*-chloroalkylamines) in inert solvents:

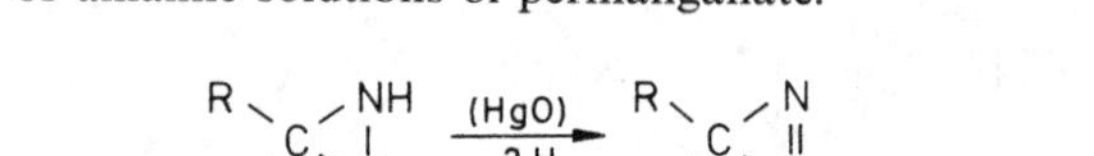

R   R   N—R'
 \   \ |
  C=N—R' + Cl—NH₂ ⟶   C   + HCl
 /   / |
R   R   NH

In the first method a ketimine (or aldimine) is the initial product. This then reacts with chloramine, formed from the ammonia and chlorine, to give the diaziridine.

Diaziridines are moderate reducing agents and can be dehydrogenated to diazirines by mercury(II) oxide or alkaline solutions of permanganate:

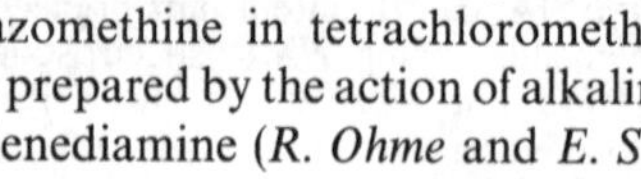

R   NH   (HgO)   R   N
 \ | ⟶   \   ‖
  C   -2 H   C
 / | /   ‖
R   NH   R   N

A dialkyldiazirine

2.17.2.2 Diazirine may be synthesised in a variety of ways. One way is from formaldehyde, chloramine and ammonia, and dehydrogenation of the initially formed diaziridine with dichromate and sulphuric acid (*E. Schmitz*, 1961). Another way is by the reaction of difluoramine with t-butylazomethine in tetrachloromethane (*W. H. Graham*, 1962). In addition, diazirine has been prepared by the action of alkaline hypochlorite or hypobromite solution on salts of methylenediamine (*R. Ohme* and *E. Schmitz*, 1964):

---

[1] See *M. Regitz*, Angew. Chem. Int. Ed. Engl. *6*, 733 (1967).
[2] See *E. Schmitz*, Three-Membered Rings with Two Hetero Atoms, in *A. R. Katritzky*, Adv. Heterocyclic Chem. Vol. 2, 83 *ff.* (Academic Press, New York 1963); *E. Schmitz*, Three-Membered Rings Containing Two Hetero-Atoms, Angew. Chem. Int. Ed. Engl. *3*, 333 (1964).

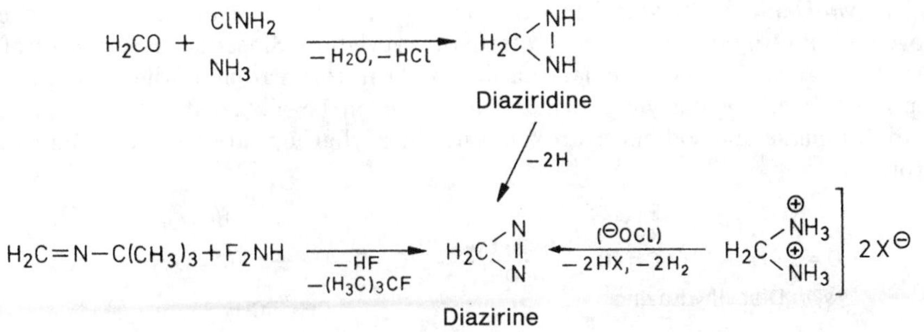

$$H_2CO + \begin{matrix} ClNH_2 \\ NH_3 \end{matrix} \xrightarrow[-H_2O,\,-HCl]{} H_2C\begin{matrix} NH \\ | \\ NH \end{matrix}$$

Diaziridine

$$\downarrow -2H$$

$$H_2C{=}N{-}C(CH_3)_3 + F_2NH \xrightarrow[-(H_3C)_3CF]{-HF} H_2C\begin{matrix} N \\ \| \\ N \end{matrix} \xleftarrow[-2HX,\,-2H_2]{(^{\ominus}OCl)} \left[H_2C\begin{matrix} \overset{\oplus}{N}H_3 \\ \underset{NH_3}{\oplus} \end{matrix} \right] 2\times^{\ominus}$$

Diazirine

Properties

Diazirine is a colourless gas. Its b.p. is $-14°C$ (259 K). It is less reactive than diazomethane, but explodes when heated. In general, diazirines differ markedly from diazoalkanes in their properties.

Curtius had earlier suggested this cyclic formula for diazomethane, and the preparation of diazirine itself thus provided chemical proof that diazomethane had a linear, rather than a cyclic, structure. When exposed to light, diazirine is converted into diazomethane.

2.18 Aliphatic Hydrazines and Azides

2.18.1 Hydrazines

There are three different types of alkyl derivatives of hydrazine:

R–NH–NH$_2$ R–NH–NH–R   R$_2$N–NH$_2$

Monoalkylhydrazine   sym-Dialkylhydrazine   unsym-Dialkylhydrazine

Preparation

2.18.1.1 As happens in the alkylation of ammonia, so in the reaction of alkyl halides or sulphates with hydrazine it is not possible to obtain monoalkyl hydrazines. The main product is the N,N,N-trialkylhydrazinium salt:

$$H_2N{-}NH_2 + 3\,R{-}I \xrightarrow[-2\,HI]{} R_3\overset{\oplus}{N}{-}NH_2]\,I^{\ominus}$$

In order to obtain **monoalkylhydrazines**, it is necessary to prevent further alkylation. To this end benzaldazine is treated with dimethyl sulphate and the resultant quaternary salt is cleaved hydrolytically; monomethylhydrazinium sulphate, benzaldehyde and methanol are the products:

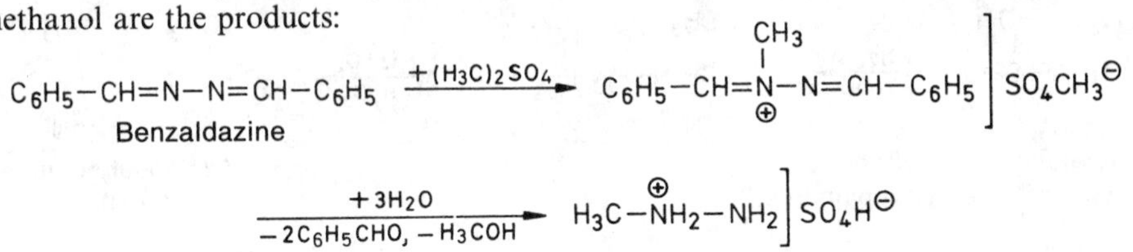

$$C_6H_5{-}CH{=}N{-}N{=}CH{-}C_6H_5 \xrightarrow{+(H_3C)_2SO_4} \left[C_6H_5{-}CH{=}\underset{\oplus}{N}{-}N{=}CH{-}C_6H_5 \right]^{\overset{CH_3}{|}} SO_4CH_3^{\ominus}$$

Benzaldazine

$$\xrightarrow[-2\,C_6H_5CHO,\,-\,H_3COH]{+3\,H_2O} H_3C{-}\overset{\oplus}{N}H_2{-}NH_2]\,SO_4H^{\ominus}$$

2.18.1.2 *sym*-**Dialkylhydrazines** can only be obtained if one hydrogen atom on each nitrogen atom of hydrazine is first protected by acylation. Since the replacement of the hydrogen by an acyl group lowers the basicity of the nitrogen atom to which it is attached, the product is a *sym*-diacylhydrazine. This can then be alkylated with, for example, dimethyl sulphate and sodium hydroxide, and the acyl groups are then removed by acid hydrolysis:

$$O=\overset{\overset{R}{|}}{C}-NH-NH-\overset{\overset{R}{|}}{C}=O \xrightarrow[\text{(NaOH)}]{\text{[(H}_3\text{C)}_2\text{SO}_4\text{]}} O=\overset{\overset{R}{|}}{C}-\overset{\underset{|}{CH_3}}{N}-\overset{\underset{|}{CH_3}}{N}-\overset{\overset{R}{|}}{C}=O$$

sym-Diacylhydrazine

$$\xrightarrow[-2\,\text{RCOOH}]{+2\,\text{H}_2\text{O}(\text{H}^\oplus)} H_3C-NH-NH-CH_3$$

sym-Dimethylhydrazine (b.p. 81°C [354 K])

2.18.1.3 *unsym*-**Dialkylhydrazines** are obtained by the reduction of dialkylnitrosamines with zinc and acetic acid:

$$R_2N-NO \xrightarrow[-H_2O]{+4\,H} R_2N-NH_2$$

Properties

Alkylhydrazines are strongly basic liquids. They are strong reducing agents, for example they reduce *Fehling's* solution. Nitrous acid deaminates unsym-dialkyl hydrazines to give a secondary amine and dinitrogen monoxide:

$$(H_3C)_2N-NH_2 + HNO_2 \rightarrow (H_3C)_2NH + N_2O + H_2O$$

unsym-Dimethylhydrazine is used as a rocket fuel.

sym-Dialkylhydrazines are oxidised by nitrous acid to the corresponding azo-compound, which contains an N=N grouping, e.g.

$$H_3C-NH-NH-CH_3 \xrightarrow[-2H]{\text{(HNO}_2\text{)}} \overset{\underset{H_3C}{\diagup}}{N=N}\overset{\diagup CH_3}{}$$

sym-Dimethylhydrazine

(*E*)-Azomethane

(*E*)-**Azomethane**, (*E*)-dimethyldiazene, is a yellow explosive gas, b.p. 1.5°C (274.7 K). At 60–100°C (335–375 K) it dissociates into nitrogen and methyl radicals, and hence can serve as a radical generator.

α,α'-**Azoisobutyronitrile** is used as an initiator for polymerisation processes and as a foaming agent in the manufacture of foam rubbers and other foam materials. It is made by oxidising the hydrazine formed from acetone cyanohydrin and hydrazine:

$$2\,H_3C-\overset{\overset{CH_3}{|}}{\underset{\underset{CN}{|}}{C}}-OH \xrightarrow[-2H_2O]{+N_2H_4} H_3C-\overset{\overset{CH_3}{|}}{\underset{\underset{CN}{|}}{C}}-NH-NH-\overset{\overset{CH_3}{|}}{\underset{\underset{CN}{|}}{C}}-CH_3 \xrightarrow[-H_2O]{+1/2\,O_2} H_3C-\overset{\overset{CH_3}{|}}{\underset{\underset{CN}{|}}{C}}-N=N-\overset{\overset{CH_3}{|}}{\underset{\underset{CN}{|}}{C}}-CH_3$$

Acetone cyanohydrin
2-Hydroxy-2-methylpropionitrile

α,α'-Azoisobutyronitrile
(AIBN)

1,1-Dimethylhydrazine induces cancers in humans. *1,2-Diethylhydrazine* and *azomethane* have been shown to be powerful carcinogens in rats.

A new class of tumor-inhibiting compounds are derivatives of *2-benzyl-1-methylhydrazine*, $H_3C-NH-NH-CH_2-C_6H_5$, with substituents in the p-position of the benzene ring. Among these *cytostatics* is *N*-isopropyl-α-(2-methyl-hydrazino)-*p*-toluamide hydrochloride (INN: **Procarbiazin**, *Natulan*).

$$\begin{array}{c} H_3C \\ \! \diagdown \\ H_3C \diagup \end{array} CH-NH-CO-\!\!\left\langle\bigcirc\right\rangle\!\!-CH_2-NH-NH-CH_3 \cdot HCl$$

2.18.2 Azides[1]

Azides are derivatives of hydrazoic acid. The simplest example is **methyl azide**, b.p. 20°C (293 K). It can be made by the reaction of sodium azide with dimethyl sulphate:

$$NaN_3 + (H_3CO)_2SO_2 \longrightarrow H_3C-N_3 + (H_3CO)SO_3Na$$
Methyl azide

Physical measurements show that the azide group has a linear structure. It is represented by the following limiting structures:

$$[R-\overset{\oplus}{\bar{N}}=N=\overset{\ominus}{\bar{N}} \longleftrightarrow R-\overset{\ominus}{\bar{N}}-N\equiv NI \longleftrightarrow R-\overset{\ominus}{\bar{N}}-N=\overset{\oplus}{N}I]$$

Alkyl azides detonate when heated rapidly. When heated slowly, they decompose to give an alkene and hydrazoic acid, e.g.

$$2\,H_3C-N_3 \longrightarrow H_2C=CH_2 + 2\,HN_3$$

2.19 Organic Compounds Containing some other Non-Metals

2.19.1 Organophosphorus Compounds[2]

These compounds are derivatives of phosphine, which, unlike ammonia, is scarcely basic; this is illustrated by the ease with which phosphonium iodide is hydrolysed to phosphine (PH_3) and hydrogen iodide in cold water. Introduction of alkyl groups into the molecule increases the basicity, and there are *primary*, *secondary* and *tertiary phosphines*[3] as well as *quaternary phosphonium bases*, e.g.

---

[1] See *S. Patai* (Ed.), The Chemistry of the Azido Group (Intersci. Publ., New York 1971).
[2] See *Houben-Weyl*, 4th edn, Vols. *12/1, 12/2, E1* (Thieme Verlag, Stuttgart 1963, 1964, 1982); *J. I. G. Cadogan* (Ed.) Organophosphorus Reagents in Organic Synthesis (Academic Press, London 1979).
[3] The name *phosphane*, which is incorrect, is also sometimes found in the literature.

H_3CPH_2	$(H_3C)_2PH$	$(H_3C)_3P$	$(H_3C)_4\overset{\oplus}{P}]OH^{\ominus}$
Methylphosphine	Dimethylphosphine	Trimethylphosphine	Tetramethylphos- phonium Hydroxide

Preparation

2.19.1.1 Aliphatic phosphines can be made analogously to amines, by the *reaction of alkyl halides with phosphine*, but the primary and secondary phosphines which are formed first react further so rapidly that only tertiary phosphines and quarternary phosphonium salts are isolable:

$$\bar{P}H_3 \xrightarrow[-HI]{+R-I} R-\bar{P}H_2 \xrightarrow[-HI]{+R-I} R_2\bar{P}H \xrightarrow[-HI]{+R-I} R_3P| \xrightarrow{+R-I} R_4P]^{\oplus}I^{\ominus}$$

Primary and secondary alkyl phosphines are obtained by treating phosphonium iodide with alkyl halides at 150°C (425 K) in the presence of zinc oxide to remove hydriodic acid, e.g.

$$PH_4]^{\oplus}I^{\ominus} + H_3C-I + ZnO \rightarrow H_3C-PH_2 + ZnI_2 + H_2O$$

$$2 H_3C-PH_2 + 2 H_3C-I + ZnO \rightarrow 2 (H_3C)_2PH + ZnI_2 + H_2O$$

2.19.1.2 Tertiary phosphines are prepared by the reaction of alkyl magnesium salts (or dialkyl zinc compounds) with phosphorus(III) chloride:

$$PCl_3 + 3 C_2H_5MgCl \rightarrow (C_2H_5)_3P + 3 MgCl_2$$

Triethylphosphine

This method is generally applicable to the synthesis of other organic derivatives of non-metals and of many metals.

In contrast to amines, tertiary phosphines with three different groups attached to phosphorus, $R^1R^2R^3P$, have stable configurations and do not invert. They can be separated into their enantiomers (*Horner* 1961). For a useful application of this property see p. 382.

2.19.1.3 To obtain *aryl* **tertiary phosphines** the reaction between aryl halides and phosphorus(III) chloride in the presence of sodium (the modified *Wurtz–Fittig* synthesis) is used, *e.g.*

$$3 C_6H_5Cl + PCl_3 + 6 Na \rightarrow (C_6H_5)_3P + 6 NaCl$$

Chlorobenzene   Triphenylphosphine (m.p. 79°C [352 K])

The same method can be used to make triaryl *arsines* and *stibines* such as triphenylstibine, $(C_6H_5)_3Sb$.

Properties

Apart from methylphosphine, which is a gas, the alkyl phosphines are liquids or low melting solids, and are insoluble in water but soluble in organic solvents. They have unpleasant smells and are strongly toxic. Particularly notable is their tendency to react with atmospheric oxygen, which results in their catching fire when exposed to air.

2.19.1.4 Derivatives of Phosphine

Controlled oxidation of primary, secondary or tertiary alkylphosphines with air or with nitric acid produces alkylphosphonic acids, dialkylphosphinic acids or trialkylphosphine oxides:

$$R-PH_2 \xrightarrow{\;+3/2\,O_2\;} R-P{\overset{\displaystyle O}{\underset{OH}{\diagdown}}}OH \qquad \text{Alkylphosphonic acids}$$

$$R_2PH \xrightarrow{\;+O_2\;} R_2P{\overset{\displaystyle O}{\underset{OH}{\diagdown}}} \qquad \text{Dialkylphosphinic acids}$$

$$R_3P \xrightarrow{\;+1/2\,O_2\;} R_3P{=}O \qquad \text{Trialkylphosphine oxides}$$

Esters of *alkylphosphonic acids* are readily obtainable from trialkylphosphites by means of the *Michaelis–Arbusov* reaction:[1]

$$|P(OR)_3 \xrightarrow{\;R'X\;} \left[R'-P{\overset{\displaystyle O-R}{\underset{OR}{\diagdown}}}OR \right]^{\oplus} X^{\ominus} \xrightarrow{\;-RX\;} R'-P{\overset{\displaystyle O}{\underset{OR}{\diagdown}}}OR$$

This is a versatile reaction and provides a good method for the formation of carbon-phosphorus bonds.

(a) Phosphine oxides are colourless, stable, crystalline substances. They differ from amine oxides in that the oxygen atom is much more strongly bound, and cannot be removed by standard reducing agents. This is because in phosphine oxides there is no semipolar bonding as in amine oxides, but rather a polar P=O double bond. Phosphorus, and also the lower members of Group 15 of the Periodic Table, can, by making use of higher orbitals, form a decet of electrons, thereby becoming pentavalent.

Wittig provided unequivocal evidence for the existence of 5-coordinate phosphorus in organic chemistry by preparing *pentaphenylphosphorane* (pentaphenylphosphorus) by the reaction of tetraphenylphosphonium iodide with phenyl lithium.

$$\left[C_6H_5-\overset{\overset{\displaystyle C_6H_5}{|\oplus}}{\underset{\underset{\displaystyle C_6H_5}{|}}{P}}-C_6H_5 \right] I^{\ominus} + LiC_6H_5 \longrightarrow \underset{C_6H_5}{\overset{C_6H_5}{C_6H_5}}\overset{\overset{\displaystyle C_6H_5}{|}}{P}\overset{C_6H_5}{\underset{C_6H_5}{}} + LiI$$

Pentaphenylphosphorane

This compound melts at 121°C (394 K) and shows no salt-like character. Its dipole moment is zero. All its properties show that all the five phenyl groups are bound to the phosphorus atom in exactly the same way.

Similar pentaphenyl derivatives of arsenic, antimony and bismuth have been made in the same way (*Wittig*).

(b) Alkylidenephosphoranes (phosphonium ylides, phosphine-alkylenes). When tetramethylphosphonium bromide is treated with methyl lithium in ether or tetrahydrofuran, instead of forming pentamethylphosphorane, methane is lost and the product is *trimethylmethylenephosphorane* (trimethylphosphonium methylide):

$$\left[(H_3C)_3\overset{\oplus}{P}-CH_3\right] Br^{\ominus} \xrightarrow[-LiBr,\ -CH_4]{+LiCH_3} \left[(H_3C)_3\overset{\oplus}{P}-\overset{\ominus}{C}H_2 \longleftrightarrow (H_3C)_3P{=}CH_2\right]$$

Phosphonium ylide Phosphorylene
Trimethyl-
methylenephosphorane

---

[1] See *A. K. Bhattacharya* et al., The *Michaelis–Arbusov* Rearrangement, Chem. Rev. *81*, 415 (1981).

Phosphonium ylides can be regarded as *carbene complexes* in which the carbene acts as σ-donor and π-acceptor (see p. 376 f.).

The reaction proceeds in the same way as that involved in the formation of an ammonium ylide, but the resultant product in this case has not the markedly ionic bonding associated with nitrogen ylides (see p. 163 ff.). The P—C bonding has more essentially double-bond character and can be considered as a polar P=C double-bond. The properties are best understood by considering the compounds to be mesomeric between the ylide and ylene structures.

If, instead of tetramethylphosphonium bromide, methyltriphenylphosphonium bromide reacts with organo-lithium compounds such as phenyl lithium, butyl lithium or methyl lithium, loss of the related hydrocarbon and formation of methylenetriphenylphosphorane (triphenylphosphonium methylide) results, e.g.

$$(C_6H_5)_3 \overset{\oplus}{P}-CH_3] \; Br^{\ominus} \xrightarrow[-LiBr,-C_6H_6]{+ LiC_6H_5} (C_6H_5)_3P = CH_2$$

Methylenetriphenylphosphorane

Wittig showed that this family of compounds is valuable for the preparation of *alkenes*, which are formed when they react with carbonyl compounds. The alkene double bond forms between the methylene group and the carbon atom of the carbonyl group.

(c) *Wittig* Reaction[1] (1949). Reaction of methylenetriphenylphosphorane with ketones, *e.g.* benzophenone, in alkaline solution at room temperature, results in the formation of *1,1-diphenylethylene* and triphenyl-phosphine oxide:

$$(C_6H_5)_3P = CH_2 + O = C(C_6H_5)_2 \rightarrow (C_6H_5)_3P = O + H_2C = C(C_6H_5)_2$$

| Methylene | Benzophenone | 1,1-Diphenylethylene |
| triphenylphosphorane | | |

If *allylenephosphoranes*, e.g. 2-propenylidenetriphenylphosphorane, are used in the *Wittig* reaction butadiene derivatives are formed, *e.g.*

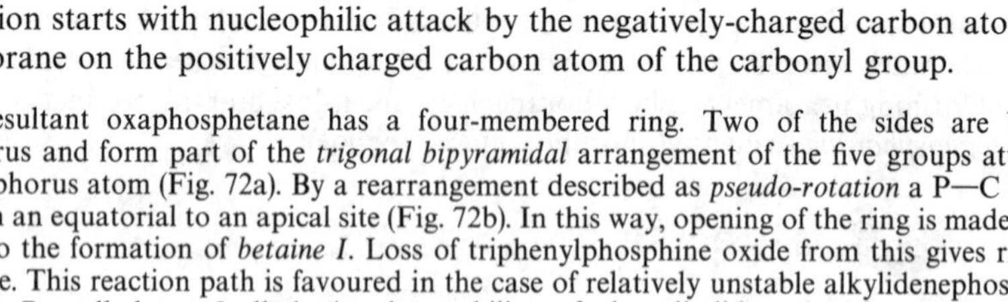

$$H_2C{=}CH{-}CH_2Cl \quad + \quad P(C_6H_5)_3 \rightarrow H_2C{=}CH{-}CH_2\overset{\oplus}{P}(C_6H_5)_3Cl^{\ominus}$$

Allyl chloride

$$\Big\downarrow LiOC_2H_5$$

$$(C_6H_5)_3PO + H_2C{=}CH{-}CH{=}CH{-}C_6H_5 \xleftarrow{C_6H_5CHO} \big[H_2C{=}CH{-}CH{=}P(C_6H_5)_3\big]$$

1-Phenylbutadiene   2-Propenylidene-
triphenylphosphorane

Reaction starts with nucleophilic attack by the negatively-charged carbon atom of the phosphorane on the positively charged carbon atom of the carbonyl group.

The resultant oxaphosphetane has a four-membered ring. Two of the sides are bonds to phosphorus and form part of the *trigonal bipyramidal* arrangement of the five groups attached to the phosphorus atom (Fig. 72a). By a rearrangement described as *pseudo-rotation* a P—C bond can shift from an equatorial to an apical site (Fig. 72b). In this way, opening of the ring is made possible, leading to the formation of *betaine I*. Loss of triphenylphosphine oxide from this gives rise to the (Z)-alkene. This reaction path is favoured in the case of relatively unstable alkylidenephosphoranes in which R = alkyl or O-alkyl. As the stability of the alkylidenephosphorane is increased (R = halogen, aryl, —CHO, —CN), so the life-time of betaine I increases, because the carbanion is stabilised by the adjacent electron-withdrawing groups. In consequence, betaine I has time to

---

[1] See A. Maercker, The *Wittig* Reaction, Org. Reactions *14*, 270 *ff*. (1965); M. Schlosser, The Stereochemistry of the Wittig Reaction, Topics in Stereochemistry *5*, 1 (1972).

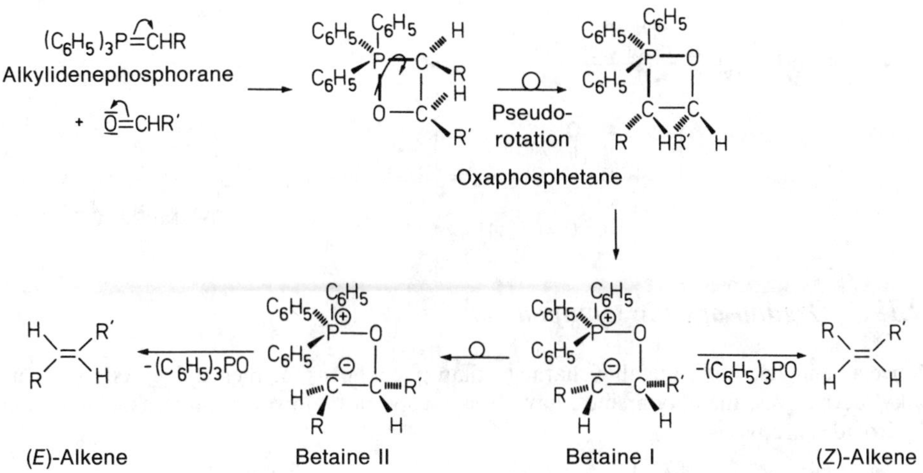

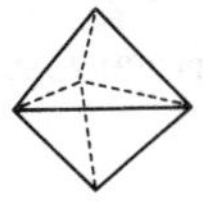

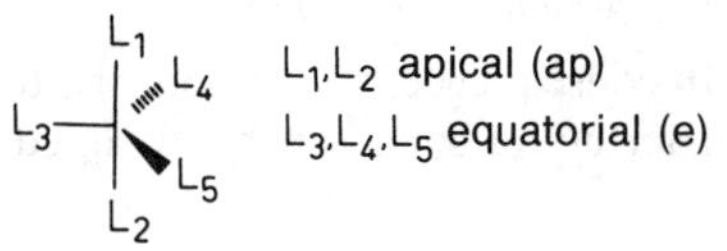

$(C_6H_5)_3P\overset{\frown}{=}CHR$

Alkylidenephosphorane

$+ \ \overset{\frown}{\underline{O}}=CHR'$

Pseudo-
rotation

Oxaphosphetane

(E)-Alkene $\xleftarrow{-(C_6H_5)_3PO}$ Betaine II $\xleftarrow{\ \ O\ \ }$ Betaine I $\xrightarrow{-(C_6H_5)_3PO}$ (Z)-Alkene

L_1,L_2 apical (ap)

L_3,L_4,L_5 equatorial (e)

Fig. 72a. Trigonal bipyramid.

Fig. 72b. Ligands attached to the 5-coordinate centre of a trigonal bipyramid.

rearrange to the thermodynamically more stable *betaine II* from which an (E)-alkene is formed. The stereochemistry of the *Wittig* reaction is thus dependent upon the identity of the substituents; it is also dependent upon the reaction conditions (solvent, presence of salts).[1]

The *Wittig* reaction can also take place with the carbonyl group of activated amides, and with other double bonds such as C=N and C=S. In the chemistry of natural products, this reaction was used in the syntheses of various carotenoids and also of vitamins A and D.[2] At present methylenephosphoranes play a vital role in organic chemistry in the synthesis of a wide variety of different compounds.[3]

(d) *Horner-Emmons* Reaction (1958).[4] Another way of obtaining alkenes from carbonyl compounds is based on their reaction with phosphonate esters or phosphine oxides which have an acidic CH group next to the phosphorus atom (at the α-position). They are more reactive than comparable alkylenephosphoranes. This is sometimes called *PO-activated* formation of alkenes.

If, for example, diethyl benzyl phosphonate reacts with benzophenone, in 1,2-dimethoxyethane as solvent and in the presence of sodium hydride, then triphenylethylene is formed together with the sodium salt of the diethyl ester of phosphoric acid.

[1] See H. J. Bestmann et al., Selected Topics of the *Wittig* reaction in the Synthesis of Natural Products, Topics in Current Chemistry *109*, 85 (1983).
[2] See H. Pommer, Synthesen in der Vitamin-A-Reihe (Syntheses in the Vitamin A Series), Angew. Chem. *72*, 811 (1960); Synthesen in der Carotinoid-Reihe (Syntheses in the Carotenoid Series); *ibid*, *72*, 911 (1960).
[3] See H. J. Bestmann et al., *Houben-Weyl*, 4th edn, Vol. *5/1b*, 383 (Thieme Verlag, Stuttgart 1972).
[4] See W. S. Wadsworth, Synthetic Applications of Phosphoryl-Stabilised Anions, Org. Reactions *25*, 73 (1977).

$$(C_2H_5O)_2\overset{\overset{|\ddot{O}|}{\|}}{P}-CH_2-C_6H_5 \xrightarrow[-H_2]{\overset{+NaH}{(H_3CO-CH_2-CH_2-OCH_3)}}$$

$$\left[(C_2H_5O)_2\overset{\overset{O}{\|}}{P}-\overset{\ominus}{\underset{\underset{\ddot{O}=C(C_6H_5)_2}{|}}{C}}H-C_6H_5\right] Na^\oplus \longrightarrow (C_2H_5O)_2\overset{\overset{O}{\|}}{\underset{\underset{O^\ominus\ Na^\oplus}{|}}{P}} + \overset{\overset{O}{\|}}{\underset{C(C_6H_5)_2}{HC}}-C_6H_5$$

2.19.2 Organo-arsenic Compounds[1]

Since arsenic has more metallic character than phosphorus, neither arsine, AsH_3, nor its alkyl derivatives, the alkylarsines, have basic properties. However quaternary arsonium hydroxides are bases.

$RAsH_2$	R_2AsH	R_3As	$[R_4\overset{\oplus}{As}]\,OH^\ominus$
primary	secondary	tertiary	quaternary Arsonium hydroxide
Arsine	Arsine	Arsine	

In consequence, and in contrast to amines and phosphines, neither primary, secondary nor tertiary arsines can be obtained by alkylation of arsine.

Preparation

2.19.2.1. Primary arsines are obtained by reduction of monoalkyl- (or aryl-) arsonic acids, which are themselves made by alkylation or arylation of arsenic acids:

$$H_3AsO_3 \xrightarrow[-HI]{+R-I} R-AsO_3H_2 \xrightarrow[-3H_2O]{+6H} R-AsH_2$$

2.19.2.2 Secondary arsines are obtained by reduction of dialkyl- or diaryl-arsenic chlorides, e.g.

$$(H_3C)_2As-Cl + 2H \longrightarrow (H_3C)_2AsH + HCl$$

Dimethylarsenic chloride   Dimethylarsine

2.19.2.3 Tertiary arsines are made in the same way as trialkylphosphines by treating arsenic(III) chloride with Grignard reagents, e.g.

$$AsCl_3 + 3H_3CMgCl \longrightarrow (H_3C)_3As + 3MgCl_2$$

Properties and Uses

Methylarsine is a gas, other arsines are highly volatile poisonous liquids which are insoluble in water; they have garlic-like odours. They are strong reducing agents. Tertiary arsines react with atmospheric oxygen to give arsine oxides, R_3AsO, in which the arsenic–oxygen bond is more polar than the P=O bond in phosphine oxides and is a hybrid of dipolar and double-bonded forms $[R_3As{=}O \leftrightarrow R_3As^\oplus{-}O^\ominus]$. Tertiary arsines react with alkyl halides to give *quaternary arsonium* salts.

---

[1] See *Houben-Weyl*, 4th edn, Vol. *13/8* (Thieme Verlag, Stuttgart 1978).

Alkylidene- (or arylidene-)arsoranes (*Arsonium Ylides*).[1] Arsonium ylides closely resemble phosphonium ylides (p. 173) in their physical and chemical properties, but they are more reactive in *Wittig*-type reactions.

Cacodyl reaction. The first known organo-arsenic compound was cacodyl oxide, which was made by *Cadet* (1760) by heating arsenic(III) oxide with potassium acetate. It is a liquid, b.p. 150°C (423 K) with a disagreeable odour and it ignites spontaneously in air. It was prepared in a pure form by *Bunsen*:

$$As_2O_3 + 4\,H_3CCOOH \longrightarrow \quad \begin{array}{c} H_3C \\ \diagdown \\ As-O-As \\ \diagup \\ H_3C \end{array} \begin{array}{c} CH_3 \\ \diagup \\ \diagdown \\ CH_3 \end{array} \quad +\,4\,CO_2\,+\,2\,H_2O$$

Cacodyl oxide

Only brief mention of stibines and bismuthines, derivatives, respectively, of SbH_3 and BiH_3, is needed. Because of the increased metallic character of the elements the stability of their alkyl derivatives is lower.

2.19.3 Organosilicon Compounds

Alkylsilanes

Since silicon comes directly underneath carbon in the Periodic Table, alkylsilanes, which are derived from the simplest silicon hydride, silane, SiH_4, by replacing one or more of the hydrogen atoms by alkyl groups, have properties similar to those of the related hydrocarbons. There are mono-, di-, tri- and tetra-alkylsilanes, *viz.*

$$SiH_3R \qquad SiH_2R_2 \qquad SiHR_3 \qquad SiR_4$$

Most alkylsilanes are colourless oils, which are sensitive to air and moisture. The tetraalkylsilanes are relatively stable to atmospheric oxygen and to acids and bases.

Preparation

2.19.3.1 *By treatment of tetrachlorosilane with dialkyl zinc derivatives* (*Friedel* and *Crafts*, 1863) e.g.

$$SiCl_4 + 2\,Zn(CH_3)_2 \longrightarrow Si(CH_3)_4 + 2\,ZnCl_2$$

Tetramethylsilane

2.19.3.2 *By the reaction between Grignard reagents and tetrachlorosilane* (*Kipping*, 1904), e.g.

$$SiCl_4 + 4\,C_2H_5MgI \longrightarrow Si(C_2H_5)_4 + 4\,MgCl\,I$$

Tetraethylsilane

2.19.3.3 *By heating tetrachlorosilane with an alkyl halide in the presence of sodium* (modified *Wurtz* reaction), e.g.

$$SiCl_4 + 4\,C_2H_5Br + 8\,Na \longrightarrow Si(C_2H_5)_4 + 4\,NaCl + 4\,NaBr$$

The same method can be used for the preparation of arylsilanes.

---

[1] See D. Lloyd, I. Gosney and R. A. Ormiston, Arsonium Ylides (with some mention also of Arsinimines, Stibonium and Bismuthonium Ylides) Chem. Soc. Rev. *16*, 45 (1987); D. *Lloyd* and I. *Gosney*, Arsonium, Stibonium and Bismuthonium, Ylides and Imines. The Chemistry of Organic Arsenic, Antimony and Bismuth Compounds, ed. S. *Patai* (Wiley, Chichester 1994).

Tetramethylsilane (TMS) is used in NMR spectroscopy as the reference compound to provide a signal to which other signals can be referred. It is a very volatile liquid; a small amount is added to the solution of the substance being tested in, for example, deuteriochloroform or carbon tetrachloride as solvent.

Alkylhalogenosilanes

Alkylhalogenosilanes are important industrially as the materials from which *silicones* are prepared.

Preparation

2.19.3.4 *By partial substitution of the halogen atoms in tetrachlorosilane by means of Grignard reagents, e.g.*

$$SiCl_4 + 2 C_2H_5MgCl \rightarrow (C_2H_5)_2SiCl_2 + 2 MgCl_2$$

Dichlorodiethylsilane

2.19.3.5 *By the action of methyl chloride in* the gas-phase at 350°C (625 K) on *powdered silicon in the presence of a copper catalyst* (*Müller–Rochow* synthesis):

$$2 H_3CCl + Si \xrightarrow{(Cu)} (H_3C)_2SiCl_2$$

Dichlorodimethylsilane

The reaction does not give just the one product, but trichloromethylsilane, H_3CSiCl_3 and chlorotrimethylsilane, $(H_3C)_3SiCl$, are by-products formed by *disproportionation* of the dichlorodimethylsilane.

In practice this method is used only with methyl chloride. Higher silanes are made by using *Grignard* reagents.

Properties and Uses

The trimethylsilyl group can activate an acidic α-CH group:

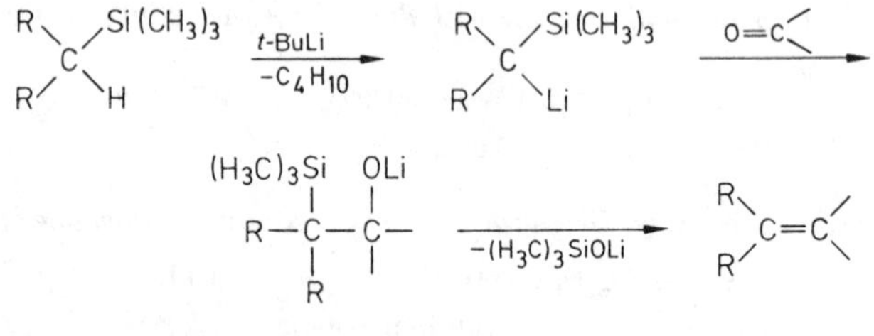

The *Peterson* reaction which is formulated here provides an alternative to the *Wittig* and *Horner–Emmons* reactions. Its mechanism is analogous to that of these two reactions. It has proved useful for the preparation of alkenes which are difficult to prepare in other ways.

Organosilicon compounds have found an increasing use in synthetic chemistry[1].

If dialkyldichlorosilanes are hydrolysed using a large amount of water or a mixture of water and an organic solvent, an exothermic reaction takes place and silanediols are formed. The hydrochloric acid which is also formed promotes the rapid polycondensation

---

[1] See E. *Colvin*, Silicon in Organic Synthesis (Butterworths, London 1981, 1988).

of these diols to form high molecular mass plastics, which are called *poly(dialkyl-siloxanes)*, or, in shorthand, *silicones*:[1]

$$n\text{Cl} - \underset{\underset{R}{|}}{\overset{\overset{R}{|}}{\text{Si}}} - \text{Cl} \xrightarrow[-2n\text{HCl}]{+2n\text{H}_2\text{O}} n\text{HO} - \underset{\underset{R}{|}}{\overset{\overset{R}{|}}{\text{Si}}} - \text{OH} \xrightarrow[-n\text{H}_2\text{O}]{} \left[\text{O} - \underset{\underset{R}{|}}{\overset{\overset{R}{|}}{\text{Si}}} \right]_n$$

Dialkyldichlorosilane	Silanediol	Poly(dialkylsiloxane) (Alkylsilicone)

Industrially these polycondensations are carried out above all using dialkyldichlorosilanes ($R=CH_3$, C_2H_5, etc.) but alkyltrichlorosilanes and trialkylchlorosilanes can also be hydrolysed and condensed.

These polycondensation products are entirely based on Si—O—Si bonding. According to the reaction conditions employed, alkysilicones of higher or lower molecular mass can be produced.

The special properties of the silicones derive from their chemical structure, which is partly inorganic and partly organic. Their stability to both heat and cold, and their marked chemical unreactivity, can be attributed to the inorganic contribution and may be compared to the properties of silica. On the other hand, the plastic-like properties of silicones, i.e. their oily or plastic or resinous nature, and above all their water-repellent character, are derived from the organic contribution. The extremely heat-resistant *silicone oils* are linear or branched or cyclic methylsilicones of lower molecular mass, and are used as heating fluids in high-temperature baths and in diffusion pumps and also in pharmacy as protective materials for the skin. *Silicone rubbers* (*siloxane elastomers*) consist of high molecular mass silicones, for the most part with linear structures. On the other hand, the more or less solid *silicone resins* are made up from three-dimensionally cross-linked silicones. Both of the latter kinds of silicones are of particular industrial importance as electrical insulators. By varying the substituents, for example by partial replacement of the methyl groups by phenyl groups, *methylphenylsilicone oils* are obtained, which are lubricants for use at high temperatures or in high vacuum systems. In the pharmaceutical industry, silicones are used in the preparation of implantable carriers for medicaments.

Among organo-silicon compounds there are some which are highly toxic and produce surprising physiological reactions (*Woronkow, Wannagat*).

2.19.4 Organoboron Compounds

Of such compounds only the *alkylboranes* (*boron trialkyls*) and *arylboranes* (*boron triaryls*) are discussed here.

Preparation

2.19.4.1 *By means of the reaction between boron trifluoride and alkyl or aryl Grignard reagents (E. Krause), e.g.*

$$BF_3 + 3\,C_2H_5MgCl \longrightarrow B(C_2H_5)_3 + 3\,MgFCl$$
$$\text{Triethylborane (b.p. 95°C [368 K])}$$

$$BF_3 + 3\,C_6H_5MgBr \longrightarrow B(C_6H_5)_3 + 3\,MgFBr$$
$$\text{Triphenylborane}$$

---

[1] See *W. Noll*, Chemie und Technologie der Silicone, 2nd edn. (Verlag Chemie, Weinheim 1968).

2.19.4.2 *Hydroboration of alkenes* (H. C. *Brown*[1]). In this process trialkylboranes are obtained by addition of diborane, B_2H_6, best prepared from boron trifluoride and sodium borohydride, to an alkene dissolved in tetrahydrofuran, ether or diethyleneglycol dimethyl ether (diglyme):

$$4\ BF_3 + 3\ Na[BH_4] \longrightarrow 2\ B_2H_6 + 3\ Na[BF_4]$$

$$6\ R{-}CH{=}CH_2 + B_2H_6 \longrightarrow 2\ (R{-}CH_2{-}CH_2)_3B$$

Alkene   Diborane   Trialkylborane

Tetrahydrofuran-
trihydroborane

Monomeric BH_3, borane or trihydroborane, is a *Lewis* acid and with *Lewis* bases forms relatively stable adducts: base$^\oplus BH_3^\ominus$, the tetrahydrofuran-trihydroborane formulated above is an example. For reasons of safety it is preferable to use such adducts rather than free diborane.

Three steps are involved in the hydroboration process, but they normally take place so rapidly that only the trialkylborane is isolated:

$$2\ R{-}CH{=}CH_2 \xrightarrow{\ +\,B_2H_6\ } 2\ R{-}CH_2{-}CH_2{-}BH_2 \xrightarrow{\ +\,2\,R{-}CH{=}CH_2\ } 2\ (R{-}CH_2{-}CH_2)_2BH$$

$$\xrightarrow{\ +\,2\,R{-}CH{=}CH_2\ } 2\ (R{-}CH_2{-}CH_2)_3B$$

In the case of unsymmetric alkenes, reaction proceeds in an anti-*Markovnikov* manner, since the boron atom acts as the electrophile. For example, propene reacts regioselectively to give tri-n-propylborane. Addition of borane to *α-pinene* provides (+)-di-3-pinanylborane [(+)-diisopino-camphenylborane, Ipc_2BH (see p. 669)], which is used as a *chiral* reagent in enantioselective synthesis.[2]

Steric effects can increase the *regio-selectivity of hydroboration* reactions; *boranes with bulky substituent groups* but retaining a B—H bond are used for this purpose, for example 2,3-dimethylbut-2-ylborane (thexylborane), bis-(3-methylbut-2-yl)borane (diisoamylborane) or *9-borabicyclo-[3.3.1]nonane (9-BBN)*, which is obtained from cycloocta-1,5-diene and diborane:

2,3-Dimethylbut-
2-ylborane
(thexylborane)   Diisoamylborane   Cycloocta-1,5-diene   9-BBN

The high regioselectivity of these reagents is illustrated by the following example of the *hydroboration of (Z)-4-methylpent-2-ene*. The arrows indicate the sites of attack of the hydroboration reagents.

43% 57%   0.1% 99.9%
B_2H_6 9-BBN

Properties and Uses

With the exception of trimethylborane, which is a gas, b.p. $-21.8°C$ (251.4 K), trialkylboranes are colourless liquids, with an odour reminiscent of onions. When exposed to

---

[1] See H. C. *Brown*, Organoborane Compounds in Organic Synthesis, in G. *Wilkinson* et al. (Ed.), Comprehensive Organometallic Chemistry (Pergamon Press, Oxford 1982).
[2] See H. C. *Brown*, Asymmetric Synthesis made easy, Chemtracts — Org. Chem. *1*, 77 (1988).

atmospheric oxygen they are very rapidly oxidised and ignite spontaneously. As a consequence trialkylboranes are generally not isolated, but used *in situ*.

Protolysis of trialkylboranes, leading to the formation of *alkanes*, takes place when they are heated with dry carboxylic acids, *e.g.*

$$(H_3C-CH_2-CH_2)_3B + 3\,R-COOH \longrightarrow 3\,H_3C-CH_2-CH_3 + B(O-CO-R)_3$$

Tri-n-propylborane   Propane

Alkaline hydrogen peroxide converts trialkylboranes into the corresponding alcohols, *e.g.*

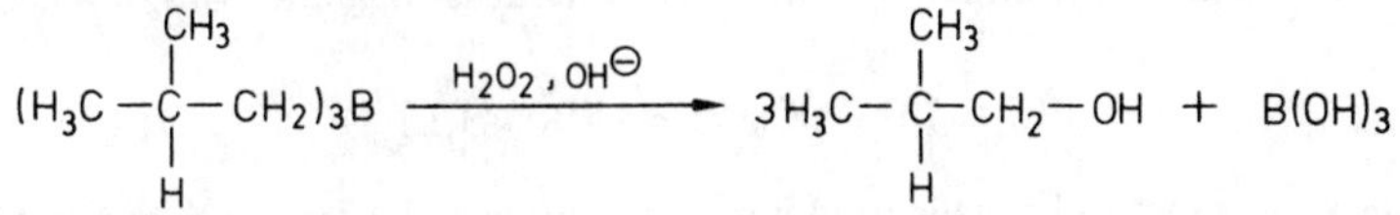

The combination of hydroboration followed by oxidation may be regarded formally as an anti-*Markovnikov* addition of water to isobutene. The whole process may be summed up by the following scheme:

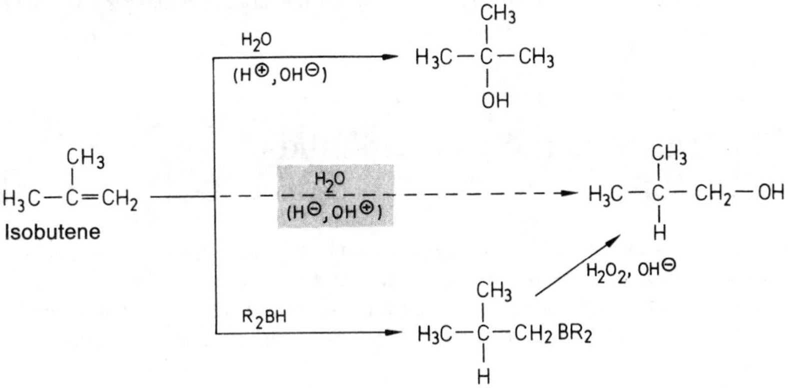

This two-step hydroxylation, *via* hydroboration, allows water to be added to isobutene in such a way that the product is what would be obtained if the alkene were to be attacked by the hypothetical $HO^\oplus$ species. This involves an inversion of the normal polarity of water ($HO^\ominus H^\oplus$). This concept of reversal of polarity is usually described by the German term for it, *umpolung*. The present example provides a simple example of umpolung, which was formulated as a new type of reaction concept by *Corey* and by *Seebach*; further examples will be provided when ketones are considered.

The formation of the products obtained in hydroboration can be explained mechanistically by a one-step concerted reaction, whose transition state takes the form of a four-membered ring (i.e. a four-centre process):

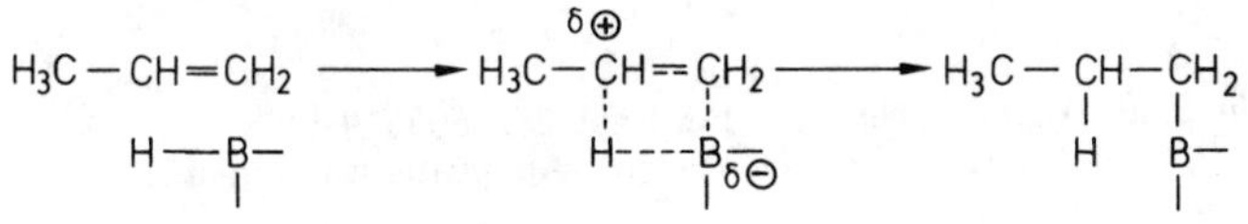

In the oxidation step the HO group goes to the *same site* as was previously occupied by the boron. As a result, *hydroboration* followed by *oxidation* converts, for example, *1-methylcyclopentene* into (*E*)-*2-methylcyclopentanol*:

1-Methylcyclopentene	(*E*)-2-Methylcyclopentanol

A compound of further interest is triphenylborane, m.p. 151°C (424 K), which is easily oxidised and fumes in air. With phenyl lithium it forms *lithium tetraphenylborate* (*Wittig*), which is a stable compound:

$$(C_6H_5)_3B + LiC_6H_5 \rightarrow Li[B(C_6H_5)_4]$$

This complex salt can also be made from boron trifluoride and 4 molecules of phenyl lithium in ethereal solution.

When tris(pentafluorophenyl)borane and xenon difluoride are mixed at low temperature and treated with acetonitrile the crystallisable compound $[CH_3CN\text{---}Xe\text{---}C_6F_5]^{\oplus}[(C_6F_5)_2BF_2]^{\ominus}$ is obtained. The presence of xenon–carbon bonding in the acetonitrile(pentafluorophenyl)xenon(II) cation was shown by an X-ray structure determination on this organic noble-gas compound.[1]

2.20 Organometallic Compounds

In organometallic compounds, both alkyl and aryl, the metal atom concerned is linked directly to one or more carbon atoms. According to the kind of bonding involved in this, these compounds are classified into two types: those with *metal to carbon σ-bonding* and those in which *π-complexes* are formed *between the metal and carbon–carbon multiple bonds*.

In the latter case the metal atom interacts with the π-electrons of an unsaturated organic molecule. While a variety of metals take part in the first type of bonding, only the transition metals participate in the formation of π-complexes. Such metal complexes will be considered at a later stage (see pp. 380 *f.*, 648 *f.*).

Because of the different electronegativities of the two atoms involved in metal–carbon σ-bonds, the bonding acquires a partial ionic character, which affects the reactivity of the particular organometallic compound. Among the most reactive are those compounds involving the alkali metals. The different reactivities of the various organometallic compounds makes them of great importance in organic synthetic work.[2] The extensive treatment available in the relevant 15 parts of *Houben-Weyl* should be mentioned.[3]

In the following section those organometallic compounds whose properties are principally determined by the metal to carbon σ-bonding are considered.

---

[1] See *H. J. Frohn* et al., Angew. Chem. Int. Ed. Engl. *28*, 1506 (1989).
[2] See *S. G. Davies*, Organotransition Metal Chemistry: Applications to Organic Synthesis (Pergamon Press, Oxford 1982).
[3] See *Houben-Weyl*, Methoden der Organischen Chemie *13* (Thieme Verlag, Stuttgart 1970–1986).

2.20.1 Alkali Metal Compounds

Organolithium compounds are becoming of increasing importance for their use in synthetic work.

Preparation

2.20.1.1 By the reaction of metallic lithium with alkyl or aryl halides (halogen–metal exchange) in an inert solvent, such as dry ether or benzene, with the exclusion of moisture, oxygen and carbon dioxide from the reaction mixture. (*K. Ziegler*, 1930), e.g.

$$H_3C-Cl + 2 Li \rightarrow H_3CLi + LiCl \qquad C_6H_5-Cl + 2 Li \rightarrow C_6H_5Li + LiCl$$

    Methyl lithium   Phenyl lithium

These reactions work best when alkyl or aryl chlorides are used. If bromides or iodides are used, the organolithium compound which is formed reacts rather easily with more of the halide in a *Wurtz–Fittig* type of reaction to form hydrocarbons.

2.20.1.2 Organolithium derivatives can be made by displacing the hydrogen atom from compounds with sufficiently reactive C—H bonds. This *metallation* can be brought about by the use of metallic lithium, lithium amides such as lithium diisopropylamide (LDA) or other accessible organolithium compounds such as those mentioned above.

$$HC \equiv CH \xrightarrow{\text{Li, fl.NH}_3} HC \equiv CLi$$

                        Lithium acetylide

$$\begin{matrix} R^1 \\ \quad \diagdown \\ \qquad CH-CN \\ \quad \diagup \\ R^2 \end{matrix} \xrightarrow{\text{LiNR}_2} \begin{matrix} R^1 \quad Li \\ \diagdown \diagup \\ C \\ \diagup \diagdown \\ R^2 \quad CN \end{matrix}$$

$$R-CH_2-\overset{\oplus}{N}\equiv\overset{\ominus}{C}l \;+\; C_4H_9Li \longrightarrow R-\underset{\underset{Li}{|}}{CH}-\overset{\oplus}{N}\equiv\overset{\ominus}{C}l$$

    Isonitrile   n-Butyl   Li
                            lithium

Tetrahydrofuran has been used as solvent for many metallation reactions. Lithium alkyls and aryls ignite spontaneously in air, but can be sublimed or distilled without decomposition in an atmosphere of nitrogen. The Li—C bond is only weakly polar and as a result organolithium compounds dissolve in a number of organic solvents such as ether, benzene and hexane. Because of their reactivity they are sometimes used instead of *Grignard* reagents for the introduction of alkyl or aryl groups into organic compounds (*Wittig, Gilman*).

It is not possible to make **organosodium compounds** from alkyl halides and sodium (see the *Wurtz* synthesis). However, some can be prepared by metallation reactions (see p. 94) and from the reaction of *dialkylmercury compounds with sodium* in the absence of air and moisture, e.g.

$$(H_3C)_2Hg + 2 Na \rightarrow 2 H_3CNa + Hg$$

    Dimethylmercury   Methylsodium

Organopotassium compounds can be obtained similarly by metal–metal exchange reactions. The corresponding aryl derivatives can be prepared by halogen–metal exchange using the method of *Ziegler* (see above).

The very reactive sodium and potassium compounds form colourless amorphous powders which in air catch fire spontaneously, and often explosively. Organosodium compounds are much more polar than their lithium counterparts.

Still more strongly polar are benzyl sodium derivatives, for example *benzylsodium* itself and especially *triphenylmethylsodium*. They are deep red in colour, and, when dissolved in sulphur dioxide or organic solvents, conduct electricity (*W. Schlenk sen.*).

2.20.2 Organomagnesium Compounds

The discovery of this group of compounds, which is of enormous importance in synthetic organic chemistry, stemmed from the work of *Cahours* (1860) and *Barbier* (1898) and from the extensive studies of *Grignard* (1900).

Preparation

If magnesium is added to a solution of an alkyl halide in completely dry ether it dissolves, forming an *alkylmagnesium halide*:

$$R-X + Mg \rightarrow RMgX$$

Grignard reagents prepared in this way are usually complexed to two molecules of ether. When the solvent ether is distilled off, these complexed ether molecules remain in the product but they can be removed by longer heating at 100°C (375 K) *in vacuo*. The ether-free magnesium compounds can also be obtained by using petrol, benzene or xylene as the solvent in their preparation. Sometimes the reactions of alkyl or aryl halides only set in at higher temperatures; it is then convenient to use butyl ether or pentyl ether as solvent.

Work by *W. Schlenk jun.* shows that in solution *Grignard* reagents take part in the following equilibrium (*Schlenk* equilibrium):

$$2\,RMgX \rightleftharpoons R_2Mg + MgX_2 \qquad (R = Alkyl, Aryl; X = Halogen)$$

More recent NMR studies indicate that in solution *Grignard* reagents exist as dimers which probably have one or other of the following complex structures.

For the sake of simplicity the original formulation as R—MgX remains the general usage.

Gilman's reagent, which is made up from *Michler's* ketone, water and iodine in solution in acetic acid, is used to test for the presence of a *Grignard* compound. When the three constituents of this reagent are added in turn to the test solution, a green-blue colour is produced if a *Grignard* compound is present.

Properties and Uses

Grignard reagents react with substances which have one or more reactive hydrogen atoms, for example HO-, H_2N-groups. The simplest example is their decomposition in the presence of *water*:

$$R-MgX + H-OH \rightarrow R-H + Mg(OH)X$$

Alcohols react similarly, leading to the formation of a hydrocarbon, *e.g.*

$$H_3C-MgI + H-OC_2H_5 \rightarrow CH_4 + Mg(OC_2H_5)I$$

The same type of reaction takes place with phenols, enols, carboxylic acids, and primary and secondary amines. Even hydrogen atoms attached to a carbon atom, in alkynes, react with *Grignard* reagents, for example in the case of hept-1-yne:

$$C_5H_{11}-C\equiv C-H \xrightarrow[-C_2H_6]{C_2H_5MgCl} C_5H_{11}-C\equiv C-MgCl$$

If *methylmagnesium iodide* is used, then each *reactive* hydrogen atom in a compound with which it reacts gives rise to the formation of 1 mol of methane; this can be measured gas-volumetrically by the method of *Zerevitinov*. This has served as a way of determining the numbers of reactive hydrogen atoms in the molecule when the structures of organic compounds were being investigated. Another use of this type of reaction is to provide a way for selective *deuteriation* in a molecule. For example, an *isotopomer*[1] of 2-methylbutane can be made by reaction of a *Grignard* reagent with deuterium oxide:

$$H_3C-CH_2-\underset{\underset{MgBr}{|}}{\overset{\overset{CH_3}{|}}{C}}-CH_3 \ + \ D_2O \ \longrightarrow \ H_3C-CH_2-\underset{\underset{D}{|}}{\overset{\overset{CH_3}{|}}{C}}-CH_3 + \ DOMgBr$$

[2-D]2-Methylbutane

This example illustrates one of the methods of nomenclature used to describe deuteriated substances; other methods are exemplified for this isotopomer as follows: 2-deuterio-2- methylbutane, 2-deutero-2-methylbutane, 2-methylbutane-2d, [2-²H]-2-methylbutane.[2]

The most important property of organomagnesium compounds is their ability to add to *polar double bonds* such as $\diagup\!\!\!C{=}O$, $\diagup\!\!\!C{=}N{-}$, ${-}C{\equiv}N$, ${-}N{=}O$ and $\diagup\!\!\!S{=}O$. In these reactions the magnesium–halogen portion of the molecule behaves like a *cation* and adds to the most electronegative element, whilst the alkyl or aryl group acts as a *carbanion* and adds to the least electronegative atom, such as carbon. If an overall view is taken of the reaction of an alkyl halide, *via* its conversion into a *Grignard* reagent, with one of the above groups, it may be seen that it provides another example of an umpolung reaction, the alkyl group being transformed from an electrophile into a nucleophile. Hydrolysis of the adduct results in the replacement of Mg–halogen by hydrogen. In this way the following products can be made:

2.20.2.1 *Secondary alcohols* are obtained from aldehydes, *e.g.*

$$R-\overset{\overset{H}{\diagup}}{C}\diagdown_O \xrightarrow{+R'-MgI} R-\overset{\overset{H}{\diagup}}{\underset{\underset{R'}{|}}{C}}\diagdown_{OMgI} \xrightarrow[-Mg(OH)I]{+H_2O} \overset{R}{\underset{R'}{\diagup}}C\overset{\diagup H}{\diagdown_{OH}}$$

In the case of formaldehyde (R=H) only, a *primary alcohol* is obtained, *e.g.*

$$C_5H_{11}-C\equiv C-MgCl \xrightarrow[2.\ H^\oplus,\ H_2O]{1.\ H_2CO} C_5H_{11}-C\equiv C-CH_2OH$$

Oct-2-yn-1-ol

---

[1] *Isotopomers* are isotopic isomers, *i.e.* compounds with the same structural formula, but with different isotopes present.
[2] An omnibus term for proton, deuteron and triton (³H) is *hydron*. See Pure Appl. Chem. *60*, 1115 (1988).

2.20.2.2 Ketones provide *tertiary alcohols*, e.g.

$$\underset{R}{\overset{R}{\diagdown}}C=O \xrightarrow{\;+R'-MgBr\;} \underset{R}{\overset{R}{\diagdown}}\underset{OMgBr}{\overset{R'}{\diagup}}C \xrightarrow[-Mg(OH)Br]{\;+H_2O\;} \underset{R}{\overset{R}{\diagdown}}\underset{OH}{\overset{R'}{\diagup}}C$$

Tertiary alcohols can also be produced from a two-step reaction between *Grignard* reagents and carboxylic acid esters. The first step is the usual addition to the C=O group; the adduct then reacts with another mol of the *Grignard* reagent:

$$R-C\underset{OC_2H_5}{\overset{O}{\diagup}} \xrightarrow{\;+R'-MgBr\;} R-\underset{R'}{\overset{OMgBr}{\underset{|}{\overset{|}{C}}}}OC_2H_5 \xrightarrow[-Mg(OC_2H_5)Br]{\;+R'-MgBr\;} R-\underset{R'}{\overset{R'}{\underset{|}{\overset{|}{C}}}}OMgBr$$

$$\xrightarrow[-Mg(OH)Br]{\;+H_2O\;} R-\underset{R'}{\overset{R'}{\underset{|}{\overset{|}{C}}}}OH$$

2.20.2.3 Introduction of dry carbon dioxide into a solution of a *Grignard* reagent leads to the formation of a *carboxylic acid*, e.g.

$$R-MgCl \xrightarrow{\;+CO_2\;} R-COOMgCl \xrightarrow[-Mg(OH)Cl]{\;+H_2O\;} R-COOH$$

Addition of a *Grignard* reagent to a carbonyl group often proceeds by a polar mechanism, which can be expressed in a simplified form as follows:

$$RMgX + \;\;\diagdown C=O \longrightarrow \overset{\delta\oplus\;\;\delta\ominus}{\diagdown C=O} \longrightarrow \diagdown\underset{R}{\overset{|}{C}}OMgX$$
$$\underset{R-MgX}{\overset{\delta\ominus\;\;\delta\oplus}{}}$$

It has been shown, however, that a mechanism involving radical intermediates can also be involved (see p. 616).

2.20.3 Organozinc Compounds

Dialkylzinc compounds (zinc dialkyls) were first made by *Frankland* (1849), by *treatment of alkyl halides with zinc* and distillation of the first-formed alkylzinc halides, e.g.

$$2\,C_2H_5-I \xrightarrow{\;+2\,Zn\;} 2\,C_2H_5ZnI \xrightarrow[-ZnI_2]{} Zn(C_2H_5)_2$$

Diethylzinc is a clear colourless liquid of b.p. 118°C (391 K) which inflames spontaneously in air. It can only be kept if all moisture is excluded; it reacts violently with water giving ethane and zinc hydroxide.

Organozinc compounds are used preparatively in the *Reformatsky* reaction (see p. 278 *ff.*) and in the *Simmons–Smith* reaction.

Simmons–Smith reaction.[1] In this reaction a solution of diiodomethane and an alkene in dry ether is treated with an excess of activated zinc; a cyclopropane derivative is formed.

---

[1] See *H. E. Simmons* et al., Cyclopropanes from Unsaturated Compounds, Methylene Iodide and Zinc-Copper Couple, Org. Reactions, *20*, 1 (1973).

The yield can be increased by the application of ultrasound to the mixture.[1] Reaction proceeds *via* the formation of *iodomethyl-zinc iodide*:

$$CH_2I_2 + Zn \xrightarrow{\text{Ether}} ICH_2-ZnI$$

It is stereospecific; *cis*- and *trans*-alkenes give, respectively, *cis*- and *trans*-cyclopropane derivatives. This suggests that the methylene group is transferred in an ordered transition state; it is a carbenoid reagent:

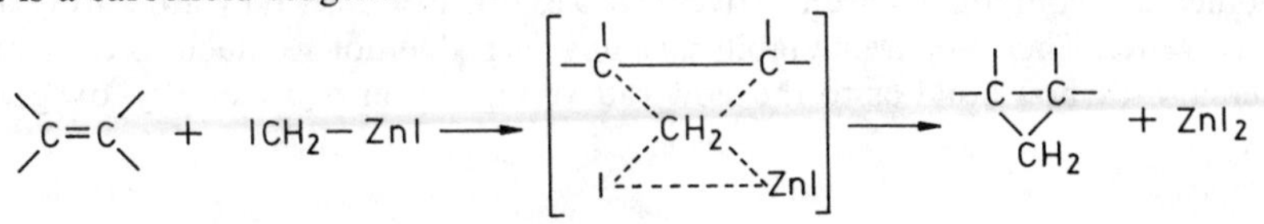

2.20.4 Organotitanium Compounds[2]

Organic titanium compounds are obtained by *transmetallation* reactions involving treatment of *organo-lithium, -magnesium* or *-zinc* compounds with titanium halides in ether, which are themselves got from the corresponding 'titanates', *e.g.*

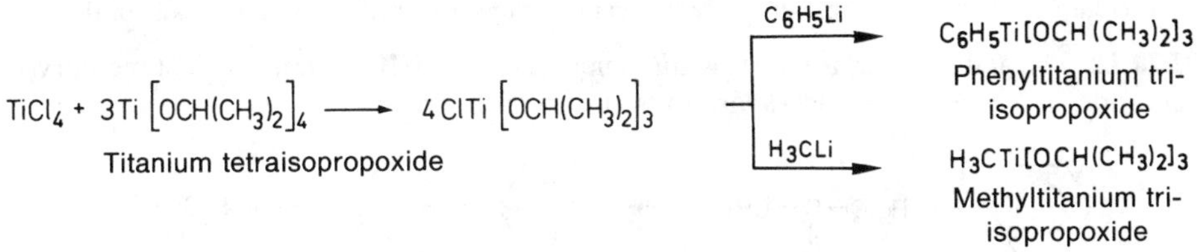

$$TiCl_4 + 3Ti\left[OCH(CH_3)_2\right]_4 \longrightarrow 4ClTi\left[OCH(CH_3)_2\right]_3$$

Titanium tetraisopropoxide

$C_6H_5Li \longrightarrow C_6H_5Ti[OCH(CH_3)_2]_3$ Phenyltitanium tri-isopropoxide

$H_3CLi \longrightarrow H_3CTi[OCH(CH_3)_2]_3$ Methyltitanium tri-isopropoxide

In contrast to the *Grignard* reagents organotitanium compounds have markedly different reactivities towards aldehydes and ketones, *e.g.*

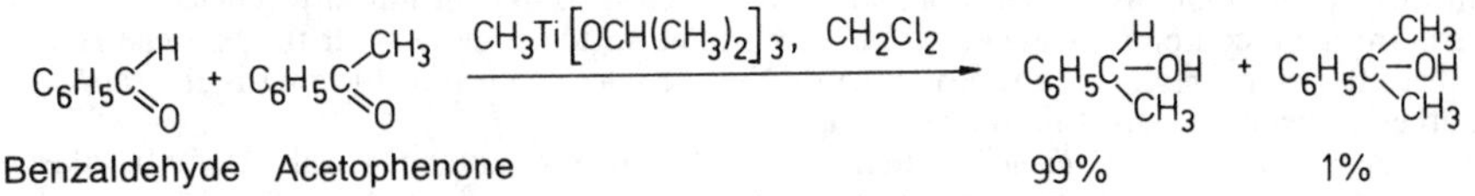

Benzaldehyde   Acetophenone   99%   1%

There is a similar *chemoselectivity* when a mixture of acetaldehyde and acetone is treated with phenyltitanium triisopropoxide. The *stereoselectivity* of organotitanium compounds as intermediates has already been mentioned in connection with *Ziegler–Natta* catalysts; it also plays a signficant role in the behaviour of *Sharpless* reagents.

The *Tebbe* reagent is prepared from dichlorotitanium cyclopentadienide $(C_5H_5)_2TiCl_2$, (see p. 649) and trimethyl-aluminium, $Al(CH_3)_3$; it has proved effective in the preparation of *enol ethers* from esters.[3] *e.g.*

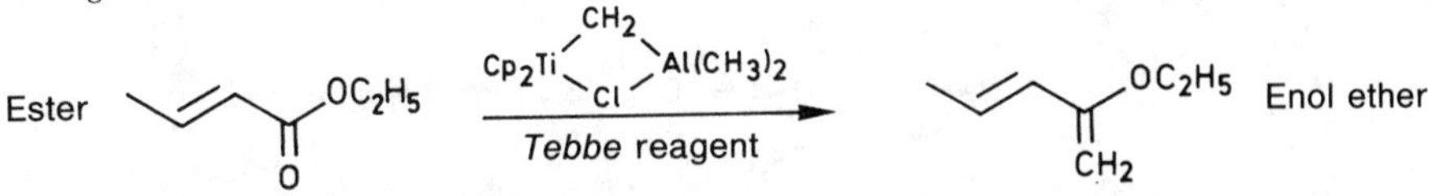

Ester   Tebbe reagent   Enol ether

[1] See K. S. *Suslick*, in R. *Scheffold* (Ed.) Modern Synthetic Methods 1986, *4*, (Springer-Verlag, Berlin 1986); S. V. *Levy* et al., Ultrasound in Chemistry (Springer-Verlag, Berlin 1989); T. J. *Mason* et al., Sonochemistry (Ellis Horwood, Chichester 1988).
[2] See M. T. *Reetz*, Organotitanium Compounds in Organic Synthesis (Springer-Verlag, Berlin 1986).
[3] See F. N. *Tebbe* et al., Olefin Homologation withTitanium Methylene Compounds, J. Amer. Chem. Soc., *100*, 3611 (1978); R. *Scheffold* p. 249, see p. 188, footnote 1.

A brief mention must also be made of organozirconium compounds which are analogues of the titanium derivatives.[1]

2.20.5 Organomercury Compounds

Compared to the other organometallic compounds which have been mentioned, organomercury compounds are relatively stable. For example, they are hardly attacked by air or water. Their apparent stability is, however, due not so much to the strength of the mercury–carbon bond as to the slow rate of formation of a mercury–oxygen bond.

Preparation

2.20.5.1 A general method is by *transmetallation of organolithium compounds or of Grignard reagents by means of mercury(II) halides*:

$$R-Li + HgBr_2 \longrightarrow R-Hg-Br + LiBr$$

$$2\ RMgI + HgCl_2 \longrightarrow R_2Hg + MgI_2 + MgCl_2$$

Mercury derivatives having two different organic groups attached can also be made in this way.

2.20.5.2 The action of *polar mercury salts*, such as mercury(II) acetate, but not mercury(II) chloride, *on aromatic hydrocarbons*, gives rise to arylmercury salts, *e.g.*

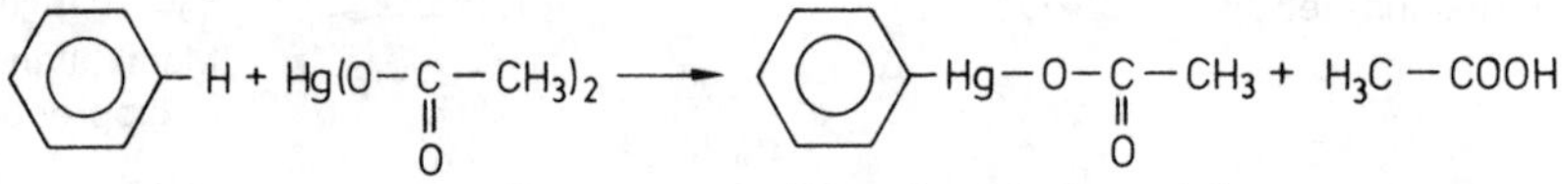

Benzene   Phenylmercuric acetate

In this *mercuriation*, hydrogen atoms directly attached to aryl groups are replaced by mercury; hetero-aromatic compounds may also be mercuriated in the same way. If the benzene ring has a substituent group present, the site of entry of the mercury atom into the ring is the same as that for other electrophilic substitution reactions.

Hydroxymercuriation of alkenes[2] with mercury(II) acetate in aqueous tetrahydrofuran takes place *regioselectively* to give an organomercury compound which can be reduced *in situ* to an alcohol, *e.g.*

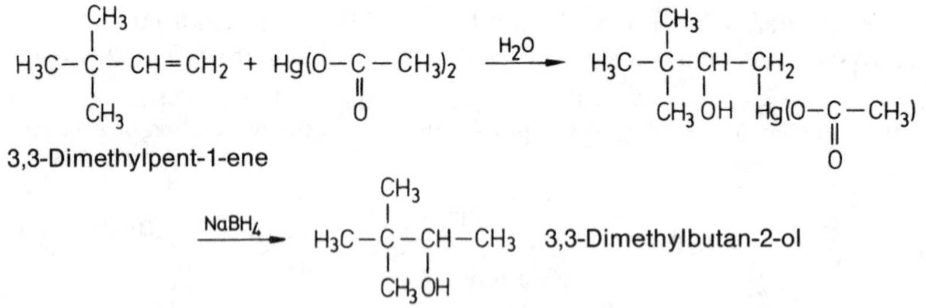

3,3-Dimethylpent-1-ene

3,3-Dimethylbutan-2-ol

---

[1] See p. 78, and *D. Seebach* et al., Titanium and Zirconium Derivatives in Organic Synthesis, Vol. 3, p. 217 in *R. Scheffold*, Modern Synthetic Methods (Otto Salle Verlag, Frankfurt a.M. 1983); *E. Negishi* et al., Organozirconium Compounds in Organic Synthesis, Synthesis 1988, 1.
[2] See *R. C. Larock*, Solvomercuriation/Demercuriation Reactions in Organic Synthesis (Springer-Verlag, Berlin 1986).

This reaction, which gives an excellent yield, is in effect an *addition of water to a C=C double bond* which has taken place according to *Markovnikov's* rule.

Properties

Alkyl mercury derivatives (mercury dialkyls) are colourless liquids which are readily volatile. They are dangerously poisonous when inhaled.[1]

Because of the different electronegativities of carbon and mercury, a C—Hg bond is polarised, but in the opposite way to a carbon–halogen bond. The carbon atom carries a partial negative charge and can react with *electrophiles* such as protons or *Lewis* acids.

2.20.6 Organoaluminium Compounds

The most important of these compounds are the *trialkylaluminium compounds* (*aluminium trialkyls*), AlR_3, which have become readily available, thanks to the work of *K. Ziegler*.

Preparation

2.20.6.1 *By reaction of an Al–Mg alloy with alkylhalides, e.g.*

$$Al_2Mg_3 + 6\ ClC_2H_5 \longrightarrow 2\ Al(C_2H_5)_3 + 3\ MgCl_2$$

Triethylaluminium (b.p. 194°C [467 K])

2.20.6.2 An alternative method for the preparation of triethylaluminium is by the reaction of *ethylene with aluminium hydride* at about 70°C [350 K]. Reaction proceeds stepwise, as follows:

$$AlH_3 \xrightarrow{+C_2H_4} Al{<}^{C_2H_5}_{H} \xrightarrow{+C_2H_4} Al{<}^{C_2H_5}_{C_2H_5}_{H} \xrightarrow{+C_2H_4} Al{<}^{C_2H_5}_{C_2H_5}_{C_2H_5}$$

Properties and Uses

Trialkylaluminium compounds are colourless, mobile liquids which can be distilled. They are sensitive to moisture and to atmospheric oxygen. The lower members of the series ignite spontaneously in air and react explosively with water. In the liquid state and also just above the boiling point they exist as *dimers*. When they come into contact with skin they cause severe and painful burns.

Industrially *trialkylaluminium compounds* are important for their use in *Ziegler–Natta* catalysts (see the polymerisation of alkenes) and in the preparation of alk-1-enes (Mülheimer Alkene Synthesis).

---

[1] A series of poisonings which occurred in the Japanese city of Minamata in 1956–1978 was caused by methylmercury chloride and methylmercury sulphide, which were carried as effluents into the Bay of Minamata. Organic derivatives of mercury are much more poisonous than inorganic mercury compounds. These compounds arose from the action of microorganisms on inorganic mercury compounds. They were present in fish, being the end-products of a natural food chain (*e.g.* in tunny fish to the extent of 0.4 mg per kg).

2.20.7 Organotin Compounds

In recent times the chemistry of organotin compounds has become of increasing importance. A C—Sn bond is much more polar than a C—C or a C—Si bond, and as a result is more easily broken. The most stable organotin compounds are the *tetraalkyl-* and *tetraaryl-tin* or *-stannane* derivatives.

These are best made by treatment of *Grignard* reagents or organolithium compounds with tin(IV) halides in an inert solvent such as ether or benzene, *e.g.*

$$SnCl_4 + 4\ RMgBr \longrightarrow R_4Sn + 2\ MgCl_2 + 2\ MgBr_2$$

The lower *tetraalkylstannanes* are colourless liquids resembling hydrocarbons (*tetramethylstannane* or *tetramethyltin*, b.p. 77°C (350 K)], the higher members are waxy materials which cannot be distilled. *Tetraarylstannanes* are solids with high melting points, *e.g.* tetraphenylstannane melts at 225°C (498 K).

Organotinhalides of general formula R_nSnX_{4-n} (n = 1–3; X = Cl, Br) are for the most part crystalline solids; they are very reactive. They can be obtained from the reactions of tetraalkyl- or tetraaryl-stannanes with tin(IV) halides at temperatures above 180°C (450 K) (*Kocheshkov*), *e.g.*

$$3\ R_4Sn + SnCl_4 \longrightarrow 4\ R_3SnCl \qquad \text{Tributyltin chloride } [R = CH_3(CH_2)_3]$$

If tributyltin chloride is treated with lithium aluminium hydride, tributyltin hydride is obtained, which has become of importance both as a selective reducing agent and as an initiator for the forming of C—C bonds by means of *radical chain reactions.*[1]

$$(n\text{-}C_4H_9)_3SnCl \xrightarrow{\text{LiAlH}_4} (n\text{-}C_4H_9)_3SnH$$

Tributyltin chloride   Tributyltin hydride

Some dialkyltin derivatives, for example dibutyltin dilaurate or maleate, which are made from dibutyltin dichloride and the sodium salt of the appropriate acid, are used as stabilisers against the action of light or heat on *PVC* and chlorinated transformer oils. They act by inhibiting elimination of HCl. Tributyltin acetate, triphenyltin acetate (Brestan) and bis(tributyltin)oxide, $(C_4H_9)_3SnOSn(C_4H_9)_3$ have use as powerful *fungicides*.

2.20.8 Organolead Compounds

Tetraethylead (*lead tetraethyl*) is of industrial importance. It is made by heating a lead–sodium alloy with ethyl chloride under pressure:

$$4\ PbNa + 4\ ClC_2H_5 \longrightarrow Pb(C_2H_5)_4 + 4\ NaCl + 3\ Pb$$

Tetraethyllead is a colourless liquid, b.p. ~ 200°C (475 K), which is highly poisonous. It has been used as an antiknock agent. *Tetramethyllead*, $Pb(CH_3)_4$, b.p. 110°C (383 K) has been used for the same purpose in petrols which are rich in arenes.

Paneth (1927) used the thermal decomposition of organolead compounds to detect for the first time the presence of very short-lived *free radicals* such as methyl, $\cdot CH_3$, and ethyl, $\cdot CH_2$—CH_3.

---

[1] See *W. P. Neumann*, Tri-n-butyltin Hydride as Reagent in Organic Chemistry, Synthesis 1987, 665; *D. P. Curran*, The Design and Application of Free Radical Chain Reactions in Organic Synthesis, Synthesis 1988, 417.

2.21 Aliphatic Aldehydes (Alkanals)

The oxidation or dehydrogenation of *primary* alcohols leads to the formation of aldehydes (see p. 112), which can be further oxidised to carboxylic acids:

$$R-CH_2-OH \xrightarrow{-2H} R-C\overset{H}{\underset{O}{\diagdown}} \xrightarrow{[O]} R-C\overset{O}{\underset{OH}{\diagdown}}$$

primary Alcohol   Aldehyde   Carboxylic acid

The name 'aldehyde' is derived from '*alcohol dehyd*rogenatus'. The characteristic feature of aldehydes is the *aldehyde* or *formyl* group:

$$-C\overset{H}{\underset{O}{\diagdown}}$$

In the traditional systems, aldehydes have been named after the carboxylic acid which they provide when they are oxidised. The first members of the series are:

		b.p.
Formaldehyde, methanal	H–CHO	−19.2 °C (253.4 K)
Acetaldehyde, ethanal	H_3C–CHO	20.8 °C (294.0 K)
Propionaldehyde, propanal	H_3C–CH_2–CHO	49.0 °C (322.2 K)
n-Butyraldehyde, butanal	H_3C–CH_2–CH_2–CHO	75.7 °C (348.9 K)
n-Valeraldehyde, pentanal	H_3C–CH_2–CH_2–CH_2–CHO	103.7 °C (376.9 K)

By the *IUPAC rules*, aldehydes are named by replacing the final 'e' of the name of the corresponding hydrocarbon by the ending 'al'; for example, 'acetaldehyde' becomes 'ethanal'. Another way of naming them involves adding -*carbaldehyde* as a final syllable, as in cyclohexanecarbaldehyde (p. 192). The names formaldehyde and acetaldehyde are still generally used, but the IUPAC names are most common for the higher aldehydes.

2.21.1 General Methods of Preparation

2.21.1.1 *By oxidation or dehydrogenation of primary alcohols*, for example using potassium dichromate in sulphuric acid or chromium trioxide in pyridine (*Collins* reagent) or pyridinium chlorochromate.[1]

As the aldehyde is formed it must be removed quickly from the solution, by distillation, to prevent it from being oxidised further. In aqueous solution a hydrate is formed which is itself oxidised faster than the original alcohol.

An oxidising agent which has proved popular for this purpose is obtained by activating *dimethyl sulphoxide* with *oxalyl chloride* (*Swern*, 1978). This is effective in solution in dichloromethane even at −60°C (213 K) and in most cases provides yields of more than 90%.[2]

---

[1] See *G. Piancatelli* et al., Synthesis 1982, 245.
[2] *T. Tidwell*, The *Moffat, Swern* and Related Oxidations, Org. Reactions *39*, 297 (1990).

$$(CH_3)_2SO + Cl-\overset{\underset{\displaystyle O}{\|}}{C}-\overset{\underset{\displaystyle O}{\|}}{C}-Cl \xrightarrow[213K]{CH_2Cl_2,} \left[(CH_3)_2\overset{\oplus}{S}-O-\overset{\underset{\displaystyle O}{\|}}{C}-\overset{\underset{\displaystyle O}{\|}}{C}-Cl \right] Cl^{\ominus} \xrightarrow[]{\substack{-CO_2 \\ -CO}}$$

Dimethyl   Oxalyl
sulphoxide   chloride

$$RCHO \xleftarrow[-H^{\oplus}, -(CH_3)_2S]{(C_2H_5)_3N} \left[(CH_3)_2\overset{\oplus}{S}OCH_2R \right] Cl^{\ominus} \xleftarrow{RCH_2OH} \left[(CH_3)_2\overset{\oplus}{S}-Cl \right] Cl^{\ominus}$$

Alkoxysulphonium salt   (I)

The oxygen-free intermediate (I) arises by spontaneous decarboxylation and decarbonyl-ation of the initial reaction product formed from dimethyl sulphoxide and oxalyl chloride. It then reacts with the alcohol to give an *alkoxysulphonium salt*, which in the presence of a base loses a proton and dissociates into an aldehyde and dimethyl sulphide.

2.21.1.2 *By partial reduction, using complex hydrides at low temperatures, of derivatives of carboxylic acids.*[1]

The earlier method of obtaining aldehydes from acid chlorides by means of the *Rosenmund reaction* has been almost entirely superseded by using lithium aluminium hydride or, with advantage, *lithium alkoxyaluminium hydrides, e.g.* LiAlH(OC$_2$H$_5$)$_3$ or LiAlH(O-t-C$_4$H$_9$)$_3$(LTBA) as well as *diisobutylaluminium hydride* (DIBAH) and other reagents such as tributyltin hydride. Aldehydes can be obtained in good yield from esters, acid chlorides, amides and nitriles, *e.g.*

$$(H_3C)_3C-C\overset{\nearrow OR}{\underset{\searrow O}{}} \xrightarrow{LTBA, 200K} (H_3C)_3C-C\overset{\nearrow H}{\underset{\searrow O}{}} \quad 67\%$$

Ester of
trimethylacetic
acid   Trimethylacetaldehyde

$$cC_6H_{11}-C\overset{\nearrow N(CH_3)_2}{\underset{\searrow O}{}} \xrightarrow{LiAlH(OC_2H_5)_3, 273K} cC_6H_{11}-C\overset{\nearrow H}{\underset{\searrow O}{}} \quad 78\%$$

N,N-Dimethyl-
cyclohexanecarboxamide   Cyclohexanecarbaldehyde

In general, carboxyl groups are not attacked. Nitriles are first reduced to an *aldimine* which can undergo acid hydrolysis to give the aldehyde:

$$R-C\equiv N \xrightarrow{+2H \ (LiAlH_4)} R-C\overset{\nearrow H}{\underset{\searrow NH}{}} \xrightarrow[-NH_3]{+H_2O(H^{\oplus})} R-C\overset{\nearrow H}{\underset{\searrow O}{}}$$

(Aldimine)

2.21.2 *Tests for Aldehydes*

Aldehydes, but *not* ketones, are *strong reducing agents*, because of their ready oxidation to carboxylic acids. This property forms the basis for the following *tests for aldehydes*.

2.21.2.1 Reduction of *ammoniacal solutions of silver nitrate* (*Tollens'* reagent) to metallic silver.

---

[1] See *J. Málek*, Reductions by Metal Alkoxyaluminium Hydrides. Org. Reactions *34*, 1 (1985).

2.21.2.2 When an aldehyde is heated with *Fehling's* solution a red or dark brown precipitate appears. The composition of this precipitate is not known with certainty. It contains traces of copper(I) oxide if an excess of *Fehling's* solution is used.

Fehling's solution is prepared by adding to a known amount of copper sulphate solution an equivalent of an alkaline solution of sodium potassium tartrate (*Rochelle* salt). The latter prevents precipitation of copper(II) hydroxide by forming a complex salt.

2.21.2.3 Reduction of *Nylander's* reagent, which is an alkaline solution of bismuth(III) hydroxide containing tartaric acid as a complexing agent, leads to the formation of black metallic bismuth.

2.21.2.4 Traces of aldehyde turn *Schiff's reagent* [an aqueous solution of fuchsin (magenta, pararosaniline) which has been decolourised with SO_2] pink-violet. This reaction is not specific and reducing agents other than aldehydes can give the same result.

2.21.2.5 Aldehydes decolourise solutions of *potassium permanganate*.

2.21.2.6 Aldehydes are often easily recognisable from their ^{1}H-NMR spectra because the hydrogen atom attached to the carbonyl group, RC*H*O, provides a signal in a region of the spectrum ($\delta 9.3$–10.0) where very few other kinds of hydrogen atoms provide signals.

2.21.3 The C=O Double Bond

Like C=C bonds, C=O bonds are described in terms of a σ-bond and a π-bond. Because of the different electronegativities of carbon and of oxygen the C=O double bond, like the C—O single bond, is *polar*. This can be seen from the dipole moments of aldehydes and ketones. The *dipole moment* associated with a carbonyl group is 9.01×10^{-30} C m, whereas that of a C—O single bond (σ-bond) in simple ethers is only about 4.00×10^{-30} C m. This shows that the π-bond is more polar than the σ-bond. The polarity of the π-bond may be demonstrated as in Fig. 73.

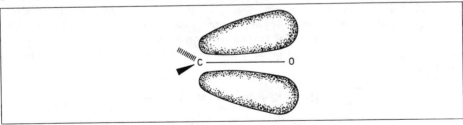

Fig. 73. The polarity of the π-Bonding in the Carbonyl Group.

In chemical formulae this can be denoted as follows:

$$\begin{array}{ccc} \searrow\!\!\!C=\underline{O} & \text{or} & \overset{\delta^+}{\searrow}\!\!\!C=\overset{\delta^-}{O} \\ \nearrow & & \nearrow \end{array}$$

The chemical properties of aldehydes and ketones are largely connected with the reactivity of the carbonyl group. This is especially apparent in the addition and condensation reactions of aldehydes. These reactions for the most part involve atoms or groups of atoms having lone pairs of electrons or anions, *i.e. nucleophilic* reagents, which attack the positively charged carbon atom of the carbonyl group, while protons, *Lewis* acids or other suitable species, *i.e. electrophiles*, react with the oxygen atom.

2.21.4 Addition Reactions of Aldehydes

2.21.4.1 Reduction to alcohols. Addition of hydrogen to aldehydes leads to the formation of *primary alcohols*:

$$R-C\overset{H}{\underset{O}{\diagup}} \quad \xrightarrow{+2H} \quad R-CH_2OH$$

Reduction can be brought about catalytically (Ni, Pd) or by 'nascent' hydrogen (sodium amalgam and water, or sodium and alcohol).

Lithium aluminium hydride, LiAlH$_4$, and *sodium borohydride*, NaBH$_4$, have become important reagents for the reduction of carbonyl groups. Markedly more reactive is lithium triethylborohydride, Li(C$_2$H$_5$)$_3$BH (Superhydride), which is also valuable because of its selectivity.[1]

Sometimes the *Meerwein–Ponndorf–Verley* reduction[2] (1925), using aluminium isopropoxide in propan-2-ol is employed. This is a *redox process* and the products are a primary alcohol and acetone:

$$R-C\overset{H}{\underset{O}{\diagup}} + H_3C-\overset{H}{\underset{OH}{\underset{|}{\overset{|}{C}}}}-CH_3 \quad \underset{}{\overset{Al(OCH(CH_3)_2)_3}{\rightleftharpoons}} \quad R-CH_2OH + H_3C-\overset{}{\underset{O}{\overset{\|}{C}}}-CH_3$$

Aldehyde   Propan-2-ol   primary Alcohol   Acetone

This method of reduction is specific for carbonyl groups. Above all, therefore, it is used if the aldehyde or ketone also has other groups present, such as C=C double bonds, nitro groups, etc., which might otherwise be reduced as well.

There is a reverse reaction in which primary alcohols can be oxidised to aldehydes by acetone in the presence of an aluminium alkoxide (see *Oppenauer* oxidation).

2.21.4.2 Formation of Cyanohydrins. A useful preparative reaction is the addition of hydrocyanic acid to aldehydes, which, especially with a base present as catalyst, leads to the formation of α-hydroxynitriles (cyanohydrins). Reaction involves attack by lone pairs of electrons on the carbon atoms of cyanide ions on the positively charged carbon atom of the carbonyl group. The initial product stabilises itself by taking a proton from water molecule.

$$H-C\equiv NI + OH^\ominus \rightleftharpoons I\overset{\ominus}{C}\equiv NI + H_2O$$

$$R-C\overset{H}{\underset{O}{\diagup}} \quad \underset{}{\overset{+I\overset{\ominus}{C}\equiv NI}{\rightleftharpoons}} \quad \left[R-\overset{H}{\underset{\underset{N}{\overset{|}{\underset{\|}{C}}}}{\overset{|}{\underset{O}{\overset{\ominus}{C}}}}} \right] \quad \underset{}{\overset{+H_2O(-OH^\ominus)}{\rightleftharpoons}} \quad R-\overset{H}{\underset{\underset{N}{\overset{|}{\underset{\|}{C}}}}{\overset{|}{\underset{OH}{C}}}}$$

Aldehyde   Cyanohydrin

In the last step of this *equilibrium reaction* the basic catalyst (HO$^\ominus$) is regenerated.

---

[1] See *H. C. Brown* et al., J. Org. Chem. *45*, 1 (1980).
[2] See *A. L. Wilds*, Reduction with Aluminium Alkoxides (The *Meerwein–Ponndorf–Verley* Reduction), Org. Reactions *2*, 178 *ff.* (1944); *N. C. Deno* et al., The Hydride-Transfer Reaction, Chem. Rev. *60*, 7 (1960); *J. Huskens* et al., Synthesis, 1994, 1007.

2.21.4.3 Addition of sodium hydrogen sulphite leads to the formation of rather insoluble crystals of the sodium salts of so-called bisulphite compounds. These either separate out spontaneously or are precipitated out (salted out) by excess sodium hydrogen sulphite. Like the formation of cyanohydrins, this reaction proceeds by nucleophilic attack of a bisulphite anion, which is followed by migration of a proton:

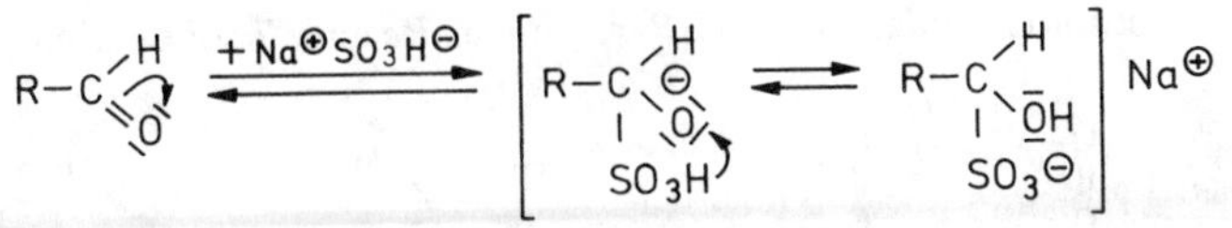

When a bisulphite compound is warmed with a dilute acid or with sodium carbonate the aldehyde is regenerated, and these compounds thus provide a way for the separation and purification of aldehydes. They react with potassium cyanide to give the corresponding cyanohydrin.

2.21.4.4 Addition of Grignard reagents to formaldehyde to give *primary* alcohols and with other aldehydes to give *secondary* alcohols.

2.21.4.5 If dry ammonia is passed into an ethereal solution of an aldehyde it first adds to the carbon atom of the carbonyl group. A proton can then transfer from the nitrogen atom to the oxygen atom. Most of the 'aldehyde-ammonias' which are formed in this way are unstable and lose water to give aldimines; the latter trimerise, forming heterocyclic compounds. As an example acetaldehyde and ammonia are converted, *via* acetaldimine (iminoethane), into *2,4,6-trimethylhexahydro-1,3,5-triazine*:

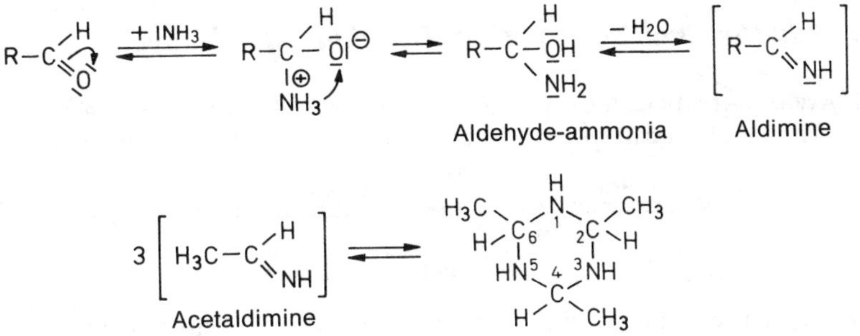

Aldehyde-ammonia   Aldimine

Acetaldimine

The reaction of ammonia with formaldehyde proceeds even further, resulting in the formation of hexamethylenetetramine.

2.21.4.6 Addition of water leads to *aldehyde hydrates* which in some cases are isolable.

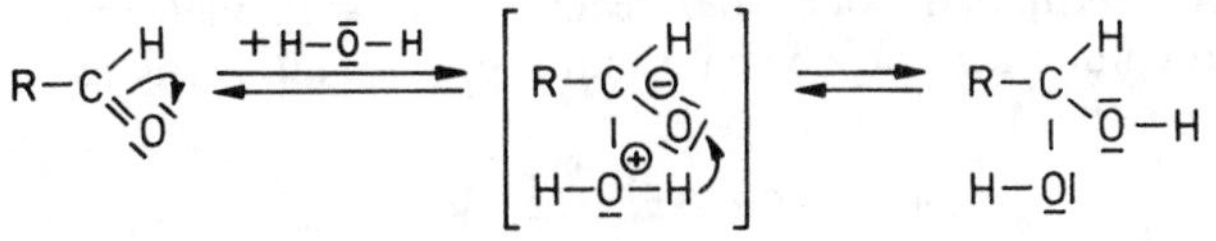

In the first step a lone pair of electrons on the oxygen atom of a water molecule makes a nucleophilic attack on the positively charged carbon atom of the aldehyde group. Migration of a proton in the intermediate so formed provides the aldehyde hydrate which is only isolable in exceptional cases (*e.g.* chloral hydrate, *q.v.*).

2.21.5 Enantiotopy and Prochirality[1]

In the course of the conversion of an aldehyde into its cyanohydrin, the achiral carbon atom of the carbonyl group becomes a centre of chirality whose configuration depends upon which face of the molecule is attacked by the cyanide ion. The two distinct faces are said to be *enantiotopic* and the reaction can be called enantiofacial. For a way of specifying two enantiotopic sides by the *CIP* system, as *Re* and *Si* configurations, see p. 277.

Just as an aldehyde can be turned through 180° to present a different face so the same thing can be done with a tetrahedral molecule of ethanol. If one of the hydrogen atoms of the CH$_2$ group of ethanol is replaced by a different group a chiral molecule results. For example, replacement by an ethyl group gives ($+$)- or ($-$)-butan-2-ol, and which of these enantiomers is formed depends on which of the two hydrogen atoms has been replaced. The two hydrogen atoms of the CH$_2$ group are also enantiotopes; the carbon atom of the CH$_2$ group is described as a *prochiral centre*. Such molecules or groups are said to be *prochiral*. Prochirality plays an important role in, for example, the building up of monosaccharides and in enzymatic reactions (see p. 893).

2.21.6 Addition and Substitution Reactions of Aldehydes

2.21.6.1 Acetal formation.
Aldehydes react with alcohols to form *hemiacetals* which, in the presence of an acid catalyst can react further, by loss of water, to give an *acetal*.

Acetals can be regarded as dialkyl ethers derived from aldehyde hydrates. As ethers, they are *very stable to alkali*, whereas they are readily converted back into an aldehyde by dilute acids. They can thus be used to 'protect' aldehyde groups while reactions involving alkali are being carried out on other parts of an aldehyde molecule, and can be readily removed again by acid when such reactions have been completed. In this way acetals are sometimes used instead of aldehydes in organic synthesis.

2.21.6.2 Thioacetal formation.
Aldehydes react readily with mercaptans (thiols) in the presence of hydrochloric acid or zinc chloride to provide *dithioacetals* (*mercaptals*).

---

[1] See K. *Mislow* et al., in Topics in Stereochemistry *1*, 1 (Wiley, New York 1967), J. Am. Chem. Soc. 106, 3319 (1984).

2.21.7 Condensation Reactions of Aldehydes

In the following condensation reactions, reaction does not end with addition of the reagent to the aldehyde, but is followed by loss of water, involving elimination of the oxygen atom which formed part of the carbonyl group.

2.21.7.1 Formation of Azomethines. The products from the reaction of *primary* amines with aldehydes, followed by elimination of water, are *azomethines* (*aldimines, Schiff's bases*).

$$R-C\underset{O}{\overset{H}{<}} + H_2N-R' \longrightarrow R-\overset{H}{\underset{}{C}}=N-R' + H_2O$$

(R,R' = Alkyl, aryl)   Azomethine

2.21.7.2 Formation of Aminals. *Secondary* amines react with aldehydes in the same way that alcohols do, giving first *hemiaminals* (*N,O-hemiacetals*), and, when heated, *aminals* (*N,N-acetals*):

$$R-C\underset{O}{\overset{H}{<}} \xrightarrow{+HNR_2'} R-\underset{OH}{\overset{H}{C}}-NR_2' \xrightarrow[-H_2O]{+HNR_2'} R-\underset{NR_2'}{\overset{H}{C}}-NR_2'$$

Aldehyde   Hemiaminal   Aminal

The aldehyde-ammonia which is mentioned on p. 195 is an unsubstituted hemiaminal.

2.21.7.3 Oxime Formation. When aldehydes are treated with hydroxylamine hydrochloride *aldoximes* are formed:

$$R-C\underset{O}{\overset{H}{<}} \xrightarrow{+H_2\bar{N}OH(H^\oplus)} \left[R-\underset{\underset{\oplus}{H_2NOH}}{\overset{H}{C}}-\bar{\underset{}{O}}I^\ominus \rightleftharpoons R-\underset{NHOH}{\overset{H}{C}}-OH \right] \xrightarrow{-H_2O} R-C\underset{NOH}{\overset{H}{<}}$$

Aldoxime

With the exception of formaldoxime, oximes are colourless, crystalline substances. They are weakly acidic and with alkalis form salts, from which the aldehyde can be regenerated by the action of carbon dioxide in aqueous solution.

$$R-C\underset{NOH}{\overset{H}{<}} + NaOH \longrightarrow R-C\underset{N-\underset{}{\bar{O}}I Na^\oplus}{\overset{H}{<}}{}^\ominus + H_2O$$

Dilute acids reconvert oximes back into aldehydes and a hydroxyammonium salt.

2.21.7.4 Formation of Semicarbazones. Condensation of aldehydes with *semicarbazide hydrochloride* provides *semicarbazones*, which often are nicely crystalline:

$$R-C\underset{O}{\overset{H}{<}} \xrightleftharpoons{\overset{1}{H_2}\overset{2}{N}-NH-\overset{O}{\overset{\|3}{C}}-\overset{4}{NH_2}(H^\oplus)}$$

$$\left[R-\underset{\underset{\oplus}{H_2N-NH-C-NH_2}}{\overset{H}{C}}-\bar{\underset{}{O}}I^\ominus \rightleftharpoons R-\underset{HN-NH-\underset{O}{\overset{\|}{C}}-NH_2}{\overset{H}{C}}-OH \right] \xrightleftharpoons{-H_2O} R-C\underset{N-NH-\underset{O}{\overset{\|}{C}}-NH_2}{\overset{H}{<}}$$

As in the case of oxime formation, the first step here is a *nucleophilic* attack by the NH_2 group on the aldehydic carbon atom, followed by migration of a proton. The next step is the rate-determining step, involving acid-catalysed loss of water; the optimum pH for the reaction is about 5.

2.21.7.5 Hydrazone Formation. Hydrazine derivatives react with aldehydes to form hydrazones. Thus *phenylhydrazine*, in solution in acetic acid, provides phenylhydrazones:

Phenylhydrazine   A phenylhydrazone

Hydrazine itself, or its hydrate, cannot be used as a typical example of this reaction since both NH_2 groups normally react with the aldehyde giving an *aldazine*:

An aldazine

When aldazines are heated for a short time with hydrazine hydrate a rather unstable *aldehydrazone* is obtained:

An aldehydrazone

Simple aldehydes are generally liquids, and the condensation reactions are used in order to provide crystalline derivatives which are of use in both the purification and the identification of aldehydes.

2.21.7.6 Aldol Addition Reactions. Aldehydes which have α-methyl or -methylene groups dimerise in the presence of base or acid catalysts (dilute hydroxides, carbonates, cyanides, sodium acetate, or dilute hydrochloric acid), giving *β-hydroxyaldehydes (aldols)*. For example, acetaldehyde gives *acetaldol* (β-hydroxybutyraldehyde), a colourless syrupy liquid, b.p. 83°C (356 K)/ 26 mbar (26 hPa).

Acetaldol

The net result is the transfer of an α-hydrogen atom from one molecule to the oxygen atom of another, with concomitant formation of a new C—C bond.

The name 'aldol' is a shortening of '*aldehydealcohol*' and brings out the dual functionality of the compound. *Aldol addition* is a more correct description of the reaction than the older term 'aldol condensation' since no loss of water is involved in the process.

The mechanism of the aldol reaction has been explained as follows (R. P. Bell, 1941). A basic catalyst such as the HO-anion removes a proton from the methyl group of acetaldehyde, which is activated by the −I-effect of the carbonyl group.

The resultant anion (which can be called the methylene component) is *stabilised by mesomerism*; a pair of electrons attacks the positively charged carbon atom of a carbonyl group in another molecule of acetaldehyde (the carbonyl component) and forms an aldol anion. This removes a proton from the water which is present, forming acetaldol.

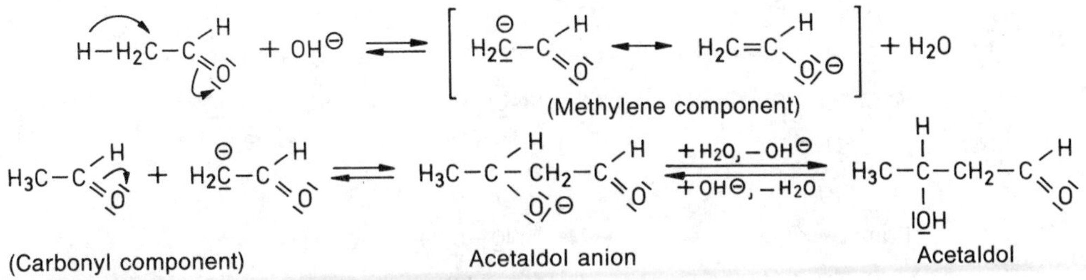

(Methylene component)

(Carbonyl component)   Acetaldol anion   Acetaldol

The addition is not restricted to reactions involving one aldehyde, or to reactions between two different aldehydes, but can also involve reaction between an aldehyde and a ketone or between two ketones. Since the carbonyl group of an aldehyde is more reactive than that of a ketone, in a reaction involving both, the aldehyde acts as the carbonyl component, unless other circumstances change the situation (see p. 207). Only hydrogen atoms which are vicinal (α) to the carbonyl group can be involved in this process. The more acidic the C—H in the methylene group is, the easier reaction proceeds.

Aldol addition is an equilibrium reaction and very dependent upon conditions such as pH and temperature. It is also an important biochemical process, for example in important steps in alcoholic fermentation and in glycolysis, in which the C_6 chain of *glucose is reversibly split* into two molecules of glyceraldehyde.

2.21.7.7 Cannizzaro Reactions of Aldehydes.[1] Aldehydes which have no reactive α-hydrogen atoms and are thus prevented from participating in the aldol reaction instead undergo the *Cannizzaro* reaction (1853) when treated with alkali. Two molecules of aldehyde give rise to one molecule of a primary alcohol and one of the salt of a carboxylic acid:

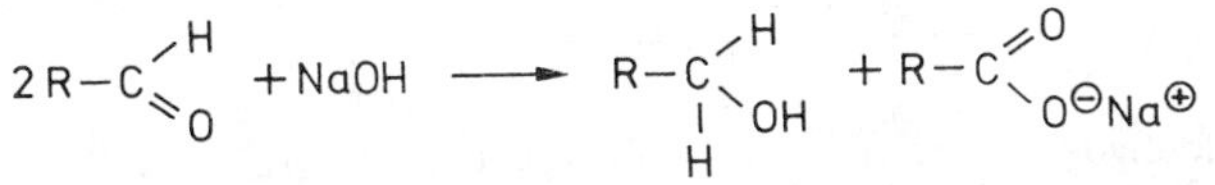

Such a coupled conversion of a substance into products of higher and lower oxidation states has been called *oxidoreduction*. Examples of aliphatic aldehydes, which easily disproportionate in this way, are formaldehyde and trimethylacetaldehyde, neither of which have α-hydrogen atoms.

If formaldehyde is mixed with another aldehyde a *crossed Cannizzaro reaction* takes place; this provides almost exclusively sodium formate and the alcohol derived from the other aldehyde.

The *mechanism of this disproportionation* is even now not completely clear. The rate of reaction depends not only on the concentrations of aldehyde and of hydroxide ion, but also on that of the cation of the base. It was shown early on that weaker bases such as calcium and barium hydroxides are better catalysts than are alkali-metal hydroxides. Thallium(I) hydroxide has proved to be particularly effective.

It is assumed that in the first place two mols of aldehyde and one of the metal hydroxide form a so-called 'primary complex' (*Pfeil*). In this complex the metal atom is co-ordinated to the oxygen atoms of the carbonyl groups. This results in an increase of positive charge on the carbon atoms of the C=O groups, which are thus made more susceptible to attack by the hydroxide anion of the metal hydroxide. When such attack takes place on one of the carbonyl groups a hydride ion migrates to the other carbonyl group, producing a carboxylic acid together with an alkoxide ion. A simple acid–base reaction between these products leads to the formation of an alcohol and a carboxylate anion.

---

[1] See *T. A. Geissman*, The *Cannizzaro* Reaction, Org. Reactions **2**, 94 *ff.* (1944).

Primary complex

The hydride ion migration was demonstrated by *Bonhoeffer* and *Fredenhagen* (1938) by carrying out the conversion of benzaldehyde into benzyl alcohol and benzoic acid in deuteriated water. They showed that there was no deuterium bonded to carbon in the resultant benzyl alcohol and thus that no deuterium exchange with the solvent had taken place. Therefore the extra hydrogen atom in the alcohol must have as its source the hydrogen atom of an aldehyde group.

2.21.7.8 *Tishchenko* Reaction. Whilst the *Cannizzaro* reaction, with just a few exceptions, only gives good yields from arylaldehydes, it was found by *Tishchenko* (1906) that when an aluminium alkoxide is used as reagent, aliphatic aldehydes also take part in such a disproportionation reaction on a preparative scale. Since the alkoxide is destroyed by water it is necessary to carry out the reaction in dry methanol. Under these conditions the resultant carboxylic acid and alcohol react together to give an ester; for example, under the appropriate conditions a quantitative yield of ethyl acetate can be obtained from a mixture of acetaldehyde and aluminium ethoxide (100 : 3–5):

Acetaldehyde   Ethyl acetate

The latter is used especially in the lacquer/varnish industries as a solvent.

2.21.7.9 Polymerisation of Aldehydes. Another typical property of aldehydes is their tendency, especially in the presence of $H^{\oplus}$ ions, to link together to form higher molecular mass compounds. This involves reactions of the aldol type, and the process is exothermic. Compounds with ring structures as well as open-chain products may result.

Formaldehyde *polymerises*[1] to give *paraformaldehyde*, which has relatively low molecular mass, either when an aqueous solution of it is evaporated, or if either acidic or basic catalysts are added to it. A *linear* product is formed which has the following constitution:

$$H\text{---}[O\text{--}CH_2]_{\overline{n}}\text{---}OH$$

Paraformaldehyde

On the other hand, water-free formaldehyde, either in the liquid state or in solution in inert solvents, polymerises at low temperatures ($-80°C$ [193 K]), in the presence of either an alkoxide or a tertiary amine as initiator, to give products which are either more or less transparent solids or powders. These are made up from *polyformaldehydes*

---

[1] See *W. Kern*, Über die Polymerisation und Copolymerisation von Trioxan und von Formaldehyde (On the Polymerisation and Copolymerisation of Trioxan and of Formaldehyde, Chemiker-Ztg. *88*, 623 (1964).

(poly(oxymethylenes)) of high molecular mass which can be processed at higher temperatures to form films or filaments (Delrin).

This polymerisation involves an *anionic chain mechanism*:

$$\underset{H_2C=\underline{\underline{O}}}{\overset{\delta^+ \quad \delta^-}{}} \quad \xrightarrow{+ \ I\underline{\underline{O}}H^{\ominus}} \quad H\underline{\underline{O}}-CH_2-\underline{\underline{O}}I^{\ominus} \quad \xrightarrow{+ nCH_2=O} \quad HO\!-\!\!\left[CH_2-O\right]_{\!n}\!\!CH_2-\underline{\underline{O}}I^{\ominus}$$

Termination of the chain reactions is brought about by cations.
Polyoxymethylenes have been detected in the tail of Halley's comet.
As well as the open-chain polymers of *formaldehyde* there is also a *cyclic* trimer, trioxan (1,3,5-trioxan, trioxymethylene). This deposits as needles (m.p. 63°C [336 K]) from gaseous formaldehyde. It is readily soluble in water and in organic solvents and is *not* a reducing agent.

$$3\,H-C\!\!\overset{H}{\underset{O}{\diagup}} \quad \longrightarrow \quad$$

Trioxan (Trioxymethylene)

Other products resulting from the *condensation* of formaldehyde are mentioned on p. 432.

When a drop of concentrated sulphuric acid is added to acetaldehyde a vigorous reaction takes place. The product is the liquid trimer *paraldehyde* (2,4,6-trimethyl-1,3,5-trioxan), b.p. 124°C (397 K):

$$3\,H_3C-C\!\!\overset{H}{\underset{O}{\diagup}} \quad \underset{(H^{\oplus})}{\rightleftharpoons} \quad$$

Paraldehyde

It may be seen from their formulae that the cyclic trimers have no free aldehyde groups, and they do not react as reducing agents or with the standard aldehyde reagents. In their properties they resemble acetals; they are *unaffected by alkalis* but when heated with acids they are reconverted into the corresponding aldehyde.

The More Important Aldehydes

2.21.8 *Formaldehyde (Methanal)*

Preparation

On the industrial scale it is largely made by *dehydrogenation of methanol*, using atmospheric oxygen in the presence of a silver or copper catalyst at about 600°C (875 K) (*A. W. v. Hofmann*, 1867):

$$H_3C-OH + \tfrac{1}{2}O_2 \xrightarrow{(Ag)} H-C\!\!\overset{H}{\underset{O}{\diagup}} + H_2O \qquad \Delta H = -159\,kJ/mol$$

Hydrogen is removed on the catalyst and the role of the oxygen is to burn it to give water. A 40% solution of formaldehyde is produced which also contains 8–10% methanol; it is known as *formalin*.

Properties

Formaldehyde is a poisonous, pungent-smelling gas which dissolves readily in water to form a hydrate, HO—CH_2—OH. It is a component of tobacco smoke. It takes part in the usual reactions of aldehydes, but differs from higher aliphatic aldehydes, having some special properties of its own. The special point about formaldehyde is that the carbonyl group has two hydrogen atoms attached to it. One consequence of this is that with dilute aqueous alkali it undergoes *oxidoreduction* to methanol and formic acid (*Cannizzaro* reaction):

$$2\,H-C\overset{H}{\underset{O}{\diagup}} + H_2O \xrightarrow{(OH^{\ominus})} H_3COH + H-C\overset{O}{\underset{OH}{\diagup}}$$

With *ammonia* it gives *hexamethylene-tetramine* (1,3,5,7-tetraazaadamantane). An aldehyde-ammonia is formed first. This loses water to give methanimine (formaldimine) which cyclises to its trimer 1,3,5-triazinane; this reacts further to provide the final product:

$$3\,H-C\overset{H}{\underset{O}{\diagup}} + 3NH_3 \rightleftharpoons \left[3\,H-C\overset{H}{\underset{NH_2}{-OH}} \xrightarrow{-3H_2O} 3\,H_2C=NH \rightleftharpoons \underset{\text{1,3,5-Triazinane}}{\text{(ring structure)}} \right]$$

Aldehyde-ammonia   Methanimine   1,3,5-Triazinane

$$\xrightarrow[+3H_2O]{+3HCHO+NH_3} \underset{\text{Hexamethylenetetramine}}{\text{(ring structure)}} \xrightarrow[-12\,H_3CCOOH]{+6(H_3CCO)_2O \; +4HNO_3+2NH_4NO_3} 2\; \underset{\text{RDX}}{\text{(ring structure)}}$$

Hexamethylenetetramine   RDX

When hexamethylenetetramine is treated with concentrated nitric acid in the presence of ammonium nitrate and acetic anhydride, it undergoes nitration and oxidative breakdown to give the high explosive RDX (Cyclonite), 1,3,5-trinitro-1,3,5-triazinane.

Hexamethylenetetramine, $(CH_2)_6N_4$, is a crystalline substance which is soluble in water. It undergoes acid hydrolysis to regenerate formaldehyde and is hence employed as a hardening agent for synthetic resins.

In medicine, as INN: *Methenamine* or *urotropine*, it is used as a urinary disinfectant.

Uses

Formaldehyde is used as a domestic disinfectant and as a preservative and hardening agent for anatomical specimens, because with proteins it forms condensation products which are insoluble in water (*cf.* tanning). The preservative effect of 'smoking' is largely due to the influence of formaldehyde. Formaldehyde condenses with phenol- or naphthalene-sulphonic acids to form synthetic tanning agents (Nerandols). It is important industrially as a starting material for the preparation of dyestuffs, for the hardening of casein and gelatine and, above all, for the preparation of plastics (phenol-formaldehyde resins). Formaldehyde reacts with sodium dithionite, $Na_2S_2O_4$, to give the sodium salt of hydroxymethanesulphinic acid, $H_2C(OH)-SO_2Na$ (Rongalit), which is used as a reducing agent in vat-dyeing and in printing.

2.21.9 Acetaldehyde (Ethanal)

Preparation

2.21.9.1 From ethanol. (a) In the laboratory by *oxidation of ethanol* with potassium dichromate and sulphuric acid:

$$3\,H_3C-CH_2OH+2CrO_3 \longrightarrow 3\,H_3C-C\overset{\nearrow H}{\underset{\searrow O}{}} + Cr_2O_3 + 3H_2O$$

(b) In countries which are rich in carbohydrates it is made by *dehydrogenation of ethanol*, using atmospheric oxygen in the presence of a silver catalyst at about 550°C (825 K). The acetaldehyde which is formed is separated from ethanol and then distilled. The yield is 85–95%.

$$H_3C-CH_2-OH + \tfrac{1}{2}O_2 \xrightarrow{(Ag)} H_3C-C\overset{\nearrow H}{\underset{\searrow O}{}} + H_2O$$

2.21.9.2 *Wacker*-Hoechst Process. By oxidation of ethylene with air or oxygen in an aqueous solution of $PdCl_2/CuCl_2$. A yield of 95% is achieved.

$$H_2C=CH_2 + \tfrac{1}{2}O_2 \xrightarrow{(PdCl_2/CuCl_2)} H_3C-C\overset{\nearrow H}{\underset{\searrow O}{}} \qquad \Delta H = -218\,kJ/mol$$

Oxidation of higher alkenes provides ketones; for example, acetone is obtained from propene and butanone from but-1-ene. Reaction proceeds according to the following equation:

$$H_2C=CH_2 + PdCl_2 + H_2O \longrightarrow H_3C-CHO + 2HCl + Pd$$

The course of the reaction is dealt with in more detail on p. 383. Regeneration of the catalyst is brought about by atmospheric oxygen in the presence of a catalytic amount of copper chloride.

$$Pd + 2\,CuCl_2 \rightleftharpoons PdCl_2 + 2\,CuCl$$

$$2\,CuCl + HCl + \tfrac{1}{2}O_2 \longrightarrow 2\,CuCl_2 + H_2O$$

This synthesis has superseded the preparation of acetaldehyde from ethanol or by hydration of acetylene (see p. 95). Since more than half of the acetaldehyde prepared in this way is further converted into acetic acid, a more economic synthesis of acetic acid (see p. 235) provides competition for the *Wacker*-Hoechst process.

Properties and Uses

Acetaldehyde is a colourless, pungent-smelling liquid, b.p. 20.2°C (293.4 K), which is completely miscible with water, ethanol or ether. Its polymerisation products have already been described (see p. 200). It plays an important role as an intermediate in the biochemical degradation of sugars (see alcoholic fermentation).

Industrially, acetaldehyde is used in the preparation of acetic acid, acetic anhydride, acetone, ethanol, acrolein, acetaldol, crotonaldehyde and butadiene.

2.21.10 Propionaldehyde (Propanal)

Propionaldehyde is a liquid with a suffocating smell. It is made industrially by *hydroformylation of ethylene.*

Oxo-synthesis.[1] This term is applied to the simultaneous addition of carbon monoxide and water to C=C double bonds to form *aldehydes*. Reaction is carried out under pressure at temperatures between 50 and 200°C (325–475 K) (*O. Roelen*, 1938). Cobalt is usually used as catalyst, in the form of $HCo(CO)_4$; an alternative is rhodium.

The simplest example of an oxo-reaction is the *synthesis of propionaldehyde* from ethylene and water gas at about 200 bar (20 MPa) and an optimal temperature of 100–115°C (375–390 K).

$$H_2C=CH_2 + CO + H_2 \xrightarrow{\text{(Co)}} H_3C-CH_2-C\overset{\displaystyle H}{\underset{\displaystyle O}{\diagup}}$$

Ethylene   Propionaldehyde

If one of the hydrogen atoms in ethylene is replaced by an alkyl group, then the carbon monoxide might add to C(1) or C(2) of the alkene. In the first case an aldehyde is formed and the carbon chain is lengthened; in the second case an aldehyde is also formed but a branching methyl group is added to the carbon chain:

$$\underset{(2)\quad(1)}{R-CH=CH_2} + CO + H_2 \xrightarrow{\text{(Co)}} \begin{cases} R-CH_2-CH_2-C\overset{H}{\underset{O}{\diagup}} \\[2ex] R-\underset{\underset{CH_3}{|}}{CH}-C\overset{H}{\underset{O}{\diagup}} \end{cases}$$

Which of the two reactions predominates depends upon both the catalyst used and the nature of the substituent group R.

Thus *hydroformylation of propene* ($R = CH_3$) over a cobalt catalyst gives rise to butyraldehyde (butanal) and isobutyraldehyde (2-methylpropanal) in the ratio 4:1, whilst in the presence of a rhodium catalyst only butyraldehyde is formed.

It is also possible to produce alcohols by means of the oxo-synthesis. If the temperature is raised to about 180°C (455 K) a cobalt catalyst can also bring about hydrogenation of the first-formed aldehyde to give an alcohol. For example, under these conditions propan-1-ol is obtained from ethylene, carbon monoxide and hydrogen.

This process is used industrially for the *preparation of higher alcohols.* Mixtures of alkenes, obtained from cracking processes, serve as starting materials.

$$R-CH=CH_2 + CO + 2H_2 \xrightarrow{\text{(Co)}} \begin{cases} R-CH_2-CH_2-CH_2-OH \\[2ex] R-\underset{\underset{CH_3}{|}}{CH}-CH_2-OH \end{cases}$$

The fraction containing alkenes with 5–11 carbon atoms provides alcohols used as solvents and plasticisers; those with 12–18 carbon atoms are used as detergents.

2.21.11 Halogenoaldehydes

The trihalogen derivatives of acetaldehyde are of particular importance.

---

[1] See *P. Pino* et al., in *J. Wender* and *P. Pino*, in Organic Synthesis *via* Metal Carbonyls, Vol. 2, 43 *ff.* (1977); *H. M. Colquhuon, D. J. Thompson* and *M. V. Twigg*, Carbonylation (Plenum Press, New York 1991).

Chloral (trichloroacetaldehyde) is made industrially by *passing chlorine into aqueous ethanol*. Ethanol is dehydrogenated giving various chlorinated intermediates which are converted into *chloral hydrate*:

$$H_3C-CH_2OH + 4\,Cl_2 + H_2O \rightarrow Cl_3C-CH(OH)_2 + 5\,HCl$$

Chloral hydrate

If chloral hydrate, m.p. 58°C (331 K), is distilled with concentrated sulphuric acid (1:1 mixture) it is converted into chloral, a pungent smelling liquid, b.p. 98°C (371 K).

In chloral hydrate, *two* HO groups are attached to the same carbon atom, providing an example of an exception to *Erlenmeyer's* rule. The hemiacetals and aldehyde-ammonia obtained from chloral and alcohols or ammonia are also relatively stable.

$$Cl_3C-\overset{\displaystyle H}{\underset{\displaystyle OH}{C}}-OH \qquad Cl_3C-\overset{\displaystyle H}{\underset{\displaystyle OR}{C}}-OH \qquad Cl_3C-\overset{\displaystyle H}{\underset{\displaystyle NH_2}{C}}-OH$$

Chloral hydrate   Chloral hemiacetal   Choral-ammonia

The stability of these adducts is a consequence of the inductive effect ($-$I-effect) of the trichloromethyl group.

Chloral hydrate is the starting material for the preparation of pure chloroform and especially of *DDT*. It is one of the oldest soporifics and is still used for this purpose.

Bromal (tribromoacetaldehyde) is made by *brominating paraldehyde* in the presence of sulphur:

$$(H_3C-CHO)_3 + 9\,Br_2 \xrightarrow{\;(S)\;} 3\,Br_3C-\overset{\displaystyle H}{\underset{\displaystyle O}{C}} + 9\,HBr$$

Paraldehyde   Bromal

It is a liquid, b.p. 174°C (447 K) which with water forms *bromal hydrate*, m.p. 53°C (326 K).

2.21.12 Unsaturated Aldehydes (Alkenals and Alkynals)

These compounds are derived from alkenes or alkynes by replacing a hydrogen atom with an aldehyde group. The commonest examples are:

		b.p.
Acrolein, Propenal	$H_2C=CH-CHO$	52,3 °C (325,5 K)
But-2-enal, Crotonaldehyde	$H_3C-CH=CH-CHO$	104 °C (377 K)
Propargaldehyde, Propynal	$HC\equiv C-CHO$	69 °C (333 K)

Acrolein, the first member of the alkenal series, is the principal dehydrogenation product from allyl alcohol. It is easily oxidised to acrylic acid.

Preparation

2.21.12.1 By *heating glycerol with dehydrating agents, e.g.* with potassium hydrogen sulphate at about 200°C (475 K).

$$
\begin{array}{ccccc}
\begin{matrix} CH_2-OH \\ | \\ CH-OH \\ | \\ CH_2-OH \end{matrix}
& \xrightarrow[-H_2O]{(KHSO_4)}
& \left[\begin{matrix} CH_2OH \\ | \\ CH \\ || \\ C \overset{H}{\underset{OH}{\diagdown}} \end{matrix} \right]
\rightleftharpoons
\begin{matrix} CH_2OH \\ | \\ CH_2 \\ | \\ C \overset{H}{\underset{O}{\diagdown}} \end{matrix}
& \xrightarrow[-H_2O]{}
& \begin{matrix} CH_2 \\ || \\ CH \\ | \\ C \overset{H}{\underset{O}{\diagdown}} \end{matrix}
\\
\text{Glycerol} & & \text{Enol form} & \text{Keto form} & \text{Acrolein}
\end{array}
$$

First of all, the secondary alcohol group is eliminated. The resultant β-hydroxyallyl alcohol (enol form) is in equilibrium with β-hydroxypropionaldehyde (keto form) (see keto–enol tautomerism), and water is eliminated from the latter to form acrolein. It is evident that the conversion of glycerol into allyl alcohol on treatment with formic acid (p. 122) is dependent upon a hydride shift, as described on p. 200.

2.21.12.2 On the industrial scale acrolein is made by *aldol addition of acetaldehyde to formaldehyde* in the gas phase at 300°C (575 K), in the presence of an alkaline dehydration catalyst (silica gel):

$$
H-C\overset{H}{\underset{O}{\diagdown}} + H_3C-C\overset{H}{\underset{O}{\diagdown}} \rightleftharpoons \underset{OH}{CH_2-CH_2-C}\overset{H}{\underset{O}{\diagdown}} \xrightarrow{-H_2O} H_2C=CH-C\overset{H}{\underset{O}{\diagdown}}
$$

2.21.12.3 By *reaction of propene with atmospheric oxygen* at 200–500°C (475–775 K) and 1–10 bar (0.1–1 MPa) over a mixed catalyst:

$$
H_2C=CH-CH_3 + O_2 \xrightarrow{[cat.], H_2O} H_2C=CH-C\overset{H}{\underset{O}{\diagdown}} + H_2O
$$

Properties

Acrolein polymerises readily. It is a colourless lachrymatory liquid with a strong pungent odour (*Lat. acer* = pungent, *oleum* = oil). It is formed from the glycerides when fats are overheated. Acrolein is an α,β-unsaturated aldehyde and has a *conjugated* double-bond system; it is a *heterodiene*. In its chemical properties it shows some resemblance to butadiene, but, because in this case there are two different sorts of double bond, the range of reactions is greater.

But-2-enal (crotonaldehyde) is a colourless liquid with similar properties to those of acrolein. There are two isomeric forms, *cis* and *trans*, which are only separated from one another with difficulty.

$$
\begin{array}{ccc}
\begin{matrix} \text{But-2-enal} \\ ((Z)\text{-form}) \\ (\text{Isocrotonaldehyde}) \end{matrix}
& \underset{H}{\overset{H_3C}{\diagdown}}C=C\underset{H}{\overset{CHO}{\diagup}}
\rightleftharpoons
\underset{H_3C}{\overset{H}{\diagdown}}C=C\underset{H}{\overset{CHO}{\diagup}}
& \begin{matrix} \text{But-2-enal} \\ ((E)\text{-form}) \\ (\text{Crotonaldehyde}) \end{matrix}
\end{array}
$$

It is best made by distilling acetaldol with acetic acid (as catalyst):

$$
\text{Acetaldol} \quad H_3C-CH(OH)-CH_2-C\overset{H}{\underset{O}{\diagdown}} \xrightarrow[-H_2O]{(H^\oplus)} H_3C-CH=CH-C\overset{H}{\underset{O}{\diagdown}} \quad \begin{matrix}\text{But-2-enal} \\ (\text{Crotonaldehyde})\end{matrix}
$$

If dimerisation of acetaldehyde is carried out in the presence of *piperidine acetate* as catalyst, dehydration of the first-formed acetaldol takes place so rapidly that the product obtained is crotonaldehyde. In this case it is appropriate to call the reaction an *aldol condensation*.

The terminal methyl group of crotonaldehyde can also take part in aldol condensation reactions. It can react with other aldehydes to provide higher polyunsaturated aldehydes, for example it reacts with acetaldehyde in the presence of *piperidine acetate* to give water and hexadienal.

$$H_3C-C{\overset{H}{\underset{O}{\lessgtr}}} + H_3C-CH=CH-C{\overset{H}{\underset{O}{\lessgtr}}} \longrightarrow H_3C-CH=CH-CH=CH-C{\overset{H}{\underset{O}{\lessgtr}}} + H_2O$$

Acetaldehyde   Crotonaldehyde   Hexadienal

2.21.13 Directed Aldol Reaction *(Wittig, 1964)*[1]

If an aldehyde is converted into a *Schiff's* base, such as *N*-ethylidenecyclohexylamine, then its reactivity as a carbonyl-type reagent is sufficiently reduced to make it resemble that of a ketone (see p. 199). The *Schiff's* base can be *metallated* with lithium diisopropylamide and serve as the *methylene component* in aldol reactions:

$$H_3CCHO + H_2N-cC_6H_{11} \xrightarrow[-H_2O]{} H_3CCH=N-cC_6H_{11} \xrightarrow[-(iPr)_2NH]{+(iPr)_2NLi}$$

Cyclohexylamine   *N*-Ethylidenecyclohexylamine

$$CH_2-CH=N-cC_6H_{11} \xrightarrow[-LiOH]{\overset{1)\ (C_6H_5)_2CO}{2)\ H_2O}} C_6H_5-\underset{OH}{\overset{C_6H_5}{\underset{|}{\overset{|}{C}}}}-CH_2-CH=N-cC_6H_{11} \xrightarrow[-H_2N-cC_6H_{11}]{\overset{+H_3O^{\oplus}}{-H_2O}} (C_6H_5)_2C=CHCHO$$

β-Phenylcinnamaldehyde
(3,3-Diphenylacrolein)

A *methylene component* for an aldol reaction can also be obtained by making a suitable *enol ether*. Trimethylsilyl enol ethers, for example 1-trimethylsiloxycyclopentene, are very useful for this purpose, and in the presence of a *Lewis* acid such as titanium tetrachloride undergo regioselective and stereoselective[2] reactions under mild conditions.

$$C_6H_5-CH_2-C{\overset{H}{\underset{O}{\lessgtr}}} + {\overset{OSi(CH_3)_3}{\bigcirc}} \xrightarrow[-(H_3C)_3SiCl]{+TiCl_4} \left[C_6H_5-CH_2-CH{\overset{O\cdots TiCl_3}{\underset{O}{\diagdown}}} \right] \xrightarrow{H_2O} C_6H_5-CH_2-\underset{OH}{\overset{|}{CH}}{\overset{O}{\diagdown}}$$

(68%)

Propargaldehyde (propynal), $HC\equiv C-CHO$ is the simplest aldehyde which also has a $C\equiv C$ triple bond. It is made by *oxidation of propargyl alcohol*, using, for example, potassium permanganate. It is a colourless liquid whose vapour irritates the mucous membranes in the same way that acrolein does.

2.22 Aliphatic Ketones (Alkanones)

The characteristic feature of ketones is the presence of a *carbonyl* or *keto* group, $C=O$, flanked by two alkyl groups. If the two alkyl groups are identical, then the compound is described as a *simple ketone*, if they are different as a *mixed ketone*:

---

[1] See *G. Wittig* et al., Directed Aldol Condensations, Angew. Chem. Int. Ed. Engl. 7, 7 (1968); *T. Mukaiyama* The Directed Aldol Reaction, Org. Reactions *28*, 203 (1982).
[2] See *S. Kobayashi* et al., Asymmetric Aldol Reaction of Silyl Enol Ethers with Aldehydes, Chem. Lett. *1989*, 297.

$$R-\underset{\underset{O}{\|}}{C}-R \qquad R-\underset{\underset{O}{\|}}{C}-R'$$

The simplest ketone is *acetone* (*propanone*), $H_3C-CO-CH_3$.

Higher ketones are named (in accord with IUPAC rules) either by replacing the '-e' ending of the related alkane from which they are derived by the ending '-one', e.g. *pentan-2-one*, $H_3C-CO-CH_2-CH_2-CH_3$, or by using the prefix 'oxo', as in *2-oxopentane*. The older method of nomenclature, which is still encountered, added the names of the two attached alkyl groups to the word ketone, e.g. *ethyl methyl ketone*, $H_3C-CO-CH_2-CH_3$ (*butanone*).

2.22.1 General Methods of Preparation

2.22.1.1 From secondary alcohols.

(a) When secondary alcohols are *oxidised* or *dehydrogenated*, for example, using stoichiometric amounts of chromium(VI) oxide in dilute sulphuric acid (*Jones reagent*), ketones are formed:

$$\underset{R}{\overset{R}{>}}C\overset{H}{\underset{OH}{<}} \quad \xrightarrow[-\,H_2O]{[O]} \quad \underset{R}{\overset{R}{>}}C=O$$

An elegant and efficient process involves heterogeneous oxidation, using a clay, for example bentonite, which has been impregnated with iron(III) nitrate or copper(II) nitrate, in a non-polar solvent.[1]

$$3\ \underset{R^1}{\overset{R}{>}}C\overset{H}{\underset{OH}{<}} \quad \xrightarrow[(-2NO, -4H_2O)]{\substack{\text{Iron(III) nitrate} \\ \text{Clayfen } (+2H^{\oplus},+2NO_3^{\ominus})}} \quad 3\ \underset{R^1}{\overset{R}{>}}C=O$$

The progress of the reaction can be followed from the evolution of the nitrogen oxides; yields commonly exceed 80%.

(b) *Oppenauer* **Oxidation (1937).**[2] When a secondary alcohol is heated with an excess of a ketone such as cyclohexanone in the presence of aluminium t-butoxide, a *redox process* ensues which provides the required ketone together with cyclohexanol:

$$\underset{R}{\overset{R}{>}}C\overset{H}{\underset{OH}{<}} + \underset{\text{Cyclohexanone}}{\hexagon\!\!=\!O} \quad \underset{\xleftarrow{\hspace{2cm}}}{\xrightarrow{\text{Al t-butoxide}}} \quad \underset{R}{\overset{R}{>}}C=O + \underset{\text{Cyclohexanol}}{\hexagon\!\!<\!\overset{H}{\underset{OH}{}}}$$

This method is used in particular with natural products such as steroids and alkaloids.

2.22.1.2 By distillation of the calcium salts of carboxylic acids.
This is a very old method; only in a few cases are good yields obtained and it is not a general method.

$$\begin{matrix} R-COO^{\ominus} \\ R-COO^{\ominus} \end{matrix} Ca^{2\oplus} \quad \longrightarrow \quad \underset{R}{\overset{R}{>}}C=O + CaCO_3$$

[1] See P. Laszlo et al., Clay-supported Copper(II) and Iron(III) Nitrates: Novel Multi-purpose Reagents for Organic Synthesis, Synthesis 1985, 909; Organic Chemistry using Clays. Reactivity and Structure Concepts in Organic Chemistry, Vol. 29 (Springer, Berlin 1993).

[2] See C. Djerassi, The Oppenauer Oxidation, Org. Reactions 6, 207 ff. (1951); J. Huskens et al., Synthesis, 1994, 1007.

2.22.1.3 By addition of Grignard reagents to nitriles, followed by hydrolysis:

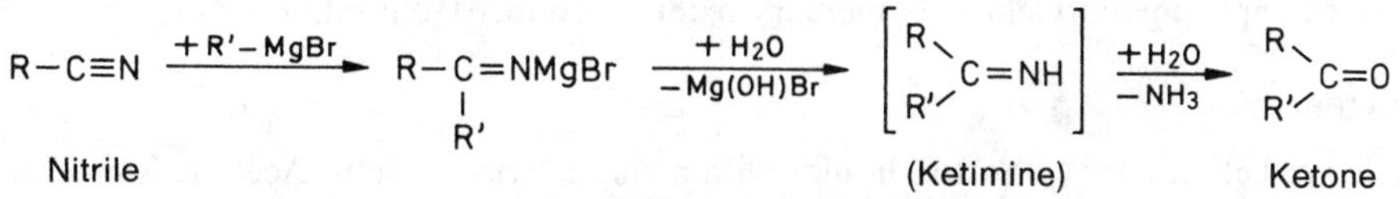

$$R-C\equiv N \xrightarrow{+R'-MgBr} R-\underset{\underset{R'}{|}}{C}=NMgBr \xrightarrow[-Mg(OH)Br]{+H_2O} \left[\underset{R'}{\overset{R}{}}C=NH\right] \xrightarrow[-NH_3]{+H_2O} \underset{R'}{\overset{R}{}}C=O$$

Nitrile   (Ketimine)   Ketone

2.22.1.4 From alkenes by the *Wacker*-Hoechst Process. (See p. 203).

2.22.1.5 From cyclopropyl derivatives by 'dry ozonolysis'.[1]

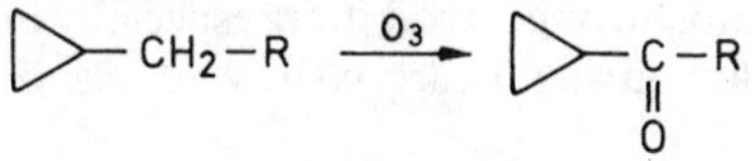

The hydrocarbon is adsorbed together with ozone on kieselgel at $-78°C$ (200 K). When it has been saturated with ozone (this can be detected by the kieselgel becoming blue), it is allowed to come slowly to room temperature.

2.22.1.6 Methods which are well adapted to the preparation of *methyl ketones* are the **alkylation of acetoacetic ester** followed by ketonic cleavage (see p. 298 *ff*), and the reaction of methylsulphinyl carbanions with esters of carboxylic acids coupled with reductive cleavage of the β-ketosulphoxides which are formed (see p. 148).

In the decisive step of forming the carbon–carbon bond, the carbonyl carbon atom of the ester group is attacked by a carbanion. In the process a C—O bond is lost and a C—C bond is formed. If it were possible to reverse the polarity of the carbon atom of a carbonyl group it would have important consequences for synthetic work.[2] An *umpolung* (dipole inversion) of this sort is involved in the following reaction.

2.22.1.7 Conversion of aldehydes into ketones by the *Corey–Seebach* method. The aim of the umpolung in this case is the generation of a species of reagent, which reacts like an *acyl anion* $[R\underline{C}{=}O]^{\ominus}$. Because of the marked ability of sulphur to stabilise anionic centres, the aldehyde group is converted, by reaction with propane-1,3-dithiol, into a dithiane derivative (a dithioacetal), which can be metallated by n-butyl lithium:

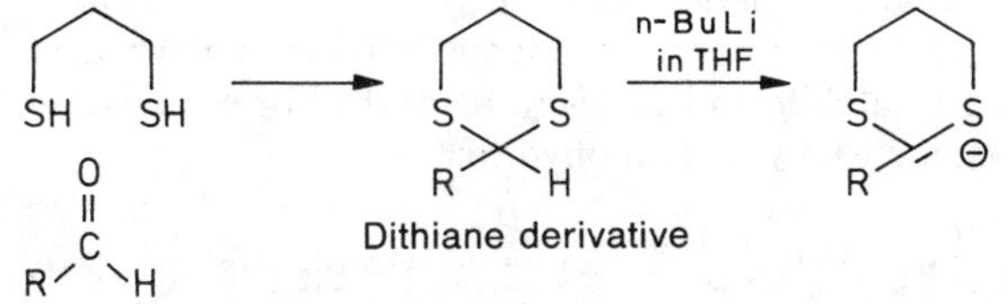

This anion is a *masked acyl anion*, which can react as a nucleophile with alkyl halides:

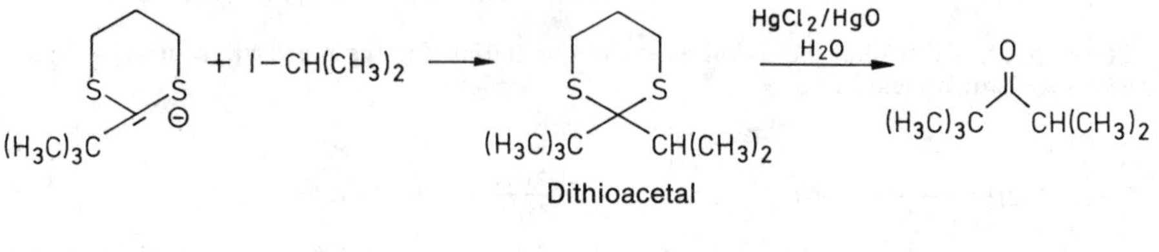

Dithioacetal

---

[1] See *A. de Meijere*, Nachr. Chem. Tech. Labor. *27*, 177 (1979).
[2] See *D. Seebach* et al., Methods of Reactivity Umpolung, Angew. Chem. Int. Ed. Engl. *18*, 239 (1979); Umpolung (dipole inversion) of Carbonyl Reactivity, Chem. Ind. (London) *1974*, 687.

The resultant *dithioacetal* (*dithioketal*) can be hydrolysed to give a ketone by the use of, for example, mercury chloride/mercury oxide or copper(II) nitrate on clay.

Properties

The lower ketones are colourless liquids with a characteristic odour. Acetone is miscible with water in all proportions, but as the number of carbon atoms in ketones increases their solubility in water decreases. Higher ketones are solids.

In contrast to aldehydes, ketones are *not reducing agents*. They are fairly stable to oxidising agents. This is because they do not have a hydrogen atom attached to the carbonyl group as aldehydes do. Stronger oxidation, especially in acidic conditions, results in splitting of the carbon chain, giving a carboxylic acid, *e.g.*

$$H_3C-\overset{\overset{\displaystyle O}{\|}}{C}-CH_2-CH_3 \xrightarrow{+\,3/2\,O_2} 2\ H_3C-COOH$$

Butanone   Acetic acid

If ketones are treated with *sodium* or *sodamide* in dry ether, salt-like compounds are formed, whose *anions are stabilised by mesomerism, e.g.*

$$H_3C-\overset{\overset{\displaystyle O}{\|}}{C}-CH_3 + NaNH_2 \longrightarrow \left[H_3C-\overset{\overset{\displaystyle \ominus}{|}}{\underset{\displaystyle |\underline{O}|}{C}}-\overset{\ominus}{C}H_2 \longleftrightarrow H_3C-\underset{\displaystyle |\underline{O}|_\ominus}{C}=CH_2 \right] Na^\oplus + NH_3$$

Acetone   Sodium acetonide

2.22.2 Addition and Condensation Reactions

Ketones undergo almost identical addition and condensation reactions to those of aldehydes, since these reactions are dependent upon the presence of the carbonyl group. In general, ketones react less readily than aldehydes. This is also evident in the fact that ketones are less prone to polymerise than aldehydes.

Ketones undergo addition reactions analogous to those of aldehydes with hydrocyanic acid, sodium hydrogen sulphite and *Grignard* reagents. For example, acetone reacts with aqueous sodium cyanide to give a cyanohydrin:

$$\overset{H_3C}{\underset{H_3C}{}}C\overset{\delta+}{=}\overset{\delta-}{\underline{O}} + |\overset{\ominus}{C}N \rightleftharpoons \overset{H_3C}{\underset{H_3C}{}}C\overset{\overset{\displaystyle |\underline{O}|^\ominus}{}}{\underset{C\equiv N}{}} \xrightarrow{+H_2O(\ominus OH)} \overset{H_3C}{\underset{H_3C}{}}C\overset{\overset{\displaystyle OH}{}}{\underset{C\equiv N}{}}$$

Acetone   Acetone cyanohydrin (2-Cyanopropan-2-ol)

The cyanohydrins can often be useful as starting materials for the preparation of unsaturated nitriles or of α-hydroxyacids, *e.g.*

$$\overset{\overset{\displaystyle CH_3}{|}}{H_2C=C-C\equiv N} \xleftarrow{-H_2O} H_3C-\overset{\overset{\displaystyle CH_3}{|}}{\underset{\underset{\displaystyle OH}{|}}{C}}-C\equiv N \xrightarrow[-NH_3]{+2H_2O(H^\oplus)} H_3C-\overset{\overset{\displaystyle CH_3}{|}}{\underset{\underset{\displaystyle OH}{|}}{C}}-COOH$$

Methacrylonitrile   Acetone   2-Hydroxy-2-methylpropionic acid
                           cyanohydrin   (α-Hydroxyisobutyric acid)

As in the case of aldehydes, condensation reactions of ketones with, for example, primary amines, hydroxylamine, semicarbazide or substituted hydrazines, lead to the formation of, respectively, *Schiff's bases* (*imines*), *ketoximes*, *semicarbazones* or *ketone hydrazones*. The tosylhydrazones are of preparative importance, since, when treated with an excess of a strong base such as butyl lithium, they give rise to *alkenes* (*Shapiro* reaction).[1]

Tosylhydrazone   Dianion (I)

Reaction proceeds *via* a dianion (I):

Aldehydes ($R^1 = H$) can be used as well as ketones.

The directed aldol reaction (p. 207) has been applied to *chiral hydrazones* and is a very efficient method for stereoselective alkylation leading to the formation of chiral α-*alkylcarbonyl compounds*. (*S*)-1-Aminomethoxy-methylpyrrolidine (SAMP), which is obtained from L-proline, serves as a chiral auxiliary. By stereoselective α-alkylation with 1-iodopropane of the chiral hydrazone (I) of pentan-3-one (diethyl ketone), (*S*)-4-methyl-heptan-3-one is obtained; this acts as an alarm pheromone in some species of ants.[2] The portion of the chiral auxiliary which remains unchanged in the synthesis is given a grey background in the formulae.

(*S*)-4-Methylheptan-3-one
Yield 60%, >99% ee

After the stereoselectively alkylated hydrazone (II) has been split by ozonolysis, the SAMP is recovered by reduction of the resultant *N*-nitroso-compound with lithium aluminium hydride.

---

[1] See *R. H. Shapiro*, Alkenes from Tosylhydrazones, Org. Reactions *23*, 405 (1976).

[2] See *D. Enders*, Alkylation of Chiral Hydrazones, in *J. D. Morrison* (Ed.), Asymm. Synthesis *3*, 275 (Academic Press, New York 1984).

The enantiomer of SAMP is known as RAMP; it is prepared from D-glucose.[1] Thus, by choice of the appropriate chiral auxiliary, the desired stereochemistry of the targeted molecule can be achieved. It is possible, in the case of enolates which are formed as intermediates in the *Birch* reduction of arylcarboxylic acid derivatives, to introduce substituents with sufficiently high stereoselectivity to provide a way of making chiral cyclohexane rings.[2]

2.22.2.1 Enamines[3]

Secondary amines react with aldehydes or ketones to form *α,β-unsaturated amines*. By analogy to *enols* they are known as *enamines*.

Cyclohexanone   1-(*N,N*-Dimethylamino)cyclohexene

Heterocyclic compounds such as pyrrolidine, piperidine or morpholine are often used as the secondary amine.

Pyrrolidine   Piperidine   Morpholine

The reaction is catalysed by acid. The mechanism is as follows:

Iminium ion   Enamine

Following addition of the amine to the carbonyl group, the HO group is protonated and water is eliminated. The iminium ion which is formed stabilises itself by losing a proton from the β-position, providing an *enamine*. Azeotropic removal of water shifts the equilibrium towards formation of the enamine.

---

[1] For the preparation and use of SAMP and RAMP see D. *Enders* et al., Org. Synth *65*, 173, 183 (1982).
[2] See *A. G. Schultz*, Acc. Chem. Res. *23*, 207 (1990).
[3] See *A. G. Cook* (Ed.), Enamines, 2nd edn (M. Dekker, New York 1988); *P. W. Hickmott*, Enamines, Tetrahedron *38*, 1975 (1982).

Enamines undergo *imine–enamine tautomerism*, similar to *keto–enol tautomerism*:

$$-\overset{|}{C}=\overset{|}{C}-\underset{H}{N}-R \ \rightleftharpoons \ -\overset{|}{C}-\underset{H}{\overset{|}{C}}=NR$$

Enamine   Imine (*Schiff*'s base)

The equilibrium usually lies very much on the side of the imine.[1] It can be appreciated therefore why secondary amines are especially useful for the preparation of enamines.

Mesomerism in the enamines results in the β-carbon atom being susceptible to *electrophilic* attack. For this reason they are of preparative importance. For example, *alkylation* or *acylation* of the enamine shown on p. 212, followed by hydrolysis, leads to the formation of *2-alkyl-* or *2-acyl-cyclohexanones* respectively, in the *Stork* reaction, e.g.

2-Methylcyclohexanone   2-Benzoylcyclohexanone

These reactions of *enamines* are a consequence of their being *electron-rich* alkenes. Enamines may be regarded as vinylogous amines.

2.22.2.2 Ketone acetals and Enol ethers

In the presence of acid catalysts, ketones form ketone acetals (ketals) only with polyhydroxy compounds (see p. 443), but not with simple alcohols with only one HO-group. However, ketone acetals derived from such alcohols can be obtained by the reaction of *ethyl orthoformate* with ketones:

| Cyclo-hexanone | Ethyl orthoformate | Cyclohexanone diethylacetal (ketal) | 1-Ethoxycyclohexene (Enol ether) |

In the example given, cyclohexanone diethylacetal is further converted into the *enol ether*, 1-ethoxycyclohexene, when it is heated with 0.5% toluene-*p*-sulphonic acid (TsOH).

Silyl enol ethers[1] are of preparative importance, since they provide reactive but *regiostable* enol derivatives. They are made by the reaction of enolate salts with chlorotrimethylsilane:

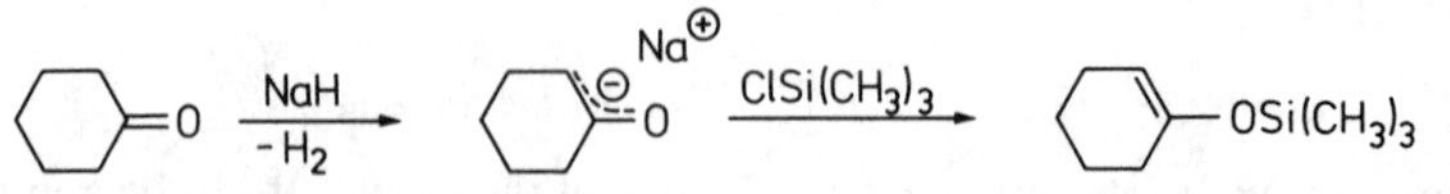

Enolate

1-Trimethylsiloxycyclohexene
(Silyl enol ether)

Silyl enol ethers show a similar regioselectivity to that of *enamines* and can be used, for example, in the directed aldol reaction (p. 207). The stereoselectivity in the formation of C—C bonds which is possible when these compounds are used is of importance.[2]

2.22.3 Reduction Products from Ketones

Various products can be obtained from the reduction of ketones, depending upon the conditions used. *Catalytic hydrogenation* (Pt or *Raney*-nickel), 'nascent hydrogen' (metal and acid) or the *Meerwein–Ponmdorf–Verley* reduction of ketones give *secondary alcohols*:

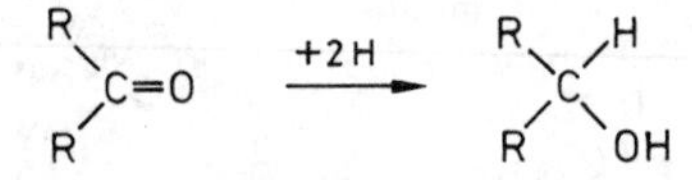

Ketones and aldehydes can be best reduced to alcohols, in high yields, by metal hydrides such as *lithium aluminium hydride*, e.g.

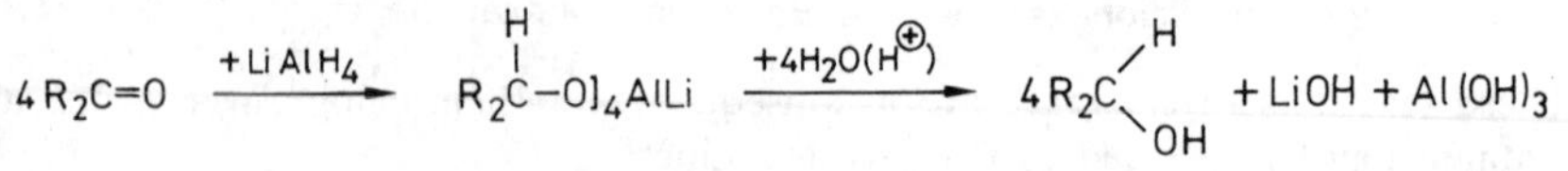

In this reaction all the hydrogen atoms of the aluminium hydride are made use of in the reduction.

The reaction mechanism resembles that followed by *Grignard* reagents, and involves *nucleophilic attack* by a hydride ion on the carbon atom of the carbonyl group:

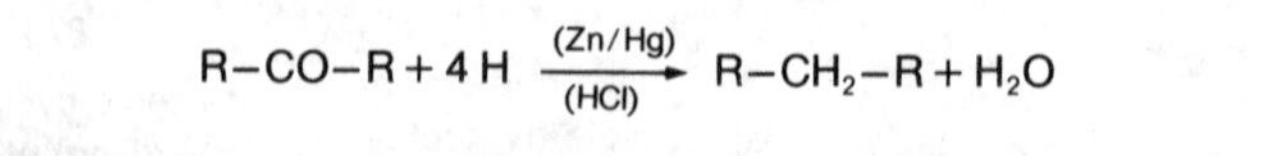

Sodium borohydride is a selective reagent for the reduction of aldehydes and ketones and, unlike $LiAlH_4$, *does not reduce* carboxylic acids or their esters.

2.22.3.1 Clemmensen Reduction.[3] When a ketone is heated with amalgamated zinc and concentrated hydrochloric acid, the carbonyl group is reduced to a methylene group:

$$R-CO-R + 4H \xrightarrow[\text{(HCl)}]{\text{(Zn/Hg)}} R-CH_2-R + H_2O$$

In a few cases a similar reaction takes place with aldehydes.

[1] See *P. Brownbridge*, Silyl Enol Ethers in Synthesis, Synthesis *1983, 1*, 85.
[2] See *D. Seebach* et al., EPC Syntheses with C,C Bond Formation via Acetals and Enamines, in *R. Scheffold* (Ed.), Modern Synthetic Methods 1986 (Springer-Verlag, Berlin 1986).
[3] See *E. Vedejs*, The *Clemmensen* Reduction of Ketones in Anhydrous Organic Solvents, Org. Reactions *22*, 401 (1975).

The detailed mechanism of this reduction of carbonyl to methylene groups is still not completely clear. It seems that an alcohol cannot be formed as an intermediate since alcohols cannot themselves be reduced to hydrocarbons in this way.

2.22.3.2 Wolff–Kishner Reduction.[1] A better method for the conversion of a carbonyl group into a methylene group is by heating a ketone hydrazone (or an aldehyde hydrazone) with sodium ethoxide at about 200°C (475 K). Nitrogen is evolved, leaving the desired hydrocarbon:

$$R_2C{=}N{-}NH_2 \xrightarrow{(C_2H_5ONa)} R_2CH_2 + N_2$$

In a modified procedure introduced by *Huang-Minlon* the hydrazone is not isolated, but instead the carbonyl compound is heated with a solution of hydrazine hydrate and sodium hydroxide in *ethylene glycol* at about 150°C (425 K). It is often possible to carry out the reaction at room temperature by using *dimethyl sulphoxide* as solvent and *potassium t-butoxide* as base.

Because the reaction conditions involved in the *Clemmensen* reduction (acidic solution) and the *Wolff–Kishner* reduction (alkaline solution) are quite different, the appropriate method can be chosen, taking account of whether the ketone to be reduced is sensitive to either acid or base.

2.22.3.3 Reductive Amination. Aldehydes and ketones are reduced by ammonia and hydrogen in the presence of nickel as a catalyst to give amines:

$$R_2C{=}O \xrightarrow[-H_2O]{+NH_3} \left[R_2C{=}NH \right] \xrightarrow[(Ni)]{+2H} R_2C(H){-}NH_2$$

Imine   Amine

2.22.3.4 Reductive Coupling. When ketones are reduced *electrolytically* or with sodium or magnesium amalgams, two ketone molecules link together and a diol is formed. Thus from acetone, in addition to the formation of some propan-2-ol, the main product is *pinacol* (tetramethylethyleneglycol), which crystallises out as a hexahydrate, in the form of plates, m.p. 46°C (319 K). *Azeotropic* distillation of the hydrate with benzene provides water-free pinacol as a liquid with b.p. 174°C (447 K).

$$(H_3C)_2C{=}O + O{=}C(CH_3)_2 \xrightarrow{+2H} (H_3C)_2C(OH){-}C(OH)(CH_3)_2$$

Acetone   Pinacol

This reductive coupling (hydrodimerisation) proceeds *via single electron transfer* (see p. 615) from the metal to the carbonyl compound, resulting in the formation of a radical anion:[2]

$$R_2C{=}\underset{\cdot\cdot}{O} \xrightarrow{+Na} R_2\overset{\cdot}{C}{-}\underset{\cdot\cdot}{O}^{\ominus} Na^{\oplus}$$

Radical anion

[1] See D. Todd, The *Wolff–Kishner* Reduction, Org. Reactions **4**, 378 (1948); H. H. Szmant, The Mechanism of the *Wolff–Kishner* Reduction, Elimination and Isomerization Reactions, Angew. Chem. Int. Ed. Engl. **7**, 120 (1968).

[2] See B. Giese, Radicals in Organic Synthesis: Formation of Carbon–Carbon Bonds (Pergamon Press, Oxford 1986).

The radical anions dimerise to give pinacolate dianions, which react with water to form pinacol:

Radical anion   Pinacolate dianion

These reactions take place on the surface of the metal.

The **McMurry reaction**[1] is preparatively useful since aldehydes or ketones can be converted in one step into alkenes, without the isolation of intermediate diols. This reaction also takes place on the surface of a metal, in this case finely divided titanium, which is generated by treatment of titanium trichloride with lithium aluminium hydride or with potassium. It is applicable to the preparation of alkenes having large and sterically crowded substituent groups, *e.g.*

Di-isopropyl   Diol   Tetraisopropylethylene
ketone   (m.p. 125°C [398 K])

As in the case of pinacol reduction, reaction starts with the formation of a radical anion. Formation of the alkene is a consequence of the great affinity of titanium for oxygen.

2.22.4 Pinacol–pinacolone rearrangement

If pinacol or its hexahydrate is distilled with conc. sulphuric acid a molecule of water is eliminated and at the same time an *intramolecular rearrangement* takes place to give pinacolone (t-butyl methyl ketone; 3,3-dimethylbutan-2-one), b.p. 106°C [379 K].

Pinacol   Pinacolone

The mechanism of the reaction is as follows (*Whitmore*, 1932):

Oxonium ion   Carbenium ion   Pinacolone

First of all, the acid protonates pinacol to form an *oxonium ion*. Loss of water from this generates a *carbenium ion*, whose lack of an octet of electrons is overcome by migration of an alkyl group, together with its bonding pair of electrons, from the neighbouring carbon atom; these two steps are

---

[1] See *J. E. McMurry*, Titanium-induced Dicarbonyl-Coupling Reactions, Accounts Chem. Res. *16*, 405 (1983).

concerted, i.e. they take place simultaneously. The new carbenium ion which is generated as a result of the migration is more stable because of interaction with the lone pairs of electrons on the adjacent oxygen atom, resulting in its having a mesomeric structure. In the final step, further stabilisation is achieved by loss of a proton to give pinacolone.

2.22.5 *Wagner–Meerwein* Rearrangement

Reduction of pinacolone gives 3,3-dimethylbutan-2-ol. If water is removed from the latter compound with the aid of an alumina catalyst the product consists almost entirely of t-butylethylene. This reaction is comparable to the preparation of 2,3-dimethyl-butadiene from pinacol.

If, however, removal of water is brought about by the action of mineral acids or of anhydrous oxalic acid, only small amounts of t-butylethylene are formed and the main product, whose formation involves an *intramolecular rearrangement*, is tetramethyl-ethylene:

Tetramethylethylene
(2,3-Dimethylbut-2-ene)

3,3-Dimethylbutan-2-ol

t-Butylethylene
(3,3-Dimethylbut-1-ene)

The mechanism of this reaction can be likened to a *reverse pinacol–pinacolone rearrangement*. First of all, the secondary alcohol is protonated to form an *oxonium ion*. Loss of water causes the formation of a carbenium ion, in which migration of an alkyl group and loss of a proton provides a stable alkene:

Oxonium ion   Carbenium ion

Wagner–Meerwein rearrangements are of particular importance in the chemistry of *terpenes*.

More Important Ketones

2.22.6 *Acetone (Propanone)*

Acetone arises as a by-product from the dry distillation of wood. It was first prepared by heating lead acetate (*Libavius*, 1606). Nowadays, acetone is obtained for the most part from the petrochemical industry.

Preparation

2.22.6.1 By dehydrogenation of propan-2-ol over a copper catalyst at 250°C (525 K):

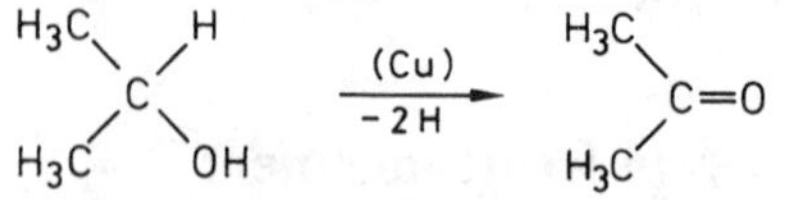

$$H_3C-CH(OH)-CH_3 \xrightarrow[-2H]{(Cu)} H_3C-CO-CH_3$$

Propan-2-ol   Acetone

2.22.6.2 By the *Wacker*-Hoechst Process. Propene is oxidised by air in the presence of aqueous $PdCl_2/CuCl_2$ at 110–120°C (385–395 K) and 10–14 bar (1–1.4 MPa) pressure. Yield = 92–94%:

$$H_3C-CH=CH_2 + \tfrac{1}{2}O_2 \xrightarrow{(PdCl_2/CuCl_2)} H_3C-CO-CH_3 \qquad \Delta H = -255\,kJ/Mol$$

Reaction involves the formation of a $PdCl_2$-π-alkene complex, which is hydrolysed to acetone, metallic palladium and hydrogen chloride. (See the synthesis of acetaldehyde). The copper(II) chloride acts as an oxygen carrier.

2.22.6.3 As a co-product in the preparation of phenol from cumyl hydroperoxide.

Properties

Acetone is a colourless liquid, b.p. 56.2°C (329.4 K), with a characteristic odour. It is miscible in all proportions with water, ethanol or ether. It is a weak proton acid, $pK_a = 20$. It arises as an anomalous metabolic product in people who are suffering from diabetes (*diabetes mellitus*), and in critical cases can be detected in their urine (*ketonuria*).

In the presence of bases such as barium hydroxide, acetone dimerises, by an aldol condensation, to give 4-hydroxy-4-methylpentan-2-one (diacetonealcohol), a colourless liquid, b.p. 165°C (438 K):

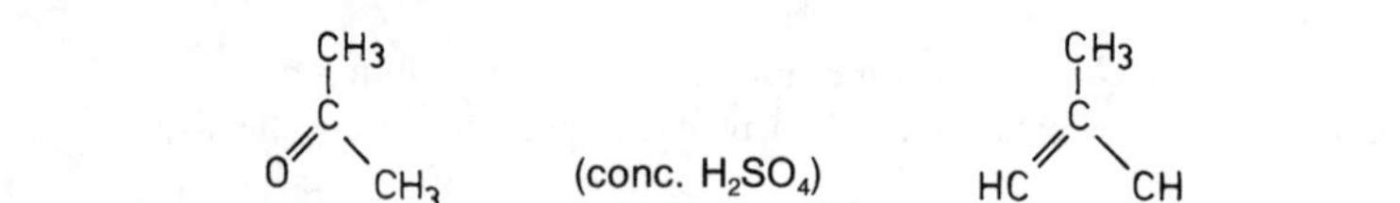

$$(H_3C)_2C=O + H-CH_2-C(=O)-CH_3 \xrightarrow{(OH^\ominus)} (H_3C)_2C(OH)-CH_2-C(=O)-CH_3$$

4-hydroxy-4-methylpentan-2-one
(Diacetonealcohol)

On treatment of acetone with conc. sulphuric acid, three molecules condense together to give mesitylene (1,3,5-trimethylbenzene):

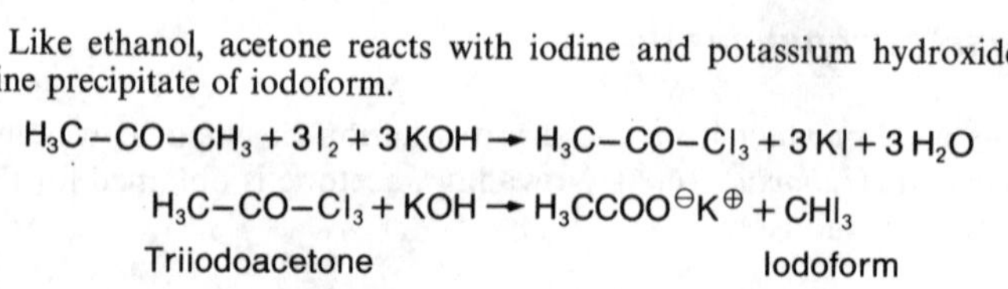

$$\xrightarrow[-3H_2O]{(conc.\ H_2SO_4)}$$

Mesitylene

Iodoform test. Like ethanol, acetone reacts with iodine and potassium hydroxide to produce a yellow crystalline precipitate of iodoform.

$$H_3C-CO-CH_3 + 3I_2 + 3KOH \rightarrow H_3C-CO-CI_3 + 3KI + 3H_2O$$

$$H_3C-CO-CI_3 + KOH \rightarrow H_3CCOO^\ominus K^\oplus + CHI_3$$

Triiodoacetone   Iodoform

Uses

Acetone is an excellent solvent for many organic substances. Industrially it is used for gelatinising nitrocellulose in the film and explosives industries, and as a solvent for acetate fibres, lacquers and acetylene. It is also an important synthetic intermediate in the preparation of, for example, ketene, halogenoforms, esters of methacrylic acid, mesityl oxide and phorone.

2.22.7 *Butanone (Ethyl methyl ketone)*

This homologue of acetone is a liquid, b.p. 79.6°C (352.8 K), with an ether-like smell:

Preparation

2.22.7.1 By catalytic dehydrogenation of butan-2-ol over copper at 250–330°C (525–575 K):

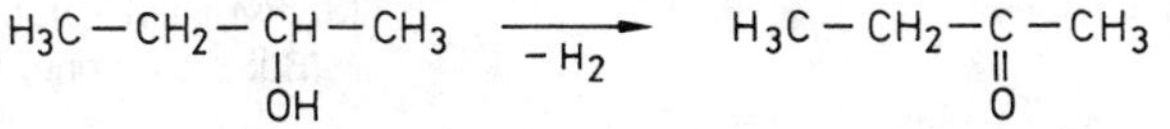

$$H_3C-CH_2-\underset{\underset{OH}{|}}{CH}-CH_3 \xrightarrow{-H_2} H_3C-CH_2-\underset{\underset{O}{\|}}{C}-CH_3$$

Butan-2-ol   Butanone (Ethyl methyl ketone)

2.22.7.2 Using the *Wacker*-Hoechst process, by atmospheric oxidation of but-1-ene in the presence of a solution of $PdCl_2/CuCl_2$ (yield ~55%).

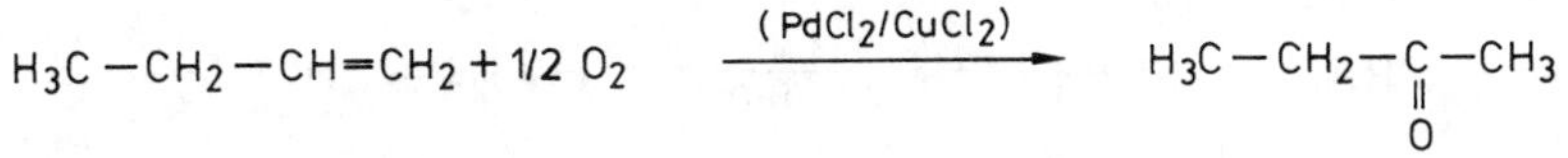

$$H_3C-CH_2-CH=CH_2 + 1/2\ O_2 \xrightarrow{(PdCl_2/CuCl_2)} H_3C-CH_2-\underset{\underset{O}{\|}}{C}-CH_3$$

Uses

Butanone is widely used as a solvent for nitrocelluloses, acetylcelluloses, vinyl resins, chlorinated rubbers and celluloid in the lacquer and adhesives industries, as well as for removing paraffin from heavy oils.

2.22.8 *Halogenoketones*

Hydrogen atoms attached to an α-carbon atom of a ketone can easily be replaced by halogen atoms. For example, *chlorination of acetone* provides *chloroacetone*, a colourless, lachrymatory liquid, b.p. 119°C (392 K).

$$H_3C-CO-CH_3 + Cl_2 \rightarrow H_3C-CO-CH_2Cl + HCl$$

Chloroacetone

Halogenation of ketones is catalysed by either bases or acids. *Base-catalysed bromination of acetone* proceeds so rapidly that only *1,1,1-tribromoacetone* (1,1,1-tribromo-propan-2-one), and no intermediate product, is formed.

$$H_3C-CO-CH_3 + 3\ Br_2 \rightarrow H_3C-CO-CBr_3 + 3\ HBr$$

1,1,1-Tribromoacetone

Because of their strong lachrymatory action, monohalogenoketones have been used as tear gases. Since the halogen atoms in α-halogenoketones are easily displaced, they are used in preparative organic chemistry in the synthesis of heterocyclic compounds.

2.22.9 Unsaturated ketones

Of these compounds, mention must be made of *methyl vinyl ketone (but-3-en-2-one)*, H_3C—CO—CH=CH$_2$, *mesityl oxide*, $(H_3C)_2C$=CH—CO—CH$_3$, and *phorone* $(H_3C)_2$ C=CH—CO—CH=C(CH$_3$)$_2$, a doubly unsaturated ketone.

Methyl vinyl ketone, but-3-en-2-one, is the simplest unsaturated ketone. It can be made by hydration of vinylacetylene in the presence of sulphuric acid and mercury(II) sulphate (*cf.* formation of acetaldehyde from acetylene).

$$H_2C=CH-C\equiv CH + H_2O \xrightarrow{(H_2SO_4, Hg^{2\oplus})} H_2C=CH-\underset{\underset{O}{\|}}{C}-CH_3$$

Vinylacetylene
(But-1-en-3-yne)

Methyl vinyl ketone
(But-3-en-2-one)

It is a pungent-smelling liquid, b.p. 81.4°C (354.6 K), which readily polymerises.

Mesityl oxide[1] is a colourless liquid, b.p. 130°C (403 K),with a peppermint-like smell. It is formed by loss of water from diacetone-alcohol when the latter is heated with a trace of acid or of iodine:

$$(H_3C)_2\underset{\underset{OH}{|}}{C}-CH_2-\underset{\underset{O}{\|}}{C}-CH_3 \xrightarrow{-H_2O} (H_3C)_2C=CH-\underset{\underset{O}{\|}}{C}-CH_3$$

Diacetonealcohol

Mesityl oxide, 4-Methylpent-3-en-2-one

Phorone is a yellow compound, m.p. 28°C (301 K), b.p. 197°C (470 K), which is obtained by the action of hydrogen chloride on acetone. Three molecules of acetone condense together to form phorone; mesityl oxide is a by-product.

$$\underset{H_3C}{\overset{H_3C}{>}}C=CH-CO-CH=C\underset{CH_3}{\overset{CH_3}{<}}$$

Phorone, 2,6-Dimethylhepta-2,5-dien-4-one

Ammonia converts mesityl oxide or phorone, respectively, into di- and tri-acetonamines. With hydrogen sulphide, mesityl oxide gives 4-mercapto-4-methylpentan-2-one, which has a repulsive odour resembling the urine of cats. It occurs sometimes in acetone-containing sewage.

4-Mercapto-4-
methylpentan-2-one

Mesityl
oxide

Diacetonamine
(4-Amino-4-methyl-
pentan-2-one)

Triacetonamine
(2,2,6,6-Tetramethyl-
piperid-4-one)

[1] See *M. Hauser*, The Reactions of Mesityl Oxide, Chem. Rev. *63*, 311 (1963).

2.22.10 Photochemistry of Carbonyl Compounds[1]

The UV spectrum of acetone is shown in Fig. 74 (p. 222). It was recorded using a solution in hexane and has a characteristic absorption band at λ_{max} 279 nm.

If acetone vapour, which has a very similar spectrum, is irradiated at about this wavelength, at 120°C (392 K), it is broken down, with *carbon monoxide* and *ethane* as the main products:

$$H_3CCOCH_3 + h\nu \text{ (313 nm)} \rightarrow CO + C_2H_6 \tag{1}$$

Methane is also formed in smaller amounts, and its yield increases at the expense of the yield of ethane as the temperature is raised higher. A better picture of the reactions involved is given by the following equations:

$$H_3CCOCH_3 + h\nu \text{ (313 nm)} \rightarrow CO + 2 CH_3\bullet \tag{2}$$

$$2 CH_3\bullet \rightarrow C_2H_6 \tag{3}$$

$$CH_3\bullet + H_3CCOCH_3 \rightarrow CH_4 + H_3CCOCH_2\bullet \tag{4}$$
Methyl radical   Acetonyl radical

Reactions (3) and (4) are secondary reactions which follow reaction (2). Reaction (3) requires almost no activation energy, whereas a considerable activation energy is involved in reaction (4). Hence (4) is favoured more by raising the temperature. Increase in the intensity of the irradiation favours (3), since more methyl radicals result from reaction (2), and they recombine according to reaction (3) much more quickly than they react with unchanged acetone as in reaction (4). Secondary products derived from acetonyl radicals are detected when acetone is photolysed.

If irradiation is carried out below 120°C (393 K), the characteristic product is *biacetyl*, $H_3CCOCOCH_3$, which is formed by the combination of acetyl radicals:

$$2 H_3CCO\bullet \rightarrow H_3CCOCOCH_3 \tag{5}$$
Acetyl radical   Biacetyl

The primary reaction of the photolysis of acetone in this case is as follows:

$$H_3CCOCH_3 + h\nu \text{ (313 nm)} \rightarrow H_3CCO\bullet + CH_3\bullet \tag{6}$$

Above 120°C (393 K) the acetyl radical decomposes:

$$H_3CCO\bullet \rightarrow CH_3\bullet + CO \tag{7}$$

By combining equations (6) and (7), equation (2) is obtained; if reaction (3) is also included, then equation (1) results. The *quantum yield* for *carbon monoxide*:

$$\left(\frac{\text{Number of CO molecules formed}}{\text{Number of quanta of light absorbed}} \right)$$

---

[1] See *N. J. Turro*, Modern Molecular Photochemistry (Benjamin, New York 1978); Adiabatic Photo-reactions of Organic Molecules, Angew. Chem. Int. Ed. Engl. *18*, 572 (1979); *Houben-Weyl*, Methoden der Organischen Chemie *4/5a, 4/5b* (Thieme Verlag, Stuttgart 1975); *H. G. O. Becker* et al., Einführung in die Photochemie (VEB Deutscher Verlag der Wissenschaften, Berlin 1983); *M. Klessinger* et al., Lichtabsorption und Photochemie Organischer Moleküle (VCH, Weinheim 1989).

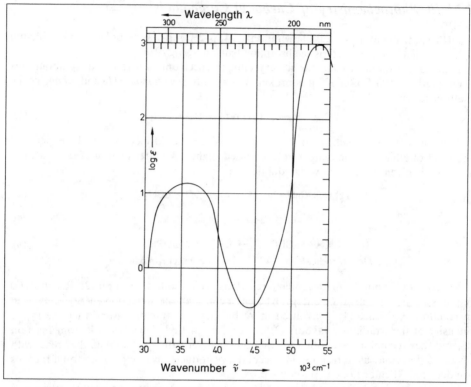

Fig. 74. UV spectrum of acetone in hexane.

for reactions (1) and (6) is equal to 1, so long as the temperature is kept above 120° (393 K). At lower temperatures it falls below 1. This is because not only the reaction (5), but also other reactions not considered here, lead to the product of the primary reaction (6) being consumed without carbon monoxide being formed.

This example illustrates how many factors have to be taken into account in order to clarify a relatively simple photochemical reaction. Further information can be obtained by making use of radical-traps. For example, in the photolysis of acetone, if sufficient oxygen is present neither methane nor ethane is formed, but instead the methyl radicals react with the oxygen. One generalisation which can be made, and which is exemplified in equation (6), is that, when a carbonyl compound is photochemically activated, the first bond which breaks is that between the carbonyl group and an adjacent carbon atom, whereby a *radical pair* is generated, consisting of an acyl radical and an alkyl radical. This is the so-called *Norrish type I cleavage* or *α-cleavage*. For example, pentan-2-one behaves as follows:

$$H_3CCOCH_2CH_2CH_3 + h\nu \ (313 \ nm) \ \longrightarrow \ H_3CCO\bullet + C_3H_7\bullet$$

Pentan-2-one   Acyl radical   Alkyl radical

If photolysis is carried out using an alkene as the solvent, the solvent also can react with the irradiated molecule. Thus, when acetone is photolysed in cyclopentene a (2+2)-

cycloaddition reaction is observed, known as the *Paterno-Büchi reaction*, and an *oxetane derivative* is formed:

$$H_3CCOCH_3 \quad + \quad \text{(cyclopentene)} \quad + \quad h\nu\,(313\,nm) \quad \longrightarrow \quad \text{(oxetane derivative)}$$

Acetone   Cyclopentene   Oxetane derivative

When pentan-2-one is photolysed, in addition to α-splitting, another reaction takes place, which is also a general type of reaction:

$$H_3CCOCH_2CH_2CH_3 + h\nu\,(313\,nm) \longrightarrow H_3CCOCH_3 + C_2H_4$$

This is the so-called *Norrish type II cleavage*. It proceeds *via* formation of an intermediate diradical, which arises from the migration of a hydrogen atom from the γ-carbon atom to the oxygen atom. This reaction is similar to the *McLafferty rearrangement* which is frequently observed in mass spectrometry.

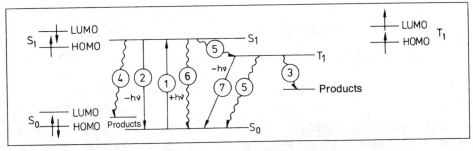

Pentan-2-one   Diradical

The enol form of acetone which is generated can be observed spectroscopically in the gas phase; it rearranges back to acetone on the surface of the vessel.

Fig. 75. Simplified *Jablonski* diagram of photochemical processes.

——▶ Radiative processes   S: Singlet (S₀ ground state)
⁓⁓⁓▶ Radiationless processes   T: Triplet

① Photochemical excitation (light absorption)
② Fluorescence (light emission, luminescence)
③ Photochemical reactions
④

⑤ Radiationless transition from a singlet to a triplet state or from a triplet to a singlet state. Spin reversal. Intersystem crossing
⑥ Radiationless transition between two states of the same multiplicity. Internal conversion
⑦ Phosphorescence (light emission, luminescence)

The diradical can also stabilise itself by ring-closure, giving 1-methylcyclobutanol:

$$
\begin{array}{ccc}
\overset{\displaystyle CH_3}{\underset{\displaystyle H_2C - \overset{\displaystyle\bullet}{C}H_2}{H_2C - C - OH}} & \longrightarrow & \overset{\displaystyle CH_3}{\underset{\displaystyle H_2C - CH_2}{H_2C - C - OH}} \\[2mm]
\text{Diradical} & & \text{1-Methylcyclobutanol}
\end{array}
$$

The chemical changes discussed hitherto form only part of a very complicated situation, which is outlined in the *Jablonski* diagram in Fig. 75.

When a molecule absorbs the appropriate radiation it is promoted ① from its ground state S_0 to the next highest singlet state (S_1, S_2, etc.), with paired spins (electronic configuration as on the left-hand side of Fig. 75). It can then, by radiationless spin reversal, transfer to an electronic triplet state, *e.g.* T_1 (electronic configuration on the right-hand side of Fig. 75); in addition, a coupling between the electron spin and the electronic orbit is necessary (spin–orbit coupling). Measurements of the fluorescence ② and the phosphorescence ⑦, which can be distinguished by their characteristic spectra, provide information about the mechanisms of photochemical reactions. For this it is advantageous to make use of substances (quenchers) which can quench or inhibit the normal photoreactions. The use of oxygen for this purpose may in particular be mentioned.

The photoreactions considered up to now correspond to arrows ③ and ④ of the *Jablonski* diagram. It is often difficult to distinguish between these two possibilities, because reactions involving the excited singlet state S_1 are unaffected by triplet quenching, while selective reactions involving the triplet state T_1 are promoted by triplet sensitisers. α-Cleavage can take place from both the S_1 and T_1 states.

By a combination of the various methods described above, it proved possible to distinguish between the reactions of the S_1 and T_1 states of acetone (*Turro* et al., 1970).

(*E*)-1,2-Dicyanoethylene also reacts with acetone in the *singlet state* in a *Paterno-Büchi* reaction to give an *oxetane* derivative with *trans*-cyano groups, *i.e.* with retention of configuration. In contrast, triplet acetone brings about an inversion of configuration to (*Z*)-1,2-dicyanoethylene.

The ability of oxygen to act as a quencher and as a radical-trap results from the fact that oxygen is a triplet ($^3O_2$) in its ground state. By provision of 94 kJ/mol of energy it can be converted, either chemically or photochemically, into its singlet state ($^1O_2$), in which it is no longer a diradical, but behaves like an alkene.

In the simplest method, singlet oxygen is generated photochemically in the presence of a sensitiser (sens), which, on the route ①–⑤, attains a state in which the conversion

$$^3O_2 \longrightarrow {}^1O_2$$

can be achieved. Singlet oxygen reacts with electron-rich alkenes to form *dioxetans*:

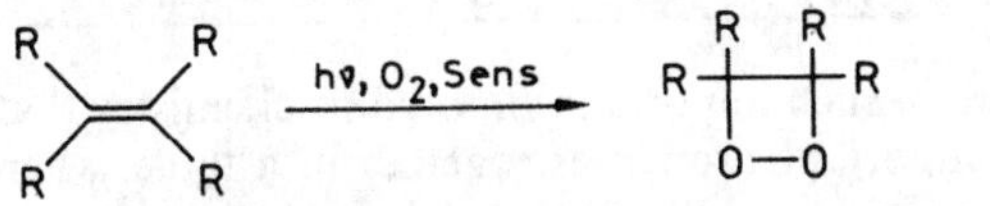

Tetramethyldioxetan can also be made in other ways. When it is heated it decomposes to form two molecules of acetone, one of which is photochemically excited and displays chemiluminescence.

If the groups R contain allylic hydrogen atoms hydroperoxides are formed in a singlet oxygen-ene reaction (see p. 666).

(*E*)-1,2-Dicyanoethylene reacts with dioxetane to give the same products which are obtained when it reacts with acetone in the presence of light, but in different proportions:

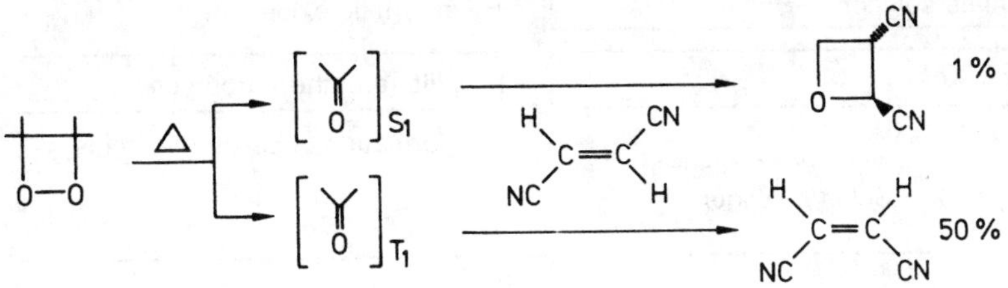

From the results of this reaction it appears that tetramethyldioxetane is a substance which can be used to carry out *photochemical processes without light*, since it attains excited states by means of a thermal process, which otherwise is only attainable by photochemical means.

2.23 Tabular Comparison of the Properties of Aldehydes and Ketones

Aldehydes and ketones are of such importance in organic chemistry that it is worthwhile to gather all their most important properties together in a table, wherein the common features and the differences between them can be brought out. Some reactions which have yet to be mentioned in the text are indicated by references to the pages where they are mentioned.

Aldehydes	Ketones
Oxidation	

Act as reducing agents, especially in alkaline solution	Not reducing agents By energetic oxidation:
$R-CHO \longrightarrow R-COOH$	Splitting of the carbon chain
	and formation of carboxylic acids

Tollen's reagent   *Nylander*'s reagent
Fehling's solution   *Schiff*'s reagent

Reduction	

$R-CHO \longrightarrow R-CH_2OH$ Primary alcohol	$R-CO-R' \longrightarrow R-\underset{\underset{OH}{\mid}}{CH}-R'$ Secondary alcohol

H_2 (catalyst)
$LiAlH_4$
Selective reagents: $NaBH_4$, *Meerwein–Ponndorf–Verley* reduction

$R-CHO \longrightarrow R-CH_3$	$R-CO-R' \longrightarrow R-CH_2-R'$

Clemmensen reduction, *Wolff–Kishner* reduction

Oxidoreduction	

$2\,R-CHO + NaOH \longrightarrow R-CH_2OH + R-COOH$	

Cannizzaro reaction: occurs if R has no α-hydrogen atoms

$2\,RCHO \xrightarrow{Al(OC_2H_5)_3} RCOOR$	

Tishchenko reaction: also takes place if R has an α-hydrogen atom

Reductive alkylation	

$R-CHO + R'MgX \longrightarrow \underset{R'}{\overset{R}{>}}C\underset{OH}{\overset{H}{<}}$ Secondary alcohol	$\underset{R}{\overset{R}{>}}C=O + R'MgX \longrightarrow \underset{R}{\overset{R}{>}}C\underset{OH}{\overset{R'}{<}}$ Tertiary alcohol

Grignard reaction Selective for aldehydes: phenyltitanium triisopropoxide	

Aldehydes	Ketones

Formation of C—O bonds

Hydrate

$$R-CHO + H_2O \rightleftharpoons R-\underset{OH}{\overset{H}{\underset{|}{\overset{|}{C}}}}-OH$$

$$R-\underset{O}{\overset{\|}{C}}-R' + H_2O \rightleftharpoons R-\underset{HO \quad OH}{C}-R'$$

Half-acetal

$$R-CHO + R'OH \rightleftharpoons R-\underset{OR'}{\overset{H}{\underset{|}{\overset{|}{C}}}}-OH$$

Of importance only in carbohydrate chemistry, see p. 435

Acetal

$$R-CHO + 2\,R'OH \overset{H^{\oplus}}{\rightleftharpoons} R-\underset{OR'}{\overset{H}{\underset{|}{\overset{|}{C}}}}-OR'$$

$$R-\underset{O}{\overset{\|}{C}}-R' \overset{H^{\oplus}}{\rightleftharpoons} R-\underset{O \qquad O}{C}-R'$$
$$+ HO-CH_2-CH_2OH \qquad H_2C-\!\!\!-\!\!\!-CH_2$$

Formation of C—S bonds

Dithioacetal

$$R-CHO + 2\,R'SH \overset{H^{\oplus}}{\rightleftharpoons} R-\underset{SR'}{\overset{H}{\underset{|}{\overset{|}{C}}}}-SR'$$

$$R-\underset{O}{\overset{\|}{C}}-R' + 2\,R''SH \rightleftharpoons R-\underset{R''S \quad SR''}{C}-R'$$

Hydrogen sulphite adducts

$$R-CHO + Na^{\oplus}SO_3H^{\ominus} \rightleftharpoons R-\underset{Na^{\oplus} \quad SO_3^{\ominus}}{\overset{H}{\underset{|}{\overset{|}{C}}}}-OH$$

This reaction only works with methyl ketones and cyclic ketones

Formation of C—N bonds

Aldehyde-ammonia

$$R-CHO + NH_3 \rightleftharpoons R-\underset{NH_2}{\overset{H}{\underset{|}{\overset{|}{C}}}}-OH$$

Leuckart – Wallach reaction, see p. 525

Azomethines (*Schiff*'s bases)
Aldimines

Secondary products

$$R-CHO + R'NH_2 \rightleftharpoons R-CH=N-R'$$

e.g. Triacetonamine

$$R-\underset{O}{\overset{\|}{C}}-R' + R''NH_2 \longrightarrow R-\underset{N\sim R''}{C}-R'$$

Ketimine

Transamination, see p. 890

Aminals

$$R-CHO + 2\,HNR'_2 \longrightarrow R-\underset{NR'_2}{\overset{H}{\underset{|}{\overset{|}{C}}}}-NR'_2$$

———

Oximes

$$R-CHO + H_2NOH \rightleftharpoons R-CH=NOH$$

$$R-\underset{O}{\overset{\|}{C}}-R' + H_2NOH \rightleftharpoons R-\underset{N\sim OH}{C}-R'$$

Hydrazones

$$R-CHO + H_2N-NH_2 \rightleftharpoons RCH=N-NH_2$$

$$R-\underset{O}{\overset{\|}{C}}-R' + H_2N-NH_2 \longrightarrow R-\underset{N\sim NH_2}{C}-R'$$

Phenylhydrazones

$$RCH=N-NH-C_6H_5$$

Semicarbazones

$$RCH=N-NH-\underset{O}{\overset{\|}{C}}-NH_2$$

t-Amines
Eschweiler– Clarke methylation

$$2\,HCHO + R-NH_2 \overset{HCOOH}{\longrightarrow} R-N\underset{CH_3}{\overset{CH_3}{<}}$$

———

Aldehydes	**Ketones**
Formation of C—C bonds	

Cyanohydrin reaction

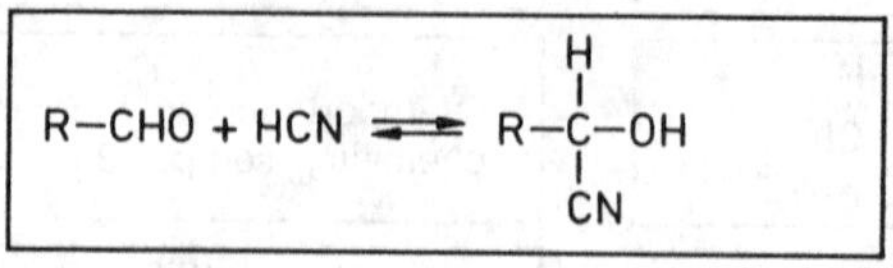

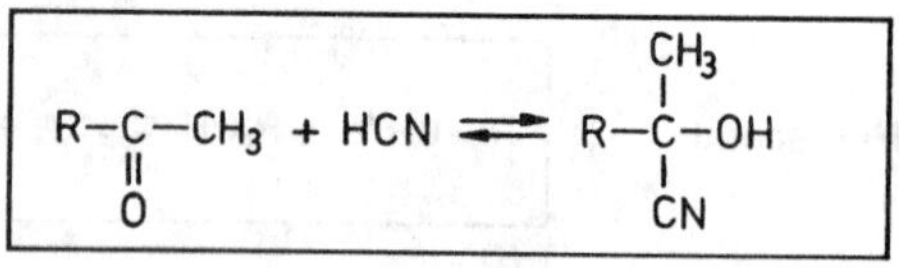

$$R-CHO + HCN \rightleftharpoons R-\underset{\underset{CN}{|}}{\overset{\overset{H}{|}}{C}}-OH$$

$$R-\underset{\overset{\|}{O}}{C}-CH_3 + HCN \rightleftharpoons R-\underset{\underset{CN}{|}}{\overset{\overset{CH_3}{|}}{C}}-OH$$

Aldol reaction

$$2\ R_2CH-CHO \underset{}{\overset{H^\oplus\ or\ HO^\ominus}{\rightleftharpoons}} R_2-CH-\underset{\underset{HO}{|}}{\overset{\overset{H}{|}}{C}}-\underset{\underset{R}{|}}{\overset{\overset{R}{|}}{C}}-CHO$$

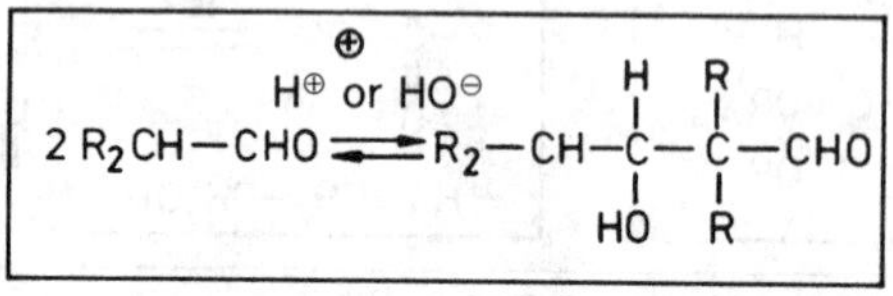

A few examples, e.g. diacetonealcohol

Directed Aldol reaction

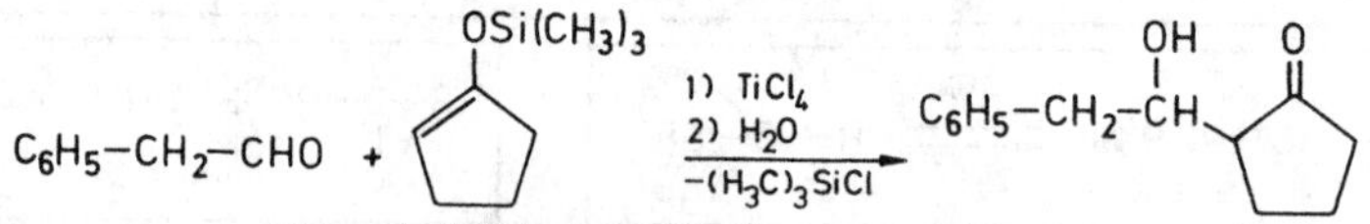

$$C_6H_5-CH_2-CHO\ +\ \text{(OSi(CH}_3)_3 \text{ cyclopentene)} \xrightarrow[{-(H_3C)_3SiCl}]{1)\ TiCl_4 \atop 2)\ H_2O} C_6H_5-CH_2-\underset{\overset{OH}{|}}{CH}.\ \text{(cyclopentanone-O)}$$

Trimethylsilylenol ether of cyclopentanone

Benzoin addition, see p. 533

$$2\ C_6H_5-CHO \xrightarrow{CN^\ominus} C_6H_5COCH(OH)C_6H_5$$

———————

Pinacol formation

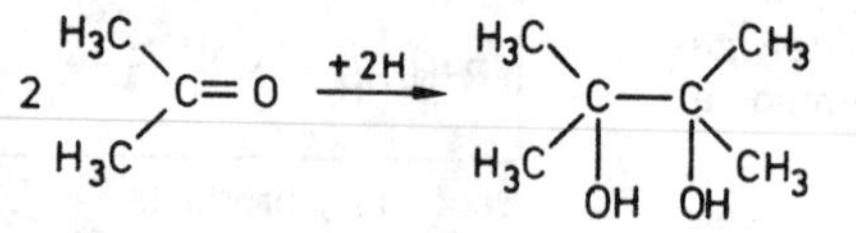

———————

$$2\ \underset{H_3C}{\overset{H_3C}{>}}C=O \xrightarrow{+2H} \underset{H_3C}{\overset{H_3C}{>}}\underset{\underset{OH}{|}}{C}-\underset{\underset{OH}{|}}{C}\overset{CH_3}{\underset{CH_3}{<}}$$

Formation of alkenes; *McMurry* reaction

———————

$$2\ \underset{R}{\overset{R}{>}}C=O \xrightarrow{TiCl_3/K,\ THF} \underset{R}{\overset{R}{>}}C=C\overset{R}{\underset{R}{<}}$$

***Wittig* reaction**

———————

$$\underset{R'}{\overset{R}{>}}C=O \xrightarrow[-(C_6H_5)_3P=O]{+(C_6H_5)_3P=CH_2} \underset{R'}{\overset{R}{>}}C=CH_2$$

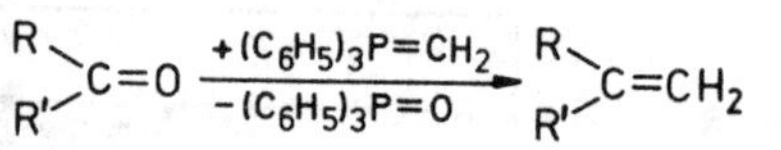

see also

Horner–Emmons reaction
Peterson reaction

| **Formation of C—C and C—N bonds** | |

Amino-methylation *Mannich* reaction see p. 543

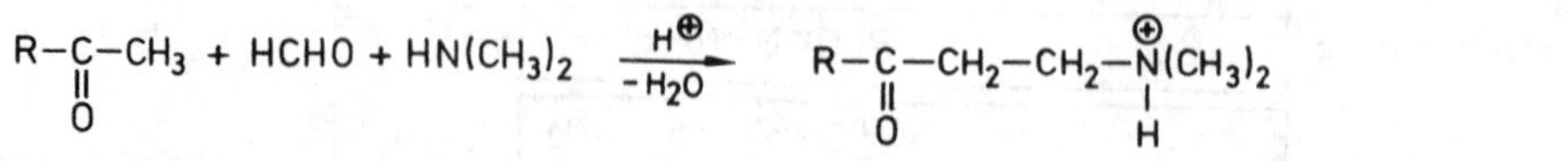

$$R-\underset{\overset{\|}{O}}{C}-CH_3 + HCHO + HN(CH_3)_2 \xrightarrow[-H_2O]{H^\oplus} R-\underset{\overset{\|}{O}}{C}-CH_2-CH_2-\underset{\underset{H}{|}}{\overset{\oplus}{N}}(CH_3)_2$$

2.24 Saturated Aliphatic Monocarboxylic Acids (Alkanoic Acids, Fatty Acids)

When *primary* alcohols are oxidised strongly, saturated monocarboxylic acids are formed. These have the general formula:

$$R-C{\overset{\displaystyle O}{\underset{\displaystyle OH}{}}} \quad \text{or} \quad C_nH_{2n+1}-C{\overset{\displaystyle O}{\underset{\displaystyle OH}{}}}$$

The characteristic feature of this class of compounds is the *carboxyl group*. Since some of the higher carboxylic acids are obtained from fats, they are also sometimes known as *fatty acids*. Such acids usually have trivial names derived from the source, animal or plant, from which they are obtained.

The trivial and the systematic names of the lower members of the *homologous series* are as follows:

		b.p.
Formic acid: methanoic acid	H–COOH	100,7 °C (373,9 K)
Acetic acid: ethanoic acid	H_3C–COOH	118 °C (391 K)
Propionic acid: propanoic acid	H_3C–CH_2–COOH	141 °C (414 K)
Butyric acid: butanoic acid	H_3C–CH_2–CH_2–COOH	164 °C (437 K)
Valeric acid: pentanoic acid	H_3C–CH_2–CH_2–CH_2–COOH	187 °C (460 K)

Of the higher carboxylic acids having unbranched chains the following are the most important:

		m.p.
Palmitic acid: hexadecanoic acid	H_3C–$(CH_2)_{14}$–COOH	63 °C (336 K)
Stearic acid: octadecanoic acid	H_3C–$(CH_2)_{16}$–COOH	69,5 °C (342,7 K)

Carboxylic acids can be thought of as derivatives of water in which a hydrogen atom has been replaced by an *acyl group* (*acyl* being derived from *acid*). The *acyl groups* are named as follows: formyl, H—CO—; acetyl H_3C—CO—; propionyl, H_3C—CH_2—CO—; butyryl, H_3C—CH_2—CH_2—CO—, *etc.* Names for the acyl groups can also be derived from the IUPAC names, for example, ethanoyl, propanoyl (also 1-oxopropyl), octadecanoyl, etc.

2.24.1 General Methods of Preparation

2.24.1.1 By oxidation of primary alcohols or aldehydes (see p. 112)

2.24.1.2 By hydrolysis of nitriles

$$R-C\equiv N + 2\,H_2O \rightarrow R-COOH + NH_3$$

2.24.1.3 By carbonylation of alkenes

Carbonylation of alkenes takes place in the presence of palladium complexes and hydrochloric acid, as follows:

$$-C=C- \xrightarrow{\overset{\oplus}{H}} -\overset{H}{\underset{|}{C}}-\overset{\oplus}{\underset{|}{C}}- \xrightarrow{\overset{\ominus}{I}\overset{\oplus}{C\equiv OI}} -\overset{H}{\underset{|}{C}}-\overset{\overset{\oplus}{C=\bar{O}}}{\underset{|}{C}}- \xrightarrow{H_2O} -\overset{H}{\underset{|}{C}}-\overset{COOH}{\underset{|}{C}}-$$

Other compounds having *acidic* hydrogen atoms can take the place of water. For example, alcohols, thiols or amines can be used and they give rise to the formation of, respectively, *esters*, *thioesters* and *amides of carboxylic acids*.

An important industrial application of carbonylation is for the preparation of *pivalic acid* (trimethylacetic acid) and its analogues (*Koch–Haaf* reaction):

$$\underset{\begin{subarray}{c}\text{2-Methylpropene}\\\text{(Isobutene)}\end{subarray}}{H_3C-\overset{CH_3}{\underset{|}{C}}=CH_2} \xrightarrow{CO(H_2SO_4),\ H_2O} \underset{\begin{subarray}{c}\text{Trimethylacetic acid}\\\text{(Pivalic acid)}\end{subarray}}{\overset{H_3C}{\underset{H_3C}{\overset{|}{\underset{|}{H_3C}}}}C-COOH}$$

2.24.1.4 By saponification of fats and oils

2.24.1.5 From the reaction of *Grignard* reagents with carbon dioxide (see p. 186).

2.24.1.6

Another laboratory method **involves the use of malonic esters** (see p. 327). Both of the hydrogen atoms attached to the methylene group of malonic esters can be replaced by alkyl groups, providing a wide variety of substituents. These substituted malonic esters can be converted, by hydrolysis and decarboxylation, into monocarboxylic acids:

$$ROOC-CH_2-COOR \xrightarrow{Base} ROOC-\overset{\ominus}{C}H-COOR \xrightarrow{H_3CBr} ROOC-\underset{CH_3}{\underset{|}{C}}H-COOR \xrightarrow{Base}$$

$$ROOC-\underset{CH_3}{\underset{|}{\overset{\ominus}{C}}}-COOR \xrightarrow{C_2H_5Br} ROOC-\underset{CH_3}{\overset{C_2H_5}{\underset{|}{\overset{|}{C}}}}-COOR \xrightarrow{H_2O,\ H^{\oplus}} HOOC-\underset{CH_3}{\overset{C_2H_5}{\underset{|}{\overset{|}{C}}}}-COOH \xrightarrow[-\ CO_2]{(heat)}$$

$$\underset{CH_3}{C_2H_5-\underset{|}{C}H-COOH} \qquad \text{2-Methylbutyric acid}$$

2.24.1.7

The reactivity of 1,3-dicarbonyl compounds can also be utilised in another way, developed by *Stetter*,[1] in order to prepare longer-chain monocarboxylic acids. In the first step, cyclohexane-1,3-dione is alkylated. The resultant product can be converted by alkali to give a long-chain δ-keto acid, which can then be reduced almost quantitatively by the *Wolff–Kishner* method, in diethyleneglycol, $HO-CH_2-CH_2-O-CH_2-CH_2-OH$, as solvent, to the corresponding carboxylic acid:

---

[1] See Angew. Chem. *67*, 769 (1955).

Cyclohexane-1,3-dione   δ-Keto acid   Carboxylic acid

By choosing the appropriate substituents R, a variety of unbranched and branched chain carboxylic acids can be readily obtained.

Properties

The first three members of the homologous series of carboxylic acids are colourless liquids with sharp irritating smells; they are completely miscible with water. Acids with four to nine carbon atoms have rancid, foetid smells. Those with ten or more carbon atoms are odourless, waxy solids, since in these cases the carboxyl group has relatively little effect on the physical properties associated with the carbon chain.

As in the case of n-alkanes, the *melting points* of n-saturated carboxylic acids alternate in value, the acids with even numbers of carbon atoms melting at higher temperatures than the neighbouring members of the series having odd numbers of carbon atoms. *X-ray analyses* of solid carboxylic acids have shown that the carbon atoms take up the shape of an extended zigzag chain.

It is noteworthy that carboxylic acids, either in their solid or liquid states, or in solution in non-polar solvents, have relative molecular masses double what would be expected from their molecular formulae. This is because they exist in a *dimeric* form. Thus the molecular association of water is reflected not only in its monoalkyl derivatives, the alcohols, but also in its monoacyl derivatives, the carboxylic acids.

Dimerisation of carboxylic acids involves intermolecular hydrogen bonding and partial sharing of the hydrogen atoms between the oxygens in a ring-shaped pairing of molecules.

The hydrogen bonding can be detected from the IR spectra. At higher temperatures formic acid provides a characteristic HO-absorption band at $3570 \, cm^{-1}$, associated with its monomeric form. As the temperature is lowered, this band shifts to lower frequencies, due to formation of the hydrogen-bonded dimeric form of formic acid. There is an energetic gain, $\Delta H = -60 \, kJ/Mol$, in the process of dimerisation.

2.24.2 *Acidity of the Carboxyl Group*

The hydroxy groups in carboxylic acids differ markedly from the hydroxy groups in alcohols, in the much greater acidity of the former. This difference is due to the presence of the neighbouring C=O double bond, which facilitates loss of the proton by means of its −*I-effect*. Furthermore, the resultant carboxylate anion is stabilised by *mesomerism*.

$$R-C{\overset{O}{\underset{O-H}{}}} + H_2O \; \rightleftharpoons \; \left[R-C{\overset{\bar{\bar{O}}|}{\underset{O|^{\ominus}}{}}} \longleftrightarrow R-C{\overset{O|^{\ominus}}{\underset{O|}{}}} \right] + H_3O^{\oplus}$$

In this mesomeric carboxylate anion the two oxygen atoms are totally equivalent and the negative charge is delocalised (Fig. 76). The four π-electrons are spread over two molecular orbitals, which are schematically represented as in Fig. 77.

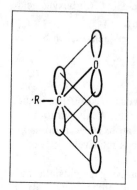

Fig. 76.

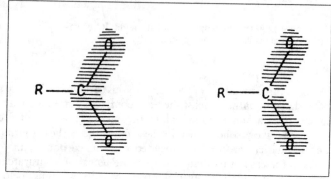

Fig. 77.

Carboxylic acids are much weaker acids than mineral acids and, in the equilibrium shown above, the balance is displaced towards the left-hand side. The strength of an acid[1] in aqueous solution at 25°C (298 K) is defined by the *acidity constant* K_a, but, to avoid the use of negative indices, this is usually expressed in the form of the corresponding pK_a value.

$$K_a = \frac{[RCOO^{\ominus}][H^{\oplus}]}{[RCOOH]} \cdot \frac{f\,RCOO^{\ominus} \cdot f\,H^{\oplus}}{f\,RCOOH} \qquad pK_a = -\log_{10} K_a$$

Making the assumption that the activity coefficients $f\,RCOO^{\ominus}$ etc. can be neglected, it follows that in sufficiently dilute solutions the following simple relationship (the *Henderson equation*) can be set up:

$$pK_a = -\log[H^{\oplus}] - \log \frac{[RCOO^{\ominus}]}{[RCOOH]} = pH - \log \frac{[RCOO^{\ominus}]}{[RCOOH]}$$

Henderson equation

In a half-neutralised solution, $[RCOO^{\ominus}] = [RCOOH]$. Hence the *Henderson* equation can be simplified to:

$$pK_a = pH \text{ (half-neutralised solution)}$$

The smaller the numerical value of pK_a, the greater is the strength of the acid in question (see p. 233).

---

[1] See *H. F. Ebel*, Die Acidität der CH-Sauren (The Acidity of CH Acids) (Thieme Verlag, Stuttgart 1969); *J. J. Christensen* et al., Handbook of Proton Ionization Heats and Related Thermodynamic Quantities (Wiley, New York 1976).

The acidity of carboxylic acids in solution is markedly influenced by the structure of the alkyl group. Whilst most substituents attached to carbon atoms exert a −I-effect and hence increase the acidity (see, for example, halogenocarboxylic acids), alkyl groups themselves act as electron-donating groups, *i.e.* they repel the bonding electrons and raise the electron density on the adjacent carbon atom. It has been found that there is an increasing +I-effect on going from primary to secondary to tertiary alkyl groups, e.g.

$$-CH_3 < -CH_2-CH_3 < -CH(CH_3)_2 < -C(CH_3)_3$$

As the hydrogen atom of formic acid is replaced by this series of alkyl groups, so the observed pK_a values for the resultant acids become greater and the acidities decrease:

Formic acid	Acetic acid	Propionic acid	Isobutyric acid	Trimethylacetic acid
H–COOH	H_3C–COOH	H_3C–CH_2–COOH	$(H_3C)_2$CH–COOH	$(H_3C)_3$C–COOH
$pK_a = 3,77$	4,76	4,88	4,86	5,05

The numerical values obtained in this way can be of use for predicting pK_a values.[1]

Aqueous solutions of salts of carboxylic acids are *alkaline* due to partial hydrolysis:

$$\left[R-C \substack{\\ \\} \longleftrightarrow R-C \substack{\\ \\} \right] Na^{\oplus} + H_2O \rightleftharpoons R-C \substack{\\ \\} + Na^{\oplus} + OH^{\ominus}$$

With the exception of formic acid, unbranched carboxylic acids are *resistant to oxidising agents*. In consequence, acetic acid (*etc.*) can be used as solvents in oxidation processes. *Hydrogenation* of higher carboxylic acids to give the corresponding alcohols only takes place at high temperatures and pressures in the presence of heavy metal catalysts (see p. 247).

The Most Important Carboxylic Acids

2.24.3 Formic Acid (Methanoic Acid)

Formic acid (Lat. *formica* = ant) occurs in the free state in, for example, ants and stinging nettles. In ants it is both a defensive material and also a pheromone.

Preparation

2.24.3.1 By oxidation of methanol or of formaldehyde.

2.24.3.2 By hydrolysis of chloroform (which is the trichloride of orthoformic acid, see p. 260) with potassium hydroxide in ethanol. This reaction was responsible for the name 'chloroform'.

$$CHCl_3 + 4\ KOH \longrightarrow \left[H-C \substack{\\ \\} \right] K^{\oplus} + 3\ KCl + 2\ H_2O$$

Potassium formate

---

[1] See *D. D. Perrin* et al., pK_a Prediction for Organic Acids and Bases (Chapman & Hall, London 1981).

2.24.3.3 Industrially, in quantitative yield, from sodium hydroxide and carbon monoxide at 120–130°C (390–400 K) and under a pressure of 6–8 bar (0.6–0.8 MPa).

$$NaOH + CO \longrightarrow \left[H-C{\overset{\displaystyle \overline{O}|}{\underset{\displaystyle \underset{\ominus}{\overline{O}|}}{}}} \right] Na^{\oplus} \qquad \text{Sodium formate}$$

Almost 100% pure formic acid can be obtained by treatment of the sodium formate under the appropriate conditions, *e.g.* by an initial addition of some high-percentage formic acid, followed by water-free sulphuric acid and finally vacuum distillation.

2.24.3.4 By alkaline hydrolysis of methyl formate which is obtained from carbon monoxide and methanol in the presence of sodium methoxide.

$$H_3COH + CO \xrightarrow{Na^{\oplus} \, (^{\ominus}OCH_3)} HCOOCH_3$$

2.24.3.5 As a by-product in the preparation of pentaerythritol (see p. 314).

Properties

Anhydrous formic acid is a colourless liquid, b.p. 100.7°C (373.9 K), with a pungent odour. Its salts are called *formates*. It is the strongest of the monocarboxylic acids and is a unique member of this series since it can react as an aldehyde (see the heavy type in the following formula) as well as as an acid.

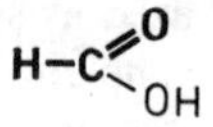

$$H-C{\overset{\displaystyle O}{\underset{\displaystyle OH}{}}} \qquad \text{Formic acid}$$

The aldehyde-like properties may be seen in the following reactions in which formic acid acts as a *reducing agent*.

(1). It reduces ammoniacal silver nitrate solutions to metallic silver.
(2). It precipitates metallic mercury from solutions of mercury(II) nitrate.
(3). It decolourises solutions of potassium permanganate.
(4). It reduces solutions of potassium dichromate to chromium(III) salts.

Formic acid is *oxidised* to carbon dioxide and water:

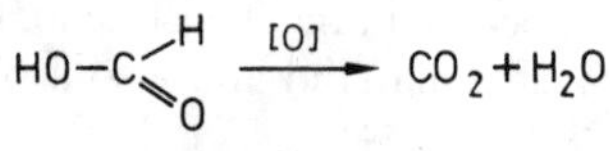

$$HO-C{\overset{\displaystyle H}{\underset{\displaystyle O}{}}} \xrightarrow{[O]} CO_2 + H_2O$$

In the presence of platinum or palladium, formic acid is *dehydrogenated*, even at room temperature, to carbon dioxide:

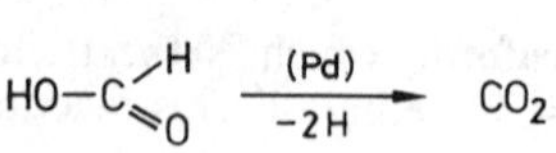

$$HO-C{\overset{\displaystyle H}{\underset{\displaystyle O}{}}} \xrightarrow[-2H]{(Pd)} CO_2$$

When *heated with concentrated sulphuric acid* formic acid loses a molecule of water and gives carbon monoxide; this is a convenient laboratory method for the preparation of this gas:

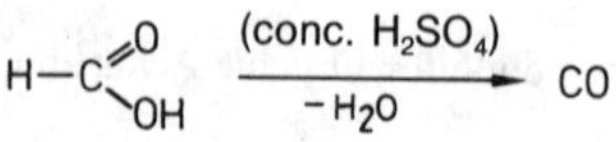

$$H-C{\overset{\displaystyle O}{\underset{\displaystyle OH}{}}} \xrightarrow[-H_2O]{(conc. \, H_2SO_4)} CO$$

Uses

Like formaldehyde, formic acid is a strong bactericide, and as such it is used in the preservation of fruit juices and in the disinfecting of wine- and beer-barrels. More important uses are as an auxiliary mordant in the dyeing of wool, and in tanning, for the removal of lime from leather, which depends on the fact that calcium formate is soluble in water.

2.24.4 *Acetic Acid (Ethanoic Acid)*

The most widely encountered member of the carboxylic acid series is acetic acid (Lat. *acetum* = vinegar). A dilute aqueous solution of acetic acid, vinegar, is produced when beer or cider or wine go sour in air, and has been known and used by man since early times. Acetic acid is one of the products obtained from the dry distillation of wood.

Preparation

2.24.4.1 By enzymatic oxidation of ethanol:

$$H_3C-CH_2OH + O_2 \xrightarrow{\text{(Acetobacter)}} H_3C-COOH + H_2O$$

2.24.4.2 By atmospheric oxidation of acetaldehyde in the presence of manganese(II) acetate:

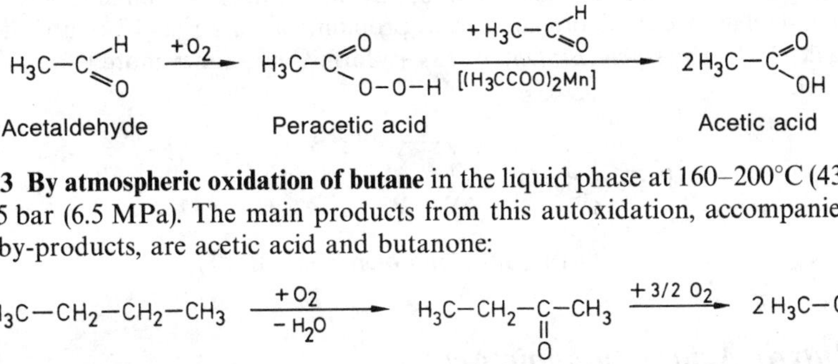

Acetaldehyde   Peracetic acid   Acetic acid

2.24.4.3 By atmospheric oxidation of butane in the liquid phase at 160–200°C (430–470 K) and 65 bar (6.5 MPa). The main products from this autoxidation, accompanied also by some by-products, are acetic acid and butanone:

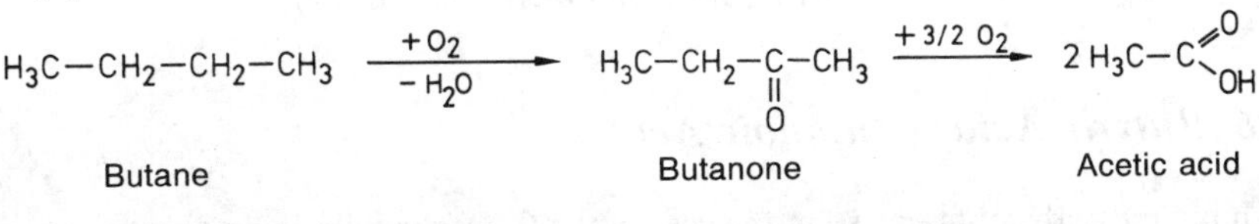

Butane   Butanone   Acetic acid

Acetic acid is also prepared industrially by oxidation of butene (Hüls).

2.24.4.4 More recently acetic acid has been made industrially **by oxidation of light petroleum** (B.P.). The principal by-product is propionic acid.

2.24.4.5 By carbonylation of methanol at about 175°C (450 K) in the presence of a rhodium catalyst and iodide ions:

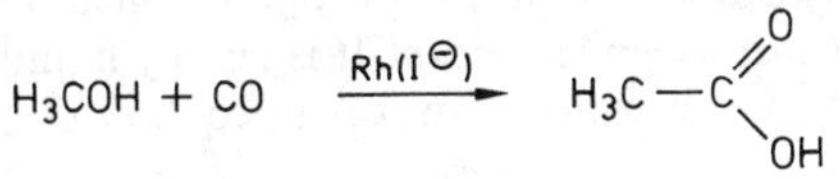

$$H_3COH + CO \xrightarrow{\text{Rh(I}^\ominus)} H_3C-C\underset{OH}{\overset{O}{\big\langle}} \qquad \Delta H = -138 \, kJ/Mol$$

Properties and Uses

Anhydrous acetic acid is a pungent-smelling liquid of b.p. 118°C (391 K), which crystallises at 16.5°C (289.7 K) to an ice-like mass, this giving rise to its common description as glacial acetic acid. Its salts are called *acetates*. With the exception of the mercury(I) and silver salts they are soluble in water.

Industrially, acetic acid is used in the preparation of a range of useful products such as pharmaceuticals. In dilute solution in water (5–8%) it is used as *vinegar*. Some of its salts have special applications, *e.g.* the sodium salt as a buffer, and the aluminium salt as a mordant in dyeing and for impregnating fabrics. Lead tetraacetate is a specific oxidising agent for 1,2-glycols (see *Criegee* oxidation).

2.24.5 Propionic Acid (Propanoic Acid)

Propionic acid is obtained in small amounts by the dry distillation of wood. It can be prepared by oxidising propan-1-ol. It is made industrially by carbonylation of ethylene with carbon monoxide and water in the presence of nickel propionate; the latter is converted *in situ* into nickel tetracarbonyl (*Reppe*):

$$H_2C=CH_2 + CO + H_2O \xrightarrow{\text{Ni propionate}} H_3C-CH_2-C\underset{OH}{\overset{O}{\lessgtr}} \qquad \Delta H = -159\,kJ/Mol$$

Ethylene   Propionic
                                                              acid

It is miscible with water, and is the lowest member of the homologous series which can be forced out of aqueous solution, as an oil, by addition of calcium chloride.

Antirheumatic agents can be prepared from propionic acid, *e.g.* 2-(4-isobutylphenyl)-propionic acid, INN **Ibuprofen** (*Brufen*); its (*S*)(+)-form is 150 times more effective than its enantiomer.

$$(H_3C)_2CH-CH_2-\!\!\bigcirc\!\!-\overset{\overset{CH_3}{\vdots}}{\underset{\underset{H}{\uparrow}}{C}}-COOH$$

INN: (*S*)(+)-*Ibuprofen* (see p. 274ff.)

2.24.6 Butyric Acids (Butanoic Acids)

As with butane and isobutane, so in the case of the C_4-carboxylic acids, two *isomers* exist:

          $H_3C-CH_2-CH_2-COOH$ $(H_3C)_2CH-COOH$

              Butyric acid   Isobutyric acid
                                              2-Methylpropanoic acid

The glycerol ester of **butyric acid** occurs in butter; *Chevreul* (1823) isolated the acid from this source. Butyric acid can be prepared by the various general methods used for making carboxylic acids. It can also be obtained by *butyric acid fermentation* of carbohydrates by bacteria such as *Clostridium butyricum*. It is an oily liquid with a rancid smell. Its calcium salt is less soluble in hot water than it is in cold.

Isobutyric acid occurs in its free state in locust beans, and as an ester in oil of camomile. It is best obtained by *atmospheric oxidation of isobutyraldehyde* (2-methyl-propionaldehyde). It boils at 154°C (427 K) and it also has a rancid smell. Because of its tertiary carbon atom it is easily oxidised to acetone and carbon dioxide; this is in marked contrast to butyric acid, which is resistant to oxidising agents.

2.24.7 *Pentanoic Acids, Valeric Acids*

There are four isomeric carboxylic acids containing five carbon atoms, as follows:

Pentanoic acid, valeric acid	$H_3C-CH_2-CH_2-CH_2-COOH$
3-Methylbutanoic acid, isovaleric acid	$\begin{matrix} H_3C \\ H_3C \end{matrix}{>}CH-CH_2-COOH$
2-Methylbutanoic acid	$H_3C-CH_2-\underset{\underset{CH_3}{\vert}}{CH}-COOH$
Trimethylacetic acid, pivalic acid, 2,2-dimethylpropionic acid	$\begin{matrix} H_3C \\ H_3C \\ H_3C \end{matrix}{>}C-COOH$

The 2- and 3-methylbutanoic acids occur as esters in valerian root (*Valeriana officinalis*).

2.24.8 *Higher Carboxylic Acids*

Mention will only be made of those higher unbranched carboxylic acids which are common in nature. They occur as esters of the lower alcohols in the essential oils of plants, where they contribute to a large extent to the taste and smell of many fruits, as esters of long-chain alcohols in waxes, and as glycerides in fats and oils.

Hexanoic acid (caproic acid), $H_3C-(CH_2)_4-COOH$, b.p. 205.5°C (478.7 K) occurs as a glyceride in butter and in coconut oil, as well as being formed as a by-product in butyric acid fermentation.

Octanoic acid (caprylic acid), $H_3C-(CH_2)_6-COOH$, m.p. 16.5°C (289.7 K) also occurs as a glyceride in butter.

Decanoic acid (capric acid), $H_3C-(CH_2)_8-COOH$, m.p. 31.5°C (304.7 K) is present in its free state in various kinds of cheeses, and also as a glycerol ester in butter and coconut oil.

The trivial names of these three acids, which are not accepted by the IUPAC rules, are derived from *capra* (Latin, goat) and came about because of their being present in goat butter.

Lauric acid (dodecanoic acid), $H_3C-(CH_2)_{10}-COOH$, m.p. 44°C (317 K) is found as its glycerol ester in the oil from the fruit of the bay tree (*Laurus nobilis* = bay tree), in coconut oil and in sperm oil.

Palmitic acid (hexadecanoic acid), $H_3C-(CH_2)_{14}-COOH$, and **stearic acid (octadecanoic acid)**, $CH_3-(CH_2)_{16}-COOH$, occur, together with oleic acid, as glyceride esters which form the major constituents of plant and animal fats.

It is noteworthy that all the naturally occurring higher unbranched carboxylic acids are exclusively compounds having even numbers of carbon atoms. *Margaric acid* (*heptadecanoic acid*), $H_3C—(CH_2)_{15}—COOH$, the acid between the last two naturally occurring acids mentioned above, is not found in nature.

In contrast, some carboxylic acids having branched chains do occur in animals and plants. For example, *tuberculostearic acid*, found in tubercular bacteria, has been shown, by oxidative breakdown products and by synthesis, to be 10-methylstearic acid.

Hexacosanoic acid, $H_3C—(CH_2)_{24}—COOH$, m.p. 88–89°C (361–362 K) is found together with palmitate esters in some natural waxes as esters of higher monohydric alcohols.

Lithium stearate is used to provide thermal stabilisation of the viscosity of greases. Further details about the preparation and properties of the higher carboxylic acids and their salts follow in section 2.26, Fats, Oils and Waxes.

2.25 Unsaturated Aliphatic Monocarboxylic Acids, Alkenoic Acids

This class of compounds differs from the saturated monocarboxylic acids in having a $C=C$ *double bond* in the molecules. The general formula for the homologous series is thus $C_nH_{2n-1}COOH$.

The simplest acid of this sort is *acrylic acid* (*propenoic acid*), which is the oxidation product of allyl alcohol or acrolein:

| Allyl alcohol | Acrolein | Acrylic acid |

The interaction of the double bond with the carboxyl group increases its acidity, thus acrylic acid ($pK_a = 4.25$) is more dissociated than the corresponding saturated acid, propionic acid ($pK_a = 4.88$).

The increased acidity is associated with the sp^2 hybridisation of the unsaturated α-carbon atom which is more electron-withdrawing than a saturated sp^3 carbon atom.

Accordingly, an sp^2-hybridised carbon atom is a weaker electron donor than an sp^3-hybridised carbon atom. This effect is even more noticeable for an sp-hybridised carbon atom of a $C≡C$ triple bond. Thus *propiolic acid* (*propynoic acid*), $HC≡C—COOH$ has a pK_a value of 1.85.

2.25.1 General Methods of Preparation

α,β-Unsaturated acids are usually made by one of the following methods:

2.25.1.1 By means of the *Knoevenagel* reaction (see p. 328).

2.25.1.2 Removal of hydrogen chloride from β-halogenocarboxylic acids by means of warm alcoholic potassium hydroxide:

$$R—\underset{\underset{Br}{|}}{CH}—CH_2—COOH + KOH \longrightarrow R—CH=CH—COOH + KBr + H_2O$$

2.25.1.3 Industrially by oxidation of acrolein, itself obtained by atmospheric oxidation of *propene* (overall yield 85–90%):

$$H_2C = CH-CHO \xrightarrow{\text{[cat.]}} H_2C = CH-COOH$$

Properties

The characteristic feature of unsaturated carboxylic acids is the readiness of the $C=C$ double bond to undergo addition reactions. They are converted into saturated carboxylic acids by *catalytic hydrogenation*:

$$R-CH = CH-COOH \xrightarrow{+2H} R-CH_2-CH_2-COOH$$

As a consequence of the effect of the carboxyl group, α,β-unsaturated carboxylic acids can be reduced by means of nascent hydrogen.

In similar fashion, both *halogens* and *hydrogen halides* add to unsaturated carboxylic acids, *e.g.*

$$\underset{\substack{\text{3-Bromopropionic}\\\text{acid}}}{\overset{\begin{array}{c}CH_2-CH_2-COOH\\ |\\ Br\end{array}}{}} \xleftarrow{+HBr} \underset{\substack{\text{Acrylic}\\\text{acid}}}{H_2C=CH-COOH} \xrightarrow{+Cl_2} \underset{\substack{\text{2,3-Dichloropropionic}\\\text{acid}}}{\overset{\begin{array}{c}CH_2-CH-COOH\\ |\quad\;\; |\\ Cl\quad Cl\end{array}}{}}$$

The $C=C$ double bond may be broken by the action of oxidising agents such as potassium permanganate or ozone.

The Most Important Unsaturated Carboxylic Acids

2.25.2 Acrylic Acid (Propenoic Acid)

Acrylic acid is derived from ethylene by replacement of a hydrogen atom by a carboxyl group.

Preparation

2.25.2.1 From 2-chloroethanol (ethylene chlorohydrin) by treating it with sodium cyanide to give *3-hydroxypropionitrile* (ethylene cyanohydrin), which is then converted into acrylic acid by means of conc. sulphuric acid, which brings about hydrolysis of the nitrile and loss of a mol of water:

$$\underset{\text{2-Chloroethanol}}{\overset{\begin{array}{c}CH_2-CH_2\\ |\quad\;\; |\\ OH\quad Cl\end{array}}{}} \xrightarrow[-NaCl]{+NaCN} \underset{\text{3-Hydroxypropionitrile}}{\overset{\begin{array}{c}CH_2-CH_2\\ |\quad\;\; |\\ OH\quad C\equiv N\end{array}}{}} \xrightarrow[-NH_3]{+2H_2O} \underset{\text{3-Hydroxypropionic acid}}{\overset{\begin{array}{c}CH_2-CH_2\\ |\quad\;\; |\\ OH\quad COOH\end{array}}{}} \xrightarrow{-H_2O} \underset{\text{Acrylic acid}}{H_2C=CH-COOH}$$

2.25.2.2 By carbonylation of acetylene.

2.25.2.3 By direct oxidation of propene without isolating the intermediate *acrolein* (see p. 206). This process is of great importance industrially.

Properties

Acrylic acid is a liquid with a pungent smell; it boils at 141°C (414 K). It is miscible with water in all proportions. When kept, and especially when heated, it polymerises giving a glassy product, poly(acrylic acid), which has the following chain structure:

$$n \ H_2C = CH \longrightarrow \left[CH-CH_2 \right]_n$$
$$\qquad \ \ | \qquad\qquad\qquad | \qquad\qquad$$
$$\qquad COOH \qquad\qquad COOH$$

Poly(acrylic acid)

$$H \diagdown\diagup COOR^* \qquad R^*OOC \diagdown\diagup H$$
$$Si \qquad\qquad\qquad Re$$

This water-soluble polymer is used as a finishing agent and as a thickening agent, and also for medical coverings and as a base for ointments. The ethyl ester of poly(acrylic acid) has proved useful as a copolymerisation partner in the preparation of weather-proof elastomers.

The ethylenic part of acrylic acid is prochiral. If it is esterified with a chiral alcohol, then the *enantiotopic* faces (see pp. 196, 277 *f.*) become *diasterotopic*. Such esters are valuable starting materials for asymmetric syntheses.

2.25.3 Acrylonitrile

Acrylonitrile, $H_2C=CH-C\equiv N$, is a colourless liquid, b.p. 78°C (351 K), which polymerises very readily[1]. It has been found to be a carcinogen in tests on animals.

Preparation

2.25.3.1 By addition of hydrocyanic acid to acetylene:

$$HC\equiv CH + H-C\equiv N \xrightarrow{\text{(CuCl, NH}_4\text{Cl, HCl)}} H_2C=CH-C\equiv N$$

This process has been superseded by the ammono-oxidation process.

2.25.3.2 Ammono-oxidation. From propene, ammonia and oxygen at 450°C (720 K) and 1.5 bar (0.15 MPa) in the presence of a bismuth–molybdenum–iron catalyst in a fluidised bed reactor (Sohio process):

$$H_2C=CH-CH_3 + NH_3 + \tfrac{3}{2}\,O_2 \rightarrow H_2C=CH-C\equiv N + 3\,H_2O$$

This process gives about 67% acrylonitrile and, as by-products, acetonitrile, $H_3C-C\equiv N$, (14–18%) and hydrogen cyanide (3–4%).

This reaction resembles the *Andrussov* process for making hydrogen cyanide. *Methacrylonitrile* (2-methylpropenenitrile), $H_2C=C-CN$, can be made in a similar way from
$$\qquad\qquad\qquad\qquad\qquad | \qquad\qquad$$
$$\qquad\qquad\qquad\qquad\quad CH_3$$
isobutene (methylpropene).

Uses

Acrylonitrile is useful for introducing *cyanoethyl groups*, $-CH_2-CH_2-CN$, into compounds with reactive methylene groups.[1] Reaction proceeds as in *Michael addition* reactions (see p. 328 *f.*).

---

[1] See *O. Bayer*, Die Chemie des Acrylnitrils (The Chemistry of Acrylonitrile), Angew. Chem. *61, 229* (1949).

Precipitation-polymerisation of acrylonitrile in water, promoted by *peroxides*, gives rise to *poly(acrylonitrile)* (PAN):

$$n\ HC=CH_2 \longrightarrow \left[\begin{array}{c} CH-CH_2 \\ | \\ C\equiv N \end{array}\right]_n \qquad n\ \begin{array}{c} CN \\ | \\ C=CH_2 \\ | \\ COOR \end{array} \longrightarrow \left[\begin{array}{c} CN \\ | \\ C-CH_2 \\ | \\ COOR \end{array}\right]_n$$

Poly(acrylonitrile)   Poly(cyanoacrylate)

Since the polymer which is produced decomposes before it melts, at about 250°C (525 K), it is not possible to spin it as it stands. Only when it was found that *N,N-dimethylformamide* was a suitable solvent for poly(acrylonitrile) did it prove possible to process it by means of dry or wet spinning techniques. The fibres of poly(acrylonitrile) (PAN-*fibre, acrylic fibre, Orlon, Dralon, Wolpryla*) obtained in this way show high strength, excellent resistance to light and to weather, and a resemblance to wool. When pyrolysed in the absence of oxygen it is converted into carbon fibre (CFK), which shows a graphite-like structure. It surpasses steel in its mechanical properties whilst being much lighter in weight. Acrylonitrile is also used as a co-monomer in the preparation of elastomers. Cyanoacrylic ester polymerises spontaneously under the influence of atmospheric moisture, giving poly(cyanoacrylate). It is hence used in large amounts as an adhesive (cyanoacrylate rapid glue).

2.25.4 *Unsaturated Carboxylic Acids with Four Carbon Atoms*

There are four *isomeric* C_4 alkenecarboxylic acids:

$$H_2C=C-COOH \atop |\atop CH_3$$ Methacrylic acid, 2-methylpropenoic acid
(b.p. 163°C [436 K])

$$H_3C-CH=CH-COOH$$ Crotonic acid, *trans*-but-2-enoic acid
(m.p. 72°C [345 K])

Isocrotonic acid, *cis*-but-2-enoic acid
(m.p. 15.5°C [288.7 K])

$$H_2C=CH-CH_2-COOH$$ Vinylacetic acid, but-3-enoic acid
(b.p. 70°C [343 K]/16 mbar [16 hPa]).

2.25.4.1 Methacrylic acid occurs in oil of camomile. Its esters are more important than the acid itself, in particular methyl methacrylate. This is made on the industrial scale by hydrolysis and removal of water from acetone cyanohydrin (α-hydroxypropionitrile), using sulphuric acid; this gives methacrylic acid which is at the same time esterified with methanol:

$$\begin{array}{c} H_3C-C=O \\ | \\ CH_3 \end{array} \xrightarrow{+HCN} \begin{array}{c} \quad OH \\ H_3C-C \\ | \quad C\equiv N \\ CH_3 \end{array} \xrightarrow[-NH_4HSO_4]{+H_2SO_4,+H_3COH} \begin{array}{c} H_2C=C-COOCH_3 \\ | \\ CH_3 \end{array}$$

Acetone   Acetone cyanohydrin   Methyl methacrylate
(b.p. 100°C [373 K])

Methyl methacrylate is also obtained industrially by dehydrogenation of methyl isobutyrate (methyl 2-methylpropionate) at about 500°C (770 K).

Radical polymerisation of methyl methacrylate (MMA) leads to the formation of a hard, transparent material, *poly(methyl methacrylate)* (PMMA). This is used as an organic glass

(*O-Glass, Piacryl, Plexiglass*) which does not splinter, in motorcars, aircraft and for optical lenses. Poly(diethyleneglycol bis[allylcarbonate]), CR39, is particularly useful for making spectacles. It is a polycarbonate, made by radical polymerisation from monomeric diethyleneglycol bis(allyl carbonate).

$$n \begin{array}{c} CH_3 \\ | \\ C{=}CH_2 \\ | \\ COOCH_3 \end{array} \longrightarrow \left[\begin{array}{c} CH_3 \\ | \\ C{-}CH_2 \\ | \\ COOCH_3 \end{array} \right]_n \qquad O(CH_2{-}CH_2{-}O{-}\underset{\underset{O}{\|}}{C}{-}O{-}CH_2CH{=}CH_2)_2$$

Poly(methyl methacrylate)   Diethyleneglycol bis(allyl carbonate)

A solution of poly(methyl methacrylate) which also contains some monomeric methyl methacrylate is used in surgery as an adhesive. On admixture with peroxide the mixture hardens and sets as a consequence of the further polymerisation which is initiated.

There are two isomeric but-2-enoic acids, $H_3C{-}CH{=}CH{-}COOH$, the *trans* form, *crotonic acid*, *E*-but-2-enoic acid, and the *cis* form, *isocrotonic acid*, *Z*-but-2-enoic acid. If isocrotonic acid is either heated or irradiated with UV light, it is isomerised to crotonic acid in an equilibrium reaction which favours the *E*-form.

$$\underset{HOOC}{\overset{H_3C}{}}\underset{H}{\overset{H}{C{=}C}} \rightleftharpoons \underset{H}{\overset{H_3C}{}}\underset{COOH}{\overset{H}{C{=}C}}$$

Isocrotonic acid, (*Z*)-form   Crotonic acid, (*E*)-form

Both acids can be reduced to butanoic acid.

2.25.4.2 Crotonic acid can be made either by *oxidation of crotonaldehyde* or by *condensation of acetaldehyde with malonic acid* in pyridine (see *Knoevenagel* reaction).

$$H_3C{-}\overset{H}{\underset{O}{C}} + H_2C(COOH)_2 \xrightarrow[+H_2O]{\text{Pyridine}} H_3C{-}CH{=}C(COOH)_2 \xrightarrow[-CO_2]{} \begin{array}{c} H_3C{-}CH \\ \| \\ HC{-}COOH \end{array}$$

Malonic acid   Crotonic acid

Isocrotonic acid is formed when *β*-hydroxyglutaric acid (3-hydroxyglutaric acid) is distilled under reduced pressure:

$$\begin{array}{c} CH_2{-}COOH \\ | \\ CHOH \\ | \\ CH_2{-}COOH \end{array} \xrightarrow{-H_2O,-CO_2} \begin{array}{c} H{-}C{-}CH_3 \\ \| \\ H{-}C{-}COOH \end{array}$$

β-Hydroxyglutaric acid   Isocrotonic acid

2.25.4.3 Vinylacetic acid (but-3-enoic acid) can be made by *acid hydrolysis of allyl cyanide*:

$$H_2C{=}CH{-}CH_2{-}C{\equiv}N \xrightarrow[-NH_3]{+2H_2O} H_2C{=}CH{-}CH_2{-}COOH$$

When vinylacetic acid is heated with alkali or with sulphuric acid it is almost totally *isomerised to crotonic acid*.

$$\underset{(2\%)}{H_2C{=}CH{-}CH_2{-}COOH} \underset{}{\overset{(OH^{\ominus})}{\rightleftharpoons}} \underset{(98\%)}{H_3C{-}CH{=}CH{-}COOH}$$

Higher *β,γ*-unsaturated carboxylic acids isomerise similarly, especially in the presence of alkali, with the double bond shifting to the *α,β*-position, where it is *conjugated* with the C=O double bond of the carboxyl group:

$$R-CH=CH-CH_2-\underset{\underset{OH}{|}}{C}=O \rightleftharpoons R-CH_2-CH=CH-\underset{\underset{OH}{|}}{C}=O$$

2.25.5 Oleic acid (Octadec-9-enoic acid)

This is the commonest of the unsaturated carboxylic acids and is the main constituent of many natural oils, and especially of olive oil and almond oil, and of fish oils. For details of its preparation see section 2.26 Fats, Oils and Waxes.

Oleic acid, $C_{17}H_{33}COOH$, is a colourless, almost odourless liquid, which is insoluble in water, but it dissolves in ethanol or ether. It melts at 13°C (286 K). On standing in air, it darkens in colour and acquires a rancid smell.

Its structure was demonstrated as follows. On catalytic hydrogenation it provided stearic acid, thus showing that the carbon chain was unbranched. The position of the double bond was investigated by ozonolysis. Hydrolysis of the resultant ozonide gave, as well as a corresponding aldehyde and carboxy-aldehyde, a monocarboxylic acid, nonanoic acid, and a dicarboxylic acid, nonanedicarboyxlic acid. Since each of these acids was made up from nine carbon atoms, the double bond of oleic acid must link the 9- and 10-carbon atoms in the chain. This procedure has industrial use for the preparation of nonanedicarboxylic acid (azelaic acid).

$$H_3C-(CH_2)_7-CH=CH-(CH_2)_7-COOH \rightarrow H_3C-(CH_2)_7-COOH + HOOC-(CH_2)_7-COOH$$

| Oleic | Nonanoic | Azelaic acid |
| acid | acid | Nonanedicarboxylic acid |

In the presence of nitrogen dioxide, oleic acid undergoes equilibration (Z–E isomerisation) with the more stable elaidic acid (m.p. 44° [317 K]):

Oleic acid
(Z)-form, 34%

$$\underset{HOOC-(CH_2)_7}{\overset{H_3C-(CH_2)_7}{}}\underset{}{\overset{}{\diagdown C}}\underset{\overset{\|}{C}}{\diagup}\overset{H}{\diagup}\underset{H}{\diagdown}$$

Elaidic acid
(E)-form, 66%

$$\underset{H}{\overset{H_3C-(CH_2)_7}{}}\overset{}{\diagdown C}\underset{C}{\overset{\|}{}}\underset{(CH_2)_7-COOH}{\diagup}\overset{H}{\diagdown}$$

2.25.6 Polyunsaturated Monocarboxylic Acids

An example of a monocarboxylic acid having two *conjugated* C=C double bonds is **sorbic acid**, (E,E)-hexa-2,4-dienoic acid, $H_3C—CH=CH—CH=CH—COOH$, m.p. 133–134°C (405–407 K), which occurs in the juice of mountain ash, or rowan, berries (*Sorbus aucuparia*) and is obtained therefrom. Since it inhibits the growth of yeasts and fungi, it is a valuable preservative for foodstuffs, and in consequence is made synthetically, for example by the reaction of crotonaldehyde with ketene. The lactone of 3-hydroxyhex-4-enoic acid is formed and this, in the presence of sulphuric acid, is hydrolysed and loses water to give sorbic acid.

$$H_3C-CH=CH-CHO \quad + \quad H_2C=C=O \longrightarrow H_3C-CH=CH-CH-CH_2-\underset{\underset{O}{|}}{C}=O$$

3-Hydroxyhex-4-enoic acid lactone

$$\xrightarrow{H_2SO_4} \quad H_3C-CH=CH-CH=CH-COOH \equiv \diagup\diagdown\diagup\diagdown COOH$$

Sorbic acid

Industrially the reaction is carried out in the presence of zinc salts. This leads to the formation of a polyester derived from 3-hydroxyhex-4-enoic acid instead of a lactone; this polyester can be hydrolysed by either acid or alkali.

In natural fats and oils, various unsaturated acids besides oleic acid occur, which have two or more C=C double bonds in their molecules. Linseed oil, walnut oil, poppy-seed oil, hemp-seed oil and fish oils all contain large amounts of polyunsaturated monocarboxylic acids. The commonest are **linoleic acid** (octadeca-9,12-dienoic acid), $C_{17}H_{31}$—COOH, **linolenic acid** (octadeca-9,12,15-trienoic acid), $C_{17}H_{29}$—COOH, and eicosa-5,8,11,14,17-pentenoic acid, $C_{19}H_{29}$—COOH:

Linoleic acid (C_{18} : 2 $n - 6$)

Linolenic acid (C_{18} : 3 $n - 3$)

Eicosapentenoic acid (C_{20} : 5 $n - 3$)

In mammals it is only possible to introduce double bonds into the carbon chain biochemically between carbon atoms 1–9. Because of this, acids with two or more double bonds which have such bonds between carbon atoms further along the carbon chain than C-9, are essential ingredients in the diet (essential fatty acids). Unsaturated C_{20}-carboxylic acids in the organism are converted by partial oxidation into versatile biologically active agents such as the prostaglandins. The notation given above has been introduced in order to clarify the biochemical modifications in the unsaturated fatty acids. For example, in the case of linoleic acid, it indicates that there are 18 carbon atoms and two double bonds. The symbol $n - 6$ shows that the double bond between C-12 and C-13 lies six carbons from the end of the carbon chain. Similarly, in this system oleic acid is represented as C_{18}:1 $n - 9$.

If the only information required is the distance of the double bond from the end of the chain, then linoleic and linolenic acids are described, respectively, as ω-6 and ω-3 carboxylic acids.

The *action of atmospheric oxygen on polyunsaturated carboxylic acids* brings about autoxidation[1] and radical polymerisation of the acids to give solid, high molecular weight resins. Hence they have been used, especially linseed oil, as 'drying oils' in the oil-paint industry. In order to accelerate the hardening process as much as possible, *drying agents* or *driers* (cobalt, manganese or leads salts of linoleic, resin and naphthenic acids) are added in small amounts (1–5%). These act as oxygen carriers and result in the formation of a solid film within a day. The mixture of drying oils and drying agents is known as *varnish*.

Saturated fatty acids also undergo slow autoxidation. The special reactivity of the drying oils depends upon the stabilisation of the radical intermediate —CH=CH—ĊH—CH=CH—, in which the unpaired electron can be delocalised over five carbon atoms.

2.26 Fats, Oils and Waxes

2.26.1 Fats and Oils

The natural fats and oils, together with carbohydrates, form an important group of replenishable raw materials. They are, without exception, glycerol esters of higher

[1] See *W. Kern* et al., Die Katalyse der Autoxidation ungesättigter Verbindungen (Catalysis of the Autoxidation of Unsaturated Compounds), Angew. Chem. **67**, 573 (1955).

carboxylic acids containing even numbers of carbon atoms (*Chevreul*, 1823). They are generally known as *glycerides* but a better description is *acylglycerols*.

Animal fats consist for the most part of mixed glycerides of three acids, palmitic, stearic and oleic. The greater the amount of oleic acid, the more easily the fat liquifies when it is heated (pork fat, goose grease, whale oil). In contrast, a preponderance of saturated acids leads to substantially higher melting fats (beef or mutton fat, tallow). Butter contains not only the three common acids, but also lower acids with even numbers of carbon atoms (see p. 236).

The following *triacylglycerol* (*triglyceride*) is an example of a fat:

$$H_2C-O-CO-(CH_2)_{14}-CH_3 \qquad \text{(Palmitic acid)}$$

$$H^*C-O-CO-(CH_2)_{16}-CH_3 \qquad \text{(Stearic acid)}$$

$$H_2C-O-CO-(CH_2)_7 \underset{}{\overset{H}{\diagup}}C=C\underset{(CH_2)_7-CH_3}{\overset{H}{\diagdown}} \quad \text{(Oleic acid)}$$

Such triglycerides contain a chiral centre (*C), but are none the less not optically active. For this reason the suitability of the widely used term *optical isomers* is open to question.

When fats are heated with aqueous sodium hydroxide they undergo hydrolysis, giving glycerol and the sodium salts of the carboxylic acids, which are *soaps*, *e.g.*

$$
\begin{array}{ll}
CH_2-O-CO-(CH_2)_{16}-CH_3 & \\
CH-O-CO-(CH_2)_{16}-CH_3 \; + \; 3\,NaOH \;\longrightarrow\; & CH-OH \; + \; 3\,H_3C-(CH_2)_{16}-COO^{\ominus}Na^{\oplus} \\
CH_2-O-CO-(CH_2)_{16}-CH_3 & CH_2-OH
\end{array}
$$

$$CH_2-OH$$

| Tristearin | Glycerol | Sodium stearate (a soap) |

The term *saponification* (Lat. *sapo* = soap), which is frequently used to describe alkaline cleavage of esters, has its origin in the alkaline hydrolysis of fats.

In contrast to animal fats, plant oils (*e.g.* olive oil, rapeseed oil, palm oil, coconut oil, linseed oil, castor oil) contain not only glycerides of palmitic, stearic and oleic acids, but above all glycerol esters of polyunsaturated acids.

Analysis of fats. Formerly the main criteria for the composition of the various fats and oils were their saponification values and their iodine values; these gave some indication of their molecular weights and of the amount of unsaturation in the molecules.

Nowadays the fat is transesterified with methanol and hydrochloric acid, giving rise to the methyl esters of the constituent carboxylic acids; these can then be *separated and estimated by means of gas chromatography*. Further information, especially about the stereochemistry of the unsaturated acids, can be derived from IR and NMR spectra.

2.26.2 Production of Fats

2.26.2.1 Animal fats are most commonly obtained by rendering (melting) the fatty tissue, less frequently by pressing it or by extracting it with ether, carbon disulphide, benzene or trichloroethylene.

2.26.2.2 Plant oils are usually accumulated in the seeds and fruits. The latter are either pressed to squeeze the oil out, or broken up and then extracted with organic solvents. The residual *oilcake* left after the material has been pressed and the residues left from the extraction process are valuable as high-protein foodstuffs.

2.26.2.3 Hardening of Fats. A majority of the naturally occurring fats and oils are mixtures of various acylglycerols which are liquid at room temperature. Since there is substantially greater demand for solid fats, especially as foodstuffs for humans, the conversion of liquid fats and oils into solids is a problem of some importance. It is achieved by *catalytic hydrogenation* in the liquid phase at 180°C (450 K) and 5 bar (0.5 MPa) in the presence of finely divided nickel. Double bonds present in the glyceryl esters of the oils are thereby hydrogenated (*Normann*, 1902). By careful choice and control of the reaction conditions, selective hydrogenation can be achieved, whereby all but one of the C=C double bonds in each of the acids are hydrogenated. The hardening is above all of importance in the making of *margarine* from vegetable oils.

2.26.2.4 Hydrolysis of Fats and Oils. The hydrolysis of fats and oils to give carboxylic acids and glycerol is an industrial process which can be carried out in a variety of ways:

(a) By using superheated steam in an autoclave at 170°C (440 K) in the presence of a basic catalyst, *e.g.* CaO, MgO or ZnO. Reaction is completed in about 8 hours. The insoluble carboxylic acids float on top of the aqueous solution of glycerol, which is run off.

(b) *Twitchell* Process. The problem about hydrolysing fats with dilute sulphuric acid is that fats and water do not mix. When fats are saponified with alkali, the soaps which are formed cause emulsification, but in acid conditions *emulsifiers* must be added. The *Twitchell* reagent has proved useful for this purpose. It is made by the action of conc. sulphuric acid on a mixture of oleic acid and naphthalene. About 0.5–1% of this reagent is added to the mixture of fats, and hydrolysis is carried out at 100°C (370 K) for 24 hours, providing a 90% yield of hydrolysis products.

(c) Enzymatic hydrolysis using the enzyme *lipase* from castor oil beans can be carried out at lower temperatures (not above 40°C [310 K]) and produces very pure acids. This method is only of limited use.

2.26.3 Waxes

Waxes are esters of long-chain carboxylic acids which are found in both plant and animal sources. They differ from fats and oils in that, whereas the latter are esters of glycerol, waxes are esters derived from long-chain *monohydric primary alcohols*. For example, in beeswax may be found myricyl palmitate ($\sim$ 75%), together with myricyl cerotate ($\sim$ 10%) and paraffin ($\sim$ 15%), (see myricyl alcohol, p.121). Cetyl palmitate, $C_{15}H_{31}COOC_{16}H_{33}$ is found in whale oil and cetyl cerotate, $C_{25}H_{51}COOC_{26}H_{53}$, in Chinese wax.

Among plant waxes carnauba wax should be mentioned. It forms scales on the leaves of the Brazilian wax pine, and consists mainly of myricyl cerotate.

Fats and *oils* are a related group of compounds which form part of the class of compounds known as **lipids**. The *lipids* also include other groups of compounds such as the *waxes*, carotenoids, steroids, glycolipids, phosphatides, prostaglandins, etc.

2.27 Soaps and Synthetic Detergents

Compounds which are to be used as washing agents must act as wetting or emulsifying agents. They do this by lowering the surface tension, thereby facilitating wetting.

They are classified into *capillary-* or *surface-active materials*, and in the technical literature *as detergents* or *surfactants*. The properties as a washing material, *i.e.* as a wetting, dispersing, emulsifying and absorbent material, depend upon the presence of two particular types of group in the molecule. A *hydrophobic* group, usually an aliphatic hydrocarbon residue with 10–18 carbon atoms or an alkylbenzene residue, is responsible for the surface activity. The term 'amphiphilic compound' describes this property. A *hydrophilic* group, which controls the solubility in water, may be one of a variety of groups such as —COOH (soaps), —OSO_3H (alkyl sulphates), —SO_3H (alkane sulphonates) and ether-type oxygen bridges (non-ionic washing materials). Surface active materials[1] are classified according to their electrochemical properties as anionic, cationic or non-ionic agents.

If untouched detergents such as washing agents are carried by rivers into the sea, plants growing on the shore can be badly damaged by them. The presence of surfactants in the spume from breakers enables sodium chloride to penetrate the leaves or needles of the plants where it destroys the chloroplasts.

2.27.1 Anionic Surfactants

2.27.1.1 Soaps

In the manufacture of soap the starting materials are either fats or free long-chain carboxylic acids which are neutralised with alkali. Soaps are a mixture of the *sodium salts of the higher carboxylic acids*, of palmitic, stearic and oleic acids as well as some polyunsaturated acids. The sodium salts are isolated from the solution resulting from hot alkaline saponification of fats by 'salting them out' with sodium chloride. The expansion of the soda industry based on the *Leblanc* process in the first half of the nineteenth century was a prerequisite for the general use of soaps both as toilet soaps and as powdered or scale-like domestic washing powders.

One disadvantage of soap is the insolubility of its calcium salt in water. In hard water this separates out to form a flocculant sediment on textile fibres, and it is unsuitable for washing purposes. The problem can be overcome by adding *water softeners* such as meta- or poly-phosphates, or zeolites. Unlike phosphates, zeolites do not lead to problems of eutrophication in waste water.

The development of washing materials not associated with natural fats has led to the use of the following *anionic surfactant* washing materials, of which the most important are the *alkyl sulphates*.

2.27.1.2 n-Alkyl sulphates

The starting materials for the manufacture of these compounds are higher primary alcohols. These are obtained either by the oxo-synthesis, or by hydrogenation of the acids obtained by the saponification of fats and oils, using a copper chromite catalyst at 250°C (520 K) and 200 bar (20 MPa). (*Schrauth* and *Bertsch*):

$$H_3C-(CH_2)_n-COOH \xrightarrow[-H_2O]{+ 4 H(CuO/Cr_2O_3)} H_3C-(CH_2)_n-CH_2OH \qquad (n = 10-16)$$

Carboxylic acid   Primary alcohol

The resultant *alcohol* is *esterified* with conc. sulphuric acid (or chlorosulphonic acid), giving an n-*alkyl sulphate*:

$$H_3C-(CH_2)_n-CH_2OH + HO-SO_3H \rightarrow H_3C-(CH_2)_n-CH_2-O-SO_3H + H_2O$$

n-Alkyl sulphate

---

[1] See *J. Falbe* and *U. Hasserodt*, Katalysatoren, Tenside and Mineralöladditive (Catalysts, Surfactants and Mineral Oil Additives) (Thieme Verlag, Stuttgart 1978).

The sodium salts of these alkyl sulphates are neutral compounds which are soluble in water and have detergent properties similar to those of soap. A special advantage is that their calcium salts are readily soluble in water, so that they can also be used as washing materials in hard water.

In recent times the unbranched primary alcohols made by the method of K. *Ziegler* have also been used as starting materials in the detergents industry. As described on p. 61 *f*, ethylene reacts with triethylaluminium to give longer-chain trialkylaluminium compounds. These are oxidised by air to aluminium alkoxides, which can be hydrolysed to give unbranched long-chain alcohols with even numbers of carbon atoms (*Alfol Process*):

$$Al[(CH_2-CH_2)_n-C_2H_5]_3 \xrightarrow{+3/2\ O_2} Al[O-(CH_2-CH_2)_n-C_2H_5]_3 \xrightarrow[-Al_2O_3]{H_2O} 3\ C_2H_5-(CH_2-CH_2)_n-OH$$

Aluminium trialkyls   Aluminium alkoxides   Primary alcohols

2.27.1.3 s-Alkyl sulphates

To prepare secondary alcohols n-alkanes are oxidised in air, in the presence of boric acid, at 140–180°C (410–450 K):

$$H_3C(CH_2)_n-CH_2(CH_2)_mCH_3 + \tfrac{1}{2}\ O_2 \xrightarrow{H_3BO_4} H_3C(CH_2)_n-\underset{\underset{OH}{|}}{CH}(CH_2)_mCH_3$$

n-Alkane   Secondary alcohol

The secondary alcohols are esterified to secondary alkyl sulphates in just the same way as the primary alcohols are (see above).

When distillates from crude oil which are rich in n-alkanes are cracked, higher monoalkenes with even numbers of carbon atoms are obtained. These can be converted into secondary alkyl sulphates by reaction with sulphuric acid:

$$H_3C-(CH_2)_{n+m}-CH=CH_2 + H_2SO_4 \longrightarrow H_3C-(CH_2)_n-\underset{\underset{O-SO_3H}{|}}{CH}-(CH_2)_m-CH_3 \ (n+m=9-15)$$

Alkene   secondary-Alkyl sulphate

If some monoalk-1-enes are also incorporated, a mixture of sulphates with the sites of the —OSO_3H group distributed over the whole hydrocarbon chain is obtained. Sodium salts of secondary alkyl sulphates are used as detergents (Teepol).

2.27.1.4 Alkanesulphonates

Because they can be obtained economically from n-paraffins, alkanesulphonates are important materials. They are made by *chlorosulphonation* or *sulphoxidation of alkanes*, in radical reactions promoted by UV light, *e.g.*

$$\underset{R'}{\overset{R}{>}}CH_2 + SO_2 + \tfrac{1}{2}O_2 \xrightarrow{h\nu,\ 300\ K} \underset{R'}{\overset{R}{>}}CH-SO_3H$$

2.27.1.5 Alkylbenzenesulphonates

Among the detergents which are produced in the largest quantities are alkylbenzenesulphonates having one alkyl substituent containing 8–14 carbon atoms.

Originally the starting material for making these compounds was tetrapropylene, produced from propylene. The disadvantage of the tetrapropylenebenzenesulphonates was their resistance to biological degradation by microorganisms. Troublesome formation of foams occurred in the waste waters; this was due to the presence of the highly branched side chain attached to the benzene nucleus. Because of this it was necessary that alkylbenzene- sulphonates having an n-alkyl substituent should be used. The problem was overcome by making unbranched monoalkenes from specific alkane fractions obtained from mineral oil. This can be done in various ways.

Alkenes obtained from cracking are possible starting materials. Alternatively n-alkanes can be chlorinated and then catalytically dehydrochlorinated to provide open-chain monoalkenes. For example, dodecenes can be made from dodecane:

$$C_{12}H_{26} \xrightarrow[-HCl]{+ Cl_2 \text{ (400 K)}} C_{12}H_{25}Cl \xrightarrow[-HCl]{Cat. \text{ (525 K)}} C_{12}H_{24}$$

Dodecane $\qquad\qquad\qquad$ Dodecenes

The monoalkenes are either converted directly into alkyl sulphates, using sulphuric acid, or into alkylbenzenes, using *hydrogen fluoride* (or aluminium chloride) as catalyst. Sulphonation of the n-alkylbenzenes with sulphur trioxide and neutralisation of the product with sodium hydroxide gives n-alkylbenzenesulphonates with alkyl chains having 8–14 carbon atoms, *e.g.* dodecylbenzene sulphonate, which are easily biodegradable.

$n + m = 8 - 14$ $\quad$ $H_3C-(CH_2)_n-CH-(CH_2)_m-CH_3$ $\qquad$ $H_3C-(CH_2)_n-CH-(CH_2)_m-CH_3$

The alkylation of benzene with an alk-1-ene is accompanied by migration of the double bond within the alkene, leading to a statistical distribution of isomeric products. Almost all synthetic detergents also contain a variety of additives such as *bleaching agents* (perborate), *perborate stabilisers* (silicates) *and optical brighteners* (whiteners).

2.27.2 Cationic Surfactants

These compounds contain an *ammonium group* attached to the *hydrophobic* part of the molecule. As a consequence of their being salts they go into solution in water. They have also been described as *ammonium amphiphiles* and as *invert soaps*.

They are not used as detergents, but they do have a number of special uses, for example the processing of textiles and of leather, in dyeing, and as emulsifying and dispersing agents; they are able to precipitate proteins. Some of them have bactericidal properties and are therefore used as disinfectants, e.g. alkylbenzyldimethylammonium chlorides, INN: *Benzalkonium chlorides*:

n = 7–17

Ammonium amphiphiles having two charged end groups separated from one another by 20–30 carbon atoms (p. 763), *e.g.*

can form spherical aggregates (vesicles) which have diameters of 10–60 nm and are filled with liquid. They are obtained by the action of ultrasound on an aqueous suspension of the amphiphilic compound. Vesicles can also form spontaneously when anionic and cationic surfactants, *e.g.* sodium dodecylbenzene-sulphonate and trimethylammonium tosylate, are present at the same time.[1] The vesicles form microemulsions[2] which are clear liquids and are able to dissolve hydrophobic compounds.

Compounds can be included in the cavities of the vesicles and in this way can be transported to places in an organism which they would not otherwise be able to reach. Lecithin vesicles (liposomes) are used in this way to transport medicaments such as penicillin or peptides and proteins. In this way the enclosed proteins are protected from setting up antibody reactions.

Transport of *neurotransmitters* to their active sites, the synaptic gaps, also takes place in vesicles in the nerve cells. In this way electrical nerve signals are transmitted from one nerve to the next or to a muscle.

2.27.3 Non-ionic Surfactants

Ethylene oxide adducts

These washing agents (detergents) are adducts of ethylene oxide to long-chain alcohols or carboxylic acids or to alkylphenols (see p. 304 *ff*).

The non-ionic detergents, which are for the most part liquids, are not affected by hard water, and can be used in either acidic or alkaline solutions.

2.28 Derivatives of Aliphatic Monocarboxylic Acids

Carboxylic acid derivatives are compounds which have modified carboxyl groups.

2.28.1 Acyl Halides (Acid Halides)

In these compounds the HO-group of a carboxyl group is replaced by a halogen atom; the *homologous series* hence has the following general formula:

$$C_nH_{2n+1}-C\underset{X}{\overset{O}{\lessgtr}}$$

In general they are known as acyl halides or acid halides. Individual names are derived from the corresponding acids, *e.g.* acetyl chloride (ethanoyl chloride), priopionyl chloride (propanoyl chloride), butyryl chloride (butanoyl chloride) *etc.*

The acyl chloride derived from formic acid is only stable at the temperature of liquid air and decomposes at higher temperatures. However for some synthetic purposes a mixture of carbon monoxide and hydrogen chloride can behave as though it were formyl chloride (methanoyl chloride).

---

[1] See *J-H. Fuhrop* et al., Angew. Chem. Int. Ed. Engl. *19*, 550 (1980); *E. W. Kaler* et al., Science *245*, 1371 (1989).
[2] See *D. Langevin*, Microemulsions, Accounts Chem. Res. *21*, 255 (1988).

The replacement of the HO- in a carboxyl group by a halogen atom can be achieved in various ways, which resemble those used to carry out a similar reaction involving the HO group of an alcohol.

2.28.1.1 Preparation

Acyl halides are prepared by treatment of carboxylic acids or their alkali-metal salts with halides of phosphorus or of sulphur, *e.g.*

$$3\,R-C\overset{O}{\underset{OH}{}} + PCl_3 \longrightarrow 3\,R-C\overset{O}{\underset{Cl}{}} + H_3PO_3$$

In the reaction of phosphorus pentachloride with an acid only one of the chlorine atoms is used in the formation of the acyl chloride; hydrogen chloride is evolved and this gives an indication of the progress of the reaction.

$$R-C\overset{O}{\underset{OH}{}} + PCl_5 \longrightarrow R-C\overset{O}{\underset{Cl}{}} + POCl_3 + HCl$$

A more convenient method involves heating the sodium salt of an acid with phosphorus pentachloride:

$$3\,R-C\overset{\bar{O}|}{\underset{\bar{O}|^{\ominus}}{}}\,Na^{\oplus} + PCl_5 \longrightarrow 3\,R-C\overset{O}{\underset{Cl}{}} + 2\,NaCl + (NaPO_3)_n$$

Phosphorus oxychloride reacts with the salts of carboxylic acids, but not with acids themselves:

$$2\,R-C\overset{\bar{O}|}{\underset{\bar{O}|^{\ominus}}{}}\,Na^{\oplus} + POCl_3 \longrightarrow 2\,R-C\overset{O}{\underset{Cl}{}} + NaCl + (NaPO_3)_n$$

Thionyl chloride is commonly used in the laboratory; it has the advantage that the other products formed are gases:

$$R-C\overset{O}{\underset{OH}{}} + SOCl_2 \longrightarrow R-C\overset{O}{\underset{Cl}{}} + HCl + SO_2$$

Acid chlorides can be obtained under very mild conditions by treating carboxylic acids with *triphenylphosphine in tetrachloromethane*.[1]

$$R-C\overset{O}{\underset{OH}{}} + (C_6H_5)_3P + CCl_4 \longrightarrow R-C\overset{O}{\underset{Cl}{}} + (C_6H_5)_3PO + CHCl_3$$

Industrially, acetyl chloride is made by heating sodium acetate with sulphuryl chloride:

$$2\,H_3C-C\overset{\bar{O}|}{\underset{\bar{O}|^{\ominus}}{}}\,Na^{\oplus} + SO_2Cl_2 \longrightarrow 2\,H_3C-C\overset{O}{\underset{Cl}{}} + Na_2SO_4$$

[1] See *J. Blee*, J. Am. Chem. Soc. **88**, 3440 (1966).

The acyl bromides, iodides and some fluorides can best be made by treating the corresponding acyl chloride with hydrogen bromide, hydrogen iodide or hydrogen fluoride (or antimony fluoride).

Properties

The lower acyl halides are colourless pungent liquids which fume in air. They have a powerful irritant effect on the mucous membrane. Higher acyl halides are crystalline solids. They have lower boiling points than the corresponding carboxylic acids, e.g. acetyl chloride boils at 52°C (325 K). They are appreciably more reactive than alkyl halides, and are used to introduce acyl groups into certain classes of organic compounds (*acylation*[1]). In particular, acetyl chloride is much used. It reacts readily, either at room temperature or on gentle heating, with water, alcohols, ammonia and primary or secondary amines to give, respectively, acetic acid, acetate esters or amides:

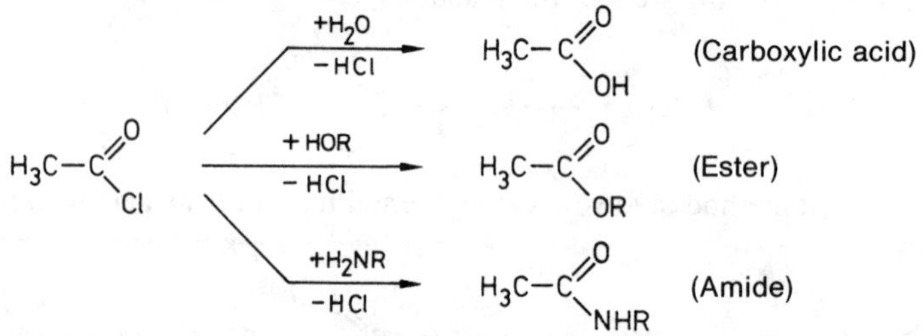

If the hydrogen chloride which is evolved is absorbed by having a *trapping agent* such as pyridine present, the formation of the acetyl derivative is promoted. Acetylation is also of use for the detection of *acidic hydrogen atoms* in hydroxy, amino, imino or mercapto groups and for the identification of these types of compounds.

2.28.1.2 Arndt–Eistert Reaction[2] (1935)

The *carbon atom of diazomethane can be acylated* by means of an acyl chloride, and the *diazonium betaine* which is formed as an intermediate at once loses hydrogen chloride to give a *diazoketone*, which is stabilised by *mesomerism*. Such a reaction may be said to proceed by an *addition–elimination mechanism*:

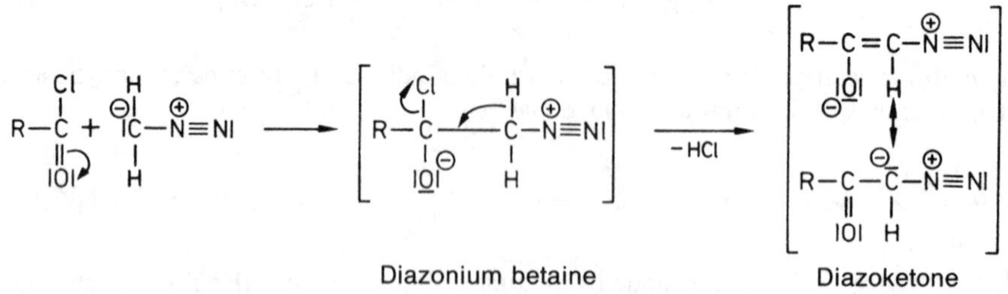

Diazonium betaine   Diazoketone

---

[1] See *D. P. N. Satchell*, An Outline of Acylation. Quart. Rev. *17*, 160 (1963); See also *D Seebach*, Methods and possibilities of Nucleophilic Acylation, Angew, Chem. Int. Ed. Engl. *8*, 639 (1969).
[2] See *W. E. Bachmann* and *W. S. Struve*, The Arndt–Eistert Synthesis, Org. Reactions, *1*, 38 *ff.* (1942).

At higher temperatures *diazoketones*[1] decompose with loss of nitrogen, especially in the presence of metal catalysts (colloidal silver or copper). The residue is a *carbene* with only six electrons associated with the carbon atom. This stabilises itself by migration of the alkyl group, in the form of an anion, from the acyl group, with concomitant formation of a double bond between the two carbon atoms. In this rearrangement, known as the *Wolff rearrangement*, the final product is thus a ketene. If the decomposition of a diazoketone is carried out in water, an alcohol or an amine, the solvent reacts with the ketene with the formation, respectively, of a carboxylic acid, an ester or an amide:

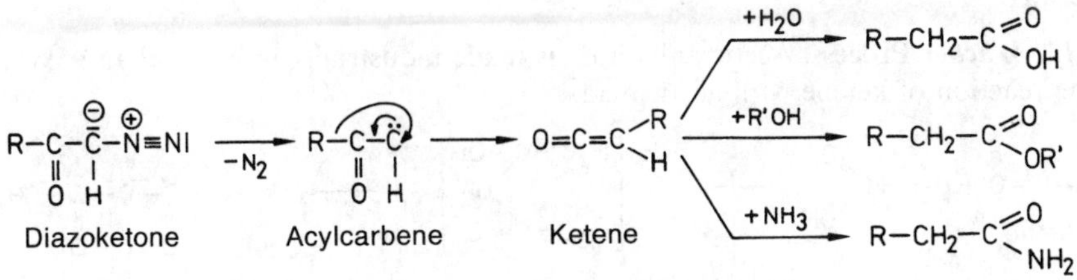

Diazoketone	Acylcarbene	Ketene

The *Arndt–Eistert* synthesis provides a method for converting an acid chloride into the next higher homologous carboxylic acid.

2.28.2 Acid Anhydrides

Acid anhydrides can be thought of as being derived by loss of water from two molecules of a carboxylic acid.

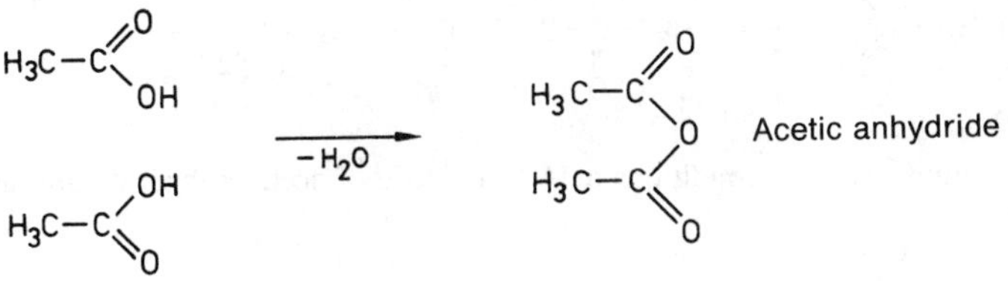

Acetic anhydride

In the same way that ethers can be considered as dialkyl derivatives of water, so acid anhydrides can be regarded as diacyl derivatives of water.

By far the most important member of this series is **acetic anhydride**; **propionic anhydride** is also being used increasingly, although on a much smaller scale.

Preparation

2.28.2.1 On the laboratory scale, acid anhydrides are made **by the reaction of acyl chlorides with the sodium salts of carboxylic acids**:

---

[1] See *F. Weygand* and *H. J. Bestmann*, Synthesen unter Verwendung von Diazoketonen, Angew. Chem. *72*, 535 (1960); *H. Meier* and *K-P. Zeller*, The *Wolff* Rearrangement of α-Diazo Compounds, Angew. Chem. Int. Ed. Engl. *14*, 32 (1975); *D. Whittaker*, Rearrangements involving the Diazo and Diazonium Groups in *S. Patai* (Ed.) The Chemistry of Diazonium and Diazo Compounds, Part 2, 593 (Wiley, New York 1978).

$$R-C\overset{O}{\underset{Cl}{\diagup}} \\ \qquad\qquad\qquad\qquad \longrightarrow R-C-O-C-R \ + \ NaCl \\ R-C\overset{\ominus}{\underset{O}{\diagup}}O^{\ominus}Na^{\oplus}$$

Mixed acid anhydrides, having two different alkyl substituents, can be made in the same way.

2.28.2.2 Wacker Process. Acetic anhydride is made industrially in better than 95% yield by the reaction of ketene with acetic acid:

$$H_2C=C=O + H_3C-C\overset{O}{\underset{OH}{\diagup}} \longrightarrow \left[\begin{array}{c} H_2C=C-OH \\ \diagdown O \\ H_3C-C=O \end{array} \right] \longrightarrow H_3C-\underset{O}{\overset{O}{\underset{\parallel}{C}}}-O-\underset{O}{\overset{O}{\underset{\parallel}{C}}}-CH_3$$

Ketene

2.28.2.3 Hoechst-Knapsack Process. In this process acetic anhydride is made by oxidising acetaldehyde, in ethyl acetate as diluent, with atmospheric oxygen over a copper and cobalt acetate catalyst at 50–60°C (320–330 K) and 4 bar (0.4 MPa).

$$2 \ H_3C-C\overset{H}{\underset{O}{\diagup}} \xrightarrow[-H_2O]{+O_2} H_3C-\overset{O}{\overset{\parallel}{C}}-O-\overset{O}{\overset{\parallel}{C}}-CH_3$$

Peracetic acid is formed as an intermediate and this reacts with acetaldehyde with elimination of water and formation of acetic anhydride:

$$H_3C-C\overset{O}{\underset{O-OH}{\diagup}} + H_3C-C\overset{H}{\underset{O}{\diagup}} \longrightarrow H_3C-\underset{O}{\overset{}{\underset{\parallel}{C}}}-O-O-\underset{OH}{\overset{}{\underset{\mid}{CH}}}-CH_3 \xrightarrow{-H_2O} H_3C-\overset{O}{\overset{\parallel}{C}}-O-\overset{O}{\overset{\parallel}{C}}-CH_3$$

Peracetic acid

Under the optimum reaction conditions a 56:44 mixture of acetic anhydride and acetic acid is formed.

Properties

Formic anhydride is unknown but a mixed formic–acetic anhydride HCO—O—COCH$_3$, exists. Most acid anhydrides are colourless pungent-smelling liquids with boiling points higher than those of the corresponding acids. Thus acetic anhydride boils at 140°C (410 K). The lower anhydrides react briskly when warmed with water, producing carboxylic acids. They are not nearly so reactive as acid chlorides, despite the fact that acetate is a good leaving group. Acetic anhydride, like acetyl chloride, is used extensively as an *acetylating agent*, e.g.

$$H_3C-\underset{O}{\overset{}{\underset{\parallel}{C}}}-O-\underset{O}{\overset{}{\underset{\parallel}{C}}}-CH_3 \quad \begin{cases} \xrightarrow{+H_2O} & 2 \ H_3C-C\overset{O}{\underset{OH}{\diagup}} \\ \xrightarrow{+HOR} & H_3C-C\overset{O}{\underset{OR}{\diagup}} + H_3CCOOH \\ \xrightarrow{+2H_2NR} & H_3C-C\overset{O}{\underset{NHR}{\diagup}} + H_3CCOO^{\ominus} \ H_3\overset{\oplus}{N}R \end{cases}$$

Whereas hydrogen chloride is evolved in reactions of acetyl chloride, reactions involving acetic anhydride result in the formation of one molecule of acetic acid. These acetylations are often activated by addition of a small amount of anhydrous sodium acetate or of conc. sulphuric acid. The appreciable solubility of acetic anhydride in water (13.6 g in 100 cm^3 cold water) makes it useful as an acetylating agent in the aqueous phase.

On the industrial scale the acetylating agent which is used is almost exclusively *acetic anhydride*, for example in the manufacture of pharmaceuticals, dyestuffs and especially of acetylcellulose.

Diacetyl peroxide, which is used as a radical source to initiate polymerisation reactions, is a colourless liquid which explodes readily if heated. It is prepared industrially from acetic anhydride and sodium metaborate peroxide hydrate:

$$(H_3C-CO)_2O + NaBO_2 \cdot H_2O_2 \cdot 3H_2O \longrightarrow H_3C-\overset{O}{\overset{\|}{C}}-O-O-\overset{O}{\overset{\|}{C}}-CH_3 + NaBO_2 + 4\,H_2O$$

Acetic anhydride   Diacetyl peroxide

2.28.3 *Ketenes*

Ketenes were discovered by *Wilsmore* and by *Staudinger* and can be regarded as inner (intramolecular) anhydrides of carboxylic acids. A distinction can be made between *aldoketenes*, R—CH=C=O, and *ketoketenes*, R$_2$C=C=O. The simplest ketene, H$_2$C=C=O (skeletal formula $= \cdot = O$) and its derivatives are of preparative importance.[1]

Preparation

2.28.3.1 An important way of making ketenes is **by the dehydrohalogenation of acyl halides.** This is usually done using a tertiary amine, *e.g.*

$$(C_6H_5)_2CH-COCl \xrightarrow[-NH^{\oplus}(C_2H_5)_3Cl^{\ominus}]{+N(C_2H_5)_3} (C_6H_5)_2C=C=O$$

Diphenylacetyl chloride   Diphenylketene

Many ketenes, and especially aldoketenes, dimerise in basic conditions. In consequence the yields obtained may be lowered.

2.28.3.2 Ketene is formed **in the pyrolytic breakdown of acetone** (*Schmidlin*). This is carried out industrially over a chrome-nickel spiral at 780°C (1050 K) and gives a good yield:

$$H_3C-\overset{O}{\overset{\|}{C}}-CH_3 \xrightarrow{(Cr/Ni)} H_2C=C=O + CH_4$$

The ketene is washed out of the mixture of gas which is formed, with strongly cooled acetone, and hence separated from methane.

2.28.3.3 *Wacker* **Process.** In this process ketene is obtained by removal of a molecule of water from acetic acid at 700°C (975 K) in the presence of a suitable catalyst, *e.g.* triethyl phosphate:

$$H_3C-C\overset{O}{\underset{OH}{\diagup}} \rightleftharpoons H_2C=C=O + H_2O \qquad \Delta H = +209\,kJ/Mol$$

---

[1] See *D. Borrmann* in *Houben-Weyl*, Methoden der Organischen Chemie 7/4, 53 (Thieme Verlag, Stuttgart 1968); *J. Streith* et al., Methylketen, Justus Liebigs Ann. Chem. *1983*, 1393.

Since the process is reversible, water must be removed quickly from the equilibrium mixture; this is done by cooling it. The cold gaseous ketene is either converted directly into acetic anhydride or *dimerised* to *diketene*, b.p. 127°C (400 K).

$$H_2C{=}C{=}O \rightleftharpoons \quad H_2C{=}C{-}O \qquad \qquad \varDelta H = -113\,kJ/Mol$$
$$+ \qquad\qquad\qquad | \quad | $$
$$H_2C{=}C{=}O \qquad\qquad CH_2{-}C{=}O$$

The IR spectrum of diketene shows it to have the structure of 4-methylene-2-oxetanone. Industrially it is converted into ethyl acetate by reaction with ethanol.

Properties and Uses

The reactivity of ketene derives from its two *cumulated* double bonds. Most ketenes dimerise very rapidly but diphenylketene is stable at room temperature.

Ketene is a colourless gas, b.p. $-56°C$ (217 K), with a sharp odour. It is very poisonous. Only at low temperatures $(-80°C$ [190 K]) is it even slightly stable. It is essential therefore that ketene is freshly made and used immediately, otherwise it dimerises to diketene. Ketene is used principally as an acetylating agent for substances which can easily provide a proton. Thus with water it gives acetic acid, with ethanol ethyl acetate and with ammonia or even more readily with primary amines the corresponding amides:

$$H_2C{=}C{=}O \qquad \begin{array}{l} +H_2O \longrightarrow H_3C{-}COOH \\ +HOR \longrightarrow H_3C{-}COOR \\ +NH_3 \longrightarrow H_3C{-}CONH_2 \end{array}$$

For the uses of ketene in the preparation of β-lactones and small-ring alicyclic compounds see p. 280 and p. 391 *ff*.

2.28.4 *Esters of Carboxylic Acids*

Esters are named simply by combining the name of the alkyl group that takes the place of the acidic hydrogen of the acid and the name of the anion derived from the acid, *e.g.* ethyl acetate, methyl propionate.

Preparation

2.28.4.1 By esterification of carboxylic acids. The formation of esters of carboxylic acids is analogous to that of esters of inorganic acids (see p. 138), and is based on an equilibrium reaction between the acid and an alcohol:

$$R{-}C{\overset{O}{\underset{OH}{\big<}}} + H{-}OR' \overset{(H^\oplus)}{\rightleftharpoons} R{-}C{\overset{O}{\underset{OR'}{\big<}}} + H_2O$$

The rapid setting up of this equilibrium requires raised temperatures and an excess of $H^\oplus$ ions. The latter is achieved by adding an inorganic acid (conc. H_2SO_4 or HCl gas) as an acid catalyst. The rate of esterification is directly proportional to the hydrogen ion concentration.

As is the case for all equilibrium reactions, the *law of mass action* applies to the esterification of carboxylic acids. The *equilibrium constant K* is derived from the concentration of the reactants as follows:

$$K = \frac{[\text{Ester}] \cdot [\text{Water}]}{[\text{Acid}] \cdot [\text{Alcohol}]}$$

The objective of any esterification is to achieve as complete a conversion of an acid into an ester as is possible; in other words [Ester] should be as large as possible:

$$[\text{Ester}] = K \cdot \frac{[\text{Acid}] \cdot [\text{Alcohol}]}{[\text{Water}]}$$

Complete esterification of an acid can be achieved, therefore, on the one hand, by increasing [Alcohol] and, on the other, by decreasing [Water], *i.e.* by removing the water which is formed in the equilibration. This can be done, for example, by the conc. sulphuric acid which is added (about 5–10%, based on the amount of carboxylic acid to be esterified), which, because of its ability to become hydrated, can bind a set amount of water to itself. In addition, a chemically inert liquid such as benzene or toluene, which forms a *binary* mixture with water and a *ternary* mixture with water/alcohol, can be added to the acid–alcohol mixture. If esterification is carried out just above the boiling point of the *azeotropic* mixture, then the water distils off at a relatively lower constant temperature and the equilbrium is displaced in favour of ester formation.

The *mechanism* of acid-catalysed esterification of carboxylic acids with *primary* or *secondary* alcohols is as follows. A proton adds to an oxygen atom of the carboxyl group to give a stabilised mesomeric cation. When this cation is formed the carbon atom bears an increased positive charge. The oxygen atom of the alcohol makes a *nucleophilic* attack on this carbon atom. Loss of water and a proton from the resultant species gives rise to the ester:

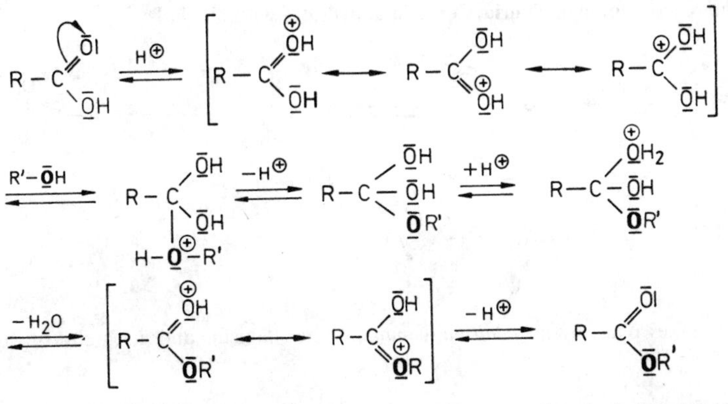

It has been shown by *isotopic labelling* that the water molecule formed is derived from an HO-group of the carboxyl group and a proton from the HO-group of the alcohol. If the esterification is carried out using an alcohol whose oxygen atom is partially labelled with ^{18}O (printed in heavy type in the reaction scheme), then this isotope is not found in the water which separates, but in the ester.

This mechanism applies, however, only to *primary* and *secondary* alcohols.

When carboxylic acids are esterified with *tertiary* alcohols, the hydroxy group of the alcohol is lost, leading to the formation of a *carbenium ion*, which is stabilised because of the $+I$-effects of the three alkyl groups. This then makes an *electrophilic attack* on an oxygen atom of the carboxyl group:

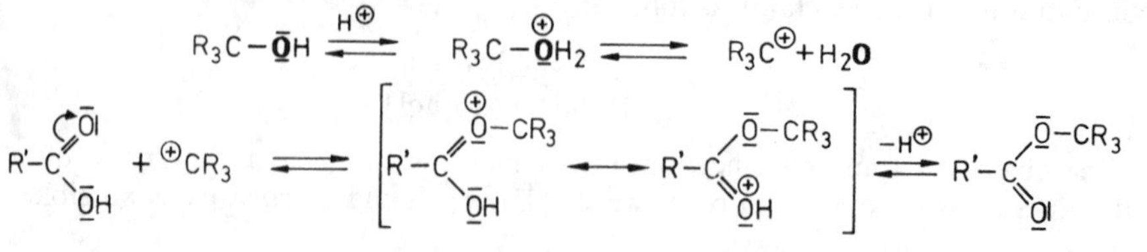

Esters can be hydrolysed to carboxylic acids either by acid or by alkali. The mechanism of *acid hydrolysis* depends upon the structure of the alkyl group which is displaced, and takes a similar course to that of the reverse reaction, esterification (p. 257). In *alkaline hydrolysis*, or *saponification*, of an ester the carbon atom of the alkoxycarbonyl group first of all undergoes nucleophilic attack by a hydroxide anion. The resultant tetrahedral adduct then stabilises itself by loss of an alcohol molecule *via* a tetrahedral transition state, leaving a mesomerically stabilised carboxylate anion. This is an example of an *addition–elimination* mechanism.

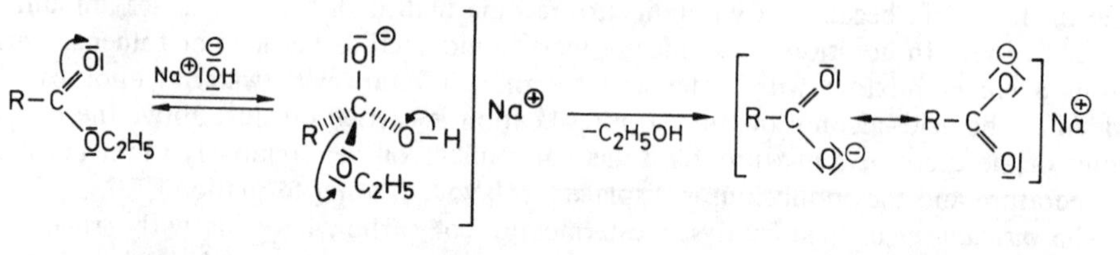

An advantage of alkaline hydrolysis lies in the fact that the last step is virtually irreversible, and, as a result, quantitative hydrolysis is achieved. Since the alkali is consumed in the reaction, it is necessary to have at least an equivalent amount of it present.

The reaction of the methylsulphinyl carbanion with carboxylic esters, which is mentioned on p. 148, proceeds in just the same way.

2.28.4.2 By the reaction of acid chlorides or acid anhydrides with alcohols:

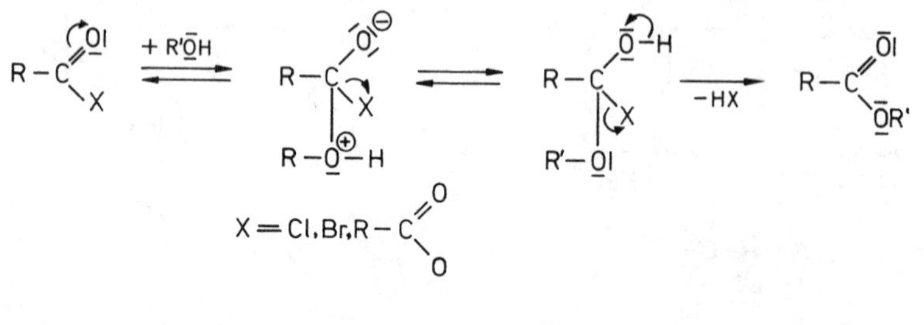

This reaction takes place by an *addition–elimination* mechanism and gives almost quantitative yields.

2.28.4.3 By the reaction of the silver salts of carboxylic acids with alkyl halides:

This method is expensive and is only used if the direct method fails or gives poor yields.

2.28.4.4 Ethyl acetate is made industrially from acetaldehyde by the *Tishchenko* reaction.

Properties

Esters are colourless liquids with fruity smells, although higher members are odourless. They are neutral, have a lower density than water and are only slightly soluble in it. In contrast to carboxylic acids, esters do not associate, since it is not possible to form hydrogen bonds. As a result they have lower boiling points than the corresponding acids, *e.g.* ethyl acetate boils at 77°C (350 K).

In a reaction which is analogous to hydrolysis, esters can be *transesterified* by other alcohols, *e.g.*

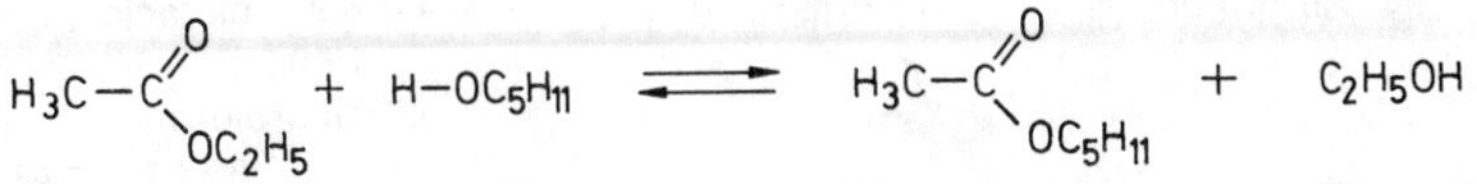

$$H_3C-C\overset{O}{\underset{OC_2H_5}{}} + H-OC_5H_{11} \rightleftharpoons H_3C-C\overset{O}{\underset{OC_5H_{11}}{}} + C_2H_5OH$$

Ethyl acetate   Pentan-1-ol   Pentyl acetate   Ethanol

2.28.4.5 Transesterification takes place at room temperature in the presence of either a mineral acid or a base as catalyst. It can also be carried out under neutral conditions in the presence of titanates, $Ti(OR)_4$; in this way, transesterification can be carried out without affecting other groups present which might react with either acids or bases.[1]

2.28.4.6 Esters which have at least one hydrogen atom attached to the α-C atom can, like ketones, form *enolate* anions from which enol ethers, which correspond to ketene acetals, can be prepared, as illustrated by the following reaction of an ester with lithium diisopropylamide and chlorotrimethylsilane:

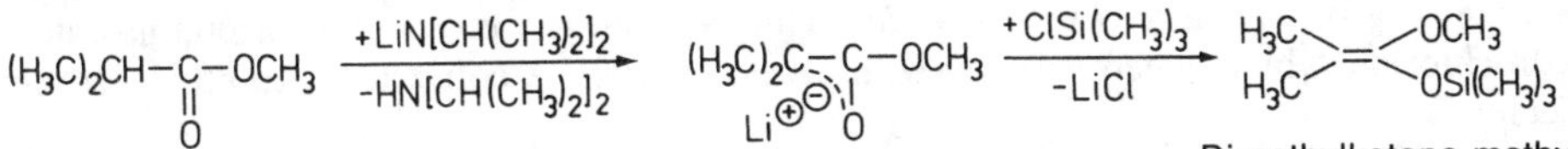

$$(H_3C)_2CH-\underset{O}{\overset{}{C}}-OCH_3 \xrightarrow[-HN[CH(CH_3)_2]_2]{+LiN[CH(CH_3)_2]_2} (H_3C)_2C\overset{}{\underset{Li^{\oplus}\overset{\ominus}{O}}{=}}C-OCH_3 \xrightarrow[-LiCl]{+ClSi(CH_3)_3} \overset{H_3C}{\underset{H_3C}{>}}=\overset{OCH_3}{\underset{OSi(CH_3)_3}{<}}$$

Dimethylketene methyl-
trimethylsilyl acetal

Dimethylketene methyl-trimethylsilyl acetal is of preparative use, for example in *group-transfer polymerisation*.

2.28.4.7 Reduction of esters to primary alcohols. This reduction can be carried out in various ways, by sodium and ethanol (*Bouveault–Blanc reduction*), with lithium aluminium hydride or catalytically over copper chromite:

$$R-C\overset{O}{\underset{OR'}{}} + 4H \longrightarrow R-CH_2OH + R'OH$$

Esters react with *Grignard* reagents to give tertiary alcohols (see p. 186).

Uses

Esters, especially ethyl acetate and butyl acetate, are used as solvents for nitrocellulose and for resins in the paint and varnish industry; they also serve as starting materials in condensation reactions of esters. Some esters are used to provide aromas or essences, *e.g.* ethyl formate (rum, arrack), isobutyl acetate (banana), methyl butyrate (apple), ethyl butyrate (pineapple) and isopentyl butyrate (pear).

---

[1] See *D. Seebach* et al., Titanate-Mediated Transesterifications with Functionalised Substrates, Synthesis *1982*, 138.

2.28.5 Orthoesters

In addition to simple esters, under certain conditions compounds known as *orthoesters* can be prepared, which have the following structure:

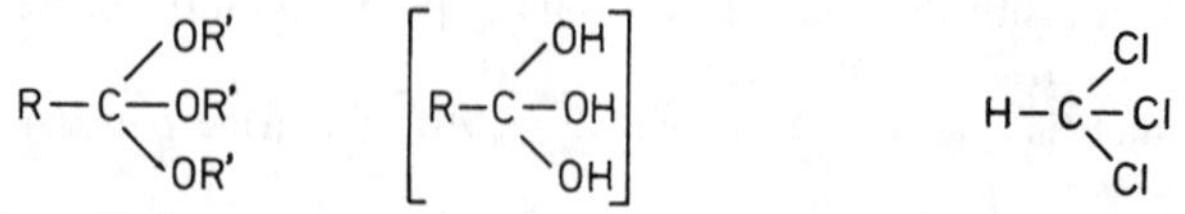

$$R-C\begin{matrix} OR' \\ -OR' \\ OR' \end{matrix} \qquad \left[R-C\begin{matrix} OH \\ -OH \\ OH \end{matrix}\right] \qquad H-C\begin{matrix} Cl \\ -Cl \\ Cl \end{matrix}$$

An orthoester   An orthoacid   Orthoformic acid trichloride (Chloroform)

They are derived from the non-existent orthocarboxylic acids, which one could imagine being formed by addition of 1 mol of water to a carboxyl group (see the *Erlenmeyer* rule). In contrast, orthoesters are very stable liquids. The most important ones are **ethyl orthoformate** (triethoxymethane), b.p. 146°C(419 K) and **ethyl orthoacetate** (1,1,1-triethoxy-ethane), b.p. 145° (418 K). The former is made by *treating chloroform* (the trichloride of orthoformic acid) *with sodium ethoxide (Williamson)*:

$$HCCl_3 + 3\ NaOC_2H_5 \rightarrow HC(OC_2H_5)_3 + 3\ NaCl$$
Ethyl orthoformate

For the most part they are made from *nitriles*, which react with ethanolic hydrogen chloride to give *iminoester hydrochlorides*. Thus acetonitrile provides an acetiminoester hydrochloride, which, when warmed with excess ethanol, is converted into ethyl ortho-acetate:

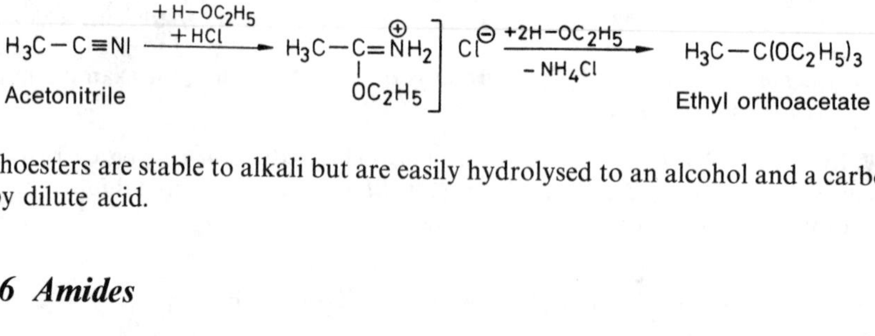

$$H_3C-C\equiv NI \xrightarrow[+HCl]{+H-OC_2H_5} \left[H_3C-C\overset{\oplus}{=}NH_2 \atop OC_2H_5\right] Cl^{\ominus} \xrightarrow[-NH_4Cl]{+2H-OC_2H_5} H_3C-C(OC_2H_5)_3$$

Acetonitrile   Ethyl orthoacetate

Orthoesters are stable to alkali but are easily hydrolysed to an alcohol and a carboxylic acid by dilute acid.

2.28.6 Amides

Replacement of the HO-group of carboxylic acids by an H_2N-group gives rise to amides, which have the general formula:

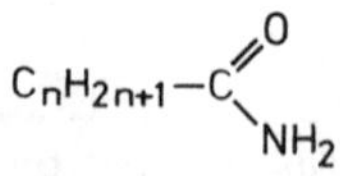

$$C_nH_{2n+1}-C\begin{matrix} O \\ NH_2 \end{matrix}$$

They can be regarded as monoacyl derivatives of ammonia. Primary or secondary amines can take the place of ammonia to provide *N-substituted amides*:

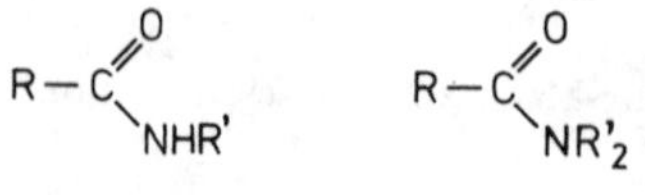

$$R-C\begin{matrix} O \\ NHR' \end{matrix} \qquad R-C\begin{matrix} O \\ NR'_2 \end{matrix}$$

The names of amides can be derived either from the acyl group of the corresponding acid, in which case the ending-*yl* is replaced by *amide*, (*e.g.* formamide, acetamide), or from the related amine. In the latter case a compound (RCO)$_2$NH (or NR), which is obtained by diacylation of ammonia (or of a primary amine), is called a *diacylamine*.

Preparation

2.28.6.1 Formamide is obtained from *carbon monoxide* and *ammonia* under pressure in the presence of sodium methoxide in methanol as a catalyst.

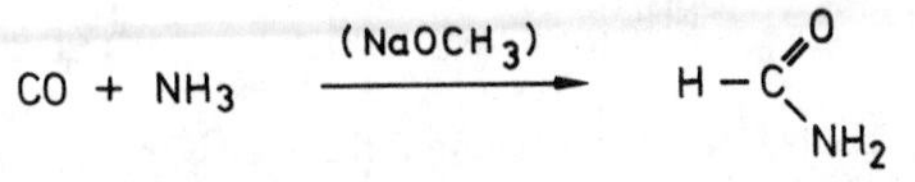

2.28.6.2 Homologous amides are obtained by treating *acid chlorides, anhydrides* or *esters* with *ammonia*. In the first two cases, reaction proceeds in the cold by an *addition–elimination* mechanism:

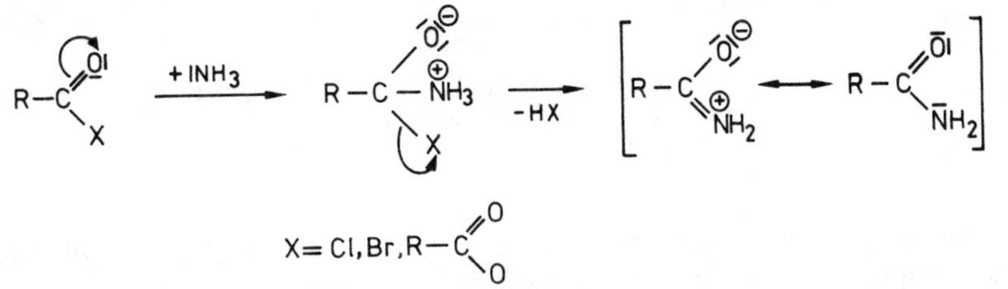

$$X = Cl, Br, R - C \overset{O}{\underset{O}{\diagup}}$$

The hydrogen halide or carboxylic acid which is also formed can be trapped by using an excess (2 mol) of ammonia or amine.

Esters usually only react at higher temperatures:

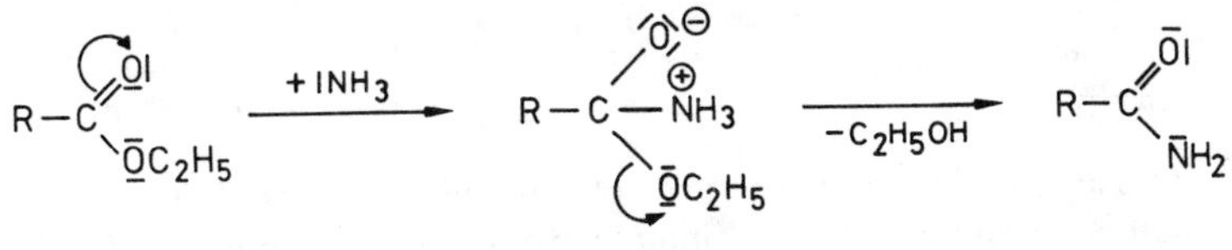

2.28.6.3 When *ammonium salts of carboxylic acids are heated* they lose water to give amides, *e.g.*

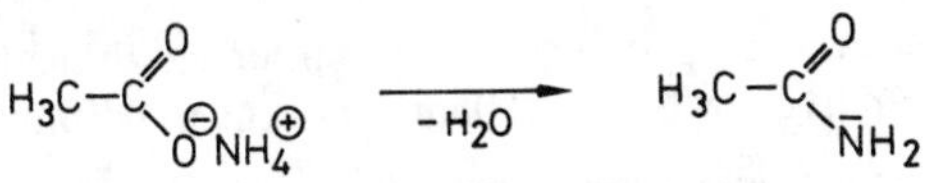

Ammonium acetate   Acetamide (m.p. 82°C [355 K])

Properties

Formamide, *N*-methylformamide and *N,N,*-dimethylformamide (DMF) are liquids at room temperature; higher amides are colourless crystalline substances which can be distilled without decomposition. They differ from amines in that, because of the adjacent C=O double bond, the C—N bond can be split by means of water or acids, or, more quickly, by alkali. There are a few amides which are difficult to hydrolyse.

$$R-C\underset{NH_2}{\overset{O}{=}} + NaOH \longrightarrow \left[R-C\underset{O^{\ominus}}{\overset{\bar{O}l}{=}}\right] Na^{\oplus} + NH_3$$

Amides are stabilised by mesomerism:

$$\left[R-C\underset{NH_2}{\overset{\bar{O}l}{=}} \longleftrightarrow R-C\underset{\overset{\oplus}{N}H_2}{\overset{\bar{\underset{.}{O}}^{\ominus}}{=}}\right]$$

Because of the partial double-bond character of the C—N bond, rotation about this bond is hindered and stereoisomers (rotamers) are observed; NMR shows the two *N*-methyl groups of DMF to be non-equivalent at room temperature. (See *Z–E* isomerism of thioamides, p. 263 *f.*).

Aqueous solutions of amides are neutral. Amides behave as *weak bases* towards acids; they only form salts with concentrated mineral acids and these salts are readily *hydrolysed*. The *cations are mesomeric*:

$$\left[R-C\underset{\underset{.}{N}H_2}{\overset{\overset{\oplus}{O}-H}{=}} \longleftrightarrow R-C\underset{\overset{\oplus}{N}H_2}{\overset{\bar{\underset{.}{O}}-H}{\diagup}}\right] X^{\ominus}$$

Amides can also act as *weak acids*. For example, *N*-arylformamides react with potassium hydride to give *anions*, which are stabilised by *mesomerism*:

$$K^{\oplus}\left[\underset{H}{\overset{l\bar{O}l}{\diagdown}}C-\bar{\underset{R}{N}}^{\ominus} \longleftrightarrow \underset{H}{\overset{\ominus\bar{\underset{.}{O}}}{\diagdown}}C=\bar{N}\underset{R}{\diagdown}\right] \rightleftharpoons \left[\underset{H}{\overset{\ominus\bar{\underset{.}{O}}}{\diagdown}}C=N\overset{R}{\diagup} \longleftrightarrow \underset{H}{\overset{l\bar{O}l}{\diagdown}}C-N\overset{R}{\underset{\ominus}{\diagup}}\right] K^{\oplus}$$

$$\qquad\qquad (E)\text{-Form} \qquad\qquad\qquad\qquad\qquad\qquad (Z)\text{-Form}$$

NMR spectroscopy indicates the presence of (*E*)- and (*Z*)-forms of the anion and it has been shown that isomerisation does not take place by rotation about the C—N bond, as happens in the case of neutral amides, but rather by inversion.[1] These stereoisomers should be differentiated from the rotamers of free amides described above, and can be called *invertomers*.

Primary amides can be reduced to primary amines either catalytically or by sodium and ethanol, lithium aluminium hydride (see p. 160) or diborane in boiling tetrahydrofuran.

With phosphorus pentoxide, amides lose water to give *nitriles*. Amides react in the same way as aliphatic amines with nitrous acid in mineral acids; nitrogen is evolved and the corresponding acid is formed.

$$R-C\underset{NH_2}{\overset{O}{=}} + NaNO_2 + HCl \longrightarrow R-C\underset{OH}{\overset{O}{=}} + NaCl + N_2 + H_2O$$

---

[1] See *C. L. Perrin* et al., Mechanism of *E/Z* Stereoisomerization of Imidate Anions, J. Org. Chem. *54*, 764 (1989).

The previously described (see p. 159) *Hofmann* conversion of amides into primary amines should also be mentioned. Recently mono- and also di-methylformamide have been used as extraction solvents in the purification of hydrocarbons.

'*Amide chlorides*' (chloromethyleneiminium salts), which are appreciably more reactive than amides, are made by the action of phosgene on tertiary amides.

Amide   'Amide Chloride'

More accurately they are called *carboxylic acid iminium halides*. Compounds with similar reactivity are the *dichloromethyleneiminium salts* ('*Phosgeneiminium Salts*'), prepared by the reaction of tertiary thioformamides with chlorine.[1]

Thioformamide   Thiocarbamoyl chloride   Phosgeneiminium salt

It is not necessary to isolate the intermediate thiocarbamoyl chloride. Both classes of compounds are valuable synthons (see p. 544) and the same is true for a large number of their initial reaction products. As an example, the reactive amidoacetals are obtained from the reaction of amide chlorides with alkoxides:

Amide chloride   Amidoacetal

2.28.7 *Thioamides*

These compounds can be made from amides, for example by heating them with tetraphosphorus decasulphide or with 2,4-bis(4-methoxyphenyl)-2,4-dithioxo-1,3,2,4-dithiadiphosphetan (*Lawesson* reagent),[2] which replace the oxygen atom by sulphur:

Lawesson reagent (LR)

[1] See *H. G. Viehe* and *Z. Janousek*, The Chemistry of Dichloromethyleneammonium Salts (Phosgeneimonium Salts), Angew. Chem. Int. Ed. Engl., *12*, 806 (1973).
[2] See *Houben-Weyl*, Methoden der Organischen Chemie, Vol. *E5*, p. 1244 *ff.* (Thieme Verlag. Stuttgart 1985).

Thioamides, like amides, are stabilised by *mesomerism*, resulting in the C—N bond having partial double-bond character

$$
\left[
R-C{\overset{\overline{S}|}{\underset{\overline{N}H_2}{}}}
\quad\longleftrightarrow\quad
R-C{\overset{S^{\ominus}}{\underset{\overset{\oplus}{N}H_2}{}}}
\right]
$$

Because of this it would be expected that different diastereomers of *N*-alkylated thioamides should exist. *W. Walter* et al.[1] succeeded in separating the two *rotation isomers* of *N*-alkylated *thioformamides*, which are diastereomers:

$$
\underset{H}{\overset{S}{\diagdown}}C-N\overset{H}{\underset{CH_3}{\diagup}}
\quad\rightleftharpoons\quad
\underset{H}{\overset{S}{\diagdown}}C-N\overset{CH_3}{\underset{H}{\diagup}}
$$

N-Methylthioformamide

(*E*)-form	(*Z*)-form
m.p. 75°C (348 K)	liquid

In cases where it is not possible to isolate the different rotamers, for example if the energy barrier to rotation is so low that at room temperature they interconvert very rapidly, it is still possible to detect them at lower temperatures by means of NMR spectroscopy. The methyl groups in the (*E*)- and (*Z*)-forms of thioformamides are in quite different chemical environments and their signals appear in their NMR spectra at different chemical shifts. This difference is also apparent for amides in which two identical groups are attached to nitrogen, as is the case in dimethylformamide:

$$
\underset{H}{\overset{O}{\diagdown}}C-N\overset{CH_3 \quad \delta=2{,}78\ ppm}{\underset{CH_3 \quad \delta=2{,}94\ ppm}{}}
$$

Here also there is hindered rotation between the two isomeric forms, both of which have the same chemical structure. Such *degenerate isomers* are known as **topomers.**[2] Ligands which show these characteristics have been classified as *heterotopic ligands*[3] (*Mislow, Raban* 1969). If, in this case, one of the methyl groups is substituted, for example by hydrogen, two *different rotamers*, which are *diastereomers*, result. Such ligands are hence known as *diastereotopic* ligands.[4] What is in evidence here is the *protostereoisomerism* of the methyl groups. Prochirality is also concerned with protostereoisomerism.

2.28.8 *Hydrocyanic Acid and Nitriles (Cyanides)*

Related to amides are the nitriles, or cyanides, R—C≡N. The two classes of compounds may be interconverted by gain or loss of water.

$$
R-C{\overset{O}{\underset{NH_2}{\diagup}}}
\quad\underset{+H_2O}{\overset{-H_2O}{\rightleftharpoons}}\quad
R-C\equiv N
$$

Amides	Nitriles (or Cyanides)

[1] See Angew. Chem. Int. Ed. Engl. *7*, 467 (1968).
[2] See *G. Binsch, E. L. Eliel* and *H. Kessler*, Nomenclature for Intramolecular Exchange Processes, Angew. Chem. Int. Ed. Engl. *10*, 570 (1971).
[3] In keeping with this, ligands which are in identical environments are described as *homotopic ligands.*
[4] See *B. Testa*, Grundlagen der Organischen Stereochemie (Foundations of Organic Stereochemistry), p. 148*ff*. (Verlag Chemie, Weinheim 1983).

The nitriles can be regarded as *alkyl derivatives of hydrocyanic acid*, which will be dealt with here, as the parent compound of this series.

The C≡N group is known as the *nitrile group*. Compounds containing this group are named either as *cyanoalkanes* or as *nitriles*. In the latter case, for nitriles related to carboxylic acids usually described by trivial names, 'nitrile' replaces 'ic acid', *e.g.* CH_3CN, acetonitrile (from acetic acid), C_6H_5CN, benzonitrile (from benzoic acid), while for other nitriles the suffix 'nitrile' is added to the name of the alkane which has the same number of carbon atoms, including the carbon atom of the nitrile group, *e.g.* $CH_3CH_2CH_2CH_2CN$, pentanenitrile. Alternatively these three compounds may be called cyanomethane, cyanobenzene or 1-cyanobutane. Nitriles were also commonly known as alkyl cyanides and names related to this were (and sometimes still are) met, *e.g.* for the above-mentioned nitriles, methyl cyanide, phenyl cyanide, n-butyl cyanide. The prefix 'cyano' is usually used for compounds also having other functional groups present, *e.g.* $NC-CH_2-COOH$, cyanoacetic acid. The suffix 'carboxylic acid' or 'acid' takes precedence, for numbering and nomenclature, over 'cyano'.

2.28.8.1 Hydrocyanic Acid, Hydrogen Cyanide

Hydrocyanic acid straddles the border between organic and inorganic chemistry. As an organic compound it can be regarded as the *nitrile derived from formic acid*.

Preparation

(a) *Andrussov* **Process.** In this process it is made by oxidation of a mixture of ammonia and methane at about 1000°C (1275 K) over a Pt/Rh catalyst:

$$CH_4 + NH_3 + \tfrac{3}{2}O_2 \xrightarrow{\text{(Pt/Rh)}} HCN + 3\,H_2O \qquad \Delta H = -474\ kJ/Mol$$

(b) By *removal of water from formamide* at 300°C (575 K) over an aluminium oxide catalyst:

$$H-C\!\!\begin{array}{c}{}^{\displaystyle O}\\{}_{\displaystyle NH_2}\end{array} \xrightarrow[-H_2O]{(Al_2O_3)} HC\equiv N$$

Properties

Pure dry hydrocyanic acid is a colourless liquid which smells like bitter almonds and boils at 26°C (299 K). In aqueous solution it is weakly acidic; its salts are known as *cyanides*. On hydrolysis it is converted into formic acid:

$$\overset{\delta^{\oplus}}{H}-\overset{\delta^{\ominus}}{C\equiv N} + 2\,H_2O \rightarrow H-COOH + NH_3$$

The carbon atom of hydrocyanic acid is ambivalent in its chemical behaviour. In its undissociated state it bears a partial positive charge and is electrophilic (see, for example, the formation of triazine), whilst, when dissociated, as a cyanide ion it is a nucleophile (see, for example, the formation of cyanohydrins). It is an extremely dangerous poison; the fatal dose for a human is 60 mg. Hydrogen cyanide can be adsorbed onto kieselguhr and released again by gentle heating. *Cyclon B* is such a preparation which is used for fumigating rooms, etc. It was used to murder the victims in the gas chambers of Auschwitz.

2.28.8.2 Fulminic Acid (Hydrocyanic Acid Oxide)

It has been shown, on the basis of its IR spectrum and microwave spectrum in the gas phase, that fulminic acid has the structure of hydrocyanic acid oxide (formonitrile oxide); its ground state is mesomeric:

$$\left[H-C\equiv\overset{\oplus}{N}-\overset{\ominus}{\underset{..}{O}}\ \longleftrightarrow\ H-\overset{\ominus}{\underset{..}{C}}=\overset{\oplus}{N}=\underset{..}{O} \right]\ \text{Fulminic acid}\qquad \overset{\ominus}{I}\overset{\oplus}{C}\equiv N-\underset{..}{O}H\ \text{ Isofulminic acid}$$

An isomer of fulminic acid is cyanic acid, HO—CN; this *isomerism* was early recognised (*Liebig* and *Wöhler*, 1824). *Isofulminic acid* can be obtained by irradiating dibromoformoxime (Br$_2$C=NOH) in an argon matrix at 12 K (*G. Maier*, 1988). It can be regarded as the oxime of carbon monoxide (carboxime). For a long time this structure was attributed to fulminic acid.

In the free state, fulminic acid is unstable and has a strong tendency to polymerise. Its salts are different from the corresponding cyanates in that, when ignited, they decompose explosively with a flash above 190° (460 K). It is from this that their name *fulminates* (Lat. *fulmen* = lightning) is derived. Fulminic acid itself does not have this property.

Mercury fulminate, Hg(CNO)$_2$, which was used as a detonator before lead azide was discovered, is obtained by a rather complicated process in which mercury is dissolved in conc. nitric acid and ethanol is added to the mixture. It is detonated by shock or on being struck, decomposing to carbon monoxide, nitrogen and mercury.

$$\text{Hg(CNO)}_2 \longrightarrow 2\,\text{CO} + \text{N}_2 + \text{Hg}\qquad \Delta H = -510\ \text{kJ/Mol}$$

2.28.8.3 Nitriles (Alkyl Cyanides)

The first members of the homologous series of nitriles are as follows:

		Sdp.
Acetonitrile (Methyl cyanide, Cyanomethane)	H$_3$C–C≡N	81,6 °C (354,8 K)
Propionitrile (Ethyl cyanide, Cyanoethane)	H$_3$C–CH$_2$–C≡N	97,1 °C (370,3 K)
n-Butyronitrile (*n*-Propyl cyanide, 1-Cyanopropane)	H$_3$C–(CH$_2$)$_2$–C≡N	117,6 °C (390,8 K)

Preparation

(a) *Kolbe* **nitrile synthesis.** When alkyl halides react with potassium cyanide in solution in aqueous ethanol the main product is a nitrile, together with smaller amounts of an isonitrile (isocyanide), *e.g.*

$$\text{H}_3\text{C–I} + \text{K}^{\oplus}\overset{\ominus}{I}\text{C}\equiv\text{NI} \xrightarrow{\ -\text{KI}\ } \begin{cases} \text{H}_3\text{C}-\text{C}\equiv\text{NI}\quad \text{Acetonitrile (Methyl cyanide)} \\[2mm] \text{H}_3\text{C}-\overset{\oplus}{\text{N}}\equiv\overset{\ominus}{\text{CI}}\quad \text{Methyl isocyanide (isonitrile)} \end{cases}$$

The alkyl isocyanide can be removed from such a mixture by shaking it with cold dilute hydrochloric acid.

Alkylation of the ambidentate cyanide ion proceeds by an S_N2 mechanism, and takes place largely at the more strongly nucleophilic carbon atom and only to a lesser extent at the nitrogen atom.

The alkyl halides decrease in reactivity from iodide to bromide to chloride; the structure of the alkyl group also affects the reactivity.

(b) By *heating amides* with dehydrating agents:

$$R-C{\overset{O}{\underset{NH_2}{\big\backslash}}} \xrightarrow[-H_2O]{(P_2O_5)} R-C\equiv N$$

(c) By *heating aldoximes* with acetic anhydride:

$$R-C{\overset{H}{\underset{NOH}{\big\backslash}}} \xrightarrow{-H_2O} R-C\equiv N$$

Properties

The lower members of the series are stable, colourless liquids with rather agreeable smells; higher members are crystalline materials. They are far less poisonous than hydrocyanic acid.

When heated with strong bases or acids, nitriles lose ammonia and are converted into carboxylic acids with the same number of carbon atoms:

$$R-C\equiv N \xrightarrow{+H_2O} R-CONH_2 \xrightarrow[-NH_3]{+H_2O} R-COOH$$

The conversion of nitrile groups into carboxyl groups is of preparative importance. Amides, which are formed as intermediates, can be isolated if the hydrolysis is carried out using 96% sulphuric acid or alkaline hydrogen peroxide.

Nitriles can be *reduced* catalytically or by means of sodium and an alcohol, or lithium aluminium hydride, giving primary amines.

$$R-C\equiv N + 4 H \rightarrow R-CH_2NH_2$$

In a reaction similar to the aldol addition of aldehydes, nitriles having α-hydrogen atoms, which are acidic, can *dimerise* in solution in ether in the presence of a base (sodium or a sodium alkoxide). For example, acetonitrile gives *β-iminobutyronitrile*, which exists to a large extent in the tautomeric *enamine* structure as *β-aminocrotononitrile*. The reaction takes place by the following mechanism:

$$H{-}CH_2-C\equiv N| \xrightarrow[-ROH]{RO^{\ominus}} {}^{\ominus}|CH_2-C\equiv N|$$

$$H_3C-C\equiv N| + |CH_2-C\equiv N| \longrightarrow H_3C-\underset{\underset{|N|^{\ominus}}{\overset{\|}{C}}}{C}-CH_2-C\equiv N \xrightarrow{+H^{\oplus}}$$

$$H_3C-\underset{NH}{\overset{\|}{C}}-CH_2-C\equiv N \rightleftharpoons H_3C-\underset{NH_2}{\overset{|}{C}}=CH-C\equiv N$$

β-Imino form   Enamine form

2.28.9 *Hydroxamic acids*[1]

Hydroxamic acids can be prepared by the reaction of *hydroxylamine* with acid chlorides or esters:

---

[1] See *L. Bauer* and *O. Exner*, The Chemistry of Hydroxamic Acids and *N*-Hydroxyimides, Angew. Chem. Int. Ed. Engl. *13*, 376 (1974).

$$R-\overset{O}{\underset{OC_2H_5}{C}} \xrightarrow[-C_2H_5OH]{+H_2NOH} R-\overset{O}{\underset{NHOH}{C}}$$

Rhodotorulinic acid, m.p. 217–218°C (490–491 K)

They are weak acids ($pK_a \sim 9$); they reduce *Fehling's* solution. With iron(III) chloride they give an intense red colour; this is a sensitive test for carboxylic acids and their derivatives.

The dihydroxamic acid *rhodotorulinic acid*, which is a derivative of 2,5-dioxopiperazine, acts as a siderophore (see p. 507). It is isolated from the red yeast *Rhodotorula pilimanae*.

2.28.10 Iminoesters, Amidines and Amidrazones

Iminoesters are ester derivatives in which the oxygen atom of the carbonyl group has been replaced by an imino group. They are made by saturating a cold solution of a nitrile in absolute ethanol with hydrogen chloride, *e.g.*

$$H_3C-C\equiv N + C_2H_5OH + HCl \longrightarrow \left[H_3C-\overset{\overset{\oplus}{N}H_2}{\underset{OC_2H_5}{C}} \right] Cl^{\ominus}$$

Acetonitrile   Ethyl acetamidate
                                  hydrochloride

When the hydrochlorides, which usually form good crystals, are treated with alkali the free iminoesters are formed; they are liquids with a characteristic odour (see also orthoesters).

Amidines are derivatives of amides in which the oxygen atom of the amide group has been replaced by an imino group. They are formed by the action of dry ammonia on iminoester hydrochlorides, *e.g.*

$$\left[H_3C-\overset{\overset{\oplus}{N}H_2}{\underset{OC_2H_5}{C}} \right] Cl^{\ominus} \xrightarrow[-C_2H_5OH]{+NH_3} \left[H_3C-\overset{\overset{\oplus}{N}H_2}{\underset{NH_2}{C}} \longleftrightarrow H_3C-\overset{\overset{-}{N}H_2}{\underset{\overset{\oplus}{N}H_2}{C}} \right] Cl^{\ominus}$$

Ethyl acetamidate   Acetamidine
hydrochloride   hydrochloride

Amidines are strong bases, which are usually only stable in the form of their salts.

Amidrazones[1] can be considered either as amide hydrazones or, in an alternative tautomeric form, as hydrazides of iminoacids. They are made by treating nitriles with, successively, sodium hydrazide (at $-20°C$ [250 K] under nitrogen) and water (*T. Kauffmann*, 1963):

$$R-C\equiv N \xrightarrow[2)+H_2O\;(-NaOH)]{1)+NaNH-NH_2} \left[R-\overset{NH_2}{\underset{N-NH_2}{C}} \rightleftharpoons R-\overset{NH}{\underset{NH-NH_2}{C}} \right]$$

Nitrile

Amidrazone

---

[1] See *D. G. Neilson* et al., The Chemistry of Amidrazones, Chem. Rev. *70*, 151 (1970).

2.28.11 Acid Hydrazides and Acid Azides

Acid hydrazides are made in a similar way to acid amides by the reaction of hydrazine with acid halides, anhydrides or esters (*Curtius*), *e.g.*

$$H_3C-C\overset{O}{\underset{Cl}{\big\langle}} + H_2N-NH_2 \longrightarrow H_3C-C\overset{O}{\underset{NH-NH_2}{\big\langle}} + HCl$$

Acetyl chloride   Acetohydrazide

The free acid hydrazides are crystalline compounds which are basic in character. They reduce both ammoniacal silver nitrate and *Fehling's* solution.

On treatment with nitrous acid, these hydrazides are converted into acid azides (acyl azides) which are, for the most part, crystalline. Many of them explode when heated.

$$H_3C-C\overset{O}{\underset{NH-NH_2}{\big\langle}} + HNO_2 \longrightarrow H_3C-C\overset{O}{\underset{N_3}{\big\langle}} + 2 H_2O$$

Acetazide

Alternatively, they can also be prepared directly from acid chlorides and sodium azide:

$$R-C\overset{O}{\underset{Cl}{\big\langle}} + NaN_3 \longrightarrow R-C\overset{O}{\underset{N_3}{\big\langle}} + NaCl$$

Curtius thought that the three nitrogen atoms were arranged in a ring, but *Angeli* and *Thiele* proposed that the nitrogen atoms have a linear structure, and this was later confirmed by physical measurements and by isotopic (^{15}N) labelling. This linear structure is stabilised by *mesomerism*.

$$\left[R-C\overset{\bar{O}|}{\underset{\underset{\oplus}{N}=\underset{}{N}=\underset{\ominus}{\bar{N}}}{\big\langle}} \longleftrightarrow R-C\overset{\bar{O}|}{\underset{\underset{\ominus}{N}-\underset{\oplus}{N}\equiv N|}{\big\langle}} \right]$$

2.29 Substituted Aliphatic Monocarboxylic Acids

Substituted carboxylic acids contain a carboxyl group but one or more of the hydrogen atoms in the alkyl group have been replaced by another atom or group of atoms.

2.29.1 Halogenocarboxylic Acids

The simplest example of this sort of carboxylic acid is *chloroacetic acid*, $Cl-CH_2-COOH$. There are two isomeric *chloropropionic acids*:

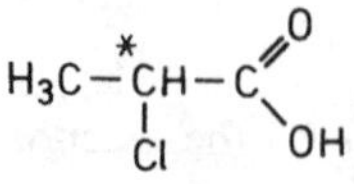

$$H_3C - \overset{*}{C}H - C\overset{\displaystyle O}{\underset{\displaystyle OH}{}}$$
$$\underset{\displaystyle Cl}{|}$$

2-Chloropropionic acid
α-Chloropropionic acid

$$CH_2 - CH_2 - C\overset{\displaystyle O}{\underset{\displaystyle OH}{}}$$
$$\underset{\displaystyle Cl}{|}$$

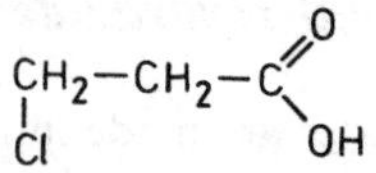

3-Chloropropionic acid
β–Chloropropionic acid

In numbering the alkyl chain of carboxylic acids, the carbon atom of the carboxyl group is assigned the number 1. An alternative approach uses the letters of the Greek alphabet to designate the alkyl carbon atoms. The carbon next to the carboxyl group is the α-carbon atom, and successive atoms are assigned successive letters of the Greek alphabet, β, γ, δ, etc. 2-Chloropropionic acid contains an *asymmetric carbon* atom (*) (see p. 277).

There is no general method available for the preparation of halogenoacids. The methods used depend upon the site in the molecule which the halogen atom is to occupy. The simplest ones to make are α-halogenoacids since an α-hydrogen atom is readily replaced by a halogen atom because the α-position is activated by the adjacent carboxyl group (−I-effect). Acids having more than one α-halogen atom can be made in the same way.

Preparation

2.29.1.1 Halogenoacetic acids. *Direct halogenation of acetic acid* with chlorine or bromine proceeds very sluggishly. It can be speeded up by either the action of light or the presence of a halogen carrier such as iodine or sulphur, and also by heating. Under these conditions acetic acid and chlorine first of all give chloroacetic acid and, if there is an excess of chlorine, dichloro- and trichloro-acetic acids:

$H_3C - COOH \ + Cl_2 \rightarrow HCl + ClH_2C - COOH$ Chloroacetic acid, m.p. 61°C (334 K)
$ClH_2C - COOH + Cl_2 \rightarrow HCl + Cl_2HC - COOH$ Dichloroacetic acid, b.p. 194°C (467 K)
$Cl_2HC - COOH + Cl_2 \rightarrow HCl + Cl_3C - COOH$ Trichloroacetic acid, m.p. 58°C (331 K)

Industrially two ways are used to make **chloroacetic acid**:

(a) *by heating trichloroethylene with 70% sulphuric acid*:

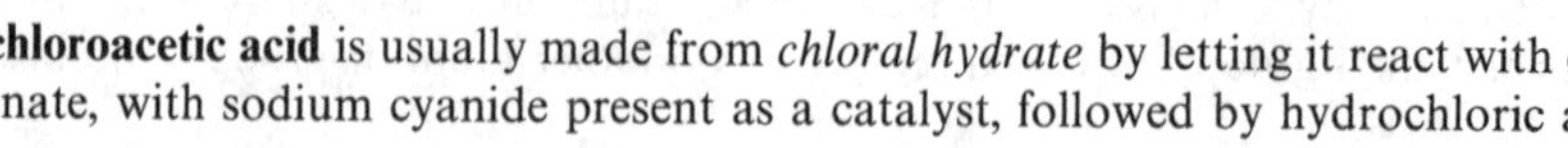

$$ClHC = CCl_2 \xrightarrow{+ H_2SO_4} ClH_2C - CCl_2 - OSO_3H \xrightarrow[-H_2SO_4, -2\,HCl]{+ 2\,H_2O} ClH_2C - C\overset{\displaystyle O}{\underset{\displaystyle OH}{}}$$

(b) *by oxidation of 2-chloroethanol* with nitric acid:

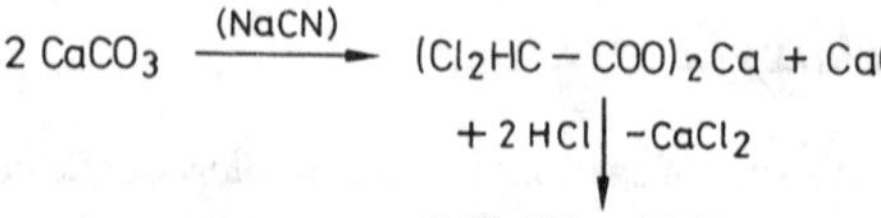

$$\underset{\displaystyle Cl \quad OH}{\overset{\displaystyle CH_2 - CH_2}{| \qquad |}} \xrightarrow[-H_2O]{[O]} Cl - CH_2 - COOH$$

Dichloroacetic acid is usually made from *chloral hydrate* by letting it react with calcium carbonate, with sodium cyanide present as a catalyst, followed by hydrochloric acid:

$$2\,Cl_3C - CHOH)_2 + 2\,CaCO_3 \xrightarrow{(NaCN)} (Cl_2HC - COO)_2Ca + CaCl_2 + 2\,CO_2 + 2\,H_2O$$

$$+ 2\,HCl \downarrow -CaCl_2$$

$$2\,Cl_2HC - COOH$$

Trichloroacetic acid is best obtained by *oxidation of chloral hydrate* with conc. nitric acid:

$$Cl_3C-CH(OH)_2 \xrightarrow[-H_2O]{[O]} Cl_3C-COOH$$

Trifluoroacetic acid is a liquid which fumes in air and is among the strongest carboxylic acids ($pK_a = 0.23$). It is obtained from *m*-trifluoromethylaniline by oxidising it with chromic acid:

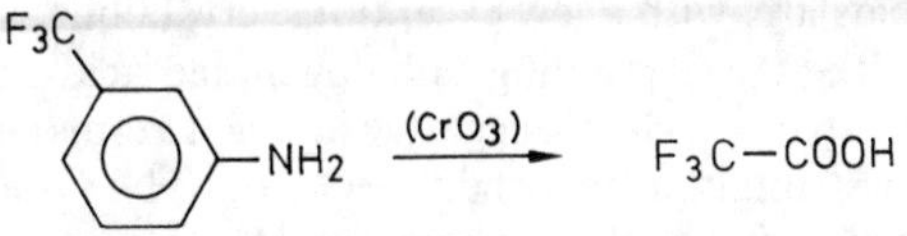

If *higher carboxylic acids are chlorinated* under the appropriate conditions, other hydrogen atoms in the alkyl chain, as well as the α-hydrogens, may be substituted by chlorine. Acid chlorides and anhydrides can be halogenated appreciably easier than can carboxylic acids.

2.29.1.2 *Hell–Volhard–Zelinsky* reaction. This is a method which provides solely **α-bromoacids**. The acid to be brominated is mixed with red phosphorus, and bromine is allowed to drip slowly on to the mixture which is heated to about 80°C (350 K). Phosphorus tribromide is generated which reacts with the acid to convert it into the acid bromide, which is readily brominated to give the α-bromoacid bromide:

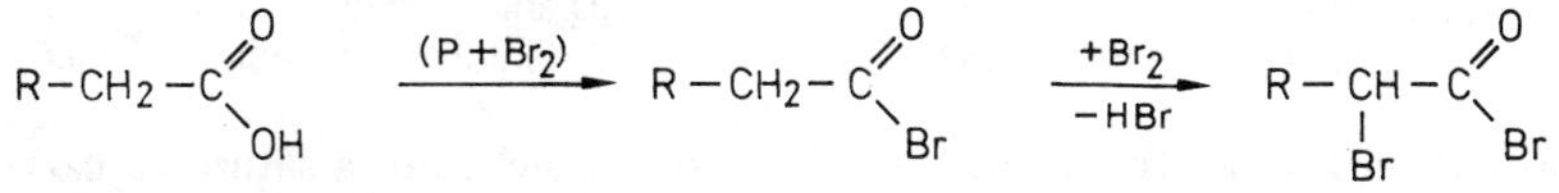

Usually only a small amount of phosphorus is necessary, because the α-brominated acid bromide which is formed reacts with excess carboxylic acid as follows:

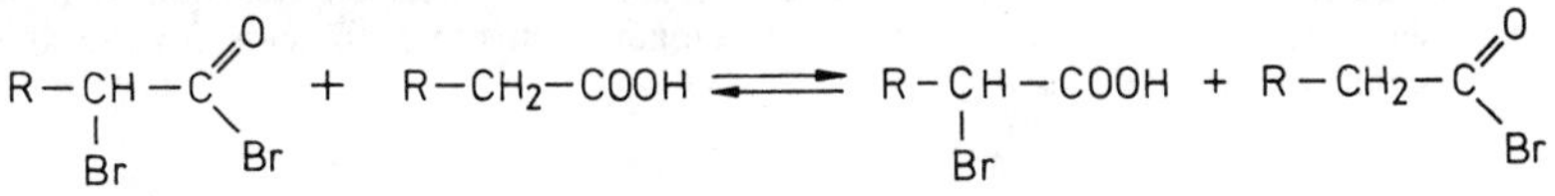

The newly generated acyl bromide is then brominated on the α-carbon atom.

α-Iodoacids are obtained from the corresponding *α-chloro-* or *α-bromo-acids* by warming them with potassium iodide in methanol or acetone:

2.29.1.3 β-Halogenoacids are obtained by *addition of a hydrogen halide to an α,β-unsaturated carboxylic acid*, e.g.

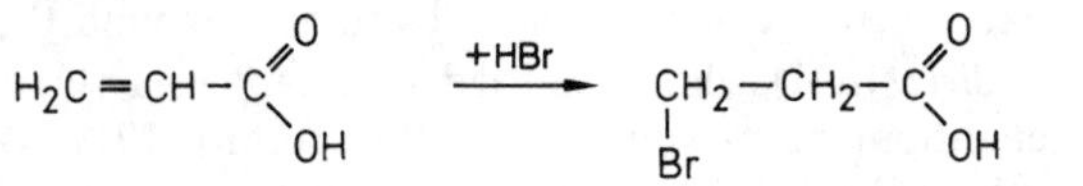

Acrylic acid   3-Bromopropionic acid

2.29.1.4 γ-Halogenoacids are formed by the *addition of hydrogen halides to β,γ- unsaturated carboxylic acids*, e.g.

$$H_2C=CH-CH_2-COOH + HBr \longrightarrow \begin{array}{c} CH_2-CH_2-CH_2-COOH \\ | \\ Br \end{array}$$

Vinylacetic acid (But-3-enoic acid)   4-Bromobutyric acid

Properties

Halogenocarboxylic acids are, with a few exceptions, crystalline materials and in solution are stronger acids than the corresponding unhalogenated acids, because of the electron-withdrawing effect of the strongly electronegative halogen atoms ($-$I-effect). This electron-withdrawing effect also strongly affects the C—O- and H—O-bonds and facilitates the removal of the hydrogen atom of the carboxyl group as a proton; in other words it is more acidic.

The $-$I-effect is felt not only at the α-position, but, although less noticeably, at the β- and γ-positions. At greater distances along the chain its effect is very slight.

The decrease in the inductive effect along the chain is evident from the increase of the pK_a values from α- to β- to γ-chlorobutyric acids. Even so, all of them are *stronger acids* than butyric acid itself.

	pK_a		pK_a
α-Chlorobutyric acid	2,84	Acetic acid	4,76
β-Chlorobutyric acid	4,06	Chloroacetic acid	2,86
γ-Chlorobutyric acid	4,52	Dichloroacetic acid	1,29
Butyric acid	4,82	Trichloroacetic acid	0,65

If more than one electron-withdrawing group is attached to a saturated carbon atom then the $-$I-effect is even greater, as shown above by the pK_a values for the chlorinated acetic acids.

Conversely, the carboxyl group in halogenoacids exerts a $-$I-effect on α-carbon–halogen bonds. This shows itself in the increased reactivity of the halogen atoms in such acids, for example in their nucleophilic replacement by HO-groups.

2.29.2 D- and L-Configuration on Asymmetric Carbon Atoms which are Stereogenic Centres

Earlier on, on p. 118 *ff*, the stereochemistry of asymmetric carbon atoms and the structural explanation of optical activity in organic compounds were discussed. Since α-*chloropropionic acid* is optically active this is an appropriate place to consider more closely the relationship between spatial configuration and optical rotation. Originally, optically active substances were denoted by the letters *d* (Lat. *dextro* = right), for the form which rotated the plane of polarisation to the right, *l* (Lat. *laevo* = left) for the form which rotated the plane to the left, and *dl* for the racemic form. This nomenclature said nothing about the actual spatial arrangement of the substituents around the tetrahedral carbon atom. It was *E. Fischer* (1891) who first tried to establish a relationship between the configuration of enantiomers, basing it on glucose. In 1917 *Wohl* suggested that *glyceraldehyde* be adopted as the reference standard for the configuration of optically active compounds and arbitrarily assigned a *d*-configuration to the form which rotated the plane to the right. The prefixes *d* and *l* thereafter specified the relative configurations in space about the asymmetric carbon atom, while the observed directions of rotation were to be indicated by the symbols (+) and (−), *e.g. d*(+)glyceraldehyde and *d*(−)glyceric acid. Since no clear distinction was made in the older literature between the attributions of *d* and *l* and the directions of optical rotation (+) and (−) a suggestion made by *Vickery* (1947) and *Hudson* (1948), that *enantiomers* should be represented by the small capitals D and L, was adopted. These symbols indicated only the *configuration of the chiral centre*, and had no connection with the direction of rotation. If it were necessary to specify the latter, this should be done by means of the symbols (+) and (−) which are inserted directly after the prefixes D or L.

The reference system was still to be based on D-glyceraldehyde. Following the proof of the absolute configuration of (+)-tartaric acid by *Bijvoet*, it followed that the configuration which had been arbitrarily assigned to D-glyceraldehyde was indeed the correct *absolute configuration*. This can be represented by a tetrahedron (Fig. 78), whose peak is occupied by the aldehyde group, while of the groups in the foreground, the HO-group is on the right and a hydrogen atom on the left. If the HO-group and the hydrogen atom are interchanged this gives the mirror-image form, which is L(−)-glyceraldehyde. These two enantiomers are not superimposible.

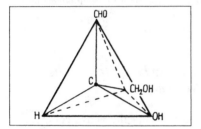

Fig. 78. D(+)-Glyceraldehyde.

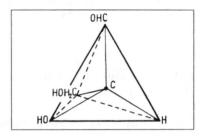

Fig. 79. L(−)-Glyceraldehyde.

The tetrahedral models of the enantiomers of glyceraldehyde can be simplified as shown in the following formulae. It is the convention that the carbon atom with the highest oxidation state is always shown at the top of the formulae.

D(+)-Glyceraldehyde   L(−)-Glyceraldehyde

In *Fischer projection formulae*, substituents shown to the left and right of the asymmetric carbon atom represent groups lying in front of the plane of the paper; those above and below this atom represent groups which are either in or behind the plane of the paper.

Every compound whose experimentally determined configuration corresponds to that of the asymmetric carbon atom in D-glyceraldehyde belongs to the D-series, irrespective of the sense in which it rotates the plane of polarisation.

In order to determine the configuration of (−)-lactic acid, the following reaction sequence can be carried out in which there is *no* change of configuration at the asymmetric centre at any step. D(+)-Glyceraldehyde is oxidised by mercury(II) oxide to give D(−)-glyceric acid, which is also available from D(+)-isoserine by the action of nitrous acid. Since D(+)-isoserine also reacts with nitrosyl bromide, followed by reduction, to give (−)-lactic acid, the latter must therefore have a D configuration.

$$
\begin{array}{cccccccc}
& \overset{H}{\underset{\parallel}{C}}{\diagdown}O & & \text{COOH} & & \text{COOH} & & \text{COOH} \\
& | & \xrightarrow{\;(\text{HgO})\;} & | & \xrightarrow{\;(\text{HNO}_2)\;} & | & \xrightarrow[\;(\text{NaHg})\;]{\;(\text{ONBr})\;} & | \\
\text{H}-\text{C}-\text{OH} & & \text{H}-\text{C}-\text{OH} & & \text{H}-\text{C}-\text{OH} & & \text{H}-\text{C}-\text{OH} \\
| & & | & & | & & | \\
\text{CH}_2\text{OH} & & \text{CH}_2\text{OH} & & \text{CH}_2-\text{NH}_2 & & \text{CH}_3
\end{array}
$$

D(+)-Glyceraldehyde   D(−)-Glyceric acid   D(+)-Isoserine   D(−)-Lactic acid

Racemic forms are designated either by DL, if the configurations of the enantiomeric forms are known unequivocally, or by (±) or the prefix *rac.*; should there be 2 or 2n centres of chirality in a molecule, the prefix *racem.* is used (see p. 346*f.*).

In the cases of optically active compounds, and especially for natural products, whose classification as D or L cannot be assigned unequivocally, then only the particular sense of rotation associated with the compound, either (+) or (−) or *d* and *l* is indicated. The first method is preferable, *e.g.* as in (−)-brucine. An unambiguous method of designating the configuration of compounds can be achieved with the help of the following system.

2.29.3 Absolute Configuration at Asymmetric Carbon Atoms. The Cahn–Ingold–Prelog (CIP) (RS) System

The *absolute configuration* or the *sense of chirality* of an optically active compound can be either calculated[1] or, as mentioned above, determined experimentally. In order to provide a completely specific system of nomenclature, *Cahn, Ingold* and *Prelog*[2] introduced the following system, often called either the CIP or the *RS* system. In this system, the four ligands attached to an asymmetric carbon atom are arranged according to their *priority*, that is in decreasing order of the atomic numbers of the vicinal atoms which are linked directly to the asymmetric centre. If two of the vicinal atoms are the same, then recourse is made to the next atom in each ligand. If need be, this may be continued along the ligand chain until a point of difference is reached. This is best explained and illustrated from the following examples.

1st Sequence Rule. Of four different ligands attached to an asymmetric centre, the highest priority goes to the one with the highest atomic number.

2nd Sequence Rule. If two of these atoms are isotopes of the same element, the higher priority goes to the atom with the higher atomic mass, *i.e.* D has higher priority than H.

As an example let us consider *bromo-chloro-iodo-methane*. The order of priority is I > Br > Cl > H.

---

[1] See *W. Kuhn*, Z. Elektrochem. *56*, 506 (1952).
[2] See *R. S. Cahn, C. K. Ingold* and *V. Prelog*, Specification of Chirality, Angew. Chem. Int. Ed. Engl. *5*, 385 (1966); *V. Prelog* and *G. Helmchen*, Basic Principles of the CIP System and Proposals for a Revision, Angew. Chem. Int. Ed. Engl. *21*, 567 (1982).

Chirality Rule. The tetrahedral model is arranged in such a way that the atom with the lowest priority — in this case H — is directed towards the rear. The remaining atoms (or groups of atoms) might now be arranged in two different ways. In one case the order of priorities decreases in a clockwise direction, in the other case in an anticlockwise direction. In the first of these cases the configuration is defined as (R) (Lat: *rectus* = right, correct) (Fig. 80), in the second case as (S) (Lat. *sinister* = left, inverted) (Fig. 81).

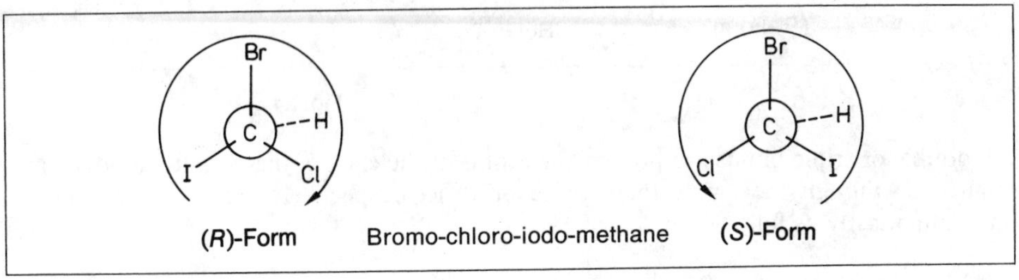

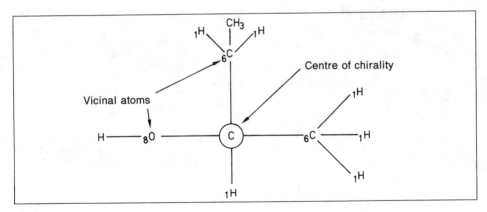

(R)-Form   Bromo-chloro-iodo-methane   (S)-Form

Fig. 80.   Fig. 81.

If the relative priorities of two vicinal atoms cannot be decided in this way, *i.e.* they are the same, then the atoms attached to the vicinal atoms are examined. *Butan-2-ol* may be taken as an example. In this case the atoms vicinal to the chiral centre consist of a hydrogen atom, an oxygen atom and two carbon atoms. Oxygen has the highest priority and hydrogen the lowest. Of the two carbon atoms, one, in the CH_3 group, has three hydrogen atoms attached to it, the other, in the CH_2CH_3 group, has one carbon and two hydrogen atoms attached to it. Since carbon has a higher priority than hydrogen, the ethyl group has higher priority than the methyl group. Thus the complete order for the ligands attached to the asymmetric centre of butan-2-ol is $HO > C_2H_5 > CH_3 > H$. This can be illustrated as follows (Fig. 82):

Fig. 82.

According to the chirality rule the enantiomers take up the following configurations:

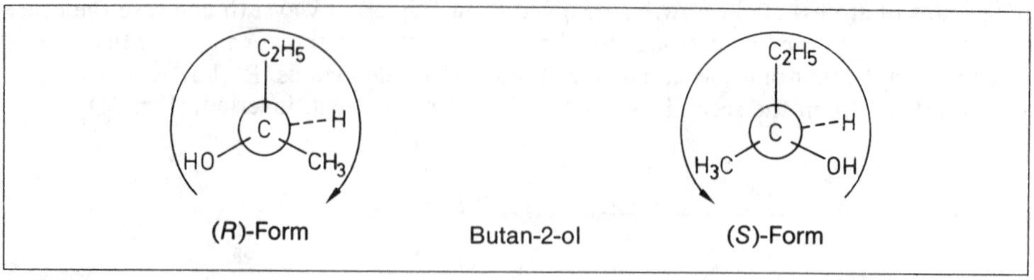

(R)-Form   Butan-2-ol   (S)-Form

Fig. 83.   Fig. 84.

If double or triple bonds are present the atoms at the ends of these multiple bonds are counted as though they were there twice or thrice respectively. This may be shown diagrammatically as follows:

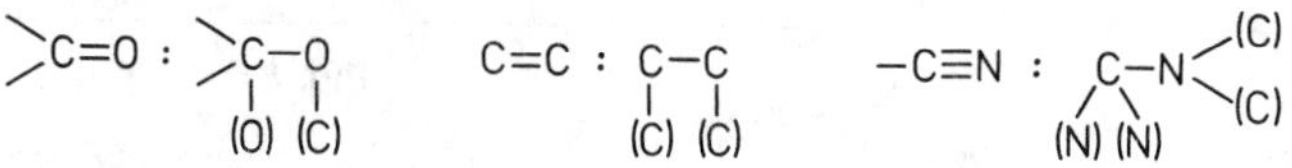

On the basis of the chirality rule the following order of priorities can be assigned to this series of C-ligands attached to an asymmetric centre:

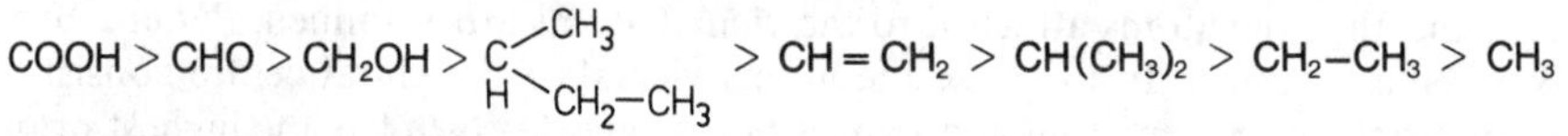

$$COOH > CHO > CH_2OH > C\underset{H}{\overset{CH_3}{\diagdown}}CH_2{-}CH_3 > CH{=}CH_2 > CH(CH_3)_2 > CH_2{-}CH_3 > CH_3$$

Glyceraldehyde may be taken as an example to illustrate the application of these rules. The HO-group (vicinal atom = O) has the highest priority, the hydrogen atom the lowest. The relative priorities of the other two groups can be settled by considering the atoms attached to the vicinal carbon atoms. Since the double bonded oxygen of the formyl group counts as two oxygens the atoms attached to its carbon atom are taken to be O, O, H, whilst in the case of the hydroxymethyl group the atoms attached to the carbon are O, H, H. Hence the formyl group takes priority and the overall ranking of the ligands is HO > CHO > CH$_2$OH > H (Fig. 85).

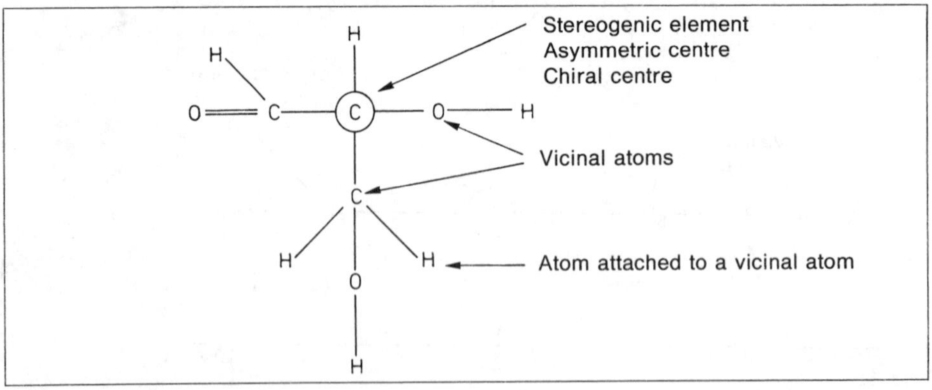

Stereogenic element
Asymmetric centre
Chiral centre

Vicinal atoms

Atom attached to a vicinal atom

Fig. 85.

If the tetrahedral models of D- and L-glyceraldehyde shown in Figs. 78 and 79 (p. 273) are turned 60° to the left and right respectively about the C—CHO axis, then in each case the lowest priority group, *i.e.* the hydrogen atom, recedes backwards behind the plane of the paper. The following arrangements in space correspond to the (R)- and (S)-forms:

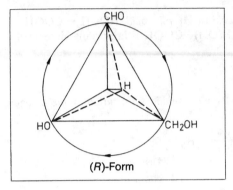

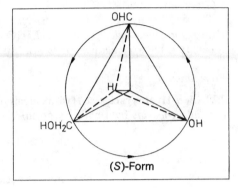

Fig. 86.   Fig. 87.

Thus for glyceraldehyde the (R)-form is identical to the D-form and the (S)-form to the L-form.

Optically active *α-chloropropionic acids* can similarly be assigned structures by the CIP system; the order of priorities is $Cl > COOH > CH_3 > H$.

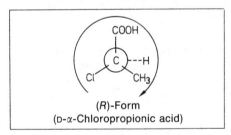

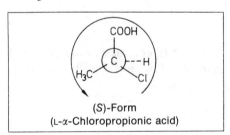

Fig. 88.   Fig. 89.

The *Cahn–Ingold–Prelog* method does not depend upon any reference system and can be applied to complicated systems, *e.g.* those found in alkaloids. In compounds having *more than one* asymmetric centre the absolute configuration of *each centre of chirality* can be assigned. Furthermore this system can be applied to stereogenic hetero-atoms, such as N, S, B, and, by the use of a few extra rules, to other classes of chiral compounds, *e.g.* allenes, spirans, biphenyl derivatives, cyclophanes and asymmetric alicyclic compounds.[1] There are cases where the *Fischer* projections (D, L) and the *CIP* system give contradictory answers, e.g. in D-ampicillin (p. 749) the starred carbon atom has an S-configuration.

The CIP system can be applied to two-dimensional enantiotopic surfaces as follows, using the symbols *Si* and *Re* instead of (S) and (R):

[1] See *W. Klyne* et al., Atlas of Stereochemistry, Absolute Configurations of Organic Molecules, 2nd edn; *J. Buckingham* et al., 2nd edn. Supplement (Chapman & Hall, London 1974, 1978, 1986).

2.29.4 Hydroxyacids

Hydroxyacids are derivatives of carboxylic acids in which one of the hydrogen atoms in the alkyl chain has been replaced by a hydroxy group. The site of the hydroxy group with respect to the carboxyl group is often indicated, as in the case of halogenoacids, by a Greek letter.

The simplest hydroxyacid is *hydroxyacetic acid* or *glycollic acid*, $HOCH_2—COOH$. α-*Hydroxypropionic acid* or *lactic acid* $H_3C—CH(OH)—COOH$ is important biologically. It can exist in two *enantiomeric forms*.

Preparation

2.29.4.1 α-Hydroxyacids, 2-Hydroxyacids. (a) By *nucleophilic substitution of the halogen in α-halogenoacids* by a hydroxy group, e.g.

$$H_3C-CH-COOH \xrightarrow[-HCl]{+H_2O} H_3C-CH-COOH$$
$$\quad\quad\; |\quad\quad\quad\quad\quad\quad\quad\quad\quad\quad\; |$$
$$\quad\quad Cl \quad\quad\quad\quad\quad\quad\quad\quad\quad OH$$

DL-α-Chloropropionic acid   DL-Lactic acid

(b) By *forming a cyanohydrin from an aldehyde* and then hydrolysing the cyano group to a carboxyl group:

$$R-C\overset{H}{\underset{O}{<}} \xrightarrow{+HCN} R-C\overset{H}{\underset{CN}{-}}OH \longrightarrow R-CH-COOH$$
$$\quad\quad\quad\quad\quad\quad\quad\quad\quad\quad\quad\quad\quad\quad\quad\quad |$$
$$\quad\quad\quad\quad\quad\quad\quad\quad\quad\quad\quad\quad\quad\quad\quad\quad OH$$

Aldehyde   Cyanohydrin   DL-α-Hydroxyacid

2.29.4.2 β-Hydroxyacids, 3-Hydroxyacids. (a) *β-Hydroxypropionic acid, 3-hydroxypropionic acid* (hydracrylic acid) is made by hydrolysis of 3-hydroxypropionitrile:

$$CH_2-CH_2-C\equiv N \xrightarrow[-NH_3]{+2H_2O} CH_2-CH_2-COOH$$
$$|\quad\quad\quad\quad\quad\quad\quad\quad\quad\quad\quad\quad |$$
$$OH \quad\quad\quad\quad\quad\quad\quad\quad\quad\quad OH$$

3-Hydroxypropionitrile   3-Hydroxypropionic acid

(b) *Reformatsky reaction*[1] (1890). Organozinc derivatives of α-halogenoesters in inert solvents (ether, benzene) add to the C=O group of aldehydes or ketones to provide β-hydroxyesters. Unlike organomagnesium compounds, organozinc derivatives do not react with ester groups. The initial adduct is hydrolysed to the β-hydroxyester, which readily loses water to give an α,β-unsaturated acid, e.g.

$$Br-CH_2-COOC_2H_5 + Zn \longrightarrow BrZnCH_2-COOC_2H_5$$

$$R-C=O + BrZnCH_2-COOC_2H_5 \longrightarrow R-C\overset{OZnBr}{\underset{R'\ CH_2-COOC_2H_5}{<}}$$
$$\quad |$$
$$\quad R'$$

$$\xrightarrow[-ZnBr(OH)]{+H_2O} R-C\overset{OH}{\underset{R'}{-}}CH_2-COOC_2H_5 \xrightarrow{-H_2O} R-C=CH-COOC_2H_5$$
$$\quad\quad\quad\quad\quad\quad\quad\quad\quad\quad\quad\quad\quad\quad\quad\quad\quad\quad\quad\quad |$$
$$\quad\quad\quad\quad\quad\quad\quad\quad\quad\quad\quad\quad\quad\quad\quad\quad\quad\quad\quad\quad R'$$

β-Hydroxyester

---

[1] See *M. W. Rathke*, The Reformatsky Reaction, Org. Reactions *22*, 423 (1975).

In general, aldehydes (R' = H) react more readily than ketones. A small amount of iodine is usually used as a catalyst. This organometallic variant of the aldol synthesis has wide synthetic uses. The carbonyl component can be alkyl or aryl aldehydes as well as alkyl, cycloalkyl or aryl ketones (see also the *Ivanov* reaction). The methylene component can be branched or unbranched aliphatic α-, β- or γ-bromoesters.

2.29.4.3 γ- and δ-Hydroxyacids, 4- and 5-Hydroxyacids.

These hydroxyacids can be obtained by hydrolysis of the corresponding halogenoacids. They can only be isolated in the form of their salts since in acid conditions they at once lose water to give γ- or δ-lactones, *e.g.*

γ-Chlorovaleric acid   γ-Hydroxyvaleric acid   γ-Valerolactone
4-Chloropentanoic Acid   4-Hydroxypentanoic acid

Properties

Hydroxyacids are more soluble in water and less soluble in ether than the parent carboxylic acids. Like halogen atoms in halogenoacids, but to a smaller extent, α-hydroxy groups increase the *acidity* of hydroxyacids (−I-effect). Their chemical properties are those associated with both the hydroxy and carboxyl groups, and they undergo the reactions associated with each of these groups.

When heated they lose water. The nature of the product depends upon the site of the hydroxy group. In the case of *α-hydroxyacids*, loss of water takes place in an intermolecular reaction leading to ring closure and formation of a lactide, *e.g.*

Lactic acid   3,6-Dimethyl-1,4-dioxan-2,5-dione

The name *lactide* comes from lactic acid (Lat. *acidum lacticum*) since the above was the first example of this type of reaction to be observed.

β-Hydroxyacids lose water intramolecularly to give α,β-unsaturated carboxylic acids, *e.g.*

β-Hydroxypropionic acid   Acrylic acid

γ- and *δ-Hydroxyacids* lose water even at room temperature, giving γ- or *δ-lactones* respectively, *e.g.*

γ-Hydroxy-   γ-Butyrolactone   δ-Hydroxy-   δ-Valerolactone
butyric acid   valeric acid

γ-Mercaptocarboxylic acids similarly cyclise spontaneously forming γ-*thiolactones*.

2.29.5 Lactones

Lactones can be regarded as intramolecular esters derived from hydroxyacids. They can be classified into α-, β-, γ-, δ- and higher-membered ring lactones. Lactones originating from aliphatic hydroxyacids can be named by adding the ending -olide to the name of the *non-hydroxylated carboxylic acid* having the same number of carbon atoms. The position along the chain at which ring-closure takes place is indicated by a number which precedes this ending, e.g. propan-3-olide = β-propiolactone, hexadec-7-en-16-olide (= ambrettolide), pentadecan-15-olide (= exaltolide).

α-Lactones can in general only be observed in solution[1] and β-lactones[2] can only be prepared under special conditions, but γ- and δ-lactones are easily obtained from the corresponding hydroxyacids. The commonest examples are *γ-butyrolactone* and *γ-* and *δ-valerolactones*.

Preparation

In addition to the general methods of preparation some lactones are made industrially in the following ways.

2.29.5.1 β-Propiolactone is made by cycloaddition of ketene to formaldehyde in the presence of zinc chloride in an anhydrous medium (yield 90%):

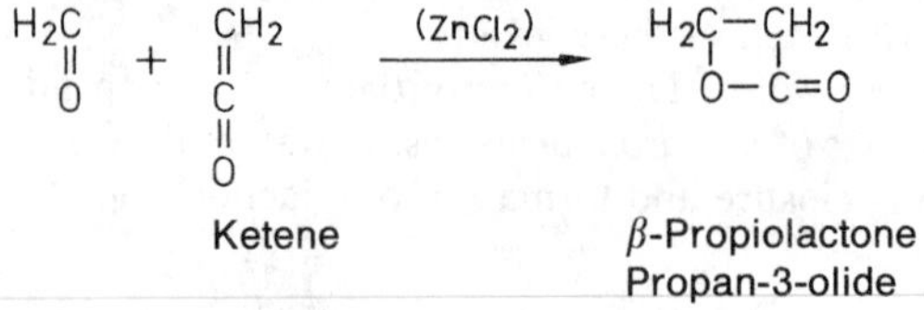

Ketene   β-Propiolactone
                       Propan-3-olide

It is extremely reactive and has wide use in synthetic work. It has been shown to be strongly carcinogenic in animals.

2.29.5.2 γ-Butyrolactone is obtained by dehydrogenation of butane-1,4-diol in the presence of a copper catalyst at 200°C (475 K) (*Reppe*):

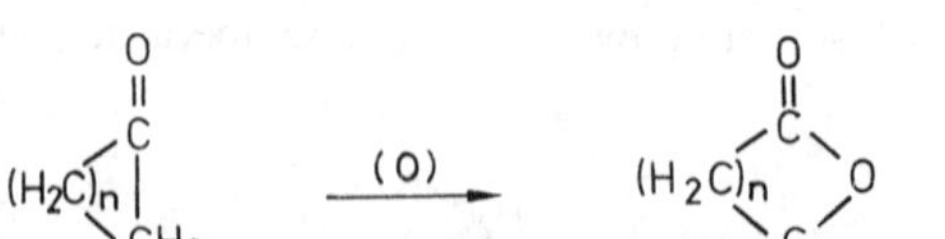

Butane-1,4-diol   γ-Butyrolactone

2.29.5.3 Larger ring lactones have been made by oxidising cyclic ketones with peroxomono-sulphuric acid (*Caro's* acid), H_2SO_5 (*Ruzicka*).

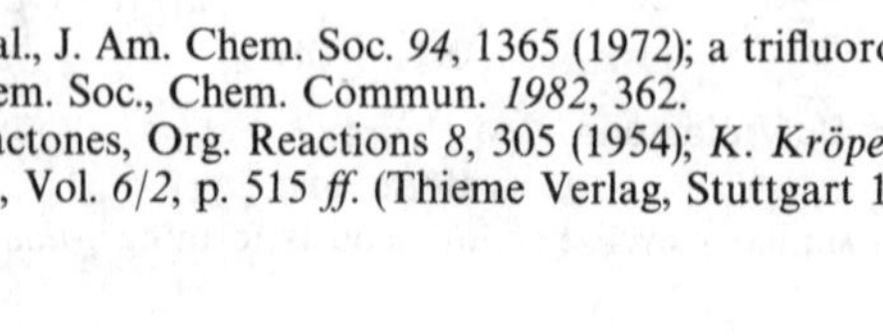

---

[1] See *O. L. Chapman* et al., J. Am. Chem. Soc. **94**, 1365 (1972); a trifluoroalkyl substituted α-lactone has been isolated: J. Chem. Soc., Chem. Commun. *1982*, 362.
[2] See *H. E. Zaugg*, β-Lactones, Org. Reactions **8**, 305 (1954); *K. Kröper, Houben-Weyl*, Methoden der organischen Chemie, Vol. 6/2, p. 515 *ff.* (Thieme Verlag, Stuttgart 1963).

Properties

The lactones boil at lower temperatures than the corresponding hydroxy acids. They have strong, usually pleasing smells. In particular, those with larger rings of 15–17 ring atoms are used in perfumery. These include *ambrettolide*, a perfume found in, among other things, the oils from musk seeds, and a lactone from angelica root, which is prepared artificially and sold under the name of *exaltolide*.

Ambrettolide   Exaltolide

In rings up to seven-membered, the lactone group is in an (*E*)-configuration; if there are more than ten atoms in the ring a (*Z*)-configuration is preferred (see p. 66).

The chemical reactivity of lactones is for the most part restricted to the lactone group. They can be reduced by sodium amalgam in acidic conditions to give the corresponding carboxylic acid:

In contrast, when the same reaction is applied to polyhydroxylactones in weakly acidic solution, aldoses are formed.

The lactone is broken and a corresponding substituted carboxylic acid is formed when they are treated with hydrogen halides, alkali-metal salts of cyanides or sodium hydrogen sulphite. With ammonia, hydroxy-amides are formed, which lose water to give lactams, *e.g.*

γ-Butyrolactone   Pyrrolidin-2-one (2-Pyrrolidinone)
                                          Butane-4-lactam

2.29.6 Commoner Hydroxyacids

2.29.6.1 Glycollic acid (hydroxyacetic acid) $HOCH_2$—COOH, m.p. 79°C (352 K), occurs in plants, for example in unripe grapes and in sugar cane.

2.29.6.2 DL-Lactic acid (α- or 2-hydroxypropionic acid), H_3C—CH(OH)—COOH, was discovered by *Scheele* in 1780 in sour milk, where it is formed by the fermentation of lactose (milk sugar) brought about by *streptococcus lactis* and *lactobacillus lactis*. It is also found in sour cucumbers, and it occurs in the gastric juices. The optically active lactic acids have the following configurations:

$$
\begin{array}{ccc}
\text{COOH} & \text{COOH} & \text{COOH} \\
| & | & | \\
\text{H–C–OH} & \text{HO–C–H} & \\
| & | & \\
\text{CH}_3 & \text{CH}_3 & \\
\text{D(\,--\,)-Lactic acid} & \text{L(\,+\,)-Lactic acid} & \text{(S)-Form}
\end{array}
$$

L(+)-**Lactic acid** occurs in various organs of the bodies of animals and also in muscle fluids and was originally called *sarcolactic acid* (Grk. *sarx* = flesh) (*Berzelius*, 1808). It has a configuration related to that of the α-amino-acids found in proteins, namely the (S)-form.

Pure DL-lactic acid is a syrupy liquid which solidifies at 18°C (291 K). D and L lactic acids melt at 25°C (298 K) (pK_a = 3.86). Their salts are called *lactates*.

2.29.6.3 Lactic fermentation. Industrially lactic acid is obtained from *carbohydrates* (largely starch from potatoes or cereals). They are converted by the enzyme *diastase* into maltose, which is fermented at 35–45°C (310–320 K) by *lactobacillus delbrueckii*, in the first place to glucose and then to lactic acid.

$$C_6H_{12}O_6 \;\rightarrow\; 2\,H_3C\text{–CH(OH)–COOH}$$

Since the fermentation of the sugars is very dependent upon the pH of the solution, the lactic acid is removed as it is formed by means of added calcium carbonate, which reacts with it to form *calcium lactate*. The free acid can be obtained from this by treating it with sulphuric acid. The mechanism of lactic fermentation is described on p. 905. The main product is DL-lactic acid but, by using specifically acting bacteria, a preponderant amount of either D(−)- or L(+)-lactic acids can be obtained. This fermentation process is also important in *silage making*.

Uses. Lactic acid is used in tanning, to remove lime from hides, in dyeing as a reducing agent for chrome-mordants, and, because it is easily digested, as an additive in alcohol-free drinks.

A material called *Royal Jelly*, which is important for the breeding of queen bees, consists of a mixture of proteins, carbohydrates and lipids and also a number of C_{10} carboxylic acids, especially (E)-10-hydroxydec-2-enoic acid. The closely related (E)-9-hydroxydec-2-enoic acid is found in the *queen bee substance*, which is secreted by queen bees.

$$
\begin{array}{cc}
\text{HO–(CH}_2\text{)}_7\text{–C}=\text{C–COOH} & \text{CH}_3\text{–CH–(CH}_2\text{)}_5\text{–C}=\text{C–COOH} \\
\text{(E)-10-Hydroxydec-2-enoic acid} & \text{(E)-9-Hydroxydec-2-enoic acid}
\end{array}
$$

2.29.6.4 Prostaglandins[1] are a group of unsaturated hydroxy- or hydroxyketo-carboxylic acids made up of 20 carbon atoms, which were first isolated in 1957 from the prostates of sheep (*Bergström*). They are formed in, among other places, leucocytes, from essential fatty acids, for example from arachidonic acid:

---

[1] See *W. Bartmann*, Prostaglandins, Angew. Chem. Int. Ed. Engl. *14*, 337 (1975); *R. Noyori* et al., Prostaglandin Synthesis by Three-Component Coupling, Angew. Chem. Int. Ed. Engl. *23*, 847 (1984).

Arachidonic acid
(Eicosa-5,8,
11,14-tetraenoic acid)
$C_{20}:4\,n-6$ (see p. 244)

Prostaglandin (PGE$_{2\alpha}$)

2 H

Prostaglandin (PGF$_{2\alpha}$)

The basic skeleton of prostaglandins is that of the hydrocarbon *prostane* and the related prostanoic acid.

Prostane

Prostanoic acid

In the names for the prostaglandins given above, the final letters in the brackets indicate the structures of the five-membered rings, the subscripts indicate the number of double bonds in the side chains, and an α or β the configuration of the HO-group on carbon atom 9.

The prostaglandins have high physiological activity and are used medicinally, *e.g.* PGF$_{2\alpha}$ (INN: *Dinoprost, Minprostin*). Also formed in the *arachidonic acid cascade* are *thromboxanes*,[1] *prostacyclins*[2] and *leucotrienes*.[3] Aspirin prevents the operation of the arachidonic acid cascade. Because of the number of carbon atoms that are present in these various compounds they are classified collectively as *eicosanoids*.

2.29.7 Amino acids

If the halogen atom of a halogenocarboxylic acid is replaced by an *amino group*, —NH$_2$, the result is the corresponding aminocarboxylic acid. The first member of this series of compounds is *glycine* (aminoacetic acid).

Glycine

As in the case of halogeno and hydroxy acids, the position of the amino group relative to the carboxyl groups is commonly indicated by the use of Greek letters, α-, β-, γ-, *etc.* Of by far the greatest importance are the α-amino acids since they are the essential

---

[1] See *F. Berti* et al. (Eds), Prostaglandins and Thromboxanes (Plenum Press, New York 1977); *D. Schinzer*, Carbacycline: Stabile Analoga der Prostacycline, Nachr. Chem. Tech. Lab. *37*, 734 (1989).
[2] See *W. Bartmann* and *G. Beck*, Prostacyclin and Synthetic Analogues, Angew. Chem. Int. Ed. Engl. *21*, 751 (1982).
[3] See *B. Samuelsson*, The Leucotrienes. Highly Biologically Active Substances Involved in Allergy and Inflammation, Angew. Chem. Int. Ed. Engl. *21*, 902 (1982); From Studies of Biochemical Mechanism to Novel Biological Mediators: Prostaglandin Endoperoxides, Thromboxanes and Leucotrienes, Angew. Chem. Int. Ed. Engl., *22*, 805 (1983).

building blocks from which proteins are built up. Except for glycine, the α-*amino acids* which are formed by enzymatic acid hydrolysis of proteins are *optically active*, and all of them, irrespective of the way in which they rotate the plane of polarised light, are members of the L-series.

2.29.7.1 The configuration of amino acids has been elucidated as follows:

(a) by converting compounds of unknown configuration into compounds of known (or the opposite) configuration, using reactions in which it is known that no change of configuration at the asymmetric carbon atom takes place;

(b) by converting compounds of unknown configuration into compounds of known configuration by means of reactions in which a *Walden inversion* (1895) is known to take place (*Hughes, Ingold*, 1935). Such inversions are represented by the special arrow $\rightarrow$ $\rightarrow$.

For example, alkaline hydrolysis involving an S_N2 *mechanism* converts (+)-α-bromopropionic acid into L(+)-lactic acid. Hence (+)-α-bromopropionic acid must have a D-configuration. This acid also reacts with sodium azide in an S_N2 *reaction* to give L(−)-α-azidopropionic acid, which is reduced catalytically in the presence of *Adams'* catalyst to (+)-alanine. Since no *Walden* inversion is involved in this reduction, (+)-alanine must also have an L-configuration.

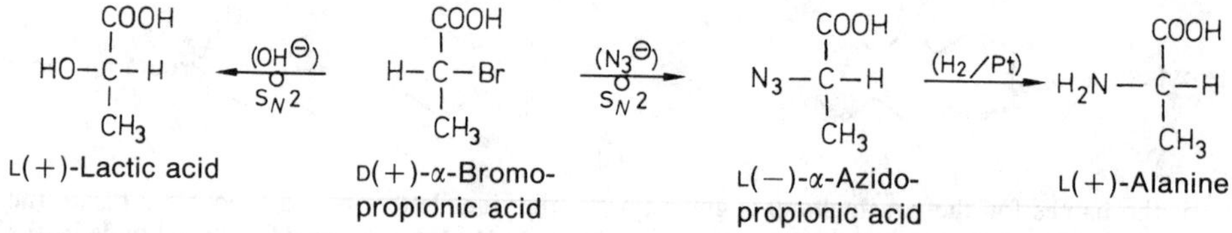

L(+)-Lactic acid   D(+)-α-Bromo-propionic acid   L(−)-α-Azido-propionic acid   L(+)-Alanine

The *enantiomers* of alanine can be represented by the following simplified spatial and projection formulae:

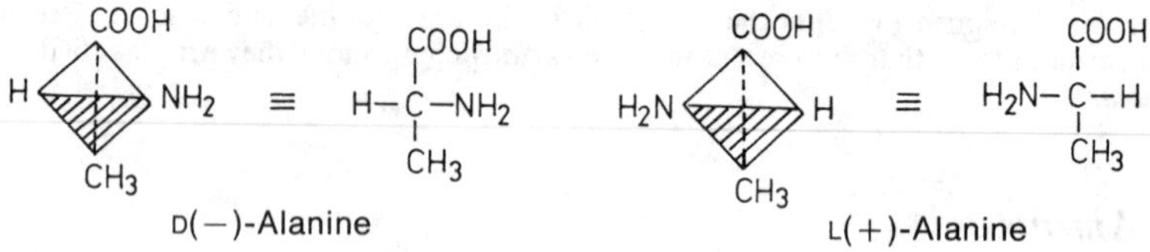

D(−)-Alanine   L(+)-Alanine

It should be noted here that by convention the carboxyl group next to the asymmetric carbon atom in amino acids is written at the top.

In the *CIP* system the priorities lie in the order $NH_2 > COOH > CH_3 > H$, and the molecules have the following three-dimensional structures:

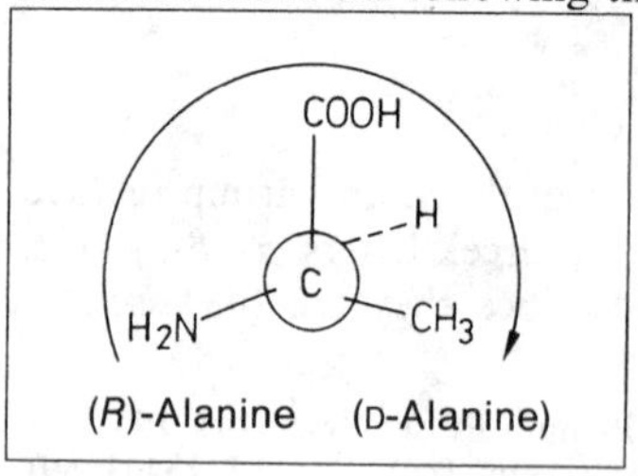

(R)-Alanine   (D-Alanine)

(S)-Alanine   (L-Alanine)

Fig. 90. 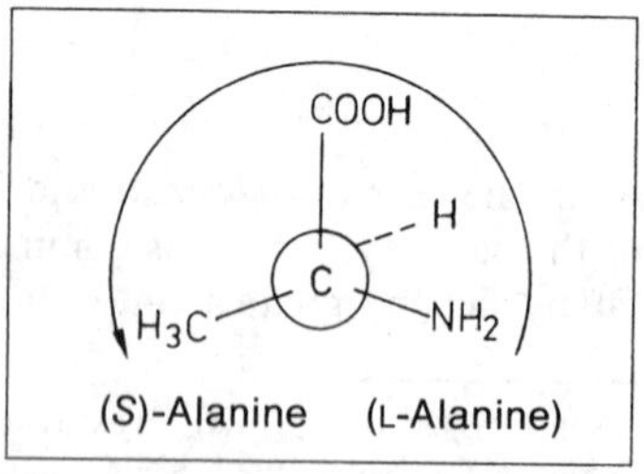   Fig. 91.

Preparation

2.29.7.2 α-Amino acids can be obtained by treating α-*halogenoacids with excess conc. ammonia*:

$$R-\underset{\underset{Cl}{|}}{CH}-C\overset{\nearrow O}{\searrow_{OH}} \;+\; 2NH_3 \longrightarrow R-\underset{\underset{NH_2}{|}}{CH}-C\overset{\nearrow O}{\searrow_{OH}} \;+\; NH_4Cl$$

2.29.7.3 *Strecker* Synthesis (1850). Addition of hydrogen cyanide to aldehydes provides *cyanohydrins*, which react with ammonia to give *aminonitriles* and these are in turn hydrolysed by conc. mineral acids to α-amino acids:

$$R-C\overset{\nearrow H}{\searrow_O} \xrightarrow{+HCN} R-\underset{CN}{\overset{H}{C}}-OH \xrightarrow[-H_2O]{+NH_3} R-\underset{CN}{\overset{H}{C}}-NH_2 \xrightarrow[-NH_3]{+2H_2O} R-\underset{\underset{NH_2}{|}}{CH}-COOH$$

Cyanohydrin   Aminonitrile

Properties

Amino acids are crystalline, relatively high-melting substances. Most have a sweetish taste. They show the properties of both bases and acids, and form salts with both mineral acids and alkalis, *e.g.*

$$\underset{\underset{NH_2}{|}}{CH_2}-C\overset{\nearrow \overline{O}|}{\searrow_{\underline{O}H}} \;+\; HCl \longrightarrow \left[\underset{\underset{H-\overset{\oplus}{N}H_2}{|}}{CH_2}-C\overset{\nearrow \overline{O}|}{\searrow_{\underline{O}H}}\right] Cl^{\ominus}$$

Glycine   Glycine hydrochloride

$$\underset{\underset{NH_2}{|}}{CH_2}-C\overset{\nearrow \overline{O}|}{\searrow_{\underline{O}H}} \;+\; NaOH \longrightarrow \left[\underset{\underset{NH_2}{|}}{CH_2}-C\overset{\nearrow \overline{O}|}{\searrow_{\underline{O}^{\ominus}}}\right] Na^{\oplus} + H_2O$$

Sodium salt of glycine

Compounds which can both accept and give up a proton, forming cations in acid solutions and anions in basic solutions, are known as *amphoteric* electrolytes or *ampholytes*.

In aqueous solutions amino acids are either neutral or very weakly acidic, and the following *equilibria* are set up:

$$\underset{\underset{NH_2}{|}}{CH_2}-C\overset{\nearrow \overline{O}|}{\searrow_{O^{\ominus}}} \underset{-H^{\oplus}}{\overset{+H^{\oplus}}{\rightleftarrows}} \underset{\underset{H-\overset{\oplus}{N}H_2}{|}}{CH_2}-C\overset{\nearrow \overline{O}|}{\searrow_{O^{\ominus}}} \underset{-H^{\oplus}}{\overset{+H^{\oplus}}{\rightleftarrows}} \underset{\underset{H-\overset{\oplus}{N}H_2}{|}}{CH_2}-C\overset{\nearrow \overline{O}|}{\searrow_{\underline{O}-H}}$$

Zwitterion

Because of this, amino acids exist predominantly as dipolar *inner salts* or *zwitterions* (Ger. *zwitter* = hybrid). These zwitterions can be formed by transferring a proton from the carboxyl group onto the lone pair of electrons of the nitrogen atom in the amino group. In the crystalline state amino acids also exist as zwitterions. The uncharged form is not found; it is just a convenient way to write the formula.

The molecules, $H_3N^{\oplus}-CHR-COO^{\ominus}$ bear no overall charge. In electrical fields they align themselves in the same way that magnets do in a magnetic field. In a pure aqueous solution in which they exist almost entirely as zwitterions, they do not migrate if an electric field is applied. This happens at their *isoelectric point*, a specific pH value, at which the concentration of the dipolar form is at a maximum, and at which the amino acids show their lowest solubility in water.

Like amines, amino acids can form quaternary ammonium salts. These only exist in the form of betaines. A prime example is trimethylglycine or betaine, which occurs in sugar beet (*Beta vulgaris*). It can also be made by warming an aqueous solution of chloroacetic acid with trimethylamine:

$$(H_3C)_3N| + Cl-CH_2-COOH \longrightarrow (H_3C)_3\overset{\oplus}{N}-CH_2-C\overset{\overline{O}|}{\underset{O_\ominus}{\diagdown}} + HCl$$

Betaine

All other compounds $R_3^{\oplus}N-CH(R')-COO^{\ominus}$, which have three alkyl groups attached to the nitrogen atom, are called *betaines*, deriving this general name from the name of betaine.

Betaine hydrazide chlorides, $(H_3C)_3^{\oplus}N-CH_2-CO-NH-NH_2[Cl^{\ominus}]$, react with carbonyl compounds to form normal hydrazones, which, because of the hydrophilic ammonium group, are soluble in water. These hydrazide chlorides are known as *Girard's* reagents. They have been used to separate naturally occurring ketones, *e.g.* certain steroids, from accompanying materials which are insoluble in water.

The salt-like character of amino acids, which was mentioned above, is evident from the following physical properties. They have relatively high melting points (230°C [500 K] and higher), they cannot be distilled without decomposing, and they dissolve at least to some extent in water, with more difficulty in ethanol, but not at all in ether or other organic solvents.

2.29.7.4 Ninhydrin Reaction. When amino acids are heated with an aqueous solution of ninhydrin an intense blue or blue-violet colour develops. This colour test is very sensitive but not specific. The mechanism of this reaction will be referred to later (see p. 624).

Amino acids can be esterified. The acid group is lost in the process so that the resultant esters are basic in character.

$$\underset{NH_2}{\overset{|}{CH_2}}-COOH + H-OC_2H_5 \underset{\longleftarrow}{\overset{(H^{\oplus})}{\longrightarrow}} \underset{NH_2}{\overset{|}{CH_2}}-COOC_2H_5 + H_2O$$

Glycine   Glycine ethyl ester

When a dry amino acid is heated with soda-lime, it undergoes decarboxylation and is converted into an amine:

$$R-\underset{\underset{NH_2}{|}}{CH}-COOH \xrightarrow[-CO_2]{(NaOH, CaO)} R-CH_2-NH_2$$

The amino groups of amino acids react with acetyl chloride or acetic anhydride to give *N-acetylamino acids*:

$$R-\underset{\underset{NH_2}{|}}{CH}-COOH + (H_3CCO)_2O \longrightarrow R-\underset{\underset{HN-COCH_3}{|}}{CH}-COOH + H_3CCOOH$$

Quantitative estimations of α-amino acids are carried out using the method of *van Slyke*, in which addition of nitrous acid leads to the evolution of nitrogen, which can be measured volumetrically. The amino acids are converted into α-hydroxyacids:

$$R-\underset{\underset{NH_2}{|}}{CH}-COOH \quad + NaNO_2 + HCl \quad \longrightarrow \quad R-\underset{\underset{OH}{|}}{CH}-COOH \quad + NaCl + N_2 + H_2O$$

2.29.7.5 Diazoacetic ester. Unlike the free amino acids, their esters react with cold nitrous acid to give relatively stable *diazoesters* (*Curtius*, 1883). For example, glycine ethyl ester reacts with sodium nitrite plus one molecule of hydrochloric acid at 0–3° (273–276 K) to give diazoacetic ester, which is *stabilised by mesomerism* (for the reaction mechanism, see p. 161):

$$\underset{\underset{NH_2}{|}}{CH_2}-COOR \quad \xrightarrow[-NaCl, \ -H_2O]{+NaNO_2 + HCl} \quad \left[\underset{\underset{N=N-OH}{}}{CH_2-COOR} \right]$$

Glycine ethyl ester (R = C$_2$H$_5$)

$$-H_2O \quad \left[\overset{\ominus}{I}\underset{\underset{\underset{N}{|||}}{\overset{\oplus}{N}}}{CH}-C\overset{\overset{\overline{O}I}{\nearrow}}{\diagdown}_{OR} \quad \longleftrightarrow \quad \underset{\underset{\underset{INI^{\ominus}}{|||}}{\overset{N}{\parallel}}}{CH}-\overset{\overset{\overline{O}I}{\nearrow}}{\underset{\oplus}{C}}\diagdown_{OR} \quad \longleftrightarrow \quad \underset{\underset{\underset{N}{|||}}{\overset{\oplus}{N}}}{CH}=C\overset{\overset{\overset{\ominus}{\overline{O}I}}{\nearrow}}{\diagdown}_{OR} \right]$$

Diazoacetic ester

Diazoacetic ester is a yellow oil which boils at 45°C (318 K)/16 mbar (16 hPa). It dissolves in ethanol and in ether but is insoluble in water. In the presence of dilute acids, the lone pair of electrons on the carbon atom of diazoacetic ester take up a proton to form a diazonium salt, which immediately loses nitrogen. The resultant carbenium ion stabilises itself by addition of a hydroxy group or of some other anion which is present, being converted thereby into an α-substituted carboxylic acid ester:

$$\underset{\underset{\oplus N\equiv NI}{}}{\overset{\ominus}{I}CH}-COOR \quad \xrightarrow{+H^{\oplus}X^{\ominus}} \quad \left[\underset{\underset{\oplus N\equiv NI}{}}{H-CH}-COOR \right] X^{\ominus} \quad \xrightarrow{-N_2} \quad \left[H-\overset{\oplus}{CH}-COOR \right] X^{\ominus} \quad \longrightarrow$$

$$\underset{\underset{X}{|}}{H-CH}-COOR \qquad R = C_2H_5, \quad X = OH, \ Cl, \ OCOCH_3 \ etc.$$

When α-, β-, γ-, δ-, ε- or other amino acids are heated, similar reactions to those of the corresponding halogeno- or hydroxy-acid ensue.

α-Amino acids lose water to form cyclic anhydrides such as *2,5-dioxopiperazines* (2,5-diketopiperazines), *e.g.*

$$\underset{O}{\overset{HO}{\diagdown}}C\diagdown_{CH_2-NH_2} \quad + \quad H_2N-CH_2\diagdown_{C}\overset{\overset{O}{\nearrow}}{\diagdown}_{OH} \quad \xrightarrow{-2\,H_2O} \quad$$

Glycine

2,5-Dioxopiperazine
(Piperazine-2,5-dione)

Higher yields are obtained by cyclising amino acid esters.

When melted, *β-amino acids* lose ammonia to form *α,β-unsaturated carboxylic acids*, *e.g.*

$$H_2N-CH_2-CH_2-COOH \quad \underset{+NH_3}{\overset{-NH_3}{\rightleftharpoons}} \quad H_2C=CH-COOH$$

β-Aminopropionic acid   Acrylic acid

The reverse process provides a way for making β-amino acids.

When γ- or *δ-amino acids* are heated they lose water intramolecularly to form lactams. By analogy with lactones these are known as γ- or *δ-lactams, e.g.*

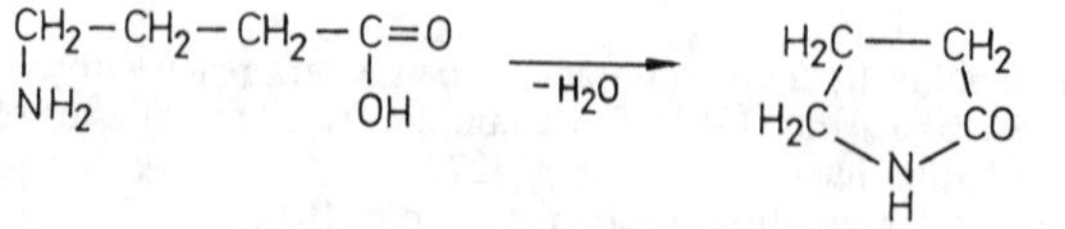

γ-Aminobutyric acid   γ-Butyrolactam

γ- and δ-lactams are reconverted into amino acids when warmed with aqueous acid or alkali.

Amino acids in which the amino group is even further away from the carboxyl group undergo polycondensation reactions giving high molecular weight polyamide fibres. The most important starting material for this process is ε-aminocaproic acid (6-aminohexanoic acid) or the corresponding ε-caprolactam (hexahydro-2*H*-azepin-2-one), which is made industrially from phenol or benzene by the following reaction sequences:

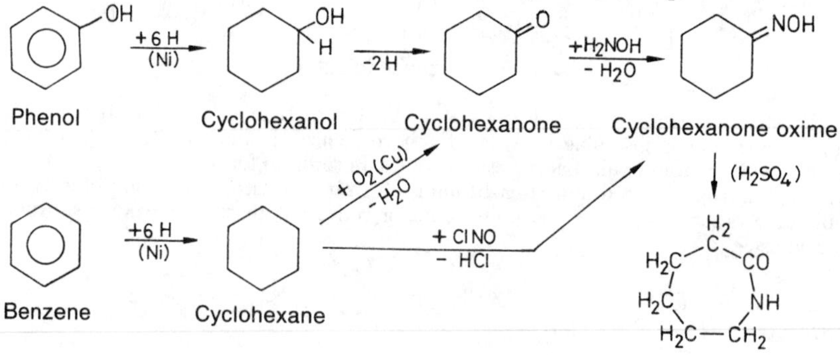

Phenol   Cyclohexanol   Cyclohexanone   Cyclohexanone oxime

Benzene   Cyclohexane

ε-Caprolactam

The ring-enlargement of cyclohexanone oxime to give ε-caprolactam is brought about by means of fuming sulphuric acid, and is an example of a *Beckmann* rearrangement (see p. 544).

The polycondensation of ε-caprolactam proceeds in the presence of a small amount of water. In the first step ε-aminocaproic acid is formed; this reacts with more of the lactam in a ring-opening chain reaction, providing an open-chain polymer (*P. Schlack*, 1938):

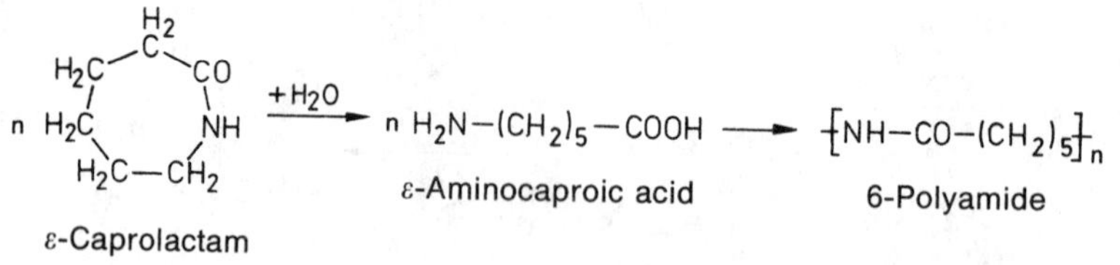

ε-Caprolactam   ε-Aminocaproic acid   6-Polyamide

These synthetic *polyamide fibres* (Perlon, Dederon, 6-Polyamide) are spun from a melt of the polymer. The NH-groups are separated from one another by six carbon atoms. Polyamides are relatively stable to alkali, but less so to acids. Like other synthetic fibres, their strength is increased by stretching them very strongly (cold-drawing). This serves to orientate the molecules along the axis of the fibre (crystallite formation).

A polyamide fibre derived from plant sources has been developed in France under the name *Rilsan.* The starting material is undecylenic acid (undec-10-enoic acid), which is obtained from castor oil. This is converted into 11-aminoundecanoic acid, $H_2N—(CH_2)_{10}—COOH$, whose polycondensation provides a polyamide fibre (11-polyamide) of high quality.

2.29.8 The Most Important Aliphatic Amino Acids

2.29.8.1 Glycine, $H_2N—CH_2—COOH$, was obtained by *Braconnot* (1819) by acid or alkaline hydrolysis of glue. It is formed as a breakdown product from almost all proteins, m.p. 232–236°C (505–509 K) (decomp.).

2.29.8.2 Sarcosine (*N*-methylglycine), $H_3C—NH—CH_2—COOH$, was obtained by *Liebig* from creatine which occurs in meat broth. If sarcosine is treated with oleyl chloride a detergent with the formula

$$H_3C—(CH_2)_7—CH=CH—(CH_2)_2—CO—\underset{\underset{CH_3}{|}}{N}—CH_2—COO^{\ominus} \quad Na^{\oplus}$$

is produced (Medialan), which does not form an insoluble calcium salt.

2.29.8.3 Hippuric acid (*N*-benzoylglycine) was first isolated by *Liebig* (1829) from horses' urine (Grk. *hippos* = horse, *uron* = urine). It can be made by benzoylating glycine.

$$C_6H_5—C\overset{\diagup O}{\diagdown Cl} \quad + \quad H_2N—CH_2—COOH \longrightarrow C_6H_5—\underset{\underset{O}{\|}}{C}—NH—CH_2—COOH \quad + HCl$$

Benzoyl chloride   Glycine   Hippuric acid

If chloroacetic acid is allowed to react with a small amount of ammonia, as well as glycine, *iminodiacetic acid*, $HN(CH_2—COOH)_2$, and *nitrilotriacetic acid*, $N(CH_2—COOH)_3$ are formed. The latter compound forms stable water-soluble complexes with calcium or magnesium cations. In the form of its sodium salt (Trilon A, NTA) it is used as a partial replacement for phosphates in washing and cleaning materials.

The corresponding complexes of *ethylenediaminetetraacetic acid* (*EDTA*) (INN: **Edetic acid**) are much more stable. This is made by alkaline hydrolysis of the related tetranitrile which is itself obtained, using the *Strecker* synthesis, from formaldehyde and hydrogen cyanide in the presence of ethylenediamine.

$$\begin{array}{c} HOOC—H_2C \\ HOOC—H_2C \end{array} N—CH_2—CH_2—N \begin{array}{c} CH_2—COOH \\ CH_2—COOH \end{array}$$

Ethylenediaminetetraacetic acid (EDTA)

Its disodium salt has industrial use.

Schwarzenbach described such acids, which have a strong tendency to form complexes, as *complexones*. They find use as titrating agents, as colorimetric reagents and in polarography. They are also used commercially for softening water and in dyeing processes for complexing undesirable metal ions.

2.29.8.4 L(+)-Alanine, α-aminopropionic acid, $H_3C—CH(NH_2)—COOH$, is obtained from the hydrolysis of silk fibres, D(−)-alanine by resolution of racemic DL-alanine.

L(+)-Alanine has the same configuration as L(+)-lactic acid. It melts, with decomposition, at about 297°C (570 K).

2.29.8.5 β-Alanine, β-aminopropionic acid, $H_2N—CH_2—CH_2—COOH$, is one of the naturally occurring β-amino acids; they are rather less common than α-amino acids. It is involved in the formation of *pantothenic acid*, which is one of the water-soluble vitamins of the vitamin B complex, and is widely distributed as a growth factor (*R. J. Williams*, 1933). Only the D(+)-pantothenic acid is biologically active. It occurs especially in yeast, and is also present in the liver; it is a precursor in the formation of *coenzyme A*.

The constitution of pantothenic acid was proved by its synthesis from α-hydroxy-β,β-dimethyl-γ-butyrolactone (pantolactone), (the lactone of 2,4-dihydroxy-3,3-dimethylbutanoic acid) and β-alanine (*R. Kuhn* and *T. Wieland*, 1940):

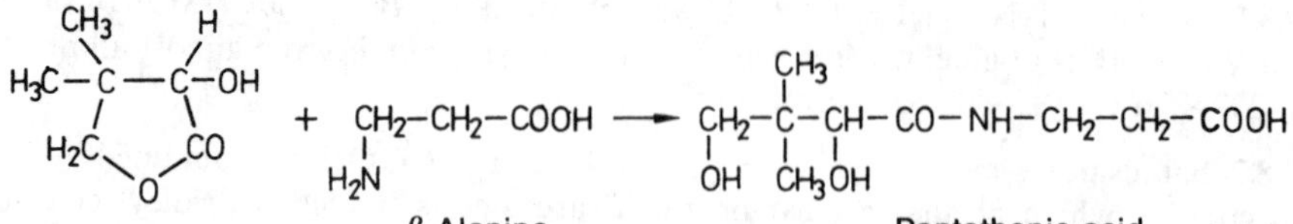

β-Alanine   Pantothenic acid

Lack of pantothenic acid leads in chicken to symptoms resembling those of pellagra, and in rats to greying of their coats; the effects on human beings are not known.

2.29.8.6 L(+)-**Valine** occurs very widely, but usually only in small amounts, in proteins. The enzymes present in yeast used in alcoholic fermentation cause any valine to be converted into isobutanol, which forms part of the fusel oil:

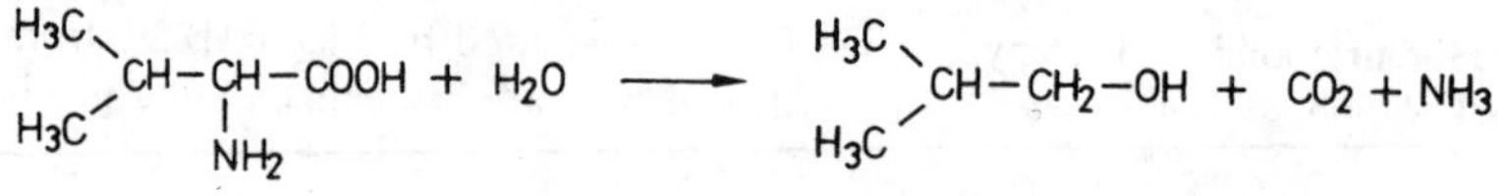

2.29.8.7 L(−)-**Leucine** is one of the most important constituents of albumens, for example of casein and of horns, and is produced when they are hydrolysed by either acid or alkali, as well as in the putrefaction of proteins.

2.29.8.8 L(+)-**Isoleucine**, (2S, 3S)-(+)-2-amino-3-methylpentanoic acid, occurs in proteins and also in molasses. It has two asymmetric carbon atoms and hence exists in four distinct optically active forms; in the racemisation of L-(+)-isoleucine, which only involves the 2-carbon atom, D(−)-alloisoleucine, (2R, 3S)-(−)-2-amino-3-methylpentanoic acid is formed.

During alcoholic fermentation L(−)-leucine and L(+)-isoleucine give rise, respectively, to *3-methylbutan-1-ol* and *2-methylbutan-1-ol*, which are constituents of so-called 'fermentation amyl alcohol':

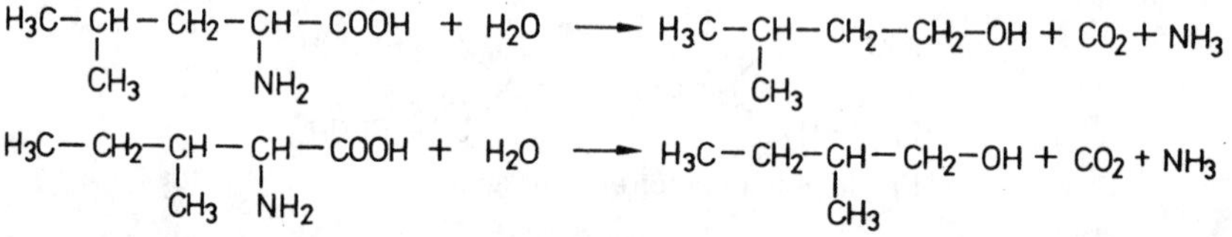

This shows that the by-products in alcoholic fermentation which make up fusel oil do not arise from sugars, but from amino acids, which form the structural units of proteins that are also present.

A few of the about 500 amino acids[1] presently known to exist in nature are dealt with in the later section 'Amino acids, Peptides and Proteins'. It should also be mentioned that carbon-containing meteorites often contain α-aminoisobutyric acid, $(H_3C)_2-C-COOH$ and isovaline C_2H_5

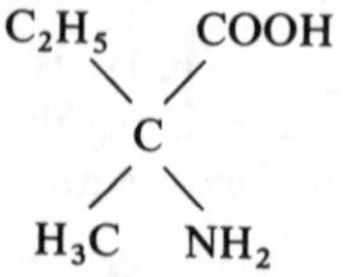

[1] See *I. Wagner* and *H. Musso*, New Naturally Occurring Aminoacids, Angew. Chem. Int. Ed. Engl. **22**, 816 (1983).

2.30 Aliphatic Aldehydoacids and Ketoacids

Ketoacids are of greater importance than aldehydoacids. Both can be considered as oxidation products of hydroxyacids.

2.30.1 Aldehydoacids

Glyoxylic acid is the simplest aldehydo acid. It is widely distributed in nature and is found especially in unripe fruit such as gooseberries and red currants; when they ripen it disappears. When it crystallises out from water it does so in the form of a hydrate (whose structure provides an exception to *Erlenmeyer's* rule).

$$
\begin{array}{cc}
\underset{O}{\overset{H}{\diagdown}}\!C-COOH & \underset{HO}{\overset{HO}{\diagdown}}\!CH-COOH \\[2pt]
\text{Glyoxylic acid} & \text{Hydrate (Dihydroxyacetic acid)}
\end{array}
$$

Preparation

2.30.1.1 The best method is by *hydrolysis of dichloroacetic acid*:

$$
Cl_2HC-COOH \xrightarrow[-2HCl]{+2\,H_2O} (HO)_2HC-COOH
$$

2.30.1.2 By electrolytic reduction of a solution of oxalic acid in sulphuric acid:

$$
\underset{\text{Oxalic acid}}{HOOC-COOH} \xrightarrow{+2H} \underset{\text{Glyoxylic acid hydrate}}{(HO)_2HC-COOH}
$$

The anhydrous form of glyoxylic acid can be obtained as a syrup by vacuum distillation of the hydrate over phosphorus pentoxide.

Properties

Glyoxylic acid behaves both as an acid and as an aldehyde. When heated with alkali it undergoes a *Cannizzaro* reaction to give glycollic acid and oxalic acid:

$$
2\;\underset{\underset{COOH}{|}}{\overset{H}{\overset{|}{C\!\diagup\!\!\diagdown O}}} + H_2O \longrightarrow \underset{\underset{COOH}{|}}{CH_2OH} + \underset{\underset{COOH}{|}}{COOH}
$$

Glyoxylic acid   Glycollic acid Oxalic acid

The next homologue, *formylacetic acid*, cannot be isolated as such. Its esters can be prepared by a *Claisen* condensation reaction between esters of formic and acetic acid in ethereal solution, in the presence of a sodium alkoxide; they are only stable as their sodium salts. The acidic character of formylacetic esters is a consequence of *keto–enol tautomerism*:

$$
H-C\!\!\overset{O}{\underset{OR}{\diagdown}} + H_3C-C\!\!\overset{O}{\underset{OR}{\diagdown}} \xrightarrow[-ROH]{(NaOR)} \left[\underset{O}{\overset{H}{\diagdown}}C-CH_2-C\!\!\overset{O}{\underset{OR}{\diagdown}} \rightleftharpoons \underset{HO}{\overset{H}{\diagdown}}C=CH-C\!\!\overset{O}{\underset{OR}{\diagdown}} \right]
$$

Formate ester Acetate ester   Formylacetic ester   (Hydroxymethylene form)

Ketoacids

According to the site of the oxo group with respect to the acid group, ketoacids may be classified as α-, β-, γ-, δ-, *etc.* ketoacids.

2.30.2 α-Ketoacids

In contrast to aldehydoacids, α-ketoacids are stable in the free state.

Pyruvic acid (α-ketopropionic acid, 2-oxopropionic acid), $H_3C-C-COOH$, is the simplest and also the most important α-ketoacid.

Preparation

2.30.2.1 Pyruvic acid was first prepared by *distilling tartaric acid with potassium hydrogen sulphate* (*Berzelius*, 1835). Reaction proceeds *via* the following intermediates:

$$\begin{array}{c} CH(OH)-COOH \\ | \\ CH(OH)-COOH \end{array} \xrightarrow[-H_2O]{(KHSO_4)} \left[\begin{array}{c} CH-COOH \\ \| \\ C(OH)-COOH \end{array} \rightleftharpoons \begin{array}{c} CH_2-COOH \\ | \\ CO-COOH \end{array}\right] \xrightarrow{-CO_2} \begin{array}{c} CH_3 \\ | \\ CO-COOH \end{array}$$

| Tartaric acid | Hydroxymaleic acid | Oxaloacetic acid | Pyruvic acid |

This still remains the best way to prepare pyruvic acid.

2.30.2.2 α-Ketoacids can also be obtained by *oxidation of α-hydroxyacids* or by *hydrolysis of acyl cyanides*, which are themselves prepared by the reaction of alkali metal cyanides with acyl chlorides, *e.g.*

$$H_3C-\overset{\overset{\displaystyle O}{\|}}{C}-CN \; + \; 2H_2O \longrightarrow H_3C-\overset{\overset{\displaystyle O}{\|}}{C}-COOH \; + \; NH_3$$

| Acetyl cyanide | Pyruvic acid |

Properties

Pyruvic acid is a colourless liquid which is miscible with water and smells like acetic acid. It boils at 165°C (438 K) with slight decomposition. Its acidity ($pK_a = 2.49$) is, like that of lactic acid ($pK_a = 3.86$), higher than that of propionic acid ($pK_a = 4.87$). It reacts not only as an acid but also as a ketone; for example, it forms an oxime and a phenylhydrazone. Its salts are known as *pyruvates*.

On reduction with 'nascent hydrogen' (sodium amalgam) pyruvic acid is converted into DL-lactic acid. When heated with dilute sulphuric acid it decomposes to give acetaldehyde and carbon dioxide:

$$H_3C-\overset{\overset{\displaystyle }{}}{\underset{\overset{\displaystyle \|}{O}}{C}}-COOH \xrightarrow{(dil.\ H_2SO_4)} H_3C-C\overset{\nearrow H}{\searrow_O} \; + \; CO_2$$

The enzyme *pyruvate-decarboxylase* promotes decarboxylation at room temperature. This biologically important reaction emphasises the central role of pyruvic acid as an intermediate product in the breakdown of carbohydrates and of fats.

A typical reaction of α-ketoacids, called *decarbonylation*, takes place when they are warmed with conc. sulphuric acid; for example, pyruvic acid loses carbon monoxide and is converted into acetic acid:

$$H_3C-\underset{\underset{O}{\|}}{C}-COOH \xrightarrow{(conc.\ H_2SO_4)} H_3C-COOH + CO$$

2.30.3 β-Ketoacids

Unlike α-ketoacids, β-ketoacids are not stable and can only be isolated in the free state if suitable precautions are taken.

Acetylacetic acid, $H_3C-CO-CH_2-COOH$, the simplest β-ketoacid, decomposes readily to acetone and carbon dioxide:

$$H_3C-\underset{\underset{O}{\|}}{C}-CH_2-COOH \longrightarrow H_3C-\underset{\underset{O}{\|}}{C}-CH_3 + CO_2$$

It occurs, together with L-β-hydroxybutyric acid and acetone, in appreciable quantity in the urine of diabetics.

Its ethyl ester, *ethyl acetoacetate*, is more important than the acid itself. It exists in two *tautomeric* forms.

2.30.4 Keto–enol Tautomerism (Oxo–enol Tautomerism)

Tautomerism (*Butlerov*, 1877, *Laar*, 1885) describes the situation whereby two isomeric compounds differ from one another by interchanging the positions of σ- and π-bonds; these compounds are in equilibrium with each other. Usually two *tautomers* also differ from one another in the position to which a hydrogen atom is attached.

The concept of tautomerism has been mentioned earlier when discussing nitroalkanes (nitro–aci–nitro–tautomerism) and formylacetic esters. Experimental support for the concept of tautomerism was provided by *ethyl acetoacetate*, which was discovered by *Geuther* (1863). Its physical properties and chemical reactivity indicated that it consisted of a mixture of two isomeric compounds. There was a heated argument between *Geuther* and *Frankland* about the 'correct' structural formula of this ester, which was resolved by L. Knorr in 1911, when he suggested that under normal conditions the following equilibrium was involved:

$$H_3C-\underset{\underset{O}{\|}}{C}-CH_2-C\overset{\diagup O}{\underset{\diagdown OC_2H_5}{}} \rightleftharpoons H_3C-\underset{\underset{OH}{|}}{C}=CH-C\overset{\diagup O}{\underset{\diagdown OC_2H_5}{}}$$

Keto-Form 92,5%   **Enol**-Form 7,5%

Ethyl acetoacetate

In consequence, ethyl acetoacetate (the *keto* form) can rearrange into ethyl β-hydroxycrotonate and *vice versa*. It has been said to have 'two-fold reactivity' (*Nesme-*

yanov). When two such forms are in equilibrium with one another, the general term used to describe the situation is *keto–enol tautomerism*.[1].

The ketoester was obtained in crystalline form by *Knorr* by cooling an ethereal solution of the normal mixture to $-78°C$ (195 K). The enol form can be isolated as an oil by passing dry hydrogen chloride through a suspension of the sodium salt of ethyl acetoacetate in dimethyl ether at $-78°C$ (195 K). At room temperature each tautomer quickly gives rise to the normal equilibrium mixture of the two forms. These rearrangements are catalysed by traces of acid or base.

A *qualitative test* for the presence of the *enol form* is provided by adding a solution of iron(III) chloride, which forms an iron(III)-enolate complex. This colour reaction is given by almost all compounds which contain the group of atoms $> C{=}C{-}OH$.

A *quantitative* estimate of the amount of enol present can be obtained by a rapid titration (*K. H. Meyer*, 1912). The equilibrium mixture is treated with an excess of bromine at $-10°C$ (260 K). After a short time, unreacted bromine is removed by addition of β-naphthol. The dibromo adduct formed by addition of bromine to the $C{=}C$ double bond of the enol loses hydrogen bromide to give ethyl α-bromoacetoacetate. α-Bromoketo compounds can be reduced by hydrogen iodide, so the solution of the α-bromoester is acidified, potassium iodide is added, and the iodine which is liberated is titrated with sodium thiosulphate solution.

$$H_3C-\underset{\underset{OH}{|}}{C}{=}CH-CO_2C_2H_5 \xrightarrow{+Br_2} H_3C-\underset{\underset{HO}{}}{\overset{}{C}}-\underset{\underset{Br}{|}}{\overset{}{C}}H-CO_2C_2H_5 \xrightarrow{-HBr} H_3C-\underset{\underset{O}{||}}{C}-\underset{\underset{Br}{|}}{C}H-CO_2C_2H_5$$

Ethyl acetoacetate (enol form)   Dibromo-adduct   Ethyl α-bromo-acetoacetate

$$H_3C-\underset{\underset{O}{||}}{C}-\underset{\underset{Br}{|}}{C}H-CO_2C_2H_5 + 2\,HI \longrightarrow H_3C-\underset{\underset{O}{||}}{C}-CH_2-CO_2C_2H_5 + I_2 + HBr$$

Since the rate of addition of bromine is markedly faster than keto–enol interchange under the above conditions, the accuracy of this method of determination is satisfactory. Keto–enol equilibration can nowadays be investigated by physical methods such as NMR spectroscopy.

Simple aldehydes and ketones containing the grouping $-CO-CH_2-R$ exist almost entirely in the keto form; for example, the enol content of acetone is only $2.5 \times 10^{-4}\%$. In 1,3-dicarbonyl compounds, $R-CO-CH_2-CO-R$, however, the hydrogen atoms attached to the central carbon atom are more acidic, thanks to the $-I$-*effects* of the adjacent carbonyl groups. Hence *prototropy* can occur readily in these compounds, although there is a significant activation energy for the process.

In these prototropic rearrangements of ethyl acetoacetate, loss of a proton leads to the formation of an *anion*, which is stabilised by *mesomerism* and in which *the negative charge is delocalised*. In the reverse reaction a proton can add either to the oxygen atom to give an enol, or to the carbon atom to give the keto form:

$$H_3C-\underset{\underset{O}{||}}{C}-CH_2-CO_2C_2H_5 \underset{+H^\oplus}{\overset{-H^\oplus}{\rightleftharpoons}} \left[\begin{array}{c} H_3C-\underset{\underset{|Ol}{||}}{C}-\overset{\ominus}{C}H-CO_2C_2H_5 \\ \updownarrow \\ H_3C-\underset{\underset{|Ol^\ominus}{|}}{C}{=}CH-CO_2C_2H_5 \end{array} \right] \underset{-H^\oplus}{\overset{+H^\oplus}{\rightleftharpoons}} H_3C-\underset{\underset{OH}{|}}{C}{=}CH-CO_2C_2H_5$$

Keto ester   Ambident anion   Enol ester

---

[1] For a general classification of tautomerism see *N. S. Zefirov* et al., J. Org. Chem. USSR (Engl. Transl.) *12*, 695 (1976).

Hence in keto–enol tautomerism one is considering the equilibria between two acids and their common ambient anion. The positions of the equilibria depend very much upon the solvent as well as on the temperature and the concentration of the solution. The enol content is higher in hydrophobic, relatively non-polar solvents than in water or in alcohols. Thus for ethyl acetoacetate at 18°C (290 K) the percentage of enol form present is 0.4% in water, 10.5% in ethanol, 16.2% in benzene, 32.4% in carbon disulphide and 46.4% in hexane.

Enolisation is energetically favoured since the enol form, but *not* the keto form, contains a conjugated system of double bonds, and in addition is stabilised by formation of intramolecular hydrogen bonding.

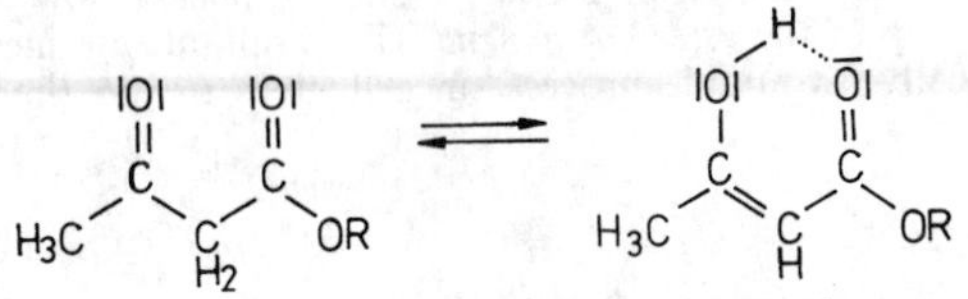

Enol

By carrying out a distillation in quartz apparatus, *K. H. Meyer* showed that the enol ester boils at a lower temperature than the keto ester. The greater volatility of the enol is due to the chelation resulting from the hydrogen bonding.

Like other acids, acetoacetic esters react with sodium with evolution of hydrogen and formation of a sodium salt, whose anions are *mesomeric*:

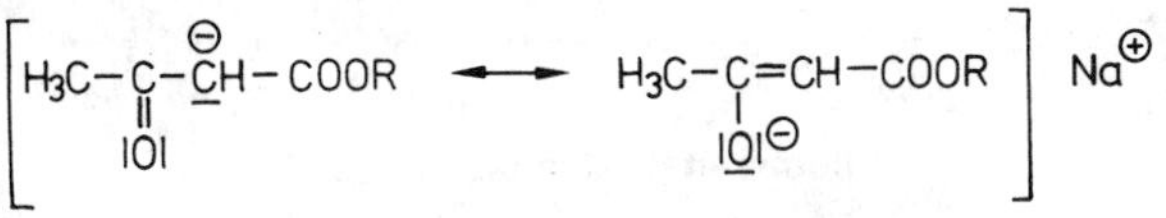

Ethyl acetoacetate, sodium salt

On addition of acid the ambient anion reacts first of all to give the enol ester, but after a short time the usual keto–enol equilibrium is set up.

2.30.5 *Preparation of Acetoacetic Esters*

2.30.5.1 Industrially by the reaction of diketene with ethanol:

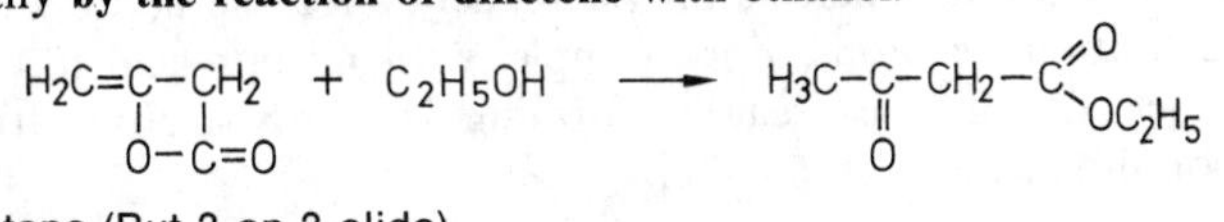

Diketene (But-3-en-3-olide)

2.30.5.2 *Claisen* Ester Condensation[1] (1887). By condensation of two molecules of ethyl acetate under the influence of alkaline reagents such as sodium ethoxide, metallic sodium or sodium amide:

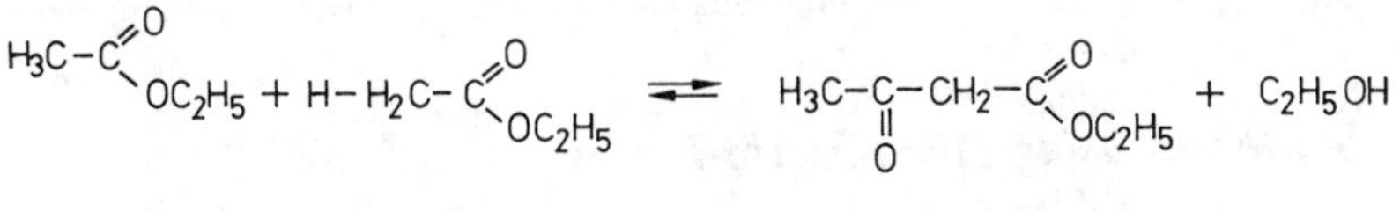

The yield of the resultant keto ester can be raised by distilling off the ethanol which is formed, thus shifting the equilibrium to favour formation of the product.

---

[1] See *S. Patai*, The Chemistry of the Carbonyl Group, pp. 567 *ff.* (Interscience Publishers, London 1966).

The mechanism of the reaction makes it clear why in ester condensation, as in aldol addition, the presence of the strongly basic ethoxide or amide anions promotes reaction by removing a proton:

$$H-H_2C-C\overset{O}{\underset{OC_2H_5}{}} + \overset{\ominus}{I}\overline{O}C_2H_5 \;\rightleftharpoons\; \left[H_2\overset{\ominus}{C}-C\overset{\overline{O}I}{\underset{OC_2H_5}{}} \;\longleftrightarrow\; H_2C=C\overset{O^{\ominus}}{\underset{OC_2H_5}{}} \right] + C_2H_5OH$$

The mesomerically stabilised anion of the *methylene component* now acts as a *nucleophile* and attacks the carbonyl group of the *ester component*. The resultant intermediate loses a molecule of ethanol in an energetically favourable process since it gives rise to the mesomeric anion of the acetoacetic ester:

$$H_3C-\overset{IOI}{\underset{OC_2H_5}{C}} \;+\; \left[\overset{\ominus}{I}\overset{H}{\underset{H}{C}}-C\overset{O}{\underset{OC_2H_5}{}} \right]Na^{\oplus} \;\rightleftharpoons\; \left[H_3C-\overset{IOI^{\ominus}}{\underset{OC_2H_5}{C}}\overset{H}{\underset{H}{-}}C-C\overset{O}{\underset{OC_2H_5}{}} \right]Na^{\oplus}$$

Ester	Methylene	Intermediate
component	component	

$$\xrightarrow{-C_2H_5OH} \left[H_3C-\overset{IOI}{C}-\overset{H}{\underset{\ominus}{C}}-C\overset{O}{\underset{OC_2H_5}{}} \;\longleftrightarrow\; H_3C-\overset{IOI^{\ominus}}{C}=\overset{H}{C}-C\overset{O}{\underset{OC_2H_5}{}} \right]Na^{\oplus}$$

Sodium salt of acetoacetic ester

The final product of the *Claisen* condensation is the sodium salt of the acetoacetic ester; acidification converts this into the ester itself.

Properties

Ethyl acetoacetate ($pK_a = 11$), b.p. 181°C (454 K), is a colourless liquid with a pleasant odour. Its keto group reacts with the usual reagents for carbonyl groups; *e.g.*, with phenylhydrazine it gives a *phenylhydrazone*, which readily cyclises to form 5-methyl-2-phenyl-(1*H*)-pyrazol-3-one. There is also a tendency for the *active methylene group* to undergo condensation reactions. For example, with nitrous acid it gives an *isonitroso* compound which decomposes readily, forming isonitrosoacetone (the monoxime of glyoxal), carbon dioxide and an alcohol:

$$H_3C-\underset{O}{\overset{\|}{C}}-CH_2-COOR \xrightarrow[-H_2O]{+\,O=NOH} H_3C-\underset{O}{\overset{\|}{C}}-\underset{NOH}{\overset{\|}{C}}-COOR \xrightarrow[-ROH,\,-CO_2]{+\,H_2O} H_3C-\underset{O}{\overset{\|}{C}}-\underset{NOH}{\overset{\|}{CH}}$$

Acetoacetic ester	Isonitrosoacetoacetic ester (Hydroxyiminoacetoacetic ester)	Isonitrosoacetone Hydroxyiminoacetone

2.30.6 Syntheses using Acetoacetic Esters

These syntheses are all based on the reactivity of the *ambident anions* of the sodium salts of acetoacetic esters, which can react at either the oxygen atom or the carbon atom with an *electrophile*.

When *alkyl halides* react with the sodium salts of acetoacetic esters α-*C-alkyl derivatives* are formed:

$$\left[\;H_3C-\underset{\underset{|\ominus}{|\text{IOI}}}{\overset{\overset{H}{|}}{C}}=\overset{\overset{H}{|}}{C}-COOR\quad\longleftrightarrow\quad H_3C-\overset{}{\underset{\underset{}{\parallel}\text{IOI}}{C}}-\overset{\overset{H}{|}}{\underset{\ominus}{C}}-COOR\;\right]\;Na^{\oplus}\quad\xrightarrow[-NaI]{+I-R'}\quad H_3C-\underset{\parallel\text{IOI}}{C}-\underset{R'}{\overset{\overset{H}{|}}{C}}-COOR$$

In this reaction, alkyl iodides react faster than alkyl bromides.

Following the substitution of one hydrogen atom of the methylene group by an alkyl group, the process can be repeated to provide a *C,C-dialkyl derivative* of acetoacetic ester.

If one adds successively 1 mol of sodium hydride and 1 mol of butyl lithium to acetoacetic ester, the *dicarbanion* of acetoacetic ester is formed (*Hauser*, 1958):

$$H_3C-\underset{\underset{O}{\parallel}}{C}-CH_2-COOR\quad\xrightarrow[\text{2) BuLi (Hexane)}]{\text{1) Natt (THF)}}\quad H_2\overset{\ominus}{C}-\underset{\underset{O}{\parallel}}{C}-\overset{\ominus}{C}H-COOR\quad\xrightarrow[\text{2) H}_2\text{O}]{\text{1) R'X}}\quad R'CH_2-\underset{\underset{O}{\parallel}}{C}-CH_2-COOR$$

When this dianion reacts with an alkyl halide a *γ*-alkyl derivative is formed.[1]

The reactions of acyl chlorides with acetoacetic ester may lead to the formation of either *C*- or *O*-acyl derivatives, depending upon the reaction conditions, for example, the use of magnesium in benzene gives the *C-acyl derivatives* (*Spassov*, 1937).

$$H_3C-\underset{\underset{O}{\parallel}}{C}-CH_2-COOR\;+\;\underset{Cl}{\overset{O}{\diagdown}}C-R'\quad\xrightarrow[-HCl]{(Mg)}\quad H_3C-\underset{\underset{O}{\parallel}}{C}-\underset{COR'}{\overset{\overset{H}{|}}{C}}-COOR$$

On the other hand, reaction of an acyl chloride with the sodium derivative of acetoacetic ester (or acetoacetic ester itself) in *pyridine* provides an *O-acyl derivative*, which is an enol ester:

$$\left[\;H_3C-\underset{\underset{\text{IOI}}{\parallel}}{C}-\overset{\ominus}{C}H-COOR\quad\longleftrightarrow\quad H_3C-\underset{\underset{\ominus}{|}\text{IOI}}{C}=CH-COOR\;\right]\;Na^{\oplus}\quad\xrightarrow[-NaCl]{+\;\underset{Cl}{\overset{O}{\diagdown}}C-R'}\quad H_3C-\underset{O-COR'}{C}=CH-COOR$$

The *pyridine* itself plays a role in the reaction by acting as a *template, e.g.*

Benzoylacetic ester

Such participation is described as a *template effect*. This sort of effect can also be observed if metal ions are present that form intermediate complexes which can favour one or other of different possible reaction paths.

Ketonic and Acid Cleavage of Esters of β-Ketocarboxylic Acids

Depending upon the reaction conditions that are used, esters of acetoacetic acid and its derivatives can be split by hydrolysis to give either ketones or carboxylic acids. Because

---

[1] See *E. M. Kaiser*, Di- and Poly-alkali Metal Derivatives of Heterofunctionally Substituted Organic Molecules, Synthesis *1977*, 509.

of the possibility of making a wide range of derivatives having a variety of substituents present, these reactions are of synthetic importance.

2.30.6.1 Ketonic cleavage of acetoacetic ester derivatives is brought about by heating them with either dilute acid or dilute alkali. The initial hydrolysis product is the corresponding β-keto acid which is unstable and breaks down into a ketone and carbon dioxide, e.g.

$$H_3C-\underset{\underset{O}{\|}}{C}-\underset{\underset{R}{|}}{\overset{\overset{H}{|}}{C}}-COOC_2H_5 + H_2O \longrightarrow H_3C-\underset{\underset{O}{\|}}{C}-CH_2-R + CO_2 + C_2H_5OH$$

Ketones prepared in this way all contain the acetyl group, H_3C-CO-. Very good yields of ketone can be obtained by carrying out the hydrolysis in aqueous dioxan with either aluminium oxide[1] or a solution of 1,4-diazabicyclo[2,2,2]octane (triethylene-diamine, DABCO) in o-xylene.

2.30.6.2 Acid cleavage of acetoacetic ester derivatives, so-called because the main product is the salt of an acid, is brought about by heating them with concentrated alkali. In this process the bond linking the acetyl group to the α-carbon atom is cleaved, and the sodium salts of acetic acid and of a mono- or di-substituted acetic acid are formed, e.g.

$$H_3C-\underset{\underset{O}{\|}}{C}-\underset{\underset{R}{|}}{\overset{\overset{H}{|}}{C}}-COOC_2H_5 + 2\,NaOH \longrightarrow H_3C\overset{\ominus}{COO}\overset{\oplus}{Na} + H-\underset{\underset{R}{|}}{\overset{\overset{H}{|}}{C}}-\overset{\ominus}{COO}\overset{\oplus}{Na} + C_2H_5OH$$

Addition of mineral acid converts the sodium salts into the carboxylic acids.

This reaction is the reverse of a *Claisen* condensation. First of all a hydroxide anion makes a nucleophilic attack on the positively charged carbon of the carbonyl group; this is followed by breaking of the central C—C bond and formation of a *carbanion*. Addition of a proton, and hydrolysis of the ester group provide the carboxylic acid

$$H_3C-\underset{\underset{|\underline{O}|}{\overset{\|}{C}}}{C}-\underset{\underset{R}{|}}{\overset{\overset{H}{|}}{C}}-COOC_2H_5 \xrightarrow{+Na\overset{\oplus\ominus}{OH}} \left[H_3C-\underset{\underset{|\underline{O}|}{|}}{\overset{\overset{|\overline{O}H}{|}}{C}}-\underset{\underset{R}{|}}{\overset{\overset{H}{|}}{C}}-COOC_2H_5 \right] Na^{\oplus}$$

$$\longrightarrow \left[H_3C-\underset{\underset{|\underline{O}|}{\|}}{\overset{\overset{|\overline{O}-H}{|}}{C}} + \overset{\ominus}{|}\underset{\underset{R}{|}}{\overset{\overset{H}{|}}{C}}-COOC_2H_5 \right] Na^{\oplus} \xrightarrow[-C_2H_5OH]{+NaOH} H_3C-\overset{\ominus}{COO}\overset{\oplus}{Na} + H-\underset{\underset{R}{|}}{\overset{\overset{H}{|}}{C}}-\overset{\ominus}{COO}\overset{\oplus}{Na}$$

Carbanion

Uses

Acetoacetic esters are important starting materials for a variety of synthetic processes, for example, industrially for the preparation of pyrazolone dyestuffs, and in the field of pharmaceuticals of *antipyrine* (INN: **Phenazone**) and 4-aminoantipyrine (*Aminophenazone*).

[1] See A. McKillop et al., Organic Synthesis using Supported Reagents, Synthesis *1979*, 401, 481
P. Laszlo (Ed.) Preparative Chemistry Using Supported Reagents (Academic Press, San Diego, Cal. 1987).

2.30.7 γ-Keto acids

Levulinic acid (4-oxopentanoic acid) is the simplest γ-keto acid.

Preparation

By heating hexoses, *e.g.* fructose (levulose), with hydrochloric acid:

$$C_6H_{12}O_6 \xrightarrow{\text{(HCl)}} H_3C-\underset{\underset{O}{\|}}{C}-CH_2-CH_2-COOH + HCOOH + H_2O$$

Levulinic acid

The *mechanism* of the breakdown of hexoses into levulinic acid is not completely understood. 5-(Hydroxymethyl)furfural is an intermediate; this is unstable to acid, and forms levulinic acid and formic acid.

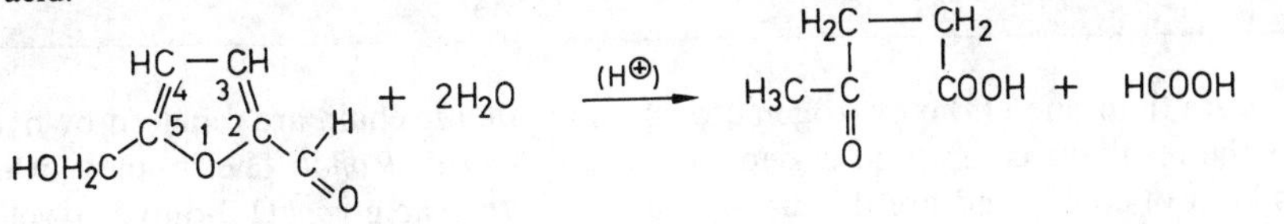

5-(Hydroxymethyl)furfural   Levulinic acid   Formic acid

Better yields can be obtained by the action of hydrochloric acid under pressure on cane sugar.

Properties

Levulinic acid forms colourless leaflets with m.p. 34°C (307 K) which are readily soluble in water, ethanol or ether. It behaves both as a ketone and as a carboxylic acid. Some of its reactions suggest that levulinic acid exists in the form of a hydroxy-lactone.

In contrast to β-keto acids, levulinic acid does not lose carbon dioxide even when heated for a long period. Instead it undergoes intramolecular loss of water to give *unsaturated lactones*, pent-3-en-4-olide and pent-2-en-4-olide:

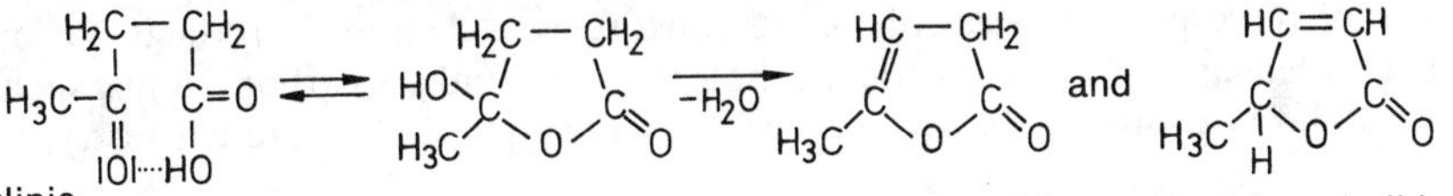

Levulinic   Hydroxy-lactone   Pent-3-en-4-olide   Pent-2-en-4-olide
acid

These unsaturated γ-lactones can also be regarded as α-ketodihydrofurans; they are abundant in nature. One, for example, found in species of lichen,[1] above all in *Letharia vulpina*, is the yellow pigment *vulpinic acid*, $C_{19}H_{14}O_5$, which has the lactone carboxylic acid methyl ester structure shown below. Alkaline hydrolysis saponifies the ester group to give *pulvinic acid* which, when heated, is converted into *pulvinic acid lactone*, $C_{18}H_{10}O_4$:

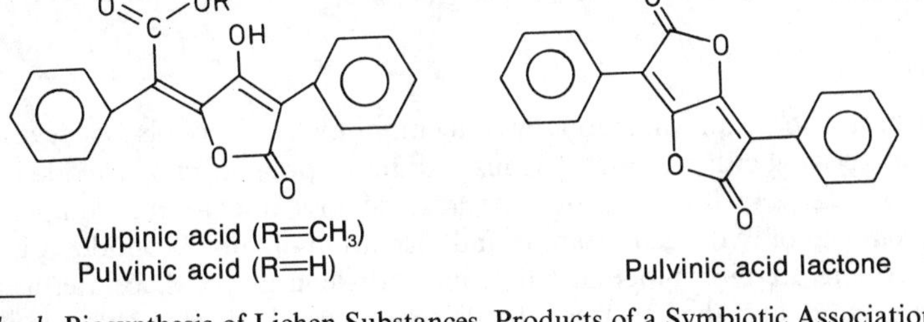

Vulpinic acid (R=CH₃)
Pulvinic acid (R=H)   Pulvinic acid lactone

[1] See *K. Mosbach*, Biosynthesis of Lichen Substances. Products of a Symbiotic Association, Angew. Chem. Int. Ed. Engl. *8*, 240 (1969).

(E)-9-Oxodec-2-enoic acid, $H_3C-\overset{\displaystyle \|}{\underset{\displaystyle O}{C}}-(CH_2)_5-\overset{\displaystyle H}{\underset{\displaystyle H}{C}}=C-COOH$, is excreted by queen bees

throughout their lives. So long as this compound is present, no new queen bee is reared. The same substance and also its *racemic* reduction product (E)-9-hydroxydec-2-enoic acid (see p. 282), as well as other odiferous materials, are released by the queen bee only just after the bees have swarmed, in order to quieten them.

2.31 Polyhydric Alcohols

When more than one of the hydrogen atoms on an alkane chain are replaced by hydroxy groups the resultant compound is known as a *polyhydric alcohol*. The simplest examples of this kind of compound are the dihydric alcohol *ethylene glycol* (1,2-dihydroxyethane), the trihydric alcohol *glycerol* (1,2,3-trihydroxypropane) and the tetrahydric alcohol erythritol (1,2,3,4-tetrahydroxybutane).

$$
\begin{array}{ccc}
& & H_2C-OH \\
& H_2C-OH & | \\
H_2C-OH & | & HC-OH \\
| & HC-OH & | \\
H_2C-OH & | & HC-OH \\
& H_2C-OH & | \\
& & H_2C-OH \\
\\
\text{Ethyleneglycol} & \text{Glycerol} & \text{Erythritol}
\end{array}
$$

As the number of hydroxy groups increases so does the solubility of the compound in water increase, whilst the solubility in ethanol or ether decreases.

Some of these polyhydric alcohols are constituents of natural products, for example glycerol in fats and oils. Polyhydric alcohols with five and six carbon atoms, such as the pentitols and hexitols, play a role as sugar alcohols in carbohydrate chemistry.

2.31.1 Dihydric Alcohols (Glycols, 1,2-Diols)

2.31.1.1 Ethylene glycol (Ethane-1,2-diol,1,2-dihydroxyethane), HOH_2C-CH_2OH, is a dihydric alcohol having two primary alcohol groups.

Preparation

(a) It can be made in an analogous way to monohydric alcohols by *hydrolysis of 1,2-dibromoethane*, using either dilute aqueous sodium or potassium hydroxide or carbonate. The yield is relatively low ($\sim$ 50%) because of loss due to the competing reaction involving removal of hydrogen bromide and formation of vinyl bromide. A better process is to convert the 1,2-dibromoethane first into ethylene glycol diacetate, using silver or potassium acetate, and then to treat the diacetate with hydrogen chloride in methanol. This provides ethylene glycol in an overall yield of $\sim$ 85%:

$$\begin{array}{l} CH_2-Br \\ | \\ CH_2-Br \end{array} \quad \xrightarrow[-2\,KBr]{+\,2\,H_3CCOO^{\ominus}K^{\oplus}} \quad \begin{array}{l} CH_2-O-COCH_3 \\ | \\ CH_2-O-COCH_3 \end{array} \quad \xrightarrow[-2\,H_3CCOOCH_3]{+\,2\,H_3COH(H^{\oplus})} \quad \begin{array}{l} CH_2-OH \\ | \\ CH_2-OH \end{array}$$

1,2-Dibromoethane   Ethylene glycol   Ethylene glycol
                                                     diacetate

1,2-Dibromoethane has been shown to be carcinogenic in some animals.

(b) Industrially ethylene glycol is made *from ethylene oxide* by heating it under pressure
with water:

$$H_2C\overset{\displaystyle\diagup\!\!\!\diagdown}{\underset{O}{}}CH_2 \; + \; H_2O \quad \xrightarrow{450\,K/2\,MPa} \quad \begin{array}{l} H_2C-CH_2 \\ | \quad\; | \\ OH\;\; OH \end{array}$$

Ethylene oxide

(c) Ditertiary glycols, such as pinacol, can be obtained by the reduction of ketones with,
for example, magnesium amalgam.

Properties

Ethylene glycol is a colourless, viscous oil with a sweetish smell, which boils at 198°C
(471 K). It is completely miscible with water or with ethanol, but almost insoluble in
ether. It is relatively toxic.

The chemical properties resemble those of monohydric alcohols, but with the additional
factor that the two hydroxy groups may react successively, one after the other. Thus it
is possible to make both *mono-* and *di-sodium salts, glycol monoethers* (cellosolves) and
diethers, and *mono-* and *di-esters*. Compounds in which one of the hydroxy groups
has been replaced by a halogen atom are known as *hydrins*, e.g. 2-bromoethanol,
$HOCH_2-CH_2Br$, is called ethylene bromohydrin.

If ethylene glycol is heated with a small amount of conc. sulphuric acid or phosphoric
acid *1,4-dioxan* (dioxan) is formed. This is a liquid of b.p. 101.5°C (374.7 K) which has an
aromatic smell. It is completely miscible, in all proportions, with water, ethanol or ether.

$$2\;HOCH_2-CH_2OH \quad \xrightarrow{H_2SO_4} \quad \text{[1,4-dioxan ring structure]}$$

1,4-Dioxan

Dioxan can also be obtained by the sulphuric acid-induced dimerisation of ethylene
oxide.

The very small dipole moment of 1,4-dioxan indicates that the molecules take up chair
conformations (see p. 399):

[chair conformation structure of 1,4-dioxan]

Analogously to the products obtained from the oxidation of monohydric primary
alcohols, ethylene glycol, as a diprimary alcohol, can give rise in turn to the following
oxidation products:

$$
\begin{array}{ccc}
\begin{array}{c} CH_2OH \\ | \\ CH_2OH \end{array} &
\begin{array}{c} C{\nwarrow}^H \\ \|{\searrow}O \\ CH_2OH \end{array} &
\begin{array}{c} C{=}O \\ |{\nwarrow}OH \\ CH_2OH \end{array} \\[2mm]
\text{Ethylene} & & \text{Glycollic acid} \\
\text{glycol} & \text{Glycolaldehyde} & \text{(Hydroxyacetic acid)}
\end{array}
$$

$$
\begin{array}{ccc}
\begin{array}{c} C{\nearrow}^H \\ |{\searrow}O \\ C{\nearrow}^H \\ {\searrow}O \end{array} &
\begin{array}{c} C{\nearrow}^H \\ |{\searrow}O \\ C{\nearrow}O \\ {\searrow}OH \end{array} &
\begin{array}{c} C{\nearrow}O \\ |{\searrow}OH \\ C{\nearrow}O \\ {\searrow}OH \end{array} \\[2mm]
\text{Glyoxal} & \text{Glyoxylic acid} & \text{Oxalic acid}
\end{array}
$$

Each of these compounds is the parent compound of a series. It is easy to isolate glycollic acid and oxalic acid when ethylene glycol is oxidised by nitric acid. The other intermediates which are in theory formed can only be obtained in small amounts because they are more readily oxidised than the glycol is. They have to be made by special methods.

The **oxidative cleavage of glycols**, in which the bond linking the two carbon atoms bearing the hydroxy groups is broken, is of preparative importance. Oxidising agents used for this purpose include lead tetraacetate in glacial acetic acid[1] (*Criegee*, 1931) and aqueous *periodic acid*[2] (*Malaprade*, 1928). The cleavage products are aldehydes or ketones:

$$
\begin{array}{c} H \\ | \\ R-C-OH \\ | \\ R-C-OH \\ | \\ H \end{array}
+ Pb(OOCCH_3)_4 \longrightarrow 2\ R-C{\nearrow}^H_{{\searrow}O} + Pb(OOCCH_3)_2 + 2\ H_3CCOOH
$$

Uses

Ethylene glycol, and also its monomethyl, monoethyl and dimethyl ethers (dimethoxy-ethane = DME), and dioxan are valuable solvents, for example for varnishes and lacquers and for acetylcelluloses. Polyester fibres are made from ethylene glycol. It is also used as an antifreeze; some insects, which are found in polar regions contain it, together with glycerol, in their body fluids, which in consequence only freeze at $-58°C$ (215 K).

The More Important Derivatives of Glycols

2.31.1.2 Ethylene oxide (oxiran), $H_2C{-}CH_2$, is the simplest epoxide.
$$\underset{O}{\diagdown\ \diagup}$$

---

[1] See Oxydationen mit Bleitetraacetat und Perjodsäure (Oxidations with Lead Tetraacetate and Periodic Acid), Angew. Chem. *53*, 321 (1940); *70*, 173 (1958); G. M. *Rubottom*, Lead Tetraacetate, in *W. S. Trahanovsky* (Ed.), Oxidation in Organic Chemistry, Part D, pp. 1–145 (Academic Press, New York 1982).

[2] See B. *Sklarz*, Organic Chemistry of Periodates, Quart. Rev. *21*, 3 (1967).

Preparation

(a) *By distilling 2-chloroethanol in the presence of potassium hydroxide* (or calcium hydroxide):

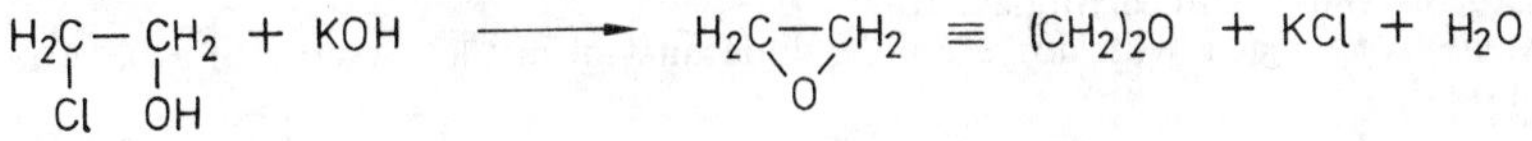

$$H_2C-CH_2 + KOH \longrightarrow H_2C-CH_2 \equiv (CH_2)_2O + KCl + H_2O$$

Ethylene oxide

(b) Industrially, by *atmospheric oxidation of ethylene*, in the presence of silver as catalyst, under pressure at about 250°C (525 K):

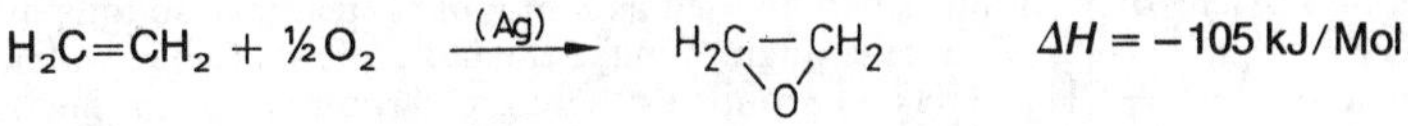

$$H_2C=CH_2 + \tfrac{1}{2}O_2 \xrightarrow{(Ag)} H_2C-CH_2 \qquad \Delta H = -105 \text{ kJ/Mol}$$

Properties

Ethylene oxide is a colourless gas which boils at 10.7°C (283.9 K); it is poisonous, and has been shown to be carcinogenic in animals. It is used as a disinfectant and for sterilising medical equipment. It reacts slowly with water, and more rapidly with dilute hydrochloric acid, to give ethylene glycol. Whereas normal ethers are rather unreactive compounds, ethylene oxide is highly reactive. This is because it has an energy-rich, strained three-membered ring, and this strain can be released when it is converted into an open-chain compound.

As a result, under the appropriate conditions ethylene oxide and its alkyl derivatives undergo addition reactions with water, alcohols, hydrogen halides, hydrogen cyanide, ammonia or primary amines, *i.e.* compounds with active hydrogen atoms, as shown in the following scheme:

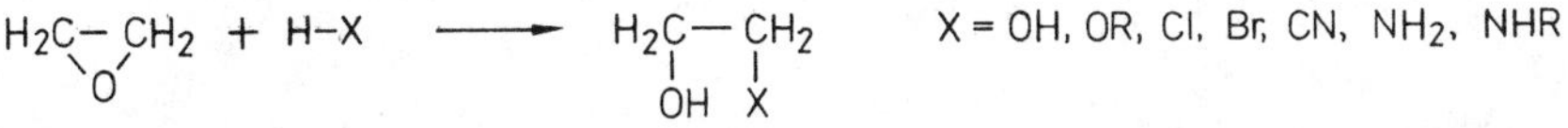

$$H_2C-CH_2 + H-X \longrightarrow H_2C-CH_2 \qquad X = OH, OR, Cl, Br, CN, NH_2, NHR$$

Ethylene oxide also has a marked tendency to polymerise. For example, it undergoes *cationic polymerisation* in the presence of tin(IV) chloride and a small amount of water; the latter adds to the enols of the polymer chains, giving *poly(ethylene glycols)*:

$$n(CH_2)_2O \xrightarrow{+H_2O} H\text{-}[O-CH_2-CH_2\text{-}]_n OH$$

Poly(ethylene glycol)

Depending upon the degree of polymerisation, either viscous liquids or waxy solids (carbowaxes) which are soluble in water are obtained. *Diethyleneglycol* ($n = 2$) is used industrially as a solvent and for the extraction of aromatic hydrocarbons (Udex process). It damages the kidneys and in consequence is poisonous. *Triethyleneglycol* ($n = 3$) is added to soaps. Higher polymers find use as plasticisers, as finishing agents for textiles and as additives in lacquers.

The reaction of ethylene oxide with ethyleneglycol monoalkyl ethers provides monoalkyl ethers of diethyleneglycol (carbitols), which have use in industry as solvents, especially for varnishes and lacquers:

$$HOCH_2-CH_2OR \xrightarrow{+ (CH_2)_2O} HOCH_2-CH_2-O-CH_2-CH_2OR$$

Diethyleneglycol monoalkyl ether

Diethyleneglycol dimethyl ether (diglyme) is a valuable solvent.

Higher molecular weight *polyethyleneglycol ethers* (polyglycol ethers) are obtained when alcohols or phenols react with an excess of ethylene oxide. When long-chain alcohols or alkylated phenols are used, the resultant *long-chain alkyl-* or *alkylphenyl-polyethyleneglycol ethers* are of value as non-ionic detergents and as emulsifying agents.

If ethylene oxide reacts with, for example, dodecan-1-ol in the presence of a base an addition reaction takes place:

$$H_3C-(CH_2)_{10}-CH_2OH + x\,(CH_2)_2O \xrightarrow{(OH^\ominus)} H_3C-(CH_2)_{10}-CH_2-(O-CH_2-CH_2)_x-OH$$

Dodecan-1-ol   α-Dodecyl-ω-hydroxypoly-(oxyethylenes)

The resultant ethylene oxide adducts can be used as non-ionic detergents, so long as the number of ($-O-CH_2-CH_2-$) groups is large enough to make the molecule soluble in water. There must be at least as many ($-O-CH_2-CH_2-$) groups as there are carbon atoms in the alcohol; in the present example the number required, n, is 12 to 15. If $n < 12$ the solubility in water can be achieved by making a sulphate ester and then converting it into its sodium salt:

$$H_3C-(CH_2)_{10}-CH_2-(OCH_2CH_2)_4OH \xrightarrow{H_2SO_4} H_3C-(CH_2)_{10}-CH_2-(OCH_2CH_2)_4OSO_2OH \xrightarrow{NaOH}$$

$$H_3C-(CH_2)_{10}-CH_2-(OCH_2CH_2)_4OSO_3{}^\ominus\,Na^\oplus$$

The products so obtained are anionic surfactants.

Ethylene oxide reacts similarly with long-chain carboxylic acids to give *polyethyleneglycol esters*, $RCO\text{+}O-CH_2-CH_2\text{+}_nOH$, which are used as specialised detergents.

2.31.1.3 Macrocyclic polyethers became generally available in 1967 as a result of the work

of *Pedersen*, and have turned out to have a quite remarkable ability to form complexes.[1] They are generally known as *crown ethers*, and are formed by the following type of S_N2 ring-closure reaction:

[18]Crown-6, [18]-O_6

The shorthand name [18]crown-6 indicates that there is an 18-membered ring which contains 6 oxygen atoms. The advantage of using this shorthand name rather than the precise but clumsy full name, 1,4,7,10,13,16-hexaoxacyclooctadecane, is self-evident. The ring-size of crown ethers, which may be thought of as cyclic glymes (*cf.* diglyme), can have any value, as can the number of oxygen atoms. The latter can be replaced either partially or completely by other hetero-atoms such as nitrogen or sulphur.[2]

---

[1] See *C. J. Pedersen* et al, Macrocyclic Polyethers and their Complexes, Angew. Chem. Int. Ed. Engl. *11*, 16 (1972); The Discovery of Crown Ethers, Science *241*, 536 (1988); *G. W. Gokel*, Crown Ethers and Cryptands (Royal Society of Chemistry, London 1991).

[2] See, for example, *J. M. Lehn*, Supramolecular Chemistry: Receptors, Catalysts and Carriers, Science *227*, 849 (1985); *D. J. Cram*, Preorganization — From Solvents to Spherands, Angew. Chem. Int. Ed. Engl. *25*, 1039 (1986); *D. J. Cram* and *J. M. Cram*, Container Molecules and their Guests (Royal Society of Chemistry, London 1994); *F. Vögtle*, Supramolekulare Chemie (Teubner, Stuttgart 1989); *F. Vögtle* et al. (Ed.) Host-Guest Chemistry/Macrocycles (Springer-Verlag, Berlin 1985); *F. P. Schmidtchen*. Wirt-Gast-Chemie, Molekulare Wirte für Anionen (Host-Guest Chemistry, Molecular Hosts for Anions), Nachr. Chem. Tech. Lab. *36*, 8 (1988).

Crown ethers and related compounds can, by means of ion-dipole interactions, form strong complexes (super-molecules) with metal cations. These are known as *cryptates*, and the crown ether itself is called a *cryptand*. Specificity for a particular cation can be achieved by suitably adjusting the ring-size, and also the number and type of hetero-atoms. For example, the *potassium ion* forms such a strong cryptate with [*18*]crown-6 that it enables potassium permanganate to dissolve in benzene in concentrations sufficient to carry out oxidations in anhydrous homogeneous solution.

Cryptates of cations can bring about results similar to those achieved by using quaternary ammonium salts as phase-transfer catalysts. With carbanions they form ion pairs which dissolve in the organic phase where they can take part in reactions. *Crown ethers* thus provide many different possibilities for use as *phase-transfer catalysts*.

S_N2 *reactions in a homogenous phase* can be greatly accelerated by *crown ethers* if the solvent used (*e.g.* acetonitrile) does not solvate anions satisfactorily. If the cation is firmly held in a cryptate, the reactivity of the accompanying anion becomes much greater because it is very weakly solvated; it is called a *naked* anion. Under such conditions, potassium fluoride becomes a strong nucleophile:

$$H_3C(CH_2)_8CH_2Br \xrightarrow[-KBr]{KF, [18]crown-6, H_3CCN} H_3C(CH_2)_8CH_2F \quad (82\%)$$

For similar reasons the hydrolysis of esters by means of potassium hydroxide is greatly accelerated by the presence of a crown ether. Crown ethers in the presence of alkali-metal ions can participate in interesting template effects[1] (see p. 297). When crown ethers are used it is important to remember their high *toxicity*.

2.31.1.4 2-Aminoethanol (ethanolamine) is the simplest aminoalcohol. It is a colourless oil, b.p. 171°C (444 K), and is obtained by the reaction of ethylene oxide with excess ammonia:

$$H_2C\!\!-\!\!CH_2 + NH_3 \longrightarrow \begin{array}{c} H_2C\!\!-\!\!CH_2 \\ | \quad\quad | \\ NH_2 \quad OH \end{array}$$

2-Aminoethanol

Also formed are *2,2′-iminodiethanol* (diethanolamine), $HN(CH_2-CH_2-OH)_2$ and *2,2′,2″-nitrilotriethanol* (triethanolamine), $N(CH_2-CH_2-OH)_3$, which is the major product if ethylene oxide is present in excess.

Commercially ethanolamines are used as basic components in soaps and creams, as wetting agents for textiles, and as softeners in the leather industry. They are also starting materials in the syntheses of heterocyclic bases.

2.31.1.5 Choline (2-hydroxyethyl-trimethyl-ammonium hydroxide) is the basic component in phospholipids of the phosphoglyceride type, and is widely distributed in both plants and animals. It is most readily made *from ethylene oxide* and *trimethylamine* in aqueous solution:

$$H_2C\!\!-\!\!CH_2 + N(CH_3)_3 + H_2O \longrightarrow \left[\begin{array}{c} H_2C\!\!-\!\!CH_2 \\ | \qquad |\oplus \\ OH \quad N(CH_3)_3 \end{array}\right] OH^\ominus$$

Choline

---

[1] *K. N. Raymond*, Molecular Recognition and Metal Ion Template Synthesis, Science *244*, 938 (1989).

Choline forms hygroscopic crystals, which, because of their strong basicity, take up carbon dioxide from the air. With hydrochloric acid it forms a crystalline hydrochloride. On oxidation it is converted into *betaine*.

Choline is an important factor in biochemical processes, for example in methylation. Lack of it in animals leads to the development of fatty livers.

The reaction of *trimethylamine* with *1,2-dichloroethene* gives rise to *chlorocholine chloride* (2-chloroethyl-trimethylammonium chloride), which restrains the height to which plants grow:

$$ClCH=CHCl \; + \; N(CH_3)_3 \longrightarrow \left[\begin{array}{c} H_2C-CH_2 \\ | \quad\;\; |\oplus \\ Cl \quad N(CH_3)_3 \end{array} \right] Cl^{\ominus}$$

Acetylcholine, $H_3CCO-O-CH_2-N(CH_3)_3^{\oplus}OH^{\ominus}$, is formed by acetylation of choline in the presence of 'activated acetic acid' and adenosine triphosphate (ATP), under the influence of the enzyme *cholinacetyltransferase* (CAT).

Physiologically acetylcholine liberates blood pressure depressants and also strongly muscle-contracting agents, which do not, however, persist for any length of time, since acetylcholine is split into choline and acetic acid by *acetylcholinesterase* which is present in the blood. Acetylcholine plays an important controlling role in organisms as a *neurohormone* (neurotransmitter) in the parasympathetic nerve system. The chemical nerve agents *Tabun*, *Sarin* and *Soman* interfere with the action of acetylcholinesterase, and this causes a lasting overstimulation of essential parts of the nervous system. *Atropine* nullifies this effect.

Carnitine, in the form of its *O*-acyl derivatives, participates in the transport of fatty acids through the mitochondrial membrane. The acylcarnitines are formed in cells from activated carboxylic acids (CoA = coenzyme A).

$$\underset{O}{\underset{\|}{R-C}}-SCoA \; + \; \underset{\underset{CH_2-COO^{\ominus}}{|}}{HO-CH}-CH_2-\overset{\oplus}{N}(CH_3)_3 \; \rightleftharpoons \; \underset{O}{\underset{\|}{R-C}}-O-\underset{\underset{CH_2-COO^{\ominus}}{|}}{CH}-CH_2-\overset{\oplus}{N}(CH_3)_3 \; + \; HSCoA$$

Carnitine   Acylcarnitine

2.31.1.6 Phosphoglycerides or **Phosphatides (Phospholipids)**. This class of natural product consists of fat-like *triglycerides* which are made up from two long-chain carboxylic acids and a phosphoric acid residue to which a base is also attached. They occur in all plant and animal cells, above all in the brain, the heart, the liver, and in egg yolks and soya beans. They play an important role in the formation and the functioning of biological membranes.

The most important phosphatides (glycerol-*P* or Gro-*P*) are *O-phosphatidylethanolamine* and *O-phosphatidylcholine* (earlier known as cephalin and lecithin), in which the base present is colamine or choline, and which both have betaine structures:

$$^3CH_2-O-CO-R$$
$$^2HC-O-CO-R'$$
$$^1CH_2-O-\overset{\overset{O}{\diagup\!\!\diagdown}}{P}\!\!\underset{O^{\ominus}}{\diagdown}O-CH_2-CH_2-\overset{\oplus}{N}H_3$$

O-Phosphatidylethanolamine (Cephalin)

$$^1CH_2-O-CO-R$$
$$R'-CO-O-^2CH$$
$$^3CH_2-O-\overset{\overset{O}{\diagup\!\!\diagdown}}{P}\!\!\underset{O^{\ominus}}{\diagdown}O-CH_2-CH_2-\overset{\oplus}{N}\!\!\underset{\diagdown CH_3}{\overset{\diagup CH_3}{-}}CH_3$$

O-Phosphatidylcholine (Lecithin)

In order to indicate the configuration of a chiral glycerine derivative a *stereospecific numbering* (*sn*) is used. In this convention, the carbon atom which comes above carbon atom 2, when the secondary hydroxy group lies to the *left* in the *Fischer* projection, is designated as carbon atom 1. According to this system, *lecithin* becomes *3-sn*-phosphatidylcholine, or, more precisely, 1,2-diacyl-*sn*-glycero-3-phosphocholine. On the CIP system it has an (*R*)-configuration. *Cephalin*, which has an (*S*)-configuration, is described in the *sn* convention as 2,3-diacyl-*sn*-glycero-1-phosphoethanolamine, because in this case the carbon atom which carries the phosphoric acid group becomes carbon atom 1 when the *Fischer* convention is applied, with the secondary *O*-acyl group on the left-hand side. The *chiral triglycerides* which were mentioned on p. 245 are also defined by the *sn* system.

O-Phosphatidylethanolamines are most readily obtained from material from brain tissue; their poor solubility in ethanol enables them to be separated from phosphatidylcholines. Both classes of compounds consist of *mixtures of phosphatides* containing various saturated and unsaturated carboxylic acids, including palmitic, stearic, oleic, linoleic and linolenic acids.

On the other hand, sphingolipids are isolable from brain tissue, spleen and the spinal cord. In them glycerol is replaced by unsaturated higher aminoalcohols, for example sphingosin, H_3C—$(CH_2)_{12}CH$=CH—$CH(OH)$—$CH(NH_2)$—CH_2OH, in which a carboxylic acid residue is always attached to the amino group. In the following formula the characteristic amide group is given a grey background. *Ceramides* (Cer.) are sphingolipids having two free hydroxy groups. *Sphingosinphosphatides* are called *sphingomyelins*.

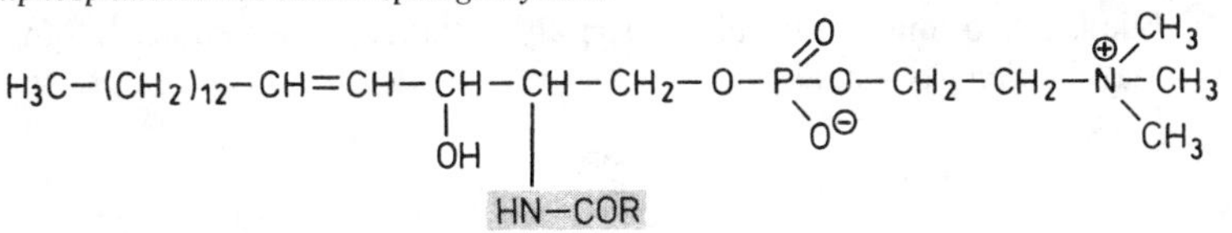

$$H_3C-(CH_2)_{12}-CH=CH-CH-CH-CH_2-O-\overset{\overset{O}{\parallel}}{\underset{\underset{O^\ominus}{\mid}}{P}}-O-CH_2-CH_2-\overset{\oplus}{N}\begin{smallmatrix}CH_3\\CH_3\\CH_3\end{smallmatrix}$$

with OH below the first CH and HN—COR below the second CH

Sphingomyelins

These *phospholipids* are also for the most part mixtures, in which the acids involved can be palmitic acid, stearic acid, nervonic acid, H_3C—$(CH_2)_7$—CH=CH—$(CH_2)_{13}$—$COOH$, and lignoceric acid, H_3C—$(CH_2)_{22}$—$COOH$ (*Klenk*).
Cerebrosides, which are closely related to phospholipids, belong to the group called *glycolipids*, which are found especially in the brain and in nerve tissues, and are made up from fatty acids, sphingosin and D-galactose, but contain no phosphoric acid residue or choline.

Amphiphilic compounds of the lecithin type can form extended bimolecular layers in which the hydrophilic groups are solvated by water and the hydrophobic chains lie within the resultant membrane. They are known as bilayer lipid membranes (BLM). Under suitable conditions the BLM form *vesicles* or *liposomes*.[1]

Neurine (trimethylvinylammonium hydroxide), H_2C=CH—$N(CH_3)_3]^\oplus OH^\ominus$, is formed from choline in the putrefaction of proteins, or by loss of water from choline by heating it with barium hydroxide. It is highly toxic and is one of the *ptomaine* poisons.

Aziridine (ethylene imine) is prepared from ethanolamine hydrochloride by treating it with thionyl chloride to give 2-chloroethylamine hydrochloride, which, when heated with aqueous alkali, is converted into aziridine:

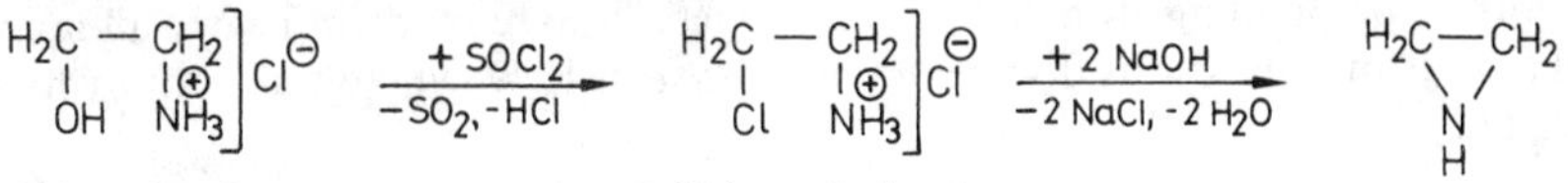

Ethanolamine hydrochloride $\xrightarrow[-SO_2,-HCl]{+SOCl_2}$ 2-Chloroethylamine hydrochloride $\xrightarrow[-2\,NaCl,\,-2\,H_2O]{+2\,NaOH}$ Aziridine

---

[1] See *H. Ti Tien*, Bilayer Lipid Membranes (BLM), (Dekker, New York 1974).

On the industrial scale aziridine is made in a similar way from *β-aminoethylsulphuric acid* (H_2N—CH_2—CH_2—OSO_3H).

It is a liquid of b.p. 56°C (329 K) which is toxic, strongly corrosive and a carcinogen. Like ethylene oxide aziridine is very reactive and is used in synthetic work, especially for the introduction of aminoethyl groups.

In both of the aziridine syntheses the hydroxy group is first replaced by a group which will be a better leaving group in the following intramolecular S_N2 reaction. This can also be achieved in one step by using dibromotriphenylphosphorane.

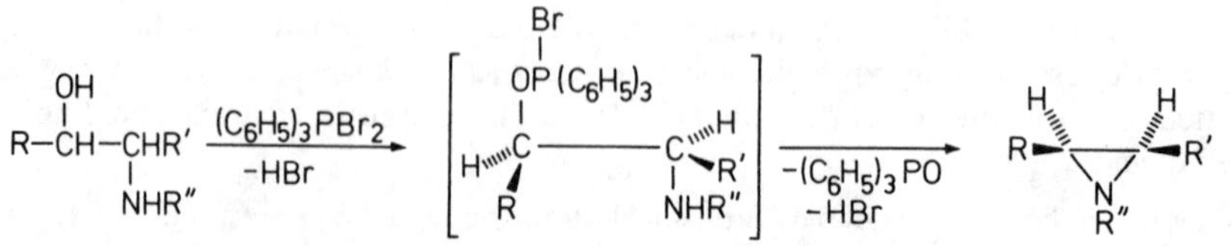

The driving force for this reaction is the formation of the thermodynamically very stable triphenylphosphine oxide.

Taurine (2-aminoethanesulphonic acid) is a constituent of taurocholic acid, which is found in the bile fluids of many animals, especially in cattle. It can be made synthetically from *aziridine* and *sulphurous acid*:

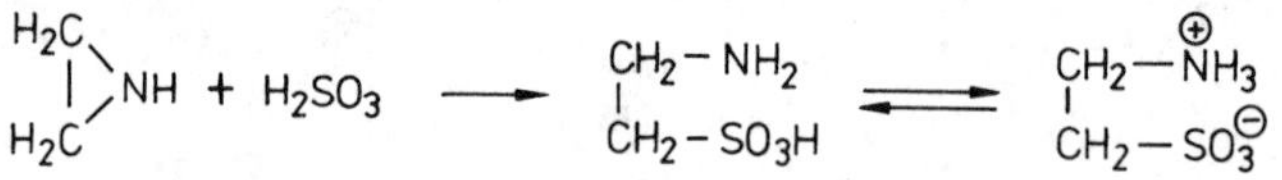

Taurine is a crystalline substance which melts above 240°C (520 K). It dissolves in water giving a *neutral* solution, but is insoluble in ethanol and ether. These facts are indicative of its salt-like character, and it is thus best represented by a *betaine* structure as shown above.

The biochemical precursor of taurine is L-cysteine. This undergoes enzymatic oxidation to give L-cysteic acid, which is then converted into taurine by a specific *cysteic acid decarboxylase*:

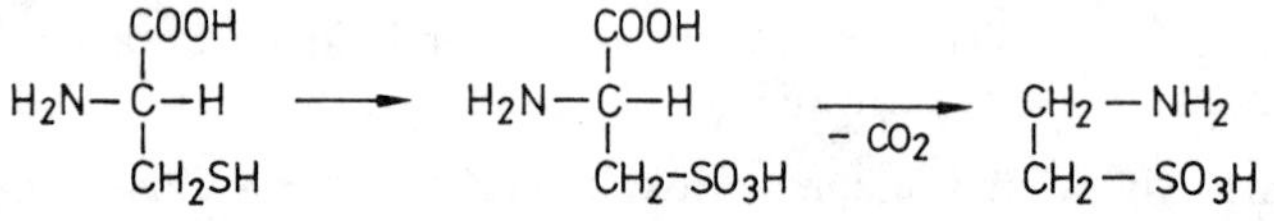

2.31.2 Trihydric Alcohols

2.31.2.1 Glycerol (INN: *Glycerol,* propane-1,2,3-triol) is the most important trihydric alcohol. It is the building stone from which glycerides are made in nearly all animal fats as well as in plant oils. It was discovered by *Scheele* in 1779, who obtained it by hydrolysis of olive oil.

Preparation

(a) By *hydrolysis of fats* using sulphuric acid, sodium hydroxide, the Twitchell reagent or the enzyme *lipase*.

(b) *From propene obtained from cracking gases.* The action of chlorine on propene at 500°C (775 K) gives allyl chloride as the major product. Treatment of this with hypochlorous acid provides a mixture of 2,3-dichloropropan-1-ol and 1,3-dichloropropan-2-ol. In order to restrict the amount of sodium hydroxide used, this mixture is first of all hydrolysed by calcium hydroxide to epichlorohydrin (1-chloro-2,3-epoxypropane), and this is then converted into glycerol by sodium hydroxide:

$$
\begin{array}{c}
\underset{\substack{\text{Allyl}\\\text{chloride}}}{\underset{\displaystyle CH_2-Cl}{\overset{\displaystyle CH_2}{\underset{\displaystyle CH}{\|}}}}
\xrightarrow{+\,HOCl}
\underset{\substack{\text{2,3-Dichloro-}\\\text{propan-1-ol}}}{\underset{\displaystyle CH_2-Cl}{\underset{\displaystyle CH-Cl}{CH_2-OH}}}
\;\text{and}\;
\underset{\substack{\text{1,3-Dichloro-}\\\text{propan-2-ol}}}{\underset{\displaystyle CH_2-Cl}{\underset{\displaystyle CH-OH}{CH_2-Cl}}}
\xrightarrow[-\,HCl]{[Ca(OH)_2]}
\underset{\text{Epichlorohydrin}}{\underset{\displaystyle CH_2-Cl}{\underset{\displaystyle HC}{\underset{\displaystyle H_2C}{}}}\!\!\diagdown\!O}
\xrightarrow[-\,NaCl]{+\,NaOH,\,+\,H_2O}
\underset{\substack{\text{Glycerol}\\\text{(Gro)}}}{\underset{\displaystyle CH_2-OH}{\underset{\displaystyle CH-OH}{CH_2-OH}}}
\end{array}
$$

In this process the chlorine is lost in the form of valueless calcium chloride. This loss can be overcome in the following process.

(c) By *hydroxylation* of allyl alcohol with hydrogen peroxide in the presence of WO_3 as a catalyst; glycidol (2,3-epoxypropan-1-ol) is formed as an intermediate:

$$
H_2C{=}CH{-}\underset{\displaystyle OH}{CH_2} + H_2O_2 \xrightarrow{(WO_3)} \underset{\text{Glycidol}}{H_2C\!-\!CH\!-\!CH_2OH} \xrightarrow{H_2O} \underset{\displaystyle OH\;\;OH\;\;OH}{CH_2-CH-CH_2}
$$

Properties

Glycerol is a colourless, syrupy liquid with a sweet taste. It boils at 290°C (563 K). It is completely miscible with water or ethanol in all proportions but is insoluble in ether. Completely dry glycerol solidifies to crystals at 18°C (291 K). Its chemical properties are those of its three alcohol groups. When it is *oxidised* either a primary or a secondary hydroxy group may be attacked to give glyceraldehyde and dihydroxyacetone (see glycerose). With stronger oxidising agents glyceric acid (2,3-dihydroxypropionic acid), $HOCH_2$—CHOH—COOH, is formed, whose phosphate ester is an intermediate in both alcoholic fermentation and in photosynthesis.

Teichoic acids are polymeric phosphate-diesters of polyhydric alcohols; they occur in bacterial cell walls. The simplest teichoic acids can be made from glycerol, for example, those isolated from *Lactobacillus casei*, which consist of a polyester chain made up from 1,3-diphosphorylated glycerol molecules also esterified at the 2-position by D-alanine:

$$
\left[-O-CH_2-\underset{\displaystyle H}{\overset{\displaystyle OR}{C}}-CH_2O-\underset{\displaystyle O}{\overset{\displaystyle OH}{\underset{\|}{P}}}-\right]_n
\qquad R = \underset{\displaystyle CH_3}{H-\overset{\displaystyle -C=O}{C}-NH_2}
$$

Not only D-alanine, but also glucose and other sugars, may take the place of R.

Uses

Glycerol has many uses. It is used in the preparation of lotions, toothpastes and cosmetics, as a finishing agent in the textile industry, and for maintaining the moisture content of tobacco. It is also used as a brake fluid (in hydraulic brakes), admixed with water as an antifreeze (in motor-cars and gas meters), as a plasticiser in films, and as a hygroscopic constituent in typewriter ribbons, copying inks and printing inks. A small amount of glycerol production (*ca* 4%) goes to explosive manufacturers for the making of *nitroglycerine* (glycerol trinitrate) and *dynamite*. In addition it takes part as an ester in the making of alkyd resins.

2.31.2.2 Alkyd resins are polyesters made by polycondensation reactions involving polyhydric alcohols and aliphatic or aryl dicarboxylic acids or their anhydrides. For example, when glycerol and phthalic anhydride are heated together at about 250°C (525 K), an adduct is formed which then, by loss of water, forms a polyester that has a chain structure and is meltable; this reacts with excess phthalic anhydride to form a cross-linked glyptal resin that cannot be melted:

Glycerol   Phthalic anhydride

Glyptal resin

These polyester resins (alkydals, glyptals) play a very important role as the raw materials used in the paint, lacquer and varnish industry. If in the preparation of the alkyd resins, non-drying oils such as castor oil, or drying oils such as linseed oil are added, an extensive network of alkyd resin modified by the presence of the long chains of the oil molecules, results. These are used, respectively, as stoving enamels and as lacquers or enamels which dry in air. They adhere well, are resistant to heat, and are weatherproof. Resopal is made from a combination of alkyd resins and melamine resins.

Unsaturated polyester resins (UP) are made from unsaturated starting materials such as maleic acid or unsaturated diols. These are dissolved in a solvent which can be polymerised, such as styrene. Polymerisation is induced by addition of a peroxide. A highly cross-linked copolymer is formed. Its mechanical properties are further improved by adding, for example, glass fibre to the reaction mixture.

2.31.2.3 The More Important Derivatives of Glycerol
Halogenohydrins, and especially chlorohydrins, are very reactive compounds which can be useful intermediates in synthetic procedures.

When glycerol is treated with hydrogen chloride under mild conditions two chlorohydrins are formed; the main product is *3-chloropropane-1,2-diol* together with a small amount of *2-chloropropane-1,3-diol*:

CH₂—Cl
CH—OH
CH₂—OH
3-Chloropropane-1,2-diol

CH₂—OH
CH—OH
CH₂—OH
Glycerol

+ HCl
−H₂O

CH₂—OH
CH—Cl
CH₂—OH
2-Chloropropane-1,3-diol

+ HOCl

CH₂
CH
CH₂—OH
Allyl alcohol

These two chloropropanediols are made industrially by addition of hypochlorous acid to an aqueous solution of allyl alcohol.

2,3-Dichloropropan-1-ol and *1,3-dichloropropan-2-ol* are made by treating glycerol with hydrogen chloride in the presence of acetic acid or by the action of hypochlorous acid on allyl chloride (see the preparation of glycerol, method (b), p. 309).

Glycidol, 2,3-epoxypropan-1-ol, is a colourless liquid, b.p. 166–167°C (439–440 K) (decomp.), which can be made by treating epichlorohydrin with potassium acetate or by the reaction of ethanolic sodium hydroxide with the mixture of chloropropanediols made from glycerol:

H₂C
HC
CH₂—Cl
Epichlorohydrin
(1-Chloro-2,3-epoxypropane)

+ H₃CCOO⁻K⁺
− KCl

H₂C
HC
CH₂—OCOCH₃

+ H₂O
−H₃CCOOH

H₂C
HC
CH₂—OH
Glycidol

(alcohol, NaOH)
−HCl
−HCl

CH₂—Cl
CH—OH
CH₂—OH
3-Chloropropane-1,2-diol

CH₂—OH
CH—Cl
CH₂—OH
2-Chloropropane-1,3-diol

Epichlorohydrin (1-chloro-2,3-epoxypropane, chloromethyloxiran) is made industrially from propene, *via* allyl chloride (see process (b), p. 309). It is a colourless liquid, b.p. 116.5°C (389.7 K), which smells like chloroform. Because it has a reactive epoxide ring and chlorine atom, it is a useful synthetic starting material, for example for the preparation of glycerol and of epoxide resins. It has been shown to be carcinogenic in some animals.

2.31.2.4 Epoxide resins are *polyethers* which are made from condensation reactions of epoxides having a vicinal reactive group with polyhydric alcohols or phenols. The most commonly used reagents are *epichlorohydrin* and di(4-hydroxyphenyl)dimethylmethane (4,4′-isopropylidenediphenol, bisphenol A, Dian), the latter compound being made from phenol and acetone.

When these two compounds are heated together in the presence of alkali at 100–150°C (375–425 K) *diepoxides* are formed whose relative molecular masses depend upon the relative molar ratio of epichlorohydrin to bisphenol A which is used. If more than two molecules of epichlorohydrin are used for every one molecule of bisphenol A, then bisphenol A-bisepoxy ether is almost the sole product:

$$CH_2-CH-CH_2 \quad + \quad HO-\langle\rangle-\overset{\overset{CH_3}{|}}{\underset{\underset{CH_3}{|}}{C}}-\langle\rangle-OH \quad + \quad CH_2-CH-CH_2$$

Epichlorohydrin   Bisphenol A

$$\xrightarrow{(OH^\ominus)} \quad CH_2-CH-CH_2-O-\langle\rangle-\overset{\overset{CH_3}{|}}{\underset{\underset{CH_3}{|}}{C}}-\langle\rangle-O-CH_2-CH-CH_2$$

Bisphenol A-bischlorohydrin

$$\xrightarrow[-2HCl]{(OH^\ominus)} \quad H_2C-CH-CH_2-O-\langle\rangle-\overset{\overset{CH_3}{|}}{\underset{\underset{CH_3}{|}}{C}}-\langle\rangle-O-CH_2-CH-CH_2$$

Bisphenol A-bisepoxy ether (Bisphenol A-diglycidyl ether)

If smaller amounts of epichlorhydrin are used the product consists of high molecular weight diepoxides with the general formula:

$$H_2C-CH-CH_2-\left[O-\langle\rangle-\overset{\overset{CH_3}{|}}{\underset{\underset{CH_3}{|}}{C}}-\langle\rangle-O-CH_2-CH-CH_2\right]_n$$

$$O-\langle\rangle-\overset{\overset{CH_3}{|}}{\underset{\underset{CH_3}{|}}{C}}-\langle\rangle-O-CH_2-CH-CH_2$$

The relative molecular masses of the diepoxides (> 3000) are low compared to those of other artificial resins, but the addition of, for example, diamines or the cyclic anhydrides of dicarboxylic acids, *e.g.* phthalic anhydride, leads to products with higher molecular weights. Their addition brings about extensive cross-linking of the chains, which leads to a marked hardening of the resins.

The epoxide resins (Araldite, Epikote) serve as the raw materials for making varnishes and lacquers, above all for stoving enamels, in conjunction with polyurethanes as adhesives, as surface protectants on metals, and in the electrical industry as resins for moulding, usually as mixtures with fillers such as powdered quartz, mineral fibres or fibreglass.

2.31.2.5 Propylene oxide (methyloxiran, 1,2-epoxypropane), the methyl derivative of ethylene oxide, is a colourless liquid of b.p. 34°C (307 K). It is obtained by the Halcon process from the reaction of *propene* with *hydroperoxides* in the presence of tungsten/molybdenum/vanadium or titanium catalysts:

$$C_6H_5-\underset{\underset{OOH}{|}}{CH}-CH_3 \quad + \quad H_3C-CH=CH_2 \quad \longrightarrow \quad H_3C-CH-CH_2 \quad + \quad C_6H_5-CH=CH_2$$

1-Phenylethyl   Propene   Propylene   Styrene
hydroperoxide   oxide

Industrially propylene oxide is converted into propane-1,2-diol by treating it with water. This diol is used in the plastics industry as a starting material for making polyesters, epoxide resins and polyurethane foams. Like poly(ethylene glycols) (see p. 303), higher molecular weight poly(propylene glycols) are used for the preparation of non-ionic detergents, while those of lower molecular weight have use as plasticisers. Propylene oxide has been shown to be carcinogenic in some animals.

2.31.2.6 Nitroglycerol (nitroglycerine, glyceryl trinitrate) is an important *explosive*. Industrially it is made by running glycerol into a strongly cooled mixture of conc. nitric and sulphuric acids, known as *nitrating acid*. The temperature must not rise above 30°C (300 K); usually it is held at 10–20°C (280–290 K) or lower. The nitroglycerol which is formed separates out as a colourless oil on the surface of the acid and is purified by washing it with water and sodium carbonate solution.

$$\begin{array}{l} CH_2-OH \\ | \\ CH-OH \\ | \\ CH_2-OH \end{array} \; + \; 3\;HO-NO_2 \quad \longrightarrow \quad \begin{array}{l} CH_2-O-NO_2 \\ | \\ CH-O-NO_2 \\ | \\ CH_2-O-NO_2 \end{array} \; + \; 3\;H_2O$$

$\qquad$ Glycerol $\qquad\qquad\qquad\qquad\qquad\qquad\qquad$ Nitroglycerol

The name 'nitroglycerol' is unsatisfactory because the compound is not in fact a nitro-compound but rather the trinitrate ester of glycerol. This incorrect name derives from the use of *nitrating acid* in its preparation; in many instances this acid is used to bring about *nitration*.

Nitroglycerol is a colourless, oily liquid; it is poisonous. In small amounts it can be burned safely, but it explodes violently if heated strongly or on impact, giving rise to large quantities of gaseous combustion products. When absorbed in kieselguhr it is insensitive to shock; this mixture is known as *dynamite (Alfred Nobel, 1886)* and can only be made to explode by the use of a detonator. *Blasting gelatine* is made by dissolving about 7% nitrated cotton (see nitrocellulose) in nitroglycerol.

Nitroglycerol causes blood vessels to dilate strongly and hence brings about a temporary fall in blood pressure. It is used for this purpose medicinally. The action of INN *Glycerol nitrate, Nitrolingual*, depends upon its reduction to the neurotransmitter nitrogen monoxide, NO. The active reducing agent in the body is cysteine (p. 821).

Instead of glycerol, alkyd resins and also the structurally related compound 2-hydroxy-methyl-2-methylpropane-1,3-diol [1,1,1-tris(hydroxymethyl)ethane] can be used for making explosives. The latter compound has three primary alcohol groups and is obtained from the reaction of propionaldehyde with three molecules of formaldehyde. In this reaction the reactive methylene group of propionaldehyde undergoes aldol reactions with two molecules of formaldehyde whilst at the same time its aldehyde group takes part in a crossed *Cannizzaro* reaction with the third molecule of formaldehyde and is reduced to a primary alcohol group; formic acid is also produced:

$$3\,H-C\!\!\overset{H}{\underset{O}{\diagdown}} \; + \; H_3C-CH_2-C\!\!\overset{H}{\underset{O}{\diagdown}} \; + \; H_2O \quad \longrightarrow \quad H_3C-\overset{\displaystyle CH_2OH}{\underset{\displaystyle CH_2OH}{\overset{|}{\underset{|}{C}}}}-CH_2OH \; + \; HCOOH$$

$\qquad\qquad$ Propionaldehyde $\qquad\qquad\qquad\qquad\qquad$ 1,1,1-Tris(hydroxymethyl) $\qquad$ Formic
$\qquad\qquad\qquad\qquad\qquad\qquad\qquad\qquad\qquad\qquad\qquad$ -ethane $\qquad\qquad\qquad\qquad$ acid

2,3-Dimercaptopropanol (1,2-dithio-glycerol), (INN: *Dimercaprol, Sulfactin*) is a highly efficient antidote 'BAL' (*British-Anti-Lewisite*) to the blistering arsenic-containing chemical warfare agents of the Lewisite-type, *e.g. Lewisite* = 2-chlorovinylarsenicdichloride. The toxicity of these arsenic compounds is due to their bonding to the mercapto groups in enzymes, which inhibits their reactivity and prevents them from operating. BAL is administered to counteract this by reacting with the *Lewisite* as follows:

$$\begin{array}{l} H_2C-SH \\ | \\ HC-SH \\ | \\ H_2C-OH \end{array} \; + \; \overset{Cl}{\underset{Cl}{\diagdown\!\!\diagup}}As-CH=CHCl \quad \xrightarrow{-2\,HCl} \quad \begin{array}{l} H_2C-S \\ | \qquad\;\; \diagdown \\ HC-S \qquad As-CH=CHCl \\ | \qquad\;\; \diagup \\ H_2C-OH \end{array}$$

$\qquad$ BAL $\qquad\qquad\qquad$ Lewisite

BAL has therapeutic use against heavy metal poisons, especially for acute arsenic or mercury poisoning.

2.31.3 Tetrahydric Alcohols

The simplest tetrahydric alcohol is **threitol**. It contains two asymmetric carbon atoms which are, however, *structurally identical*. Because of this, the number of optically active forms (2^n) is reduced to two, and in addition there is an optically inactive *meso* form, known as *erythritol*, which cannot be resolved. Thus there are three isomers, as follows:

CH₂OH	CH₂OH	CH₂OH
HO—C—H	H—C—OH	H—C—OH
H—C—OH	HO—C—H	H—C—OH
CH₂OH	CH₂OH	CH₂OH
(2R,3R)-Threitol	(2S,3S)-Threitol	Erythritol
D-Threitol	L-Threitol	(*meso*-Erythritol)

Of these, only erythritol occurs naturally, in algae and as esters in lichens.

The two threitols are *enantiomers*, although they show a two-fold axis of symmetry (C_2-symmetry). If the *Fischer* projection formulae of the threitols are rotated in the plane of the paper by 180°, in each case the resultant formula is identical to the initial one. In order to describe satisfactorily the existence of enantiomers of molecules which have axes of symmetry, it is necessary to describe them as *chiral* (see p. 120). In contrast, there is a plane of symmetry which lies through a molecule of erythritol, since the ligands attached to each of the two asymmetric carbon atoms are mirror images of each other. This compound is *achiral*. Because of their symmetry properties, molecules of threitol (m.p. 119°C [392 K]) cannot be unequivocally assigned to either a D or an L configuration. On the other hand, a precise (R) or (S) configuration can be assigned to each chiral centre. (2R,3R)-Threitol is obtained by reducing D-threose and is therefore described as D-threitol.

Pentaerythritol is a branched-chain tetrahydric alcohol. It is prepared by the action of calcium hydroxide on a mixture of acetaldehyde and formaldehyde (the latter in excess). In this aldol-like addition reaction the *reactive methyl group* of acetaldehyde reacts with the carbonyl groups of formaldehyde molecules and all three of its hydrogen atoms are replaced by hydroxymethyl groups. The intermediate 3-hydroxy-2,2-bis(hydroxymethyl)-propanal which is formed is then reduced by more formaldehyde in a crossed *Cannizzaro* reaction, giving pentaerythritol and formic acid.

3-Hydroxy-2,2-bis-(hydroxymethyl)propanal   Pentaerythritol

Pentaerythritol (m.p. 262°C [535 K]) is used in the preparation of alkyd- and other polyester resins, and to improve the quality of drying oils. Its tetranitrate is the explosive pentaerythritol tetranitrate (PETN, nitropenta). The plastic explosive *Semtex* is made up from a mixture of PETN and RDX which is finely dispersed in a partially organic matrix material.

2.31.4 Pentahydric Alcohols (Pentitols)

The asymmetric carbon atoms 2 and 4 of pentitols are structurally identical.

$$HOCH_2-CHOH-CHOH-CHOH-CH_2OH$$
$$(2) \quad (3) \quad (4) \quad (5)$$

In consequence, as in the case of tetritols, one *meso* and two *optically active* pentitols should exist. As well as the two enantiomers D(+)-**arabitol** and L(−)-**arabitol**, there are in fact *two* optically inactive diastereomeric pentitols, **adonitol (ribitol)** and **xylitol**, *i.e.* in all there are four pentitols:

CH$_2$OH	CH$_2$OH	CH$_2$OH	CH$_2$OH
HO–C–H	H–C–OH	H–C–OH (S)	H–C–OH (R)
H–C–OH	HO–C–H	------ H–C–OH (3)|r ·	⋯⋯ HO–C–H (3)|s ⋯⋯
H–C–OH	HO–C–H	H–C–OH (R)	H–C–OH (S)
CH$_2$OH	CH$_2$OH	CH$_2$OH	CH$_2$OH
D(+)-Arabitol,	L(−)-Arabitol,	Adonitol	Xylitol
D(+)-Arabinitol,	L(−)-Arabinitol,		
(2R,4R)-Arabitol	(2S,4S)-Arabitol		

Adonitol occurs in pheasant's eye plants (*Adonis vernalis*), and D-arabitol in fungi and in lichens. They can be made synthetically by reduction of the corresponding pentoses (arabinose, ribose or xylose).

Teichoic acids (see p. 309), which consist of polyester chains made up from adonitol linked 1,5 to phosphoric acid, form the greater part of the cell walls of gram-positive bacteria. The hydroxy groups at the 2,3 and 4 positions may be linked to D-alanine or to sugars.

As in the case of erythritol, there is a plane of symmetry in molecules of adonitol and xylitol, which passes through carbon atom 3. Both compounds are therefore *meso* forms. A carbon atom such as C(3) in adonitol or xylitol, which has two different and two similar but mirror-image substituents attached to it, is described as a *pseudoasymmetric* carbon atom. The symbols for the absolute configuration at a pseudoasymmetric centre are *r* and *s*. They are assigned using the CIP system, with the extra proviso that a ligand with an (*R*)-configuration takes precedence over a similar ligand with an (*S*)-configuration.

Problems[1] associated with the definition of pseudoasymmetric centres can be elucidated by using *Mislow's* analytical approach (see p. 120). The 3-carbon atoms of the *meso* forms are stereogenic, since diastereomers arise by exchanging the positions of the H and OH (permutation isomerism); they are achirotopic as far as the local symmetry is concerned. (See the plane of symmetry in adonitol and xylitol.) In contrast, exchange of the H and OH on C(3) of D-arbitol or L-arabitol does not provide stereoisomers; the 3-carbon atom is *not stereogenic*. It is however chirotopic, because it has four different ligands attached to it.

2.31.5 Hexahydric Alcohols (Hexitols)

Hexitols contain *four* asymmetric carbon atoms, consisting of two similar pairs (2 and 5; 3 and 4).

$$HOCH_2–CHOH–CHOH–CHOH–CHOH–CH_2OH$$
    (1)   (2)   (3)   (4)   (5)   (6)

---

[1] See *J. K. O'Loane*, Chem. Rev. *80*, 41 (1980).

In consequence, the number of stereoisomers is 10, and not 16. Only the three most important of the ten known isomeric hexitols are shown here:

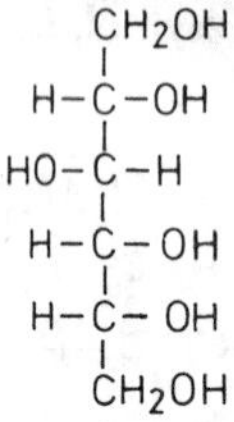

D-Sorbitol
D-Glucitol

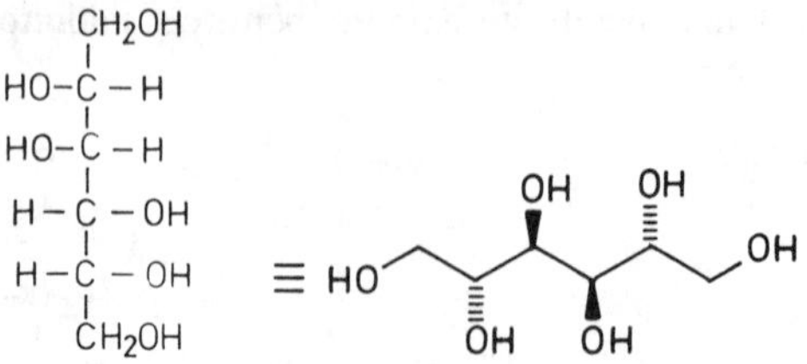

Fischer projection   Skeleton formula
D-Mannitol

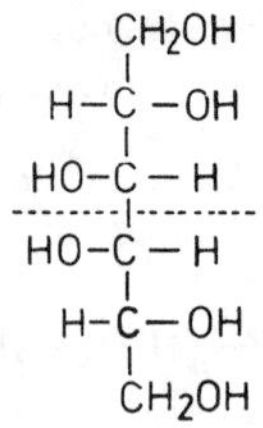

Dulcitol,
Galactitol

D-Sorbitol occurs plentifully in mountain ash (rowan) berries, D-mannitol in the exudates from mannas, and dulcitol also in exudates from mannas. Mannitol is a normal constituent of silage, where it is formed by bacterial reduction of fructose. They can be made synthetically by catalytic reduction, using nickel, of the appropriate aldo- or keto-hexoses. They form good crystals and taste sweet. D-Sorbitol is a sweetening agent which can be used by diabetics. It is stable up to 150°C (423 K), and hence can be used in cooking and baking. Its viscous aqueous solutions are used instead of glycerol, for example, in hectographic inks, printer's ink and to make papers more flexible.

2.32 Aliphatic Hydroxy-Aldehydes and Hydroxy-Ketones

Both of these classes of compounds arise as oxidation products of polyhydric alcohols.

2.32.1 Hydroxy-aldehydes

Hydroxy-aldehydes contain both an aldehyde group and an alcohol group. The simplest examples are glycolaldehyde and glyceraldehyde.

2.32.1.1 Glycolaldehyde, hydroxyacetaldehyde, $HOCH_2$—CHO, is the simplest *aldehyde sugar* or *aldose*, with the characteristic grouping

$$—CH(OH)—C\overset{\displaystyle H}{\underset{\displaystyle O}{<}}$$

Preparation

(a) *By oxidation of ethylene glycol with hydrogen peroxide* in the presence of iron(II) salts (*Fenton's* reagent).

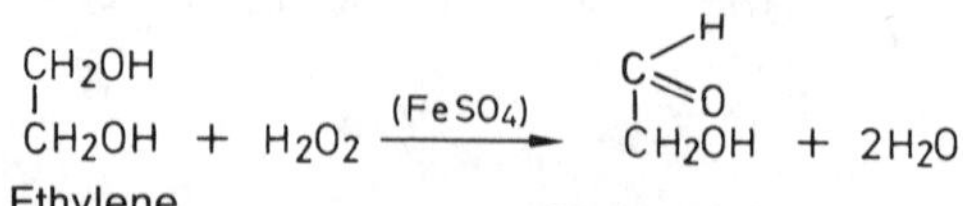

Ethylene
glycol

Glycolaldehyde

(b) From *vinyl acetate* (or divinyl ether) by *oxidation,* using hydrogen peroxide or a peracid in t-butanol in the presence of *osmium(VIII) oxide*:

$$H_2C=CH-O-COCH_3 + H_2O_2 \xrightarrow{(OsO_4)} \underset{\underset{OH}{|}}{CH_2}-C\underset{O}{\overset{H}{\lessgtr}} + H_3CCOOH$$

Properties. Glycolaldehyde is a syrupy liquid with a sweet taste. It boils at 96–97°C (369–370 K) and very readily polymerises. There is also a *dimeric* form, which is crystalline, m.p. 96°C (369 K). It dissolves readily in water and gradually changes into the monomeric form. With water, glycolaldehyde forms a stable *hydrate* $HOCH_2$—$CH(OH)_2$ (*cf.* chloral hydrate). It reduces ammoniacal silver nitrate and *Fehling's* solution.

A characteristic reaction of α-hydroxy-carbonyl compounds is that with *phenylhydrazine,* in which not only is the carbonyl group converted into its phenylhydrazone derivative, but in addition the neighbouring hydroxy group is converted into a carbonyl group by a second molecule of phenylhydrazine, ammonia and aniline also being formed. This newly formed carbonyl group reacts with a third molecule of phenylhydrazine to give a bisphenylhydrazone or *osazone*:

$$\underset{\underset{CH_2OH}{|}}{\overset{}{C}}\overset{\diagup H}{\underset{\lessgtr O}{}} \xrightarrow[-H_2O]{+C_6H_5NHNH_2} \underset{\underset{CH_2OH}{|}}{HC}=N-NHC_6H_5 \xrightarrow[-C_6H_5NH_2,\,-NH_3,\,-H_2O]{+2\,C_6H_5NHNH_2} \begin{array}{l} HC=N-NHC_6H_5 \\ |\\ HC=N-NHC_6H_5 \end{array}$$

$$\text{osazone}$$

The *mechanism* of this oxidation is not unambiguously settled but may proceed *via* a pericyclic reaction involving the neighbouring hydrazone and hydroxy groups. Formation of osazones has played an important role in determining the constitution of sugars (see p. 429).

2.32.1.2 Glyceraldehyde, dihydroxypropanal, $HOCH_2$—CHOH—CHO, is a dihydroxyaldehyde. Because it is closely related structurally to the sugars, it was chosen as the *reference compound* in assigning optically active compounds to either the D or L series.

DL-Glyceraldehyde crystallises from aqueous methanol in colourless needles, m.p. 142°C (415 K). This form is dimeric.

It can be prepared by the *partial oxidation of glycerol,* using for example, hydrogen peroxide in the presence of iron(II) salts or an alkaline solution of bromine (sodium hypobromite). In these processes the secondary alcohol group or the primary alcohol group may be oxidised, giving rise to a mixture of glyceraldehyde and dihydroxyacetone (1,3-dihydroxypropan-2-one). These two products are in *equilibrium* with one another, and in the presence of dry pyridine are interconvertible, reaction proceeding *via* proton migration and formation of an intermediate *enediol* (*Lobry de Bruyn-van Ekenstein rearrangement,* 1895):

| Glyceraldehyde | *Enediol* | Dihydroxyacetone |

(The directions of the arrows in these formulae signify the migration of hydrogen atoms and not of electrons)

The mixture of the two oxidation products, in which the major component is glyceraldehyde, is known as *glycerose*.

Both glyceraldehyde and dihydroxyacetone provide the same *osazone* on reaction with phenylhydrazine:

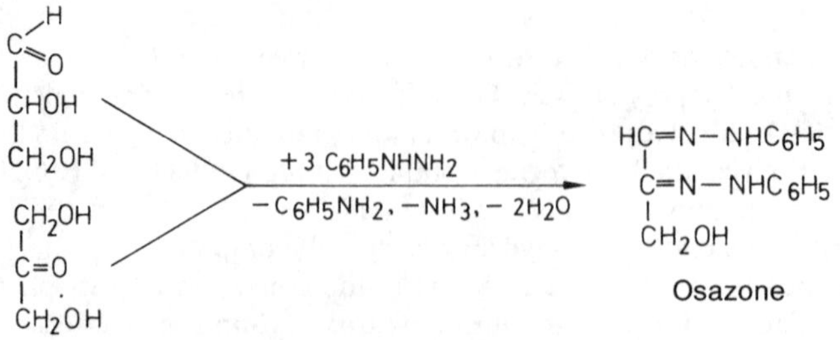

2.32.2 Hydroxyketones (Keto-alcohols)

The simplest hydroxyketone, *hydroxyacetone* or *acetol*, $H_3C—CO—CH_2OH$, is derived from acetone by replacement of a hydrogen atom with a hydroxy group. In general, compounds having the grouping $—CO—CH_2OH$ are known as *ketols*.

Acetol is obtained by treating chloro- or bromo-acetone with potassium formate to give acetol formate (2-oxopropyl formate), and then letting this react with methanolic potassium hydroxide:

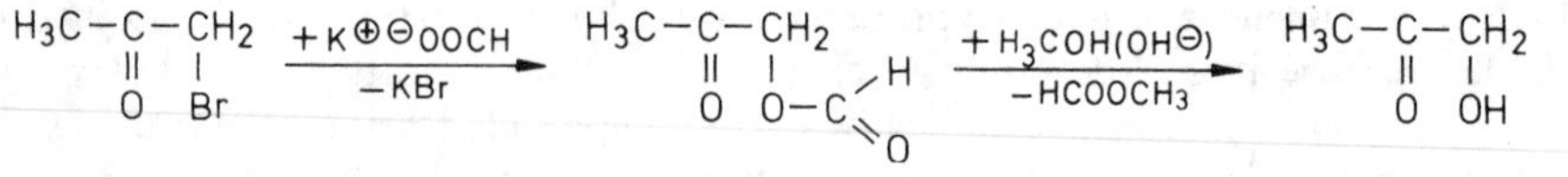

Bromoacetone   2-Oxopropyl formate   Hydroxyacetone

Hydroxyacetone is a colourless liquid, b.p. 145°C (418 K), which has a sweet taste. Like hydroxyaldehydes it reduces *Fehling's* solution. With phenylhydrazine it gives the same *osazone* as that obtained from methylglyoxal, $H_3C—CO—CHO$.

Acetoin (3-hydroxybutan-2-one), $H_3C—CHOH—CO—CH_3$, is a liquid with b.p. 142°C (415 K); it readily dimerises. It is obtained either by reduction of diacetyl, $CH_3—CO—CO—CH_3$, or by fermentation of sugars using, for example, *Bacillus polymyxa*.

Acyloins[1] are hydroxyketones which have identical hydrocarbon or heterocyclic residues attached to the two ends of the $—CH(OH)—CO—$ grouping.

Dihydroxyacetone, $HOCH_2—CO—CH_2OH$, is derived from glycerol and is the simplest keto-sugar. It is formed together with glyceraldehyde by mild oxidation of glycerol. Enzymatic dehydrogenation of glycerol by either *Acetobacter suboxydans* or *A. xylinum* provides exclusively dihydroxyacetone, since these bacteria attack only secondary alcohol groups.

---

[1] See S. M. McElvain, The Acyloins, Org. Reactions 4, 256 *ff.* (1948); K. T. Finlay, The Acyloin Condensation as a Cyclisation Method, Chem. Rev. 64, 573 (1964).

2.33 Aliphatic Dialdehydes, Ketoaldehydes and Diketones

Because they have two carbonyl groups these three classes of compounds are particularly reactive and they serve as starting materials for the preparation of a range of cyclic compounds.

2.33.1 Dialdehydes

2.33.1.1 Glyoxal (oxaldehyde, diformyl, ethanedial), OHC—CHO is the parent compound of this homologous series.

Preparation
(a) *By oxidation of paraldehyde with selenium dioxide*[1] or *nitric acid* and trapping the glyoxal as its sodium hydrogen sulphite adduct:

$$H_3C - C\overset{H}{\underset{O}{\diagdown}} \quad \xrightarrow[-H_2O]{[O]} \quad \overset{H}{\underset{O}{\diagup}}C - C\overset{H}{\underset{O}{\diagdown}}$$

Glyoxal

(b) *Industrially by oxidation of ethylene glycol with atmospheric oxygen* in the presence of a silver or copper catalyst at 300°C (575 K):

$$HOCH_2 - CH_2OH \;+\; O_2 \quad \xrightarrow{(Ag)} \quad \overset{H}{\underset{O}{\diagup}}C - C\overset{H}{\underset{O}{\diagdown}} \;+\; 2\,H_2O$$

Properties. Glyoxal polymerises readily, giving colourless *polyglyoxal*. This can be reconverted into the monomeric form by heating it with phosphorus pentoxide. Glyoxal forms a green vapour which has a pungent smell and which, when cooled, condenses to yellow prisms, m.p. 15°C (288 K). It is the simplest organic compound which is coloured; the colour is due to the presence of the two adjacent carbonyl groups. In aqueous solution glyoxal forms a dihydrate which is isolable.

Aqueous alkali brings about an *intramolecular disproportionation* reaction leading to the formation of a salt of glycollic acid. This is a special case of the *Cannizzaro* reaction:

$$\begin{array}{c} C\overset{H}{\underset{O}{\diagdown}} \\ | \\ C\overset{H}{\underset{O}{\diagdown}} \end{array} \quad \xrightarrow{+NaOH} \quad \begin{array}{c} CH_2OH \\ | \\ C\overset{O}{\diagup}\underset{O^{\ominus}\;Na^{\oplus}}{} \end{array}$$

Glyoxal undergoes the customary addition and condensation reactions of aldehydes twice, for example, with 2 molecules of hydroxylamine it gives the simplest example of a

---

[1] See *N. Rabjohn*, Selenium Dioxide Oxidation, Org. Reactions 5, 331 *ff.* (1949).

1,2-dioxime, *glyoxime*. Phenylhydrazine gives the same *osazone* as that obtained from glycolaldehyde.

2.33.1.2 Malonaldehyde is best obtained from β-ethoxyacrolein acetal by shaking the latter with water; precise neutralisation of the resultant yellow solution with sodium hydroxide provides the mesomerically stabilised sodium monoenolate (*Hüttel*, 1941):

$$CH=CH-C\begin{smallmatrix}H\\|\\OC_2H_5\\\diagdown OC_2H_5\end{smallmatrix} \quad \xrightarrow[-3\,C_2H_5OH]{+2\,H_2O} \quad CH=CH-C\begin{smallmatrix}\diagup H\\\diagdown O\\|\\OH\end{smallmatrix} \quad \xrightarrow[-\,H_2O]{+\,NaOH} \quad \left[\begin{array}{c}{}^{\ominus}O-CH=CH-CH=O\\ \updownarrow\\ O=CH-CH=CH-O^{\ominus}\end{array}\right] Na^{\oplus}$$

$$(I)$$

Treatment of this sodium salt with a solution of hydrogen chloride in ether at 0°C (273 K) gives the enol form of free malonaldehyde. This is not stable and has a strong tendency to polymerise. Because of this, in synthetic work it is common practice to use instead a *diacetal of malonaldehyde*. The bis(dimethyl acetal), 1,1,3,3-tetramethoxy-propane, is made by the reaction of methyl vinyl ether with trimethyl orthoformate at 25–40°C (300–315 K) in the presence of boron trifluoride:

$$H_2C=CH-OCH_3 + HC(OCH_3)_3 \quad \xrightarrow{(BF_3)} \quad \begin{smallmatrix}H_3CO\\\diagdown\\H_3CO\diagup\end{smallmatrix}CH-CH_2-CH\begin{smallmatrix}\diagup OCH_3\\\diagdown OCH_3\end{smallmatrix}$$

Methyl vinyl ether

Malonaldehyde
bis(dimethylacetal)

2.33.1.3 Succinaldehyde (butanedial), OHC—CH$_2$—CH$_2$—CHO, is a pungent smelling oil, b.p. 170°C (443 K) which readily polymerises to a glassy material, especially when heated. It exists in two tautomeric forms.

Preparation. *By the reaction of carbon monoxide and hydrogen with acrolein diacetate* (1,1-diacetoxyprop-2-ene) *at* 120°C (390 K) *and* 200 bar (20 MPa) *in the presence of a dicobaltoctacarbonyl catalyst* (*Adkins*, 1949), *followed by hydrolysis with dilute hydrochloric acid* (see oxo-synthesis):

$$H_2C=CH-C\begin{smallmatrix}\diagup H\\|\\\diagdown O-COCH_3\end{smallmatrix}O-COCH_3 \quad \xrightarrow[{[Co_2(CO)_8]}]{+CO,\,+H_2} \quad \begin{smallmatrix}H\\\diagdown\\O\diagup\end{smallmatrix}C-CH_2-CH_2-C\begin{smallmatrix}\diagup H\\|\\\diagdown O-COCH_3\end{smallmatrix}O-COCH_3$$

$$\xrightarrow[-2\,H_3CCOOH]{+H_2O} \quad \begin{smallmatrix}H\\\diagdown\\O\diagup\end{smallmatrix}C-CH_2-CH_2-C\begin{smallmatrix}\diagup H\\\diagdown O\end{smallmatrix} \quad \rightleftharpoons \quad \begin{smallmatrix}H\\\diagdown\\HO\diagup\end{smallmatrix}C=CH-CH=C\begin{smallmatrix}\diagup H\\\diagdown OH\end{smallmatrix}$$

Dioxo form   Dienol form

Succinaldehyde

2.33.2 Ketoaldehydes

Methylglyoxal (pyruvic aldehyde, 2-oxopropanol), H$_3$C—CO—CHO is the first member of the series of ketoaldehydes. It is derived from glyoxal by replacing a hydrogen atom with a methyl group.

It can be made by oxidising acetol with hydrogen peroxide, by oxidising acetone with selenium dioxide, or by hydrolysis of *hydroxyiminoacetone* (isonitrosoacetone):

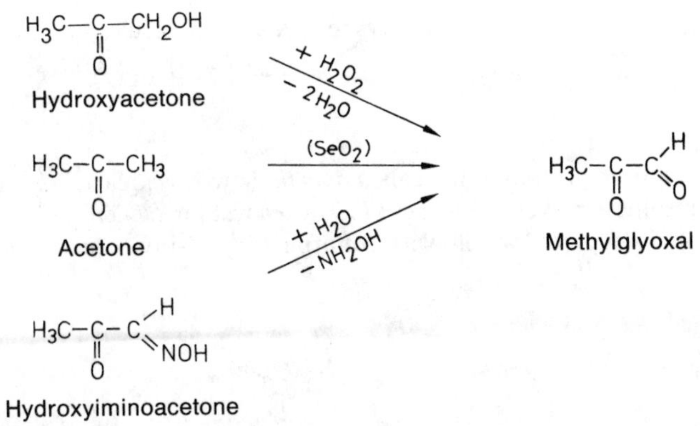

H₃C—C—CH₂OH
‖
O
Hydroxyacetone

$+ H_2O_2$
$- 2H_2O$
(SeO₂)

H₃C—C—CH₃
‖
O
Acetone

$+ H_2O$
$- NH_2OH$

H₃C—C—C
‖   ‖
O   O

H

H₃C—C—C
          ⟍H
‖   ⟍O
O
Methylglyoxal

H₃C—C—C
‖   ⟍NOH
O
Hydroxyiminoacetone

Methylglyoxal is a yellow liquid, b.p. 72°C (345 K) which readily polymerises. By the action of either alkalies or enzymes it undergoes *intramolecular disproportionation* to form lactic acid, in a special case of the *Cannizzaro* reaction:

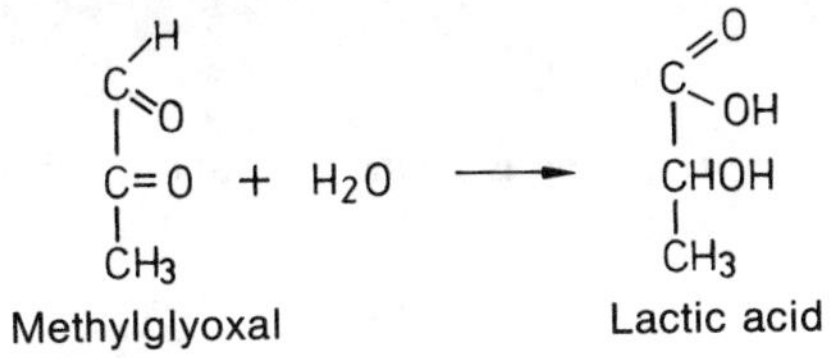

C
⟋H
⟍O
|
C=O + H₂O ⟶
|
CH₃

Methylglyoxal

C
⟍OH
⟋O
|
CHOH
|
CH₃

Lactic acid

2.33.3 Diketones

Diketones are classified as follows, based on the relative sites of the two carbonyl groups:

α or 1,2-Diketones   R—CO—CO—R,
β or 1,3-Diketones   R—CO—CH₂—CO—R,
γ or 1,4-Diketones   R—CO—CH₂—CH₂—CO—R.

2.33.3.1 1,2-Diketones

Diacetyl, butane-2,3-dione, *biacetyl*, (dimethylglyoxal), is the simplest 1,2-diketone. Like glyoxal and methylglyoxal it absorbs light in the visible region.

Preparation
(a) *By oxidation of butanone with selenium dioxide*:

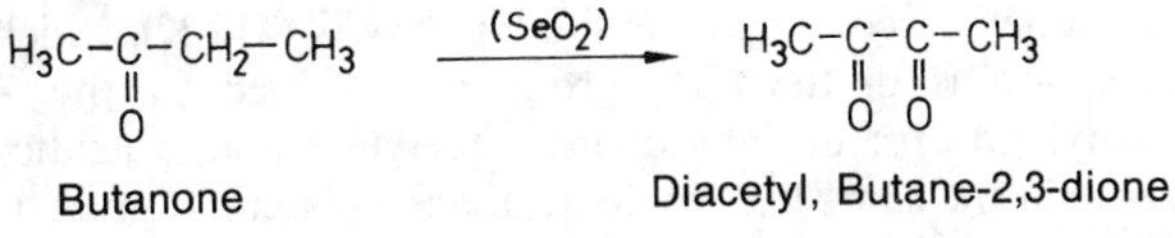

H₃C—C—CH₂—CH₃   (SeO₂)   H₃C—C—C—CH₃
‖   ⟶   ‖   ‖
O   O   O

Butanone   Diacetyl, Butane-2,3-dione

(b) *By acid hydrolysis of 3-hydroxyiminobutan-2-one* (itself made from butanone):

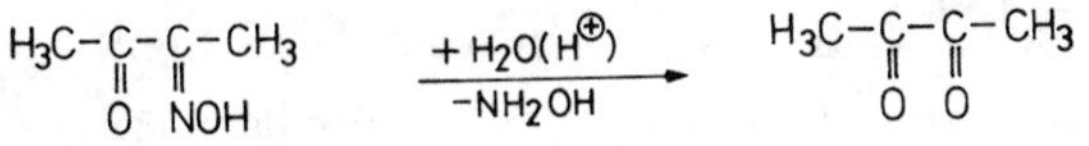

H₃C—C—C—CH₃ $+ H_2O (H^\oplus)$ H₃C—C—C—CH₃
‖   ‖ $- NH_2OH$ ‖   ‖
O   NOH   O   O

Properties. Diacetyl is a greenish yellow liquid, b.p. 88°C (361 K), which has a smell resembling that of *p*-benzoquinone. Together with acetoin, it provides the characteristic odour of butter (in which it occurs to the extent of 0.5 mg per kg).

Hydrogen peroxide breaks down diacetyl to give two molecules of acetic acid:

$$H_3C-\overset{\overset{O}{\|}}{C}-\overset{\overset{O}{\|}}{C}-CH_3 \ + \ H_2O_2 \ \longrightarrow \ 2\,H_3C-COOH$$

The dioxime of diacetyl, commonly called *dimethylglyoxime*, reacts with nickel salts to form red, very insoluble nickel complexes (*Tschugaev* test for nickel). The *chelate complex* has the following structure, which was confirmed by X-ray structure determination (*Rundle*, 1953):

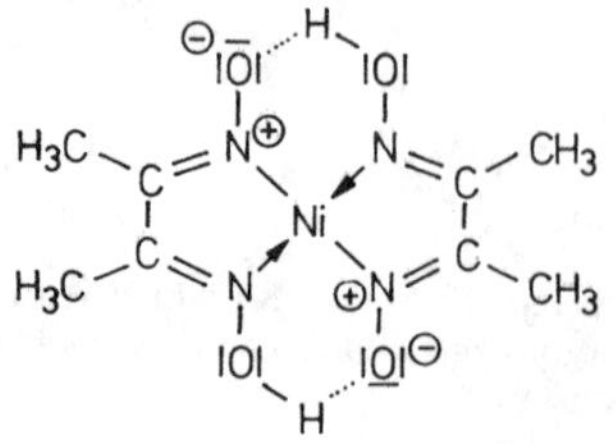

2.33.3.2 1,3-Diketones

The first compound in this series is **acetylacetone** (pentane-2,4-dione), $H_3C-CO-CH_2-CO-CH_3$.

Preparation

(a) A general method for the preparation of 1,3-diketones is by the *Claisen condensation of esters with ketones* in the presence of a *sodium alkoxide* or *sodamide*. Acetylacetone is made in this way from ethyl acetate and acetone:

$$H_3C-\overset{\overset{O}{\diagup\!\!\|}}{C}\diagdown_{OC_2H_5} \ + \ H_3C-\overset{\overset{O}{\|}}{C}-CH_3 \ \xrightarrow{\text{(NaOR)}} \ H_3C-\overset{\overset{O}{\|}}{C}-CH_2-\overset{\overset{O}{\|}}{C}-CH_3 \ + \ C_2H_5OH$$

Acetylacetone, Pentane-2,4-dione

(b) Acetylacetone is obtainable in very good yield *by the reaction of acetic anhydride with acetone* in the presence of boron trifluoride:

$$H_3C-\overset{\overset{O}{\|}}{C}-CH_3 + (H_3CCO)_2O \ \xrightarrow{\text{(BF}_3)} \ H_3C-\overset{\overset{O}{\|}}{C}-CH_2-\overset{\overset{O}{\|}}{C}-CH_3 \ + \ H_3CCOOH$$

Properties. Acetylacetone is a colourless liquid, b.p. 139°C (412 K), which has a pleasant smell. The hydrogen atoms of the CH_2 group are affected by the $-$I-effect of the neighbouring carbonyl groups; in consequence, acetylacetone is acidic and undergoes *keto–enol tautomerism* in the same way as do acetoacetic esters. It gives a red colour with iron(III) chloride. The following equilibrium lies very much on the side of the enol form:

$$H_3C-\overset{\overset{\text{|}\overset{\cdot\cdot}{O}\text{|}}{\|}}{C}-CH_2-\overset{\overset{\text{|}\overset{\cdot\cdot}{O}\text{|}}{\|}}{C}-CH_3 \ \rightleftharpoons \ H_3C-\overset{\overset{\text{|}\overset{\cdot\cdot}{O}\text{|}}{\|}}{C}-CH=\overset{\overset{\text{|}\underline{O}H}{\text{|}}}{C}-CH_3$$

Keto form (15%)   Enol form (85%)

Acetylacetone

As in the case of acetoacetic esters, enolisation is favoured because it results in the formation of a conjugated system and intramolecular hydrogen bonding. Furthermore, in this case there is marked *mesomeric stabilisation*, much more so than in acetoacetic esters:

Acetylacetone

Various light and heavy metal ions, such as Be, Al, Cr, Fe, Cu, *etc.* form *acetylacetonates* with acetylacetone. These are chelate complexes and differ greatly from the normal metal salts in their properties. They are soluble in organic solvents such as ether, benzene or chloroform, and can be distilled. They are not hydrolysed in aqueous solution.

2.33.3.3 1,4-Diketones

Acetonylacetone, hexane-2,5-dione, is the simplest 1,4-diketone. It is a colourless liquid, b.p. 194°C (467 K) with an aromatic smell. It is made by *alkaline hydrolysis of diacetylsuccinic acid* and pyrolysis of the resultant acid (ketonic hydrolysis):

Diacetylsuccinic acid
(Diethyl 2,5-dioxohexane-3,4-dicarboxylate)

Acetonylacetone
Hexane-2,5-dione

1,4-Diketones are not soluble in aqueous alkali and with metals form neither salts nor chelate complexes. One characteristic feature of 1,4-diketones is the ease with which they can be converted into five-membered ring heterocyclic compounds.

2.34 Saturated Aliphatic Dicarboxylic Acids

Dicarboxylic acids are the final products obtained when diprimary alcohols are oxidised. They contain two carboxyl groups which may be either adjacent to each other or separated from one another by one or more methylene groups. The names of the first members of the series are as follows:

| Oxalic acid | Malonic acid | Succinic acid | Glutaric acid | Adipic acid |

These names are for the most part derived from the natural sources from which the acids were obtained, and are the everyday names of these compounds; higher members of the series are described by their systematic names. Systematic nomenclature resembles that of monocarboxylic acids; the ending *dioic acid* is appended to the name of the alkane having the same number and arrangement of carbon atoms, *e.g. heptane-1,7-dioic acid*, $HOOC-(CH_2)_5-COOH$.

Dicarboxylic acids are generally crystalline substances. Their solubility in water decreases with increasing relative molecular mass. Except for oxalic acid they are stable towards oxidising agents. Their melting points alternate, those with even numbers of carbon atoms melting higher than those with odd numbers of carbon atoms (see Fig. 92).

As in the case of monocarboxylic acids, the first member of the series is a stronger acid than the others. This is because the two adjacent carboxyl groups exert a $-I$-effect on each other. Dissociation takes place in *two steps*. The acidity of the first step is notably higher than that of monocarboxylic acids, whereas that of the second is somewhat lower. For oxalic acid, which is the strongest dicarboxylic acid, the first dissociation constant $pK_a^1 = 1.46$, which is notably stronger than formic acid ($pK_a = 3.77$) and acetic acid ($pK_a = 4.76$). As might be expected, the strength of the acids decreases as the two carboxyl groups get further away from each other. Thus malonic acid has a first $pK_a^1 = 2.80$, whilst for succinic acid $pK_a^1 = 4.17$, about equal to the acidity of monocarboxylic acids.

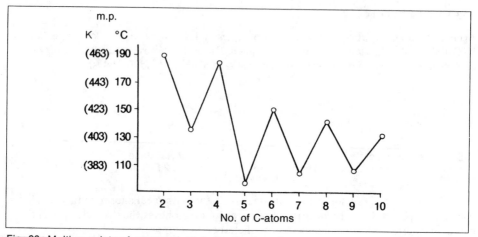

Fig. 92. Melting points of saturated aliphatic dicarboxylic acids

The chemical properties of dicarboxylic acids are those of the two carboxyl groups, which resemble those of the monocarboxylic acids.

There are some special reactions, which depend upon the positions of the two carboxyl groups relative to one another. Among these are the *effect of heat* on dicarboxylic acids. Thus, when *oxalic acid* is heated above its melting point for some time, it decomposes completely, giving carbon dioxide, carbon monoxide and water; formic acid is an intermediate in this process:

$$HOOC-COOH \xrightarrow[-CO_2]{} HCOOH \rightarrow CO + H_2O$$

Malonic acid, in which the two carboxyl groups are separated by one carbon atom, is easily decarboxylated by heating it to 140–150°C (410–420 K):

$$HOOC-CH_2-COOH \rightarrow H_3C-COOH + CO_2$$

Malonic acid   Acetic acid

When *succinic acid* or *glutaric acid* are heated for some time, best in the presence of acetic anhydride, *intramolecular loss of water* ensues with formation of a *cyclic anhydride*:

Succinic anhydride   Glutaric anhydride

In complete contrast, when *adipic acid* or *pimelic acid* (heptane-1,7-dioic acid) are heated with acetic anhydride, a cyclic ketone is distilled out of the mixture, respectively *cyclopentanone* and *cyclohexanone*, e.g.

Adipic acid   Cyclopentanone

These experimental findings form the basis of *Blanc's* rule (1905), which says that dicarboxylic acids whose carboxyl groups are separated from one another by 2 or 3 carbon atoms give cyclic anhydrides when they are heated with acetic anhydride, while under the same conditions dicarboxylic acids whose carboxyl groups are separated from one another by 4 or 5 carbon atoms give cyclic ketones.

Under these conditions dicarboxylic acids whose carboxyl groups are separated apart by more than 5 carbon atoms do not give cyclic products but instead undergo *intermolecular* loss of water with concomitant formation of linear polymeric anhydrides.

When the *calcium* or *barium* salts of dicarboxylic acids whose carboxyl groups are separated from one another by 4 or more carbon atoms are heated, cyclic ketones are formed, e.g.

Calcium pimelate   Cyclohexanone

The yields obtained from higher dicarboxylic acids are extremely low, but *Ruzicka* employed a similar method, using instead thorium or cerium salts, to prepare cyclic ketones with up to 34 carbon atoms.

The Most Important Saturated Dicarboxylic Acids

2.34.1 Oxalic acid

Oxalic acid is the most widely distributed acid in plants and, above all, is found as its mono-potassium salt in wood sorrel (*Oxalis acetosella*), in common sorrel (*Rumex acetosa*) and in rhubarb. It occurs in some cell fluids as the calcium salt, especially in algae and lichens. The bars of certain species of eucalyptus contain as much as 20% of it. Small amounts of calcium oxalate exist in the urine of animal organisms; in pathological cases (*oxaluria*) this can lead to the formation of bladder and kidney stones. Larger amounts of oxalic acid are poisonous because it upsets the calcium balance in the body.

Preparation

2.34.1.1 *By acid hydrolysis of cyanogen*, which is the dinitrile of oxalic acid, *oxalonitrile* (*Wöhler*, 1824):

$$
\begin{array}{c} C\equiv N \\ | \\ C\equiv N \end{array}
\;+\; 4\,H_2O \;\longrightarrow\;
\begin{array}{c} COOH \\ | \\ COOH \end{array}
\;+\; 2\,NH_3
$$

Cyanogen   Oxalic acid

This process is now only of historic interest.

2.34.1.2 *By the oxidation of cane sugar* with concentrated nitric acid in the presence of vanadium(V) oxide:

$$
C_{12}H_{22}O_{11} + 18\,[O] \xrightarrow[\;(V_2O_5)\;]{(HNO_3)} \; 6 \begin{array}{c} COOH \\ | \\ COOH \end{array} + 5\,H_2O
$$

2.34.1.3 Industrially oxalic acid is made by *heating sodium formate rapidly* to 360°C (630 K), best in the presence of sodium hydroxide, when evolution of hydrogen is accompanied by the formation of sodium oxalate:

$$
\begin{array}{c} H-COONa \\ \\ H-COONa \end{array}
\;\longrightarrow\;
\begin{array}{c} COONa \\ | \\ COONa \end{array}
\;+\; H_2
$$

The sodium oxalate is treated with calcium hydroxide to convert it into the almost insoluble calcium oxalate which is then converted into free oxalic acid by reaction with sulphuric acid. A genetic relationship exists between formic acid and oxalic acid, each being convertible into the other.

Properties. Oxalic acid crystallises out from aqueous media as colourless crystals, m.p. 101.5°C (374.7 K), which contain two molecules of water of crystallisation. If the dihydrate is heated, best in tetrachloromethane, anhydrous oxalic acid, m.p. 189.5°C (462.7 K), is obtained. It is readily soluble in water or ethanol but almost insoluble in ether. It forms two series of salts, the normal and the acid *oxalates*. Among its derivatives the following may be mentioned: *dimethyl oxalate*, m.p. 54°C (327 K), *diethyl oxalate*, b.p. 185.4°C (458.6 K), *oxalyl chloride*, ClOC—COCl, b.p. 63.5°C (336.7 K), *oxamide*, H_2NOC—$CONH_2$, m.p. 419°C (692 K, decomp.) and *oxamic acid* (oxalic acid monoamide), HOOC—$CONH_2$, m.p. 210°C (483 K, decomp.); they are often used as starting materials in synthetic work.

Uses. Oxalic acid dihydrate is used analytically as a *standard* in titrations involving either alkali or permanganate, and also for the quantitative determination of calcium as calcium oxalate. It is also used in the separation of the rare earths.

2.34.2 Malonic acid

Malonic acid occurs in the juice of sugar beet, but otherwise is not commonly found in plants. It was first made by the oxidation of malic acid, and it is from this that its name was derived.

Preparation. When potassium chloroacetate is treated with potassium cyanide, potassium cyanoacetate is formed. Cyanoacetic acid is the mononitrile of malonic acid, and acid hydrolysis of the potassium cyanoacetate gives malonic acid in about 85% yield.

$$Cl-CH_2-COO^{\ominus}K^{\oplus} \xrightarrow[-KCl]{+KCN} N\equiv C-CH_2-COO^{\ominus}K^{\oplus} \xrightarrow[-NH_4Cl,-KCl]{+2H_2O,\ +2HCl} H_2C\!\!\begin{array}{c}\diagup COOH\\ \diagdown COOH\end{array}$$

Properties. Malonic acid forms crystals which melt at 135.6°C (408.8 K); at slightly higher temperatures it loses carbon dioxide and acetic acid is formed. Its salts are called *malonates.*

If malonic acid is heated with phosphorus pentoxide at about 150°C (425 K) intramolecular loss of two molecules of water takes place giving, in poor yield, *carbon suboxide (Diels).*

$$H_2C\!\!\begin{array}{c}\diagup COOH\\ \diagdown COOH\end{array} \xrightarrow[-2H_2O]{(P_2O_5)} O=C=C=C=O$$

Malonic acid   Carbon suboxide

This is a poisonous gas, b.p. 7°C (280 K), with a pungent smell. It burns with a sooty blue-edged flame. Even at room temperature it polymerises to a red, amorphous material. It can be regarded as a diketene with four *cumulated* double bonds.

Penta-1,2,3,4-tetraen-1,5-dione (C_5O_2 = O=C=C=C=C=C=O) contains six cumulated double bonds. It is a stable yellow solid at temperatures below about −90°C (183 K).[1]

2.34.2.1 Malonic Esters

Esters of malonic acid, and especially the diethyl ester, diethyl malonate, are of more importance than the free acid. Diethyl malonate is made from potassium cyanoacetate by treating it with absolute ethanol and conc. sulphuric acid or hydrogen chloride. Reaction proceeds *via* the iminoester, which is hydrolysed to the diester:

$$N\equiv C-CH_2-COO^{\ominus}K^{\oplus} \xrightarrow[-KX,\ -H_2O]{+C_2H_5OH,\ +H^{\oplus}X^{\ominus}} N\equiv C-CH_2-COOC_2H_5 \xrightarrow{+C_2H_5OH(H^{\oplus})}$$

$$HN=\underset{\underset{OC_2H_5}{|}}{C}-CH_2-COOC_2H_5 \xrightarrow[-NH_3]{+H_2O} C_2H_5OOC-CH_2-COOC_2H_5$$

Diethyl malonate

Diethyl malonate is a colourless liquid, b.p. 199°C (472 K), with a pleasant smell. Its importance in synthetic work is due to its *reactive methylene group.* The hydrogen atoms are acidic because of the −I-effect of the neighbouring ethoxycarbonyl groups, and may be replaced by other groups. Unlike ethyl acetoacetate, diethyl malonate *does not spontaneously enolise.* Hence it is not soluble in aqueous alkali. A proton is removed from the methylene group only by the action of a base such as a sodium alkoxide or sodamide in an anhydrous solvent. The resultant sodio-malonic ester serves as a starting material for so-called *malonic ester syntheses.* Reaction with alkyl halides gives exclusively *C-alkylmalonic esters:*

$$\left[\underset{RO}{\overset{O}{\diagdown}}C-\underset{\underset{\ominus}{|}}{\overset{H}{\underset{}{C}}}-\underset{OR}{\overset{O}{\diagup}}C\right]Na^{\oplus} + I-R' \longrightarrow \underset{RO}{\overset{O}{\diagdown}}C-\underset{\underset{R'}{|}}{\overset{H}{\underset{}{C}}}-\underset{OR}{\overset{O}{\diagup}}C + NaI$$

Sodio-malonic ester   *C*-Alkylmalonic ester

---

[1] See *G. Maier* et al., Angew Chem. Int. Ed. Engl., **27**, 566 (1988).

If the sodium derivative of the monoalkyl-malonic ester is treated with a further molecule of an alkyl iodide a *dialkylmalonic ester* results. In contrast to acetoacetic ester, dialkylation of a malonic ester can be carried out in one step, using two molecules of sodium ethoxide and two molecules of alkyl halide to one molecule of malonic ester.

When monoalkyl- or dialkyl-malonic esters are hydrolysed by heating them with alkali and then with acid, carbon dioxide is evolved and a substituted acetic acid is the product (*acid splitting*), e.g.

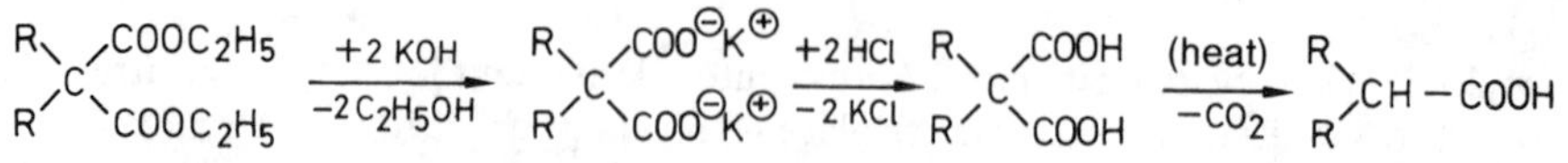

2.34.2.2 Knoevenagel reaction[1] (1898).

The *reactive methylene group* of malonic esters reacts with aldehydes or ketones in the presence of a secondary base as a catalyst (piperidine and diethylamine are often used) to form an intermediate aldol-like product which loses water to give an unsaturated dicarboxylic acid. For example, a *benzylidenemalonic ester* is obtained from diethyl malonate and benzaldehyde:

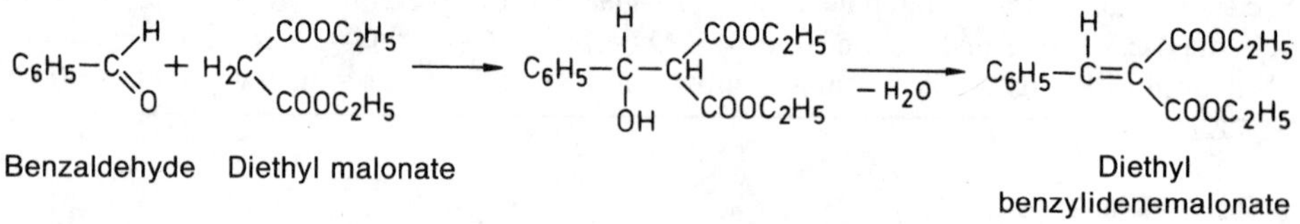

Benzaldehyde   Diethyl malonate   Diethyl
                                                                benzylidenemalonate

Acid hydrolysis of benzylidenemalonic ester leads to loss of carbon dioxide and formation of cinnamic acid, C_6H_5—CH=CH—COOH.

The *reaction mechanism*, which is similar to that of aldol addition, is as follows. First of all the basic catalyst acts as a proton acceptor and removes a proton from the methylene reactant converting it into a *carbanion*. This then makes a *nucleophilic* attack on the positively charged carbon atom of the aldehyde or ketone (the carbonyl reactant). The resultant adduct then loses water to provide an unsaturated diester:

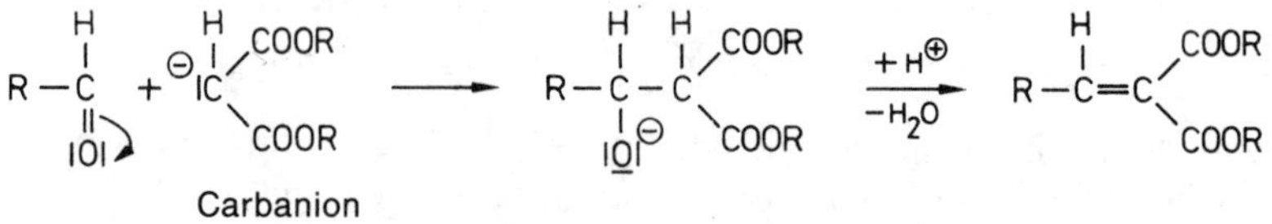

Carbanion

An analogous reaction takes place when a cyanoacetic ester, $N{\equiv}C$—CH_2—COOR, is used in place of a malonic ester.

2.34.2.3 Michael addition reaction[2] (1887).

This reaction involves the *nucleophilic addition* of compounds with *reactive methylene groups*, such as malonic esters or acetoacetic esters, to activated C=C double bonds (e.g. α,β-unsaturated carbonyl compounds, carboxylic acid esters or nitriles; described as *Michael* acceptors), in the presence of a base catalyst such as piperidine, diethylamine or sodium alkoxides, e.g.

[1] See G. *Jones*, The *Knoevenagel* Condensation, Org. React. *15*, 204 *ff.* (1967).
[2] See E. D. *Bergmann* et al., The *Michael* Reaction, Org. React. *10*, 179 *ff.* (1959).

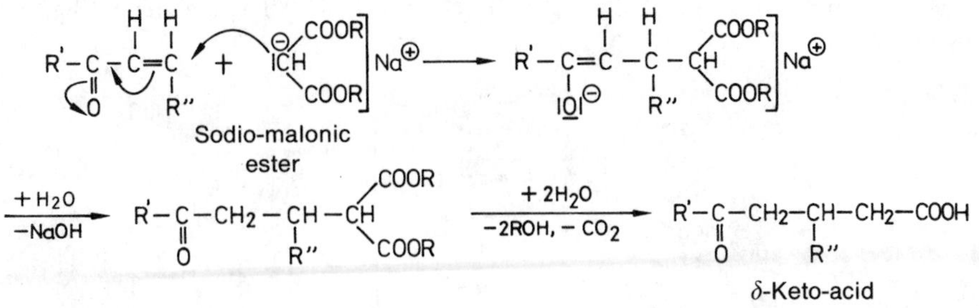

δ-Keto-acid

Since the product of such a reaction can be converted by hydrolysis, accompanied by decarboxylation, into a ketocarboxylic acid, the *Michael* addition is frequently of use in synthesis.

The activation of the C=C double bond results from the properties of the carbonyl group being passed to the β-carbon atom of the unsaturated system.

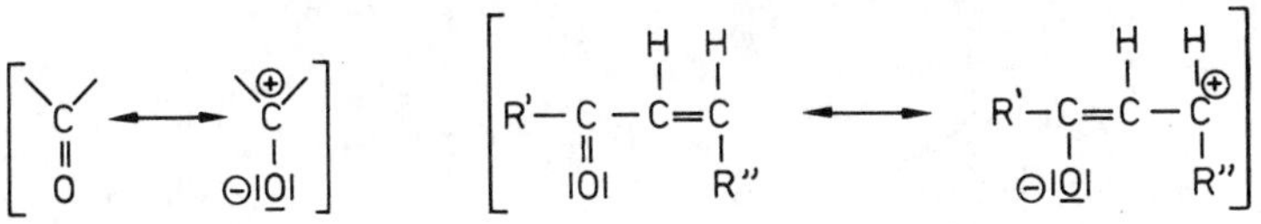

Such systems are described as *vinylogues* of the corresponding carbonyl compound; thus according to this principle of *vinylology*,[1] acrolein can be thought of as the *vinylogue* of formaldehyde.

It is also possible to add hydrogen cyanide to vinylogous systems.[2] The competing reaction to give a cyanohydrin has only a minor effect on the yield since the reverse reaction to regenerate hydrogen cyanide is much faster for the cyanohydrin than it is for the β-cyanoketone.

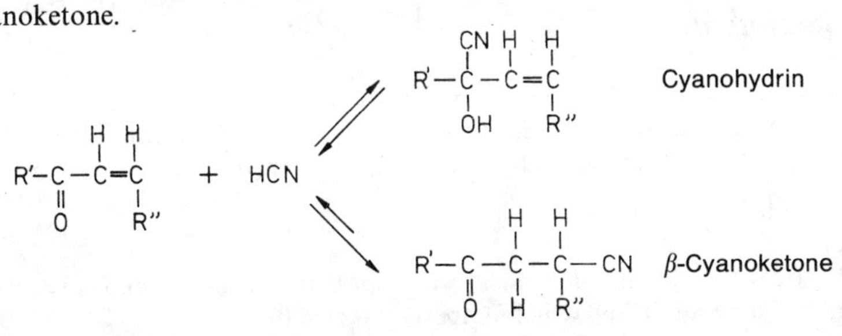

Indeed, innocuous acetone cyanohydrin can take the place of toxic hydrogen cyanide as a reagent for this reaction.

A very promising industrial use for a *Michael* addition reaction is in *Group-transfer Polymerisation* (GTP), by means of which it is possible to make polymers with relatively uniform relative molecular masses. This is achieved by a base-catalysed *Michael* addition of *dimethylketene methyl(trimethylsilyl)acetal* to *methyl methacrylate*; at each polymeris-

---

[1] See *R. C. Fuson*, Chem. Rev. *16*, 1 (1935).
[2] See *W. Nagata* et al., Hydrocyanation of Conjugated Carbonyl Compounds, Org. React. *25*, 255 ff. (1977).

ation step the trimethylsilyl group migrates to the end of the chain as it grows, so that there is always a ketene methyl(trimethylsilyl)acetalgroup (shown with a grey background in the following scheme) at the end:

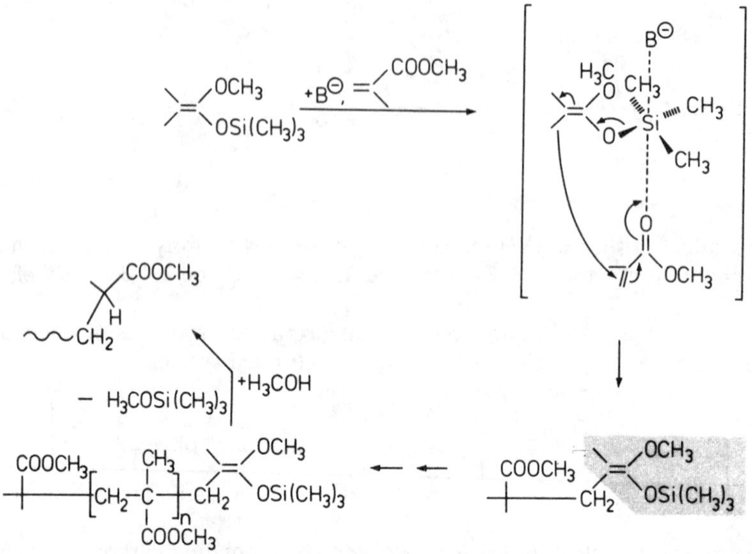

Polymerisation proceeds rapidly at room temperature; it can be terminated by, for example, adding methanol.

2.34.3 *Succinic acid*

Succinic acid is found in amber (*succinum*) and in some other resins, in much brown coal, and in numerous plants, for example, algae, fungi and lichens, in rhubarb, and in unripe grapes and tomatoes. Physiologically it is formed as an intermediate in the citric acid cycle.

Preparation

2.34.3.1 By treating *1,2-dibromoethane with potassium cyanide* and hydrolysing the succinonitrile (butanedinitrile), which is thereby formed:

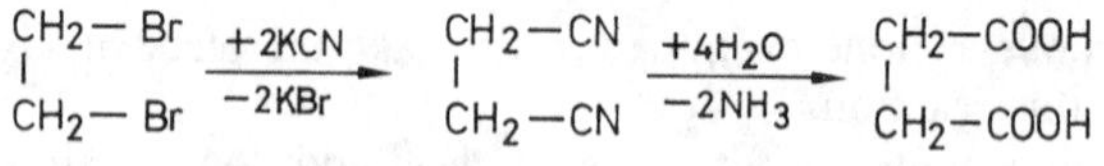

1,2-Dibromoethane Succinonitrile Succinic acid

2.34.3.2 Industrially by *catalytic hydrogenation of maleic acid*:

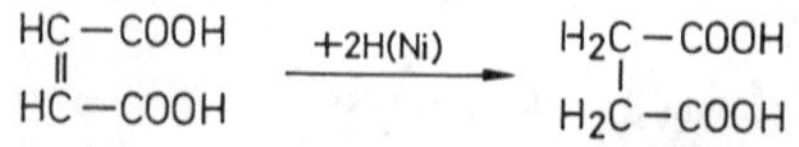

Maleic acid Succinic acid

Properties. Succinic acid crystallises as plates with m.p. 185°C (458 K). Its salts are called *succinates*. When heated it loses water and is converted into succinic anhydride (m.p. 120°C [393 K]), which is sometimes useful in synthesis. When ammonium succinate is distilled rapidly it is converted into succinimide (m.p. 126°C [399 K]).

N-Bromosuccinimide, NBS[1], is obtained by treating succinimide with bromine at 0°C (273 K) in the presence of sodium hydroxide.

$$
\begin{array}{ccc}
\text{Succinimide} & & \text{N-Bromosuccinimide}
\end{array}
$$

It is used as a *brominating agent*, especially for compounds containing allyl groups, since it does *not* attack the C=C double bond (*Wohl-Ziegler reaction*), e.g.

$$R-CH_2-CH=CH-CH_3 + \text{(N-Br)} \longrightarrow R-CH(Br)-CH=CH-CH_3 + \text{(N-H)}$$

The α-methylene group is brominated more rapidly than the methyl group.

In addition, *N*-bromosuccinimide has become of preparative importance as a *dehydrogenating* agent.

2.34.3.3 *Stobbe* condensation[2] (1893). When ketones or aldehydes react with succinic esters in the presence of a base such as potassium t-butoxide (*Johnson*, 1944), an aldol-type addition occurs and the product is the potassium salt of an alkylidenesuccinic acid half-ester. Acid hydrolysis of the latter material, best carried out using conc. hydrobromic acid in acetic acid, brings about decarboxylation and formation of a γ-lactone and also the corresponding β,γ-unsaturated mono-carboxylic acid. Both of these compounds can be reduced catalytically to the related saturated carboxylic acid, e.g.

$$
\begin{array}{l}
\underset{\text{Ketone}}{R-C=O} + \underset{\text{Diethyl succinate}}{\begin{array}{l}H_2C-COOC_2H_5\\H_2C-COOC_2H_5\end{array}} \xrightarrow[-(H_3C)_3COH]{\substack{+KOC(CH_3)_3\\-C_2H_5OH,}} \underset{}{R-C=C-COOC_2H_5} \xrightarrow[\substack{-C_2H_5OH,-CO_2,\\-KBr}]{HBr(H_3CCOOH)}
\end{array}
$$

γ-Lactone   or $R-C(CH_3)=CH-CH_2-COOH$ $\xrightarrow{+2H}$ $R-CH(CH_3)-CH_2-CH_2-COOH$

4-Alkyl(aryl)-carboxylic acid

The method is especially used to displace the oxo-group of aryl ketones by a propionic acid group.

---

[1] See *L. Horner* et al., N-Bromosuccinimid, Eigenschaften und Reaktionsweisen (*N*-Bromosuccinimide, Properties and Reactions), Angew. Chem. *71*, 349 (1959); *R. Filler*, Chem. Rev. *63*, 21 (1963).
[2] See *W. S. Johnson* et al., The *Stobbe* Condensation, Org. React. *6*, 1 *ff.* (1951).

2.34.4 *Higher Dicarboxylic acids*

2.34.4.1 Glutaric acid ($HOOC-(CH_2)_3-COOH$ is a crystalline substance which melts at 98°C (371 K). It occurs with malonic acid in the juice of sugar beet. It is obtained industrially by *oxidative ring-opening of cyclopentanone* using 50% nitric acid in the presence of vanadium(V) oxide:

$$\text{Cyclopentanone} \xrightarrow[(V_2O_5)]{(HNO_3)} \begin{array}{l} H_2C-COOH \\ | \\ H_2C-CH_2-COOH \end{array}$$

Cyclopentanone   Glutaric acid

2.34.4.2 Adipic acid, $HOOC-(CH_2)_4-COOH$, m.p. 153°C (426 K) is also found in the juice of sugar beet. It is made industrially by *oxidative ring-opening of cyclohexanol or cyclohexanone* with 65% nitric acid at 30–40°C (300–310 K):

$$\text{Cyclohexanol} \longrightarrow \text{Cyclohexanone} \longrightarrow \begin{array}{l} COOH \\ | \\ (CH_2)_4 \\ | \\ COOH \end{array}$$

Cyclohexanol   Cyclohexanone   Adipic acid

Adipic acid and hexamethylenediamine (1,6-diaminohexane), $H_2N-(CH_2)_6-NH_2$ are used as starting materials for the preparation of *nylon fibres* which are notable for their tensile strength and elasticity.

In the preparation of this first *polyamide fibre* (Nylon-6,6; 6,6-polyamide) equimolar amounts of the *dicarboxylic acid* and *diamine* are melted together at about 280°C (550 K) and the resultant polycondensed material is spun out of the melt (*Carothers*, 1935). This *polyamide* has the following chain structure:

$$+NH-CO-(CH_2)_4-CO-NH-(CH_2)_6 \rightarrow_n: \quad \text{Poly(hexamethyleneadipamide)}$$

A characteristic feature of such polycondensations is that at each chain-lengthening step a small molecule, for example water, is formed, which must be taken out of the equilibrium. *Polycondensation* and *polyaddition* are thus fundamentally different modes of *polymerisation* (see p. 73 *ff.*).

The name 'Nylon' originally served as the generic name for all long-chain synthetic polyamides. Nowadays it is usually accompanied by numbers, as in Nylon-6,6 or 6,6-Polyamide, which in this case indicate that, of the starting materials, the diamine (first digit) had six carbon atoms and so had the dicarboxylic acid (second digit).

2.34.4.3 Hexamethylenediamine, 1,6-diamino-hexane, m.p. 44°C (317 K), is usually prepared by catalytic hydrogenation of adiponitrile, which is itself obtained from butadiene, as follows:

$$H_2C=CH-CH=CH_2 \xrightarrow{+Cl_2} \underset{\begin{array}{cc} | & | \\ Cl & Cl \end{array}}{H_2C-CH=CH-CH_2} \xrightarrow[-2\,NaCl]{+2\,NaCN}$$

$$\underset{\begin{array}{cc} | & | \\ CN & CN \end{array}}{H_2C-CH=CH-CH_2} \xrightarrow[(Ni)]{+2H} \underset{\text{Adiponitrile}}{NC-(CH_2)_4-CN} \xrightarrow[(Ni,NH_3)]{+8H} \underset{\text{Hexamethylenediamine}}{H_2N-(CH_2)_6-NH_2}$$

Adiponitrile (hexanedinitrile) can also be made electrochemically, in one step, from acrylonitrile:

$$2\,H_2C\!=\!CH\!-\!CN \xrightarrow{\;2\ominus,\,2\,H\oplus\;} NC\!-\!(CH_2)_4\!-\!CN$$

Acrylonitrile   Adiponitrile

When this is carried out in aqueous solution, with a quaternary ammonium salt as a supporting electrolyte, a chemical yield of 95% and an electrical efficiency of 90% is achieved. This effective industrial process is an example of electrolytic **hydrodimerisation** (EHD), which can be applied to many compounds which have double bonds conjugated with electron-withdrawing groups.[1]

The reaction does not involve neutral radicals, as in the case of the *Kolbe* reaction, but rather a *radical anion*, generated at the cathode; the following mechanism has been suggested for the hydrodimerisation:

Radical anion

Michael-donor

Cyanoethyl anion

Michael reaction

Adiponitrile

The radical anion is protonated and reduced to give the cyanoethyl anion, which reacts with acrylonitrile in a *Michael* reaction. Of the two participants in the dimerisation, one of them undergoes a *redox-umpolung* in being converted into a *carbanion* which then acts as the donor in the *Michael* reaction. The carbanion arises from a so-called ECE process in which an *Electrochemical* step is followed by a *Chemical* step and then by a further *Electrochemical* step.[2]

Pimelic acid, heptane-1,7-dioic acid, $HOOC\!-\!(CH_2)_5\!-\!COOH$, m.p. 105°C (378 K), is obtained by *catalytic hydrogenation of salicyclic acid* to tetrahydrosalicyclic acid, which is in *tautomeric equilibrium* with 2-oxocyclohexanecarboxylic acid, and when heated with conc. potassium hydroxide under pressure at 300°C (575 K) undergoes so-called acid cleavage to give pimelic acid:

Salicylic acid

Tetrahydro-salicylic acid

2-Oxocyclohexane-carboxylic acid

Pimelic acid

---

[1] Concerning the large developments in organic electrochemistry see *F. Beck*, Elektroorganische Chemie (Verlag Chemie, Weinheim 1974); *M. R. Rifi* and *F. H. Covitz*, Introduction to Organic Electrochemistry (Dekker, New York 1974); *M. M. Baizer*, Recent Developments in Organic Synthesis by Electrolysis, Tetrahedron *40*, 935 (1984).

[2] See *H. J. Schäfer*, Anodic and Cathodic C—C Bond Formation, Angew. Chem. Int. Ed. Engl. *20*, 911 (1981).

Suberic acid, octane-1,8-dioic acid, m.p. 140°C (413 K) is formed in small amounts when *cork is oxidised* with nitric acid (*Quercus suber* = cork oak). It is more satisfactorily obtained, together with the two next highest homologues, azelaic acid and sebacic acid, by oxidising castor oil with nitric acid.

2.35 Unsaturated Aliphatic Dicarboxylic Acids

2.35.1 Ethylenedicarboxylic Acids (Maleic and Fumaric Acids)

The two commonest members of this series are maleic acid and fumaric acid, which were involved in the first studies of *cis–trans* isomerism by *Wislicenus*. The stereochemical relationship of these two acids was evident from the fact that, because of the spatial arrangement of its two carboxyl groups, only *maleic acid*, the (Z)-form, was converted, by loss of water, into a cyclic anhydride when it was heated quickly, especially in the presence of acetic anhydride:

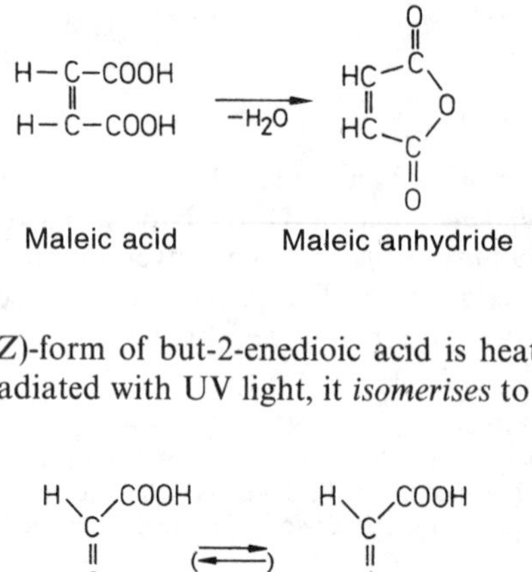

$$\begin{array}{ccc}
\text{H-C-COOH} \\
\quad\| \\
\text{H-C-COOH}
\end{array}
\xrightarrow{-\text{H}_2\text{O}}
\text{(Maleic anhydride)}$$

Maleic acid   Maleic anhydride

When maleic acid, the (Z)-form of but-2-enedioic acid is heated for a longer time at 150°C (425 K), or if it is irradiated with UV light, it *isomerises* to the more stable *fumaric acid*, the (E)-form:

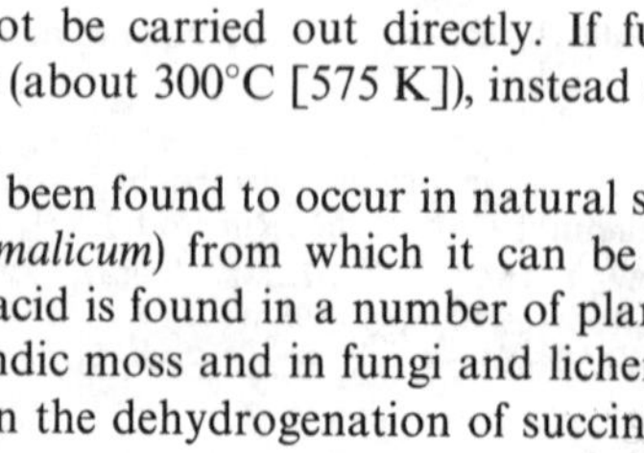

Maleic acid   Fumaric acid
(Z)-form   (E)-form

The reverse change cannot be carried out directly. If fumaric acid is heated to a sufficiently high temperature (about 300°C [575 K]), instead of isomerising it loses water and forms maleic anhydride.

Maleic acid has not as yet been found to occur in natural sources. Its name came from that of malic acid (*acidum malicum*) from which it can be prepared by removal of a molecule of water. Fumaric acid is found in a number of plants, for example in fumitory (*Fumaria officinalis*), in Icelandic moss and in fungi and lichens. In the citric acid cycle it appears as an intermediate in the dehydrogenation of succinic acid.

Preparation

2.35.1.1 *When malic acid is heated rapidly* at about 250° (525 K) it loses water to give maleic anhydride, which can be converted into maleic acid by treatment with water:

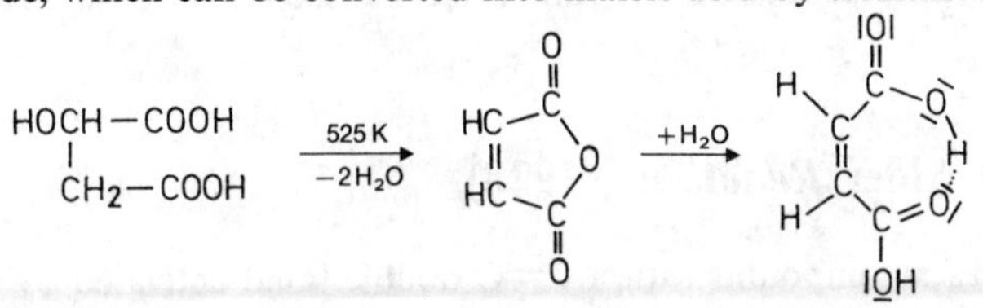

| Malic acid | Maleic anhydride | Maleic acid |

2.35.1.2 Industrially *maleic anhydride* is obtained as follows:

(a) By *catalytic oxidation of benzene* with atmospheric oxygen at 400–450°C (675–725 K) in the presence of a catalyst based on V_2O_5, yield 50%.

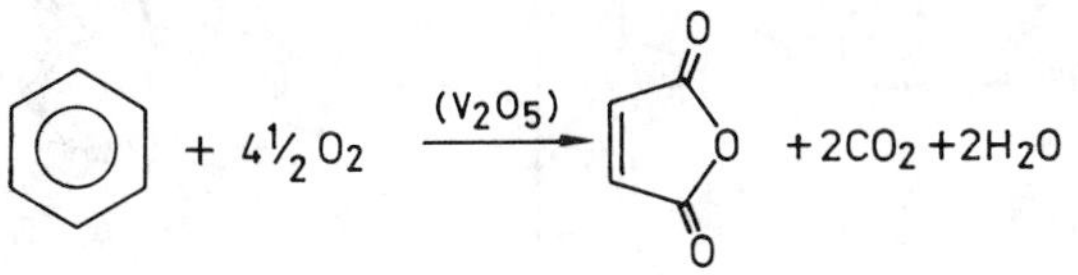

Benzene   Maleic anhydride

Fumaric acid and *p*-benzoquinone are formed as by-products.

(b) It is also made by a similar *oxidation of butenes*:

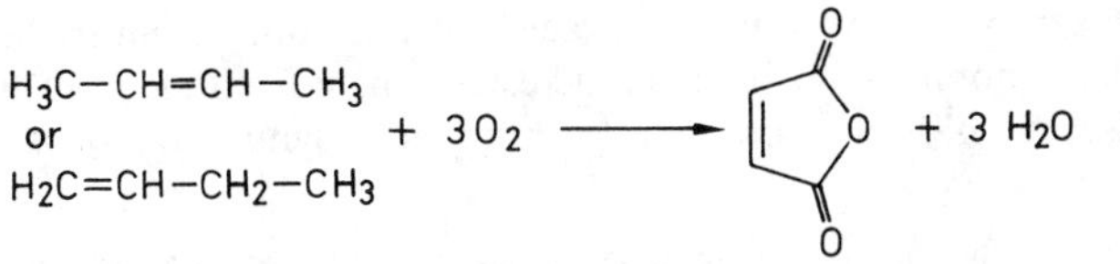

$$H_3C-CH=CH-CH_3$$
or
$$H_2C=CH-CH_2-CH_3$$

$+ 3O_2 \longrightarrow$ [maleic anhydride] $+ 3 H_2O \quad \Delta H = -1315 \text{ kJ/Mol}$

2.35.1.3 *Fumaric acid* can be made by removing hydrogen bromide from monobromo-succinic acid by heating it with dilute alkali:

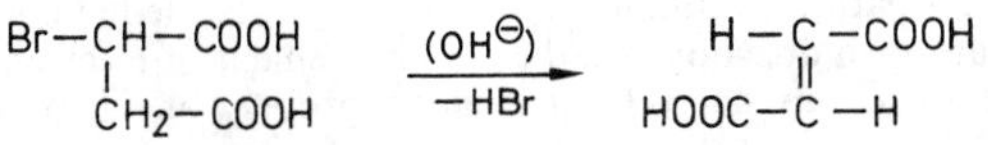

$$Br-CH-COOH \xrightarrow[-HBr]{(OH^{\ominus})} \begin{array}{c} H-C-COOH \\ \| \\ HOOC-C-H \end{array}$$
$$CH_2-COOH$$

Monobromosuccinic acid   Fumaric acid

Industrially it is prepared by *catalytic rearrangement of maleic acid* at 100°C (375 K) in the presence of thiourea.

Properties

Maleic acid forms colourless prisms of mp. 130.5°C (403.7 K); it is readily soluble in water. Fumaric acid is only slightly soluble in water; it sublimes at 200°C (475 K) and melts in a closed tube at 287°C (560 K). The first dissociation constant, pK_a^1, of maleic acid is 1.92. It is a much stronger acid than fumaric acid, $pK_a^1 = 3.02$, because the anion of maleic acid, like that of salicyclic acid, is stabilised by *intramolecular hydrogen bonding* (see p. 551). Maleic acid is catalytically reduced to succinic acid appreciably faster than fumaric acid is under identical reaction conditions.

Uses

Synthetically, maleic acid is used as a component in the making of alkyd resins and unsaturated polyester resins. Maleic anhydride is much used as a reactant in 'Diene syntheses'.

2.35.2 Diels–Alder *Reaction*[1] (1928)

In *diene syntheses* a compound with a C=C double bond activated by a neighbouring conjugated electron-withdrawing group such as COOH, CN, NO_2 undergoes cyclo-addition to a hydrocarbon having conjugated double bonds as part of its structure. For example, *maleic anhydride* readily takes part in a *[4+2]cyclo-addition reaction with butadiene*:

Butadiene	Maleic anhydride		1,2,3,6-Tetrahydro-phthalic
(Diene, *s-cis* form)	(Dienophile)		anhydride (*cis* form)

The two starting materials in diene syntheses are described as the *diene* and the *dienophile*, and the resultant cyclic product as the *adduct*. The rate of reaction is very little influenced by the polarity of the solvent, and no intermediates have been observed. Nevertheless the reaction can be catalysed by *Lewis* acids such as aluminium trichloride.

By using diastereotopic dienophiles, such as acrylic esters made from alcohols having a chiral alcohol group (see p. 240), it is possible to carry out asymmetric *Diels–Alder* reactions.[2]

In bicyclic systems formed by reactions involving maleic anhydride the two rings are *cis*-linked. The *Diels–Alder* reaction can be described as a *1,4-cyclo-addition* reaction of the activated C=C double bond of the dienophile to the conjugated system of the diene. In the course of the reaction in most cases two new σ-bonds are formed in a *cyclic transition state* (depicted in the above reaction scheme), but the reaction path is continuous, without any intermediate. The reaction is described as *concerted*, but it is an open question whether the movement of electrons indicated by means of the dashes in the circle is a simultaneous (synchronous) process, or if there is a minute time interval between the two steps[3].

Cyclic dienes such as cyclopentadiene, cyclohexa-1,3-diene, furan or anthracene (see p. 636) can be used instead of butadiene and its derivatives, but compounds such as 1,2,3,5,6,7-hexahydronaphthalene in which the two conjugated double bonds are fixed in an *s-trans* form cannot.

Cyclopentadiene	Cyclohexa-1,3-diene	Furan	1,2,3,5,6,7-Hexahydro naphthalene

[1] See *M. C. Kloetzel*, The *Diels–Alder* reaction with Maleic Anhydride, Org. React. *4*, 1 *ff.* (1948); *J. Sauer* and *R. Sustmann*, Mechanistic Aspects of the *Diels–Alder* Reactions — A Critical Survey, Angew. Chem. Int. Ed. Engl. *19*, 779 (1980).
[2] See *P. Welzel*, Nachr. Chem. Tech. Lab. *31*, 979 (1983).
[3] See *M. J. S. Dewar*, Multibond Reactions Cannot Normally be Synchronous, J. Am. Chem. Soc. *106*, 209 (1984).

The following compounds are reactive dienophiles:

Tetracyanoethylene, $(NC)_2C{=}C(CN)_2$ Acrylonitrile, $H_2C{=}C{-}CN$
But-2-enal, $H_3C{-}CH{=}CH{-}CHO$ Acrolein, $H_2C{=}CH{-}CHO$
1-Nitropropene, $H_3C{-}CH{=}CH{-}NO_2$ Ethyl acrylate, $H_2C{=}CH{-}COOC_2H_5$
β-Nitrostyrene, $C_6H_5{-}CH{=}CH{-}NO_2$ Maleimide

All of these compounds have electron-withdrawing substituents attached to the alkene group; in consequence they are 'electron-poor' alkenes. The dienes which have been mentioned have no such substituents and in contrast may be called 'electron-rich'. The electronic character of the two reactants thus provides a reason, based on normal electronic requirements, for a rapid reaction to take place between them.

Other examples of dienophiles include acetylenedicarboxylic esters, which have electron-poor $C{\equiv}C$ triple bonds, and nitroso compounds.[1]

Functionalised dienes have proved to be very useful starting materials in syntheses of complex compounds.[2]

When cyclic dienes are used, two different cyclo-addition products are possible, for example, in the reaction of furan with maleimide:

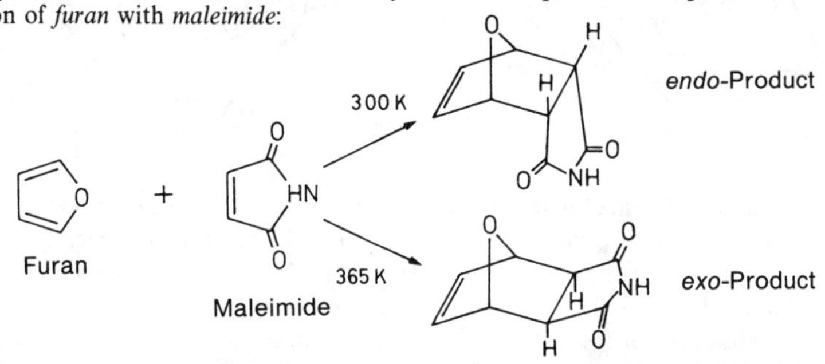

In the case of the endo-product, the dienophile portion lies beneath the newly formed six-membered ring. In the example given, this is the main product when the reaction is carried out at 25°C (300 K). At 90°C (365 K) the exo-isomer is the main product; at this temperature the endo-compound rearranges to the exo-form.

This is a further example of kinetic and thermodynamic effects producing different results (cf. p. 102). A smaller activation energy is required to form the endo-isomer than the exo-isomer, but the latter compound is the more thermodynamically stable of the two. Hence formation of the endo-compound is favoured at lower temperatures, whilst at higher temperatures the amount of exo-compound formed increases.

In the case of dienophiles such as ethyl acrylate, the resultant cyclic transition state (see equation p. 336) can have either an exo- or an endo-configuration. Thus in the reaction with cyclopentadiene stereoisomeric addition products are formed:

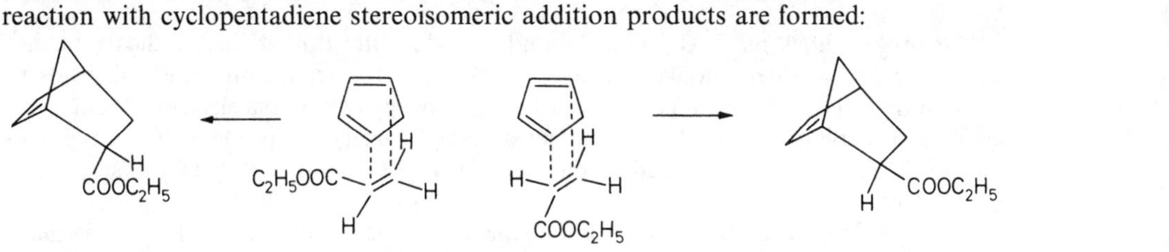

endo-form   Transition states   exo-form

[1] See G. Kresze et al., Diensynthesen mit Nitrosoverbindungen, Fortschr. Chem. Forsch. 11, 245 (1968/69).
[2] See S. Danishefsky, Cyclo-addition and Cyclocondensation Reactions of Highly Functionalised Dienes: Applications to Organic Synthesis, Chemtract-Organic Chemistry 2, 273 (1989).

In this case also, formation of the *endo*-form has the lower activation energy.

Diene synthesis is an example of a *pericyclic reaction*, whose mechanism can be explained by the principle of *retention of orbital symmetry* (*Woodward–Hoffmann* rules), as can the mechanisms of *electrocyclic* and *sigmatropic* reactions. Their common feature is that the stereochemical course of the reactions depends on the number of electrons involved in the process and not on the number of atoms involved.

In the case of the *Diels–Alder* reaction a 4-π-electron system (the diene) reacts with a 2-π-electron system of the dienophile, giving a [4π + 2π] reaction. The principle of retention of orbital symmetry now requires that the transition state, which is formed in a thermal reaction, arises from mutual interaction of the HOMO of one of the reactants with the LUMO of the other. This comes about if the overlapping orbitals at the termini of the reacting systems have the same sign.[1]

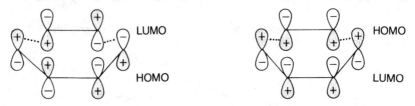

It is evident that in the places where the two new σ-bonds are being formed a *suprafacial* (s) (same-sided) interaction is possible; therefore the transition state is sterically favourable. The *Diels–Alder* reaction is thus better described as a [4πs + 2πs] process. It follows that the newly formed σ-bonds will have a *cis* configuration.

The kinetic preference for formation of the *endo*-product in the reaction between *furan* and *maleimide* is evident from a consideration of the frontier orbitals in the transition state:

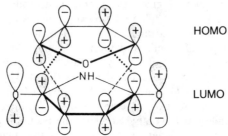

The broken lines mark the new σ-bonding, while the dotted lines indicate further secondary orbital interaction which only takes place in a transition state, and leads to formation of the *endo*-product. Previously the concept of highest electron density was invoked to explain the result. The higher activation energy required to form the *exo*-transition state can be attributed to the fact that a less favourable transition state must be involved.

A [2πs + 2πs] reaction between butadiene and maleic anhydride, which is formally conceivable, is not in fact observed:

[1] See *T. L. Gilchrist* and *R. C. Storr*, Organic Reactions and Orbital Symmetry (Cambridge University Press 1972, and later editions); *I. Fleming*, Frontier Orbitals and Organic Chemical Reactions (Wiley, London 1976).

Application of the principle of retention of orbital symmetry shows that the transition state of a synchronous reaction would not be favourable, since it would involve an antibonding interaction between the HOMO and LUMO (shown by a wavy line).

2.35.2.1 Ene-reaction.[1] This term is used to describe the reaction of an *alkene* which has a hydrogen atom at its allyl position (3-position), the *ene-component*, with an *activated π-bond, the enophile*.

In this the *ene-component* plays an analogous role to that of the *diene* in the *Diels–Alder* reaction; the same situation also obtains in both reactions as regards the electronic requirements of the two reactants. As a result it is not surprising that both reactions may proceed at the same time, if the structures of the reactants are suitable.

The *transition state for an ene-reaction* is boat-shaped:

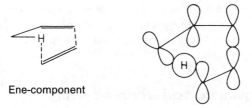

Ene-component

Because of the mobility of the system in the vicinity of the allylic hydrogen atom a pericyclic suprafacial transition state is possible, in which six electrons, two from each π-bond and a σ-pair from the C—H bond, participate. In the general terminology of the

[1] *H. M. R. Hoffmann*, The Ene Reaction, Angew. Chem. Int. Ed. Engl. *8*, 556 (1969); *W. Oppolzer* and *V. Snieckus*, Intramolecular Ene Reactions in Organic Synthesis, Angew. Chem. Int. Ed. Engl. *17*, 476 (1978).

Woodward–Hoffmann rules, the reaction can therefore be described as a $[\sigma 2s + \pi 2s + \pi 2s]$ process. An example is the reaction of propene with a dialkyl acetylenedicarboxylate:

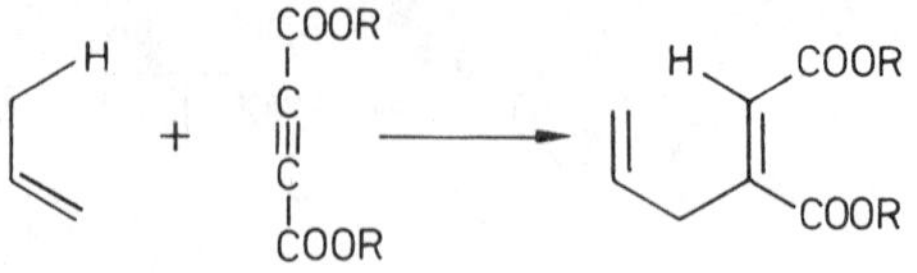

2.35.3 Acetylenedicarboxylic Acid

This acid is prepared by the action of *methanolic potassium hydroxide* on D,L-α,β-dibromosuccinic acid (itself obtained by addition of bromine to disodium fumarate); two molecules of hydrogen bromide are eliminated.

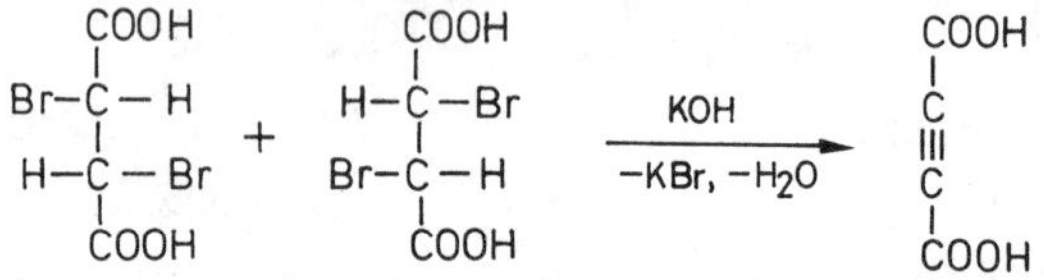

D,L-α,β-Dibromosuccinic acid   Acetylenedicarboxylic acid
(m.p. 177°C [450 K])

Acetylenedicarboxylic acid is a relatively stable compound with m.p. 177°C (450 K) (decomp.). The carboxyl groups are activated by the C≡C triple bond, making them strongly acidic. The dialkyl esters have much use in synthesis, for example in 1,3- and 1,4-dipolar cyclo-addition reactions.

2.36 Aliphatic Hydroxy-Dicarboxylic and -Tricarboxylic Acids

This class of compounds includes some important plant acids such as *malic acid, tartaric acid* and *citric acid*. All three of these acids were first isolated by *Scheele* from fruit juices, making use of their almost insoluble calcium salts. These were treated with equivalent amounts of sulphuric acid to obtain and isolate the free acids.

2.36.1 Hydroxymalonic Acid (Tartronic Acid)

The simplest hydroxydicarboxylic acid, hydroxymalonic acid, tartronic acid, is not found in nature. It is made either by hydrolysis of bromomalonic acid with silver hydroxide or by reduction of mesoxalic acid with sodium amalgam:

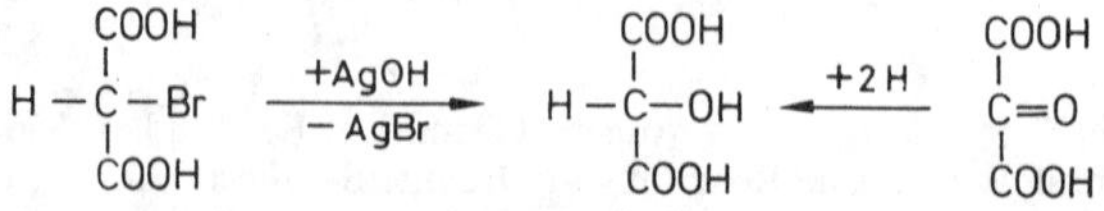

Bromomalonic acid   Tartronic acid   Mesoxalic acid

Its hemi-hydrate forms colourless prisms which melt at 160°C (433 K) with evolution of carbon dioxide.

2.36.2 Malic Acid (Hydroxysuccinic Acid)

This molecule contains an asymmetric carbon atom, and thus exists in two *enantiomeric forms* and also as a racemic form.

$$
\begin{array}{cc}
\text{COOH} & \text{COOH} \\
| & | \\
\text{H}-\text{C}-\text{OH} & \text{HO}-\text{C}-\text{H} \\
| & | \\
\text{CH}_2 & \text{CH}_2 \\
| & | \\
\text{COOH} & \text{COOH} \\
\text{D}(+)\text{-Malic acid} & \text{L}(-)\text{-Malic acid}
\end{array}
$$

It may be seen from these projection formulae that, as in the case of lactic acid, the carboxyl group attached to the asymmetric carbon atom is written above it. The naturally occurring laevorotatory malic acid belongs to the L-series and has the same configuration as L(+)-lactic acid and L(+)-alanine. Using the *Cahn–Ingold–Prelog* system this is the (S)-configuration.

L(−)-Malic acid is found in unripe apples, gooseberries and other fruits and can be obtained from the juice of rowan (mountain ash) berries and barberries (*Scheele*, 1785).

D(+)-Malic acid has not been found in plants. It is obtained either by resolution of DL-malic acid or from L(−)-malic acid by means of a *Walden* inversion. The salts of malic acid are known as malates or hydroxysuccinates.

Preparation

2.36.2.1 By treating DL-*monobromosuccinic acid with alkali* or with silver oxide and water.

$$
\begin{array}{ccc}
\text{Br}-\text{CH}-\text{COOH} & & \text{HO}-\text{CH}-\text{COOH} \\
| & +\text{KOH} \longrightarrow & | \quad +\text{KBr} \\
\text{CH}_2-\text{COOH} & & \text{CH}_2-\text{COOH} \\
\text{DL-Monobromosuccinic acid} & & \text{DL-Malic acid}
\end{array}
$$

2.36.2.2 The best method is by *heating maleic acid with dilute sulphuric acid* under pressure:

$$
\begin{array}{ccc}
\text{HC}-\text{COOH} & & \text{HO}-\text{CH}-\text{COOH} \\
\| \quad\quad +\text{H}_2\text{O} & \xrightarrow{\text{(H}_2\text{SO}_4)} & | \\
\text{HC}-\text{COOH} & & \text{CH}_2-\text{COOH}
\end{array}
$$

Properties

L(−)-Malic acid forms deliquescent needles, m.p. 100–101°C (373–374 K). The specific rotation of its aqueous solutions depends upon the concentration. Solutions containing less than 25% of the acid are laevorotatory, but this rotation decreases as the concentration increases and eventually inverts to become dextrorotatory. The origins of this unusual behaviour in aqueous solution stem from the previously mentioned association of acid molecules and their solvation. Solutions in acetone are all laevorotatory, irrespective of their concentration. DL-Malic acid melts at 130°C (403 K).

2.36.3 Walden *Inversion* (1895)

This concerns substitution reactions which take place at an asymmetric carbon atom wherein *epimerisation* occurs, *i.e.* a D-configuration becomes an L-configuration or *vice versa*, usually accompanied by a reversal of the optical rotation. This was first noticed by *Walden* in work on malic acid. When laevorotatory L-malic acid reacts with phosphorus pentachloride, dextrorotatory D-chlorosuccinic acid is formed. If this is now treated with moist silver oxide, no change of configuration takes place and it is converted into dextrorotatory D-malic acid. In this way it is possible to get D(+)-malic acid from L(−)-malic acid.

In an analogous way *Walden* obtained laevorotatory L-malic acid from dextrorotatory D-malic acid, *via* laevorotatory L-chlorosuccinic acid, thus completing a cycle of interchanges. He also observed that formation of L(−)-malic acid from L(−)-chloro-succinic acid depended upon the identity of the *cation* of the base. When potassium hydroxide was used instead of silver hydroxide, a *Walden* inversion took place and D(+)-malic acid was obtained from the L(−)-chlorosuccinic acid.

The whole situation can be summed up as follows:

The arrow $\overset{\circ}{\rightarrow}$ indicates that a *Walden* inversion takes place.

The cycle below provides an example involving amino acids. The action of nitrosyl chloride on L(+)-alanine provides L(−)-α-chloropropionic acid and similarly D(−)-alanine gives D(+)-α-chloropropionic acid. In these processes the direction of rotation changes but there is no change of configuration. In contrast, when either L(−)- or D(+)-α-chloropropionic acids react with ammonia, there is a reversal of configuration and D(−)- or L(+)-alanines are formed, respectively; in this case, however, there is no change in the direction of rotation.

Reactions in which a change in configuration takes place at the asymmetric carbon atom in most cases involve bimolecular reactions of the S_N2 type. In reactions where no *Walden* inversion occurs, no precise statement can be made about the mechanism. The configuration at a chiral centre can also remain the same in the product if interaction by a neighbouring *nucleophilic group* leads to the involvement of two successive *Walden* inversions.

2.36.3.1 Neighbouring Group Participation. This term is used to describe the action of certain functional groups which carry either a negative charge or a lone pair of electrons, for example a carboxylate anion, on a neighbouring reaction centre. An example is provided by the hydrolysis of optically active α-halogenocarboxylic acids. Whilst the replacement of the halogen atom by a hydroxy group follows an S_N2 mechanism with *Walden* inversion in strongly alkaline conditions, in dilute alkali the configuration remains the same.

Winstein showed that in the latter case, the first step involves *nucleophilic* attack by the carboxylate anion on the rear of the α-carbon atom with loss of a bromide ion and formation of an unisolable α-lactone with an opposite configuration from that of the bromo-acid. This lactone then undergoes rearward attack by water, again involving reversal of configuration, and formation of the α-hydroxyacid, e.g.

D-α-Bromo-   α-Lactone   D-Lactate anion
propionic acid   (L-form)

Both substitution reactions proceed with *inversion*, and the two steric changes cancel one another out, leading to an overall retention of configuration.

Neighbouring group effects can have not only steric consequences, but may in some cases greatly increase the rate of reactions.[1]

2.36.3.2 Racemisation. Substitution reactions at asymmetric carbon atoms are often accompanied by partial or complete racemisation. Sometimes even warming of an *optically active* compound in the presence of a base or acid catalyst can lead to the formation of the racemic form; reaction proceeds through an intermediate anion, cation or enolised form. Thus when optically active 2-methylbutyric acid is warmed it loses its activity; the intermediate in this process is probably a tautomeric optically inactive ketene hydrate. When this reverts back to the acid, migration of a proton may lead to the formation of either enantiomer, thus bringing about racemisation.

(+)-2-Methylbutyric acid, (R)-form   (−)-2-Methylbutyric acid, (S)-form

[1] See *W. Lwowski*, Nachbargruppen-Effekte in der Organischen Chemie (Neighbouring Group Effects in Organic Chemistry), Angew. Chem. *70*, 483 (1958).

2.36.4 Asymmetric Synthesis (Marckwald, 1904), Stereoselective Synthesis, Asymmetric Induction[1]

When optically active compounds are prepared from optically inactive reagents, a racemic mixture is usually obtained, i.e. 50% of each of the D- and L-forms. This is because a reagent may attack from either direction; for example, in the formation of mandelonitrile from benzaldehyde the cyanide ion may approach either the upper or the lower side of the aldehyde molecule:

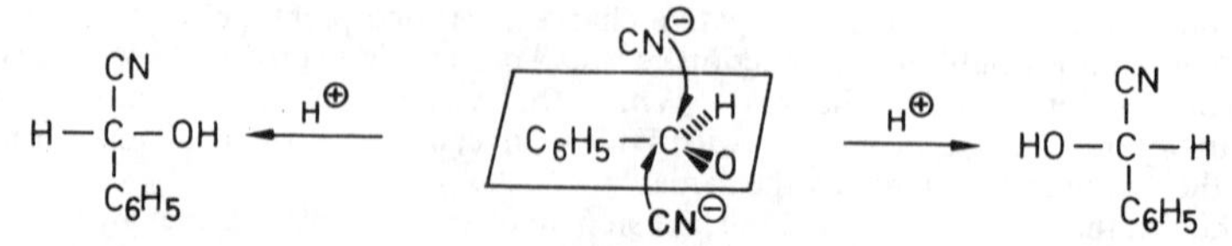

D-Mandelonitrile   Benzaldehyde   L-Mandelonitrile

On the other hand, some reactions which lead to the formation of a new asymmetric chiral centre may, owing to the influence of an asymmetric carbon atom already present in the molecule, lead to preferential formation of one of the diastereoisomers (E. Fischer, 1894, see p. 433). These are called 'asymmetric syntheses' or diastereoselective syntheses. This preferential generation of one form is especially marked if the existing chiral centre is near to the site of the newly formed one, e.g. in the reduction of chiral ketones with lithium aluminium hydride:

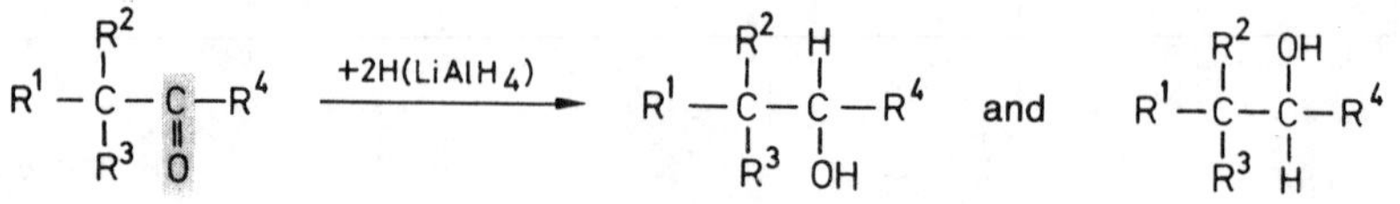

In this 1,1-asymmetric induction there is preferred formation (up to 75%) of the diastereoisomer in which the attacking reagent approaches the least sterically hindered side of the ketone (Cram's rule of asymmetric induction[2]).

An asymmetric synthesis may also result if the asymmetric centre already present in a molecule is further away from the newly formed centre (1-n-asymmetric induction) or if a reaction takes place in the presence of an asymmetric catalyst. The diastereoisomeric excess is defined as de[%] = $(D_1 - D_2) \cdot 100 / (D_1 + D_2)$; see ee, p. 121.

A valuable method of this type, discovered in 1980 by Sharpless, is the stereoselective epoxidation of allyl alcohols. The chiral catalyst is made from an ester of (+)- or (−)-tartaric acid (DET = diethyl tartrate) and titanium tetraisopropoxide, and the oxidising agent is t-butyl hydroperoxide (TBHP — care! explosive!).[3] An enantioselective reaction using the Sharpless reagent is illustrated by the following example, involving tridec-2-enol:

[1] See M. Nógrádi, Stereo-selective Synthesis (VCH Verlagsgesellschaft, Weinheim 1987); W. Bartmann and K. B. Sharpless, Stereochemistry of Organic and Bioorganic Transformations (VCH Verlagsgesellschaft, Weinheim, 1987); E. Winterfeldt, Stereo-selektive Synthese, Prinzipien und Methoden (Vieweg, Braunschweig 1988); R. A. Aitken and S. N. Kilényi, Asymmetric Synthesis (Blackie Academic, Glasgow 1992).
[2] See K. N. Houk et al., Theory of Selectivity of Nucleophilic Additions to Carbonyl Compounds, in W. Bartmann and K. B. Sharpless (Ed.), Stereochemistry of Organic and Bioorganic Transformations, p. 247 (VCH Verlagsgesellschaft, Weinheim 1987).
[3] See K. B. Sharpless et al., J. Am. Chem. Soc. 109, 5765 (1987).

(R,R)-(+)-Diethyl tartrate

(S) — H O H — (R)   82% ee

HO

R

R (H₃C)₃COOH Ti(OiPr)₄

R=n—C₈H₁₇
(Z)-Tridec-2-enol

(S,S)-(−)-Diethyl tartrate

(R) — H O H — (S)   80% ee

HO

R

When the complex of (+)diethyl tartrate with titanium is used, attack by the oxidising agent takes place predominantly from below; when the complex of (−)diethyl tartrate is used, attack is predominantly from above.[1] Making use of molecular sieves, ee of more than 90% can be achieved.[2]

Reactions catalysed by enzymes lead almost exclusively to the formation of one enantiomer. Thus, for example, when the previously mentioned addition of hydrogen cyanide to benzaldehyde is carried out in the presence of the mixture of enzymes known as *emulsin* found in bitter almonds, it leads to the formation of only the L(+)-mandelonitrile. In this particular enantioselective synthesis, only one chiral centre is introduced.

In general, when a reaction could possibly lead to the formation of more than one stereoisomer, but in fact only one reaction path is followed, leading to one product exclusively, then such a reaction is termed *stereoselective*. This may be distinguished from *stereospecific* reactions, in which one particular stereoisomeric starting material specifically provides only one stereoisomeric product, whilst a different stereoisomer gives a different stereoisomeric product[3] (see p. 71). *Stereospecificity* is indeed a special case of *stereoselectivity*.

A consequence of the concerted nature of the *ene-reaction* is that chirality associated with the 3-position of the ene can be transferred to an atom in the reaction product which originally formed part of the enophile, *e.g.*

H₃C
C₆H₅ —H
3-Phenylbut-1-ene

+

→

H₃C C₆H₅ H

This type of process is also called *asymmetric induction*.

If a chiral compound contains other appropriate centres, such as the CO group surrounded by grey in the formula above (p. 344), which might be converted into chiral centres, creation of this new centre results in the formation of diastereomers (see p. 240). There is *diastereofacial selectivity* in such a reaction, since the two sides which the reagent might approach are *diastereotopic*. The CH₂ group of malic acid contains a prochiral centre and is prochiral, and the two hydrogen atoms of this group

---

[1] See E. J. Corey, On the Origin of Enantioselectivity in the *Katsui-Sharpless* Procedure, J. Org. Chem. *55*, 1693 (1990).
[2] See K. B. Sharpless et al., J. Am. Chem. Soc. *109*, 5765 (1987).
[3] For a discussion 'Selectivity versus Specificity' see R. S.Ward, Chem. Brit. *1991*, 803.

are diastereotopic substituents. If one or other of these hydrogens is replaced, for example by an OH group, then this results in the formation of, respectively, a diastereomeric molecule of tartaric acid, or a molecule of *meso*-tartaric acid.

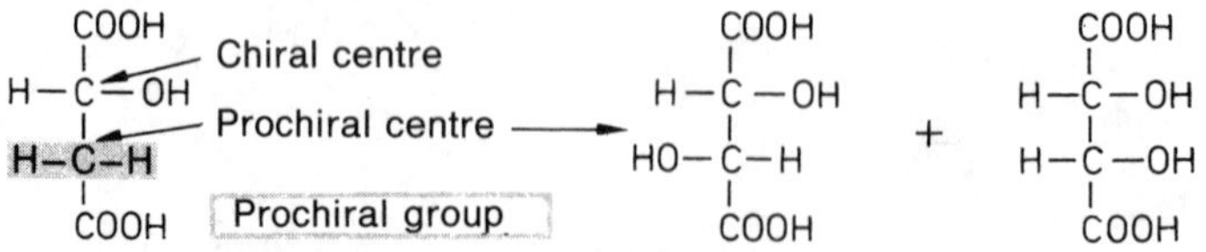

D(+)-Malic acid   (2R,3R)-(+)-   *meso*-Tartaric acid
                         Tartaric acid

2.36.5 *Tartaric Acid (Dihydroxysuccinic Acid)*

A molecule of tartaric acid contains two chiral centres, which, as is the case in the tetritols, are structurally the same. Hence there are only three different tartaric acids, namely the two *enantiomers*, (+)-tartaric acid and (−)-tartaric acid, and the optically inactive *meso*-tartaric acid, which cannot be resolved and whose molecule has a plane of symmetry.

The *sense of chirality or the absolute configuration* of sodium-rubidium tartrate was determined (*Bijvoet*, 1949) making use of anomalous diffraction of X-rays (zirconium K_α X-rays).[1]

The assignment of a configuration to the D- or the L-series in the case of optically active tartaric acid is not consistent in the literature, but depends on whether it is based on carbon atom 2 or carbon atom 3, so it has proved expedient in this case to use the *Cahn–Ingold–Prelog* system.

In (+)-tartaric acid, both asymmetric carbon atoms belong to the R-series, and in (−)-tartaric acid to the S-series, whilst *meso*-tartaric acid contains one R- and one S-carbon atom. Thus (+)-tartaric acid is described as (2R,3R)-(+)-tartaric acid and (−)-tartaric acid as (2S,3S)-(−)-tartaric acid.

The configurations of the tartaric acids are shown by the following *Fischer* projection formulae:

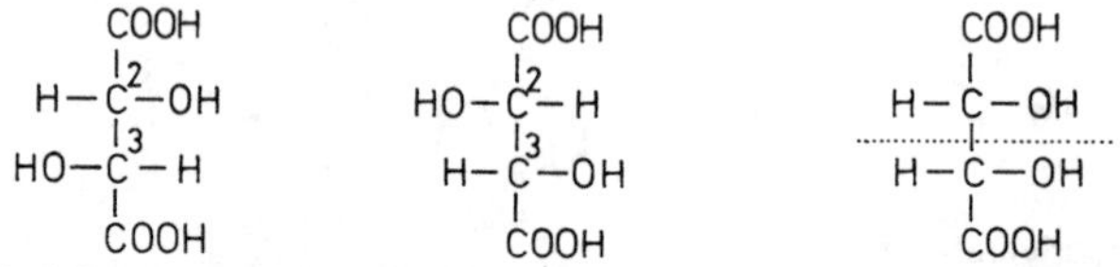

(2R,3R)-(+)-Tartaric acid   (2S,3S)-(−)Tartaric acid   *meso*-Tartaric acid
                                                          (inactive, achiral)

racemic Tartaric acid

(2R,3R)-(+)-Tartaric acid occurs both as the free acid and as salts in many fruits. The salts are known as *tartrates*. In the preparation of wine it partially separates out as potassium hydrogen tartrate (cream of tartar). In order to isolate the free acid this is treated with calcium hydroxide to give calcium tartrate, which is converted into tartaric acid by adding sulphuric acid (*Scheele*, 1769). (+)-Tartaric acid forms translucent prisms

---

[1] See R. *Parthasarathy*, The Determination of Relative and Absolute Configurations or Organic Molecules by X-Ray Diffraction Methods, in H. B. *Kagan*, Stereochemistry, Vol. *1*, 181 (Thieme Verlag, Stuttgart 1977).

with m.p. 170° (443 K). In aqueous solution the optical rotation changes from right to left as the concentration of the solution is increased. It is used as a preservative in food; potassium sodium tartrate (Rochelle salt) is a constituent of *Fehling's* solution.

(2*S*,3*S*)-(−)-**Tartaric acid** occurs in the leaves of the indigenous Central African bush *Bauhinia* (see p. 451, Footnote).

(±)-**Tartaric acid** ((±)-2,3-dihydroxysuccinic acid) is not found in nature, but can be made synthetically in a number of ways, for example, by oxidation of maleic anhydride with hydrogen peroxide, from glyoxal *via* a cyanohydrin, or by the reaction of dibromo-succinic acid with silver oxide. It crystallises from water as a monohydrate, m.p. 206°C (479 K). In 1844 *Pasteur* separated the sodium ammonium salt of racemic tartaric acid into its two *enantiomeric forms*.

meso-**Tartaric acid** does not occur in nature. It can be made by oxidising maleic acid with potassium permanganate or by oxidising D(−)-erythrose with nitric acid. It crystallises as a monohydrate, m.p. 140°C (413 K).

meso-Tartaric acid has an intramolecular plane of symmetry (....); because of this it is achiral.

2.36.6 Methods for the Resolution of Racemic Forms[1]

2.36.6.1 Spontaneous Crystallisation (*Pasteur*, 1848). Racemic mixtures almost always crystallise out of solution as *racemates*; each unit cell of the crystals contains the same number of molecules of both the D- and L-forms and is thus made up of two molecules of different chirality. Only in rare cases does recrystallisation at a certain temperature give rise to mixtures of crystals of the D- and L-forms in the form of a *conglomerate*. *Pasteur* observed that, at temperatures below 28°C (300 K), sodium ammonium tartrate crystallised out, not as a racemate, but as a conglomerate, or mixture, of the (+)- and (−)-salts.

In this case the individual crystals are made up from molecules of the same chirality; the (+)- and (−)-forms have hemihedral facets of opposite symmetry, and are known as *enantiomorphic* forms. These mirror-image forms can be mechanically sorted out from one another; in solution they are optically active. By adding a suitable chiral auxiliary, it is possible to obtain crystals all having the same chirality from a solution of the racemate. The crystals formed have a stereochemistry which depends upon that of the auxiliary, and they have chirality which is the reverse of that of the auxiliary.[2] As an example, if a solution of DL-glutamic acid hydrochloride is allowed to crystallise in the presence of a small amount of L-lysine, D-glutamic acid with an ee of 100% results. This method represents a kinetic method for the resolution of enantiomers and is a result of hindrance of the crystal growth of one of the enantiomeric forms.

2.36.6.2 Biochemical Separation (*Pasteur*, 1858). When some moulds or bacteria are provided with the appropriate nutrient media they only consume as food one optically active form of a racemic mixture. Making use of this, *Pasteur* was able to isolate (−)-tartaric acid from racemic tartaric acid by the action of the mould *penicillium glaucum*; in the process the (+)-acid was destroyed. The disadvantage of this method is that one of the enantiomers is lost.

2.36.6.3 Chemical Separation making use of Optically Active Materials (*Pasteur*, 1858). This is the commonest method of resolution. Racemic forms of organic acids can be converted into *diastereomers*

---

[1] See *S. H. Wilen* et al., Strategies in Optical Resolutions, Tetrahedron *33*, 2725 (1977); *J. Jacques* et al., Enantiomers, Racemates and Resolutions (Wiley, New York 1981); *P. Newman* Optical Resolution Procedures for Chemical Compounds Vols 1–3 (Optical Resolution Information Center, Manhattan College, Riverside, N.Y. 1979–84); *C. J. Sih* and *S. H. Wu*, Resolution of Enantiomers by Biocatalysis, Top. Stereochem. *19*, 63 (1989).
[2] See *L. Addadi* et al., Growth and Dissolution of Organic Crystals with Tailor-made Inhibitors — Implications in Stereochemistry and Materials Science, Angew. Chem. Int. Ed. Engl. *24*, 466 (1985).

by the action of optically active bases such as (+)-cinchonine, and this provides a means for separating the enantiomers. The two salts which are formed are not mirror image forms:

$$\text{Racemic form} \begin{cases} (+)\text{-Acid} \\ (-)\text{-Acid} \end{cases} (+)\text{-Base} \rightarrow \begin{Bmatrix} (++)\text{-Salt} \\ (-+)\text{-Salt} \end{Bmatrix} \text{Diastereomers}$$

Stereoisomers which are not mirror images of each other are called *diastereomers*. They have different energies and differ from one another in both their physical and chemical properties. They have different solubilities and because of this it is possible to achieve a separation of diastereoisomeric salts by fractional crystallisation. After separation, the (+)-base can be removed by hydrolysis with an inorganic acid, leaving samples of the pure (+)- and (−)-acids. Similarly, racemic bases can be resolved by use of an optically active acid such as (S)-(+)-camphor-10-sulphonic acid or (+)-3-bromocamphor-10-sulphonic acid.

Racemic forms of *Lewis* bases which contain no functional group can be resolved by forming *charge-transfer complexes* with *Newman*'s reagent, TAPA, 2-(2,4,5,7-tetranitro-7-fluorenylideneamino-oxy)-propionic acid.

2.36.6.4 A further development of the original method of *Pasteur* involves the use of chiral stationary phases on which gas-chromatographic separation of racemic forms can be achieved.[1] This enantiospecific separation process was realised by *Gil-Av* in 1966, who carried out a quantitative separation of the racemic forms of derivatives of amino acids. The chiral stationary phases, such as polyorganosiloxanes, contain regularly distributed chiral units (*E. Bayer*).

The commercial product *Chirasil-Val* formulated above uses a diamide of L-valine as a chiral unit. The chiral stationary phase is described as the *selector* and the separated racemic form as the *selectand*.

2.36.6.5 Very pure enantiomers (ee > 96%) can be obtained from racemic secondary allyl alcohols by the *stereoselective generation of chiral epoxides*.[2] For this purpose the *rac.-sec.*-allyl alcohol is treated with a *Sharpless* reagent using (+)- or (−)-isopropyl tartrate. In most cases, the one enantiomer reacts so much more quickly that the other slower-reacting enantiomer effectively remains unchanged and can be isolated directly.

Such a *kinetic separation of enantiomers* depends upon the highly selective *diastereofacial reaction* which takes place.

2.36.6.6 Separation using Urea Inclusion Compounds (*W. Schlenk jun.*, 1952). Urea forms 'inclusion compounds' (clathrates) with a variety of organic substances, for example with alkanes, whose hexagonal crystal lattices can take up two *enantiomorphic* forms, corresponding to the right- and left-handed pitches of screws. By using the appropriate experimental conditions, *e.g.* by seeding with

---

[1] See *V. Schurig*, Gas Chromatographic Separation of Enantiomers on Optically Active Metal-Complex-free Stationary Phases, Angew. Chem. Int. Ed. Engl. *23*, 747 (1984); (see also p. 460).
[2] See *K. B. Sharpless* et al., J. Amer. Chem. Soc. *103*, 6237 (1981).

suitable crystals, crystallisation can be induced in which only crystals of one particular pitch separate out.

Using this method the *racemic form of 2-chlorooctane* can be separated into its enantiomers. When the racemic form is confronted with, say, the right-handed form of urea, then inclusion leads to the formation of two different adducts, which are not mirror images of each other, and hence have different solubilities.

Right-handed urea + (*R*)-2-chlorooctane
Right-handed urea + (*S*)-2-chlorooctane

The two adducts can be separated by fractional crystallisation and the two enantiomers obtained by releasing them from the inclusion compounds. This process enables racemic forms to be resolved without the involvement of any other optically active agent.

2.36.7 *Citric Acid*

This is the main acid present in lemons, oranges, pineapples, strawberries, red currants, cranberries and other fruit. It can be isolated from lemon juice, which contains about 5–7% of it (*Scheele*, 1784). Citric acid is also found in milk, and in small amounts in blood and in urine; it plays a vital role as an intermediate in metabolic processes.

Preparation

2.36.7.1 From *carbohydrates* (glucose, cane sugar, molasses or starch), in about 50–70% yield, by a fermentation process brought about by certain varieties of *penicillium* in the presence of air.

2.36.7.2 Synthetically by the following process, starting from *1,3-dichloroacetone*:

$$
\begin{array}{l}
CH_2-Cl \\
|\quad \\
C=0 \\
|\quad \\
CH_2-Cl
\end{array}
\xrightarrow{+HCN}
\begin{array}{l}
CH_2-Cl \\
|\_OH \\
C \\
|^{\diagdown}CN \\
CH_2-Cl
\end{array}
\longrightarrow
\begin{array}{l}
CH_2-Cl \\
|\_OH \\
C \\
|^{\diagdown}COOH \\
CH_2-Cl
\end{array}
\xrightarrow[-2\,KCl]{+2\,KCN}
\begin{array}{l}
CH_2-CN \\
|\_OH \\
C \\
|^{\diagdown}COOH \\
CH_2-CN
\end{array}
\longrightarrow
\begin{array}{l}
CH_2-COOH \\
|\_OH \\
C \\
|^{\diagdown}COOH \\
CH_2-COOH
\end{array}
$$

1,3-Dichloroacetone   Citric acid

Properties

Citric acid crystallises from aqueous solutions as rhombic prisms of its monohydrate; they are very readily soluble in water. The anhydrous acid melts at 153°C (426 K). Its salts are called *citrates*. Calcium citrate is one of the few salts which are less soluble in hot water than in cold water.

When heated at 175°C (448 K) *intramolecular* loss of water takes place and citric acid is converted into aconitic acid:

$$
\begin{array}{l}
H-CH-COOH \\
|\quad \\
HO-C-COOH \\
|\quad \\
CH_2-COOH
\end{array}
\xrightarrow{-H_2O}
\begin{array}{l}
CH-COOH \\
\|\quad \\
C-COOH \\
|\quad \\
CH_2-COOH
\end{array}
\xrightarrow{Na\,(Hg)}
\begin{array}{l}
CH_2-COOH \\
|\quad \\
CH-COOH \\
|\quad \\
CH_2-COOH
\end{array}
$$

Citric acid   Aconitic acid   Tricarballylic
                                                                      acid

Aconitic acid occurs in some plants, *e.g.* in monkshood (*aconitum napellus*). When it is reduced, using sodium amalgam, the product is tricarballylic acid, propane-1,2,3-tri-carboxylic acid, which occurs in sugar beet.

When heated with conc. sulphuric acid citric acid, as an α-hydroxyacid, loses water and carbon monoxide to give *acetonedicarboxylic acid*. When this is heated more strongly it decomposes to acetone and two molecules of carbon dioxide. Esters of citric acid are more stable.

$$
\begin{array}{ccc}
\text{CH}_2\text{—COOH} & \text{CH}_2\text{—COOH} & \text{CH}_3 \\
| & | & | \\
\text{C}\substack{-\text{OH} \\ \searrow\text{COOH}} \xrightarrow{-\text{CO},-\text{H}_2\text{O}} & \text{C}=\text{O} \xrightarrow{-2\,\text{CO}_2} & \text{C}=\text{O} \\
| & | & | \\
\text{CH}_2\text{—COOH} & \text{CH}_2\text{—COOH} & \text{CH}_3
\end{array}
$$

Acetonedicarboxylic acid

2.37 Aliphatic Ketodicarboxylic Acids

2.37.1 Mesoxalic Acid (oxomalonic acid)

The simplest ketodicarboxylic acid, mesoxalic acid (oxomalonic acid), is only isolable as its hydrate (which provides an exception to the *Erlenmeyer* rule). It is related structurally to tartronic acid:

$$
\begin{array}{ccc}
\text{COOH} & \text{COOH} & \text{COOH} \\
| & | & | \\
\text{H—C—OH} \xrightarrow[-\text{H}_2\text{O}]{[\text{O}]} & \text{C}=\text{O} \xrightarrow{+\text{H}_2\text{O}} & \text{C}\substack{\nearrow\text{OH} \\ \searrow\text{OH}} \\
| & | & | \\
\text{COOH} & \text{COOH} & \text{COOH}
\end{array}
$$

Tartronic acid   Mesoxalic acid   Mesoxalic acid hydrate
(m.p. 121°C [394 K])

When heated in aqueous solution mesoxalic acid loses carbon dioxide and is converted into glyoxylic acid.

2.37.2 Oxaloacetic Acid

Oxaloacetic acid (oxosuccinic acid) undergoes not only keto–enol tautomerism but also (Z)–(E) isomerism; the following equilibria are involved:

$$
\begin{array}{ccc}
\text{O}=\text{C—COOH} & \text{HO—C—COOH} & \text{HO—C—COOH} \\
| \rightleftharpoons & \| \rightleftharpoons & \| \\
\text{H}_2\text{C—COOH} & \text{H—C—COOH} & \text{HOOC—C—H}
\end{array}
$$

Oxaloacetic acid   Hydroxymaleic acid   Hydroxyfumaric acid

Keto–enol tautomerism   (Z)–(E) Isomerism

Free oxaloacetic acid itself is not known; rather, depending upon the reaction conditions, either *hydroxymaleic acid*, m.p. 152°C (425 K) or *hydroxyfumaric acid*, m.p. 184°C (457 K) are isolated. Aqueous solutions give colours with iron(III) chloride which are typical of *enols*. Oxaloacetic acid is an important biochemical intermediate in the citric acid cycle.

Its esters are more stable. **Diethyl oxaloacetate** can be made by a *Claisen* condensation reaction between diethyl oxalate and ethyl acetate in the presence of sodium ethoxide:

$$C_2H_5OOC-CO\ OC_2H_5 + H-CH_2-COOC_2H_5 \longrightarrow C_2H_5OOC-\underset{\underset{O}{\|}}{C}-CH_2-COOC_2H_5 + C_2H_5OH$$

| Diethyl oxalate | Ethyl acetate | Diethyl oxaloacetate |

The liquid diethyl ester exists largely (up to 80%) in the *enol* form. It can be distilled under reduced pressure (b.p. 131°C [404 K]/32 mbar [32 hPa], but when distilled under atmospheric pressure it loses carbon monoxide (decarbonylation) and becomes diethyl malonate:

$$C_2H_5OOC-\underset{\underset{O}{\|}}{C}-CH_2-COOC_2H_5 \longrightarrow H_2C(COOC_2H_5)_2 + CO$$

Oxaloacetic esters and their C-alkyl derivatives are useful synthetically, since, like acetoacetic esters, they can undergo ketonic cleavage (with acids) or acid cleavage (with alkalis):

Ketonic cleavage

$$C_2H_5OOC-\underset{\underset{R}{|}}{CH}-CO-COOC_2H_5 \xrightarrow[-2\,C_2H_5OH]{+2\,H_2O(H^\oplus)} R-CH_2-CO-COOH + CO_2$$

α-Keto acid

Acid cleavage

$$C_2H_5OOC-\underset{\underset{R}{|}}{CH}-CO-COOC_2H_5 \xrightarrow[-2\,C_2H_5OH]{+3\,H_2O(OH^\ominus)} R-CH_2-COOH + (COOH)_2$$

2.38 Carbonic Acid Derivatives

Carbonic acid, HO—CO—OH, can be regarded either as a hydrate of carbon dioxide or as hydroxyformic acid. It is incapable of existence in solution, but has been detected in the gas phase.[1]

The following equilibria, which are largely displaced towards the left, take place in aqueous solutions of carbon dioxide:

$$CO_2 + H_2O \rightleftarrows [H_2CO_3] \rightleftarrows H^\oplus + HCO_3^\ominus$$

While carbon dioxide and carbonates are usually dealt with as inorganic materials, derivatives of carbon dioxide are usually treated as organic compounds.

The amount of carbon dioxide in the atmosphere has risen since the industrial revolution from about 280 ppmv:volume to over 380 ppmv (more than 576 ppm by weight); it increases by about 1.5 ppm *per annum* and contributes up to about 50% of the greenhouse effect in the earth's atmosphere.[2]

---

[1] See *J. K. Terlouw* et al., Thermolysis of NH_4HCO_3 — A Simple Route to the Formation of Free Carbonic Acid (H_2CO_3) in the Gas Phase. Angew. Chem. Int. Ed. Engl. 26, 354 (1987).

[2] See *S. H. Schneider*, The Greenhouse Effect: Science and Policy. Science *243*, 771 (1989). If the carbon dioxide content of the atmosphere were doubled it would probably lead to an increase in temperature of about 3 ± 1.5 K.

By producing an atmosphere of carbon dioxide in grain silos it is possible to protect them against rodents and insects. At temperatures above the critical solution temperatures, liquid carbon dioxide dissolves organic compounds and this is made use of industrially for the extraction of caffeine from coffee beans and of nicotine from tobacco.

2.38.1 Halogen Derivatives of Carbonic Acid

In theory, both of the following chloro-derivatives of carbonic acid are available, but in the case of chloroformic acid only its esters, and not the free acid, are stable and isolable.

$$\left[O=C \begin{matrix} Cl \\ OH \end{matrix} \right] \qquad O=C \begin{matrix} Cl \\ Cl \end{matrix}$$

Chloroformic acid   Phosgene

2.38.1.1 Phosgene (carbonyl chloride) was prepared by *Davy* (1812) by the action of sunlight on a mixture of *carbon monoxide and chlorine* (*phosgene* = generated by light):

$$CO + Cl_2 \xrightarrow{h\nu} COCl_2 \qquad \Delta H = -96\ kJ/Mol$$

The energy provided by the radiation leads to splitting of the chlorine molecules into chlorine atoms and a radical chain reaction is set up.

Industrially phosgene is obtained by passing chlorine and carbon monoxide over activated charcoal as a catalyst at about 100–120° (375–395 K). In the laboratory it can be made from *tetrachloromethane and fuming sulphuric acid* (45% oleum) at 78°C (350 K); chlorosulphonic acid is also formed:

$$CCl_4 + SO_3 + H_2SO_4 \rightarrow COCl_2 + 2\ ClSO_3H$$

Phosgene is a colourless gas, which liquefies at 8°C (280 K). It is transported in compressed form in steel cylinders. It reacts only slowly with cold water, but more rapidly when heated:

$$COCl_2 + H_2O \rightarrow CO_2 + 2\ HCl$$

Phosgene is used as the starting material in some synthetic processes, *e.g.* for making isocyanates and in the dyestuffs industry. Physiologically it is a powerful *respiratory poison* (see also under diphosgene).

2.38.2 Esters of Carbonic Acid

When phosgene reacts with alcohols the products are, depending upon the reaction conditions, either *esters of chloroformic acid* or *diesters of carbonic acid* (*dialkyl carbonates*) e.g.

$$O=C \begin{matrix} Cl \\ Cl \end{matrix} \xrightarrow[-HCl]{+H-OCH_3} O=C \begin{matrix} Cl \\ OCH_3 \end{matrix} \xrightarrow[-3\ HCl]{+3\ Cl_2,\ h\nu} O=C \begin{matrix} Cl \\ OCCl_3 \end{matrix}$$

Methyl chloroformate
(b.p. 71.5°C [344.7 K])   Diphosgene

$$O=C \begin{matrix} Cl \\ Cl \end{matrix} \xrightarrow[-2\ HCl]{+2H-OCH_3} O=C \begin{matrix} OCH_3 \\ OCH_3 \end{matrix} \xrightarrow[-6\ HCl]{\substack{+6\ Cl_2,\ h\nu \\ 200\ K}} O=C \begin{matrix} OCCl_3 \\ OCCl_3 \end{matrix}$$

Dimethyl carbonate (b.p. 91° [364 K])   Triphosgene

Esters of chloroformic acid are liquids with suffocating smells. As acid chloride-esters they are very reactive. In consequence the ethyl ester, b.p. 94.5°C (367.7 K), for example, is used in order to introduce the *ethoxycarbonyl group*, —$COOC_2H_5$, into compounds which have reactive hydrogen atoms, *e.g.* alcohols, ammonia, primary and secondary amines. Photo-chlorination leads to the formation of trichloromethyl chloroformate (diphosgene).

Dialkyl carbonates are liquids with ethereal odours; they are readily soluble in water. Photo-chlorination of dimethyl carbonate provides crystalline bis(chloromethyl) carbonate (triphosgene, m.p. 80°C [313 K]). The chemical properties of diphosgene and triphosgene resemble those of phosgene but, compared to phosgene, they are less harmful substances.

Orthocarbonate esters are derivatives of the non-existent orthocarbonic acid, $(C(OH)_4)$, and are formed by the reaction of trichloro-nitromethane (chloropicrin) with sodium alkoxides, *e.g.*

$$Cl_3CNO_2 + 4\,NaOC_2H_5 \longrightarrow C(OC_2H_5)_4 + 3\,NaCl + NaNO_2$$

Orthocarbonic esters are colourless liquids which are fairly stable to water; they are useful in synthesis. Tetraethyl orthocarbonate boils at 158°C (431 K).

2.38.3 Amides of Carbonic Acid

The following are amides derived from carbonic acid:

$$\left[O{=}C\!\!\begin{array}{c} \diagup NH_2 \\ \diagdown OH \end{array} \right] \qquad\qquad O{=}C\!\!\begin{array}{c} \diagup NH_2 \\ \diagdown NH_2 \end{array}$$

Carbamic acid   Urea

2.38.3.1 Carbamic acid is the mono-amide of carbonic acid. It is unknown in the free state because the carboxyl group attached to the nitrogen atom immediately splits off as carbon dioxide. However, its salts, the *carbamates*, and its esters, the alkyl carbamates or *urethanes*, are stable substances. Ammonium carbamate can be made from dry carbon dioxide and gaseous ammonia:

$$CO_2 + NH_3 \;\rightleftharpoons\; \left[O{=}C\!\!\begin{array}{c} \diagup NH_2 \\ \diagdown OH \end{array} \right] \xrightarrow{+NH_3} \left[O{=}C\!\!\begin{array}{c} \diagup NH_2 \\ \diagdown \overset{\ominus}{O} \end{array} \right] \overset{\oplus}{NH_4}$$

Carbamic acid   Ammonium carbamate

It is a colourless, crystalline substance which is readily soluble in water. If an aqueous solution of this salt is heated above 60°C (330 K) it is hydrolysed to ammonium carbonate which decomposes into ammonia and carbon dioxide:

$$\left[O{=}C\!\!\begin{array}{c} \diagup NH_2 \\ \diagdown \overset{\ominus}{O} \end{array} \right] \overset{\oplus}{NH_4} \xrightarrow{+(H_2O)} \left[O{=}C\!\!\begin{array}{c} \diagup \overline{O}{}^{\ominus} \\ \diagdown \overline{O}{}^{\ominus} \end{array} \right] 2\,\overset{\oplus}{NH_4} \xrightarrow{-(H_2O)} 2\,NH_3 + CO_2$$

2.38.3.2 Urethanes or alkyl carbamates can be made from chloroformic esters and ammonia at low temperatures, *e.g.*

$$O=C\overset{Cl}{\underset{OC_2H_5}{\big<}} \xrightarrow[-HCl]{+NH_3} O=C\overset{NH_2}{\underset{OC_2H_5}{\big<}}$$

Urethane (m.p. 50°C [323 K])

$$H_3C-CH_2-CH_2\overset{CH_2-O-\overset{O}{\overset{\|}{C}}-NH_2}{\underset{CH_2-O-\underset{\|}{\underset{O}{C}}-NH_2}{\big<}}C\overset{}{\underset{H_3C}{\big<}}$$

INN: *Meprobamate*

The generic name *urethane* is commonly used to denote the specific compound ethyl carbamate; it has been shown to act as a carcinogen in some animals. 2-Methyl-2-propyl-propane-1,3-diyl dicarbamate (INN: *Meprobamate*) is a general sedative which also brings about relaxation of muscles.

N-Substituted carbamate esters are made by *heating isocyanate esters with alcohols* (see *Curtius* decomposition of acid azides) or *phenols*, *e.g.*

O=C=NCH$_3$ +

Methyl
isocyanate

1-Naphthol (with OH)

$\longrightarrow$

N-Methyl-1-naphthyl carbamate (with O−C(=O)−NHCH$_3$)

The *N*-methyl-1-naphthyl carbamate which is produced is used as a pesticide (Carbaryl, Sevin).

Aryl isocyanates also give crystalline adducts with anhydrous alcohols, for example *N-phenylurethane* (m.p. 53°C [326 K]) is obtained from ethanol and phenyl isocyanate:

$$C_6H_5-N=C=O \ + \ C_2H_5OH \ \longrightarrow \ C_6H_5-NH-C\overset{\nearrow O}{\underset{OC_2H_5}{\big<}}$$

Phenyl isocyanate   *N*-Phenylurethane

Because of this, phenyl, *p*-nitrophenyl and 1-naphthyl isocyanates are often used to prepare crystalline derivatives to assist in the identification of alcohols.

2.38.3.3 Polyurethanes are of industrial importance in the plastics industry.[1] They were discovered by *O. Bayer* and are made by the *diisocyanate polyaddition process*,[2] for example by polyaddition of butane-1,4-diol to hexamethylenediisocyanate:

$$n \ O=C=N-(CH_2)_6-N=C=O \ + \ n \ HO-(CH_2)_4-OH \ \longrightarrow$$

Hexamethylene-diisocyanate   Butane-1,4-diol

$$\left[\!\!-O-(CH_2)_4-O-\underset{\underset{O}{\|}}{C}-NH-(CH_2)_6-NH-\underset{\underset{O}{\|}}{C}-\!\!\right]_n$$

Polyurethane

---

[1] See *G. Woods*, The ICI Polyurethane Book, 2nd Edtn. (Wiley, Chichester 1990).
[2] See *O. Bayer*, Das Di-Isocyanat-Polyadditionsverfahren (Polyurethane) (The Di-isocyanate Polyaddition Process), Angew. Chem. *59*, 257 (1947).

The units within the polyurethane chain are linked together by means of *carbamic acid groups*, —O—CO—NH—.

The *diisocyanates* are made from the reaction of phosgene with the appropriate diamine, e.g.

$$H_2N-(CH_2)_6-NH_2 + 2\,OCCl_2 \;\longrightarrow\; O=C=N-(CH_2)_6-N=C=O + 4\,HCl$$

Industrially the most commonly used reactants are *aryldiisocyanates* such as phenylene-1,4-diisocyanate, 4-methylphenylene-1,3-diisocyanate (tolylene-2,4-diisocyanate), 2-methylphenylene-1,3-diisocyanate (tolylene-2,6-diisocyanate), biphenyl-4,4'-diisocyanate, bis(4-isocyanatophenyl)methane or naphthylene-1,5-diisocyanate.

Among the *polyurethane elastomers* is *Vulkollan*[1] which has rubber-like properties, and is made from poly(adipic acid glycol ester) and naphthylene-1,5-diisocyanate.

If the polyaddition is carried out in such a way that the isocyanate groups undergo a reaction with water, catalysed by the presence of triethylenediamine (DABCO), this brings about evolution of carbon dioxide and a *polyurethane foam* is produced. This foam has been used as filling and padding in furniture and in vehicles and also as an insulating material in refrigerators.

2.38.3.4 Urea

2.38.3.4 Urea (carbamide), which is the diamide of carbonic acid, was discovered in urine by *Rouelle* (1773) and is the end-product from the degradation of proteins in human beings and in land mammals. An adult human excretes about 20 g of urea per day. Urea has also been detected in many plants and especially in fungi.

Preparation

(a) By *Wöhler's* synthesis, which was by evaporating an aqueous solution of ammonium cyanate:

$$|N\equiv C-\underline{\bar{O}}|^{\ominus}\; NH_4^{\oplus} \;\rightleftharpoons\; O=C\diagup^{NH_2}_{\diagdown NH_2}$$

urea

(b) In the laboratory by the usual methods for the preparation of amides, namely by the action of ammonia on phosgene, chloroformate esters, urethanes or diesters of carbonic acid. The yields are appreciably lower, however, than those achieved for amides of normal carboxylic acids:

$$O=C\diagup^{Cl}_{\diagdown Cl} \xrightarrow[-\,2\,HCl]{+\,2\,NH_3} O=C\diagup^{NH_2}_{\diagdown NH_2} \xleftarrow[-\,2\,ROH]{+\,2\,NH_3} O=C\diagup^{OR}_{\diagdown OR}$$

(c) Industrially from *carbon dioxide and ammonia, via* ammonium carbamate. In the presence of a threefold excess of ammonia at 135–150°C (410–425 K) and 35–40 bar (3.5–4 MPa) this is converted into urea:

$$O=C\diagup^{NH_2}_{\diagdown O^{\ominus}} \Bigg]\; NH_4^{\oplus} \;\rightleftharpoons\; O=C\diagup^{NH_2}_{\diagdown NH_2} + H_2O$$

Ammonium carbamate

Under these conditions the competing reaction of hydrolysis of ammonium carbamate to ammonium carbonate, ammonia and carbon dioxide is supressed.

---

[1] See *O. Bayer* and *E. Müller*, Das Aufbauprinzip der Urethan-Elastomeren „Vulkollan" (The 'Principal of Constructing Urethane Elastomers), Angew. Chem. 72, 934 (1960); *H. Rinke*, Elastomere Fasern auf Polyurethanbasis (Elastomeric Fibres from Polyurethanes) Angew. Chem. 74, 612 (1962).

(d) *N-Alkylated ureas* are made by the reaction of primary or secondary amines with potassium cyanate or isocyanate esters.

Properties. Urea forms tetragonal prisms, m.p. 132.7°C (405.9 K), which are soluble in water and ethanol. In aqueous solution it denatures proteins and nucleic acids. With mineral acids urea behaves as a very weak base. Its salts are extensively hydrolysed and give an acid reaction. Its nitrate only dissolves sparingly in conc. nitric acid and this has been utilised to separate urea out of urine.

Like the amides of carboxylic acids urea is *stabilised by mesomerism*, as follows:

$$\left[\overline{O}=C\underset{\overline{N}H_2}{\overset{\overline{N}H_2}{\diagup}} \longleftrightarrow \overset{\ominus}{\overline{|O}}-C\underset{\overline{N}H_2}{\overset{\oplus}{=}NH_2} \longleftrightarrow \overset{\ominus}{\overline{|O}}-C\underset{\overset{\oplus}{N}H_2}{\overset{\overline{N}H_2}{\diagup}} \right]$$

Urea undergoes hydrolysis when heated with either aqueous acids or bases:

$$H_2N-CO-NH_2 + H_2O \longrightarrow CO_2 + 2\,NH_3$$

A similar reaction results in the breakdown of urea by the enzyme *urease*.

When *urea reacts with nitrous acid*, as in the case of other amides, nitrogen is evolved and the quantity can be measured volumetrically (*van Slyke*):

$$H_2N-CO-NH_2 + 2\,HNO_2 \longrightarrow 2\,N_2 + CO_2 + 3\,H_2O$$

Uses. Urea has been used increasingly as a fertiliser having a high nitrogen content, and also serves as a nitrogen-rich feedstock. Industrially it is the starting material from which *melamine* is prepared and it is also used for the preparation of urea-formaldehyde resins. These arise from the condensation of urea with formaldehyde which react together to form long-chain molecules:

$$n\,H_2N-\underset{O}{\overset{}{C}}-NH_2 + n\,CH_2O \xrightarrow[-n\,H_2O]{} (\!-NH-CH_2-NH-CO-\!)_n$$

If there is excess formaldehyde it cross-links the chains, giving a thermosetting plastic.

Another important use of urea is in the manufacture of effective soporifics of the ureide type. **Ureides** are amide-like derivatives obtained from organic acids and urea; they have an —NH—CO—NH$_2$ group instead of the —NH$_2$ group found in simple amides, *e.g.* acetylurea, $H_3CCO-NH-CO-NH_2$. They are prepared by the reaction of urea with acid chlorides or esters and are crystalline materials which, in contrast to urea, do not form salts with acids.

The ureides which are used as soporifics are cyclic ureides; they are heterocyclic compounds and as such are considered further in a later section.

Tetramethylurea, $(H_3C)_2N-CO-N(CH_3)_2$, which is a liquid of b.p. 176.5°C (449.7 K), is used as a solvent for organic substances and also as a reaction medium in which to carry out base-catalysed isomerisations, alkylation and acylation reactions, and other condensation reactions.

Derivatives of Urea

The most important of these are *semi-carbazide* and *guanidine*:

$$O=C\begin{array}{c}{}^{NH_2}\\{}_{NH-NH_2}\end{array} \qquad HN=C\begin{array}{c}{}^{NH_2}\\{}_{NH_2}\end{array}$$

Semicarbazide   Guanidine

The commonly used name semicarbazide does not strictly tally with the structure, since no azide group is present.

2.38.3.5 Semicarbazide, the hydrazide of carbamic acid, can be made by a method analogous to *Wöhler's* synthesis of urea, from *potassium cyanate* and *hydrazine sulphate*; the following reaction between isocyanic acid and hydrazine occurs:

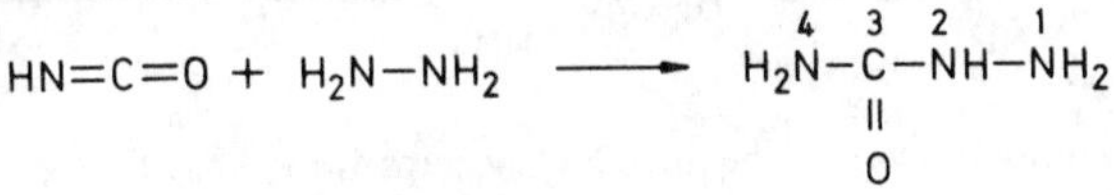

$$HN=C=O + H_2N-NH_2 \longrightarrow \overset{4}{H_2N}-\overset{3}{\underset{\overset{\|}{O}}{C}}-\overset{2}{NH}-\overset{1}{NH_2}$$

Semicarbazide

If acetic acid is used as solvent and there is an excess of the potassium cyanate, reactions proceed further to give hydrazine-1,2-dicarbonamide (bis-urea), $H_2N-CO-NH-NH-CO-NH_2$.

Semicarbazide is a crystalline base, m.p. 96°C (369 K). Its hydrochloride salt is often used in the isolation of carbonyl compounds, since it reacts with the latter to give *semicarbazones* which are rather insoluble.

$$\begin{array}{c}R\\R\end{array}C=O + H_2N-NH-\underset{\overset{\|}{O}}{C}-NH_2 \longrightarrow \begin{array}{c}R\\R\end{array}C=N-NH-\underset{\overset{\|}{O}}{C}-NH_2 + H_2O$$

Semicarbazone

The semicarbazones can be hydrolysed using dilute mineral acids, regenerating the carbonyl compound. Semicarbazide can be used instead of hydrazine in carrying out *Wolff–Kishner* reductions.

2.38.3.6 Guanidine occurs in the juice of sugar beet and in vetch seeds. It was first obtained by oxidative decomposition of guanine derived from *guano* (*Strecker*, 1861). It also forms part of the structure of *streptomycin*.

Preparation

(a) Its hydrochloride is made by *heating cyanamide (or dicyanodiamide) with a solution of ammonium chloride*:

$$N\equiv C-NH_2 + NH_4Cl \longrightarrow \left[\overset{\oplus}{H_2N}=C\begin{array}{c}{}^{NH_2}\\{}_{NH_2}\end{array} \right] Cl^{\ominus}$$

Cyanamide   Guanidine hydrochloride

(b) Guanidine nitrate is made industrially in 85% yield from dicyanodiamide (cyanoguanidine) and ammonium nitrate:

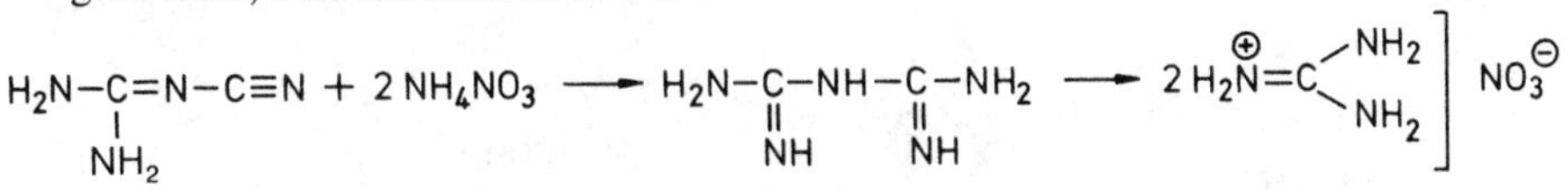

$$H_2N-\underset{\overset{|}{NH_2}}{C}=N-C\equiv N + 2NH_4NO_3 \longrightarrow H_2N-\underset{\overset{\|}{NH}}{C}-NH-\underset{\overset{\|}{NH}}{C}-NH_2 \longrightarrow 2\left[\overset{\oplus}{H_2N}=C\begin{array}{c}{}^{NH_2}\\{}_{NH_2}\end{array} \right]NO_3^{\ominus}$$

Dicyanodiamide   Biguanide   Guanidinium nitrate

Biguanide[1] is formed as an intermediate in the process.

---

[1] See *F. Kurzer* et al. The Chemistry of Biguanides, Fortschr. Chem. Forsch. *10*, 375 *ff.* (1968).

Properties. It is difficult to isolate pure guanidine. It forms crystals m.p. *ca* 50°C (325 K) which are very deliquescent. Its basicity is comparable to that of alkali metal hydroxides, and in air it takes up carbon dioxide and water to form *guanidinium hydrogen carbonate*:

$$HN=C(NH_2)_2 + CO_2 + H_2O \rightarrow H_2N=C(NH_2)_2]^{\oplus} HCO_3^{\ominus}$$

Guanidine is most commonly isolated as its nitrate, which is sparingly soluble in water.

The *guanidinium ion* is a very low energy species, because of its mesomeric stabilisation:

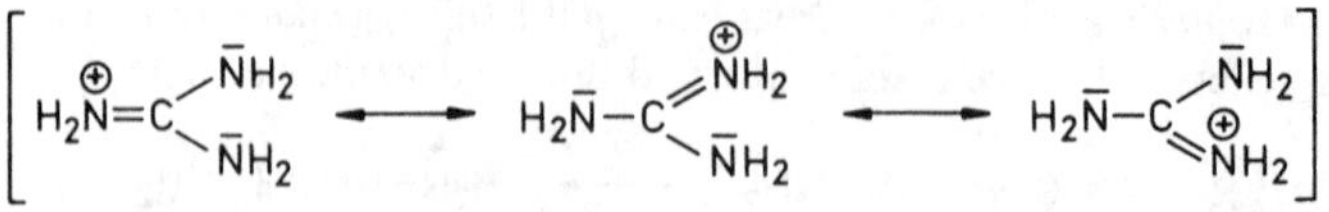

$$\left[H_2\overset{\oplus}{N}=C\overset{\bar{N}H_2}{\underset{\bar{N}H_2}{}} \longleftrightarrow H_2\bar{N}-C\overset{\overset{\oplus}{N}H_2}{\underset{\bar{N}H_2}{}} \longleftrightarrow H_2\bar{N}-C\overset{\bar{N}H_2}{\underset{\overset{\oplus}{N}H_2}{}} \right]$$

An X-ray crystal structure shows that the three NH_2 groups are arranged completely symmetrically about the central carbon atom and are all at the same distance from it (118 pm) (*Theilacker*, 1935).

Cold conc. sulphuric acid converts guanidinium nitrate into *nitroguanidine*:

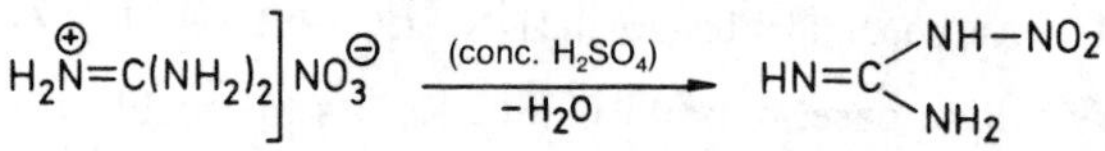

$$H_2\overset{\oplus}{N}=C(NH_2)_2 \Big] NO_3^{\ominus} \xrightarrow[-H_2O]{(conc.\ H_2SO_4)} HN=C\overset{NH-NO_2}{\underset{NH_2}{}}$$

Nitroguanidine (m.p. 246–247°C [519–520 K])

The latter compound is explosive and is used as a constituent of explosives. Zinc dust and acetic acid reduce it to *aminoguanidine*, which is usually isolated by means of its insoluble hydrogen carbonate salt. It can be used for detecting the presence of carbonyl groups with which it forms *amidinohydrazones*:

$$\overset{R}{\underset{R}{}}C=O + H_2N-NH-C\overset{NH_2}{\underset{NH}{}} \longrightarrow \overset{R}{\underset{R}{}}C=N-NH-C\overset{NH_2}{\underset{NH}{}} + H_2O$$

Aminoguanidine   Amidinohydrazone

Biologically important guanidines include L(+)-arginine, creatine and creatinine.

L(+)-**Arginine**, 2-amino-5-guanidinopentanoic acid, is a building block in proteins.

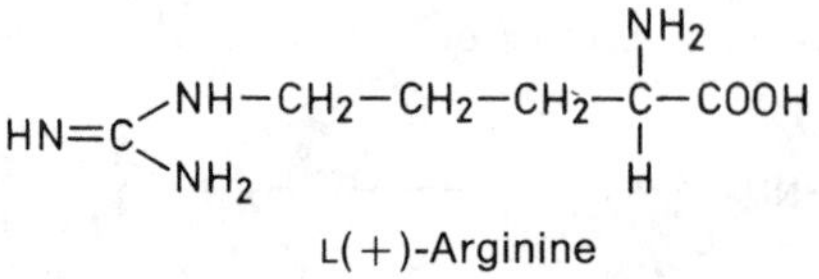

$$HN=C\overset{NH-CH_2-CH_2-CH_2-\overset{\overset{NH_2}{|}}{\underset{\underset{H}{|}}{C}}-COOH}{\underset{NH_2}{}}$$

L(+)-Arginine

Creatine, *N*-amidinosarcosine, *N*-methylguanidinoacetic acid, occurs in the muscle fluids of vertebrates, and in smaller quantity in the brain and in blood. It was first discovered by *Chevreul*, in meat broth. It is made synthetically from *sarcosine* (*N*-methylglycine) and *cyanamide*:

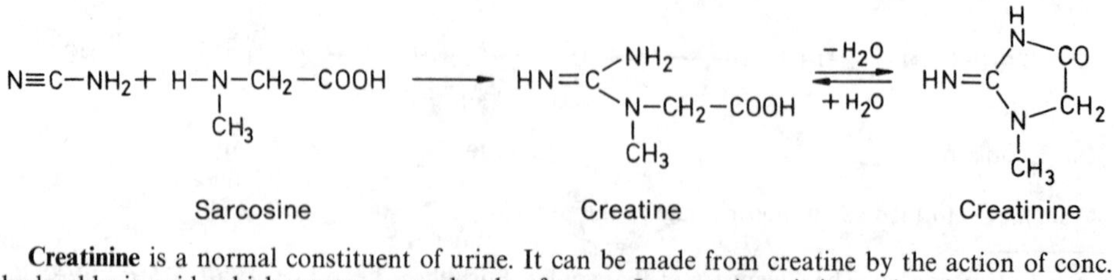

$$N\equiv C-NH_2 + \underset{\underset{CH_3}{|}}{H-N-CH_2-COOH} \longrightarrow HN=C\overset{NH_2}{\underset{\underset{\underset{CH_3}{|}}{N-CH_2-COOH}}{}} \underset{+H_2O}{\overset{-H_2O}{\rightleftharpoons}} HN=C\overset{\overset{\overset{H}{|}}{N}-CO}{\underset{\underset{\underset{CH_3}{|}}{N}-CH_2}{}}$$

Sarcosine   Creatine   Creatinine

Creatinine is a normal constituent of urine. It can be made from creatine by the action of conc. hydrochloric acid, which removes a molecule of water. In organisms it is produced from creatine phosphate, the energy-rich phosphate store of muscles.

$$HN=C\overset{\displaystyle NHPO_3H_2}{\underset{\displaystyle N(CH_3)CH_2COOH}{}}$$ Creatine phosphate

2.38.4 *Amides of Orthocarbonic Acid*[1]

Stepwise substitution of the alkoxy groups of esters of orthocarbonic acid by amino groups leads to amides of orthocarbonic acid.

$$R_2^1N\!\!\diagdown\!\!\underset{R^2O}{\overset{}{C}}\!\!\diagup\!\!OR^2 \qquad R_2^1N\!\!\diagdown\!\!\underset{R_2^1N}{\overset{}{C}}\!\!\diagup\!\!OR^2 \qquad R_2^1N\!\!\diagdown\!\!\underset{R_2^1N}{\overset{}{C}}\!\!\diagup\!\!NR_2^1 \qquad R_2^1N\!\!\diagdown\!\!\underset{R_2^1N}{\overset{}{C}}\!\!\diagup\!\!NR_2^1$$

Orthocarbamic acid esters   Orthocarbonic acid
                                                       octaalkylamides

The first three are esters of orthocarbamic acids.

The key substances for their synthesis are carbenium-iminium ions, which are obtained by alkylating urethanes and ureas, *e.g.*

$$(H_3C)_2N\!\!\diagdown\!\!\underset{(H_3C)_2N}{\overset{}{C}}\!\!=\!\!O \quad \xrightarrow{+(C_2H_5O)_3O^{\oplus}\,BF_4^{\ominus}} \quad \left[\,(H_3C)_2N\!\!\diagdown\!\!\underset{(H_3C)_2N}{\overset{\oplus}{C}}\!\!-\!\!OC_2H_5\,\right]\,BF_4^{\ominus}$$

Tetramethylurea

$$(H_3C)_2N\!\!\diagdown\!\!\underset{(H_3C)_2N}{\overset{}{C}}\!\!\diagup\!\!\overset{OC_2H_5}{\underset{OC_2H_5}{}} \quad \xrightarrow[-\,NaBF_4]{+NaOC_2H_5} \quad \left[\,(H_3C)_2\overset{\oplus}{N}\!\!=\!\!\underset{(H_3C)_2N}{\overset{}{C}}\!\!-\!\!OC_2H_5\,\right]$$

Tetramethylurea diethylacetal   Carbenium-iminium ion
(b.p. 35°C [308 K]/3 mbar [3 hPa])

Compounds of this series are useful as synthetic starting materials (see amide acetals).

2.39 Thio-Derivatives of Carbonic Acid

By replacing the oxygen atoms of carbonic acid systematically with sulphur atoms, the following types of compounds can be postulated:

$$O=C\overset{\displaystyle SH}{\underset{\displaystyle OH}{}} \qquad\qquad S=C\overset{\displaystyle OH}{\underset{\displaystyle OH}{}} \qquad\qquad S=C\overset{\displaystyle SH}{\underset{\displaystyle OH}{}}$$

Thiolocarbonic acid   Thionocarbonic acid   Thiolothionocarbonic acid
Thiocarbonic-*S*-acid   Thiocarbonic-*O*-acid   Dithiocarbonic-*O,S*-acid
                                                                (Xanthic acid)

$$O=C\overset{\displaystyle SH}{\underset{\displaystyle SH}{}} \qquad\qquad S=C\overset{\displaystyle SH}{\underset{\displaystyle SH}{}}$$

Dithiolocarbonic acid   Trithiocarbonic acid
Dithiocarbonic-*S*-acid

---

[1] See *W. Kantlehner*, Die präparative Chemie der *O*- und *N*-funktionellen Orthokohlensäurederivate (The preparative chemistry of *O*- and *N*-functionalised derivatives of orthocarbonic acid), Synthesis *1977*, 73.

Of these, only the trithiocarbonic acid is known in the free state. Important derivatives of thiolothionocarbonic acid are its anhydride, *carbon disulphide*, and its *O*-esters, known as xanthate esters.

Originally the *O*-ethyl ester of thiolothionocarbonic acid was known as xanthic acid. Since, however, it was desirable to allow for a variety of *O*-esters, the name xanthic acid was instead allotted to the hypothetical thiolothionocarbonic acid, and its *O*-esters were called xanthates.

2.39.1 Carbon Disulphide

Carbon disulphide, CS_2, is a colourless, strongly refracting liquid of b.p. 46°C (319 K) which, when pure, has an ether-like smell. It does not mix with water but is miscible with ethanol and ether. Its vapour is poisonous.

Carbon disulphide is obtained by passing sulphur vapour over red-hot charcoal at 900°C (1200 K), or from methane and sulphur vapour at 700°C (1000 K) in the presence of an aluminium oxide catalyst or from methane and pyrites:

$$C + 2S \rightleftharpoons S=C=S \qquad \Delta H = +88\,kJ/Mol$$

$$CH_4 + 4S \xrightarrow{(Al_2O_3)} CS_2 + 2H_2S$$

Carbon disulphide is an outstandingly good solvent for sulphur, white phosphorus and iodine as well as for organic materials, especially fats, oils and resins. Its disadvantage is its high inflammability. Industrially it is used for making *viscose* and in the preparation of ammonium thiocyanate and tetrachloromethane.

Xanthates, the alkali metal salts of xanthic acid, can be made by adding sodium alkoxides to carbon disulphide, *e.g.*

$$S=C=S + NaOC_2H_5 \longrightarrow S=C\begin{smallmatrix}SNa\\OC_2H_5\end{smallmatrix}$$

If this sodium salt is treated with copper(II) sulphate, yellow copper(I) ethyl xanthate separates out (Gr. *xanthos* = yellow). The reduction of copper(II) ions to copper(I) ions is balanced by an oxidation of xanthate anions to a dixanthogen, bis(ethoxythiocarbonyl)disulphide.

$$4\,S=C\begin{smallmatrix}\overset{\ominus}{\underline{S}}\,\overset{\oplus}{Na}\\OC_2H_5\end{smallmatrix} + 2\,CuSO_4 \longrightarrow 2\,S=C\begin{smallmatrix}SCu\\OC_2H_5\end{smallmatrix} + S=C\begin{smallmatrix}S\text{———}S\\OC_2H_5 \quad C_2H_5O\end{smallmatrix}C=S + 2\,Na_2SO_4$$

Sodium ethyl xanthate   Copper(I) ethyl xanthate   Dixanthogen
Sodium *O*-ethyldithiocarbamate

Derivatives of Thionocarbonic Acid

Among these are the following compounds:

$$S=C\begin{smallmatrix}NH_2\\NH_2\end{smallmatrix} \qquad S=C\begin{smallmatrix}NH_2\\NH-NH_2\end{smallmatrix} \qquad S=C\begin{smallmatrix}NH-NH_2\\NH-NH_2\end{smallmatrix}$$

Thiourea   Thiosemicarbazide   3-Thiocarbazide
                                            Thiocarbonohydrazide

2.39.2 Thiourea

The chemical properties of thiourea closely resemble those of urea.

Preparation

2.39.2.1 In a manner analogous to *Wöhler's* synthesis of urea, by *melting ammonium thiocyanate* at 140–180°C (410–450 K); yield 15–20%.

$$\left[{}^{\ominus}\overline{\underline{S}}-C\equiv N| \longleftrightarrow \overline{\overline{S}}=C=\overline{\underline{N}}{}^{\ominus} \right] NH_4^{\oplus} \rightleftharpoons S=C \begin{matrix} \diagup NH_2 \\ \diagdown NH_2 \end{matrix}$$

Thiourea

2.39.2.2 Industrially by the reaction of *hydrogen sulphide with calcium cyanamide* at 150–180°C (410–450 K):

$$CaN-C\equiv N + 2 H_2S \longrightarrow CaS + H_2N-\underset{\underset{S}{\|}}{C}-NH_2$$

Thiourea

Properties and Uses

Thiourea forms colourless, rhombic prisms, m.p. 180–182°C (453–455 K). They dissolve in water giving a neutral solution. Like urea, thiourea is *stabilised by mesomerism*. Consideration of the contributing structures indicates that there should be increased electron density on the sulphur atom.

$$\left[\overline{\overline{S}}=C \begin{matrix} \diagup \overline{N}H_2 \\ \diagdown \overline{N}H_2 \end{matrix} \longleftrightarrow {}^{\ominus}\overline{\underline{S}}-C \begin{matrix} \diagup {}^{\oplus}NH_2 \\ \diagdown \overline{N}H_2 \end{matrix} \longleftrightarrow {}^{\ominus}\overline{\underline{S}}-C \begin{matrix} \diagup \overline{N}H_2 \\ \diagdown {}^{\oplus}NH_2 \end{matrix} \right]$$

Thiourea

As a result of this, for example, thiourea reacts readily with alkyl halides to give *S-alkylisothiouronium salts*, and is oxidised by hydrogen peroxide at low temperatures in neutral solution to give what was called 'formamidinesulphinic acid', whose structure has been shown by X-ray studies to be *thiourea-S-dioxide*.

$$\left[R-S-C \begin{matrix} \diagup {}^{\oplus}NH_2 \\ \diagdown NH_2 \end{matrix} \right] X^{\ominus} \qquad O_2S=C \begin{matrix} \diagup NH_2 \\ \diagdown NH_2 \end{matrix}$$

S-Alkylisothiouronium salt   Thiourea-S-dioxide

S-Alkyl- and S-aryl-isothioureas, and especially *S-benzylisothiourea*, C_6H_5—CH_2 S—C(=NH)—NH_2, react with either carboxylic acids or sulphonic acids to give highly crystalline isothiouronium salts.

Some thiourea derivatives are used for protecting plants, and are effective against both insects and rodents.

2.39.3 Thiosemicarbazide[1]

Thiosemicarbazide can be regarded as the hydrazide of thiocarbamic acid.

Preparation

By *heating 85% hydrazine hydrate with excess ammonium thiocyanate under nitrogen* (yield *ca* 60%) (*Audrieth*, 1954):

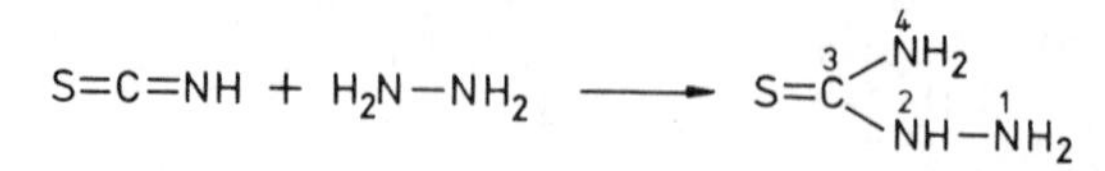

$$S=C=NH + H_2N-NH_2 \longrightarrow S=C\overset{4}{\underset{2}{\overset{3}{\diagup}}}\overset{NH_2}{\underset{NH-NH_2}{}}{}^1$$

Thiosemicarbazide

In addition *hydrazine-1,2-bis(thiocarbonamide)* (bis-thiourea) is formed:

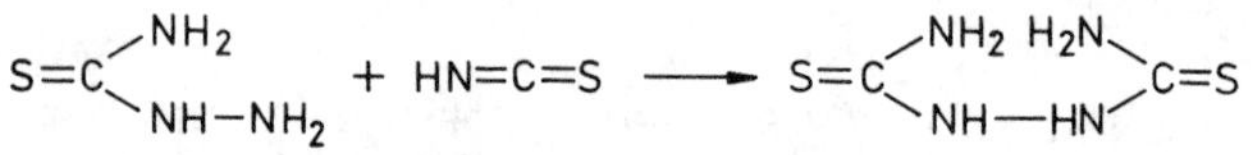

$$S=C\overset{NH_2}{\underset{NH-NH_2}{}} + HN=C=S \longrightarrow S=C\overset{NH_2}{\underset{NH-HN}{}}\overset{H_2N}{\underset{}{}}C=S$$

Hydrazine-1,2-bis(thiocarbonamide)

If an aqueous solution of hydrazine sulphate and potassium thiocyanate is heated to 90°C (365 K) and then acetone is added to it, acetone thiosemicarbazone crystallises out in very good yield, and this can be easily hydrolysed to provide thiosemicarbazide.

Properties and Uses

Thiosemicarbazide forms crystals with m.p. 181–183°C (454–456 K). It is a base and forms salts with acids. The numbering system shown above is used in naming substituted derivatives of thiosemicarbazide.

1-Acetylthiosemicarbazide, $H_2N-CS-NH-NH-COCH_3$, is formed by heating thiosemicarbazide with acetic acid. Like thiosemicarbazide itself, this acetyl derivative has much use as a starting material for the preparation of heterocyclic compounds.

The most important reaction of thiosemicarbazide is its condensation with either aldehydes or ketones in solution in water or ethanol to give the corresponding *thiosemicarbazones*:

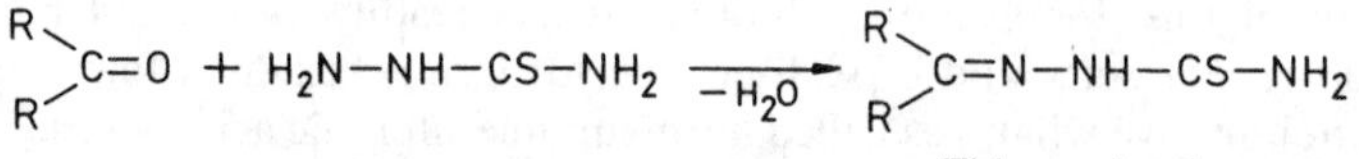

$$\overset{R}{\underset{R}{\diagup}}C=O + H_2N-NH-CS-NH_2 \xrightarrow{-H_2O} \overset{R}{\underset{R}{\diagup}}C=N-NH-CS-NH_2$$

Thiosemicarbazone

Thiosemicarbazones have achieved some importance because of the work of *Domagk*. It was shown that certain of them have the ability to arrest the growth of tubercular bacilli. Among those which have proved effective in the treatment of tuberculosis may be mentioned *p-acetylaminobenzaldehyde thiosemicarbazone*, although this one is not now made commercially.

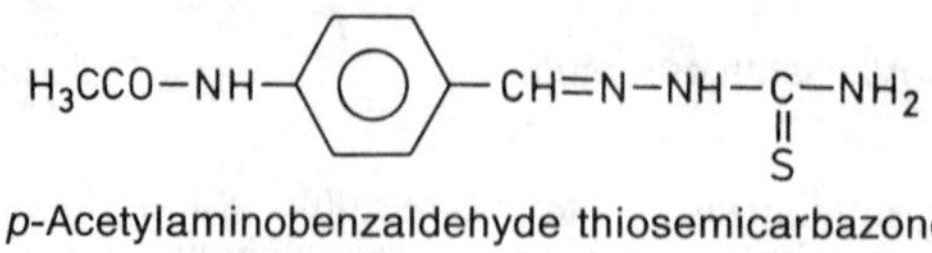

$$H_3CCO-NH-\langle\bigcirc\rangle-CH=N-NH-\underset{\underset{S}{\|}}{C}-NH_2$$

p-Acetylaminobenzaldehyde thiosemicarbazone

[1] See *F. Kurzer* et al., The Chemistry of Carbohydrazide and Thiocarbohydrazide, Chem. Rev. *70*, 111 (1970).

2.39.4 3-Thiocarbazide (Thiocarbonohydrazide)

This compound is made by *heating carbon disulphide with an excess of aqueous hydrazine* (*Audrieth, 1954*).

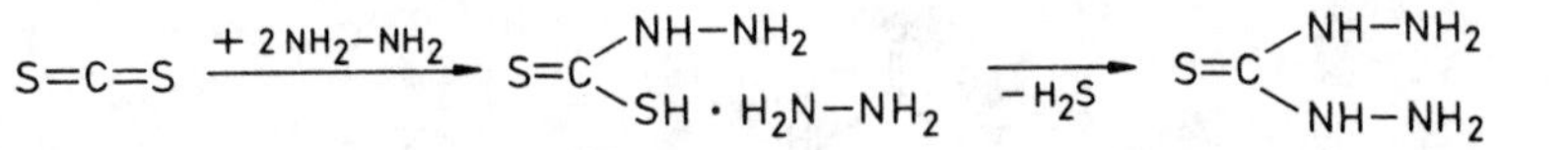

Hydrazine salt of
dithiocarbazinic acid

3-Thiocarbazide
Thiocarbonohydrazide

It is only slightly soluble in water and in the common organic solvents. It melts at 171°C (444 K) and forms a hydrochloride.

A diphenyldehydro-derivative of 3-thiocarbazide called *Dithizone* (*3-mercapto-1,5-diphenylformazan*) is used in analytical work as a reagent for the quantitative estimation of traces of heavy metals such as Ag, Hg, Cu, Pb, Zn and Cd (*H. Fischer*). Intensely coloured chelate complexes are formed, which can be determined colorimetrically.

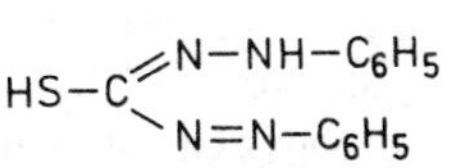

Dithizone

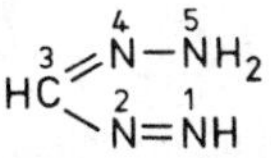

Formazan

2.40 Cyanic Acid and its Derivatives

Cyanic acid can exist in two *tautomeric* forms. Under normal conditions it exists exclusively as *isocyanic acid*, $\overline{O}=C=\overline{N}-H$. *Cyanic acid*, $HO-C\equiv N|$, has been detected when isocyanic acid is irradiated with low-wavelength UV light (224 nm) in an argon matrix at 20 K.

Isocyanic acid is a *heterocumulene* with the following structure:

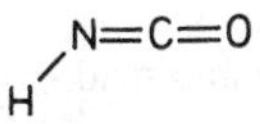

The angle N=C is about 128°.

Alkali metal salts of cyanic acid have a *mesomeric anion*:

$$[^{\ominus}\,|\overline{O}-C\equiv N| \longleftrightarrow \overline{O}=C=\overline{N}^{\ominus}]\ Me^{\oplus}$$

Potassium cyanate is prepared by oxidising potassium cyanide with either lead(IV) oxide or potassium dichromate:

$$K^{\oplus}\,|\overset{\ominus}{C}\equiv N| + \tfrac{1}{2}\,O_2 \rightarrow K^{\oplus}\,|\overset{\ominus}{\overline{O}}-C\equiv N|$$

When urea is heated above its melting point, *isocyanic acid* is formed, together with ammonia. It polymerises very readily to a mixture of *cyanuric acid* ($\sim 70\%$) and *cyamelide* ($\sim 30\%$). The former is a trimer of isocyanic acid; it is a 1,3,5-triazine derivative and its formal name is 2,4,6-trihydroxy-1,3,5-triazine.

| Isocyanic acid | Cyanuric acid | 2,4,6-Trihydroxy-1,3,5-triazine |

The triketo structure of cyanuric acid follows from an X-ray crystal structure determination which shows that it exists predominantly in the solid state as 1,3,5-triazine-2,4,6(1H,3H,5H)-trione.

When cyanuric acid is heated it depolymerises, regenerating isocyanic acid which, if rapidly cooled below 0°C (273 K), can be isolated as a colourless liquid with a pungent odour (*Wöhler*).

The other product obtained from the polymerisation of isocyanic acid, cyamelide $(H—N=C—O)_n$, a white product resembling porcelain, is probably a linear poly[oxy(iminomethylene)]:

In the presence of water, isocyanic acid is rapidly hydrolysed to carbon dioxide and ammonia:

Carbamic acid

With alcohols it forms *urethanes*, which with excess isocyanic acid are converted into *allophanates*, the esters of the unknown allophanic acid (urea-1-carboxylic acid):

| Isocyanic acid | A urethane | An allophanate |

2.40.1 Cyanogen Halides

Cyanogen halides can be regarded either as the acid halides derived from cyanic acid or as halogen-substituted derivatives of hydrocyanic acid. They are formed by the action of halogens on hydrogen cyanide or metal cyanides, *e.g.*

$$N \equiv C^{\ominus}K^{\oplus} + Br_2 \longrightarrow KBr + Br—C \equiv N \quad \text{Cyanogen bromide}$$

Cyanogen chloride, $Cl—C \equiv N$ is a poisonous, lachrymatory gas which liquefies at 13°C (286 K). It readily *trimerises* to cyanuric chloride (2,4,6-trichloro-1,3,5-triazine). With sulphur trioxide it gives chlorosulphonyl isocyanate, b.p. 106–107°C (379–380 K).

$$CI-C \equiv N + SO_3 \rightarrow CISO_2N=C=O \quad \text{Chlorosulphonyl isocyanate}$$

Cyanogen bromide melts at 52°C (325 K) and **cyanogen iodide** sublimes, without melting, above 45°C (318 K). Both are used in organic synthesis for the introduction of *cyano groups* into molecules.

Cyanogen bromide reacts with tertiary amines to form quaternary salts which lose a molecule of alkyl bromide to give, for example, a dialkylcyanamide (*von Braun* reaction).

$$R_3NI \xrightarrow{+BrCN} R_3N-C \equiv NI]^{\oplus} Br^{\ominus} \xrightarrow{-RBr} R_2N-C \equiv N$$

Usually the group which is lost from the amine is the one that gives the most reactive halide, *i.e.* the probability of cleavage is in the order $C_6H_5CH_2 > CH_3 > C_2H_5 >> C_6H_5$. Hydrolysis of the dialkylcyanamide provides the related secondary amine:

$$R_2N-C \equiv N \xrightarrow[-NH_3]{+2 H_2O} R_2N-COOH \xrightarrow{-CO_2} R_2NH$$

This method provides a means of obtaining secondary amines from tertiary amines.

The ease of formation of the quaternary salt depends on the identity of the substituent groups R. Reaction proceeds faster the more nucleophilic the tertiary amine, *i.e.* alkylamines react more readily than arylamines. If there are two or more phenyl substituents no reaction takes place with the cyanogen bromide. Ring-opening of cyclic tertiary amines is brought about by the *von Braun* reaction.

2.40.2 Cyanate Esters

In contrast to alkyl and aryl esters of isocyanic acid, which have been known for a long time, cyanate esters (*alkyl* and *aryl cyanates*) have only recently been isolated in their monomeric forms. The following methods were developed quite independently from one another.

Preparation

2.40.2.1 In general, treatment of alcohols or phenols with a cyanogen halide in alkaline conditions does *not* provide the desired cyanate ester, but rather a diester of iminocarbonic acid which slowly transforms into a triester of cyanuric acid:

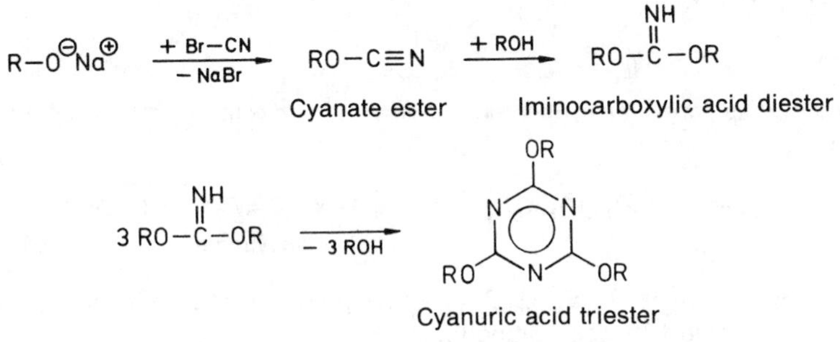

The cyanuric acid triester is derived from the tautomeric iminol form of cyanuric acid (R=H), which is not detectable, as mentioned in the earlier discussion of cyanuric acid.

On the other hand, aryl esters of cyanic acid, which are formed as intermediates, can be obtained in high yields by changing the reaction conditions; a base such as triethylamine is dropped into a solution of *equimolecular* amounts of a phenol and a cyanogen halide in acetone at about 0°C (273 K) (*Grigat*).

Only in special cases, *e.g.* from trihalogenoethanols and from some enols, can alkyl cyanates be obtained in this way.

2.40.2.2 When *O*-phenyl thiocarbonate chloride is treated with sodium azide at about 0°C (273 K) unstable 5-phenoxy-1,2,3,4-thiatriazole is formed which, even at room temperature, loses nitrogen and sulphur to give phenyl cyanate in 93% yield (*Martin*). (Caution — *danger of explosions*).

$$C_6H_5O-\underset{\underset{S}{\|}}{C}-Cl \xrightarrow[-NaCl]{+NaN_3} C_6H_5-O-\underset{S}{\overset{N-N}{\underset{}{C^5_{~1}~^2N}}} \xrightarrow[-N_2,-S]{\text{Room temperature}} C_6H_5O-C\equiv N$$

Phenyl cyanate

In a similar way *K. A. Jensen* obtained alkyl cyanates (yield *ca* 90%) from the reaction of *O*-alkyl thiocarbonate hydrazides and nitrous acid at about 0°C (273 K), which provides 5-alkoxy-1,2,3,4-thiatriazoles as intermediates. These alkyl cyanates rearrange to the corresponding alkyl isocyanates or *N,N,N*-trialkyl cyanurates in 1–2 days at room temperature.

Properties

Aryl cyanates are colourless liquids with pungent smells. They can be distilled under reduced pressure and when pure can be kept for a long period. In the presence of acid they readily undergo trimerisation to form the related triaryl cyanurates. They react with *nucleophiles*; for example, they add water to form *O*-arylurethanes and add hydrogen sulphide to give *O*-aryl thiocarbamates:

$$\underset{\text{O-Arylurethane}}{ArO-\underset{\underset{}{\overset{\|}{O}}}{C}-NH_2} \xleftarrow{+H_2O} \underset{\text{Aryl cyanate}}{ArO-C\equiv N} \xrightarrow{+H_2S} \underset{\text{O-Aryl thio-carbamate}}{ArO-\underset{\underset{}{\overset{\|}{S}}}{C}-NH_2}$$

Alkyl cyanates undergo similar reactions.

2.40.3 Isocyanate Esters

Preparation

2.40.3.1 *By heating potassium cyanate with dialkyl sulphates in the presence of dry sodium carbonate*, *e.g.*

$$KO-C\equiv N+(H_3C)_2SO_4 \xrightarrow{(Na_2CO_3)} H_3C-N=C=O+H_3COSO_3K$$

Methyl isocyanate

Methyl isocyanate (b.p. 39.1°C; 321.3 K) is very poisonous; the unfortunate accident in Bhopal was associated with this compound.

2.40.3.2 A generally applicable method involves the *reaction of phosgene with a primary amine*, e.g.

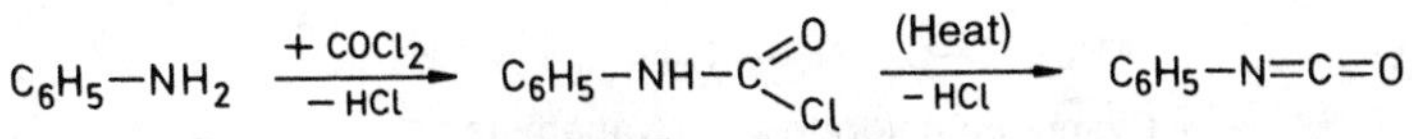

$$C_6H_5-NH_2 \xrightarrow[-HCl]{+COCl_2} C_6H_5-NH-C\underset{Cl}{\overset{O}{<}} \xrightarrow[-HCl]{(Heat)} C_6H_5-N{=}C{=}O$$

Aniline   Phenylcarbamoyl chloride   Phenyl isocyanate

Properties

Alkyl isocyanates are stable liquids which have pungent smells and are poisonous. Their *cumulated* double-bond system makes them reactive, and alcohols, ammonia and primary and secondary amines readily add to them, *e.g.*

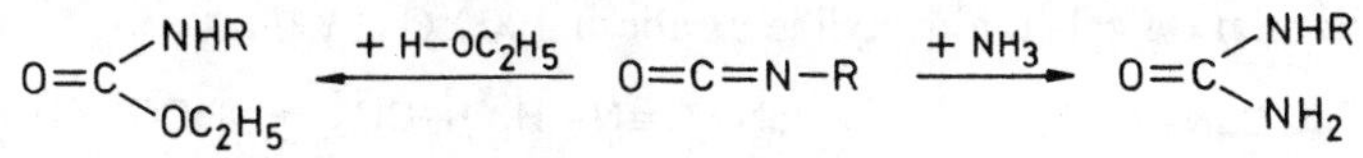

$$O{=}C\underset{OC_2H_5}{\overset{NHR}{<}} \xleftarrow{+H-OC_2H_5} O{=}C{=}N-R \xrightarrow{+NH_3} O{=}C\underset{NH_2}{\overset{NHR}{<}}$$

N-Alkylurethane   Alkyl isocyanate   *N*-Alkylurea

Since good crystalline products are obtained, isocyanates, and especially *phenyl isocyanate*, b.p. 166° (439 K), are used in the characterisation of these types of compounds.

In the presence of bases, alkyl and aryl isocyanates trimerise to *N,N,N-trialkyl-* or *-triaryl-cyanuric acids*:

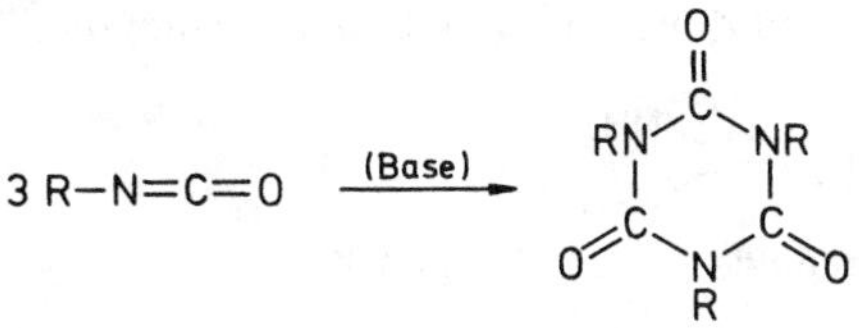

$$3\ R-N{=}C{=}O \xrightarrow{\text{(Base)}} \begin{array}{c} O \\ \| \\ RN{\diagup}^{C}{\diagdown}NR \\ | \quad\quad | \\ O{=}C{\diagdown}_{N}{\diagup}C{=}O \\ | \\ R \end{array} \qquad \Delta H = -226\ \text{kJ/Mol (for } R = CH_3)$$

N,N,N-Trisubstituted cyanuric acids

When heated, isocyanate esters can be hydrolysed to primary amines and carbon dioxide.[1]

$$R-N{=}C{=}O + H_2O \rightarrow R-NH_2 + CO_2$$

Alkyl isocyanates are formed as intermediates in the decomposition of amides, azides and hydroxamic acids.

2.40.4 Cyanamides

Cyanamide, $H_2N-C{\equiv}N$, can be regarded either as the amide of cyanic acid or as the nitrile derived from carbamic acid. Its *Raman* spectrum indicates that there is a tautomeric equilibrium as shown:

$$H_2N-C{\equiv}N \ \rightleftharpoons\ HN{=}C{=}NH$$

Cyanamide   Carbodiimide

---

[1] This strongly exothermic reaction was probably responsible for the misfortune in Bhopal in which 3500 people lost their lives.

Preparation

2.40.4.1 From *cyanogen halides and ammonia*, e.g.

$$NH_3 + Cl-C \equiv N \rightarrow H_2N-C \equiv N + HCl$$

Cyanogen chloride   Cyanamide

2.40.4.2 From *disodium cyanamide*, $Na_2N—C \equiv N$, by careful acidification with strong sulphuric acid whose concentration is so adjusted that all the water becomes bound to the sodium sulphate as water of crystallisation. Cyanamide is extracted from the reaction mixture with either absolute ethanol or ether.

2.40.4.3 Industrially *by passing carbon dioxide into an aqueous suspension of calcium cyanamide*, which is obtained by the *Frank–Caro* process (1897) from the action of a stream of nitrogen on calcium acetylide at about 1000°C (1300 K).

$$CaC_2 + N_2 \rightleftharpoons CaN-C \equiv N; \qquad CaN-C \equiv N + H_2O + CO_2 \rightarrow CaCO_3 + H_2N-C \equiv N$$

Properties

Cyanamide is a colourless crystalline substance, m.p. 43–44°C (316–318 K). It is very readily soluble in water, ethanol and ether. It is *amphoteric*, forming metal salts, of which the calcium is the most important, with strong bases, and also salts with acids, which are readily hydrolysed. Cyanamide is stable in aqueous solution at pH < 5, but at pH 8–9.5 it dimerises rapidly to the industrially important dimer *dicyanodiamide* (cyanoguanidine).

$$2\ H_2N-C \equiv N \longrightarrow H_2N-C \underset{N-C \equiv N}{\overset{NH_2}{<}}$$

Cyanamide   Dicyanodiamide (m.p. 209°C [482 K])

A series of other compounds can be made from dicyanodiamide. For example, if it is heated with 3N sodium hydroxide a 50% yield of *cyanourea* is obtained. If it is heated for a short time with 50% sulphuric acid, addition of water leads to the formation of *amidinourea* in almost quantitative yield. Hydrogen sulphide reacts similarly with dicyanodiamide to give, depending upon the reaction conditions, *amidinothiourea* or *dithiobiuret*.

Uses

Calcium cyanamide is a valuable nitrogen-containing fertiliser (nitrochalk) for soil which is deficient in lime. It is hydrolysed, *via* cyanamide and urea as intermediates, to ammonia and carbon dioxide:

$$CaN-C\equiv N \xrightarrow[-Ca(OH)_2]{+2H_2O} H_2N-C\equiv N \xrightarrow{+H_2O} O=C\overset{NH_2}{\underset{NH_2}{\diagup}} \xrightarrow{+H_2O} 2NH_3 + CO_2$$

In addition, calcium cyanamide is the starting material from which a number of other products such as cyanamide, dicyanodiamide, urea, thiourea, guanidine, nitroguanidine and aminoguanidine are made industrially.

2.40.5 Carbodiimides[1]

Carbodiimides are made from ureas or from thioureas by treating them with *triphenylphosphine* and *tetrachloromethane* in the presence of base (*Appel*).

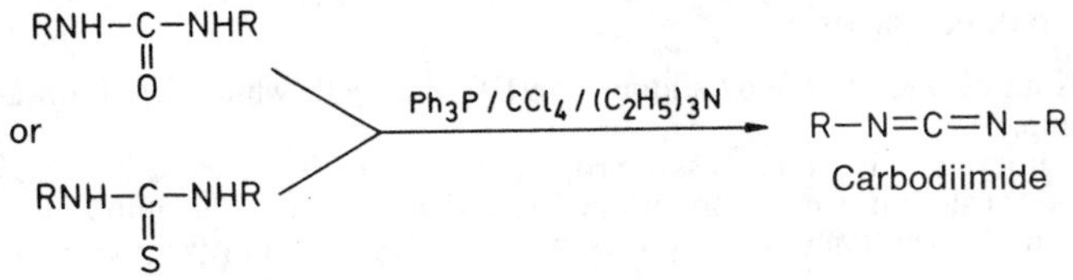

$$\begin{array}{c} RNH-\overset{O}{\underset{\|}{C}}-NHR \\ \text{or} \\ RNH-\overset{S}{\underset{\|}{C}}-NHR \end{array} \xrightarrow{Ph_3P\,/\,CCl_4\,/\,(C_2H_5)_3N} \begin{array}{c} R-N=C=N-R \\ \text{Carbodiimide} \end{array}$$

They are important as reactants in cycloaddition reactions. *Dicyclohexylcarbodiimide* ($R=C_6H_{11}$) (DCC) has proved useful as a dehydrating agent in the formation of amide and peptide links.

2.41 Thiocyanic Acid and its Derivatives

If a sulphur atom takes the place of the oxygen atom in cyanic acid the resultant compound is *thiocyanic acid*, which has the following tautomeric forms:

$$H-\bar{S}-C\equiv N| \rightleftharpoons \bar{S}=C=\bar{N}-H$$

Thiocyanic acid   Isothiocyanic acid

In contrast to cyanic acid, thiocyanic acid exists almost entirely as such, rather than as the iso-tautomer. Its salts, the *thiocyanates*, have *mesomeric anions*:

$$[{}^{\ominus}|\bar{S}-C\equiv N| \leftrightarrow \bar{S}=C=\bar{N}^{\ominus}]\overset{|}{Me}{}^{\oplus}$$

2.41.1 Preparation

2.41.1.1 Thiocyanate salts can be prepared in an analogous way to cyanates, by *melting cyanides with sulphur*, e.g.

$$K^{\oplus}[{}^{\ominus}|C\equiv N| + S \rightarrow K^{\oplus}[{}^{\ominus}|\bar{S}-C\equiv N|$$

Potassium thiocyanate

---

[1] See *M. Mikolajczyk*, Recent Developments in the Carbodiimide Chemistry, Tetrahedron *37*, 233 (1981).

2.41.1.2 Ammonium thiocyanate is made industrially by the reaction of carbon disulphide with excess ammonia:

$$S=C=S \xrightarrow{+\ 2\,NH_3} \left[S=C\begin{smallmatrix} NH_2 \\ S^{\ominus} \end{smallmatrix} \right] NH_4^{\oplus} \xrightarrow[-(NH_4)_2S]{+\ 2\,NH_3} NH_4^{\oplus} \left[^{\ominus}|\underline{S}-C\equiv N| \right]$$

Ammonium dithiocarbamate   Ammonium thiocyanate

2.41.1.3 Thiocyanic acid is obtained from its salts by treating them with sulphuric acid, the reaction mixture meantime being cooled with liquid air.

Properties

Free thiocyanic acid only exists at very low temperatures; above $-40°C$ (230 K) polymerisation sets in. In aqueous solution it is more stable; its acidity is then comparable to that of the halogen acids.

Cyanate ions may be tested for by adding iron(III) ions, with which they form a deep red iron thiocyanate complex.

Mercury(II) thiocyanate has the unusual property that when it is burned it swells up strongly to form a voluminous ash. If it is made up into pellets with a little gum, on ignition it forms a curious snake-like structure called '*Pharaoh's Serpents*' which is, appropriately, very poisonous.

2.41.2 Thiocyanate and Isothiocyanate Esters

As in the case of cyanate esters, both thiocyanate and isothiocyanate esters are isolable.

Preparation

Thiocyanate esters (alkyl thiocyanates) are formed by alkylation of thiocyanate anions:

$$K^{\oplus}[^{\ominus}|\underline{S}-C\equiv N| + I-R \longrightarrow R-\underline{S}-C\equiv N| + KI$$

Isothiocyanate esters (alkyl isothiocyanates) are best prepared by the reaction between primary amines and carbon disulphide in the presence of sodium hydroxide. The sodium *N*-alkyldithiocarbamide which is formed reacts with ethyl chloroformate to give an alkyl isothiocyanate, carbonyl sulphide and ethanol:

$$S=C=S \xrightarrow[-\ H_2O]{+\ RNH_2 + NaOH} \left[S=C\begin{smallmatrix} NHR \\ S^{\ominus} \end{smallmatrix} \right] Na^{\oplus} \xrightarrow[-\ NaCl]{+\ Cl-COOC_2H_5} \left[S=C\begin{smallmatrix} NHR \\ S-COOC_2H_5 \end{smallmatrix} \right]$$

(unstable)

$$\longrightarrow S=C=N-R\ +\ COS\ +\ C_2H_5OH$$

Alkyl isothiocyanate

Properties

Alkyl thiocyanates are rather unstable oils which have garlic-like odours. They are immiscible with water. They are oxidised to *sulphonic acids* by conc. nitric acid, and reduced to *thiols* by zinc and sulphuric acid.

$$R-SH + HCN \xleftarrow{+\ 2H} R-S-C\equiv N \xrightarrow{(HNO_3)} R-SO_3H$$

Some alkyl thiocyanates are used as *insecticides*.

Alkyl isothiocyanates are also insoluble in water. They have pungent, lachrymatory odours and raise blisters on the skin. When heated with hydrochloric acid at 100°C (373 K) they are hydrolysed to primary amines and carbonyl sulphide which is transformed further into carbon dioxide and hydrogen sulphide:

$$R-N{=}C{=}S + 2 H_2O \xrightarrow[-(COS)]{(HCl)} R-NH_2 \cdot HCl + CO_2 + H_2S$$

The main constituent of the mustard oil obtained from black mustard seed (*Brassica nigra*) is the thioglucoside *sinigrin* (allyl glucosinolate), which is hydrolysed either by acid or by the enzyme *Myrosinase* into allyl isothiocyanate, $H_2C{=}CH{-}CH_2{-}N{=}C{=}S$, b.p. 152°C (425 K), glucose and potassium hydrogen sulphate. The *thiohydroximic acid grouping* outlined with grey in the following formulae forms part of many *glucosinolates*.[1] This group of compounds consists of a number of substances which impair or hinder the utilisation of foodstuffs (antinutritional factors), *e.g.* *glucobrassicine* (R = 3-indolylmethyl), in various members of the cabbage family.

Sinigrin $CH_2{=}CH{-}CH_2{-}\underset{\substack{\| \\ N-O-SO_3^{\ominus}K^{\oplus}}}{C}{-}S{-}(C_6H_{11}O_5)$ $R{-}\underset{\substack{\| \\ N-O-SO_3^{\ominus}K^{\oplus}}}{C}{-}S{-}(C_6H_{11}O_5)$ Glucosinolate

Chlorination of isothiocyanate esters provides the reactive isocyanidedichlorides.

$$Ar{-}N{=}C{=}S \xrightarrow{Cl_2} Ar{-}N{=}\underset{Cl}{\overset{}{C}}{-}SCl \xrightarrow[-SCl_2]{Cl_2} Ar{-}N{=}C\begin{smallmatrix}Cl\\ \\Cl\end{smallmatrix}$$

Isothiocyanate   Isocyanidedichloride

Isocyanidedichlorides can also be made by pyrolysing phosgeneiminium salts, *e.g.*

$$\left[\begin{smallmatrix}H_3C\\ \\H_3C\end{smallmatrix}\overset{\oplus}{N}{=}C\begin{smallmatrix}Cl\\ \\Cl\end{smallmatrix}\right] Cl^{\ominus} \xrightarrow[-H_3CCl]{\Delta\ (>375\,K)} H_3C{-}N{=}C\begin{smallmatrix}Cl\\ \\Cl\end{smallmatrix}$$

2.42 Dicyanogen and Di(thiocyanogen)

2.42.1 Dicyanogen

Dicyanogen (cyanogen), $N{\equiv}C{-}C{\equiv}N$, may be regarded as the dinitrile derived from oxalic acid (oxalonitrile). It has a symmetrical linear structure and its dipole moment is equal to zero. It has been called a *pseudohalogen* because it has properties resembling those of the halogens.

Preparation

2.42.1.1 *By heating mercury(II) cyanide (Gay–Lussac, 1815):*

$$Hg(CN)_2 \rightarrow Hg + N{\equiv}C{-}C{\equiv}N \quad \text{Dicyanogen}$$

---

[1] See *F. Hoffmann*, Chemie in unserer Zeit **12**, 182 (1978); *E. Block*, Scientific American *1985* Vol. 5, p. 66.

2.42.1.2 When a *solution of potassium cyanide and copper(II) sulphate* is heated, copper(II) cyanide is formed, which spontaneously decomposes into copper(I) cyanide and cyanogen radicals. Some of these radicals combine to form dicyanogen:

$$2\,CuSO_4 + 4\,KCN \rightleftarrows 2\,Cu(CN)_2 + 2\,K_2SO_4 \qquad\qquad 2\,Cu(CN)_2 \rightarrow 2\,CuCN + (CN)_2$$

Others of the cyanogen radicals react with water, forming hydrocyanic acid and isocyanic acid:

$$2\cdot CN + H_2O \rightarrow HCN + HOCN$$

Properties

Dicyanogen is a highly poisonous colourless gas, b.p. $-21°C$ (252 K). It burns in air with a flame which has the colour of peach blossom, with a blue edge. It is moderately soluble in water. Partial hydrolysis, using conc. hydrochloric acid, gives *oxamide*:

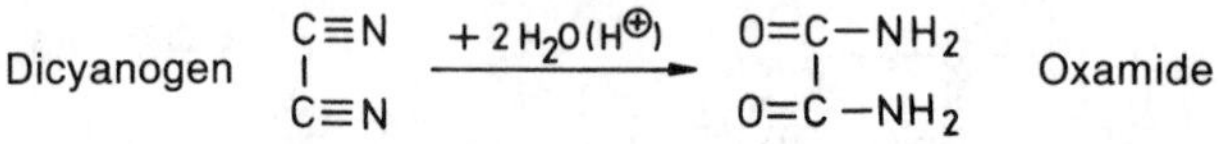

Stepwise addition of hydrogen sulphide to the $C\equiv N$ triple bonds of dicyanogen leads to the formation of *cyanothioformamide* and *dithiooxamide* as, respectively, yellow and orange crystals:

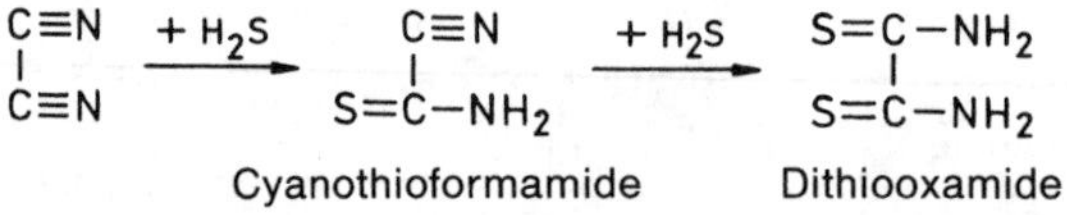

Of the two other possible isomers of dicyanogen, namely *cyanoisocyanogen*, NCNC, and *diisocyanogen*, CNNC, the former has been prepared[1] by an elimination reaction from norbornadienone azine:

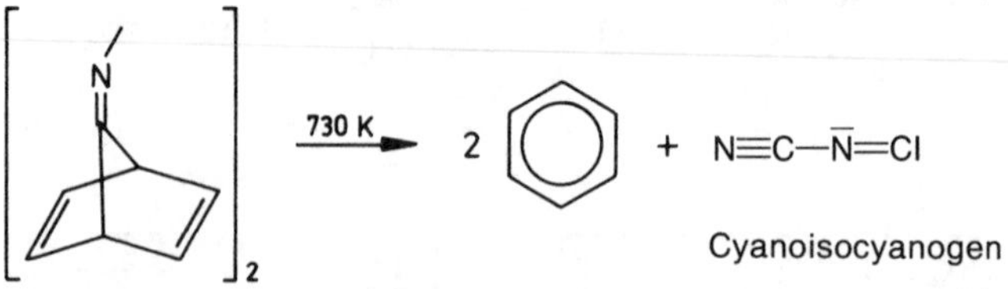

Cyanoisocyanogen polymerises at $-30°C$ (243 K) but is stable in the gas phase up to 500°C (730 K). Diisocyanogen is also formed as a by-product; it is far less stable and is only isolable in an argon matrix at 12K when made by an alternative route.[2]

2.42.2 Di(thiocyanogen)

Di(thiocyanogen), $N\equiv C-S-S-C\equiv N$, (thio-cyanogen) is prepared by the *action of bromine on heavy metal thiocyanates* in an inert solvent, such as carbon disulphide (*Söderbäck*):

$$2\,Ag-S-C\equiv N + Br_2 \rightarrow 2\,AgBr + N\equiv C-S-S-C\equiv N$$

Di(thiocyanogen)

It is an unstable substance, m.p. $-3°C$ (270 K). Chemically it behaves as a pseudo-halogen, *e.g.* with heavy metals it forms thiocyanates.

[1] See *F. Bickelhaupt* et al., Angew. Chem. Int. Ed. Engl. *27*, 936 (1988).
[2] See *G. Maier* et al, Angew. Chem. Int. Ed. Engl. *31*, 1218 (1992).

2.43 Carbon Monoxide and its Derivatives

2.43.1 Carbon Monoxide[1]

Preparation

Carbon monoxide is evolved as a decomposition product of formic acid, of α-ketoacids and of oxalic acid (*carbon monoxide extrusion* or *decarbonylation*). A cooled mixture of formic acid and sulphuric acid is useful for making it *in situ*.

2.43.1.2 Industrially it is prepared from *synthesis gas* by a *low temperature separation process* (*Linde* process). As an alternative to its preparation from methane (see p. 54), due to the increasing prices of natural oil and gas the gasification of coal is again becoming of importance. The gasification of coal involves the exothermic *partial combustion* of coal (1) and the endothermic *formation of water gas* (2):

$$2\,C + O_2 \;\rightleftharpoons\; 2\,CO \qquad \Delta H = -222\,kJ/Mol \qquad (1)$$
$$C + H_2O \;\rightleftharpoons\; CO + H_2 \qquad \Delta H = +130\,kJ/Mol \qquad (2)$$

The ratio of carbon monoxide to hydrogen can be adjusted by means of the *water-gas equilibrium* (3) and the *Boudouard equilibrium* (4):

$$CO + H_2O \;\rightleftharpoons\; CO_2 + H_2 \qquad \Delta H = -\;42\,kJ/Mol \qquad (3)$$
$$C + CO_2 \;\rightleftharpoons\; 2\,CO \qquad \Delta H = +\;172\,kJ/Mol \qquad (4)$$

Properties

Because of the 1:1 ratio of carbon to oxygen in carbon monoxide, CO, it occupies a special position among carbon compounds, for the principle of the tetravalency of carbon is not kept. Carbon monoxide is formulated with triple bonding between the constituent atoms, with a superimposed ionic character:

$$|\overset{\ominus}{C} \equiv \overset{\oplus}{O}|$$

A similar type of bonding is found in *isocyanides* (isonitriles), which can thus be regarded as isoelectronic derivatives of carbon monoxide; they have been described as *isosteres*.

Evidence for this type of bonding comes from the *Raman* spectrum of carbon monoxide, which clearly indicates a triple bond and a complete lack of carbonyl character. The above formulation also explains why carbon monoxide can act as a *Lewis* base.

Uses

Together with the alkenes and acetylenes, carbon monoxide is a very important starting material for the synthesis of a variety of aliphatic compounds. In many instances, synthesis gas can be used directly for this purpose. According to the conditions employed,

---

[1] See *J. Falbe*, Reactivity and Structure/Concepts in Organic Chemistry. Vol. 11 — New Syntheses with Carbon Monoxide (Springer-Verlag, Berlin 1980) *H. M. Colquhoun, D. J. Thompson* and *M. V. Twigg* Carbonylation (Plenum Press, New York 1991).

hydrogenation can provide alkanes, methanol or higher alcohols. Carbonylation of alkenes gives carboxylic acids and their derivatives, and alkynes give α,β-unsaturated carboxylic acids. Reaction of alk-1-enes with carbon monoxide and hydrogen provides aldehydes or alcohols (see oxo-synthesis). Further examples include the synthesis of sodium formate, the addition of chlorine to carbon monoxide giving phosgene, $OCCl_2$, and of sulphur to give carbonyl sulphide, COS, an odourless, poisonous gas with b.p. $-50°C$ (233 K).

A characteristic reaction of carbon monoxide is its tendency to form metal carbonyls. The most important reactions of carbon monoxide are gathered together in Table 12.

Table 12

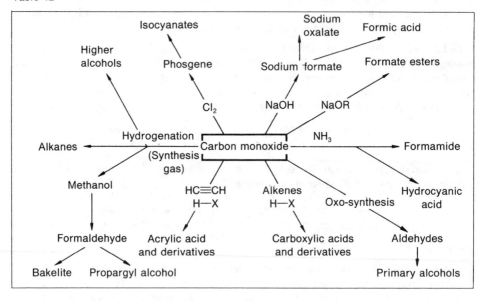

2.43.2 Alkyl Isocyanides (Isonitriles)

Alkyl isocyanides are formally derived from carbon monoxide by substituting an N—R group for the oxygen atom:

$$\left[\overset{\ominus}{|}C\equiv\overset{\oplus}{N}-R \leftrightarrow |C=\bar{N}-R \right]$$

They are isomeric with nitriles (alkyl cyanides) and hence are also known as *isonitriles*, but contain a quite different bond system. This may be represented, as in the above formula, as a hybrid of the two structures shown, with the dipolar form making the greater contribution.

Preparation[1]

2.43.2.1 When alkyl halides are treated with solid silver cyanide the sole products are isocyanides (*Gautier*, 1868):

---

[1] See *I. Ugi* et al., Isonitrile Syntheses, Angew. Chem. Int. Ed. Engl. *4*, 472 (1965).

$$R-I \xrightarrow{+ Ag-C \equiv NI} Ag-C \equiv N-R]^{\oplus} I^{\ominus} \xrightarrow[-AgI]{} I\bar{C} \equiv \overset{\oplus}{N}-R$$

If copper(I) cyanide is used instead of silver cyanide, 56% of isocyanides are obtained; sodium or potassium cyanides give only small amounts of isocyanide. At the same time the yield of nitriles increases. This trend is explained if it is assumed that the lone pair of electrons on the negatively charged carbon atom is prevented from reacting with the alkyl halide by being linked to a heavy metal, and in consequence alkylation takes place instead at the nitrogen atom.

2.43.2.2 Isocyanides which are free from any accompanying cyanides are obtained by treating primary amines with chloroform and potassium hydroxide (*A. W. von Hoffmann*, 1867; see *isonitrile reaction*).

When chloroform reacts with a strong base it is first of all transformed into unstable and very reactive *dichloromethylene* (*dichlorocarbene*), which has a sextet rather than an octet of electrons, and this associates with the lone pair of electrons of the nitrogen atom of the amine. The resultant betaine, which structurally resembles an ylide, reacts with the base and loses two molecules of hydrogen chloride, being converted thereby into an isocyanide:

$$CHCl_3 + KOH \longrightarrow :CCl_2 + KCl + H_2O$$
$$\text{Dichlorocarbene}$$

$$R-\bar{N}H_2 \xrightarrow{+ ICCl_2} R-\overset{\oplus}{N}H_2 \xrightarrow{+ 2 KOH} R-\overset{\oplus}{N}\equiv\overset{\ominus}{C}I + 2 KCl + 2 H_2O$$
$$\qquad\qquad\qquad \overset{\ominus}{I}CCl_2 \qquad\qquad\qquad \text{Isocyanide}$$

2.43.2.3 The best way of obtaining isocyanides is by the removal of a molecule of water from formylated primary amines with phosphorus oxychloride (or phosgene) in the presence of a base such as pyridine or triethylamine (*Ugi* et al., 1960):

$$2\,R-NH-C\overset{H}{\underset{O}{\diagdown}} + POCl_3 \xrightarrow{(Base)} 2\,R-\overset{\oplus}{N}\equiv\overset{\ominus}{C}I + 3\,HCl + (HPO_3)_n$$

Properties

Isocyanides are usually liquids which have highly repulsive smells and, in contrast to nitriles, are highly toxic. They have boiling points about 20° lower than the corresponding nitriles; for example, acetonitrile (methyl cyanide), $H_3C-C\equiv N$, boils at 81.6°C (354.8 K), whereas methyl isocyanide (methyl isonitrile), $H_3C-\overset{\oplus}{N}\equiv\overset{\ominus}{C}I$, boils at 60°C (333 K). Isocyanides are stable to alkali, but are hydrolysed at room temperature by dilute mineral acids to formic acid and a primary amine.

$$R-\overset{\oplus}{N}\equiv\overset{\ominus}{C}I + 2\,H_2O \longrightarrow R-NH_2 + HCOOH$$

Catalytic reduction of isocyanides converts them into secondary amines:

$$R-\overset{\oplus}{N}\equiv\overset{\ominus}{C}I \xrightarrow[\text{(Pt or Ni)}]{+4H} R-NH-CH_3$$

This reaction is reminiscent of the hydrogenation of carbon monoxide to give methanol. It provides a way of making secondary amines in which one of the alkyl substituents is a methyl group.

Isocyanides react with mercury(II) oxide to add oxygen and form isocyanates. Isothiocyanates are formed similarly by addition of sulphur:

$$R-\overset{\oplus}{N}\equiv\overset{\ominus}{C}I + HgO \longrightarrow R-N=C=O + Hg$$

$$R-\overset{\oplus}{N}\equiv\overset{\ominus}{C}I + S \longrightarrow R-N=C=S$$

The similarity of the structure of isocyanides to that of carbon monoxide is particularly evident in the way that they both form *metal complexes*. For example, the four CO groups in nickel tetracarbonyl can be displaced by reaction with phenyl isocyanide, leading to the formation of tetrakis(phenyl isocyanide) nickel as a yellow crystalline material (*Hieber,*[1] *Klages*):

$$Ni(CO)_4 \xrightarrow[-4\,^\ominus C \equiv O^\oplus]{+4\,C_6H_5-N^\oplus \equiv C^\ominus} Ni(CN-C_6H_5)_4$$

Nickel tetracarbonyl has been shown to be a carcinogen in some animals.

A hydrogen atom on the carbon atom next to the nitrogen atom is somewhat acidic; therefore this carbon atom is readily metalated and alkylated.[2]

$$R^1-\underset{\underset{R^2}{|}}{\overset{\overset{H}{|}}{C}}-N^\oplus \equiv C^\ominus \xrightarrow{n-Bu\,Li} R^1-\underset{\underset{R^2}{|}}{\overset{\ominus}{C}}-N^\oplus \equiv C^\ominus \xrightarrow{R^3X} R^1-\underset{\underset{R^2}{|}}{\overset{\overset{R^3}{|}}{C}}-N^\oplus \equiv C^\ominus$$

In this way umpolung of an α-carbon atom of an amine can be achieved. In the normal way such a carbon atom acts as an electron acceptor (electrophile), but by converting the amine group into an isonitrile group, this carbon atom becomes an electron donor (nucleophile).[3]

2.44 Carbenes and Nitriles as Reactive Intermediates

Many reactions involve the formation of short-lived reactive intermediate products, for example aliphatic carbon radicals, carbenium ions and carbanions. In the following section the substances to be considered are uncharged, reactive, usually not isolable intermediates which have sextets of electrons, *i.e.* derivatives of methylene, CH_2, (carbene) and of nitrene $\overline{N}H$. Studies of their chemical reactivity have opened up new aspects of both theoretical and preparative importance. Carbenium ions are the protonated derivatives of carbenes.

2.44.1 Carbenes[4]

Carbenes have a divalent carbon atom with a sextet of electrons.

Two possible structures can be considered, one with the two free electrons on the carbon atom paired (singlet carbene) and the other with them unpaired (triplet carbene, a diradical). In the first case the carbon atom has a pair of electrons and a vacant orbital.

[1] See Neuere Forschungsergebnisse auf dem Gebiet der Metallcarbonyle (Newer Research Results concerning Metal Carbonyls), Angew. Chem. *64*, 465 (1952).
[2] See *D. Hoppe*, α-Metalated Isocyanides in Organic Synthesis, Angew. Chem. Int. Ed. Engl. *13*, 789 (1974).
[3] See *D. Seebach*, Methods of Reactivity Umpolung, Angew. Chem. Int. Ed. Engl. *18*, 239 (1979).
[4] See *Houben-Weyl*, Methoden der Organischen Chemie, E *19b* (Thieme Verlag, Stuttgart 1989); *H. F. Schaefer III*, Methylene: A Paradigm for Computational Quantum Chemistry, Science *231*, 1100 (1986); *T. L. Gilchrist* and *C. W. Rees*, Carbenes, Nitrenes and Arynes (Nelson, London 1969).

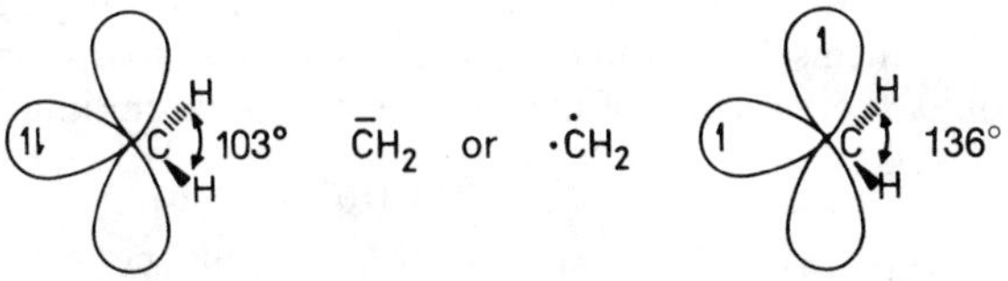

Singlet carbene   Triplet carbene

The triplet state of methylene represents the ground state, while the singlet state represents the first excited state. The energy difference between these two states is about 36 kJ/mol. In the case of some halogenocarbenes the singlet state is the ground state. Singlet carbenes can be regarded as examples of *electron-poor* (electron deficient) compounds, and this is in accord with their chemical reactivity. Because of the participation of both paired and unpaired electrons they have the character of 1,1-dipoles. The free electrons are denoted by the symbol · ·, *e.g.* :CH$_2$, methylene.

Carbenes are formed as intermediates in the following types of reactions.

2.44.1.1 Carbenes are generated by *thermal or photochemical decomposition of diazoalkanes or ketenes*:

$$R-\overset{\ominus}{C}H-\overset{\oplus}{N} \equiv N| \xrightarrow[-N_2]{} R-\ddot{C}H \xleftarrow[-CO]{} R-CH=C=O$$

This method can also be used to prepare substituted carbenes such as acylcarbenes and alkoxycarbonylcarbenes.

$$\underset{\text{Diazoketone}}{R-\overset{O}{\overset{\|}{C}}-CHN_2} \xrightarrow[-N_2]{} \underset{\text{Acylcarbene}}{R-\overset{O}{\overset{\|}{C}}-\ddot{C}H}$$

$$\underset{\text{Diazoacetic ester}}{ROOC-CHN_2} \xrightarrow[-N_2]{} \underset{\text{Alkoxycarbonylcarbene}}{ROOC-\ddot{C}H}$$

2.44.1.2 Dichlorocarbene, :CCl$_2$, is obtained by the action of alkali on chloroform (see α-elimination, p. 132).

The ensuing reactions of the unstable carbenes all result from their tendency to convert their sextets of electrons into octets. They can be classified into the three following groups:

2.44.1.3 Isomerisation. In this case the carbene stabilises itself by means of a hydride-ion shift, with concomitant formation of an alkene, *e.g.*

$$R-CH_2-\ddot{C}H \rightarrow R-CH=CH_2$$

In *branched-chain carbenes* alkyl or aryl groups can also *migrate in the form of anions*, *e.g.*

$$\underset{\substack{\text{Triphenylmethylcarbene}\\ \text{(Tritylcarbene)}}}{(C_6H_5)_3C-\ddot{C}H} \rightarrow \underset{\text{Triphenylethylene}}{(C_6H_5)_2C=CH-C_6H_5}$$

Acylcarbenes rearrange to *ketenes* (see *Wolff* rearrangement):

$$R-\underset{\overset{\|}{O}}{C}-\ddot{C}H \longrightarrow O=C=CH-R$$

2.44.1.4 Insertion reactions. *Branched* alkylcarbenes can not only rearrange to alkenes, but can also stabilise themselves by undergoing ring-closure to cyclopropane derivatives. This process involves the insertion of the carbene carbon atom into a C—H bond, *e.g.*

It is not possible at present to make any certain statements about the mechanism of this insertion process.[1]

2.44.1.5 Addition reactions. With nucleophiles carbenes react smoothly as *electrophiles*; for example, they add to alkenes to give cyclopropane derivatives. Thus methylene adds to cyclohexene to form bicyclo[4.1.0]heptane (norcarane):

Cyclohexene   Norcarane   Cycloheptatriene

This type of reaction is described as a *cis-1,1-cycloaddition* reaction.

Benzene derivatives can take part in similar reactions, leading to ring-expansion; for example, methylene adds to benzene to give cycloheptatriene (tropilidene).

2.44.1.6 Carbene complexes have advantages in the preparative use of carbenes, both making the process easier and extending its range.[2]

2.44.2 Nitrenes[3]

The nitrogen analogues of carbenes are known as nitrenes. They have a *monovalent* nitrogen atom with a sextet of electrons and thus, like carbenes, are electron-deficient compounds.

Nitrenes are normally formed, as non-isolable intermediates, by photochemical decomposition of azides; this method is analogous to the formation of carbenes from diazo-compounds. Aryl nitrenes can be made from aryl nitro- or nitroso-compounds by treating them with phosphorus(III) compounds, *e.g.*

$$Ar-NO + PCl_3 \rightarrow ArN + POCl_3$$

It has been possible to investigate alkyl and aryl nitrenes in their ground state, dispersed in solid matrices at $-269°C$ (4 K); ESR spectra show this to be a triplet state. Singlet nitrenes are generally formed under the usual photolytic conditions.

---

[1] See *R. A. Moss*, Carbenic Reactivity Revisited, Acc. Chem. Res. *22*, 15 (1989).
[2] See *K. H. Dötz*, Carbene Complexes in Organic Synthesis, Angew. Chem. Int. Ed. Engl. *23*, 587 (1984).
[3] See *L. Horner* et al., Angew. Chem. Int. Ed. Engl. *2*, 599 (1963); *W. Lwowski*, Nitrenes (Interscience Publishers, New York 1970).

$$R-\underline{\ddot{N}} \qquad\qquad R-\dot{\underline{N}}\cdot \qquad\qquad R-\overset{\cdot\cdot}{\underline{N}}$$

Singlet nitrene   Triplet nitrene   General formula
(see p. 377)

Like carbenes, nitrenes try to stabilise themselves by attaining an octet of electrons instead of a sextet. Again this can be achieved by means of isomerisation, insertion reactions or addition to a nucleophilic reactant. In the case of nitrenes, the first two of these are the preferred modes of reaction. Thus they may undergo *isomerisation* to an imine by migration of a hydride ion from the α-positions or formation of a cyclic product by an *intramolecular insertion reaction*:

$$H_3C-(CH_2)_3-\overset{\cdot\cdot}{\underline{N}}$$
n-Butylnitrene

$$H_3C-(CH_2)_2-CH=NH$$
Butan-1-imine (59%)

Pyrrolidine (12%)

A characteristic reaction of arylnitrenes is their *dimerisation* to azobenzenes:

$$2\,Ar-\overset{\cdot\cdot}{\underline{N}} \longrightarrow Ar{\sim}\bar{N}=\bar{N}{\sim}Ar$$

The symbol ~ indicates that there is the possibility of *Z–E* isomers.

2.44.2.1 Addition reactions to nucleophiles are successful with, for example, ethoxycarbonylnitrene, and, as in the case of carbenes, reaction with cyclohexene gives a bicyclic product and with benzene leads to ring-expansion and formation of a seven-membered ring, which to some extent reacts further and rearranges to an N-phenylurethane.

Cyclohexene $+\; \overset{\cdot\cdot}{\underset{\cdot\cdot}{N}}-COOC_2H_5 \longrightarrow$ Ethyl 7-azanorcarane-7-carboxylate

Benzene $+\; \overset{\cdot\cdot}{\underset{\cdot\cdot}{N}}-COOC_2H_5 \longrightarrow$ $N-COOC_2H_5$ Ethyl azepine-1-carboxylate

$\longrightarrow$ $N-COOC_2H_5$
              H
N-Phenylurethane, Ethyl phenyl carbamate

Ethoxycarbonylnitrene is generated *in situ*, either by photolysis of ethyl azidoformate or from *p*-nitrobenzenesulphonyl oxyurethane by deprotonation with triethylamine followed by solvolysis:

$$
\underset{\text{Ethyl azidoformate}}{N_3-\overset{\overset{\displaystyle O}{\|}}{C}-OC_2H_5} \xrightarrow[-N_2]{h\nu} \underset{\text{Ethoxycarbonylnitrene}}{\overset{\displaystyle ..}{\underset{\displaystyle ..}{N}}-COOC_2H_5} \longleftarrow
$$

$$
-p-O_2N-C_6H_4SO_3^{\ominus}
$$

$$
\underset{\substack{\text{\textit{p}-Nitrobenzenesulphonyl}\\ \text{oxyurethane}}}{p-O_2N-C_6H_4-SO_2O\overset{\displaystyle -}{\underset{\displaystyle H}{N}}-COOC_2H_5} \xrightarrow[-BH^{\oplus}]{+B} p-O_2N-C_6H_4-SO_2O-\overset{\ominus}{N}-COOC_2H_5
$$

2.45 Organic Transition-Metal Complexes[1]

In addition to the organometallic compounds which have already been mentioned, whose properties can for the most part be dealt with in terms of metal–carbon σ-bonds, there are large numbers of metal complexes in which the organic ligands can each provide *two* electrons for the metal–carbon bonding.

Because of their importance in many reactions which have been described, in which they serve as heavy-metal catalysts, it seems appropriate to consider here some of these transition-metal complexes.

2.45.1 The 18-Electron Rule

The *18-electron rule* (see p. 22) is of value in defining the influence of a metal atom on the structure and properties of its complexes. It makes clear whether a complex is co-ordinatively saturated or unsaturated. For example, it may be deduced as follows that the nickel tetracarbonyl, $Ni(CO)_4$, which is formed by carbonylation of nickel in the *Reppe* synthesis, is a co-ordinatively saturated complex.

From the Periodic Table it can be seen that nickel has, in addition to the electrons of the argon shell, ten further electrons. The eight electrons required to build up a noble-gas-type electronic structure are provided by lone pairs of electrons from four molecules of carbon monoxide, which acts as a *Lewis* base and as an electron-donating species. Nickel behaves as a *Lewis* acid, *i.e.* as an electron acceptor. The hexacyanoferrate(II) anion, $[Fe(CN)_6]^{4\ominus}$ is also a co-ordinatively saturated molecule since $Fe^{2\oplus}$ has six electrons and requires twelve electrons, which are provided by the six cyanide ions, to achieve a noble-gas-type structure.

---

[1] See *S. G. Davies*, Organotransition Metal Chemistry Applications to Organic Synthesis (Pergamon Press, Oxford 1982); *R. Scheffold* (Ed.) Modern Synthetic Methods 1983, Transition Metals in Organic Chemistry (Salle Verlag, Frankfurt am Main 1983); *A. J. Pearson*, Metallo-organic Chemistry (Wiley, New York 1985), *J. P. Collman* et al., Principles and Applications of Organotransition Metal Chemistry, 2nd edn (University Science Books, Mill Valley, California 1987); *G. Wilkinson*, Comprehensive Coordination Chemistry (Pergamon Press, Oxford 1988).

An example of a co-ordinatively saturated complex of iron with an oxidation state $= 0$ is $Fe(CO)_5$. Since co-ordinatively unsaturated complexes play an important role in the reactions of organo-complexes of transition metals, it is essential to know the electronic configuration of the metal. It is thus important to specify both the oxidation state and the number of electrons in excess of the noble-gas structure which are available. Thus for the complexes just discussed we have:

$$Ni(CO)_4 \qquad [Fe(CN)_6]^{4\ominus} \qquad Fe(CO)_5$$
$$0\,(18) \qquad\qquad +2\,(18) \qquad\qquad 0\,(18)$$

Carbenes (*Fischer*, 1964) and phosphines form similar transition-metal complexes; an example is chloro-tris(triphenylphosphine)-rhodium(I), which as *Wilkinson's* catalyst, has become very important both in the laboratory and in industry. This is a planar co-ordinatively unsaturated complex:

$$\left[(C_6H_5)_3P\right]_3RhCl \;\equiv\;$$

$$+1 \;\; (16)$$

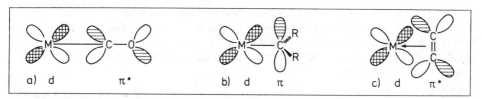

$(C_6H_5)_3P$ — Rh — Cl

$(C_6H_5)_3P$ — $P(C_6H_5)_3$

Wilkinson's catalyst

η^2-Complexes are compounds in which ligands, acting as *Lewis* bases, donate pairs of π-electrons in order to form complexes. Similarly alkyl groups covalently bonded to a transition metal can be described as η^1-ligands. Benzene rings, which can form complexes in which all their six π-electrons are involved, are called η^6-ligands.

2.45.2 Back Bonding

Carbon monoxide, carbenes and alkenes have vacant orbitals of low energy and the correct symmetry to overlap occupied d orbitals of the metal atom. Hence electrons can be fed back to the ligands (back donation). The extent of this back bonding influences the stability and properties of complexes.

a) d π^* b) d π c) d π^*

Fig. 93. Back bonding in various transition-metal complexes: (a) carbon monoxide, (b) a carbene, (c) an alkene.

Figure 93a shows that the $C{=}O$ double bond makes an important contribution to the complex by the participation of its π^*-orbital in the MO. This can be seen, for example, from their IR spectra, in the lowering of the wavenumber of the stretching vibrations of the $\overset{\ominus}{C}{\equiv}\overset{\oplus}{O}$ group. The metal–carbon bond in the carbene ligand (Fig. 93b) is shorter than that in an η^1-alkyl ligand. The π-electron density in alkenes forming part of transition-metal complexes depends upon the extent of the back bonding; it may be greater or less than that in the unco-ordinated alkene. The distribution of electrons between the metal and ligands is decisive in determining the catalytic performance of an organic transition-metal complex. The back bonding with carbon monoxide is very marked, and in consequence the electron density in the d orbital of the metal is lowered. In the case of phosphines there is little back bonding, so that in phosphine complexes the electron density on the metal is large.

2.45.3 Oxidative Addition and Reductive Elimination

The two processes of oxidative addition and reductive elimination play an important role in many reactions of organic transition-metal complexes. As an example, catalytic hydrogenation making use of *Wilkinson's* catalyst will be described in a simplified manner.

2.45.3.1 Oxidative addition is often carried out by the splitting off of a ligand. In the case of *Wilkinson's* catalyst with $L = P(C_6H_5)_3$:

Oxidative addition of hydrogen takes place with *raising of the co-ordination number* and formation of a *trigonal bipyramid*. The oxidation number of the metal is raised from $+1$ to $+3$, and both hydrogen atoms are bonded to it (a). Next the alkene adds on as a π-complex (b), and then inserts itself in the rate-determining step (c) into the *cis*-aligned metal–hydrogen bond. In this way a trigonal bipyramid with an η^1-ligand (σ-complex) is formed from an octahedral complex with an η^2- alkene-ligand. When the alkane splits off (d), the catalyst is regenerated and its oxidation state is lowered from $+3$ to $+1$; this is the *reductive elimination*.

2.45.3.2 By using chiral ligands *Wilkinson's* catalyst can bring about stereoselective hydrogenation. For this purpose either *chiral tertiary phosphines* (*Knowles, Horner*, 1968) or *biphosphines* with a chiral carbon framework (*Kagan*, 1971) can be used:

(R)-2-Methoxyphenyl-
methyl-phenylphosphine (PAMP)

(−)(4R,5R)-2,3,-O-isopropylidene-2,3-dihydroxy-
1,4-bis(diphenylphosphino)butane
(−)(R,R)-DIOP

$(-)$-(R,R)-(DIOP) is made by a four-stage synthesis from tartaric acid.[1] The chiral *Wilkinson's* catalyst is then made *in situ* by a double decomposition reaction with the pre-catalyst [RhCl(cyclooctene)$_2$]$_2$. An example of its use is for the reduction of (Z)-2-acetamidocinnamic acid to (R)-N-acetylphenylalanine:

2.45.4 Wacker-Hoechst Process

When ethylene is oxidised by the *Wacker*-Hoechst process, the following transformations to the first-formed π-complex (I) take place:

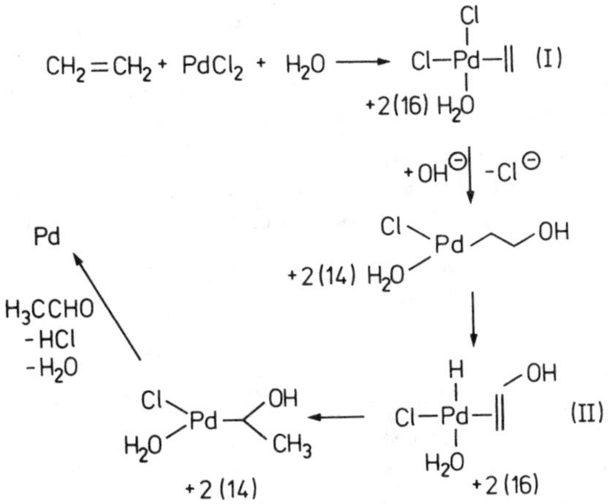

Nucleophilic attack by hydroxide ion on the π-complex (I) converts it into a σ-complex, which undergoes β-elimination to form another π-complex (II). This is the reverse of the insertion seen in the previous example. This palladium complex provides an exception to the 18-electron rule.

2.45.5 Oxo-synthesis

The action of the cobalt catalyst which is most commonly used in the oxo-synthesis depends upon the formation of tetracarbonylcobalt hydride, which is produced by oxidative addition to cobaltoctacarbonyl:

---

[1] See *H. B. Kagan* et al., *J. Am. Chem. Soc.* **94**, 6429 (1972).

The course of the reaction is governed by the two insertions, which result in the conversion of the initially formed π-complex into the σ-complex (I). Further take-up of carbon monoxide is then followed by CO-insertion. Interaction with another molecule of tetracarbonylcobalt hydride results in a reductive elimination and liberation of propionaldehyde. The corresponding rhodium catalyst is:

$$HRh(CO)_2L_2 \quad cf. \quad HCo(CO)_4$$
$$+1\,(18) \quad\quad\quad +1\,(18)$$

The carbonylation of acetylene (p. 98) or methanol (p. 235) and alkylations using lithium alkyl cuprates (p. 55) involve similar mechanisms, which, as in the example given, are characterised by interchanges between π- and σ-complexes. Among the compounds discussed in section 2.20 are some which are derivatives of transition metals. In such cases also π-complexes may be formed, and this is more likely the higher the atomic number of the metal.

3 Alicyclic Compounds

This class of compounds consists of cyclic compounds in which the atoms making up the rings are all carbon atoms. If the rings are made up entirely of saturated carbon atoms, the compounds are called *cycloalkanes*; if they contain one or more double bonds they are described as *cycloalkenes*. Both cycloalkanes and cycloalkenes resemble their aliphatic acyclic analogues in their chemical and physical properties and in their reactivity and this had led to their description as *alicyclic compounds*.

Because they take part in very different types of reactions, a group of compounds known as 'aromatic compounds', which are derivatives of benzene (cyclohexatriene), are treated separately in section 5.

Cycloalkanes

The saturated cyclic hydrocarbons form a homologous series with the general formula C_nH_{2n}. They are named after the open-chain alkanes having the same number of carbon atoms, but with the prefix 'cyclo' added. The simplest cycloalkanes are cyclopropane, cyclobutane, cyclopentane, cyclohexane, cycloheptane, *etc.*

Cyclopropane Cyclobutane Cyclopentane Cyclohexane Cycloheptane

There are marked differences, depending upon the size of the ring, both in the ease of preparation and the stabilities of the different cycloalkanes. The easiest to obtain are the stable 5- and 6-membered rings. In consequence, these are commonly found in natural sources, for example cyclopentane and cyclohexane derivatives in the Caucasian oil deposits. The most important family of alicyclic compounds found in natural products comprises the cyclic terpenes, which also include some 3- and 4-membered ring compounds.

Synthetically prepared compounds having 3- and 4-membered rings have proved to be reactive compounds which are sometimes unstable. These compounds are described as 'small-ring' compounds.

Baeyer's **Strain Theory**. *A. von Baeyer* (1885) first tried to explain the variations in ease of formation and stability of cycloalkanes with ring-size. He made the assumption that all the rings are flat. If this were so, then the bond angles at the carbon atoms must deviate to a greater or lesser extent from the normal tetrahedral angle (109° 28'). Deformation of the bond angles must therefore take place in closing the rings, introducing 'ring strain'; this results in the formation of an energy-rich system. The more the deformation of the bond angles the greater should be the strain; this could be defined as in the following equation:

$$d = \tfrac{1}{2}(109°28' - \text{bond angle in a planar ring})$$

The calculated values of d for the first eight alicyclic hydrocarbons, in each case assuming that the rings are flat, and that all the angles within any one ring are identical, are given in Table 13. Also included is ethylene, whose C=C double bond was regarded by *Baeyer* as a 2-membered ring (see p. 64). It is seen that the compression of the normal tetrahedral bond angles in cyclopropane and cyclobutane results in a considerable 'positive' strain. Their chemical properties are in accord with this. It was as a result of these examples that the concept was first introduced that bond angles have an effect on the stability and reactivity of organic compounds. A cyclobutane ring is both more stable and less reactive than a cyclopropane ring. The formation of 5- and 6-membered rings is especially favoured; ring strain is at a minimum in their cases. In the case of larger rings, an expansion of the bond angle would be necessary if they are planar and this could result in a 'negative' strain.

Table 13

Hydrocarbon C_nH_{2n}	Bond angles in a planar form	Deformation of tetrahedral angle	Heat of combustion per CH_2 group kJ/mol	Strain energy per CH_2 group kJ/mol
Ethylene (2-ring)	0°	+ 54° 44′	711,3	52,3
Cyclopropane (3-ring)	60°	+ 24° 44′	697,1	38,5
Cyclobutane (4-ring)	90°	+ 9° 44′	685,8	27,2
Cyclopentane (5-ring)	108°	+ 0° 44′	664,0	5,4
Cyclohexane (6-ring)	120°	− 5° 16′	658,6	0,0
Cycloheptane (7-ring)	128° 34′	− 9° 34′	662,3	−3,8
Cyclooctane (8-ring)	135°	− 12° 46′	663,6	−5,0

The dependence of the energy content on ring size can be seen from their *heats of combustion* per CH_2 group. The values for the 3- and 4-rings are appreciably higher than those for the 5- and 6-rings which have little or no strain (see the last but one column in Table 13). The strain energy per CH_2 group in a planar ring in the case of the two small rings is 27–38 kJ/mol (see last column in Table), as compared to nil, which is assumed to be the contribution of strain to the heats of combustion for n-alkanes, which are 658.1 kJ/mol per CH_2 group. This is the crucial feature of the *Baeyer* strain theory, which historically was important for stimulating interest in the stereochemistry of alicyclic compounds, and for enriching the whole outlook of organic chemistry. On the other hand, its application to the larger rings, for which it proposed a 'negative' strain, was not borne out. Their heats of combustion per CH_2 group are effectively the same, irrespective of their ring size, and are little greater than the values for acyclic n-alkanes. Cyclohexane and larger rings are *not* normally *planar*, and thus the whole basis of strain theory is not applicable to these compounds. In all rings with six or more members there is no classical Baeyer strain, and all of them are stable.

Classification of Cycloalkanes

On account of the differences in the ease of making cycloalkanes of different ring sizes, and because of other structural and chemical distinctions between rings of different sizes, it is convenient to classify 3- and 4-membered rings as 'small' rings, 5-,6- and 7-membered rings as 'common' rings, 8–12-membered rings as 'medium-sized rings' and rings with 13 or more members as 'large rings' (*H. C. Brown*, 1951).

3.1 Small Ring Compounds[1]

3.1.1 Cyclopropane

As mentioned above, the tendency for 3-membered rings to be formed is relatively small, and this was discussed in terms of *Baeyer's* strain theory. The carbon atoms of cyclopropane must form the corners of an equilateral triangle which has angles of 60°. More recent calculations indicate that the best orbital overlap between the three carbon atoms is achieved if the C—C—C bonds are at an angle of about 101°. Furthermore, the electron diffraction studies of *Bastiansen* and *Hassel* show that the H—C—H angle is almost 120°, with the plane of the carbon atoms running through the middle of it.

On the basis of these results the carbon atoms in cyclopropane are sp^3-hybridised, as in saturated aliphatic compounds, but it appears that the overlapping of the sp^3-hybrid orbitals of the carbon atoms in the ring is incomplete, and that they lie not in a straight but a bent line between the atoms. They are therefore known as 'bent' bonds or 'banana' bonds (see Fig. 94a).

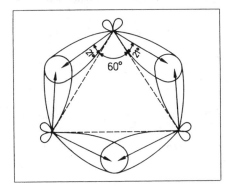

Fig. 94a.

Fig. 94b.

Walsh proposed an alternative model involving sp^2-hybridised orbitals (see Fig. 94b) in which the three carbon-carbon bonds are formed by overlap between three p orbitals, with an sp^2 orbital from each atom directed towards the centre of the ring; as a result there is some anti-bonding (+ −) interaction. This model provides a better description of the chemical reactivity and of the results of physical measurements such as the coupling constant $^1J_{(13\text{CH})} = 162$ Hz; it expresses better the π-character of the bonding. The shortening of the C—C bonds to 152 pm from 153.4 pm as found for bonds between two sp^3-hybridised carbon atoms is also explained by either model, which, with appropriate postulates, are interconvertible.[2]

---

[1] See *Houben-Weyl*, Methoden der organischen Chemie, Vol. 4/3 and 4/4 (Thieme Verlag, Stuttgart 1971).

[2] See *M. Klessinger*, Elektronstruktur organischer Moleküle, p. 103 (Verlag Chemie, Weinheim 1982).

Preparation

3.1.1.1 By *dehalogenation of 1,3-dihalogeno-alkanes,* using either sodium, in an intramolecular *Wurtz* reaction (*Freund*, 1882), or zinc (*Gustavson*, 1887). For example, cyclopropane is best made by the reaction of zinc with 1-bromo-3-chloropropane, obtained by addition of hydrogen bromide to allyl chloride in the presence of peroxide:

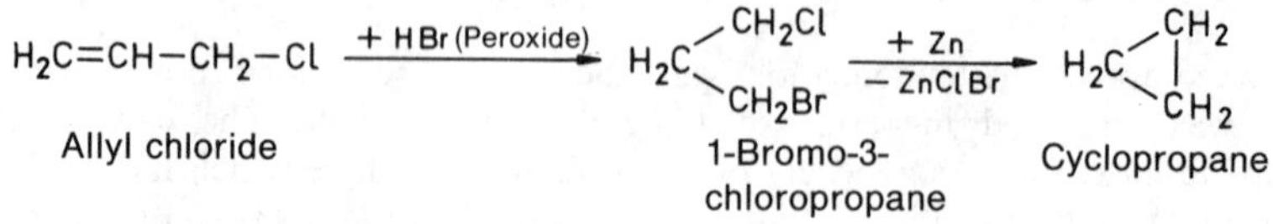

$$H_2C=CH-CH_2-Cl \xrightarrow{+ HBr\,(Peroxide)} H_2C\begin{smallmatrix}CH_2Cl\\ \\CH_2Br\end{smallmatrix} \xrightarrow[- ZnClBr]{+ Zn} H_2C\begin{smallmatrix}CH_2\\|\\CH_2\end{smallmatrix}$$

Allyl chloride   1-Bromo-3-   Cyclopropane
                               chloropropane

3.1.1.2 By *addition of diazo-compounds to* C=C *double bonds,* e.g.

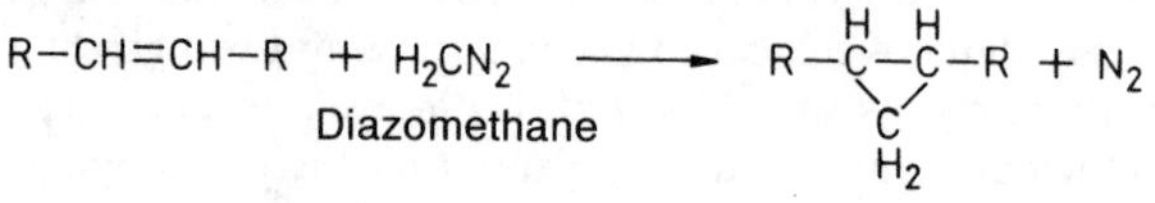

$$R-CH=CH-R + H_2CN_2 \longrightarrow R-\underset{\underset{H_2}{C}}{\overset{\overset{H\ \ H}{|\ \ |}}{C-C}}-R + N_2$$

Diazomethane

There are two possible ways in which the reaction might proceed:

(a) A 1,3-dipolar cycloaddition or (3 + 2)-cycloaddition reaction gives rise to a dihydropyrazole derivative, which, when heated, loses nitrogen and is transformed into a cyclopropane derivative.

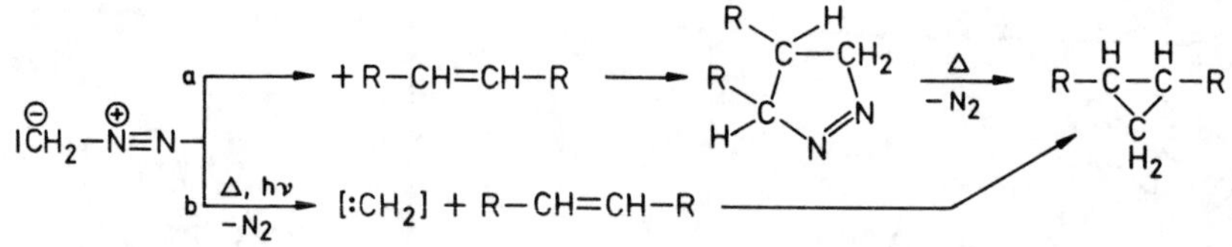

(b) Thermal or photochemical decomposition of diazomethane generates nitrogen and a carbene, which undergoes a (2 + 1) cycloaddition reaction with the C=C double bond.

3.1.1.3 By means of the *Simmons–Smith reaction* (see p. 186 f.).

Properties

Cyclopropane, C_3H_6, is a gas (b.p. $-32.8°C$ [241.4 K]), which is sometimes used as an anaesthetic. It isomerises to propene when it is heated at 100–200°C (375–475 K) in the presence of a catalyst (Pt). In its chemical properties cyclopropane shows some resemblance to ethylene. Thus it undergoes addition reactions which involve cleavage of the ring with, for example, conc. sulphuric acid, hydrobromic acid or bromine; it can also be catalytically hydrogenated:

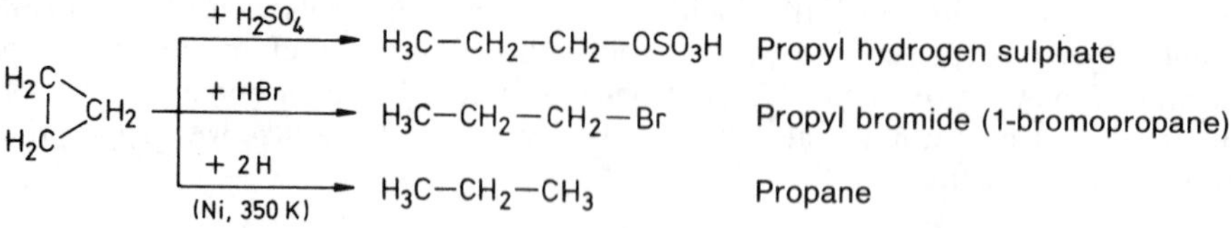

$$H_2C\begin{smallmatrix}\\|\\\end{smallmatrix}CH_2\begin{cases}\xrightarrow{+ H_2SO_4} H_3C-CH_2-CH_2-OSO_3H & \text{Propyl hydrogen sulphate}\\ \xrightarrow{+ HBr} H_3C-CH_2-CH_2-Br & \text{Propyl bromide (1-bromopropane)}\\ \xrightarrow[\text{(Ni, 350 K)}]{+ 2H} H_3C-CH_2-CH_3 & \text{Propane}\end{cases}$$

This reactivity of cyclopropane can be explained in terms of the structural models in Figs. 94a and 94b, in which partial π-bond character leads to there being some kinship to ethylene.

In contrast to ethylene, cyclopropane is unaffected by cold aqueous potassium permanganate or ozone. It only reacts with bromine under the influence of light or a

Lewis acid, and 1,3-dibromopropane is only one of a number of brominated propanes which are formed.

3.1.2 The Stereochemistry of Carbocyclic Compounds

1,2-Di- and 1,2,3-tri-substituted cyclopropanes can exist not only as *cis–trans* isomers, but also, in some cases, as *enantiomers*. This can be illustrated in the case of **cyclopropane-1,2-dicarboxylic acids**.

cis Form   trans Form

In the case of the *cis* form, both carboxyl groups lie on the same side of the plane of the ring; they are adjacent to one another and can readily form an anhydride by loss of a molecule of water. In contrast, the *trans* form, in which the carboxyl groups are attached to opposite sides of the ring, does not form an anhydride.

The *cis* form has a plane of symmetry lying through the CH_2 group, but the *trans* form has neither a plane of symmetry nor a centre of symmetry, and is resolvable into *enantiomers* (*E. Buchner*).

3.1.3 Cyclopropene

Cyclopropene, C_3H_4, is a gas (b.p. $-36°C$ [237 K]). It can be made from cyclopropylamine by exhaustive methylation and dry distillation of the resultant cyclopropyl-trimethyl-ammonium hydroxide (*Demjanov*):

Cyclopropylamine   Cyclopropene

Cyclopropene is very labile and readily polymerises. It reacts vigorously with bromine, forming 1,2-dibromocyclopropane.

The *cyclopropenium* (cyclopropenylium) *cation* has a delocalised two π-electron system, which is of theoretical interest (see p. 655).

Long-chain carboxylic acids having cyclopropane or cyclopropene rings in their structure occur naturally. Examples include lactobacillic acid (*K. Hofmann*, 1950) in the lipids of *lactobacillus arabinosus*, and sterculic acid (*Nunn*, 1952) in the kernel oil from *sterculia foetida*. Sterculic acid is poisonous, but saturated dihydrosterculic acid occurs together with lactobacillic acid in products obtained from sour milk and is not poisonous.

$$H_3C-(CH_2)_5 \quad \overset{H}{\underset{H}{\triangle}} \quad (CH_2)_9-COOH \qquad H_3C-(CH_2)_7-C\!=\!C-(CH_2)_7-COOH$$

Lactobacillic acid (m.p. 30°C [303 K])
(11R,12S)-Methylene-octadecanoic acid

Sterculic acid (m.p. 18°C [291 K])
9,10-Methyleneoctadec-9-enoic acid

1-Aminocyclopropane-1-carboxylic acid is the precursor of the phytohormone ethylene in plants, where it arises by the action of auxins on L-methionine.[1]

1-Aminocyclopropane-
1-carboxylic acid

$$H_2N \underset{\triangle}{\overset{COOH}{\diagup}}$$

$$\overset{\cdot\cdot}{\underset{\triangle}{}} \quad \text{Cyclopropenylidene}$$

The cyclic carbene cyclopropenylidene (see p. 419) is one of the most abundant interstellar molecules.

3.1.4 Cyclobutane and its Derivatives

Until a few years ago cyclobutane was a very difficult hydrocarbon to obtain. It was first made by *Willstätter* from cyclobutanecarboxylic acid. Unlike the ready preparation of cyclopropane from 1,3-dihalogenopropanes, reaction of sodium with 1,4-dibromobutane gives only a small amount (7%) of cyclobutane. Only more recently has a reasonable method for the preparation of cyclobutane and its derivatives been described.[2]

Preparation

3.1.4.1 Treatment of 1,3-dibromopropane with diethyl malonate in the presence of a sodium alkoxide provides the diethyl ester of cyclobutane-1,1-dicarboxylic acid, which can be converted by alkaline hydrolysis and decarboxylation into cyclobutanecarboxylic acid (see malonic ester syntheses, p. 327 *ff.*):

H_2C-Br	$H_2C-C(COOR)_2$	$H \atop H_2C-C-COOH$
H_2C-CH_2-Br	H_2C-CH_2	H_2C-CH_2
1,3-Dibromo propane	Diethyl cyclobutane-1,1-dicarboxylate	Cyclobutane-carboxylic acid

$\xrightarrow[-2\,NaBr]{+\,Na_2C(COOR)_2}$ $\xrightarrow[-CO_2]{(Hydrolyse)}$

This method for the preparation of alicyclic carboxylic acids, which was discovered by *Perkin*, is also applicable to higher ω,ω'-dihalogenoalkanes.

3.1.4.2 *Hunsdiecker* reaction (*Borodin* reaction) (*A.P. Borodin*, 1861). When the silver salt of cyclobutanecarboxylic acid is treated with bromine, bromocyclobutane is formed; this can be converted *via* formation of a *Grignard* reagent into cyclobutane (*Cason*, 1949).

[1] See *K. Lürssen*, Das Pflanzenhormon Ethylen (The Plant Hormone Ethylene), Chemie in unserer Zeit *15*, 122 (1981); *E. W. Ainscough* et al., Ethylene. An Unusual plant Hormone, J. Chem. Ed. *69*, 315 (1992).
[2] See *D. Seebach*, *Houben-Weyl*, Methoden der Organischen Chemie, Vol. 4/4 (Thieme Verlag, Stuttgart 1971).

$$\begin{array}{c} \text{COOAg} \\ | \\ H_2C-CH \\ | \quad\quad | \\ H_2C-CH_2 \end{array} \quad \xrightarrow[- \text{AgBr}, -CO_2]{+ Br_2} \quad \begin{array}{c} \text{Br} \\ | \\ H_2C-CH \\ | \quad\quad | \\ H_2C-CH_2 \end{array} \quad \xrightarrow{+ Mg} \quad \begin{array}{c} \text{MgBr} \\ | \\ H_2C-CH \\ | \quad\quad | \\ H_2C-CH_2 \end{array} \quad \xrightarrow[- \text{Mg(OH)Br}]{+ H_2O} \quad \begin{array}{c} H_2C-CH_2 \\ | \quad\quad | \\ H_2C-CH_2 \end{array}$$

Bromocyclobutane   Cyclobutane

3.1.4.3 The best way to obtain cyclobutane is by means of a *Wolff–Kishner reduction of cyclobutanone* (yield 65%):

$$\begin{array}{c} H_2C-C=O \\ | \quad\quad | \\ H_2C-CH_2 \end{array} \quad \xrightarrow[-H_2O]{+ N_2H_4} \quad \begin{array}{c} H_2C-C=N-NH_2 \\ | \quad\quad | \\ H_2C-CH_2 \end{array} \quad \xrightarrow[-N_2]{(NaOC_2H_5)} \quad \begin{array}{c} H_2C-CH_2 \\ | \quad\quad | \\ H_2C-CH_2 \end{array}$$

3.1.4.4 1,2-Cycloaddition or (2 + 2)cycloaddition reactions. A large number of cyclobutane derivatives can be made by *cycloaddition* of compounds which contain *activated* double bonds, including ketenes, allenes, cinnamic acids, and fluoro- or chloro-ethylenes.

(a) Whilst ketene and aldoketenes dimerise rapidly to give *β*-lactones (see p. 256), *ketoketenes dimerise* to form substituted cyclobutane-1,3-diones:

$$2\ R_2C=C=O \quad\longrightarrow\quad \begin{array}{c} R_2C-C=O \\ {}^2|_3 \quad {}^1| \\ O=C-CR_2 \end{array}$$

(b) *Allene dimerises* in a sealed tube at about 500°C (775 K), forming *1,2-dimethylenecyclobutane* (*Blomquist*, 1956):

$$2\ H_2C=C=CH_2 \quad\longrightarrow\quad \begin{array}{c} H_2C-C=CH_2 \\ | \quad\quad | \\ H_2C-C=CH_2 \end{array}$$

Allene   1,2-Dimethylenecyclobutane

(c) *Irradiation of cinnamic acid* derivatives makes them *dimerise* to *truxillic acid* (2,4-diphenylcyclobutane-1,3-dicarboxylic acid) and *truxinic acid* (3,4-diphenylcyclobutane-1,2-dicarboxylic acid) derivatives, *e.g.*

$$2\ C_6H_5-CH=CH-COOH \quad\xrightarrow{h\nu}\quad \begin{array}{c} \quad H \quad H \\ C_6H_5-C-C-COOH \\ {}^{|4}\quad {}^{1|} \\ {}_{|3}\quad {}_{2|} \\ HOOC-C-C-C_6H_5 \\ \quad H \quad H \end{array} \quad \text{and} \quad \begin{array}{c} \quad H \quad H \\ C_6H_5-C-C-COOH \\ {}^{|4}\quad {}^{1|} \\ {}_{|3}\quad {}_{2|} \\ C_6H_5-C-C-COOH \\ \quad H \quad H \end{array}$$

Truxullic acid   Triuxinic acid

Both the truxillic and truxinic acids can exist in different *stereoisomeric* forms. Some cinnamic acids which are substituted in the benzene ring also undergo similar photochemical dimerisation.

(d) Cyclobutane derivatives are obtained, usually in very good yields, by *thermal dimerisation of fluoro- or chlorofluoro-ethylenes* such as $F_2C=CF_2$, $F_2C=CFCl$ and $F_2C=CCl_2$ (see p. 137 f.) (*Harmon*, 1946 and *Barrick*, 1947), *e.g.*

$$2\ F_2C=CF_2 \quad\longrightarrow\quad \begin{array}{c} F_2C-CF_2 \\ | \quad\quad | \\ F_2C-CF_2 \end{array} \quad\quad 2\ F_2C=CCl_2 \quad\longrightarrow\quad \begin{array}{c} F_2C-CCl_2 \\ {}^{|4}\quad {}^{1|} \\ {}_{|3}\quad {}_{2|} \\ F_2C-CCl_2 \end{array}$$

Octafluorocyclobutane   1,1,2,2-Tetrachloro-3,3,4,4-tetrafluorocyclobutane

As a result of this *1,2-cycloaddition* of fluorinated alkenes, by using different starting materials a variety of cyclobutane derivatives can readily be prepared.

Cyclobutane itself cannot be made by dimerisation of ethylene (see p. 338 *f.*), but 1,1,2,2-tetrafluorocyclobutane can be obtained by heating *ethylene and tetrafluoroethylene* together (*Coffman*, 1949):

$$H_2C{=}CH_2 + CF_2{=}CF_2 \longrightarrow \begin{array}{c} H_2C{-}CF_2 \\ | \quad\;\; | \\ H_2C{-}CF_2 \end{array}$$

Ring-closure proceeds *via* either a radical or a dipolar intermediate; the principle of conservation of orbital symmetry is not applicable to the reaction.

Properties

Cyclobutane, C_4H_8, b.p. $12°C$ (285 K), is more stable than cyclopropane. It does not react at room temperature with either sulphuric acid, hydrobromic acid or bromine. Its catalytic reduction to butane requires stronger reaction conditions than the corresponding reduction of cyclopropane:

$$\begin{array}{c} H_2C{-}CH_2 \\ | \qquad | \\ H_2C{-}CH_2 \end{array} \xrightarrow[\text{(400K)}]{+\,2\,H(Ni)} H_3C{-}CH_2{-}CH_2{-}CH_3$$

Cyclobutane   Butane

The C—C bonds of cyclobutane are longer (157 pm) than those in cyclopropane. This can be ascribed to steric hindrance between the carbon atoms which are not directly bonded to one another. Their distance apart, 222 pm, is much shorter than the distance which non-bonded carbon atoms normally take up. The cyclobutane ring is not planar, but is folded (see Fig. 95, p. 394).

The special nature of the bonding in small rings causes certain reactions, not associated with common rings, to take place, in which there is interchange between cyclopropyl and cyclobutyl structures. One of the best-known examples involves the reactions of cyclobutylamine or cyclopropylmethylamine with *nitrous acid*. Both amines provide a mixture of cyclobutanol and cyclopropylmethanol (*Demjanov* reaction, 1903):

$$\underset{\text{Cyclobutylamine}}{\begin{array}{c} H \\ H_2C{-}C{-}NH_2 \\ | \qquad | \\ H_2C{-}CH_2 \end{array}} \xrightarrow[-N_2,-H_2O]{+HNO_2} \underset{\text{Cyclobutanol}}{\begin{array}{c} H \\ H_2C{-}C{-}OH \\ | \qquad | \\ H_2C{-}CH_2 \end{array}} + \underset{\substack{\text{Cyclopropyl-}\\\text{methanol}}}{\begin{array}{c} H_2C\;\;\; H \\ \quad>C< \\ H_2C\;\; CH_2OH \end{array}} \xrightarrow[-N_2,-H_2O]{+HNO_2} \underset{\text{Cyclopropylmethylamine}}{\begin{array}{c} H_2C\;\;\; H \\ \quad>C< \\ H_2C\;\; CH_2{-}NH_2 \end{array}}$$

Both *ring-contraction* amd *ring-expansion* reactions are involved.

3.1.4.5 Cyclobutanone, C_4H_6O, b.p. $98{-}100°C$ (371–373 K), is obtained by the action of *diazomethane on ketene* at $-70°C$ (200 K) (*Kaarsemaker*, 1951):

$$H_2C{=}C{=}O \xrightarrow[-N_2]{+H_2CN_2} \left[\underset{\text{Cyclopropanone}}{\begin{array}{c} H_2C \\ \quad>C{=}O \\ H_2C \end{array}}\right] \begin{array}{l} \xrightarrow[-N_2]{+H_2CN_2} \underset{\text{Cyclobutanone}}{\begin{array}{c} H_2C{-}C{=}O \\ | \qquad | \\ H_2C{-}CH_2 \end{array}} \\[2ex] \xrightarrow{+H_2O} \underset{\substack{\text{Cyclopropanone}\\\text{hydrate}}}{\begin{array}{c} H_2C\;\;\; OH \\ \quad>C< \\ H_2C\;\;\; OH \end{array}} \end{array}$$

Ketene

Cyclopropanone is formed as an intermediate in this reaction. In anhydrous conditions it then reacts with excess diazomethane undergoing a ring-expansion reaction to form cyclobutanone (*P. Lipp*). Cyclopropanone can only be isolated, and then in the form of its hydrate or hemi-acetal, if water or an alcohol are present.

3.1.4.6 Cyclobutanol, C_4H_7OH, b.p. 125°C (398 K), can be obtained in almost quantitative yield by *reduction of cyclobutanone with lithium aluminium hydride* (*J.D. Roberts*, 1949):

Cyclobutanol   INN: **Carboplatin**

3.1.4.7 The organoplatinum compound INN: **Carboplatin** (*Carboplat*) which is used in the treatment of testicular cancer is a derivative of cyclobutane-1,1-dicarboxylic acid (see p. 390).

3.1.4.8 Cyclobutene, C_4H_6, is the simplest unsaturated four-membered ring compound and is best made by thermal decomposition of cyclobutyl-trimethylammonium hydroxide:

Cyclobutene

3.1.4.9 Squaric acid[1] (3,4-dihydroxy-cyclobut-3-ene-1,2-dione) was first prepared by the reaction of chlorotrifluoroethylene with zinc. The product is perfluorocyclobutene which reacts with ethanol to give 1,2-diethoxy-3,3,4,4-tetrafluorocyclobutene; this is converted by acid hydrolysis into squaric acid (*S. Cohen* et al., 1959):

Chlorotrifluoroethylene   Perfluorocyclobutene   Squaric acid

It is a colourless, crystalline dibasic acid ($pK_a^1 = 0.59$, $pK_a^2 = 3.9$) which decomposes when heated to about 293°C (565 K). It forms salts, esters of lower alcohols, acid chlorides and amides; its derivatives can cause allergic reactions. The mycotoxin *moniliformin* is closely related to squaric acid. The *dianion* of squaric acid is a member of a novel series of oxocarbon dianions with the general formula $(C_nO_n)^{2\ominus}$, which appear to have some aromatic character (see p. 647 *ff*). If an oxygen atom is replaced by another atom X, *e.g.* N,S, $= C\!\!\stackrel{\diagup}{\diagdown}$, the resultant compounds, $C_nO_{n-m}X_m^{x\ominus}$, are called *pseudo-oxocarbons*.[2]

The biosynthesis of moniliformin (p. 394) probably starts from acetoacetic acid, the first member of a series of *polyketides* or *acetogenins*, which play a role in the formation of natural products such as flavonoids, naphthoquinones and tetracyclins in organisms. Acetoacetic acid, as the first member of the polyketide series ($n=0$) is also known as diketide, since it arises metabolically from two molecules of activated acetic acid (see p. 897).[3]

[1] See *G. Maahs* et al., Syntheses and Derivatives of Squaric Acid, Angew. Chem. Int. Ed. Engl. 5, 888 (1966).

[2] *G. Seitz* and *P. Imming*, Oxocarbons and Pseudo-oxocarbons, Chem. Rev. *92* 1227 (1992); *A. H. Schmidt*, Oxokohlenstoffe, Chemie in unserer Zeit *16*, 57 (1982).

[3] See *B. Franck*, Mycotoxins from Mould Fungi, Angew. Chem. Int. Ed. Engl. *23*, 493 (1984).

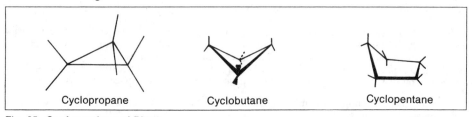

Squaric acid   Moniliformin   Polyketide
dianion

3.2 Common Ring Compounds

The parent compounds of this group of alicyclic compounds are cyclopentane, cyclohexane and cycloheptane.

3.2.1 Cyclopentane and its Derivatives

For a long time cyclopentane was thought to have a planar structure. A comparison of the *heats of combustion* of cyclopentane and of cyclohexane indicate that, although there is no distortion of the bond angles in cyclopentane, there is a strain energy of 5.4 kJ per CH_2 group and also that the first ring to be strainless is cyclohexane (see Table 13, p. 386). *Electron diffraction studies* by *Hassel* showed that only four of the ring carbon atoms in cyclopentane are coplanar, and that *one* carbon is out of this plane. This arrangement is a consequence of the fact that the hydrogen atoms cannot take up an energetically preferable staggered conformation. The resultant *conformational strain* is also known as *Pitzer strain*. It is also present in cyclopropane and in cyclobutane (see Fig. 95) and in medium-sized rings.

Cyclopropane   Cyclobutane   Cyclopentane

Fig. 95. Conformations of Rings

3.2.1.1 Preparation. Cyclopentane is best prepared by *Clemmensen reduction of cyclopentanone*, which is readily available from adipic acid:

Cyclopentanone (b.p. 130.5°C [403.7 K])   Cyclopentane

Properties. Cyclopentane, C_5H_{10}, is a liquid which boils at 50°C (323 K). Like cyclopropane and cyclobutane it is stable to permanganate and to ozone. Catalytic hydrogenation over nickel takes place at temperatures above 300°C (575 K), leading to opening of the ring and formation of n-pentane (*Zelinsky*). Cyclopentane resembles normal alkanes in its chemical behaviour.

1,2-Substitued and 1,3-substituted derivatives of cyclopentane exist as *cis* and *trans* isomers, since the two substituents can be either on the same side or on opposite sides of the ring. **Cyclopentane-1,2- and -1,3-dicarboxylic acids** serve as examples:

cis Form
(m.p. 135°C [408 K])

trans Form
(m.p. 182°C [455 K])

Cyclopentane-1,2-dicarboxylic acids

cis Form
(m.p. 121°C [394 K])

trans Form
(m.p. 188° [461 K])

Cyclopentane-1,3-dicarboxylic acids

cis-Cyclopentane-1,2-dicarboxylic acid, but not the *trans* isomer, readily forms an anhydride, m.p. 72°C (345 K). Neither of the two *trans* acids has a plane of symmetry and both are resolvable into *enantiomers*.

Alicyclic carboxylic acids are found in crude oil; they can be extracted out with alkali and obtained from this extract by acidifying it. A mixture of acids is present, of which the larger part consists of alkylated cyclopentane- and cyclohexane-carboxylic acids.

The crude acids, and especially their metal salts, have acquired a number of industrial uses. Thus the lead, manganese, zinc and iron salts are used as drying agents in varnishes (see p. 244). Copper salts serve as *fungicides* which kill moulds, and are impregnated into wood and in natural fibres in, for example, sandbags and ships' ropes. A mix of their aluminium salts with the aluminium salts of fatty acids is used in the preparation of some incendiary bombs (Napalm).

3.2.1.2 Cyclopentane,

C_5H_8O, can be obtained in good yield by the *Dieckmann* **condensation**[1] (1901) from esters of adipic acid and sodium alkoxide or metallic sodium in benzene.

This is an example of an *intramolecular* ester condensation reaction; its mechanism is as follows (*cf.* *Claisen* ester condensation p. 295):

Adipic acid diester

Cyclopentanone (b.p. 130°C [403 K])

[1] See *J. P. Schaefer* and *J. J. Bloomfield*, The Dieckmann Condensation, Org. React. *15*, 1 (1967).

Cyclohexanone and cycloheptanone can be prepared in the same way, starting from diesters of, respectively, pimelic and suberic acids.

3.2.1.3 Cyclopentene, C_5H_8, b.p. 45°C (318 K) can be made by *removal of a molecule of water from cyclopentanol* or by *heating bromocyclopentane with an alcoholic solution of potassium hydroxide*.

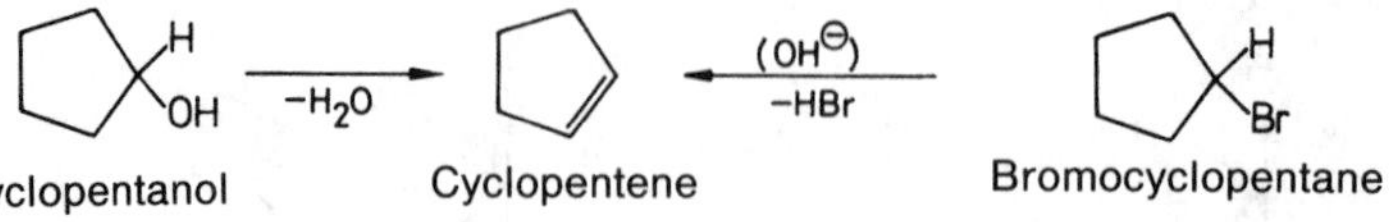

Cyclopentanol $-H_2O$ Cyclopentene $(OH^\ominus)$ / $-HBr$ Bromocyclopentane

Like open-chain alkenes, cycloalkenes can be hydroxylated. Depending upon whether the *oxidising agent* is potassium permanganate or performic acid, either the *cis* or *trans* isomer is formed; for example, *cyclopentene* gives the following *isomers*:

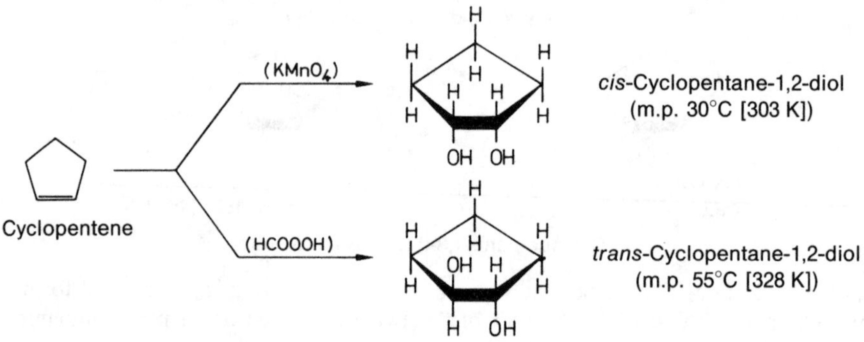

Cyclopentene

$(KMnO_4)$ → *cis*-Cyclopentane-1,2-diol (m.p. 30°C [303 K])

$(HCOOOH)$ → *trans*-Cyclopentane-1,2-diol (m.p. 55°C [328 K])

The *cis* isomer is a *meso* form, whilst the *trans* form can be resolved into *enantiomers*.

An extremely powerful natural *insecticide*, given the trade name '*PYR*', can be extracted from the flower-heads of African plants of the *pyrethrum* family. Its active ingredients are esters formed from a cyclopropanecarboxylic acid ((+)-*trans*-chrysanthemic acid) with a *ketocyclopentenyl alcohol* (rethrolone); for example, **Pyrethrin I**[1] has the following structure:

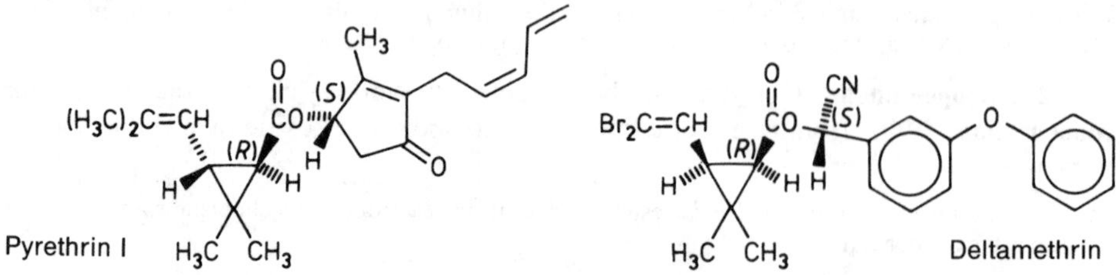

Pyrethrin I   Deltamethrin

Even more powerful pyrethroid insecticides can be produced by planned modification of the pyrethrin system, for example, by having the ester group and the side-chain *cis* to one another in the three-membered ring. *Deltamethrin*, which is a compound of this sort, is 3000 times more active than DDT.

3.2.1.4 Cyclopentadiene, C_5H_6, b.p. 41°C (314 K), is a diene which is found in the gasoline fraction from coal tars and in the C_5-fraction obtained from steam-cracking. Even at

---

[1] See *L. Crombie* et al., Chemistry of the Natural Pyrethrins, Fortschr. Chem. Org. Naturst. *19*, 120 ff. (1961).

room temperature it *dimerises* to dicyclopentadiene(tricyclo[$5.2.1.0^{2,6}$]deca-3,8-diene, (see p. 413 *f.*), m.p. 33°C (306 K), which can be *depolymerised* back to cyclopentadiene by distillation.

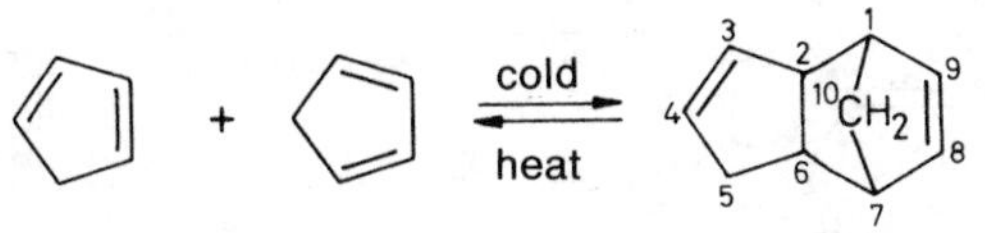

Cyclopentadiene   Dicyclopentadiene

This dimerisation is a *Diels–Alder* reaction in which cyclopentadiene acts both as a *diene* and as a *dienophile* (see p. 336 *ff.*). Cyclopentadiene reacts as a dienophile with butadiene to give 5-vinylbicyclo[2.2.1]hept-2-ene, which in the presence of alkali rearranges to 5-ethylidenebicyclo [2.2.1]hept-2-ene (5-ethylidene-norbornene):

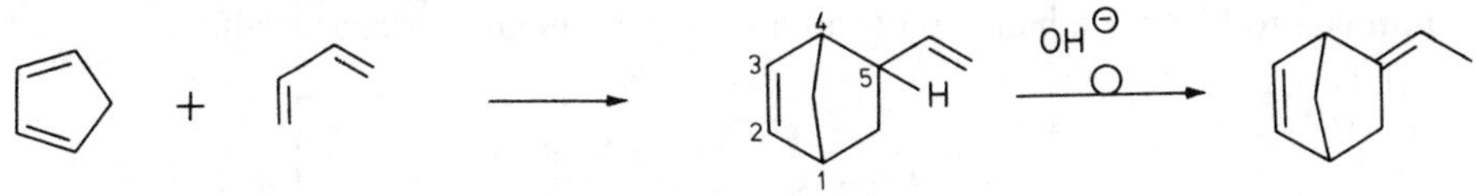

Cyclopentadiene   Butadiene   5-Vinylbicyclo   5-Ethylidene-
                                          [2.2.1]hept-2-ene   norbornene

Another feature of the chemistry of cyclopentadiene is the reactivity of its methylene group, which is activated by the two adjacent C=C double bonds. It condenses with aliphatic aldehydes or ketones in the presence of alkali to give yellow or orange coloured products called **fulvenes**[1] (Lat. *fulvus* = red-yellow), many of which are oils which are readily autoxidised and polymerise. Aryl aldehydes or ketones, such as benzaldehyde or benzophenone, give red crystalline fulvenes:

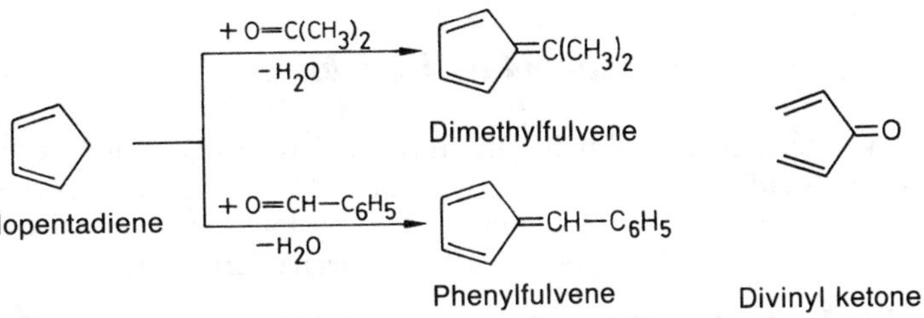

Cyclopentadiene   Dimethylfulvene   Phenylfulvene   Divinyl ketone

As derivatives of 5-methylene-1,3-cyclopentadiene, fulvenes contain *cross-conjugated double bonds*, which are particularly reactive. In cross-conjugated systems, three groups are present, two of which may not be conjugated with each other although each is conjugated with the third group, as, for example, in divinyl ketone, wherein the two alkene groups are separately conjugated with the keto group. In the case of fulvenes, the two ring double bonds are each independently conjugated with the exocyclic double bond, although in this case they are in fact also conjugated with one another.

A more fully developed system of this sort is found in *fulvalene*, which is very unstable. Some stable substituted derivatives are known, for example hexaphenylfulvalene.

---

[1] See *P. Yates*, Advances in Alicyclic Chemistry, Vol. II, 59 *ff.* (Academic Press, New York, 1968); *E. D. Bergmann*, Chem. Rev., *68*, 41 (1968).

Derivatives of *tetrathiafulvalenes* have become of importance as constituents of organic conductors, wherein they act as electron donors in charge-transfer complex salts.[1]

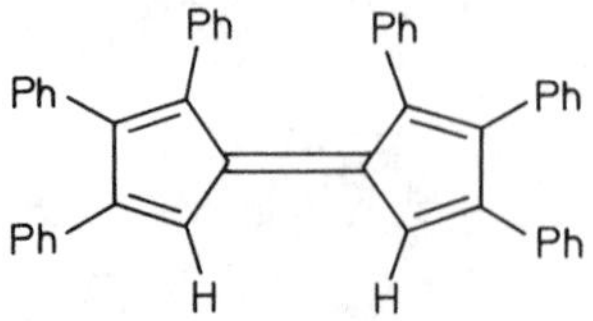

Hexaphenylfulvalene
(m.p. 279–282°C [552–555 K])

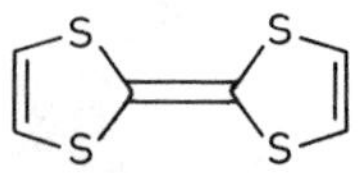

Tetrathiafulvalene

The hydrogen atoms of the methylene group in cyclopentadiene are sufficiently acidic to form salts; for example, in solution in dry benzene cyclopentadiene gives a yellow potassium salt on addition of potassium metal. It also reacts with ethyl magnesium bromide to liberate ethane and form a cyclopentadienyl Grignard salt.

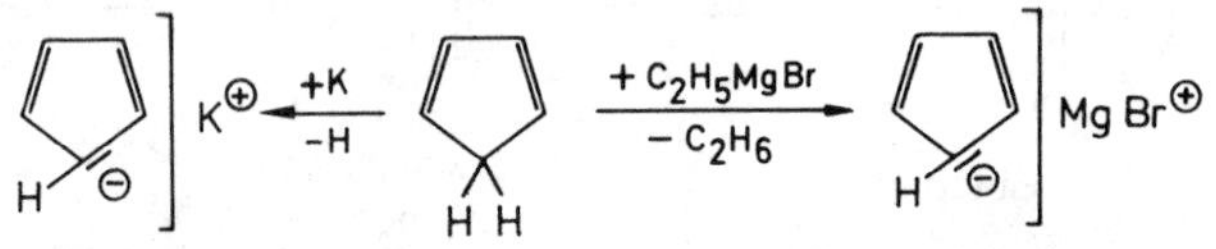

Potassium   Cyclopentadiene   Cyclopentadienyl-
cyclopentadienide   magnesium bromide

Addition of carbon dioxide to either of these salts leads to the formation of salts of cyclopentadienecarboxylic acid.

The *cyclopentadienide anion* resembles benzene in having a sextet of π-electrons and is hence of theoretical interest.

3.2.2 Cyclohexane and its Derivatives

Cyclohexane, and also its unsaturated derivatives cyclohexene and the two isomeric cyclohexadienes, are the parent compounds of a whole series of natural products, above all the terpenes and camphors. Most cyclohexane derivatives are made by catalytic hydrogenation of the corresponding benzene derivatives.

3.2.2.1 Cyclohexane (hexahydrobenzene), C_6H_{12}, occurs in mineral oil. It is a colourless liquid, b.p. 81° (354 K), which smells rather like benzene.

Industrially cyclohexane is made by *catalytic hydrogenation of benzene* in the presence of a nickel catalyst at about 200°C (475 K) and 20–40 bar (2–4 MPa):

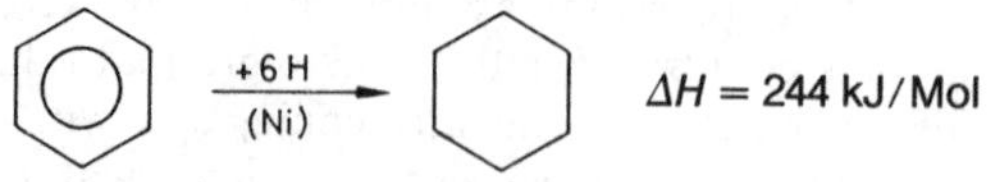

Benzene   Cyclohexane

$\Delta H = 244 \text{ kJ/Mol}$

Cyclohexane is a starting material from which nylon and perlon are made, *via* cyclohexanone.

---

[1] See E. *Amberger* et al., Angew. Chem. Int. Ed. Engl. *24*, 968 (1985); J. M. *Williams* et al., Prog. Inorg. Chem. *35*, 51 (1987).

3.2.2.2 Conformations[1] of Cyclohexane Derivatives. As long ago as 1890 *Sachse* expounded the hypothesis that the carbon atoms in the ring tried, as far as possible, to retain their tetrahedral structure. On this basis he postulated two strain-free shapes for cyclohexane, the *chair* and *boat* (*tub*) forms. In the chair form (Fig. 96a) the six carbon atoms are arranged in two parallel planes each containing three carbon atoms; atoms 1 and 4 point away from one another. On the other hand, in the boat form (Fig. 96b) atoms 1 and 4 lie on the same side of a plane containing the other four carbon atoms. All attempts, however, to isolate the two isomers proved unsuccessful. *Mohr* (1918) added the suggestion that the boat form could easily be converted into the chair form and *vice versa* by inverting one corner of the ring, and that a small amount of strain would be introduced into the ring in the course of the process.

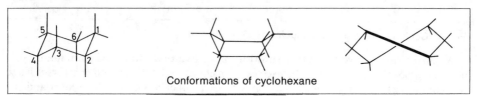

Conformations of cyclohexane

Fig. 96a. Chair form   Fig. 96b. Boat form   Fig. 96c. Twist form

In more recent years more has been learned about the conformation of cyclohexane and its derivatives (see p. 57 *ff.*). The chair and boat forms of cyclohexane are seen as two conformations in which the normal tetrahedral angles at carbon are maintained. Physical measurements carried out on a variety of different cyclohexane derivatives, using electron diffraction (*Hassel*), and NMR and IR spectroscopy, and theoretical calculations on the interaction between atoms in the molecule which are not bonded to each other (*Pitzer*), have led to the conclusion that *the chair form is more stable than the boat form and the twist form* (Fig. 96c). The latter forms differ only slightly in their energy content and can rapidly interchange. This is responsible for the flexibility of these two forms. The boat form can be regarded as a transition state between two twist forms while the twist form is higher in energy by 23 kJ/mol than the chair form.[2]

The reason for this can be traced to non-bonded interactions between the hydrogen atoms (*Pitzer* or conformational strain). The *energy barrier* which separates the chair and twist forms is about 45 kJ/mol. Consequently cyclohexane and its derivatives, and also related six-membered heterocyclic rings (see conformation of 1,4-dioxan, p. 301) exist predominantly in *chair forms*. *Barton* pointed out that in these chair forms it is possible to distinguish between two sorts of C—H bonds. Six C—H bonds, one attached to each carbon atom of a cyclohexane ring, are arranged parallel to the three-fold axis of symmetry of the ring, three above and three below the plane of the ring. These are described as *axial* bonds; often shortened to *a*-bonds. The other six C—H bonds lie more or less in the plane of the ring (they are alternately directed a little above and below the plane) and are called *equatorial* (*e*) bonds. Since the hydrogen atoms in the chair form of cyclohexane are completely staggered, and the C—H bonds do not eclipse one another, this represents an energetically favourable situation and an especially stable conformation.

---

[1] See *M. Hanack*, Organic Chemistry, A Series of Monographs, Vol. 3, Conformation Theory (Academic Press, New York 1965).

[2] See *F. G. Riddell* et al., Non Chair Conformations of Six-membered Rings, Topics in Stereochemistry *8*, 225 (1974).

It may be noticed that bonds linking e-substituents to the ring are parallel to the next-but-one C—C bonds of the ring, $e.g.$, in Fig. 86a, e-C^1 and C^2—C^3 or C^5—C^6.

If a hydrogen atom in cyclohexane is replaced by another atom or group of atoms, the substituent (R) can occupy either an axial or an equatorial position. Hence two *conformational isomers* (conformers) should be possible for monosubstituted cyclohexane derivatives in the chair form (Fig. 97):

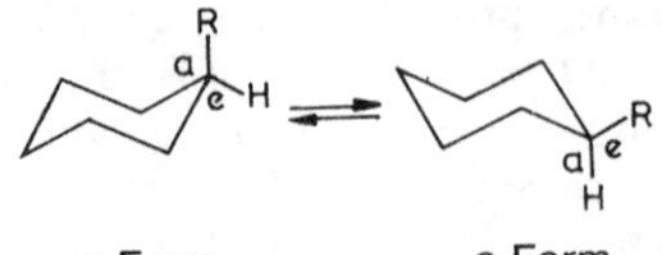

a-Form   e-Form

Fig. 97 Conformational isomers (conformers) of cyclohexane

When the ring is inverted, either substituents or hydrogen atoms which initially were in an axial position instead find themselves in an equatorial position and *vice versa*. These two conformers, which are *in equilibrium* with each other are called the a-form and the e-form (see Fig. 97). This type of e-form should be differentiated from the e-position of ligands attached to pentacoordinate phosphorus (see p. 175). *Electron diffraction* studies on monoalkyl- and monohalogeno-cyclohexanes in the gas phase indicate that single substituents prefer to take up an equatorial conformation, for example the equilibrium mixture of chlorocyclohexanes contains more of the e-conformer than of the a-conformer (*Hassel*).

For cyclohexane derivatives with more than one substituent the question of the ratios of different conformers present in an equilibrium mixture becomes more complicated. Introduction of a second identical substituent into a cyclohexane ring first of all presents the possibility of positional isomers, *i.e.* either 1,2-, 1,3- or 1,4-derivatives may result. Each of these may be present as either *cis* or *trans* forms (Figs. 98–100). In the *cis* isomer both substituents are on the same side of the ring, in the *trans* isomer they are on opposite sides of the plane of the ring. If the chair forms of cyclohexanes are projected onto the plane of a piece of paper then classical configuration formulae for the six stereoisomeric disubstituted cyclohexane derivatives are produced.[1] As was done above for monosubstituted cyclohexane derivatives, so in Figs. 98–100 the possible conformations for these disubstituted derivatives are shown.

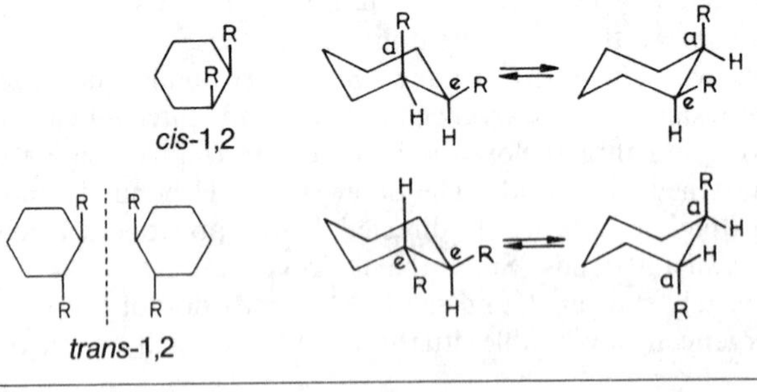

cis-1,2

trans-1,2

Fig. 98. Conformations of 1,2-disubstituted cyclohexanes

---

[1] For a justification of these projections see *J. E. Leonard* et al., The Apparent Symmetry of Cyclohexane, J. Am. Chem. Soc. *97*, 5052 (1975).

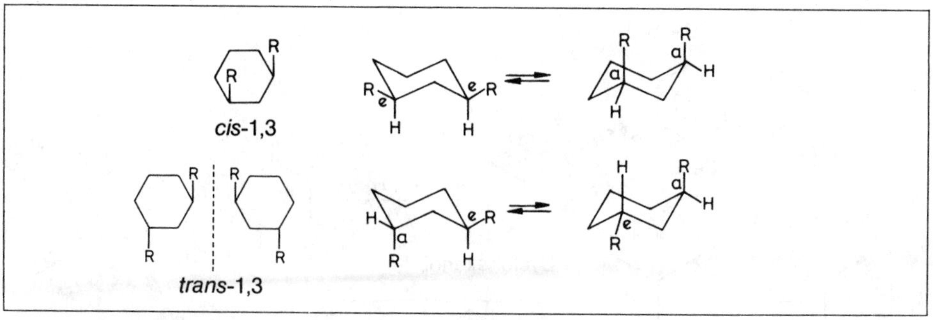

Fig. 99. Conformations of 1,3-disubstituted cyclohexanes

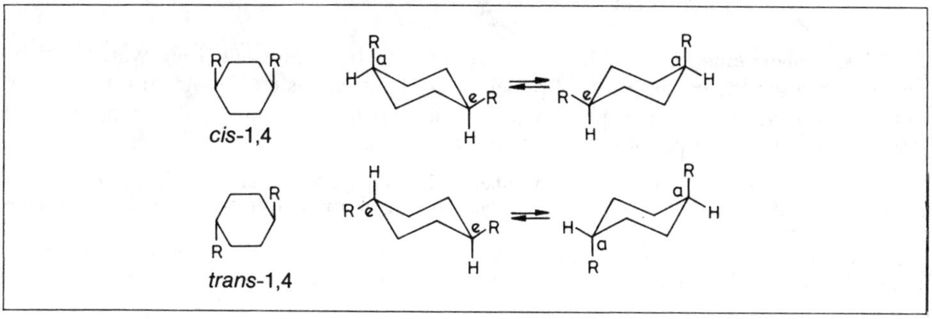

Fig. 100. Conformations of 1,4-disubstituted cyclohexanes

From *Hassel's* observations, the *cis*-1,2-, *trans*-1,3- and *cis*-1,4-derivatives must have one of the substituents in an equatorial and one in an axial position. If one of the chair forms is inverted into the alternative form, then the former axial substituent becomes equatorial and *vice versa*. Therefore the two chair conformations are equivalent and cannot be differentiated, one from the other. On the other hand, in the case of the *trans*-1,2-, *cis*-1,3- and *trans*-1,4-derivatives there are two energetically non-equivalent conformations, in one of which both substituents are axial whilst in the other they are both equatorial.

Since both the 1,2-*trans*- and 1,3-*trans*-derivatives are chiral, this increases the number of possible isomers by a further two, since both derivatives are resolvable into *enantiomers* (see Figs. 98, 99).

As was the case for monosubstituted cyclohexanes, so with di- and poly-substituted cyclohexanes, the substituents usually prefer to take up equatorial positions, so long as there are no strong repulsive electrostatic effects, as is the case when, for example, halogeno or carboxyl groups are present; then the picture is completely changed. With increasing numbers of substituents some of them may be forced to go into axial positions. To date, no compound is known in which three axial substituents are present on the same side of the plane of the ring.

Since the equatorial positions are less subject to steric hindrance than axial positions, some reactions take place more readily at these positions. For example *e*-carboxyl groups are more easily esterified than *a*-carboxyl groups. In contrast, secondary alcohols having an *a*-hydroxy group are more easily oxidised to ketones by chromic acid than are those with *e*-hydroxy-groups. The reason for this is that reaction actually takes place at the adjacent *e*-C—H bond. A consideration of the conformation of a cyclohexane derivative thus provides a deeper insight into the progress of the reaction than the configuration does.

3.2.2.3 Cyclohexanol, $C_6H_{11}OH$, is an oil with a camphor-like smell which boils at 161°C (434 K) (m.p. 25°C [298 K]). It is made industrially by *catalytic hydrogenation of phenol* at about 160°C (430 K), or by oxidation of cyclohexane at about 150°C (420 K). It has the same properties as an aliphatic secondary alcohol. Cyclohexanol is oxidised to cyclohexanone by dilute nitric acid and to adipic acid by conc. nitric acid (see p. 332).

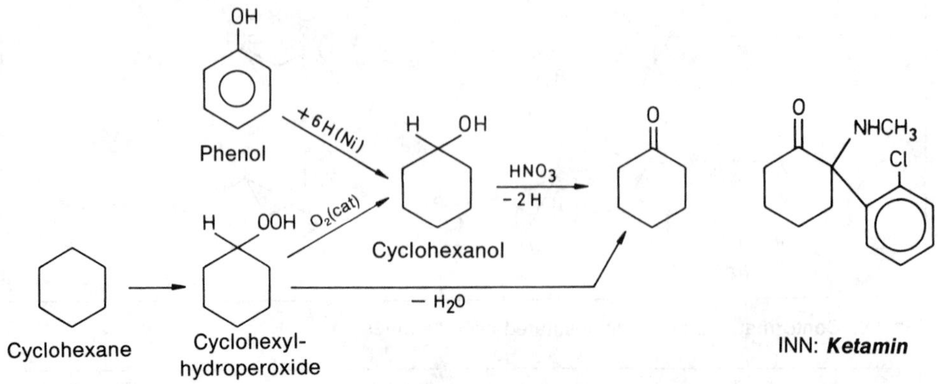

3.2.2.4 Cyclohexanone, $C_6H_{10}O$, is a colourless oil, b.p. 156°C (429 K), which can be made by heating the calcium salt of pimelic acid. Its oxime is an important intermediate in the manufacture of ε-caprolactam which is the starting material for the preparation of *Perlon* or *Dederon* (see p. 288).

The cyclohexanone derivative 2-(2-chlorophenyl)-2-aminoxycyclohexanone, INN: *Ketamin*, is a powerful anaesthetic that brings about a change in consciousness rather than making a person unconscious (dissociative anaesthesia).

Reduction of cyclohexanones by lithium aluminium hydride leads to the preferential formation of cyclohexanols which have equatorial hydroxy groups. For example, when R = t-butyl:

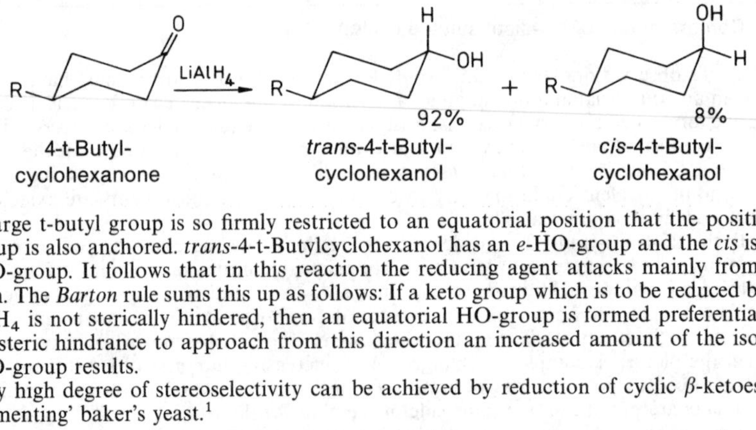

4-t-Butyl-cyclohexanone	*trans*-4-t-Butyl-cyclohexanol	*cis*-4-t-Butyl-cyclohexanol
	92%	8%

The large t-butyl group is so firmly restricted to an equatorial position that the position of the HO-group is also anchored. *trans*-4-t-Butylcyclohexanol has an *e*-HO-group and the *cis* isomer has an *a*-HO-group. It follows that in this reaction the reducing agent attacks mainly from an axial direction. The *Barton* rule sums this up as follows: If a keto group which is to be reduced by LiAlH$_4$ or NaBH$_4$ is not sterically hindered, then an equatorial HO-group is formed preferentially, but if there is steric hindrance to approach from this direction an increased amount of the isomer with an *a*-HO-group results.

A very high degree of stereoselectivity can be achieved by reduction of cyclic β-ketoesters with 'non-fermenting' baker's yeast.[1]

3.2.2.5 Cyclohexane-1,4-diol (quinitol), $C_6H_{10}(OH)_2$ is made by *catalytic hydrogenation of hydroquinone*:

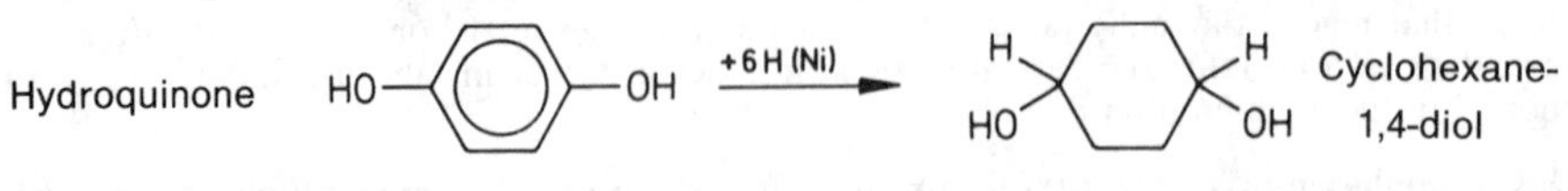

There are *cis* and *trans* isomers of cyclohexane-1,4-diol, both of which are achiral, since each has a plane of symmetry; the *trans* isomer also has a centre of symmetry.

---

[1] See *D. Seebach* et al., Helv. Chim. Acta *70*, 1605 (1987).

3.2.2.6 Inositols (hexahydrocyclohexanes), $C_6H_6(OH)_6$, are related to the sugar alcohols; they are soluble in water and have a sweet taste. There are eight possible *cis–trans* isomers, only one of which is resolvable into *enantiomers*. The other seven are all *meso* forms. The two optically active inositols, (+)- and (−)-*chiro*-inositol, *myo*-inositol (formerly called *meso*-inositol) and *scyllo*-inositol (formerly scyllitol), a constituent of streptidine (see p. 456), occur in nature.

(+)-1-D-*chiro*-Inositol	(−)-1-L-*chiro*-Inositol	*myo*-Inositol	*scyllo*-Inositol
(1,2,4/3,5,6)	(1,2,4/3,5,6)	(1,2,3,5/4,6)	(1,3,5/2,4,6)

myo-Inositol is found both in the free state and as part of larger molecules in various organs (liver, brain) and in the body fluids. In plants it exists partially as a hexaphosphoric acid ester (phytic acid, INN: *fytic acid*), whose calcium–magnesium salt is used therapeutically under the name *Phytin*. In addition *myo*-inositol acts as an essential *growth factor* in yeasts and other microorganisms. It is achiral since it has a mirror plane which passes perpendicularly through the molecule, indicated in the formula above by a dashed line.

Quinic acid and shikimic acid should be mentioned as examples of polyhydroxycyclohexane-carboxylic acids.

3.2.2.7 Quinic acid, 1-L-1(OH), 3,4/5-tetrahydroxycyclohexanecarboxylic acid,[1] is a plant acid which is very widely distributed in nature. It was first found in cinchona bark; it is also present in coffee beans and in the leaves of many plants.

3.2.2.8 Shikimic acid, 3/4,5-trihydroxycyclohexene-1-carboxylic acid, was first isolated from the fruits of a Japanese anise (*Illicium anisatum*) (1885).

Quinic acid   Shikimic acid

Investigations using biochemical mutants of *Escherichia coli* (Davis, 1953) have shown that shikimic acid should be seen as an important intermediate in the biosynthesis of aromatic amino acids (phenylalanine, tyrosine and tryptophane) from carbohydrates. The formation of benzene rings in lignin also occurs *via* these plant acids.

---

[1] The configuration at carbon atom 1 is apparent from the middle formula of quinic acid: the oblique lines between the numerals in the names of quinic and shikimic acids and shown in brackets below the formulae of the inositols separate the hydroxy groups which lie above the plane of the ring from those below the plane. See Pure Appl. Chem. *37*, 285 (1974).

The following unsaturated hydrocarbons are derivatives of cyclohexane:

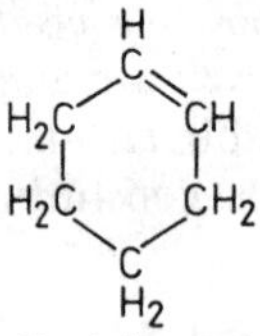

Cyclohexene   Cyclohexa-1,3-diene   Cyclohexa-1,4-diene   Benzene (see p. 467)
(b.p. 83°C [356 K]) (b.p. 80.4°C [355 K])   (b.p. 86°C [359 K])   (b.p. 80.1°C [353 K])

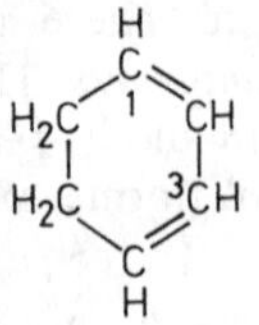

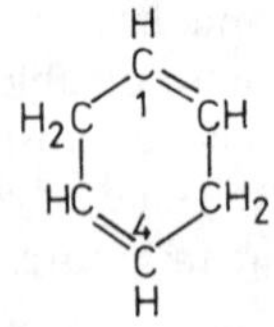

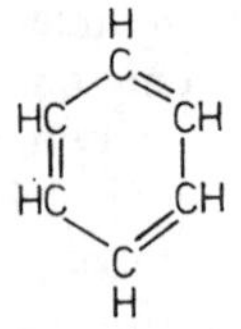

3.2.2.9 Cyclohexene (tetrahydrobenzene), C_6H_{10}, is prepared by removal of a molecule of water from cyclohexanol by treating it with conc. sulphuric acid. It exists in a half-chair conformation:

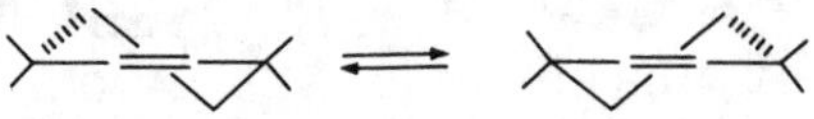

Half-chair conformation

3.2.2.10 Cyclohexa-1,3-diene and **cyclohexa-1,4-diene**, C_6H_8, can also be regarded as isomeric dihydrobenzenes. They have a strong tendency to polymerise. The former shows the typical properties of a conjugated diene; for example, it reacts with tetracyanoethylene (TCNE) in a *Diels–Alder* reaction at 25°C (330 K), giving a *cis* adduct:

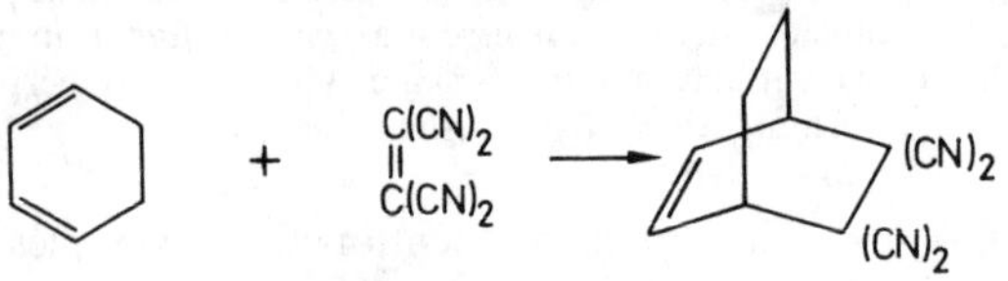

Cyclohexa-1,3-diene   Tetracyanoethylene   *cis* Adduct

3.2.3 *Cycloheptane and its Derivatives*

3.2.3.1 Cycloheptane, C_7H_{14}, occurs in mineral oil. It can be made by *Clemmensen* reduction of cycloheptanone, which is made by dry distillation of calcium suberate (calcium octanedioate):

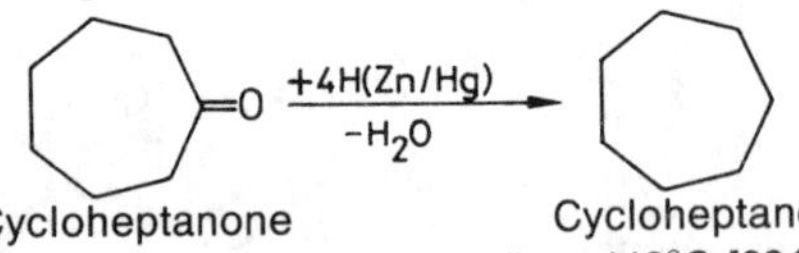

Cycloheptanone   Cycloheptane
(b.p. 180°C [453 K])   (b.p. 118°C [391 K])

3.2.3.2 Cycloheptatriene (tropilidene), C_7H_8, is obtained as a decomposition product of the alkaloids atropine and cocaine.

Preparation

(a) By irradiating a mixture of benzene and diazomethane with ultra-violet light (see p. 378).

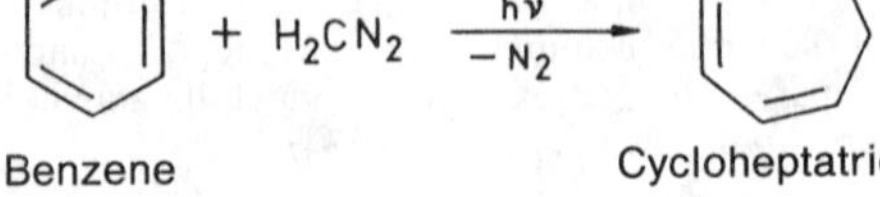

Benzene   Cycloheptatriene

(b) Cyclopentadiene and ketene react together readily to form *bicyclo[3.2.0]hept-2-en-6-one*. Reduction of this provides a secondary alcohol, from which water can be removed by acetolysis of its methylsulphonate ester. The resultant bicyclodiene isomerises to cycloheptatriene and to 6-methylfulvene (*Dryden*, 1954).

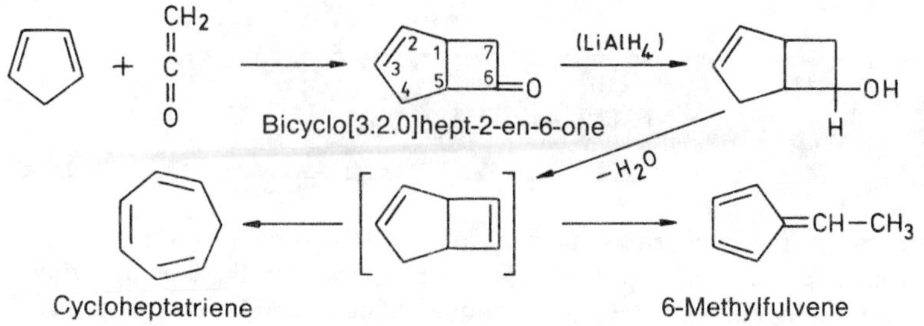

Bicyclo[3.2.0]hept-2-en-6-one

Cycloheptatriene   6-Methylfulvene

(For the nomenclature of bicyclic compounds see p. 414).

The initial cyclo-addition step in the above reaction scheme merits some comment, for a system in which a $[4\pi s + 2\pi s]$ reaction should be possible, in fact takes part in a $[2\pi s + 2\pi a]$ cyclo-addition, which is not observed with normal alkenes. A situation like this, in which a choice of behaviour appears to be possible but in which, in the event, only some of the electrons of a conjugated system take part in the reaction, has been described as *periselectivity*.

The reaction is stereoselective and with an unsymmetrically substituted ketene leads to formation of the sterically less favourable product:

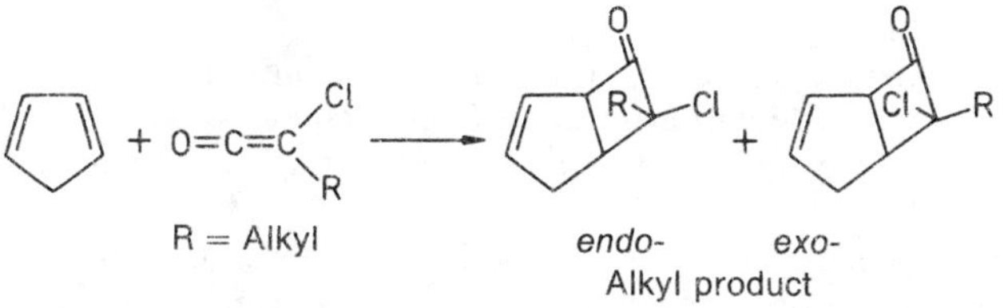

R = Alkyl   *endo-*   *exo-*
                                    Alkyl product

The *endo/exo* ratio varies as follows as the steric requirements of the alkyl group increase: $R = CH_3$:4.3; $R = C_2H_5$:5.3; $R = CH(CH_3)_2$:10.[1] The reaction thus appears to be masochistic as far as the stereochemistry of the product is concerned.

The results can be accommodated within the *Woodward–Hoffmann* rules if the following antarafacial (a) transition state is assumed:

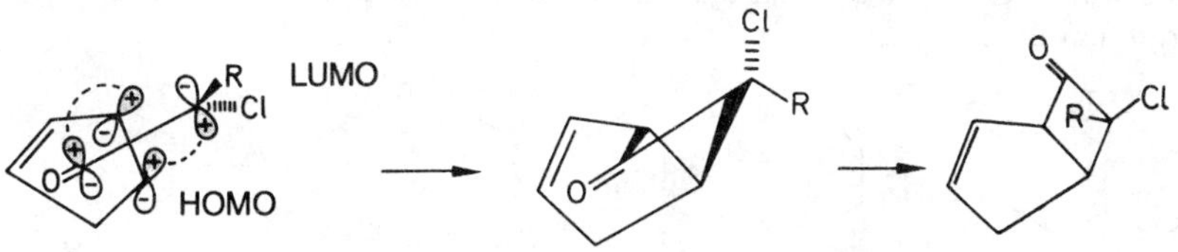

However, the sterically unfavourable arrangement in which two bonding interactions are possible (see dashed lines) is a result of the CO-group of the ketene. This needs only a small amount of space, and can therefore find room for itself over the ring. The resultant arrangement becomes more favourable the greater the size of R, because in the situation shown there is no interference between R and the ring which lies well beneath it. Realignment of the four-membered ring then brings R into the *endo* position.

This synchronous mechanism is described as a $[2\pi s + 2\pi a]$ process.

---

[1] See *W. T. Brady* et al., J. Am. Chem. Soc. 92, 4618 (1970).

Cycloheptatriene is a liquid, b.p. 115.5°C (388.7 K), which solidifies at −79.5°C (193.7 K). It readily undergoes both atmospheric oxidation and polymerisation.

The so-called 'aromatic' tropylium salts, $C_7H_7^{\oplus}X^{\ominus}$, and also tropones and tropolones, are derivatives of cycloheptatriene (tropilidene).

3.3 Medium-Sized Rings

The medium-sized alicyclic rings made up from 8–11 carbon atoms form a special group, whose properties are often markedly different from those of the common-ring and large-ring compounds as well as those of their aliphatic analogues. Most striking of all, perhaps, is the difficulty with which the rings are closed. This was originally attributed solely to the conformational (*Pitzer*) strain in these molecules. With the study of space-filling models of medium-sized rings it was realised that they had unfavourable energy-rich conformations in which there must be steric repulsive forces between adjacent hydrogen atoms which would result in an increase of energy. A further factor which had to be taken into account for these medium-sized rings was repulsive interaction between non-neighbouring hydrogen atoms across the ring, which leads to further strain, called *transannular strain* or *Prelog* strain, in the molecules. *X-ray studies* have shown that in the case of a 10-membered ring this latter effect is of overriding importance, but there is very little conformational strain.

Many of the anomalous reactions of medium-sized rings are associated with transannular interactions between hydrogen atoms. In particular, account must be taken of its effect upon reactions in which the nature of the bonding at the reacting carbon atoms changes.

For example, hydrogen cyanide does not add to cyclodecanone, because such a reaction involves a change of hybridisation of the carbon atom from sp^2 to sp^3 and this would result in an increase in transannular strain. On the other hand, *acetolysis of cyclodecyl tosylate* proceeds more than 500 times faster than the corresponding reaction involving cyclohexyl tosylate, since in the *carbenium ion* which is formed as an intermediate, the transannular strain decreases because of its sp^2 hybridisation. Furthermore, isotopic labelling with ^{14}C (black dot) has shown that in this reaction transannular migration of a hydride ion takes place:

$$Ts = SO_2-C_6H_4-CH_3-p \qquad Ac = COCH_3$$

3.3.1 Cyclooctatetraene

Cyclooctatetraene has been prepared industrially by cyclisation of acetylene (*Reppe*) (see p. 97). It was first prepared by *Willstätter* (1911) in a many-step synthesis based on the degradation of the alkaloid pseudopelletierine. It can also be prepared from cycloocta-

1,5-diene by reaction with butyl lithium and tetramethylenediamine (TMEDA) (1,4-diaminobutane) and atmospheric oxidation of the intermediate dianion.[1]

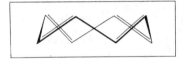 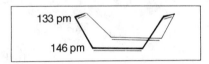

While this compound resembles benzene in having a complete cyclic system of conjugated double bonds, it is in fact a typically *unsaturated* compound in its properties. Cyclooctatetraene (COT) is a yellow liquid with b.p. 142–143°C (415–416 K). It reacts readily with bromine or with hydrogen halides, and is oxidised in the cold by permanganate. Reactions may involve rearrangement of the carbon skeleton. *X-ray analysis* shows that single and double bonds occur alternately around the 8-membered ring (*Kaufman*, 1948); the valence angles in the ring are about 126°C. Of the two conformations which are possible, one with only *trans* double bonds (crown form, Fig. 101), and the other with only *cis* double bonds (tub form, Fig. 102), the latter is the form taken up by the molecules.

Fig. 101. Cyclooctatetraene, crown form

Fig. 102. Cyclooctatetraene, tub form

133 pm

146 pm

NMR spectroscopy indicates that the tub form may be in equilibrium with planar cyclooctatetraene. This allows the double bonds to shift; this must take place through a planar delocalised transition state.

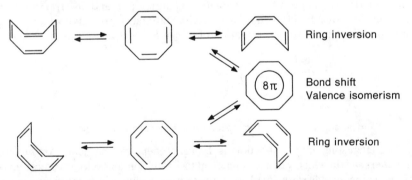

Ring inversion

Bond shift
Valence isomerism

Ring inversion

The energy barrier to bond migration is higher than that for ring inversion.

Such reversible rearrangements, which, like the bond shifts, are entirely dependent upon reorganisation of the bonding electrons, are described by the terms *valence isomerism* or valence tautomerism. In complete contrast to mesomerism, in valence isomerism there is a genuine *equilibrium* between two different compounds which often have different

[1] See *G. Wilke*, Angew. Chem. Int. Ed. Engl. *17*, 371 (1978); *L. A. Paquette*, The Renaissance in Cyclooctatetraene Chemistry, Tetrahedron *31*, 2855 (1975); *G. I. Fray* and *R. G. Saxton*, Chemistry of Cyclooctatetraene and its Derivatives (Cambridge University Press, Cambridge 1978).

topologies. Another example of valence isomerism is the equilibrium between cyclooctatetraene and bicyclo[4.2.0]octa-2,4,7-triene:

Cyclooctatetraene   Bicyclo[4.2.0]octa-
2,4,7-triene

3.3.2 Cyclooctane, Cyclooctene

Catalytic hydrogenation of cyclooctatetraene leads to the formation of cyclooctene and cyclooctane. Oxidation of cyclooctene with conc. nitric acid converts it into suberic acid:

Cyclooctane   Cyclooctatetraene   Cyclooctene   Suberic acid

Some reactions of cyclooctatetraene (catalytic hydrogenation, addition of alkali metals, oxidation with peracids) result in retention of the 8-membered ring, but others lead to its conversion into a benzene ring with formation of derivatives of ethylbenzene and *p*-xylene (*p*-dimethylbenzene). If an aqueous suspension of cyclooctatetraene is warmed with mercury(II) sulphate, metallic mercury separates out and *phenylacetaldehyde* is formed. On the other hand, oxidation of cyclooctatetraene with hypochlorous acid in alkaline conditions leads to the formation of *terephthalaldehyde*:

Phenylacetaldehyde   Cyclooctatetraene   Terephthalaldehyde

Another interesting ring-contraction is brought about by the action of *bromine on cyclooctatetraene*, in which valence isomerism to 7,8-dibromobicyclo[4.2.0]octa-2,4-diene occurs. Reaction in this case involves formation of a dibromocyclooctatetraene which then undergoes isomerism to the isolated product. The product of the reaction of cyclooctatetraene with maleic anhydride has likewise undergone valence isomerism, but in this case rearrangement to the bicyclic isomer precedes the addition reaction and it is the valence isomer of cyclooctatetraene rather than cyclooctatetraene itself which adds to the maleic anhydride.

3.3.3 Cyclododecatriene

By making use of catalysts, especially nickel, *Wilke* succeeded in converting *butadiene*, *via* formation of *metal-π-complexes*, into an excellent yield of the dimer, cycloocta-1,5-diene, or into a mixture of the trimers, **all-*trans*** and ***trans-trans-cis*-cyclododeca-1,5,9-triene**:

| Cycloocta-1,5-diene | all-*trans* | *trans-trans-cis* |
| | Cyclododeca-1,5,9-trienes | |

trans,trans,cis-Cyclododeca-1,5,9-triene is converted into the corresponding ketone, *via* initial formation of a monoepoxide. This ketone is then hydrogenated to give cyclododecanone. When the oxime of the latter compound is made and treated with sulphuric acid a *Beckmann* rearrangement (see p. 544) takes place and the product is a lactam, *laurolactam*, which is used in the manufacture of C_{12}-polyamides (Vestamid, Hüls).

3.4 Large Rings

Whilst the medium-sized rings with 8–11 carbon atoms are anomalous in their physical and chemical properties, the large rings show the normal expected behaviour.

The first large-ring ketones to be isolated, by *Ruzicka* in 1926, were muscone and civetone, from the scent glands of, respectively, the musk deer and the civet cat. Civetone has the same perimeter as that of androst-16-ene-3-one (see p. 692).

| Muscone | Civetone |
| 3-Methylcyclopentadecanone | Cycloheptadec-9-enone |

These 15- and 17-membered-ring ketones, which were used as perfumes, provided an impetus to the search for ways to make large- and medium-ring ketones.

3.4.1 Preparation

3.4.1.1 By *intramolecular cyclisation of long-chain dinitriles* in the presence of ether-soluble alkali-metal alkylanilides such as lithium ethylanilide, $C_6H_5N(C_2H_5)Li$, as condensing agents (*K. Ziegler*, 1933).

This method is a modified version of the *Dieckmann* condensation, using dinitriles instead of diesters. The best results are only achieved, however, by applying the *Ruggli–*

Ziegler high dilution principle to the reaction. In this method, a dilute ethereal solution of the appropriate dinitrile is slowly dropped into a very dilute ethereal solution of the condensing agent. The effectiveness of this method depends upon the stationary low concentration of the reactant and the rate of reaction. The high dilution suppresses the competing intermolecular polymerisation process and favours intramolecular cyclisation.

The reaction mechanism is as follows:

Dinitrile   Lithium salt of dinitrile

β-Ketonitrile   β-Keto acid

By using this method it is possible to make cyclic ketones with 14–33-membered rings in yields of 60–85%, but yields of medium-sized rings are exceedingly small.

3.4.1.2 The best method available for the preparation of large rings was found, quite independently of one another, by both *Prelog* and *M. Stoll* (1947). In this method an ester of a long-chain dicarboxylic acid is dropped into a heated suspension of sodium in xylene which is under an atmosphere of nitrogen. Reaction takes place on the surface of the metal and leads to the formation, after acid hydrolysis, of *cyclic acyloins* (see hydroxyketones, p. 318).[1]

Enediol   Acyloin

This acyloin cyclisation method is also successful for the preparation of medium-sized rings so long as the reaction is carried out in very dilute solutions; for example the 9- and 10-membered rings can be obtained in 30–60% yield. For larger rings, yields are 80% or more.[2] The corresponding cycloalkanes can be made by reducing the cyclic acyloins.

Properties

Compounds with large rings containing 13 or more carbon atoms differ very little in their properties from their acyclic long-chain analogues. X-ray studies have shown that in large rings made up of more than 20 carbon atoms, the normal tetrahedral angles are maintained and the atoms are arranged in two almost parallel zigzag carbon chains (Fig. 103).

---

[1] See *R. Brettle*, Acyloin Coupling Reactions, Comprehensive Organic Synthesis Vol. 3, p. 613 (Pergamon, Oxford 1991).
[2] See *J. J. Bloomfield* et al., The Acyloin Condensation, Org. React. **23**, 259 (1976).

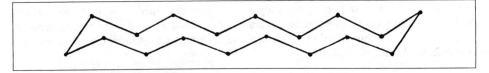

Fig. 103. The conformation of larger rings

3.5 Bi- and Poly-Cyclic Hydrocarbons

This class of compounds is subdivided into spiro-compounds (spiranes), condensed-ring compounds and bridged-ring compounds. In these three classes there are, respectively, one, two, or three or more atoms common to two rings.

3.5.1 Spiro-alkanes (Spiranes)

Spiranes are characterised by having just one carbon atom common to two rings. The planes of the two rings are perpendicular to one another.

Their steric structure resembles that of unsymmetrically substituted allenes, with two alicyclic rings taking the place of the two double bonds. As in the case of allenes (see p. 621), suitable substitution in spiranes can lead to molecular asymmetry. This gives rise to enantiomeric molecules which do not have an asymmetric carbon atom; rather the whole molecule is chiral. They have neither a plane nor a centre of symmetry, but they do, however, have an *axis of chirality* (see p. 603 *ff.*). An example is provided by **spiro[3.3]-heptane-2,6-dicarboxylic acid** (*H. J. Backer*, 1928).

Spiro[3.3]heptane-2,6-dicarboxylic acid   4-Methylcyclohexylidene-acetic acid

The numbers in the square brackets indicate that, starting from the central quaternary carbon atom, there are in addition three atoms in each of the rings (see also footnote,[1] p. 414).

An alicyclic compound which has an exocyclic double bond can provide isomers which are half-way houses between spirane- and allene-type isomers; an example is 4-methylcyclohexylideneacetic acid, which is indeed chiral (*Pope*, 1909).

3.5.2 Condensed Rings

If two cyclohexane molecules are joined together in such a way that two adjacent carbon atoms are shared by the two rings, the resultant molecule is the hydrocarbon *decalin*, $C_{10}H_{18}$.

In general this way of joining rings together may be described as 'ortho-condensation', and the resulting di- or poly-cyclic structures are known as condensed ring-systems. In like fashion, ortho-condensation of a cyclohexane ring and a cyclopentane ring provides a molecule of hydrindane, C_9H_{16}.

Decalin   Hydrindane

These two saturated bicyclic hydrocarbons can be made by complete catalytic hydrogenation of the corresponding arenes, naphthalene and indene.

3.5.2.1 The Stereoisomerism of cis- and trans-decalin

As long ago as 1918 Mohr pointed out that the two non-planar cyclohexane rings in decalin could be joined together in cis or trans forms, both of which would be strain-free. Separation of the two isomers was achieved by careful fractional distillation of the reaction mixture obtained from the complete reduction of naphthalene (W. Hückel, 1925). Both rings in both isomers have chair conformations. The two isomers differ in the way their rings are joined together. In the trans form, the rings are joined together by equatorial bonds; the two hydrogen atoms attached to the atoms common to the two rings are both axial. In cis-decalin, the rings are joined by one axial and one equatorial bond, and of the two hydrogen atoms at the ring junctions, one is equatorial and one axial. This may be seen in Figs. 104a and 104b.

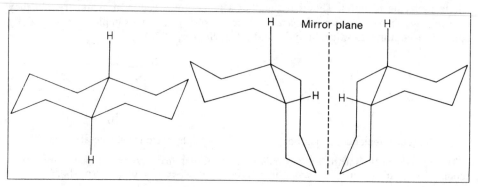

Fig. 104a. trans-Decalin (b.p. 185°C [458 K])   Fig. 104b. cis-Decalin (b.p. 194°C [467 K])

The trans isomer is the more stable of the two; the energy difference measured from their heats of combustion is, however, only 8.4 kJ/Mol. The trans form is rigid, but the cis form is flexible; it can be converted into its mirror-image form by inverting the conformation.

Instead of being shown by the conformational representation they are usually drawn in simple projection formulae, in which the trans or cis link is shown by either a dot or by wedge-shaped and dashed lines to the attached substituents (or hydrogen atoms). The dashed line indicates that the hydrogen atom or substituent lies below the plane of the projection formula.

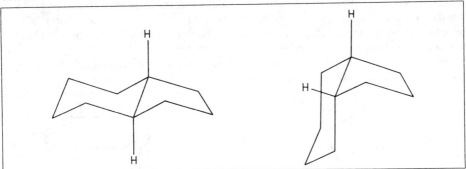

trans-Decalin   *cis*-Decalin

This type of representation is used extensively for polycyclic terpenes and steroids. Analogous stereochemical considerations apply to the *hydrindanes*, which also exist in *cis* and *trans* forms, for which the following conformations (Figs. 105a and 105b) can be drawn:

Fig. 105a. *trans*-Hydrindane   Fig. 105b. *cis*-Hydrindane

From this it may be seen that, as in the case of the cyclopentane molecule (see p. 394), one carbon atom of the 5-membered ring lies out of the plane of the other four. Heats of combustion show that *trans*-hydrindane is only 7.5 kJ/mol lower in energy than *cis*-hydrindane.

Decalin and hydrindane play an important role in the structure of a large number of natural products, for example the sesqui-, di- and tri-terpenes, as well as in steroids; these compounds will be dealt with separately in section **6.**

3.5.3 *Bridged-ring Compounds*

There is a further type of bicyclic ring-system, typified by the examples in the formulae below, in which a cyclohexane ring is linked between its 1- and 4-carbon atoms by a bridge containing one or more methylene groups. Such systems have three or more ring atoms common to both rings. These two bridged-ring compounds are simple examples of this class of compound; the first ring-system is found in a number of naturally occurring terpenes.

Bicyclo [2.2.1]heptane
(m.p. 86–87°C [359–360 K])
(8,9,10-Trinorbornane; trivial name: Norbornane)

Bicyclo[2.2.2]octane
(m.p. 170°C [443 K])

The nomenclature used for this class of compounds[1] can be traced back to a suggestion made by *A. von Baeyer*. The name starts with the prefix 'Bicyclo' and ends with the name of the corresponding alkane having the same number of carbon atoms as go to make up the bicyclic system. Between the two, not hyphenated, are a pair of square brackets containing the numbers of carbon atoms in each of the bridges which link the two tertiary carbon atoms. These numbers are arranged in decreasing order, and are separated by dots. In the dicyclopentadiene (p. 397) there is also a secondary bridge, containing no carbon atoms, between atoms 2 and 6. Its location is indicated by superscripts, separated by a comma, denoting the positions of the atoms it joins ($0^{2.6}$).

In contrast to strain-free condensed ring-systems, the stability of bridged-ring compounds is affected by the ring strain present as a consequence of the boat form rather than the chair form of cyclohexane being involved. The two carbon atoms in the 1,4-positions of cyclohexane can only be sufficiently close to one another to enable them to be bridged by a methylene or ethylene group if the ring takes up a boat rather than a chair form. Whilst the ethylene bridge in **bicyclo[2.2.2]octane** is effectively free of strain, **bicyclo[2.2.1]-heptane** has a strain energy of 38 kJ/mol, which is due to the bridgehead carbon atoms of the boat being forced inwards towards one another. These two bridged-ring compounds have the following structures (Figs. 106a and 106b).

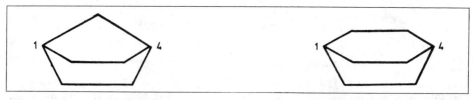

Fig. 106a. Bicyclo[2.2.1]-heptane   Fig. 106b. Bicyclo[2.2.2]-octane

It may be seen from these structures that, overall, the first compound contains two 5-membered and one 6-membered ring, whereas the second contains three strain-free rings, all in boat form.

In such systems with endomethylene or endoethylene bridges it is not possible to have a double bond involving the bridgehead carbon atoms (*Bredt*'s rule, 1924).[2] For example, because of the high ring strain that it would possess, bicyclo[2.2.2]oct-1-ene is incapable of existence. Orbital theory provides an explanation for the non-existence of such compounds since it requires that the four substituents at the ends of a C═C double bond must be effectively coplanar (see p. 63 *ff.*). Consideration of the strain-free model of bicyclo[2.2.2]octane shows clearly that the bridgehead atoms cannot take up planar arrangements and hence the formation of planar double bonds is not possible.

Substitution reactions at bridgehead carbon atoms[3] are sometimes difficult For example, *nucleophilic* replacement of a halogen atom by a hydroxyl group in an S_N1 reaction becomes increasingly difficult the smaller a bicyclic molecule is. Thus the chlorine atom of 1-chloro-7,7-dimethylnorbornane does *not* undergo substitution under conditions which are ideal for S_N1 reactions (treatment with an aqueous ethanolic solution of silver nitrate). The reason for this is obviously the pyramidal structure of the bridgehead atom, which would inevitably be retained in a transition state and this prevents the formation of a planar carbenium ion.

---

[1] See: Handbook for Chemical Society Authors, Ed. *R. S. Cahn*, p. 74 *ff.* (The Chemical Society, London 1960); Definitive Rules for Nomenclature of Organic Chemistry, J. Am. Chem. Soc. *82*, 5557 (1960).
[2] See *G. Köbrich*, Bredt Compounds and Bredt's rule, Angew. Chem. Int. Ed. Engl. *12*, 464 (1973).
[3] See *R. C. Fort jun.* and *P. von R. Schleyer*, Adv. Alicyclic Chem. *1*, 284 (1966); *R. C. Bingham* et al., Recent Developments in the Chemistry of Adamantane and Related Polycyclic Hydrocarbons, Fortsch.Chem.Forsch. *18*, 52 (1971); *R. C. Fort jun.* in Carbonium Ions *4*, 1783, *G. A. Olah* et al. (Ed.) (Wiley, New York 1973).

In S_N2-reactions the entering substituent attacks from the rear and this is obviously completely ruled out at a bridgehead carbon atom. However, *radical or electrophilic substitution reactions* at a bridgehead carbon atom are possible since the pyramidal arrangement of the intermediate concerned, be it radical or carbanionic, exerts little influence on the course of the reaction. Examples of such reactions are the gas-phase radical nitration of norbornane (*Hass*) and the S_E reaction between 4-bornyllithium and carbon dioxide:

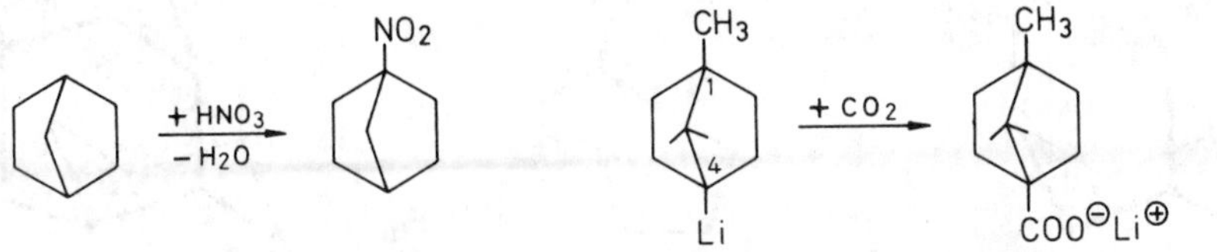

| 8,9,10-Trinorbornane Norbornane | 1-Nitro-8,9,10-trinorbornane | 4-Bornyl lithium | Lithium salt of bornane-4-carboxylic acid |

Bredt's rule does not apply to bicyclic ring-systems which have greater numbers of ring atoms, in which ring strain, the cause of the non-formation of bridgehead double bonds, is lower. In the border region of ring sizes where stable '*anti-Bredt*' compounds having bridgehead double bonds might become feasible, (S)-(−)-bicyclo[3.3.1]non-1(2)-ene has been isolated as a chiral compound. Its structure has a clear resemblance to that of *trans*-cyclooctene, which was resolved into its enantiomeric forms by means of a chiral platinum(II) complex (*Cope*, 1962).

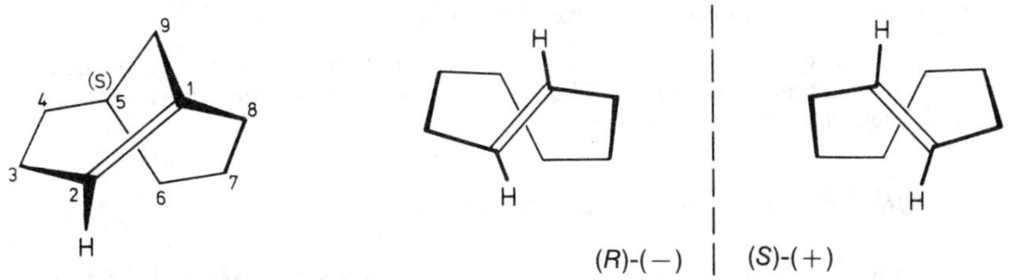

(S)-(−)-Bicyclo[3.3.1]-non-1(2)-ene

(R)-(−) | (S)-(+)
Cyclooctene

3.5.4 Ring-Systems Related to Diamond

It is appropriate to mention here some polycyclic ring-systems which are tetracyclic and have the same type of structure as urotropine[1] (see p. 202). **Adamantane**, tricyclo[3.3.1.1^{3,7}]decane, $C_{10}H_{16}$, m.p. 268°C (541 K), may be taken as the parent member of this series of hydrocarbons. It was first isolated in 1933 from mineral oil, by *Landa* and *Macháček*, and later was made synthetically by *Prelog* (1941).

If one thinks of a tetrahedral arrangement of ten carbon atoms as the smallest unit of the diamond lattice, then its carbon framework is the same as that of adamantane. Ring-systems of this sort have sometimes been called 'diamondoid' or 'adamantoid' compounds.

It may be seen from Fig. 107a that the four 6-membered rings of adamantane are strain-free and arranged identically, in *chair* forms. The high symmetry of the adamantane structure results in the molecules being almost spherical in form and determines both the physical and chemical properties of these compounds. For example, they have relatively

[1] R. C. Fort jun. and P. von R. Schleyer, Adamantane: Consequences of the Diamondoid Structure, Chem. Rev. *64*, 277 (1964); J. C. Angus et al., Low-Pressure, Metastable Growth of Diamond and 'Diamondlike' Phases, Science *241*, 913 (1988).

high melting points and are volatile. Chemically they are not very reactive and are very stable. The introduction of a double bond into an adamantane ring system is sterically impossible; *Bredt*'s rule is applicable here to each one of the carbon atoms.

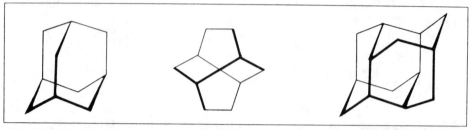

Fig. 107a. Adamantane   Fig. 107b. Twistane   Fig. 107c. Congressane (Diamantane)

Twistane, $C_{10}H_{16}$, m.p. 163–164.8°C (436–437 K) is an isomer of adamantane which is also constructed from four cyclohexane rings, but in this case they are in twisted boat or *twist* forms instead of chair forms (Fig. 107b). Like adamantane, twistane is very volatile.

Another adamantane-like hydrocarbon is **congressane** or **diamantane**, m.p. 236–237°C (509–510 K), which was synthesised by *Schleyer* in 1965. Its carbon skeleton represents a larger unit out of the diamond lattice (Fig. 107c).

Empirical force field or molecular mechanics (MM) calculations have proved of value in planning the syntheses of this type of molecule.[1] They provide information about which of a large number of isomers is the most stable (see p. 422).

3.5.5 Small Bi- and Poly-cyclic Systems

Among alicyclic compounds there are a number of ring-systems which had at one time been thought to be non-existent, but which were later prepared, in particular making use of *photochemical reactions*. Examples of these are the three non-planar valence isomers of benzene,[2] called 'Dewar benzene', *benzvalene* and *prismane*:

'Dewar benzene' (*J. Dewar*, 1867), Bicyclo-[2.2.0]hexa-2,5-diene
Synthesis: *E. E. van Tamelen* (1963)

Benzvalene (*E. Hückel*, 1937), Tricyclo-[3.1.0.0.^{2,6}]hex-3-ene
Synthesis: *T. J. Katz* (1971)

Prismane (*A. Ladenburg*, 1869), Tetracyclo-[2.2.0.0^{2,6}.0^{3,5}]hexane
Synthesis: *T. J. Katz* (1973)

---

[1] See *S. A. Godleski* et al., J. Org. Chem. *41*, 2596 (1976); *E. Osawa* et al., Molecular Mechanics Calculations in Organic Chemistry, Angew. Chem. Int. Ed. Engl. *22*, 1 (1983).
[2] See *E. E. van Tamelen*, Valence Isomers of Aromatic Systems, Ang.Chem. Int. Ed. Engl. *4*, 738 (1965); *A. H. Schmidt*, Valenzisomere des Benzols (Valence Isomers of Benzene), Chemie in unserer Zeit *11*, 118 (1977).

The *Dewar* benzene formula was originally postulated as an alternative planar structure for benzene. It is completely misnamed, because *Dewar* did not present it as an alternative structure for benzene, but rather listed it as one of a number of alternative structures for C_6H_6 and rejected it in favour of *Kekulé*'s formula.[1] The alternative name *tectadiene* has been suggested (Lat. *tectum* = roof), since the synthesised compound has a bent form resembling a roof.

Derivatives of all three of the benzene isomers have been made. They have for the most part been made by photochemical methods and their identification has depended on the use of gas chromatography and NMR spectroscopy.

The syntheses of these three compounds have been achieved as follows:

3.5.5.1 *Dewar* benzene. *Phthalic acid* was reduced by means of sodium amalgam to *dihydrophthalic acid*. This was converted into its anhydride by treating it with acetic anhydride and irradiation of the resultant anhydride isomerised it into the anhydride of bicyclo[2.2.0]hex-5-ene-2,3-dicarboxylic acid. This was hydrolysed to the corresponding acid, which was decarboxylated by electrolysis to give *Dewar* benzene:

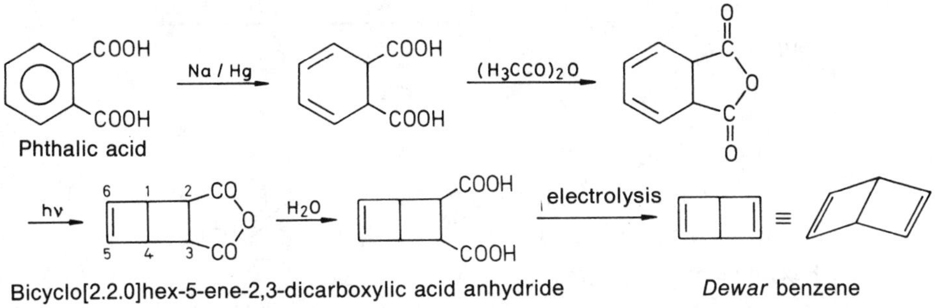

Bicyclo[2.2.0]hex-5-ene-2,3-dicarboxylic acid anhydride   *Dewar* benzene

Dewar benzene has a half-life of 37 hours at room temperature in tetrachloroethylene.

3.5.5.2 Benzvalene was obtained by the reaction of *lithium cyclopentadienide* with a *carbene* which was prepared *in situ* from dichloromethane and methyllithium:

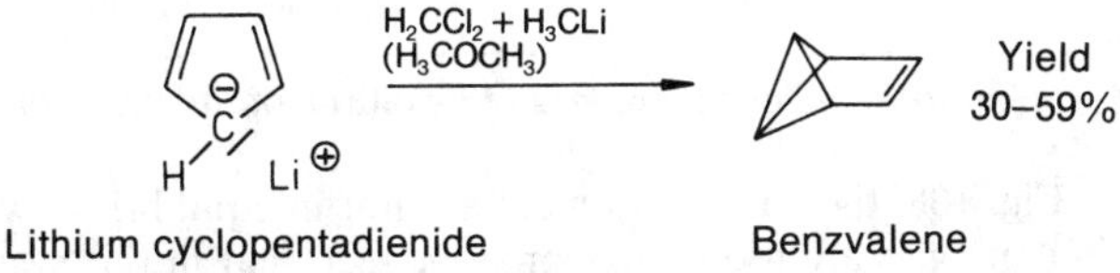

Lithium cyclopentadienide   Benzvalene

Pure benzvalene is highly explosive. It has a putrid odour. At room temperature in isohexane it has a half-life of about 10 days.

3.5.5.3 Prismane was made from benzvalene by means of a cycloaddition reaction with the highly reactive compound *4-phenyl-3H-1,2,4-triazoline-3,5-dione* (PTAD). Hydrolysis and decarboxylation converts the adduct into an azo-compound which, on irradiation, gives a 6% yield of prismane:

---

[1] See *J. Dewar*, Proc. Roy. Soc. Edinburgh, *84* (1866–67).

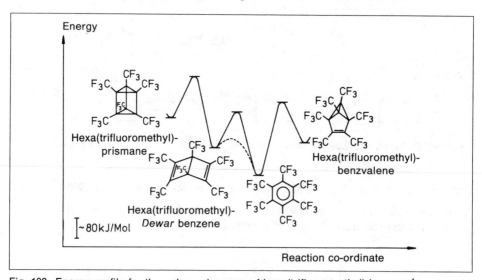

4-Phenyl-3*H*-1,2,4-
triazoline-3,5-dione

Prismane is a highly explosive, colourless liquid, with a half-life in toluene at 90° (363 K) of 11 hours.

When *benzene* is photolysed, the main product is *fulvene*; *benzvalene* and *prismane* are also formed. When hexa(trifluoromethyl)-benzene was photolysed in perfluoropentane at room temperature all three valence isomers were formed in a total yield of 82%.

3.5.5.4 These rearrangements are also examples of **valence isomerisation reactions**.

Fig. 108. Energy profile for the valence isomers of hexa(trifluoromethyl)-benzene[1]

As indicated in Fig. 108, the valence isomers are not in equilibrium with one another since the activation energies are too high. It may be seen that in this system the benzene derivative is the most stable compound thermodynamically, but that all the others show some kinetic stability. The *lifetime* (*half-life*) of a chemical compound under specified external conditions depends upon its *kinetic stability*. The kinetic stability of *Dewar* benzene would be significantly lower if the reaction co-ordinate drawn in Fig. 108 as a dotted line were operative.

The activation energies in Fig. 108 are surprisingly high, and to account for this, not only the perfluoroalkyl, or R_F, effect due to the CF_3-groups,[2] but also a kinetic stabilisation is involved, which arises from the principle of *retention of orbital symmetry*.

[1] See *D. M. Lemal* et al., J. Am. Chem. Soc. *94*, 6562 (1972).
[2] See *A. Greenberg* et al., On the Nature of the Perfluoroalkyl (R_F) Effect, Tetrahedron, *36*, 1161 (1980).

If the principles governing electrocyclic reactions which were outlined on p. 104 *ff.* are applied to *Dewar* benzene, the following considerations apply. To convert *Dewar* benzene into benzene a *disrotatory ring-opening* is necessary to achieve the required (Z,Z,Z)-configuration:

Dewar benzene   Benzene (Z,Z,Z)   (E,Z,Z)-Cyclo-
hexatriene

Under thermal conditions the preferred mode of ring-opening of a cyclobutene ring is *conrotatory*, but this would lead to the formation of a highly strained (E,Z,Z)-cyclo-hexatriene, which is indeed too strained to exist.

Thermal ring-opening of *Dewar* benzene can thus not take place by a synchronous mechanism, but must proceed *via* either a radical or an ionic transition state, either of which would require a large activation energy. On the other hand, the *Woodward–Hoffmann* rules indicate that a photochemical disrotatory electrocyclic ring-open reaction is energetically favourable, and *Dewar* benzene is readily photolysed to benzene. The importance of photochemical reactions in this field is evident.

The conversion of *norbornadiene* (or, more precisely, 8,9,10-trinorbornadiene) into *quadricyclane* (tetracyclo[3.2.0.0^{2,7}.0^{4,6}]heptane) is another example of a photochemical electrocyclic reaction:

$\Delta H = +109\ kJ/Mol$ Cyclopropenylidene

The reverse reaction can be brought about with the aid of catalysts, and in the process the energy taken up in the endothermic photo-reactions is released again. Norbornadiene is thus a compound which might be made use of for storing the sun's energy.

Flash pyrolysis of a cyclopropene derivative of quadricyclane generates cyclopropenylidene (see p. 390).

3.5.5.5 Propellanes are compounds in which a bond between two carbon atoms forms the axis of three linked bridges; their structure can be compared to that of a propeller (although one without pitch). For a long time there was debate as to whether the parent compound [1.1.1]propellane (tricyclo[1.1.1.0^{1,3}]pentane) would be stable or not. Its synthesis by *Wiberg* (1982) showed that it is. It is now accessible on a preparative scale by a two-step synthesis:[1]

Methallyl dichloride   [1.1.1]Propellane

---

[1] See *G. Szeimies* et al., Brückenkopfgekoppelte Bicyclo[1.1.1]pentane: Synthesis und Struktur (Bridgehead-coupled bicyclo[1.1.1]-pentanes: Syntheses and Structure), Chem. Ber. *121*, 1785 (1988).

Reaction of methallyl dichloride [3-chloro-2-(chloromethyl)propene] with dibromocarbene, generated from bromoform and 50% aqueous sodium hydroxide in the presence of a phase-transfer catalyst, gives 1,1-dibromo-2,2-bis(chloro-methyl)cyclopropane in 45% yield. Treatment of this with methyllithium brings about an intramolecular *Wurtz* reaction and provides an ethereal solution of [1.1.1]propellane. The four bonds from carbon atoms 1 and 3 are all directed into the tricyclic ring system: C≣. They have been called *inverted carbon atoms*.

An X-ray crystal structure determination of a [3.1.1]propellane derivative shows that the bonds linking the two inverted carbon atoms are only slightly longer than normal C—C bonds between sp^3-hybridised carbon atoms. MO calculations suggest that [1.1.1]propellane has a *non-bonding HOMO*, and that the C^1—C^3 bonds are three-centre two-electron orbitals[1]. These bonds can take part in addition reactions under conditions appropriate for radical chain reactions. In this way it is possible to make bicyclo[1.1.1]pentane-1,3-dicarboxylic acid, which is useful for preparative purposes, from [1.1.1]propellane and diacetyl:

| [1.1.1] Propellane | 1,3-Diacetylbicyclo[1.1.1]-pentane | Bicyclo[1.1.1]pentane-dicarboxylic acid |

3.5.5.6 It is possible to obtain the **cubane system**,[2] which is of steric interest, starting from 2-*bromocyclopentadienone*. This dimerises in a *Diels–Alder* reaction, and, through a series of four steps, this dimer is converted finally into *dimethyl cubane-1,4-dicarboxylate*, m.p. 161–162°C (434–435 K).

Unsubstituted **cubane**, C_8H_8, can be obtained similarly. It is formed as an extremely volatile hydrocarbon which melts at 130–131°C (403–404 K) and begins to decompose at about 200°C (470 K). If 1,1-diiodocubane is treated with t-butyl lithium, cubene, C_8H_6, is formed and can be trapped as a *Diels–Alder* adduct.

3.5.5.7 It can be shown by NMR spectroscopy that the 3-membered ring in **homotropylidene**, bicyclo[5.1.0]octa-2,5-diene, is not fixed, but that there is a temperature-dependent *equilibrium* between *two identical structures*. This sort of *valence isomerism* has been described by *Doering* as a *fluxional system*.

Homotropylidene

[1] See *J. E. Jackson* et al., J. Am. Chem. Soc. *106*, 591 (1984).
[2] See *P. E. Eaton*, Cubanes: Starting Materials for the Chemistry of the 1990s and the New Century, Angew. Chem. Int. Ed. Engl. *31*, 1421 (1992).

The rate of interchange between the two structures depends very much upon the temperature. At $-50°C$ (223 K) it comes to a standstill.

3.5.5.8 An ideal example of a fluxional molecule is the hydrocarbon **bullvalene**, $C_{10}H_{10} = (CH)_{2n}$, $n = 5$. It is prepared from cyclo-octatetraene, which, when heated at $100°C$ (375 K) is converted into a *dimer*, m.p. $76°C$ (349 K). UV-irradiation of the dimer gives an 80% yield of bullvalene, m.p. $96°C$ (369 K).

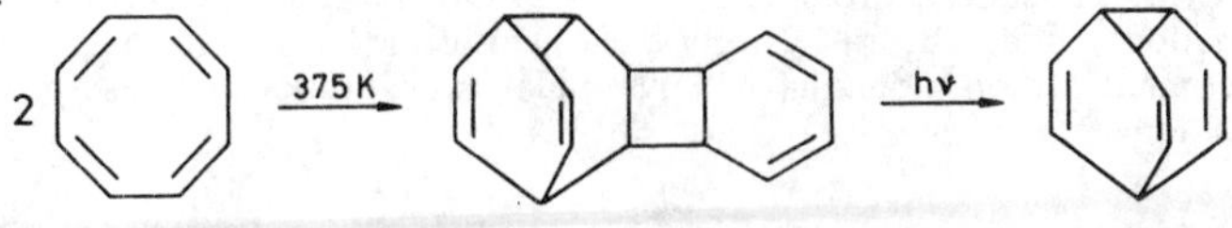

In bullvalene also, the position of the 3-membered ring is not fixed. Its bonds can open up, whilst at the same time a new cyclopropane ring is formed elsewhere in the molecule, creating another bullvalene structure. Each carbon atom and each hydrogen atom can exist in four different kinds of chemical environment, namely two vinyl positions, an allyl position and a position in a 3-membered ring. Altogether there are $\dfrac{10!}{3} = 1,209,600$ possible identical structures, all of which can change into one another. This can be observed by means of the temperature-dependent NMR spectra. Some of the equilibria of this degenerate valence isomerism are given in Fig. 109, with the fluxional bonds indicated. A much simpler system involving fluxional bonds has already been encountered in the case of cyclooctatetraene (p. 407).

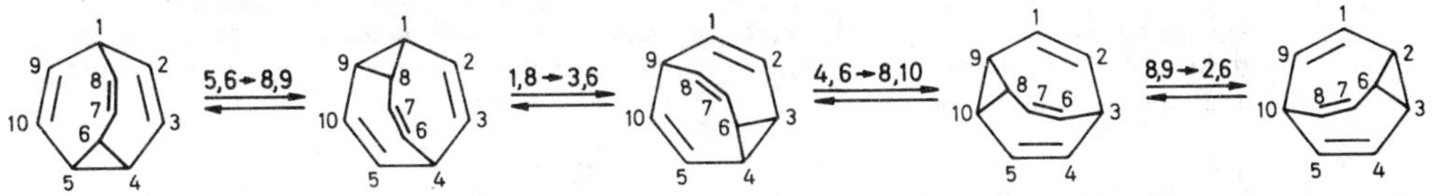

Fig. 109. Valence isomerism of bullvalene

3.5.5.9 Among the structurally stable hydrocarbons encompassed by the general formula $(CH)_{2n}$, **dodecahedrane**. $C_{20}H_{20}$, is of great significance because of its high symmetry (I_h). Following tetrahedrane (see p. 656) and cubane, it is the third of the five platonic solids of which a hydrocarbon equivalent has been prepared.[1]

This *spherically annelated polyquinane* can be made on a preparative scale from the isomeric hydrocarbon [1.1.1.1]pagodane.[2]. The isomerisation to dodecahedrane proceeds as shown in the following formulae, in which the two nearest of the fifteen cyclopentane rings are shaded in grey from the starting material through to the product.

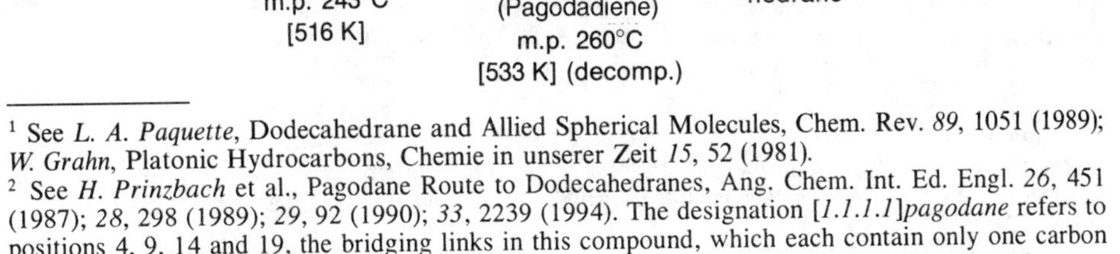

| $\Delta H°f$ [kJ/Mol] | 269 | 259 | 167 | 92 |

| Isodrin | [1.1.1.1]Pagodane m.p. 243°C [516 K] | Bissecododeca-hedradiene (Pagodadiene) m.p. 260°C [533 K] (decomp.) | Bissecododeca-hedrane | Dodecahedrane |

[1] See *L. A. Paquette*, Dodecahedrane and Allied Spherical Molecules, Chem. Rev. *89*, 1051 (1989); *W. Grahn*, Platonic Hydrocarbons, Chemie in unserer Zeit *15*, 52 (1981).
[2] See *H. Prinzbach* et al., Pagodane Route to Dodecahedranes, Ang. Chem. Int. Ed. Engl. *26*, 451 (1987); *28*, 298 (1989); *29*, 92 (1990); *33*, 2239 (1994). The designation *[1.1.1.1]pagodane* refers to positions 4, 9, 14 and 19, the bridging links in this compound, which each contain only one carbon atom. Its preparation from the insecticide *isodrin* involved 45 functional and structural transformations yet was carried out in an overall yield of 24%. J. Am. Chem. Soc. *109*, 4626 (1987).

From the MM-calculated enthalpies of formation ($\Delta H°f$) it appears that of the above series of compounds *dodecahedrane* is the most stable *isomer*, and it is obtained in 8% yield by gas-phase isomerisation of [1.1.1.1]pagodane. The latter compound can readily be functionalised at the methylene groups 4, 9, 14 and 19. When the above series of reactions is applied to such a functionalised compound the substituents end up at the sites marked with arrows in the resultant dodecahedrane.

A seven-step synthesis starting from dimethyl 14,19-dioxopagodane-*syn-syn*-4,9-dicarboxylate leads to the formation of dodecahedra-1,2;16,17-diene. This diene has two strongly distorted double bonds, whose terminal carbon atoms have a pyramidal configuration. Because of this it is very sensitive to oxidation.

Dodecahedra-1,2;16,17-diene   Dication (I)

The distortion of the two double bonds of dodecahedradiene is characterised by the angle Φ, which specifies how far the substituents deviate from the planar form which is normal for C=C double bonds.

When [1.1.1.1]pagodane is oxidised in superacid medium the non-classical dication (I) is formed which has been described as the first example of a σ-homoaromatic compound.

3.5.5.10 The hexabenzo-derivative of the highly symmetric *centrohexaquinane* (Td), called **centrohexaindane** (hexabenzohexacyclo[5.5.2.2$^{4.10}$.1$^{1.7}$.0$^{4.17}$.0$^{10.17}$]heptadecane) is made surprisingly easily from tetrabromofenestrindane (I).[1]

Centrohexaquinane   Tetrabromo-fenestrindane (II)   Centrohexaindane

The precursor of (II), fenestrindane, is made on the gram scale by an eleven-step convergent synthesis with dimethyl phthalate, ethyl acetate, acetone, benzaldehyde and perchlorobutadiene as the starting materials.[2] Its bromination, to give (II), and the subsequent ring-closures, accompanied by folding-in of the benzene rings, in which all four new C—C bonds are formed at the same time, take place with an overall yield of 50%. Centrohexaindane is obtained as colourless needles which do not melt below 420°C (693K). Its spectra, which indicate its high degree of symmetry, do not give any hint of any interaction between the six aromatic π-electron systems.

---

[1] See *D. Kuck* et al., Centrohexaindane, the First Hydrocarbon with Topologically Non-planar Molecular Structure, Angew. Chem. Int. Ed. Engl. *27*, 1192 (1988); J. Chem. Soc., Perkin Trans. 1, *1995*, 721.

[2] For the parent compound, *fenestrane*, see *B. R. Venepalli* et al., Fenestranes and the Flattening of the Tetrahedral Carbon Atom, Chem. Rev. *87*, 339 (1987).

4 Carbohydrates

Included in this group of compounds are the various sugars which are found in animals and in plants, which, together with fats and proteins, provide the major constituents of food.[1] The term 'carbohydrates' originally described compounds which contained carbon together with hydrogen and oxygen, the latter being present in the same ratio as in water (2:1), and which had the general formula $C_nH_{2n}O_n$ (*Karl Schmidt*, 1844).

This definition is no longer applied strictly since many natural sugars are known in which the hydrogen to oxygen ratio is not 2:1. In addition, there are some sugars which contain nitrogen or sulphur as well as carbon, hydrogen and oxygen. Despite this, the overall term 'carbohydrates' is still used for this large class of compounds, although it conveys no indication concerning their constitution. Names of individual sugars are characterised by the word-ending '*ose*'.

It is convenient to divide the family of carbohydrates into the three following groups, based on their constitution and on their chemical and physical properties.

1. Monosaccharides (simple sugars)
The most important monosaccharides (Grk. *sakkharon* = sugar) are the *pentoses*, $C_5H_{10}O_5$, *e.g.* ribose, and the *hexoses*, $C_6H_{12}O_6$, which include glucose and fructose.

2. Oligosaccharides
These sugars are made up from two to six monosaccharide units which are joined together by acetal-like links.

They can be subdivided into:
(a) *disaccharides*, *e.g.* sucrose (cane sugar), maltose (malt sugar), lactose (milk sugar);
(b) *trisaccharides*, *e.g.* raffinose;
(c) *tetra-*, *penta-* and *hexa-saccharides*.

3. Polysaccharides
If the linking together of simple sugar units is continued further then polysaccharides are obtained. Their properties are essentially different from those of mono- and oligo-saccharides. The most important of these high molecular weight natural products are starch, glycogen and cellulose.

Examples of all three of these groups (*e.g.* glucose, sucrose, starch) are important starting materials for industrial processes, and are replenishable feedstocks. Because of this, efforts to determine the

---

[1] See *R. D. Guthrie* and *J. Honeyman*, An Introduction to the Chemistry of Carbohydrates, 3rd edn (Clarendon Press, Oxford 1968); *R. J. McIlroy*, Introduction to Carbohydrate Chemistry (Butterworths, London, from 1967); *W. W. Pigman* et al., The Carbohydrates, 2nd edn (Academic Press, New York 1970); *P. M. Collins*, Carbohydrates (Chapman and Hall, London 1987).

essential structures of carbohydrates (*e.g.* their chirality) have intensified, so that use may be made of them, rather than just breaking them down to simple molecules such as ethanol for use as fuels.

4.1 Monosaccharides[1]

The clarification of the constitution and the steric structure of carbohydrates derived in the first place from the work of *Emil Fischer, Kiliani* and *Tollens*. This showed the simplest sugars to be *oxidation products of polyhydric alcohols*, in which either a primary or a secondary alcohol group had been oxidised to provide, respectively, an aldehyde or a keto group. Thus they may be classified respectively as either **aldoses** or **ketoses**.

The individual aldoses or ketoses with the general formula $C_nH_{2n}O_n$ are derived from the polyhydric alcohols glycol, glycerol, tetritols, pentitols and hexitols. Oxidation of glycol provides the simplest aldose, glycolaldehyde, $HOCH_2$—CHO. Since it has just two carbon atoms directly linked together it is described as **aldobiose**. Similarly the next highest sugar, glyceraldehyde, HO—CH_2—CHOH—CHO, is called an **aldotriose** ($n = 3$).

The simplest ketose, which is formed by removal of hydrogen from glycerol, is dihydroxyacetone, $HOCH_2$—CO—CH_2OH, a **ketotriose** ($n = 3$).

4.1.1 Configuration of Sugars

Glyceraldehyde (see p. 272 *ff.*) is used as the reference compound in assigning simple sugars to the D- or L-series. The *absolute configuration* of monosaccharides is then derived from that of the asymmetric carbon atom which is furthest away in the chain from the carbonyl group, *i.e.* that has the highest number in the numbering scheme (CHO = 1) (*Wohl* and *Freudenberg*).

$$
\begin{array}{cc}
\overset{\displaystyle C\!\!\nwarrow^{H}_{\searrow O}}{} & \overset{\displaystyle C\!\!\nwarrow^{H}_{\searrow O}}{} \\
H-C-OH & HO-C-H \\
CH_2OH & CH_2OH \\
\text{D(+)-Glyceraldehyde} & \text{L(−)-Glyceraldehyde}
\end{array}
$$

Aldotetroses have two asymmetric carbon atoms and can occur in four stereoisomeric forms ($2^2 = 4$). If the 3-dimensional model of a sugar molecule is projected onto the plane of the paper using the same convention as described for glyceraldehyde (see p. 273), then the *enantiomers* of *erythrose* and *threose* have the following *Fischer* projection formulae:

$$
\begin{array}{cccc}
(1) & C\!\!\nwarrow^{H}_{\searrow O} & C\!\!\nwarrow^{H}_{\searrow O} & C\!\!\nwarrow^{H}_{\searrow O} & C\!\!\nwarrow^{H}_{\searrow O} \\
(2) & H-C-OH & HO-C-H & HO-C-H & H-C-OH \\
(3) & H-C-OH & HO-C-H & H-C-OH & HO-C-H \\
(4) & CH_2OH & CH_2OH & CH_2OH & CH_2OH \\
 & \text{D(−)-Erythrose} & \text{L(+)-Erythrose} & \text{D(−)-Threose} & \text{L(+)-Threose}
\end{array}
$$

[1] See *J. Lehmann*, Chemie der Kohlenhydrate, Monosaccharide und ihre Derivate (Chemistry of Carbohydrates, Monosaccharides and their Derivatives) (Thieme Verlag, Stuttgart 1976).

(+)-Erythrose and (−)-erythrose are a pair of enantiomers, as are (+)- and (−)-threose, and in each case the (+)- and (−)-forms rotate the plane of linearly polarised light to the same extent but in opposite directions. D(−)-Erythrose and D(−)-threose are *diastereomers*, as are L(+)erythrose and L(+)-threose.

Compounds having two asymmetric carbon atoms are often of use in the study of reaction mechanisms, and the prefixes *erythro-* and *threo* are frequently used to denote *diastereomers* which have at least two identical or similar substituents on the two vicinal (neighbouring) optically active carbon atoms. This designation has been applied to all systems of the following type:[1]

$$
\begin{array}{ccc}
\text{R} & & \text{R} \\
| & & | \\
\text{a}-\text{C}-\text{b} & & \text{b}-\text{C}-\text{a} \\
| & \text{or} & | \\
\text{a}-\text{C}-\text{c} & & \text{a}-\text{C}-\text{c} \\
| & & | \\
\text{R}' & & \text{R}'
\end{array}
$$

erythro Form   *threo* Form

If the two identical ligands lie on the same side of the *Fischer* projection formula, as the HO-groups do in erythrose, then this isomer is described as the *erythro* form. If, on the other hand, the two identical ligands are on opposite sides, as in threose, then it is called the *threo* form. The symbols (⇄) and (⇅) have been used to indicate, respectively, *erythro* and *threo* forms. When a *Newman* projection is used, the substituents b and c are antiperiplanar in the *erythro* form and synclinal (see p. 58) in the *threo* form (see p. 526).

The naturally occurring monosaccharides are almost all *pentoses* or *hexoses* with unbranched chains of carbon atoms. The *Fischer* projection formulae of the most important naturally occurring **aldopentoses** and **aldohexoses** and of the **ketohexose** fructose are as follows:

L(+)-Arabinose   D(−)-Ribose   D(+)-Xylose

D(+)-Glucose   L(−)-Glucose   D(+)-Mannose   L(−)-Mannose

---

[1] This system of assignment has not proved to be unequivocal and it is recommended that definition of a configuration should rather be on the basis of the *RS* system. See D. *Seebach* and V. *Prelog*, The Unambiguous Specification of the Steric Course of Asymmetric Syntheses, Angew. Chem. Int. Ed. Engl. *21*, 654 (1982).

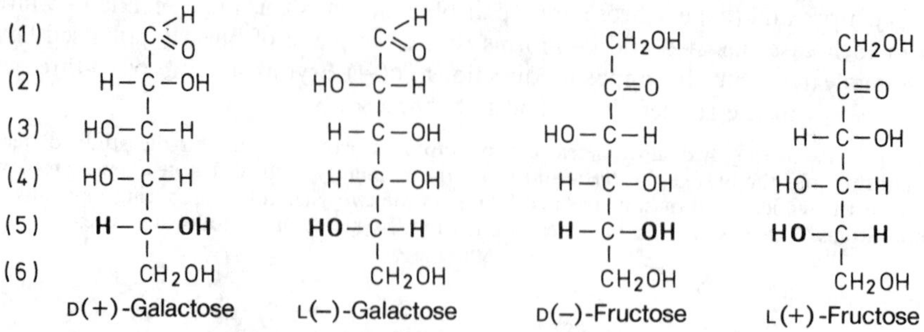

(1)	C≤⁰ / H	C≤⁰ / H	CH₂OH	CH₂OH
(2)	H—C—OH	HO—C—H	C=O	C=O
(3)	HO—C—H	H—C—OH	HO—C—H	H—C—OH
(4)	HO—C—H	H—C—OH	H—C—OH	HO—C—H
(5)	H—C—OH	HO—C—H	H—C—OH	HO—C—H
(6)	CH₂OH	CH₂OH	CH₂OH	CH₂OH
	D(+)-Galactose	L(−)-Galactose	D(−)-Fructose	L(+)-Fructose

It may be seen from these formulae how the arrangement of the substituents on carbon atom 4 in the pentoses and on carbon atom 5 in the hexoses shows whether the sugar belongs to the D- or the L-series. The arrangement of substituents on the other asymmetric carbon atoms plays no part in this assignment. If the arrangement of the substituents on any one of the other asymmetric carbon atoms is inverted then a different stereoisomeric sugar is produced.

In general, a compound having n asymmetric carbon atoms can exist as 2^n different optical isomers, so long as there is no plane of symmetry within the molecule. For example, there are thus $2^4 = 16$ *stereoisomers* ($= 8$ enantiomeric pairs) of the *aldohexoses* and $2^3 = 8$ *stereoisomeric ketohexoses* ($= 4$ enantiomeric pairs).

Monosaccharides which have *identical configuration except at carbon atom 2* are known as **epimers**. They are diastereomers, differing in configuration from one another at only *one* of their chiral centres. It is characteristic of epimeric sugars, *e.g.* D-glucose and D-mannose that they react with phenylhydrazine to give the *same osazone* (see p. 429 *ff*.).

Table 14 includes a *family tree of the D-aldoses*, showing in a simplified way how they are all related to D-glyceraldehyde.

Similarly Table 15 shows the relationship of the various other D-ketoses besides D-fructose, which has already been mentioned. They are all derived from the optically inactive dihydroxyacetone (ketotriose). Allocation of ketotetroses, ketopentoses and ketohexoses to either the D- or L-series is also based on the *absolute configuration* of glyceraldehyde. In systematic nomenclature their names have the suffix '-ulose'.

The corresponding aldoses and ketoses in the L-series are mirror-image isomers of the monosaccharides listed in Tables 14 and 15.

Compared to the hexoses and pentoses, *tetroses* play only a minor role; they are only obtained synthetically.

The formulae above indicate that the simple sugars can be thought of as *polyhydroxyaldehydes* or *polyhydroxyketones*. In fact, as will be seen later, they exist in the form of *cyclic hemiacetals*, which sometimes can be in α- or β-forms. These are in equilibrium with one another, interchange taking place *via* the *open-chain aldehyde* or *ketone forms*. Therefore solutions of monosaccharides can react as either hydroxyaldehydes or hydroxyketones and their functional groups can be detected under the appropriate conditions by the usual reagents. In order to understand them more easily, the following reactions are therefore treated on the basis of the open-chain forms.

4.1.2 Reactions of Monosaccharides

A noteworthy feature is the ease with which the hydroxy groups which are on the carbon atom next to the oxo-group can be oxidised. As a result of this, ketoses, unlike simple

Table 14

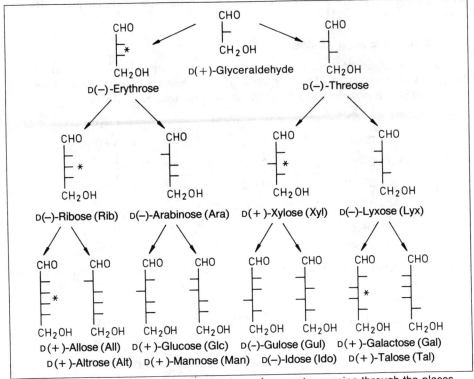

For some of the compounds in Table 14 a plane of symmetry passing through the places marked * is introduced when those sugars are either reduced to the corresponding sugar alcohol or oxidised to a dicarboxylic acid.

Table 15

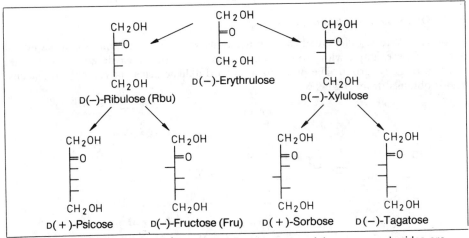

The abbreviation in brackets alongside the names of some of the monosaccharides are used in formulating oligosaccharides and polysaccharides.

ketones, can act as reducing agents. This means that the difference in chemical properties between aldoses and ketoses is less marked than in the case of simple aldehydes and ketones.

4.1.2.1 Reduction to Sugar Alcohols

Monosaccharides are reduced to polyhydric alcohols either by sodium borohydride or by catalytic hydrogenation. Thus D-glucose and L-gulose are converted into D-sorbitol, D-mannose into D-mannitol, and D-galactose into dulcitol (see p. 315 *f.*).

D-Glucose   →   D-Sorbitol   ←   L-Gulose

When the carbonyl group in a ketose is reduced, a *new* asymmetric centre is created, so that a pair of *diasteromers* results. For example, D-sorbitol and D-mannitol are obtained from D-fructose.

D-Mannitol   +2 H ⇌   D-Fructose   +2 H →   D-Sorbitol

4.1.2.2 Oxidation to Polyhydroxyacids

Aldoses and ketoses differ greatly in their reactions with oxidising agents. *Gentle oxidation of aldoses*, for example with bromine-water, affects only the aldehyde group, converting it into a carboxyl group. The products are called **aldonic acids**; for example, D-*gluconic acid* is obtained from D-glucose.

D-Glucose   +[O] →   D-Gluconic acid (GlcA)   −H$_2$O →   D-Gluconic acid-γ-lactone

Like other γ- and δ-hydroxyacids, aldonic acids readily lose water to give γ- or δ-lactones. γ-Lactones are formed preferentially.

When *aldoses* are *oxidised more strongly*, for example with conc. nitric acid, then the primary alcohol group as well as the aldehyde group are transformed into carboxyl groups. The products are polyhydroxydicarboxylic acids known as **aldaric acids**; for example, D-glucose is oxidised to D-glucaric acid (also called D-saccharic acid), D-mannose to D-mannaric acid (or D-mannosaccharic acid), and D-galactose to galactaric acid (or mucic acid). The latter acid has a plane of symmetry through the molecule and hence is a *meso* form; its salts are known as *mucates*.

Four of the sixteen aldohexoses, those marked with an asterisk in Table 14, like D-galactose give rise to *meso*-dicarboxylic acids.

COOH	COOH	COOH	C⟨H⟩O
H – C – OH	HO – C – H	H – C – OH	H – C – OH
HO – C – H	HO – C – H	HO – C – H	HO – C – H
H – C – OH	H – C – OH	HO – C – H	H – C – OH
H – C – OH	H – C – OH	H – C – OH	H – C – OH
COOH	COOH	COOH	COOH
D-Saccharic acid	D-Mannaric acid	Mucic acid (*meso* form)	D-Glucuronic acid (GlcU)

All the sugar acids readily form *monolactones* (lactonic acids) and sometimes also *dilactones*.

The **uronic acids**, which are obtained by oxidising the terminal primary alcohol group of simple sugars to carboxyl groups, are of physiological significance. For example, D-glucuronic acid is obtained from D-glucose.

In organisms, D-glucuronic acid acts as a detoxifying agent towards invasive substances such as, for example, phenols. It serves this purpose in blood and urine, where it is present as part of a glucuronide.

Strong oxidation of ketoses brings about cleavage of the carbon chain, giving rise to carboxylic acids with smaller numbers of carbon atoms, for example oxalic acid.

4.1.2.3 Osazone formation

The *reaction of monosaccharides*, both aldoses and ketoses, *with phenylhydrazine* has been one of the most important reactions in sugar chemistry (*E. Fischer*, 1884). In the first place the carbonyl group undergoes a normal reaction with 1 molecule of phenylhydrazine to give the corresponding *phenylhydrazone*, e.g. glucose gives glucose phenylhydrazone. In weakly acidic solution (acetic acid) reaction proceeds further if excess phenylhydrazine is present, and yellow 'osazones', which are almost insoluble in water, separate out, the other products being aniline and ammonia.

The second step was explained by *E. Fischer* as follows. An alcohol group on a carbon atom next to the carbonyl group is dehydrogenated by a second molecule of phenylhydrazine, providing as an intermediate an *osonehydrazone*. In removing the hydrogen atoms the phenylhydrazine is cleaved to form aniline, $C_6H_5NH_2$, and ammonia. The newly generated carbonyl group then reacts with a third molecule of phenylhydrazine to form an osazone which is stabilised as a chelate by intramolecular hydrogen bonding (*Fieser*, 1944):[1]

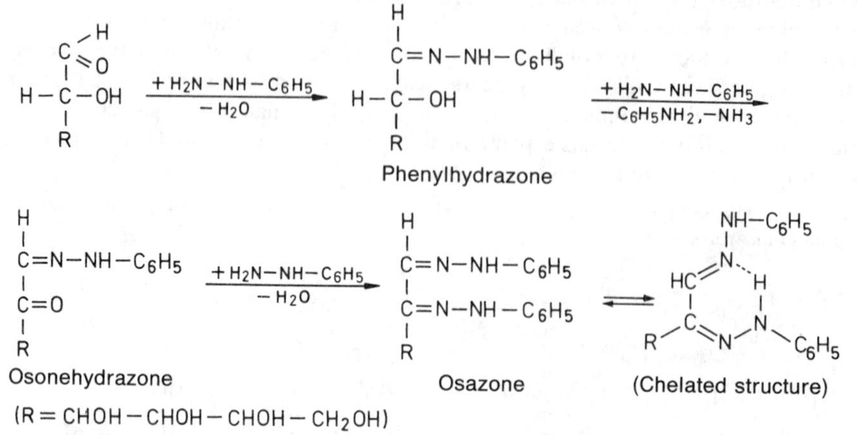

Phenylhydrazone

Osonehydrazone   Osazone   (Chelated structure)

(R = CHOH − CHOH − CHOH − CH$_2$OH)

This reaction scheme does not explain why phenylhydrazine, which is itself a strong reducing agent, is so readily reductively cleaved by the two hydrogen atoms from the CH(OH) group on carbon atom 2 to form aniline and ammonia. It is also not clear why the secondary hydroxyl group on carbon atom 3 of the osazone should be resistant to further attack by phenylhydrazine. Thus far, experiments have failed to clarify the *reaction mechanism*. At present it is thought that the formation of osazones probably involves the mechanism suggested by F. *Weygand*.[2]

Properties of Osazones. Whilst the colourless monophenylhydrazones of sugars often show no great tendency to be crystalline, the yellow osazones usually crystallise readily, providing an excellent means of separating monosaccharides out of dilute or impure solutions. They are also useful for *identifying* simple sugars (melting point, specific rotation). Furthermore the osazones may assist in determining the configuration of particular sugars, for example, the two *epimeric* sugars D-glucose and D-mannose, and also D-fructose, all give the same osazone. Hence all three must have the same configurations at the asymmetric carbon atoms 3, 4 and 5. Such observations have played a vital role in determining the structures of sugars (by E. *Fischer*) (see * in Table 14).

Osazones are hydrolysed by conc. hydrochloric acid to **osones** (α-ketoaldehydes). A better way of achieving this conversion is by heating the osazone with excess benzaldehyde:

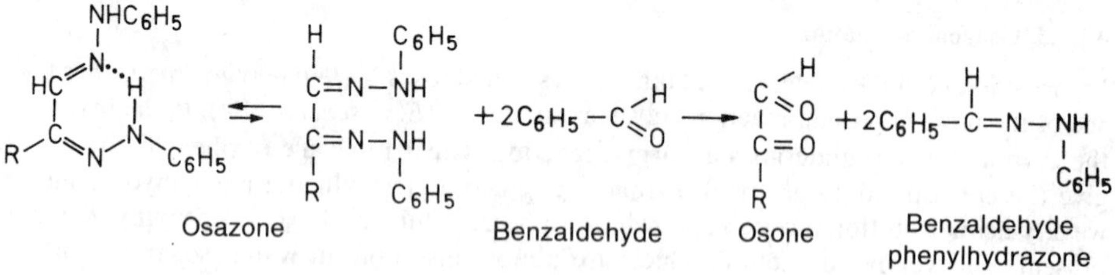

Osazone   Benzaldehyde   Osone   Benzaldehyde
                                                            phenylhydrazone

This interconversion takes place due to the greater stability of benzaldehyde phenylhydrazone.

[1] See L. *Mester* et al., J. Chem. Soc. C *1970*, 2567.
[2] Chem. Ber. *73*, 1284 (1940) and *95*, 17 (1962).

4.1.2.4 *Maillard* Reaction (Non-enzymatic browning)

Reducing sugars react with the amino groups of amino acids, peptides or proteins to give brown-coloured products (*L. C. Maillard*, 1912).[1] This gives rise to, among other things, the materials which provide the characteristic smells of boiled or roast foodstuffs. In the first step of this reaction a derivative of an *N-glycosylamine* (see glycosides, p. 439 *f.*) is formed, which undergoes an *Amadori rearrangement* to a derivative of a 1-amino-1-deoxy-2-ketose.

This transformation is shown using lysine, because this essential amino acid is irreversibly made ineffective by inappropriate treatment of foodstuffs, owing to this reaction which is called non-enzymic glycosylation or 'glycation'. The later steps of the *Maillard* reaction, in which separation of water, ring-closure and *Strecker* synthesis are all involved, are not totally clear. It is worth noting that the *Maillard* reaction is involved in such diverse processes as the cooking of meat, the increasing opacity of lenses of the eye in cataracts, and very widely in the ageing of organisms.

4.1.3 *Interconversion of Monosaccharides*

The formation of osones provides a route for the conversion of aldohexoses into related ketoses. If the *osone is reduced with sodium amalgam* in weakly acidic solution, then the aldehyde group is hydrogenated appreciably faster than the keto group. In this way D-glucose can be converted *via* the osazone and osone into D-fructose:

E. *Fischer* (1890) found another method for stereochemical interconversion of sugars, by heating an *aldonic acid* with pyridine, quinoline or dilute alkali. Under these conditions racemisation takes place only at the asymmetric carbon atom next to the carboxyl group, and an *equilibrium mixture* of the two *epimeric* aldonic acids results. Thus D-gluconic acid is partially converted into D-mannonic acid and *vice versa*:

---

[1] See F. *Ledl*, New Aspects of the *Maillard* Reaction in Foods and in the Human Body, Angew. Chem. Int. Ed. Engl. *29*, 565 (1990).

The epimeric acids are isolated as lactones, from which the related simple sugars are obtained by reduction with *sodium amalgam*. In this way an aldose can be converted into the epimeric sugar, *e.g.* D-glucose → D-gluconic acid → D-mannonic acid → D-mannonolactone → D-mannose (see Epimerisation).

4.1.4 Test Reactions for Monosaccharides

These depend for the most part on the ready oxidation of monosaccharides in alkaline solution and are used, for example, in testing for glucose in urine and in blood.

Tollens **reaction**. When warmed with an ammoniacal solution of a silver salt, monosaccharides cause reduction of the silver ion and precipitation of metallic silver.

Fehling and *Trommer* **reactions**. When monosaccharides are heated with *Fehling's* solution (see p. 193), a yellow to red precipitate is formed. Because the simple sugars, as polyhydroxy compounds, form complexes with copper, the precipitation of copper(II) hydroxide from alkaline solution is prevented and, according to *Trommer*, this test can also be carried out in the absence of tartrate.

Nylander **reaction**. When warmed with an alkaline solution of a bismuth salt (*Nylander's* reagent) monosaccharides produce a black precipitate of metallic bismuth.

Osazone reaction. When monosaccharides are heated with excess *phenylhydrazine* in dilute acetic acid, yellow crystalline precipitates of their osazones precipitate out.

Fermentation. The commonest hexoses can be fermented by means of yeast to give ethanol and carbon dioxide. Pentoses are not attacked, thus providing a method for their separation.

4.1.5 Syntheses, Chain-lengthening and Chain-shortening of Monosaccharides

4.1.5.1 Syntheses

The first synthesis of a sugar from simple starting materials was carried out by *Butlerov* (1861), by the action of a warm dilute solution of calcium hydroxide on paraformaldehyde. The resultant syrup had a sweet taste and gave the normal reactions of sugars but it was *not optically active*. *Loew* (1886) repeated this work with formaldehyde at room temperature and obtained a mixture of sugars ('Formose') which was extremely difficult to separate. *Loew* also noticed that the reaction depended upon the cation of the chosen base and proceeded faster the more dilute the solution of formaldehyde. The rate decreased with increasing concentration of formaldehyde. It is now known that when a large excess of formaldehyde is present, a *Cannizzaro* reaction takes place preferentially instead. *E. Fischer* (1889) was the first person to isolate individual sugars, namely small amounts of DL-fructose (α-acrose) and DL-sorbose (β-acrose), from the mixture of hydroxyaldehydes and hydroxyketones obtained from condensation reactions of formaldehyde; this was carried out by means of their osazones. *E. Fischer* also achieved the first successful *synthesis of an optically active hexose*. He reduced the DL-fructose he had obtained, oxidised the resultant DL-mannitol to DL-mannonic acid, resolved the racemic mixture and reduced the lactone of D-mannonic acid to D-mannose. This D-mannose was converted by means of the methods described earlier into D-fructose and D-glucose.

Since there is no carbon atom α to the carbonyl group of formaldehyde, potassium or sodium hydroxides bring about a *Cannizzaro* reaction. If alkali or alkaline-earth hydroxides whose cations show a strong tendency to form complexes, and more dilute solutions of formaldehyde are used, then condensation reactions of formaldehyde increasingly take place.

This is an *autocatalytic* reaction. In the first phase of the reaction the carbonyl group of a formaldehyde molecule is activated by the formation of a 'primary complex' with the cation of the metal hydroxide. It can then dimerise with a second molecule of formaldehyde to form *glycolal-*

dehyde.[1] This now accelerates reaction by interacting further with the formaldehyde–metal hydroxide complex thereby stabilising it.

The next step must involve the formation of DL-glyceraldehyde, which is in *equilibrium* with dihydroxyacetone (see *Lobry de Bruyn–van Ekenstein* rearrangement, p. 317) with which it reacts in an *aldol-like reaction* to give DL-fructose and DL-sorbose. *H. O. L. Fischer* (1936) provided evidence for a reaction of this sort when he obtained D-fructose and D-sorbose in about 45% yield by the reaction of D-glyceraldehyde with a dilute solution of barium hydroxide. This aldol-like reaction proceeds even faster if a mixture of D-glyceraldehyde and dihydroxyacetone is used as starting material:

D-Fructose   D-Sorbose

4.1.5.2 Chain Lengthening

(a) Cyanohydrin synthesis. In this method an aldose with *n* carbon atoms is converted into one with (*n* + 1) carbon atoms (*Kiliani*; *E. Fischer*). It is carried out by adding hydrogen cyanide to the aldehyde to form a cyanohydrin, which is hydrolysed to an aldonic acid. The lactone of this acid is then reduced by means of *sodium amalgam* in a weakly acidic solution to provide the next highest aldoses. Since cyanohydrin formation involves the creation of a *new* chiral centre, two isomeric compounds are formed which are not however enantiomers but *diastereomers*, and thus, because of their different solubilities, can be separated by fractional crystallisation.

The preparation of D-gluconitrile and D-mannonitrile from D-arabinose may be taken as an example. Starting from L-arabinose *E. Fischer* obtained, after hydrolysis of the L-mannononitrile, about 50% de of mannononitrile; this was the first example of asymmetric induction *in vitro.*

D-Gluconitrile   D-Arabinose   D-Mannononitrile
(epimeric cyanohydrin)   (epimeric cyanohydrin)

The equilibrium between the two cyanohydrins is dependent upon the pH. In addition to the method described above it is also possible to convert the nitriles directly into D-glucose and D-mannose by hydrogenation, using $PdO/BaSO_4$ as a catalyst.

---

[1] See *E. Pfeil* et al., Kinetik und Reaktionsmechanismus der Formaldehyde-Kondensation, Chem. Ber. *85*, 293 (1952); *R. Mayer* et al., Zur Umwandlung von Formaldehyd in Kohlenhydrate und der gegenwärtige Stand der Arbeiten, Z. Chem. *3*, 134 (1963).

By repeated cyanohydrin synthesis and reduction of the lactones of the aldonic acids it is possible to carry out a stepwise preparation of aldoses, at each step producing aldoses with one more carbon atom. Using this method aldoses with up to ten carbon atoms (decoses) have been obtained.

(b) Nitroalkane synthesis. In basic media the carbonyl groups of aldoses or ketoses react with nitroalkanes. In this way, for example, the anions of two *epimeric 1-nitropentitols* are obtained from D-*arabinose* and *nitromethane*. They can be separated by fractional crystallisation and converted into D-glucose and D-mannose by means of the *Nef* reaction.

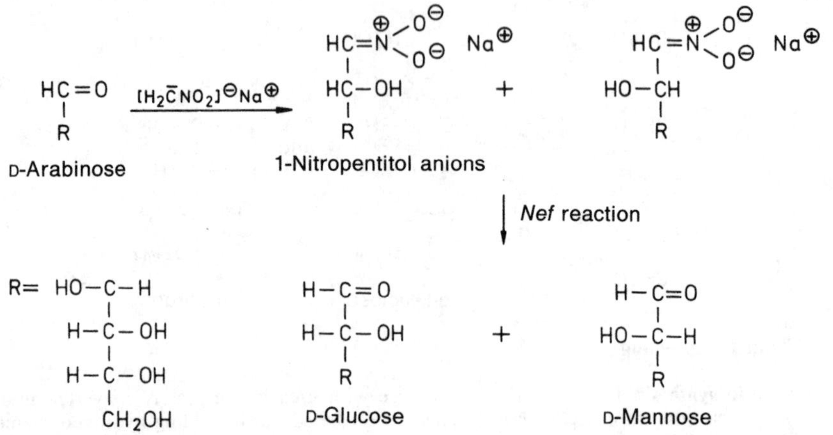

$$
\begin{array}{ccc}
\underset{R}{\overset{HC=O}{|}} & \xrightarrow{[H_2\bar{C}NO_2]^{\ominus}Na^{\oplus}} &
\begin{array}{c}
HC=\overset{\oplus}{N}\diagdown{}^{O^{\ominus}}_{O^{\ominus}}\ Na^{\oplus} \\
| \\
HC-OH \\
| \\
R
\end{array}
\quad + \quad
\begin{array}{c}
HC=\overset{\oplus}{N}\diagdown{}^{O^{\ominus}}_{O^{\ominus}}\ Na^{\oplus} \\
| \\
HO-CH \\
| \\
R
\end{array}
\end{array}
$$

D-Arabinose   1-Nitropentitol anions

Nef reaction ↓

$$
R=\ \begin{array}{c} HO-C-H \\ | \\ H-C-OH \\ | \\ H-C-OH \\ | \\ CH_2OH \end{array}
\qquad
\begin{array}{c} H-C=O \\ | \\ H-C-OH \\ | \\ R \end{array}
\quad + \quad
\begin{array}{c} H-C=O \\ | \\ HO-C-H \\ | \\ R \end{array}
$$

D-Glucose   D-Mannose

4.1.5.3 Chain-shortening

The following processes are especially useful for the step-wise shortening of the carbon chain in aldoses:

(a) *Ruff's* method (1898). The aldose is first oxidised to an aldonic acid and the calcium salt of the latter is then treated with hydrogen peroxide in the presence of iron(III) acetate. Carbon dioxide and water are lost, leaving an aldose with one less carbon atom. For example, D-arabinose can be obtained in this way from D-glucose.

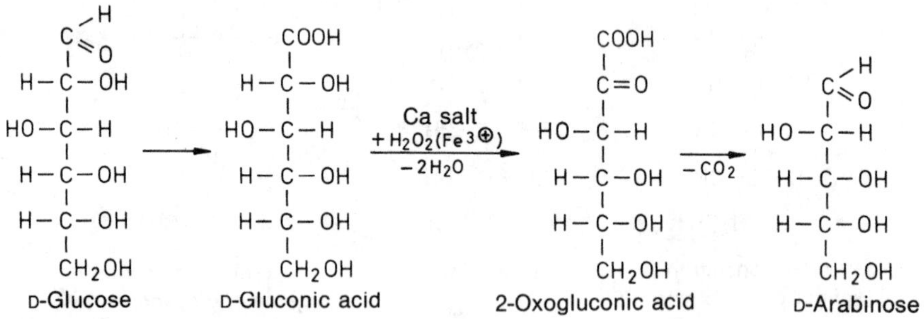

$$
\begin{array}{c} C\diagup^{H}_{\diagdown O} \\ | \\ H-C-OH \\ | \\ HO-C-H \\ | \\ H-C-OH \\ | \\ H-C-OH \\ | \\ CH_2OH \end{array}
\ \longrightarrow\
\begin{array}{c} COOH \\ | \\ H-C-OH \\ | \\ HO-C-H \\ | \\ H-C-OH \\ | \\ H-C-OH \\ | \\ CH_2OH \end{array}
\ \xrightarrow[-2H_2O]{\substack{Ca\ salt \\ +H_2O_2(Fe^{3\oplus})}}\
\begin{array}{c} COOH \\ | \\ C=O \\ | \\ HO-C-H \\ | \\ H-C-OH \\ | \\ H-C-OH \\ | \\ CH_2OH \end{array}
\ \xrightarrow{-CO_2}\
\begin{array}{c} C\diagup^{H}_{\diagdown O} \\ | \\ HO-C-H \\ | \\ H-C-OH \\ | \\ H-C-OH \\ | \\ CH_2OH \end{array}
$$

D-Glucose   D-Gluconic acid   2-Oxogluconic acid   D-Arabinose

It is assumed that the calcium salt of 2-oxogluconic acid is an intermediate and that it is easily decarboxylated. The epimer of D-glucose, D-mannose, is similarly converted into D-arabinose.

(b) *Wohl's* method (1893). Reaction of an aldose with hydroxylamine gives the corresponding aldoxime. When the latter is heated with acetic anhydride the oxime group loses a molecule of water and the product is a fully acetylated aldononitrile. This is heated with ammoniacal silver oxide; hydrogen cyanide is eliminated and at the same time the acetyl groups are hydrolysed, leaving an aldose with one less carbon atom, *e.g.*

D-Glucose oxime   Penta-acetyl-D-gluconitrile   D-Arabinose

Appreciably better yields (60–70%) are obtained if the modified method of *Zemplén* (1927) is used. This involves removal of the nitrile group by sodium methoxide in chloroform.

4.1.6 The Ring Structure of Monosaccharides

The reactions of aldoses and ketoses which have been described so far can all be satisfactorily explained in terms of the *open-chain* formulae, for example the fact that they undergo the reactions typical of aldehyde and ketone groups. Experience shows, however, that the aldehyde and ketone formulae are not adequate to explain all the properties of simple sugars. For example, the carbonyl groups do *not* undergo addition reactions with either sodium hydrogen sulphite or with ammonia. In addition, aldoses do not give the usual colour test that other aldehydes do with *Schiff's* reagent.

Because of these differences *Tollens* (1883) came to the conclusion that the simple sugars do *not* exist in the aldehyde or keto forms. The explanation for this was that the carbonyl groups formed *intramolecular* bonds with a hydroxy group of the same sugar molecule, resulting in ring formation.

It was known that aldehydes react with alcohols to form hemiacetals and acetals:

Aldehyde   Hemiacetal   Acetal

In the case of monosaccharides it remained to be decided which of the hydroxy groups was involved in the ring closure and which proton shifted to the carbonyl group. Later researches showed that in the case of aldohexoses, the proton from the hydroxy group on carbon atom 5 was the one predominantly involved in forming a cyclic hemiacetal. A heterocyclic 6-membered ring which was a hydrogenated pyran ring was formed. There were also instances of simple sugars in which the hydroxy group on carbon atom 4 was the one which took part in ring formation, giving a 5-membered heterocyclic ring, namely a hydrogenated furan. *Haworth* therefore made a distinction between monosaccharides having *tetrahydropyran* or *tetrahydrofuran* rings, calling them **pyranoses** and **furanoses**.

Tetrahydropyran   Tetrahydrofuran

The ring structures of simple sugars are further explained in terms of *Fischer projection formulae* using D-glucose as an example. Ring closure to D-glucopyranose involves migration of a proton from the hydroxy group on carbon atom 5 to the oxygen atom of the carbonyl group. In the process the carbon atom which originally formed part of the carbonyl group becomes a *new* asymmetric carbon atom, so that *two diastereomeric* D-*glucopyranoses* (D-Glc*p*) are produced, which are known as the α- and β-forms. Both of these, **α-D-glucose** and **β-D-glucose**, are in *equilibrium* with each other in solution.

4.1.6.1 In the *formulae for these rings* involving D-sugars, the HO-group on carbon atom 1 is drawn to the right for the α-form and to the left for the β-form. In the case of L-sugars the reverse applies. Interchange between the α- and β-forms proceeds *via* the open-chain *aldehydo* form (*al* form) in a system of *oxo-cyclo tautomerism*. The isomeric α- and β-forms, which have identical configurations except for their difference at the chiral centre on carbon atom 1, are called *anomers*. (The anomeric carbon atoms 1 and the anomeric protons are printed in heavy type in the following formulae. The rounding off of the corners in the printed formulae indicates that no change of the skeletal formula is involved; were corners used rather than bends it would imply the introduction of an extra CH_2 group at each corner).

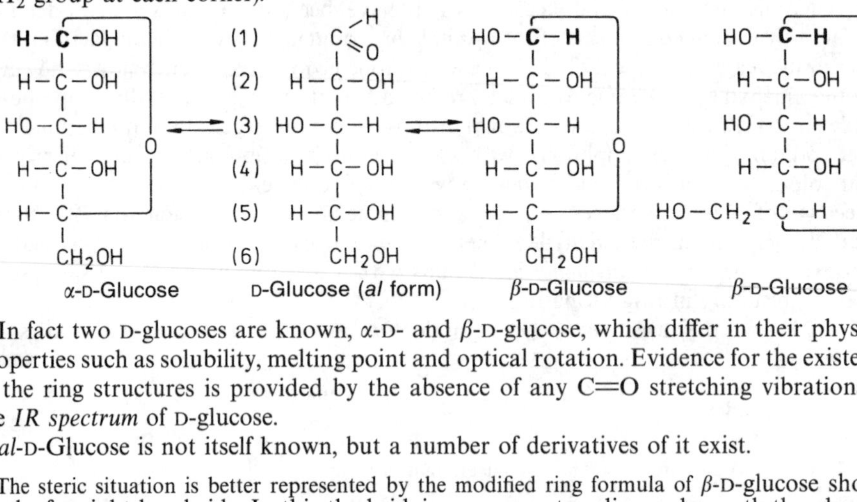

```
    H—C—OH           (1)        C      HO—C—H              HO—C—H
        |                        ‖\H       |                   |
    H—C—OH           (2)   H—C—OH    H—C—OH            H—C—OH
        |                        |         |                   |
   HO—C—H        ⇌   (3)  HO—C—H  ⇌  HO—C—H            HO—C—H
        |    O                   |         |     O             |
    H—C—OH           (4)   H—C—OH    H—C—OH            H—C—OH   O
        |                        |         |                   |
    H—C———           (5)   H—C—OH    H—C———      HO—CH₂—C—H
        |                        |         |
     CH₂OH           (6)      CH₂OH     CH₂OH

   α-D-Glucose      D-Glucose (al form)   β-D-Glucose         β-D-Glucose
```

In fact two D-glucoses are known, α-D- and β-D-glucose, which differ in their physical properties such as solubility, melting point and optical rotation. Evidence for the existence of the ring structures is provided by the absence of any C=O stretching vibrations in the *IR spectrum* of D-glucose.

al-D-Glucose is not itself known, but a number of derivatives of it exist.

The steric situation is better represented by the modified ring formula of β-D-glucose shown on the far right-hand side. In this the bridging oxygen atom lies underneath the plane of the paper, and it is immediately comparable with *Haworth's* ring formula. It shows the *trans* configuration of the primary hydroxy group with respect to the hydroxy groups at the 2- and 4-positions as well as to the anomeric proton.

4.1.6.2 Further support for *Tollens'* ring formula for monosaccharides is provided by the phenomenon of **mutarotation** (Lat. *mutare* = to change), which was first observed by *Dubrunfaut* in 1846. If the optical rotation of a freshly made-up solution of a sugar is recorded at intervals of time, it shows a continual change until it finally settles down to a constant value. This end-value is dependent upon the nature of the solvent. If, for example, aqueous solutions of α-D-glucose and β-D-glucose are made up, the initial optical rotation of these solutions are, respectively, $[\alpha] = +112.2$ and $+18.7°$. Eventually the rotations for each of the solutions become $+52.7°$.[1]

---

[1] The use of the traditional symbol of degrees (°) for the numerical value of $[\alpha]$ is not strictly correct. As the equation on p. 121 shows, $[\alpha]$ has the dimensions [degree.10^{-1}.cm^2.g^{-1}]. IUPAC recommends that the numerical value of $[\alpha]$ should either be given *without* ° or with the correct dimensions.

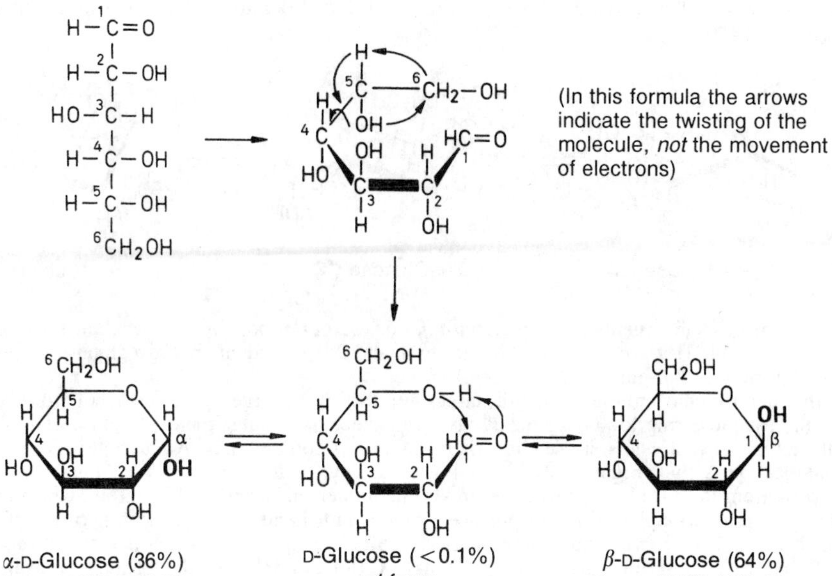

$\begin{array}{c} H-\overset{1}{C}=O \\ H-\overset{2}{C}-OH \\ HO-\overset{3}{C}-H \\ H-\overset{4}{C}-OH \\ H-\overset{5}{C}-OH \\ \overset{6}{C}H_2OH \end{array}$

(In this formula the arrows indicate the twisting of the molecule, *not* the movement of electrons)

α-D-Glucose (36%)
(m.p. 146°C [419 K],
$[\alpha]_D = +112.2°$)

D-Glucose (<0.1%)
al-form

β-D-Glucose (64%)
(m.p. 150°C [423 K],
$[\alpha]_D = +18.7°$)

Equilibrium in water $[\alpha]_D = +52.5$

The *rate* at which the equilibrium between α- and β-D-glucose is set up depends very much on the temperature and on the pH of the solution. The presence of small amounts of alkali leads to mutarotation taking place at a rate too fast to be measured. The final value is not the arithmetic mean of the rotation values for α- and β-D-glucose. The two anomers are diastereomers and are not both present to the same extent. In aqueous solution there is more β-D-glucose (64%) than α-D-glucose (36%). In order to illustrate the various factors which affect this situation it is necessary to look further into the ring structure of monosaccharides (*Böeseken, Haworth, Hirst*). The *ring* or *configurational* formulae present a perspective picture of the sugar molecules. To simplify these formulae, the ring atoms are represented by a flat pyranose (or furanose) ring lying in one plane (perpendicular to the plane of the paper and with the half of the ring lying above this plane drawn in thick type), and the substituents on the different carbon atoms are attached by perpendicular lines.

In order to arrive at the *Haworth* formulae from the *Fischer* projection formulae it is necessary to change the carbon chain into the shape of a hexagon, doing this in such a way that carbon atom 1 is placed in front at the right-hand corner. All the hydroxy groups which lay on the right-hand side of the C—C axis in the *Fischer* projection now appear below the plane of the ring, while those that lay to the left-hand side of the axis appear above the plane of the ring. In order that the HO-group on carbon atom 5 may be in the correct position to participate in ring closure it is necessary that the substituents on this carbon atom must be twisted in the manner indicated by the arrows.

Mutarotation depends upon the fact that the cyclic hemiacetal can easily open and then ring-close again.

4.1.6.3 Conformation of sugars: Since the tetrahydropyran ring in hexoses is *not planar but, like cyclohexane rings, takes up an energetically favoured chair form* (see p. 399), α-D-glucose and β-D-glucose (α-D-Glc$p$ and β-D-Glc$p$) take up the following conformations (*Reeves*, 1950):

α-D-Glucose (^{1}C$_4$)   α-D-Glucose (^{4}C$_1$)   β-D-Glucose (^{4}C$_1$)

The various substituents take up either *axial* or *equatorial* positions in the chair form. From the positions of the HO-groups on carbon atoms 1 and 2 it is evident that α-D-glucose represents the *cis*-1,2 form and β-D-glucose the *trans*-1,2 form.

In general, conformations having the largest number of substituents located in equatorial positions on the pyranose ring are energetically favoured, since axial substituents, because of 1,3-repulsive interactions (⌒), are less stable than their equatorial counterparts. A characteristic of the *chair conformation* is that, because of the folding of the pyranose ring, carbon atoms 2, 3, 5 and the oxygen atom form a plane (marked in grey). The numerical site of the ring atom, which lies above this plane is indicated by a superscript placed *before* a shorthand symbol indicating ring conformation (in this case $\mathbf{C}$ = chair), whilst the numerical site of the ring atom which lies below this plane is indicated by a subscript placed *after* the symbol $\mathbf{C}$ (for examples see the formulae above).

The $^1\mathbf{C}_4$-conformation of α-D-glucose is 13.4 kJ/mol higher in energy than the $^4\mathbf{C}_1$-conformation. The difference in energy between α- and β-D-glucose is in contrast only 1 kJ/mol.

4.1.6.4 The **anomeric effect**[1] provides a model to explain the fact that, especially in non-polar solvents, the α-anomer is usually energetically favoured over the β-anomer. It appears to be due to a stereoelectronic effect in which electrostatic repulsion between the dipoles of the ring oxygen atom and the anomeric HO-group play an important role.[2] The following *Newman* projections (from C-1 to the ring oxygen atom) illustrate this.

α-Anomer   β-Anomer

The two dipoles, of the HO-group and of the ring oxygen atom, are parallel and in the same direction in the *β-anomer* (*equatorial HO-group*) and it is therefore disfavoured compared to the α-anomer in which the dipole of the *axial HO-group* partially neutralises that of the ring oxygen atom.

The preponderance of β-D-glucose in aqueous solution can be related to the polar solvent, which lowers the effects of the dipole moments by solvation.

Like aldoses, *ketoses* also form cyclic hemiacetal structures, but with the difference that in this case the tetrahydropyran ring arises by migration of a proton from carbon atom 6 to the keto group on carbon atom 2. The following equilibration, involving D-fructose, may be taken as an example:

---

[1] See A. J. Kirby, The Anomeric Effect and Related Stereoelectronic Effects in Oxygen. Reactivity and Structure, Concepts in Organic Chemistry *15* (Springer-Verlag, Berlin 1983).
[2] See P. Deslongchamps, Stereoelectronic Effects in Organic Chemistry (Pergamon Press, Oxford 1983).

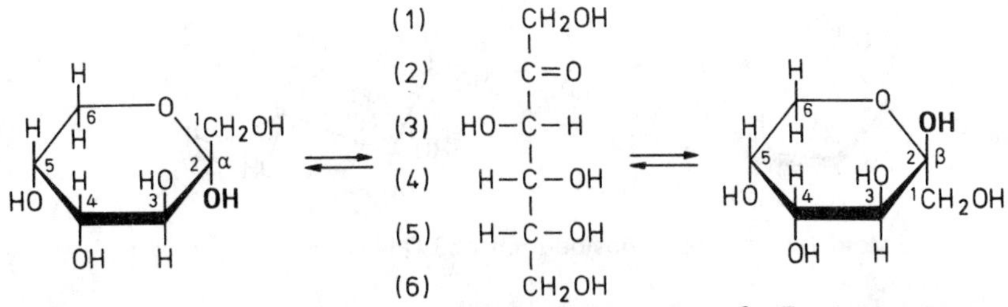

(1)	CH_2OH
(2)	$C=O$
(3)	$HO-C-H$
(4)	$H-C-OH$
(5)	$H-C-OH$
(6)	CH_2OH

α-D-Fructopyranose
(α-Anomer)

D-Fructose
(*keto* form)

β-D-Fructopyranose
(β-Anomer)

4.1.6.5 To date, only one form of fructose has been isolated, namely **β-D-fructopyranose**, which has the following conformational formula:

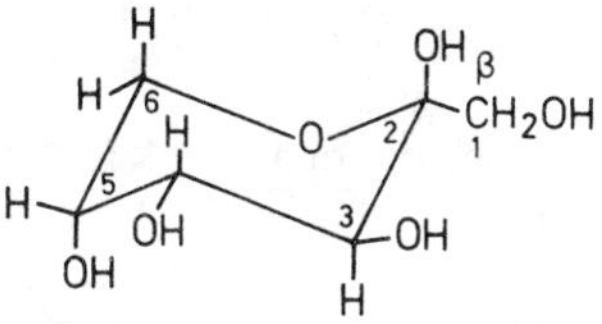

From the above examples it may be seen that the structures of monosaccharides or their anomers can be represented in three different ways, namely by the *Tollens* ring formulae, the ring formulae of *Haworth* and the conformational formulae of *Reeves*; the two latter ways are preferable. The *conformational formulae* reflect the real stereochemical structures of sugars. In order to simplify the picture it is sometimes the practice, as in the case of formulae of benzene derivatives, to omit the hydrogen atoms and show only the substituent groups.

Not only carbon atom 1 in the anomeric glucopyranoses but also carbon atom 2 in the fructopyranoses can be regarded as *latent carbonyl groups*, and in their respective tautomeric *al* and *keto* forms the *carbonyl reactions* already mentioned can take place.

In contrast to β-D-fructopyranose, the **fructofuranoses** are not known as separate substances, but they do exist as *furanoside* units in *glycosides*. D-Fructofuranose (D-fru *f*) occurs in, for example, sucrose and in the polysaccharide inulin. The configurations of α-D- and β-D-fructofuranoses are known from the resultant methyl fructofuranosides.

4.1.7 Glycosides

The HO-group which appears in the cyclic hemiacetal forms of aldo- or keto-hexoses on carbon atom 1 or 2 respectively is a particularly reactive hydroxy group and readily reacts with an alcohol (or a phenol) in the presence of a small amount of hydrogen chloride to form a real *acetal*. These compounds are known as 'glycosides' (*Gerhardt*, 1852) and specifically in the case of individual sugars as *glucosides, mannosides, fructosides, etc.*, or, if heterocyclic 5- or 6-membered heterocyclic rings are involved, as *furanosides* or as *pyranosides*.

Methyl-D-glucopyranoside may be taken as the simplest example of this large class of compounds. It is formed by treatment of D-glucose with methanol/HCl and can occur as **methyl-α-D-** and **methyl-β-D-gluco-pyranosides** (*E. Fischer*, 1893). The two anomers have different melting points and different specific rotations:

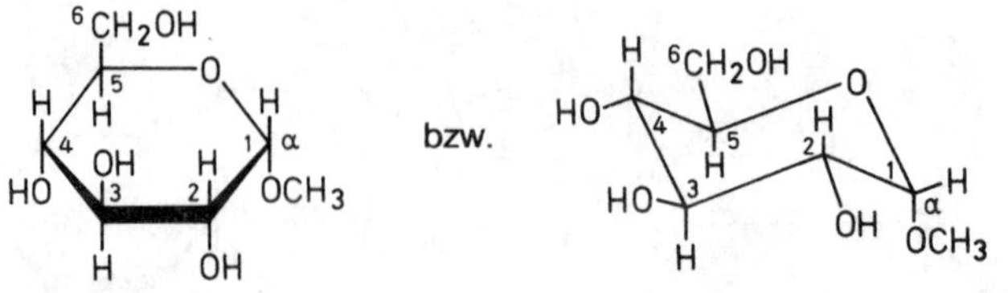

Methyl-α-D-glucopyranoside (m.p. 166°C [439 K], $[\alpha]_D = +158°$)

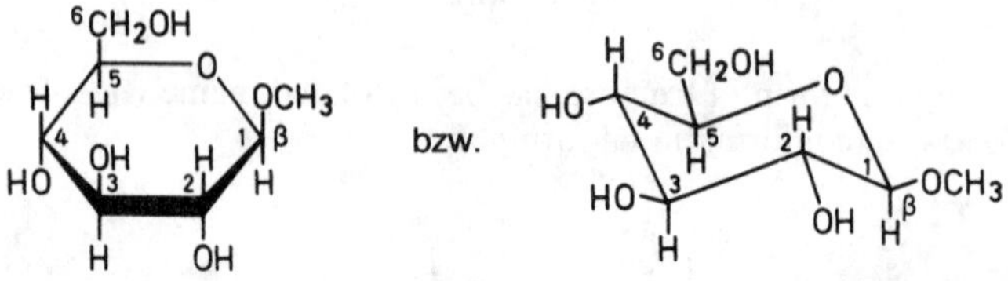

Methyl-β-D-glucopyranoside (m.p. 107°C [380 K], $[\alpha]_D = -33°$)

In acidified methanol an equilibrium is set up between the α- and β-methylglucosides which at 35°C [308 K] contains 66% of the α-anomer; this is a consequence of the anomeric effect.

Haworth (1927) also isolated **α-** and **β-methyl-D-glucofuranosides** as by-products from the preparation of α- and β-methyl-D-glucopyranosides:

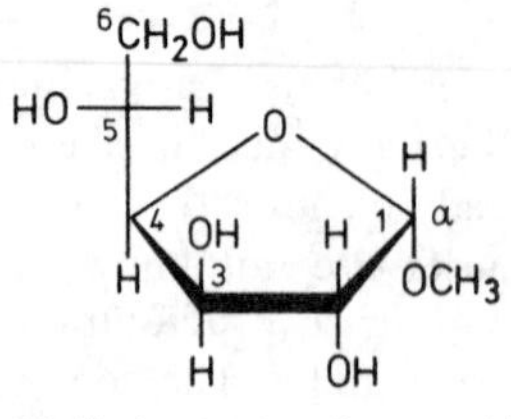

Methyl-α-D-glucofuranoside

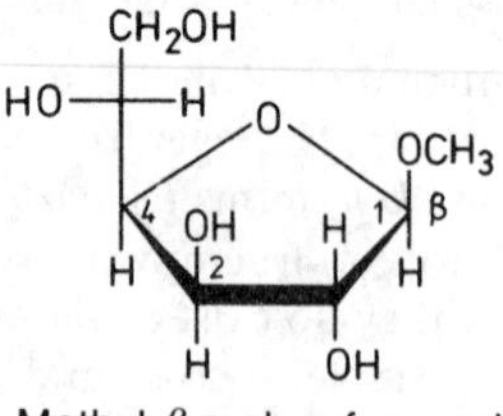

Methyl-β-D-glucofuranoside

As already mentioned, *ketofuranoses* are only stable in the form of *furanosides, e.g.* **methyl-α-D-** and **methyl-β-D-fructofuranosides** have the following ring formulae:

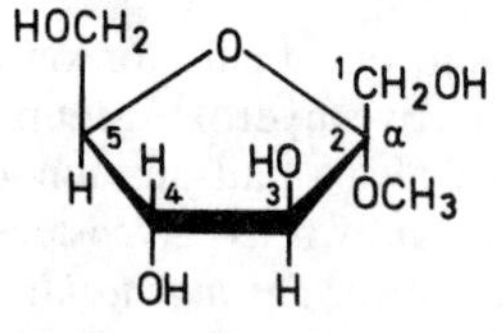

Methyl-α-D-fructofuranoside

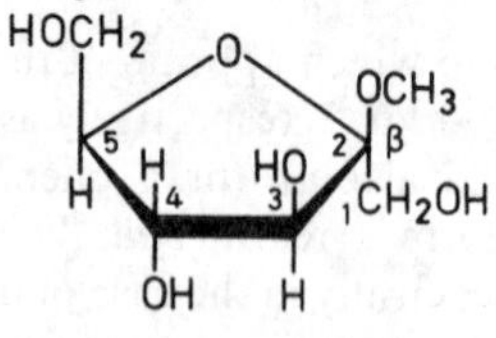

Methyl-β-D-fructofuranoside

Glycosides are of great importance biochemically. They occur in a very wide range of natural products, especially in plants. Examples are amygdalin, salicin and coniferin. The non-sugar components of glycosides are known as *aglycones*. In the enzymatic cleavage of glycosides, depending upon the specificity of the enzymes involved, either the α- or the β-forms of the sugar in question may be hydrolysed.

Glycosides are crystalline materials which are soluble in water. They show the typical properties of *acetals* in that they are stable to alkali but are readily hydrolysed by hot, dilute mineral acids to the sugar and the aglycone. Since they do not have a latent aldehyde group they do *not reduce Fehling's* solution.

Oligo- and poly-saccharides are also glycosides in which the glycosidic hydroxy group of the sugar has formed an acetal, not with an alcohol (or a phenol), but rather with a glycosidic or alcohol group of another monosaccharide molecule. In teichonic acids the sugar component has a glycosidic link to glycerol or to ribitol.

The Most Important Monosaccharides

4.1.8 Pentoses, $C_5H_{10}O_5$

Pentoses occur in plants, principally as components of polysaccharides, in the *pentosans*, *e.g.* in the structural material in wood, as well as in plant gums and mucilages. In animals they form part of glycosides, *e.g.* in the *nucleoproteins* of the pancreas and the liver. All naturally occurring pentoses are aldoses. They are not fermented by yeast. They include arabinose, xylose, lyxose and, above all, ribose (for formulae see p. 427).

L(+)-Arabinose is obtained from *araban* (*cherry gum*) by boiling it with dilute sulphuric acid. It forms needles which melt at 160°C (433 K).

D(−)-Arabinose occurs as a glycoside in *aloes* and forms the same osazone as its *epimer* D(−)-ribose.

D(+)-Xylose is found in *xylan* (*wood gum*), in bran, in maize cobs and in straw, and can be obtained from these sources by hydrolysing them. It melts at 145°C (418 K).

D(−)-Ribose, m.p. 95°C (368 K), is present as the *N*-glycoside of purine and pyrimidine bases in nucleic acids.

When pentoses are warmed with dilute mineral acids *furfural* (2-furaldehyde) is formed.

4.1.9 Hexoses, $C_6H_{12}O_6$

Of the sixteen possible *optically active* aldohexoses only four have been found in nature, namely D(+)-glucose, D(+)-mannose, D(+)-galactose and D(+)-talose. Their main source is in polysaccharides in which the hexose units are joined together by glycosidic links. Unlike pentoses, some of the hexoses, *e.g.* D-glucose, D-mannose, D-galactose and D-fructose, are fermented by yeast.

D(+)-Glucose (grape sugar, dextrose) occurs in large amounts in sweet fruits such as grapes, and together with fructose is the main constituent of honey. In living organisms D-glucose is found in blood (the normal blood sugar content is about 0.1%) and in other body fluids. It is found in the urine of diabetics. In both plants and animals it plays a key role in biogenetic processes.

Industrially D-glucose is made by hydrolysing potato or maize starch with dilute hydrochloric acid under pressure. It is used sometimes as a quick-acting tonic.

It crystallises from cold aqueous ethanol as **α-D-glucose**, m.p. 146°C (419 K). It is only half as sweet as cane sugar (sucrose). **β-D-Glucose**, m.p. 150°C (423 K) is obtained by dissolving glucose in hot pyridine and subsequent crystallisation. D-Glucose can be partially converted into D-fructose with the help of immobilised enzymes. The resultant mixture (isomerose) is as sweet as sucrose.

D(+)-**Mannose**, m.p. 132°C (405 K) occurs in the polysaccharide mannan which is found, for example, in ivory nuts and in the seeds of locust beans. It is obtained by hydrolysing pieces of ivory nut with hydrochloric acid.

D(+)-**Galactose**, m.p. 165.5°C (438.7 K), is a component of lactose (milk sugar) and is obtained by hydrolysis of the latter sugar. It is also found in some gums, the *galactans*.
The following are the stable conformations of D-mannose and D-galactose:

β-D-Mannopyranose   α-D-Galactopyranose

D(−)-**Fructose** (fruit sugar, laevulose) was discovered by *Dubrunfaut* (1847) who prepared it by hydrolysing cane sugar. It occurs abundantly in sweet fruits and in honey. The sweetness depends upon the structure. β-D-Fructopyranose is about twice as sweet as sucrose, which is made up from one molecule of D-glucose and one molecule of D-fructose. β-D-Fructofuranose is nearly tasteless. D-Fructose does not crystallise readily and melts at 102–104°C (375–377 K). It readily reduces *Fehling's* solution and ammoniacal silver nitrate (*cf.* hydroxyketones, p. 318).

L-**Sorbose** is another ketohexose, which is formed by removal of hydrogen from carbon atom 5 of sorbitol by the enzyme *acetobacter xylinum*. It melts at 165°C (438 K), tastes sweet, and is *not* fermented by yeast. It is an intermediate in the industrial production of vitamin C (see p. 448).

4.1.10 Reactions of the Hydroxy Groups

Methylation of the alcoholic hydroxy groups of sugars can be achieved using either methyl iodide and silver oxide (*Purdie* and *Irvine*) or methyl sulphate and alkali (*Haworth*). Methyl fluorosulphonate ('magic methyl') (CARE—carcinogenic) is an especially reactive methylating agent. It is usual to protect the glycosidic hydroxy group by first *converting it into an acetal*, i.e. the sugar is made into a methyl glycoside which is stable to alkali. Methylation then converts the free hydroxy groups into *ether groups*, e.g. if methyl β-D-glucopyranoside is treated with dimethyl sulphate in alkaline solution the product is methyl 2,3,4,6-tetra-O-methyl-β-D-glucopyranoside. If this is heated with dilute hydrochloric acid *only* the glucosidic ether bond is broken and 2,3,4,6-*tetra-O-methyl-β-D-glucopyranoside* is formed. The latter compound undergoes *mutarotation* and is a *reducing* sugar.

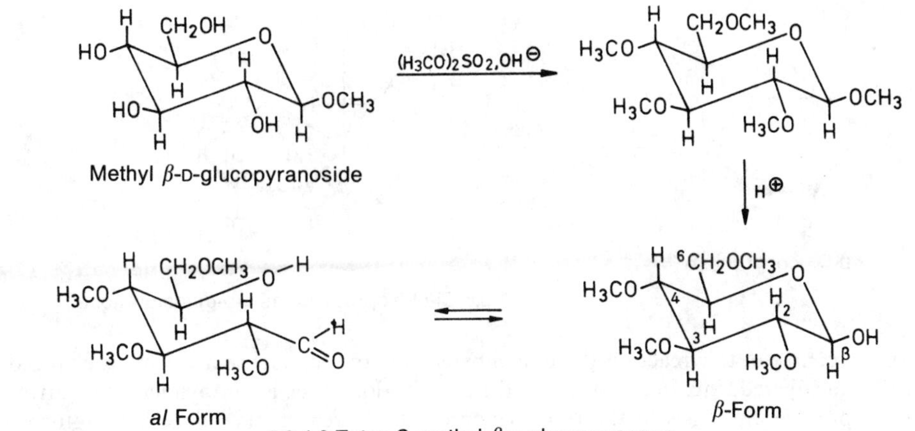

Methyl β-D-glucopyranoside

(H₃CO)₂SO₂,OH⊖ →

2,3,4,6-Tetra-O-methyl-β-D-glucopyranose

al Form

β-Form

4.1.10.1 Acetylation. In the acetylation of sugars by acetic anhydride an acid or base catalyst, such as zinc chloride, conc. sulphuric acid, anhydrous sodium acetate or pyridine, is employed. In this way hexoses provide penta-acetates while pentoses give tetra-acetates; for example, penta-acetylglucose is formed from glucose. These acetates do not reduce Fehling's solution, since they have no free aldehyde group. They can exist in α- or β-forms. Which anomer is preponderant depends upon the catalyst used, as the following examples demonstrate:

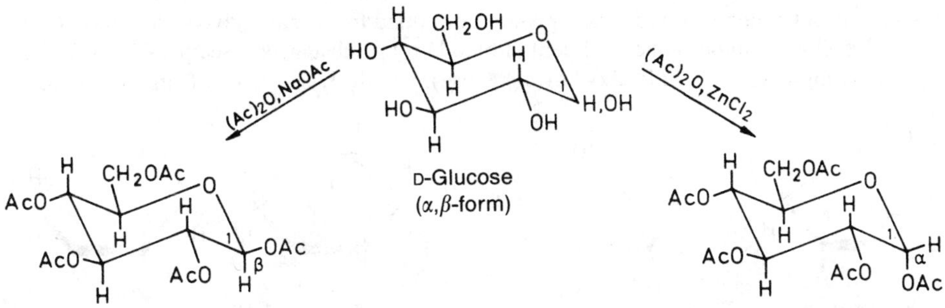

D-Glucose
(α,β-form)

(Ac)₂O,NaOAc

(Ac)₂O, ZnCl₂

β-D-Glucopyranose penta-acetate

α-D-Glucopyranose penta-acetate

The *acetoxy group* at carbon atom 1 in either an α- or a β-hexose penta-acetate can generally be replaced by a bromine atom. This is done by treating the penta-acetate with acetic acid which has been saturated with hydrogen bromide. For example, 'acetobromoglucose', which is an important starting material in the synthesis of sugar derivatives, is made in this way from penta-acetyl-D-glucose.

4.1.10.2 Isopropylidene Derivatives of Sugars. Monosaccharides which have two vicinal hydroxy groups *cis* to one another react with aldehydes or ketones in the presence of sulphuric acid, zinc chloride or phosphorus(V) oxide to give *acetals* or *ketals* (E. Fischer, 1895). As an example, in the presence of sulphuric acid, *acetone reacts with D-glucose* at room temperature to give *1,2:5,6-di-O-isopropylidene-D-glucofuranose* (diacetone-α-D-glucose):

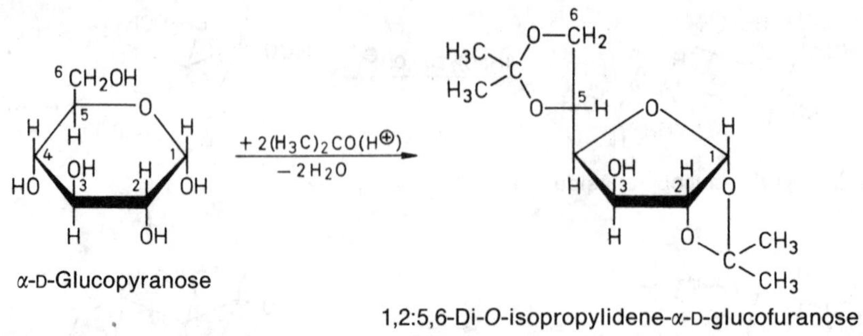

α-D-Glucopyranose

1,2:5,6-Di-O-isopropylidene-α-D-glucofuranose

In the course of this reaction the 6-membered ring of the glucose molecule is changed into a 5-membered ring. In other words, the constitution of the isopropylidene derivatives do not provide any clues to the ring structures of the free sugars. Diacetone-D-glucose has a free hydroxy group on C-3, which can be methylated or acetylated. It does *not* reduce *Fehling*'s solution. It is easy to hydrolyse off the isopropylidene groups with dilute acid.

4.1.11 Deoxysugars

Some naturally occurring aldopentoses or aldohexoses have one or more of the hydroxy groups replaced by hydrogen atoms. These compounds are known as *deoxysugars*. Among them are the biologically important sugar **2-deoxy-D-ribose**, which forms part of the deoxyribonucleic acids, and also the very commonly occurring L(+)-**rhamnose** (6-deoxy-L(+)-mannose) which is a hydrolysis product obtained from many glycosides, and L(−)-**fucose**, which is an important constituent of blood group substances. (See p. 845). A further example is **digitoxose** (2,6-dideoxy-D-ribohexose), a hydrolysis product from digitoxin.

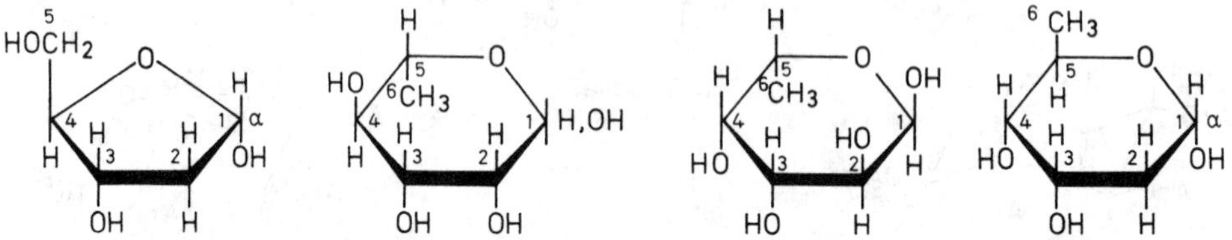

2-Deoxy-α-D-
-ribofuranose (2d Rib*f*)

L(+)-Rhamnose (Rha)
(6-Deoxy-L-mannose)

L(−)-Fucose (Fuc)
(6-Deoxy-L-galactose)

Digitoxose
(2,6-Dideoxy-α-D-
-ribohexose)

4.1.12 Aminosugars[1]

Aminosugars are derivatives of monosaccharides in which one of the alcohol groups (*other than* the glycosidic group) has been replaced by an amino group. The two commonest examples are D-**glucosamine** and D-**galactosamine**. In particular, in the form of their *N*-acetyl derivatives they contribute to the structure of *N*-containing

---

[1] See R. *Kuhn*, Aminozucker (Aminosugars), Angew. Chem. *69*, 23 (1957); E. A. *Balasz* et al., The Amino Sugars, 3 Vols (Academic Press, New York 1966).

polysaccharides. Their structures resemble those of the cyclic hemiacetal forms of D-glucose and D-galactose.

D-Glucosamine (β-form)	D-Galactosamine (β-form)	4-Amino-4,6-dideoxy-α-D-
(2-Amino-2-deoxy-D-	(2-Amino-2-deoxy-D-	glucopyranose
glucopyranose, GlcN)	galactopyranose, GalN)	

The aminosugars are derived from the corresponding deoxysugars, with one of the hydrogen atoms of the CH_2 group replaced by an amino group. They can exist in α- and β-forms. The 4-amino derivative of 6-deoxy-D-glucose, 4-amino-4,6-dideoxy-α-D-glucopyranose, forms part of the pseudo-oligosaccharides found in actinomycetin.

In particular, D-*glucosamine* (chitosamine) forms the constituent unit of the polysaccharide *chitin*. It was the first aminosugar to be isolated, being obtained from acid hydrolysis of lobster shells (*Ledderhose*, 1876).

D-*Galactosamine* (chondrosamine), in the form of its *N*-acetyl derivative, occurs in *mucopolysaccharides*.

Both of these aminosugars form crystalline hydrochlorides. They *reduce Fehling's* solution and undergo mutarotation.

Rational preparations of D-glucosamine (and D-galactosamine) have been achieved using the *Strecker* synthesis (see p. 285). Treatment of D-arabinose (or D-lyxose) with ammonia and hydrogen cyanide gives an aminonitrile, which can be reduced catalytically in dilute acid solution to give the aminosugar (*R. Kuhn*, 1956), *e.g.*

| D-Arabinose (*al* form) | Aminonitrile | D-Glucosamine hydrochloride | Muraminic acid (Mur) |

The yield of the hydrochlorides of D-glucosamine or D-galactosamine was about 70–74%.

N-**Methyl-L-glucosamine** forms part of *streptomycin*. It can be prepared in the same way as glucosamine, starting from L-arabinose, methylamine and hydrogen cyanide. If the 3-hydroxy group of D-glucosamine is made into an ether derived from the hydroxy group of lactic acid, the resultant compound, *muraminic acid* (3-*O*-(1'-carboxyethyl)-D-glucosamine) is important as a constituent of bacterial cell-walls.

Neuraminic acid plays a central role in the structure of membrane-forming glycolipids. It occurs there in acetylated forms; the *N*-acetyl and *O*-acetyl derivatives are classed together as *sialic acids*. They are among the most important constituents of the mucins (mucoids) which have been separated from the mucous membranes of the respiratory and urogenital tracts. They play an important role in the specific recognition of glycoproteins (see p. 844).

Neuraminic acid arises from an aldol addition reaction between *pyruvic acid* and D-*mannosamine*.

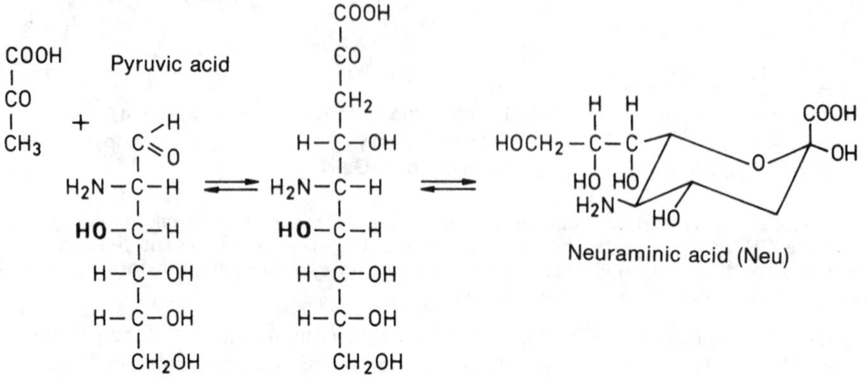

Neuraminic acid (Neu)

D-Mannosamine (D-Man N)

The newly formed HO-group takes up a position *trans* to the amino group, and the carbonyl group derived from the pyruvic acid forms a pyranose ring in the $^1C_4$ conformation with the 3-HO-group of the original mannosamine (shown in bold type).

(+)-**Muscarine**, the poisonous constituent of toadstools (*Amanita muscaria*), has been shown to be a tetrahydrofuran derivative (*Eugster*, 1956). 260 mg of pure (+)-muscarine chloride could be isolated from 125 kg of fresh toadstools. It is hygroscopic, has an m.p. 178–179° (451–452 K), $\alpha[D] = +6.7°$, and the molecular formula $C_9H_{20}O_2N^{\oplus}Cl^{\ominus}$. *Kögl* (1957) showed that it is the *trimethylammonium chloride derivative of 3-hydroxy-2-methyl-5-(aminomethyl)-tetrahydrofuran*.[1]

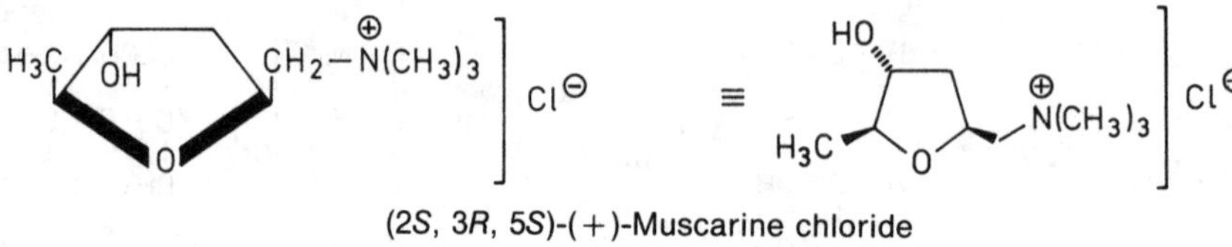

(2S, 3R, 5S)-(+)-Muscarine chloride

Structurally it can be regarded as a derivative of a deoxyaminosugar. L(+)-Muscarine chloride has been synthesised from D-glucosamine (*Hardegger*, 1957); it had the same effect as the natural product on a frog's heart.

4.1.13 Sugar Mercaptals and Thiosugars[2]

Mention was made on p. 435 that when sugars react with alcohols they do not form open-chain acetals, but instead form glycosides in which *only one* oxygen atom of the hydrated

[1] See C. H. *Eugster* et al., Helv. Chim. Acta *54*, 2704 (1971).
[2] See H. *Horton* and D. H. *Hutson*, Developments in the Chemistry of Thio Sugars in M. L. *Wolfrom* and R. S. *Tipson*, Adv. Carbohydrate Chem. *18*, pp. 123 *ff.* (Academic Press, New York 1963).

carbonyl group is alkylated whilst the other is involved in intramolecular acetal formation leading to the usual ring structure. In contrast, mercaptans (thiols) react with sugars in the presence of protic acids to give *sugar mercaptals* (thioacetals), which exist exclusively in an *open-chain* form. For example, D-glucose reacts with ethanethiol to give first of all *ethylthioglucoside* and finally D-**glucose diethylmercaptal**:

D-Glucose   Ethylthioglucoside   D-Glucose diethylmercaptal

Because of their relative insolubility and because they crystallise well, sugar mercaptals are of use in the purification and characterisation of sugars. Like the mercaptals of aldehydes and ketones they are stable to alkali.

Thiosugars are derivatives of sugars in which an oxygen atom has been replaced by a sulphur atom. Among them are the 1-thioaldoses which have an HS-group instead of an HO-group at position 1. They are best prepared starting from 'acetobromoglucose' which is got from penta-acetyl-D-glucose and hydrogen bromide. When this is treated with potassium disulphide an acetylated sugar disulphide is formed, which is converted into **1-β-D-thioglucose** by hydrolysis and reductive cleavage of the disulphide bond. Treatment of the thioglucose with alkyl halides in the presence of alkali provides 1-*β*-D-*thioglucosides*:

Acetobromoglucose   Disulphide

hydrolysis, reduction

1-*β*-D-Thioglucose   1-*β*-D-Thioglucoside

The importance of the thioglucoside group in the mustard oil glucosides has already been mentioned (p. 371). The preparative use of thiosugars lies in their ready conversion into deoxysugars, which is brought about by catalytic removal of sulphur, using *Raney* nickel.

Thiosugars in which the ring oxygen is replaced by sulphur are made from the appropriate derivatives of 4- or 5-thiosugars,[1] which, after acid hydrolysis, cyclise to give sulphur analogues of furanoses or pyranoses.

4.1.14 L(+)-Ascorbic Acid (Vitamin C)

Vitamins are organic reagents which are produced in very small amounts in plants and which are essential metabolites in human beings and in animals. If these organic catalysts or reactants are lacking in the food uptake then severe vitamin-deficiency diseases result.

The name 'vitamin' dates from the first researches on vitamin B_1 carried out by *C. Funk* in 1911, who established that it was essential for life (Lat. *vita* = life) and that it reacted like an amine. Although chemically most vitamins are not amines, this name has become established generally for this class of compounds.

INN: L(+)-*Ascorbic acid* belongs to the water-soluble vitamins and is found especially in fresh fruits (rose hips, black currants, oranges, lemons) as well as in red peppers and greenstuffs (brassica). Absence of ascorbic acid causes the disease scurvy, which results in bleeding from the skin and the gums and in severe cases leads to teeth falling out. Vitamin C has therefore been called the *antiscorbutic* vitamin. It was first isolated by *Szent-Györgi* and its structure was elucidated above all by *Micheel* and *Hirst* (1932/33). It is the γ-lactone of a ketohexonic acid whose keto group is in the enol form.

INN: L-Ascorbic acid

It may be seen from this formula that there are two adjacent hydroxy groups attached to either end of a C=C double bond. This *enediol* or *reductone* group is responsible for both the *acidity* of L-ascorbic acid and its strongly reducing character. Another interesting feature is the *chelating stabilisation* illustrated in the right-hand formula.

Tillmann's reagent (2,6-dichlorophenol-indophenol) is used as a test for ascorbic acid; in acid solution it is reduced to a colourless leuco-compound.

4.1.14.1 Preparation

In the industrial preparation of L-ascorbic acid, D-glucose is used as the starting material. This is catalytically hydrogenated to D-sorbitol which is then biochemically dehydrogenated, using *acetobacter*, to L-sorbose, whose carbon atom 5 has the same

---

[1] See *H. Paulsen*, Angew. Chem. Int. Ed. Engl. *5*, 495 (1966); Adv. Carbohydrate Chem. *23*, 115 (1968).

configuration as that of naturally occurring ascorbic acid. L-Sorbose is oxidised to 2-keto-L-gulonic acid, either directly, using oxygen and a platinum catalyst, or by treating its di-isopropylidene derivative with potassium permanganate and then removing the isopropylidene groups by acid hydrolysis. When 2-keto-L-gulonic acid is heated with dilute acid it undergoes γ-lactonisation to give L-ascorbic acid.

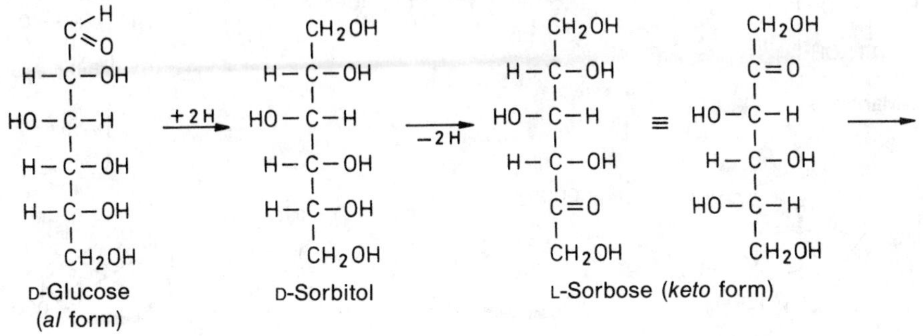

D-Glucose (al form) — D-Sorbitol — L-Sorbose (keto form)

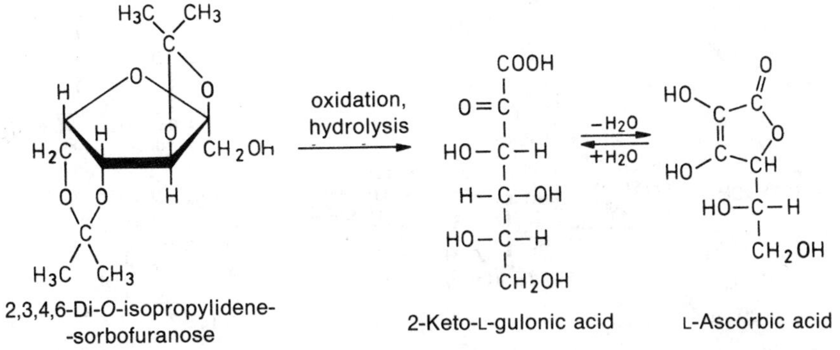

2,3,4,6-Di-O-isopropylidene-
-sorbofuranose

2-Keto-L-gulonic acid

L-Ascorbic acid

4.1.14.2 The 'Chiral Pool'

Together with other readily available chiral compounds, such as (S)-lactic acid, (S)-malic acid, (R,R)-tartaric acid, (S,S)-tartaric acid and (S)-glutamic acid, carbohydrates are of use in syntheses of *natural products* where the aim is to bring in *chiral centres* with the correct configuration at the very start of the synthesis. The synthesis of (+)-biotin from D-mannose is described as an example of this sort of methodology. This is called a *linear synthesis* since the target molecule is prepared in a series of consecutive steps from the starting material, mannose, in contrast to a *convergent synthesis* such as that used for the preparation of vitamin B_{12} (see p. 715).

The carbon atoms which are given a grey background in the formulae of D-mannose and (+)-biotin signify those which are already disposed in the *correct configuration in the starting material.* This can be appreciated from the fact that in the di-isopropylidene compound I, all the H-atoms shown in heavy type in the formula of mannose lie on the *same side* of the furanose ring (see p. 437 f.). In the following steps, partial hydrolysis to II, glycol cleavage to give III, *Wittig* reaction and hydrogenation to give IV, as well as in the opening of the furanose ring to V, the configuration at these centres remains unchanged. Ring-closure to the thiolan VI involves an S_N2 reaction; hence a *Walden* inversion takes place at C-4 of the original mannose (HMPT = hexamethylphosphoric triamide: $[(H_3C)_2N]_3P{=}O$).[1] In the introduction of the acetylamino groups in the step VI → VII, the initially introduced methanesulphonyl groups are replaced by azide groups in an S_N2 reaction; these are then reduced and acetylated. In the S_N2 reaction *Walden* inversions occur at C-2 and C-3 of the original mannose. The three chiral centres thus once again occupy the same relative configurations, because in the course of the synthesis each centre has undergone *one Walden* inversion.

---

[1] HMPT has been shown to be a carcinogen in animals; it can be replaced by the non-mutagenic tetrahydro-1,3-pyrimidin-2-one (DMPU). See Nach. Chem. Tech. Lab. *33*, 396 (1985).

Tartaric acid is a particularly valuable starting material for the synthesis of homochiral compounds (EPC: Enantiomeric Pure Compound)[1] because of the ready availability of both of its enantiomers and the many possibilities available for their conversion into reactive derivatives (functionalisation).

4.2 Oligosaccharides

The most important oligosaccharides are the di- and tri-saccharides in which, respectively, two or three monosaccharide units are joined together by glycosidic links. Being glycosides they can be split by acid into the component monosaccharides.

4.2.1 Disaccharides, $C_{12}H_{22}O_{11}$

There are only a few naturally occurring disaccharides; examples are sucrose, lactose and maltose. In cyclic monosaccharides there are two different sorts of hydroxy groups, the *hemiacetal* HO (on carbon atom 1 in aldoses and on carbon atom 2 in ketoses), and the others, which are alcohol groups. Glycosidic linkage of two monosaccharides can take place in two different ways: loss of a molecule of water involves either a hemiacetal HO-group from each monosaccharide or a hemiacetal group of the one monosaccharide and an alcoholic HO-group of the other. As a result of these alternative modes of linkage, it is possible to have both *non-reducing* and *reducing* disaccharides. Examples of these two types are, respectively, sucrose and lactose.

4.2.1.1 Non-reducing disaccharides. Sucrose (cane sugar) is made up from one molecule of D-glucose and one molecule of D-fructose. An *acetal-like* linkage, involving the hemiacetal HO-groups of both molecules, joins these two units together. Carbon atom 1 of glucose is thus linked to carbon atom 2 of fructose by means of an oxygen bridge. In consequence the *latent* carbonyl groups of both of the monosaccharides are blocked and sucrose does *not* give the reactions typical of monosaccharides such as reduction of *Fehling's* solution, osazone formation or mutarotation. It also cannot form a simple alkyl glycoside. Further investigations showed that sucrose is made up from α-D-glucopyranoside and the otherwise seldom found β-D-fructofuranose. Sucrose can thus be called either *α-D-glucopyranosyl-β-D-fructofuranose* or *β-D-fructofuranosyl-α-D-gluco-pyranose*. The second is in fact the more correct name. It has the following structure which has been confirmed both by *X-ray structure determination* and by *synthesis* (*Lemieux*, 1956).

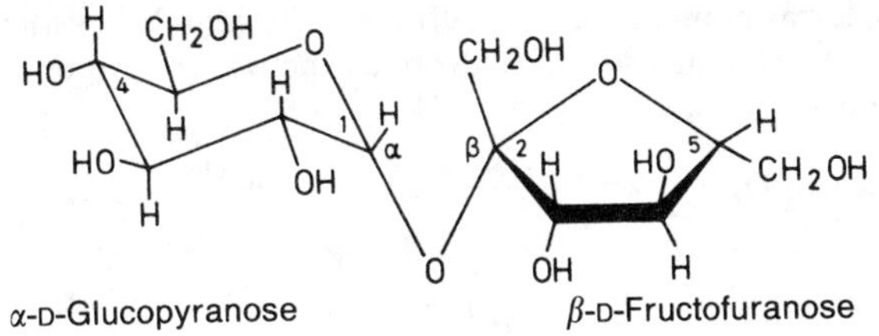

α-D-Glucopyranose   β-D-Fructofuranose

Sucrose

---

[1] See *D. Seebach* et al., Chiral Reagents from Tartaric Acid, Angew. Chem. Int. Ed. Engl. *18*, 958 (1979); *J. W. Scott*, Readily Available Chiral Carbon Fragments and Their Use in Synthesis; *J. D. Morrison* et al. (Ed.), Asymmetric Synthesis *4*, 1 (Academic Press, New York 1984).

In this conformational formula, the β-D-fructofuranose (see p. 440) is twisted at 180° to an axis which passes through the ring oxygen atom and the middle of the $C_{(3)}$—$C_{(4)}$ bond. When sucrose is hydrolysed the products are D-glucopyranose and D-*fructopyranose*, *i.e.* in the process the 5-membered ring of the D-fructofuranose is converted into a more stable pyranose ring.

4.2.1.2 Reducing disaccharides. Lactose is assembled from one molecule of D-galactose and one of D-glucose. They are joined by an *acetal-like* link between the hydroxy groups on carbon atom 1 of galactose and on carbon atom 4 of D-glucose. Lactose has the following conformational formula:

D-Galactose   D-Glucose

Lactose (β-form)

Lactose, 4-*O*-(β-D-*galactopyranosyl*)-D-*glucopyranose*, can exist in an α-form as well as in the β-form. Since there is a hemiacetal hydroxy group on carbon atom 1 of D-glucose, lactose is a reducing agent; it forms an osazone and undergoes mutarotation. The same applies for *allolactose* [6-*O*-(α-D-*galactopyranosyl*)-D-*glucopyranose*].

The Most Important Disaccharides

4.2.1.3 Sucrose (saccharose, cane sugar) is the most abundantly occurring disaccharide in plants. It occurs in almost all fruits and in many vegetable juices, especially in sugar beet (16–20%) and in sugar cane (14–16%). The latter is cultivated particularly in India, Brazil, China, the Caribbean, southern Africa and the Philippines.

Production
(a) A syrup is obtained by pressing sugar cane (*Saccharum officinarum*); this is neutralised with lime and evaporated *in vacuo* until crystallisation sets in.
(b) Sugar beet (*Beta vulgaris*) is sliced, and sugar is leached out by water in a counter-current type of operation. The solution of sugar is treated with lime-water in order to remove plant acids such as oxalic acid, tartaric acid and other hydroxyacids, as well as protein material. Excess lime which is present as soluble calcium saccharate is removed as calcium carbonate by passing carbon dioxide through the solution. The solid residue is separated off by pressing it, and the resultant 'thin juice' is concentrated *in vacuo* to 'thick juice'. Sugar crystallises out from the syrup and is separated off by centrifuging. The residual brown syrup is used as cattle-feed or is subjected to alcoholic fermentation. The crystalline sugar, which is still mixed with some syrup, is refined by recrystallisation, filtration through active charcoal and vacuum drying.[1]

Properties. Sucrose forms monoclinic prisms. It is readily soluble in water but only slightly soluble in ethanol. Dilute mineral acid hydrolyses sucrose to give one molecule each of D-*glucopyranose* and D-*fructopyranose*.

$$C_{12}H_{22}O_{11} + H_2O \xrightarrow{H^\oplus} C_6H_{12}O_6 + C_6H_{12}O_6$$

Sucrose   D-Glucose   D-Fructose

$[\alpha]_D = +66.5°$ $[\alpha]_D = +52.5°$ $[\alpha]_D = -92°$

Final Value in Water $[\alpha]_D = -20°$

[1] World production of sucrose (from sugar cane and sugar beet) amounts to about 100×10^6t per year. See *W. Walter*, Süsse and Macht (Sweetness and Power), Naturwiss. Rundschau *43*, 1 (1990).

Fructose rotates the plane of rotation to the left more than glucose rotates it to the right. As a result, when sucrose, which is dextrorotatory, is hydrolysed, the rotation changes from right- to left-handed. This has led to the hydrolysis of sucrose being described as the *inversion of sucrose* and the product mixture being called *invert sugar*. Artificial honey consists of inverted sucrose whereas real honey is a natural invert sugar.

Inversion of sucrose by dilute acid proceeds much faster than the hydrolysis of other disaccharides, and the rate is *proportional* to the hydrogen ion concentration. Sucrose is also hydrolysed by the specific enzyme invertase (saccharase) and the product is then fermented by yeast.

When sucrose is heated above its melting point a brown mass is obtained which is a mixture of a number of decomposition products and is called '*caramel*'. Conc. sulphuric acid reacts violently with sucrose to produce nearly pure carbon.

4.2.1.4 (+)-Trehalose is found in mushrooms and in other more primitive plants. It is, like sucrose, a *non-reducing* disaccharide. Acid hydrolysis provides two molecules of D-glucose. Trehalose has the structure of α-D-*glucopyranosyl-α-D-glucopyranose*, *i.e.* the two glucose residues are joined together by an acetal-like link between their 1-carbon atoms.

α,α-Trehalose

4.2.1.5 (+)-Lactose (milk sugar), accompanied by (+)-*allolactose*, is the most important sugar in milk. It occurs to the extent of 5–7% in human milk and of 4–5% in cows' milk. Lactose is produced from animal milk, from which the emulsified fats, in the form of cream, and casein have first been removed. Evaporation of the residual whey leads to the separation, first of lactalbumin, and then of lactose.

Lactose is a colourless, crystalline substance which dissolves readily in water and tastes faintly sweet. It is a strong *reducing* agent. It undergoes *mutarotation* and the final value for the specific rotation of the mixture of α- and β-lactoses in water, $[\alpha]_D$, is equal to +52.3°. It is hydrolysed by acid or by the enzyme *lactase* (β-galactosidase) to give one molecule each of D-glucose and D-galactose. Large numbers of the world's adult population have no lactase; because of this, for such people milk powder is only of help to children and is of no value to hungry adults. When milk is curdled or turned into yoghurt, bacterial action converts the lactose into lactic acid.

4.2.1.6 (+)-Maltose (malt sugar) occurs especially in germinating seeds, for example in malted barley, and is a water-soluble intermediate in the biochemical formation of starch. Maltose is made by enzymatic degradation of starch by the enzyme *diastase*. It crystallises as needles which are readily soluble in water but only dissolve with difficulty in ethanol. Either dilute acids or the enzyme *maltase* (α-glucosidase), which is present in yeast, split maltose into two molecules of D-glucose. Maltose reduces *Fehling's* solution, forms an osazone and undergoes *mutarotation*; its chemical properties thus resemble those of lactose. It is a *reducing* disaccharide; the linking oxygen bridge joins carbon atom 1 of one glucose unit to carbon atom 4 of the other glucose unit, thus leaving the half-acetal hydroxy group on carbon atom 1 of the second glucose molecule untouched. Maltose is 4-O-(α-D-glucopyranosyl)-D-glucopyranose and it can exist in α- and β-forms. The final value for the optical rotation of the anomeric equilibrium in water is $[\alpha]_D = +128.5°$.

Maltose (α-form)

A general method for the elucidation of the constitution of disaccharides may be illustrated using *maltose* as an example. First of all, maltose was oxidised by bromine water to maltobionic acid. This was then methylated, using dimethyl sulphate in alkaline solution, giving a *methyl ester of octa-O-methylmaltobionic acid*. When hydrolysed, this ester is split into 2,3,4,6-*tetra-O-methyl-D-glucose* and 2,3,5,6-*tetra-O-methyl-D-gluconic acid*:

The hydrogen atoms have been omitted from the above formulae for the sake of clarity.

The first fragment obtained from the octa-O-methylmaltobionic acid methyl ester is derived from the 'half-acetal component' and the second fragment from the 'alcohol component' of maltose. In this way, by successive oxidation, methylation and hydrolysis reactions it was possible to establish the structural formula of maltose.

4.2.1.7 (+)-Cellobiose is a breakdown product derived from cellulose and, like maltose, is made up from two molecules of D-glucose. The two disaccharides differ in their structure, in that maltose is the α-glucoside, whereas cellobiose is the corresponding β-glucoside. Its structure is that of 4-O-(β-D-glucopyranosyl)-D-glucopyranose.

Cellobiose (β-form)

Cellobiose can be obtained by *acetolysis of pure cellulose* with acetic anhydride in the presence of conc. sulphuric acid, which provides the *octa-acetate* of cellobiose. This is hydrolysed to cellobiose using either potassium hydroxide or sodium methoxide. It is a colourless, crystalline substance which is soluble in water. It gives the usual reactions of a *reducing* disaccharide. It is dextrorotatory, $[α]_D = +34.6°$. The enzyme *emulsin* (β-glucosidase) splits cellobiose into two molecules of D-glucose.

4.2.1.8 Gentiobiose and **melibiose** *do not occur as free disaccharides* in nature, but arise from hydrolysis of, respectively, the trisaccharides gentianose or raffinose. Gentiobiose also occurs in the glycoside *amygdalin*. Both of these disaccharides differ from the others in that glycoside formation involves the HO-group on carbon atom 6. Gentiobiose is 6-O-(β-D-glucopyranosyl)-D-glucopyranose, while melibiose is 6-O-(α-D-galactopyranosyl)-D-glucopyranose:

Melibiose (β-form)   Gentiobiose (β-form)

4.2.1.9 *Streptomycin* was isolated from a culture of *Streptomyces griseus* (*Waksman*, 1944). It has attained great importance in chemotherapy as an *antibiotic* active against tuberculosis, meningitis and pneumonia.

Antibiotics are specific chemical compounds, originating from fungi or microorganisms, which interfere with the growth or the propagation of certain bacteria.

INN: *Streptomycin* (p. 456) includes in its structure the disaccharide *streptobiose*, which is joined by a glycosidic link to *streptidine* (*Folkers, Wolfrom, Wintersteiner* and others). Streptidine is a derivative of *scyllo*-inositol, which is the *all-trans* isomer of inositol (see p. 403), in which two of the hydroxy groups are replaced by guanidino groups.

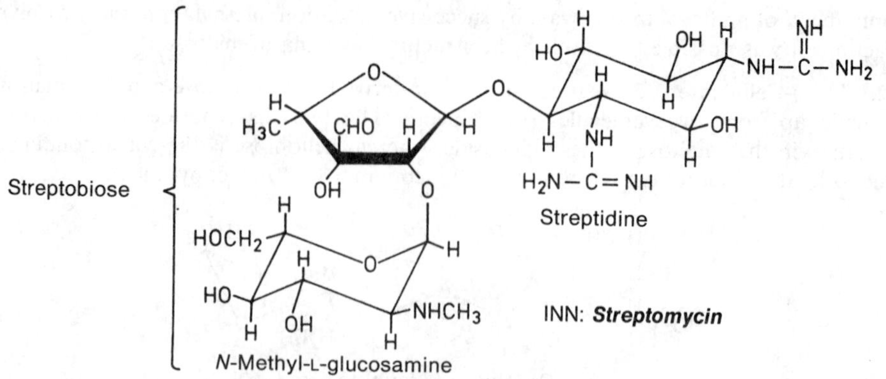

Streptobiose

Streptidine

INN: **Streptomycin**

N-Methyl-L-glucosamine

It is of interest that two sugars from the L-series, *i.e.* N-methyl-L-glucosamine (see p. 445) and a dialdehyde-deoxysugar, L-streptose, are present in streptomycin (*Wintersteiner*).

4.2.2 *Trisaccharides*, $C_{18}H_{32}O_{16}$

The few known trisaccharides resemble disaccharides in their structure. **Raffinose**, which occurs in sugar beet, may be taken as an example. In this molecule D-glucose, D-fructose and D-galactose are joined together by glycosidic links. It is a non-reducing sugar and has the following conformational formula:

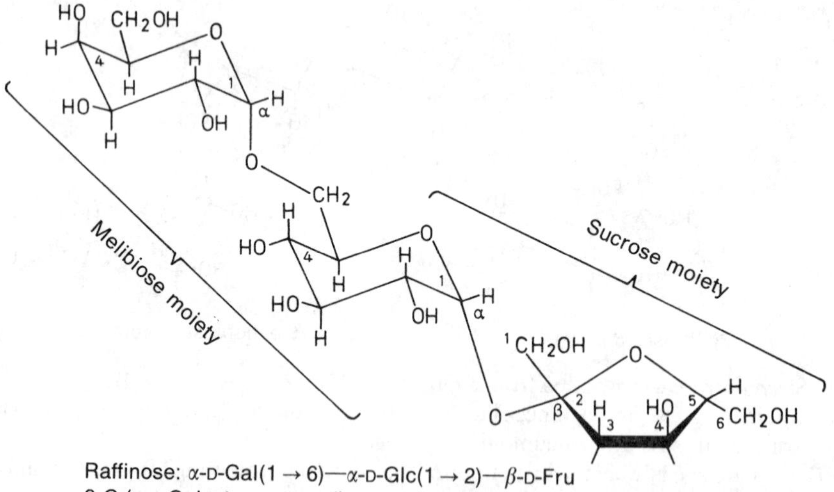

Melibiose moiety

Sucrose moiety

Raffinose: α-D-Gal(1 → 6)—α-D-Glc(1 → 2)—β-D-Fru
6-*O*-(α-D-Galactopyranosyl)—α-D-glucopyranosyl—β-D-fructofuranoside

The enzyme *invertase* (β-fructosidase) hydrolyses raffinose to D-fructose and melibiose whereas the enzyme *emulsin* (α-galactosidase) hydrolyses it to sucrose and D-galactose. This proves that the three monosaccharide units are linked in the order shown.

4.2.3 Pseudo-oligosaccharides[1]

Pseudo-oligosaccharides differ from the oligosaccharides in that one of the monosaccharide units is replaced by a cyclitol unit, as shown in the accompanying formula of *acarbose*.

Acarbose

Acarbose is found in actinomyces cultures. It is a pseudotetrasaccharide, in which an unsaturated cyclitol (hydroxymethylconduritol) is joined by an α-pseudo-*N*-glycosidic link to 4-amino-4,6-dideoxy-D-glucose, which in turn is joined by a glucosidic link to maltose.

The two pseudoglycidically linked units make up the core of a series of pseudo-oligosaccharides, to which acarbose belongs. Pseudo-oligo- and pseudopoly-saccharides with different structural cores are widely distributed in cereals, and in bacteria and fungi. They inhibit α-glucosidases and are of interest in view of their ability to regulate carbohydrate metabolism. INN: *Acarbose* is used as an oral medicine in the treatment of diabetes.

4.3 Polysaccharides (Glycans)[2]

The most important polysaccharides are *starch*, *glycogen* and *cellulose*, which are all built up from D-glucose units, almost entirely joined together by bonds from the hydroxy groups attached to carbon atom 1 of one glucose unit to the alcoholic HO-group at carbon atom 4 of the next glucose unit, with loss of a molecule of water for each link. These *acetal-like* chains involve large numbers, hundreds or even thousands, of monosaccharide units, resulting in the formation of *macro-molecules*. These high-polymer sugars have quite different physical and chemical properties from those of the monosaccharides and oligosaccharides. The glycosidic coupling of monosaccharide units may provide unbranched or branched chains of these units. Most polysaccharides with the overall formula $(C_6H_{10}O_5)_n$ are either insoluble in water or form colloidal suspensions. The colloidal form appears to have a high relative *molecular mass*, which can be anywhere between 17,000 and several million.

---

[1] See *E. Truscheit* et al., Chemistry and Biochemistry of Microbial α-Glucosidase Inhibitors, Angew. Chem. Int. Ed. Engl. *20*, 744 (1981).
[2] See *G. O. Aspinall* (Ed.), Polysaccharides, *1–3* (Academic Press, New York 1982–1985).

Chemical, enzymatic and physical methods have been used in the determination of the structures and the relative molecular masses of polysaccharides, for example osmotic pressure, viscosity measurements, behaviour in an ultracentrifuge, electron microscopy and X-ray structure analysis. In addition, further information about the molecular size of these high polymers can be obtained from end-group analysis. In this a quantitative estimate is made of the amount of a specific end-group relative to that of the entire molecule (*Staudinger, Haworth*). Accurate information on the chain length of single polysaccharides has not as yet been achieved.

4.3.1 Starch (Amylum)[1]

Starch is formed by photosynthesis in the cells of green plants.

Under the influence of the green plant pigment chlorophyll and the action of sunlight (photosynthesis), carbon dioxide from the air is converted into starch. In the course of this process, which is not as yet completely understood, the plants give off molecular oxygen. In this way almost 10^{11} tons of carbohydrate are produced every year on the earth.

Biochemically formed starch is deposited in the chloroplasts in the form of small colourless starch granules, which have a laminated structure. Transport of starch within plants is not possible since the plant cell membranes are impervious to colloidal starch. However, plants have the ability to break down photochemically formed starch into maltose and then glucose. Glucose can migrate in the plant and at certain storage points, especially in tubers (*e.g.* potatoes) and in seeds (*e.g.* cereals), can be reconverted into starch to form reserve stores of it.

Starch granules consist of *amylopectin* ($\sim 80\%$) in the hulls and *amylose* ($\sim 20\%$) in the interiors. The two polysaccharides differ from each other in their chemical and physical properties.

4.3.1.1 Amylose is made up largely of unbranched chains of glucose units, which are joined together by *α(1,4)-glucosidic* links: (α-1,4-glucan):

Amylose
(1 → 4)-α-D-Glucopyranan

Sterically, amylose forms a *helix* (screw) with six D-glucose units per turn. The *relative molecular masses* range from 17,000 to 225,000, which represents a chain length of about 100–1400 glucose units.

---

[1] See *T. Gaillard* (Ed.), Starch: Properties and Potential (Wiley, New York 1987).

Useful information concerning the structure and the relative molecular mass of amylose was obtained, as in the case of disaccharides, by methylation followed by hydrolysis. It may be seen from the above formula that each glucose unit in the amylose chain has three free HO-groups. In fact, methylation and subsequent hydrolysis of amylose provides for the most part *2,3,6-tri-O-methyl-D-glucose* together with a small amount (0.5%) of *2,3,4,6-tetra-O-methyl-D-glucose*, which is derived from the units at the ends of the amylose chain (**end-group analysis**).

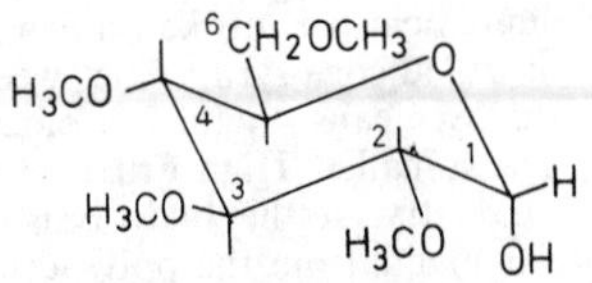

2,3,4,6-Tetra-O-methyl-
-D-glucose (α-form)

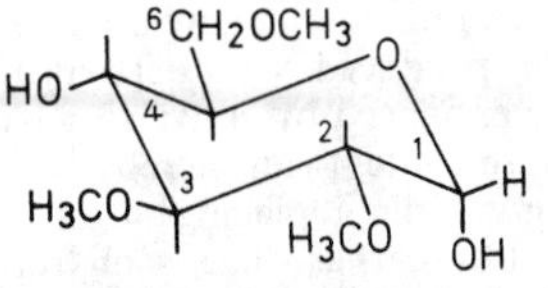

2,3,6-Tri-O-methyl-
-D-glucose (α-form)

4.3.1.2 Amylopectin is the main constituent of starch. Like amylose it is constructed from D-glucose units, but in the case of amylopectin they are assembled in shorter, rather bush-like, branched chains, containing only 20–25 units. The links in the chain are *α(1,4)-glucosidic*, while the branching points involve *α(1,6)-glucosidic* bonds. The *relative molecular masses* range from 200,000 to a million or sometimes even higher.

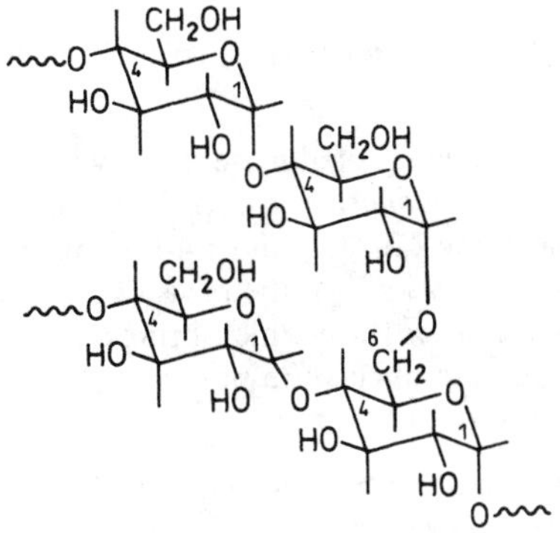

Portion of the branched
amylopectin chain

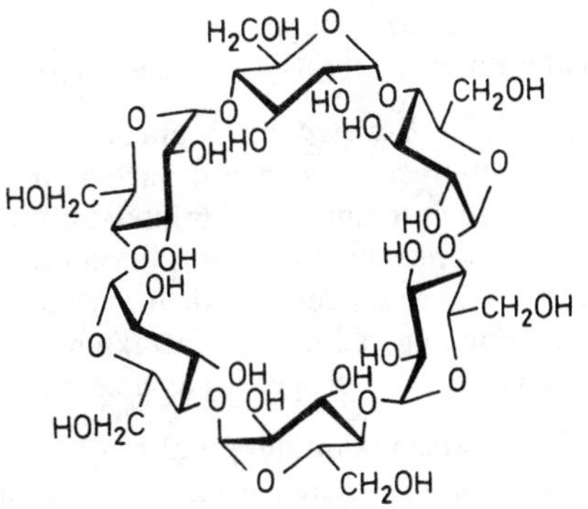

α-Cyclodextrin

Production

The main sources of starch are potatoes, grain, maize and rice. Finely ground meal from cereals is almost pure starch; protein-rich binding material is separated from the heavier starch granules by elution. The cell walls of potatoes are broken down by grinding and the starch granules are washed out. The coarse debris from the cells is sieved out, and the starch granules separate out from the milky suspension. They are dried at 25–30°C (300–305 K) and milled to provide *flour*.

Properties

Starch is a white hygroscopic powder which is insoluble in cold water. When heated with water to about 90°C (365 K) it forms *starch paste*; the gelatinous nature of paste is a result of the amylopectin swelling up. This contrasts with the formation of a colloidal suspension by amylose. If the filtrate is treated with alcohol, *soluble starch* separates out; this gives the blue colouration typical of amylose when an iodine–potassium iodide solution is added to it. This colouration stems from the intercalation of the iodine molecules in the amylose to form an *inclusion compound*. The *iodine–starch reaction* serves as a very sensitive test for amylose or for iodine. When warmed, the blue colour disappears, but it reappears when the mixture is cooled again. Amylopectin gives a red-violet colour with iodine.

When starch (amylose + amylopectin) is heated with dilute acid it is broken down into glucose and a syrup is produced. That derived from maize starch is called corn syrup. The sweetness of this syrup can be enhanced by converting part of the D-glucose, by means of glucoseisomerase, into D-fructose, whose content can thereby be more than 50%. The so-called 'High Fructose Corn Syrup' (HFCS) is gradually replacing sugar in the soft-drink industry. If the hydrolysis of starch is interrupted at an earlier stage, a mixture of breakdown fragments from the polysaccharide chain, which is extremely difficult to separate, is obtained. This is called *dextrin*, which is used, among other things, as an adhesive and as a size. It is an amorphous substance which forms a colloidal suspension in water. It has free aldehyde groups and reduces *Fehling's* solution fairly well. Starch can also be broken down enzymatically.

Cyclodextrins[1] (p. 459) are formed by the action of *B. macerans* on starch. They contain 6, 7 (or 8) α-1,4-linked glucose units (α-, β- or γ-cyclodextrins have internal diameters of 500–800 pm). Because of their well-defined hydrophobic inner surface they serve as useful models for the character of cavities (pockets) in enzymes. By introducing n-pentyl groups onto the HO-groups of the glucose units in α- and β-cyclodextrins, amorphous waxy compounds are obtained, which are used as chiral stationary phases in capillary gas chromatography[2] (see p. 348).

Cyclic *glucans* made up from 1,2-linked β-D-glucose units play a role in osmotic regulation in bacteria in varying surroundings.

4.3.1.3 INN: *Dextrans* are branched polysaccharides made up of glucose units, which are formed by micro-organisms, particularly those of the *leuconostoc* group. For the most part the glucose units are 1,6-linked but there is also some 1,2-, 1,3- and 1,4-linking. They are used as plasma-substitutes in blood transfusions and also form the basis of *Sephadex* gels, which are made from them by reaction with *epichlorohydrin* which brings about 3-dimensional cross-linking. The extent of the cross-linking affects the porosity of the gel and therewith its properties as a separating agent (see p. 7).

4.3.1.4 Xanthan (xanthan gum) is a polysaccharide with a relative molecular mass of one to ten million, isolated from the bacterium *Xanthomonas campestris*. There is a β-1,4-glucan main chain and, on average, a trisaccharide with the structure β-D-Man$(1 \rightarrow 4)$-β-D-GlcU$(1 \rightarrow 2)$-α-D-Man is attached to each 3-position.

About 50% of the terminal D-mannose units in each side chain form acetals at their 4,6-positions with pyruvic acid, while the α-D-mannose unit which is joined to a glucose unit in the main chain is acetylated at its 6-position.[3] Xanthan forms highly viscous aqueous solutions which are used on a large scale industrially, and in the manufacture of foodstuffs, as emulsifiers, stabilisers and gelling agents.

---

[1] See *W. Saenger*, Cyclodextrin Inclusion Compounds in Research and Industry, Angew. Chem. Int. Ed. Engl. *19*, 344 (1980); *A. P. Croft* et al., Synthesis of Chemically Modified Cyclodextrins, Tetrahedron *39*, 1417 (1983).

[2] See *W. A. König*, Eine neue Generation chiraler Trennphasen für die Gaschromatographie (A New Generation of Chiral Stationary Phases for Use in Gas Chromatography), Nachr. Chem. Tech. Lab. *37*, 471 (1989).

[3] See *P. E. Janssen* et al., Structure of the Extracellular Polysaccharide from Xanthomonas Campestris, Carbohdr. Res. *45*, 275 (1975).

4.3.2 Glycogen

While starch is produced exclusively in plants, glycogen serves as the 'reserve carbohydrate' in animal organisms. It arises from conversion of part of the carbohydrate food uptake and is stored in the liver (up to 18% of liver weight) as well as in muscles. It can be isolated by, for example, heating rabbit liver with aqueous potassium hydroxide.

When glycogen is hydrolysed by acid, D-glucose is formed; with β-amylase maltose is the product. The chemical structure of glycogen resembles that of amylopectin, but the chains are more highly branched and of higher molecular mass. With iodine it gives a red-brown colour. The *relative molecular mass* of glycogen varies in different organs in the range 10^5 to 10^7. There is a branch chain attached to the main chain at intervals of from 3 to 12 glucose units. The enzymatic breakdown of glycogen to L-lactic acid will be discussed later (p. 905).

4.3.3 Inulin

Another plant reserve carbohydrate is inulin, which is found especially in dahlia and artichoke tubers, from which it is prepared. It also has a high molecular mass. In contrast to the two polysaccharides mentioned previously, inulin is built up from D-*fructose* units which are almost all joined together by β(1,2)-*glycosidic* links, *i.e.* (2 → 1)-β-D-fructofuranan. Inulin forms a colloidal suspension with warm water. It is stable to alkali and does *not* reduce *Fehling's* solution. Enzymatic hydrolysis of inulin provides also a small amount of D-glucopyranose and it is concluded from this that sucrose molecules form the end-groups of inulin molecules. The relative *molecular mass* of inulin is about 5200, *i.e.* it consists of a chain of about 30 fructose units with a glucose unit at the ends of the chain.

4.3.4 Chitin

Chitin is found in the shells of crustaceans, in the exo-skeletons of insects (*e.g.* cockchafers) and in fungi (*e.g. boletus edulis*). The whole molecule is made up of a chain of *N-acetyl-D-glucosamine units* (Glc NAc), joined together by β(1,4)-*glycosidic* links. X-ray structure determination has shown that the structure of chitin resembles that of cellulose, with the difference that the HO-groups on the 2-carbon atoms are replaced by acetylamino groups.

Portion of a chitin chain, 2-acetamido-2-deoxy-D-glycan

When chitin is heated with dilute hydrochloric acid, it is split up into D-glucosamine and acetic acid.

4.3.5 Pectins

In plants, and especially in fruits, tubers and stalks, other polysaccharides are formed, which readily gelatinise; these are the *pectins*. They are polygalacturonans in which some of the carboxyl groups of the D-galacturonic acid units are in the form of their methyl esters.

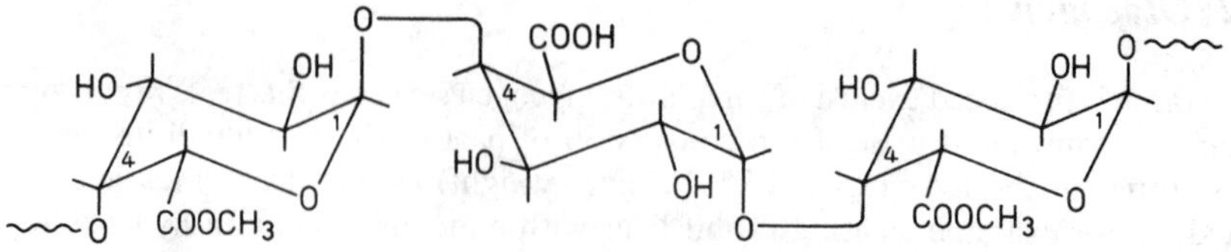

Portion of a pectin chain

The characteristic property of the pectins is their ability in the presence of acids and sugar to produce gels. As a result of this they are used in the preparation of jellies and jams and other condiments. Fruit brandies, such as raspberry brandy, contain varying amounts of methanol which is derived from hydrolysis of the methyl ester groups; in the formation of fruit wines small amounts of methanol are also formed; the permitted quantity is not more than 0.02%.

4.3.6 Cellulose

Cellulose forms the structural material for the greater part of plant cell walls, and is the most commonly occurring polysaccharide. Plant fibres such as cotton, jute, flax and hemp are more or less pure cellulose. Wood from coniferous and deciduous trees is a composite material containing 40–50% cellulose together with about 25–30% of hemicelluloses and lignins. Straw contains about 30% cellulose.

Pure cellulose can be obtained from cotton by first of all removing fat and oil with dilute sodium hydroxide. Acid or enzymatic hydrolysis of cellulose converts it into D-glucose; if suitable reaction conditions are used, cellobiose, cellotriose, cellotetrose and so on can be isolated as intermediate products. Hydrolysis of fully methylated cellulose gives in high yield 2,3,6-tri-O-methyl-D-glucose; cellulose thus has a similar structure to that of starch. Extensive research has shown that cellulose has the following structure (Haworth, Freudenberg):

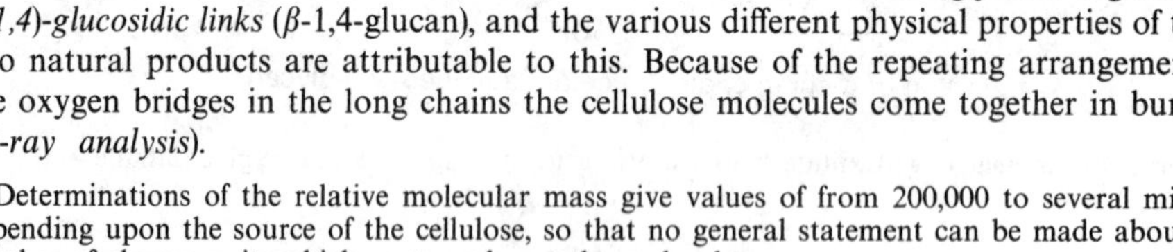

Portion of a cellulose chain

The D-glucose units in cellulose differ from those in starch in being joined together by β(1,4)-glucosidic links (β-1,4-glucan), and the various different physical properties of these two natural products are attributable to this. Because of the repeating arrangement of the oxygen bridges in the long chains the cellulose molecules come together in bundles (X-ray analysis).

Determinations of the relative molecular mass give values of from 200,000 to several million, depending upon the source of the cellulose, so that no general statement can be made about the number of glucose units which go to make up the molecules.

Cellulose is a colourless substance which is insoluble in water and in most organic solvents. It dissolves readily in ammoniacal copper(II) hydroxide solution (Schweizer's reagent), forming a complex in the process. It also dissolves in concentrated solutions of hydrochloric acid; this is accompanied by some decomposition to products of lower molecular mass. Cellulose can be partially digested by ruminants; if it is treated with hydrogen peroxide in dilute alkaline solution its ease of digestion is appreciably increased.

Cellulose is very important industrially. For the most part it is obtained from wood or straw. It is known commercially as *woodpulp*.

4.3.6.1 Woodpulp.

(a) In the most commonly applied *acid extraction process* (*sulphite extraction*) wood chips (spruce, pine or beech), of about the size of nuts, are heated in pressure chambers, at about 4 bar (0.4 MPa), with calcium hydrogen sulphite solution, $Ca(HSO_3)_2$. In the course of this treatment the lignins (as sulphonic acids), hemicelluloses and resinous materials go into solution and become the sulphite waste liquors. They contain fermentable sugars from which ethanol can be made. The remaining sulphited pulp is washed and treated with bleaching powder; it then forms white plates which are about 85–90% cellulose.

(b) *Alkali process* (*sulphate process*). Pulverised wood or straw is heated with aqueous sodium hydroxide to which a specific amount of sodium sulphide and sodium sulphate has been added, at 7–10 bar (0.7–1 MPa). This largely separates the cellulose from other accompanying materials. The work-up of this sodium woodpulp is done in the same way as in the case of that obtained from the sulphite process.

4.3.6.2 Paper.

In order to make paper, the woodpulp is stirred with water to make a paste, moulded and then dried. The resultant paper is porous and closely resembles filter paper (unsized paper) which consists of almost pure cellulose. In order to prepare high-grade paper, various materials such as barium or calcium sulphate and also china clay (kaolin) and especially resin soap are added to the woodpulp paste. This provides *sized paper*. In making cheaper paper, especially that for use in newspapers, crude cellulose is treated with up to 90% finely powdered wood, whose high lignin content results in the yellowing of the paper with time. When unsized paper, *e.g.* filter paper, is dipped briefly into 75% sulphuric acid and then at once washed with water, it is turned into *parchment paper*.

Since the middle of the last century many different kinds of paper have been produced which contain acid and are therefore unstable and do not last. For example, a combination of resin acids and alum provides a size whose pH can approach 4.8. In combination with the water which is present absorbed in the paper (4–6%), this slowly hydrolyses the cellulose. Such paper becomes brittle with the passage of time and ultimately disintegrates. This process, which seriously threatens the durability of books and journals, can be stopped if the acid is neutralised by treatment with gaseous diethyl zinc under reduced pressure. This process works satisfactorily for the present but it is not without its problems.

4.3.7 Hemicelluloses

Hemicelluloses are always associated with cellulose and are found especially in cell walls. The main examples are the *plant gums*, which consist of a mixture of various polysaccharides which on hydrolysis are converted into D-xylose (from xylan), D-mannose (from mannan), D-galactose (from galactan) and D-glucuronic acid; they also contain methoxy groups. The *pentosans* (xylan, araban, arabinan) are abundant in straw (20–30%) and in bran. **Xylan** contains D-xylopyranose units joined together by *β(1,4)-glycosidic links*:

Portion of a xylan chain

4.3.8 Cellulose Ethers

The three free hydroxy groups in each of the glucose units in cellulose can be converted into ethers, using alkyl halides such as ethyl chloride or sodium chloroacetate. Ethyl chloride provides *ethylcellulose* (EC), which is insoluble in water; sodium chloroacetate gives the sodium salt of *carboxymethylcellulose* (CMC, CMCellulose), which is soluble in water. Usually only a portion of the available hydroxy groups are converted into ether groups. Depending upon the extent of etherification and the degree of polymerisation, they are used variously as detergents, in varnishes and paints, as adhesives, and also for thickening and gelling foodstuffs. In the pharmaceutical industry they are used as the base of ointments and as coating for solid pharmaceuticals.

4.3.9 Cellulose Nitrates

The three hydroxy groups of each glucose unit in cellulose can be esterified by a mixture of nitric acid and sulphuric acid (*nitrating acid*) to give *mono-*, *di-* and *tri-nitrocellulose*. As in the case of glycerol, the products are not nitro-compounds but rather are esters, *i.e.* nitrates. The name nitrocellulose is chemically incorrect. Highly nitrated cellulose, containing about 13% nitrogen, is known as *guncotton* (*C. F. Schönbein*, 1846); it burns harmlessly in open air, but when moulded it explodes on initiation by a detonator.

When guncotton is converted into a gel by the action of an ethanol–ether mixture, *smokeless powder* results, which is used as a propellant for projectiles.

Cellulose which is less nitrated ($\sim$ 10% N) dissolves in a mixture of ethanol and ether to give *collodion*, which is used to plasticise nitroglycerol and also in medicine for sealing wounds. Thorough kneading of collodion with an ethanolic solution of camphor produces the elastic horn-like material *celluloid* (*J. W. Hyatt*, 1868). It was among the first plastics to be used, and even now it has some use as a thermoplastic material. The camphor acts as a *plasticiser*. Cellulose nitrates are especially useful in the preparation of nitrocellulose lacquers.

4.3.10 Cellulose-based Synthetic Fibres (Half-synthetic Fibres)

The manufacture of half-synthetic fibres is based largely on cellulose. A distinction is made between *regenerated cellulose* fibres (hydrated cellulose) and *cellulose esters*.

Hydrated cellulose is not a hydrate of cellulose but cellulose whose structure has been altered by degradation and depolymerisation.

Originally, fibres which were made by modifying natural biopolymers were known as 'artificial fibres' or as 'artificial silks'. Textile fibres (staple fibres) of definite length are differentiated from fibres of undefined length, which were formerly called filaments or 'silks'. They are made either by dissolving woodpulp or cottonwool linters, or cellulose esters, in a suitable solvent, followed by reprecipitation through fine spinnerets in special precipitation baths (*solution-spun fibres*) or by evaporating off the solvent (*dry-spun fibres*).

4.3.10.1 Viscose fibres

If woodpulp is treated with 15–20% aqueous sodium hydroxide and allowed to age or 'ripen', the cellulose swells up; in the course of the process there is a shortening of the average chain-length of the cellulose molecules. The product is known as *alkali cellulose*. This reacts with carbon disulphide to give *cellulose xanthate*:

$$\text{—OH} + CS_2 + NaOH \longrightarrow \text{—O—C—S}^{\ominus}Na^{\oplus} \xrightarrow{(H^{\oplus})} \text{—OH} + CS_2 + Na^{\oplus}$$

| Cellulose | Cellulose xanthate (soluble) | Regenerated cellulose (insoluble) |

The cellulose xanthate contains on average about one molecule of CS_2 for two glucose units. After a further period of 'ripening' it is spun into a bath of dilute sulphuric acid where the xanthate decomposes into cellulose, carbon disulphide and sodium sulphate. The regenerated cellulose is stretched and forms viscose fibre. Uncut viscose fibres were formerly called *viscose-rayon*, while cut fibres were called staple rayon. Viscose fibres, which are built up from cellulose molecules, resemble *cotton* both in their structure and in their properties.

Alternatively, the cellulose xanthate can be cast into the form of a broad thin film in the acid precipitation bath. Addition of glycerol as a plasticiser to this regenerated cellulose provides *cellophane*.

4.3.10.2 Cellulose fibres

Cellulose can also be spun into fibres without any preceding chemical treatment. This process, which is now superseded, depends upon the solubility of cellulose in tetra-aminecopper(II) hydroxide, $[Cu(NH_3)_4](OH)_2$ (*Schweizer's* reagent). In this process the cuprammonium rayon was obtained as fibres of indeterminate length; they are soluble in 4-methylmorpholine-1-oxide.

4.3.10.3 Acetate- and Triacetate-fibres

In contrast to viscose fibres which consist of regenerated cellulose, acetate fibres consist of acetate esters of cellulose. If cellulose (usually cotton linters) is treated with acetic anhydride together with a small amount of sulphuric acid it is acetylated, forming cellulose triacetate (primary acetate). Since this product poses problems because of its poor solubility, it is partially hydrolysed by addition of a calculated amount of water to provide a 'mixed' cellulose acetate in which, on average, there are $2–2\frac{1}{2}$ acetyl groups present for each glucose unit (secondary acetate). The resultant *acetylcellulose* is soluble in acetone and is dry-spun to produce acetate fibres (formerly called acetate silk). Acetate fibres are also used in the preparation of cigarette filters.

With the help of plasticisers, commonly phthalate esters, acetylcellulose can be made into materials which only burn with difficulty (*Cellon, Ecarit*) and can take the place of the easily ignited celluloid. Foil and nonflammable *cellitfilm* (safety film) are similarly made from acetylcellulose. Acetylcellulose varnishes are characterised by a special gloss and by their resistance to light and heat. Mixed esters of cellulose, especially acetyl-butyryl-cellulose, also have special uses.

The chemical fibres which are based on naturally occurring macromolecules such as cellulose and modified cellulose should be distinguished from the synthetic fibres, such as polyamides and polyesters. The synthetic fibres are of wider importance than the chemical fibres derived from cellulose[1].

---

[1] See *B. v. Falkai*, Synthesefasern (Synthetic Fibres) (VCH Verlagsgesellschaft, Weinheim, 1981).

5 Aromatic Compounds. Derivatives of Benzene

This class of compounds consists of *benzene* and its derivatives, including compounds in which more than one benzene ring are condensed together, *e.g.* naphthalene, anthracene. The term 'aromatic compounds' originated from the fact that a number of the earliest known benzene derivatives, which were isolated from plant material, had 'aromas' or 'aromatic' smells. This name originally carried no information about their chemical structure, but the name came to be used to describe this extensive and important group of compounds.[1] Benzenoid hydrocarbons are known as *arenes* and the groups derived by removal of a hydrogen atom from arenes are called *aryl groups*, *e.g.* a halogeno-derivative of benzene is an example of an *aryl halide*. Thus *aryl* is derived from *arene* in the same way that *alkyl* is derived from *alkane*.

Some special chemical properties and a series of typical reactions are associated with the presence of a benzene ring in a molecule. One consequence of these special properties has been the difficulty they engendered in providing a suitable formula to express them, and this has occupied the attention of chemists right down to the present day. Mention has already been made to some valence isomers of benzene (see p. 416), but these only represent the tip of the iceberg, for in all there are 217 possible compounds having the molecular formula C_6H_6.

Structure of Benzene

Benzene has the molecular formula C_6H_6; it is made up from six functionally identical CH-groups. *Kekulé* (1865/67) first suggested a triply-unsaturated 6-membered ring formula for benzene; in it the six CH-groups were joined together by alternate C—C single and C=C double bonds.[2] If specific numbers are allocated to the six carbon atoms it becomes evident that there are two ways of writing the formula of benzene, each having D_{3h} symmetry:

---

[1] See *H-G. Franck*, Industrielle Aromatenchemie (Springer-Verlag, Berlin 1987); *J. P. Snyder*, Aromaticity: Preelectron Events, Non-benzenoid Aromatics Vol. *1*, p. 1 (Academic Press, New York 1969).
[2] There has been much discussion recently concerning the priorty of *Loschmidt* in suggesting a cyclic formula for benzene; see, for example, *C. R. Noe* and *A. Bader*, Chem. Brit. *29*, 126 (1993); *G. P. Schiemenz*, Goodbye Kekulé? Joseph Loschmidt und die monocyclishe Struktur des Benzols, Naturwiss, Rundschau, *46*, 85 (1993); Chem. Ind. (London) *1993*, 522.

Whilst these two *Kekulé* formulae provide satisfactory explanations for the bonding in the ring, they do not adequately explain some of the properties and reactions of benzene. If, for example, two hydrogen atoms on adjacent carbon atoms are replaced by other atoms or groups of atoms (X), then in theory there should be two *isomeric compounds* in one of which the two CX-groups are linked by a single bond, and in the other by a double bond:

However, two different '*ortho* isomers' (see p. 471) have never been discovered. This led *Kekulé* to postulate that the bonding in the benzene ring was not fixed but was in a permanent state of exchange between the two possible forms (*oscillation theory*).

It might be expected that benzene should have similar chemical properties to those of alkenes, which undergo a characteristic set of reactions, and especially to those of dienes. But benzene is in fact a rather unreactive molecule. For example, *addition of bromine* to benzene takes place very much more slowly than it adds to an alkene. Similarly, a weakly alkaline solution of potassium permanganate is not decolourised by benzene, even when the mixture is heated. Furthermore, benzene shows no tendency to polymerise, a characteristic property of unsaturated compounds.

On the other hand, and in contrast to this apparent anomalous behaviour of benzene, there are some reactions which do serve to show the presence of three double bonds in the molecule. For example, ozone reacts with benzene to give a *triozonide*, which can be either hydrolysed or reduced by zinc and acetic acid to provide three molecules of *glyoxal*:

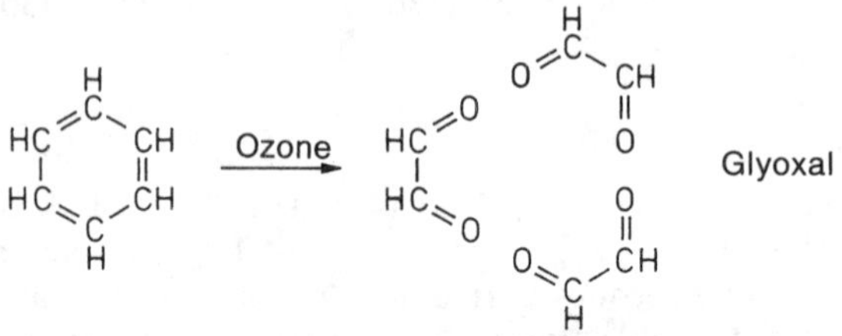

In addition, *catalytic hydrogenation of benzene*, which proceeds stepwise to give dihydro-, tetrahydro- and hexahydro-benzene (cyclohexane), C_6H_{12}, provides evidence in support of *Kekulé*'s formula. It is noteworthy, however, that a number of dihydrobenzene (cyclohexadiene) derivatives readily lose two hydrogen atoms to regenerate a benzene ring. Whilst formation of a C=C double bond by loss of two hydrogen atoms generally requires an expenditure of about 117–126 kJ/mol, *1,2-dihydrobenzene* is converted into benzene in a very rapid weakly exothermic reaction:

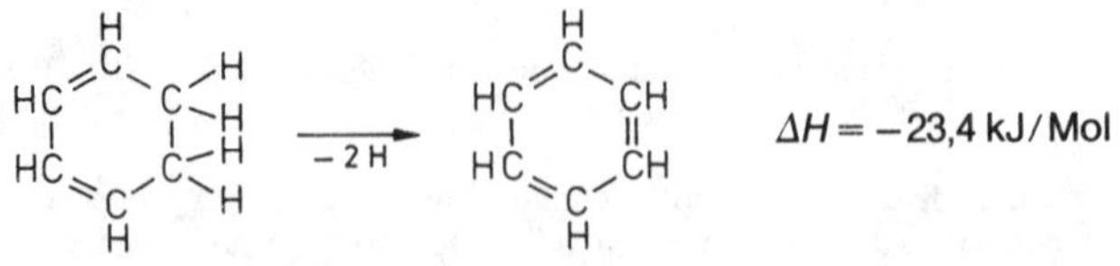

$\Delta H = -23,4\ kJ/Mol$

This release of energy suggests that the system of bonding in the benzene ring must be especially stable.

In order to overcome the incompatibility of the *Kekulé* formula with the chemical properties of benzene, a variety of other formulae were put forward, such as the *Dewar*-benzene formula (1867) (so called although *J. Dewar* himself did not favour it; see p. 417), the diagonal formula of *Claus* (1867) and the centric formula of *Armstrong* and of *Baeyer* (1887), which is a variant of the diagonal formula:

<table>
<tr>
<td>

H
C
HC | CH
‖ | ‖
HC | CH
C
H

'Dewar'
benzene

</td>
<td>

H
C
HC CH
| ✕ |
HC CH
C
H

(Claus)

</td>
<td>

H
C
HC CH
| ✳ |
HC CH
C
H

(Armstrong, Baeyer)

</td>
<td>

▷◁

Dicyclopropenyl

</td>
</tr>
</table>

In addition to *Dewar benzene*, *prismane*, *fulvene* and *benzvalene*, a derivative of another type of cyclic compound having the molecular formula C_6H_6 has been prepared, namely the hexaphenyl derivative of dicyclopropenyl.[1] All of these systems are less stable than the corresponding benzene derivatives, and are converted into benzene or derivatives of benzene when heated.

The first quantum mechanical investigations on the structure of benzene were carried out by *E. Hückel* (1931) and *L. Pauling* (1933), with the help of approximation methods, and led to a more precise concept of the structure of benzene.

X-ray structural analysis indicates that the carbon atoms of benzene are in the form of a flat, regular hexagon, which shows a very good approximation to D_{6h} symmetry, and whose edges are 139.7 pm in length. This bond distance lies between those of a single bond (147.6 pm, see p. 103) and those of a double bond (133.8 pm) (in each case for bonds between sp^2-hybridised carbon atoms). Thus this suggests that in benzene—in contrast to butadiene—there must be an *ideal interchange of bonds*, in other words, benzene is a typical *mesomeric system*, which can be suitably represented as an average of the two *Kekulé* forms. In order to simplify the following formulae, the carbon and hydrogen atoms are not included.

In benzene and its derivatives the π-electrons form a characteristic *π-electron sextet*, from which the stability and other properties of this class of compound are derived. The π-electron sextet can be represented by an inscribed circle in the ring, following a suggestion by *Robinson* (1925). Since this formula more and more appears as a symbol

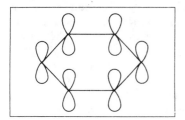

Fig. 110.

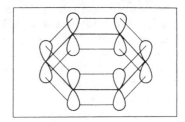

Fig. 111.

[1] See *R. Breslow*, J. Am. Chem. Soc. *81*, 4747 (1959).

for aromatic character it is frequently used in succeeding formulae for benzene. The so-called *Kekulé* formula is, however, also an accepted formula for benzene; it must be read as a symbol and not as a representation of the structure of benzene.

According to orbital theory, after all the σ-bonding system of benzene has been constructed, there remains on each carbon atom a $2p_z$ orbital. All six of these orbitals are arranged parallel to one another and perpendicular to the plane of the ring (Fig. 110), with the result that they can overlap one another (Fig. 111). The resultant molecular orbital is spread over the whole system, above and below the plane of the ring, which itself forms a nodal plane.

Since the *Pauli* exclusion principle limits the number of electrons in any molecular orbital to two, the six π-electrons associated with the benzene ring must be distributed between three molecular orbitals. These are shown schematically, viewed from above the ring, in Figs. 112a–c. The orbital of lowest energy (Fig. 112a) is illustrated again in Fig. 112d in a more detailed representation (see Fig. 20, p. 41). The other two orbitals each have a nodal plane perpendicular to the plane of the benzene ring; the sign of the wave function changes across this nodal plane. (See Table 11, p. 105.) These two orbitals are of identical energy; because of this they are described as *degenerate orbitals*. There are also two degenerate anti-bonding orbitals; it is in these orbitals that the unpaired electron in the radical anion of benzene, which was mentioned on p. 46 f., is delocalised.

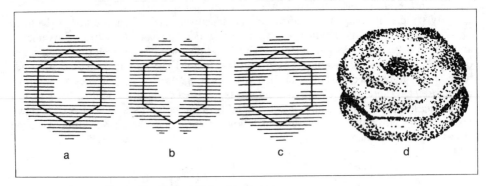

Fig. 112. The bonding molecular orbitals of benzene

Benzene, as a mesomeric system, has considerable stability. It is appreciably lower in energy than would be expected for a compound having three double bonds.

The difference in energy between a *Kekulé* structure and the actual molecular structure can be estimated from the theoretically calculated and the experimentally determined heat of hydrogenation. Hydrogenation of cyclohexene to cyclohexane is an exothermic reaction involving the loss of 120 kJ/mol. Since additivity of heats of hydrogenation has been confirmed experimentally it would be expected that, if benzene had a cyclohexatriene structure, its hydrogenation to cyclohexane should involve the loss of $3 \times 120 = 360$ kJ/mol. It was shown, however, by *Kistiakowski* (1936) that the heat of hydrogenation of benzene, when it is reduced to cyclohexane, is only 209 kJ/mol. The difference between these two values indicates that benzene is 151 kJ/mol lower in energy than the hypothetical cyclohexatriene molecule. This figure of 151 kJ/mol has up to now been variously called the *mesomeric energy*, the *resonance energy* and the *delocalisation energy*. M. J. S. Dewar has proposed that, instead, the term *stabilisation energy* should be used, since only a fraction of this energy is attributable to the delocalisation of the π-electrons.

In recent years, however, it has been pointed out that critical appraisal of the experimental data on the structure of benzene suggests that rigorous proof of the generally accepted D_{6h} structure for benzene does not in fact exist, and that X-ray and spectroscopic studies are equally interpretable in terms of very rapidly interconverting D_{3h} (*Kekulé*) structures.[1] It has also been suggested (*Hiberty, Shaik, Vollhardt*) that the high symmetry and delocalisation of the π-electrons in benzene follows from the symmetric nature of the σ-skeleton.[2]

Nomenclature of Benzene Derivatives

As more of the hydrogen atoms of benzene are replaced by other atoms or groups of atoms, so the number of possible structural isomers increases; they are positional isomers, differing in the relative positions to which identical substituents are attached. To indicate the positions where the substituents are present, the atoms of the ring are numbered consecutively 1 to 6 in such a way that the lowest numbers possible are given to the substituents. In the following example, if a second methyl group is introduced into a toluene ring (wherein the carbon atom carrying the initial methyl group must be taken as C-1), it may be at either the 2-, 3- or 4-positions. Hence there are three *positional isomers* of dimethylbenzene, more commonly called *xylene*, the 1,2-, 1,3- and 1,4-dimethylbenzenes. An alternative form of nomenclature, introduced by *Körner*, uses the prefixes *ortho-, meta-* and *para-* (abbreviated to *o-, m-* and *p*, respectively) to describe 1,2-, 1,3- and 1,4-disubstituted benzenes.

| Toluene | *o*- or 1,2- | *m*- or 1,3- | *p*- or 1,4- |

Dimethylbenzenes (xylenes)

In toluene the 2- and 6-positions are equivalent to one another, as are the 3- and 5-positions; but numbering is always such as to use the lowest numbers, *i.e.* 1,2- rather than 1,6-. Many benzene derivatives (for example, the xylenes) have trivial names, often derived from the natural sources from which they were originally obtained.

For trisubstituted derivatives of benzene the number of isomers depends upon the types of substituents present. If all three are identical then there are three possible isomers. The substituents may either all be adjacent to one another or *vicinal* (*vic.*), or *symmetric* (*sym*), or *asymmetric* (*asym.*). These three types are, respectively, 1,2,3-, 1,3,5- and 1,2,4-trisubstituted benzene derivatives, *e.g.*

1,2,3- or *vic.*-   1,3,5- or *sym.*-   1,2,4- or *asym.*-

Trimethylbenzenes

(Hemimellitene)   (Mesitylene)   (Pseudocumene)

---

[1] See *O. Ermer*, Angew. Chem. Int. Ed. Engl. *26*, 782 (1987); *B. Y. Simkin, V. I. Minkin* and *M. N. Glukhovtsev*, Adv. Het. Chem. 56, 303 (1993).
[2] See *S. Shaik, P. C. Hiberty* et al., J. Org. Chem. *50*, 4659 (1985); *A. Stanger* and *K. P. C. Vollhardt*, J. Org. Chem. *53*, 4889 (1988); J. Am. Chem. Soc. *107*, 3089; *E. Jug* and *A. M. Köster*, J. Am. Chem. Soc. *112*, 6772 (1990); *A. Streitwieser, K. P. C. Vollhardt, F. Weinhold* et al., J. Am. Chem. Soc. *115*, 10952 (1993).

If the three substituents are all different then the number of possible isomers is increased to ten. If there are more than three substituents their sites are normally defined by numbers.

Removal of a hydrogen atom from an *arene* leaves an *aryl group* (shortened in formulae to Ar); examples are *phenyl*, $-C_6H_5$, and *tolyl*, $-C_6H_4-CH_3$. Removal of a hydrogen atom of the methyl side-chain of toluene leaves the *benzyl* group $-CH_2-C_6H_5$.

In writing out formulae, phenyl groups are frequently represented by the symbol Ph, or sometimes by ϕ (Grk. phi).

5.1 Arenes. Aromatic Hydrocarbons

The fundamental source of arenes and their derivatives is *coal tar*,[1] a black viscous liquid with a characteristic smell, which is a valuable by-product from the coking of coal in coke-works and town gasworks. Coal tar, which is carcinogenic, consists of a mixture of neutral, acidic and basic substances; up to the present about 350 different compounds have been definitely identified or isolated.

In recent years coal and coal tar have decreased in importance as sources of benzene and its derivatives. The so-called BTX aromatics, *benzene, toluene* and the *xylenes* (*o-, m-* and *p*-xylene) have increasingly been obtained from mineral oil; this had risen to more than 85% in Europe and North America by 1985.

The first step in the industrial treatment of coal tar is to fractionate it by distillation. About 40% of the original tar comes over as a distillate, which may be classed into the following five main fractions:

(a) 80–170°C (355–445 K) *Light oil* (1–2%). This consists principally of *benzene, toluene* and the three isomeric *xylenes* (*o-, m-* and *p*-xylene), as well as some trimethylbenzenes (mesitylene, pseudocumene and hemimellitene), durene and other tetramethylbenzenes and also *styrene, cyclopentadiene*, dicyclopentadiene, *indene*, hydrindene and other hydrocarbons. These may be separated by further fractional distillation in fractionating towers.

Nitrogen-containing bases such as *pyridine* and its methyl derivatives and aniline are also found in light oil. They can be separated out as salts by treatment with acid. In addition, there are sulphur-containing compounds such as *thiophene*.

(b) 170–215°C (445–490 K) *Middle oil* (2–3%). The main constituents are *naphthalene* and *phenol* together with the various cresol and xylenol isomers and nitrogenous bases such as toluidines, quinolines and isoquinolines. The acidic phenols can be separated as salts with aqueous sodium hydroxide and the bases as salts with sulphuric acid.

(c) 215–240°C (490–515 K) *Naphthalene oil* (8–10%). As well as naphthalene, phenol and their homologues (alkylnaphthalenes, cresols and xylenols) this contains higher hydrocarbons such as *acenaphthylene* and *fluorene*, and also dibenzofuran. There are also bases belonging to the indole, quinoline and isoquinoline series. The majority of the naphthalene solidifies and is separated off.

---

[1] See *K. F. Lang* et al., Im Steinkohlenteer nachgewiesene organische Verbindungen (Organic Compounds Found in Coal Tar), Fortschr.chem. Forsch. *8*, 91 (1967).

(d) 230–290°C (505–565 K) *Wash oil* (8–10%). This contains the same substances found in (c) and also *biphenyl* and *dibenzofuran*. Because all the constituents mutually lower each other's melting points, no solid residue forms down to 0°C (273 K). Wash oil is used to wash benzene out of the coke-oven gases.

(e) 270–350°C (545–625 K) *Anthracene oil* (18–25%). In the cold this solidifies to a soft, crystalline paste, from which higher hydrocarbons such as *anthracene, phenanthrene* and *fluorene*, and nitrogen-containing substances such as *carbazole, acridine* and *phenanthridine* are isolable. It is used in the preparation of carbon black.
The remaining oily material is used as *creosote*, for impregnating wood, and as fuel oil.

The residual *pitch* ($\sim$ 55%) has a softening point of about 60–75°C (335–350 K). Chemically it is a very complex mixture made up largely of polycyclic arenes with smaller amounts of high-molecular-weight, soot-like compounds. These are insoluble and cannot be distilled and are thus only separated chemically with great difficulty. Among the more important constituents which have been isolated are *pyrene, chrysene, fluoranthrene* and homologues of carbazole.

Most of the pitch is used in the manufacture of electrodes and for conversion into coke.[1]

5.1.1 Benzene

Benzene was first isolated by *Faraday* (1825) from a compressed illuminating gas which had been made commercially by decomposing fish oils at a red heat. It was later obtained (E. *Mitscherlich*, 1833) by decarboxylation of benzoic acid, C_6H_5—COOH. The cyclisation of acetylene to give benzene has already been mentioned (see p. 97). The increasing demand for benzene has resulted in its being produced not only from coal tar distillation and from gases from coke ovens but also, in ever-increasing amount by the petrochemical industry, from processes based on mineral oil.

Benzene is produced in differing amounts in reforming processes (see p. 85), for example by the dehydrogenative ring-closure of *n-alkanes* to cycloalkanes and their subsequent aromatisation:

Hexane   Cyclohexane   Benzene

In a similar way benzene is obtained by dehydrogenation of methylcyclopentane–cyclohexane mixtures:

Methylcyclopentane   Cyclohexane   Benzene

---

[1] See *H. G. Franck* et al., Steinkohlenteer, Chemie, Technologie und Verwendung (Coal Tar, Chemistry, Technology and Uses) (Springer-Verlag, Berlin 1968).

Pyrolysis benzene obtained from *steam cracking* must be freed from polymerisable alkenes by hydrogenating them.

Propane, obtained from oil refining, is converted into benzene and hydrogen in the presence of a suitable catalyst.[1]

Alkylbenzenes can be dealkylated thermally to provide benzene; for example, if toluene is heated to 600–700°C (875–975 K) under pressure in the presence of hydrogen, benzene and methane are the products.

$$\langle\!\langle O \rangle\!\rangle\!-\!CH_3 + H_2 \longrightarrow \langle\!\langle O \rangle\!\rangle + CH_4 \qquad \Delta H = -126 \text{ kJ/Mol}$$

The arenes are concentrated by extraction with suitable solvents, *e.g.* with *glycols* in the Udex-*process*, by tetrahydrothiophene dioxide in the *sulfolan process*, or with *N-methyl-2-pyrrolidone* in the *Arosolvan process*. They are then isolated by distillation from the mixtures.

Properties

Benzene is a colourless refracting liquid, b.p. 80.1°C (353.3 K), with a characteristic odour. It burns with a very sooty flame. It is miscible in all proportions with most organic solvents, *e.g.* ether, alcohols, acetone or acetic acid. Water dissolves in benzene to the extent of 1%.

When continuously inhaled, benzene acts as a strong *poison*, bringing about vertigo, vomiting and unconsciousness. This action results from the ready solubility of fats and lipids in benzene. Chronic exposure leads to damage to the bone marrow, the liver and kidneys, and causes leukemia. Benzene is also a carcinogen.

Uses

Above all, benzene is used as an important starting material in the chemical industry, for example for the preparation of chlorobenzene, phenol, nitrobenzene, aniline, styrene, alkyl benzenes, cyclohexane and maleic anhydride, as well as in the preparation of many dyestuffs, insecticides and pharmaceuticals.

5.1.2 Homologues of Benzene. Alkylbenzenes

Alkylbenzenes can be obtained not only by the industrial methods from mineral oil, coke-oven gas and coal tar, but also by synthetic methods.

Preparation

5.1.2.1 Wurtz–Fittig Synthesis (1863). *Wurtz's* method for the preparation of alkanes was applied to the preparation of alkylbenzenes by *Fittig*. When a halogenobenzene is treated with an alkyl halide in solution in ether in the presence of sodium, a two-step reaction ensues: a first-formed organosodium derivative reacts with the alkyl halide to give an alkylbenzene.

$$\langle\!\langle O \rangle\!\rangle\!-\!Br \xrightarrow[-\text{NaBr}]{+\,2\,Na} \langle\!\langle O \rangle\!\rangle\!-\!Na \xrightarrow[-\text{NaCl}]{+\,Cl-R} \langle\!\langle O \rangle\!\rangle\!-\!R$$

Bromobenzene   Phenyl sodium   Alkylbenzene

[1] See *P. C. Doolan* and *P. R. Pujado*, Hydrocarbon Processing, *1989, September*, 72

In general, the alkylbenzene is not the sole product of this reaction, biphenyl and thé corresponding bialkyl being formed as by-products. The yields of alkylbenzenes depend very much upon the aryl halide reacting much faster with sodium than the alkyl halide does. In addition, the phenyl sodium must react faster with the alkyl halide than it does with the aryl halide, *i.e.* the carbon–halogen bond of the alkyl halide must be polarised more readily than that of the aryl halide; this is usually the case.

5.1.2.2   *Friedel–Crafts* Alkylation[1] (1877). When alkyl halides react with arenes in the presence of catalytic amounts of anhydrous *aluminium chloride*, alkylated arenes are formed, often accompanied by vigorous evolution of hydrogen halide, *e.g.*

The mechanism of alkylation is as follows. The aluminium chloride molecule is a *Lewis* acid with a sextet of electrons and has the tendency to bind to the halogen atom of the alkyl halide to form a complex anion such as $AlCl_4^\ominus$. In doing so it does not lead to a complete cleavage of the carbon–halogen bond. The carbon atom of this bond bears a strong positive charge and makes an *electrophilic* attack on the π-electron sextet of the benzene ring:

π-Complex   Carbenium ion

π-Complex

The initially formed π-complex can stabilise itself by conversion into a carbenium ion, which is known as a σ-complex; its positive charge is delocalised.[2] This mechanism resembles that of electrophilic additions to C=C double bonds (see p. 67 *f*). While in the case of alkenes and alkynes an addition reaction takes place, in the case of the benzene ring a proton is eliminated from the π-complex which is formed, resulting in the reformation of the energetically favoured delocalised system of benzene.

A real disadvantage of the *Friedel–Crafts* alkylation of arenes stems from the fact that the entering alkyl group activates the benzene ring, with the result that reaction is not limited to the introduction of only one substituent group. The mono-alkyl derivative reacts faster with alkyl halides than benzene itself does, so that polyalkylated derivatives, especially *m*-derivatives, are also formed. In addition, under the reaction conditions used, aluminium chloride can also bring about dehydrogenation and isomerisation reactions. For example, the main product obtained from 1-halogenopropanes and benzene is isopropylbenzene. *Friedel–Crafts* alkylation is also reversible so that isomerisation, both intra- and inter-molecular, can take place, leading ultimately to a thermodynamically controlled mixture of products.

Friedel–Crafts reactions, including also the *acylation* of arenes, are usually carried out using aluminium chloride, but iron(III) chloride, tin(IV) chloride, zinc chloride, boron trifluoride, anhydrous hydrogen fluoride, conc. sulphuric acid or phosphoric acid are also sometimes used.

---

[1] See *C. C. Price*, The Alkylation of Aromatic Compounds by the Friedel–Crafts Method, Org. React. *3*, 1 *ff.* (1946); *G. A. Olah* et al., Friedel–Crafts Alkylations, Comprehensive Organic Synthesis, Vol. 3, p. 293, ed. *B. M. Trost* (Pergamon, Oxford 1991).

[2] It is still uncertain whether or not the π-complex is formed as a precursor of the σ-complex. See *D. V. Banthorpe*, Chem. Rev. *70*, 295 (1970).

5.1.2.3 Instead of alkyl halides, *alkenes* can be used to react with arenes in the presence of *aluminium chloride* to provide alkylbenzenes. For example, ethylbenzene is made industrially in this way from benzene and ethylene.

Further alkylation to diethyl- and polyethyl-benzenes is suppressed if benzene is present in excess.

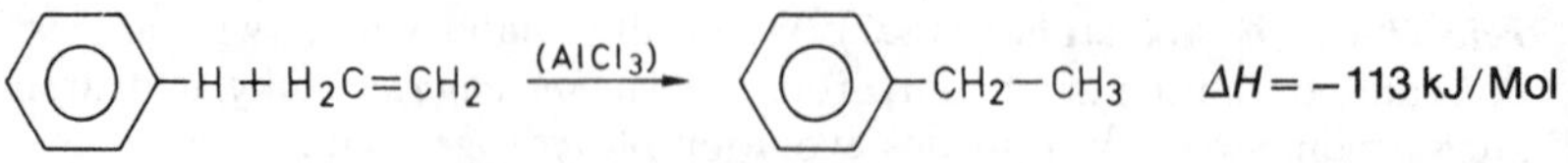

$$\text{[benzene]}-H + H_2C=CH_2 \xrightarrow{(AlCl_3)} \text{[benzene]}-CH_2-CH_3 \qquad \Delta H = -113 \text{ kJ/Mol}$$

Ethylbenzene

The Most Important Alkylbenzenes

In general, the homologues of benzene differ very little from benzene itself in their chemical and physical properties. Introduction of a methyl substituent on the ring raises the boiling point by about 30 K.

5.1.3 Toluene

Toluene (methylbenzene), C_6H_5—CH_3, was discovered by *Pelletier* and *Walther* (1836), who obtained it by dry-distilling pine needles. Its name is derived from Tolu balsam from which it was also isolated by distillation.

Preparation

5.1.3.1 From the light oil fraction from coal-tar distillation.

5.1.3.2 From the C_7-*fraction* of catalytically cracked oil or from mineral oil by *aromatisation* at about 500°C (775 K) under pressure (catalytic reforming process, see p. 473).

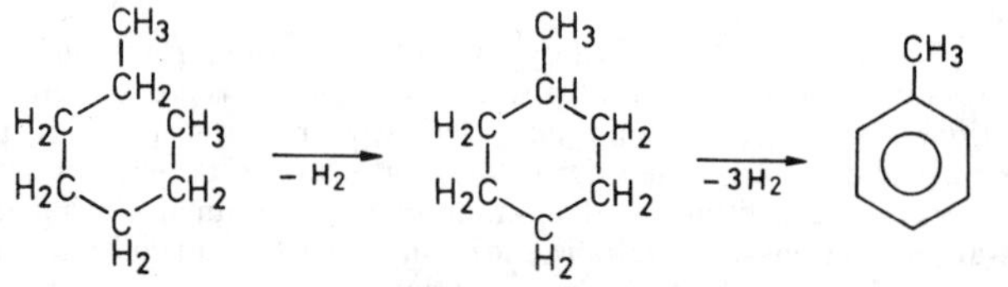

n-Heptane   Methylcyclohexane   Toluene

Properties. Toluene is a colourless liquid, b.p. 110.8°C (384.0 K), with a smell resembling that of benzene. Strong oxidation, using sodium dichromate and sulphuric acid, potassium permanganate or catalytically in air, converts toluene into benzoic acid:

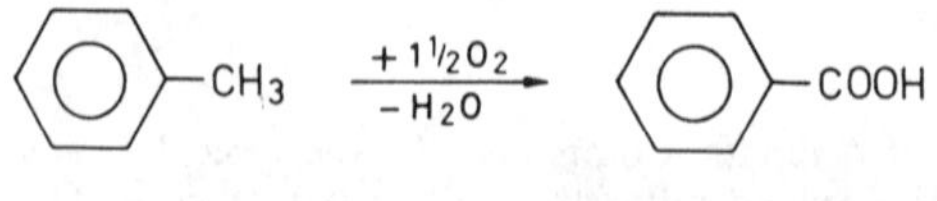

$$\text{[benzene]}-CH_3 \xrightarrow[-H_2O]{+1\frac{1}{2}O_2} \text{[benzene]}-COOH \qquad \text{Benzoic acid}$$

The oxidation of a methyl group to a carboxyl group is commonly used as a method for converting alkylarenes into arylcarboxylic acids. Longer side-chains attached to a benzene ring are similarly degraded oxidatively to carboxyl groups.

Uses. Toluene is used as a solvent, and also as an intermediate in the industrial preparation of trinitrotoluene, vinyltoluene, benzoic acid, saccharin, dyestuffs, *etc.*

5.1.4 Ethylbenzene

Ethylbenzene, C_6H_5—C_2H_5, is a colourless liquid, b.p. 136°C (409 K) which is important as a starting material for the manufacture of styrene. For its preparation see p. 476.

5.1.5 Xylenes

The xylenes (dimethylbenzenes) are colourless liquids (Grk. *xylon* = wood). Xylene obtained from coal tar is a mixture of the three *isomeric* dimethylbenzenes (for formulae see p. 471) in the proportions *m*-xylene (50–60%), *o*-xylene and *p*-xylene (each 20–25%). The three xylenes have closely similar boiling points and for a long time their separation was only achieved by sulphonating them. Their boiling points are *o-xylene*, 143.6°C (416.8 K) (m.p. −28°C [245 K]); *m-xylene*, 139.2°C (412.4 K) (m.p. −49°C [224 K]); *p-xylene* 138.4°C (411.6 K) (m.p. 13°C [286 K]). On the industrial scale, the higher-boiling *o-xylene* is now separated out in a fractionating column, and *p-xylene* is separated from *m*-xylene by freezing it out.

In more recent times, the xylenes have been obtained by *aromatisation* of the petrol fraction and by conversion of butane, obtained from oil refining, in the presence of a suitable catalyst, into xylenes and hydrogen;[1] *p*-xylene is separated out of the resultant mixture by using molecular sieves and is used for making terephthalic acid.

5.1.6 Trimethylbenzenes

There are three isomeric trimethylbenzenes which are found in the light oil fraction from coal tar distillation, namely *mesitylene* (1,3,5- or *sym.*-trimethylbenzene), b.p. 165°C (438 K), *pseudocumene* (1,2,4- or *asym.*-trimethylbenzene), b.p. 170°C (443 K) and *hemimellitene* (1,2,3- or *vic.*-trimethylbenzene), b.p. 176°C (449 K) (for formulae see p. 471).

5.1.7 Cumene

Cumene (isopropylbenzene) is a liquid, b.p. 153°C (426 K), which is best made industrially from *propene* and *benzene* in the presence of *phosphoric acid* as a catalyst at 250°C (525 K) and 400 bar (40 MPa):

	Propene		Cumene

$\Delta H = -113\ kJ/Mol$

First of all an *isopropyl cation* is formed, which, as a *secondary* carbenium ion, is more stable than the *primary* propyl cation. This cation then makes an *electrophilic* attack on the benzene ring. This reaction is a special type of *Friedel–Crafts* alkylation.

---

[1] See *P. C. Doolan* and *P. R. Pujado*, Hydrocarbon Processing, *1989*, September, 72.

Cumene is used as the starting material for the preparation of phenol and of acetone.

p-Cymene (*p*-isopropyltoluene) is found in oil of thyme, eucalyptus oil and oil of carraway, and is structurally related to the terpenes.

p-Cymene (b.p. 177°C [450 K]

It can be obtained by the dehydrogenation of some terpenes, and from wood resins as a by-product in the manufacture of woodpulp by the sulphite process.

5.2 Halogen Derivatives of Arenes

Mention has already been made of the differing reactivities of aliphatic and of aryl hydrocarbons towards halogens (chlorine and bromine). Nevertheless, under the appropriate conditions arenes may take part in both addition and substitution reactions involving the benzene ring.

5.2.1 Addition of Halogens to the Benzene Ring

Addition of chlorine or bromine to benzene only occurs under the influence of UV light. Under these conditions the first step is the splitting of halogen molecules into very reactive halogen atoms. These then add to the benzene ring in *radical chain reactions*, which lead to the stepwise formation of hexachlorocyclohexane, HCH, (benzene hexachloride), $C_6H_6Cl_6$, or hexabromocyclohexane (benzene hexabromide), $C_6H_6Br_6$, *e.g.*

Hexachlorocyclohexane

Hexachlorocyclohexane can exist in eight different stereoisomeric forms. The γ-isomer has the aaaeee conformation and is an effective insecticide (*Gammexane, Lindane*, BHC), m.p. 112°C (385 K). This compound does not decompose readily and can cause harm to the environment. It is possible that 'dioxin' (see p. 789) is formed from it. The α-isomer, which has an aaeeee conformation, is chiral. Estimations of the concentrations of the two enantiomers in sea water provides some insight into the biological breakdown and transformation processes of the hexachlorocyclohexanes (*Hühnerfuss*, 1990).

Halogeno-substituents on the Benzene Ring

Substitution reactions involving the benzene ring are much more important than the addition reactions, in particular direct *halogenation, nitration* and *sulphonation*, most of which involve *ionic electrophilic mechanisms.*[1]

5.2.2 Halogenobenzenes

Direct chlorination or bromination of benzene only takes place in the presence of a so-called *halogen-carrier* such as iron, iron(III) chloride, aluminium chloride or the corresponding bromides, all of which are capable of taking up a halide anion. The actual reagent involved can be regarded as a halonium cation which makes an *electrophilic* attack on the π-electron sextet of the benzene ring (see p. 475). A *π-complex* results which in turn gives a *carbenium ion*; the benzene system is regenerated by loss of a proton, this step proceeding *via* formation of another π-complex, *e.g.*

| π-Complex | Carbenium ion | π-Complex | Chloro benzene |

$$H^{\oplus} + FeCl_4^{\ominus} \longrightarrow FeCl_3 + HCl$$

The main products, chloro- or bromo-benzene, are formed in 80–90% yield; the by-products are *p*- and *o*-dihalogenobenzenes.

Reaction of two molecules of chlorine with one molecule of benzene gives *p*-dichlorobenzene (1,4-dichlorobenzene) as the main product together with some *o*-dichlorobenzene (1,2-dichlorobenzene) and a trace of *m*-dichlorobenzene:

p-Dichlorobenzene   *o*-Dichlorobenzene

Industrially *chlorobenzene* is made by the *Raschig* process, in which a gaseous mixture of benzene, hydrogen chloride and air is passed over a copper catalyst (*oxychlorination*):

$$C_6H_6 + HCl + \tfrac{1}{2} O_2 \xrightarrow{(Cu)} C_6H_5-Cl + H_2O$$

Bromination of toluene in the presence of iron gives a mixture of *o*- and *p*-bromotoluene:

---

[1] See *R. Taylor*, Electrophilic Aromatic Substitution (Wiley, Chichester 1990).

o-Bromotoluene   p-Bromotoluene

Arenes cannot be directly iodinated since the reaction *equilibrium* is strongly on the side of the starting material. Reaction proceeds only in the presence of nitric acid or mercury(II) oxide, which remove the hydrogen iodide, thereby displacing the equilibrium, *e.g.*

Iodobenzene

$$4HI + 2HNO_3 \longrightarrow 2I_2 + N_2O_3 + 3H_2O$$

$$2HI + HgO \longrightarrow HgI_2 + H_2O$$

Direct fluorination of arenes can be achieved using *xenon difluoride*. Iodo- and fluoro-arenes are usually made from the corresponding arylamines, *via* aryldiazonium salts. More recently a method for the preparation of highly fluorinated arenes has been developed.[1]

Properties

Monohalogenobenzenes are colourless liquids whose physical properties, like those of alkyl halides, vary regularly, depending upon the atomic weight of the halogen atom. For example, the *boiling points* are as follows: *fluorobenzene*, C_6H_5—F, 85°C (358 K); *chlorobenzene*, C_6H_5—Cl, 132°C (405 K); *bromobenzene*, C_6H_5—Br, 156°C (429 K); *iodobenzene*, C_6H_5—I, 188.5°C (461.7 K). It is interesting to note how small an increase in boiling point results when a hydrogen atom in benzene (b.p. 80.1°C [353.3K]) is replaced by a fluorine atom.

Like alkyl halides, aryl bromides and aryl iodides react readily with *magnesium* in ether to form *Grignard* reagents, which undergo the usual reactions of such reagents, *e.g.*

Phenyl magnesium bromide

In the aryl series lithium derivatives are frequently used instead of *Grignard* reagents (see p. 183 *f.*). Related to this, the reaction of aryl halides with sodium in the *Wurtz–Fittig* reaction should be noted.

In contrast to alkyl halides, the halogen atoms in aryl halides are rather unreactive towards *nucleophiles*. For example, aryl halides may be heated with alkali, ammonia, potassium cyanide, potassium sulphide or alcoholic silver nitrate without any exchange or loss of halogen taking place. Reaction does take place if the halogen atom is activated by neighbouring polar electron-withdrawing groups of atoms such as —NO_2, —CN or —COOH. This will be discussed more fully later (see p. 503 *f.*).

---

[1] See R. D. *Chambers*, Fluorine in Organic Chemistry (Wiley, New York 1974).

5.2.2.1 Treatment of an aryl halide with an alkali-metal amide in liquid ammonia, does, however, bring about *nucleophilic aromatic substitution*. This takes place by means of an *elimination–addition mechanism* involving a short-lived intermediate *dehydrobenzene* (aryne, benzyne) which has been detected in an argon matrix. *J. D. Roberts* (1953) showed that when [1-^{14}C]-chlorobenzene reacts with potassium amide in liquid ammonia both [1-^{14}C]-aniline and [2-^{14}C]-aniline are formed in approximately equal amounts. Ammonia adds equally well to either end of the triple bond of benzyne:

[1-^{14}C]-Chlorobenzene   Benzyne   [1-^{14}C]-Aniline   [2-^{14}C]-Aniline

This explains why in halogen-exchange reactions of *o*- and *p*-substituted aryl halides, *m*-substituted products are often formed.

o-Bromofluorobenzene reacts with lithium amalgam to form benzyne as an intermediate which in the presence of furan undergoes a *Diels–Alder* reaction. The product is 1,4-dihydro-1,4-epoxynaphthalene, which readily isomerises to α-naphthol in the presence of methanolic hydrogen chloride (*Wittig*, 1955).

α-Bromofluorobenzene   Benzyne   1,4-Dihydro-1,4-   α-Naphthol
epoxynaphthalene

Unlike the other halogens, the iodine atom in *aryl iodides* has a tendency to increase the number of groups attached to it, thereby gaining an *onium-like* character. *Willgerodt* showed that if chlorine is passed through a cold solution of iodobenzene in chloroform yellow crystalline *phenyliododichloride* (iodobenzene dichloride), $C_6H_5ICl_2$, is formed. In light, or when heated to 110°C (383 K) it is converted into *p*-chloroiodobenzene. Aqueous hydroxide hydrolyses the chlorine atoms of phenyliododichloride and loss of water then gives yellow, amorphous *iodosobenzene*, C_6H_5—IO:

$C_6H_5I + Cl_2 \longrightarrow C_6H_5ICl_2$
Iodobenzene   Phenyliodo-
dichloride

Cl—⟨ ⟩—I + HCl
p-Chloroiodobenzene

$C_6H_5IO + 2NaCl + H_2O$
Iodosobenene

Iodosobenzene can also be prepared by oxidising iodobenzene with ozone. It is an unstable substance which, slowly at room temperature, but more rapidly when heated with boiling water, *disproportionates* to iodoxybenzene and iodobenzene:

$$2 C_6H_5IO \longrightarrow C_6H_5IO_2 + C_6H_5I$$
Iodosobenzene   Iodoxybenzene   Iodobenzene

Iodoxybenzene can also be made by oxidising iodobenene, using permonosulphuric acid (*Caro's* acid), or by the action of bleaching powder on iodosobenzene. Iodosobenzene, and even more so iodoxybenzene, explode when they are heated. With acid solutions of potassium iodide they liberate equivalent amounts of iodine, being themselves reduced to iodobenzene.

The structures of iodosobenzene and iodoxybenzene resemble those of sulphoxides and sulphones (see p. 147 *ff.*).

Uses. Some aryl halides are of use in industry, for example, chlorobenzene is an important intermediate in the preparation of aniline, phenol, *DDT*, and of dyestuffs. *p*-Dichlorobenzene has use as a moth repellent, and *o*-dichlorobenzene, to which some of the *p*-isomer has been added, as a filler for heat-exchangers in the range 150–250°C (425–525 K).

5.2.3 *Mechanisms of Further Electrophilic Substitution in the Benzene Ring*

The main influence on the introduction of a second substituent into a benzene ring is first and foremost the nature of the substituent group which is already present. Dependent upon this, the second substituent may enter the ring either predominantly at the *p*- and *o*-positions or predominantly at the *m*-position. The following rules concerning the *orientation* of the second substituent apply:

5.2.3.1 Substituents of type 1 direct the second substituent mainly to the *o*- and *p*-positions, usually giving more of the *p*-derivative. The following substituents act in this way:

$$-NR_2, \; -NHR, \; -NH_2, \; -OH, \; -OR, \; -NHCOR, \; -I, \; -Br, \; -Cl, \; -F, \; -CH_3,$$
$$-CH=CH_2$$

5.2.3.2 Substituents of type 2, which direct a second substituent principally to the *m*-position, include the following:

$$-\overset{\oplus}{N}R_3, \; -\overset{\oplus}{N}H_3, \; -NO_2, \; -C\equiv N, \; -SO_3H, \; -CF_3, \; -CCl_3, \; -CHO, \; -CO-R,$$
$$-COOH, \; -COOR, \; -CONH_2$$

These classifications do not provide hard and fast rules of behaviour but rather reflect empirical findings, and there are exceptions to them. For example, chlorination of nitrobenzene gives *m*-chloronitrobenzene as the main product but it is accompanied by 17.6% of the *o*- and 1.5% of the *p*-isomers. Nitration of ethyl benzoate gives mainly the *m*-nitro-derivative but also 23.8% of the of the *o*- and 3.3% of the *p*-derivatives.

Both the rate and site of substitution of a second electrophilic substitution reaction depend upon the reciprocal interactions between the benzene ring and the first substituent. This can be described in terms of two effects which can each either raise or lower the *electron density on the benzene ring*.

5.2.3.3 I-Effect

As described on p. 28 *f*, alkyl groups exert a $+I$-*effect* and hence raise the electron density on a benzene ring. In contrast, most other substituents, such as $NR_2^{\oplus}$, halogen atoms, —OH, —OR, —NH$_2$, —NR$_2$, —SO$_3$H, *etc.* act in the opposite way; because they are more electronegative than the carbon atoms to which they are attached they have an electron-withdrawing $-I$-*effect*. The push or *pull effects on the electrons* in the bonds can be symbolised as follows:

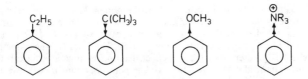

5.2.3.4 M-Effect

In the majority of cases the I-effect is overshadowed by a second effect, in which a lone pair of electrons, or the π-electrons of a double bond which forms part of the substituent, interact mesomerically with the π-electrons of the benzene ring. This interaction is described as a *mesomeric* effect (M-effect). Depending upon the nature of the substituent this can bring about either an increase or a decrease in the electron density on the benzene ring, these being designated, respectively, as $+M$ or $-M$ effects.

5.2.3.5 +M-Effect

If there is a lone pair of electrons on an atom directly attached to the benzene ring as, for example, in the case of *aniline*, mesomeric interaction with the ring takes place as follows:

From the contributing structures it may be seen that the nitrogen atom partially gives up its lone pair of electrons to the benzene ring, which as a result acquires an *increased electron density* ($+M$-effect). Similar contributing structures can be drawn for *phenol* or for *chlorobenzene*. The greater the electronegativity of the atom attached to the benzene ring the more reluctant it is to bear a positive charge, and in consequence the $+M$-effect is smaller. The effect decreases in the order $NH_2 > OH > Cl$.

5.2.3.6 −M-Effect

If the substituent contains a double bond which is conjugated with the benzene ring and which has a strongly electronegative atom at its end furthest from the ring, then electrons are withdrawn from the benzene ring, thus *lowering the electron density* on the ring ($-M$-effect). An example is provided by the nitro group:

Similar contributing structures can be drawn in the cases of, for example, alkyl phenyl ketones, benzenesulphonic acid and benzonitrile.

The interpretation of second electrophilic substitution in a benzene ring is made difficult by the fact that usually both *mesomeric* and *inductive* effects operate. These can act either in the same direction or oppose one another, and can in combination produce either a lowering or a raising of the electron density on the benzene ring, which in turn is responsible for the *rate at which electrophilic substitution takes place*. It is not easy to predict which effect will contribute most in any particular case; for example, electrophilic substitution in chlorobenzene takes place more slowly than it does in benzene. This can be explained by the fact that, because of its high electronegativity, the chlorine atom causes a strong −I-effect, which is only opposed by a weak +M-effect, so that the overall effect is a withdrawal of electrons from the benzene ring. However, in the cases of *aniline* and of *phenol*, the +M-effect is greater than the −I-effect, leading to increased electron density on the benzene ring and an enhanced rate of reaction. In the *phenoxide anion*, both effects (+I and +M) activate the benzene ring and further electrophilic substitution takes place very readily. In contrast, in the case of nitrobenzene both effects are electron-withdrawing (−I and −M-effects), leading to deactivation of the benzene ring and decreased reactivity towards electrophiles.

The decisive factor for the *orientating effect of a substituent* is the *distribution of electrons in the transition state*; the best approximation to this can be obtained from the structure of the *carbenium ion, i.e.* the σ-complex, which is formed as an intermediate. Whether the newly introduced second substituent goes to the *o*-, the *m*- or the *p*-position depends above all upon which intermediate or which transition state is more favourable energetically. Because of their polarity σ-complexes are much more sensitive than the ground states to the effects of polar substituent groups.

When phenol is chlorinated, the first-formed π-complex can give rise to three different carbenium ions, dependent upon the site, *o*-, *m*- or *p*-, to which the chlorine atom attaches itself. The most favoured site will be the one that gives the most stable (lowest energy) product. In this case, this means the *cation* in which the positive charge can be most effectively delocalised. The hydroxy group can only contribute to the mesomerism in the cation if the chlorine is attached to an *o*- or *p*-position, and hence there is much better delocalisation of positive charge in these two products and they are energetically favoured compared to the *m*-product.

In the two reaction schemes which follow, the intermediate step involving formation of the second π-complex, which comes between the carbenium ions and the final products, is omitted.

It is thus clear why about 50% each of the *o*- and *p*-chlorophenols are formed, but no *m*-chlorophenol. Similar situations arise when further substitution reactions are carried out on aniline, chlorobenzene or styrene.

In the further nitration of *nitrobenzene* the situation is completely different. In this case the first nitro group cannot contribute to the mesomerism in the intermediate carbenium ion, irrespective of where the σ-complex is formed, and thus cannot help to delocalise the positive charge in any of them. This is so because the nitrogen atom attached to the ring itself bears a positive charge. Mesomerism in each of the cations is limited to the ring.

It can be seen that in the case of *o*- and *p*-substitution, in both cases one of the possible contributing structures has positive charges on directly adjacent carbon and nitrogen atoms. This results in an appreciable raising of the energy content of these structures which therefore make only a small contribution to the overall mesomerism of these *carbenium ions*, in turn lowering the extent of possible delocalisation of the positive charge over the ring. In the case of *m*-substitution there is not the same limitation of delocalisation in the ring and hence the *m*-derivative is energetically favoured compared to the *o*- and *p*-derivatives. In this particular case the reaction gives 93% *m*-, 6% *o*- and 1% *p*-dinitrobenzene. Similar considerations apply to electrophilic substitution reactions of benzene derivatives which have —SO$_3$Na, —COOH, —CN, —$\overset{\oplus}{N}R_3$, —CF$_3$ or —CCl$_3$ substituents.

Further detail will not be considered here. In certain cases other factors such as temperature, solvent, concentration, use of catalysts, and the polarity of the new substituent which is being introduced, can influence the rate of reaction and the site at which it takes place as well as the proportions of the *o*-, *m*- and *p*-derivatives which are formed.

5.2.4 *Side-chain Halogenation of Alkylbenzenes*

Like aliphatic hydrocarbons, the alkyl side-chains of alkylbenzenes can be halogenated under the appropriate conditions. Unlike halogenation of the benzene ring, halogenation of side-chains takes place without a halogen-carrier being needed, at higher temperatures and under the influence of short-wavelength light. It proceeds *via* a radical chain mechanism, and is a faster reaction than that of addition and aromatic substitution reactions.

As an example, if a mixture of chlorine and toluene is heated and exposed to sunlight, or in the presence of *tetraethyl-lead* as a radical generator, all three hydrogen atoms of the methyl group are in turn replaced by chlorine atoms, forming, successively, *benzyl chloride, benzylidene chloride* (benzal chloride) and *benzotrichloride*:

Toluene   Benzyl chloride   Benzylidene chloride   Benzotrichloride
            (α-chlorotoluene)   (α,α-dichlorotoluene)   (α,α,α-trichlorotoluene)

In the presence of a radical generator some chlorine atoms are formed, which react with toluene to form a benzyl radical that is mesomerically stabilised. This radical reacts with chlorine to give benzyl chloride, also generating another chlorine atom to continue the chain reaction.

Industrially *benzyl chloride* is made by irradiating a mixture of chlorine and toluene at 130–140°C (405–415 K) with UV light. This temperature is above the b.p. of toluene (110.8°C [384.0 K]) but below that of benzyl chloride (179°C [452 K]). The latter separates out and is thus protected from further chlorination.

5.2.4.1 Chloromethylation. A chloromethyl group, —CH_2Cl, may be introduced into a benzene ring by an *electrophilic* reaction involving the use of formaldehyde and hydrogen chloride in the presence of zinc chloride, aluminium chloride or phosphoric acid as catalyst (*Blanc* reaction).

Benzyl chloride (α-Chlorotoluene)

Unlike halogen atoms bonded directly to aryl rings, those attached to alkyl side-chains are very reactive.

Benzyl chloride, C_6H_5—CH_2Cl, is a colourless liquid, b.p. 179°C (452 K). Its vapour is lachrymatory. Its chemical properties resemble those of an alkyl halide; for example, when heated with dilute alkali it gives *benzyl alcohol*, C_6H_5—CH_2OH, b.p. 205°C (478 K), with aqueous potassium cyanide it gives *benzyl cyanide* (phenylacetonitrile), b.p. 223°C (506 K) and with alcoholic ammonia *benzylamine*, b.p. 184.5°C (457.7 K).

Benzylidene chloride (benzal chloride, α,α-dichlorotoluene), C_6H_5—$CHCl_2$, m.p. 205°C (478 K) is the starting material in the industrial preparation of benzaldehyde.

Benzotrichloride (α,α,α-trichlorotoluene), C_6H_5—CCl_3, b.p. 221°C (494 K) is converted into benzoic acid by hydrolysis.

5.3 Nitro-Arenes. Aromatic Nitro-Compounds

While aliphatic nitro-compounds have been made on an industrial scale only since 1940, nitroarenes have been of considerable importance for a much longer time, above all as starting materials for the manufacture of dyestuffs, explosives and pharmaceuticals. This is a consequence of the fact that arenes, and especially some of their substituted derivatives, can be readily nitrated with conc. nitric acid. Most aromatic nitro-compounds are colourless or yellow solids; only a very few mononitro-derivatives are liquid at room temperature.

Arenes are usually nitrated by heating them gently with a mixture of conc. nitric acid and conc. sulphuric acid (nitrating acid).

Other nitrating agents used include fuming nitric acid (with 6–12% NO_2), nitronium tetrafluoroborate, $NO_2^{\oplus}BF_4^{\ominus}$ and acetyl nitrate, $H_3CCOONO_2$, which itself readily explodes.[1]

5.3.1 Nitrobenzenes

Preparation

5.3.1.1   When *benzene is warmed with conc. nitric acid or nitrating acid* the main product is nitrobenzene (yield 85%).

Nitrobenzene

Like halogenation of the benzene ring, nitration is an *electrophilic substitution* reaction. The nitrating agent is the *nitronium ion*, $NO_2^{\oplus}$, which is obtained from nitric acid as follows (*Hughes* and *Ingold*, 1946):

$$HNO_3 + 2\,H_2SO_4 \rightleftharpoons NO_2^{\oplus} + H_3O^{\oplus} + 2\,HSO_4^{\ominus}$$

The role of the sulphuric acid in this is not to remove water as suggested by the above equation, but to protonate nitric acid to give *nitracidium ions*, $H_2NO_3^{\oplus}$, which lose water to provide nitronium ions.

If nitration is carried out using moderately strong nitric acid in an organic solvent, the nitronium ion is generated as follows:

$$2\,HNO_3 \rightleftharpoons NO_2^{\oplus} + H_2O + NO_3^{\ominus}$$

The mechanism of nitration involves electrophilic attack by a nitronium ion on the benzene ring, forming a *π-complex*. The *carbenium ion* which is thus formed loses a proton, regenerating the delocalised system of the benzene ring and giving nitrobenzene:

π-Complex   Carbenium ion   Nitrobenzene

5.3.1.2   When nitronium tetrafluoroborate is used as nitrating agent, the nitronium ion similarly makes an *electrophilic* attack on the benzene ring (*Olah*):

---

[1] See *K. Schofield*, Aromatic Nitration (Cambridge University Press 1980); *L. Eberson*, Electron Transfer Mechanisms in Electrophilic Aromatic Nitration, Acc. Chem. Res. *20*, 53 (1987).

5.3.1.3   Nitrobenzene can be further nitrated using excess nitrating acid at a higher temperature. The product is *m*-dinitrobenzene (yield ~ 80%):

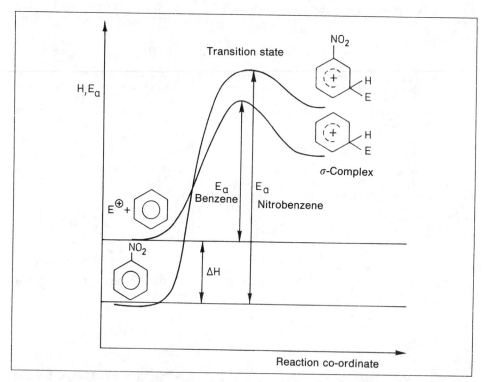

m-Dinitrobenzene

Introduction of a third nitro group into the benzene ring is much more difficult.

Under identical conditions *m-dinitrobenzene* is formed 10000 times more slowly than is *nitrobenzene*.

This can be explained as follows in terms of earlier discussion of such reactions. The contributing structures b, c and d of nitrobenzene on p. 483 indicate that the benzene ring is relatively electron-poor. As a consequence the chance of electrophilic attack on the ring is lowered compared to that on benzene itself: the inductive effect of the nitro group reinforces this result. The combination of these two effects means that nitrobenzene is lower in energy than benzene (see Fig. 113).

The σ-complex derived from nitrobenzene and the transition state leading to its formation are, however, destabilised compared to those derived from benzene, because of the inability of the nitro group to assist in the delocalisation of the positive charge on the ring and of its destabilising –I-effect (see pp. 484, 485). Thus the activation energy for nitration of nitrobenzene is greater than that for benzene.

Fig. 113 illustrates the results of these effects for the two systems and the resultant difference in the energies of activation, and qualitatively explains the differences in reactivity.

Fig. 113. Results of inductive and mesomeric effects on the transition states of electrophilic aromatic substitution (ignoring the π-complexes, see p. 475)

Properties and Uses

Nitrobenzene, C_6H_5—NO_2, is a yellow liquid, b.p. 211°C (484 K), with a strong smell of bitter almonds. It is scarcely soluble in water, but is volatile in steam. Its vapour is poisonous. The reduction products obtained from nitrobenzene will be discussed in the section on arylamines. Nitrobenzene has a dipole moment of $13.8.10^{-30}$ Cm, resulting from the distribution of the electrons in the ground state of the molecule, which derives from the contributing structures a–d shown on p. 483.

The main use of nitrobenzene is as a precursor in the manufacture of aniline. It is also used as a solvent and as an oxidising agent in preparative chemistry.

m-Dinitrobenzene is a pale yellow substance, m.p. 90°C (363 K) which is quite insoluble in water but volatile in steam.

1,3,5-Trinitrobenzene is not made by direct nitration of benzene (see p. 487 *f*.) but instead by *decarboxylation of 2,4,6-trinitrobenzoic acid*, which is itself made by oxidising 2,4,6-trinitrotoluene with sodium dichromate in conc. sulphuric acid (yield *ca* 45%).

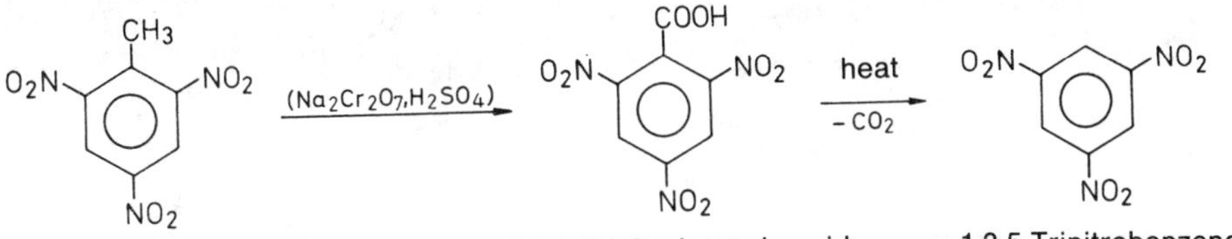

2,4,6-Trinitrotoluene   2,4,6-Trinitrobenzoic acid   1,3,5-Trinitrobenzene

1,3,5-Trinitrobenzene is a more powerful explosive than 2,4,6-trinitrotoluene, but it is much more difficult to prepare and is therefore not used as an explosive.

5.3.2 *Nitrotoluenes*

Alkylbenzenes are more easily nitrated than benzene itself. *Nitration of toluene* at 0°C (273 K) gives *o*-nitrotoluene (*ca* 65%) and *p*-nitrotoluene (30%) together with a small

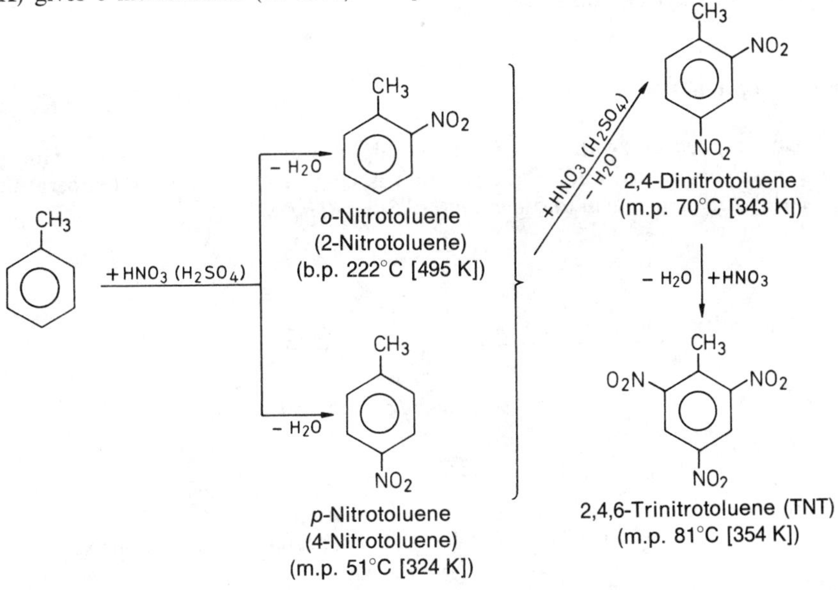

o-Nitrotoluene
(2-Nitrotoluene)
(b.p. 222°C [495 K])

p-Nitrotoluene
(4-Nitrotoluene)
(m.p. 51°C [324 K])

2,4-Dinitrotoluene
(m.p. 70°C [343 K])

2,4,6-Trinitrotoluene (TNT)
(m.p. 81°C [354 K])

amount of *m*-nitrotoluene (5%). The *o*- and *p*-isomers can be separated by fractional distillation. Further nitration provides *2,4-dinitrotoluene* and finally *2,4,6-trinitrotoluene*, which is used as an explosive under the name *TNT*. It is a safe explosive to handle and requires a detonator to initiate explosion.

A by-product which is obtained in the nitration of toluene is *phenylnitromethane* (ω-nitrotoluene), which results from nitration of the side-chain. It undergoes the same reactions as a primary aliphatic nitroalkane, and can be isolated in both the neutral form, b.p. 225°C (498 K) or in the *aci*-form, m.p. 84°C (357 K) (*Hantzsch*). Dinitrotoluene has been shown to be carcinogenic in some animals.

Some trinitrated alkylbenzenes containing a t-butyl group have musk-like smells and are used in perfumery. Among such synthetic products are so-called toluene-musk and xylene-musk, which have the following formulae:

Toluene-musk   Xylene-musk

5.4 Arenesulphonic Acids

A particularly characteristic reaction of arenes and their derivatives is that involving the direct substitution of one or more hydrogen atoms by *sulphonic acid* groups, —SO_3H, using conc. or fuming sulphuric acid. As a result arylsulphonic acids are much more easy to obtain than alkylsulphonic acids.

5.4.1 Preparation

Sulphonation of benzene, with conc. or fuming sulphuric acid (oleum) leads to the formation of *benzenesulphonic* acid and, if the reaction is carried out at higher temperatures, to *benzene-1,3-disulphonic* acid and *benzene-1,3,5-trisulphonic acid*:

Benzenesulphonic acid

Benzene-1,3-disulphonic acid   Benzene-1,3,5-trisulphonic acid

Alkylbenzenes are sulphonated mainly at the o- and p-positions. The rate of reaction depends largely upon the substituents already present in the benzene ring. The usual rules normally apply.

Mechanistically sulphonation is an *electrophilic substitution* reaction, but the details are not fully understood. Kinetic studies by *Ingold* indicate that *sulphur trioxide*, either in free form, or in a disguised form, *e.g.* as $H_3SO_4^{\oplus}$ or as pyrosulphuric acid ($H_2S_2O_7$), is involved.[1]

$$2\,H_2SO_4 \rightleftharpoons HSO_4^{\ominus} + H_3SO_4^{\oplus} \qquad H_3SO_4^{\oplus} \rightleftharpoons SO_3 + H_3O^{\oplus}$$

Unlike other electrophilic substitution reactions, sulphonation is an equilibrium reaction.

Properties

Arylsulphonic acids are colourless, crystalline substances which are strongly hygroscopic. They are readily soluble in water and are completely dissociated in aqueous solution:

$$Ar-SO_3H + H_2O \rightarrow Ar-SO_3^{\ominus} + H_3O^{\oplus}$$

The *acidity* of benzenesulphonic acid ($pK_a = 0.70$) is similar to that of sulphuric acid. Electron-withdrawing substituents at the o- or p-positions increase the acidity, for example, 2,4-dinitrobenzenesulphonic acid is a stronger acid than sulphuric acid. It is very difficult to obtain sulphonic acids completely free from water, and they are usually isolated as *sodium sulphonates* by salting them out with saturated sodium chloride solution:

$$C_6H_5-SO_3H + NaCl \rightarrow C_6H_5-SO_3^{\ominus}Na^{\oplus} + HCl$$

Sodium benzenesulphonate

In contrast to the corresponding sulphates, calcium, barium and lead sulphonates are soluble in water.

Desulphonation

If benzenesulphonic acid is heated to 150–180°C [420–450 K] with dilute hydrochloric acid in a sealed tube the sulphonic acid group is replaced by hydrogen. In other words, under suitable conditions sulphonation is *reversible*:

$$C_6H_5-SO_3H + H_2O \overset{(HCl)}{\rightleftharpoons} C_6H_6 + H_2SO_4$$

Examples of *nucleophilic substitution* of the sulphonic acid group which are of practical importance include its replacement by hydroxy or nitrile groups. If sodium benzenesulphonate is melted with sodium hydroxide sodium phenoxide is formed; with sodium cyanide the product is *benzonitrile* (cyanobenzene):

[1] See *H. Cerfontain*, Mechanistic Aspects in Aromatic Sulfonation and Disulfonation. (Interscience, New York 1968).

$$C_6H_5-SO_3Na + NaCN \rightarrow C_6H_5-C\equiv N + Na_2SO_3$$
Benzonitrile

Uses

Because they are only weak oxidising agents, free arylsulphonic acids are sometimes used as strong acids instead of sulphuric acid where acid catalysis is required. The ready replacement of the sulphonic acid group by other groups leads to their being used industrially as synthetic intermediates.

Sodium sulphonate groups are often introduced into the molecules of dyestuffs, pharmaceuticals and other products in order to make them soluble in water. The sodium salts of alkylbenzenesulphonates are of importance in the detergents industry since the calcium and magnesium salts of these acids are soluble in water and they can hence be used where the water is hard.

5.4.2 *Derivatives of Arenesulphonic Acids*

Like carboxylic acids and alkylsulphonic acids, arenesulphonic acids form sulphonyl chlorides, esters, amides, *etc.*

5.4.2.1 Sulphonyl Chlorides

The most important of the sulphonyl chlorides are those derived from benzene- and toluene-sulphonic acids.

Preparation.

(a) *By the action of phosphorus(V) chloride on arenesulphonic acids* or, better, on their sodium salts, *e.g.*

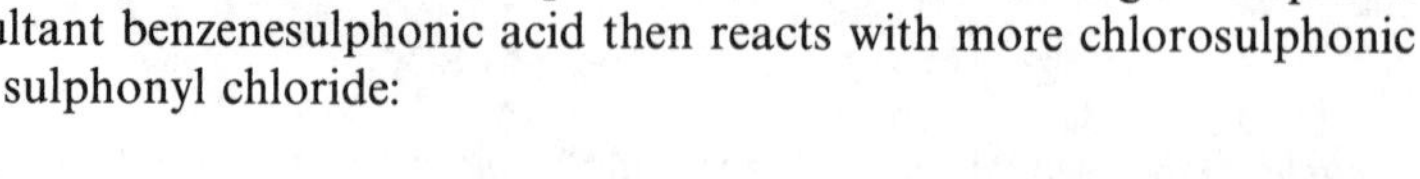

$$\langle\bigcirc\rangle-SO_2ONa + PCl_5 \longrightarrow \langle\bigcirc\rangle-SO_2Cl + NaCl + POCl_3$$

Sodium benzenesulphonate   Benzenesulphonyl chloride (m.p. 14.5°C [287.7K])

(b) Industrially benzenesulphonyl chloride is made by *chlorosulphonation of benzene* using chlorosulphonic acid. In this process benzene first undergoes sulphonation, and the resultant benzenesulphonic acid then reacts with more chlorosulphonic acid to form the sulphonyl chloride:

$$C_6H_6 + Cl-SO_3H \rightarrow C_6H_5-SO_2-OH + HCl$$
$$C_6H_5-SO_2-OH + Cl-SO_3H \rightleftharpoons C_6H_5-SO_2Cl + H_2SO_4$$
Benzenesulphonyl chloride

Diphenylsulphone is formed as a by-product.

Chlorosulphonation of toluene leads to a mixture of *o*- and *p*-toluenesulphonyl chlorides:

CH$_3$

$\xrightarrow[{-H_2SO_4,\,-HCl}]{+2Cl-SO_3H}$

CH$_3$ — SO$_2$Cl

and

CH$_3$ — SO$_2$Cl

Toluene

o-Toluenesulphonyl chloride
(m.p. 10°C [283 K])

p-Toluenesulphonyl chloride
(m.p. 68°C [341 K])

Properties. Arenesulphonyl chlorides boil at lower temperatures than the corresponding sulphonic acids and can be purified either by distillation under reduced pressure or by recrystallisation. They react with water only very slowly because of their insolubility in it; it requires lengthy heating to reconvert them into acids, *e.g.*

CH$_3$ — SO$_2$Cl + H$_2$O $\xrightarrow{heat}$ CH$_3$ — SO$_3$H + HCl

Tosyl
chloride

p-Toluenesulphonic acid
(m.p. 160.5°C [433.7 K])

The name of the *p*-toluenesulphonyl group, H$_3$C—C$_6$H$_4$—SO$_2$—, is often abbreviated to *tosyl* group (*Ts*), *e.g.* *TsCl* = *tosyl chloride*. The esters of these acids are sometimes described as *alkyl* or *aryl tosylates* instead of the IUPAC names *alkyl* or *aryl p-toluenesulphonates*. Introduction of a tosyl group into a molecule is called *tosylation*.

Reduction of arenesulphonyl chlorides, depending upon the reaction conditions, leads to the formation of either *arenesulphinic acids* or *thiophenols* (arylmercaptans), *e.g.*

$$2\,C_6H_5-SO_2Cl \xrightarrow[-ZnCl_2]{+2\,Zn\,(in\,Ether)} (C_6H_5-SO_2)_2Zn \xrightarrow[-ZnCl_2]{+2\,HCl} 2\,C_6H_5-SO_2H$$
Benzenesulphinic acid

$$C_6H_5-SO_2Cl + 6\,H \xrightarrow{(LiAlH_4)} C_6H_5-SH + HCl + 2\,H_2O$$
Thiophenol, Phenylmercaptan

Arenesulphinic acids are unstable compounds and are readily oxidised to sulphonic acids in air.

Thiophenol is a colourless, poisonous liquid, b.p. 170°C (443 K) with a disagreeable smell. It is acidic (pK_a = 6.50) and forms metal salts. In air it undergoes ready oxidation to *diphenyl disulphide*:

$$2\,C_6H_5-SH + \tfrac{1}{2}\,O_2 \rightarrow C_6H_5-S-S-C_6H_5 + H_2O$$
Diphenyl disulphide

Benzenesulphonyl chloride takes part in a *Friedel–Crafts* reaction with benzene to give *diphenylsulphone*, m.p. 128°C (401 K):

$$C_6H_5-SO_2-Cl + C_6H_6 \xrightarrow{(AlCl_3)} C_6H_5-SO_2-C_6H_5 + HCl$$
Diphenylsulphone

5.4.2.2 Esters of Arenesulphonic Acids

Arenesulphonyl chlorides react with alcohols or phenols in the presence of dilute alkali or, better, of pyridine, to form arenesulphonate esters, *e.g.*

$$H_3C-\langle\bigcirc\rangle-SO_2Cl + ROH + NaOH \longrightarrow H_3C-\langle\bigcirc\rangle-SO_2-OR + NaCl + H_2O$$

Tosyl chloride   *p*-Toluenesulphonate
                                         ester

Alkyltosylates are often used instead of alkyl halides as *alkylating agents, e.g.*

$$R'COOH + ROTs \rightarrow R'COOR + TsOH$$

5.4.2.3 Sulphonamides

When sulphonyl chlorides react with aqueous ammonia *arenesulphonamides* are formed, *e.g.*

$$C_6H_5-SO_2-Cl + 2NH_3 \rightarrow C_6H_5-SO_2-NH_2 + NH_4Cl$$
Benzenesulphonamide
(m.p. 151°C [424 K])

In the same way sulphonyl chlorides react with primary or secondary amines to give *N*-substituted benzenesulphonamides, *e.g.*

$$C_6H_5-SO_2-N\langle^H_R \qquad\qquad C_6H_5-SO_2-N\langle^R_R$$

N-Alkylbenzenesulphonamide   *N,N*-Dialkylbenzenesulphonamide
(soluble in alkali)   (insoluble in alkali)

The monoalkyl derivatives dissolve in alkali; the hydrogen on the nitrogen atom is lost and a metal salt is formed. This difference in properties between mono- and di-alkyl derivatives provides a method for the separation of primary and secondary amines (see *Hinsberg* reaction, p. 158). It is better to use *p*-toluenesulphonyl chloride (tosyl chloride) for this purpose than benzenesulphonyl chloride.

Arenesulphonamides are well-crystalline compounds. They are only hydrolysed with difficulty. This is best achieved by heating them with 30% hydrobromic acid in acetic acid with some phenol also present:

$$Ar-SO_2-NHR + 2HBr \rightarrow Ar-SO_2-Br + R-NH_3]^{\oplus} Br^{\ominus}$$

Tosyl chloride occurs as a by-product in the preparation of saccharin and is converted into tosyl amide, which is treated with sodium hypochlorite solution to give the *sodium salt of N-chloro-p-toluenesulphonamide*:

$$H_3C-\langle\bigcirc\rangle-SO_2-NH_2 + NaOCl \longrightarrow H_3C-\langle\bigcirc\rangle-SO_2-\overset{\ominus}{N}-Cl\Big] Na^{\oplus} + H_2O$$

p-Toluenesulphonamide   Chloramine-T (*N*-chloro-*p*-toluene-
                                 sulphonamide, sodium salt)

This salt is stable when dry. In aqueous solution it behaves like hypochloric acid and it is used as an *antiseptic* under the name *Chloramine-T* (not to be confused with chloramine, see p. 168).

When *N*-methyl-*p*-toluenesulphonamide is treated with nitrous acid a stable *N*-nitroso-derivative is obtained, which is used in the preparation of diazomethane (see p. 167).

$$H_3C-\underset{}{\bigcirc}-SO_2-\underset{\underset{H}{|}}{N}-CH_3 + HNO_2 \longrightarrow H_3C-\underset{}{\bigcirc}-SO_2-\underset{\underset{NO}{|}}{N}-CH_3 + H_2O$$

N-Methyl-   N-Methyl-N-nitroso-
-p-toluenesulphonamide   -p-toluenesulphonamide

5.5 Phenols

The name of this class of compounds stems from an early suggested name for benzene, namely 'phene' (*Laurent*, 1837). In phenols, one or more hydroxy groups are directly attached to the benzene ring. They might be regarded as the tertiary alcohols of the aromatic series of compounds, but the ways of making them are completely different from those used for making aliphatic alcohols. According to the number of hydroxy groups present they are known as monohydric, dihydric or trihydric phenols.

5.5.1 Monohydric Phenols

Phenol (hydroxybenzene), C_6H_5OH, is the parent compound of this class. It was first isolated from coal tar by *Runge* (1834) and became known as carbolic acid. Although a considerable amount of phenol is still obtained in this way, increasing demands for phenol have led to a number of industrial synthetic processes.

Preparation

5.5.1.1   By *melting sodium benzenesulphonate with alkali* at about 300°C (570 K):

$$C_6H_5-SO_3Na + 2\,NaOH \xrightarrow{\text{Melt}} C_6H_5-ONa + Na_2SO_3 + H_2O$$

Sodium phenoxide

This is the oldest method for the synthesis of phenol; it involves displacement of the sulphonate group by the strongly basic hydroxide ion (nucleophilic aromatic substitution, see p. 481).

Free phenol is obtained from sodium phenoxide by treating it with acid:

$$C_6H_5-\overline{\underline{O}}|]^{\ominus}Na^{\oplus} \xrightarrow[-NaHSO_3]{+\,SO_2,\,H_2O} C_6H_5-OH$$

Phenol

5.5.1.2   By *hydrolysis of chlorobenzene* with steam over tricalcium phosphate or silica as catalyst (*Raschig* process):

$$C_6H_5-Cl + H_2O \xrightarrow{Ca_3(PO_4)_2} C_6H_5-OH + HCl$$

Among the by-products are diphenyl ether and o- and p-phenylphenols.
Replacement of chlorine by a hydroxy group takes place much more easily when the chlorine is activated by electron-withdrawing substituents at the o- or p-positions (see p. 503).

5.5.1.3 *Cumyl hydroperoxide* (α,α-dimethylbenzyl hydroperoxide), which is obtained by atmospheric oxidation of cumene, is split by dilute acid into *phenol* and *acetone* (*Hock*,[1] 1945);

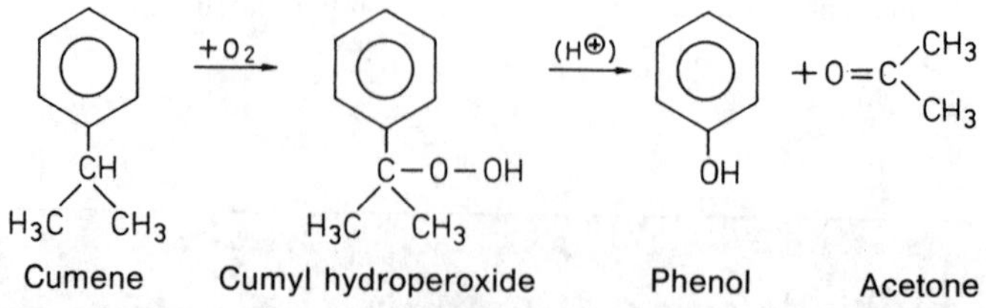

Cumene   Cumyl hydroperoxide   Phenol   Acetone

Most phenol is made by this method nowadays. There is no waste of material since the other product is acetone.

This acid-catalysed reaction proceeds by an *ionic mechanism*. First a proton from the acid removes the hydroxy group from the hydroperoxide. The phenyl group migrates to the resultant electron-deficient oxygen atom, thereby generating a carbenium ion which, by addition of water from the solvent, forms an oxonium salt of a hemiacetal. This breaks down into phenol, acetone and a proton:

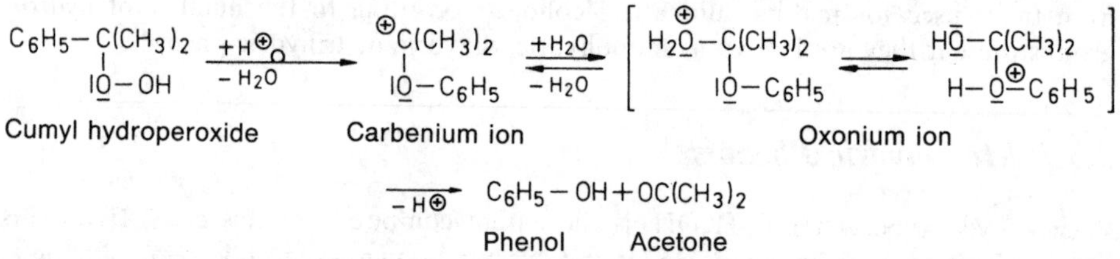

Cumyl hydroperoxide   Carbenium ion   Oxonium ion

$$C_6H_5-OH + OC(CH_3)_2$$

Phenol   Acetone

Properties. Pure phenol forms colourless needles, m.p. 41°C (314 K) which have a typical 'phenolic' smell. They become reddish when exposed to air. Phenol is moderately soluble in water, less so in ethanol or ether. Like *enols*, in which the hydroxy group is directly attached to an unsaturated carbon atom, phenol is acidic ($pK_a = 10$), and in aqueous solution gives a characteristic colour reaction with iron(III) chloride. This is due to the formation of a violet *iron complex*.

The *greater acidity of phenol* as compared to aliphatic alcohols can be explained by *mesomeric stabilisation* in the phenoxide anion. One of the lone pairs of electrons on the oxygen atom contributes to the mesomerism in the ring and the negative charge is delocalised thereby.

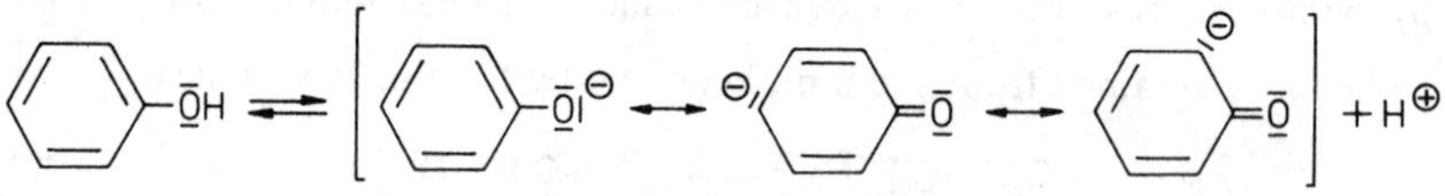

Dilute alkali reacts with phenol to form a solution of a phenoxide salt, *e.g.*

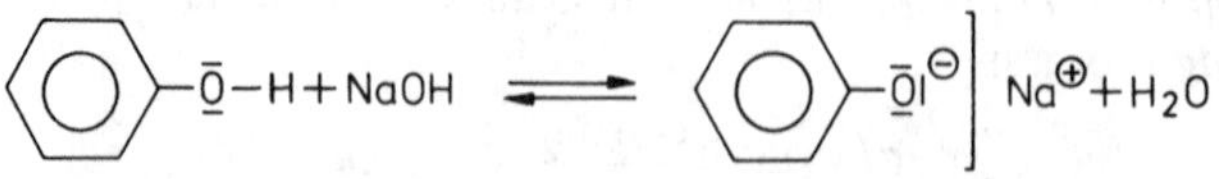

---

[1] See H. *Kropf*, Moderne technische Phenol-Synthesen, Chem. Ing. Technik 36 (759) (1964); F. R. *Mayo*, Free-radical Autoxidations of Hydrocarbons, Acc. Chem. Res. 1, 193 (1968).

Because of hydrolysis such solutions give an alkaline reaction. If carbon dioxide is passed into the solution, phenol is regenerated:

$$C_6H_5-\overline{\underline{O}}|]^{\ominus} Na^{\oplus} + CO_2 + H_2O \longrightarrow C_6H_5-OH + NaHCO_3$$

Uses. Phenol is an important starting material in industrial processes and is used in the manufacture of dyestuffs, plastics, anti-oxidants, and artificial tanning agents. It also plays a vital role in many organic syntheses.

Phenol Derivatives

5.5.2 Alkyl Phenyl Ethers (Phenol Ethers)

These ethers are made either by the *Williamson* synthesis, by heating sodium phenoxide with an alkyl halide in ethanolic or aqueous solution, or industrially by the reaction of an alkaline solution of phenol with a dialkyl sulphate, *e.g.*

$$C_6H_5-\overline{\underline{O}}|]^{\ominus} Na^{\oplus} + X-R \longrightarrow C_6H_5-O-R + NaX$$

$$C_6H_5-\overline{\underline{O}}|]^{\ominus} Na^{\oplus} + (H_3CO)_2SO_2 \longrightarrow \underset{\text{Anisole}}{C_6H_5-O-CH_3} + (H_3CO)SO_3]^{\ominus} Na^{\oplus}$$

Because of its acidity, phenol is also methylated by diazomethane to form anisole:

$$C_6H_5-O-H + |\overset{\ominus}{C}H_2-\overset{\oplus}{N}\equiv N| \longrightarrow C_6H_5-O-CH_3 + N_2$$

Alkyl phenyl ethers are stable to alkali. They are insoluble in water; some of them have pleasing odours. Among the most important are anisole (methyl phenyl ether, methoxybenzene), b.p. 154°C (427 K); phenetole (ethyl phenyl ether, ethoxybenzene), m.p. 170°C (443 K) and anethole (4-[1-propenyl]anisole), b.p. 235°C (508 K), the major constituent of oil of aniseed.

Anisole   Phenetole   Anethole

It has commonly been taught that phenyl esters (phenol esters) *cannot* be made by esterifying phenols with carboxylic acids; this has been shown to be incorrect.[1] They are usually made by the reactions of phenols with acid chlorides or acid anhydrides in the presence of potassium carbonate or pyridine, in the so-called *Schotten–Baumann* reaction, *e.g.*

$$C_6H_5-OH + Cl-COR \longrightarrow C_6H_5-O-COR + HCl$$

$$C_6H_5-OH + (H_3CCO)_2O \longrightarrow C_6H_5-O-COCH_3 + H_3CCOOH$$

---

[1] See *M. B. Hocking*, J. Chem. Educ. *57*, 527 (1980).

5.5.2.1 Fries rearrangement[1] (1908).

5.5.2.1 *Fries* **rearrangement**[1] (1908). When heated with *aluminium chloride* (boron trifluoride, zinc chloride or titanium(IV) chloride) in dry nitrobenzene, either alkyl or aryl phenyl esters rearrange into *p*- or *o*-acylphenols. The reaction is *partly intramolecular*. At temperatures below 100°C (373 K) the acyl group migrates from the oxygen atom principally to the *p*-position, but above 100°C (373 K) usually to the *o*-position:

Phenyl ester   *p*-Acylphenol   *o*-Acylphenol

The *Fries* rearrangement is useful for the preparation of acylphenols.

5.5.2.2 *Claisen* **rearrangement of phenyl allyl ethers** (1912)[2]. *Allyl phenyl ether*, which can be made from phenol and allyl bromide in acetone in the presence of potassium carbonate, rearranges when heated at about 200°C (475 K) to give *o-allylphenol* in almost quantitative yield:

Allyl phenyl ether   *o*-Allylphenol

Making use of isotopic labelling, for example with [γ-^{14}C]allyl phenyl ether, it has been shown that these *o*-rearrangements are *intramolecular* and proceed with inversion of the allyl side-chain.

[γ-^{14}C]-Allyl phenyl ether   Cyclic intermediate

Cyclohexadienone derivative   *o*-[α-^{14}C]-Allylphenol

[1] See *H. Henecka*, in *Houben-Weyl*, Methoden der organischen Chemie, Vol. *7/2a*, 379 *ff.*(Thieme Verlag, Stuttgart 1973); *H. Kwart* et al. in *S. Patai*, The Chemistry of Carboxylic Acids and Esters, p. 347 *ff.* (Wiley, New York 1969).
[2] See *D. S. Tarbell*, The *Claisen* Rearrangement, Org. React. *2*, 1 *ff.* (1944); *F. E. Ziegler*, Stereo- and Regio-chemistry of the *Claisen* Rearrangement: Applications to Natural Product Synthesis, Acc. Chem. Res. *10*, 227 (1977); *P. Wipf*, Claisen Rearrangements, Comprehensive Organic Synthesis, Vol. 5, p. 827, ed. *B. M. Trost* (Pergamon, Oxford 1991).

The initially formed cyclic intermediate rearranges to a cyclohexadienone derivative which rapidly isomerises to the o-[α-^{14}C]-allylphenol.

If both the o-positions are occupied by substituents, then the p-allylphenol is formed instead. Crossover experiments, using radioactively labelled compounds, have demonstrated that the p-rearrangement is an intramolecular reaction. The following *mechanism* shows that the initial step is an o-rearrangement giving a cyclohexadienone derivative; this quickly stabilises itself by rearranging to the corresponding p-allylphenol. In this process the allyl side-chain undergoes inversion twice.

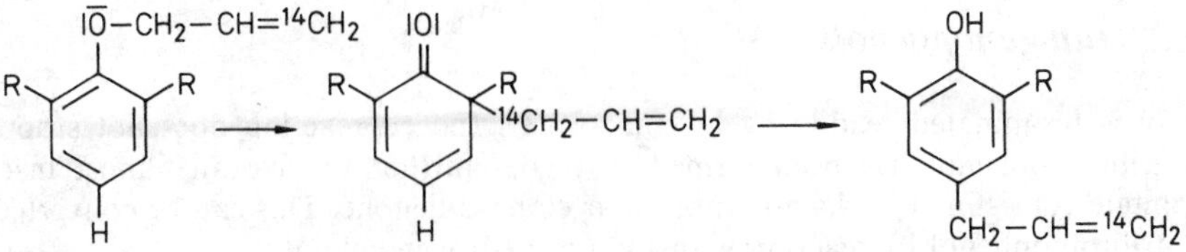

This type of reaction is not limited to benzenoid compounds, but also occurs in the aliphatic series where it takes place in the case of *enol ethers*, and is a sigmatropic [3,3]-shift analogous to the *Cope* rearrangement:

Thus *cyclohex-2-enyl vinyl ether* rearranges at 150–160°C (425–435 K) to form the γ,δ-unsaturated *cyclohex-2-enylacetaldehyde*.

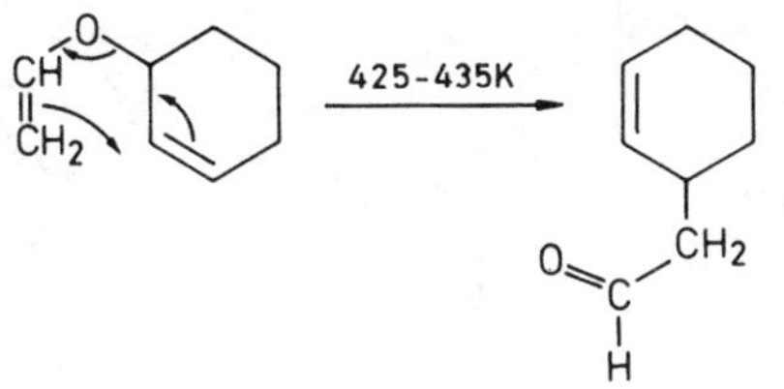

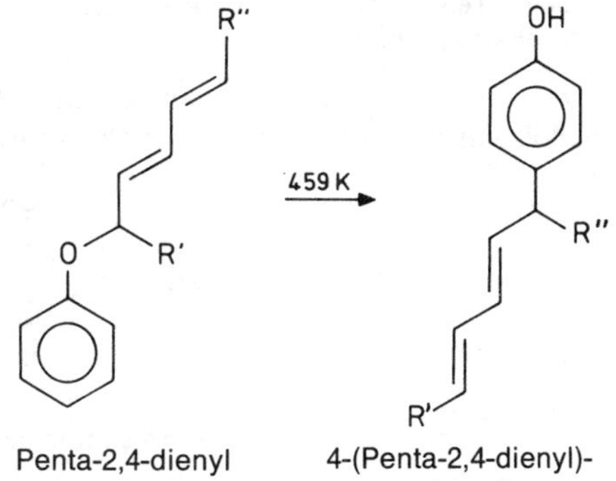

| Cyclohex-2-enyl vinyl ether | Cyclohex-2-enyl-acetaldehyde | Penta-2,4-dienyl phenyl ether | 4-(Penta-2,4-dienyl)-phenol |

Sulphur (in a thio-*Claisen* rearrangement) can take the place of oxygen (oxo-*Claisen* rearrangement).[1] *Penta-2,4-dienyl phenyl ether* rearranges in a sigmatropic [5,5]-shift to *4-(penta-2,4-dienyl)-phenol*.[2]

[1] *P. Metzner*, The Use of Thiocarbonyl Compounds in Carbon – Carbon Bond Forming Reactions, Synthesis *1992*, 1185.
[2] See *G. Fráter* and *H. Schmid*, Helv. Chim. Acta *53*, 269 (1970).

Substituted Phenols

The important factor in the following reactions is that the sites *o*- and *p*- to a phenolic hydroxy group are strongly activated towards *electrophilic substitution* (see p. 484).

5.5.3 *Halogenophenols*

Phenol is brominated readily by bromine-water. Indeed reaction does not stop when 2,4,6-tribromophenol has been formed, but goes further to give an almost insoluble precipitate consisting of 2,4,4,6-tetrabromocyclohexadienone. This can be converted into 2,4,6-tribromophenol by reaction with sodium hydrogen sulphite.

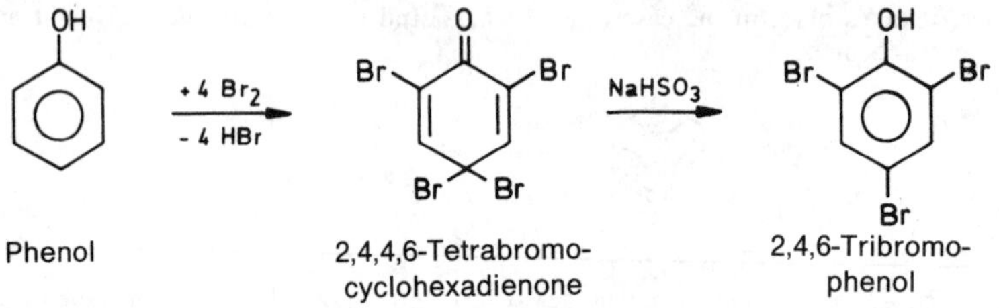

| Phenol | 2,4,4,6-Tetrabromo-cyclohexadienone | 2,4,6-Tribromo-phenol |

2,4,4,6-Tetrabromocyclohexadienone can be used in analogous fashion to *N*-bromosuccinimide as a *dehydrogenating agent* (*Messmer*).

o- and *p*-Bromophenols can be obtained by brominating phenol in non-aqueous media, such as carbon disulphide, under the appropriate conditions:

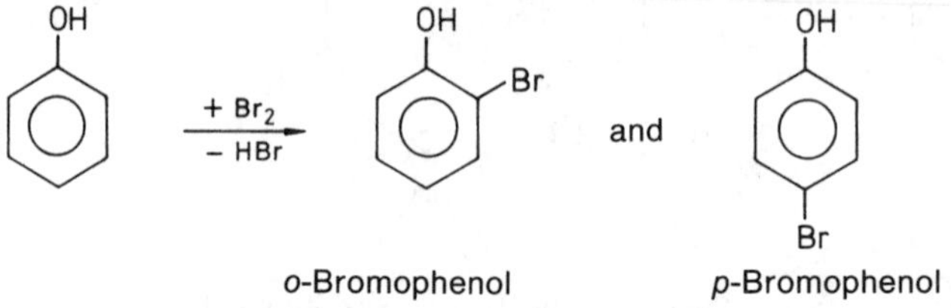

o-Bromophenol   *p*-Bromophenol

2,4-Dichlorophenoxyacetic acid (2,4-D) is an analogue of the *auxins* (see p. 726) and is used as a herbicide (weed-killer).[1] It is made by treating sodium 2,4-dichlorophenoxide with sodium chloroacetate.

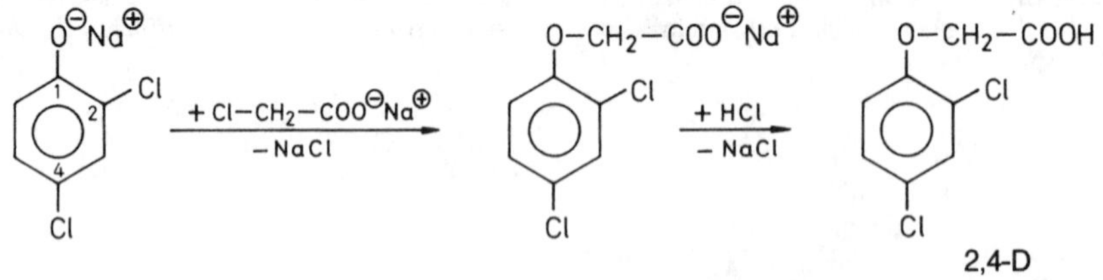

2,4-D

---

[1] *Herbicides*, together with *fungicides*, *insecticides* and *rodenticides* are all plant-protecting agents (*pesticides*, *biocides*). They act in helping the development of useful plants by partially or completely destroying the wild plants which grow alongside them. These biocides are thus better described as protectors of plants which are needed, rather than just as plant-protecting agents.

Mixtures of the n-butyl ester of 2,4-dichlorophenoxyacetic acid (2,4-D) and of 2,4,5-trichloro-phenoxyacetic acid (2,4,5-T, 'Agent Orange') were used as defoliating agents. 2,4,5-T is a *teratogen*, that is it leads to deformities in developing embryos.

5.5.4 Phenolsulphonic acids

Sulphonation of phenol with conc. sulphuric acid proceeds very readily. At room temperature the main product is phenol-*o*-sulphonic acid, and at about 100°C (373 K) the *p*-isomer. Further sulphonation leads to the formation of phenol-2,4-disulphonic acid:

Phenol-*o*-sulphonic acid

Phenol-*p*-sulphonic acid

Phenol-2,4-disulphonic acid
(4-Hydroxybenzene-1,3-disulphonic acid)

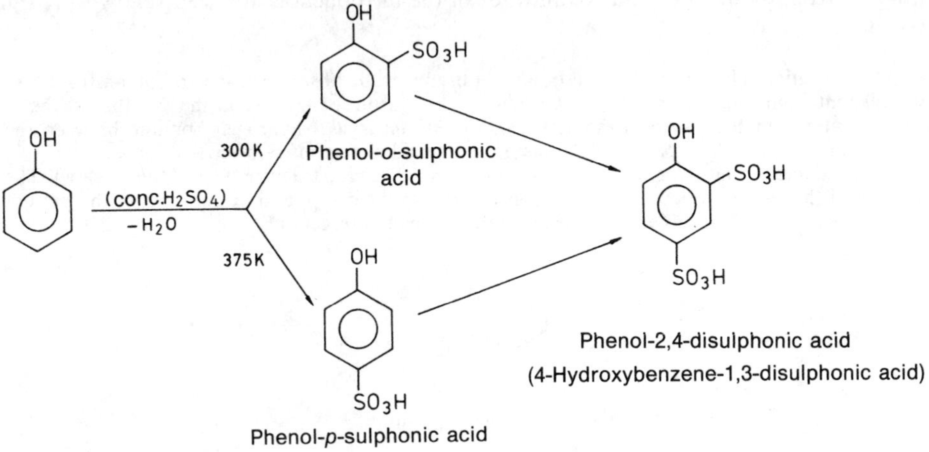

5.5.5 Nitrophenols

Nitration of phenol with dilute nitric acid at room temperature leads to the formation of a mixture of *o*-nitrophenol (~35%) and *p*-nitrophenol (~15%):

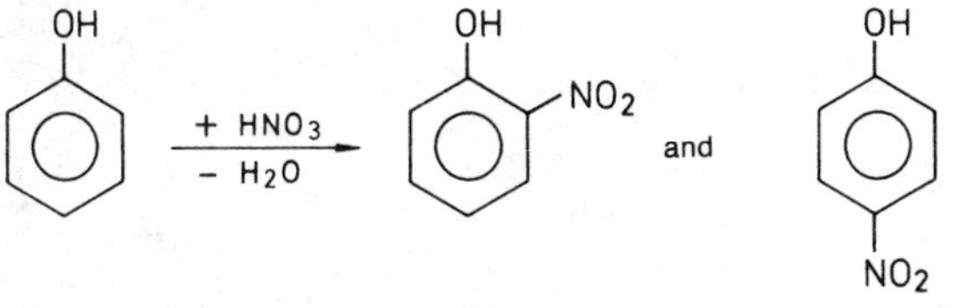

o-Nitrophenol   *p*-Nitrophenol

Separation is achieved by steam distillation; only the *o*-nitrophenol is volatile in steam.

m-Nitrophenol is prepared by *diazotising* *m*-nitroaniline and heating the aqueous solution of the resultant diazonium salt:

m-Nitroaniline   Diazonium salt   m-Nitrophenol

o-Nitrophenol is a yellow substance, m.p. 45°C (318 K), with a pungent smell. m-Nitrophenol and p-nitrophenol are colourless and melt, respectively, at 93°C (366 K) and 114°C (387 K). o-Nitrophenol and p-nitrophenol ($pK_a \sim 7.15$) are more strongly acidic than m-nitrophenol ($pK_a = 8.35$); all three are stronger acids than phenol itself ($pK_a = 10.0$). Alkali-metal salts of all three of the nitrophenols are deep yellow to red in colour.

5.5.5.1 Chelation. The physical properties (volatility in steam, absorption spectra) of o-nitrophenol are different from those of the m- and p-isomers. The reason for this lies in the fact that in the o-isomer, but not in the m- and p-isomers, there is *intramolecular* hydrogen bonding between the hydroxy group and a lone pair of electrons on the neighbouring nitro group.

This produces a ring-structure which is known as a *chelate* structure (Grk. *khele* = claw). The strength of the chelation depends on the constitution of the compound in question; in the present case the geometry of the molecule is ideally suited to such interaction:

o-Nitrophenol

Conc. nitric acid reacts with o- and p-nitrophenols to produce 2,4-dinitrophenol, m.p. 114°C (387 K), which has a pK_a of 4.02.

5.5.5.2 Trinitrophenol (Picric acid). This can only be made with difficulty by nitrating phenol under forcing conditions because much of the material is destroyed by oxidation.

On the industrial scale picric acid is made from phenol by first converting the latter compound into phenol-2,4-disulphonic acid (4-hydroxybenzene-1,3-disulphonic acid), which is not readily oxidised, and then treating this with nitric acid. This brings about both *electrophilic substitution* of the sulphonic acid groups by nitro-groups and direct introduction of a third nitro-group.

Phenol-2,4-disulphonic acid   Picric acid

In another process chlorobenzene is treated with nitrating acid, giving 1-chloro-2,4-dinitrobenzene. The nitro groups activate the chlorine atom, which in alkali readily undergoes *nucleophilic substitution* and is replaced by a hydroxy group. Further nitration leads to the formation of picric acid:

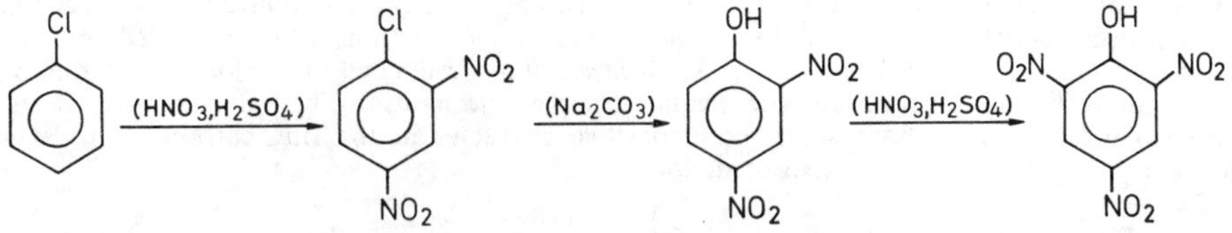

1-Chloro-2,4-dinitrobenzene   2,4-Dinitro-phenol   Picric acid

Picric acid gets its name from its relatively high acidity ($pK_a = 1.02$) and bitter taste (Grk. *pikros* = bitter). It crystallises from water in deep yellow leaflets of m.p. 122°C (395 K). It is poisonous. Picric acid and its ammonium salt can be detonated by impact and have been used as *explosives*. In the laboratory, picric acid has been used for characterising organic bases, most of which form highly crystalline and nearly insoluble *picrates*. Picric acid also forms, in high yield, coloured *molecular complexes* with condensed aromatic hydrocarbons such as naphthalene.

Analogous adducts are formed by other polynitro compounds, *e.g.* 1,3,5-trinitrobenzene, in which the $-I$- and $-M$-effects of the nitro groups markedly lower the electron density on the benzene ring. This electron deficiency is partly compensated by interaction with another suitable π-electron system, such as that in naphthalene. The resultant π-complex is called a *charge-transfer* complex.

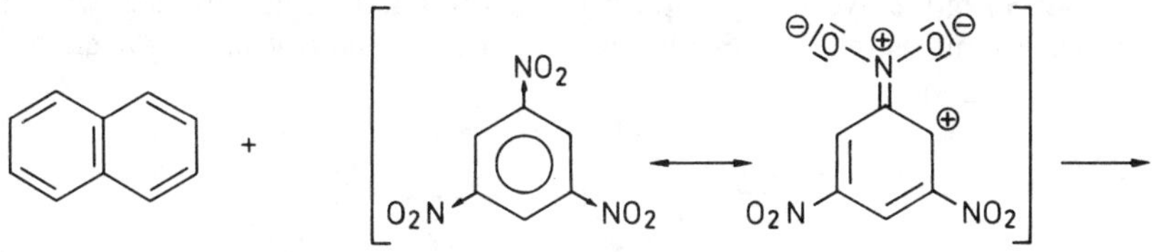

Naphthalene   +   1,3,5-Trinitrobenzene

$-I$-Effect $-M$-Effect

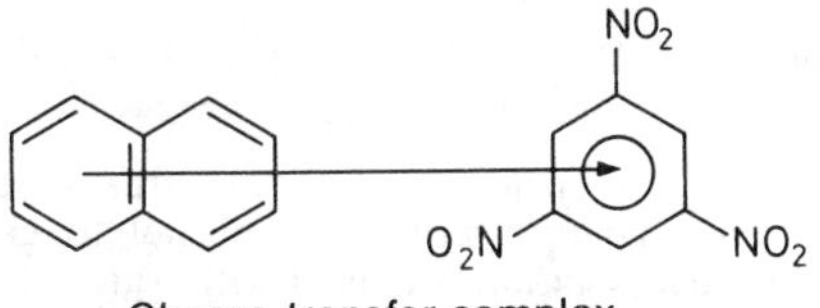

Charge-transfer complex

In this complex the naphthalene molecule is acting as a *Lewis* base.

5.5.5.3 Picryl chloride. When picric acid is treated with phosphorus(V) chloride the hydroxy group, which is activated by the three nitro groups, is replaced by a chlorine atom. The product, picryl chloride, behaves like an acid chloride and reacts with ammonia to form *picramide*, m.p. 190°C (463 K).

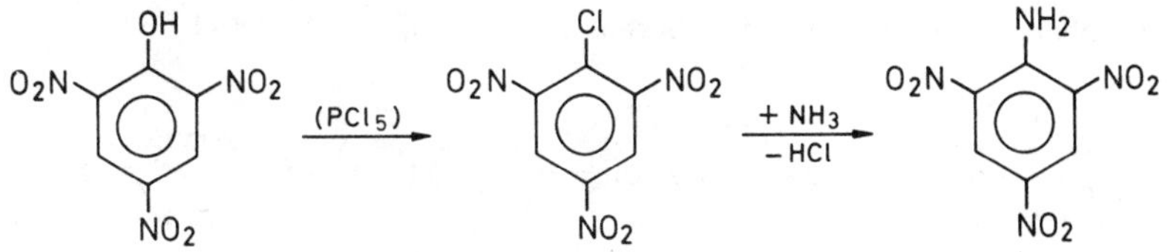

Picryl chloride   Picramide

5.5.5.4 This reaction, and the previously mentioned conversion of 2,4-dinitrochlorobenzene into 2,4-dinitrophenol belong to a class of *nucleophilic substitution* reactions, which activated derivatives of benzene undergo. They take place by a S_N-*Ar mechanism*, designated by the following terms: S_N2-Ar, addition–elimination reaction, intermediate-complex mechanism. The $-I$- and $-M$-effects of the nitro groups bring about some electropositive character in the ring carbon atoms, which facilitates *nucleophilic* attack by hydroxide ions:

1-Chloro-2,4-dinitrobenzene   2,4-Dinitrophenol

In some cases the initial adduct can be isolated; it then loses a chloride ion to regenerate the delocalised system in the ring.

5.5.6 Nitrosophenols

The $+M$-effect of the hydroxy group is sufficient to allow phenol to react with *nitrous acid*, producing *p*-nitrosophenol together with a small amount of the *o*-isomer.

p-Nitrosophenol   p-Benzoquinone
monoxime

p-Nitrosophenol crystallises from hot water in the form of yellow needles, m.p. 133°C (406 K). In the air these soon turn brown. It is *tautomeric* with *p*-benzoquinone monoxime.

If the reaction of phenol with *nitrous acid* is carried out in the presence of sulphuric acid, the *p*-nitrosophenol which is formed condenses with more phenol to give a blue-green solution of *indophenol monosulphate*. If the acid solution is diluted with water, the colour changes to red, and on addition of alkali to deep blue:

Indophenol monosulphate (blue-green)   red

deep blue

This colour reaction serves as a *test* for the presence of nitrite and nitroso groups (*Liebermann* reaction).

5.5.7 *Homologues of Phenol*

The three isomeric hydroxytoluenes are often called cresols; the hydroxy derivatives of xylene are sometimes described as *xylenols*. They occur in coal-tar.

o-Cresol   *m*-Cresol   *p*-Cresol   3,5-Dimethylphenol (a xylenol)

Crude cresol is used as a disinfectant; a solution of it in aqueous soap solution is called *Lysol*. m-*Cresol* has the strongest antiseptic action.

p-Cresol is formed from tyrosine in the putrefaction of proteins, and occurs as its sulphate ester, $H_3C-C_6H_4-O-SO_3H$ in human urine. Its esterification by sulphuric acid may be thought of as a detoxification process.

Tricresyl phosphate, which is manufactured from sodium cresolate and phosphoryl chloride, is used as a plasticiser in *PVC* plastics. Because of its toxicity alternative plasticisers are used in articles for everyday use.

Higher homologues of phenol include *thymol*[1] (2-isopropyl-5-methylphenol), which occurs in oil of thyme and is an *antiseptic*, and its isomer carvacrol[1], (5-isopropyl-2-methylphenol), which is found in many essential oils:

Thymol (m.p. 51°C [324 K])   Carvacrol (b.p. 237°C [510 K])

5.5.7.1 Uses. *Polycondensation products* formed from phenol and formaldehyde are of great importance as plastics and resins. They have the overall names of *phenolic plastics*, or *bakelite*, the latter after their discoverer *Baekeland* (1909).

Phenol, and also its homologues, the cresols and xylenols, are important industrial starting materials. They can be obtained from coal-tar or from low-temperature carbonisation of brown coal, but this does not provide sufficient to meet the needs for them, and phenol is also made from benzene or from cumene (see p. 496).

Phenol–formaldehyde resins can be prepared in either acid or alkaline media.

(a) The *alkaline condensation* of phenol ($\sim \frac{2}{3}$) and its homologues ($\sim \frac{1}{3}$) with a 30–40% solution of formaldehyde in the ratio phenol:formaldehyde = 1:1.2 to 1:1.5, using ammonia, sodium

---

[1] The numbering system used in these formulae is a trivial one used in terpene chemistry (see p. 665).

hydroxide or sodium carbonate as catalyst, takes place in three steps. In the first step an easily melting resin A (*resol*) is formed, which is soluble in organic solvents and in aqueous alkali. The resols are low-molecular-weight linear condensation products which are formed as follows by addition (to the carbonyl group of formaldehyde) and condensation reactions:

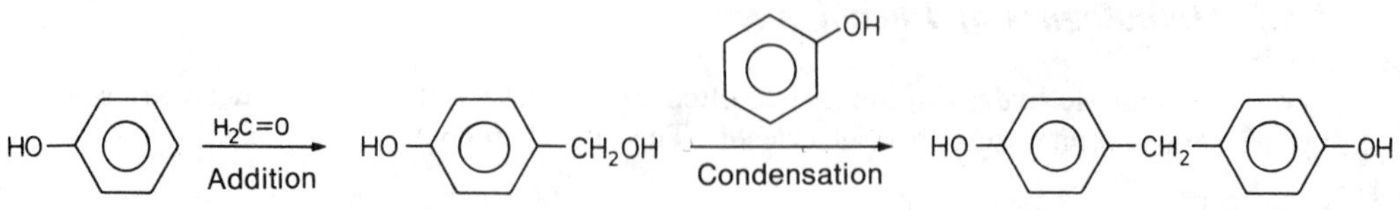

p-Hydroxybenzyl alcohol   4,4′-Dihydroxy-phenylmethane

In addition to the *p*-hydroxybenzyl alcohol shown, some of the *o*-isomer is also formed and then undergoes condensation similarly.

(b) In the *acid-catalysed condensation* a small excess of phenol is used (about 1.2 mol phenol to 1 mol formaldehyde) and hydrochloric acid is the catalyst. In this case also the first products are linear condensates, which melt and are soluble; they are known as *novolakes*. Both the resols and the novolakes can be hardened. This is done by addition of *hexamethylenetetramine*, which leads to the formaldehyde forming methylene bridges which cross-link the linear chains. When the products, which are *thermoplastic*, are heated they are converted into infusible *thermosetting* plastics.

In the preparation of hard fibreboard, wood shavings are steeped in an aqueous alkaline solution of resol; after acidification it is hardened under pressure at 140–170°C (410–440 K).

Special phenol–formaldehyde condensates, which were first discovered in Great Britain in 1935 (*Adams* and *Holmes*), called *Wofatite* and *Levatite* are used as *ion-exchange resins*.[1] These gel-like synthetic resins contain phenolic hydroxy groups, whose normal weak acidity has been increased by incorporating carboxyl or especially sulphonyl groups into the phenol nuclei. Like inorganic zeolites or organic humus material they act as *cation exchangers* and are used as such in various industries, for example in the purification of molasses and for softening water. In the latter case there is an exchange on the ion-exchange resin between $H^{\oplus}$ or $Na^{\oplus}$ ions and the $Ca^{2\oplus}$ ions in the water:

$$(H_2\text{-resin}) + Ca(HCO_3)_2 \rightarrow (Ca\text{-resin}) + 2\,CO_2\!\uparrow + 2\,H_2O \quad \Big\}$$

$$\text{or} \quad (Na_2\text{-resin}) + CaSO_4 \rightleftharpoons (Ca\text{-resin}) + Na_2SO_4 \qquad \Big\} \text{ Softening}$$

$$(Ca\text{-resin}) + 2\,NaCl \rightleftharpoons (Na_2\text{-resin}) + CaCl_2 \qquad \text{Regenerating}$$

Phenol–formaldehyde resins containing methylenesulphonic acid groups (Amberlite) or sulphonic acid groups attached to the benzene rings (Dowex 50) have proved effective for the separation of lanthanides by the ion-exchange method and in this way individual rare-earth compounds have been obtained pure in larger amounts.

In addition to cation exchange, anion exchange has been achieved with the help of certain *amino resins*. Anion exchangers are condensation products which are basic in character and contain reactive aryl- or alkyl-amino groups. They have been made by *condensation reactions* of either *urea* or *aromatic diamines*, e.g. *m*-phenylenediamine, with *formaldehyde*. They can also take part in HO-ion exchange. These exchange reactions can be presented in a similar way to the equilibria shown above for cation exchange:

$$(\text{Resin-OH}) + HCl \rightleftharpoons (\text{Resin-Cl}) + H_2O$$

$$\text{or} \quad (\text{Resin-Cl}_2) + Na_2SO_4 \rightleftharpoons (\text{Resin-SO}_4) + 2\,NaCl$$

---

[1] See *K. Dorfner*, Ionenaustauscher, 3rd edn. (Verlag W. de Gruyter & Co., Berlin 1970); *R. Kunin*, Ion Exchange Resins, 3rd edn (Krieger, Huntington 1972).

Derivatives of urea such as dicyandiamide, melamine and thiourea have also been used for the preparation of amino resins.

If acidic and basic ion-exchange resins are used one after the other, it is possible to obtain a completely salt-free solution. In addition, ion-exchange resins are employed in the recovery of precious metals or scarce natural products from strongly diluted solutions. Finally, ion-exchange resins can be made use of to promote organic reactions which are catalysed by $H^\oplus$ or $HO^\ominus$ ions, and for removing radioactive fission materials.

Biochemically, exchange resins are used successfully to obtain enzymes, vitamins, hormones and other active substances, as well as for quantitative chromatographic separation of mixtures of amino acids (ion-exchange chromatography).

Resins which can provide or take up electrons or protons are called *electron exchangers*.[1]

5.5.8 Dihydric Phenols

The three isomeric dihydroxybenzenes are:

Catechol   Resorcinol   Hydroquinone

5.5.8.1   Catechol (*o*-dihydroxybenzene, 1,2-dihydroxybenzene) was first obtained from the dry distillation of catechins, which are tannin-like substances found in some Asian plants. In general the syntheses of dihydric phenols resemble those of monohydric phenols.

Preparation
(a) *By melting phenol-o-sulphonic acid with alkali.*
(b) *Industrially by heating o-chlorophenol with 20% aqueous sodium hydroxide* under pressure in the presence of copper(II) sulphate at about 200°C (475 K).

Catechol

Properties. Catechol is a crystalline compound, m.p. 104°C (377 K), which is readily soluble in water. In alkaline solution it is a strong *reducing agent*; for example, it reduces ammoniacal solutions of silver salts or alkaline solutions of copper salts, and, as a result, is used as a photographic developer. With iron(III) chloride it gives a characteristic *emerald green colour* which turns to red-violet on addition of alkali or sodium acetate; this is due to chelate formation. Catechol derivatives take part in the metabolic transport of iron and act as *siderophores* (Grk., iron bearer) in living creatures; an example is enterobactin in *Escherichia coli*:

---

[1] See B. *Sansoni*, Über Elektronenaustauscher und Redox-Ionenaustauscher (Electron Exchangers and Redox Ion Exchangers), Angew. Chem. *66*, 143 (1954).

Enterobactin
N,N′,N″-(2,6,10-Trioxo-1,5,9-trioxa-
cyclododecane-3,7,11-triyl)-tris(2,3-
dihydroxy)-benzamide

The central trilactone is made up from L-serine units, and only this particular configuration is biologically active. The three catechol groups of this siderophore act together to form an octahedral iron(III) complex.

Catechol can be methylated by dimethyl sulphate in alkaline solution to give, in turn, *guaiacol* and *veratrole*.

Guaiacol   Veratrole

Guaiacol (2-methoxyphenol) was first obtained from the distillation products of guaiac, a resin obtained from trees of the genus *guaiacum* (*Unverdorben*, 1826). It is also abundant in beech-wood tar. It melts at 32°C (305 K) and has a very characteristic odour; it is an effective *disinfectant*. With iron(III) chloride it gives a blue colour.

When melted, the allyl ether of guaiacol undergoes an *intramolecular rearrangement* to form *o*-eugenol. The allyl chain takes the place of an *o*-hydrogen atom and is *inverted* in the course of this *Claisen* rearrangement:

Guaiacyl allyl ether   *o*-Eugenol

Veratrole has an agreeable odour and melts at 22.5°C (295.7 K).

5.5.8.2   Resorcinol (*m*-dihydroxybenzene, 1,3-dihydroxybenzene) is obtained from the distillation of natural resins. It is prepared by *melting benzene-1,3-disulphonic acid with alkali*:

Benzene-1,3-disulphonic acid   Resorcinol

Resorcinol, m.p. 110°C (383 K), is a colourless crystalline substance which is readily soluble in water, ethanol or ether. It reduces ammoniacal silver nitrate solution on heating. With iron(III) chloride an aqueous solution of resorcinol gives a violet colour.

The ease with which resorcinol can take up two hydrogen atoms to form *dihydroresorcinol* (cyclohexane-1,3-dione), *e.g.* by the action of sodium amalgam and water, is of interest. This is due to the fact that, energetically, the formation of two keto groups largely compensates for the loss of stabilisation associated with delocalisation of electrons in the benzene ring, and keto–enol tautomerism results.

Dienol form   *Diketo* form

Resorcinol   Dihydroresorcinol

A consequence of this tautomerism is that resorcinol forms a diacetyl derivative when treated with acetic anhydride, and a dioxime with hydroxylamine.

Resorcinol is used as an *antiseptic* in dermatology and in the dyestuffs industry as an important starting material.

A derivative of resorcinol, 5,5-dimethylcyclohexane-1,3-dione or *dimedone*, is readily prepared from mesityl oxide and diethyl malonate, and is used as a sensitive *reagent* for the gravimetric estimation of *formaldehyde*:

Dimedone   Formaldehyde

Nitration of resorcinol provides *styphnic acid* (2,4,6-trinitroresorcinol), m.p 175°C (448 K) which, like picric acid, with organic bases gives salts, *styphnates*, which form good crystals.

A homologue of resorcinol, orcinol (5-methylresorcinol, 3,5-dihydroxytoluene), m.p. 107°C (380 K), can be isolated from lichens, and is the parent compound of the dyestuffs *archil* and *litmus*.

Styphnic acid   Orcinol

5.5.8.3 Hydroquinone (quinol, *p*-dihydroxybenzene, 1,4-dihydroxybenzene) is prepared by *reducing p-benzoquinone* with sulphurous acid or with nascent hydrogen:

p-Benzoquinone   Hydroquinone

It forms colourless needles of m.p. 172°C (445 K). Because of its reducing properties it is used in photography as a *developer*. Hydroquinone and catechol, but *not* resorcinol, quickly turn red-brown in air when moist, due to autoxidation. The oxidation products obtained from these two dihydric phenols are dealt with in the section on benzoquinones.

5.5.9 Cyclophanes and Catenanes

When dihydric phenols such as hydroquinone or resorcinol form ethers by reaction with ω,ω'-dihalogenoalkanes under high dilution conditions (*Ruggli–Ziegler*), *cyclic ethers* with large rings are formed (*K. Ziegler, Lüttringhaus*). The mono-ether, which is formed first, cyclises in the presence of a sodium alkoxide to give a *cyclophane*; these compounds were formerly described as ansa-compounds (Lat. *ansa* = handle), e.g.

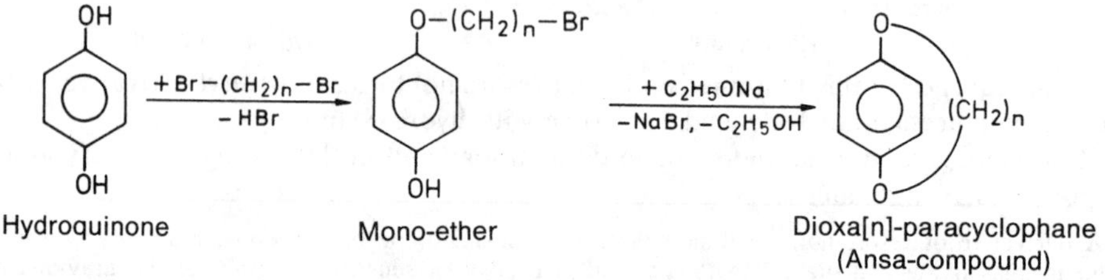

Hydroquinone   Mono-ether   Dioxa[n]-paracyclophane
(Ansa-compound)

In order to form such cyclophanes, at least eight methylene groups in the case of hydroquinone, and at least seven in the case of resorcinol, must be present in the bridge, *i.e.* dioxa[8]-paracyclophane and dioxa[7]-metacyclophane, respectively. If larger substituents are present in the benzene ring of the cyclophane, then free rotation of the benzene ring about its *p*-axis is sterically hindered by 'handles' of appropriate size and this leads to *molecular asymmetry* and optical activity; a large substituent cannot pass through the 'handle'. The following cyclophanes have been resolved into enantiomers:

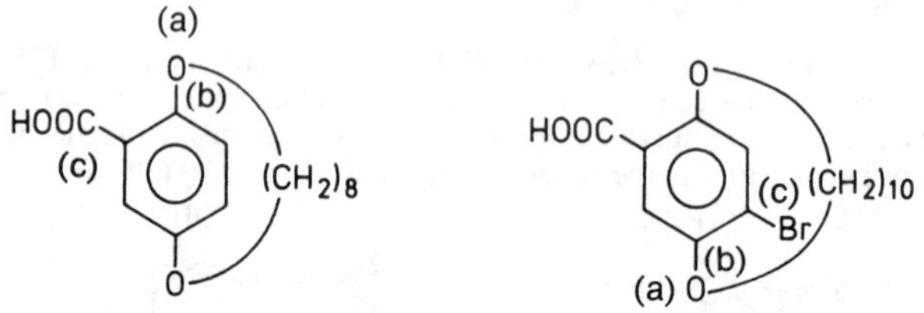

There is a problem concerning the configuration of molecules having a chiral plane which lies in the plane of the benzene ring. In order to classify such a molecule according to the CIP system, a *pilot atom* is selected, which is directly attached to an atom in the chiral plane and which is attributed the highest priority (a): starting from here, one moves to the in-plane atom to which it is attached; this is assigned priority (b). From (b) one moves to the in-plane atoms next to it and assigns priority (c) to whichever of these has the higher priority according to the standard priority rules. If going from (a) to (b) to (c) involves a clockwise route then there is a (R_p)-configuration.[1] This is the case in the two examples above. If an anticlockwise route is involved there is a (S_p)-configuration.

The cyclophenylalkanes, given the name *cyclophanes*[2] by *Cram*, are bridged derivatives of benzene. Their nomenclature, structure and numbering are illustrated by the following three examples.

---

[1] The subscript p indicates that the designation R refers to a plane.
[2] See *P. M. Keehn* et al. (Ed.), Cyclophanes, 2 vols. (Academic Press, New York 1983); *F. N. Diederich*, Cyclophanes (Royal Society of Chemistry, London 1992).

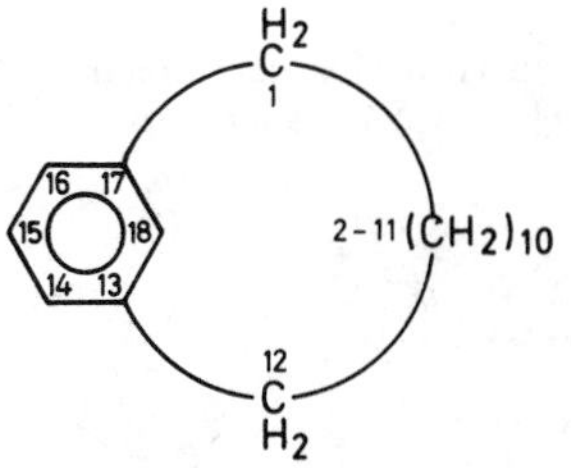

[12]Metacyclophane 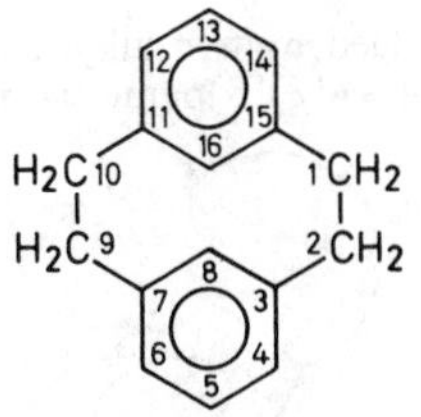   [2.2]Metacyclophane 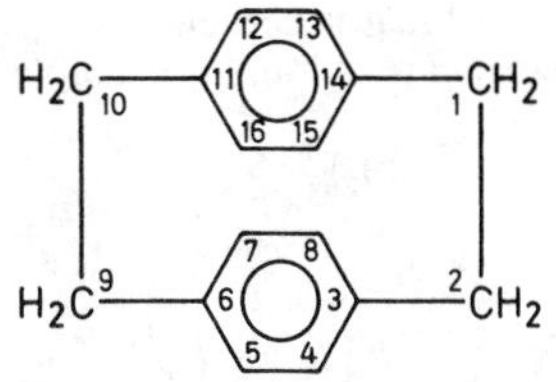   [2.2]Paracyclophane

5.5.9.1 [2.2]-Paracyclophane is an intermediate in the industrial preparation of poly-p-xylylene (Parylene); it can be made from p-xylene by dehydrogenation in the gas phase. Thermal polymerisation ($\sim 500°C$ [775 K]), followed by cooling to 50°C (325 K) gives rise to products with relative molecular masses of more than 500,000.

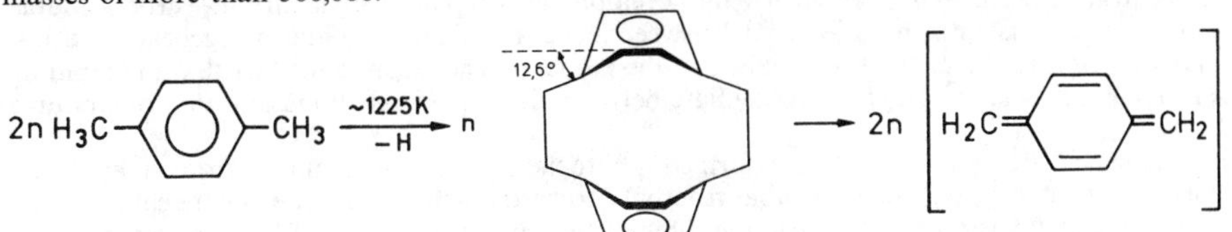

p-Xylene   [2.2]-Paracyclophane

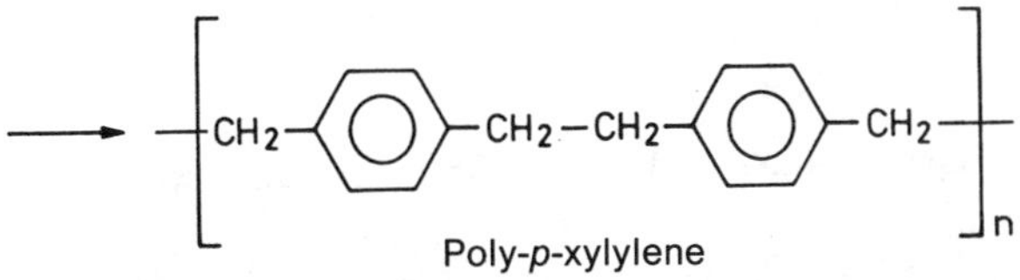

Poly-p-xylylene 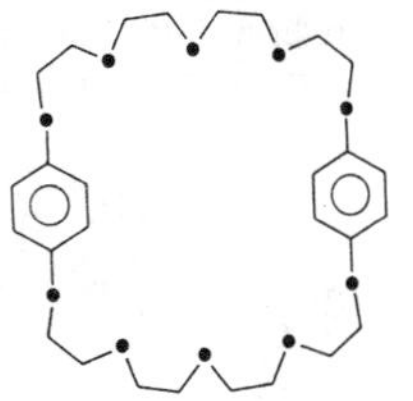   Decaoxa[13.13]paracyclophane

In the laboratory, [2.2]-paracyclophane has been prepared by a *Hofmann* elimination reaction on p-methylbenzyl-trimethylammonium hydroxide, $p-CH_3-C_6H_4-CH_2\overset{\oplus}{N}(CH_3)_3\ OH^{\ominus}$. The two benzene rings in [2.2]-paracyclophane are boat-shaped, but this does not lead to any marked interference with the delocalisation of the π-electrons.

1,4,7,10,13,20,23,26,29,32-Decaoxa[13.13]-paracyclophane is a π-electron-rich macrocycle, whose role as a starting material in a one-stage synthesis of a catenane is mentioned on p. 513. A large step forward in cyclophane chemistry resulted from using the pyrolysis of sulphones, which leads to the extrusion of SO_2:[1]

$$R-SO_2-R' \xrightarrow[-SO_2]{\Delta} R-R'$$

The first synthesis of a poly-bridged cyclophane (*Boekelheide*, 1973) will serve as an example:

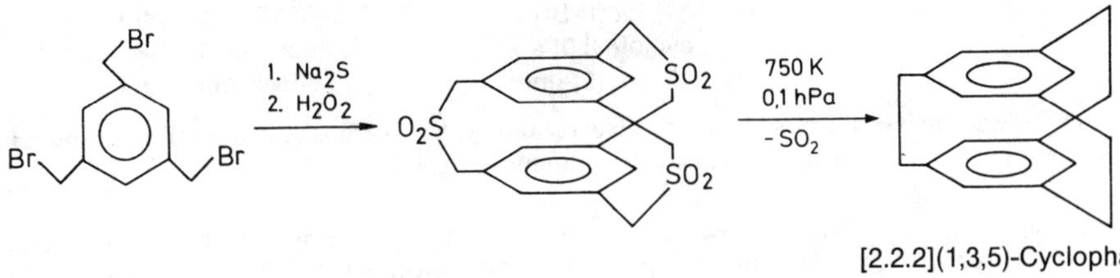

[2.2.2](1,3,5)-Cyclophane

---

[1] See *F. Vögtle* et al., Pyrolysis of Sulphones as a Synthetic Method, Angew. Chem. Int. Ed. Engl. *18*, 515 (1979).

5.5.9.2   When a benzene ring is bridged by two alkyl chains, as in [m][n]paracyclophanes, the molecules are chiral, for example (S)-$(+)$-[8][8]paracyclophane, $[\alpha]_D^{20}$ $-5.4°$ (*Nakazaki*).

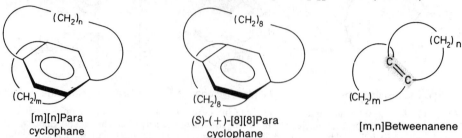

[m][n]Para
cyclophane

(S)-$(+)$-[8][8]Para
cyclophane

[m,n]Betweenanene

If the central benzene ring of an [m][n]paracyclophane is replaced by a carbon–carbon double bond, the molecule is known as an [m,n]betweenanene; the ligands at the bridgeheads are not coplanar with one another. The last syllables of the name, -anene, express the fact that the bonding in the *trans*-bridge (shaded grey) is intermediate between that of a single bond and that of a double bond.

Betweenanenes are chiral. They are formed photochemically, for example from achiral *cis*-bicyclo[8.8.0]octadecenene. Their variable reactivity towards dichlorocarbene is explicable if only the *cis*-compound takes part in the reaction. One of the two enantiomers of [8,8]betweenanene is formed preferentially if a suitable derivative, such as (Z)-2-oxobicyclo-[8.8.0]octadec-1(10)-ene (I) is irradiated in the presence of $(R.R)$-$(+)$-diethyl tartrate.

$(CH_2)_8$ $(CH_2)_7$

$\xrightarrow[\text{(+)-Diethyl tartrate}]{h\nu,}$ I +

$(CH_2)_7$ $(CH_2)_8$ (II)

C
‖
O   (I)

(Z)-2-Oxobicyclo[8.8.0]
octadec-1(10)-ene

I : II = 1 : 5,5

Wolff–Kishner reduction

$(CH_2)_8$ $(CH_2)_8$ $\xleftarrow[\text{NaOH}]{CHCl_3,}$ Cl Cl $(CH_2)_8$ $(CH_2)_8$ + $(CH_2)_8$ $(CH_2)_8$

1rC²,10cC⁹-Bi-
cyclo[8.8.0]
octadec-1(10)-ene[1]

1rC²,10tC⁹-Bicyclo
[8.8.0]octadec-1(10)-ene.[1]
[8.8]Betweenanene

The [8.8]betweenanene which is obtained is laevorotatory, but there is only a small enantiomeric excess of about 1%, which probably has an (R)-configuration.[2]

---

[1] Designation as (Z) or (E) is not applicable here, since the two bridges each have the same constitution. An unequivocal description is possible if one indicates whether a substituent on the bridge is *cis* (c) or *trans* (t) to a prescribed substituent defined by the symbol (r). See p. 533, and *Beilsteins* Handbuch der Organischen Chemie, 4th edn, 4th Supplement Vol. *1*, p. X *ff*.

[2] See M. *Nakazaki* et al., *J. Org. Chem.*, **45**, 3229 (1980). [8.8]Betweenane has D_2-symmetry; its unequivocal name is derived from this and is (R)-$(-)$-D_2-bicyclo[8.8.0]octadec-1(10)-ene.

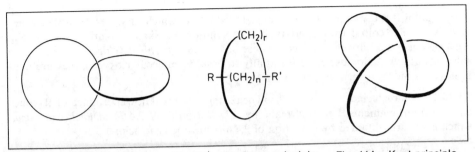

Fig.114a. Catenane principle   Fig.114b. Rotaxane principle   Fig. 114c. Knot principle

Going a stage further from the [n]-cyclophanes, there are compounds in which two macrocyclic rings are linked through each other, like links on a chain (Fig. 114a); these were the first molecules to be prepared whose parts were *not* held together by chemical bonding (*Lüttringhaus* and *Schill*, 1964). Such compounds are described as catenanes (Lat. *catena* = chain). In rotaxanes (Fig. 114b) a linear unit is threaded through a ring and cannot escape from this position because the end groups, R,R' are too bulky.[1] An example of the knot principle illustrated in Fig. 114c is the cloverleaf knot.[2] *Wasserman* has proposed the name *topological isomers* for these classes of compounds.

A single-step synthesis of a [2]-catenane[3] has been achieved, using 1,4,7,10,13,20,23,26,29,32-decaoxa[13.13]paracyclophane (I, see p. 511) as a template. In the process I forms the one ring, whilst the other arises from a reaction between 1,1'-*p*-xylylene-bis(4,4'-bipyridinium) hexafluorophosphate and 1,4-bis(bromomethyl)benzene.[4]

5.5.10 Trihydric Phenols

There are three isomeric trihydric phenols as follows:

Pyrogallol
(vic)

Phloroglucinol
(sym.)

Hydroxyhydroquinone
(asym.)

5.5.10.1 Pyrogallol (1,2,3-trihydroxybenzene) was prepared in 1786 by *Scheele* by heating gallic acid, and is still made by this process:

Gallic acid

$-CO_2$

Pyrogallol

[1] See G. Schill, Catenanes, Rotaxanes and Knots (Academic Press, New York 1971).
[2] See J. P. Sauvage et al., A Synthetic Molecular Trefoil Knot, Angew. Chem. Int. Ed. Engl. 28, 189 (1989); Acc. Chem. Res. 23, 319 (1990).
[3] [2]-Catenane indicates that two rings in a molecule are interlinked according to the catenane principle.
[4] See J. F. Stoddart et al. A [2]-Catenane Made to Order, Angew. Chem. Int. Ed. Engl. 28, 1396 (1989).

Pyrogallol is a crystalline powder, m.p. 133° (406 K) which dissolves readily in water and gives a blue colour with iron(III) chloride solution. In alkaline solution it is a strong *reducing agent*, and in consequence it is used as a *photographic developer*. Since alkaline solutions of pyrogallol also have the ability to bond to molecular oxygen, they are used in gas analysis to absorb oxygen.

5.5.10.2   Phloroglucinol (1,3,5-trihydroxybenzene) is a breakdown product of flavones and flavylium pigments. It is best made by *acid hydrolysis of 2,4,6-triaminobenzoic acid*, which is in turn obtained by reduction of 2,4,6-trinitrobenzoic acid:

2,4,6-Trinitrobenzoic acid   2,4,6-Triaminobenzoic acid   Phloroglucinol

Phloroglucinol crystallises out from water with two molecules of water of crystallisation. The anhydrous form melts at 218°C (491 K). It is used for *testing for lignins* in wood; in the presence of hydrochloric acid a red-purple colour results.

While on the one hand phloroglucinol gives a blue-violet colour with iron(III) chloride and reacts as expected with diazomethane to give a trimethyl ether, and with acetic anhydride to give a triacetyl derivative, on the other hand with hydroxylamine it forms a trioxime. This is because phloroglucinol (like resorcinol) can exist in two different *tautomeric* forms, a tri-enol and a tri-keto form, each of which can take part in their characteristic reactions:

Tri-enol form   Tri-keto form
Phloroglucinol

5.6   Benzoquinones[1]

Strictly speaking, benzoquinones are diketo derivatives of cyclohexadiene and could thus be classified as partially hydrogenated benzene rings. However, for teaching purposes it is more convenient to consider them together with the phenols.

---

[1] See S. *Patai* et al. (Ed.) The Chemistry of the Quinonoid Compounds, 2 vols. (Wiley, Chichester 1974, 1988).

5.6.1 The Most Important Benzoquinones

Benzoquinones can be regarded as oxidation products derived from dihydric phenols, but only catechol and hydroquinone can be dehydrogenated to provide benzoquinones. Loss of two hydrogen atoms is accompanied by loss of two π-electrons from the benzene ring. In the process the aromatic character of the ring is lost and the following quinonoid structures arise:

Catechol   o-Benzoquinone   Hydroquinone   p-Benzoquinone

It is necessary to distinguish between o- and p-quinonoid systems. In both of them the doubly-bonded oxygen atom can be replaced by other groups such as $=NH$, $=NR$, $=CH_2$ or $=CR_2$.

o-Quinonoid system   p-Quinonoid system

In contrast, a m-benzoquinone cannot be obtained by oxidising resorcinol. It is not possible to construct m-quinonoid systems using the normal requirements for bonding.

5.6.1.1 **o-Benzoquinone** is an unstable odourless substance which is decomposed by water. It is obtained by careful dehydrogenation of catechol by silver oxide in absolute ether in the presence of anhydrous sodium sulphate. In the first place a colourless intermediate is formed, which at once changes into red o-benzoquinone (see above).

o-Benzoquinone is a strong oxidising agent; for example, it liberates free iodine from acidic solutions of potassium iodide.

5.6.1.2 **p-Benzoquinone** (1,4-benzoquinone, sometimes called just quinone) was first obtained by oxidising quinic acid, a constituent of cinchona bark (*Woskressensky*, 1838).

Derivatives of p-benzoquinone are used as biological defence weapons by the bombardier beetle. When needed they are formed from the corresponding hydroquinones and hydrogen peroxide, which, in the presence of the enzyme oxidoreductase (see p. 884), react explosively to spray the products onto their target.

Preparation. p-Benzoquinone is made by *oxidising aniline* with chromic acid or manganese(IV) oxide (manganese dioxide) in sulphuric acid.

The *reaction mechanism* is rather complicated, but can be explained as follows (*Willstätter, Green*): The first product, N-phenylquinonediimine, which is yellow, is formed by loss of hydrogen from two molecules of aniline:

N-Phenylquinonediimine

Further oxidation provides a derivative of quinonediimine, which is hydrolysed by mineral acid to *p*-benzoquinone, *p*-phenylenediamine, aniline and ammonia:

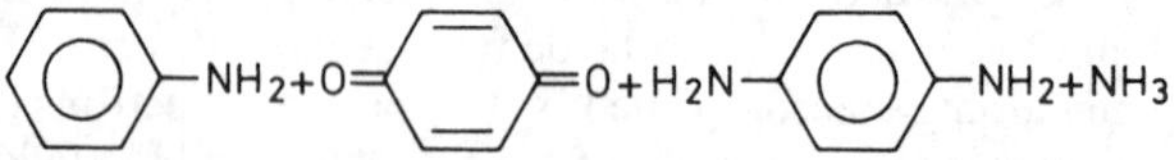

Aniline   *p*-Benzoquinone   *p*-Phenylenediamine

Under the reaction conditions used, the *p*-phenylenediamine is dehydrogenated to *p*-benzoquinone diimine and this is hydrolysed to *p*-benzoquinone:

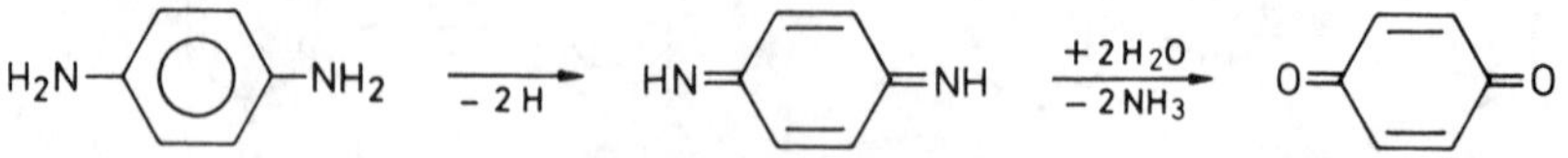

p-Benzoquinone diimine

Properties. *p*-Benzoquinone forms yellow crystals, m.p. 116°C (389 K), which have a sharp smell; it is volatile in steam.

The colour is due to the presence of the quinonoid system; this system is one which leads to colour in a molecule; such groups are known as *chromophores*.

Chemically *p*-benzoquinone behaves as a 2,3-unsaturated-1,4-diketone. Thus with hydroxylamine hydrochloride it gives both a mono- and a di-oxime; the former is a tautomer of *p*-nitrosophenol. The *oxime* is more stable than the *nitroso form*; in solution there is an *equilibrium* between the two tautomeric forms:

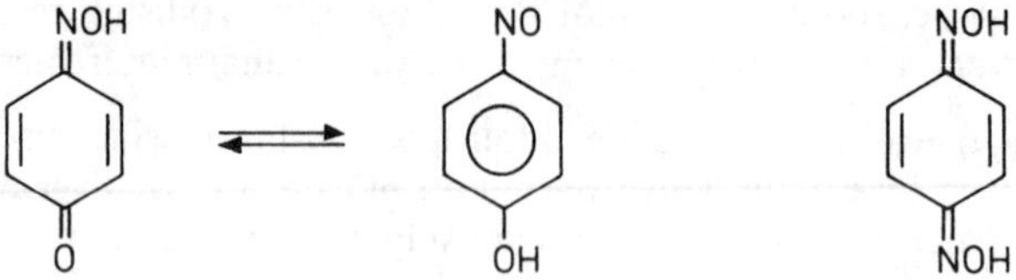

p-Benzoquinone monoxime   *p*-Nitrosophenol   *p*-Benzoquinone dioxime

Another diketo derivative is *p*-benzoquinone-amidinohydrazone-thiosemicarbazone (INN: **Ambazone**) which is made by treating *p*-benzoquinone monoguanylhydrazone with thiosemicarbazide in acid solution, and is present in *Iversal*, which is used as a disinfectant in the mouth and throat.

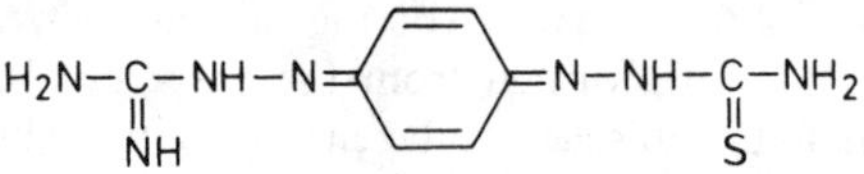

INN: Ambazone

This compound is strongly basic and forms copper-brown crystals which melt at 193°C (466 K).

Only one of the keto groups of *p*-benzoquinone reacts with *Grignard* reagents to form hydroxycyclohexadienones (quinoles, *p*-quinols), e.g.

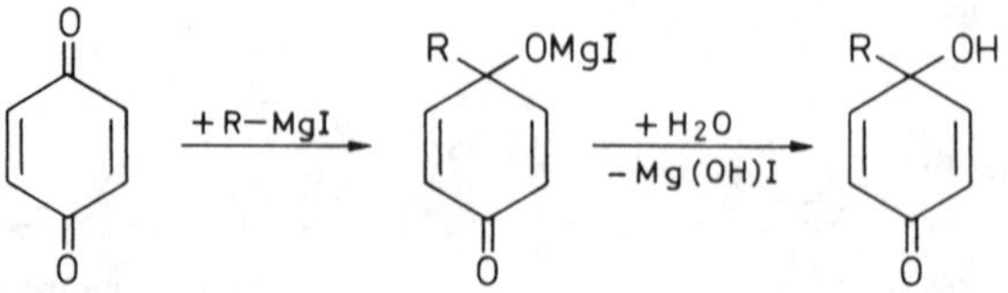

A quinole

5.6.2 Redox Reactions of p-Quinones

An important redox reaction is the reduction of *p*-benzoquinone to hydroquinone or the oxidation of hydroquinone to *p*-benzoquinone:

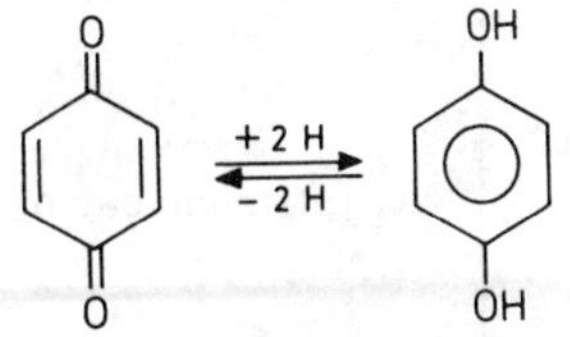

In both reactions an intermediate product, quinhydrone, is formed which is an iridescent deep green substance, only slightly soluble in water. The same adduct is obtained in the form of dark brown crystals with a green lustrous surface by mixing equimolecular solutions of *p*-benzoquinone and hydroquinone (*Wöhler*, 1844). It consists of one molecule each of *p*-benzoquinone and hydroquinone and was once thought to be a molecular compound held together by hydrogen bonding between the two constituents. However, it became apparent that other compounds which were incapable of undergoing hydrogen bonding could form quinhydrones. Phenol, hydroquinone ethers or hexamethylbenzene can all take the place of hydroquinone.

It is now thought that formation of these adducts is the result of *reciprocal interaction* between the two π-electron systems. This takes place between two compounds because of mesomeric or inductive effects which cause the one compound to be π-electron-rich (an electron donor D), whilst the other is π-electron-deficient (an electron acceptor A). As a result the adducts, described variously as π-complexes, electron donor–acceptor complexes (EDA complexes) or *charge-transfer* complexes, are formed, *e.g.*

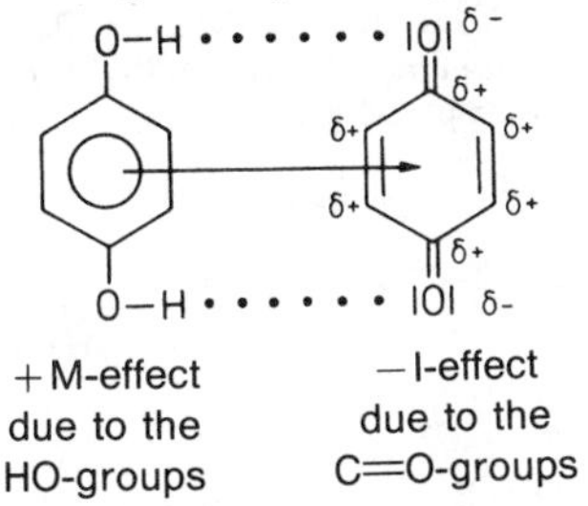

The intense colour of quinhydrones is also associated with the mutual interactions between the π-electrons of the two rings. In addition, formation of two hydrogen bonds may contribute to the stability of the complexes.

In alkaline media the quinhydrones are unstable and they dissociate, one electron from the donor D being transferred to the acceptor A to produce *semiquinone anions*. These are *radical anions* and are *stabilised by mesomerism*. The unpaired electron is delocalised over all the carbon atoms and the oxygen atoms:

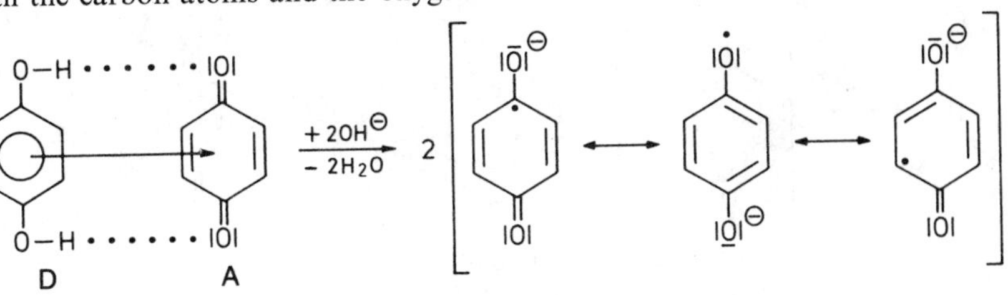

Semiquinones are also formed by the *reduction of p-benzoquinones in alkaline media*, but in the case of *p*-benzoquinone itself only the quinhydrone can be isolated. In contrast, from *p*-duroquinone (2,3,5,6-tetramethyl-*p*-benzoquinone), the semiquinone semiduroquinone can be isolated as its brown sodium salt:

Na⁺   Sodium salt of semiduroquinone

The radical character of this semiquinone is evident from its *paramagnetism*. Such monomolecular semiquinones are only stable in alkaline solution. In neutral or acid solution they convert into the corresponding quinone and hydroquinone or quinhydrone.

The oxidative capacity of quinones is directly associated with the ease with which they are reduced. The *oxidation potential* of a quinone is strongly affected by substituents in the ring. Electron-donating groups such as NH_2, NR_2, OH, OR or CH_3 lower the oxidation potential whereas electron-withdrawing groups such as NO_2, SO_3Na, CN and halogen atoms raise it. They also have a marked effect on the reactions of quinones with reagents of the type H—X, which take part in 1,4-addition reactions.

In connexion with the redox character of *p*-quinones, the compounds known as *ubiquinones* and *plastoquinones*, which are found in the cells of both plants and animals, are of interest. Because of this redox behaviour and the conversion of ubiquinones into dianions, they are involved in electron transport and hydrogen exchange in the respiratory processes. The *p*-quinone rings of ubiquinones have as substituents a methyl group at the 2-position, methoxy groups at the 5- and 6-positions, and at position 3 an isoprenoid side-chain which may be of different lengths ($n = 6$–10). Similar side chains occur in vitamin K_2 and Vitamin E (α-tocopherol). Farnesyl diphosphate is involved in the biosynthesis of ubiquinones. *Plastoquinones* have methyl groups at their 5- and 6-positions and the side-chain contains at least nine isoprene groups.

Ubihydroquinone dianion   Ubiquinone ($n = 6$–10)

5.6.3 1,4-Addition Reactions of p-Quinones

Hydrogen chloride reacts with *p*-benzoquinone to give a 1,4-adduct which at once spontaneously rearranges, generating a benzenoid ring and forming chlorohydroquinone. This can be dehydrogenated to chloro-*p*-benzoquinone:

1,4-Adduct   Chlorohydroquinone   Chloro-*p*-benzoquinone

If this addition of hydrogen chloride and subsequent oxidation is repeated several times the ultimate product is chloranil (tetrachloro-*p*-benzoquinone). This forms yellow leaflets which sublime at 80°C (353 K) and melt in a sealed tube at 90°C (363 K).

Chloranil   2,3-Dichloro-5,6-dicyano-*p*-benzoquinone (DDQ)

Chloranil is a *fungicide*. Because of its high oxidation potential it finds use in preparative chemistry as a *dehydrogenating agent*. The even more powerful reagent 2,3-dichloro-5,6-dicyano-*p*-benzoquinone (DDQ) is used for the same purpose.

Ethyleneimino-*p*-benzoquinones are readily converted under biological conditions into the corresponding hydroquinones; they probably act *in vivo* in both the oxidised and reduced forms.

Methanol or *aniline* undergo 1,4-addition reactions with *p*-benzoquinone analogous to that of hydrogen chloride. The resultant hydroquinone derivative is oxidised to the related quinone derivative by excess *p*-quinone present. In this case, the oxidation potential of the substituted quinone is lower than that of *p*-benzoquinone itself.

Anilinohydroquinone   Anilino-*p*-benzoquinone

2,5-Dianilinohydroquinone   2,5-Dianilino-*p*-benzoquinone

2,5-Dianilino-*p*-benzoquinone dianil

Since the substituted *p*-benzoquinone can react further, for example with more aniline, the product contains more than one substituent group, *e.g.* 2,5-dianilino-*p*-benzoquinone; its keto group can condense with excess aniline to give a *quinonediimine*, which is known traditionally as a dianil.

If *p*-benzoquinone is treated with bromine in acetic acid, a normal addition reaction ensues, involving addition, in turn, of two and four bromine atoms:

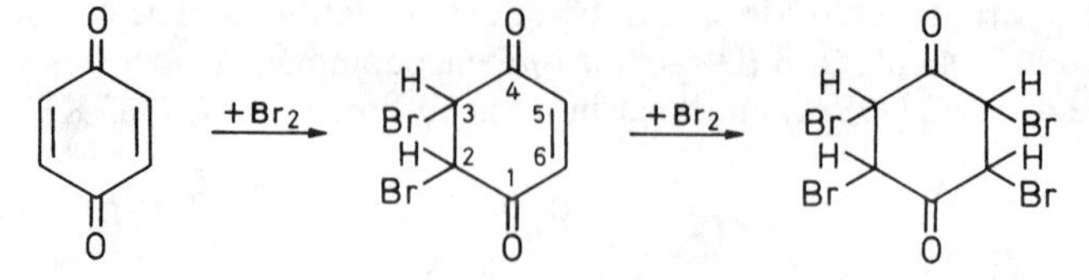

2,3,5,6-Tetrabromocyclohexane-1,4-dione

5.6.4 Quinoid Dyestuffs

The simplest quinoid dyestuffs are derivatives of *N*-phenyl-*p*-benzoquinone monoimine, containing one quinonoid and one benzenoid ring. Examples are the indophenols and indoanilines, whose parent compounds have the following constitutions:

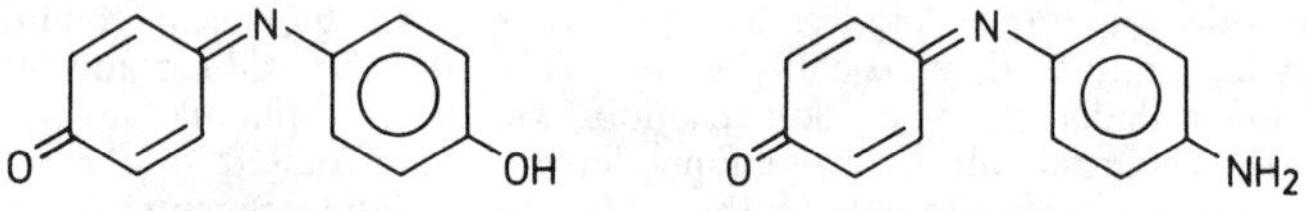

Indophenol   Indoaniline

Indophenol, m.p. 160°C (433 K), forms brown leaflets which have a metallic lustre. It is made by oxidising equimolecular amounts of *p*-aminophenol and phenol with sodium hypochlorite in alkaline solution (*cf.* p. 504).

Phenol blue, m.p. 162°C (435 K), is an example of an indoaniline dyestuff. It is made, *via* its colourless *leuco-base*, by oxidising a mixture of *N,N*-dimethyl-*p*-phenylenediamine and phenol with potassium hexacyanoferrate(III).

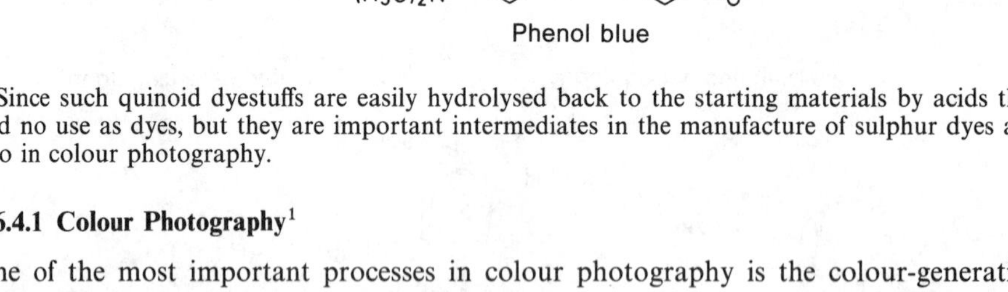

Leuco-base

Phenol blue

Since such quinoid dyestuffs are easily hydrolysed back to the starting materials by acids they find no use as dyes, but they are important intermediates in the manufacture of sulphur dyes and also in colour photography.

5.6.4.1 Colour Photography[1]

One of the most important processes in colour photography is the colour-generating (chromogenic) developing process, whose principle was discovered by *Rudolf Fischer* in 1911. It depends upon the fact that, when a silver halide emulsion is developed with *N,N-dialkyl-p-phenylenediamine*, the silver ions in the exposed grains are transformed into

---

[1] See *W. Püschel*, Die Farbphotographie (Colour Photography), Chemie in unserer Zeit *4*, 9 (1970); *M. Schellenberg* et al. Die Silberfarbbleich-Farbphotographie (Silver colour-bleach Colour Photography), *ibid. 10*, 131 (1976).

metallic silver and the *p*-phenylenediamine into the corresponding quinone diimine. The latter then reacts with a colour-developer, *i.e.* an organic compound containing a reactive *methylene* or *methine* group, to produce yellow, purple or blue-green dyestuffs in the emulsion layer, *e.g.*

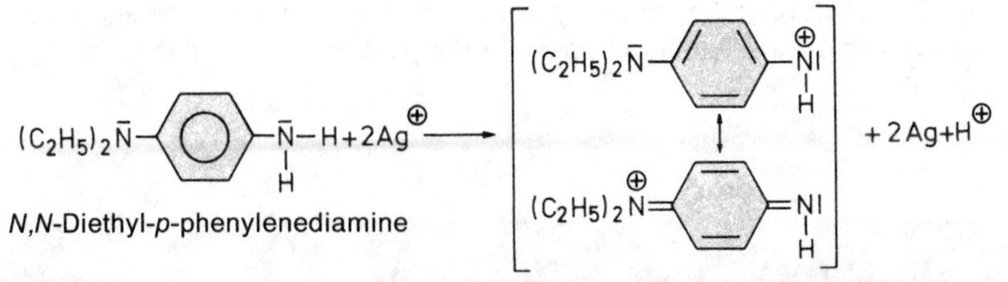

N,N-Diethyl-*p*-phenylenediamine

N,N-Diethylquinone diimine

Open-chain or cyclic compounds may be used as 'methylene couplers'. For example, with *β-ketoanilides yellow azomethine dyestuffs* are formed as follows:

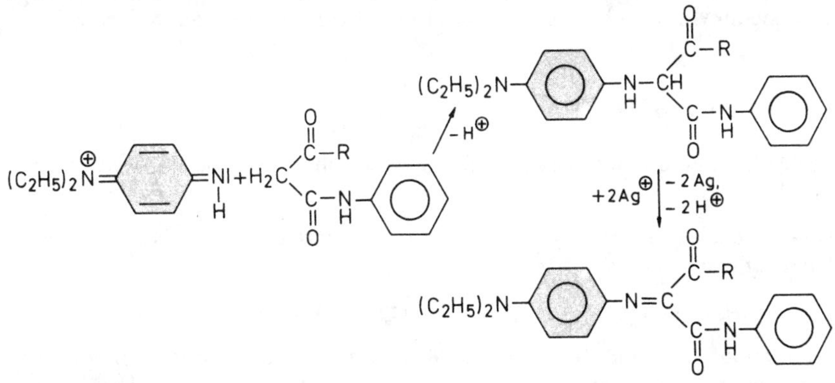

Azomethine dyestuffs (yellow)

In the same way, if *pyrazolones* are used as 'cyclic methylene couplers', oxidative developing gives rise to *purple azomethine dyestuffs*, and if 'methine couplers', which are almost exclusively phenols or α-naphthols, are used, *blue-green indoaniline dyestuffs* are formed.

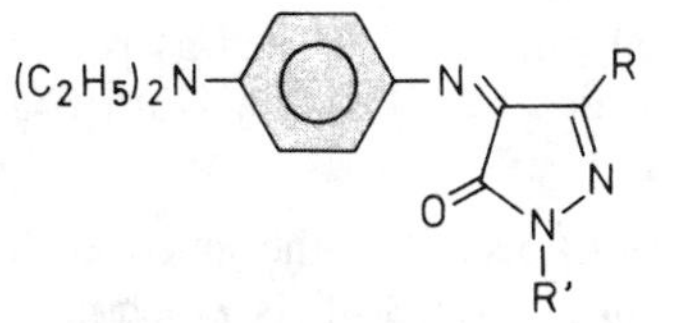

Azomethine dyestuffs (purple)

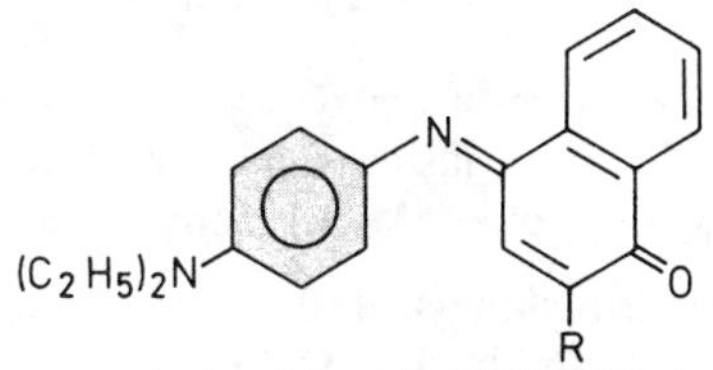

Indoaniline dyestuffs (blue)

The *three-layer colour film* (Agfacolour Process, 1936) is based upon these three kinds of colours, in all of which the benzene ring, picked out in grey in the above formulae, is part of an identical *N,N*-dialkyl-*p*-phenylenediamine building block.

In colour photography it is important that the coupling components should be soluble, and the resulting dyestuffs, as far as is possible, insoluble and resistant to diffusion. The solubility in water or alkali can be increased by the presence of sulphonyl groups or carboxyl groups, the resistance to diffusion by the introduction of a long-chain alkyl group.

In modern subtractive *many-layered colour films*, each of the three coupling components is included in separate layers of differently sensitised silver bromide, one on top of the other, so that on developing with only one developer, all three of the subtractive colour mixtures are formed at the correct places in the picture.

5.7 Aryl Alcohols and Arylalkylamines

These two classes of compounds are derivatives of alkylbenzenes in which a hydrogen atom of the alkyl group has been replaced by a hydroxy or an amino group.

5.7.1 Aryl Alcohols

Of particular interest are benzyl and salicyl alcohols, α- and β-phenylethanols and cinnamyl alcohol.

Benzyl alcohol, C_6H_5—CH_2OH, is the simplest aryl alcohol. It is formed by *alkaline hydrolysis of benzyl chloride*.

$$C_6H_5-CH_2-Cl + Na_2CO_3 + H_2O \longrightarrow C_6H_5-CH_2-OH + NaHCO_3 + NaCl$$

It is a colourless liquid, b.p. 205°C (478 K), with a pleasant odour. It is only slightly soluble in water. Like primary aliphatic alcohols it can be oxidised to benzaldehyde and benzoic acid.

$$C_6H_5-CH_2OH \longrightarrow C_6H_5-C{\overset{H}{\underset{O}{}}} \longrightarrow C_6H_5-C{\overset{O}{\underset{OH}{}}}$$

| Benzyl alcohol | Benzaldehyde | Benzoic acid |

Some esters of benzyl alcohol occur in essential oils and are used as perfumes. For example, oil of jasmine contains *benzyl acetate*, which can be made synthetically from benzyl alcohol and acetic anhydride.

o-**Hydroxybenzyl alcohol** (salicyl alcohol, saligenin) occurs as the glucoside *salicin* in the leaves and bark of willow trees (*Salix*). The alcohol forms leaflets which melt at 86°C (359 K).

CH$_2$OH
OH

o-Hydroxybenzyl alcohol (Salicyl alcohol)
(m.p. 86°C [359 K])

α-Phenylethanol, C_6H_5—CH(OH)—CH_3, is *optically active* and is an important intermediate in the perfume industry. It is obtained in its racemic form, b.p. 204°C (477 K), by *reduction of acetophenone*, C_6H_5—CO—CH_3.

β-Phenylethanol (phenethyl alcohol), C_6H_5—CH_2—CH_2—OH, b.p. 220°C (493 K), is the main constituent of rose-oil. Together with its esters it plays an important role in perfumery. It can be prepared by *reduction of ethyl phenylacetate* with sodium and alcohol (Bouveault–Blanc *reduction*):

$$C_6H_5\text{-}CH_2\text{-}COOC_2H_5 + 4Na + 3C_2H_5OH \rightarrow C_6H_5\text{-}CH_2\text{-}CH_2OH + 4C_2H_5ONa$$

Cinnamyl alcohol, [(*E*)-3-phenylallyl alcohol, 3-phenylprop-2-en-1-ol], C_6H_5—CH=CH—CH_2OH, forms colourless needles, m.p. 34°C (307 K), b.p. 257.5°C (530.7 K). It has a hyacinth-like odour, and both it and its esters are used in perfumery. Industrially it is made by *Meerwein–Ponndorf–Verley* reduction of cinnamaldehyde, [(*E*)-3-phenylpropenal], C_6H_5—CH=CH—CHO.

5.7.1.1 Unsaturated Alcohols derived from Phenol Ethers

K. Freudenberg identified a number of unsaturated alcohols derived from phenyl ethers, which act as building blocks in the structure of lignins. Among them are *coniferyl alcohol* (4-hydroxy-3-methoxycinnamyl alcohol) and *sinapinyl alcohol* (3,5-dimethoxy-4-hydroxycinnamyl alcohol):

Coniferyl alcohol   Sinapinyl alcohol

Both of these unsaturated alcohols can be regarded as either *phenylpropenol* derivatives of *guaiacyl* or *syringyl* type, or as derivatives of hydroxycinnamyl alcohol. Coniferyl alcohol is found in the cambial sap of various conifers as part of the glucoside coniferin and is oxidised to vanillin (yield 25%). Deciduous trees contain not only coniferin but, especially, *syringin* and when oxidised this gives syringaldehyde as well as vanillin.

Coniferin (R = H)
Syringin (R = OCH_3)

Vanillin (R = H)
Syringaldehyde (R = OCH_3)

5.7.1.2 Lignin[1]

Together with cellulose and hemicellulose, lignin forms a major constituent (30%) of wood. An insight into the structure of lignin first became possible when vanillin was

---

[1] See *K. Freudenberg* and *A. C. Neish*, Constitution and Biosynthesis of Lignin (Springer-Verlag, Berlin 1968); *K. V. Sarkanen* et al. (Ed.) Lignins, Occurrence, Formation, Structure and Reactions (Wiley, New York 1971); *K. B. G. Torssell*, Natural Product Chemistry, pp. 96 *ff.* (Wiley, Chichester 1983).

isolated (see above) from conifer lignin, which has a considerably simpler structure than lignin from deciduous trees. As a result it was concluded that lignin was an oxidation product of coniferyl alcohol. *Freudenberg* produced an amorphous powder, which had similar properties to those of naturally occurring lignin from spruce, by the action of *phenoldehydrogenase* (from mushroom juices) on coniferyl alcohol.

When lignin is heated with a hydrogen sulphite solution the ether links are broken and *ligninsulphonic acids* are formed, which behave as surfactants. Lignin is the source of the methanol which is obtained by the dry distillation of wood.

The precise structure of lignin has not yet been elucidated. Its biosynthesis proceeds *via* shikimic acid (see p. 403), phenylalanine and tyrosine as intermediates to give ferulic acid, which can be reduced enzymatically to coniferyl alcohol:

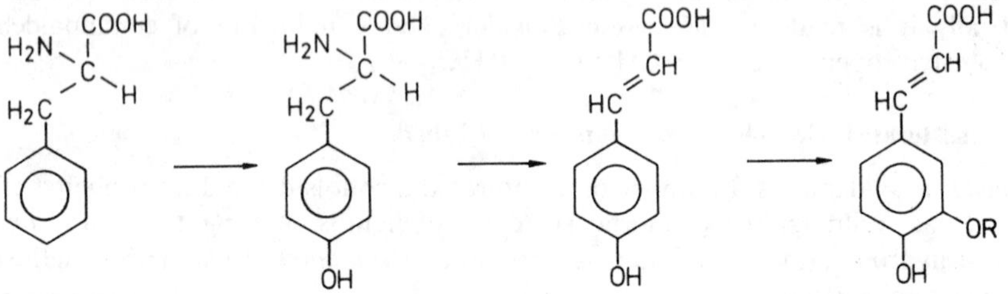

L-Phenylalanine   L-Tyrosine   p-Coumaric acid   Caffeic acid (R = H)
                                                     Ferulic acid (R = CH₃)

Freudenberg fed radioactive L-[^{14}C]phenylalanine to young spruce trees and after a few days small amounts of labelled coniferin and syringin could be detected.

5.7.2 *Arylalkylamines*

Among the parent compounds of this series are benzylamine and α- and β-phenylethylamine. In all these compounds the NH_2-group is attached to the side chain.

Benzylamine, C_6H_5—CH_2—NH_2, is made from benzyl chloride by means of the *Gabriel* synthesis, or by *catalytic reduction* of *benzonitrile*:

$$C_6H_5-C\equiv N \xrightarrow[\text{(Ni)}]{+4H} C_6H_5-CH_2-NH_2$$

Benzonitrile   Benzylamine

It is a liquid, b.p. 184°C (457 K), which fumes and has an ammoniacal smell. Unlike aniline, it gives a strongly basic reaction in aqueous solution.

5.7.2.1 α-Phenylethylamine, b.p. 186°C (459 K), exists in two *enantiomeric forms*.

Preparation
(a) By *reduction of acetophenone oxime* with sodium and ethanol or with sodium amalgam in dilute acetic acid.

$$C_6H_5-\underset{\underset{NOH}{\|}}{C}-CH_3 \xrightarrow[-H_2O]{+4H} C_6H_5-\underset{\underset{NH_2}{|}}{CH}-CH_3$$

(±)-α-Phenylethylamine

(b) By *reductive amination of acetophenone*, using ammonium formate at 150–180°C (425–455 K) (*Leuckart–Wallach* reaction):

$$C_6H_5-\overset{\underset{\|}{O}}{C}-CH_3+HCOONH_4 \longrightarrow C_6H_5-\overset{\underset{|}{NH_2}}{CH}-CH_3+CO_2+H_2O$$

Formic acid acts as a reducing agent. The advantage of this process over the previous one lies in the fact that it may also be used if halogen atoms or nitro groups are present as substituents in the benzene ring, since they remain unaffected by this reagent.

5.7.2.2 β-Phenylethylamine (2-Phenethylamine), b.p. 186°C (459 K), raises the blood pressure and is the parent compound of a number of physiologically active *biogenic amines*.[1] It is best prepared by *reducing benzyl cyanide* in liquid ammonia in the presence of *Raney* nickel at 100–130°C (375–400 K):

$$C_6H_5-CH_2-C\equiv N \xrightarrow{+4H} C_6H_5-CH_2-CH_2-NH_2$$

Benzyl cyanide   β-Phenylethylamine

By working in liquid ammonia, formation of secondary di-β-phenylethylamine, $(C_6H_5-CH_2-CH_2)_2NH$, which otherwise is a by-product, is suppressed. *Lithium aluminium hydride* has also proved to be a good reagent for this reduction.

Tyramine, 2-(4-hydroxyphenyl)ethylamine, 4-(2-aminoethyl)phenol, is found in rotting flesh, and also in ergot. It arises from the enzymatic decarboxylation of the amino acid L-tyrosine:

$$HO-C_6H_4-CH_2-\overset{\underset{|}{H}}{\overset{\overset{NH_2}{|}}{C}}-COOH \xrightarrow{-CO_2} HO-C_6H_4-CH_2-CH_2-NH_2$$

L-Tyrosine   Tyramine

It can be made synthetically by the *reduction of p-hydroxyphenylacetonitrile*:

$$HO-C_6H_4-CH_2-C\equiv N \xrightarrow{+4H} HO-C_6H_4-CH_2-CH_2-NH_2$$

Tyramine crystallises as prisms, m.p. 164–165°C (437–438 K). Like β-phenylethylamine it raises blood pressure.

Hordenine, *N,N*-dimethyltyramine, m.p. 118°C (391 K) has been isolated from barleygerm and from the cactus *Ariocarpus fissuratus*.

$$HO-C_6H_4-CH_2-CH_2-N(CH_3)_2$$

Hordenine

Mescaline forms needles which melt at 35–36°C (308–309 K); it has been obtained from peyotl cactus (*Lophora williamsii*). Its structure has been proved by synthesis. Condensation of 3,4,5-trimethoxybenzaldehyde with nitromethane provides 3,4,5-trimethoxy-β-nitrostyrene, which can be reduced to mescaline (*E. Späth*):

---

[1] See *J. W. Daly* and *B. Witkop*, Neuere Untersuchungen über zentral wirkende endogene Amine (Recent Investigations on Centrally Acting Endogenous Amines), Angew. Chem. *75*, 552 (1963); *P. B. Molinoff* et al., Biochemistry of Catecholamines, Ann. Rev. Biochem. *40*, 465 (1971).

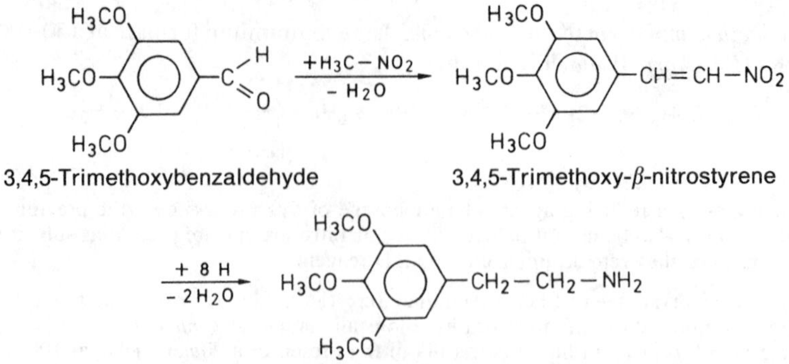

$$3,4,5\text{-Trimethoxybenzaldehyde} \xrightarrow[- H_2O]{+H_3C-NO_2} 3,4,5\text{-Trimethoxy-}\beta\text{-nitrostyrene}$$

H₃CO / H₃CO — C=O / H₃CO   →   H₃CO / H₃CO — CH=CH—NO₂ / H₃CO

3,4,5-Trimethoxybenzaldehyde   3,4,5-Trimethoxy-β-nitrostyrene

$$\xrightarrow[- 2H_2O]{+\ 8\ H}$$

H₃CO / H₃CO — CH₂—CH₂—NH₂ / H₃CO

Mescaline, 3,4,5-Trimethoxyphenylethylamine

5.7.2.3 L-(−)-**Ephedrine** was found (*Nagai*, 1887) in the Chinese drug 'Ma-Huang', which was obtained from various members of the species *Ephedra*. It has two asymmetric carbon atoms: in addition to a pair of mirror-image isomers of ephedrine, there is also another enantiomeric pair, called the *pseudoephedrines* (ψ-ephedrines). The two naturally occurring forms differ only in the configuration of the secondary alcohol group:

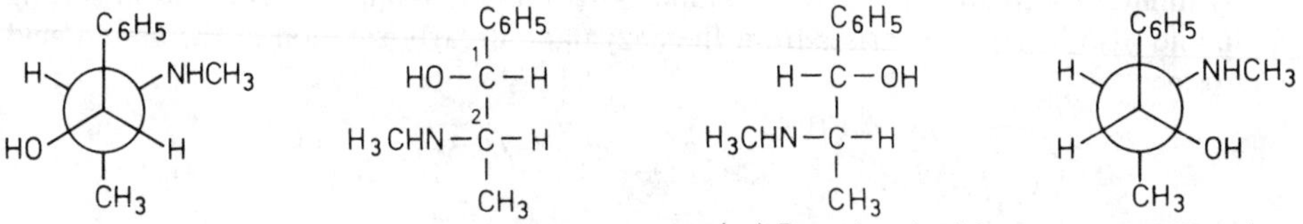

(1R,2S)(erythro)   L(−)-Ephedrine   D(+)-Pseudoephedrine   (1S,2S)(threo)
                         (m.p. 40°C [313 K])   (m.p. 118°C [391 K])

Naturally occurring L(−)-ephedrine (L-erythro-2-methylamino-1-phenylpropan-1-ol) has a (1R,2S) configuration.

The synthesis of DL-*ephedrine* (ephetonin) starts from propiophenone, which is first brominated, then treated with methylamine and finally reduced to DL-ephedrine:

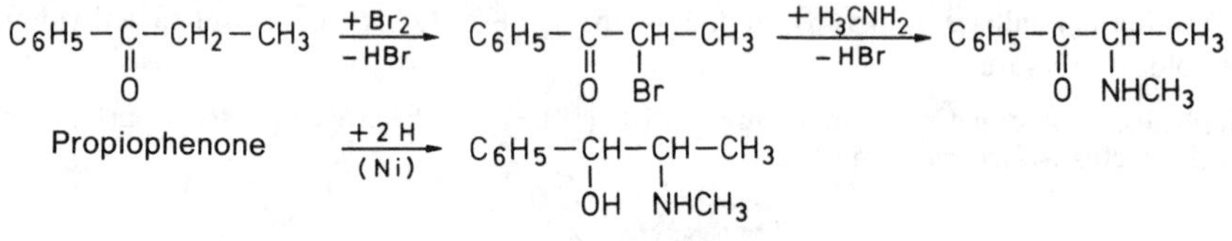

$$C_6H_5-\underset{O}{\underset{\|}{C}}-CH_2-CH_3 \xrightarrow[-HBr]{+Br_2} C_6H_5-\underset{O}{\underset{\|}{C}}-\underset{Br}{\underset{|}{CH}}-CH_3 \xrightarrow[-HBr]{+H_3CNH_2} C_6H_5-\underset{O}{\underset{\|}{C}}-\underset{NHCH_3}{\underset{|}{CH}}-CH_3$$

Propiophenone

$$\xrightarrow[(Ni)]{+2\ H} C_6H_5-\underset{OH}{\underset{|}{CH}}-\underset{NHCH_3}{\underset{|}{CH}}-CH_3$$

DL-Ephedrine

Pharmacologically, ephedrine raises blood pressure, and at the same time stimulates the sympathetic nervous system; it is used to treat persons who are susceptible to asthma. β-Phenylethylamine derivatives having no free hydroxy group are stimulants. The first compound of this sort to be prepared was (±)-α-methyl-β-phenylethylamine (amphetamine). It was later replaced by (±)-N,α-dimethyl-β-phenylethylamine (INN: *Methamphetamine*, *pervitin*). The observed side-

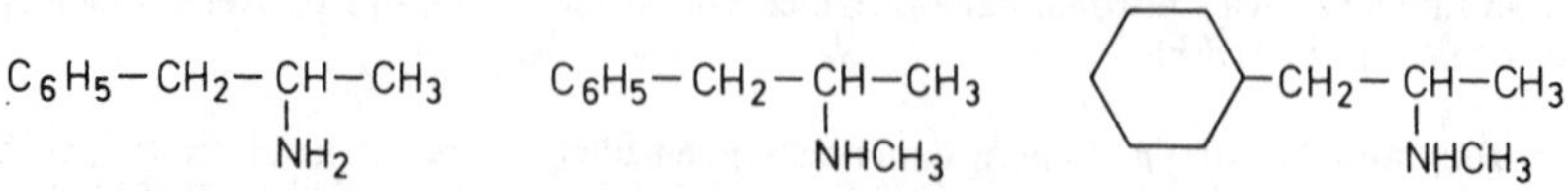

$$C_6H_5-CH_2-\underset{NH_2}{\underset{|}{CH}}-CH_3 \qquad C_6H_5-CH_2-\underset{NHCH_3}{\underset{|}{CH}}-CH_3 \qquad \text{⬡}-CH_2-\underset{NHCH_3}{\underset{|}{CH}}-CH_3$$

Amphetamine   INN: **Methamphetamine**   INN: **Levopropylhexedrine**

effect that they caused loss of appetite led to the development of hunger suppressants (anorectics), e.g. 1-cyclohexyl-*N*-methyl-2-propylamine (INN: **Levopropylhexedrine**, *eventine*).

5.7.2.4 Adrenaline was the first hormone to be isolated from the adrenal medulla (*Takamine*, 1901).

Hormones are active substances which are formed in the endocrine glands of certain organs in the bodies of humans and animals and which are transported in the blood vessels. In this way they can act to maintain the normal indispensable metabolic functions at places distant from where they are formed.

Adrenaline is another derivative of β-phenylethylamine and contains an asymmetric carbon atom. In nature it occurs in the *laevorotatory* form. Its constitution, as 1-(3,4-dihydroxyphenyl)-2-methylaminoethanol was demonstrated as long ago as 1904 (*F. Stolz*) as follows:

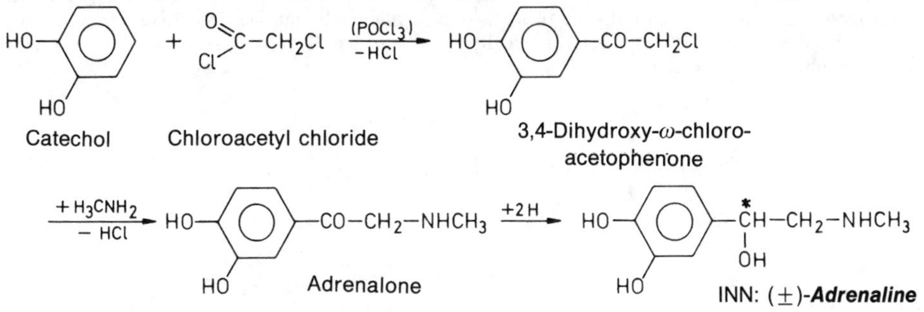

Catechol   Chloroacetyl chloride   3,4-Dihydroxy-ω-chloro-
                                                            acetophenone

                  Adrenalone   INN: (±)-**Adrenaline**

Catechol and chloroacetyl chloride react in the presence of phosphorus oxychloride to give an *O*-chloroacetyl derivative which undergoes a *Fries* rearrangement to form 3,4-dihydroxy-ω-chloroacetophenone. Treatment of this with methylamine provides adrenalone, which is reduced by either sodium or aluminium amalgam to INN: (±)-**Adrenaline (epinephrin)**. The resultant racemic form can be resolved into its enantiomers by means of its tartrate. Because adrenaline is related to catechol it is classified as a *catecholamine*.

Blood pressure is raised markedly by (*R*)-(−)-adrenaline and it speeds up the breakdown of glycogen in the liver and in muscles, leading to an increased sugar content in the blood (hyperglycemia) and in the urine (glucosuria). Andrenaline acts particularly on two types of receptors, the α- and the β-receptors. If they are specifically blocked (by α- or β-blockers) then its action is prevented.[1]

The *biosynthesis of adrenaline* starts from tyrosine and proceeds enzymatically *via* the following steps:

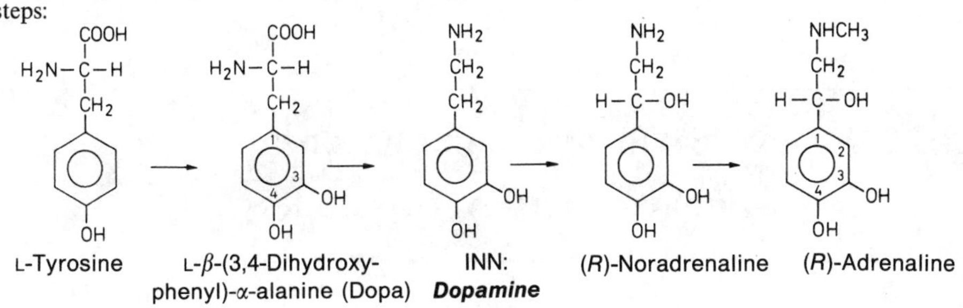

L-Tyrosine   L-β-(3,4-Dihydroxy-   INN: (*R*)-Noradrenaline   (*R*)-Adrenaline
              phenyl)-α-alanine (Dopa) **Dopamine**

---

[1] See E. *Mutschler*, Arzneimittelwirkungen (Medical Effects), 6th edn. p.257 *ff.* (Wissenschaftliche Verlagsgesellschaft, Stuttgart 1991).

5.7.2.5   Noradrenaline (norepinephrin) is another catecholamine obtained from the adrenal medulla. It differs from adrenaline by having no methyl group attached to the nitrogen atom. It raises the blood pressure and, like adrenaline, occurs at the nerve endings where it acts as a *neurotransmitter*. Dopamine is also a neurotransmitter; *Parkinson's* disease is associated with lack of it in the brain. If the HO-group at the 3-position of adrenaline is missing, *sympatol* (octopamine, α-(aminomethyl)-4-hydroxybenzyl alcohol) is available, which is used to counter weakening of the circulation.

Recently, L-α-methyl-β-(3,4-dihydroxyphenyl)alanine (INN: *Methyldopa*, *Presinol*, *Aldometil*, *Sembrina*), which is metabolised by the organism to normethyladrenaline, has been approved as a treatment for high blood pressure (antihypertonic). (*R,S*)-*N*-Isopropyl-noradrenaline sulphate (*Isoprenalin*, *aludrin*) has been used to treat bronchial asthma.

β-Blockers lower blood pressure and are effective against cardiac arrhythmia and angina pectoris. They arrest the action of the *catecholamines* at their active sites, the β-receptors; they resemble catecholamines in their chemical constitution, *e.g.* INN: *Propranolol* (*Dociton*, *Indobloc*). The (*S*)-enantiomer which is shown below is a hundred times stronger as a blocking agent than the (*R*)-enantiomer is.

INN: *Chloroamphenicol* (*Chloromycetin*, *Leucomycin*, *Paraxin*), which is an *antibiotic* produced by *Streptomyces venezuelae*, can also be thought of as a phenylethylamine derivative. It crystallises in yellow needles, m.p. 150°C (423 K), is optically active and has the structure shown below, which has been confirmed by synthesis:

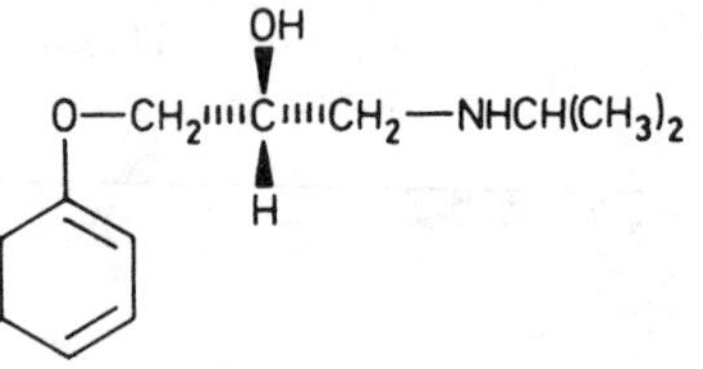

INN: *Propranolol* 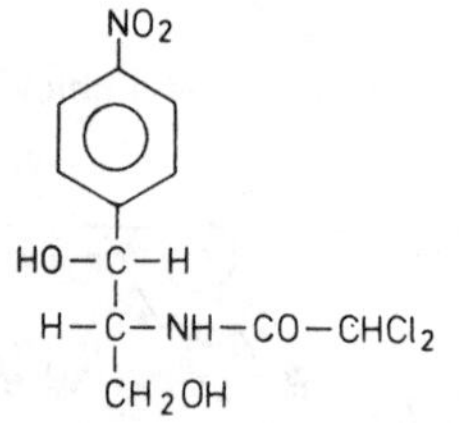   INN: *Chloroamphenicol*

Its chemical name is D-*threo-2,2-dichloro-N-[β-hydroxy-α(hydroxymethyl)-p-nitrophenylethyl]ace-tamide*; its configuration is related to that of D(−)-threose (see p. 424). This was the first time that a nitro group or a dichloroacetic acid residue had been found in a natural product. Therapeutically, *chloroamphenicol* is active, not only against bacteria, but also against rickettsia and large viruses, such as those which cause *psittacosis* and *typhus*.

5.8 Aromatic Aldehydes and Ketones. Aryl Aldehydes and Aryl Ketones

The reactions and properties of these classes of compounds differ very little from those of their aliphatic analogues. Whenever their aromatic character results in some modification of their chemical properties, this will be stressed.

Aryl Aldehydes

In aryl aldehydes the aldehyde group is joined directly to the benzene ring.

5.8.1 Benzaldehyde

Benzaldehyde, C_6H_5—CHO, was originally called *oil of bitter almonds*. It is the simplest and also the most important member of this class of compounds. It occurs in some plant oils and as a constituent of the 'cyanogenic' glycoside *amygdalin* in bitter almonds. In amygdalin the cyanohydrin of benzaldehyde (mandelonitrile) is linked glycosidically to the disaccharide gentiobiose. It is broken down by dilute acid or by the enzyme *emulsin* into two molecules of glucose and one each of benzaldehyde and hydrogen cyanide:

$$C_6H_5-\overset{\underset{|}{O-C_{12}H_{21}O_{10}}}{\underset{}{C}}\overset{H}{\underset{CN}{\diagdown}} + 2H_2O \longrightarrow 2C_6H_{12}O_6 + C_6H_5-C\overset{H}{\underset{O}{\diagdown}} + HCN$$

$$\qquad\qquad\qquad\qquad\qquad\qquad\qquad\qquad Glucose \quad\ Benzaldehyde$$

Preparation

5.8.1.1   By *hydrolysis of benzylidene chloride* with potassium carbonate at 130°C (400 K) in an atmosphere of carbon dioxide:

$$C_6H_5-CHCl_2 + H_2O \longrightarrow C_6H_5-C\overset{H}{\underset{O}{\diagdown}} + 2HCl$$

Benzaldehyde is made industrially by heating benzylidene chloride and water slowly to 100°C (323 K) in the presence of iron powder or iron(III) benzoate, and then neutralising the mixture with alkali and steam-distilling off the benzaldehyde (yield = 80–90%). About 10% of benzoic acid is also formed as a by-product.

Particularly pure benzaldehyde is made industrially by catalytic atmospheric oxidation of toluene.

5.8.1.2   *Sommelet* **Reaction**[1] (1913). By *heating benzyl chloride with hexamethylenetetramine* in aqueous ethanol or dilute acetic acid at pH 3.0–6.5 (yield ~70%).

$$C_6H_5-CH_2-(CH_2)_6N_4\big]^{\oplus}Cl^{\ominus} + 6H_2O \longrightarrow C_6H_5-C\overset{H}{\underset{O}{\diagdown}} + H_3C\overset{\oplus}{N}H_3\big]Cl^{\ominus} + 3NH_3 + 5H-C\overset{H}{\underset{O}{\diagdown}}$$

The reaction amounts to a dehydrogenation of benzylamine to an aldimine and hydrolysis of the latter to an aldehyde. Methyleneimine functions as a hydrogen acceptor, being reduced to methylamine in the course of the reaction.

The *reaction mechanism* is as follows. First of all the benzyl chloride reacts with the amine to form a quaternary ammonium salt, which is hydrolysed to benzylamine. The ammonia and formaldehyde which arise from the breakdown of hexamethylenetetramine react together to form methyleneimine, which is protonated to give a cation. Isotopic labelling with deuterium has shown that a hydrogen atom of the methylene group of benzylamine migrates *as a hydride ion* to the methyleneiminium cation, resulting in the formation of methylamine and a benzyliminium cation; the latter cation is then hydrolysed to benzaldehyde:

---

[1] See *S. J. Angyal*, The *Sommelet* Reaction, Org. React. *8*, 197 *ff*. (1954).

$$C_6H_5-CH_2-C_6H_{12}N_4]^{\oplus}Cl^{\ominus} \xrightarrow[-2NH_3,-NH_4Cl,-6HCHO]{+6H_2O} C_6H_5-CH_2-NH_2$$

Benzylamine

$$NH_3 + HCHO \xrightarrow{-H_2O} HN=CH_2 \xrightleftharpoons{+H^{\oplus}} H_2\overset{\oplus}{N}=CH_2$$

Methyleneimine

$$H_2\overset{\oplus}{N}=CH_2 + C_6H_5-\underset{H}{\underset{|}{CH}}-NH_2 \rightleftharpoons H_2N-CH_3 + C_6H_5-CH=\overset{\oplus}{NH_2}$$

Aldimine

$$C_6H_5-CH=\overset{\oplus}{NH_2} + H_2O \xrightleftharpoons{-H^{\oplus}} C_6H_5-C\overset{H}{\underset{O}{\diagdown}} + NH_3$$

5.8.1.3 By *reduction of benzoyl chloride* at $-75°C$ (200 K) by means of lithium tri-t-butoxyaluminium hydride (LTBA), $LiAlH[OC(CH_3)_3]_3$ (yield = 78%). The reagent can be obtained from t-butanol and lithium aluminium hydride in ethereal solution at room temperature.

$$C_6H_5-C\overset{O}{\underset{Cl}{\diagdown}} \xrightarrow{LTBA} C_6H_5-C\overset{H}{\underset{O}{\diagdown}}$$

p-Nitrobenzaldehyde can be obtained in the same way, in 69% yield, from *p*-nitrobenzoyl chloride (see p. 192).

5.8.1.4 *Stephen* Reduction. If hydrogen chloride is passed through an ethereal solution of an aryl nitrile in the presence of anhydrous tin(II) chloride, an aldiminium hexachlorostannate is formed as an intermediate, thanks to the reducing action of tetrachlorostannic(II) acid, and is readily hydrolysed to an arylaldehyde. In this way benzaldehyde is obtained from benzonitrile in a 97% yield.

$$C_6H_5-C\equiv N \xrightarrow{+2HCl, +H_2[SnCl_4]} \left[C_6H_5-C\overset{H}{\underset{\overset{\oplus}{NH_2}}{\diagdown}}\right] HSnCl_6^{\ominus} \xrightarrow[-NH_4H[SnCl_6]]{+H_2O} C_6H_5-C\overset{H}{\underset{O}{\diagdown}}$$

Benzonitrile

Benzaldehyde

Other recommended solvents for this reaction are ethyl formate or ethyl acetate (*Stephen*, 1956).

Properties

Benzaldehyde is a colourless liquid, b.p. 179°C (452 K), which smells like bitter almonds. It takes part in most of the reactions associated with aliphatic aldehydes; it reduces ammoniacal silver nitrate solution but not *Fehling*'s solution. It gives the corresponding adducts with either sodium hydrogen sulphite or hydrogen cyanide. It is remarkably easily oxidised in air to benzoic acid, in an autocatalytic reaction, which is accelerated by light, by radical sources or by metal ions (Fe, Ni, Cu).

If *antioxidants*,[1] *e.g.* hydroquinone, are added, autoxidation is stopped or curtailed (*Moureu, Dufraisse*).

Bäckstrom (1934) showed that autoxidation involves a *radical chain reaction*. A benzoyl radical is formed first and is converted into benzoic acid as follows:

---

[1] See L. R. Mahoney, Antioxidants, Angew. Chem. Int. Ed. Engl. *8*, 547 (1969).

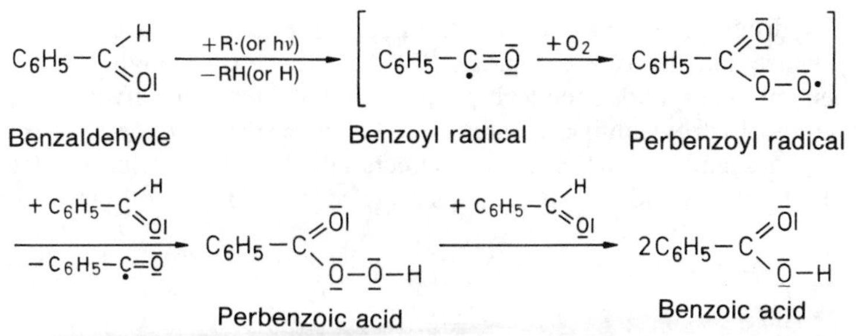

It is possible to detect the presence of the intermediate perbenzoic acid in the reaction mixture.

A reaction which is of use preparatively is the oxidation, by peracids, of ketones to esters; there is no further breakdown:

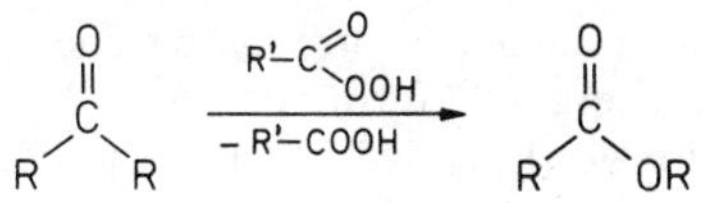

5.8.1.5 This reaction forms the basis of the *Baeyer-Villiger* reaction,[1] whose mechanism is formulated as follows:

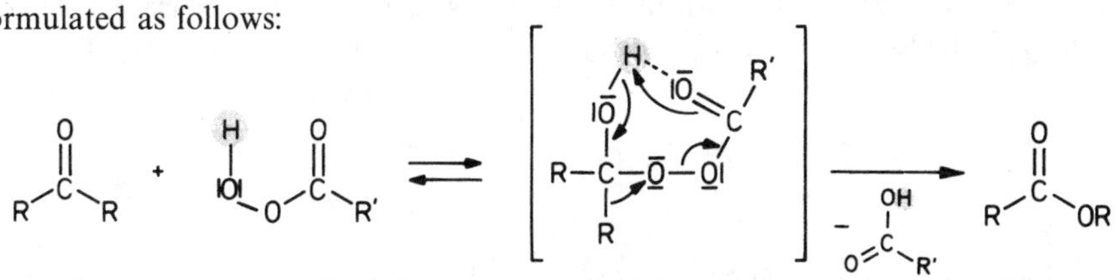

In the first step, the peracid adds to the carbonyl group of the ketone. In the process, the proton which has a grey background in the reaction scheme above formally changes its site. Nucleophilic displacement of R within this adduct leads to formation of an ester as the end-product of the reaction.

The ease with which group R undergoes displacement is in the following order, H(in aldehydes) $> CR_3 >$ cyclohexyl $> CHR_2 \sim$ phenyl $> CH_2R > CH_3$.

Uses

The main use of benzaldehyde is as a starting material in the dyestuffs industry and for a large number of organic syntheses. It is also used to enhance the odour and taste of products, and in perfumery.

Reactions of Benzaldehyde

5.8.1.6 When benzaldehyde is treated with hydroxylamine, *α-benzaldehyde oxime*, m.p. 35°C (308 K), is formed. If dry hydrogen chloride is passed into an ethereal solution of

---

[1] See *C. H. Hassall*, Org. React. *9*, 73 (1957).

this oxime a hydrochloride separates out. On addition of sodium carbonate solution to the latter, β-benzaldehyde oxime, m.p. 132°C (405 K) is precipitated. The two oximes differ in their physical and chemical properties, but both are hydrolysed to give benzaldehyde and hydroxylamine. *Hantzsch* (1890) suggested that the two oximes were stereoisomers, analogous to the Z and E isomers encountered in alkenes. The relative position of the H-atoms and the HO-groups are evident from the following formulae:[1]

$$C_6H_5\diagdown C \diagup H \qquad (UV) \qquad C_6H_5 \diagdown C \diagup H$$

(Z)-Form (β)   (E)-Form (α)

Benzaldehyde oxime

When a solution of the more stable (Z)-form in benzene is irradiated with UV light an equilibrium with the higher energy (E)-form is set up. (E)-Benzaldehyde oxime gives a normal acetyl derivative with acetic anhydride, but the (Z)-oxime loses water to form benzonitrile, C_6H_5—C≡N.

5.8.1.7 Aryl aldehydes react with primary amines to form *Schiff*'s bases (*azomethines*). Thus benzaldehyde and aniline give *benzylideneaniline* (benzalaniline), m.p. 54°C (327 K):

$$C_6H_5-C\diagup^H_{\diagdown O} + H_2N-C_6H_5 \longrightarrow C_6H_5-C^H=N-C_6H_5 + H_2O$$

Benzylideneaniline

Benzaldehyde reacts in a different way from aliphatic aldehydes with ammonia. The benzylideneimine (benzaldehyde imine), which is formed first of all, reacts with more benzaldehyde to give **hydrobenzamide** (N,N'-benzylidenebis[benzylideneamine]):

$$C_6H_5-C\diagup^H_{\diagdown O} + H_2NH \longrightarrow C_6H_5-C^H=NH + H_2O$$

Benzylideneimine

$$\begin{matrix} C_6H_5-C^H=NH \\ C_6H_5-C=NH_{\diagdown H} \end{matrix} + \begin{matrix} H_{\diagdown} \\ O^{\diagup}C-C_6H_5 \end{matrix} \longrightarrow \begin{matrix} C_6H_5-CH=N_{\diagdown} \\ C_6H_5-CH=N^{\diagup} \end{matrix} CH-C_6H_5 + H_2O$$

Hydrobenzamide (m.p. 103°C [376 K])

5.8.1.8 In the presence of dilute alkali aryl aldehydes condense with aldehydes or ketones which have at least two *reactive* hydrogen atoms on a carbon atom next to the carbonyl group to form *benzylidene derivatives* (*benzal derivatives*). The initially formed aldols lose water much more readily then their aliphatic analogues do. This is energetically favoured because the C=C double bond which is formed is conjugated not only with the carbonyl group but also with the benzene ring. For example, benzaldehyde reacts with acetaldehyde or acetone to give, respectively, **(E)**-*cinnamaldehyde* or **(E)**-*benzylideneacetone* (benzalacetone):

---

[1] In the older literature the (Z)-form was described as 'anti' and the (E)-form as 'syn'.

$$C_6H_5-C\overset{H}{\underset{O}{\diagdown}} + H_3C-C\overset{H}{\underset{O}{\diagdown}} \longrightarrow \left[C_6H_5-\overset{H}{\underset{OH}{C}}-CH_2-C\overset{H}{\underset{O}{\diagdown}} \right] \xrightarrow{-H_2O} C_6H_5-C\overset{H}{\diagdown}\overset{}{\underset{H}{\diagup}}C-C\overset{H}{\underset{O}{\diagdown}}$$

Benzaldehyde   Acetaldehyde   (Aldol)   (E)-Cinnamaldehyde (b.p. 253.5°C [526.7 K])

$$C_6H_5-C\overset{H}{\underset{O}{\diagdown}} + H_3C-\overset{CH_3}{\underset{O}{C}} \longrightarrow \left[C_6H_5-\overset{H}{\underset{OH}{C}}-CH_2-\overset{CH_3}{\underset{O}{C}} \right] \xrightarrow{-H_2O} C_6H_5-C\overset{H}{\diagdown}\overset{}{\underset{H}{\diagup}}C-\overset{}{\underset{O}{C}}-CH_3$$

Acetone   (E)-Benzylideneacetone (m.p. 42°C [315 K])

Treatment of the latter with benzaldehyde leads to the all-trans form of *dibenzylideneacetone* (1t,5t-diphenylpenta-1,4-dien-3-one [*t = trans*]):

$$C_6H_5-CH=CH-\overset{}{\underset{O}{C}}-CH_3 + \overset{H}{\underset{O}{\diagdown}}C-C_6H_5 \xrightarrow{-H_2O} C_6H_5-C\overset{H}{\diagdown}C-C-C\overset{H}{\diagdown}C-C_6H_5$$

Benzylideneacetone   (E,E)-Dibenzylideneacetone
                                       (m.p. 113°C [386 K])

Because benzaldehyde cannot undergo self-condensation, it can participate in a number of reactions which are difficult or impossible to achieve using aliphatic aldehydes.

5.8.1.9 Cannizzaro Reaction. It is typical of aryl aldehydes that they undergo this reaction very smoothly; its mechanism was outlined on p. 199 *f.* for aliphatic aldehydes.[1] In the presence of strong alkali, *oxidoreduction* of benzaldehyde to benzyl alcohol and a salt of benzoic acid takes place:

$$2C_6H_5-C\overset{H}{\underset{O}{\diagdown}} + NaOH \longrightarrow C_6H_5-CH_2OH + C_6H_5-COO^{\ominus}Na^{\oplus}$$

                                    Benzyl alcohol   Sodium benzoate

In the presence of *aluminium ethoxide* as a catalyst benzaldehyde is converted into benzyl benzoate (see *Tishchenko* reaction):

$$2C_6H_5-C\overset{H}{\underset{O}{\diagdown}} \xrightarrow{[Al(OC_2H_5)_3]} C_6H_5-\overset{O}{\overset{\parallel}{C}}-O-CH_2-C_6H_5$$

                                         Benzoyl benzoate

5.8.1.10 Benzoin Condensation. This is a typical reaction of aryl aldehydes in which, under the catalytic influence of *potassium cyanide* in aqueous alcohol, two molecules of benzaldehyde condense to form *benzoin* (in its racemic form):

$$2C_6H_5-C\overset{H}{\underset{O}{\diagdown}} \xrightarrow{(CN^{\ominus})} C_6H_5-CO-\overset{*}{C}H(OH)-C_6H_5$$

                                          (±)-Benzoin

---

[1] ESR spectroscopy has indicated the presence of a radical anion ArCHO$^{\cdot\ominus}$ in *Cannizzaro* reactions, for example of *p*-chlorobenzaldehyde, and in consequence a SET-mechanism (see p. 128) has been proposed. See *E. C. Ashby* et al., J. Org. Chem. *52*, 4079 (1987).

The *mechanism* of the reaction can be formulated as follows:

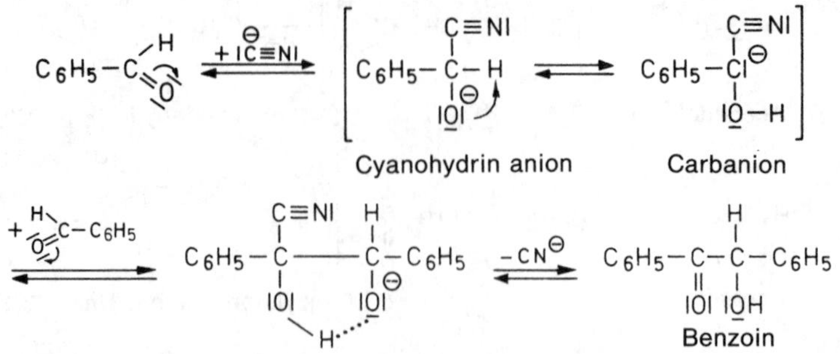

Cyanohydrin anion   Carbanion

Benzoin

The first step is a *nucleophilic* addition of a cyanide anion to the carbonyl group of benzaldehyde. The hydrogen atom next to the cyano group of the cyanohydrin anion is somewhat acidic and migrates to the oxygen atom, resulting in the formation of a carbanion. This makes a *nucleophilic* attack on a second molecule of benzaldehyde, giving rise to another anion which, by loss of a cyanide ion and concomitant migration of a proton, provides benzoin. More recent studies have shown that the benzoin condensation involves an equilibrium reaction in which the decisive step is an *umpolung* in which the normally *electrophilic* benzaldehyde becomes a *nucleophilic* reactant.

Protected cyanohydrins provide great opportunities for making use of this umpolung. Treatment of benzaldehyde with cyanotrimethylsilane, which is readily obtained from chlorotrimethylsilane and potassium cyanide in *N*-methylpyrrolidone, provides a protected cyanohydrin (carbon atom on a grey background) that can be converted by means of LDA into a versatile and readily alkylated carbanion, *e.g.*

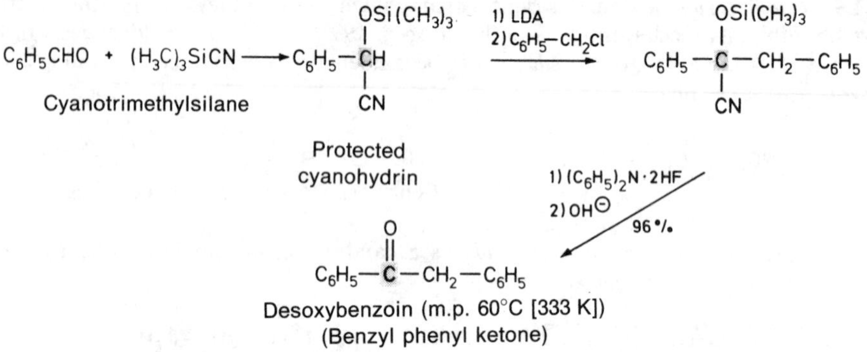

Cyanotrimethylsilane

Protected
cyanohydrin

Desoxybenzoin (m.p. 60°C [333 K])
(Benzyl phenyl ketone)

Benzyl propen-1-yl ketone can be obtained similarly in 76% yield by using crotonaldehyde instead of benzaldehyde:[1]

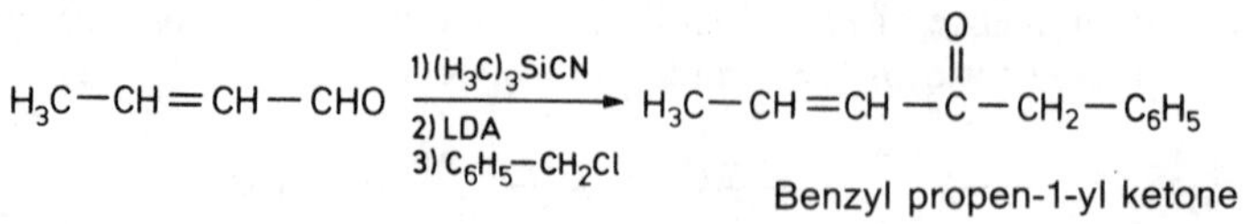

Benzyl propen-1-yl ketone

Protected cyanohydrins of many aldehydes (but not formaldehyde) can be made by treating them with ethyl vinyl ether (ethoxyethylene), thus readily providing the umpolung mentioned above (*Stork*, 1971).

---

[1] See *S. Hünig*, Synthesen mit Trimethylsilylcyanid, Chimia *36*, 1 (1982).

$$R-\underset{\underset{CN}{|}}{\overset{\overset{OH}{|}}{CH}} \xrightarrow{C_2H_5OCH=CH_2} R-\underset{\underset{CN}{|}}{\overset{\overset{O-CH-CH_3}{\overset{|}{OC_2H_5}}}{CH}} \xrightarrow[\substack{3)H_2O,\,H^{\oplus}}]{\substack{1)LDA\\2)R'X}} R-\underset{\underset{O}{\|}}{\overset{}{C}}-R'$$

The acetal group can be removed under very mild conditions. In order to alkylate a carbanion derived from formaldehyde, diethylaminoacetonitrile, $[(C_2H_5)_2N—CH_2CN]$ is used.

Benzoin, $C_6H_5—CO—CH(OH)—C_6H_5$, is a crystalline substance, m.p. 137°C (410 K), which combines the properties of an alcohol and of a ketone. Like aliphatic α-hydroxyketones and sugars, with phenylhydrazine it gives an osazone, m.p. 225°C (498 K). Benzoin is oxidised to benzil (a 1,2-diketone), m.p. 95°C (368 K), by nitric acid:

$$C_6H_5-\underset{\underset{OH}{|}}{\overset{\overset{H}{|}}{C}}-\underset{\underset{OH}{|}}{\overset{\overset{H}{|}}{C}}-C_6H_5 \xleftarrow[(Na/Hg)]{+2\,H} C_6H_5-\underset{\underset{O}{\|}}{\overset{}{C}}-\underset{\underset{OH}{|}}{\overset{}{CH}}-C_6H_5 \xrightarrow[-H_2O]{+[O]} C_6H_5-\underset{\underset{O}{\|}}{\overset{}{C}}-\underset{\underset{O}{\|}}{\overset{}{C}}-C_6H_5$$

ms-Hydrobenzoin   Benzoin   Benzil

(m.p. 137°C [410 K])

Reduction of benzoin by sodium amalgam gives *ms*-hydrobenzoin. The racemic form (isohydrobenzoin), m.p. 119°C (392 K) crystallises from ether as a conglomerate of the two *enantiomorphic* forms which can be separated out into (+)- and (−)-hydrobenzoin. (See separation of racemic forms.)

With inorganic acid chlorides such as thionyl chloride, benzoin is converted into desyl chloride (2-chloro-2-phenylacetophenone), m.p. 68°C (341 K).

$$C_6H_5-\underset{\underset{O}{\|}}{\overset{}{C}}-\underset{\underset{OH}{|}}{\overset{}{CH}}-C_6H_5+SOCl_2 \longrightarrow C_6H_5-\underset{\underset{O}{\|}}{\overset{}{C}}-\underset{\underset{Cl}{|}}{\overset{}{CH}}-C_6H_5+SO_2+HCl$$

Benzoin   Desyl chloride

5.8.1.11 Reductive Coupling. The *meso*-form of hydrobenzoin is obtained directly from benzaldehyde in ethereal solution by letting it react with sodium (*cf.* pinacols):

$$2C_6H_5-\overset{}{C}\overset{H}{\underset{O}{\diagdown\!\!\diagup}} \xrightarrow{+2Na} \left[C_6H_5-\underset{\underset{|\underset{\cdot\cdot}{O}|^{\ominus}}{}}{\overset{}{CH}}-\underset{\underset{|\underset{\cdot\cdot}{O}|^{\ominus}}{}}{\overset{}{CH}}-C_6H_5 \;\; 2Na^{\oplus} \right] \xrightarrow[-2NaOH]{+H_2O} C_6H_5-\underset{\underset{HO}{|}}{\overset{\overset{H}{|}}{C}}-\underset{\underset{OH}{|}}{\overset{\overset{H}{|}}{C}}-C_6H_5$$

ms-Hydrobenzoin

5.8.2 *Homologues of Benzaldehyde*

Aryl aldehydes can be prepared as follows by *direct formylation*.

5.8.2.1 *Gattermann–Koch* synthesis[1] (1897). If dry hydrogen chloride and carbon monoxide are passed into, for example, dry toluene, in the presence of aluminium chloride and copper(I) chloride, then this mixture of gases behaves as though it were formyl chloride, which is stable only at the temperature of liquid air. The processes involved in this reaction may be promoted by the formation of both a copper(I) chloride–carbon monoxide adduct and an aluminium chloride complex $HCO[AlCl_4]$.

---

[1] See *N. N. Crounse*, The *Gattermann–Koch* Reaction, Org. React. **5**, 290 *ff*. (1949).

The aldehyde group is introduced for the most part at the site *para* to the substituent group already present.

This reaction is a special case of *Friedel–Crafts* acylation (see p. 539 *f.*). Because of the variable yields obtained it is only seldom used.

5.8.3 Phenol and Phenol Ether Aldehydes

The following processes are usually used for the preparation of these aldehydes.

5.8.3.1 *Gattermann* aldehyde synthesis[1] (1907). The *Gattermann–Koch* synthesis is only applicable to phenols or phenol ethers if anhydrous hydrogen cyanide is used instead of carbon monoxide. Copper(I) chloride is not necessary, and aluminium chloride can be replaced by less reactive zinc chloride. In the case of polyhydric phenols, formylation proceeds in the absence of a *Friedel–Crafts* catalyst. Ether, chloroform, benzene, chlorobenzene or 1,1,2,2-tetrachloroethane are used as solvents.

It is assumed that in this reaction a non-isolable intermediate formimidoyl chloride is involved. This reacts with the phenol or phenol ether to give an *aldiminium chloride* which, when warmed with dilute aqueous acid, is hydrolysed to the aldehyde, *e.g.*

In this case also the main product is the *p*-derivative.

This reaction can also be regarded as a special case of a *Friedel–Crafts* acylation. It has been largely superseded by the Vilsmeier–Haack synthesis.

5.8.3.2 *Vilsmeier–Haack* synthesis[2] (1927) This has proved to be a good method for the formylation of benzene derivatives, especially of *phenol ethers* and of *dialkylanilines*. As an example, the reaction of an equimolecular mixture of *N*-methylformanilide (*N*-formylmethylaniline) and phosphorus oxychloride with anisole at about 10°C (285 K), followed by hydrolysis, provides an excellent yield of *p*-anisaldehyde:

---

[1] See *W. E. Truce*, The *Gattermann* Synthesis of Aldehydes, Org. React. *9*, 37 *ff.* (1957)
[2] See *C. Jutz*, The *Vilsmeier–Haack–Arnold* Acylations, Adv. Org. Chem. *9*, I, 225 (1976).

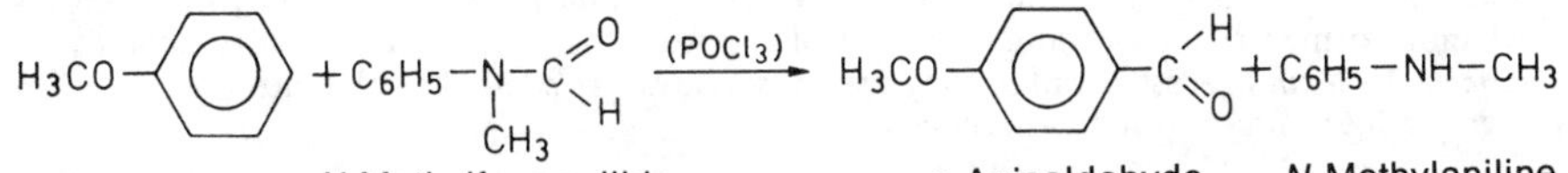

Anisole   N-Methylformanilide   p-Anisaldehyde   N-Methylaniline

Again the reaction mechanism can be regarded as a special case of *Friedel–Crafts* acylation. Reaction of phosphorus oxychloride with a formyl group gives a '*Vilsmeier–Haack* complex' which, as a chloromethyliminium salt, makes an *electrophilic* attack, usually at the *p*-position, on the benzene ring. The resultant adduct is hydrolysed to methylaniline and *p*-anisaldehyde:

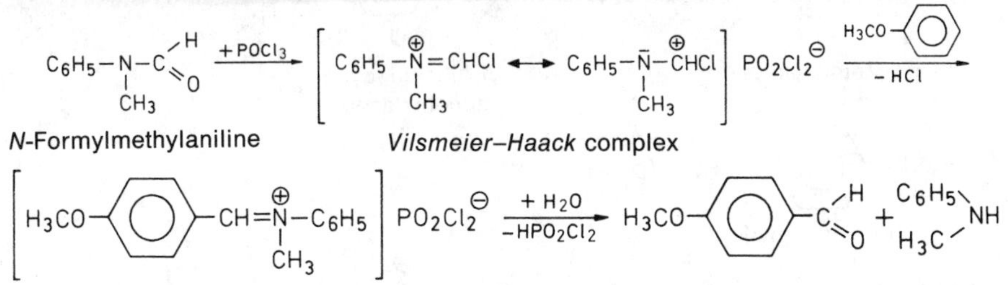

N-Formylmethylaniline   Vilsmeier–Haack complex

N-Methylformamide or *N,N*-dimethylformamide can be used instead of *N*-methylformanilide.

5.8.3.3 Reimer–Tiemann synthesis.[1] (1876). In this synthesis, which is used for the preparation of phenol aldehydes, phenol, for example, is treated with chloroform and alkali at 65–70°C (340–345 K). The main product is the *o*-hydroxyaldehyde together with a small amount of the *p*-isomer. Yields are relatively low, since considerable tar-formation occurs.

The *reaction mechanism* involves initial generation of *dichlorocarbene* (dichloromethylene), $:CCl_2$, from the reaction of alkali with chloroform (*Hine*, 1950; *Wynberg*, 1954):

$$CHCl_3 \xrightarrow[-H_2O]{+OH^\ominus} :\overset{\ominus}{C}Cl_3 \longrightarrow :CCl_2 + Cl^\ominus$$

This species, which lacks an octet of electrons on the carbon atom, attacks the *mesomeric* phenoxide anion. A proton then migrates leading to regeneration of a delocalised sextet of electrons in the benzene ring and formation of a dichloromethyl group; the latter is hydrolysed to an aldehyde group in the alkaline medium.

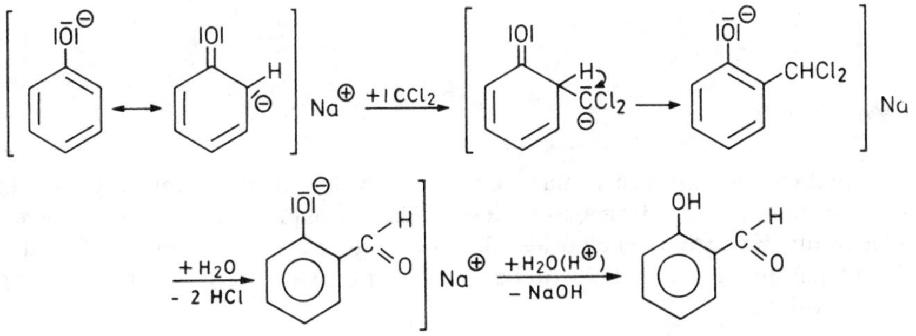

Salicylaldehyde

---

[1] See *H. Wynberg* et al., The *Reimer–Tiemann* Reaction, Org. React. *28*, 1 (1982).

After acidification, the salicylaldehyde is steam-distilled out of the mixture; some *p*-hydroxybenzaldehyde which is formed as a by-product can be isolated from the residue that remains behind.

The *Reimer–Tiemann* process is only rarely used nowadays, for example if the introduction of an aldehyde group into an *o*-position is necessary.

5.8.3.4 Electro-oxidation of *p*-substituted toluenes is used in the industrial preparation of the corresponding aldehydes (BASF, 1979), *e.g.*

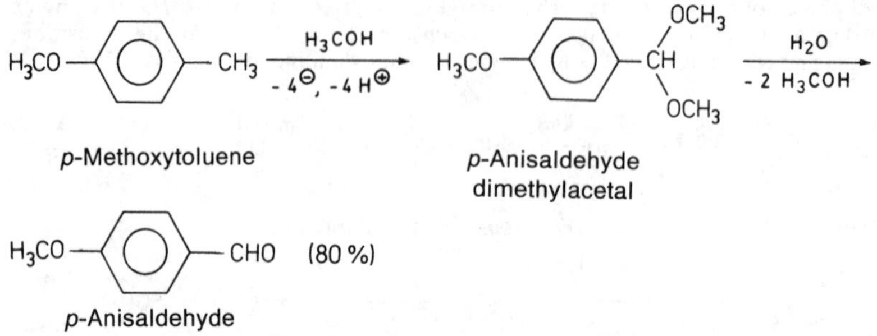

p-Methoxytoluene

p-Anisaldehyde
dimethylacetal

p-Anisaldehyde

p-**Anisaldehyde** (*p*-methoxybenzaldehyde), b.p. 248°C (521 K) occurs in various essential oils, *e.g.* in the oils of cassia flowers, aniseed and fennel. It is used as a perfume.

Salicylaldehyde (*o*-hydroxybenzaldehyde) also occurs in essential oils, especially from members of the *Spiraea* family. It is an oily liquid, b.p. 196.5°C (469.7 K), which is much less susceptible than benzaldehyde to autoxidation. It gives a deep red-violet colour with iron(III) chloride.

As in the case of *o*-nitrophenols, because of the vicinal hydroxy and carbonyl groups, both *o*-hydroxyaldehydes and *o*-hydroxyketones form hydrogen-bonded chelates, and this affects their physical properties.

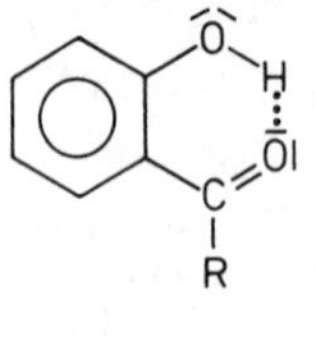

(R = H, alkyl or aryl)

Unlike benzaldehyde and cinnamaldehyde, phenol ether aldehydes are stable to atmospheric oxygen.

5.8.4 Vanillin

Vanillin (4-hydroxy-3-methoxybenzaldehyde) occurs widely in the plant kingdom, and especially in vanilla pods. It forms colourless needles of m.p. 81°C (354 K) and gives a blue-violet colour with iron(III) chloride. Because of its use as an aroma for foodstuffs and in the preparation of synthetic perfumes, a number of synthetic methods of preparing it have been evolved.

Preparation

5.8.4.1 In particular, *eugenol*, which is obtained from oil of cloves (*Eugenia caryophyllata* = *Syzygium aromaticum*), serves as a starting material. In the presence of alkali it *isomerises* to isoeugenol, in which the C=C double bond in the side-chain becomes conjugated with the benzene ring (allyl-propenyl rearrangement). Oxidation of isoeugenol with ozone gives vanillin (84% yield):

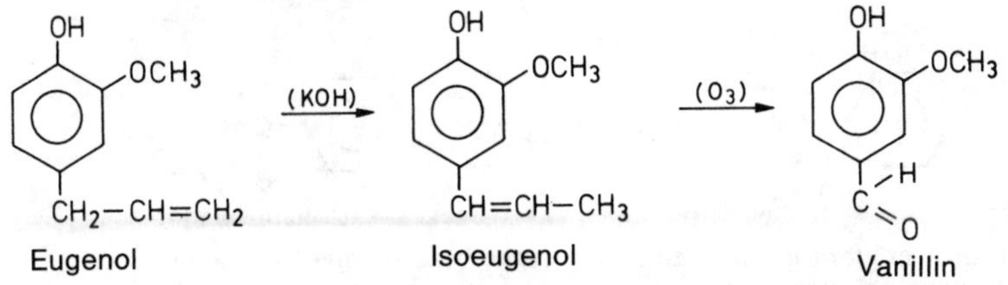

| Eugenol | Isoeugenol | Vanillin |

5.8.4.2 It is prepared in 70% yield from *guaiacol* by the *Vilsmeier–Haack* synthesis at about 10°C (285 K):

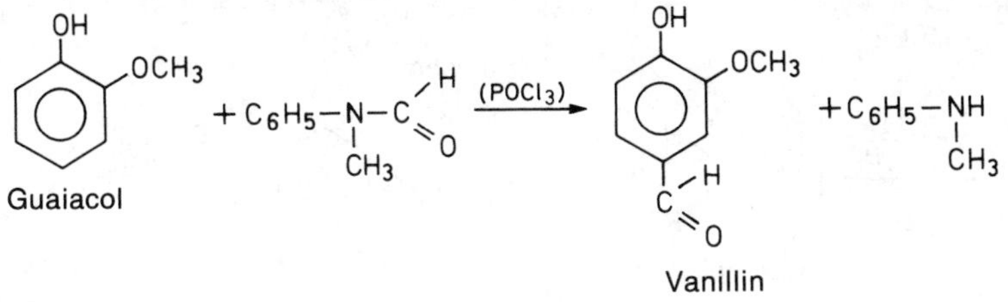

Guaiacol

Vanillin

5.8.4.3 Vanillin is obtained from the *ligninsulphonic acids* contained in the sulphite waste liquors in the breakdown of wood by the sulphite process. The yield is from 7 to 25%, depending on the kind of wood.

Piperonal (3,4-methylenedioxybenzaldehyde, 1,3-benzodioxole-5-carboxaldehyde), is known as a perfume under the name *heliotropin*. It is prepared synthetically from the natural product *safrole*. This is isomerised by alkali to isosafrole and then oxidised with ozone (*cf.* synthesis of vanillin, section 5.8.4.1).

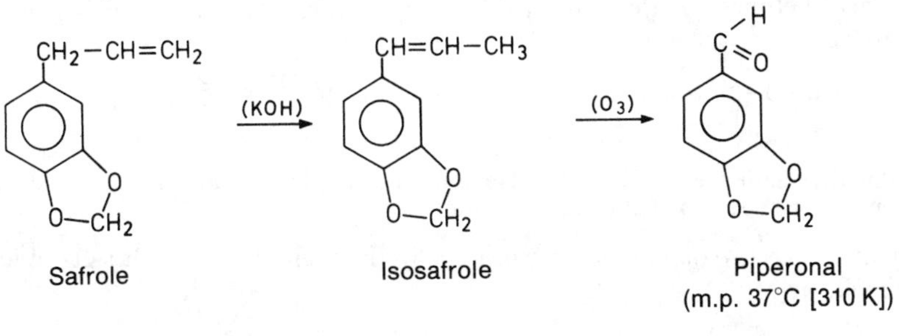

| Safrole | Isosafrole | Piperonal (m.p. 37°C [310 K]) |

5.8.5 Aryl Ketones

This class of compounds can be subdivided into alkyl aryl ketones and diaryl ketones. The simplest examples are *acetophenone* (methyl phenyl ketone), C_6H_5—CO—CH_3, and *benzophenone* (diphenyl ketone), C_6H_5—CO—C_6H_5.

Preparation

5.8.5.1 *Friedel–Crafts* **Acylation**. By treating arenes with aliphatic or aryl *acid chlorides* in the presence of *aluminium chloride, e.g.*

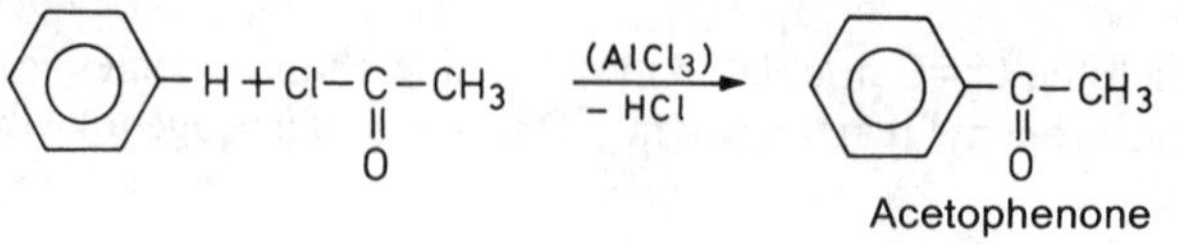

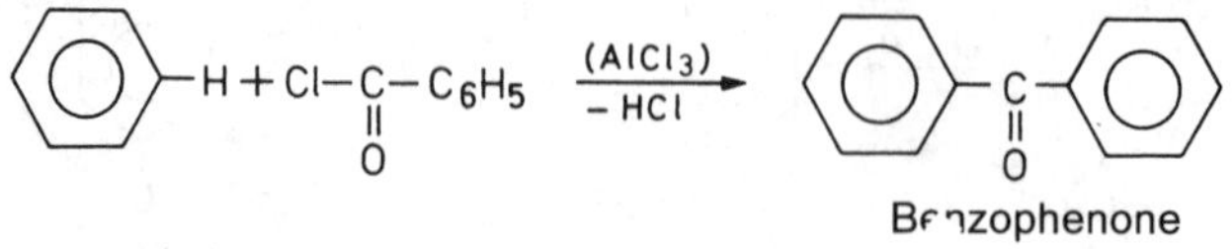

Acetophenone

Benzophenone

Reaction involves *electrophilic substitution* on the benzene ring.[1] In the presence of a *Lewis* acid such as aluminium chloride the acid chloride forms a complex with it, which is in equilibrium with an *acyl cation*.[2] Electrophilic attack by this cation on a benzene ring gives rise to a carbenium ion, which loses a proton to regenerate the aromatic system:

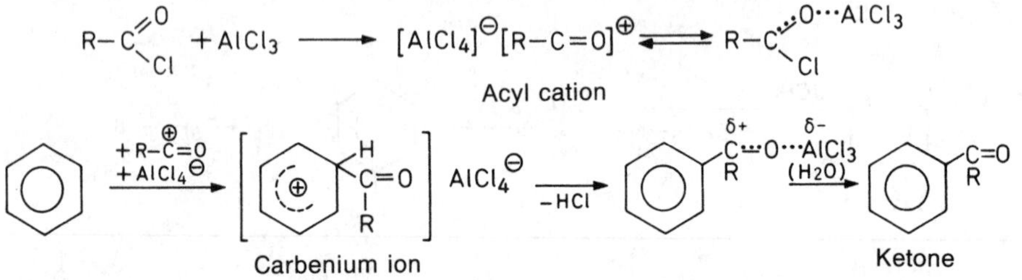

Acyl cation

Carbenium ion   Ketone

The ketone which is produced forms a complex with 1 mol of aluminium chloride; this is split by hydrolysis, consuming the aluminium chloride. In consequence, for *Friedel–Crafts* acylation (in contrast to alkylation) it is necessary to have more than 1 mol of aluminium chloride for the reaction to be successful.[3]

The acyl substituent, as a deactivating group, makes introduction of a further keto group more difficult and, again in contrast to *Friedel–Crafts* alkylation, electrophilic disubstitution does not take place.

Like acid chlorides, *acid anhydrides* react with arenes in the presence of *aluminium chloride* to give ketones. In this case also an *acyl cation* is formed, which reacts as described above.

Since in this case both the acid and the ketone are bonded to the aluminium chloride, it is necessary to use at least 3 mols of aluminium chloride.

5.8.5.2   By *atmospheric oxidation of ethylbenzene* in the presence of cobalt salts of long-chain carboxylic acids.

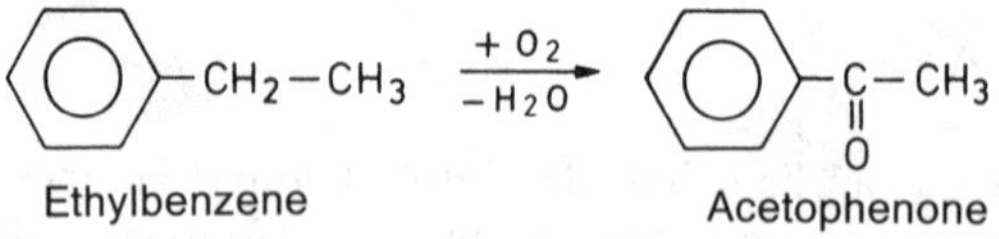

Ethylbenzene   Acetophenone

---

[1] See R. *Taylor*, Electrophilic Aromatic Substitution (Wiley, Chichester 1990).
[2] See B. *Chevrier* et al., Angew. Chem. Int. Ed. Engl. *13*, 1 (1974).
[3] But see D. E. *Pearson* et al., *Friedel–Crafts* Acylation with little or no catalyst, Synthesis *1972*, 533.

5.8.5.3 *Hoesch* **Ketone Synthesis**[1] (1927). From *polyhydric* phenols and nitriles in ethereal solution in the presence of hydrogen chloride, *e.g.*

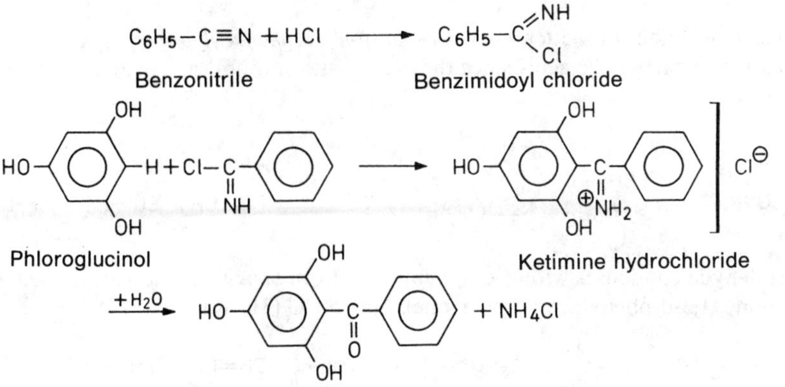

$$C_6H_5-C \equiv N + HCl \longrightarrow C_6H_5-C\begin{smallmatrix} \nearrow NH \\ \searrow Cl \end{smallmatrix}$$

Benzonitrile   Benzimidoyl chloride

Phloroglucinol + Cl−C(=NH)−C₆H₅ ⟶ [Ketimine hydrochloride] Cl⁻

Phloroglucinol   Ketimine hydrochloride

$$\xrightarrow{+H_2O} \text{2,4,6-Trihydroxybenzophenone} + NH_4Cl$$

2,4,6-Trihydroxybenzophenone

The benzimidoyl chloride which is formed reacts with phloroglucinol to give as an intermediate a ketimine hydrochloride, which, when heated with water, loses ammonia and is hydrolysed to 2,4,6-trihydroxybenzophenone (2,4,6-trihydroxyphenyl phenyl ketone, 2-benzoylphloroglucinol) (*cf. Gattermann* aldehyde synthesis).

Under these conditions monohydric phenols do not form ketimines but instead give imidoesters, *e.g.*

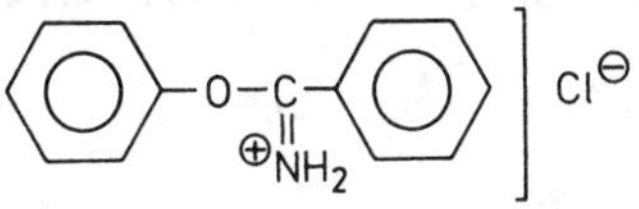

[structure: C₆H₅−O−C(=⊕NH₂)−C₆H₅] Cl⁻

Properties

The chemical properties of aryl ketones are very similar to those of aliphatic ketones. Some *butyrophenones* (peridols) are apparently effective against psychological disorders (schizophrenia). INN: **Haloperidol** (*Haldol, Sigaperidol*) has become of particular interest.

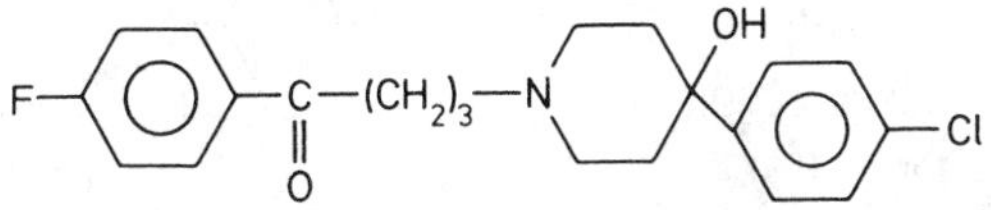

$$F-C_6H_4-C(=O)-(CH_2)_3-N\,\langle\text{piperidine ring}\rangle-C_6H_4-Cl \;(OH)$$

INN: *Haloperidol*

5.8.5.4 Acetophenone, $C_6H_5-CO-CH_3$, forms crystals which melt at 20°C (293 K). As in acetone, the hydrogen atoms of the methyl group are reactive.

If acetophenone is chlorinated or brominated the products are *α-chloroacetophenone* (phenacyl chloride), m.p. 60°C (333 K) or *α-bromoacetophenone* (phenacyl bromide), m.p. 51°C (324 K).

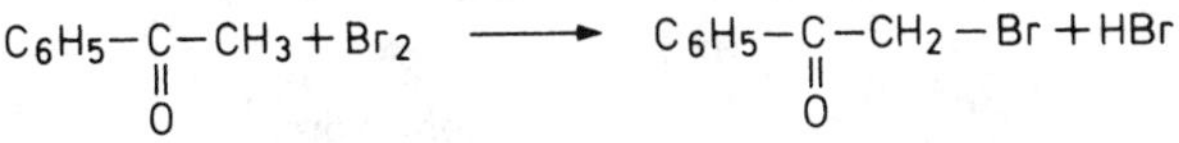

$$C_6H_5-\underset{O}{\underset{\|}{C}}-CH_3 + Br_2 \longrightarrow C_6H_5-\underset{O}{\underset{\|}{C}}-CH_2-Br + HBr$$

Acetophenone   α-Bromoacetophenone

---

[1] See *P. E. Spoerri* and *A. S. du Bois*, The *Hoesch* Synthesis, Org. React. *5*, 387 *ff.* (1949).

Both of these α-halogenoketones have wide synthetic use. Physiologically both are strongly *lachrymatory*. Solutions of phenacyl cyanide (CN) are used in sprays as tear-gases.

Phenacyl bromide, or better, *p-bromo-* or *p-nitro-phenacyl bromides* are used in the identification of carboxylic acids since they react with sodium salts of acids to form nicely crystalline esters, *e.g.*

$$R-C\underset{O^\ominus Na^\oplus}{\overset{O}{\diagdown}} + Br-CH_2-\overset{O}{\underset{O}{\overset{|}{C}}}-\bigcirc-NO_2 \xrightarrow{-NaBr} R-C\overset{O}{\diagdown}_{O-CH_2-CO}-\bigcirc-NO_2$$

Benzaldehyde condenses with acetophenone to form *benzylideneacetophenone* (benzala-cetophenone, 1,3-diphenylprop-2-en-1-one), m.p. 58°C (311 K):

$$C_6H_5-\overset{O}{\underset{||}{C}}-CH_3 + \overset{H}{\underset{O}{\diagdown}}C-C_6H_5 \longrightarrow C_6H_5-\overset{O}{\underset{||}{C}}-CH=CH-C_6H_5 + H_2O$$

Benzylideneacetophenone (Chalcone)

α,β-Unsaturated aryl ketones are sometimes called *chalcones* after a trivial name for the simplest member of the series.

Acetic anhydride reacts with acetophenone in the presence of boron trifluoride to give a good yield of benzoylacetone, m.p. 59°C (332 K) (*Meerwein*, 1934):

$$C_6H_5-\overset{O}{\underset{||}{C}}-CH_3 + (H_3C\,CO)_2O \xrightarrow{(BF_3)} C_6H_5-\overset{O}{\underset{||}{C}}-CH_2-\overset{O}{\underset{||}{C}}-CH_3 + H_3C\,COOH$$

Like acetylacetone (see p. 322), benzoylacetone undergoes *keto–enol tautomerism*; it contains up to 99% of the *enol* form, which is stabilised by *intramolecular hydrogen bonding*:

$$\bigcirc-\overset{O}{\underset{|||}{C}}-CH_2-\overset{O}{\underset{|||}{C}}-CH_3 \rightleftharpoons \bigcirc-C=CH-\overset{O}{\underset{||}{C}}-CH_3$$

Keto form (1%)   *Enol* form (99%)

Benzoylacetone

5.8.5.5 Benzophenone, C_6H_5—CO—C_6H_5, exists in an unstable and a stable modification which melt, respectively, at 26°C (299 K) and 48°C (321 K). It is insoluble in water but readily soluble in most organic solvents. Sodium amalgam reduces it to **benzhydrol**, m.p. 69°C (342 K):

$$C_6H_5-\overset{O}{\underset{||}{C}}-C_6H_5 \longrightarrow C_6H_5-\overset{|}{\underset{OH}{C}}H-C_6H_5$$

Benzhydrol

Benzophenone forms only *one* oxime, m.p. 143°C (416 K), but mixed ketones form two isomeric oximes (*cf.* benzaldehyde oximes).

5.8.5.6 *Mannich* **reaction**[1] (1917). This name is given to the condensation reactions of compounds having acidic C—H groups with formaldehyde in the presence of secondary or primary amines, usually present as their hydrochlorides.

A ketone, *e.g.* acetophenone, condenses with formaldehyde and an equimolecular amount of an ammonium salt to form β-aminoketones. In the process, a reactive hydrogen atom of the ketone is replaced by a dialkyl- or monoalkyl-*aminomethyl* group; it is described as α-aminomethylation. The two compounds which react with formaldehyde are known as the *methylene component* and the *amino component*, *e.g.*

$$\left[\text{C}_6\text{H}_5-\underset{\overset{\|}{\text{O}}}{\text{C}}-\text{CH}_3 + \text{H}-\text{C}\overset{\text{H}}{\underset{\text{O}}{}} + \text{H}_2\overset{\oplus}{\text{N}}\overset{\text{CH}_3}{\underset{\text{CH}_3}{}} \right] \text{Cl}^\ominus \xrightarrow[-\text{H}_2\text{O}]{} \left[\text{C}_6\text{H}_5-\underset{\overset{\|}{\text{O}}}{\text{C}}-\text{CH}_2-\text{CH}_2-\overset{\oplus}{\underset{\text{H}}{\text{N}}}(\text{CH}_3)_2 \right] \text{Cl}^\ominus$$

| Acetophenone (Methylene component) | Dimethylammonium chloride (Amino component) | ω-Dimethylamino-propiophenone hydrochloride |

Basically, the methylene components can only be compounds which under the reaction conditions can act as *nucleophiles*. The successful outcome of a *Mannich* reaction also depends upon the amine being a stronger nucleophile towards formaldehyde than the methylene compound is.

The *reaction mechanism* depends very much upon the reaction conditions. In neutral or acidic conditions, the first step is an attack by the lone pair of electrons of the amine nitrogen atom on the positively charged carbon atom of formaldehyde. The adduct takes up a proton and loses a molecule of water thereby forming a *cation* containing a positively charged carbon atom. This methyleneiminium ion is mesomerically stabilised (a carbenium–iminium ion) and acts as the aminomethylating agent, making an electrophilic attack on a suitable nucleophilic reagent, *e.g.* the *enol* form of acetophenone:

$$\underset{\text{H}}{\text{R}_2\overset{|}{\text{N}}\text{I}} + \underset{\underset{\text{I}\overset{\ominus}{\text{O}}\text{I}}{\overset{\|}{}}}{\overset{\oplus}{\text{CH}_2}} \longrightarrow \text{R}_2\overset{\oplus}{\underset{\text{H}}{\text{N}}}-\underset{\text{I}\underline{\text{O}}\text{I}^\ominus}{\overset{|}{\text{CH}_2}} \longrightarrow \text{R}_2\overset{\oplus}{\text{N}}-\underset{\text{OH}}{\overset{|}{\text{CH}_2}} \xrightarrow[-\text{H}_2\text{O}]{+\text{H}^\oplus} \left[\text{R}_2\overset{\oplus}{\text{N}}=\text{CH}_2 \longleftrightarrow \text{R}_2\overset{\ominus}{\text{N}}-\overset{\oplus}{\text{CH}_2} \right]$$

Methyleneiminium ion

$$\text{C}_6\text{H}_5-\underset{\underset{\text{I}\underline{\text{O}}-\text{H}}{|}}{\overset{\delta^-}{\text{C}}}=\overset{\delta^+}{\text{CH}_2} + \overset{\oplus}{\text{CH}_2}=\overset{\oplus}{\text{NR}_2} \xrightarrow[-\text{H}^\oplus]{} \text{C}_6\text{H}_5-\underset{\overset{\|}{\text{O}}}{\text{C}}-\text{CH}_2-\text{CH}_2-\text{NR}_2$$

Mannich base

The mono- or di-substituted β-aminoketones obtained in this way can be reduced to physiologically active β-aminoalcohols.

$$\text{R}-\underset{\overset{\|}{\text{O}}}{\overset{\overset{\text{R}'}{|}}{\text{C}}}-\overset{\text{R}'}{\underset{}{\text{N}}}-\text{H} + \text{CH}_2\text{O} \longrightarrow \text{R}-\underset{\overset{\|}{\text{O}}}{\overset{\overset{\text{R}'}{|}}{\text{C}}}-\text{N}-\text{CH}_2\text{OH} \xrightarrow[-\text{H}_2\text{O}]{\text{H}^\oplus\text{X}^\ominus}$$

Hydroxymethylamide

$$\left[\text{R}-\underset{\overset{\|}{\text{O}}}{\overset{\overset{\text{R}'}{\overset{|\oplus}{}}}{\text{C}}}-\text{N}=\text{CH}_2 \longleftrightarrow \text{R}-\underset{\overset{\|}{\text{O}}}{\overset{\overset{\text{R}'}{|}}{\text{C}}}-\overset{\oplus}{\text{N}}-\text{CH}_2 \right] \text{X}^\ominus \xrightarrow[-\text{H}^\oplus]{\text{ArH}} \text{R}-\underset{\overset{\|}{\text{O}}}{\overset{\overset{\text{R}'}{|}}{\text{C}}}-\text{N}-\text{CH}_2\text{Ar}$$

Acylmethyleneiminium salt

[1] See *F. F. Blicke*, The *Mannich* Reaction, Org. React. 1, 303 *ff.* (1942); *M. Tramontini*, Advances in the Chemistry of *Mannich* Bases, Synthesis *1973*, 703.

If primary amines or ammonia are used the nitrogen atom of the resultant *Mannich* base has either one or two hydrogen atoms attached to it; these can also take part in the reaction and the overall picture becomes more complicated. Hence secondary ammonium salts are usually used.

Amides react with formaldehyde to give *hydroxymethylamides*. In the presence of acid they provide *acylmethyleneiminium salts*, which can also act as *amidomethylating agents* (see p. 543):

Other aldehydes besides formaldehyde can be used for preparing iminium salts; the reactions are then α-amidoalkylations.[1]

Acylmethyleneiminium salts can be used synthetically in a variety of ways, especially for making C—C bonds.[2] Compounds of this sort have been described as *synthons*.[3] Other synthons include amide chlorides, *Vilsmeier–Haack* complexes and dichloromethyleneiminium salts.

5.8.5.7 *Willgerodt–Kindler* Reaction[4] (1887, 1923).

When alkyl aryl ketones are heated with sulphur in morpholine they are converted into carboxylic acid thioamides (in this case thiomorpholides) having the same number of carbon atoms; the original carbonyl group has been reduced to a methylene group:

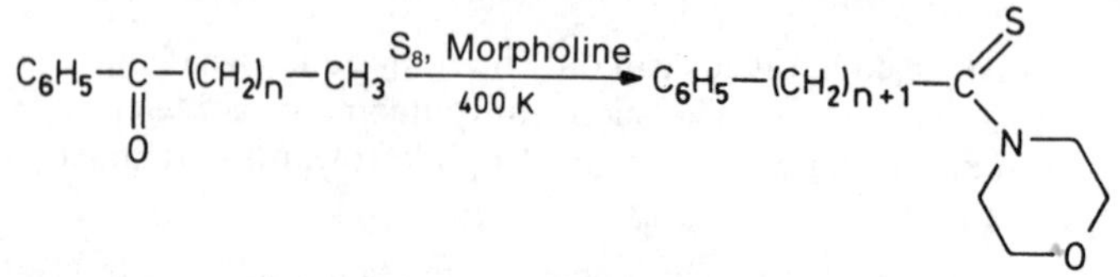

$$C_6H_5-\underset{\underset{O}{\|}}{C}-(CH_2)_n-CH_3 \xrightarrow[400\ K]{S_8,\ Morpholine} C_6H_5-(CH_2)_{n+1}-C\underset{N}{\overset{S}{\diagup}}$$

Alkaline hydrolysis of the thiomorpholide results in the formation of aryl-substituted aliphatic carboxylic acids.

The mechanism of the reaction is still not fully understood, but the suggestion that it proceeds *via* a series of reductions and oxidations along the carbon chain is widely accepted.[5]

5.8.5.8 *Beckmann* Rearrangement[6] (1886).

When a ketoxime is treated with acid chlorides such as phosphorus(V) chloride in ether, benzenesulphonyl chloride in pyridine, or with concentrated mineral acids (conc. H_2SO_4, gaseous HCl), a rearrangement reaction takes place giving rise to a carboxylic acid amide or anilide. For example, benzophenone oxime is converted into benzanilide:

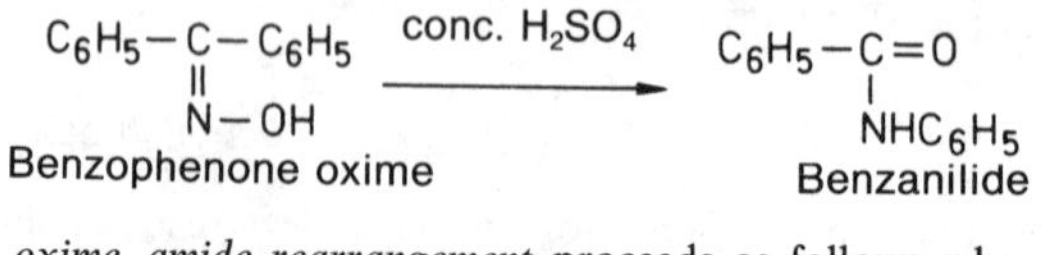

$$\underset{\text{Benzophenone oxime}}{C_6H_5-\underset{\underset{N-OH}{\|}}{C}-C_6H_5} \xrightarrow{\text{conc. } H_2SO_4} \underset{\text{Benzanilide}}{C_6H_5-\underset{\underset{NHC_6H_5}{|}}{C}=O}$$

In most cases this *oxime–amide rearrangement* proceeds as follows when catalysed by mineral acids. First of all, a proton from the acid bonds to a lone pair of electrons on either the nitrogen or oxygen atoms of the ketoxime thereby generating an iminium or an oxonium ion. Loss of a

---

[1] See *H. E. Zaugg*, Recent Synthetic Methods Involving Intermolecular α-Amidoalkylation at Carbon, Synthesis *1970*, 49.

[2] See *H. Böhme* and *H. G. Viehe*, Iminium Salts in Organic Chemistry, Adv. Org. Chem. *9*, (1976); *D. Matthies* et al., Lineare N-Acylimine, Chemiker-Zeitung, *111*, 181, 253 (1987).

[3] The originally suggested definition of a synthon by *E. J. Corey* [Pure Appl. Chem. *14*, 19 (1967)] was expressed more generally and read 'Structural units within a molecule, which are related to possible synthetic operations.' In this sense the role of synthons in retrosynthesis (see p. 549), which was associated with the idea of disconnection of bonds, was conceived, and they were differentiated from the reagents actually employed. Thus CH_3I is the reagent for the synthon $CH_3^{\oplus}$.

[4] See *M. Carmack* et al., The *Willgerodt* Reaction, Org. React. *3*, 83 (1946).

[5] *S. W. Schneller*, Name Reactions in Sulfur Chemistry, Part II, Int. J. Sulfur Chem. *8*, 579 (1976).

[6] See *L. G. Donaruma* et al., The *Beckmann* rearrangement, Org. React. *11*, 1 (1960); *R. E. Gawley* et al., The *Beckmann* Reactions, Org. React. *35*, 1 (1988).

molecule of water causes a deficiency of electrons on the nitrogen atom and the group R migrates as an anion with a pair of electrons to the nitrogen atom, generating a carbenium ion in the process. A water molecule, acting as a nucleophile, attaches itself to this carbenium atom, giving rise to an N-substituted carboxylic acid amide or anilide.

As long ago as 1921 *Meisenheimer* showed that in the *Beckmann* rearrangement the group which migrated was the one in the *anti*-position to the HO-group of the oxime. The reaction is stereospecific and only a few exceptions are known.

This rearrangement is of industrial importance for the conversion of cyclic ketoximes *via* ring-expansion into cyclic carboxylic acid amides. As an example, when cyclohexanone oxime is treated with conc. sulphuric acid the product is ε-caprolactam (see p. 288).

5.9 Arenecarboxylic Acids

A distinction is made between arenecarboxylic acids in which one or more carboxyl groups are attached directly to the benzene ring and other acids wherein the carboxyl group is attached to a side-chain. Depending upon the number of carboxyl groups attached to the ring these acids are divided into mono-, di-, *etc.* to hexa-carboxylic acids.

5.9.1 Arenemonocarboxylic Acids

The first member of this series is benzoic acid. This was isolated at the end of the 16th century from the dry distillation of gum benzoin.

5.9.1.1 Benzoic acid, C_6H_5—COOH, is found in plants, and in the urine of herbivorous mammals as benzoylglycine (hippuric acid), C_6H_5CO—NH—CH_2-COOH.

Preparation
(a) *By oxidising a side-chain attached to a benzene ring and converting it into a carboxyl group.* Among oxidising agents commonly used are potassium permanganate, chromium trioxide, nitric acid or atmospheric oxidation in the presence of cobalt or manganese naphthenates. The easiest process is the oxidation of toluene which provides benzoic acid in very good yields.

$$C_6H_5\text{--}CH_3 \xrightarrow[-H_2O]{+1\frac{1}{2}O_2} C_6H_5\text{--}COOH$$

Benzoic acid is made on a large scale industrially in this way.

(b) *By acid or alkaline hydrolysis of benzonitrile:*

$$C_6H_5{-}CN \xrightarrow[-NH_3]{+2\,H_2O} C_6H_5{-}COOH$$

Properties. Benzoic acid crystallises in the form of glistening leaflets, m.p. 122°C (395 K) and can be sublimed. It is soluble in ethanol and in ether and in hot water. It is volatile in steam; its vapours induce coughing. When sodium benzoate is melted with sodium or calcium hydroxides *decarboxylation* ensues and benzene is formed:

$$C_6H_5COONa + NaOH \rightarrow C_6H_6 + Na_2CO_3$$

Benzoic acid ($pK_a = 4.21$) is more strongly dissociated than is the corresponding saturated acid, cyclohexanecarboxylic acid ($pK_a = 4.86$). From this it appears that the phenyl group exerts an electron-withdrawing effect on the carboxyl group. On the other hand, benzoic acid is a weaker acid than formic acid ($pK_a = 3.77$), *i.e.* compared to hydrogen the phenyl group acts as an *electron donor*. Consequently, the π-electron sextet can function as either an electron acceptor or an electron donor with respect to the adjacent carboxyl group; on this depends to a large extent the effects of different substituents on the o-, m- and p-positions in the benzene ring.

The effects of the methyl group (an electron donor) and of the halogens and of the nitro group (electron acceptors) on the acidity of different substituted benzoic acids may be taken as an example:

	pK_a		pK_a
p-methylbenzoic acid	4.35	m-nitrobenzoic acid	3.45
p-bromobenzoic acid	4.18	p-nitrobenzoic acid	3.42
p-chlorobenzoic acid	4.01	o-nitrobenzoic acid	2.17

The presence of *bulky* substituents adjacent to the carboxyl group appears to have a big influence on its ease of esterification. Benzoic acid itself is readily esterified, for example with methanol and hydrogen chloride, but the rate of esterification of 2,6-dimethylbenzoic acid is much slower, while in the case of 2,6-di-isopropylbenzoic acid the rate of reaction is almost zero (*V. Meyer*, 1874). In general, 2,6-dinitrobenzoic acid is not esterified under normal conditions. The failure of these esterifications is due first and foremost to *steric hindrance* of the carboxyl group by the 2,6-substituents; this has been called an *F-strain effect* (*front-strain effect*). Steric effects on the fission of bonds can be detected quantitatively by means of the *strain enthalpy*.[1]

Uses. Benzoic acid is used as a preservative for foodstuffs.

Derivatives of Benzoic Acid

5.9.1.2   Benzoyl chloride, $C_6H_5{-}COCl$, is the most important derivative of benzoic acid.

Preparation

(a) In the laboratory it is made by *distilling a mixture of benzoic acid and phosphorus(V) chloride or thionyl chloride, e.g.*

---

[1] See *C. Rüchardt* et al., Towards the Understanding of the Carbon–Carbon Bond, Angew. Chem. Int. Ed. Engl. *19*, 429 (1980); Consequences of Strain for the Structure of Aliphatic Molecules, Angew. Chem. Int. Ed.Engl., *24*, 529 (1985).

$$C_6H_5-C\overset{O}{\underset{OH}{\diagup}} + SOCl_2 \longrightarrow C_6H_5-C\overset{O}{\underset{Cl}{\diagup}} + HCl + SO_2$$

Benzoic acid   Benzoyl chloride

(b) Industrially it is made from *benzotrichloride* and *benzoic acid* (1:1) in the presence of a silver catalyst:

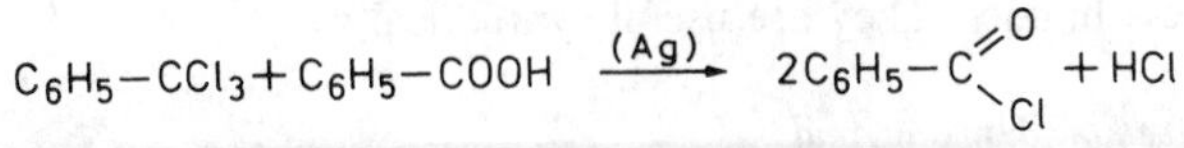

$$C_6H_5-CCl_3 + C_6H_5-COOH \xrightarrow{(Ag)} 2C_6H_5-C\overset{O}{\underset{Cl}{\diagup}} + HCl$$

(c) By the reaction of *chlorine* with *benzaldehyde* in the cold:

$$C_6H_5-C\overset{H}{\underset{O}{\diagup}} + Cl_2 \longrightarrow C_6H_5-C\overset{O}{\underset{Cl}{\diagup}} + HCl$$

Properties. Benzoyl chloride is a colourless liquid, b.p. 197°C (470 K), with an irritating smell. It is only slightly soluble in water and consequently it is hydrolysed much more slowly than aliphatic acid chlorides. Its main use is for *benzoylating* alcohols, phenols and amines; this is carried out either in aqueous alkali (*Schotten–Baumann* reaction) or in pyridine (*Einhorn*), e.g.

$$RO-H + Cl-CO\ C_6H_5 \longrightarrow RO-COC_6H_5 + HCl$$

In this reaction pyridine acts not only as a weak base to take up the hydrogen chloride which is formed as pyridine hydrochloride, but also forms a complex with the acid chloride, thereby increasing the rate of reaction. 4-Dimethylaminopyridine (DMAP) is especially effective for this purpose.

Alcohols are often characterised by letting them react with *3,5-dinitrobenzoyl chloride*, since the 3,5-dinitrobenzoate esters which are formed are usually crystalline materials.

5.9.1.3 If benzoyl chloride is shaken with an alkaline solution of hydrogen peroxide *dibenzoyl peroxide* is formed. This is used as a *bleaching agent* and, above all, as a *radical source* in polymerisation processes:

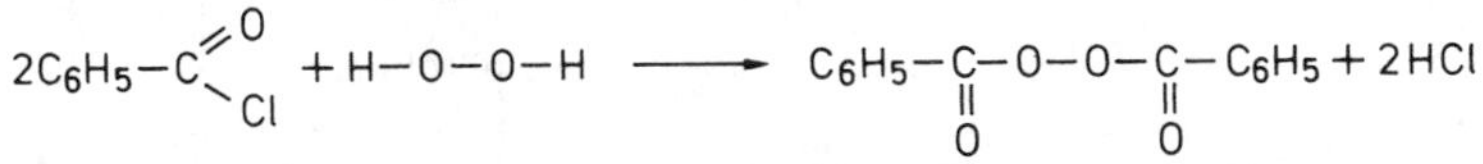

$$2C_6H_5-C\overset{O}{\underset{Cl}{\diagup}} + H-O-O-H \longrightarrow C_6H_5-\underset{\underset{O}{\|}}{C}-O-O-\underset{\underset{O}{\|}}{C}-C_6H_5 + 2HCl$$

Benzoyl chloride   Dibenzoyl peroxide

If 1 mol of dibenzoyl peroxide in ether is treated with 1 mol of sodium ethoxide in ethanol, it is cleaved to give the sodium salt of *perbenzoic acid* and ethyl benzoate:

$$C_6H_5-\underset{\underset{O}{\|}}{C}-O-O-\underset{\underset{O}{\|}}{C}-C_6H_5 + NaOC_2H_5 \longrightarrow \left[C_6H_5-C\overset{O}{\underset{O-O^{\ominus}}{\diagup}}\right]Na^{\oplus} + C_6H_5COOC_2H_5$$

Sodium
perbenzoate

5.9.1.4 Perbenzoic acid, m.p. 42°C (315 K), has a pungent smell. It is a versatile oxidising agent. For preparative purposes *m-chloroperbenzoic acid* (MCPBA), m.p. 92–94°C (365–367 K), has proved valuable. It can explode, so care must be taken in handling it.

5.9.1.5 Benzoyl cyanide, C_6H_5COCN, m.p. 34°C (307 K) is obtained by heating either benzoyl chloride or benzoyl bromide with copper(I) cyanide:

$$C_6H_5-C\diagdown{}_{Br}^{O} + CuCN \longrightarrow C_6H_5-C\diagdown{}_{C\equiv N}^{O} + CuBr$$

Benzoyl bromide   Benzoyl cyanide

Acyl cyanides or α-ketonitriles occupy an intermediate position between simple carbonyl compounds and acyl halides. They are useful synthetically.

Substituted Derivatives of Benzoic Acid

Electrophilic substitution reactions, e.g. halogenation, nitration, sulphonation, give rise to *m*-substituted derivatives. Nitrobenzoic acids and, especially, aminobenzoic acids are of importance.

5.9.1.6 *o*-Nitrobenzoic and *p*-nitrobenzoic acids (for pK_a values, see p. 546) are prepared by oxidising the corresponding nitrotoluenes with chromic acid, *e.g.*

p-Nitrotoluene   *p*-Nitrobenzoic acid
                              (m.p. 240°C [513 K])

3,5-Dinitrobenzoic acid, $pK_a = 2.82$, can be obtained by direct nitration of benzoic acid:

3,5-Dinitrobenzoic
acid (m.p. 205°C [478 K])

The three isomeric *aminobenzoic acids* are made by reducing the corresponding nitrobenzoic acids.

5.9.1.7 Anthranilic acid (*o*-aminobenzoic acid) is made industrially from phthalimide by treating it with alkaline sodium hypochlorite solution (or bleaching powder and alkali); a salt of phthalamic acid is formed as an intermediate (see *Hofmann* degradation):

Phthalimide   Sodium   Anthranilic acid
                     phthalamate

Anthranilic acid is a colourless, crystalline substance of m.p. 146°C (419 K); it is soluble in water, ethanol and ether.

5.9.1.8 p-Aminobenzoic acid (PABA) forms yellow crystals, m.p. 188°C (461 K). It is an important bacterial metabolite. It is made either by reduction of p-nitrobenzoic acid or by oxidising 4'-methylacetanilide (the acetyl group is present to protect the amino group) to p-acetylaminobenzoic acid and then removing the acetyl group.

p-Toluidine   4'-Methylacetanilide

p-Aminobenzoic acid

Some derivatives of p-aminobenzoic acid, usually in the form of their hydrochloride salts commercially, are used as *local anaesthetics*, e.g.

| Ethyl p-Aminobenzoate (*Benzocaine, Anaesthesin*) | β-Diethylaminoethyl p-aminobenzoate (INN: **Procaine**, *Novocaine*) | β-Diethylaminoethyl p-butylaminobenzoate (INN: **Tetracaine**, *Pantocaine*) |

A further group of synthetic alternatives to cocaine are *2-aminoacetanilides* e.g. 2-diethylamino--2',6'-dimethylacetanilide hydrochloride (INN: *Lidocaine*) and 2-n-butylamino-(2'-chloro-6'-methyl) acetanilide hydrochloride (INN: *Butanilicaine*). Because the former and its analogues have shorter times of onset of analgesia than *Procaine*, they are widely used in minor surgical procedures. *Lidocaine* has also proved successful in the treatment of cardiac arrythmias which are often associated with acute heart disease.

INN: **Lidocaine**, *Xylocaine*   INN: **Butanilicaine**, *Hostacaine*

5.9.1.9 Sulphobenzoic acids. Whilst *m*-sulphobenzoic acid can be obtained by direct sulphonation of benzoic acid, the *o*- and *p*-isomers are prepared by oxidising the corresponding toluenesulphonic acids. *o*-Sulphamidobenzoic acid is an important intermediate in the manufacture of the sweetener saccharin.

Saccharin may be used as an example to demonstrate the use of the idea of *retrosynthesis* (disconnection) in the planning of synthesis.[1] One starts from the *target molecule* and

---

[1] See S. Warren, Organic Synthesis, The Disconnection Approach (Wiley, New York 1982); J. H. Winter, Chemische Syntheseplanung in Forschung und Industrie (The Planning of Syntheses in Research and in Industry) (Springer-Verlag, Berlin 1982); E. J. Corey et al., The Logic of Chemical Synthesis (Wiley, New York 1989).

disconnects those bonds which would need to be formed in a synthesis based on this route. The retrosynthetic steps are signified by means of open arrows ($\Rightarrow$); modification of substituents are indicated in the same way.

Saccharin
(Target molecule)

Disconnections
S—N; C—N

Modification by oxidation
Disconnection C—S

Toluene

The first step of a restrosynthesis represents the final step of the actual synthesis. After the last step of a retrosynthesis, the compound which remains, in this example toluene, is the starting material for the synthesis, and saccharin is made from toluene industrially by the following route.

Chlorosulphonation of toluene is carried out using chlorosulphonic acid at about 0°C (273 K). o-Toluenesulphonyl and p-toluenesulphonyl chlorides are formed and, after they have been separated, the o-isomer is treated with ammonia to give o-toluenesulphonamide. This is oxidised by chromic acid to 2-carboxybenzenesulphonamide (o-sulphamoylbenzoic acid) which loses water intramolecularly when heated. Saccharin precipitates out; it is rather insoluble and is converted into its water-soluble sodium salt by sodium hydroxide.

o-Toluenesulphonyl chloride   o-Toluenesulphonamide o-Sulphamoylbenzoic acid   Saccharin

The oxidising agent is continuously regenerated electrochemically by anodic oxidation during the course of the process. Processes like this, which do not involve direct electrolysis of the reagents, but in which electrolytic methods are indirectly involved, in the regeneration of reagents, are known as *indirect electrochemical reactions*.

Saccharin, which was first synthesised in 1879 (*Fahlberg, Remsen*), is a colourless, crystalline powder, m.p. 228°C (501 K), which is rather insoluble in water but dissolves readily in ethanol. It is about 300 times as sweet as cane sugar, but has no nutritional value.[1]

Hydroxybenzoic acids. Of the three isomeric hydroxybenzoic acids the o- and p-isomers are of importance.

5.9.1.10   Salicylic acid (o-hydroxybenzoic acid) occurs in various essential oils, both as the free acid and as its methyl ester. It can be made by oxidising either salicyl alcohol (saligenin, o-hydroxybenzyl alcohol) or salicyl aldehyde (o-hydroxybenzaldehyde).

Preparation. Kolbe–Schmitt synthesis (1885). Salicylic acid is made on the industrial scale by the reaction of carbon dioxide with dry sodium phenoxide at 120–140°C (395–415 K) and 5–6 bar (0.5–0.6 MPa). It has since been shown that this involves a *direct carboxylation* at the o-position of the mesomeric phenoxide anion by means of an electrophilic substitution reaction. It can be formulated in a greatly simplified way as follows:

---

[1] For the many-faceted problem of the admissability of saccharin as a sweetening agent see *L. Caglioti*, The Saccharin Mess in the Two Faces of Chemistry, p.36 (MIT Press, Cambridge, Mass. 1983).

Sodium phenoxide   Sodium salicylate

If potassium phenoxide is used instead of the sodium salt the main product is *p-hydroxybenzoic acid*, whose esters are used as preservatives.

Properties. Salicylic acid crystallises from hot water as colourless needles, m.p. 159°C (432 K). Its $pK_a = 2.98$, i.e. it is an appreciably stronger acid than benzoic acid ($pK_a = 4.21$) or *p*-hydroxybenzoic acid ($pK_a = 4.56$). It differs from *p*-hydroxybenzoic acid, m.p. 215°C (488 K) in its greater volatility and by the blue-violet colour it gives with iron(III) chloride.

These properties, plus its higher acidity, can be attributed to *intramolecular hydrogen bonding* (see chelates, p. 502) and the colour obtained with iron(III) chloride is associated with the formation of an inner complex.

Salicylic acid

Uses. Salicylic acid has a wide range of uses, for example, in the preparation of dyestuffs and perfumes (salicylate esters) and of pimelic acid, and in medicine as an *antiseptic*. It is a permitted preservative for foodstuffs, but only for domestic use and not for commercial food products. Consequently it has not been brought into use on a commercial scale.

The acetyl derivative of salicyclic acid, *aspirin* (acetylsalicylic acid), is a valuable *antipyretic* and *antineuralgic. It stops the formation of prostaglandins* and is important as a non-steroidal inhibitor of inflammation (*cf.* p. 695). Aspirin prevents the irreversible aggregation of erythrocytes. *Salol*, phenyl salicylate, is used as an antiseptic in mouth washes.

Acetylsalicylic
acid (Aspirin)   Salol   *p*-Aminosalicylic
acid   INN: *Niclosamine*

p-Aminosalicylic acid (PAS) has been found to be particularly effective in the chemotherapy of tuberculosis (*J. Lehmann*, 1946); its action is nullified by *p*-aminobenzoic acid.

2′,5-Dichloro-4′-nitrosalicylanilide, INN: *Niclosamine*, is an effective anthelmintic used against tapeworms.

5.9.1.11 Gallic acid (3,4,5-trihydroxybenzoic acid) occurs as the free acid in, among other things, tea, oak bark and pomegranate roots, and as a glucoside in the *tannins* from galls. Commercially it is obtained from the tannin-rich aqueous extract of galls, either by hydrolysis, using hot dilute sulphuric acid, or enzymatically by the action of *tannase* (from a mould fungus). It is a crystalline substance, which in the anhydrous form melts at 253°C (526 K) and is a strong reducing agent. When heated it loses carbon dioxide to form pyrogallol. With iron(III) chloride it gives a black precipitate of iron gallate (iron gallate ink). Propyl gallate is used in foods as an antioxidant.

Gallic acid, together with other hydroxyacids which occur in plants, and sugars form important building blocks for a series of natural products, especially the turgorins, as well as for tannins and for lichens.

Turgorins[1] are plant hormones (phytohormones) which influence the internal sap pressure (turgor) in plants and regulate the hydrodynamic methods of plant movement, such as the regular daily movements of leaves, or the reaction of *Mimosa pudica* when its leaves are touched.

Turgorinic acid PLMF 1 was the first example of this class of substances to be isolated and have its structure determined (*Schildknecht*, 1981). PLMF stands for 'periodic leaf movement factor'.

Turgorinic acid PLMF 1

The compound is a β-glucoside of the *p*-hydroxy group of gallic acid, namely 4-*O*-(6-*O*-sulpho-β-D-glucopyranosyl)gallic acid and is found in other plants such as the legumes. A number of slightly modified turgorinic acids are known, *e.g.* a 4-*O*-(β-D-glucopyranouronyl)gallic acid with a COOH group instead of the —CH₂OSO₃H group in the carbohydrate part of the molecule, (4-carboxy-2,6-dihydroxyphenyl)-β-D-glucopyranosideuronic acid.

In the materials from tannins and lichens the carboxyl group of the gallic acid is esterified by the hydroxy group of a second molecule which is either the same or some other hydroxybenzoic acid. These hydroxybenzoate esters are called *depsides* (Grk. *depsein* = to tan) and are subdivided into didepsides, tridepsides, *etc.* according to the number of esterified units present.

Some such esters were synthesised by *E. Fischer*, including the important didepside *m*-digallic acid (3-*O*-galloylgallic acid), whose acid chloride reacted with glucose to form a pentadigalloylglucose.

m-Digallic acid (Didepside)   Pentadigalloylglucose

(X = *m*-Digalloyl = 3-Galloyloxy-4,5-dihydroxy-benzoyl)

---

[1] See *H. Schildknecht*, Turgorins, Hormones of the Endogenous Daily Rhythms of Higher Organised Plants, Angew. Chem. Int. Ed. Engl. *22*, 695 (1983).

The natural *tannins*[1] or tannic acids are for the most part colourless, amorphous powders which are readily soluble in water. They are able to precipitate proteins and this is made use of in the tanning of leather. The chemical processes which take place during tanning are not as yet known precisely. In practice it is standard that the tannins render the animal hide resistant to decay, make the leather retain its pliability and prevent it from swelling when exposed to water.

5.9.2 Arylaliphatic Monocarboxylic Acids (Arylalkanoic Acids)

In these compounds the substituent carboxyl group is on a side-chain attached to a benzene ring.

5.9.2.1 Phenylacetic acid, $C_6H_5-CH_2-COOH$, is a colourless, crystalline substance, m.p. 78°C (351 K). It is obtained by hydrolysis of benzyl cyanide which is itself made by the reaction of benzyl chloride with potassium cyanide:

$$C_6H_5-CH_2-Cl \xrightarrow[-KCl]{+KCN} C_6H_5-CH_2-CN \xrightarrow[-NH_3]{+2H_2O} C_6H_5-CH_2-COOH$$

Benzyl chloride   Benzyl cyanide   Phenylacetic acid

Compared to purely aliphatic carboxylic acids, the hydrogen atoms on the carbon atom next to the carboxyl group are much more *acidic*. This is apparent from the reaction of isopropyl magnesium chloride with phenylacetic acid in solution in ether:

$$C_6H_5-CH_2-COOH \ + \ 2\,(H_3C)_2CH-MgCl \ \longrightarrow \ \left[C_6H_5-CH = C \begin{smallmatrix} \ominus \\ \diagup \diagdown \end{smallmatrix} \begin{smallmatrix} O \\ \underline{O}l^\ominus \end{smallmatrix} \right] 2\,MgCl^\oplus \ + \ 2\,C_3H_8$$

If this solution of *Grignard* compound is treated with either an aldehyde or an unsymmetrically substituted ketone, a substituted *β-hydroxycarboxylic acid* results; for example, with benzaldehyde it gives 2,3-diphenyl-3-hydroxypropionic acid, which can be isolated in two diastereomeric forms (*Ivanov* reaction, 1931).[2]

$$\left[C_6H_5-C\begin{smallmatrix} \diagup H \\ \diagdown O \end{smallmatrix} + C_6H_5-\overset{\ominus}{C}H-C\begin{smallmatrix} \diagup O \\ \diagdown O \ominus \end{smallmatrix} \right] 2MgCl^\oplus \xrightarrow{(H^\oplus)} C_6H_5-\underset{\underset{OH}{|}}{CH}-\underset{\underset{C_6H_5}{|}}{CH}-COOH$$

Benzaldehyde   2,3-Diphenyl-3-hydroxy-propionic acid

This addition reaction follows a similar course to that of the aldol addition and *Reformatsky* reactions. It can also be carried out using butyl lithium instead of a *Grignard* reagent.

5.9.2.2 Mandelic acid (α-hydroxyphenylacetic acid), $C_6H_5-CH(OH)-COOH$, was first obtained by extracting bitter almonds with dilute hydrochloric acid and heating the extract. It contains an asymmetric carbon atom and can be resolved into *enantiomers*. DL-Mandelic acid can be made synthetically from benzaldehyde and hydrogen cyanide, which react to form an α-hydroxynitrile (cyanohydrin); this is then hydrolysed to give the acid:

---

[1] See G. T. *Schmidt* et al., Natürliche Gerbstoffe (Natural Tannins), Angew. Chem. *68*, 103 (1956).
[2] See D. *Ivanov* et al., Syntheses with Polyfunctional Organomagnesium Compounds, Synthesis *1970*, 615.

$$C_6H_5-C\overset{H}{\underset{O}{\lessdot}} \xrightarrow{+HCN} C_6H_5-C\overset{H}{\underset{CN}{-}}OH \xrightarrow[-NH_3]{+2H_2O} C_6H_5-CH-COOH$$
$$\underset{OH}{|}$$

Benzaldehyde

DL-Mandelic acid
(m.p. 120°C [395 K])

The racemic form can be resolved by means of (+)-cinchonine into D(−)- and L(+)-mandelic acids, m.p. 133°C (406 K). The former occurs in amygdalin.

5.9.2.3 Benzilic acid (hydroxy-diphenylacetic acid), $(C_6H_5)_2C(OH)-COOH$, can be regarded as a phenyl derivative of mandelic acid and can be prepared in the following way.

Benzilic acid rearrangement. When benzil is heated with an aqueous-ethanolic solution of potassium hydroxide an addition product is first formed. This rearranges almost quantitatively to potassium benzilate, which with acid gives free benzilic acid:

$$\begin{array}{c} C_6H_5-C=O \\ | \\ C_6H_5-C=O \end{array} \underset{}{\overset{+KOH}{\rightleftharpoons}} \begin{array}{c} C_6H_5-C\overset{O^\ominus K^\oplus}{\underset{OH}{\lessgtr}} \\ | \\ C_6H_5-C=O \end{array} \xrightarrow{} \begin{array}{c} C_6H_5 \\ \diagdown \\ C \\ \diagup \diagdown \\ C_6H_5 COO^\ominus K^\oplus \end{array}$$

Benzil   Potassium benzilate

The *mechanism* of this reaction resembles that of the *Cannizzaro* reaction (see p. 199 f.). In the first step a *complex* is formed between benzil and the metal hydroxide, in which the carbonyl groups become more strongly polarised, thereby making them more susceptible to *nucleophilic* attack by a hydroxide ion. In the intermediate adduct which is formed, the negatively charged oxygen atom exerts a strong +I-effect which is greater than the −I-effect of the hydroxy group. The resultant increase in electron density on the carbon atom weakens the bond to the phenyl group, and the latter group, together with its bonding electrons, migrates to the neighbouring positively charged carbon atom of the carbonyl group.

Initial complex (Me = metal)

The benzilic acid rearrangement, which was discovered by *J. Liebig* (1838), also takes place with aliphatic, alicyclic and heterocyclic 1,2-diketones.

Benzilic acid crystallises in needles, m.p. 150°C (423 K), which dissolve in conc. sulphuric acid to give a carmine-red solution. When 1 mol of benzilic acid reacts with 2 mol of phosphorus(V) chloride, chlorodiphenylacetyl chloride is formed; when the latter compound is treated with zinc the two chlorine atoms are removed, leaving diphenylketene (*Staudinger*):

$$\begin{array}{c} C_6H_5 \\ \diagdown Cl \\ C \diagdown \\ \diagup C-Cl \\ C_6H_5 \diagup\diagdown \\ O \end{array} + Zn \longrightarrow \begin{array}{c} C_6H_5 \\ \diagdown \\ C=C=O \\ \diagup \\ C_6H_5 \end{array} + ZnCl_2$$

Chlorodiphenylacetyl chloride   Diphenylketene

5.9.2.4 Diphenylketene, $(C_6H_5)_2C{=}C{=}O$ (see ketenes, p. 255 *f.*) is relatively stable. It is best made from benzil monohydrazone (*G. Schroeter*). This is oxidised by mercury(II) oxide to azibenzil (1-diazonio-2-oxo-1,2-diphenylethan-1-ide) which, when heated in the absence of air and moisture, decomposes to form an unstable ketocarbene. This stabilises itself by the migration of a phenyl group and its bonding electrons to the electron-deficient carbene carbon atom, thus forming diphenylketene (*cf. Wolff* rearrangement).

Benzil monohydrazone   Azibenzil   Diphenylketene

Diphenylketene is an orange-yellow liquid which can be distilled *in vacuo* (b.p. 146°C [419 K]/16 mbar [16 hPa]) and which gives yellow crystals when cooled in a freezing-bath. It is readily soluble in unreactive solvents such as ether, benzene or petrol. In contrast to dialkylketenes, diphenylketene shows little tendency to polymerise.

5.9.3 Unsaturated Arylaliphatic Monocarboxylic Acids

5.9.3.1 Cinnamic acid (β-phenylacrylic acid, 3-phenylpropenoic acid),

$$C_6H_5{-}CH{=}CH{-}COOH$$

is the simplest arylaliphatic acid with an unsaturated side-chain. It occurs as the free acid or in the form of esters in essential oils, and in resins and balsams (Peru balsam, Tolu balsam).

Preparation. *Perkin* synthesis.[1] When benzaldehyde reacts with acetic anhydride in the presence of a condensing agent, *e.g.* sodium acetate or pyridine, an aldol-like intermediate is formed. This can lose a molecule of water and be hydrolysed to cinnamic acid and acetic acid:

$$C_6H_5{-}CH{=}CH{-}COOH + H_3CCOOH$$

Cinnamic acid   Acetic acid

The *reaction mechanism* is the same as that involved in condensations of esters. First of all, acetic anhydride and sodium acetate react to set up an *equilibrium* involving the formation of an anion of the anhydride (the *methylene component*). This makes a *nucleophilic* attack on benzaldehyde (the *carbonyl component*). The resultant anion is protonated by the acetic acid present to give an intermediate which, in the presence of acetic anhydride, loses a molecule of water forming a conjugated system in the process. Finally this is hydrolysed to cinnamic acid and acetic acid:

---

[1] See *J. R. Johnson*, the *Perkin* Reaction and Related Reactions, Org. React. 1, 210 *ff.* (1942).

$$H_3C-\underset{\underset{O}{\|}}{C}-O-\underset{\underset{O}{\|}}{C}-CH_3 \underset{-H_3CCOOH}{\overset{+H_3CCOO^{\ominus}}{\rightleftharpoons}} H_3C-\underset{\underset{O}{\|}}{C}-O-\underset{\underset{O}{\|}}{C}^{\ominus}CH_2$$

$$C_6H_5-\overset{H}{\underset{\underset{\cdots}{C}}{\diagdown}} + \underset{\underset{O}{\|}}{\overset{\ominus}{I}CH_2}-\underset{\underset{O}{\|}}{C}-O-\underset{\underset{O}{\|}}{C}-CH_3 \longrightarrow C_6H_5-\underset{\underset{\cdots}{\underset{I\ominus}{I}}}{CH}-CH_2-\underset{\underset{O}{\|}}{C}-O-\underset{\underset{O}{\|}}{C}-CH_3$$

Carbonyl component   Methylene component

$$\overset{+H_3CCOOH}{\underset{-H_3CCOO^{\ominus}}{\longrightarrow}} C_6H_5-\underset{\underset{OH}{|}}{CH}-CH_2-\underset{\underset{O}{\|}}{C}-O-\underset{\underset{O}{\|}}{C}-CH_3 \underset{-H_2O}{\longrightarrow} C_6H_5-CH=CH-\underset{\underset{O}{\|}}{C}-O-\underset{\underset{O}{\|}}{C}-CH_3$$

$$\overset{+H_2O}{\underset{-H_3CCOOH}{\longrightarrow}} C_6H_5-CH=CH-COOH$$

Properties. The resultant cinnamic acid has an (*E*)-configuration. If a solution of it in benzene is irradiated with UV light, it is partly converted into the higher energy (*Z*)-form (allocinnamic acid). The two acids are in equilibrium with each other.

Cinnamic acid (*E*) (*trans*)   Allocinnamic acid (*Z*) (*cis*)
(m.p. 135°C [408 K])   (m.p. 68°C [341 K])

5.9.3.2 If salicylaldehyde is used instead of benzaldehyde in a *Perkin* synthesis, the product is *o-hydroxycinnamic* acid (*coumarinic acid*), which undergoes intramolecular loss of water to form the cyclic ester *coumarin*. Coumarinic acid must therefore have a *Z*-configuration.

(*Z*)-*o*-Hydroxycinnamic   Coumarin (the lactone of
acid   *o*-hydroxycinnamic acid)

Coumarin is responsible for the perfume of woodruff. It melts at 71°C (344 K) and has been shown to be a carcinogen.

The (*E*)-*o*-hydroxycinnamic acid (coumaric acid), m.p. 208°C (481 K) is also found naturally. It does not form a lactone.

The extremely toxic *aflatoxins*, which have been isolated as metabolites (mycotoxins) from various genera of fungi, *e.g. Aspergillus flavus*, and can hence appear in foodstuffs, are derivatives of coumarin. Aflatoxin B_1 is one of the strongest known carcinogens.[1] Other furocoumarins, such as *psoralen* (7*H*-furo[3,2-g][1]benzopyran-7-one), which occurs in figs, celery and parsley, become carcinogenic if they are activated by light.[2]

---

[1] See. *B. Franck*, Mycotoxins from Mold Fungi, Angew. Chem. Int. Ed. Engl. *23*, 493 (1984).
[2] For a consideration of the cancer risks from synthetic and natural toxins, see *B. N. Ames* et al., Misconceptions on Pollution and the Causes of Cancer, Ang. Chem. Int. Ed. Engl. *29*, 1197 (1990).

| Aflatoxin B$_1$ | Psoralen | INN: **Phenprocoumen** |

Coumarin derivatives such as Phenprocoumen (*Marcumar*) are Vitamin K-antagonists. They inhibit the coagulation of blood and have therapeutic use to this end.

5.9.4 Arenedicarboxylic Acids

There are three isomeric arenedicarboxylic acids.

| Phthalic acid | Isophthalic acid | Terephthalic acid |

Phthalic acid and terephthalic acid are of industrial importance.

Each of the benzenedicarboxylic acids can be made by oxidising corresponding dialkylbenzenes, *e.g.*

m-Xylene   Isophthalic acid

5.9.4.1 Phthalic acid (Benzene-*o*-dicarboxylic acid)

Preparation
(a) *By oxidation of o-xylene.*
(b) *By oxidation of naphthalene* with atmospheric oxygen at 380–450°C (650–725 K) in the presence of vanadium(V) oxide as catalyst.

The naphthalene is oxidised in turn to 1,4-naphthoquinone, phthalic acid and phthalic anhydride, from which phthalic acid is obtained by solution in water and reprecipitation by hydrochloric acid (yield 85–88%).

Naphthalene   1,4-Naphthoquinone   Phthalic acid   Phthalic anhydride

Maleic anhydride and p-benzoquinone are formed as by-products.

Properties. Phthalic acid forms monoclinic plates. They do not have a sharp melting point, since they lose water before they melt. If phthalic acid is kept in a molten state for a few days it loses water to give a quantitative yield of phthalic anhydride, which crystallises in needles, m.p. 131°C (404 K).

Uses. Both phthalic anhydride and phthalic acid are used as starting materials in the dyestuffs industry as well as for the preparation of anthraquinone and phthalonitrile (for making phthalocyanins). Esters of phthalic acid and higher alcohols are used as plasticisers for *PVC* (*Palatinole*) and those made from polyhydric alcohols find use as raw materials for the preparation of lacquers and varnishes (see alkyd resins, p. 310).

Phthalimide is made by heating phthalic anhydride and ammonia together under pressure. The two adjacent C=O double bonds lead to a weakening of the N—H bond, and as a result phthalimide is *acidic*. Thus it dissolves in ethanolic potassium hydroxide to form the potassium derivative of phthalimide, which finds use in the *Gabriel synthesis*.

Phthalic anhydride   Phthalimide   Potassium salt
                                                 of phthalimide

5.9.4.2 Terephthalic acid (Benzene-*p*-dicarboxylic acid)

The term 'tere' came from oil of turpentine (Ger. *Terpentinöl*) from which terephthalic acid was first obtained by oxidation.

It is of great commercial importance for the preparation of polyester fibres.

Preparation
(a) *By oxidation of p-xylene.*
(b) By *carboxylation of benzoic acid*, in 60% yield, according to the *Kolbe–Schmitt* synthesis (see p. 550), at 340°C (615 K) and 300 bar (30 MPa) in the presence of potassium hydrogen carbonate:

Benzoic acid   Terephthalic acid

(c) By thermal isomerisation of dipotassium phthalate at 20 bar (2 MPa) and 400°C (675 K) in carbon dioxide as a protective gas and in the presence of a Zn/Cd catalyst (*Henkel process*). Yield = 95%.

Properties. Terephthalic acid condenses in colourless needles which sublime at about 300°C (575 K). It does not form an anhydride. It is not very soluble in water or in organic solvents.

5.9.4.3   Polyester fibres. Both terephthalic acid itself and its dimethyl ester are used as starting materials for the preparation of synthetic fibres. Dimethyl terephthalate undergoes a transesterification reaction with ethylene glycol to give as a first product bis(2-hydroxyethyl) terephthalate (I). Polycondensation of this provides poly(ethylene terephthalate), (PETP), KRE: [oxyethyleneoxyterephthaloyl], which is drawn from a melt as *polyester fibres* (Terylene, Dacron, Trevira, Grisulten, Diolene).

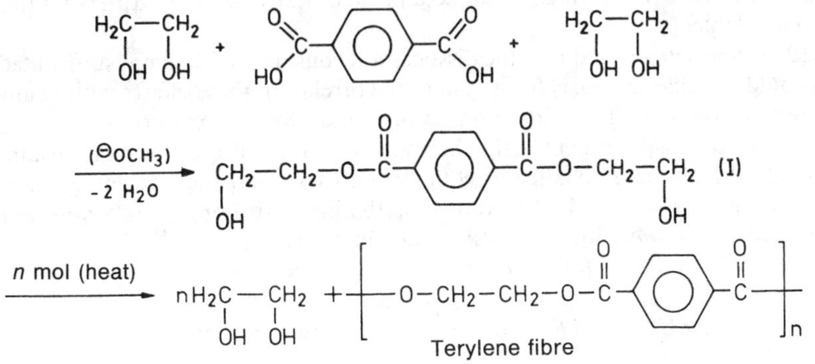

Terylene fibre

These hydrophobic polyester fibres, which were discovered in Great Britain in 1941 (*Whinfield, Dickson*) are at present the most widely produced synthetic fibres. Textiles made from them do not readily crease or need ironing. Moreover the fibres have outstanding tensile strength.

If a diester of carbonic acid, or phosgene, replaces dimethyl terephthalate in the transesterification process, a **polycarbonate** is produced. For example, if, in addition, *bisphenol A* is used instead of ethylene glycol, the following polycarbonate, KRE: [oxy(1,4-phenylene)dimethylmethylene(1,4-phenylene)oxycarbonyl] results:

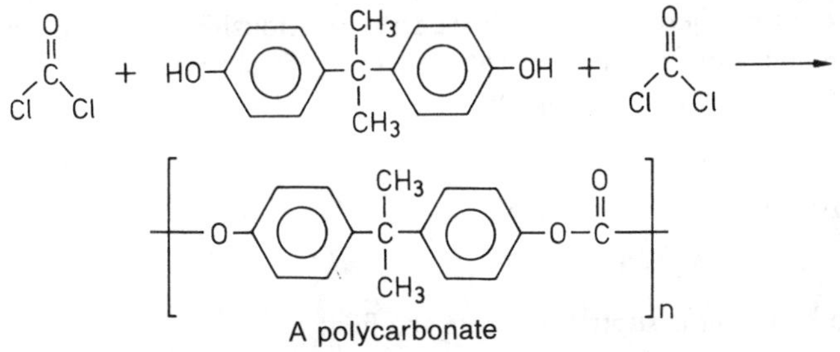

A polycarbonate

Polycarbonates are synthetic materials which have valuable properties (Makrolon); table cutlery, protective coverings and finely precisioned parts can all be made from them.

Self-strengthening liquid crystal (LC) polymers. The preparation of straight-chain polycarbonates can lead to the formation of liquid crystals if their melts are cooled. The macromolecules in them are aligned parallel to one another. This arrangement is preserved by further cooling and causes particular strength in one defined direction whilst the normal strength is retained in other directions.

Polycondensation of terephthaloyl dichloride with *p*-phenylenediamine gives the aryl polyamide poly(*p*-phenyleneterephthalamide).

$$\left[NH-\!\!\bigcirc\!\!-NH-\overset{\displaystyle}{\underset{\displaystyle O}{C}}-\!\!\bigcirc\!\!-\overset{\displaystyle}{\underset{\displaystyle O}{C}} \right]_n$$

Aramid

If this is wet-spun from conc. sulphuric acid the Aramid fibre (Kevlar) is obtained, which surpasses steel wire in its low density, strength and rigidity. Because of this it is used for space travel as well as in the construction of motorcars.

5.9.4.4   The *Hammett* equation. On p. 546 *f.* the pK_a values of benzoic acids were considered in connection with a qualitative consideration of the interaction of substituent groups with the benzene ring. A simple and quantitative way of expressing these effects is also available.

In 1935 *Hammett* showed that the dissociation constants of different substituted benzoic acids could provide the basis for a system of correlating these effects with equilibria and rates of reaction as well as with spectroscopic and physical properties.

The basis for this treatment is the discovery that there is a linear relationship between the standard free energy change at equilibrium (ΔG°) and the free energy of activation ($\Delta G^{\neq}$) of many reactions. This is known as the linear free energy relationship (LFER).[1] It is directly obtainable from the following equations.

$$\Delta G^\circ = -RT \cdot \ln K \quad (K = \text{equilibrium constant})^{[2]}$$

$$\Delta G^{\neq} = -RT \cdot \ln k + \frac{k_B \cdot T}{h} \quad (k = \text{rate constant})^{[2]}$$

In order to test whether a linear free energy relationship holds, the logarithms of the equilibrium constants, or the rate constants, for a series of compounds involved in one type of reaction are plotted against the logarithms of the equilibrium constants of a standard reaction. *Hammett* suggested that the dissociation constants of substituted benzoic acids at 25°C (298 K) should be taken as the standard reaction.

For example, Fig. 115 shows the correlation with the rates of alkaline hydrolysis of substituted ethyl benzoates at 30°C (303 K).

It can be seen that there is a 'linear free energy relationship' for the *m*- and *p*-substituted derivatives (**o**), but not for the *o*-derivatives ($\times$). If the slope of the line is represented by ρ, then the following equation applies:

$$\log k = \rho.\log K + C$$

For the benzoic acids:

$$\log k_0 = \rho.\log K_0 + C$$

If the second equation is subtracted from the first:

$$\log \frac{k}{k_0} = \rho.\log \frac{K}{K_0}; \quad \log \frac{K}{K_0} = \sigma$$

This gives rise to the *Hammett* equation:

$$\log \frac{k}{k_0} = \rho.\sigma; \quad \log \frac{K'}{K'_0} = \rho.\sigma$$

---

[1] Strictly this should be a free *enthalpy* relationship.
[2] R = general gas constant = 8.314 J K^{-1} mol; k_B = *Boltzmann* constant = R/N_A.

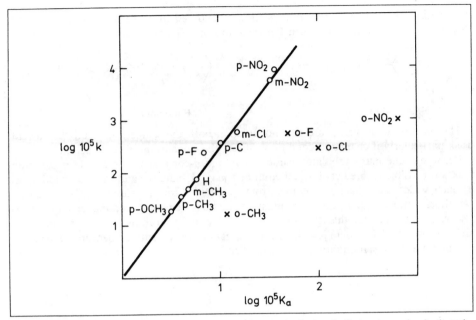

Fig. 115. Correlation between log k for the alkaline hydrolysis of ethyl esters of substituted benzoic acids at 30°C (303 K) in 85% ethanol/water and the dissociation constants of the corresponding benzoic acids in water at 25°C (298 K)

Table 16 lists some σ-values, which are described as *substituent constants*, because they characterise the effects of substituents at the *m*- or *p*-positions on reactions. By definition, hydrogen is assigned a value $= 0$. Substituents which, compared to hydrogen, have electron-donating effects, have negative values $(-\sigma)$, whilst electron-withdrawing substituents have positive values $(+\sigma)$. This relates to the standard reaction, where electron-withdrawing substituents increase the acidity of a benzoic acid and lead to a positive value for $\log\dfrac{K}{K_0}$.

Table 16. *Hammett* substituent constants

Substituent	σ_p	σ_m	Substituent	σ_p	σ_m
NH$_2$	− 0.66	− 0.16	Cl	+ 0.23	+ 0.37
OH	− 0.37	+ 0.12	COOH	+ 0.41	+ 0.36
OCH$_3$	− 0.27	+ 0.12	CHO	+ 0.45	+ 0.36
CH$_3$	− 0.17	− 0.07	COCH$_3$	+ 0.50	+ 0.38
C$_6$H$_5$	− 0.01	+ 0.06	NO$_2$	+ 0.78	+ 0.71
H	0.00	0.00	$^\oplus$N(CH$_3$)$_3$	+ 0.88	+ 0.90
F	+ 0.06	+ 0.34	N$_2^\oplus$	+ 1.91	+ 1.76

Any reaction that can be correlated with the *Hammett* equation provides a characteristic slope which is defined by the reaction constant ρ. ρ is taken to be 1 for the standard reaction. All reactions which are promoted by the presence of electron-withdrawing groups have positive ρ-values. In the case of the alkaline hydrolysis of esters, illustrated in Fig. 115, $\rho = 2.43$. The reaction is more sensitive to the effects of substituents than the dissociation of benzoic acid is. If a reaction is favoured by there being a high electron density at the site of reaction, then the value of ρ is negative:

| Phenoxide anion | | Phenetole |

The reaction constants ρ are dependent upon both the temperature and the solvent used, particularly if change of conditions affects the mechanism of the reaction.

Because of the many tabulated values for ρ- and σ-constants,[1] the *Hammett* equation makes it possible to predict states of equilibria (*e.g.* pk_a-values[2]) and rates of reaction on a much wider scale than is possible from qualitative considerations based on known substituent effects. None the less, many predictions remain uncertain, and one must always be on guard for this. It is not possible here to do more than mention that there are many more systems than just those involving *m*- and *p*-substituted benzene derivatives, to which the *Hammett* equation can be applied.[1,3]

5.10 Reduction Products from Aryl Nitro-compounds

Like aliphatic nitro-compounds, aryl nitro-compounds can be reduced in solution in a mineral acid to give primary amines. For example, nitrobenzene is converted, *via* a series of intermediate compounds, into aniline:

| Nitrobenzene | Nitrosobenzene | *N*-Phenylhydroxylamine | Aniline |

In contrast, if the reduction is carried out in either weakly acidic or neutral conditions, it comes to a halt at the *N*-phenylhydroxylamine stage. In alkaline media the end-product is a bimolecular reduction product, hydrazobenzene, C_6H_5—NH—NH—C_6H_5. Thus the extent to which reduction of aryl nitro compounds proceeds depends very much upon the pH of the solution.

---

[1]See *N. B. Chapman* et al. (Eds), Correlation Analysis in Chemistry, Recent Advances in Linear Free Energy Relationships (Plenum Press, London 1978).
[2] See *D. D. Perrin*, pk_a Predictions for Organic Acids and Bases (Chapman and Hall, London 1981).
[3] See *J. Hine*, Structural Effects on Equilibria in Organic Chemistry (Wiley, New York 1975); *J. Shorter*, Die *Hammett*-Gleichung — und was daraus in fünfzig Jahren wurde (Fifty years of the *Hammett* equation), Chemie in unserer Zeit *19*, 197 (1985).

5.10.1 Reduction in Solution in Mineral Acid

5.10.1.1 Aniline, $C_6H_5-NH_2$, which is the simplest arylamine, was first obtained by *reducing nitrobenzene* with tin and hydrochloric acid (*Zinin*, 1841). The present-day method of manufacturing aniline is based on this process. Industrially nitrobenzene is reduced by iron and water with one fortieth of the calculated required amount of hydrochloric acid present. The hydrogen needed comes for the most part from the water; the iron becomes Fe_3O_4:

$$4\,C_6H_5-NO_2 + 9\,Fe + 4\,H_2O \xrightarrow{\text{(HCl)}} 4\,C_6H_5-NH_2 + 3\,Fe_3O_4$$

At the end, the reaction mixture is neutralised with calcium oxide and the aniline is separated out from it by steam distillation. The yield is to all intents quantitative.

In the laboratory it is still customary to use tin and hydrochloric acid to bring about the reduction of aryl nitro compounds.

In recent times, *catalytic hydrogenation of nitrobenzene* in the gas phase at normal pressure over a fluidised-bed copper catalyst has partially taken the place of reduction using iron:

$$C_6H_5-NO_2 \xrightarrow[-2\,H_2O]{+6\,H(Cu)} C_6H_5-NH_2 \quad \Delta H = -493\,\text{kJ/Mol}$$

Because of their great technical importance arylamines are considered in detail in a later section.

5.10.2 Reduction in Weakly Acidic or Neutral Solution

5.10.2.1 *N*-**Phenylhydroxylamine**, C_6H_5-NHOH, is the first isolable product which can be obtained from the reduction of nitrobenzene.

Preparation
(a) By *reduction of nitrobenzene* with zinc dust in aqueous ammonium chloride solution:

$$C_6H_5-NO_2 + 4\,H \longrightarrow C_6H_5-NHOH + H_2O$$
Nitrobenzene   *N*-Phenylhydroxylamine

(b) By *cathodic reduction of nitrobenzene* in acid solution (10% sulphuric acid) using lead, mercury, copper or graphite electrodes.

This method is also applicable to derivatives of nitrobenzene and provides the corresponding substituted phenylhydroxylamines in high yields. The first intermediate, formed at the electrode by uptake of an electron, is the radical anion of nitrobenzene, $[C_6H_5NO_2]^{\cdot\ominus}$.

Properties and Uses. *N*-Phenylhydroxylamine forms colourless, lustrous needles, m.p. 83°C (356 K), which dissolve in water, ethanol or ether. It is a neutral compound which cannot be kept for any length of time. Like hydroxylamine itself, it reduces ammoniacal silver nitrate and *Fehling's* solution.

If an ethereal solution of *N*-phenylhydroxylamine is treated with dry ammonia and n-butyl nitrite, the product is the ammonium salt of *N*-nitroso-*N*-phenylhydroxylamine, which, under the name 'Cupferron', is used both to test for the presence of, and to estimate quantitatively the amounts of, metal ions such as copper, iron and titanium with which it forms *metal complexes*:

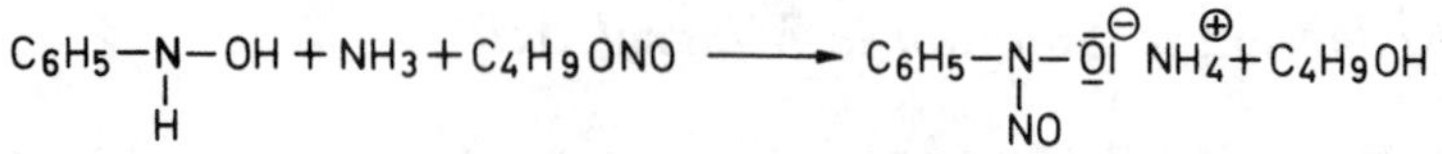

$$C_6H_5-\underset{\underset{H}{|}}{N}-OH + NH_3 + C_4H_9ONO \longrightarrow C_6H_5-\underset{\underset{NO}{|}}{N}-\overset{\ominus}{\underset{}{O|}}\ \overset{\oplus}{N}H_4 + C_4H_9OH$$

The complex with copper has the following structure:

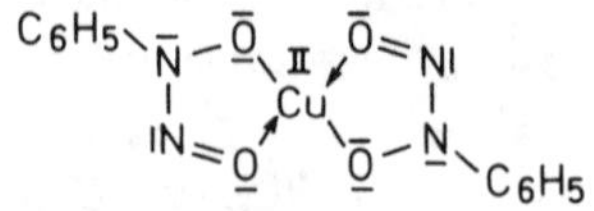

In solution in strong mineral acid *N*-phenylhydroxylamine undergoes a rearrangement reaction to form *p*-aminophenol (*Bamberger*, 1894):

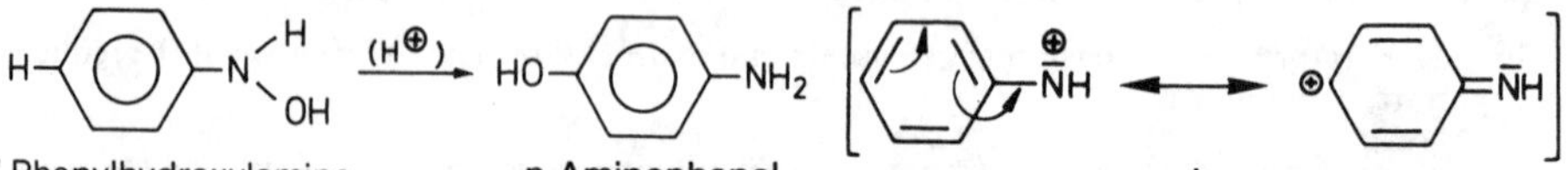

N-Phenylhydroxylamine   *p*-Aminophenol   I

The rearrangement is an intermolecular reaction and proceeds *via* the intermediate cation I which is formed by loss of water from protonated *N*-phenylhydroxylamine and then undergoes nucleophilic attack by water at the *p*-position.

5.10.2.2   *p*-Aminophenol, m.p. 186°C (459 K), is prepared industrially by electrolytic reduction of nitrobenzene in conc. sulphuric acid. Under the name *Rodinal* it, and also *p*-methylaminophenol sulphate (Metol), are used as *photographic developers*.

Aryl Nitroso Compounds

5.10.2.3   Nitrosobenzene, $C_6H_5-N=O$, cannot be isolated directly from the reduction of nitrobenzene, since it is at once reduced further. It is best obtained by *oxidising N-phenylhydroxylamine* with potassium dichromate and sulphuric acid at 0°C (273 K):

$$C_6H_5-NHOH \xrightarrow[-H_2O]{[O]} C_6H_5-N=O$$

N-Phenylhydroxylamine   Nitrosobenzene

Nitrosobenzene forms colourless crystals, m.p. 68°C (341 K). When dissolved, melted or vapourised it becomes green. The colourless solid form is a *dimer*, $(C_6H_5NO)_2$.

X-ray crystal structure determination has shown that the *dimeric* form of nitrosobenzene contains an N—N bond, and has the following structure, which is mesomerically stabilised:

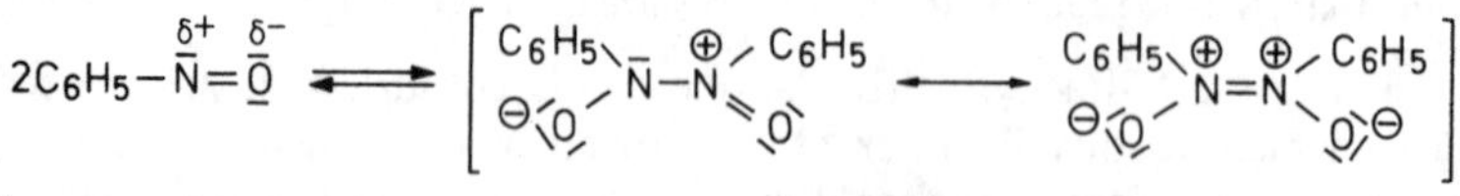

(green or blue)   (colourless or yellow)

It is the nitroso group which is responsible for the colour of the *monomeric form* of nitrosobenzene.

Nitrosobenzene reacts with phenylmagnesium bromide to give *N,N-diphenylhydroxy-lamine*, m.p. 60°C (333 K):

$$C_6H_5-N=O \xrightarrow{+C_6H_5MgBr} C_6H_5-N \overset{OMgBr}{\underset{C_6H_5}{}} \xrightarrow[-Mg(OH)Br]{+H_2O} (C_6H_5)_2N-OH$$

N,N-Diphenylhydroxylamine

5.10.3 Reduction in Alkaline Solution

Reduction of aryl nitro compounds in alkaline conditions leads eventually to the corresponding hydrazo compounds. The bimolecular reduction products arise from a secondary reaction involving the formation of azoxybenzene from a condensation reaction between nitrosobenzene and *N*-phenylhydroxylamine:

$$C_6H_5-N=O+H-\underset{OH}{N}-C_6H_5 \longrightarrow C_6H_5-\overset{\oplus}{N}=\underset{\underset{\ominus}{IOI}}{N}-C_6H_5+H_2O$$

Azoxybenzene

5.10.3.1 Azoxy-compounds

Preparation

(a) By *heating nitrobenzene with methanolic potassium hydroxide*; the methanol is oxidised to formic acid:

$$4C_6H_5-NO_2+3H_3COK \longrightarrow 2C_6H_5-\overset{\oplus}{N}=N-C_6H_5+3HCOOK+3H_2O$$

Azoxybenzene

(b) By *oxidising azobenzene* with hydrogen peroxide in acetic acid:

$$C_6H_5-N=N-C_6H_5+H_2O_2 \longrightarrow C_6H_5-\overset{\oplus}{N}=N-C_6H_5+H_2O$$

Azobenzene   Azoxybenzene

When this reaction was carried out using *p*-bromoazobenzene (*Angeli*, 1906), two *isomeric* azoxy compounds were isolated, in which the oxygen atom was attached by means of a *semipolar* bond (*cf.* amine oxides) to different nitrogen atoms:

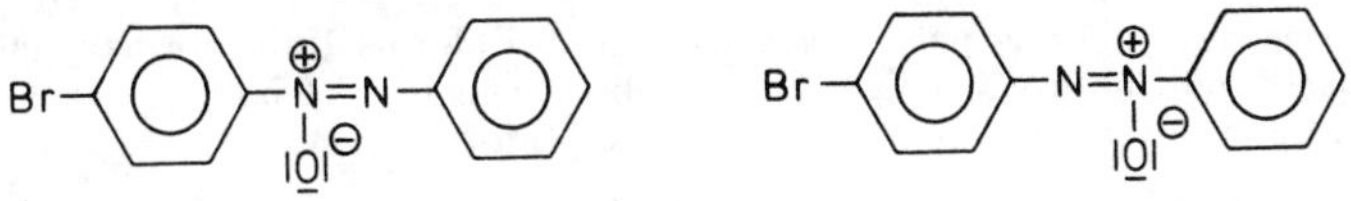

Thus two different azoxy compounds arise when unsymmetric azo compounds are oxidised with hydrogen peroxide.

Properties. Azoxybenzene crystallises in pale yellow needles, m.p. 36°C (309 K). It is a neutral compound and is stable to oxidising agents.

Analogously to the *cis–trans* isomerism associated with compounds having C=C double bonds, or to the *Z*—E isomerism of C=N double bonds, for example in oximes, so compounds having N=N double bonds could be expected to exist in *stereoisomeric* forms. In 1909 *Reissert* isolated two azoxybenzenes with m.p.s 36°C (309 K) and 84°C (357 K). It was shown later, by means of UV spectroscopy (*Szegö*) and by dipole moment measurements (*Eu. Müller*) that these were *E* and *Z* isomers. The two forms are in *equilibrium* with one another; the (*Z*)-form is the more stable:

(*Z*)- or *trans* Form (m.p. 36°C [309 K]
$\mu = 5.67 \times 10^{-30}$ C m)

(*E*)- or *cis* Form (m.p. 84°C [357 K]
$\mu = 15.58 \times 10^{-30}$ C m)

Azoxybenzene

5.10.3.2 Azo-compounds

In these compounds an alkyl or an aryl group is attached to each end of an *azo group*, —N=N—. The simplest aromatic azo compound is azobenzene, the parent compound of the series.

Azobenzene, C_6H_5—N=N—C_6H_5, can be obtained by reduction of nitrobenzene using an appropriate reducing agent. It is formed as a result of secondary reactions involving intermediates in the reduction, for example from the oxidation of hydrazobenzene with excess nitrobenzene:

$$C_6H_5-NH-NH-C_6H_5 + C_6H_5NO_2 \rightarrow C_6H_5-N=N-C_6H_5 + C_6H_5NO + H_2O$$

Preparation

(a) By *reduction of nitrobenzene* with either sodium amalgam or lithium aluminium hydride:

$$2\,C_6H_5-NO_2 + 8\,H \rightarrow C_6H_5-N=N-C_6H_5 + 4\,H_2O$$

(b) By *oxidising hydrazobenzene* with sodium hypobromite solution:

$$C_6H_5-NH-NH-C_6H_5 + NaOBr \rightarrow C_6H_5-N=N-C_6H_5 + NaBr + H_2O$$

(c) By *condensing nitrosobenzene with aniline* in acetic acid:

$$C_6H_5-N=O + H_2N-C_6H_5 \rightarrow C_6H_5-N=N-C_6H_5 + H_2O$$

This latter method is useful because it can be used to prepare unsymmetric azobenzene derivatives.

Properties. Azobenzene forms orange-red leaflets which melt at 68°C (341 K) and boils, without decomposition, at 293°C (566 K). It is fairly volatile in steam. It is almost insoluble in water but dissolves readily in organic solvents.

Like azoxybenzene, azobenzene also exists in (*E*)- and (*Z*)-forms. If the normally more stable (*E*)-form is irradiated with ultra-violet light an equilibrium mixture of both forms is set up which, depending upon the solvent, contains 15–40% of the (*Z*)-form:

(*E*)- or *trans* Form (m.p. 68°C [341 K],
$\mu = 0$)

(*Z*)- or *cis* Form (m.p. 71.4°C [344.6 K],
$\mu = 10.0 \times 10^{-30}$ C m)

Azobenzene

The isomers can be separated by means of adsorption chromatography (*A. H. Cook*).

(*Z*)-Azobenzene can be trapped by passing ketene into an irradiated solution of azobenzene in hexane at 15°C (290 K). An adduct is formed which quickly decomposes again when it is heated (*G. O. Schenk*, 1956):

$$C_6H_5\underset{N}{\overset{N}{\|}}C_6H_5 + \underset{O}{\overset{CH_2}{\underset{\|}{C}}} \longrightarrow C_6H_5\overset{N-CH_2}{\underset{C_6H_5}{\underset{|}{N-C=O}}}$$

(*Z*)-Azobenzene   Ketene   Adduct

A similar addition reaction takes place with diphenylketene.

5.10.3.3 Hydrazo compounds

Hydrazobenzene (1,2- or *sym*-diphenylhydrazine), C_6H_5—NH—NH—C_6H_5, is the end-product of the reduction of nitrobenzene under alkaline conditions.

Preparation

(a) By *reduction of nitrobenzene* with zinc dust and sodium hydroxide:

$$2\,C_6H_5\text{–}NO_2 + 10\,H \longrightarrow C_6H_5\text{–}NH\text{–}NH\text{–}C_6H_5 + 4\,H_2O$$

Industrially nitrobenzene is reduced first to azobenzene, using 8 reduction-equivalents of sodium amalgam, the latter derived from the chlorine-alkali electrolysis process, and then with zinc powder (2 reduction-equivalents) to hydrazobenzene.

Reduction of nitrobenzene can also be carried out electrochemically in aqueous sodium hydroxide.

(b) By *reduction of azoxybenzene*:

$$C_6H_5 - \overset{\oplus}{\underset{\underset{\ominus}{|O|}}{N}}{=}N - C_6H_5 + 4H \longrightarrow C_6H_5 - NH - NH - C_6H_5 + H_2O$$

Properties. Hydrazobenzene forms colourless plates, m.p. 131°C (404 K); their surfaces turn red in air, due to formation of azobenzene. It is insoluble in water but dissolves in alcohols and in ether. Reduction with a powerful reducing agent, namely tin(II) chloride and hydrochloric acid, gives aniline. Oxidation using, for example, iron(III) chloride provides azobenzene.

When heated, hydrazobenzene disproportionates to give azobenzene and aniline:

$$2\,C_6H_5\text{–}NH\text{–}NH\text{–}C_6H_5 \longrightarrow C_6H_5\text{–}N = N\text{–}C_6H_5 + 2\,C_6H_5\text{–}NH_2$$

The free hydrogen atoms which are formed in the dehydrogenation of hydrazobenzene attack a second molecule of hydrazobenzene and split it into two molecules of aniline.

5.10.3.4 Benzidine Rearrangement.[1] The most important reaction of hydrazobenzene is its rearrangement to benzidine (4,4'-biphenyldiamine, 4,4'-diaminobiphenyl), which takes place in acid:

---

[1] See *D. V. Banthorpe*, Top Carbocycl Chem. *1*, 1 (Logos Press, London 1969).

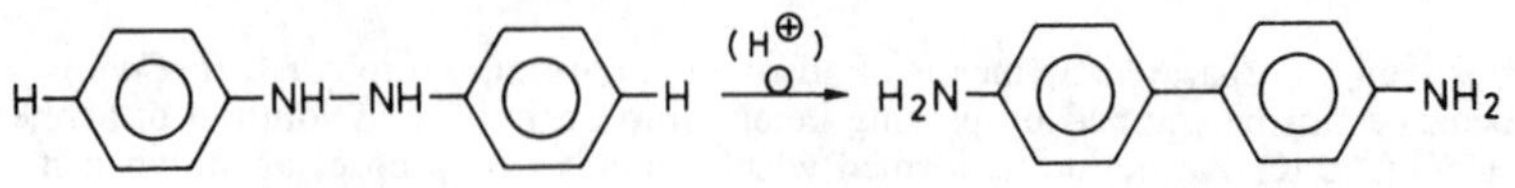

Hydrazobenzene

Benzidine

The *mechanism* of the benzidine rearrangement has been clarified with the help of kinetic isotope effects.[1] As early as 1933 *Ingold* showed that it did not involve splitting of the molecule into two pieces, but that the reaction was *intramolecular*. Protonation first leads to the formation of a diprotonated species which is converted into a *p*-quinonoid structure *via* a transition state involving concerted cleavage of the N—N bond and formation of a C—C bond [(5+5) sigmatropic rearrangement, see p. 499]. The quinonoid structure then loses two protons, leaving benzidine.

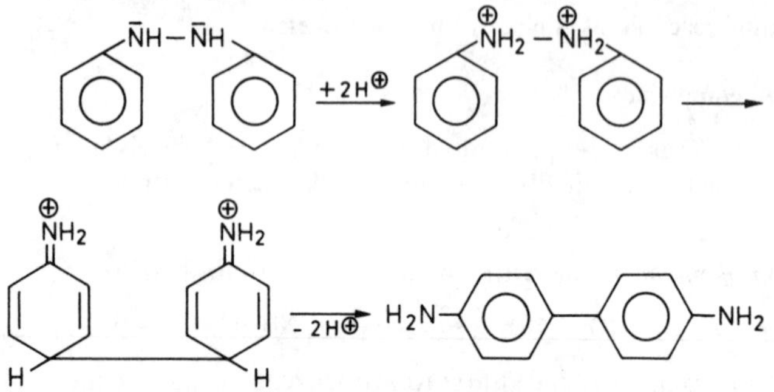

Benzidine was used as the starting material for the preparation of direct dyes (substantive dyes) for cotton (see p. 596 *ff*). It is a strong carcinogen.

The benzidine is accompanied by small amounts of 2,4′-diaminobiphenyl (2,4′-biphenyldiamine) formed in a similar rearrangement involving the *o*-position of one of the rings:

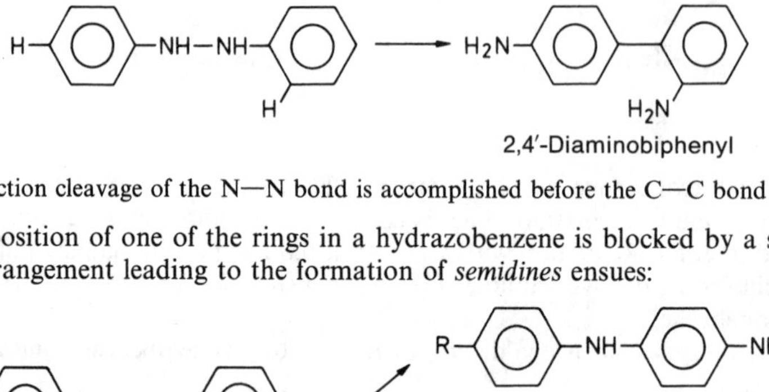

2,4′-Diaminobiphenyl

In this reaction cleavage of the N—N bond is accomplished before the C—C bond is formed[1].

If the *p*-position of one of the rings in a hydrazobenzene is blocked by a substituent, then a rearrangement leading to the formation of *semidines* ensues:

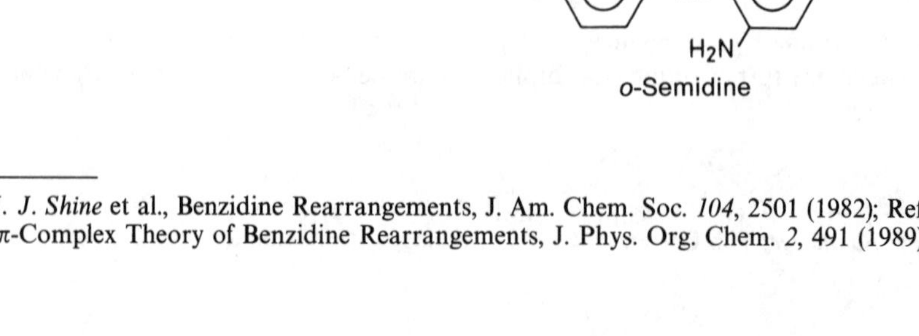

p-Semidine

o-Semidine

---

[1] See *H. J. Shine* et al., Benzidine Rearrangements, J. Am. Chem. Soc. *104*, 2501 (1982); Reflections on the π-Complex Theory of Benzidine Rearrangements, J. Phys. Org. Chem. *2*, 491 (1989).

5.11 Arylamines

In this class of compounds an amino group, or a monoalkyl- or dialkyl- (or aryl-)*amino* group is directly attached to a carbon atom in a benzene ring. As in the case of aliphatic amines, arylamines are classified as *primary, secondary* or *tertiary* amines; they may include mixed alkyl-arylamines or purely arylamines. The following derivatives of aniline serve as examples:

Primary amine	Secondary amines	Tertiary amines
$C_6H_5-NH_2$	$C_6H_5-NH-CH_3$	$C_6H_5-N\begin{smallmatrix}CH_3\\CH_3\end{smallmatrix}$
Aniline	N-Methylaniline	N,N-Dimethylaniline
	$C_6H_5-NH-C_6H_5$	$C_6H_5-N\begin{smallmatrix}C_6H_5\\C_6H_5\end{smallmatrix}$
	Diphenylamine	Triphenylamine

There are three isomeric *aminotoluenes*, toluidines:

o-Toluidine	m-Toluidine	p-Toluidine
(b.p. 200°C [473 K])	(b.p. 203°C [476 K])	(m.p. 44°C [317 K])

They are used as coupling components in the preparation of azo-dyestuffs. The three primary amines derived from xylenes are known as xylidines (or dimethylanilines). o-Toluidine has been shown to be carcinogenic to some animals.

Primary Amines

Almost all primary arylamines are made by reducing the corresponding nitro compounds.

5.11.1 Aniline

Aniline, aminobenzene, $C_6H_5-NH_2$, was first made by *Unverdoben* in 1826 by distilling natural indigo with lime. In 1834 *F.F. Runge* detected it in coal-tar by its reaction with bleaching powder. In 1841 it was rediscovered by *Fritsche* as a decomposition product of indigo and he named it aniline from *anil* (Port, = indigo).

Preparation

5.11.1.1 By *reduction of nitrobenzene in acid media* (see p. 563).

5.11.1.2 Recently it has been made industrially by *ammonolysis of phenol* (Halcon process, Japan).

Properties

Freshly distilled aniline is a colourless liquid, b.p. 184°C (457 K), with an unpleasant smell. It quickly turns brown in air due to autoxidation. It is almost insoluble in water but it is volatile in steam. Its vapour is poisonous and causes feelings of vertigo.

Runge's reaction with bleaching powder serves as a *test for aniline*. On addition of bleaching powder, even trace amounts of aniline (but *not* its salts) give rise to a red-violet colouration. The chemical constitution of this coloured material, which is formed by oxidation of aniline, is still not known.

Like the HO-groups in phenols, amino-groups attached to a benzene ring exert a +M-effect, due to mesomeric interaction of the lone pair of electrons on the nitrogen atom with the π-electrons of the benzene ring:

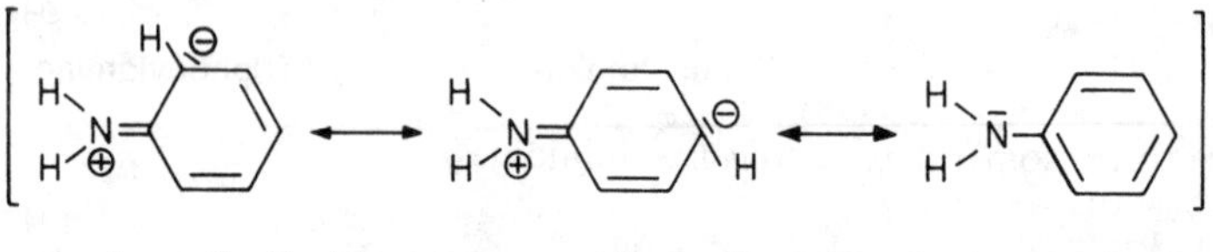

It can be seen from the limiting structures for aniline that the lone pair of electrons on the nitrogen atom, which are responsible for salt formation, are not as freely available to take up a proton as are those in an aliphatic amine. A consequence of this is that arylamines are weaker bases; for example, an aqueous solution of aniline does *not* turn red litmus paper blue.

Aniline and its *N*-methyl derivatives have the following pK_a values:

	Aniline	*N*-Methylaniline	*N,N*-Dimethylaniline
	$C_6H_5-NH_2$	$C_6H_5-NH-CH_3$	$C_6H_5-N(CH_3)_2$
pK_a	4.63	4.85	5.15

Aniline only forms salts with strong mineral acids, and even these are extensively hydrolysed in aqueous solution.

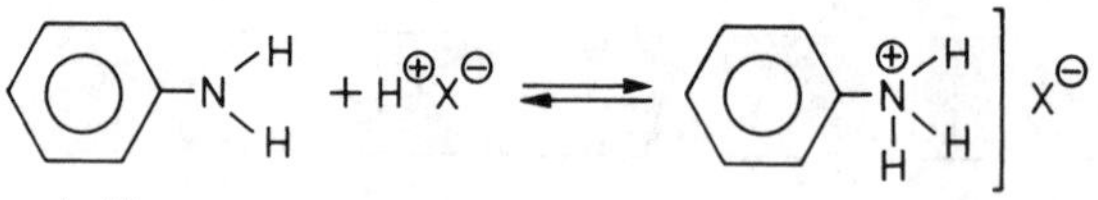

Aniline   Anilinium salt

Conversely, and in the same way that the phenyl group increases the acidity of phenols, it makes the hydrogen atoms of the amino group *more acidic*:

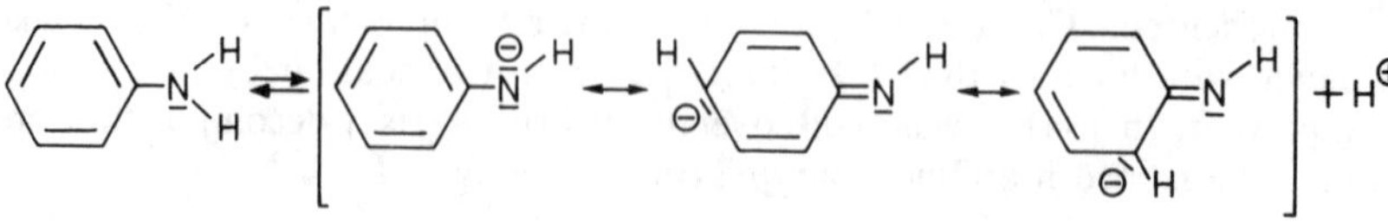

This mesomerism is responsible for the fact that aniline forms stable alkali-metal derivatives, *e.g.*

$$C_6H_5-\bar{N}H_2 + K \rightarrow [C_6H_5-\underset{..}{N}-H]^{\ominus}K^{\oplus} + \tfrac{1}{2}H_2$$

Catalytic hydrogenation of aniline provides *cyclohexylamine*. If this is sulphonated, using chlorosulphonic acid and sodium hydroxide, the product is the sodium salt of cyclohexylsulphamic acid, which is about 35 times as sweet as cane sugar and has been used as a sweetening agent (Cyclamate, Assugrin, Sucaryl). When tested on animals in extremely high doses it was shown to be carcinogenic and for this reason its use was forbidden in the USA,[1] despite the fact that the results did not appear to be reproducible.

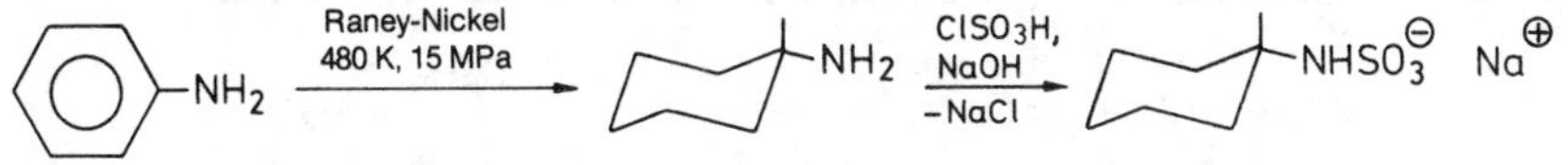

Cyclohexylamine   Cyclamate

5.11.2 Derivatives of Aniline

The many-faceted reactivity of aniline is evident from the following reactions it undergoes:

5.11.2.1 *Acetylation of aniline* by heating it with acetic anhydride (or acetyl chloride or glacial acetic acid) produces acetanilide, m.p. 115°C (388 K):

$$C_6H_5-NH_2 + (H_3CCO)_2O \rightarrow C_6H_5-NH-COCH_3 + H_3CCOOH$$

Acetanilide

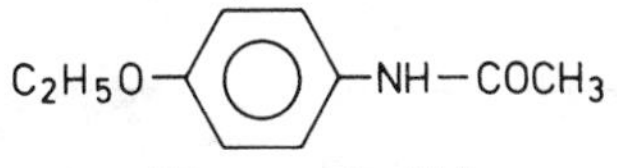

p-Ethoxyacetanilide, INN: **Phenacetin**, is used medicinally because of its analgesic and antipyretic properties. It is increasingly being replaced for this purpose by *p*-hydroxyacetanilide (INN: **Paracetamol**).

p-Ethoxyacetanilide   *p*-Hydroxyacetanilide
(INN: **Phenacetin**)   (INN: **Paracetamol**)

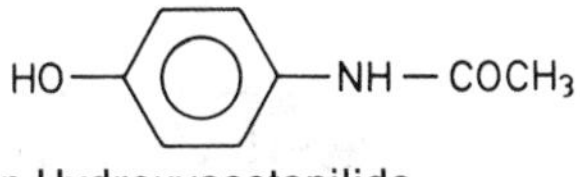

5.11.2.2 Aniline undergoes condensation reactions very readily with most arylaldehydes or arylnitroso compounds. Loss of water provides a *Schiff's* base (azomethine) or an azobenzene derivative respectively. The azomethines are useful for characterising primary arylamines.

5.11.2.3 **Isonitrile reaction**. Like aliphatic primary amines, primary arylamines react with chloroform and potassium hydroxide to form isonitriles which have extremely unpleasant odours, *e.g.*

$$C_6H_5-\bar{N}H_2 + CHCl_3 + 3KOH \rightarrow C_6H_5-\overset{\oplus}{N}\equiv\overset{\ominus}{C}| + 3KCl + 3H_2O$$

Phenylisonitrile (b.p. 166°C [439 K])

(For the reaction mechanism see p. 375).

---

[1] See *F. J. Stare* et al. Food Additives and Health. A Challenge for Nutrition Educators, Food and Nutrition 2, 2 (1976).

5.11.2.4 Like aliphatic primary amines, aniline reacts with carbon disulphide in the cold in the presence of ammonia, giving ammonium N-phenyldithiocarbamate. Treatment of this with lead(II) salts leads to loss of hydrogen sulphide and formation of **phenylisothiocyanate**:

$$S=C=S \xrightarrow{+C_6H_5NH_2, +NH_3} \left[S=C{\overset{\displaystyle \nearrow NHC_6H_5}{\underset{\displaystyle \searrow \underset{|}{S}^{\ominus}}{}}} \right] NH_4^{\oplus} \xrightarrow[-H_2S, -NH_3]{(Pb^{2\oplus})} C_6H_5-N=C=S$$

Ammonium   Phenylisothiocyanate
N-phenyldithiocarbamate   (b.p. 221°C [494 K])

If aniline and carbon disulphide are heated together in ethanolic alkali, hydrogen sulphide is evolved and N,N'-*diphenylthiourea* (thiocarbanilide), m.p. 155°C (428 K) is formed:

$$2C_6H_5-NH_2 + S=C=S \xrightarrow{(OH^{\ominus})} S=C{\overset{\displaystyle \nearrow NH-C_6H_5}{\underset{\displaystyle \searrow NH-C_6H_5}{}}} + H_2S$$

N,N'-Diphenylthiourea

N,N'-Diphenylthiourea is used as an *accelerator* in the *vulcanisation* of rubber.
When heated with conc. hydrochloric acid it loses a mol of aniline to give **phenylisothiocyanate**:

$$S = C(NHC_6H_5)_2 + HCl \longrightarrow C_6H_5-N=C=S + C_6H_5-NH_3]^{\oplus}Cl^{\ominus}$$

Phenylisothiocyanate

Phenylisothioscyanate reacts with ammonia to give N-phenylthiourea, m.p. 154°C (427 K):

$$C_6H_5-N=C=S + NH_3 \longrightarrow C_6H_5-NH-\underset{\underset{\displaystyle S}{\|}}{C}-NH_2$$

N-Phenylthiourea

When solutions of primary arylamines in acid are treated with nitrous acid *diazonium salts* are formed; because of their industrial importance they are dealt with in a separate section.

Substituted Derivatives of Aniline

It has already been mentioned that the hydrogen atoms *o* and *p* to the amino group are readily substituted. In halogenation reactions all three of these hydrogen atoms are replaced by halogens, giving rise to 2,4,6-trisubstituted anilines.

5.11.3 Nitroanilines

All three of the isomeric nitroanilines are used industrially in the preparation of dyestuffs.

o-Nitroaniline   m-Nitroaniline   p-Nitroaniline

Preparation

When arylamines are treated with conc. nitric acid nitration occurs but it is overshadowed by oxidative reactions. Because of this it is customary to protect the amino groups; this is most simply achieved by *acetylating* them.

5.11.3.1 *Nitration of acetanilide* provides *p*-nitroacetanilide as the main product (90%) together with a small amount of *o*-nitroacetanilide. The acetylated amine still directs substitution to the *p*- and *o*-positions. After the two products have been separated the protecting acetyl groups are removed by alkaline hydrolysis:

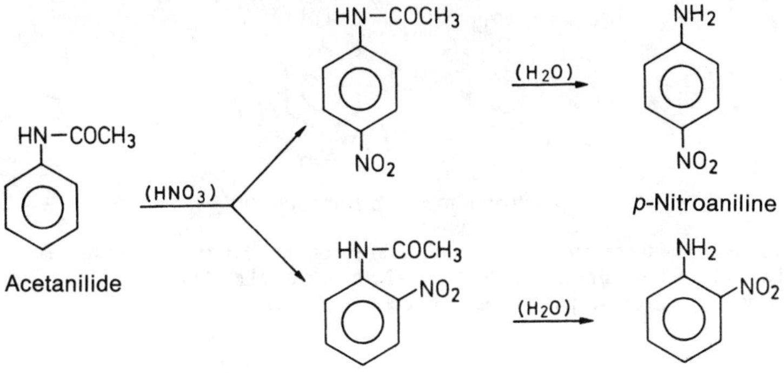

Acetanilide

(HNO₃)

p-Nitroaniline

o-Nitroaniline

5.11.3.2 *m*-Nitroaniline is made industrially by partial reduction, using aqueous sodium sulphide, of the readily available *m*-dinitrobenzene:

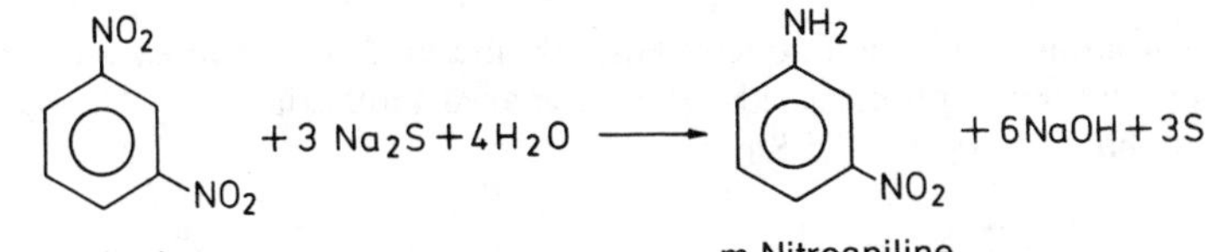

m-Dinitrobenzene $+ 3\,Na_2S + 4\,H_2O \longrightarrow$ *m*-Nitroaniline $+ 6\,NaOH + 3\,S$

5.11.3.3 *o*-Nitroaniline and *p*-nitroaniline are made industrially by heating the corresponding *o*- or *p*-chloronitrobenzene with ethanolic ammonia under pressure, *e.g.*

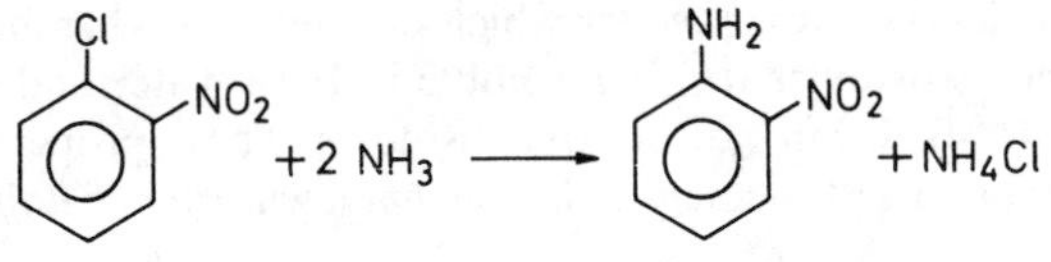

o-Chloronitrobenzene $+ 2\,NH_3 \longrightarrow$ *o*-Nitroaniline $+ NH_4Cl$

Properties

The three nitroanilines are weaker bases than aniline. They dissolve, forming salts, only in excess acid. The basicity increases from the *o*- to the *p*- to the *m*-nitroaniline, as may be seen from the following pK_a values:

	o-Nitroaniline	*p*-Nitroaniline	*m*-Nitroaniline
pK_a	−0.28	0.98	2.45

o-Nitroaniline forms yellow leaflets, m.p. 71°C (344 K), *m*-nitroaniline forms yellow needles, m.p. 114°C (387 K) and *p*-nitroaniline forms pale yellow needles, m.p. 148°C (421 K). The first two of these are volatile in steam. Whilst the nitroanilines have intense yellow colours, their salts are colourless. The deepening of colour (bathochromic effect) caused by the amino group, as compared to nitrobenzene itself, which is almost colourless in its pure form, is lost on salt formation.

o-Nitroaniline and *p*-nitroaniline, but *not m*-nitroaniline, differ from aniline in that, when they are warmed with alkali, the amino group undergoes nucleophilic substitution by a hydroxy group to form the corresponding nitrophenols, *e.g.*

p-Nitroaniline   p-Nitrophenol

The hydrolysis of 2,4,6-trinitroaniline (picramide) to picric acid and ammonia takes place even more readily. This behaviour is due to the −M-effect of the nitro groups. It illustrates the susceptibility of such nitroaryl derivatives to nucleophilic attack.

5.11.4 Anilinesulphonic acids

5.11.4.1 Sulphanilic acid (*p*-aminobenzenesulphonic acid) is the most important of the three isomeric anilinesulphonic acids. It is prepared industrially by 'baking' anilinium hydrogen sulphate at 200°C (475 K):

Sulphanilic acid

Sulphanilic acid is a crystalline powder which decomposes when heated to 280–300°C (550–575 K). It is noteworthy that it is insoluble in both water and organic solvents. It is soluble in aqueous alkali but not in aqueous acids. These properties derive from the fact that a special type of salt, namely a *dipolar ion* (*zwitterion, betaine*) is involved.

Sulphanilic acid (a betaine)   Sodium sulphanilate
insoluble in water   soluble in water

Sulphanilic acid is one of the components used in the manufacture of dyestuffs.

Sulphanilamide (*p*-aminobenzenesulphonamide) is made industrially by treating *p*-acetylamino-benzenesulphonyl chloride (which is prepared from acetanilide and chlorosulphonic acid) with aqueous ammonia, followed by hydrolysis of the acetyl group, usually under alkaline conditions.

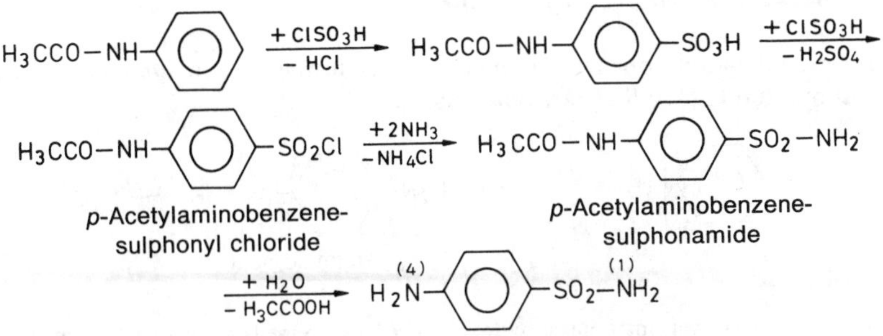

p-Acetylaminobenzene-
sulphonyl chloride

p-Acetylaminobenzene-
sulphonamide

INN: **Sulphanilamide**

INN: *Sulphanilamide*, but especially other related sulphonamides, are *chemotherapeutic agents* which are active against various sorts of bacterial infections; they were first discovered by *Domagk* et al. The sulphonamides are N^1-substituted derivatives of sulphanilamide. The substituents are for the most part heterocyclic rings.

There are some important sulphonamides which have pyrimidine rings as substituents, for example INN: *Sulphisomidine*, also called *Aristamide*. If the methyl groups attached to its pyrimidine ring are replaced by methoxy groups, then the longer-acting material INN: *Sulphadimethoxin* (*Madribon*) results. A derivative which is not reabsorbed and is therefore active against intestinal infections is INN: *Sulphaguanidine*.

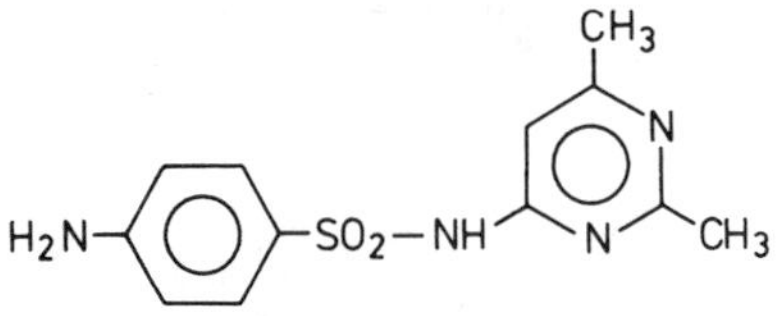

INN: **Sulphisomidine**

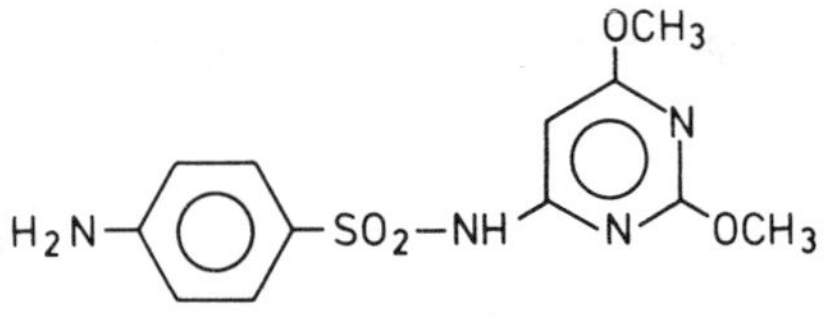

INN: **Sulphadimethoxin**

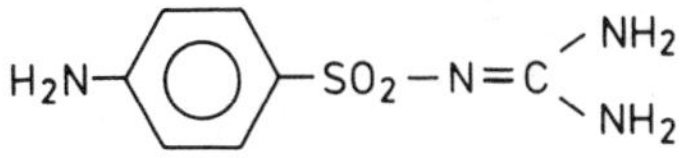

INN: **Sulphaguanidine**

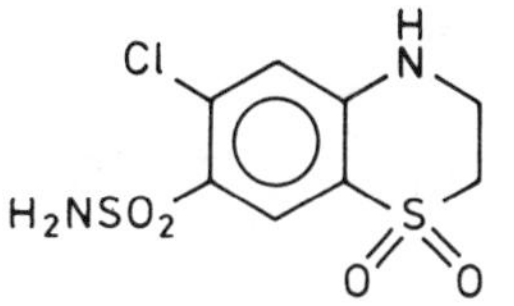

INN: **Hydrochlorothiazide**

The bacteriostatic action of the sulphonamides depends upon the sulphonamide displacing *p*-aminobenzoic acid in reactions involved in the metabolism of the bacteria. The sulphonamide thus acts as a competitive antagonist to *p*-aminobenzoic acid, *i.e.* as an antimetabolite, thereby inhibiting the growth of the bacterial organisms (*Wood–Fildes* theory).[1] The process is called competitive inhibition of an enzyme-catalysed reaction. The sulphonamides can thus also be regarded as 'bacterial antivitamins', see p. 800.

Some sulphonamides, *e.g.* INN: *Hydrochlorothiazide* (*Esidrix*) cause increased excretion of sodium and chloride ions into the urine and are used as *saluretics*.

5.11.4.2 Orthanilic acid (aniline-2-sulphonic acid, 2-aminobenzenesulphonic acid) and metanilic acid (3-aminobenzenesulphonic acid) are used industrially in the manufacture of dyestuffs.

---

[1] See *P. G. Sammes*, Comprehensive Medicinal Chemistry, Vol. 2, 258 (Pergamon, Oxford 1990).

5.11.5 Arsenic Derivatives of Aniline

p-Arsanilic acid (4-aminobenzenearsonic acid) is formed when *anilinium arsenate is heated* to about 200°C (470 K) (*Béchamp*, 1863):

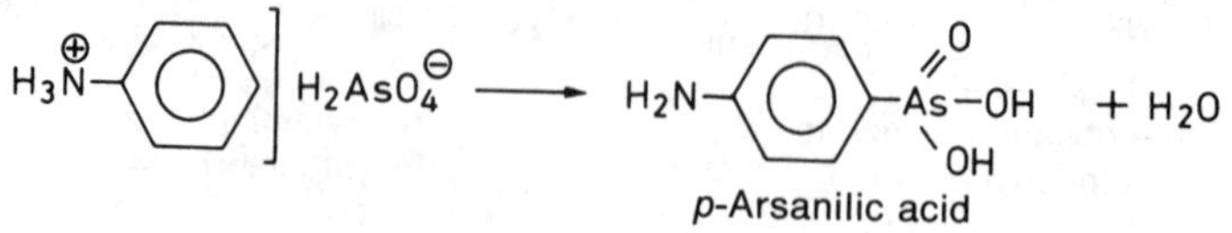

p-Arsanilic acid

Its sodium salt (Atoxyl) was introduced by *Robert Koch* as a remedy for tropical sleeping-sickness. A necessary detoxification of atoxyl is achieved by acetylating it to form *Arsacetin*.[1]

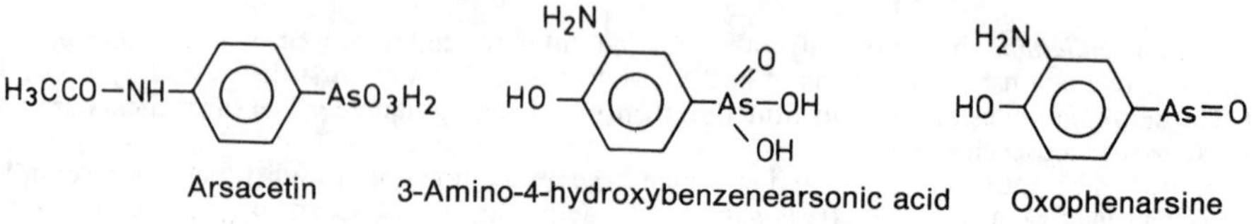

Arsacetin   3-Amino-4-hydroxybenzenearsonic acid   Oxophenarsine

3-Amino-4-hydroxybenzenearsonic acid is the starting material from which *salvarsan* (*Paul Ehrlich*, 1910) is made. Salvarsan is effective against spirochetes and trypanosomes and has been used above all in the treatment of syphilis. It exists in a cyclic trimeric form as shown below.

The action of salvarsan is not due to the compound itself, but to an oxophenarsine, 3-amino-4-hydroxy-oxo-phenylarsine, which is formed from it in the organism. Salvarsan is no longer used now.

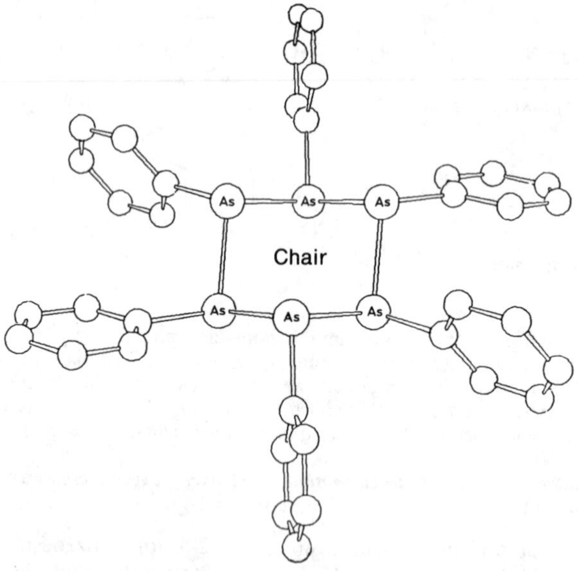

Chair

Trimeric salvarsan, X-ray structure of the σ-skeleton

---

[1] See *O. Dann*, Die Entwicklung der Arzneimittel gegen Schlafkrankheit und andere Trypanosomen-Infektionen (The Development of Remedies for Sleeping-Sickness and other Trypanosome Infections), Dtsch. Apoth. Ztg. *108*, 1595 (1968).

Secondary and Tertiary Amines

5.11.6 N-*Alkylanilines (Alkylarylamines)*

Preparation

5.11.6.1 The hydrogen atoms of the amino group in aniline can be *alkylated*[1] in the same way as those in alkylamines, *e.g.*

$$C_6H_5-NH_2 + ClCH_3 \rightarrow C_6H_5-NH-CH_3 + HCl$$
N-Methylaniline

N,N-Dimethylaniline and the quaternary salt trimethylanilinium chloride are also formed.

5.11.6.2 Industrially aniline and methanol are heated in an autoclave at 180°C (450 K) with hydrochloric acid or at 230°C (500 K) with sulphuric acid as catalyst. *N*-Methylaniline and *N,N*-dimethylaniline are formed:

$$C_6H_5-NH_2 \xrightarrow[-H_2O]{+ H_3COH} C_6H_5-NHCH_3 \xrightarrow[-H_2O]{+ H_3COH} C_6H_5-N(CH_3)_2$$

N-Ethylaniline and *N,N*-diethylaniline are made in the same way, using ethanol instead of methanol. If a large excess of the alcohol is present, the *N,N*-dialkylaniline is obtained as the main product.

Properties. *N*-Methylaniline and *N,N*-dimethylaniline are liquids, b.p., respectively, 196°C (469 K) and 194°C (467 K). They cannot be separated by distillation. However, *N*-methylaniline can be acetylated and separation achieved in this way. Both amines are slightly more basic than aniline (for pK_a values, see p. 570).

5.11.6.3 If a mixture of *N*-methylaniline and formic acid in toluene is heated, a very good yield of *N*-methylformanilide results. It is a liquid, b.p. 253°C (526 K) which is used as a *formylating agent* in *Vilsmeier–Haack* reactions:

$$C_6H_5-\underset{\underset{CH_3}{|}}{N}-H + HO-C\overset{\diagup H}{\underset{\diagdown O}{}} \longrightarrow C_6H_5-\underset{\underset{CH_3}{|}}{N}-C\overset{\diagup H}{\underset{\diagdown O}{}} + H_2O$$

N-Methylformanilide

5.11.6.4 When *N*-methylaniline is treated with *nitrous acid* a yellow *nitrosamine* is formed, which is reduced by sodium amalgam to 1-methyl-1-phenylhydrazine:

1-Methyl-1-phenylhydrazine

N,N-Dimethylaniline undergoes *electrophilic substitution* at the *p*-position when treated with nitrous acid, the product being *p*-nitrosodimethylaniline:

---

[1] See *R. Stroh* et al., Alkylierung aromatischer Amine (Alkylation of Aromatic Amines), Angew. Chem. *69*, 124 (1957); *S. Patai*, The Chemistry of the Amino Group, p. 290 *ff.* (Interscience, London 1968).

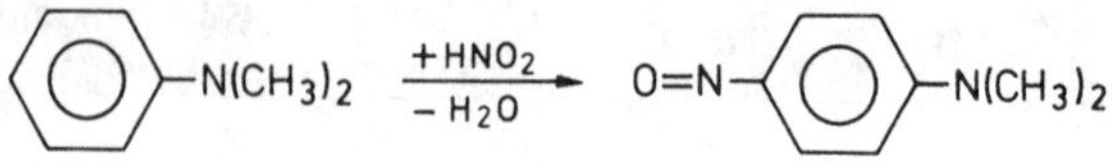

p-Nitrosodimethylaniline

p-Substituted dimethylanilines are nitrosated at an *o*-position.

The hydrochloride of *p*-nitrosodimethylaniline forms orange-red needles while the free nitroso compound forms lustrous green leaflets, m.p. 88°C (361 K). The dimethylamino group is activated by the nitroso substituent; if it is heated with dilute aqueous sodium hydroxide it is hydrolysed to dimethylamine and *p*-nitrosophenol:

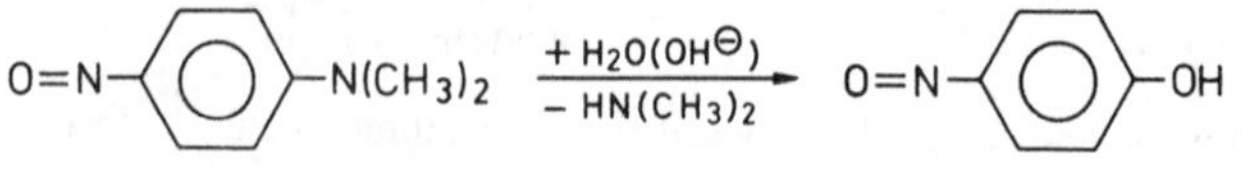

p-Nitrosophenol

Reduction of p-nitrosodimethylaniline provides *p*-aminodimethylaniline, m.p. 41°C (314 K):

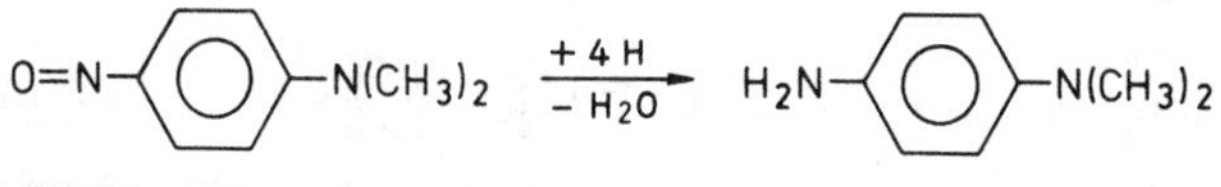

p-Nitrosodimethylaniline   *p*-Aminodimethylaniline

p-Nitrosodimethylaniline is *oxidised* by potassium permanganate to *p*-nitrodimethylaniline, which forms yellow crystals with a steel-like lustre, m.p. 164°C (437 K):

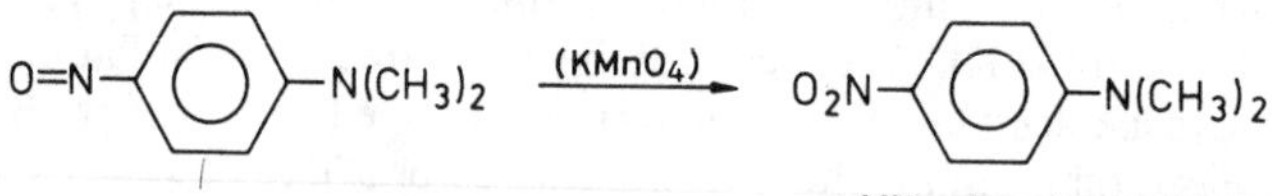

p-Nitrodimethylaniline

Phosgene reacts with *N,N*-dimethylaniline in the presence of zinc chloride to form *Michler's ketone* (4,4'-bis-(dimethylamino)benzophenone), m.p. 179°C (452 K):

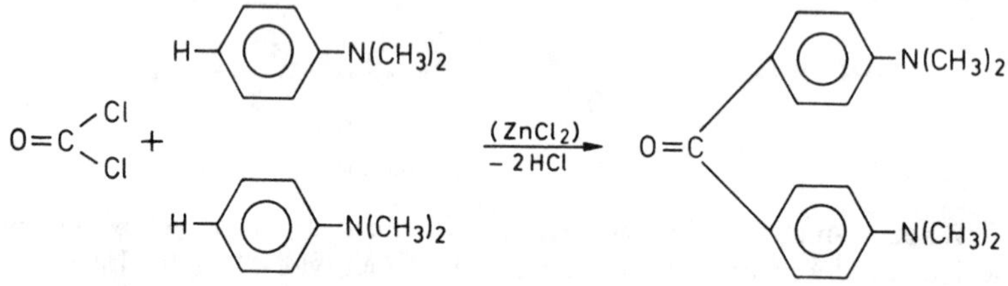

Phosgene   *N,N*-Dimethylaniline   *Michler's ketone*

5.11.7 *Secondary and Tertiary Amines having only Aryl Substituents*

5.11.7.1 Diphenylamine, C_6H_5—NH—C_6H_5, is made by heating together equimolar amounts of aniline and anilinium chloride at about 200°C (475 K) (*phenylation*):

$$C_6H_5-NH_3]^{\oplus}Cl^{\ominus} + C_6H_5-NH_2 \rightarrow C_6H_5-NH-C_6H_5 + NH_4Cl$$

Introduction of a second phenyl substituent onto an ammonia molecule brings about a further lowering of the basicity ($pK_a = 0.8$). Diphenylamine only forms salts with strong acids, and these salts revert completely to the free amine in the presence of water.

Diphenylamine forms colourless leaflets, m.p. 53°C (326 K), which have a pleasant smell. In solution in conc. sulphuric acid it gives an intense blue colour with either *nitric acid* or *nitrous acid*.

In this test reaction, diphenylamine is first of all oxidised to tetraphenylhydrazine. Under the influence of the sulphuric acid this undergoes a *benzidine rearrangement*. The initial product from this rearrangement is then oxidised to *N,N'*-diphenyl-diphenoquinonediimine sulphate (*Kehrmann*):

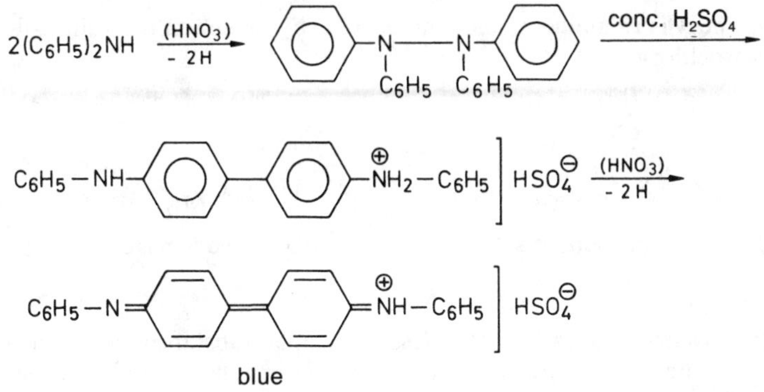

blue

Diphenylamine is used as an anti-oxidant and as a starting material in the preparation of thiazine dyestuffs.

5.11.7.2 Triphenylamine, $(C_6H_5)_3N$, is made by heating diphenylamine with iodobenzene and potassium carbonate in nitrobenzene with some copper-bronze present (*Ullmann reaction*):

$$(C_6H_5)_2NH + C_6H_5I + K_2CO_3 \xrightarrow{(Cu)} (C_6H_5)_3N + KI + KHCO_3$$

It is a colourless crystalline substance, m.p. 127°C (400 K), and is hardly basic at all.

5.11.8 Phenylenediamines

There are three isomeric phenylenediamines:

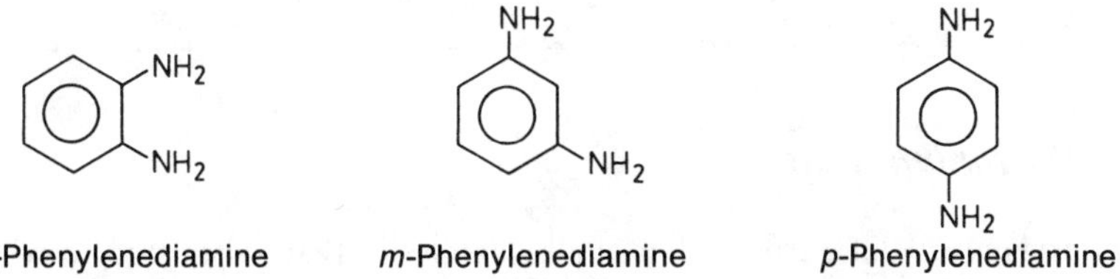

o-Phenylenediamine
(m.p. 104°C [377 K])

m-Phenylenediamine
(m.p. 63°C [338 K])

p-Phenylenediamine
(m.p. 142°C [415 K])

The methods of preparation of phenylenediamines do not differ appreciably from those used for making simple arylamines. They are usually made by reduction of the corresponding dinitro compounds or nitroanilines.

Preparation

5.11.8.1 *o*-Phenylenediamine is best made by *reducing o-nitroaniline* with zinc dust and ethanolic sodium hydroxide:

o-Nitroaniline → o-Phenylenediamine

5.11.8.2 p-Phenylenediamine is made similarly from p-nitroaniline.

5.11.8.3 m-Phenylenediamine is made industrially by reduction of m-dinitrobenzene with iron and hydrochloric acid:

m-Dinitrobenzene → m-Phenylenediamine

Properties

The three phenylenediamines are colourless, crystalline substances which dissolve readily in ethanol or ether. They resemble polyhydroxy-phenols in gradually becoming discoloured in air due to oxidative decomposition, but their salts are stable.

The chemical properties of o-phenylenediamine are particularly affected by the o-arrangement of the two amino groups. This vicinal arrangement of the two groups enables them to participate in condensation reactions which lead to the formation of *heterocyclic* compounds.

Uses

m-Phenylenediamine is a starting material for the preparation of azo-dyestuffs. Asymmetrically dialkylated p-phenylenediamine derivatives, and above all, p-diethylaminoaniline, are used in colour photography as developers.

5.12 Aryl Diazo Compounds

5.12.1 Diazonium Salts

Primary arylamines differ from primary alkylamines in that, when treated with nitrous acid under the appropriate reaction conditions, *i.e.* with sodium nitrite in mineral acid at temperatures about 0°C (273 K), a diazonium salt is formed (*P. Griess*, 1858):

$$Ar-NH_2 + 2HX + NaNO_2 \rightarrow Ar-N_2^{\oplus}]X^{\ominus} + NaX + 2H_2O$$

Diazonium salt (X = Cl, Br, HSO_4, NO_3 etc.)

The name-ending 'onium' is derived from their similarity to ammonium salts, in that they contain a quadruply bonded nitrogen atom which bears a positive charge.

This reaction is known generally as diazotisation. A specific example is the conversion of aniline dissolved in hydrochloric acid into *benzenediazonium chloride*.

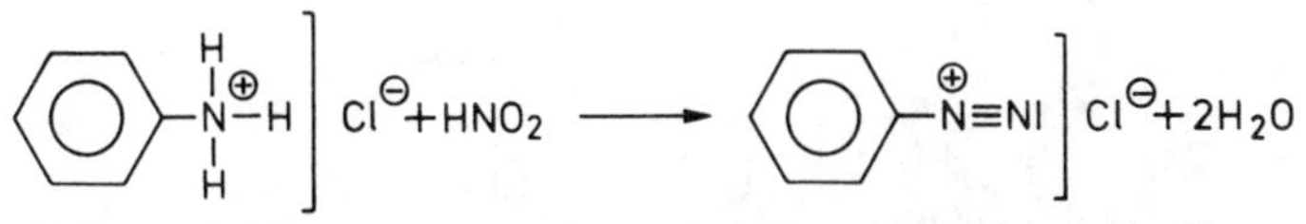

Anilinium chloride   Benzenediazonium chloride

Because of their instability, diazonium salts are not usually isolated, but are used *in situ* in aqueous solution at 0°C (273 K).

Extensive researches by *J. H. Ridd* on the mechanism of diazotisation have shown that the required reaction conditions must be such that a sufficient amount of free amine is present. It was shown that, in contrast to nitration, in which the nitrating agent is the nitronium ion (see p. 487), in diazotisation, depending on the mineral acid present, a different agent is involved. This can be understood from the following equilibria:

$$NO_2^{\ominus} + H_3O^{\oplus} \rightleftharpoons HNO_2 + H_2O$$

$$HNO_2 + H_3O^{\oplus} \rightleftharpoons H_2\overset{\oplus}{N}O_2 + H_2O$$

$$H_2\overset{\oplus}{N}O_2 + X^{\ominus} \rightleftharpoons ONX + H_2O \quad (X^{\ominus} = NO_2^{\ominus}, Cl^{\ominus}, Br^{\ominus}, HSO_4^{\ominus}, SCN^{\ominus} \text{ etc.})$$

In aqueous perchloric acid it is *dinitrogen trioxide*, $ON-NO_2$ ($X = NO_2$), in aqueous hydrogen halides it is the corresponding *nitrosyl halide* ($X = Cl, Br$), and in sulphuric acid, depending upon the water content, either the *nitrosoacidium ion* H_2NO_2, or the *nitrosylsulphate ion*, $ON-OSO_3^{\ominus}$, and only in strongly acidic solution does the *nitrosyl cation*, $NO^{\oplus}$, play a significant role (see p. 161).

Thus diazotisation proceeds as follows:

$$Ar-NH_2 + H_3O^{\oplus} \rightleftharpoons Ar-\overset{\oplus}{N}H_3 + H_2O$$

$$Ar-NH_2 + ONX \rightleftharpoons Ar-\overset{\oplus}{N}H_2-NO + X^{\ominus}$$

$$Ar-\overset{\oplus}{N}H_2-NO + H_2O \rightleftharpoons Ar-NH-NO + H_3O^{\oplus}$$

$$Ar-NH-NO \rightleftharpoons Ar-\bar{N}=\bar{N}-OH \underset{+ H_3O^{\oplus}}{\overset{}{\rightleftharpoons}} Ar-\overset{\oplus}{N}\equiv N| + 2 H_2O$$

An especially active catalyst for *nitrosation* and *diazotisation* is the *thiocyanate ion*, $SCN^{\ominus}$, which is present in human saliva as well as in the gastro-intestinal tract, and could play a part in the formation of *nitrosamines* in the human organism.

Properties

Diazonium salts are colourless, crystalline substances, which darken on exposure to air, and when dry they explode either on being heated or on being struck; this is especially so for diazonium nitrates and perchlorates. Diazonium salts of strong acids are completely dissociated in water and provide a neutral aqueous solution. They thus behave like salts of strong bases.

Double salts made from diazonium salts and zinc chloride, and also diazonium tetrafluoroborates are stable at room temperature:

$$Ar-\overset{\oplus}{N}\equiv N|]_2 ZnCl_4^{2\ominus} \qquad Ar-\overset{\oplus}{N}\equiv N| \, BF_4^{\ominus}$$

They are used as Fast Dye Salts in the manufacture of naphthol-AS dyestuffs.

A further way of stabilising diazonium ions is by their adducts with arylsulphonic acids, especially naphthalenesulphonic acids. From their *infra-red spectra* it appears that they form electron donor–acceptor complexes (*charge transfer complexes*) with the following structure:

5.12.2 Diazotates

Treatment of diazonium salts with alkali changes the diazonium ion into a *diazotate* anion:

$$Ar-\overset{\oplus}{N}\equiv N| \quad \underset{}{\overset{+\,OH^{\ominus}\,(slow)}{\rightleftharpoons}} \quad [Ar-N=N-OH] \quad \underset{}{\overset{+\,OH^{\ominus}(-H_2O,\,fast)}{\rightleftharpoons}} \quad Ar\sim N=N\sim \overline{\underline{O}}|^{\ominus}$$

Diazonium ion   Diazohydroxide   Diazotate anion

First of all, the diazonium ion adds a hydroxide anion, giving a diazohydroxide; however, because there is an equilibrium involving its reaction with a second hydroxide ion, it is only an unstable intermediate and is rapidly converted into a diazotate anion. The properties of diazotate ions depend very much upon the reaction conditions under which they are prepared. The products which precipitate out immediately after mixing the reagents are extremely reactive; in particular they can undergo coupling reactions. Diazotates which only precipitate out after long exposure to excess alkali, sometimes only when heated, behave differently.

These experimental findings led to the suggestion that there are two isomeric diazotates, which, according to *Hantzsch*, could be thought of as *cis(syn)* and *trans(anti)* isomers or, according to *Angeli* as structural isomers:

syn-Diazotate (labile)   *anti*-Diazotate (stable)   *n*-Diazotate (labile)   *iso*-Diazotate (stable)

IR spectra and X-ray structure analysis have shown that the diazotates have the structures suggested by *Hantzsch*.[1]

5.13 Reactions of Arenediazonium Salts

Diazonium salts can react in many different ways and only the most typical reactions are dealt with here. A distinction can be made between reactions in which the nitrogen atoms of the diazonium group are lost and those in which they are retained in the product.

---

[1] See *W. Lüttke* et al., Ber. Bunsenges, Phys. Chem. *67*, 2 (1963).

5.13.1 *Reactions Involving Loss of the Diazo Group*

Whilst diazonium salts are relatively stable at lower temperatures, when heated they lose nitrogen and in the presence of water are converted into *phenols, e.g.*

$$C_6H_5-\overset{\oplus}{N}\equiv N|]\ Cl^{\ominus} + H_2O\ \longrightarrow\ C_6H_5-OH + |N\equiv N| + HCl$$

Phenol

The *course of the reaction* is as follows: The diazonium ion loses nitrogen in an equilibrium reaction and the resultant *phenyl cation* reacts with an available *nucleophile*, in this case a water molecule:

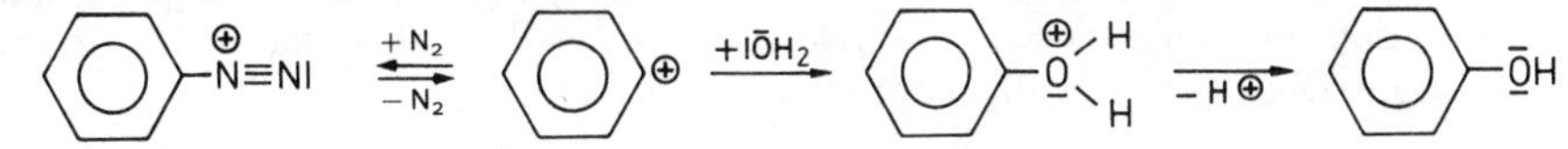

Together with nucleophilic substitution reactions of aryl halides, which take place by an elimination–addition mechanism involving transient formation of an aryne (see p. 481), and substitution reactions of activated benzene derivatives which involve an $S_N Ar$ mechanism (see p. 504), the reactions of diazonium salts now being considered provide another method of achieving nucleophilic substitution in an aryl derivative, in this latter case by *monomolecular* $S_N 1$ mechanism.

The diazonium group can be replaced by a number of other substituents, *e.g.* $I^{\ominus}$, $N_3^{\ominus}$, $H_2AsO_3^{\ominus}$ or $H_2SbO_3^{\ominus}$, but some such substitution reactions in fact involve rather different mechanisms.[1]

$$C_6H_5-\overset{\oplus}{N}\equiv N|]\ Cl^{\ominus} + KI\ \longrightarrow\ C_6H_5-I + N_2 + KCl$$

Iodobenzene

$$C_6H_5-\overset{\oplus}{N}\equiv N|]\ Cl^{\ominus} + NaN_3\ \longrightarrow\ C_6H_5-N_3 + N_2 + NaCl$$

Phenyl azide

$$C_6H_5-\overset{\oplus}{N}\equiv N|]\ Cl^{\ominus} + Na_3AsO_3\ \longrightarrow\ C_6H_5-AsO_3Na_2 + N_2 + NaCl$$

Sodium
benzenearsonate

5.13.1.1   The reaction leading to the formation of benzenearsonic acid is known as the *Bart* reaction (1910).

In concentrated acid solution, chloride and bromide ions can also replace a diazonium group attached to a benzene ring; however this is usually accomplished by means of the *Sandmeyer* reaction.

If benzenediazonium salts are heated in solution in *methanol*, anisole is formed, but in higher alcohols the latter act as reducing agents and the main product is an arene rather than the corresponding aryl ether:

$$C_6H_5-\overset{\oplus}{N}\equiv N|]\ Cl^{\ominus} + R-CH_2-OH \underset{}{\overset{}{\bigg\langle}} \begin{array}{l} C_6H_5-O-CH_2-R + N_2 + HCl \\ C_6H_6 + N_2 + HCl + R-CHO \end{array}$$

[1] See *E. S. Lewis* et al., J. Am. Chem. Soc. *84*, 3847 (1962).

Sulphite or cyanide ions do not react by displacing the diazonium group; instead competing reactions, not involving loss of nitrogen, result in the formation of diazosulphonates or diazocyanides:

$$C_6H_5 - \overset{\oplus}{N} \equiv N|\,] \; Cl^{\ominus}$$

$$\xrightarrow[- \text{NaCl}]{+ \text{Na}_2\text{SO}_3} \quad C_6H_5 - N = N - SO_3^{\ominus} \, Na^{\oplus}$$
Diazosulphonate

$$\xrightarrow[- \text{NaCl}]{+ \text{NaCN}} \quad C_6H_5 - N = N - C \equiv N$$
Diazocyanide

Physical studies carried out by *Le Fèvre* on aryl-substituted diazocyanides indicate that they can exist as *Z–E* (*syn–anti*) isomers.

The replacement of diazonium groups by chlorine or bromine atoms or by cyanide and other groups takes place more satisfactorily if the reaction takes place in the presence of either copper(I) salts or metallic copper.

5.13.1.2   *Sandmeyer* reaction (1884). If, for example, an aqueous solution of benzene-diazonium chloride is run into a warm solution of copper(I) chloride in hydrochloric acid, then *chlorobenzene* is formed:

$$C_6H_5 - \overset{\oplus}{N} \equiv N|\,] \; Cl^{\ominus} \xrightarrow{\text{(CuCl)}} C_6H_5 - Cl + N_2$$

In this reaction the copper(I) chloride acts as a catalyst, forming a 'primary complex' with the diazonium salt; when warmed, this loses nitrogen.[1] In the process, an electron is transferred from copper to nitrogen, which leads to the copper attaining an oxidation state of +2. A *phenyl radical* is formed which reacts with the copper(II) chloride to give chlorobenzene thus regenerating copper(I) chloride.

$$Ar - \overset{\oplus}{N} \equiv N| \xrightarrow{+ \text{CuCl}} Ar - \overset{\oplus}{N} \equiv N| \cdots Cu^I Cl \xrightarrow{-N_2} Ar\cdot + [Cu^{II} Cl]^{\oplus}$$

$$Ar\cdot + Cl - Cu^{II} - Cl \rightarrow Ar - Cl + Cu^I Cl$$

Azo compounds and biphenyl derivatives are always formed as side products.

Bromobenzene can be obtained in the same way by using copper(I) bromide.

The replacement of a diazonium group by a *cyano* group is a special case of the *Sandmeyer* reaction. The diazonium salt solution is treated with copper(I) cyanide which is in solution in the form of a complex with potassium cyanide, *e.g.*

$$C_6H_5 - \overset{\oplus}{N} \equiv N| \xrightarrow[-N_2]{K_3[Cu(CN)_4]} C_6H_5 - C \equiv N$$
Diazonium ion   Benzonitrile

5.13.1.3   If a solution of a diazonium salt is treated with mercury(II) chloride a double salt precipitates out. When a suspension of this in acetone is heated with copper powder it is transformed into an arylmercury chloride (*Nesmeyanov* reaction), *e.g.*

$$C_6H_5 - \overset{\oplus}{N} \equiv N|\,]C^{\ominus} \xrightarrow{+ \text{HgCl}_2} C_6H_5 - \overset{\oplus}{N} \equiv N|\,]HgCl_3^{\ominus} \xrightarrow[-2\text{CuCl}, -N_2]{+2\text{Cu}} C_6H_5 - HgCl$$

5.13.1.4   *Fluorobenzene* cannot be made by a *Sandmeyer* reaction but is obtained by heating dry benzenediazonium tetrafluoroborate (*Balz–Schiemann* reaction, 1934).

---

[1] See *Houben-Weyl*, Methoden der organischen Chemie *5/4*, p.438 *ff.* (1960); *5/3*, p.846 *ff.* (1962) (Thieme Verlag, Stuttgart).

$$C_6H_5-\overset{\oplus}{N}\equiv N|]BF_4^{\ominus} \xrightarrow{\text{heat}} C_6H_5-F + N_2 + BF_3$$

Better yields are obtained if the benzenediazonium tetrafluoroborate is decomposed in acetone at room temperature in the presence of a small amount of copper powder (*E. D. Bergmann*, 1956).

Reduction of diazonium salts to hydrocarbons is best accomplished using either phosphinic acid, phosphonic acid, sodium stannite solution or formic acid, *e.g.*

$$Ar-\overset{\oplus}{N}\equiv N|]HSO_4^{\ominus} + H_3PO_2 + H_2O \longrightarrow Ar-H + N_2 + H_2SO_4 + H_3PO_3$$

Diazonium sulphate

5.13.1.5 *Arylation* of the benzene ring in a diazonium salt can be achieved by stirring it with, for example, benzene to provide an intimate mixture of the two, and then dropping in dilute sodium hydroxide solution (***Gomberg–Bachmann* reaction**). The yield of biphenyl is about 20%.

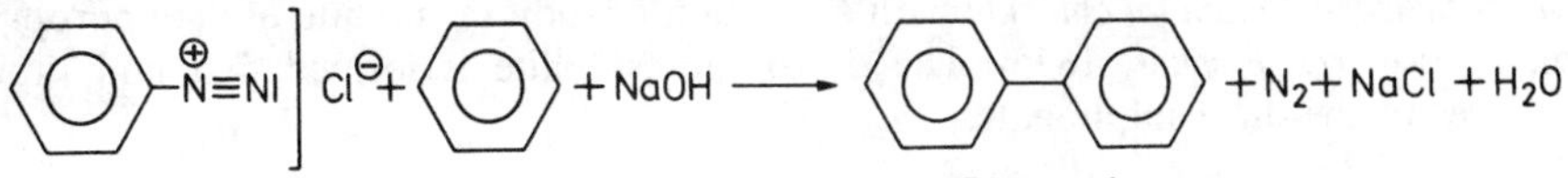

Biphenyl

Reaction proceeds *via* the diazohydroxide, which reacts further by a radical mechanism:

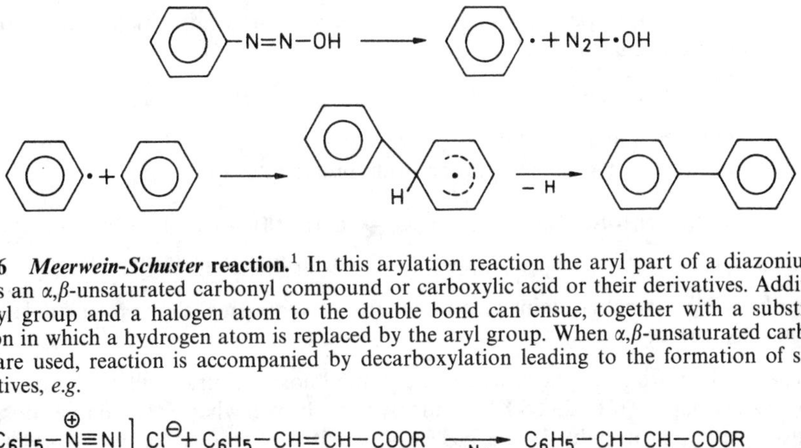

5.13.1.6 ***Meerwein-Schuster* reaction.**[1] In this arylation reaction the aryl part of a diazonium salt attacks an α,β-unsaturated carbonyl compound or carboxylic acid or their derivatives. Addition of the aryl group and a halogen atom to the double bond can ensue, together with a substitution reaction in which a hydrogen atom is replaced by the aryl group. When α,β-unsaturated carboxylic acids are used, reaction is accompanied by decarboxylation leading to the formation of stilbene derivatives, *e.g.*

$$C_6H_5-\overset{\oplus}{N}\equiv N|]Cl^{\ominus} + C_6H_5-CH=CH-COOR \xrightarrow{-N_2} C_6H_5-\underset{\underset{Cl}{|}}{C}H-\underset{\underset{C_6H_5}{|}}{C}H-COOR$$

Cinnamic acid ester   2,3-Diphenyl-3-chloropropionate ester

$$C_6H_5-\overset{\oplus}{N}\equiv N|]Cl^{\ominus} + C_6H_5-CH=CH-CN \xrightarrow{-N_2,-HCl} C_6H_5-CH=\underset{\underset{C_6H_5}{|}}{C}-CN$$

Cinnamonitrile   2,3-Diphenylacrylonitrile

$$C_6H_5-\overset{\oplus}{N}\equiv N|]Cl^{\ominus} + C_6H_5-CH=CH-COOH \xrightarrow{-N_2,-HCl,-CO_2} C_6H_5-CH=CH-C_6H_5$$

Cinnamic acid   Stilbene

---

[1] See *C. S. Rondestvedt*, jun., Arylation of Unsaturated Compounds by Diazonium Salts (The *Meerwein* Arylation Reaction), Org. React. *11*, 189 *ff.* (1960).

These reactions proceed by *radical mechanisms* and are catalysed by both copper(II) and copper(I) salts.

5.13.2 Reactions in which the Nitrogen of the Diazo Group is retained

This group of reactions of arenediazonium salts includes the preparation of phenylhydrazines, phenyl azides and the industrially important coupling reactions (*P. Griess*, 1858).

5.13.2.1 Phenylhydrazine, C_6H_5—NH—NH_2, has already been mentioned several times as a valuable reagent for the identification of aldehydes, ketones and sugars.

Preparation. Phenylhydrazine is prepared industrially as well as in the laboratory by *reducing benzenediazonium chloride* with a solution of sodium sulphite of the appropriate concentration (*E. Fischer*, 1875). The diazonium sulphite is formed first and rapidly rearranges to the diazosulphonate:

$$C_6H_5 - \overset{\oplus}{N} \equiv N|] \ Cl^\ominus \xrightarrow[-NaCl]{+ Na_2SO_3} C_6H_5 - \overset{\oplus}{N} \equiv N|] \ NaSO_3{}^\ominus \longrightarrow C_6H_5 - N = N - SO_3{}^\ominus Na^\oplus$$

Diazonium sulphite   Diazosulphonate

Addition of hydrochloric acid liberates free sulphurous acid, which reduces the azo group, producing sodium phenylhydrazine sulphonate:

$$C_6H_5 - N = N - SO_3^\ominus Na^\oplus + 2\,H \longrightarrow C_6H_5 - NH - NH - SO_3^\ominus Na^\oplus$$

Finally, under the influence of conc. hydrochloric acid and heat, the sulphonyl group is removed and phenylhydrazine hydrochloride precipitates out:

$$C_6H_5 - NH - NH - SO_3^\ominus Na^\oplus + HCl + H_2O \longrightarrow C_6H_5 - NH - \overset{\oplus}{N}H_3]\ Cl^\ominus + NaHSO_4$$

Phenylhydrazine hydrochloride

Free phenylhydrazine is obtained from the salt by treating it with the calculated amount of sodium hydroxide.

Properties. In its pure state phenylhydrazine is an almost colourless oil, which in the cold sets to crystals, m.p. 19.6°C (292.8 K). It quickly turns brown when exposed to atmospheric oxidation, especially in light. It is a base whose salts are readily soluble in water. As a hydrazine derivative it is a strong reducing agent. It can itself be reduced by strong reducing agents, being split in the process to aniline and ammonia.

Phenylhydrazine is a blood toxin; it causes eczema in many persons.
A series of phenylhydrazine derivatives which occur in fungi have been shown to have carcinogenic effects on animals.

Uses. As well as being a reagent for carbonyl groups, phenylhydrazine is an important starting material for the preparation of dyestuffs, and also of indole derivatives, especially in the pharmaceutical industry. (See, for example, antipyrine and pyramidon.)

5.13.2.2 Phenyl azide, $C_6H_5N_3$, is a yellow oil, b.p. 59°C (332 K)/18 mbar (18 hPa), which has a bitter almond-like smell. If it is heated at atmospheric pressure it explodes violently. *Clusius* showed, making use of isotope (^{15}N) labelling, that the *azide group* has the same structure as it does in acid azides. The following limiting structures can be assigned to phenyl azide.

$$[C_6H_5-\overset{\oplus}{N}=N=\overset{\ominus}{\underline{\overline{N}}} \longleftrightarrow C_6H_5-\overset{\ominus}{\underline{\overline{N}}}-N\equiv N|]$$

Its *preparation* from benzenediazonium salts and aqueous sodium azide has already been mentioned (p. 583). It is also available from the *reaction of benzenediazonium perbromide with ammonia*:

$$C_6H_5-\overset{\oplus}{N}\equiv N|]\,Br^{\ominus} \xrightarrow{+Br_2} C_6H_5-\overset{\oplus}{N}\equiv N|]\,[BrBr_2]^{\ominus} \xrightarrow[-3\,NH_4Br]{+4\,NH_3} C_6H_5-\overset{\oplus}{N}=N=\overset{\ominus}{\underline{\overline{N}}}$$

Benzenediazonium bromide   Perbromide   Phenyl azide

Yet another synthesis starts from phenylhydrazine hydrochloride, which, on treatment with nitrous acid at 0°C (273 K), first of all gives nitrosophenylhydrazine which loses water to form phenyl azide:

$$C_6H_5-NH-NH_2 \xrightarrow[-2\,H_2O]{+HNO_2} C_6H_5-\overset{\oplus}{N}=N=\overset{\ominus}{\underline{\overline{N}}}$$

5.13.3 Coupling Reactions

Coupling reactions are those in which an arenediazonium compound is linked or *coupled* to another molecule. The simplest reactions of this sort involve the formation of diazotates and of diazocyanides and diazosulphonates as described above. The most important coupling reactions are those with arylamines and phenols to give *diazoamino* or *azo compounds* (*N-* or *C-coupling*). The course of these reactions depends very much on the pH-value of the solution.

5.13.3.1 Coupling with Arylamines

Coupling of diazonium salts with primary amines in weakly acidic solution leads to the formation of *diazoamino compounds* (**N-coupling**).

Diazoaminobenzene results from the reaction of aniline with benzenediazonium chloride in acetic acid solution:

$$C_6H_5-\overset{\oplus}{N}\equiv N|]\,Cl^{\ominus} + H_2N-C_6H_5 \rightleftharpoons C_6H_5-N=N-NH-C_6H_5 + HCl$$
$$\text{Diazoaminobenzene}$$

In practice, a solution of aniline hydrochloride at 0°C (273 K) is treated with as much sodium nitrite as is necessary to diazotise half of the aniline, and sodium acetate is then added:

$$C_6H_5-\overset{\oplus}{N}\equiv N|]\,Cl^{\ominus} + H_2N-C_6H_5 \xrightarrow[-NaCl, -H_3CCOOH]{+H_3CCOONa} C_6H_5-N=N-NH-C_6H_5$$

Diazoaminobenzene forms golden yellow leaflets, m.p. 100°C (373 K), which explode when heated rapidly. It is insoluble in water, fairly soluble in hot ethanol and readily soluble in benzene or ether. It is weakly basic but with acids does not form stable salts. When heated in dilute mineral acids, it is broken down into phenol, aniline and nitrogen:

$$C_6H_5-N=N-NH-C_6H_5 + H_2O \xrightarrow{(H^{\oplus})} C_6H_5-OH + N_2 + H_2N-C_6H_5$$

A characteristic reaction of diazoamino compounds is their ready *rearrangement* into aminoazo compounds. For example, when diazoaminobenzene (*N,N'*-diphenyltriazine) is heated at about 40°C (310 K) in aniline containing a small amount of aniline hydrochloride, it rearranges to *p*-aminoazobenzene, which is a stronger base:

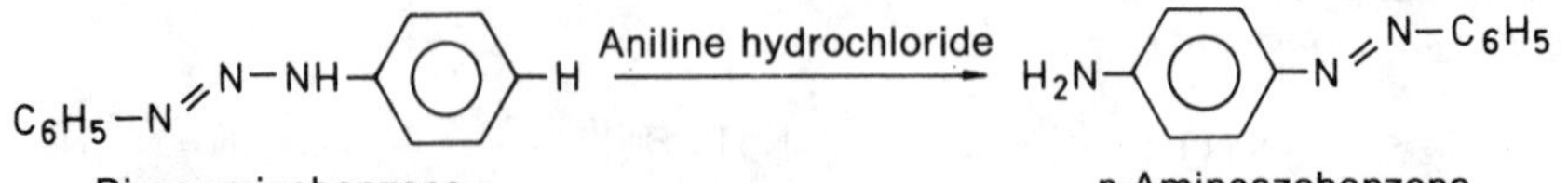

| Diazoaminobenzene | | *p*-Aminoazobenzene |

This is an *intermolecular* rearrangement, *i.e.* the first step is a dissociation into a diazonium salt and aniline, followed by a *C*-coupling reaction.

If the *p*-position of the primary amine already bears a substituent then the product is an *o*-aminoazo compound.

p-Aminoazobenzene crystallises in yellow needles, m.p. 127°C (400 K). It is sparingly soluble in water, ethanol or ether, but readily soluble in boiling benzene.

Strong reduction with tin and hydrochloric acid or, better, with titanium(III) chloride leads to cleavage of the molecule into aniline and *p*-phenylenediamine:

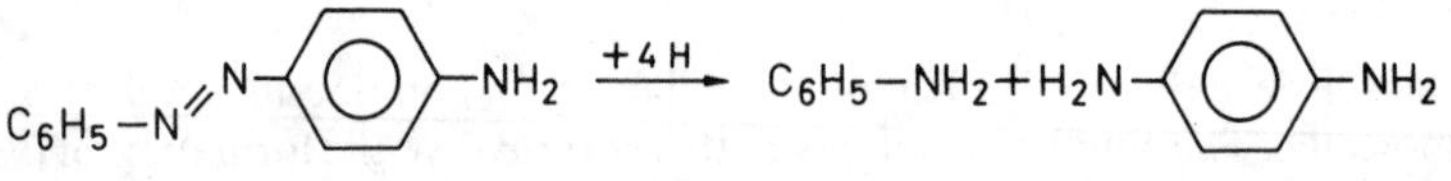

p-Aminoazobenzene   Aniline   *p*-Phenylenediamine

o-Aminoazobenzene forms red prisms, m.p. 59°C (332 K).

When a *secondary* amine, *e.g.* *N*-methylaniline, reacts with benzenediazonium chloride the first-formed diazoamino compound is produced but some of it has already rearranged to the azo compound:

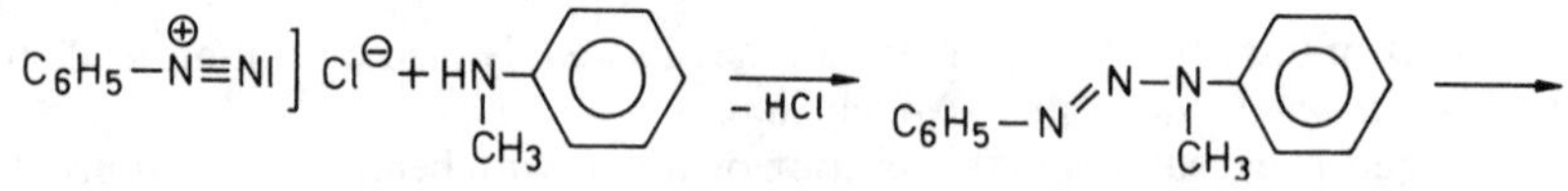

N-Methyldiazoaminobenzene

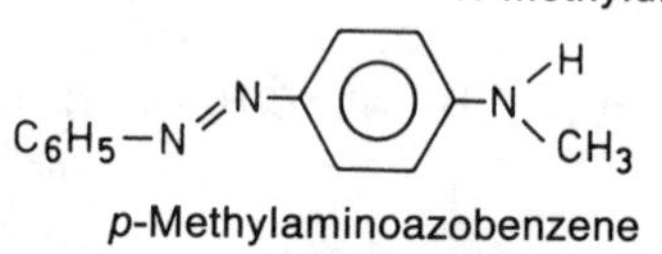

p-Methylaminoazobenzene

In contrast to the reactions just described, *tertiary* amines do not give diazoamino compounds, but rather couple with a diazonium salt in weakly acidic solution directly at the *p*-position to form *p*-aminoazo compounds (**C-coupling**), *e.g.*

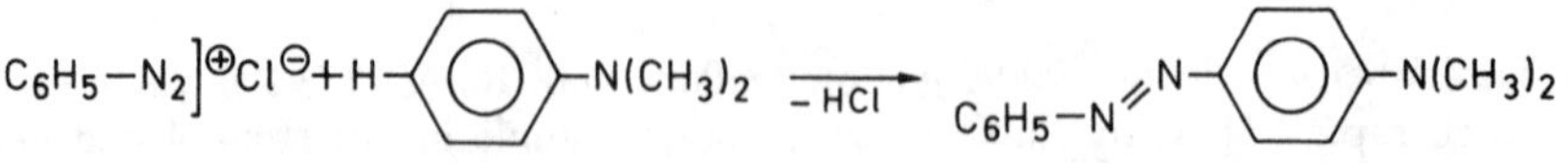

p-Dimethylaminoazobenzene

If the *p*-position in the amine is blocked by a substituent, then coupling takes place instead at the *o*-position.

p-Dimethylaminoazobenzene forms yellow leaflets, m.p. 117°C (390 K). It has been shown to be carcinogenic in some animals.

N-Coupling of diazonium salts to form diazoamino compounds is restricted to only a few primary arylamines, such as aniline, some of its *o*- and *p*-substituted derivatives, and *p*-toluidine. With most primary arylamines, *e.g. m*-toluidine, *m*-phenylenediamine and *β*-naphthylamine, *C*-coupling takes place at once.

5.13.3.2 Coupling with Phenols

The coupling of diazonium salts with phenols (or naphthols) in *alkaline* solution proceeds appreciably faster than do the reactions with comparable amines in acid solution. With phenols also the diazonium group attacks the ring at the *p*-position to the hydroxy group, so long as it is not already substituted, if it is substituted reaction takes place at an *o*-site. The products are *p*- or *o*-hydroxyazo compounds, *e.g.*

$$C_6H_5-\overset{\oplus}{N}\equiv N| \; Cl^\ominus + H-\!\!\!\left\langle\bigcirc\right\rangle\!\!\!-O^\ominus Na^\oplus \longrightarrow C_6H_5-N\overset{N}{\diagup}\!\!\!\left\langle\bigcirc\right\rangle\!\!\!-OH + NaCl$$

Sodium phenoxide	*p*-Hydroxyazobenzene

p-Hydroxyazobenzene crystallises in orange prisms, m.p. 152°C (425 K). Because of its phenolic character it is readily soluble in aqueous ammonia and in aqueous alkali. Certain hydroxyazo-compounds can also be made by a condensation reaction between the appropriate quinone and phenylhydrazine (or its derivatives); thus, *p*-hydroxyazobenzene, for example, could be thought of as a *tautomeric* form of *p*-benzoquinone monophenyl-hydrazone.

$$C_6H_5-NH-N=\!\!\!\left\langle\bigcirc\right\rangle\!\!\!=O$$

o-Hydroxyazobenzene forms orange-red needles with a blue reflex, m.p. 83°C (356 K).

5.13.3.3 Coupling with Reactive Methylene Group[1]

Diazonium salts also couple with compounds such as acetoacetic ester, malonic ester, *etc.*, which have reactive methylene groups. The initially formed azo compounds straightway *tautomerise* to arylhydrazones, *e.g.*

$$Ar-\overset{\oplus}{N}\equiv N| \; Cl^\ominus + H_2C\overset{COOR}{\underset{COOR}{\diagdown}} \;\xrightarrow{-HCl}\; Ar-N\overset{N-CH\overset{COOR}{\diagup}}{\underset{COOR}{\diagdown}} \longrightarrow Ar-NH-N=C\overset{COOR}{\underset{COOR}{\diagdown}}$$

Malonic ester	Arylhydrazone of mesoxalic ester

Further examples of this type of reaction involve coupling reactions with 1-aryl-3- methylpyrazol-5-ones (1-aryl-3-methyl-2-pyrazolin-5-ones) and their derivatives (see pyrazolone dyestuffs).

5.13.3.4 Mechanism of Coupling Reactions

Because of the −I-effect caused by the positively charged nitrogen atom, the terminal nitrogen atom of the diazonium group is electron-deficient. The diazonium group, like a

---

[1] See *S. M. Parmerter*, The Coupling of Diazonium Salts with Aliphatic Carbon Atoms, Org. React. *10*, 1 *ff.* (1959); *E. Enders*, Houben-Weyl, Methoden der Organischen Chemie *10/3*, p. 490 *ff.* (Thieme Verlag, Stuttgart 1965).

$Br^\oplus$ or an $NO_2{}^\oplus$, can hence make an electrophilic attack on an aryl ring. For example, the coupling reaction with a tertiary arylamine can be formulated as follows:

Diazo compound   Coupling partner

p-Dimethylaminobenzene

The coupling of diazonium ions with phenols in alkaline solution follows a similar course.

The rates of these coupling reactions depend very much on two factors, which are strongly affected by substituents attached to the benzene ring. First, the *coupling partner* must be activated by substituent groups such as $-NH_2$, $-NR_2$ or $-O^\ominus$. Second, the electropositive character of the terminal nitrogen atom of the diazo compound is increased by the presence of electron-withdrawing substituents such as $-NO_2$, $-CN$ or SO_3H attached to its aryl group, and this also increases the rate of its coupling reactions. For example, diazotised *p*-nitroaniline and 2,4-dinitroaniline couple much faster than benzenediazonium chloride itself does. Conversely, electron-donating substituents, such as alkyl or alkoxy groups, lower the rate of coupling of diazonium ions.

These coupling reactions find much use in the dyestuffs industry.

5.14 Azo-Dyestuffs[1]

The large group of azo-dyestuffs comprises various derivatives of azobenzene and is characterised by the presence of one or more *azo groups*, $-N=N-$.

5.14.1 Structure and Colour[2]

A range of coloured compounds has already been mentioned, but it is necessary now to distinguish between the concept of *colour* and that of an actual *dyestuff*. A chemical compound appears to us as coloured if it has an absorption band in the visible part of the spectrum (400–800 nm). The colour of the compound which we see represents the corresponding complementary colours to those in the absorption region of the spectrum and is derived from the remaining portion of the spectrum between 400 and 800 nm (see

---

[1] See *H. Zollinger*, Color Chemistry, Syntheses, Properties and Applications of Organic Dyes and Pigments, 2nd edn (VCH Verlagsgesellschaft, Weinheim 1991); *P. F. Gordon* and *P. Gregory*, Organic Chemistry in Colour, 2nd edn (Springer-Verlag, Berlin 1987).
[2] See *J. Fabian* and *H. Hartmann*, Light Absorption of Organic Colorants (Springer-Verlag, Berlin 1980); *J. Fabian* and *R. Zahradnik*, The Search for Highly Coloured Organic Compounds, Ang. Chem. Int. Ed. Engl. *28*, 677 (1989).

Table 17, the cover of the book, and Fig. 116). Thus the impression of 'red' is aroused either by light with a wavelength of 700 nm or by sunlight from which the green component has been filtered out.

In general, saturated organic compounds appear colourless to our eyes because they absorb light in the ultra-violet region, outwith the visible portion of the spectrum. If a π-bonding system is present, e.g. $>C=O$, $>C=N-$, $>C=C<$, $-N=O$ and especially if two or more of these groups are *conjugated* with one another (*e.g.* in glyoxal or in osazones) the absorption bands are shifted more or less into the visible region of the spectrum. This poses the question as to whether there is a relationship between colour and chemical constitution. A unified interpretation of this long-debated problem was provided by the theory of *mesomerism*. According to this, the π-electrons of the multiple bond are responsible for the selective absorption of light and hence for the colour (*Bury, Arndt* and *Eistert*). The greater the delocalisation of the π-electrons, the longer the wavelength of the light which a compound absorbs.

Application of quantum chemical models provides more accurate and more reliable descriptions. With the help of the schematic wave functions in Table 11 (p. 105) and the free electron (FE) model, it is possible to calculate with satisfactory accuracy the wavelengths at which symmetric azomethines absorb (λ_{max}). Fig. 116 (see also cover) illustrates the visible portion of the electromagnetic spectrum as it is defined in Table 4 (p. 30).[1]

When absorption of light takes place an electron is raised from the HOMO to the LUMO; the smaller the energy difference between the two states, the longer the wavelength of the absorption.

Groups of atoms which have a decisive effect on the selective absorption are known as *chromophores* (Grk. *chroma* = colour, *phoron* = bearer) or *chromophoric groups* (*Witt*, 1876). Compounds which contain chromophoric groups are called *chromogens*. Usually a chromophoric group in a molecule is not by itself sufficient to produce absorption of light in the visible region and the absorption is instead in the ultra-violet region. Rather it is necessary to have a number of chromophores, above all conjugated with one another, to bring about a deepening of colour (a *bathochromic effect*), *i.e.* a shift of absorption maxima to longer wavelengths in the visible spectrum (see Fig. 116).

Table 17. Absorption of light and colours

Absorbed light		Colour of compound
Wavelength (nm)	Colour	
400–440	Violet	Yellow-green
440–480	Blue	Yellow
480–490	Green-blue	Orange
490–500	Blue-green	Red
500–560	Green	Purple
560–580	Yellow-green	Violet
580–595	Yellow	Blue
595–605	Orange	Green-blue
605–750	Red	Blue-green
750–800	Purple	Green

---

[1] See *M. Klessinger*, Konstitution und Lichtabsorption organischer Farbstoffe, (Constitution and Light Absorption of Organic Pigments), Chemie in Unserer Zeit *12*, 1 (1978).

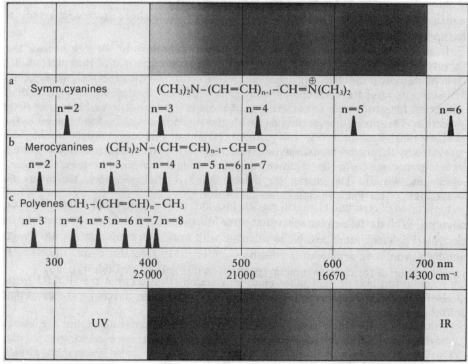

Fig. 116. Relationship of site of the long-wavelength absorption bands to the number of conjugated double bonds in some polymethine compounds. The lower line indicates colours which are absorbed out of the spectrum, the top line the related complementary colours. which are those that we see. This is illustrated in colour on the book cover.

A typical chromophoric group is the *azo group* —N=N—, which is thought of as the essential source of colour in azo-dyestuffs. The chromophoric groups, which are responsible for the essential colour of a substance, do not, however make that substance a dyestuff. Thus red azobenzene, which is the parent compound of the azo-dyestuffs, has a good colour, but it is not a dyestuff. To be a *dyestuff* it must be able to colour other materials such as natural fibres (cotton, wool, silk) or cellulosic or synthetic chemical fibres, or leather, and be resistant to light and not removed by washing. This is first achieved by the introduction of *auxochromes* (Grk. *auxesis* = increase) or *auxochromic groups* (*Witt*) into the aryl part of the coloured molecule. Among the most important auxochromes are the following electron-donating groups: —NH_2, —NHR, —NR_2, —OH, —OCH_3. In this way the simple azo-dyestuffs, *e.g.* p-aminoazobenzene (aniline yellow), p-hydroxyazo-benzene, are derived from azobenzene.

A general distinction can be made between *basic* and *acidic* dyestuffs, depending upon whether H_2N— or HO— groups are substituted in the benzene nucleus of the molecule. The acidic character of an azo-dyestuff is also often the result of the presence of a sulphonyl or a carboxyl group; in addition, this affects its solubility in water. Such groups withdraw electrons and are known as electron-withdrawing or electron-accepting groups.

These groups function as binding groups to attach the dyestuff to the material. Many also cause a *deepening of the colour*. The extent of this *bathochromic effect* depends on the electron-donating ability of the auxochromic group. For example, if the hydrogen atoms of the NH_2-group are replaced by alkyl or aryl groups then the colour is deepened.

On the other hand, the presence of some other substituents, *e.g.* alkyl groups, on the benzene ring of azo-dyestuffs brings about a *hypsochromic effect*, that is, the absorption maxima are shifted to shorter wavelengths.

In an important group of dyestuffs there is electronic interaction between a donor and an acceptor group (donor–acceptor chromogen). In such compounds it is often found that *formation of a salt* leads to deepening of the colour (*halochromism, halochromic salts*). For example, *p*-hydroxyazobenzene is orange but treatment with alkali forms a salt which is red. The deeper colour of the anion is associated with the fact that the lone pair of electrons on the oxygen atom interacts more strongly with the mesomeric system in the case of the salt:

p-Hydroxyazobenzene (orange)

Anion (red)

The conversion of a phenolic hydroxy group into a phenoxide group makes it more strongly electron-donating. The resultant interaction with the otherwise unchanged acceptor group (the benzene nuclei plus the azo group) results in a *bathochromic* shift.

As a result of such colour changes, some azo-dyestuffs, *e.g.* methyl orange, act as *indicators*.

Solvatochromism. The interaction between donor and acceptor groups can be influenced by the solvent, for example, *4-diethylamino-4'-nitroazobenzene* is yellow-orange in cyclohexane (λ_{max} 470 nm) but deep red in ethanol (λ_{max} 510 nm).

4-Diethylamino-4'-nitroazobenzene

The ground state of the compound (governed by the HOMO) is less polar than the excited state (governed by the LUMO); a polar solvent thus has more stabilising effect on the excited state than on the ground state. This decreases the difference in energy between the HOMO and LUMO and results in the absorption maximum shifting to longer wavelength.

Methyl orange (Helianthin, Orange III), the sodium salt of 4-dimethylaminoazobenzene-4'-sulphonic acid, is made by coupling diazotised sulphanilic acid with *N*,*N*-dimethylaniline and converting the acid into its sodium salt. Its pale yellow solution turns red when mineral acid is added to it:

Methyl orange (yellow)

red

Because of this, methyl orange has been used as an *indicator* in acidimetry; it is, however, sensitive to acid and thus has no use as a dyestuff.

In this case the halochromism stems from the fact that protonation increases the electron-withdrawing effect of the acceptor group.

5.14.2 *Dyestuff Technology*[1]

Because of the wide range of diazo compounds and of the compounds with which they can couple, the number of azo-dyestuffs greatly exceeds that of all other classes of dyestuffs. They can provide almost any shade of colour and can be used to dye or to print most textiles.

Azo-dyestuffs containing anionic sodium sulphonate groups are added to protein fibres (wool, silk) in *acid solution* in which the amino-acid units of the fibres are protonated; this also applies to the amino-end groups in polyamide fibres. Under these conditions there is strong ionic bonding between the acid groups of the azo-dyestuff and the basic groups of the fibres.

Direct dyes (*substantive dyes*) play a prominent role in the processing of cotton. In the dyeing they are directly linked to the plant fibre. They have completely taken the place of *mordant dyes* which only adhere to the fibres in the form of an insoluble coloured *lake* if the material is first treated with a metal salt.

Direct dyes consist of large planar molecules (anionic bis- and poly-azo-dyestuffs (see p. 618)) which diffuse into the fibres and hence are not easily washed out. This process is known in textile chemistry as *absorbing* the dyestuff onto the fibres; the permanent attachment of dyestuffs is called *fixing*.

In the case of chemical fibres it is possible to add insoluble *pigments* to the bulk material before it is spun and in this way to produce colour throughout the fibres.

In another process, the dyestuff is actually generated on the fibres (see naphthol-AS-dyestuffs, p. 597).

Industrially, *aniline* and its derivatives, together with *naphthylamines* and substituted derivatives, and also various diamines are used to provide the *diazo components*. The diamines provide the direct dyes.

As coupling components, not only *phenols* but above all *naphthols* and *naphtholdisulphonic acids* such as R- and G-acids, and also *aminonaphtholsulphonic acids*, e.g. γ-, I- and H-acids, are used.

---

[1] See *H. Balli* and *G. Ebner*, Färberei und Farbstoffe (Dyeing and Dyestuffs) (Springer-Verlag, Berlin 1988).

R-acid
(2-Naphthol-3,6-disulphonic acid)

G-acid
(2-Naphthol-6,8-disulphonic acid)

In the case of *aminonaphtholsulphonic acids,* coupling depends largely on the medium in which it is carried out. The orientating influence of the *sulphonyl groups* also comes into play. In *acid* conditions the diazo group attaches itself to the site next to the amino group, in *alkaline* conditions to the site next to the hydroxy group, as may be seen in the following examples:

γ-Acid
(2-Amino-8-naphthol-6-sulphonic acid)

I-Acid
(2-Amino-5-naphthol-7-sulphonic acid)

H-Acid
(8-Amino-1-naphthol-3,6-disulphonic acid)

The sites of acid coupling are shown by the solid arrows and those of alkaline coupling by the dotted arrows. γ-Acid is used for the most part as a coupling component in wool dyes, and I- and H-acids are used in cotton dyes.

In dyestuffs chemistry the diazotisation and coupling process is represented simply by an arrow:

D → K (D = diazocomponent, K = coupling component)

According to the number of azo groups present in a dyestuff they are known as mono-, bis-, tris- or poly-azo-dyestuffs.

Out of the very large number of azo-dyestuffs, only a few typical examples are mentioned here.

5.14.3 Basic Azo-dyestuffs

These dyestuffs have either NH_2- or NR_2-groups as auxochromes.

Chrysoidine (2,4-diaminoazobenzene) is made by coupling diazotised aniline with *m*-phenylenediamine. It was the first synthetic mono-azo-dyestuff to be prepared (*Caro*, 1875; *Witt*, 1876) and is still used today for making coloured paper.

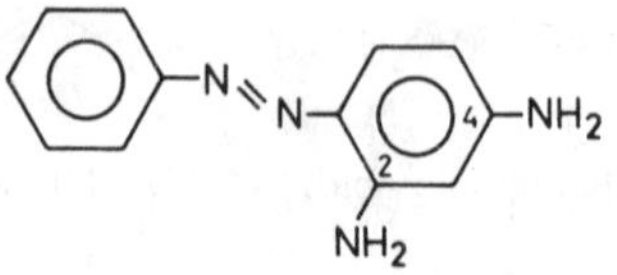

Chrysoidine (orange coloured)

5.14.4 Acid Azo-dyestuffs

Most of these dyestuffs contain sulphonyl groups.

Naphthol Blue Black 6B (Amidoblack 10B) is the most important acid wool dye which gives a reasonably fast black colour. It is a *bis-azo-dyestuff*, made by coupling H-acid with 1 mol *p*-nitrobenzenediazonium chloride in acid solution and then with 1 mol benzenediazonium chloride in alkaline solution.

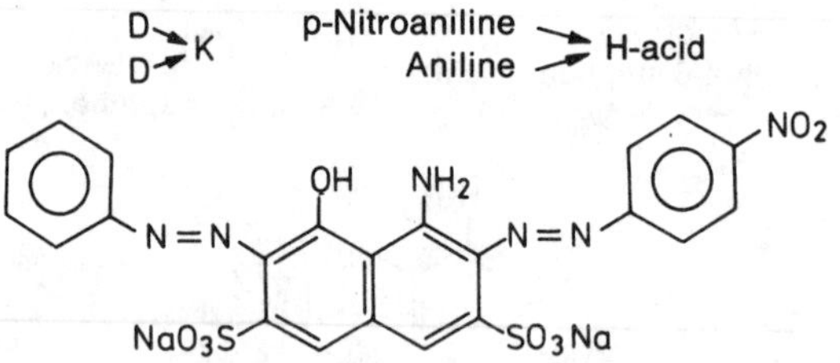

Naphthol Blue Black 6B

This process involves the sequential coupling of two different diazo compounds in different media with the same coupling component (*M. Hoffmann*, 1891).

5.14.5 Direct Dyestuffs (Substantive Dyestuffs)[1]

The azo-dyestuffs described so far are incapable of dyeing cellulose fibres directly, without the help of a mordant. A new group of dyestuffs which would colour cellulose (cotton, *etc.*) as well as regenerated cellulose (*e.g.* viscose fibres) without the aid of a mordant was first disclosed by the discovery of Congo Red (*Böttiger*, 1884). They are called *direct* or *substantive* cotton dyes. Nowadays they represent the largest group of azo-dyestuffs.

Congo Red is made by coupling doubly diazotised benzidine with 2 mol of naphthionic acid. A blue acid dye is formed first and is salted out with sodium chloride as the red disodium salt (Congo Red):

---

[1] See *E. Pfeil* et al., Molekülbau und Haftung substantiver Farbstoffe auf Cellulose (Molecular Structure and Adsorption of Substantive Dyestuffs on Cellulose), Angew. Chem. 75, 407 (1963).

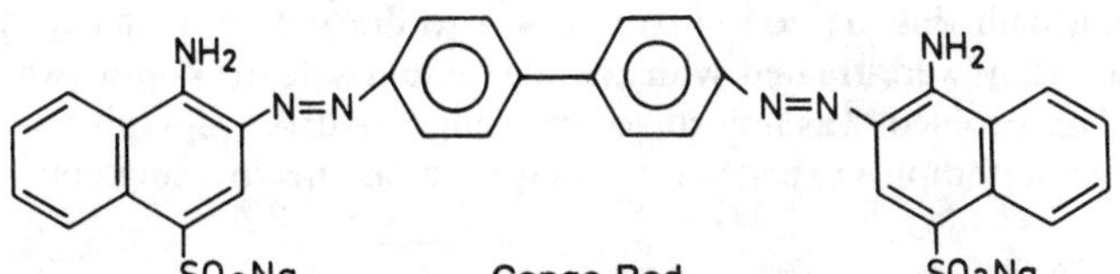

Congo Red

Treatment with mineral acid regenerates the blue acid dye, and hence Congo Red can be used as an *indicator for mineral acids.*

The deepening of colour is associated with *mesomerism* in the compound:

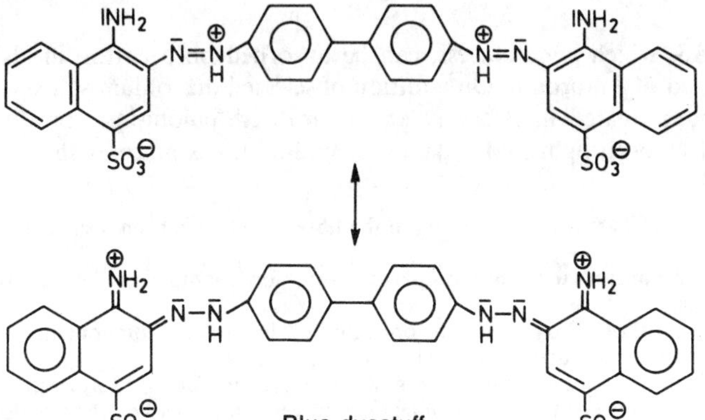

Blue dyestuff

Benzidine and its derivatives formerly played an important role in the preparation of dyestuffs for textiles. However, they are suspected of being carcinogens and are therefore no longer used. Large, planar dyestuff molecules are made in alternative ways, for example from 4,4'-diaminostilbene-2,2'-disulphonic acid, providing dyestuffs which, like those derived from benzidine, can be relied upon to be coplanar (see chrysophenine, p. 618).

5.14.6 Naphthol-AS-dyestuffs[1]

This class of compound started with the 'ice colours', so-called because they were produced on the fibres at 0°C (237 K). An important compound of this sort is Para Red (*p*-nitroaniline red, 2-hydroxynaphthalene-1-azo(4'-nitrobenzene)), which is made as an insoluble dyestuff on the cotton fibres by impregnating them with an alkaline solution of β-naphthol, drying them, and then treating them with a solution of diazotised *p*-nitroaniline. This process was used especially in the manufacture of printed cloth.

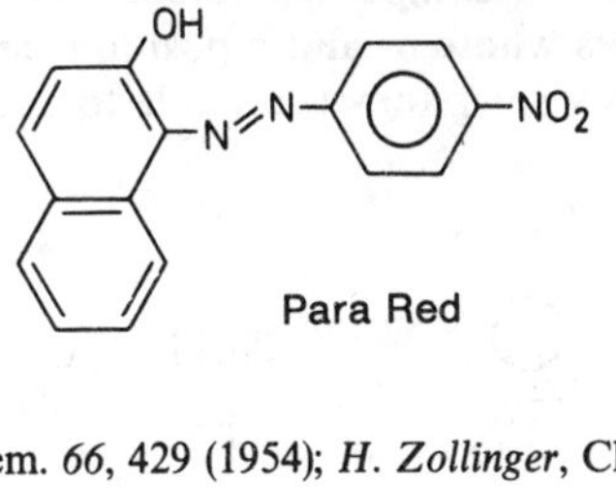

Para Red

---

[1] See *W. Kirst* et al., Angew. Chem. *66*, 429 (1954); *H. Zollinger*, Chimia *22*, 9 (1968).

The problem with this method of dyeing was to dry the impregnated fibres in such a way that when they were treated with the diazonium salt the colour was even through the product. This problem was first solved by using 3-hydroxy-2-naphthoic acid or, better still, its anilide, naphthol AS, instead of β-naphthol as the coupling constituent.

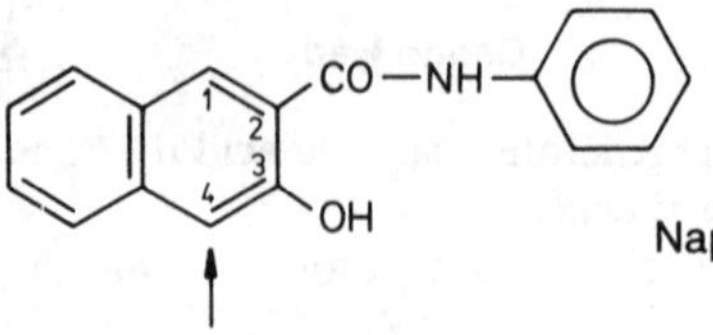

Naphthol AS

Naphthol AS, which is colourless, can be absorbed onto cotton in alkaline solution without the need of a mordant. On addition of selected diazonium salt solutions to fibres which have been treated in this way and then dried, colours are produced which are extremely fast to both light and washing. Coupling takes place at the 4-position of the naphthol ring.

Since these dyestuffs are formed directly on the fibres, they have been termed 'developed dyes' or 'ingrain dyes'.

The diazo components used are prepared from, for example, *p*- or *m*-nitroaniline or *p*-aminoazobenzene.

Because the naphthol-AS-dyestuffs are only formed from their components in the actual dye works, two points were important for their economic success.

First an industrially practicable process for the *preparation of 3-hydroxy-2-naphthoic acid arylamides* had to be worked out. This can be done in very good yield by the reaction of 3-hydroxy-2-naphthoic acid with an arylamine, *e.g.* aniline, in the presence of phosphorus(III) chloride:

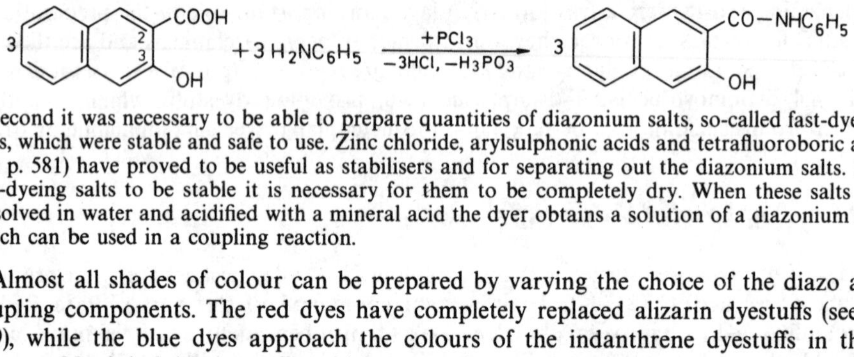

Second it was necessary to be able to prepare quantities of diazonium salts, so-called fast-dyeing salts, which were stable and safe to use. Zinc chloride, arylsulphonic acids and tetrafluoroboric acid (see p. 581) have proved to be useful as stabilisers and for separating out the diazonium salts. For fast-dyeing salts to be stable it is necessary for them to be completely dry. When these salts are dissolved in water and acidified with a mineral acid the dyer obtains a solution of a diazonium salt which can be used in a coupling reaction.

Almost all shades of colour can be prepared by varying the choice of the diazo and coupling components. The red dyes have completely replaced alizarin dyestuffs (see p. 639), while the blue dyes approach the colours of the indanthrene dyestuffs in their fastness. Naphthol-AS-dyestuffs, which were discovered by *Laska* and *Zitscher* (1913) have worldwide importance.

Rapidogen dyestuffs. Diazoamino compounds, made from diazonium salts by coupling them with primary arylamines whose *o*- and *p*-positions are substituted, have been used in order to stabilise diazonium salts; they are stable to alkali:

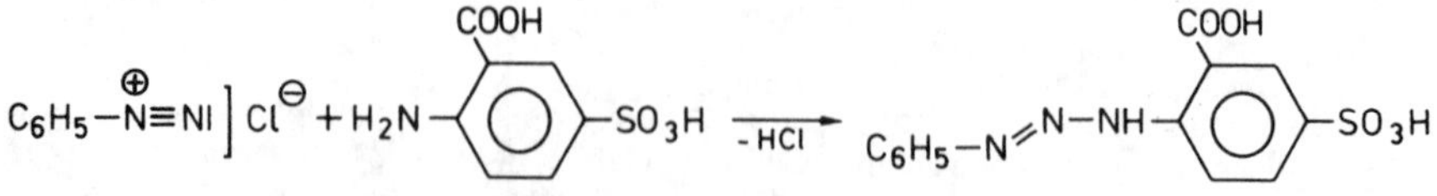

2-Amino-5-sulphobenzoic acid

When these diazoamino compounds are mixed with naphthoxides of the naphthol-AS series they can be kept without any dyestuff being formed. If cotton fabric is printed with this mixture and then exposed to formic acid or acetic acid vapour, the diazoamino compound is split and coupling then takes place to give an insoluble azo-dyestuff (*rapidogen-dyestuff*).

5.14.7 Reactive Dyestuffs[1]

A newer industrially important type of dyestuff, first made by ICI, are the *Procion dyestuffs*, made by condensing water-soluble aminoazo-dyestuffs containing *two* sulphonyl groups with 1 mol cyanuric chloride (2,4,6-trichloro-1,3,5-triazine). The three chlorine atoms of cyanuric chloride react at different rates, depending upon the temperature and the pH of the medium, so it is possible to exchange *only one* of them by the amino group of the azo-dyestuff. For example the azo-dyestuff *Procion brilliant orange GS* is obtained by coupling diazotised orthanilic acid (*o*-aminobenzenesulphonic acid) with I-acid and using the product to substitute a chlorine atom of cyanuric chloride:

Procion brilliant orange GS

The dyestuffs are fixed on the fibres because the two remaining chlorine atoms can be displaced by nucleophilic groups occurring in the fibres, *e.g.* the hydroxy groups in cellulose or the amino groups of proteins; thus covalent bonds are formed between the dyestuff and the fibres. This dyeing process produces very fast colours. Similar bonding principles underlie the action of the *Cibacron dyestuffs* (CIBA), based on monochlorotriazine, *Remazol dyestuffs*[2] (Hoechst), which utilise either a sulphate ester of β-hydroxyethylsulphone ($-SO_2-CH_2-CH_2-O-SO_3H$) or vinyl sulphones ($-SO_2-CH=CH_2$), and *Levafix* dyestuffs[3] (Leverkusen), based on the grouping $-SO_2-NH-CH_2-CH_2-O-SO_3H$.

5.14.8 Disperse Dyestuffs[4]

Disperse dyestuffs are characterised by the facts that they have no ionisable groups, are only soluble with difficulty in water, and are suitable for being deposited from colloidal dispersions onto hydrophobic fibres, especially *polyester fibres* and *acetate fibres*. Readily

[1] See H. *Zollinger*, Chemismus der Reactivfarbstoffe (Chemistry of Reaction Dyestuffs), Angew. Chem. *73*, 125 (1961); E. *Siegel*, Chemie der Reactivfarbstoffe, Suppl. Chimia *1968*, 100.

[2] See J. *Heyna*, Reaktivfarbstoffe mit Vinylsulfongruppen, Angew. Chem. *74*, 966 (1962).

[3] See K. G. *Kleb*, Levafix-Fabstoffe Chemismus und Praxis, Angew. Chem. *74*, 698 (1962); K. G. *Kleb* et al., Uber neue Reaktivfarbstoffe, *ibid* 76, 423 (1964).

[4] See C. *Müller*, Neuere Entwickelung auf dem Gebiet der Dispersionsfarbstoffe und ihrer Zwischenprodukte (Recent Developments in Disperse Dyestuffs and their Intermediates), Suppl. Chimia *1968*, 69.

sublimable disperse dyestuffs are particularly useful for transfer printing on fabrics made from synthetic fibres.[1] Among these compounds monoazo- and bisazo-dyestuffs are important.

Both aryl and heterocyclic compounds are used as diazo and coupling components. They contain suitable substituents which influence the shade of the dye and their fastness to light and heat. Among the most commonly used sources of diazo components are the following:

| 2-Amino-5-nitro benzonitrile | 2-Amino-3,5-dinitro benzophenone | 2-Amino-5-nitro thiazole |

Typical coupling components are:

| 4-(N-Ethylanilino)-butan-2-one | 3[N-(2-Hydroxyethyl)-anilino]propionitrile |

Many other coupling components are heterocyclic compounds (see azo-dyestuffs derived from pyrazole, p. 739).

5.14.9 Metal-complex azo-dyestuffs[2]

There is a group of azo-dyestuffs which form complexes with salts and, in so doing, the fastness to light and to washing is increased. In particular, chromium and cobalt complexes are used as dyes for *protein fibres* (wool, silk) and for *polyamide fibres*. As long ago as 1892 *Erdmann* observed that when *o,o'*-dihydroxyazo-dyestuffs were heated with dichromate and also, usually, with a reducing agent, the chromium is reduced to Cr(III) and a chromium lake is formed on the protein fibre. These are called *azo-chrome mordant dyes*. The metal-complex dyestuffs obtained in this way, which usually contained sulphonyl groups to enhance their solubility, had the disadvantage that they were only absorbed on the protein fibre in strongly acidic solution.

In the 1950s new metal-complex azo-dyestuffs were made from azo compounds having *hydrophilic* groups such as sulphonamide or alkanesulphonyl groups and chromium or cobalt salts. In these complexes each metal atom is bonded to one or two dyestuff molecules. Among the 1:2 metal-complex azo-dyestuff derivatives are *Cibalan* (CIBA), *Irgalan* (Geigy) and *Ortalan* (BASF) dyestuffs. These complexes are anionic and dye wool in neutral solution.

[1] See *A. Haus*, Der Transferdruck (Transfer Printing), Chemie in unserer Zeit *12*, 41 (1978).
[2] See *H. Baumann* and *H. R. Hensel*, Neue Metallkomplexfarbstoffe, Struktur und färberische Eigenschaften (New Metal Complex Dyestuffs. Structure and Dyeing Properties). Fortschr. Chem. Forsch. *7*, 643 (1967).

More recently, particularly fast chromium- and cobalt-complex dyestuffs, based on reactive dyestuffs having vinylsulphonyl groups, have been introduced commercially (*Remalan fast dyestuffs*). Some of these metal-complex dyestuffs are also used for dyeing leather (*Coranil fast dyestuffs*).

The metal ions in the waste water from these factories can seriously affect the environment.

5.14.10 Diazotype printing and related copying processes[1]

Before the invention of electrophotographic duplicating processes *diazotype paper* was of considerable importance.

Diazotype was based on the sensitivity of arenediazonium compounds to light; when irradiated they decompose with loss of nitrogen and consequently lose the ability to act as coupling components in the formation of dyestuffs. There are two types of sensitive paper, which differ in their make-up: one has *two components*, the other just *one*. In the case of the first sort, there is a diazo component, a coupling component and a stabiliser in the light-sensitive layer. After exposure to light it is developed with a gaseous alkaline reagent such as ammonia in a dry process (dry developing). In the case of one-component papers, the light-sensitive layer contains only a diazo compound. For the copying process the requisite coupling component is added in an alkaline or neutral developer solution ('half-wet process'). Both processes provide a positive copy of the original.

Whilst irradiation of simple diazo compounds converts them into phenols, photolysis of diazonium salts derived from *o*-aminophenols or *o*-aminonaphthols leads to more complicated reactions. When irradiated, these *o*-benzo- and *o*-naphtho-quinonediazides (also known as *o*-diazoanhydrides or *o*-diazoquinones) lose nitrogen to form carbenes which act as intermediates in reactions analogous to the *Wolff* rearrangement, forming ketenes that add water to give cyclopentadiene- or indene-carboxylic acids (*Süs* reaction). Because of the electronegative substituents present these acids bond to undecomposed *o*-benzoquinonediazide molecules which are still present to form azo-dyestuffs, *e.g.*

o-Benzoquinonediazide derivative   a Carbene   a Ketene

3-Sulphocyclopentadiene-1-carboxylic acid   Azo-dyestuffs

In practice, the *o*-naphthoquinonediazide derivatives which also have sulphonyl groups present to make them water-soluble only couple with compounds having electronegative substituents to give dyestuffs in alkaline conditions. In neutral or acid media only colourless *indenecarboxylic acids* are formed. They are therefore only suitable for *positive*-producing two-component type diazotype papers.

o-Quinonediazides of the naphthalene series which contain an ester or an amide group instead of the sulphonyl group are soluble in organic solvents. They have been of great industrial importance in copying technology for the preparation of presensitised print films.

---

[1] See *O. Süs* et al., Neue Entwicklungen auf dem Gebiet der vorsensibilisierten Druckfolein (Recent Advances in the Area of Presensitised Films), Angew. Chem. *74*, 985 (1962).

5.15 Biphenyl and Arylmethanes

5.15.1 Biphenyl

In biphenyl two benzene rings are directly joined to one another by a C—C bond. The following numbering system is used to designate the positions of substituents in the two rings:

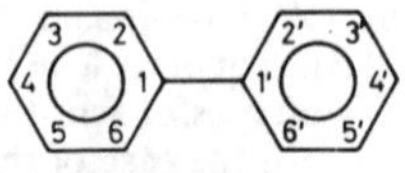

Biphenyl

As an example, benzidine is 4,4'-diaminobiphenyl.
A small amount of biphenyl is present in coal tar.

Preparation

5.15.1.1   Industrially, biphenyl is made by *pyrolytic dehydrogenation of benzene* in red-hot iron tubes filled with pumice:

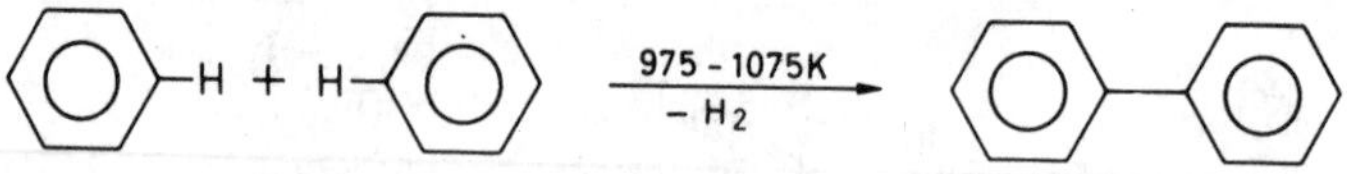

5.15.1.2   By means of the *Ullmann* reaction, by *heating iodobenzene with copper-bronze.*

5.15.1.3   Substituted biphenyls can be obtained by an electrolytic process, in which a radical cation is formed at the anode; this dimerises giving a substituted biphenyl. For example, mesitylene reacts as follows:

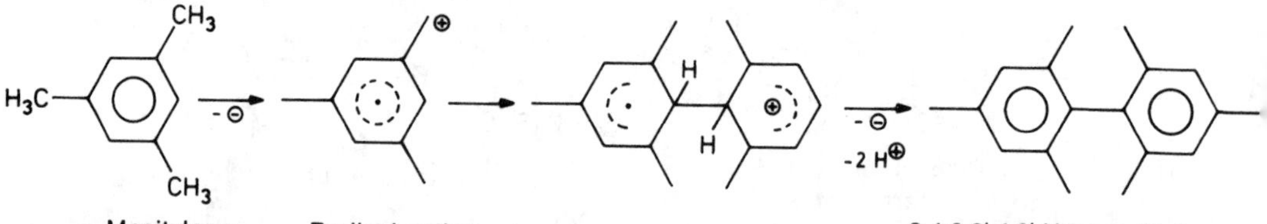

| Mesitylene (1,3,5-trimethyl-benzene) | Radical cation | 2,4,6,2',4,6'-Hexamethyl-biphenyl (Bimesityl) (m.p. 101°C [374 K]) |

In this case a C—C bond is formed at the anode, in contrast to the electrochemical synthesis of adiponitrile, wherein a C—C bond is formed at the cathode.

Properties

Biphenyl forms colourless, lustrous leaflets, m.p. 70°C (343 K). It is insoluble in water but soluble in benzene, ether or ethanol. Its chemical properties are those of an arene.

Uses

Biphenyl is used as a *preservative* for citrus fruit, and, as a eutectic mixture with phenol ethers (*e.g.* dibenzofuran), as a filling for heat-exchangers. The most important derivatives of biphenyl are benzidine, *o*-tolidine (3,3'-dimethylbenzidine) and *o*-dianisidine (3,3'-dimethoxybenzidine), all of which provide diazo components for the preparation of direct dyestuffs. Polychlorobiphenyls were used particularly in the construction of transformers, because of their good electrical properties and their stability. However, since they have a deleterious effect on the environment they are no longer used.

o-Tolidine
(3,3'-Dimethylbenzidine)

o-Dianisidine
(3,3'-Dimethoxybenzidine)

Both *o*-tolidine and *o*-dianisidine are suspected of being carcinogens.

5.15.1.4 Optical Isomerism due to Restricted Rotation ('Atropisomerism') in Biphenyl Derivatives

The two phenyl groups in biphenyl can rotate about the axis of the interannular C—C bond but this rotation may be restricted by the presence of *large substituents* at the 2,2'- and 6,6'-positions (*W. H. Mills*). If rotation is prevented the two rings cannot be coplanar, as is possible in biphenyl itself, but must be at an angle to each other. If each of the rings is unsymmetrically substituted, such a conformation has no centre of chirality (see p. 120) but an axis of chirality, and such biphenyls must be resolvable into enantiomers. A number of chiral biphenyl derivatives have been resolved, for example 2,2'-dinitro-diphenic acid (6,6'-dinitrobiphenyl-2,2'-dicarboxylic acid) (*Kenner*). The two enantiomers cannot be superimposed but are related to one another as object and mirror image (the heavily printed benzene rings are perpendicular to the plane of the paper).

Mirror plane

This type of stereoisomerism is said to be due to *restricted rotation* or *atropisomerism* (see cyclophanes). To produce such enantiomers it is necessary that the 2- and 6-substituents differ from one another and that the 2'- and 6'-substituents also differ from one another. The stability of the enantiomers depends on the size of these substituents. The less the steric hindrance the more easily racemisation takes place.

5.15.2 Diphenylmethane

The simplest arylalkane is toluene, C_6H_5—CH_3. If a second hydrogen atom of methane is replaced by a phenyl group the resultant compound is diphenylmethane.

Preparation

5.15.2.1 From *dichloromethane and benzene* by means of a *Friedel–Crafts* reaction:

$$C_6H_6 + Cl-CH_2-Cl + C_6H_6 \xrightarrow[-2\,HCl]{(AlCl_3)} C_6H_5-CH_2-C_6H_5$$

Diphenylmethane

5.15.2.2 In a similar way from benzyl chloride and benzene:

$$C_6H_5-CH_2Cl + C_6H_6 \xrightarrow[-HCl]{(AlCl_3)} C_6H_5-CH_2-C_6H_5$$

5.15.2.3 By *Clemmensen reduction of benzophenone* (*see p. 214*):

$$C_6H_5-\underset{\underset{O}{\|}}{C}-C_6H_5 \xrightarrow[-H_2O]{+4\,H} C_6H_5-CH_2-C_6H_5$$

Properties

Diphenylmethane crystallises in colourless needles, m.p. 26°C (299 K), which have an orange-like smell. The hydrogen atoms of the methylene group are rather reactive; for example, diphenylmethane is readily oxidised by chromic acid to benzophenone.

Among important derivatives of diphenylmethane are the *germicide* and *insecticide* bis(3,5,6-trichloro-2-hydroxyphenyl)methane (2,2′-methylene-bis[3,4,6]trichlorophenol, hexachlorophene) and *1,1-bis(4-chlorophenyl)-2,2,2-trichloroethane* (1,1,1-trichloro-2,2-bis(4-chlorophenyl)ethane) (DDT, chlorophenothane). The latter is made by condensing 2 mols of chlorobenzene with 1 mol of chloral hydrate in the presence of sulphuric acid:

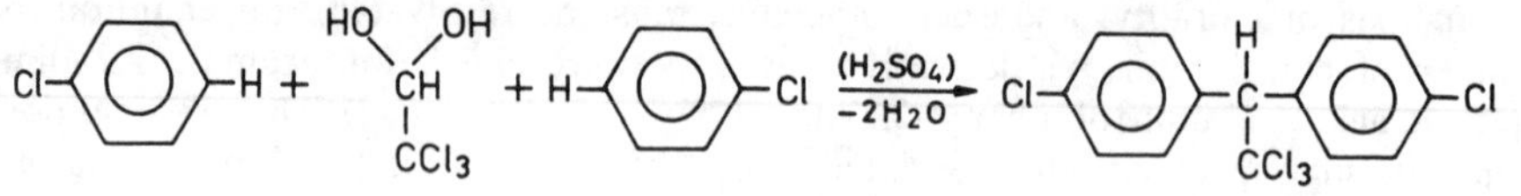

DDT

DDT is very toxic to a variety of insects and was much used in the fight against malaria. Its use in agriculture is forbidden because it is not broken down biologically. Readily hydrolysable insecticides of the *parathion* type are used instead for agricultural purposes.

5.15.3 Triphenylmethane

This hydrocarbon provides the basic structure for the triphenylmethane dyestuffs.

Preparation

5.15.3.1 By means of the *Friedel–Crafts* reaction from either *chloroform* and *benzene* or from *benzylidene chloride* and *benzene*:

$$3C_6H_6 + Cl_3CH \xrightarrow[-3HCl]{(AlCl_3)}$$

$$(C_6H_5)_3CH$$

$$2C_6H_6 + Cl_2CHC_6H_5 \xrightarrow[-2HCl]{(AlCl_3)}$$ Triphenylmethane

5.15.3.2 In appreciably better yield by the *reduction of triphenylmethanol* (see below) with zinc powder and acetic acid or by boiling it with formic acid.

Properties

Triphenylmethane forms colourless crystals, m.p. 93°C (366 K). The presence of the three phenyl groups increases the *acidity* of the hydrogen atom attached to the tertiary carbon atom. One result of this is that triphenylmethane is easily oxidised to triphenylmethanol. If sodamide is added to a solution of triphenylmethane in ether, a bluish-red solution of *triphenylmethylsodium (trityl sodium)* is formed:

$$(C_6H_5)_3C-H + NaNH_2 \longrightarrow (C_6H_5)_3C]^{\ominus} Na^{\oplus} + NH_3$$
$$\text{Triphenylmethylsodium}$$

5.15.3.3 Triphenylmethanol (triphenylcarbinol), $(C_6H_5)_3C-OH$, can be prepared by the reaction of phenyl magnesium bromide with ethyl benzoate (or benzophenone) in dry ether:

$$C_6H_5-C\overset{O}{\underset{OR}{\diagdown}} \xrightarrow[-Mg(OR)Br]{+2C_6H_5MgBr} (C_6H_5)_3C-OMgBr \xrightarrow[-Mg(OH)Br]{+H_2O} (C_6H_5)_3C-OH$$

Benzoate ester   Triphenylmethanol

Triphenylmethanol crystallises in colourless prisms, m.p. 164°C (437 K). It dissolves in conc. sulphuric acid to give an orange-red solution. Although it has the structure of a tertiary alcohol its properties differ from those of most alcohols. The hydroxy group *cannot* be esterified with either acid chlorides or anhydrides; with acetyl chloride, instead of forming triphenylmethyl acetate, it is converted into chlorotriphenylmethane (trityl chloride):

$$(C_6H_5)_3C-OH + H_3CCOCl \rightarrow (C_6H_5)_3C-Cl + H_3CCOOH$$

5.15.3.4 Chlorotriphenylmethane   (triphenylmethyl   chloride,   trityl   chloride), $(C_6H_5)_3C-Cl$, can be obtained from *benzene* by means of a *Friedel–Crafts* reaction using *tetrachloromethane*:

$$CCl_4 + 3\ C_6H_6 \xrightarrow[-3\ HCl]{(AlCl_3)} (C_6H_5)_3C-Cl$$

The fourth chlorine atom does *not react* and no tetraphenylmethane is formed.

Chlorotriphenylmethane forms colourless crystals, m.p. 112°C (385 K). Its reactivity resembles that of acid chlorides. When dissolved in liquid sulphur dioxide it forms a yellow solution which conducts electricity (*Walden*, 1902). This indicates that in this solution it exists in a dissociated form as a carbenium ion salt. Furthermore, when chlorotriphenylmethane is hydrolysed with hot water a coloured *carbenium ion* is formed rapidly in the first place before colourless, undissociated triphenylmethanol, the carbinol base, precipitates out. Hydrolysis is prevented if hydrochloric acid is added, and the following *equilibria* can be assumed to be involved:

$$(C_6H_5)_3C-Cl \rightleftharpoons (C_6H_5)_3C^{\oplus} + Cl^{\ominus} \underset{+HCl\,(-H_2O)}{\overset{+H_2O\,(-HCl)}{\rightleftharpoons}} (C_6H_5)_3C-OH$$

| Chlorotriphenyl-
methane
(colourless) | Carbenium
ion (yellow) | Carbinol base
(colourless) |

In the presence of pyridine, chlorotriphenylmethane reacts with only the *primary* alcohol groups of sugars to give trityl ethers (*B. Helferich*, 1923):

$$R-CH_2OH + Cl-C(C_6H_5)_3 \xrightarrow[-HCl]{(Pyridine)} R-CH_2-O-C(C_6H_5)_3$$

The remaining hydroxy groups of the sugar can now be methylated or acetylated and then the trityl group can be selectively removed again.

Tetraphenylmethane, $(C_6H_5)_4C$, can be made by *Friedel–Crafts* alkylation of triphenylmethanol with anilinium chloride, followed by diazotisation and deamination of the resultant amine. It forms stable, colourless needles, m.p. 283°C (556 K) (yield 82%).[1]

$$(C_6H_5)_3COH + C_6H_5NH_3^{\oplus} Cl^{\ominus} \longrightarrow (C_6H_5)_3CC_6H_5NH_2 \xrightarrow{C_5H_{11}ONO} (C_6H_5)_3CC_6H_4N_2 \xrightarrow[-N_2]{H_3PO_2} (C_6H_5)_4C$$

5.16 Triphenylmethane Dyestuffs

If *at least* two amino or hydroxy groups are introduced as *auxochromes* in the *p*-positions of the benzene rings in triphenylcarbinol or chlorotriphenylmethane, then this provides the parent compounds of the triphenylmethane dyestuffs. These dyestuffs have bright colours but they are not fast either to light or to washing.

5.16.1 *Aminotriphenylmethane dyestuffs*

These dyestuffs include malachite green and fuchsin.

5.16.1.1 **Malachite green** (*O. Fischer*, 1877).

If 1 mol of benzaldehyde is condensed with 2 mols of *N,N*-dimethylaniline in the presence of zinc chloride, the colourless 'leuco-base' of malachite green is formed. This is converted into the dyestuff by oxidation with lead(IV) oxide or manganese(IV) oxide in acid solution:

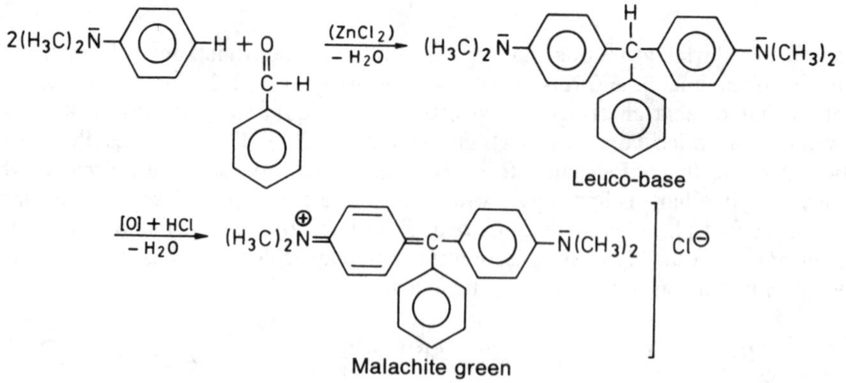

Leuco-base

Malachite green

[1] See *H. Kurreck* et al., Ang. Chem. Int. Ed. Engl. *25*, 1097 (1986).

The lone pairs of electrons on the dimethylamino groups, which act as *auxochromes*, interact mesomerically with the benzene rings (+ M-effect), providing a *mesomerically stabilised carbenium-iminium cation*:

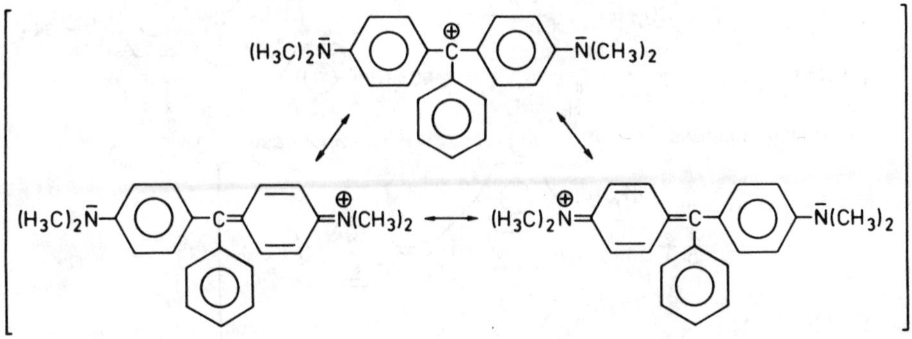

5.16.1.2 Fuchsin (rosaniline) is prepared by heating an equimolar mixture of aniline, *o*-toluidine and *p*-toluidine with nitrobenzene in the presence of hydrochloric acid:

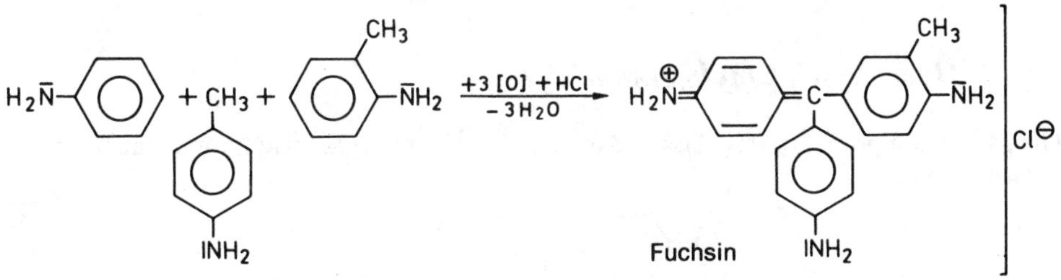

It is probable that *p*-aminobenzaldehyde is formed from *p*-toluidine as an intermediate.

Fuchsin forms green crystals with a metallic sheen. They dissolve in water to give a red-purple solution and can be used for dyeing wool or silk in this colour. In this case the dyestuff molecules contain three amino groups; this leads to increased mesomerism in the cation.

Schiff's reagent (fuchsin-aldehyde reagent) is an aqueous solution of fuchsin which has been decolourised by sulphur dioxide. With *aldehydes* it gives a *purple colour* (H. Schiff, 1865); the structure of this coloured product is uncertain.

5.16.1.3 Pararosaniline (parafuchsin) resembles fuchsin but lacks the methyl substituent ('Para' has no structural significance in this name).

If the amino groups of fuchsin are alkylated the colour changes to violet.

5.16.1.4 Crystal violet (hexamethylpararosaniline) is made by condensing *Michler's* ketone (4,4'-bisdimethylaminobenzophenone, see p. 578) with *N,N*-dimethylaniline in the presence of phosphorus oxychloride (see top of p. 608).

This dyestuff dissolves in water to give an intensely violet solution. It colours wool, silk or mordanted cotton giving a bright violet shade, but washes out readily, and hence it has only been used for copying pencils, copying ink and typewriter ribbons.

With iodine, crystal violet forms a charge-transfer complex which bonds selectively to certain kinds of bacteria and makes possible their classification into gram-positive and gram-negative bacteria (*Gram*-colouring).

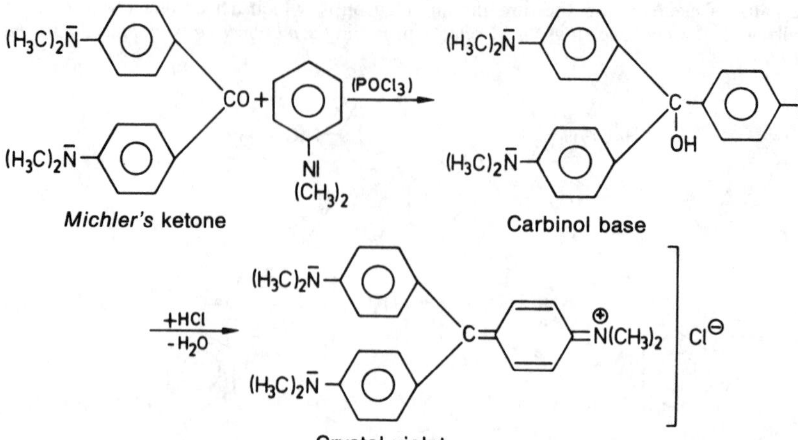

Michler's ketone

Carbinol base

Crystal violet

5.16.2 Hydroxytriphenylmethane dyestuffs

The parent compound of this class of dyestuffs is *fuchsone* (diphenylquinomethane) which is itself colourless.

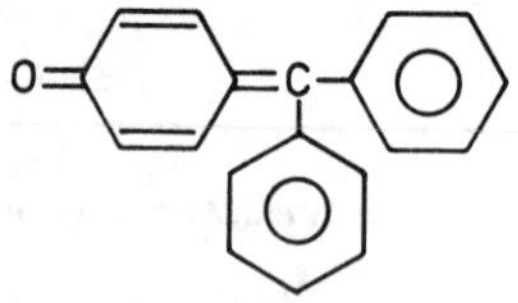

Fuchsone

Benzaurine (*p*-hydroxyfuchsone) is the simplest dyestuff of this sort and arises when 1 mol benzotrichloride and 2 mols of phenol are melted together:

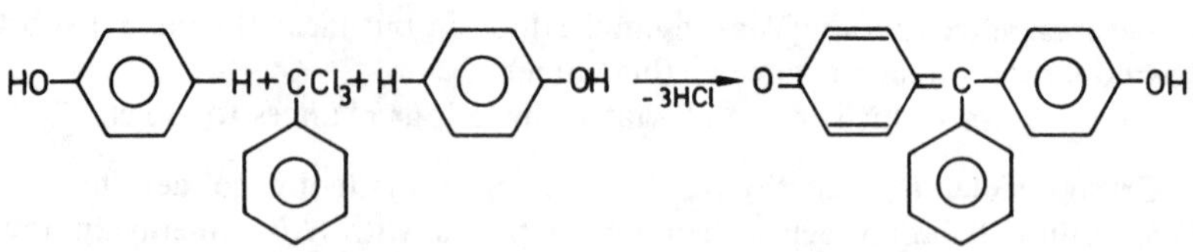

Benzaurine

As a phenol derivative, yellow-brown benzaurin is *acidic*; with dilute alkali it forms violet-coloured salts.

The strong +M-effect of the oxygen atoms results in there being extensive *mesomerism in the anion*:

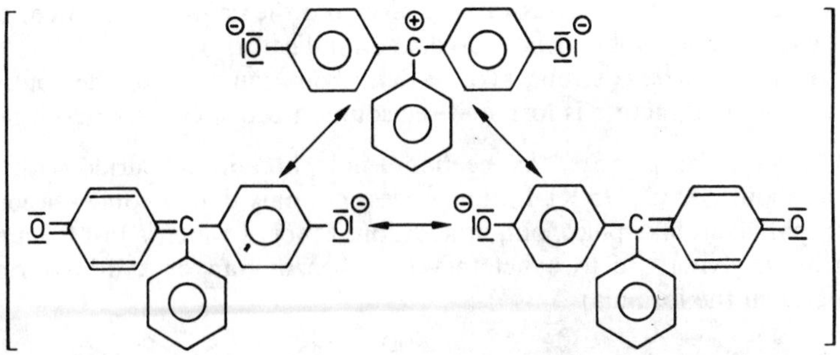

5.16.3 Phthaleins

This class of dyestuffs is derived from *o*-carboxy derivatives of triphenylmethanol; they are therefore related to the triphenylmethane dyestuffs.

5.16.3.1   Phenolphthalein (3,3-bis(4-hydroxyphenyl)phthalide) is made by heating 1 mol phthalic anhydride with 2 mols of phenol in the presence of conc. sulphuric acid. A condensation reaction, with loss of water, takes place and a lactone is formed. Addition of alkali results in opening of the lactone ring. The deep red colour (λ_{max} 553 nm) is due to *mesomerism* in the *quinonoid dianion*.

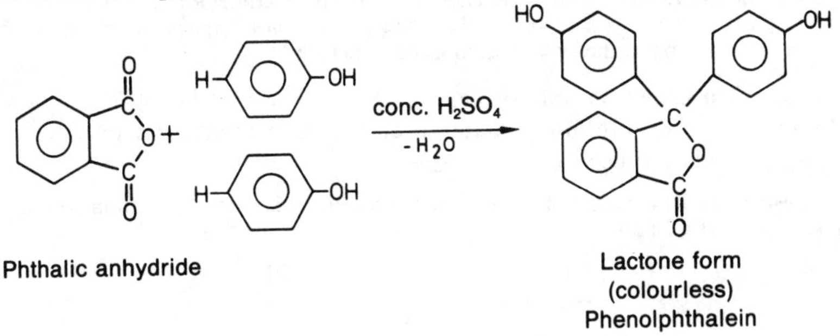

Phthalic anhydride

Lactone form
(colourless)
Phenolphthalein

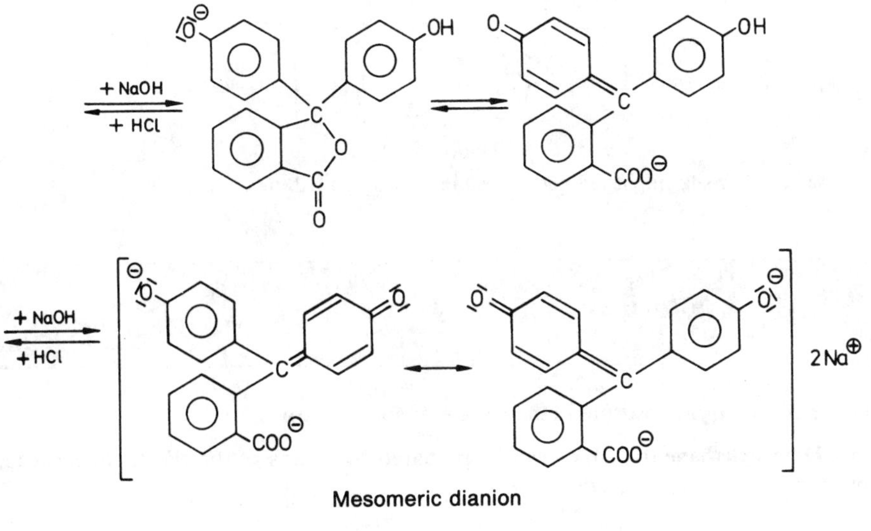

Mesomeric dianion

Addition of mineral acid reverses this reaction and the colour is restored. Because of this colour change, phenolphthalein is used as an indicator.

In the presence of *excess* strong alkali a hydroxide ion adds to the central carbon atom, the quinonoid structure is lost, and the solution becomes colourless again.

5.16.3.2 Fluorescein is prepared by heating 1 mol phthalic anhydride with 2 mols of resorçinol at about 200°C (475 K). In this process not only does a condensation reaction leading to a phthalein take place, but, in addition, 1 mol of water is lost from the two *o*-hydroxy groups, giving rise to a heterocyclic dibenzopyran or xanthene ring system (shaded in grey in the formula).

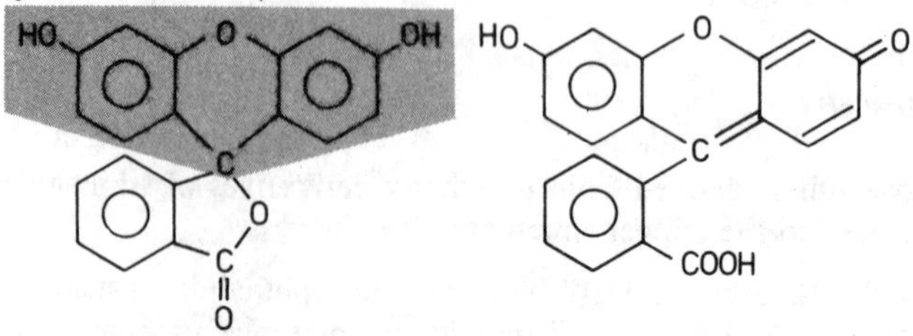

Lactone form   Quinonoid form

Fluorescein

The colour of fluorescein cannot be explained by the lactone form, which has no suitable chromophore, and its structure is represented by the *quinonoid form*. Like phenolphthalein, fluorescein forms an anion which has a mesomeric quinonoid structure.

Fluorescein is a brownish-red powder which dissolves in ethanol to give a yellowish-red solution. In dilute solution in alkali, even traces of fluorescein produce an intense yellowish-green *fluorescence*.

5.16.3.3 Eosin is the sodium salt of tetrabromofluorescein; it is made by bromination of a solution of fluorescein in acetic acid.

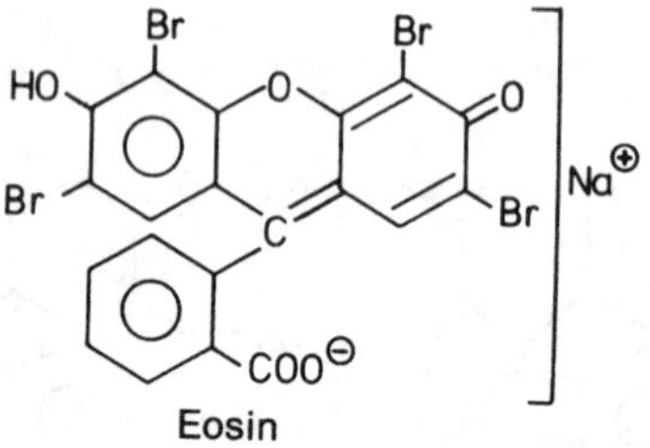

Eosin

It dyes wool or silk bright red and is used in making up red ink.

5.17 Arylethanes

Symmetrically aryl-substituted ethanes are of some importance.

1,2-Diphenylethane (bibenzyl) can be prepared by means of the *Wurtz–Fittig* reaction from *benzyl chloride and sodium:*

$$2\ C_6H_5-CH_2-Cl + 2\ Na \rightarrow C_6H_5-CH_2-CH_2-C_6H_5 + 2\ NaCl$$

Benzyl chloride   1,2-Diphenylethane

It forms colourless prisms, m.p. 53°C (326 K).

Tetraphenylethane is made in a similar way from *chlorodiphenylmethane* and *sodium or copper*:

$$2(C_6H_5)_2\underset{H}{C}-Cl + 2Cu \longrightarrow (C_6H_5)_2\underset{H}{C}-\underset{H}{C}(C_6H_5)_2 + 2CuCl$$

Chlorodiphenylmethane   1,1,2,2-Tetraphenylethane

It crystallises in colourless needles, m.p. 211°C (484 K).

The accumulation of phenyl groups in tetraphenylethane results in a weakening of the ethane C—C bond. As a result, if a solution of tetraphenylethane in ether is treated with potassium, this C—C bond is split and two molecules of diphenylmethyl potassium are formed:

$$(C_6H_5)_2\underset{H}{C}-\underset{H}{C}(C_6H_5)_2 \xrightarrow{+2K} \left. 2(C_6H_5)_2\overset{\ominus}{\underset{H}{C}l} \right] K^{\oplus}$$

From this it could be expected that the central C—C bond of hexaphenylethane should be even more readily broken. Experimentally, when an attempt was made to prepare hexaphenylethane by shaking a solution of chlorotriphenylmethane in benzene with silver powder, the product obtained was triphenylmethyl peroxide, $(C_6H_5)_3C-O-O-C(C_6H_5)_3$. Hence, some intermediate product must have reacted with atmospheric oxygen. When the same reaction was repeated in the absence of air a yellow solution resulted, from which colourless crystals, m.p. 145–147°C (418–420 K) were precipitated by addition of acetone (*Gomberg*, 1900). When this colourless material was dissolved a yellow colour was again produced, which suggested the presence of *triphenylmethyl radicals*.

It turned out that the product obtained by shaking chlorotriphenylmethane with silver in tetrachloromethane was not the α,α-dimer, hexaphenylethane, but 1-diphenylmethylene-4-tritylcyclohexa-2,5-diene, the α,p-dimer. Its structure has been shown unequivocally by its *UV* and *NMR spectra* (*Lankamp*, 1968).[1] The following equilibrium is taking place in the yellow solution:

1-Diphenylmethylene-4-tritylcyclohexa-2,5-diene   Triphenylmethyl
(Trityl)

[1] See *J. M. McBride*, The Hexaphenylethane Riddle, Tetrahedron, *30*, 2009 (1974).

The position of the equilibrium depends upon the solvent, the temperature and the concentration of the solution. When heated the degree of dissociation increases; similarly with increased dilution the colour deepens. The anomalous trivalency of the central carbon atom in triphenylmethyl characterises this compound as a *free radical* (mono-radical) and at the same time explains its special reactivity.

5.18 Free Radicals[1]

5.18.1 Carbon Radicals

5.18.1.1 Monoradicals. Whilst aliphatic free radicals are very short-lived and react spontaneously either with another radical of the same sort or some other radical to form products bound in the normal way, the triphenylmethyl radical mentioned above is more or less stable (persistent).[2] Since the spin of the unpaired electron on the trivalent carbon atom is not magnetically coupled, triphenylmethyl, like all free radicals, is *paramagnetic*. The stability of triphenylmethyl is largely associated with the fact that the unpaired electron of the central carbon atom is very much sterically shielded by the phenyl groups.

Quite aside from the mesomerism associated with each of the phenyl rings, ten limiting formulae, all, to a first approximation, of about the same energy, can be drawn for triphenylmethyl, namely one with the free electron located on the central carbon atom (a), three with it on *p*-sites of the phenyl groups (b), and six with it on *o*-sites of the phenyl groups (c):

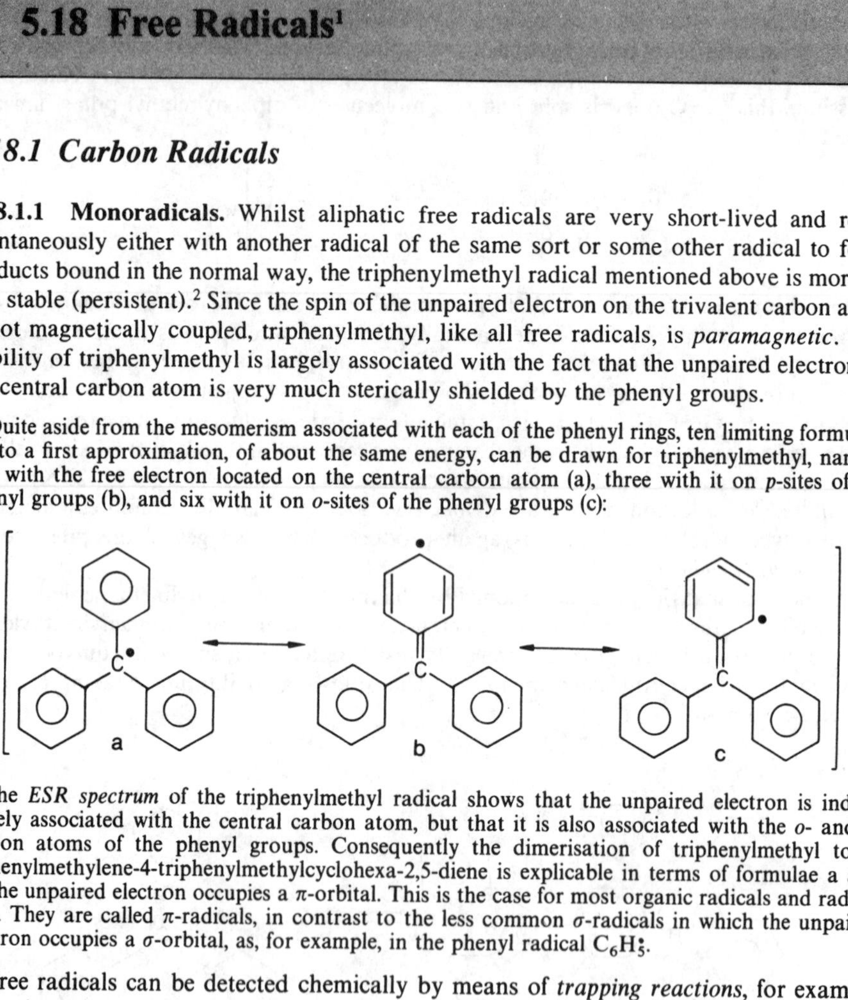

The *ESR spectrum* of the triphenylmethyl radical shows that the unpaired electron is indeed largely associated with the central carbon atom, but that it is also associated with the *o*- and *p*-carbon atoms of the phenyl groups. Consequently the dimerisation of triphenylmethyl to 1-diphenylmethylene-4-triphenylmethylcyclohexa-2,5-diene is explicable in terms of formulae a and b. The unpaired electron occupies a π-orbital. This is the case for most organic radicals and radical ions. They are called π-radicals, in contrast to the less common σ-radicals in which the unpaired electron occupies a σ-orbital, as, for example, in the phenyl radical $C_6H_5^{\cdot}$.

Free radicals can be detected chemically by means of *trapping reactions*, for example with nitrogen monoxide or sodium:

---

[1] See *A. R. Forrester* et al., Organic Chemistry of Stable Free Radicals (Academic Press, New York 1968); *J. K. Kochi*, Free Radicals in Solution, 2 Vols. (Wiley, New York 1975).
[2] See *D. Griller* and *K. U. Ingold*, Persistent Carbon-centred Radicals, Acc. Chem. Res. *9*, 13 (1976).

$$(C_6H_5)_3C \cdot + \cdot NO \rightarrow (C_6H_5)_3C-N=O$$

$$(C_6H_5)_3C \cdot + \cdot Na \rightarrow (C_6H_5)_3C \vert]^{\ominus} Na^{\oplus}$$

Triphenylmethyl sodium (trityl sodium)

5.18.1.2 Biradicals. Biradicals are characterised by having *two* unpaired electrons, each centred on a different carbon atom. The question which now arises is to what extent the two electrons can interact with one another; such interaction could ultimately lead to the formation of an electron pair. In that circumstance the molecule would no longer be paramagnetic. Thus more detailed investigations showed that *Chichibabin's hydrocarbon*, a deep violet compound first prepared in 1907, and which for a long time was thought to be a radical, is in fact *diamagnetic*. The two electrons must interact with one another through the two benzene rings to form a π-electron pair, in the process setting up a quinonoid structure. *ESR spectroscopy* has shown that the contribution of the biradical form is only of the order of 4%:

(4%)   (96%)

If the four *o*-hydrogen atoms of this biphenyl system are replaced by *larger groups, e.g.* by chlorine atoms, the result is an orange-yellow compound which is strongly *paramagnetic, i.e.* it is a *true biradical (Eu. Müller)*.

The chlorine atoms force the two phenyl groups out of coplanarity and thereby interfere with interaction between the two unpaired electrons.

5.18.2 Nitrogen Radicals

These are free radicals based on nitrogen, comparable to carbon free radicals; in them the nitrogen atom is divalent.

If diphenylamine is oxidised by potassium permanganate in acetone the product is tetraphenylhydrazine:

$$2\,(C_6H_5)_2N-H \xrightarrow[-H_2O]{[O]} (C_6H_5)_2N-N(C_6H_5)_2$$

Diphenylamine   Tetraphenylhydrazine

It is a colourless, crystalline substance, m.p. 149°C (422 K) which dissolves in conc. sulphuric acid to give a deep blue solution (see p. 579). Dissociation of tetraphenylhydrazine into *diphenylaminyl* (diphenylnitrogen) is observable only when it is heated in xylene to about 80°C (355 K) and even then the following equilibrium lies largely to the left-hand side (*H. Wieland*).

$$(C_6H_5)_2\bar{N}-\bar{N}(C_6H_5)_2 \rightleftharpoons 2\,(C_6H_5)_2\bar{N}\cdot$$

Tetraphenylhydrazine Diphenylaminyl
(colourless)   (olive-green)

The presence of diphenylaminyl radicals in the heated solution can be demonstrated by *trapping reactions* involving other free radicals, such as nitrogen monoxide. In this way *diphenylnitrosamine*, $(C_6H_5)_2N—N=O$, is formed in quantitative yield for, as the free diphenylaminyl radicals are removed by trapping, the equilibrium is disturbed, and they are replaced by further decomposition of tetraphenylhydrazine.

In general terms, diphenylaminyl radicals with their divalent nitrogen atoms are less stable than carbon radicals; this can be related to the greater electronegativity of nitrogen as compared to carbon. No sooner is an electron septet formed on a nitrogen atom than it attempts to gain another electron as quickly as possible (*Eistert*). Here again the unpaired electron plays a role in the mesomerism of the whole compound. *p*-Substitution by groups having a +M-effect, *e.g.* OCH_3 or $N(CH_3)_2$, leads to a more ready dissociation to free radicals, for example, even in solution at room temperature *tetra-anisylhydrazine* is partially dissociated into two molecules of dianisylaminyl (di-(*p*-methoxyphenyl)nitrogen), which is light green in colour:

Tetra-anisylhydrazine   Dianisylaminyl

When warmed, or diluted, the *equilibrium* is shifted towards the right-hand side and a blue-green colour develops. When cooled the solution becomes paler again. Nitrogen radicals are stable to molecular oxygen.

Another type of radical containing divalent nitrogen consists of the *hydrazyls* (S. Goldschmidt, 1922). Dehydrogenation of triphenylhydrazine leads to the formation of colourless hexaphenyltetrazanes, which exist in equilibrium with strongly coloured *tetraphenylhydrazyl* radicals:

Triphenylhydrazine   Hexaphenyltetrazane   Triphenylhydrazyl

Whilst in this case the dissociation equilibrium lies largely to the left-hand side, replacement of a phenyl group by a picryl group provides *N,N-diphenyl-N'-trinitrophenyl-hydrazyl*, which is blue-black and stable. In the crystalline state it consists exclusively of the *monomeric radical form*.

A further example is provided by *diphenylaminyl oxide*, obtained as red crystals by dehydrogenating N,N-diphenylhydroxylamine with silver oxide in ether (H. Wieland):

Diphenylaminyl oxide

The diphenylaminyl oxide radical exists as a monomer. In its properties it resembles nitrogen dioxide and, like it, is *paramagnetic*.

5.18.3 Aroxyls (Phenoxy Radicals)

Yet another kind of radicals consists of the *aroxyls*, in which the unpaired electron can reside on an oxygen atom as well as on a carbon atom; they are hence often described as *oxygen radicals*. Their existence depends on there being large *bulky groups* at both of the *o*-positions and at the *p*-position of a phenolic compound.

For example, if 3,3',5,5'-tetra-t-butyl-4,4'-dihydroxydiphenylmethane is dehydrogenated using either lead(IV) oxide or, better, potassium hexacyanoferrate(III), the solution becomes deep blue, and the radical *galvinoxyl* is formed:

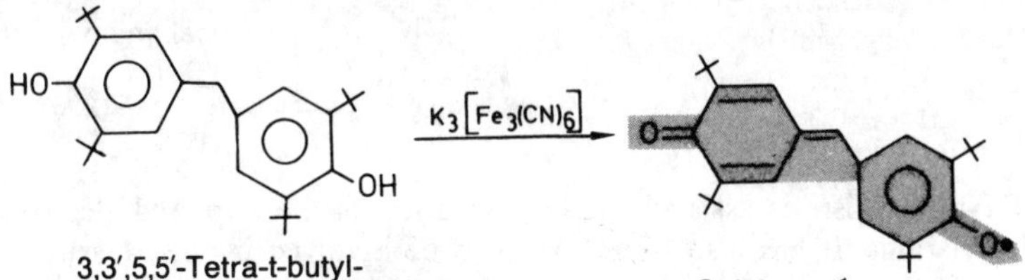

3,3',5,5'-Tetra-t-butyl-
4,4'-dihydroxydiphenylmethane   Galvinoxyl[1]

The ESR spectrum of galvinoxyl shows that the unpaired electron is delocalised over all the parts of the molecule with a grey background in the formula. It is used as a radical trap.

5.18.4 Radical ions

If a solution of benzophenone in ether in an atmosphere of nitrogen is shaken with potassium powder an ion-pair is formed consisting of $K^{\oplus}$ and the deep blue radical anion derived from the ketone ('Benzophenone potassium', benzophenone ketyl, potassium salt). Their discoverer, *W. Schlenk sen.* called these ion-pairs *metal ketyls*.

$$(C_6H_5)_2C=\underline{\overline{O}} \quad + \quad \bullet K \longrightarrow (C_6H_5)_2\overset{\bullet}{C}-\underline{\overline{O}}| \Big]^{\ominus} \quad K^{\oplus}$$

Benzophenone   'Benzophenone potassium' (blue)

Further studies showed that 'benzophenone potassium' consists of a mixture of the ion-pair (77%) and the dimeric form, the pinacolate (23%). (See pinacol.)

The radical character of metal ketyls is evident from their ESR spectra.

Metal ketyls are extremely reactive. Atmospheric oxygen regenerates the ketone; with water disproportionation into benzophenone and benzhydrol takes place in the example shown below. In the presence of dilute acid the ketyls are hydrolysed and transformed into pinacols.

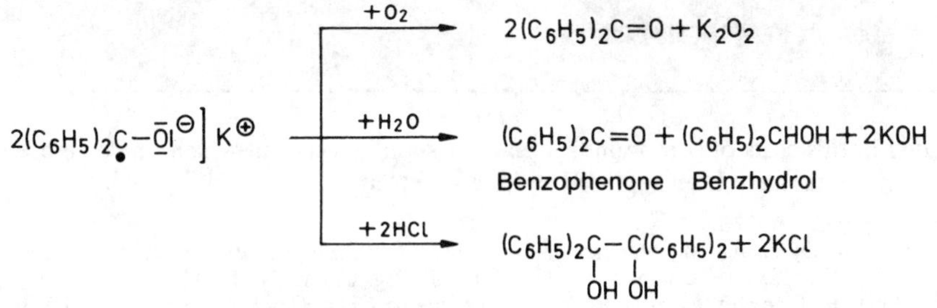

$$2(C_6H_5)_2\overset{\bullet}{C}-\underline{\overline{O}}|\Big]^{\ominus} K^{\oplus} \quad \begin{array}{l} \xrightarrow{+O_2} \quad 2(C_6H_5)_2C=O+K_2O_2 \\[2ex] \xrightarrow{+H_2O} \quad (C_6H_5)_2C=O + (C_6H_5)_2CHOH + 2KOH \\ \qquad\qquad\text{Benzophenone   Benzhydrol} \\[2ex] \xrightarrow{+2HCl} \quad (C_6H_5)_2C-C(C_6H_5)_2 + 2KCl \\ \qquad\qquad\quad |\quad\ | \\ \qquad\qquad\ \ OH\ \ OH \\ \qquad\qquad\text{Benzopinacol} \end{array}$$

[1] Named after the first name of its discoverer, *Galvin Coppinger*; see J. Am. Chem. Soc. 79, 501 (1957).

In the presence of excess of the alkali metal the central carbon atom of 'benzophenone potassium' takes up a further valence electron from another potassium atom and forms a red-violet dianion, benzophenone dipotassium:

$$(C_6H_5)_2\overset{\bullet}{\underset{}{C}}-\bar{\underset{}{O}}|\ ^{\ominus}]\ K^{\oplus} + \bullet K \longrightarrow (C_6H_5)_2\overset{\ominus}{C}-\bar{O}|\ ^{\ominus}]\ 2K^{\oplus}$$

Radical anions also arise when *Grignard* reagents react with aryl ketones by a single electron transfer (SET) mechanism:[1a]

The SET-mechanism is assisted by delocalisation of the unpaired electron over both of the phenyl rings. It has also been shown to be involved in the *Meerwein–Ponndorf–Verley* reduction of aryl ketones[1b] and in aldol reactions of aryl ketones.[1c]

Single electron transfer processes have been detected in a number of reactions which up to now had been explained exclusively in terms of transfer of pairs of electrons.[2]

A similar structure containing a *radical cation*, known as *Wurster's red salt* is obtained by oxidising N,N-dimethyl-p-phenylenediamine with bromine. The following constitution has been attributed to this salt (*Weitz*):

Wurster's red salt

This persistent radical has been examined by ESR spectroscopy. Its stability is due to the combined effects of the electron-donating NH_2-group and the electron-withdrawing $^{\oplus}N(CH_3)_2$-group. Other long-lasting radicals, such as the *semiquinone anions* are similarly stabilised by *captodative substitution* (*Viehe*).[3]

5.19 Phenyl Derivatives of Unsaturated Hydrocarbons

Included in this class of compounds are alkenes and alkynes in which one or more of the hydrogen atoms have been replaced by phenyl groups.

---

[1] See [a]*E. C. Ashby* et al., The Mechanisms of *Grignard* Reagent Addition to Ketones, Acc. Chem. Res. *7*, 272 (1974); [b]J. Org. Chem. *51*, 3593 (1986); [c]J. Org. Chem. *51*, 472 (1986).
[2]See *L. Eberson*, Electron Transfer Reactions in Organic Chemistry (Springer-Verlag, Berlin 1987); *A. Pross*, The Single Electron Shift as a Fundamental Process in Organic Chemistry: The Relationship between Polar and Electron-Transfer Pathways, Acc. Chem. Res. *18*, 212 (1985).
[3] See *H. G. Viehe* et al., The Captodative Effect, Acc. Chem. Res. *18*, 148 (1985); *C. Rüchardt* et al., A Test for Synergetic Captodative Radical Stabilisation, Ang. Chem. Int. Ed. Engl. *26*, 573 (1987).

5.19.1 Arylalkenes

The most important phenyl derivatives of ethylene are *styrene* and *stilbene*:

Styrene   Stilbene

The reactivity of the C=C double bonds of these compounds in addition and polymerisation reactions resembles that of aliphatic alkenes.

5.19.1.1 Styrene (Vinylbenzene). Styrene was first isolated in 1831 by distillation of *liquid storax*, a fragrant balsam; it is also present in coal tar. It is the most important of the arylalkene hydrocarbons.

Preparation. Industrially *by addition of benzene to ethylene* in the presence of *aluminium chloride*. This gives ethylbenzene which is converted into styrene by dehydrogenation over zinc oxide at 600°C (875 K):

$$C_6H_6 + H_2C = CH_2 \xrightarrow{(AlCl_3)} C_6H_5-CH_2-CH_3 \xrightarrow[-2H]{(ZnO)} C_6H_5-CH = CH_2$$

Ethylbenzene   Styrene

Properties. Styrene is a liquid which boils at 146°C (419 K). Under the influence of light, even at room temperature, it sets to a hard glass-like mass of polystyrene [poly(1-phenylethylene)]. Industrially, polymerisation of styrene is carried out in the presence of a *radical-equivalent* such as hydrogen peroxide, giving head-to-tail linkages:

Head   Tail   Polystyrene

Uses. Polystyrene (PS) is the oldest polymerisation product and has been used as a material for die-casting and for making electrical insulators. If an aerosol propellant is introduced into the polymerisation process, foamed or expanded polystyrenes are produced (EPS, Styropor), which have wide usage, for example in making light-weight concretes, as packaging material, as heat insulators, and, in the form of floating balls, to cover the surface of water in mosquito-rife areas, in order to prevent the propagation of the insects, thereby curtailing the spread of malaria. Statistical copolymerisation of butadiene and styrene (see also p. 682) provides synthetic styrene-butadiene rubber SBR (formerly known as *Buna S*). If *divinylbenzene*, $C_6H_4(CH=CH_2)_2$, is polymerised together with styrene, products are obtained with defined amounts of cross-linking, which show one of the typical behaviours due to cross-linking, namely they swell in solvents. They are sulphonated or converted into amines for use as *ion-exchange resins* for anions or cations (see p. 506).

5.19.1.2 Stilbene [(*E*)-1,2-diphenylethylene]. Stilbene is prepared from the reaction between *benzaldehyde* and *benzyl magnesium bromide*. 1,2-Diphenylethanol is the initial product but loses water when heated, giving (*E*)-stilbene (*trans*-stilbene):

$$C_6H_5-C\overset{H}{\underset{O}{\diagdown}} \quad \xrightarrow[- \text{Mg(OH)Cl}]{+C_6H_5CH_2MgCl(+H_2O)} \quad \underset{\underset{OH}{|}}{C_6H_5-CH-CH_2-C_6H_5} \quad \xrightarrow{-H_2O} \quad \underset{H-C-C_6H_5}{\overset{C_6H_5-C-H}{\|}}$$

$$\text{1,2-Diphenylethanol} \qquad\qquad\qquad \text{(E)-Stilbene}$$

Stilbene forms colourless crystals, m.p. 124°C (397 K). It is readily soluble in benzene and is volatile in steam. On irradiation with UV light, the stable (E)-isomer is transformed into the higher-energy labile (Z)-isomer, isostilbene, in an equilibrium process:

$$\underset{H}{\overset{C_6H_5}{\diagdown}}C=C\underset{C_6H_5}{\overset{H}{\diagup}} \quad \underset{\xrightarrow{}}{\overset{hv}{\rightleftharpoons}} \quad \underset{H}{\overset{C_6H_5}{\diagdown}}C=C\underset{H}{\overset{C_6H_5}{\diagup}}$$

Stilbene, (E)-form   Isostilbene, (Z)-form

Isostilbene is an oil with a flowery smell and can be made by decarboxylating 1-phenylcinnamic acid:

$$\underset{\underset{C_6H_5}{|}}{C_6H_5-CH=C-COOH} \quad \longrightarrow \quad \underset{C_6H_5-C-H}{\overset{C_6H_5-C-H}{\|}}$$

Isostilbene

Unlike simple alkenes, stilbene reacts with sodium to give a brown-violet coloured *disodium derivative of 1,2-diphenylethane*:

$$C_6H_5-CH=CH-C_6H_5 \quad \xrightarrow{+2\,Na} \quad C_6H_5-\overset{\ominus}{\underset{}{C}}H-\overset{\ominus}{\underset{}{C}}H-C_6H_5]\,2\,Na^{\oplus}$$

In this reaction both of the sodium atoms give up a valence electron to the C=C double bond of stilbene. A similar reaction takes place even more readily in the case of tetraphenylethylene.

The metal adducts behave like organometallic compounds; for example, with water or ethanol 1,2-diphenylethane (dihydrostilbene) is formed, and with carbon dioxide the disodium salt of diphenylsuccinic acid:

$$C_6H_5-\overset{\ominus}{C}H-\overset{\ominus}{C}H-C_6H_5]\,2\,Na^{\oplus}\Bigg\{\quad\begin{array}{l}\xrightarrow[-2\,NaOH]{+2H_2O}\quad C_6H_5-CH_2-CH_2-C_6H_5\\[2em]\xrightarrow{+2CO_2}\quad \underset{\underset{COO^\ominus}{|}}{C_6H_5-CH}-\underset{\underset{COO^\ominus}{|}}{CH-C_6H_5}\Big]\,2\,Na^{\oplus}\end{array}$$

5.19.1.3 Stilbene dyes.

These dyes are derivatives of 4,4'-diaminostilbene and are *bis-azo dyestuffs*. An important example is the yellow dye *Chrysophenine G* (Direct Yellow 12), which is made by diazotising 4,4'-diaminostilbene-2,2'-disulphonic acid and coupling the resultant diazonium salt with 2 mols of phenol; the phenolic groups are then converted into ethoxy groups by reaction with ethyl bromide. The latter is carried out to provide a product which is more stable to alkali.

Chrysophenine G

Chrysophenine G is a direct dyestuff for wool, silk and cotton.

Treatment of 4,4'-diaminostilbene-2,2'-disulphonic acid with phenyl isocyanate leads to the formation of a reddish-blue fluorescent dyestuff, (*Blankophore R*), which is used as an *optical brightener* or *whitener* in washing powders.

$$C_6H_5-NH-\underset{\underset{O}{\|}}{C}-NH-\langle O \rangle-CH=CH-\langle O \rangle-NH-\underset{\underset{O}{\|}}{C}-NH-C_6H_5$$

$$SO_3Na \quad NaO_3S$$

Blankophore R

5.19.1.4 Diphenylpolyenes. Molecules which contain a long conjugated chain of double bonds are known as polyenes. Diphenylpolyenes have the following general structure:

$$C_6H_5-(CH=CH)_n-C_6H_5$$

1,4-Diphenylbuta-1,3-diene can exist in three *stereoisomeric* forms, as follows:

(*E–E*), m.p. 152°C [425 K]   (*Z–E*), liquid   (*Z–Z*), m.p. 70°C [343 K]

Of these three isomers, only the higher melting one is stable. In sunlight the others change into this (*E–E*)-form.

The stable hydrocarbon is made by condensing phenylacetic acid with cinnamaldehyde; this is done by heating them with acetic anhydride:

$$C_6H_5-\underset{\underset{COOH}{|}}{CH_2}+O=\overset{\overset{H}{\diagup}}{C}-CH=CH-C_6H_5 \xrightarrow[-CO_2,-H_2O]{(H_3CCO)_2O} C_6H_5-CH=CH-CH=CH-C_6H_5$$

Cinnamaldehyde   1,4-Diphenylbutadiene

It forms yellowish crystals and in solution shows a blue-violet fluorescence.

As the number of double bonds increases in the higher diphenylpolyenes there is a regular rise in their melting points and their *colour deepens*. The longest diphenylpolyene prepared so far has fifteen double bonds. It is greenish-black in colour and gives violet-red solutions.

5.19.2 Arylalkynes

The simplest members of this series are *phenylacetylene* and *diphenylacetylene* (tolan).

5.19.2.1 Phenylacetylene, $C_6H_5-C\equiv CH$, is a liquid which boils at 142°C (415 K). It can be prepared from cinnamic acid. This is first converted into 2,3-dibromo-3-phenylpropionic acid. When warmed with sodium carbonate solution this gives β-bromostyrene. If this is melted with a hydroxide, phenylacetylene distils out of the mixture:

$$C_6H_5-CH=CH-COOH \xrightarrow{+Br_2} C_6H_5-\underset{\underset{Br}{|}}{CH}-\underset{\underset{Br}{|}}{CH}-COOH$$

Cinnamic acid

$$\xrightarrow[-HBr, \ -CO_2]{(Na_2CO_3)} C_6H_5-CH=CHBr \xrightarrow[-HBr]{(KOH)} C_6H_5-C\equiv CH$$

β-Bromostyrene   Phenylacetylene

As an acetylene derivative still having one CH-group, phenylacetylene reacts with solutions of silver or copper(I) salts to form the corresponding *phenylacetylide salts*.

5.19.2.2   Diphenylacetylene (tolan), $C_6H_5-C\equiv C-C_6H_5$, forms colourless leaflets, m.p. 62°C (335 K). It is made from 1,2-dibromo-1,2-diphenylethane (stilbene dibromide) by heating the latter with alcoholic potassium hydroxide:

$$C_6H_5-\underset{\underset{Br}{|}}{CH}-\underset{\underset{Br}{|}}{CH}-C_6H_5 \xrightarrow[-2HBr]{(OH^\ominus)} C_6H_5-C\equiv C-C_6H_5$$

1,2-Dibromo-1,2-diphenylethane   Diphenylacetylene

Four halogen atoms will add to the C≡C triple bonds of either phenylacetylene or diphenylacetylene.

5.19.2.3   Diphenylpoly-ynes. Whereas butadiyne is highly unstable, its diphenyl derivative is stable to both light and air for quite a long time at room temperature.

Diphenylbutadiyne is made by the reaction between the *Grignard* derivative of phenylacetylene and 2-iodo-1-phenylacetylene in the presence of *cobalt(II) chloride (Schlubach)*:

$$C_6H_5-C\equiv C-MgBr + I-C\equiv C-C_6H_5 \xrightarrow[-MgBrI]{(CoCl_2)} C_6H_5-C\equiv C-C\equiv C-C_6H_5$$

Diphenylbutadiyne (m.p. 88°C [361 K])

Catalytic perhydrogenation converts diphenylpoly-ynes into the corresponding diphenylalkanes. *Tetraphenylcumulenes* are related structurally to the diphenylpoly-ynes and are therefore considered here.

5.19.3 Cumulenes

Cumulenes are compounds in which a number of double bonds are directly linked to one another. Simple examples are allene (see p. 99) and carbon suboxide (see p. 327).

5.19.3.1   The first cumulene having more than two cumulated double bonds was made by *F. Straus* (1905) by treating diphenylbutadiyne with 1 mol bromine. More recent work has shown that this *electrophilic* addition reaction gives preponderantly (*E*)-1,4-dibromo-1,4-diphenyl-butatriene, m.p. 136–137°C (409–410 K). The *reaction mechanism* is as follows:

$$C_6H_5-C\equiv C-C\equiv C-C_6H_5 \xrightarrow{+Br^\oplus} \left[\underset{Br}{\overset{C_6H_5}{>}}C=C=C=\overset{\oplus}{C}-C_6H_5\right]$$

Diphenylbutadiyne

$$\xrightarrow{+Br^\ominus} \underset{Br}{\overset{C_6H_5}{>}}C=C=C=C\underset{C_6H_5}{\overset{Br}{<}}$$

(*E*)-1,4-Dibromo-1,4-diphenyl-
butatriene (yellow)

Higher tetraphenylcumulenes have been synthesised in a variety of ways.

5.19.3.2 Tetraphenylbutatriene forms yellow needles, m.p. 240°C (513 K) and was first made by *K. Brand* (1921), by the reaction of alcoholic potassium hydroxide with 2,3-dichloro-1,1,4,4-tetraphenylbut-2-ene:

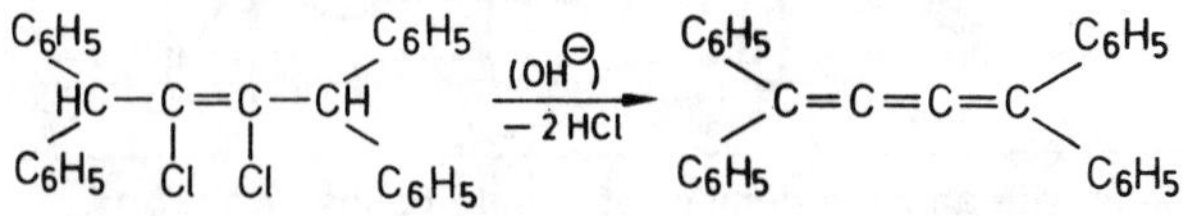

Tetraphenylbutatriene

5.19.3.3 Tetraphenylhexapentaene is a scarlet-red substance, m.p. 302°C (575 K). It was synthesised as shown below by *R. Kuhn* and *Wallenfels* (1938) by treating the dimagnesium derivative of butadiyne with benzophenone.

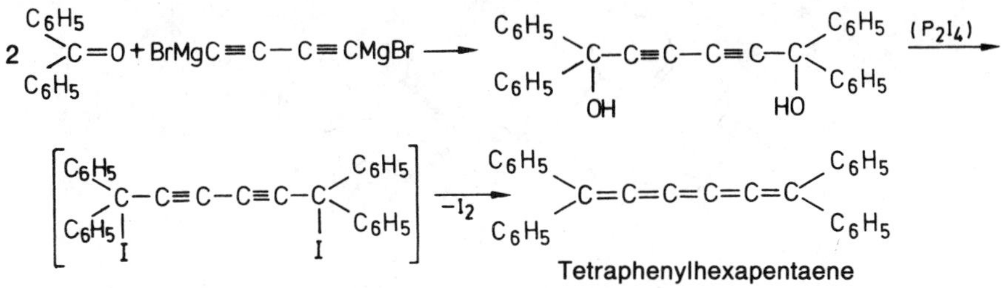

Tetraphenylhexapentaene

Tin(II) chloride and conc. hydrochloric acid can be used instead of diphosphorus tetraiodide (P_2I_4) in pyridine.

As the number of cumulated double bonds increases, so the colour of the cumulenes deepens

5.19.4 Allenic Isomerism (Molecular Asymmetry)

Stereochemical studies of cumulated bonding systems led *van't Hoff* (1874) to the conclusion that substituted allenes of the following type

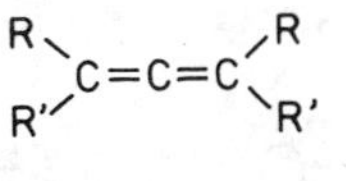

had neither a plane of symmetry nor a centre of symmetry and therefore must be resolvable into *enantiomers*.

The optical isomerism of such cumulenes also follows from the orbital structure of the C=C double bond. In an allene, the carbon atoms at the ends of the system have trigonal sp^2 hybridisation, but the central carbon atom is digonal and sp-hybridised. As a result the carbon atoms are collinear. The central carbon atom forms part of two π-bonding systems whose p orbitals are perpendicular to one another (Fig. 117). The overlap between the orbitals is represented in the diagram by lines joining the lobes, and the planes in which the substituents lie are sketched. There is an *axis of chirality* (see p. 603) and hence such a molecule is resolvable into enantiomers.

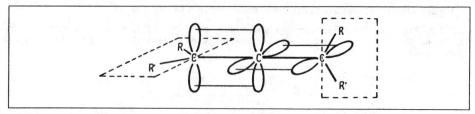

Fig. 117. A cumulene with an even number of double bonds.

Experimental confirmation of *van't Hoff's* prediction was first obtained by *Mills* and *Maitland* (1935) who synthesised 1,3-di-α-naphthyl-1,3-diphenylallene and resolved it into *enantiomers*.

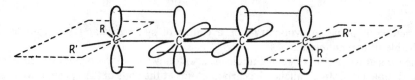

In 1952 the antibiotic *mycomycin* was isolated from the mould *Norcadia acidophilus*. It has the structure shown below (*Celmer* and *Solomons*, 1953) and its optical activity is entirely due to the allene group which is present:

$$H-C \equiv C-C \equiv C-CH = C = CH-CH = CH-CH = CH-CH_2-COOH$$

This carboxylic acid has a most interesting structure in that it contains a 1,3-diyne-, an allene- and a 1,3-diene-system.

Studies of models led *van't Hoff* to make the hypothesis that all compounds which have an odd number of cumulated double bonds must form *Z–E* isomers. If the allene chain shown in the example above is extended by a further double bond then another *sp*-hybridised carbon atom is introduced, providing a three-dimensional structure as shown in Fig. 118. In such a structure, the substituted groups at the two ends of the cumulated chain all lie in the same plane, and *Z* and *E* isomers can exist.

Fig. 118. A cumulene with an odd number of double bonds.

It thus follows that unsymmetrically substituted cumulenes fall into two types, depending upon the number of double bonds in the system. Those with *even* numbers of double bonds are *chiral*; those with *odd* numbers show *geometric* (*E–Z*) *isomerism*.

The presence of *Z* and *E* isomers of a cumulene, as predicted by *van't Hoff*, was first demonstrated in 1954,[1] when the isomers of 1,4-bis-(4-nitrobiphenyl-2,2'-diyl)butatriene were separated chromatographically.

NO₂   O₂N   NO₂

C=C=C=C   C=C=C=C

                                                O₂N

(Z)-form   (E)-form

1,4-Bis-(4-nitrobiphenyl-2,2'-diyl)butatriene (red)

5.20 Condensed Aromatic Ring-systems

This large group of compounds consists of molecules containing two or more carbon rings, the individual rings being linked to their neighbours by having two carbon atoms in common. The hydrocarbons indene and fluorene, both of which are found in coal tar, will be considered first. Both are derivatives of cyclopentadiene, to which one, or two, benzene rings, respectively, have been condensed.

Cyclopentadiene   Indene   Fluorene

The carbon atoms are numbered as shown.

5.20.1 *Indene*

Pure indene is isolated from industrial indene obtained from coal tar by heating it with metallic sodium (approximately the calculated amount) at about 150°C (425 K) or with molten sodium hydroxide. This results in the formation of **sodium indenide**, C_9H_7Na:

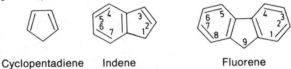

Sodium indenide

[1] See *R. Kuhn* et al., Chem. Ber., *87*, 598 (1954).

The residual hydrocarbon material is distilled off *in vacuo*. The sodium indenide is hydrolysed and pure indene separated off by steam distillation.

Indene (1*H*-cyclopentabenzene), C_9H_8, is a colourless liquid, b.p. 182°C (455 K) which crystallises when cooled (m.p. −2°C [271 K]). It polymerises on standing, especially if heated. The double bond in the five-membered ring can be reduced by nascent hydrogen (*e.g.* sodium and ethanol) *to indane (hydrindene)*:

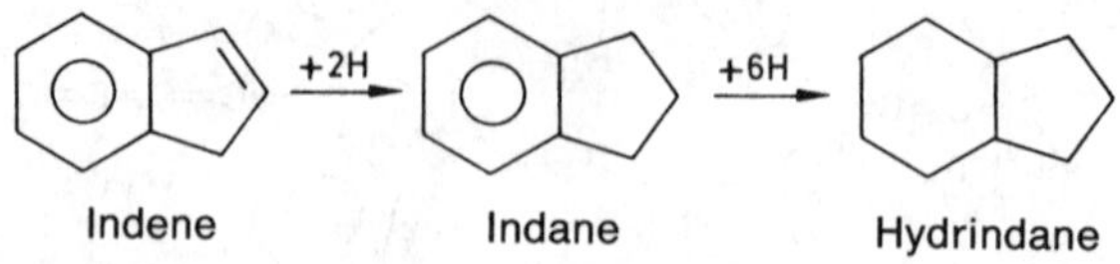

Indene   Indane   Hydrindane

Catalytic hydrogenation of indene provides *hydrindane* which can exist in *cis* and *trans* forms. (See conformation, p. 413.)

The methylene group in indene, as in cyclopentadiene, is *reactive* and in alkaline conditions condenses with aldehydes or ketones to form *benzofulvenes* (*cf.* fulvenes).

5.20.1.1 Ninhydrin, the hydrate of indane-1,2,3-trione, is used to *test* for amino acids, polypeptides and proteins, with which it produces a blue or violet colour (*Ruhemann*, 1910).

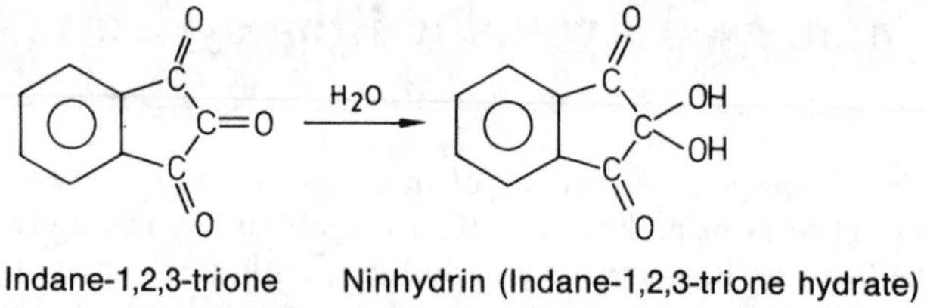

Indane-1,2,3-trione   Ninhydrin (Indane-1,2,3-trione hydrate)

A simplified version of the mechanism of the reaction can be formulated as follows:[1]

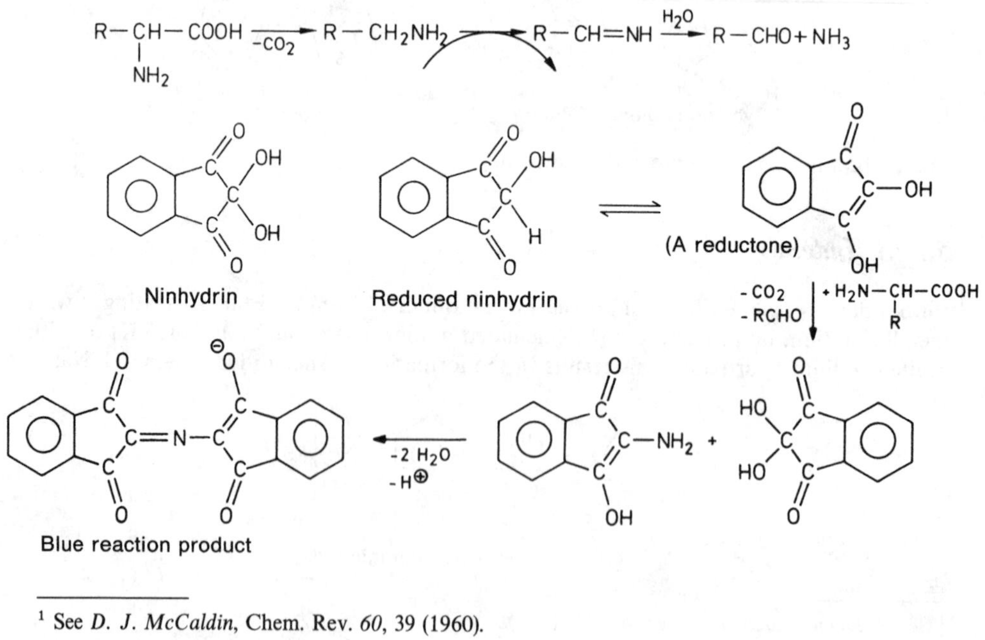

$$R-CH-COOH \xrightarrow{-CO_2} R-CH_2NH_2 \rightleftharpoons R-CH=NH \xrightarrow{H_2O} R-CHO + NH_3$$

Ninhydrin   Reduced ninhydrin   (A reductone)

Blue reaction product

---

[1] See *D. J. McCaldin*, Chem. Rev. *60*, 39 (1960).

The *amino acid is decarboxylated to an amine*. This undergoes an *oxidative deamination* to an *aldehyde* (*Strecker* degradation) in which the ninhydrin is reduced. The first isolable intermediate is an amino derivative of the reductone, which is itself a product derived by tautomerism of the *reduced ninhydrin*. The amino derivative reacts with a further molecule of ninhydrin to give a blue reaction product.

5.20.2 Fluorene

Fluorene is obtained either from coal tar, separating it by means of its sodium salt, or synthetically by passing diphenylmethane vapour through a red-hot tube where it is dehydrogenated to give fluorene:

Diphenylmethane   Fluorene

Fluorene (2,2'-methylenebiphenyl), $C_{13}H_{10}$, is a stable colourless compound which, unlike cyclopentadiene and indene, does *not* polymerise. It crystallises in lustrous leaflets, m.p. 116°C (389 K). Its name arose from the fact that impure samples of fluorene showed a blue *fluorescence*, due in fact to the contaminants. Halogenation, nitration and sulphonation lead to stepwise substitution in the 2- and 7-positions; this is noteworthy in that these sites are *meta* to the methylene group. It is oxidised to **fluorenone**, m.p. 85°C (358 K), by chromic acid in acetic acid.

Fluorenone   Benzylidenefluorene

Fluorenone is deep yellow; the colour is associated with the cross conjugated system which is present (*cf.* fulvenes). Compounds with similar structure, which are also for the most part yellow, are obtained from condensation reactions of the reactive methylene group of fluorene with aldehydes or ketones in the presence of alkoxides; for example, either in solution or in a melt, benzaldehyde and fluorene condense to form 9-benzylidenefluorene (6-phenyldibenzofulvene) (see above).

As well as indene and fluorene a number of condensed aromatic hydrocarbons made up solely from benzenoid rings are found in coal tar. The most important are naphthalene, anthracene and phenanthrene.

5.20.3 Naphthalene

Naphthalene, $C_{10}H_8$, was discovered, as a constituent of coal tar, by *Garden* in 1819. It occurs there to the extent of about 5–6% and is obtained industrially from this source.

It forms colourless shiny leaflets, m.p. 80°C (353 K), with a characteristic odour. It is insoluble in water but is volatile in steam. It dissolves readily in most common organic solvents, especially when heated.

5.20.3.1 Structure of Naphthalene. Making use of *Kekulé*-type benzene formulae, three limiting structures can be drawn for naphthalene which only differ from one another in the disposition of the C=C double bonds:

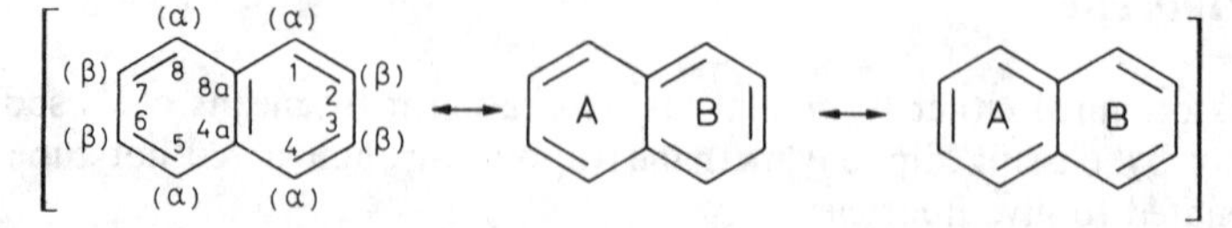

The numbering shown is used to designate the positions of substituents. In older literature it was commonplace to describe mono-substituted products by the Greek letters α and β. In addition it is customary to indicate individual rings in condensed systems by the capital letters A, B, C, D, *etc.* as shown.

Physical properties, and above all the *Raman* spectrum, show the symmetric arrangement of the π-electrons in the naphthalene molecule. The two rings are entirely equivalent and the π-electrons form a 10-π-electron system. The decet of electrons can be represented either by *Kekulé*-style formulae or as shown in (a). Formula (b), which is often used, is incorrect since it implies a 12-π-electron system. The latter is in fact present in the sodium derivative of naphthalene (c).

(a)   (b)

In contrast to benzene, naphthalene can be reduced by sodium and ethanol. Stepwise reduction goes first to 1,4- and 1,2-dihydronaphthalene and then to 1,2,3,4-tetrahydronaphthalene (tetralin). In addition to this method, hydrogen can also be added to the 1,4- or 1,2-positions of naphthalene by reaction with an alkali metal in ether. The alkali metal adds to naphthalene to give a metal derivative. The two additional electrons are not

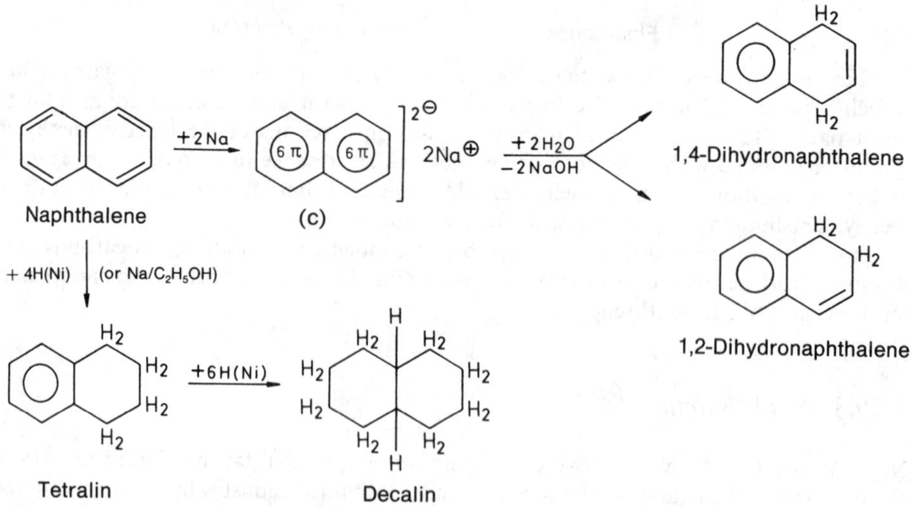

Naphthalene   (c)

+ 4H(Ni)   (or Na/C₂H₅OH)

Tetralin   Decalin

localised on any particular carbon atoms but rather form part of a system of 12 π-electrons. The anion is very readily protonated; thus, on hydrolysis it gives 1,4- or 1,2-dihydronaphthalene (see previous page).

Catalytic hydrogenation of naphthalene gives first of all tetralin which, because of its benzenoid character, is only reduced further to decalin (decahydronaphthalene) under more vigorous conditions (*G. Schroeter*).

The main product obtained in this way is *cis*-decalin (see conformation, p. 412).

5.20.3.2   *Birch* reduction. A regioselective hydrogenation of naphthalene to 1,4-dihydro-naphthalene, m.p. 25°C (298 K), is achieved by using *sodium in liquid ammonia* and alcohol. Naphthalene is dissolved in, for example, ether, and ethanol and liquid ammonia are added, followed by sodium. If excess sodium is used, 1,4,5,8-tetrahydronaphthalene (m.p. 58°C [331 K]) is formed.

1,4-Dihydro-
naphthalene

1,4,5,8-Tetra-
hydronaphthalene

(I)

(II)

(III)

Reaction is initiated by transfer of an electron from the metal to the hydrocarbon, generating a radical anion (I). This extracts a proton from the alcohol to form the radical (II), which accepts another electron from a sodium atom to give the carbanion (III), which is in turn converted into the product by adding another proton from an alcohol molecule.[1]

Birch reduction is a very valuable synthetic process and has been much developed,[2] including participation in a stereochemically important reaction, see p. 212.

5.20.3.3   The number of isomeric substitution products derived from naphthalene is inevitably greater than that from benzene. There are *two* monosubstitution products, frequently described as the α- and β-(or 1- and 2-)isomers. Oxidation of the two methylnaphthalenes provides, respectively, the α- and β-naphthalenecarboxylic acids, usually called, by analogy with benzoic acid, the α- and β- or *1*- and *2-naphthoic acids*.

α-Methylnaphthalene   1(α)-Naphthoic acid β-Methylnaphthalene   2(β)-Naphthoic acid

Correspondingly there are a larger number of possible isomers of *disubstitution products*. If both substituents are the same, there are ten possible isomers; if they are different, fourteen. Historically 1,8-disubstituted naphthalenes were described by the prefix '*peri*' and 2,6-derivatives by the prefix '*amphi*', *e.g.*

---

[1] See *A. J. Birch* et al., J. Am. Chem. Soc. *103*, 284 (1981).
[2] See *E. M. Kaiser*, A Comparison of Methods using Lithium/Amine and *Birch* Reduction Systems, Synthesis *1972*, 391.

peri-Naphthalene-
dicarboxylic acid
(Naphthalic acid)

amphi-Dihydroxy-
naphthalene

5.20.3.4   Naphthalic acid, which is used as a starting material in the dyestuffs industry, is made by atmospheric oxidation of acenaphthene in the gas phase. When heated to 180°C (450 K) it forms an intramolecular anhydride, since the two carboxyl groups, being in the *peri*-positions, are adjacent to one another.

Substitution Products of Naphthalene

Naphthalene undergoes similar substitution reactions to those of benzene but the orientation involved when a second or further substituents are introduced becomes complicated. Monosubstitution in naphthalene takes place chiefly at an α-position, with the exception of sulphonation at higher temperatures and *Friedel–Crafts* reactions carried out in nitrobenzene as solvent; these two processes result in β-substitution.

The second substituent may enter either the ring which already bears a substituent or the unsubstituted ring. As a guideline, depending upon the substituent present the following rules apply (see p. 482):

(a) When *substituents of type 1*, especially —OH, —NH$_2$, are present at position 1, they direct the next substituent mainly to the *o*- and *p*-positions of the same ring, namely the 2- and 4-positions (and also, to a limited extent, to positions 5 and 7).
    If such substituents are present at the 2-position, then further substitution takes place primarily at the 1-position, with some also occurring at the 8- and 6-positions.
(b) If *substituents of type 2*, e.g. —NO$_2$, —SO$_3$H, —COOH, are present at position 1, there is no tendency for substitution to take place at the *m*- (4-)position, but rather it takes place in the other ring at the 5-, 6- or 8-positions. If such substituents are present at position 2, then further reaction takes place at the 5-, 6- or 7-positions.

5.20.3.5   *Chlorination of naphthalene in the presence of iron gives 95% 1-chloronaphthalene and only 5% of 2-chloronaphthalene:*

$$+Cl_2(Fe) \quad -HCl$$

and

1-Chloronaphthalene   2-Chloronaphthalene

5.20.3.6   In similar fashion, *nitration* of naphthalene with conc. nitric acid provides almost exclusively *1-nitronaphthalene* and only a small amount of *2-nitronaphthalene*. *1,5-* and *1,8-dinitronaphthalenes* are formed by further nitration of 1-nitronaphthalene using nitrating acid. In tests on animals 2-nitronaphthalene has been shown to be a carcinogen.

5.20.3.7   Depending upon the temperature, *sulphonation* of naphthalene may provide either *naphthalene-1-sulphonic acid* (at temperatures below 80°C [350 K]) or *naphthalene-2-sulphonic acid* (at temperatures above 120°C [390 K]) as the main product. Under stronger conditions these acids are converted, respectively, into *naphthalene-1,5-* and *-1,6-disulphonic acids* and *naphthalene-2,6-* and *-2,7-disulphonic acids*:

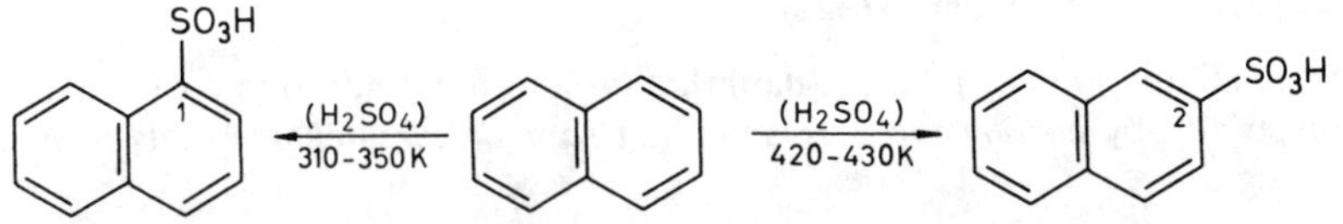

Naphthalene-1-sulphonic acid   Naphthalene   Naphthalene-2-sulphonic acid

(H_2SO_4) (H_2SO_4)

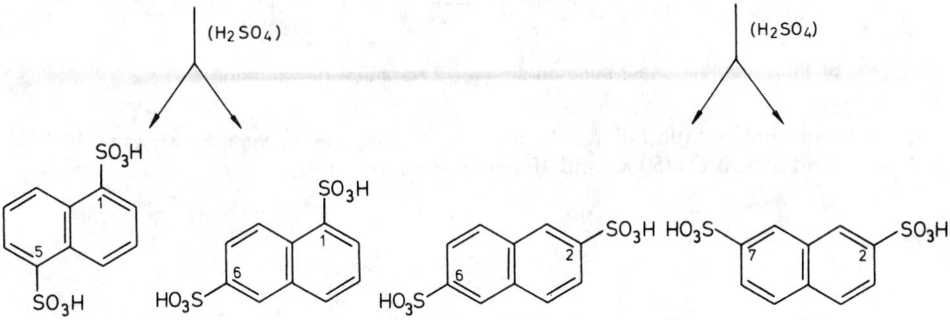

| 1,5- | 1,6- | 2,6- | 2,7- |

Naphthalenedisulphonic acid   Naphthalenedisulphonic acid

At higher temperatures naphthalene-1-sulphonic acid is converted, by means of an equilibrium reaction with free naphthalene as an intermediate, into naphthalene-2-sulphonic acid. Although the 1-acid is the product at lower temperatures it is also, on the other hand, the more readily desulphonated. At 160°C (435 K) *equilibrium* between the two monosulphonic acids takes place:

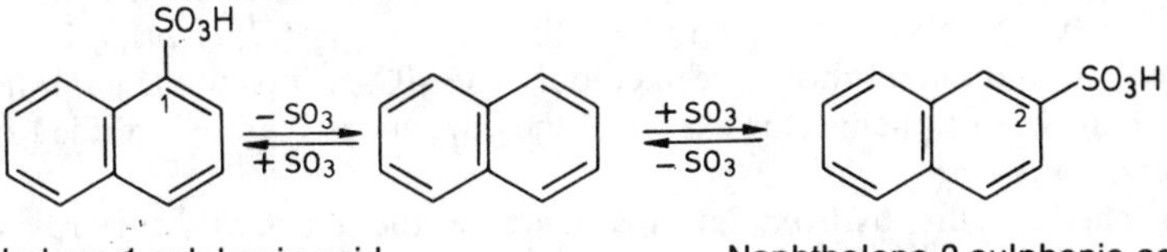

Naphthalene-1-sulphonic acid   Naphthalene-2-sulphonic acid
(15%)   (85%)

Naphthalenesulphonic acids are important intermediates in the preparation of naphthols and their sulphonic acid derivatives, which are used as *coupling components* in the dyestuffs industry.

5.20.3.8 *Friedel–Crafts acylation* of naphthalene provides a mixture of the 1- and 2-acyl derivatives, the relative amounts of the two depending upon the reaction conditions. Use of acetyl chloride in carbon disulphide gives largely the 1-substituted product and only about 25% of the 2-isomer. In contrast, in *nitrobenzene* as solvent the product is almost exclusively the 2-substituted derivative:

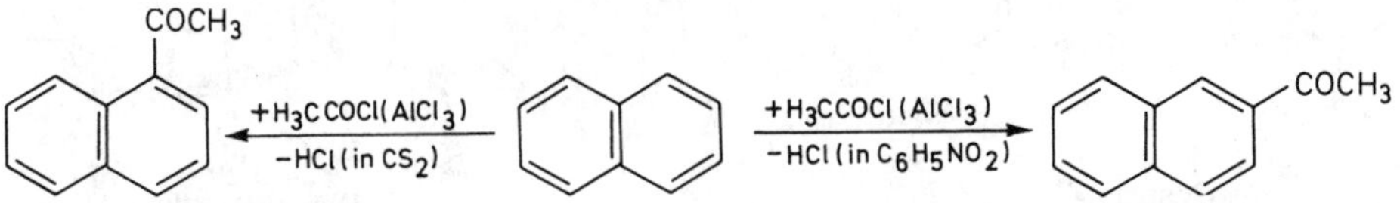

1-Acetylnaphthalene   Naphthalene   2-Acetylnaphthalene
(1-Acetonaphthone,   (2-Acetonaphthone,
Methyl 1-naphthyl ketone)   Methyl 2-naphthyl ketone)

5.20.3.9 Naphthols (Hydroxynaphthalenes)

There are small amounts of 1- and 2-naphthol (α- and β-naphthol) in coal tar. They are made industrially by *melting* the corresponding *naphthalenesulphonic acids with sodium hydroxide, e.g.*

Sodium naphthalene-2-sulphonate   2-Naphthol (β-Naphthol)

Pure 1-naphthol (α-naphthol) can be made by *hydrolysis of 1-naphthylamine* with 10% sulphuric acid at 180°C (450 K) and 10 bar (1 MPa) (yield ∼ 85%):

1-Naphthylamine   1-Naphthol

1-Naphthol crystallises in needles, m.p. 95°C (368 K), **2-naphthol** in glistening rhombic crystals m.p. 123°C (396 K). Both are only very slightly soluble in cold water. Being phenols, they are weakly acidic and dissolve in alkali to form naphthoxide salts which are extensively hydrolysed in aqueous solution. With *iron(III) chloride* 1-naphthol gives a violet colour and 2-naphthol a green colour. The HO-groups of the naphthols are more reactive than that of phenol; for example, they are more readily esterified or converted into ethers.

As in phenols, the hydroxy groups increase the reactivity towards electrophilic substitution; thus they *couple* readily with diazonium salts, in the case of 1-naphthol at the 2- and 4-positions and in the case of 2-naphthol at the 1-position.

The poisonous compound *gossypol*, which is found in cottonseed, is an anti-nutrient factor. It has an axis of chirality and is atropisomeric (see p. 603), $[\alpha]_D^{19} = +445°C$ (CHCl$_3$).

Gossypol   INN: *Rifampicin*

Once it was discovered that gossypol could be excluded from cottonseed oil, the latter could be used as a foodstuff. This is a typical example of the importance of biotechnology.

The chemically modified macrolide antibiotic *INN*: **Rifampicin** (*Rifa, Rimactan*) is effective against tuberculosis and leprosy. Its action is associated with arresting the action of the DNA-dependent RNA-polymerase.

5.20.3.10 Naphthylamines

The ready replacement of the hydroxy group in 2-naphthol by an amino group is made use of industrially for the preparation of *2-naphthylamine*.

Bucherer reaction.[1] In this process 2-naphthol is heated in an autoclave at 150°C (425 K) with an aqueous solution of ammonium hydrogen sulphite and conc. ammonia (1 : 1).

According to *Rieche* and *Seeboth* the reaction mechanism is as follows:

2-Naphthol

2-Naphthylamine

First of all a molecule of ammonium hydrogen sulphite adds across the 3- and 4-positions of 2-naphthol. Ammonia then reacts with the resultant ammonium salt of 2-tetralone-4-sulphonic acid and loss of hydrogen sulphite provides 2-naphthylamine.

In the same way 1-naphthol is converted, *via* intermediate formation of 1-tetralone-3-sulphonic acid, into 1-naphthylamine.

These reactions are reversible. They do not take place if a sulphonyl group is present at the *m*-position to the hydroxy group of the naphthol.

1-Naphthylamine (α-naphthylamine) is more satisfactorily prepared by reducing 1-nitronaphthalene (in the same way that aniline is obtained from nitrobenzene). It forms needles, m.p. 50°C (323 K). **2-Naphthylamine** (β-naphthylamine) forms leaflets, m.p. 110°C (383 K). Both amines form stable crystalline salts with mineral acids. Both can be diazotised in the normal way and are used in the preparation of dyestuffs. β-Naphthylamine is strongly *carcinogenic*.

1,8-Bis(dimethylamino)naphthalene has a very high basicity, which is associated with the fact that the lone pairs of electrons on the two nitrogen atoms are crowded into a relatively small space; it has been described as a 'proton sponge'.[2]

Proton sponge

Hydride sponge

---

[1] See *N. L. Drake*, The *Bucherer* Reaction, Org. React. *1*, 105 *ff.* (1942).
[2] See *H. A. Staab* et al., 'Proton Sponges'and the Geometry of Hydrogen Bonds: Aromatic Nitrogen Bases with Exceptional Basicities, Angew. Chem. Int. Ed. Engl. *27*, 865 (1988).

1,8-Naphthalenediyl-bis(dimethylborane) has a similar affinity for hydride ions and has been called a 'hydride sponge'.

5.20.3.11 Naphthol and Naphthylamine Sulphonic Acids

Both of these classes of compounds are of importance as coupling components in the manufacture of azo-dyestuffs. The simplest sulphonic acid derived from 1-naphthylamine is naphthionic acid which is made by 'baking' the amine hydrogen sulphate (see sulphanilic acid).

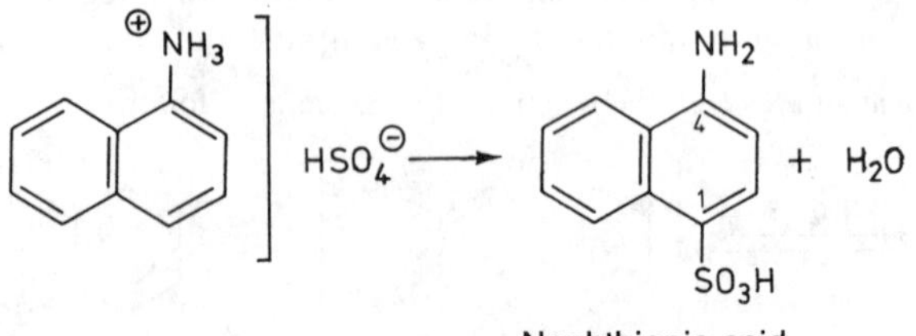

Naphthionic acid

It is used as a coupling component in making the bisazo-dyestuff Congo Red.

5.20.3.12 Hydroxynaphthoic Acids

3-Hydroxy-2-naphthoic acid is used as a starting material for the industrial preparation of naphthol-AS dyestuffs. It is obtained by the action of carbon dioxide on the sodium salt of 2-naphthol at elevated temperature (240°C [513 K]) (see *Kolbe–Schmitt* synthesis). At lower temperatures (150°C [423 K]) almost the only product is the *isomer*, 2-hydroxy-1-naphthoic acid. When heated at 240°C (513 K) the latter is converted into 3-hydroxy-2-naphthoic acid:

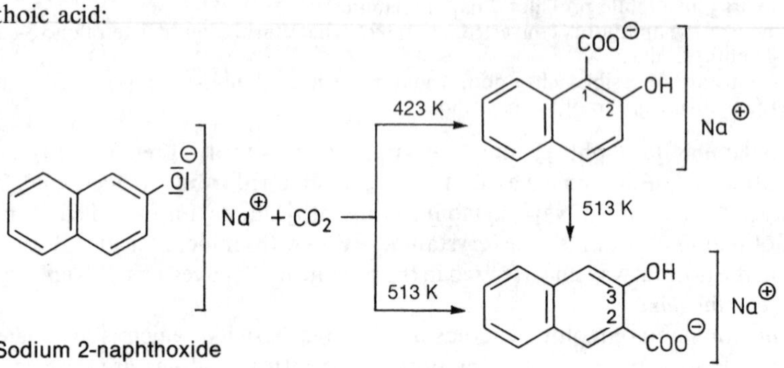

Sodium 2-naphthoxide

5.20.3.13 Naphthoquinones

There are three possible naphthoquinones:

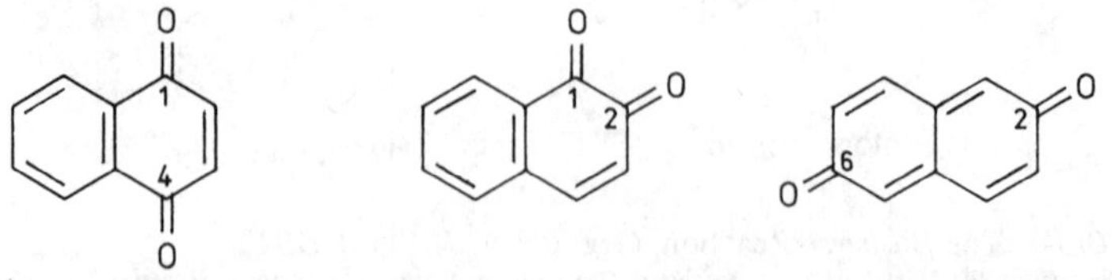

1,4-Naphthoquinone   1,2-Naphthoquinone   2,6-Naphthoquinone

1,4-Naphthoquinone is formed when either naphthalene or 4-amino-1-naphthol is oxidised with chromic acid in acetic acid. It forms yellow crystals, m.p. 126°C (399 K) and resembles 1,4-benzoquinone in its physical properties.

1,2-Naphthoquinone is obtained by oxidising 1-amino-2-naphthol. It forms yellow needles which melt with decomposition at 145–147°C (418–420 K). In contrast to 1,4-naphthoquinone it is odourless and involatile in steam.

2,6-Naphthoquinone is the only naphthoquinone having one carbonyl group in each of the two rings. It is formed as yellow-red prisms, m.p. 130–135°C (403–408 K) (decomp.) when 2,6-dihydroxynaphthalene is oxidised by lead(IV) oxide in benzene. It is a stronger oxidising agent than the other two naphthoquinones.

Some derivatives of 1,4-naphthoquinone are found occurring naturally. Examples are *juglone* (5-hydroxy-1,4-naphthoquinone), which is found in walnut shells and stains the skin brown, and the coagulating vitamins K_1 and K_2, which are derivatives of 2-methyl-1,4-naphthoquinone (menadione). Lack of vitamin K leads to interference with the normal coagulating processes of blood.

Vitamin K_1 (2-methyl-3-phytyl-1,4-naphthoquinone, phylloquinone, 3-phytylmenadione, INN: *phytomenadione*, konakion), $C_{31}H_{46}O_2$, is found in parts of green plants and was isolated by *Karrer* (1939) from *alfalfa* (lucerne) by extracting it with petrol. It is a yellow, viscous oil.

Vitamin K₁

Its structure was confirmed by synthesis from 2-methyl-1,4-naphthoquinone and phytol in the presence of anhydrous oxalic acid and oxidation of the dihydrovitamin K_1 thereby formed (*Fieser*, 1939).

Vitamin K_2, $C_{46}H_{64}O_2$, m.p. 54°C (327 K), is one of the bacterial vitamins and was obtained from rotting fishmeal (*Doisy*, 1939). In the 3-position it has a side-chain made up of seven isoprene units which are arranged in an *all-trans* configuration (*Isler*).[1] Bacteria also produce an isoprenoid vitamin K_2, $C_{41}H_{56}O_2$, m.p. 50°C (323 K), containing only six isoprene units, *i.e.* 30 carbon atoms, in the side-chain. The two vitamins are now known as $K_{2(35)}$ and $K_{2(30)}$.

Vitamin K₂₍₃₅₎

Work by *Martius* indicates that the K-vitamins play an important role in oxidative phosphorylation in the respiratory chain. Vitamin K_1 is indispensible for bringing about coagulation of blood.

2-Methyl-1,4-naphthoquinone (INN: *menadione*) and some related compounds bring about similar, and, in part, greater physiological action than the K-vitamins.

---

[1] See *O. Isler* Über die Vitamine K_1 and K_2 (About Vitamins K_1 and K_2), Angew. Chem. *71*, 7 (1959); *R. A. Morton*, Quinones as Biological Catalysts, Endeavour, *24*, 81 (1965).

INN: *Menadione*   (*R*)-BINAP   (*S*)-BINAP ruthenium(II)dicarboxylate complex

The isomerism due to restricted rotation (atropisomerism) associated with some biphenyl derivatives can also occur in binaphthyl derivatives. Thus 2,2'-bis(diarylphosphino)-1,1'-binaphthyl (BINAP) has an axis of chirality. It can be converted into chiral metal complexes which are easily prepared and which are very useful as asymmetric catalysts. The BINAP-Ru(II)dicarboxylate complex has proved to be especially useful for enantioselective hydrogenations.[1]

5.20.4 Acenaphthene

Acenaphthene (1,8- or *peri*-ethylene-naphthalene), $C_{12}H_{10}$, was isolated from the coal-tar fraction boiling at about 300°C (575 K); it forms colourless needles, m.p. 95°C (368 K). When oxidised with chromic acid in acetic acid the first product is *acenaphthenequinone* (acenaphthoquinone) which crystallises in yellow needles, m.p. 262°C (535 K). It does not behave like a quinone, but rather as a *diketone*. It is used for the preparation of indigoid vat dyes. Further oxidation of acenaphthenequinone provides *naphthalic acid*.

Acenaphthene   Acenaphthenequinone   Naphthalic acid

5.20.5 Anthracene

Anthracene, $C_{14}H_{10}$, was first discovered in 1832, as a component of the anthracene oil obtained from the distillation of coal tar; it is still produced commercially from this source. Pure anthracene forms colourless leaflets, m.p. 218°C (491 K) which show a blue-violet fluorescence, especially in UV light. They dissolve readily in boiling benzene.

5.20.5.1 Structure of Anthracene

The anthracene molecule is made up from three benzene rings fused together in a linear arrangement. Because it could be prepared by a *Friedel–Crafts* synthesis involving tetrabromomethane, benzene and aluminium chloride (*R. Anschütz*), anthracene was for a long time ascribed a structure with a bridged central ring:

---

[1] See *R. Noyori* et al., BINAP: An Efficient Chiral Element for Asymmetric Catalysis, Acc. Chem. Res. *23*, 345 (1990); Enantioselective Catalysis with Metal Complexes, an Overview, in *R. Scheffold* (Ed.), Modern Synthetic Methods 5 (Springer-Verlag, Berlin 1989).

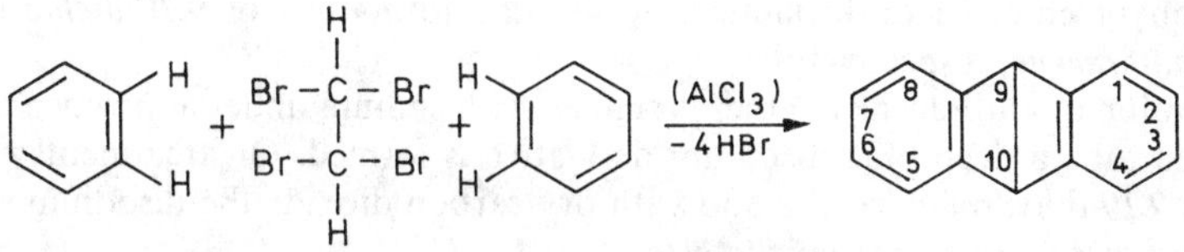

Later on, the presently accepted formula was suggested by *Armstrong* and *Hinsberg*:

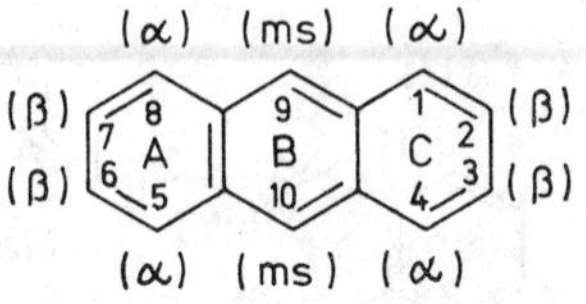

Anthracene

The numbering system for anthracene resembles that used for naphthalene and is shown above. The 9,10-positions have sometimes been called the *meso*-positions, but the preferred system is that using the numerals.

Similar considerations apply to the *14-π-electron system of anthracene* as to the 10 π-electron system of naphthalene.[1]

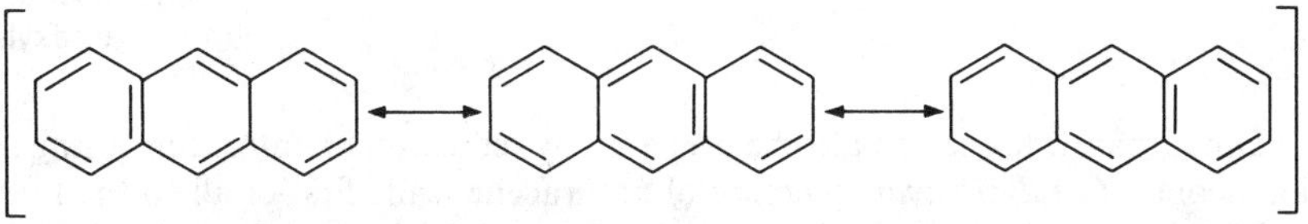

Since there are three rings in anthracene there is the possibility for larger numbers of isomeric substitution products. *Three* monosubstituted isomers are possible, namely at the α-, β- or 9,10-positions. If two identical substituents are present there is the possibility of no fewer than 15 isomers; if they are not identical the number is even greater.

5.20.5.2 Addition Reactions of Anthracene

Whereas the two rings in naphthalene have identical reactivity this is not so for the three rings of anthracene, wherein the central ring differs from the other two. The 9,10-positions in particular undergo addition reactions; this is favoured by the fact that in the process two *benzenoid systems*, each with a sextet of π-electrons, are generated in the outer rings. If a solution of anthracene exposed to oxygen is irradiated with UV light, *anthracene peroxide* is formed; crystals of this compound decompose explosively when heated to 120°C (400 K) (*Dufraisse* and *Gérard*).

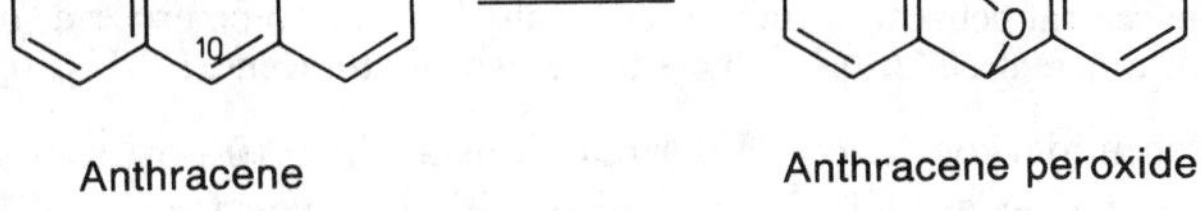

Anthracene   Anthracene peroxide

[1] See *C. Glidewell* and *D. Lloyd*, Tetrahedron *40*, 4455 (1984); J. Chem. Educ. *63*, 306 (1986).

Addition of chlorine or bromine leads to the formation of *9,10-dichloro-* or *9,10-dibromo-anthracene*, respectively.

If a solution of anthracene in ether is shaken with sodium under a nitrogen atmosphere in a sealed tube a deep blue disodium derivative is formed. On treatment with alcohol this gives *9,10-dihydroanthracene* and with dry carbon dioxide the disodium salt of *9,10-dihydroanthracene-9,10-dicarboxylic acid:*

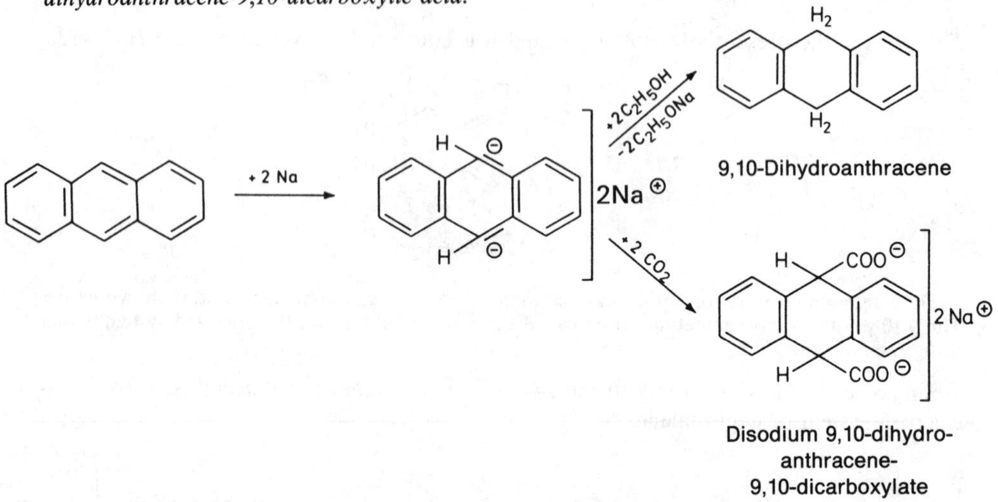

9,10-Dihydroanthracene

Disodium 9,10-dihydro-
anthracene-
9,10-dicarboxylate

9,10-Dihydroanthracene can also be prepared by reduction of anthracene using sodium and an alcohol. *Catalytic hydrogenation* of anthracene leads first of all to the formation of the 9,10-dihydro-derivative, which can be reduced further to give *1,2,3,4-tetrahydro-* and *1,2,3,4,5,6,7,8-octahydro-anthracenes.* In this process the 9,10-hydrogen atoms must shift into the neighbouring rings (*G. Schroeter*):

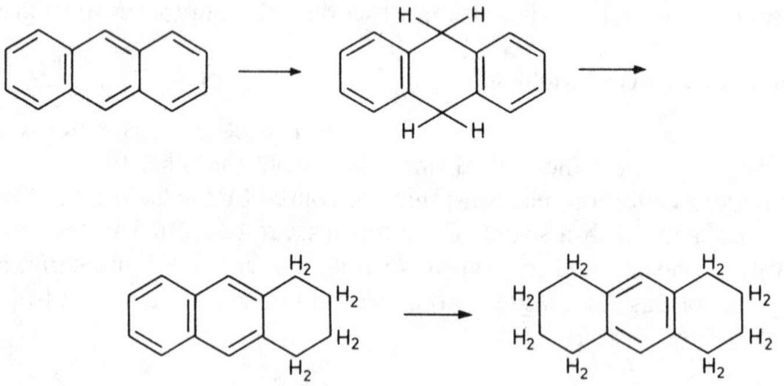

Reduction of the central benzenoid ring in the octahydroanthracene proceeds much more slowly, giving a mixture of different *cis–trans* isomers of perhydroanthracene, $C_{14}H_{24}$ (*cf.* decalin).

Anthracene also undergoes *Diels–Alder* reactions at the 9,10-positions with *dienophiles*, for example with maleic anhydride, and also with benzyne (see p. 481). Benzyne was generated by diazotising anthranilic acid in an inert solvent with n-amyl nitrite and trapped by anthracene to form *triptycene* (9,10-*o*-benzenoanthracene) (yield ~60%):

Anthranilic acid   Benzenediazonium-2-carboxylate   Benzyne

Anthracene   Triptycene

Triptycene is a colourless stable substance, m.p. 255°C (528 K).

5.20.5.3 Derivatives of anthracene

Oxidising agents readily attack anthracene at the 9,10-positions to give 9,10-anthraquinone (anthracene-9,10-dione). In the process the π-electronic system of anthracene is destroyed and two benzenoid rings are provided in its place. Anthraquinone is used as the starting material in the preparation of alizarin and indanthrene dyestuffs.

Anthraquinone

Preparation

(a) Anthraquinone was first obtained by the action of nitric acid on anthracene (*Laurent*, 1840). No nitration occurs, but a fast oxidation reaction takes place instead:

Anthracene   Anthraquinone

(b) In the laboratory *sodium dichromate* is used as the *oxidising agent*. On the industrial scale, *chromium trioxide* is used, either at 50–100°C (325–375 K) in the *liquid phase*, or at 340–400°C (615–675 K) in the *gas phase*. Air can also be used as the oxidising agent, with iron vanadate as catalyst.

(c) Anthraquinone can be obtained synthetically from a *Friedel–Crafts* reaction between *phthalic anhydride* and *benzene* in the presence of *aluminium chloride*. o-Benzoylbenzoic acid is formed first; in the presence of conc. sulphuric acid this loses water intramolecularly to give anthraquinone:

Phthalic anhydride   o-Benzoylbenzoic acid   Anthraquinone

This reaction helped to confirm the constitution of anthraquinone. By varying the reactants, substituted anthraquinones can be made in the same way. It is of industrial importance, as is the following process.

(d) Anthraquinone can also be made by a *Diels–Alder* reaction of 1,4-naphthoquinone with butadiene followed by dehydrogenation of the initial cyclo-adduct:

1,4-Naphthoquinone   Butadiene

Naphthoquinone is the first substance produced in the oxidative conversion of naphthalene into phthalic anhydride.

Properties. Anthraquinone crystallises in yellow needles, m.p. 286°C (559 K); at higher temperatures it sublimes. It dissolves best in boiling benzene. In contrast to *p*-benzoquinone it is very stable and only slightly affected by oxidising agents. It has no real quinonoid character but rather falls somewhere between a quinone and a ketone. It thus has none of the properties normally associated with quinones such as being strong oxidising agents and being volatile in steam. This divergence in properties can be attributed to the fact that the central ring is fused to the flanking benzene rings.

5.20.5.4   The most important reaction of anthraquinone is its ready reduction by sodium dithionite to **9,10-anthrahydroquinone** which, as a phenol, is soluble in alkali. In air it is oxidised back to the insoluble anthraquinone:

$(Na_2S_2O_4, NaOH)$

(O_2)

This redox process forms the basis of the vat dyeing by anthraquinone dyestuffs.

When anthraquinone is reduced by tin and hydrochloric acid in acetic acid the product is anthrone [9(10*H*)-anthracenone], m.p. 155°C (428 K). Anthrone, which is colourless, is a *tautomer* of the brownish-yellow metastable phenol *anthranol*. Of these two tautomers anthrone contains two benzene rings and is therefore the more stable form.

$+4H$
$-H_2O$

Anthraquinone   Anthrone (89%)   Anthranol (11%)

If anthraquinone is heated with zinc dust, it is reduced to anthracene which can be distilled out from the mixture.

5.20.5.5 Substitution Products From Anthraquinone. Substitution of the hydrogen atoms in the outer rings of anthraquinone only takes place with difficulty; the rings are deactivated by the two keto groups.

Direct halogenation is very difficult and does not provide a homogeneous product. Halogenated derivatives are usually obtained by using appropriately substituted reactants in method (c) for making anthraquinones (p. 637).

Anthraquinone can be nitrated by nitrating acid under strong conditions. The main product is **1-nitroanthraquinone**, accompanied by small amounts of 1,5- and 1,8-dinitroanthraquinones.

Sulphonation of anthraquinone is an industrially important process. Fuming sulphuric acid (40–50% SO_3) is used and the product is *anthraquinone-2-sulphonic acid*. If a small amount of mercury(II) sulphate is added to the sulphonating mixture the product is almost entirely *anthraquinone-1-sulphonic acid*. The catalytic function of the mercury(II) ion has not been fully established.

Anthraquinone-1-sulphonic acid   Anthraquinone   Anthraquinone-2-sulphonic acid

Further sulphonation of either of these monosulphonic acids takes place in the unsulphonated ring and leads to the formation of 2,6- and 2,7- or 1,5- and 1,8-disulphonic acids respectively.

Nucleophilic attack on the sulphonyl groups of anthraquinonesulphonic acids can lead to their replacement by either hydroxy or amino groups, and also by halogen atoms. *1-Hydroxyanthraquinone*, m.p. 193°C (466 K) and *2-hydroxyanthraquinone*, m.p. 306°C (579 K), both yellow crystals, are made in this way.

Aminoanthraquinones are made industrially by reducing the corresponding nitroanthraquinones. 1-Amino- and 2-amino-anthraquinones crystallise in red needles, m.p., respectively, 252°C (525 K) and 302°C (575 K).

5.20.5.6 Anthraquinone Dyestuffs

(a) **Mordant dyes. Alizarin** (1,2-dihydroxyanthraquinone) is a member of this class of dyestuffs. It was used in antiquity, being obtained from madder root (*Rubia tinctorum*), in which it occurs as the glucoside *ruberythric acid*. Alizarin was prepared synthetically in 1869 and soon displaced the natural product for commercial use. When anthraquinone-2-sulphonic acid is melted with alkali, not only does the sulphonyl group undergo *nucleophilic substitution* by a hydroxy group, but, especially if the mixture is exposed to the air or if some sodium chlorate is added as an oxidising agent, a second hydroxy group is introduced at the 1-position. Acidification with sulphuric acid converts the sodium salt which is obtained into alizarin (*Caro, Graebe* and *Liebermann*):

Sodium anthraquinone-2-sulphonate   Alizarin

Alizarin crystallises out from acetic acid in orange-red needles, m.p. 289°C (562 K); it dissolves in alkali giving a blue-violet solution.

Polyhydroxyanthraquinones can be made from simple hydroxyanthraquinones by oxidising them with fuming sulphuric acid in the presence of boric acid with traces of mercury or selenium present to act as catalysts (*Bohn–Schmidt* reaction). For example, the mordant dye **Alizarin cyanine R** (1,2,4,5,6-pentahydroxyanthraquinone) can be made in this way from 2-hydroxyanthraquinone:

2-Hydroxyanthraquinone   Alizarin cyanine R

The dyeing technique used with alizarin or other polyhydroxyhydroquinones requires a special preliminary treatment, continuous over several days, of the fabric to be dyed (*Turkey red dyeing*). In this process the fibres are first of all steeped in sulphonated castor oil and then, after drying, immersed in a bath of aluminium sulphate and steamed. The resultant aluminium lake is bright red.

Whilst alizarin and higher hydroxy derivatives were for centuries important mordant dyestuffs for cotton, wool and silk, they are now no longer in use. On the other hand, the sulphonated oils, which are used as *wetting agents*, are still called 'Turkey red oil'.

(b) Acid dyestuffs. The starting materials for the preparation of these dyestuffs are anthraquinones having a number of amino or hydroxy groups. Introduction of sulphonyl groups into the molecules provides compounds which are used as *direct acid dyes* for use on *wool*.

Important examples of this class of dyestuffs are *Alizarin saphirole B*, which is made from 1,5-dihydroxyanthraquinone by disulphonating it, followed by nitration and reduction of the nitro groups to amino groups, and *Alizarin cyanine green G extra*. The latter is made by condensing *p*-toluidine with *quinizarin* (1,4-dihydroxyanthraquinone), which is prepared from phthalic anhydride and chlorobenzene in the presence of sulphuric acid and boric acid (see preparation (c), p. 637), and sulphonating the product. Both of these dyes give green or blue shades and are noted for their brilliant colours and fastness to light.

Alizarin saphirole B   Alizarin cyanine green G extra

(c) Disperse dyestuffs. The dyestuffs derived from anthraquinone can scarcely be surpassed for the brilliance of their colours. 1,4-, 1,4,5- and 1,4,5,8-substituted anthraquinones and their β-substituted derivatives are dyestuffs of outstanding character. If suitable substituent groups are introduced they are applicable as violet, blue or turquoise *disperse dyes*. They have marked fastness to heat and to light, especially for use on *polyester* fibres.

As an example, one brilliant blue dyestuff has the following chemical structure.

(d) **Vat dyes.** Among these are the oldest vat dye, indigo, and also *indanthrone* and *flavanthrone*, which were discovered in 1901 by *Bohn* and are made from aminoanthraquinones.

Indanthrone (Indanthrone blue RS) is obtained by melting 2-aminoanthraquinone with potassium hydroxide, together with potassium nitrate, at 150–200°C (425–475 K). If the melt is taken to a higher temperature (270–300°C [550–580K]) with an acidic condensing agent such as antimony(V) chloride or aluminium chloride also present then the product is *flavanthrone* (indanthrone yellow G).

Indanthrone

Flavanthrone

Vat dyeing involves the following specific transformations. The dyestuff, which is insoluble in water, is converted by reduction (vatting) in alkaline solution, especially by means of sodium dithionite (hydrosulphite vat), into a water-soluble dihydro or leuco compound whose anion has a good enough affinity for the fibre (cotton, viscose). The dihydro compound is reoxidised, either by exposure to air, or more rapidly by means of an oxidising agent. This results in the dyestuff being generated as very fine insoluble particles actually within the fibre; this leads to its being very fast to both washing and to light.

When the two dyestuffs mentioned above are vatted, only one quinonoid ring is reduced to its dihydro form. In the case of indanthrone, the dyestuff and the vat look the same; in the case of flavanthrone, the yellow dyestuff gives a deep blue coloured vat.

5.20.6 Phenanthrene, $C_{14}H_{10}$

Phenanthrene is an isomer of anthracene. It too was isolated from the anthracene oil obtained from the distillation of coal tar. The three benzene nuclei in phenanthrene have an *angular* arrangement, *i.e.* a line joining the centres of the three rings is not a straight line but has a kink in it. Phenanthrene can therefore be regarded as 2,2′-vinylenebiphenyl and it is in fact formed in small amounts if a mixture of gaseous biphenyl and ethylene is passed through a red-hot tube:

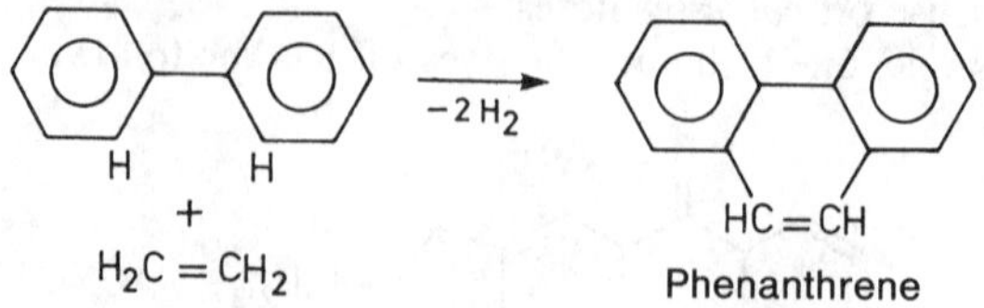

Phenanthrene

Like anthracene, phenanthrene contains a *14-π-electron system*.

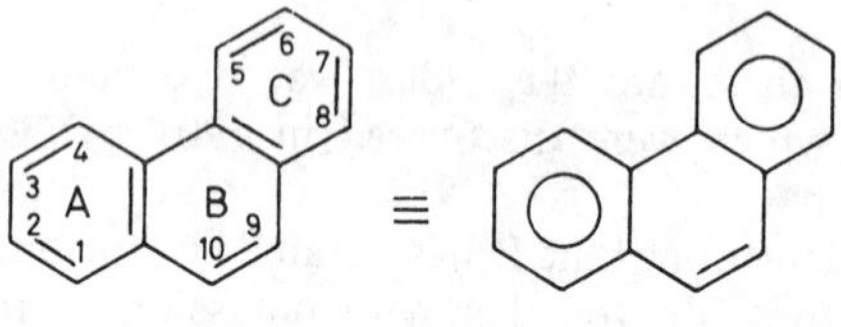

(a)

Preparation

As well as being obtained industrially from coal tar, the following synthetic processes have provided important ways of preparing phenanthrene:

5.20.6.1 *Pschorr* synthesis (1896).

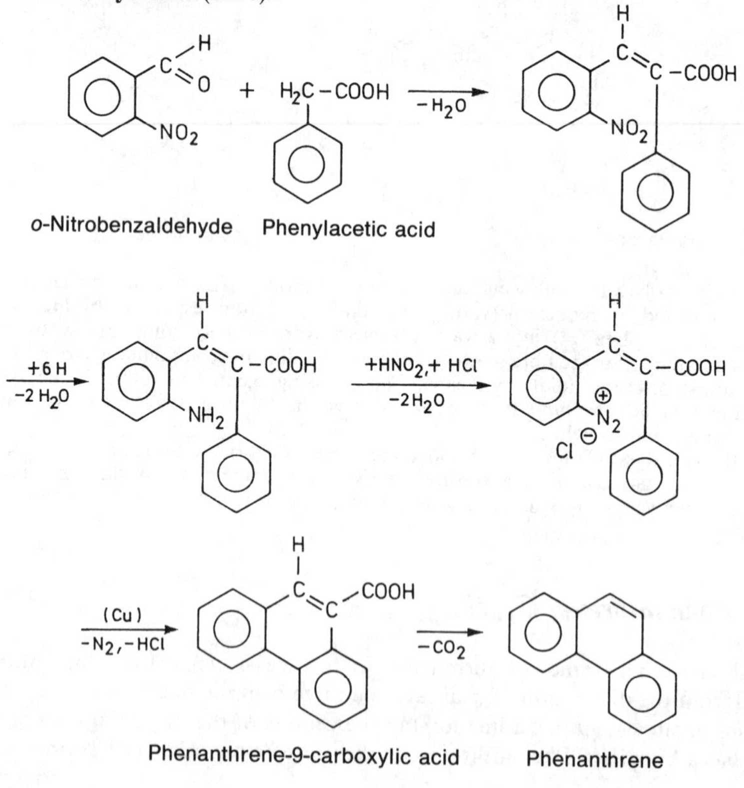

o-Nitrobenzaldehyde   Phenylacetic acid

Phenanthrene-9-carboxylic acid   Phenanthrene

In the first step *o*-nitrobenzaldehyde is condensed with phenylacetic acid to give α-phenyl-*o*-nitrocinnamic acid, in a reaction which is analogous to the *Perkin* reaction (see p. 555 *f.*). Because of repulsion between the carboxyl and the nitrophenyl groups, the two aryl groups are arranged *cis* to one another and hence there is the possibility of linking the two rings together. The nitro group is next reduced to an amino group which is diazotised, giving a diazonium salt; in the presence of copper powder this undergoes loss of nitrogen and ring-closure to form phenanthrene-9-carboxylic acid. Distillation at low pressure results in loss of carbon dioxide to give phenanthrene.

Substituted phenanthrenes can also be prepared in this way by using suitably substituted starting materials.

5.20.6.2   *Haworth* synthesis (1932). Acylation of naphthalene with succinic anhydride under *Friedel–Crafts* conditions leads to the formation of 3-(1-naphthoyl)propionic acid. *Clemmensen* reduction of this keto acid provides 4-(1-naphthyl)butyric acid which, on treatment with conc. sulphuric acid, loses water and undergoes ring-closure at the 2-position of the naphthalene ring. *Clemmensen* reduction of the resultant ketone followed by dehydrogenation of the product with selenium gives phenanthrene:

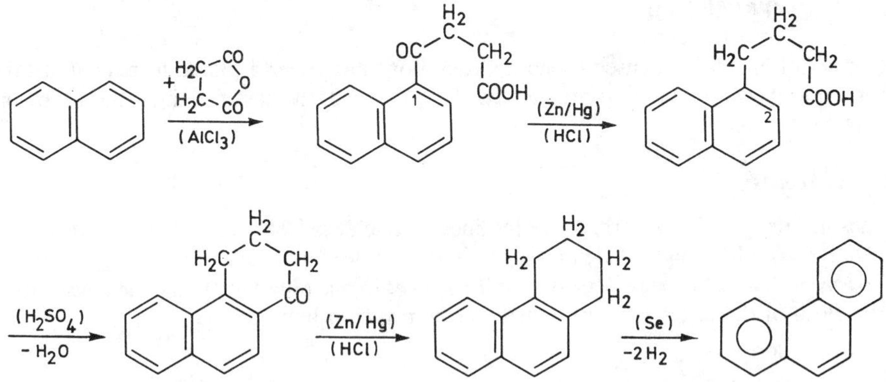

Under different conditions acylation of naphthalene can take place at the 2- as well as the 1-position, but the present reaction occurs solely at the 1-position and, as a result, phenanthrene is formed and not anthracene. The *Haworth* synthesis has been particularly useful for the preparation of alkylphenanthrenes.

Properties

Phenanthrene crystallises in colourless, shining leaflets, m.p. 101°C (374 K). It is readily soluble in benzene or ether; the solutions show a blue fluorescence. Most reactions take place at the 9- and 10-positions. Catalytic hydrogenation in the presence of a copper-chromium oxide catalyst leads to saturation of the 9,10-double bond and formation of *9,10-dihydrophenanthrene*, which crystallises from methanol in clusters of needles, m.p. 35°C (308 K). Further reduction resembles that of anthracene, leading in turn to *tetranthrene* (1,2,3,4-tetrahydrophenanthrene) and *octanthrene* (1,2,3,4,5,6,7,8-octahydrophenanthrene).

Stepwise oxidation of phenanthrene with chromic acid provides *phenanthroquinone* and then *diphenic acid* (biphenyl-2,2'-dicarboxylic acid):

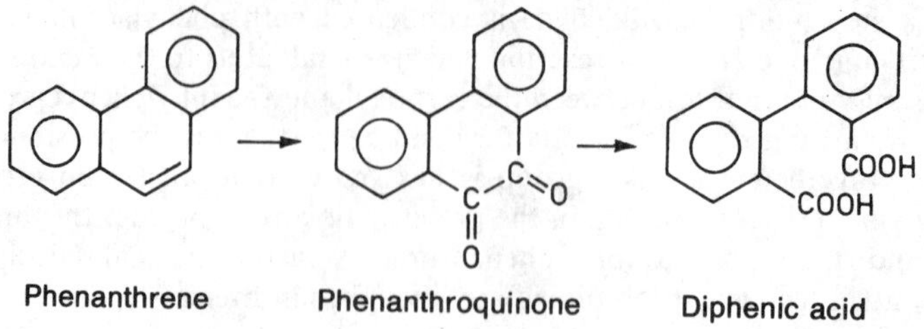

Phenanthrene   Phenanthroquinone   Diphenic acid

Phenanthroquinone forms odourless orange needles, m.p. 207°C (480 K), which are involatile in steam. It has the character of a 1,2-diketone.

Phenanthrene has not as yet found any industrial application, but the carbon skeleton of phenanthrene is present in some important natural products, such as some of the opium alkaloids (morphine) and especially in the extensive group of compounds known as the steroids. 7-Isopropyl-1-methylphenanthrene (retene) is formed when conifer wood is burned (see p. 675); its existence in sediments is thought to stem from its formation in forest fires.

5.20.7 Polycyclic Arenes

A number of highly condensed arene hydrocarbons have been isolated from coal tar and others have been prepared synthetically. They have been classified into the following groups by *Clar* et al.

5.20.7.1 Acenes[1]

Acenes are derived from anthracene by linear 'annelation' of further benzene rings and are characterised by higher chemical reactivity and deeper colours.

Tetracene (naphthacene) is orange-yellow, *pentacene*, blue-violet and *hexacene* deep green. Both of the latter are very sensitive to air and to light.

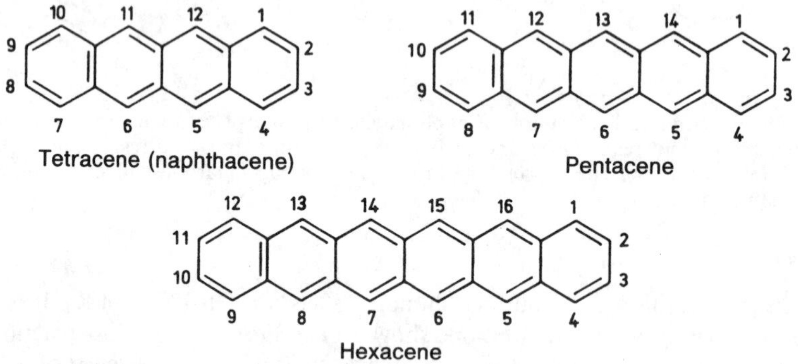

Tetracene (naphthacene)   Pentacene

Hexacene

Addition reactions such as, for example, *Diels–Alder* reactions with maleic anhydride, occur across *meso*-positions, *i.e.* *p*-positions in the same ring, and result in the acene system being converted into two smaller but more stable π-electron systems. The resultant adducts are colourless and appreciably more stable than the acenes themselves.

---

[1] See E. Clar, Polycyclic Hydrocarbons, 2 Vols (Academic Press, New York and Springer-Verlag, Berlin 1964); The Aromatic Sextet (Wiley, London 1972).

Some acenes can also be stabilised by migration of a proton to a *meso*-position. As an example, *6-hydroxypentacene* exists only in the tautomeric keto form (*cf.* anthranol ⇌ anthrone, p. 638):

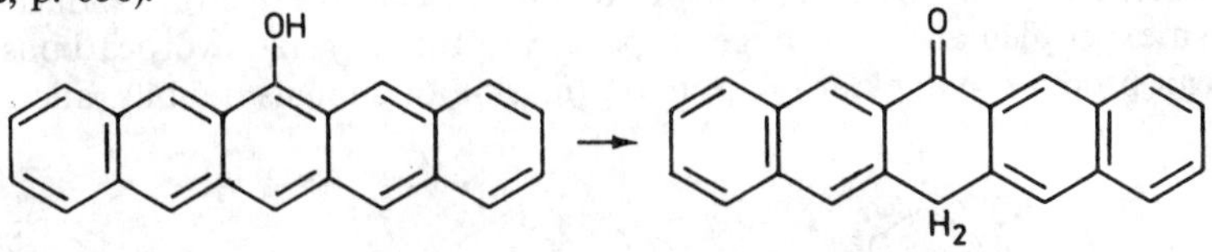

Enol form	Keto form

6-Hydroxypentacene

Perhaps more strikingly, methylacenes are converted into *tautomeric methylene isomers*; the ratio of the tautomers depends upon the temperature:

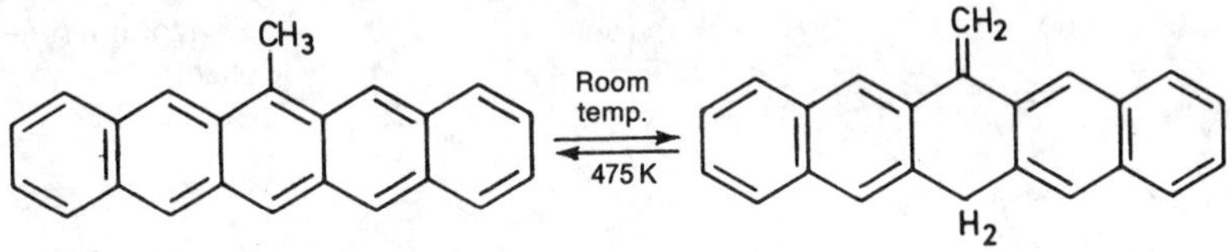

blue (unstable)	pale yellow (stable)

6-Methylpentacene

The following antibiotics are structurally related to *tetracene*:

Aureomycin was isolated from cultures of *Streptomyces aureofaciens*. It is active against many bacteria as well as against Rickettsiae and many viruses. It forms golden yellow crystals, m.p. 173°C (446 K) and is strongly laevorotatory.

Terramycin (Macocyn) is obtained from cultures of *Streptomyces rimosus* and resembles aureomycin in many of its properties.

The parent compound of these antibiotics is INN: *Tetracycline*[1], which also acts as an antibiotic and has the following structure:

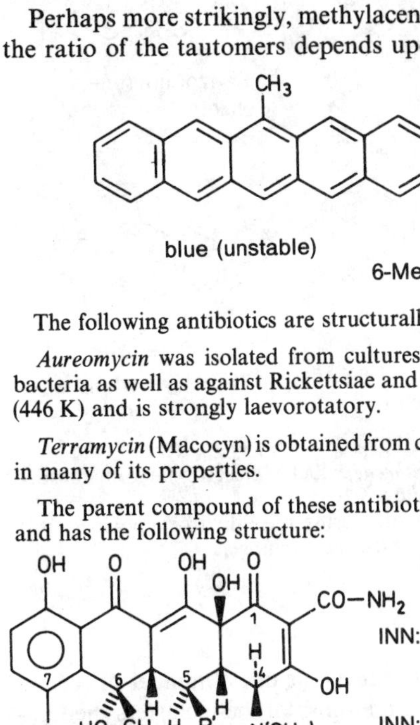

INN: *Tetracycline* (Achromycine, Hostacycline)
$R = R' = H$

INN: *Chlorotetracycline* (7-Chlorotetracycline, *Aureomycin*)
$R = Cl, \quad R' = H$

INN: *Oxytetracycline* (5-Hydroxytetracycline, *Terramycin*)
$R' = H, \quad R' = OH$

The tetracycline group of antibiotics displays a wide range of activity against infective illnesses, for example cholera. Their action depends upon their interfering with protein synthesis in the mitochondria of the bacteria (see p. 867). The *absolute configuration* shown in the formula was worked out by *Shemyakin* (1962).

5.20.7.2 Carcinogenic Hydrocarbons[2]

There is a series of carcinogenic hydrocarbons derived from *chrysene* and *pyrene* which can be regarded as arising from phenanthrene by fusing another benzene ring on to it.

[1] See H. *Muxfeldt*, Syntheses in the Tetracycline Series, Angew. Chem. Int. Ed. Engl., *1*, 372 (1962); D. L. J. *Clive*, Chemistry of Tetracyclines, Quart. Rev. *22*, 435 (1968).
[2] See P. *Rademacher*, Chemische Carcinogene, Chemie in unserer Zeit, *9*, 79 (1975); R. G. *Harvey*, Activated Metabolites of Carcinogen Hydrocarbons, Acc. Chem. Res. *14*, 218 (1981).

Benzo[a]pyrene (1,2-benzpyrene, benzo[def]chrysene), which occurs in coal tar, has been shown to be particularly carcinogenic (*J. W. Cook*, 1932); it was held to be responsible for the incidence of skin carcinomas in workers in the coal-tar dyestuffs industry. It has been shown experimentally that longer exposure to carcinogenic hydrocarbons such as chrysene can produce malignant tumours on the skin of animals (rabbits, mice).

Chrysene

Pyrene

Benzo[a]pyrene,
Benzo[def]chrysene[1]

3-Methylcholanthrene

7,8-Dihydroxy-9,10-epoxy-7,8,9,10-
tetrahydrobenz[a]pyrene

Buckminsterfullerene

A highly active hydrocarbon of this sort is *3-methylcholanthrene*, which is closely related chemically to the steroids (cholesterol, the bile acids, sex hormones, etc.).

In the course of the biosynthesis of prostaglandins in organisms the dihydroxyepoxide (I) of benzo[a]pyrene is formed and this is considered to be the intrinsic carcinogen.

5.20.7.3 Carbon Clusters

When soot is formed from vapourised graphite the products include a number of carbon clusters (fullerenes) containing even numbers of carbon atoms. Of these clusters the C_{60}-compound appears to be especially stable. It has the structure of a truncated icosahedrane and was called **buckminsterfullerene**[2] and also, from its appearance, *footballene*. The formula above shows just one of the 12500 possible limiting structures for this compound[3]. The 60 carbon atoms of this hollow structure are arranged in a mosaic of 20 6-membered and 15 5-membered rings. The compound can be obtained in crystalline form. It provides the third allotropic form of carbon, the others being diamond and graphite, and in this context has also been called *fullerite*.

---

[1] According to IUPAC rules (21.3), the second name is the correct one.
[2] *Richard Buckminster Fuller* was an American engineer, who designed dome-like constructions making use of a similar structural network.
[3] See *H. Kroto*, Ang. Chem. Int. Ed. Engl. *31*, 111 (1992); *H. W. Kroto* et al., Chem. Rev. *91*, 1213 (1991); *J. F. Stoddart*. The Third Allotropic Form of Carbon, Angew. Chem. Int. Ed. Engl. *30*, 70 (1991); *A. Hirsch*, The Chemistry of Fullerenes (Thieme Verlag, Stuttgart 1994).

5.21 Nonbenzenoid Aromatics[1]

Members of this class of compounds include some carbocyclic compounds which, in contrast to benzene, have rings containing odd numbers of ring atoms. Yet, despite this, they appear to have *aromatic character*. They exist either as cations or anions or as dipolar compounds and consequently they have chemical properties which differ from those of benzene.

In order to extend the definition of aromaticity which was used in the case of benzenoid compounds, it is hence unsatisfactory to rely solely on chemical properties, and physical properties must also be taken into account. In the broadest sense, aromatic compounds have cyclic unsaturated systems with all the ring atoms taking part in a mesomeric system, and are characterised by a high stabilisation energy. Further criteria include a planar or almost planar structure and an anomalously high diamagnetic susceptibility which results in a characteristic type of NMR spectra.

The *sextet* of *π-electrons*, which has been taken as a criterion for aromatic compounds, is also found in these non-benzenoid systems, for example in the cyclopentadienide anion and the cycloheptatrienium cation.

5.21.1 Cyclopentadienides

The ease with which cyclopentadiene forms alkali metal salts (see p. 398) is due to the *mesomeric stabilisation* of the *cyclopentadienide anion*, $C_5H_5^\ominus$ which is formed. This can be illustrated by the following limiting structures (or canonical forms):

$$\left[\begin{array}{ccccccccc} & \leftrightarrow & & \leftrightarrow & & \leftrightarrow & & \leftrightarrow & \end{array} \right]$$

It is usually written as follows:

The simple alkali-metal cyclopentadienides are organometallic compounds with ionic character. In addition there are cyclopentadienylides, in which the cyclopentadienide ion is stabilised intramolecularly. Compounds of this sort are *diazocyclopentadiene* (diazoniocyclopentadienide) (*W. v. E. Doering*, 1953), which is a mesomeric compound whose extreme limiting structures can be described as fulvenoid and dipolar aromatic, and pyridinium cyclopentadienylide (pyridiniocyclopentadienide) (*Lloyd and Sneezum*, 1955):

[1] See *J. P. Snyder* et al., Nonbenzenoid Aromatics, 2 Vols (Academic Press, New York 1970, 1971); *D. Lloyd*, Nonbenzenoid Conjugated Carbocyclic Compounds (Elsevier, Amsterdam 1984); *D. Lloyd*, The Chemistry of Conjugated Cyclic Compounds: To Be or Not To Be Like Benzene? (J. Wiley & Sons, Chichester 1989).

Diazocyclopentadiene   Pyridinium cyclopentadienylide

In the synthesis of diazocyclopentadiene, which is a stable red compound, the diazo group is introduced without the involvement of a normal *diazotisation* reaction. Instead it arises from the reaction between *lithium cyclopentadienide* and *p-toluenesulphonyl azide*. The *reaction mechanism* involves a diazo transfer reaction and is as follows:

p-Toluenesulphonyl azide

Diazocyclopentadiene

p-Toluenesulphonyl azide is made from *p*-toluenesulphonyl chloride and sodium azide. It can be regarded as an *N*-diazo-compound, and, in reactions which are analogous to azo-coupling, it undergoes coupling reactions with suitable reactive partners.

Diazocyclopentadiene has served as a model compound for the study of the chemical properties of cyclopentadienide systems. It can be nitrated, brominated, mercuriated and coupled with diazonium salts (*Cram*, 1963).

5.21.2 Aromatic complexes

5.21.2.1 Ferrocene[1]

Other examples of stabilised cyclopentadienides (Cp) are ferrocene (dicyclopentadienyl-iron) and numerous other metal π-complexes or metallocenes, for example, the *Tebbe* reagent (see p. 187).

A very stable orange crystalline substance of m.p. 173°C (446 K) (*Kealy* and *Pauson*, 1951) was obtained by the reaction of iron(II) chloride with a cyclopentadienide Grignard reagent. The same compound was also formed when cyclopentadiene was passed over activated iron at 300°C (575 K) in an atmosphere of nitrogen (*Miller*, 1952). It was given the name 'ferrocene' by *Woodward*.

$$2C_5H_5]^\ominus MgBr^\oplus + FeCl_2 \rightarrow Fe(C_5H_5)_2 + MgBr_2 + MgCl_2$$

$$2C_5H_6 + Fe \rightarrow Fe(C_5H_5)_2 + H_2$$

Ferrocene,0(18), can be sublimed. It is insoluble in water and can be recrystallised from organic solvents.

---

[1] See *P. L. Pauson*, Ferrocene and Related Compounds, Quart. Rev., *9*, 391 (1955); *K. Plesske*, Ring Substitution and Secondary Reactions of Aromatic Metal π-Complexes, I & II, Angew. Chem. Int. Ed. Engl., *1*, 312 and 394 (1962); *D. W. Slocum* et al., Metalation of Metallocenes, J. Chem. Educ., *46*, 144 (1969).

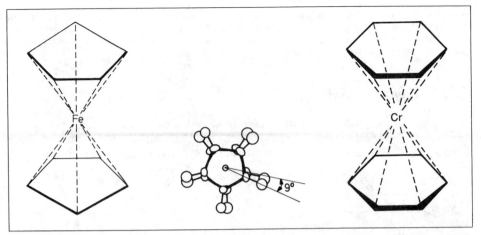

Fig. 119a. Ferrocene, 0(18)   Fig. 119b   Fig. 119c. Dibenzenechromium, 0(18)

Ferrocene is an η^5 *metal-π-complex* (see pp. 380 *ff.*), *which has a sandwich* structure. X-ray analysis shows that the iron is at the centre of a distorted antiprism, whose bases are two regular pentagons (Fig. 119a). The carbon atoms of the two rings are in a slightly staggered arrangement, with the two rings twisted at an angle of 9° from an eclipsed conformation (Fig. 119b). At room temperature the two rings oscillate about this conformation by about 8°.[1]

5.21.2.2 Dibenzenechromium

Following the discovery of ferrocene it was realised that the benzene–chromium compound which had been prepared by *Fr. Hein* in 1919 could have a similar type of structure (*L. Onsager, H. Zeiss* and *M. Tsutsui*, 1954). Benzene and a number of other aromatic compounds have been found to form π-complexes with transition metals. They are best prepared by the so-called reductive *Friedel–Crafts* reaction, for example, by heating benzene with chromium(III) chloride, aluminium chloride and aluminium powder at 180°C (450 K).

In the resulting cation $(C_6H_6)Cr^\oplus$ chromium has an oxidation level of +1. On reduction with, for example, hypophosphoric acid it provides dark brown crystalline η^6-dibenzenechromium(0), $(C_6H_6)_2Cr,0(18)$ (*E.O. Fischer*, 1955).

Dibenzenechromium is soluble in organic solvents. It can be sublimed. It has a *dipole moment* of zero, and is stable in the absence of air up to 300°C (575 K). Its diamagnetic properties show that the π-electron sextets of both benzene rings occupy vacant electron shells of the central metal atom, giving thereby a krypton-like electronic structure. Dibenzenechromium is thus another example of a compound having a *sandwich* structure (Fig. 119c). *Neutron diffraction* studies show that the benzene rings in dibenzenechromium(0) are not planar; both take up a slightly distorted chair form (*E. Förster et al.*, 1969).

Despite the similarities in their structures ferrocene and dibenzenechromium show differences in their chemical properties. In contrast to dibenzenechromium, ferrocene undergoes a series of

---

[1] The completely staggered conformation, as shown in Fig. 119a, has since been shown to be incorrect. See *J. D. Dunitz* et al., Acta Cryst., *B 35*, 1063, 2020 (1979).

reactions such as sulphonation, mercuriation, *Friedel–Crafts* alkylation and acylation, and the *Vilsmeier–Haack* reaction. With alkali metals ferrocene gives alkali-metal derivatives, from which carboxylic acids can be obtained by reaction with carbon dioxide. Halogenoferrocenes can be prepared from the mercury derivatives.

5.21.3 *Tropylium Salts*[1]

In contrast to the formation of the cyclopentadienide *anion*, from cycloheptatriene (or tropilidene, see p. 378) a stable *cation* $C_7H_7^{\oplus}$, the tropylium cation, can be obtained, which has a sextet of π-electrons, and can be considered as a hybrid of seven limiting structures:

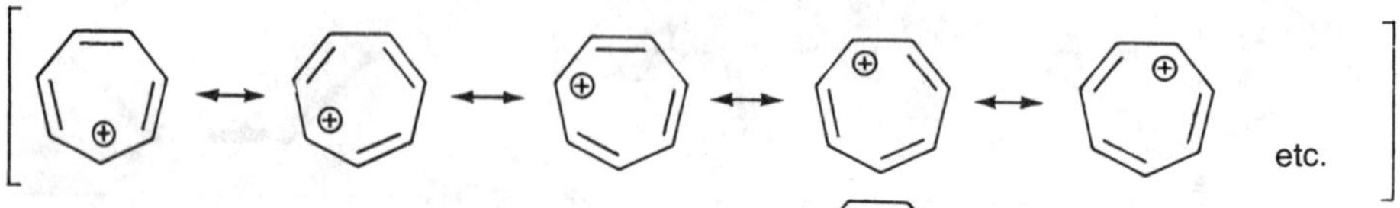

etc.

It is usually represented by the following formula:

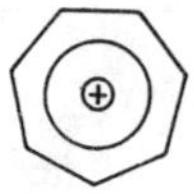

The first tropylium salt to be prepared was *tropylium bromide*, $C_7H_7^{\oplus}Br^{\ominus}$, by the reaction of 1 mol bromine with cycloheptatriene; when the product was heated hydrogen bromide was eliminated to give tropylium bromide. *Merling* had prepared this salt in 1891, but did not realise its structure. Sixty-three years later, *W. v. E. Doering* and *Knox* (1954) repeated this work and explained it correctly:

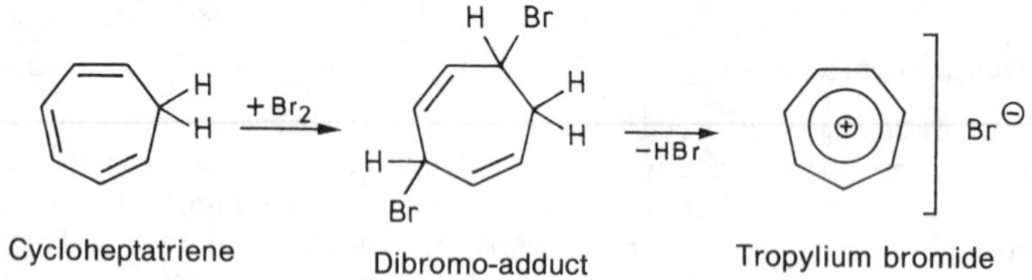

Cycloheptatriene   Dibromo-adduct   Tropylium bromide

Nowadays a variety of methods are available for the preparation of tropylium salts, starting from either cycloheptatriene or benzene. For example, reaction of benzene with a diazoacetic ester gives an ester of 8,9,10-trinorcaradienecarboxylic acid (*cf.* carane, p. 667; 'nor' signifies the absence of the methyl groups at these positions) which is normally shortened to norcaradienecarboxylic acid;[2] this is converted *via* the free acid and acid chloride into the acid azide. When this azide is heated in benzene it loses nitrogen (see *Curtius* reaction) and is converted into an isocyanate, from which tropylium bromide is obtained by the action of hydrogen bromide (*Dewar* and *Pettit*, 1955):

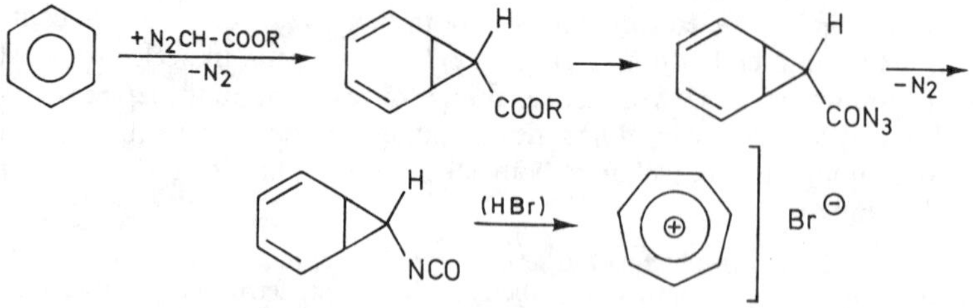

---

[1] See *W.v.E. Doering* and *H. Krauch*, Das Tropylium Ion, Angew. Chem., *68*, 661 (1956).
[2] See *G. Maier*, The Norcaradiene Problem, Angew. Chem. Int. Ed. Engl. *6*, 402 (1967).

Tropylium bromide forms yellow crystals of m.p. 203°C (476 K). It is readily soluble in water. The bromide ion is completely precipitated by silver nitrate in the cold. Catalytic hydrogenation provides cycloheptane. As a cation, the tropylium ion reacts with *nucleophilic* reagents such as hydroxide, alkoxide and cyanide ions and amines.

5.21.3.1 Homotropylium Ions[1]

When *cyclooctatetraene* is treated with concentrated sulphuric acid or with antimony pentachloride and hydrochloric acid in nitromethane a salt of the *homotropylium cation* is obtained (*Pettit*, 1962).

Homotropylium
hexachloroantimonate

Fe(CO)$_3$

The ^{1}H-NMR spectrum indicates the presence of this non-classical ion because the signal for the *endo*-proton Ha occurs at δ −0.6 p.p.m., whilst the *exo*-proton Hb gives a signal at δ 5.2 p.p.m. The difference, $\Delta\delta$, between these two chemical shifts thus amounts to 5.8 p.p.m. The bicyclic tricarbonyliron complex shows a $\Delta\delta$ of only 0.2 p.p.m.; in this case the cation is an η^4 ligand. However, the significance of this large value for $\Delta\delta$ as evidence for a diamagnetic ring current in the *homoaromatic homotropylium cation* has sometimes been overrated[2].

The tropylium cation is the parent compound of a series of 7-membered ring derivatives among which are tropone and tropolone.

5.21.4 Tropone

Tropone (cycloheptatrienone) is most simply obtained from the UV light catalysed reaction of bromobenzene with diazomethane (*Doering* and *Mayer*). The bromocycloheptatriene that is obtained reacts with bromine to give bromotropylium bromide, which is then hydrolysed to tropone:

Bromocycloheptatriene   Bromotropylium bromide

Tropone

---

[1] See *L. A. Paquette*, The Realities of Extended Homoaromaticity, Angew. Chem. Int. Ed. Engl., *17*, 106 (1978); *R. F. Childs*, The Homotropylium Ion and Homoaromaticity, Acc. Chem. Res., *17*, 347 (1984); *R. F. Childs* et al., Pure Appl. Chem. *58*, No. 1, 111 (1986).
[2] See *R. F. Childs* et al., Ring Currents, Chemical Shifts and Homoaromaticity: The Homotropylium Ion Revisited, J. Amer. Chem. Soc., *106*, 5974 (1984).

Tropone is miscible with water. It boils at 113°C (386 K)/20 mbar (20 hPa), which is appreciably higher than isomeric benzaldehyde (68°C [341 K]/20 mbar [20 hPa]). It has a high *dipole moment* (14.2×10^{-30} cm). It is basic and forms a hydrochloride, but has an unreactive carbonyl group.

5.21.5 *Tropolone and its Derivatives*[1]

α-Tropolone is a hydroxy derivative of tropone. There are also β- and γ-tropolones, which are respectively 3- and 4-hydroxytropones, but they are of lesser importance.

α-Tropolone can be obtained by oxidation of cycloheptatriene with potassium permanganate. Two *tautomeric* forms are possible; in each of them the hydroxy and keto groups are linked by *hydrogen bonding*. Because of the electronic interaction between the hydroxy group and the carbonyl group, *via* the unsaturated chain, tropolone can be regarded as a vinylogue (see p. 329) of a carboxylic acid, and is acidic.

α-Tropolone

Since the proton is rapidly exchanged between the two oxygen atoms, the 3- and 7-positions in tropolone are in effect identical, as are the 4- and 6-positions.

α-Tropolone crystallises in colourless needles of m.p. 50–51°C (323–324 K). It is readily soluble in both water and organic solvents. With iron(III) chloride it gives a green colour. In keeping with its structure it shows no carbonyl group properties. Its acidity is intermediate between that of phenols and that of acetic acid. It reacts with hydrochloric acid to give a hydrochloride (dihydroxytropylium chloride). It resembles phenol in its reactivity towards *electrophiles*; it is easily brominated and nitrated and couples with diazonium salts. When heated, it is isomerised to give benzoic acid.

The α-tropolone ring is found in a number of natural products,[2] for example in *stipitatic acid*, which is obtained from the mould *Penicillium stipitatum*, and in the heartwood of red cedar trees (*Thuja plicata*) from which *Erdtman* isolated the three isomeric thujaplicins (isopropyltropolones).

| Stipitatic acid | α-Thujaplicin (m.p. 39°C[312 K]) | β-Thujaplicin (m.p. 52–53°C [325–326 K]) | γ-Thujaplicin (m.p. 79–80°C [352–363 K]) |

[1] See *P. L. Pauson*, Chem. Rev., *55*, 9 (1955); *T. Asao* and *M. Oda*, *Houben-Weyl*, Methoden der Organischen Chemie, 4th edtn. *5/2c*, p. 710 *ff*. (Thieme Verlag, Stuttgart 1985).
[2] See *T. Nozoe*, Natural Tropolones and Some Related Troponoids, Fortschr. Ch. Org. Naturst., *13*, 232 *ff* (1956).

The thujaplicins are powerful *fungicides*; for example they prevent cedar wood from rotting.

5.21.5.1   Purpurogallin was first obtained by the oxidation of pyrogallol (1,2,3-trihydroxybenzene) (*Girard* 1869) and later identified (*Haworth*) as trihydroxybenzo-α-tropolone. It occurs as a glucoside in *Dryophantin* from red pea gall. It is best made by oxidation of pyrogallol with sodium iodate (*Evans*), and forms brick-red needles of m.p. 275° (548 K). It is rather insoluble in most solvents.

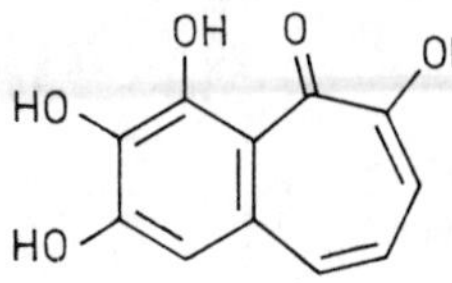

Purpurogallin 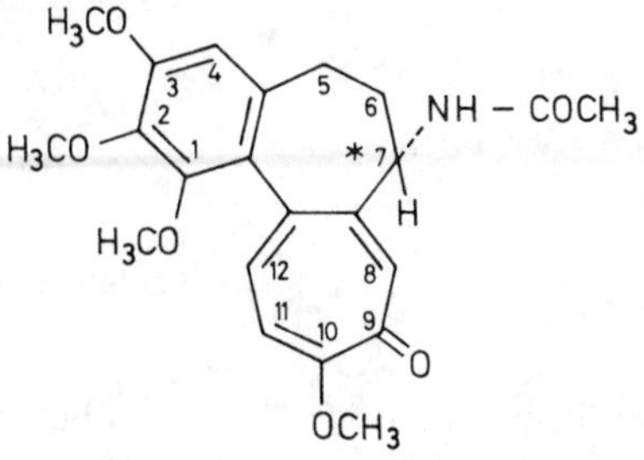   Colchicine

5.21.5.2   Colchicine, which was shown by *Pelletier* and *Caventou* in 1819 to be the principal alkaloid of the autumn crocus (*Colchicum autumnale*), is a complicated α-tropolone derivative. Following investigations by *Dewar* (1945), colchicine was assigned the above structure, which has been confirmed by a synthesis[1] starting from purpurogallin. It is a colourless crystalline substance of m.p. 155–157°C (428–430 K) and is optically active. It dissolves in ethanol or water to give neutral solutions.

Colchicine interferes with mitosis and with the separation of chromosomes at nuclear division, so that by using suitable doses of colchicine strains of plants with increased numbers of chromosomes (*polyploids*) can be bred. It is used commercially in plant breeding to produce new varieties.

5.21.6 Azulenes[2]

As two benzene rings are fused together to form naphthalene so formal fusion of a cyclopentadienide ring with a tropylium ring provides *azulene*. Azulene has a stabilisation energy of 176 kJ/mol. Derivatives of azulene, which are blue or violet coloured, occur in natural essential oils, for example oil of camomile, from *Matricaria camomilla*. A range of azulenes has been made synthetically. Some act as inflammation inhibitors.

Chemical and physical studies show that azulene can be regarded as a *non-benzenoid aromatic* compound which is slightly dipolar, with higher electron density in the 5-ring than in the 7-ring, although the dipole moment is small (2.668×10^{-30} cm). The ground state of azulene has been explained in terms of the limiting structures shown at the top of p. 654.

Because of this electronic distribution, *electrophilic* reactions, either substitution or addition reactions, take place readily at the 1(=3) position.

In strong mineral acids a proton adds to give a yellow **azulenium** cation.

Azulene can be halogenated, nitrated, sulphonated, mercuriated, and acylated under *Friedel–Crafts* conditions. It couples with diazonium salts. *Nucleophilic* reactions take

[1] See *A. Eschenmoser* et al., Synthese des Colchicins, Angew. Chem., *71*, 637 (1959).
[2] See *K. Hafner*, Neuere Ergebnisse der Azulenchemie, Angew. Chem., *70*, 419 (1958); *T. Nozoe* et al., Recent Advances in the Chemistry of Azulenes and Natural Hydroazulenes, Fortschr. Chem. Org. Naturst., *19*, 32 *ff* (1951); *T. Nozoe*, Pure Appl. Chem. *28*, 239 (1971); *V. B. Mochalin* and *Y. N. Porshnev*, Russ. Chem. Rev. (Engl.) *46*, 530 (1977) ref.[1], p. 647.

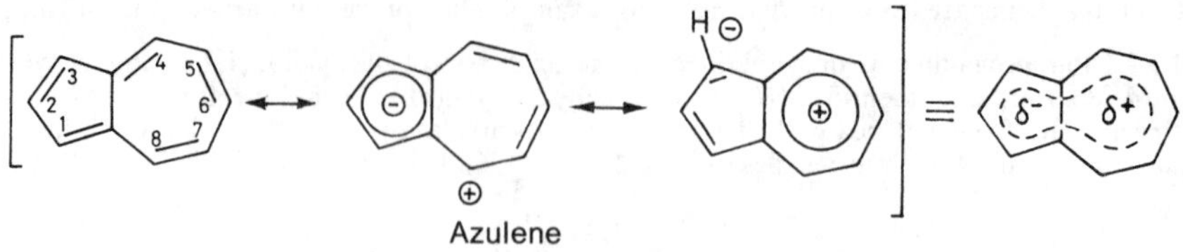

Azulene

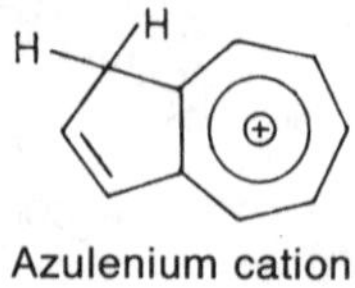

Azulenium cation

place in the 7-ring at the 4(= 8) position but, because of the sensitivity of azulene towards bases, they have been little studied.

König and *Rösler*, and *Ziegler* and *Hafner* (1955), quite independently found 5-*N*-methylanilino-2,4-pentadienal, the so-called ***Zincke*-aldehyde**, a derivative of glutacondial-dehyde, to be a useful starting material for the synthesis of azulene. It condenses with cyclopentadiene in the presence of alkali to give a *fulvene* derivative which, when heated *in vacuo* at 200–250°C (475–535 K), loses *N*-methylaniline and cyclises to give azulene in 60% yield.

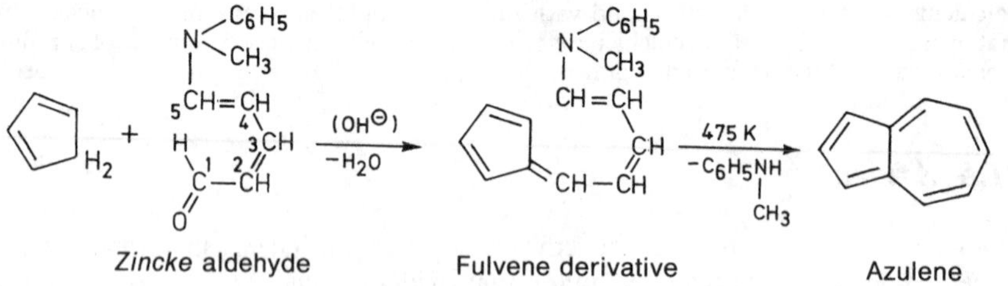

Zincke aldehyde   Fulvene derivative   Azulene

Substituted azulenes have been prepared similarly by this elegant method and variations of it.

When two 5-rings or two 7-rings are fused together the resultant systems are known as pentalenes and heptalenes respectively. In contrast to azulene they have no aromatic properties. Their physical and chemical properties indicate that they are polyenes which lack any mesomeric stabilisation.

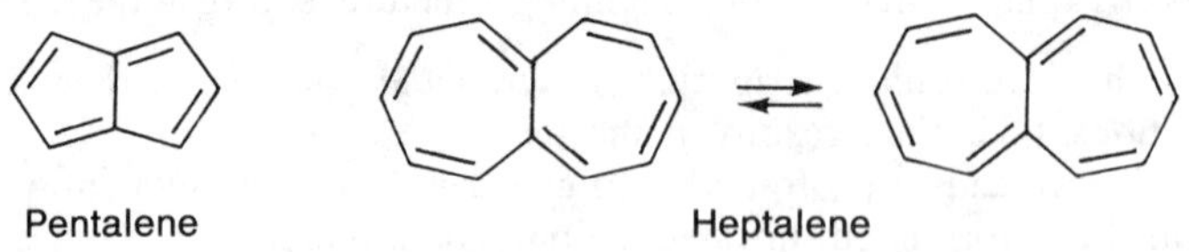

Pentalene   Heptalene

NMR spectra show that heptalene has alternate single and double bonds and that there is rapid interchange between the two valence isomeric forms.[1]

---

[1] See *E. Vogel* et al., Heptalene Syntheses from 1,6-Methano[10]annulenes: Evidence for Rapid π-Bond Interchange, Angew. Chem. Int. Ed. Engl., *13*, 732 (1974); Pentalenes and Heptalenes, Pure Appl. Chem. *28*, 153 (1971).

5.21.7 *Hückel's* Rule

Quantum mechanical calculations of the π-electron systems of aromatic compounds lead to the conclusion that fully conjugated cyclic polyenes with $(4n + 2)$ π-electrons should have a special stability (*E. Hückel*, 1931). The rule applies in particular to molecules which have an unbranched cyclic π-electron system. The π-electron sextets of benzene, and of the cyclopentadienide and tropylium ions, thus represent only a special case with closed π-electron systems in which $n = 1$. From *Hückel's* rule it follows that, in addition, carbocyclic compounds with 2,10,14,18, etc. π-electrons in the ring should also be *mesomerically stabilised* and hence can be considered as aromatic compounds. In contrast, cyclic systems with 4,8,12, etc. π-electrons in their perimeters show *antiaromatic* character, and the double bond system does not have delocalised π-electrons and is destabilised. The basic difference between *aromatic* and *antiaromatic* compounds is shown by a comparison between the energy diagrams for benzene and cyclobutadiene.

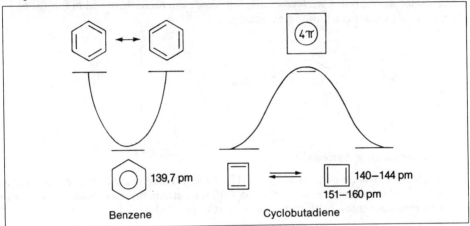

Fig. 120.

In the case of **benzene** the ground state with delocalised electrons is energetically stabilised compared to the *Kekulé* forms with localised double bonds (see p. 470), but in the case of **cyclobutadiene** the situation is reversed, with the delocalised form being of higher energy. It is the transition state through which the two localised forms must pass in order to interchange. This is shown by the bond lengths in various substituted derivatives of cyclobutadiene which have been examined by X-ray analysis. This provides evidence that there is a distinction between the *mesomerism* of benzene and the *valence isomerism* of cyclobutadiene.[1]

5.21.7.1 2-π-Electron systems

The smallest $(4n + 2)\pi$ ring system $(n = 0)$ is found in the *cyclopropenium cation*, $C_3H_3^{\oplus}$, which has two π-electrons in a 3-membered ring.[2]

---

[1] On the problem of aromaticity see *P. J. Garratt*, Aromaticity (Wiley, New York 1986); *D. Lloyd*, Nonbenzenoid Conjugated Carbocyclic Compounds (Elsevier, Amsterdam, 1984); *D. Lloyd*, The Chemistry of Conjugated Cyclic Compounds: To Be or Not To Be Like Benzene? (Wiley, Chichester 1989).
[2] *R. Breslow*, Angew. Chem. Int. Ed. Engl. *7*, 565 (1968); *S. V. Krivun, O. F. Alferova* and *S. V. Sayapina*, Russ. Chem. Rev. (Engl.) *43*, 835 (1974).

The first compound of this kind to be described was *triphenylcyclopropenium tetrafluoroborate*. It was made from diphenylacetylene and phenyldiazoacetonitrile, and the initially formed covalent cyanide was treated with boron trifluoride in moist ether to give a salt. (*Breslow*, 1957):

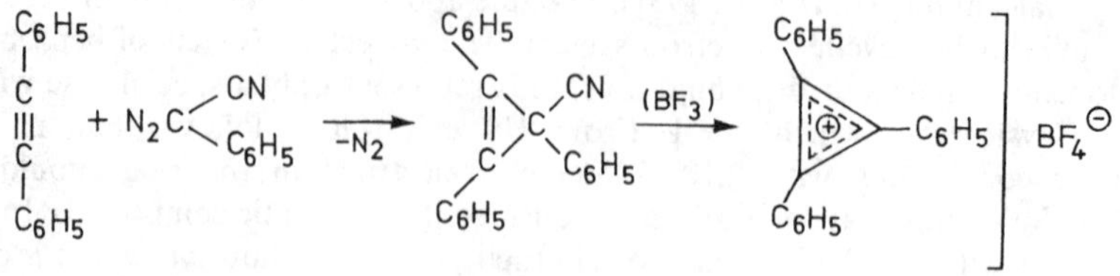

This salt is markedly more stable than other cyclopropene derivatives. Its stability is not dependent on the substituent phenyl groups. If tetrachlorocyclopropene is treated with *aluminium chloride*, a *trichlorocyclo-propenium cation* is formed, and in this the C—C bonds are stronger than those in benzene.

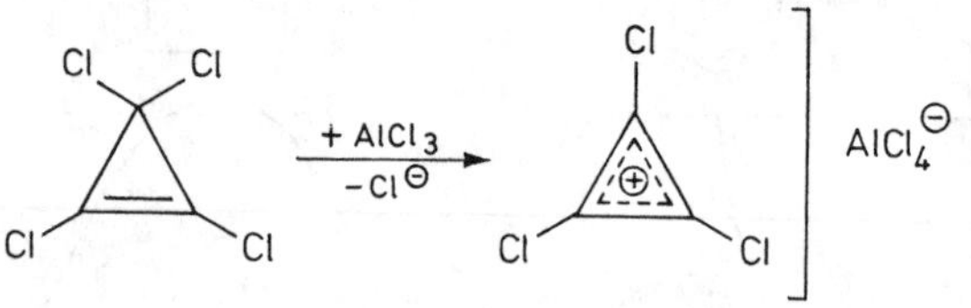

5.21.7.2 4-π-Electron Systems[1]

All attempts to isolate *cyclobutadiene*, C_4H_4, have hitherto proved unsuccessful. It has been obtained in an argon matrix at $-265°C$ (8 K). Tetramethylcyclobutadiene forms an isolable *metal π-complex* with nickel chloride (*Criegee*, 1959).

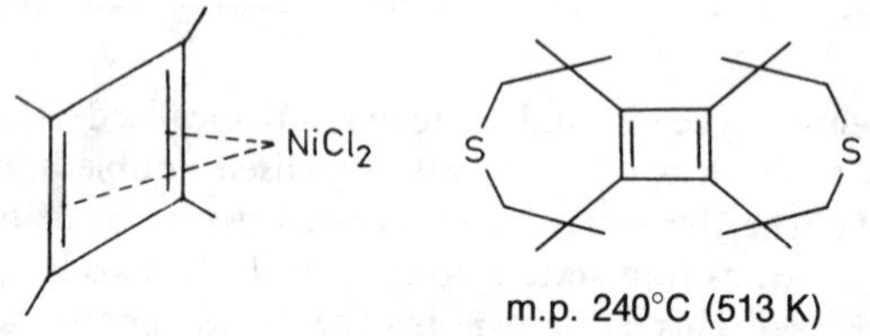

m.p. 240°C (513 K)

The kinetic stability of sensitive systems can be increased by introducing steric hindrance into the molecule. Making use of this principle, *A. W. Krebs* succeeded in isolating a stable derivative of cyclobutadiene. The substituent groups occupy similar space to four t-butyl groups. This compound forms yellow crystals of m.p. 240°C (513 K) which are very sensitive to oxygen. The single bonds are markedly longer than the double bonds.

A possible isomer of cyclobutadiene is tetrahedrane, in which four CH groups are arranged at the corners of a tetrahedron. When tetra-t-butylcyclopentadienone was irradiated at 254 nm in a matrix cooled in liquid nitrogen, tetra-t-butyltetrahedrane was produced. It decomposes when it is heated to 135°C (408 K).

---

[1] See *G. Maier*, The Cyclobutadiene Problem, Angew. Chem. Int. Ed.Engl., *13*, 425 (1974); *T. Bally* and *S. Masamune*, Cyclobutadiene, Tetrahedron, *36*, 346 (1980); *G. Maier*, Angew. Chem. Int. Ed. Engl. *27*, 309 (1988).

If tetra-t-butyltetrahedrane is heated in cyclosilane [1,1,3,3,5,5-hexakis(trideuterio-methyl)-1,3,5-trisilacyclohexane] at 130°C (400 K) it is isomerised to tetra-t-butylcyclo-butadiene.[1]

5.21.7.3 8-π-Electron Systems

The parent compound, cyclooctatetraene, C_8H_8, has already been discussed in detail (see p. 406). In the case of this ring system, and in contrast to cyclobutadiene, a largely strain-free non-planar arrangement of C-atoms linked by alternate single and double bonds is possible, and this is the reason why this compound can be isolated (see Fig. 102, p.407). It has been shown that by adding two further electrons to this system, it can be converted into an aromatic system; this is achieved in the *cyclooctatetraenide dianion*.

5.21.7.4 10-π-Electron Systems[2]

The *cyclooctatetraenide dianion*, $C_8H_8^{2\ominus}$, is obtained from the reaction of cyclooctatetra-ene with alkali metals. For example, with potassium, dipotassium cyclooctatetraenide, $C_8H_8^{2\ominus}\,2K^{\oplus}$, is produced. Its physical properties indicate that this dianion is an *aromatic system*. The ring is flat and all the C-C bonds are of the same length. It gains mesomeric stabilisation from the 10 π-electron system.

The same considerations apply to the *cyclononatetraenide anion*, $C_9H_9^{\ominus}$. As well as the all-cis form, there is also a mono-trans isomer.

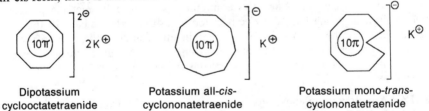

| Dipotassium cyclooctatetraenide | Potassium all-*cis*-cyclononatetraenide | Potassium mono-*trans*-cyclononatetraenide |

5.21.7.5 Annulenes

The *Hückel* rule has thus been shown to hold true for ionic $(4n + 2)$-π-electron systems. Neutral 10-π-electron systems would also be expected to show aromatic character, so long as a planar ring shape is achieved by introducing bridges across the ring. Thus, if two hydrogen atoms in *cyclodecapentaene*, $C_{10}H_{10}$([*10*]*annulene*), which would crowd one another, are replaced by a CH_2 group, then a marked, although not complete, flattening of the 10-ring is brought about. The resultant compound, *1,6-methanocyclo-decapentaene*, $C_{11}H_{10}$, has an *aromatic* character (*Vogel*, 1964).

[1] See G. *Maier* et al, Tetra-t-butyl-tetrahedrane, Angew. Chem. Int. Ed. Engl., **17**, 520 (1978).
[2] See E. *Vogel*, Aromatic 10- and 14-π-Electron Systems, Suppl. Chimia, 1968, **21**; S. *Masamune* and N. *Darby*, [10]Annulenes, Acc. Chem. Res. **5**, 272 (1972).

[10]Annulene

1,6-Methano[10]annulene

Higher homologues of this sort, which satisfy *Hückel*'s rule, are the macrocycles $C_{14}H_{14}$ and $C_{18}H_{18}$. They were first made by *Sondheimer* (1959) and have been called *annulenes*.[1] **[14]Annulene** is nearly planar, and **[18]annulene** has a sufficiently planar shape to permit an equalising of the bond lengths due to mesomerism. However, an X-ray structure determination of [18]annulene shows that its ring-bonds are of various lengths, between 138 and 142 pm. Like [14]annulene it sustains a ring-current in the same way that benzene does (*cf*. Fig. 20, p. 41). As a result of this, the signals from the hydrogen atoms within the ring are strongly shifted to higher field i.e. these protons are shielded. Compounds showing such diamagnetic shifts are described as *diatropic*.

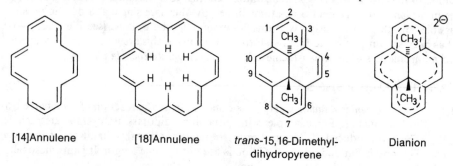

[14]Annulene   [18]Annulene   *trans*-15,16-Dimethyl-   Dianion
                                                dihydropyrene

A further example of a diatropic compound is ***trans*-15,16-dimethyldihydropyrene**, which has a [14]annulene perimeter. The signals from the hydrogens of the methyl groups appear at $\delta = -4.2$ p.p.m. When this compound is reduced by metallic potassium, a *dianion* is formed which has 16 π-electrons (*i.e.* 4×4) in its perimeter. In this case the ring-current is in the opposite sense to that in an aromatic system, and the signals from the hydrogen atoms of the methyl groups are strongly shifted to lower field ($\delta = +21$ p.p.m.). Compounds which show this behaviour, *i.e.* paramagnetic shifts, are described as *paratropic*.

5.21.7.6 Annulenoannulenes[2] are ring systems made up from two *ortho*-condensed annulene rings. The simplest [4n]annuleno[4n]annulene is *octalene*, the 8-membered ring analogue of *naphthalene*.

Octalene was first prepared by *E. Vogel*.[3] It is a lemon-yellow liquid (b.p. 50–52°C [323–325K]/1 mbar [1 hPa]), which is sensitive to air.

In their ground state the molecules are not planar. One of the possible conformations is shown in Fig. 121. Valence isomerisation of the double bonds can take place, and this involves a transition state with a flattened perimeter, which is, however stabilised by mesomerism, having 14 π-electrons. Its stabilisation energy is estimated to be about 60 kJ/mol.

---

[1] See *F. Sondheimer*, The Annulenes, Acc. Chem. Res. *5*, 81 (1972)
[2] See *M. Nakagawa*, Angew. Chem. Int. Ed. Engl. *18*, 202 (1979).
[3] Angew. Chem. Int. Ed. Engl., *16*, 871, 872 (1977).

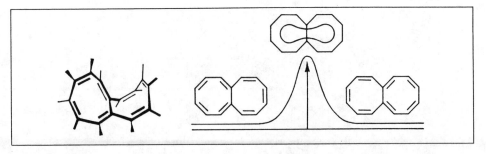

Fig. 121. Octalene.

This provides a striking contrast to the destabilisation of the transition state which is involved in the migration of the double bonds in cyclobutadiene.

In addition to the two valence isomers shown above, another structure is possible which lacks full peripheral conjugation.

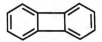

This isomer has not yet been observed. Obviously it is sufficiently disfavoured energetically that it does not participate to any worthwhile extent.

5.21.7.7 Biphenylene

Biphenylene takes up a double-bond structure as depicted above. The two single bonds linking the six-membered rings are long and there is incomplete delocalisation of the π-electrons in the benzenoid rings. This arrangement inhibits the setting-up of a destabilising cyclobutadiene ring or of a peripheral 12-π-electron system while retaining some benzenoid character in the outer rings. Biphenylene is a stable crystalline compound.

5.21.7.8 The *Evans* Principle

Pericyclic reactions have transition states which are electronically analogous to the perimeters to which *Hückel's* rule applies. This connection was noted by *Evans* in 1939. It can be expressed in its simplest form as the following rule, which was formulated by *Dewar* and described as the *Evans principle*:[1]

Thermal pericyclic reactions take place preferentially via aromatic transition states.

A simple example is provided by the *Diels–Alder* reaction. As *Dewar* has pointed out, the Evans principle provides an interesting alternative to the *Woodward–Hoffmann* rules in many cases.

[1] See *M. J. S. Dewar*, Aromaticity and Pericyclic Reactions, Angew. Chem. Int. Ed. Engl., *10*, 761 (1971).

6 Isoprenoids (Terpenes and Steroids)

The two classes of natural products known respectively as terpenes and steroids are closely similar in their molecular make-up.[1] As far back as 1922 *Ruzicka* postulated that the different terpenes were all made up from assemblies of isopropene (C_5H_8) units. This 'isoprene rule' was later extended to encompass not only the addition of isoprene units but also other modifications they might undergo, such as rearrangement reactions. More recent work has shown that (R)-*mevalonic acid* functions as the 'isoprene building block' in the construction of terpene and steroid molecules in plants and in animals. This biogenetic relationship led to both types of compound being classed together as isoprenoids.

The naturally occurring terpenes[2] were investigated especially by *Wallach, v. Baeyer, Semmler* and *Tiemann*. They include not only hydrocarbons but also derivatives such as alcohols, ethers, aldehydes and ketones. They occur in plants, especially in flowers and fruits, and are obtained as 'essential oils' by steam distillation or extraction of crushed plants. Because of their pleasant odours many terpene derivatives have been used as *perfumes*.

In addition to monoterpenes, $C_{10}H_n$, which can be regarded as dimers of isoprene and which are found in the essential oils and also in resins and in other plant materials, there are other terpenes containing more than two isoprene units. These have been described as polyterpenes or as *polyprenes* (= polyisoprenes) (*Ruzicka*). Within this overall description they can be grouped according to the number of isoprene units they contain as *sesquiterpenes* ($1\frac{1}{2}$ monoterpenes), $C_{15}H_n$; diterpenes, $C_{20}H_n$; *triterpenes*, $C_{30}H_n$; *tetraterpenes*, $C_{40}H_n$; and high molecular weight *polyterpenes* (higher terpenoids, polyprenes), $(C_5H_n)_x$; naturally occurring rubber and gutta-percha are examples of the latter. As well as these unsaturated hydrocarbons there are also naturally occurring hydrogenated and dehydrogenated derivatives and oxygen-containing derivatives. In addition there are naturally occurring compounds in which subsequent breakdown of first-formed polyprenes results in the formation of products whose constitution does not correspond to that of polyisoprenes. All these compounds are included among terpenes or terpenoids.

When a second isoprene unit is joined to an initial isoprene this may lead either to the formation of an aliphatic chain or to ring-formation. Hence monoterpenes can be divided into three main classes:

I.   acyclic (or open-chain) terpenes
II.   monocyclic terpenes

---

[1] See *W. Templeton*, An Introduction to the Chemistry of Terpenoids and Steroids (Butterworths, London 1969).
[2] See *A. A. Newman* (Ed.). Chemistry of Terpenes and Terpenoids (Academic Press, New York 1972).

III.   bicyclic terpenes.

Only a few of the more important terpenes will be mentioned in this book.

6.1 Acyclic Monoterpenes

6.1.1 Monoterpene Hydrocarbons

Of this group of compounds only two isomeric hydrocarbons need be mentioned, *ocimene* and *myrcene*, $C_{10}H_{16}$, each of which contains three C=C double bonds. They occur in particular in oil of basil (*Osimum basilicum*) and in laurel (bay oil). Both include two conjugated double bonds; they differ from one another in the sites of the double bonds.

Ocimene   Myrcene

6.1.2 Monoterpene Alcohols

The following naturally occurring alcohols are of greater importance than the above hydrocarbons.

Geraniol, $C_{10}H_{18}O$, is the main constituent of both geranium oil and rose oil. The hydroxymethyl group at carbon atom 1 is *trans* to the methylene group 4 of the main carbon chain.

If geraniol is heated with an alkoxide it is converted into the (Z) isomer, *nerol*, which occurs naturally in a number of natural oils including those from orange blossom and bergamot. Both the (Z) and (E) isomers are *perfumes* having a rose-like smell.

Linalool is an *optically active* tertiary alcohol, which is isomeric with geraniol and nerol. In particular it is found in oil of linaloe and lavender oil; it has a smell like that of lilies of the valley.

Geraniol   Nerol   Linalool   Citronellol   (S)-(−)-Ipsenol
(E) isomer   (Z) isomer

Citronellol contains two more hydrogen atoms than the previously described alcohols. It is *optically active* and occurs in rose oil in the L(−) form, in lemon oil as the D(+) form, and in geranium oil in the racemic form.

(S)-(−)-Ipsenol is a pheromone of bark-beetles of the genus *Ips*, where it is formed from *myrcene*. It becomes evident that the action of this and other chiral pheromones is increased if a mixture of the (R)- and (S)-enantiomers is present rather than just the one isomer. This provides a notable example of the mutual interaction (*synergism*) of enantiomers.

6.1.3 Monoterpene Aldehydes and Monoterpene Ketones

The two monoterpene aldehydes *citral a* (geranial) and *citral b* (neral), $C_{10}H_{16}O$, each have two $C=C$ double bonds. They are (E) and (Z) isomers derived, respectively, from geraniol and nerol. They are yellowish oils with a lemon-like odour and are isolated from lemon oil and lemon-grass oil; they are the principal constituents of the latter.

Aldol-condensation of citral with acetone in alkaline conditions leads to the formation of *pseudoionone*, $C_{13}H_{20}O$, which, when heated with acid, undergoes ring-closure to give α and β-**ionone**:

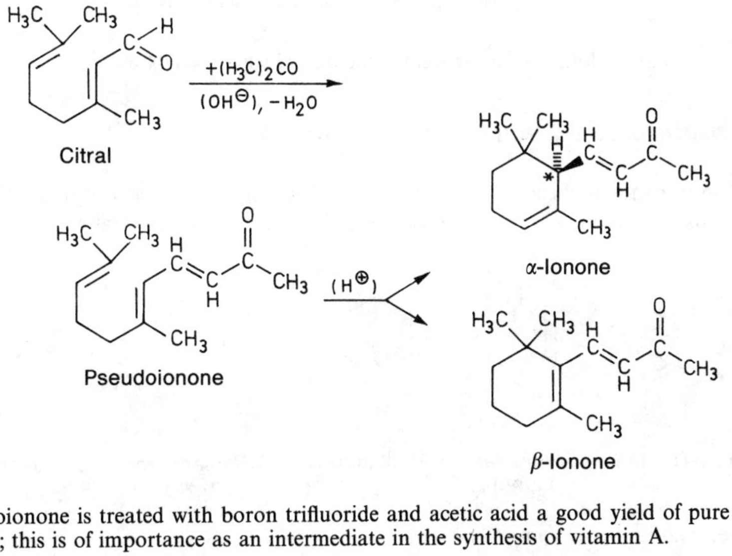

If pseudoionone is treated with boron trifluoride and acetic acid a good yield of pure β-ionone is obtained; this is of importance as an intermediate in the synthesis of vitamin A.

The ionones differ only in the site of the double bond in the ring. α-Ionone has an asymmetric carbon atom (*), which in plants has an (R)-configuration. They smell like violets and because of this are used as *perfumes*.

The ionones provide a way of obtaining monocyclic terpene derivatives and are important as building blocks for making carotenes and the A-vitamins. The odiferous properties are lost when the molecular mass exceeds 300, because the substances then become insufficiently volatile.

6.2 Monocyclic Monoterpenes

This series of monoterpenes is based on *trans-p-menthane*, 4-isopropyl-1-methylcyclohex-ane, $C_{10}H_{20}$, and also includes the mono-unsaturated derivative *menthene*, $C_{10}H_{18}$, and the doubly unsaturated *menthadiene*, $C_{10}H_{16}$.

6.2.1 p-*Menthane*

p-Menthane is a liquid smelling of peppermint which can be prepared by catalytic hydrogenation of *p-cymene* (see p. 478). It occurs in two optically inactive forms, the *cis* and *trans* isomers (*cf.* quinitol):

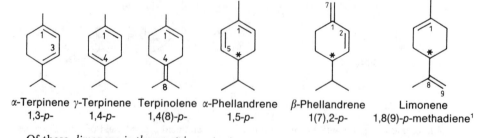

cis-*p*-Methane   trans-*p*-Methane
(b.p. 168.5°C [441.7 K])   (b.p. 169–170°C [442–443 K]))

The shorthand linear formula on the right is the one which is usually used.

6.2.2 *Unsaturated Monoterpene Hydrocarbons*

The most important of these are the naturally occurring derivatives of menthadiene. Of the numerous possible *isomers* the following six are found in essential oils:

α-Terpinene γ-Terpinene Terpinolene α-Phellandrene   β-Phellandrene   Limonene
   1,3-*p*-   1,4-*p*-   1,4(8)-*p*-   1,5-*p*-   1(7),2-*p*-   1,8(9)-*p*-methadiene[1]

Of these, *limonene is the most important*.

The numbering of these monocyclic monoterpenes is based on that of *p-menthane*. If there is an *exocyclic* double bond, *i.e.* one linking an exocyclic group to the ring, or a double bond totally outside the ring, it is customary to number both ends of the bond.

---

[1] Before the introduction of the IUPAC nomenclature as used here, the numbers referring to the double bond were designated by superscripts following the symbol Δ, *e.g.* 1,8(9)-*p*-menthadiene was called $\Delta^{1,8(9)}$-menthadiene.

(R)-(+)- and (S)-(−)-Limonene, $C_{10}H_{16}$, and also the racemic form *dipentene* occur in essential oils, (+)-limonene in oil of caraway, (−)-limonene in the oil from pine needles, and dipentene in oil of turpentine. (S)-(−)-limonene smells of lemons, and (R)-(+)-limonene like oranges.

Dipentene can be prepared by the *dimerisation of isoprene* in a *Diels–Alder* reaction:

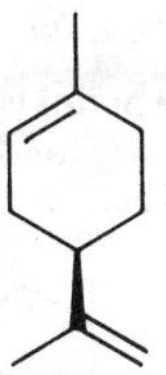

(R)-(+)-Limonene 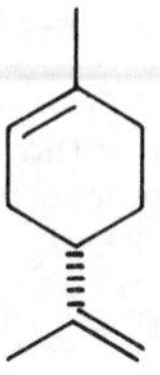   (S)-(−)-Limonene 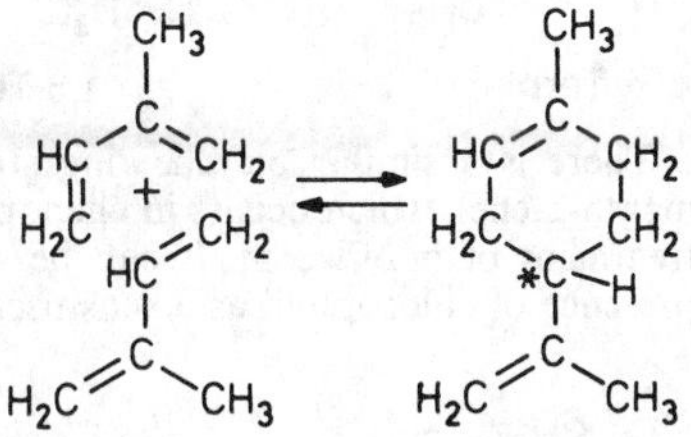   Isoprene   Dipentene

6.2.3 Monoterpene Alcohols and Monoterpene Thiols

Menthol or menthan-3-ol, $C_{10}H_{20}O$, is the most important saturated alcohol of this series. There are three chiral centres in the molecule, and in consequence there are eight stereoisomers which have all been made synthetically.

The racemic form of menthol can be obtained from the catalytic hydrogenation of thymol (isopropylmethylphenol) which is a constituent of oil of thyme.

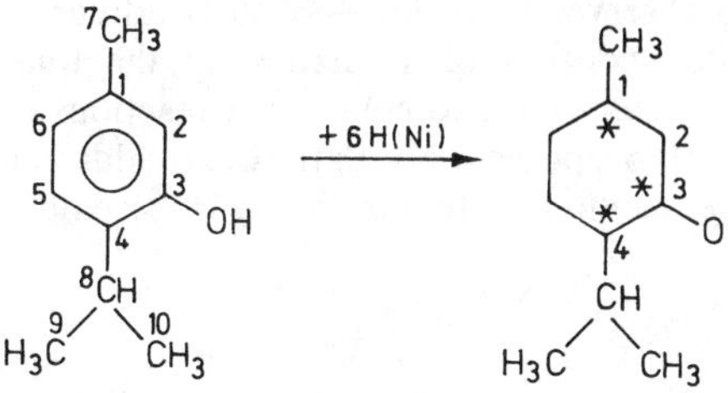

Thymol   (±)-Menthol   (1R,3R,4S)-(−)-Menthol

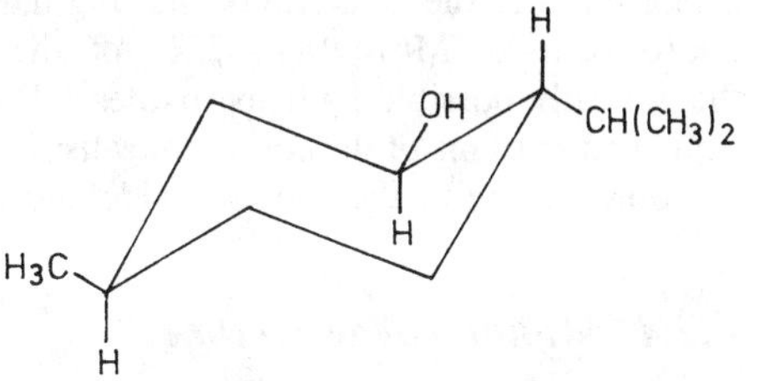

The main constituent of peppermint oil is only the (−)-menthol with m.p. 43°C (316 K) whose conformation is shown above.

α-Terpineol, $C_{10}H_{16}O$, one of the unsaturated alcohols derived from *p*-menthane, occurs in both its dextro- and laevo-rotatory forms in essential oils, for example in oil of cardamom. It smells of elderflower. It is related to *1,8-terpin*, $C_{10}H_{20}O_2$, which is obtained in the form of a hydrate by shaking α-terpineol with dilute sulphuric acid. 1-p-Menthene-8-thiol is a component of the aroma of fresh grapefruit juice; in its racemic form it can be tasted in water in concentrations down to 10^{-4} ppb, *i.e.* 1 mg per 10000 t water!

Another dehydration product derived from 1,8-terpin is *1,8-cineole*, (*eucalyptol*), $C_{10}H_{18}O$, which has a camphor-like smell and is the main constituent of eucalyptus oil from *Eucalyptus globulus*. It can be prepared from 1,8-terpin by heating the latter with dilute mineral acid which brings about intramolecular loss of water. The bicyclic ether cineole contains the completely strain-free 2-oxabicyclo[2.2.2]octane ring system.

| α-Terpineol | 1,8-Terpin | 1,8-Cineole | 1-p-Menthene-8-thiol |

There is a similar bicyclic ring structure in the peroxide derivative *ascaridole* (1,4-epidioxy-*p*-menth-2-ene), which occurs in chenopodium oil and was at one time used as an *anthelmintic* for the treatment of hookworm. It can be synthesised from α-terpinene by the action of oxygen in the presence of chlorophyll as a sensitiser (*G.O. Schenk*, 1953):

α-Terpinene   Ascaridole   (I) Endo-peroxide

This *photo-oxidation* or *photo-oxygenation* is brought about by *singlet oxygen* (see p. 225), generated by excitation from the ground (triplet) state; it reacts with the diene system of α-terpinene in a reaction analogous to the *Diels–Alder* reaction. A concurrent reaction of singlet oxygen with one of the two allylic hydrogen atoms (shown with a grey background in the formula) of the ring has been observed when R-(−)-α-phellandrene is photo-oxidised. More than 25% of the product consists of a mixture of the four theoretically possible hydroperoxides[1]. The existence of the two concurrent reactions is explained in terms of the common transition state (I) proposed for both the cyclo-addition reaction leading to the endoperoxide and the *ene*-reaction leading to the hydroperoxide.

6.2.4 Monoterpene Ketones

The three most important monocyclic ketones are:

Menthone   Pulegone   Carvone

---

[1] See *R. Matusch* et al., Competition of Endoperoxide and Hydroperoxide Formation in the Reaction of Singlet Oxygen with Cyclic Conjugated Dienes, Angew. Chem. Int. Ed. Engl. *27*, 217 (1988).

Menthone, p-menthan-3-one, $C_{10}H_{18}O$, is formed when menthol is oxidised with chromic acid. It contains two asymmetric carbon atoms. (+)-Menthone occurs in geranium oil and (−)-menthone in peppermint oil

Pulegone, $C_{10}H_{16}O$, an unsaturated ketone, is found in a dextrorotatory form as the main constituent of oil of pennyroyal.

Carvone, p-mentha-6,8(9)-dien-2-one, $C_{10}H_{14}O$, contains two C=C double bonds. Its (S)-(+)-form is found in caraway oil and dill oil, and smells of caraway. The enantiomeric (R)-(−)-form has a mint-like smell.

The position of the ketone group in carvone was shown as follows. When carvone is heated with hydrochloric acid at 125°C (400 K), the double bond in the side chain migrates into the ring, and enolisation of the ketone to a phenol provides carvacrol, an isomer of thymol:

Carvone   Carvacrol

6.3 Bicyclic Monoterpenes

This kind of monoterpene is also derived from p-menthane. The isopropyl group at position 4 is joined through carbon atom 8 (see p. 664) to the cyclohexane ring, thus forming a second ring. In the same way that o-, m- and p-positions may be involved in a benzene ring, so here loss of two hydrogen atoms can result in the isopropyl group being linked to the cyclohexane ring in three different ways, leading to the formation of, respectively, cyclopropane, cyclobutane or cyclopentane rings. In this way the three parent hydrocarbons of the bicyclic monoterpenes, carane, pinane and bornane (camphane) are produced.

Carane   Pinane   Bornane (Camphane)

All of them have the molecular formula $C_{10}H_{18}$ and are constructed in accordance with the *isoprene rule.*

The numbering of the carbon atoms in the above formulae follows IUPAC rules, which favour the use of bornane rather than camphane.

Among the oxo-derivatives of these bicyclic monoterpenes are the following:

Carone   Verbenone   Camphor

6.3.1 Carane Group

The most important of these compounds is the *optically active* carone, caran-2-one, $C_{10}H_{16}O$. It is a liquid, b.p. 101°C (374 K)/ 20 mbar (20 hPa) with a camphor-like smell, and can be prepared by adding hydrogen bromide to dihydrocarvone [*p*-menth-8(9)-en-2-one] and treating the adduct with ethanolic potassium hydroxide. The 3-membered ring in carone is readily opened, for example, when heated it is converted into carvenone, *p*-menth-3-en-2-one, b.p. 233°C (506 K), which is more stable.

Dihydrocarvone   "Dihydrocarvone   Carone   Carvenone
p-Menth-8(9)-en-2-one   hydrobromide"

6.3.2 Pinane Group

The principal derivatives are the unsaturated hydrocarbons α- and *β-pinene*. The two of them are the main constituents (80–90%) of oil of turpentine, which is obtained by steam distillation of the resin from various kinds of pine tree (*pinus*). The resins consist of a complicated mixture of alicyclic compounds. The involatile residue, *rosin*, consists almost entirely of the so-called resin acids. Because of its outstanding ability as a solvent for oils and resins, oil of turpentine is an almost unsurpassable solvent for use in the varnish and lacquer industries.

α-Pinene (b.p. 156°C [429 K])   Ipc₂BH   β-Pinene (b.p. 164°C [437 K])

α-Pinene, $C_{10}H_{16}$, is the main constituent ($\sim 60\%$) of oil of turpentine a. present in either its dextrorotatory, laevorotatory or racemic forms. Treatment of o₊ active α-pinene with borane provides an asymmetric substituted borane, for exam₊ (+)(1R,2S,3R,5S)-diisopinocampheylborane (Ipc$_2$BH) or, better, its tetramethylethylene-diamine adduct, which are used to achieve enantioselective hydroboration.

β-Pinene is isomeric with α-pinene, having an exocyclic double bond.

Dilute mineral acid opens the four-membered ring of α-pinene to give α-terpineol or 1,8-terpin or its hydrate:

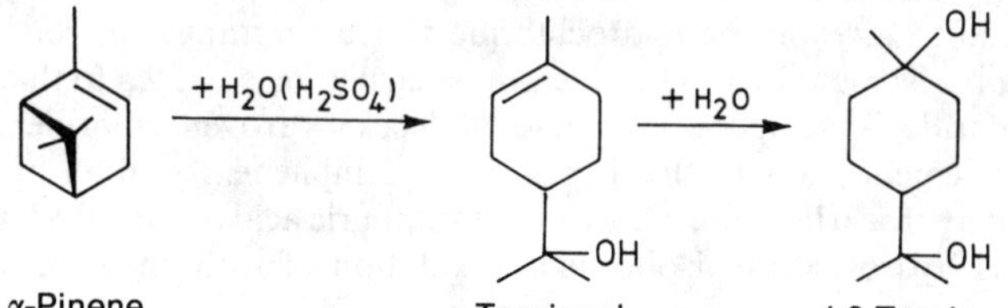

α-Pinene   α-Terpineol   1,8-Terpin

α-Pinene reacts differently, however, with anhydrous mineral acids. (See the synthesis of camphor).

6.3.3 Bornane Group

The most important member of this group of compounds is camphor, $C_{10}H_{16}O$, a bicyclic ketone with the following structure (*Bredt*, 1893):

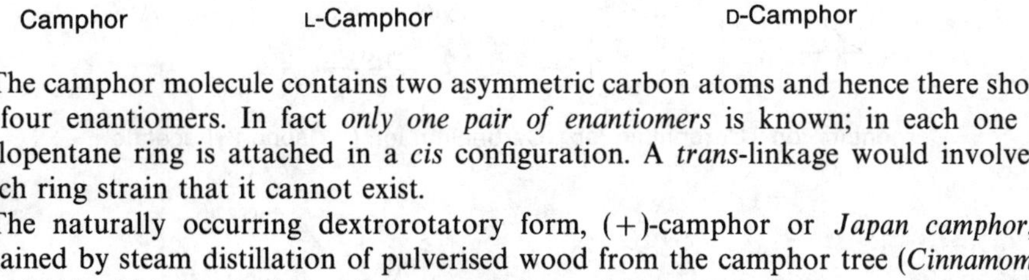

Camphor   L-Camphor   D-Camphor

The camphor molecule contains two asymmetric carbon atoms and hence there should be four enantiomers. In fact *only one pair of enantiomers* is known; in each one the cyclopentane ring is attached in a *cis* configuration. A *trans*-linkage would involve so much ring strain that it cannot exist.

The naturally occurring dextrorotatory form, (+)-camphor or *Japan camphor*, is obtained by steam distillation of pulverised wood from the camphor tree (*Cinnamomum camphora*) which has its home in Japan, China and Taiwan. Laevorotatory camphor is found in some essential oils, *e.g.* oil of feverfew, and has been given the name *matricaria camphor* from its occurrence in oil of *Matricaria parthenium* (camomile).

Freudenberg determined the absolute configuration of (+)-camphor, relating it to that of D-glucose. If the carbonyl group is drawn at the top right-hand corner of the 6-membered ring, in D-camphor the isopropyl group is above the plane of this ring and in L-camphor beneath it.

On the *Cahn–Ingold–Prelog* system carbon atoms 1 and 4 of (+)-camphor have R-configurations and the molecule is (1R,4R)-bornan-2-one.

Since the supply of naturally occurring camphor is insufficient to meet the need for it as a *plasticiser* in the manufacture of celluloid, the industrial preparation of racemic camphor is of importance.

Preparation

α-Pinene serves as the starting material. In the presence of titanium dioxide hydrate as an acid catalyst it rearranges at 180°C (450 K) into camphene. Alternatively the *Wagner–Meerwein*[1] rearrangement of the bicyclic terpene can be brought about in a stepwise fashion, starting with the addition of dry hydrogen chloride at 0°C (273 K) to the double bond of α-pinene to give pinene hydrochloride which rearranges at room temperature into its more stable isomer bornyl chloride, whose structure is related to that of camphor[2].

If bornyl chloride is heated with a base it loses hydrogen chloride and a further molecular arrangement leads to the formation of camphene. Addition of acetic acid to this unsaturated hydrocarbon in the presence of sulphuric acid as a catalyst gives isobornyl acetate which is hydrolysed to isoborneol. Oxidation of isoborneol provides *optically inactive* camphor.

The overall course of these reactions can be illustrated by the following reaction scheme, in which the three *Wagner–Meerwein* rearrangements are indicated by curved arrows:

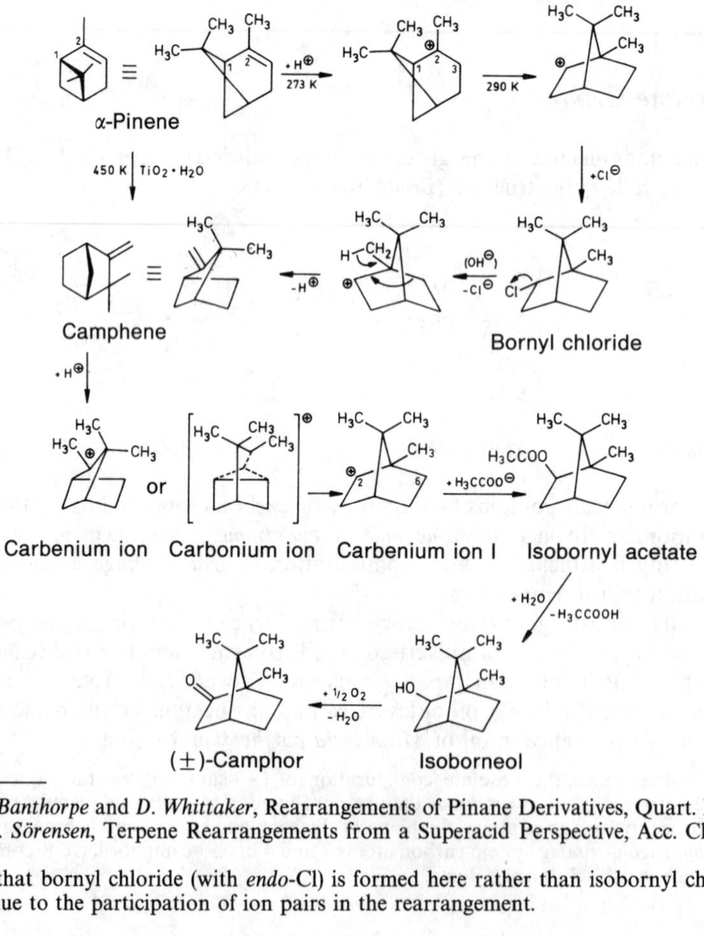

α-Pinene

Camphene

Bornyl chloride

Carbenium ion   Carbonium ion   Carbenium ion I   Isobornyl acetate

(±)-Camphor   Isoborneol

---

[1] See *D. V. Banthorpe* and *D. Whittaker*, Rearrangements of Pinane Derivatives, Quart. Rev. *20*, 373 (1966); *T. S. Sörensen*, Terpene Rearrangements from a Superacid Perspective, Acc. Chem. Res. *9*, 257 (1976).

[2] The fact that bornyl chloride (with *endo*-Cl) is formed here rather than isobornyl chloride (with *exo*-Cl) is due to the participation of ion pairs in the rearrangement.

In order to make the scheme clearer, *carbenium ions* are largely used in the above reaction scheme. However *carbonium ions* (non-classical cations, see p. 27) play an important role in this system of rearrangements.[1] They also make it easier to understand the reactions which lead to the formation of (±)camphor; the positive charge on carbenium ion (I) can rest on either carbon atom 2 (as drawn) or carbon atom 6. It is recommended that this series of rearrangements be followed with the help of molecular models.

Properties

(+)Camphor forms colourless crystals, m.p. 177°C (450 K). It has a strong, very characteristic odour. Its specific rotation is $[\alpha]_D = +44°$ (in benzene). Camphor is used medicinally as a disinfectant and local counter-irritant (camphorated oil).

The ketone group of (+)-camphor reacts with the characteristic ketone reagents such as hydroxylamine, phenylhydrazine, etc. When it is reduced to a secondary alcohol group a third asymmetric carbon atom is created. As a result, two stereoisomeric secondary alcohols, namely (+)-borneol and, in smaller amounts, (−)-isoborneol are formed, in which the hydroxy groups are, respectively, *endo* and *exo*.

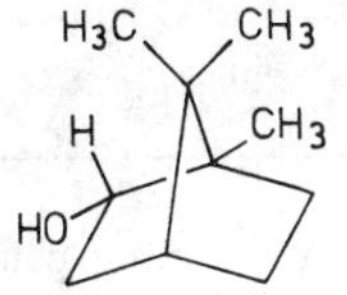

(+)-Borneol (*endo*) 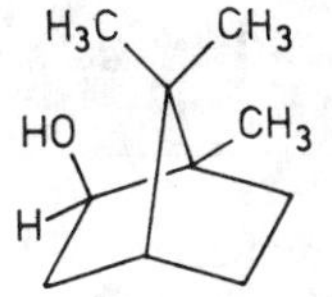   (−)-Isoborneol (*exo*)

(+)-Borneol (which, because of its relationship to camphor, has also been called Borneo camphor) occurs in the wood of the camphor tree (*Dryobalanops aromatica*) from Borneo and Sumatra and in many essential oils, for example in oil of lavender and oil of rosemary.

(−)-Borneol (Ngai camphor) occurs in valerian.

The *methylene group* at carbon atom 3, which is next to the keto group of camphor, is reactive; on treatment with isoamyl nitrite and sodium alkoxide *isonitrosocamphor* (3-hydroxyiminocamphor) is formed. When this is hydrolysed the product is golden yellow (+)-*camphorquinone*, the 2,3-diketone of camphane, m.p. 198°C (471 K):

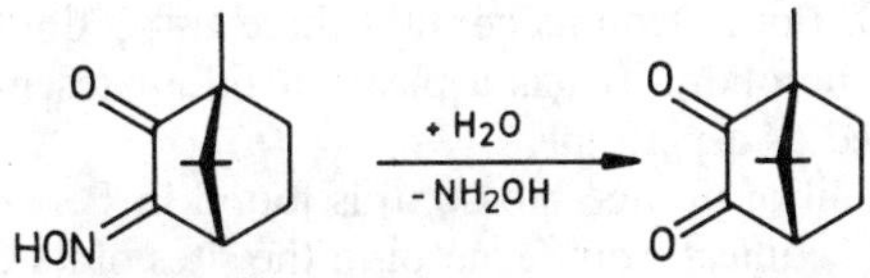

Isonitrosocamphor   (+)-Camphorquinone

Bromination of (+)-camphor at 100°C (375 K) provides 3-bromocamphor, which, when warmed with sulphuric acid and acetic anhydride, is converted into (+)-*3-bromocamphor-10-sulphonic acid*. Sulphonation of (+)-camphor with sulphuric acid and acetic anhydride gives (S)-(+)-*camphor-10-sulphonic acid*, m.p. 194°C (467 K), $[\alpha]_D = +24°$ (in water). These two readily available optically active sulphonic acids have frequently been used to *resolve racemic bases* into their enantiomers.

---

[1] See *G. A. Olah*, The σ-Bridged 2-Norbornyl Cation and its Significance to Chemistry, Acc. Chem. Res. **9**, 41 (1976); *H. C. Brown* and *P. v. R. Schleyer*, The Non-classical Ion Problem (Plenum Press, New York 1977).

(+)-3-Bromocamphor-   (+)-Camphor-
10-sulphonic acid   10-sulphonic acid

Camphoric acid

Oxidative cleavage of camphor by nitric acid leads to the formation of *camphoric acid*, which has two asymmetric carbon atoms and, unlike camphor, does form *two pairs of enantiomers*. In the first pair, namely (+)- and (−)-camphoric acids, the two carboxyl groups are *cis* to each other; they readily lose water to form anhydrides. In (+)- and (−)-isocamphoric acids, the two carboxyl groups are *trans* to one another. Since only the single 5-membered ring is present in the camphoric acids there is no steric inhibition to the formation of the *trans* isomers (*i.e.* isocamphoric acids).

6.4 Sesquiterpenes

When three isoprene units are joined together to form sesquiterpenes, $C_{15}H_n$, this may lead, as in the case of the simple terpenes, to either open-chain aliphatic compounds or to cyclic compounds. Three types of the latter are known, namely mono-, bi- and tricyclic sesquiterpenes.

Ruzicka was responsible for determining the constitutions of many cyclic sesquiterpenes.

More recent studies have led to the discovery of sesquiterpenes containing 9-, 10- and 11-membered carbon rings.[1]

6.4.1 Acyclic Sesquiterpenes

The most important examples of this class are the two sesquiterpene alcohols *farnesol* and *nerolidol*, both $C_{15}H_{26}O$. Both alcohols contain three C=C double bonds.

Farnesol is prepared from nerolidol. It has a pleasant odour resembling that of lilies of the valley and is hence used as a perfume.

Nerolidol (from neroli oil, oil of orange blossom) is found in Peru balsam and in the essential oil of many flowers. It differs from farnesol in the sites of its C=C double bond and hydroxy group. It has a chiral centre (marked *) and also exists as a mixture of *Z* and *E* forms.

Farnesol   Nerolidol

[1] See *G. Rücker*, Sesquiterpenes, Angew. Chem. Int. Ed. Engl. *12*, 793 (1973).

The regular arrangement of methyl substituents separated in each instance by three unsubstituted carbon atoms exemplifies how these alcohols are assembled from isoprene units. These two alcohols are related to one another in the same way that geraniol and linalool are to each other.

6.4.2 *Monocyclic Sesquiterpenes*

The hydrocarbon *bisabolene* is an example of this type of compound. It is widely distributed in nature and is found in lemon oil and in the oil from pine needles. It is made synthetically from farnesol (or from nerolidol) by the action of formic acid which removes a molecule of water from the alcohol:

| Farnesol | Bisabolene | Hernandulcin |

Hernandulcin, which is a *naturally occurring sweetening agent* derived from the Mexican plant *Lippia dulcis Trev.*, is a derivative of bisabolene; it is about a thousand times as sweet as saccharose.

6.4.3 *Bicyclic Sesquiterpenes*

Two hydrocarbons of this sort are *cadinene* (from the essential oils of juniper species and cedars) and *β-selinene* (from celery), both $C_{15}H_{24}$. They both have reduced naphthalene rings and can be thought of as being derived in a similar way as bisabolene is from farnesol.

Cadinene   *β*-Selinene

Both hydrocarbons can be dehydrogenated to naphthalene derivatives by heating them with sulphur or selenium, for example cadinene is converted into 4-isopropyl-1,6-dimethylnaphthalene (cadalene).

Cadinene   Cadalene

When *β*-selinene is dehydrogenated the *angular methyl group* at one of the carbon atoms common to both rings is lost as methane (*Ruzicka*).

The blue or violet **azulenes**, $C_{15}H_{18}$, (*cf. azulene*, p. 653 *f.*) represent another kind of bicyclic sesquiterpene derivative. They are fully unsaturated and are formed by dehydrogenation and/or dehydration of naturally occurring hydroazulenes.

Azulenes isolated from essential oils include especially **vetivazulene** (from oil of vetiver) and **guaiazulene** (from geranium oil); biochemically they are derived from farnesol:

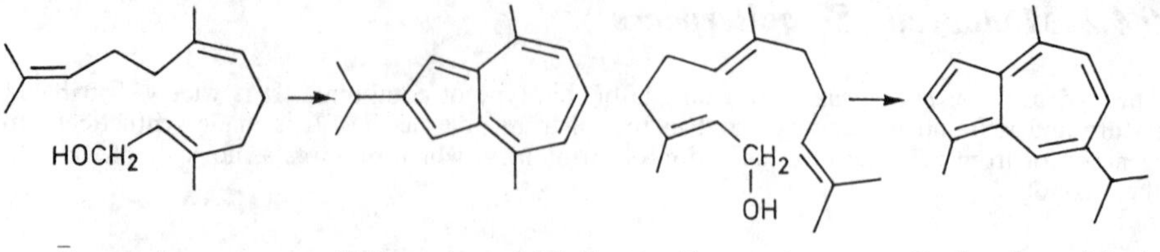

Farnesol   Vetivazulene (violet)   Farnesol   Guaiazulene (blue)
(m.p. 33°C [306 K])   (m.p. 31°C [304 K])

6.4.4 Tricyclic Sesquiterpenes

Mention will be made here only of the hydrocarbons *α-santalene*, $C_{15}H_{24}$, from sandalwood oil, and *modhephene*, $C_{15}H_{24}$, a derivative of [3.3.3]propellane isolated from golden rod (*Isocome wrightii*):

α-Santalene   Modhephene

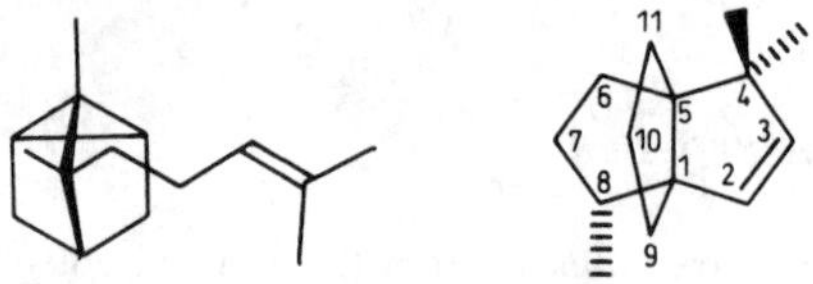

6.5 Diterpenes

6.5.1 Acyclic Diterpenes

The biologically important diterpene alcohol phytol is an example of this group.

Phytol, $C_{20}H_{40}O$, was first isolated by *Willstätter* (1909) from the hydrolysis of chlorophyll. It is a colourless oil, b.p. 202° (475 K)/13 mbar (13 hPa). The molecules of phytol contain a single C=C double bond and two chiral centres. The phytyl residue is found in vitamins E and K_1 as well as in chlorophyll.

Earlier syntheses of phytol provided the racemic form; the absolute configuration of phytol, [(2E)-(7R,11R)-3,7,11,15-tetramethylhexadec-2-enol], was demonstrated by means of a *stereospecific synthesis* (*Burrell*, 1959):

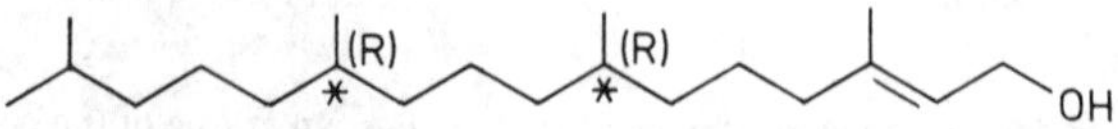

Phytol

6.5.2 Monocyclic Diterpenes

Especially important members of this group of compounds are vitamin A and its derivatives, known as *retinoids*.

Vitamin A, also known as vitamin A_1 or INN: **Retinol**, $C_{20}H_{30}O$, is an unsaturated diterpene primary alcohol. It is found especially in fish-liver oils, egg yolk and milk. It was isolated by *Karrer* in 1931 from halibut-liver oil and has the following structure, which has been confirmed by synthesis:[1]

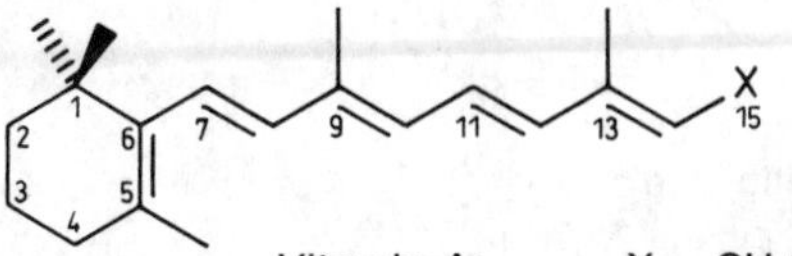

	Vitamin A:	X = CH₂OH
(β-Ionone ring)	Vitamin A acid:	X = COOH

Vitamin A: $X = CH_2OH$
Vitamin A acid: $X = COOH$ 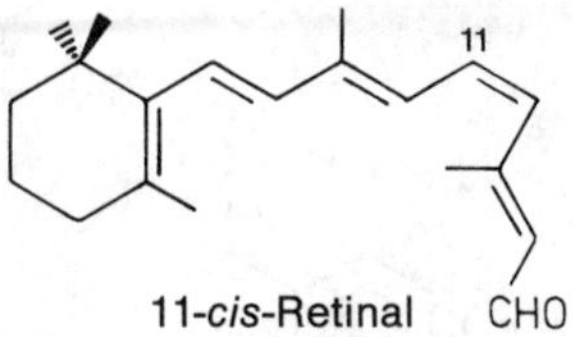
11-*cis*-Retinal CHO

A lack of the fat-soluble growth-vitamins A in the food of mammals causes a decrease in weight and the onset of degeneration of the mucous membranes, above all those of the conjunctiva and cornea of the eyes. Vitamin A is therefore a remedy for *xerophthalmia* (drying out of the conjuctiva) and as a result is known as the antixerophthalmic vitamin (axerophthol). Lack of vitamin A leads to night blindness since it plays a role in the formation of visual purple (rhodopsin) in the retina. Rhodopsin is formed from the protein opsin and 11-*cis*-retinal, which is an oxidation product of vitamin A. In the visual process 11-*cis*-retinal is changed *via* a series of intermediates into all-*trans*-retinal.[2]

Nowadays synthetic vitamin A has largely taken the place of natural liver oils.[3] Some retinoids, *e.g.* vitamin A acid, are used in the treatment of acne.

6.5.3 Tricyclic Diterpenes

Abietic acid, $C_{20}H_{30}O_2$, is the most important tricyclic diterpene. It is the major constituent of colophony, a resin obtained from the sap of various species of pine (Latin *abies* = fir). This resin consists of a mixture of resin acids and terpenes; the latter are distilled off as *oil of turpentine*. Abietic acid forms part of the residue; structurally it is an isopropyldimethyldecahydrophenanthrenecarboxylic acid. It undergoes dehydrogenation and decarboxylation to form *retene* (7-isopropyl-1-methylphenanthrene) when heated with sulphur:

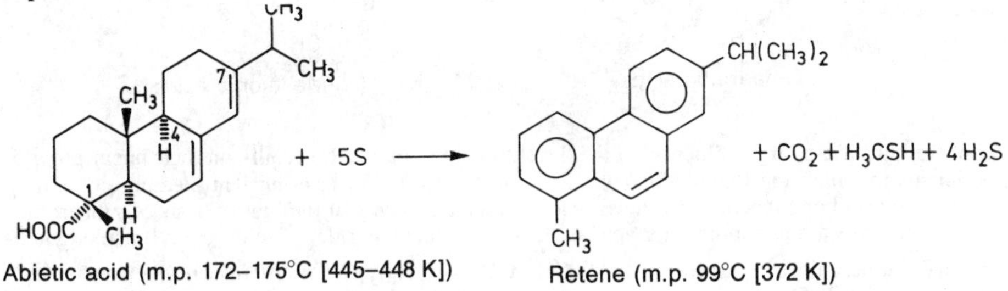

Abietic acid (m.p. 172–175°C [445–448 K])   Retene (m.p. 99°C [372 K])

---

[1] See *O. Isler*, Über das Vitamin A und die Carotenoide, Angew. Chem. *68*, 547 (1956); *O. Isler* et al., Chem. Soc. Special Publ., 1956, No. 4, p.47.

[2] See *A. Mannschreck*, Photoisomeriserung und Sehvorgang (Photoisomerisation and Visual Processes), Chemie in unserer Zeit, *2*, 149 (1968).

[3] See *H. Pommer*, Synthesen in der Vitamin-A-Reihe (Syntheses in the Vitamin A Series), Angew. Chem. *72*, 811 (1960).

If colophony is heated with palladium/active charcoal at 150–250°C (425–525 K), *disproportionation* takes place, the abietic acid being converted into dehydroabietic acid, which has a benzene ring, and di- and tetra-hydroabietic acids:

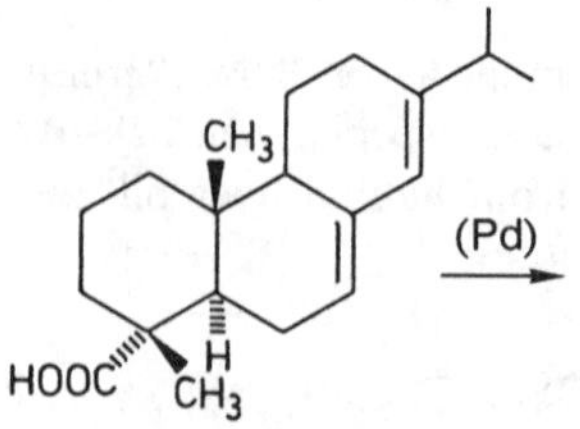

Abietic acid

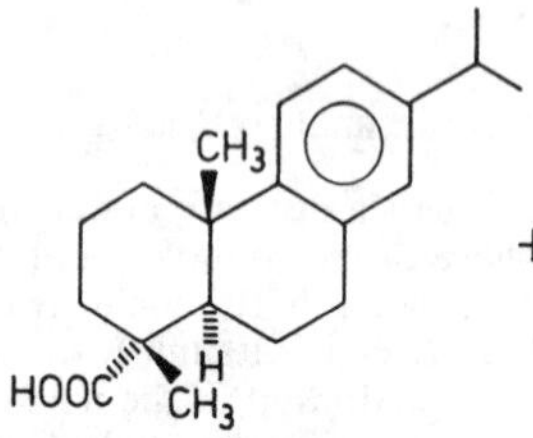

Dehydroabietic acid

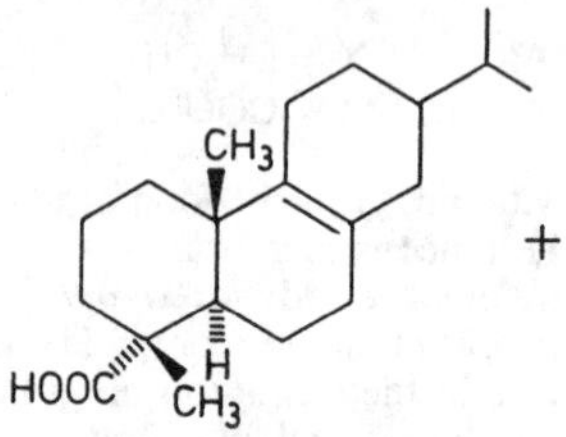

Dihydroabietic acid

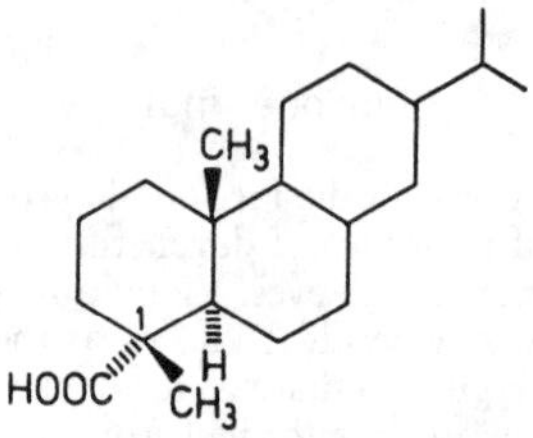

Tetrahydroabietic acid

The sodium salts of the disproportionated colophony are used as an *emulsifier* (Dresinate) in the preparation of *low temperature rubber*.

Among structurally related derivatives of the diterpenes are a group of *phytohormones*, the *gibberellins*[1], which are present as metabolites in the phytopathogenic fungi *Gibberella fujikoroi* (*Kurusawa*, 1926; *Yabuta* and *Hayashi*, 1939). Later on they were also isolated from higher plants, *e.g.* unripe beans and maize.

Of the gibberellins so far known, only gibberellic acid (gibberellin A₃), $C_{19}H_{22}O_6$, m.p. 233–235°C (506–508 K), is available in quantity (*Curtis* and *Cross*, 1954).

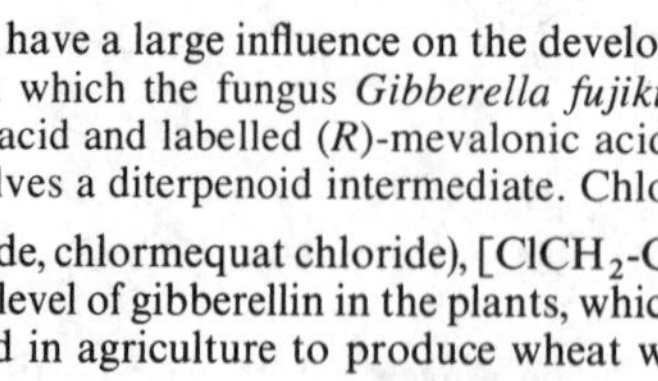

Gibberellic acid

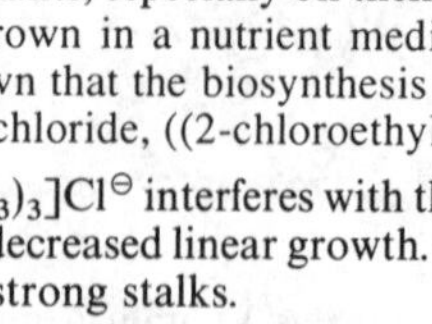

(R)-Mevalonic acid

Gibberellins have a large influence on the development of plants, especially on their linear growth. Experiments in which the fungus *Gibberella fujikuroi* was grown in a nutrient medium containing labelled acetic acid and labelled (R)-mevalonic acid have shown that the biosynthesis of gibberellins obviously involves a diterpenoid intermediate. Chlorocholine chloride, ((2-chloroethyl) trimethylammonium chloride, chlormequat chloride), $[ClCH_2\text{-}CH_2\text{-}\overset{\oplus}{N}(CH_3)_3]Cl^\ominus$ interferes with the biosynthesis and lowers the level of gibberellin in the plants, which leads to decreased linear growth. Chlorocholine chloride is used in agriculture to produce wheat with short, strong stalks.

---

[1] See *H. Kaufmann*, Die Gibberelline, *Chimia* 17, 120 (1963); *J. MacMillan*, Gibberellin Metabolism, Pure Appl. Chem. 50, 995 (1978).

6.5.4 Tetracyclic Diterpenes

An important compound of this type is phorbol, $C_{20}H_{28}O_6$, which is found as a 12,13-dicarboxylate ester in croton oil from the seeds of *Croton tiglium L., Euphorbiaceae*, and is a co-carcinogen. The latter means that the compound itself is not carcinogenic but it can greatly accelerate the action of carcinogenic hydrocarbons in promoting cancer formation (tumour promotion).

Phorbol

The group of compounds $C_{25}H_n$ intermediate between the diterpenes and the triterpenes are known as the *sesterterpenes* (see footnote 2, p. 661).

6.6 Triterpenes (Squalenoids)

The hydrocarbon *squalene*, $C_{30}H_{50}$, is an acyclic triterpene. It was first obtained from whale-liver oil, and it also occurs in small amounts in other mammals. It is made up from six isoprene units and contains six isolated C=C double bonds, all of which have (E) *configurations*.

The first synthesis of squalene was carried out by *Karrer* in 1931, from farnesyl bromide and magnesium, in a type of *Wurtz* reaction. The isoprene units are arranged symmetrically about the middle of the squalene molecule (marked by a dashed line; see p. 678). At this point there is therefore a head-to-head linkage; in the rest of the chain the units are joined by the usual head-to-tail linkage.

All the mono- to penta-cyclic triterpenes and other compounds derived from them such as the steroids are closely related structurally to squalene. In the biosynthesis of cholesterol from squalene an intermediate stabilised protolanosterol carbenium ion, whose hydroxy-groups are formed oxidatively *via* 2,3-epoxysqualene, is formed on the surface of the enzyme. The final product is the triterpene alcohol *lanosterol* (lanosta-8,24-dien-3β-ol), $C_{30}H_{50}O$, which occurs together with cholesterol in the unsaponifiable portion of wool fat. Because of their relationship to squalene triterpenoids are also known as *squalenoids*. Both compounds are strong emulsifiers; they are important ingredients of *Eucerite*, which is used in the preparation of Nivea cream.

Squalene   Lanosterol

A number of microorganisms, *e.g. acetobacter*, can bring about a different cyclisation of squalene to give *hopanoids*. These are derived from *hopane*, which is built up from four condensed cyclohexane rings and a cyclopentane ring. In addition to the *biohopanoids* which occur widely in bacteria, there are *geohopanoids* which are found in sediments, and have acquired importance as 'molecular fossils' in the search for mineral oil.[1]

Among the *pentacyclic* triterpenes found in plants are the *sapogenins*, which are often bonded to sugars as glycosides (saponins). Examples of the numerous compounds of this sort are α- and β-*amyrin*, $C_{30}H_{50}O$, which occur in the resins from many plants, and *oleanolic acid*, $C_{30}H_{48}O_3$, which is derived from β-amyrin. They are found as saponins in sugar beet and in the free state in olive leaves.

Hopane   α-Amyrin   β-Amyrin (R = CH₃)
                                  Oleanolic acid (R = COOH)

6.7 Tetraterpenes

The most important members of this series are the carotenoids, which are widely distributed in plants and in animals. The name is derived from the multiply unsaturated hydrocarbon *carotene*, $C_{40}H_{56}$, which was isolated from carrots by *Wackenroder* in 1831. The colour of carotenoids is due to the presence of a number of conjugated double bonds, which act as *chromophores*.

---

[1] See *G. Ourisson* and *P. Albrecht*, Geohopanoids, Acc. Chem. Res. *25*, 398 (1992); *G. Ourisson* and *M. Rohmer*, Biohopanoids, Acc. Chem. Res. *25*, 403 (1992).

In addition to the polyene hydrocarbons there are also naturally occurring derivatives containing alcohol, carboxylic acid and ester groups. An important property of carotenoids is their solubility in fats, which has led to them being called *lipochromes*.

6.7.1 *Carotenoids*[1] (Polyene pigments)

The most important carotenoids are the following:

6.7.1.1   Carotene, $C_{40}H_{56}$, is found together with chlorophyll and xanthophylls in green leaves as well as in flowers and fruits. In animals it is found in milk, fat and blood serum.

Carotene isolated from carrots is a mixture of *three isomers*, namely α-, β- and γ-carotene. In the first two both the terminal isoprene units of the chain have been ring-closed to form ε- or β-ionone rings (indicated in the formulae by ε or β inscribed in the ring). β-Carotene contains two β-ionone rings, α-carotene has an ε- and a β-ionone ring. These two carotenes thus differ only in the position of a double bond in an ionone ring. In γ-carotene only one end of the chain is closed to form a β-ionone ring. There is a head-to-head linkage in the middle of the carotene chains.

α-Carotene is optically active since it has an asymmetric carbon atom (*) in the ε-ionone ring; the two other carotenes are not optically active. Separation of the three carotenes was achieved by means of adsorption chromatography. Their constitutions were elucidated mainly from the work of *Willstätter, Karrer, R. Kuhn, Heilbron* and *Zechmeister*.

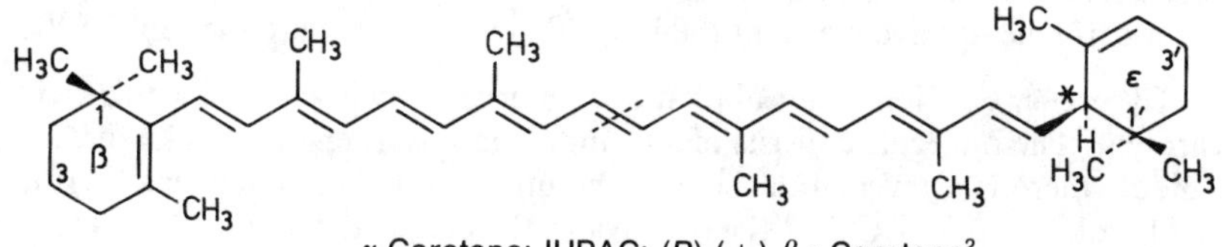

α-Carotene; IUPAC: (R)-$(+)$-β,ε-Carotene[2]

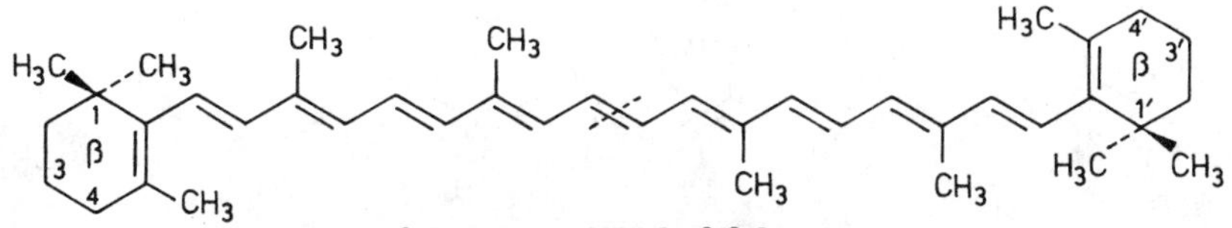

β-Carotene; IUPAC: β,β-Carotene

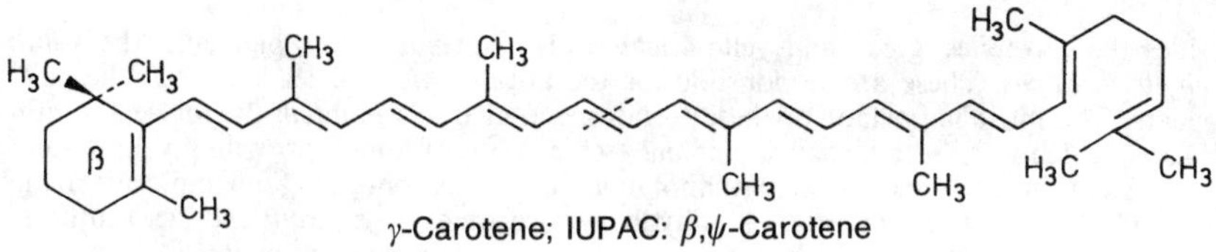

γ-Carotene; IUPAC: β,ψ-Carotene

[1] See *E. Jucker*, Zur Entwicklung der Carotinoid-Chemie (The Development of Carotenoid Chemistry), Angew. Chem. *71*, 253 (1959); *A. Winterstein*, Neuere Ergebnisse der Carotinoid-Forschung (Recent Results of Carotenoid Research), Angew. Chem. *72*, 902 (1960).
[2] According to IUPAC rules the end groups are denoted by β,ε (ionone rings) and ψ (open-chain); see Pure Appl. Chem. *41*, 405 (1975).

The presence of the large number of double bonds means that in theory there should be numerous Z–E isomers. In nature the (E) forms predominate. It is possible to bring about partial change to higher energy (Z) forms by the action of heat, light or catalysts, especially iodine. They differ from the (E) isomers in the positions of their absorption maxima.[1]

6.7.1.2 β-Carotene forms dark red plates, m.p. 184°C (457 K). As a polyene it is very reactive and sensitive to atmospheric oxygen. Industrially it is obtained from red palm oil, lucerne or carrots and used for the colouring of foodstuffs. β,β'-Carotene-4,4'-dione is present in the plumage of flamingoes and is also made industrially and used as a permitted colouring agent for foodstuffs (canthaxanthin).

The molecule of β-carotene is fully symmetrical about the central C=C double bond, i.e. the isoprene units are not all joined to each other in the same way but reverse in the middle, where there are four carbon atoms between two substituted carbon atoms instead of the usual three (cf. squalene).

β-Carotene was synthesised by *Karrer, Inhoffen* and *Isler.*

In animals, β-carotene is enzymatically split into two molecules of vitamin A and hence acts as a provitamin A.

6.7.1.3 α-Carotene forms crystals which melt at 188°C (461 K). Compared to β-carotene it is only half as effective as a provitamin A. The position of the double bond in the β-ionone ring is decisive for activity as a vitamin; the half of the molecule having the ε-ionone ring is inactive.

6.7.1.4 γ-Carotene melts at 178°C (451 K). Since in this case there is only one β-ionone ring and, unlike β-carotene, the other end of the chain is open, γ-carotene is intermediate between β-carotene and lycopene; in the latter there is no ring at either end of the chain.

6.7.1.5 Lycopene, $C_{40}H_{56}$, is an aliphatic tetraterpene hydrocarbon which is isomeric with carotene. The molecule consists of a chain of eight isoprene units; like the carotenes, at its middle there is a reversal in the way the units are linked. Lycopene crystallises in deep red needles, m.p. 174°C (447 K). It provides the deep red colour of tomatoes, of hips and of many other fruits. It contains thirteen double bonds; catalytic hydrogenation converts it into a colourless saturated hydrocarbon $C_{40}H_{82}$.

Lycopene

Besides the carotenes, green and yellow leaves also contain yellow pigments, the *xanthophylls* and *phylloxanthines*. These are responsible for the autumn tints of leaves, when the chlorophyll disappears. The most important of these tetraterpene alcohols is *lutein* (luteol, leaf xanthophyll), $C_{40}H_{56}O_2$, which is 3,3'-dihydroxy-α-carotene (see p. 679). Although providing yellow solutions in organic solvents, it crystallises from methanol in violet prisms containing one molecule of methanol, m.p. 193°C (466 K). The isomeric 3,3'-dihydroxy-β-carotene is *zeaxanthine* (zeaxanthol), yellow crystals m.p. 215°C (488 K), which is the main pigment of maize grain. Both pigments are also present in egg yolk.

As well as the carotenoids there are also other plant pigments having shorter polyene chains and carboxyl groups at each end of the chains. An important compound of this sort is crocetin.

---

[1] See L. *Zechmeister, Cis-trans* Isomeric Carotenoid Pigments, Fortschr. Ch. org. Naturst. *18*, 223 ff (1960).

Crocetin, $C_{20}H_{24}O_4$, is formed by hydrolysis of the yellow saffron pigment crocin (from *Crocus savitus*). Crocin is the digentiobiose ester of crocetin (carotidinic acid) which is made up from four isoprene units. Crocetin crystallises from acetic anhydride as red rhombs, m.p. 285°C (558 K).

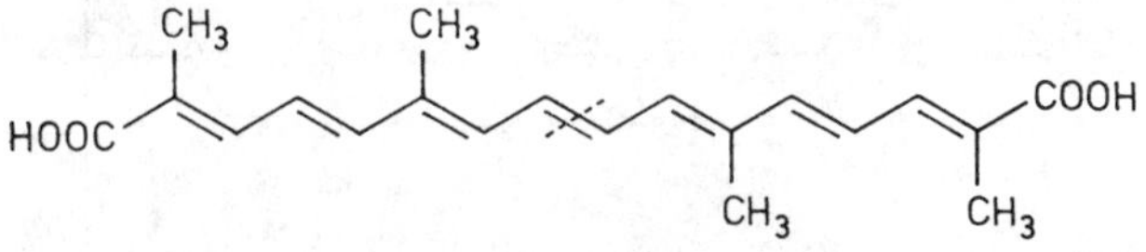

Crocetin (Carotidinic acid)

6.8 Polyprenes

6.8.1 Polyprenols

Polyprenols have the general formula:

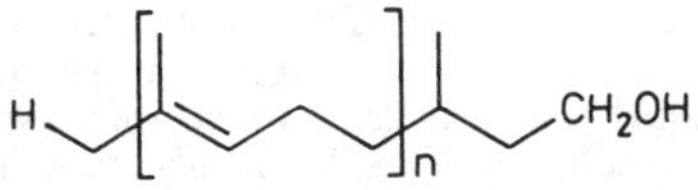

They are widely distributed in plants ($n = 8$–12) and in animals: dolichols ($n = 16$–21). For the most part their double bonds have (Z) configurations (see rubber), but there are also some (E) configurations which are specified in the formula. Dolichols have a metabolic role in the transport of monosaccharides through lipid membranes.

6.8.2 Natural Rubber

Rubber is an important natural product which is obtained from the milky liquid (*latex*) extracted from a variety of tropical trees, especially from *Hevea brasiliensis*. The amount of rubber in latex varies between 30 and 50%. The particles of rubber are suspended in an aqueous emulsion which also contains protein material. They are readily coagulated by acids such as formic or acetic acid, and for commercial use are then formed into sheets by treatment in a heated press.

Rubber consists of high molecular weight hydrocarbons, $(C_5H_8)_n$, assembled according to the isoprene rule. Pyrolysis of rubber leads to the formation of *isoprene* and *dipentene* as decomposition products. The location of the double bonds in rubber was determined by letting it react with ozone; hydrolytic work-up of the resultant ozonides $(C_5H_8O_3)_n$ gave almost quantitative yields of *levulinic acid*, CH_3—CO—CH_2—CH_2—$COOH$, and *levulinaldehyde* (4-oxopentanal) (*Harries*, 1905). Since ozonide formation always takes place at the C=C double bonds, the identity of the decomposition products shows that the rubber molecules must be constructed from isoprene units. Later work showed that all the double bonds had a (Z) configuration and that the saturated parts of the chains take up an *anti*-conformation (shown with a grey background in the formula). The mean relative molecular mass is about 350 000.

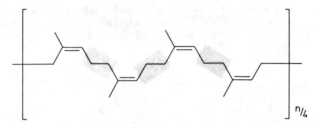

Rubber (a portion of the (Z)-1,4-polyisoprene chain of the all-*cis* polymer

6.8.2.1 Vulcanisation of Rubber

The importance of natural rubber depends on the many ways in which it can be processed to give a variety of rubber products. In its pure form it is initially a plastic material, but on keeping crystallisation and cross-linking of the linear molecules takes place and it becomes hard and brittle.

In order to impart to natural rubber the valuable properties of elasticity and resistance to the atmosphere and to chemical attack the process of *vulcanisation* was introduced by *Goodyear* in 1838. In this process the rubber is heated and kneaded in the presence of air to bring it into a plastic form (mastication) and mixed with sulphur and a filler (carbon black, zinc oxide) as well as with various materials which assist the vulcanisation process and stabilisers. The mixture is subsequently pressed into the required shape and heated to about 120°C (400 K) (hot vulcanisation); this brings about cross-linkage of the extended rubber molecules by sulphur bridges. The sulphur content varies between 2% for soft rubber to 30% for hard rubber (ebonite).

In the preparation of thin-walled rubber articles a *cold vulcanisation* process may be used, in which the rubber is immersed in a solution of disulphur dichloride in carbon disulphide.

Since the rate of reaction of the unassisted sulphur vulcanisation reaction is slow, vulcanisation-accelerators (Vulkazites) are mixed in, for example salts of N-substituted dithiocarbamic acid such as zinc dimethyldithiocarbamate or 2-mercaptobenzothiazole. *Foam rubber* is prepared by including a substance which decomposes with evolution of nitrogen at the vulcanisation temperature, *e.g.* benzenesulphonyl hydrazide, α,α'-azoisobutyronitrile or N,N'-dinitrosoterephthalic acid-N,N'-dimethyldiamide.

It has now become customary to describe all rubber-like polymers as *elastomers*.

6.8.3 Synthetic Elastomers

The first attempts at industrial preparation of synthetic elastomers were made by *Fritz Hofmann* in 1909. As his main starting material he used *2,3-dimethylbutadiene*, which had already been shown by *I. Kandakov* (1900) to polymerise spontaneously at room temperature to give a product which resembled rubber.[1] In succeeding years butadiene, which is more readily available, came more and more into use; originally *polymerisation* was induced by the presence of metallic sodium. Because of the ingredients used in the process the product was given the name *Buna*. In further developments statistical *copolymerisation of butadiene with styrene* in an emulsion with a peroxide catalyst was carried out giving *styrene-butadiene rubber*, and also statistical *copolymerisation of butadiene and acrylonitrile* (known as nitrile rubber, formerly perbunan).

Potassium or ammonium persulphate are the first choice as peroxide activating agents for these copolymerisations. Alkali metal salts of alkylbenzenesulphonic acid or of dialkylnaphthalenesul-

---

[1] See *E. Konrad*, Über die Entwicklung der Synthetischen Kautschuks in Deutschland (The Development of Synthetic Rubber in Germany), Angew. Chem. *62*, 491 (1950); *E. V. Anderson*, Rubber, Chem. Eng. News, Vol. *47*, No. *29* (1969).

phonic acids (Nekale) and especially of disproportionated abietic acid (dresinate, see p. 676) have proved useful as anionic surfactants. The latter have the advantage that they can remain in the polymerised material without impairing its properties.

A further important discovery was that the decomposition of peroxides to give radicals could be brought about not only thermally, but also at low temperatures if certain *metal ions*, which were themselves converted to a higher oxidation state in the process, were present, *e.g.*

$$R-\overset{\displaystyle O}{\underset{\displaystyle \|}{C}}-\bar{O}-\bar{O}-\overset{\displaystyle O}{\underset{\displaystyle \|}{C}}-R \; + \; Fe^{2\oplus} \longrightarrow R-\overset{\displaystyle O}{\underset{\displaystyle \|}{C}}-\bar{O}\bullet \; + \; R-CO_2^{\ominus} \; + \; Fe^{3\oplus}$$

Diacyl peroxide

If a suitable *reducing agent* such as Rongalit (sodium formaldehyde bisulphite), glucose or a sulphinic acid is added at the same time *redox activation* brings about a rapid polymerisation (cold polymerisation) even at 5°C (280 K).[1] A particularly valuable low temperature rubber (cold rubber or TT-rubber) is obtained in this way. It is characterised by high resistance to oxidation, high tensile strength and less dust, and is produced in increasing amounts to be vulcanised for use in the manufacture of car tyres.

The newest developments in this field have led to the stereoregular polymerisation of butadiene and of isoprene, which is nowadays available from the petrochemical industry, in solution, using lithium catalysts or organometallic mixed catalysts of the *Ziegler–Natta* type. *cis*-1,4-*polybutadiene* (Buna CB) and *cis*-1,4-*polyisoprene* (IR, Natsyn) are made in this way.

2-*Chlorobutadiene* (chloroprene), which can be made cheaply and in good yield by adding chlorine to butadiene to give 3,4-dichlorobut-1-ene and then removing a molecule of hydrogen chloride, is polymerised in increasing amounts for making polymers with special uses. The resultant polymer (Neoprene, Bayprene, Sowprene) is resistant to oil and to ageing. It is used, among other things, for the manufacture of adhesives.

6.8.3.1 Block polymers. In addition to the statistical copolymerisation of butadiene (B) with styrene (S), macromolecules can be prepared in which the two constituents are arranged in consecutive blocks, B_n–S_m. Such polymers can be provided with special properties, depending on the method of making them. An example is the *styrene-butadiene-three-block polymer* (SBS):

| Polystyrene | *cis*-1,4-Poly-butadiene | Polystyrene |

The rubbery elastic *cis*-1,4-polybutadiene blocks make up about 60% of the relative molecular mass and each of the hard elastic polystyrene blocks about 20%. As a result a material which holds its shape is produced, in which the more pliable B-blocks are held in place by the more rigid S-blocks, so that one can think of it as a three-dimensional physical network.

[1] See E. Konrad et al., Zur Geschichte des bei tiefer Temperatur polymerisierten synthetischen Kautschuks (The Tale of Low Temperature Polymerised Synthetic Rubbers) Angew. Chem. *62*, 423 (1950).

The yield point of the S-block is reached at 180°C (450 K) and at this temperature the SBS is thermoplastic and can be moulded, because the rigidity of the cross-linking is lowered. This is restored when it is cooled again and with it the properties of the initial three-block polymer.

6.8.3.2   Graft Polymers. Side-chains such as styrene can be attached to the double bonds which are present in poly(butadiene) and these side-chains can then be polymerised to polystyrenes. To do this, *cis*-1,4-poly(butadiene) is dissolved in monomeric styrene and polymerisation is brought about by a radical initiator. In this way a sufficient number of double bonds from the poly(butadiene) are statistically copolymerised to produce a macromolecule of the following type:

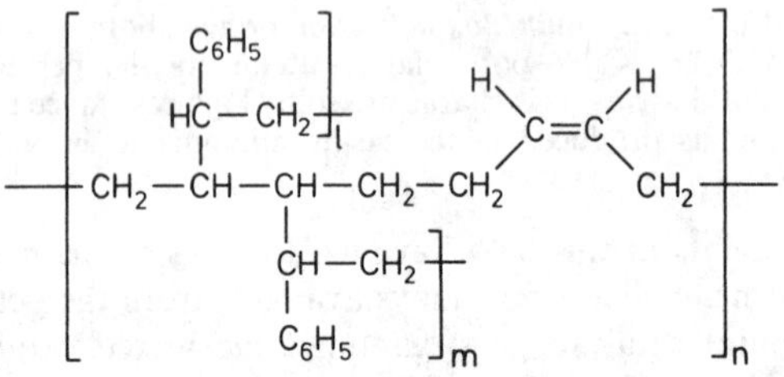

The polymeric side-chains are thus grafted onto the backbone polymer, in this case *cis*-1,4-poly(butadiene), giving rise to their description as *graft polymers* (g-polymers).

The abbreviated name for the above polymer is *cis*-1,4-polybutadiene-g-polystyrene. It combines the properties of polystyrene with those of poly(butadiene) but in a different manner from in block polymers, and it is a tough but not brittle material.

A real disadvantage of elastomers derived from dienes lies in the fact that even after vulcanisation a considerable number of double bonds remain, leading to them being sensitive to oxygen and ozone and also to light. In order to counteract this an anti-oxidant, *e.g.* N-phenyl-β-naphthylamine, is added to the rubber.

Synthetic elastomers which have been prepared by polymerising alkenes or vinyl derivatives have significantly higher stabilities and life-times. Among these are the copolymers from ethyl acrylate and acrylonitrile (acrylan rubber) and combinations of ethylene with propene and dicyclopentadiene (EPDM) or vinyl acetate as comonomers.

An especially high-grade material is obtained by simultaneously chlorinating and sulphochlorinating polyethylene (Hypalon); it can be vulcanised by the addition of either metal oxides, diamines or polyalcohols. Mention should be made here of the *thiokols*, which are obtained from the *polycondensation* reaction of 1,2-dichloroethane with sodium tetrasulphide. Other synthetic rubbers such as polyester rubbers, polyurethane rubbers and silicone rubbers have already been mentioned earlier.

6.8.4 *Gutta-percha*

Gutta-percha $(C_5H_8)_n$ is obtained by evaporating down the latex of various tropical trees, *e.g. Palaquium gutta*. It is a *structural isomer* of natural rubber, having an all-*trans*-1,4-polyisoprene chain, i.e. all the double bonds have an (E) configuration. Gutta-percha is used in the electrical industry as an insulator, for example in undersea cables.

Steroids

Biochemically active *steroids*[1] include the sterols, bile acids, sex hormones, corticoids, cardenolides and bufadienolides as well as steroid-sapogenins and steroid alkaloids.

The carbon skeleton of steroids consists of four carbocyclic rings fused together to form *perhydrocyclopenta[a]phenanthrene*, $C_{17}H_{28}$, which has also been given the names *gonane* or sterane.

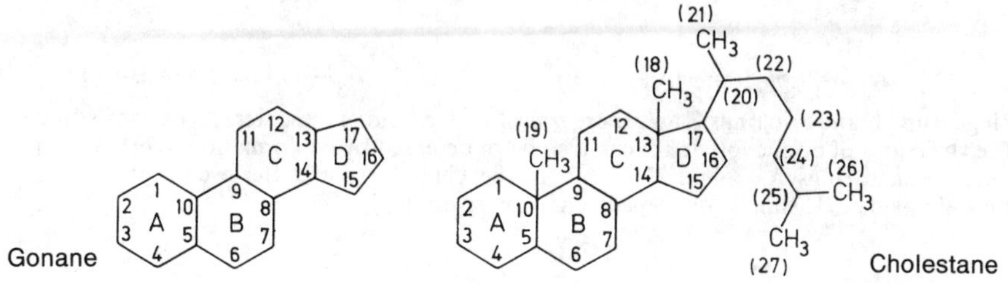

Gonane   Cholestane

These alicyclic hydrocarbons are numbered as shown above and the rings are designated A, B, C and D, as indicated. Together with their various substituted derivatives they are known collectively as the steroids. The elucidation of their constitution was based very largely on the pioneering work of *A. Windaus* and *H. Wieland*.

Nomenclature and Stereochemistry of Steroids

The *cholestanes*, $C_{27}H_{48}$, (see above), which are the parent compounds of the cholesterols, will be used to explain the numbering and the stereochemistry of steroids.

In theory the four rings of cholestane can each be fused together in either *cis*- or *trans*-configurations, as has been previously explained in the cases of decalin and hydrindane. In fact only certain methods of ring fusion are found in the steroids, which greatly reduces the number of configurations and conformations (*Barton*). Rings A & B may be either *cis*- or *trans*-fused to each other, rings B and C are always *trans*-fused, rings C and D are usually *trans*-fused but can also have a *cis*-arrangement, for example in the cardenolides and bufadienolides.

In characterising the steroids, not only the mode of fusion of the rings, but also the steric arrangement of substituents and of the hydrogen atoms, is important. The angular methyl group attached to carbon atom 10 serves as the reference point. Substituents and hydrogen atoms which are located on the same side of the molecule as this group are designated as *β*-atoms (or groups of atoms) and their bonding is represented by an unbroken line; substituents (or hydrogen atoms) on the opposite side of the ring are *α*-groups and their bonding is represented by a dashed line (*Fieser*).

In cholestane the angular methyl groups at C-10 and C-13, the side chain at C-17 and the hydrogen atom at C-8 are all *β*-groups; the hydrogen atoms at C-9, C-14 and C-17 are *α*-hydrogen atoms. If the rings A and B are *trans*-fused, the hydrogen atom at C-5 is an *α*-group, if these rings are *cis*-fused it is a *β*-group. Arising from this the two stereoisomers are called *5-α-cholestane* and *5-β-cholestane*. The following *configurational formulae* show not only the relative steric arrangement of the angular groups but also

---

[1] See *Rodd's* Chemistry of Carbon Compounds, 2nd ed, *S. Coffey* (Ed.) 2, D, E Steroids (Elsevier, Amsterdam 1970/71).

the absolute configurations of the asymmetric carbon atoms involved (see also the formula of cholesterol).

5-α-Cholestane (A:B-*trans*)   5-β-Cholestane (A:B-*cis*)

Rings B and C and also rings C and D are *trans*-fused in both of these stereoisomeric hydrocarbons. The structures of both cholestanes can also be represented by conformational formulae, in which the cyclohexane rings A, B and C are all in the chair form and the cyclopentane ring is not completely planar. (R stands for the side chain at position 17).

5-α-Cholestane   5-β-Cholestane

The α-hydrogen atom on C-3 in 5-α-cholestane is *axial*, the β-hydrogen atom is equatorial (see p. 399 *ff*.). The two β angular methyl groups in these steroids give rise to stereoselectivity with preferential reactivity at α-positions in many reactions.

The above arrangements apply only to saturated rings. If a double bond is introduced into a 6-membered ring it takes up a half-chair form as illustrated on p. 404.

Sterols have a hydroxy group at C-3, which may be either α or β to the angular methyl group at C-10. As a result, *each* cholestane gives rise to *two cholestanols* which have the following *configurational formulae*:

5-α-Cholestan-3β-ol   5-β-Cholestan-3β-ol

5-α-Cholestan-3α-ol   5-β-Cholestan-3α-ol

(R at position 17 indicates the same side chain as in cholestane).

In total syntheses of the steroid residue the *Robinson* annelation reaction (1935) is often used; it is a combination of a *Michael* reaction and an aldol reaction, *e.g.*

Under the reaction conditions methyl vinyl ketone, which acts as the *Michael*-acceptor, itself polymerises, and this can result in unsatisfactory yields. This can be overcome by using instead a reactant which under the alkaline conditions decomposes to provide methyl vinyl ketone. This can be done by using the quaternary ammonium salt $[H_3C-CO-CH_2-CH_2-N(C_2H_5)_2CH_3]^{\oplus}I^{\ominus}$, which is readily obtainable from the *Mannich* base $H_3C-CO-CH_2-CH_2-N(C_2H_5)_2$.[1]

The product from the *Robinson* annelation in this case is an octalindione, in which rings A and B of the cholestane system together with a β-angular methyl group are assembled.[2]

6.9 Sterols

In earlier times the true sterols were classified into zoo-sterols, phyto-sterols and myco-sterols, depending on the source in which they were found. As the chemical study of sterols developed it became apparent that none of them could be regarded as belonging uniquely to either of these sub-groups, and a structural classification was more desirable. In general, all sterols contain a 3β-hydroxy group. There are appreciable differences in the side-chain at C-17(20), but it always has a β-configuration.

The most important sterols are *cholesterol, ergosterol, stigmasterol* and *β-sitosterol*.

In German, sterols are known as *sterins*.

Cholesterol, $C_{27}H_{45}OH$, was the first sterol to be known. It is the main constituent of gall stones and was isolated from them by *Conradi* in 1775. The name is also derived from this source (Greek *cholesterin* = solid gall) (*Chevreul*, 1815). As a primary constituent of cells cholesterol occurs either as itself or as part of an ester in all animal cells and body fluids, but especially in the brain and spinal cord as well as in the suprarenal gland, in fish-liver oils and in wool fat (lanolin). 5α- and 5β-cholestan-3β-ol also occur together with it.

Cholesterol is a secondary alcohol which crystallises in needles, m.p. 150°C (423 K). It is insoluble in water but dissolves in ethanol, ether or chloroform. Cholesterol has eight asymmetric carbon atoms (*) and is *optically active*. There is a double bond in ring B which adds bromine and is hydrogenated to give 5α-cholestan-3β-ol.

[1] See *M. E. Jung*, A Review of Annulation, Tetrahedron *32*, 3 (1976).
[2] For further uses of this method see *L. Velluz* et al., Recent Advances in the Total Synthesis of Steroids, Angew. Chem. Int. Ed. Engl. *4*, 181 (1965).

The most important contribution to the *structure determination* of cholesterol was provided by its dehydrogenation by means of selenium to give *3'-methyl-1,2-cyclopenten-ophenanthrene*, $C_{18}H_{16}$, (*Diels*), which was also made synthetically for comparison. In the reaction with selenium, in addition to dehydrogenation, the hydroxy-group at C-3 was eliminated as water, the angular methyl group C-19 was lost as methane and, as the side chain was lost, the methyl group C-18 migrated to the neighbouring atom C-17, thus enabling the aromatic phenanthrene ring system to be formed.

Cholesterol   3'-Methyl-1,2-cyclopentenophenanthrene

The structural formula proposed by *Rosenheim* and *Wieland* in 1932 was confirmed by an X-ray structure of cholesteryl iodide (*D. Crowfoot*, 1945). *J. W. Cornforth* and *G. Popjak* determined the orientation of all the hydrogen atoms of cholesterol (cholest-5-en-3β-ol).[1]

Cholesterol is involved in a number of essential processes in the body. Derangement of the cholesterol metabolism can lead, especially in older persons, to deposition of cholesterol on the arterial walls, leading to *arteriosclerosis*.

5-β-Cholestan-3β-ol, $C_{27}H_{47}OH$, is a derivative of 5-β-cholestane; it forms needles with m.p. 101°C (374 K). It is formed from cholesterol by the action of bacteria in the gut and is hence found in faeces.

Ergosterol, (22E)-(24R)-24-methylcholesta-5,7,22-trien-3β-ol, $C_{28}H_{43}OH$, is found particularly in yeast; it forms shiny needles, m.p. 168°C (441 K) and is *optically active*. It has three double bonds, two of which in ring B are conjugated and the third is in the side-chain between carbon atoms 22 and 23. There is a methyl group attached to C-24. The constitution of the side-chain was determined by hydrolysis of the ozonide, when *isopropyl-methylacetaldehyde* (2,3-dimethylbutanal) was isolated as a cleavage product. Irradiation of ergosterol converts it into vitamin D_2.

Stigmasterol, (22E)-24(S)-24-ethylcholesta-5,22-dien-3β-ol, $C_{29}H_{47}OH$, occurs in soya beans and differs from cholesterol only in its side-chain. Like ergosterol it has a double bond between carbon atoms 22 and 23. Whereas ergosterol has a methyl substituent at C-24, in stigmasterol there is an ethyl group. As a result the ozonide cleavage product from stigmasterol is *ethyl-isopropylacetaldehyde* (2-ethyl-3-methylbutanal).[2]

Stigmasterol has been used as an important starting material for the partial syntheses of steroid hormones. Simple steroids have been valuable not only as starting materials for the preparation of other steroids but also in theoretical organic chemistry, especially in connection with the relationship between the stereochemistry and reactivity of functional groups.

---

[1] See *J. W. Cornforth*, Asymmetry and Enzyme Action, Science *193*, 121 (1976).
[2] The numbers other than 28 and 29 which are shown correspond to recent IUPAC rules; see Pure Appl. Chem. *61*, 1783 (1989).

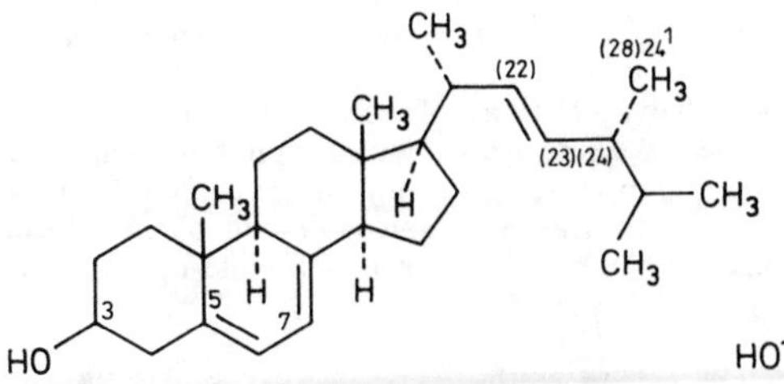

Ergosterol

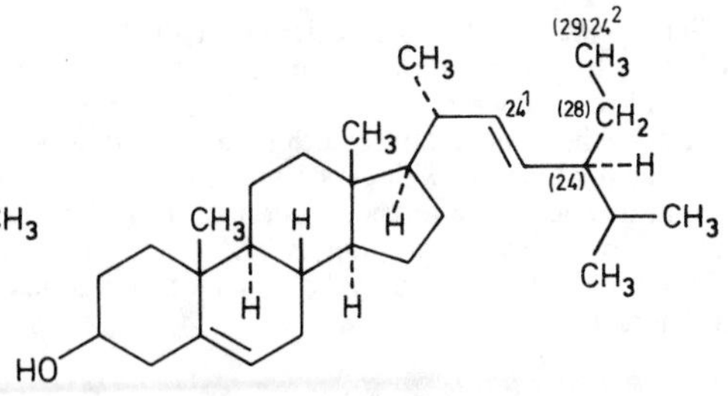

Stigmasterol

β-Sitosterol, $C_{29}H_{49}OH$, is the dihydro-derivative of stigmasterol, having a saturated side-chain. It is obtained, together with other phytosterols, in large quantities as a by-product from the saponification of *soya bean oil*, part of which is used as a starting material for the preparation of *steroid hormones* such as *oestrone*. Sitosterol has been of use for the treatment of hypertrophy (enlargement) of the prostate-gland.

6.10 Bile Acids

Sterically the bile acids are derivatives of *5-β-cholestane*. They do not occur in bile in the free state but rather are linked by peptide links to amino-acids. For example, cholic acid, which occurs very widely, is bound as *glycocholic acid* or *taurocholic acid* to either glycine or taurine, respectively. They may be hydrolysed to give cholic acid and the amino-acid (*Strecker*, 1848).

H. Wieland recognised that *cholanic acid*, $C_{29}H_{39}COOH$, which has not yet been found in nature, is the parent compound of the bile acids. It can be prepared from 5β-cholestane, and has a *cis*-fusion between rings A and B. Compared to 5β-cholestane cholic acid has three less carbon atoms in its side-chain and the terminal carbon atom forms part of a carboxyl group.

The bile acids are hydroxy derivatives of cholanic acid (cholane-24-carboxylic acid). The most important examples are *cholic acid* (3α,7α,12α-trihydroxy-5β-cholane-24-carboxylic acid), *deoxycholic acid* (3α,12α-dihydroxy-5β-cholane-24-carboxylic acid), *chenodeoxycholic acid* (3α,7α-dihydroxy-5β-cholane-24-carboxylic acid) and *lithocholic acid* (3α-hydroxy-5β-cholane-24-carboxylic acid). All four of these acids are *optically active*.

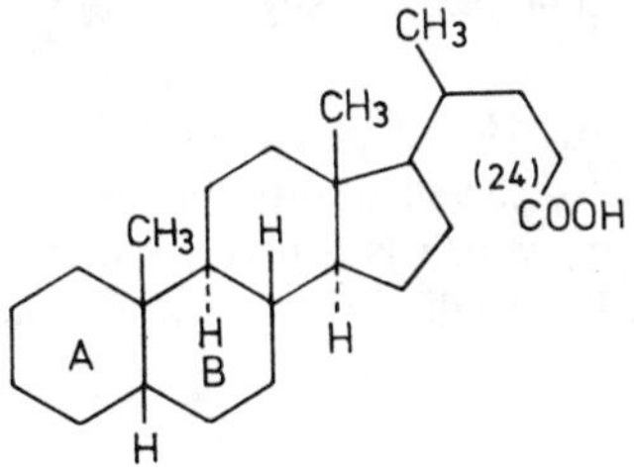

Cholanic acid

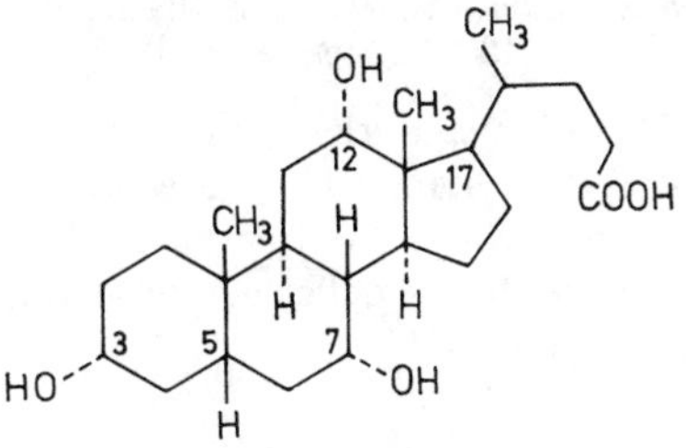

Cholic acid

The principal physiological role of the bile acids is in the digestion of fats, but it is not clear whether they further the absorption of fats by emulsifying them or activate the enzymes, *lipases*, which break down the fats.

Deoxycholic acid can form 8:1 addition compounds with fatty acids such as palmitic, stearic or oleic acid which are well crystalline and soluble in water. These are known as choleic acids and are found in bile. Their molecular constitution depends upon the chain length of the carboxylic acid. They are inclusion compounds and other compounds such as esters, alcohols and phenols form similar adducts with deoxycholic acid. Sodium deoxycholate is used to help solubilise lipids (*i.e.* as a detergent).

6.11 Steroid Vitamins

The following vitamins of the D-series[1] are closely related structurally to ergosterol.

Vitamin D$_2$ (calciferol, ergocalciferol, *vigantol*), $C_{28}H_{43}OH$, was first obtained by *Windaus* in 1932 by *photolysing ergosterol*. The processes which take place on irradiation with UV-light are rather complicated and involve a number of intermediates, including lumisterol and tachysterol. The net reaction is an electrocyclic ring-opening of ring B between carbon atoms 9 and 10, and formation of a new double bond between carbon atoms 10 and 19 to form a methylene group; the reaction is in accord with the *Woodward–Hoffmann* rules. The final result is only an *isomerisation* of the steroid structure. Ergosterol can be looked on as *provitamin D$_2$*.

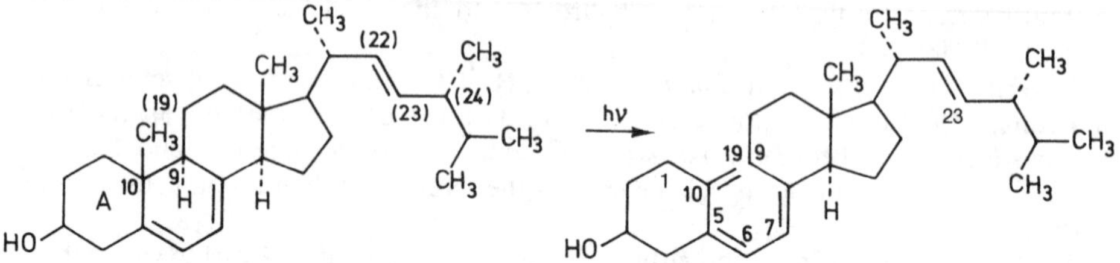

Ergosterol
Ergosta-5,7,22-trien-3-ol

Vitamin D$_2$

Evidence for the above structure of vitamin D$_2$ included the uptake of eight hydrogen atoms on catalytic hydrogenation and the addition of maleic anhydride to the conjugated system (carbon atoms 6 and 19) in a *Diels–Alder* reaction to give an octahydro-naphthalene derivative. An *X-ray structure determination* (*Crowfoot* and *Dunitz*, 1948) confirmed the structure as (5Z,7E,22E)-(3S)-9,10-secocholesta-5,7,10(19),22-tetraen-3-ol.

Vitamin D$_2$ forms colourless prisms, m.p. 121°C (394 K) and is *optically active*. Before it was realised that vitamin D$_3$ is the precursor of the substance that is actually active, vitamin D$_2$ was used for the treatment of *rickets*, which is caused by disruption of the metal ion balance due to deficiency of vitamin D$_2$, leading to insufficient calcium being deposited in the bones.

---

[1] See *H. H. Inhoffen*, Aus der Chemie der antirachitischen Vitamine (About the Antirachitic Vitamins), Angew. Chem. **72**, 875 (1960).

Vitamin D$_3$ (cholecalciferol, (5Z,7E)-(3S)-9,10-secocholesta-5,7,10(19)-trien-3-ol) occurs together with vitamin A in the liver oils of tuna, halibut and cod, and was first isolated by *Brockmann* (1936) from tuna-liver oil. It differs from vitamin D$_2$ only in the structure of the side-chain, which lacks a double bond and a methyl substituent on carbon 24. Its biological activity is like that of vitamin D$_2$. The substance which is actually the active member of this group of compounds has been shown to be 1,25-dihydroxycholecalciferol, INN *Calcitol*, which is formed from Vitamin D$_3$ in the kidneys.

Vitamin D$_3$ has been made synthetically by irradiating 7-dehydrocholesterol (Provitamin D$_3$) with UV light (*Windaus*):

7-Dehydrocholesterol   Vitamin D$_3$ (m.p. 84°C [357 K])

6.12 Steroid Hormones (Sex Hormones and Adrenal Hormones)[1]

Steroidal hormones consist of two important groups of hormones associated with internal secretory processes, namely the sex hormones and the adrenal hormones. The latter are called *corticoids*.

The group of sex hormones includes substances which may be produced in either the male or female gonads (testicles or ovaries), and which are responsible for the development and the normal functioning of the sexual organs and for the formation of secondary sexual characteristics. It is necessary to make a distinction between male and female sex hormones.

The terms 'male' and 'female' sex hormones refer only to the physiological action of these materials, and not to their sources.

These sex hormones resemble cholesterol in their constitution; some can be obtained by a partial synthesis starting from cholesterol. They are all *optically active*.

---

[1] See *L. Träger*, Steroidhormone (Springer-Verlag, Berlin 1977).

6.12.1 Male Sex Hormones (Androgens)

The first hormones of this sort were isolated by *Butenandt* in 1934 from male urine, namely *androsterone*, $C_{19}H_{30}O_2$, and *5-dehydroandrosterone*, $C_{19}H_{28}O_2$. Both crystallise well; they melt at, respectively, 185°C (458 K) and 148°C (421 K). The parent hydrocarbon of both of these androgens is androstane (10,13-dimethylgonane), $C_{19}H_{32}$ (see p. 685). Sterically, androsterone is related to 5α-cholstan-3α-ol, whilst 5-dehydroandrostane is related to cholesterol. Both compounds have a keto-group at carbon atom 17. INN: **Testosterone**, $C_{19}H_{28}O_2$, which was isolated from bulls' testicles (*E. Laqueur*, 1935), is an active androgen. It is regarded as the intrinsic male hormone because not only does it affect secondary sexual characteristics (*e.g.* the development of the comb in capons), but also brings about the growth and normal functioning of specific sexual glands (testicular glands, prostate glands).

Testosterone forms needles, m.p. 155°C (428 K). Like all steroids it is insoluble in water, but dissolves in organic solvents and oils.

The structure of testosterone was proved at just about the same time by *Butenandt* and by *Ruzicka*, who carried out partial syntheses starting from 5-dehydroandrosterone (3β-hydroxyandrost-5-en-17-one). It differs from 5-dehydroandrosterone in having a keto group at C-3, and the double bond in the 4,5-position instead of at the 5,6-position. In addition, at C-17 there is a secondary alcohol group, which is pointing towards the methyl group attached to C-13.

Androsterone
3-α-Hydroxy-5α-androstan-
17-one

INN: **Testosterone**
17-β-Hydroxyandrost-
4-en-3-one

16-Androsten-3-one

16-Androsten-3-one, m.p. 140–141°C (413–414 K), has an identical perimeter to that of civetone (see p. 409) and smells strongly of urine, α- and β-16-androsten-3-ols, which are obtained from it by reduction of the keto-group, have musk-like smells, that of the α-compound being the stronger of the two; both have been isolated from the testes of pigs.

6.12.2 Female Sex Hormones (Oestrogens (Estrogens) and Gestagens)

Female sex hormones have an essential influence on the periodic *sexual cycle* of humans and of other mammals, and on pregnancy. Other features of the sexual cycle are menstruation and the coming into season of animals (*oestrus*). Depending upon their specific functions in this cycle a distinction is made between *follicular hormones* and *pregnancy hormones*.

6.12.2.1 Follicular Hormones (Oestrogens, Estrogens)[1]

In 1927 *Aschheim   and Zondek* established that the hormone content in the urine of pregnant women was particularly high. Two years later a number of investigators (*Butenandt, Doisy, Marrian*), working quite independently of each other, isolated a crystalline follicular hormone, later called *oestrone* (*estrone*), from urine from pregnant women. The urine of pregnant mares is a rich source of hormones; however, the urine of stallions contains, in addition to male hormones, more female sexual hormones even than the urine of mares.

The follicular hormones are derivatives of the hydrocarbon *oestrane* (estrane), $C_{18}H_{30}$; the essential difference between this and androstane is the absence of a methyl group on carbon atom 10. The most important follicular hormones are *oestrone* (estrone), $C_{18}H_{22}O_2$, *17β-oestradiol* (17β-estradiol), $C_{18}H_{24}O_2$ and *oestriol* (estriol), $C_{18}H_{24}O_3$.

Oestrone   17β-Oestradiol   Oestriol

It is evident from these formulae that, in contrast to male hormones, the *oestrogens* have a benzenoid ring A and thus the hydroxy group attached to it has phenolic character. Carbon atom 17 is the site either of a keto group or a secondary alcohol group.

The most active oestrogen hormone is *17β-oestradiol* (oestra-1,3,5(10)-trien-3,17β-diol), m.p. 178°C (451 K), which was isolated by *Doisy* from pig ovaries.

In connection with the relationship between chemical constitution and oestrogenic activity it has been found that some synthetically produced stilbene derivatives surpass the follicular hormones in oestrogenic activity (*Dodds, R. Robinson*). Among these (*E*)-α,β-diethyl-4,4′-stilbenediol is particularly noteworthy; its activity is 2-3 times that of oestrone. However this compound entails risks as a serious carcinogen and teratogen. The bis-phosphoric ester of diethylstilboestrol (INN: *Fosfestrol*) is in use for the treatment of carcinomas of the prostate.

Diethylstilboestrol (R = H)
(Diethylstilbestrol)

INN: *Fosfestrol*

$$\left(R = \; \underset{\underset{O}{\|}}{P} \begin{matrix} \diagup OH \\ \diagdown OH \end{matrix} \right)$$

INN: *Clomiphene*

By the use of such compounds in calf-feed, the slaughter-weight is increased by up to 15% compared to that of untreated animals. The consumption of flesh which contains unnatural oestrogens can cause problems and the use of any oestrogens in cattle-feed is forbidden in some countries. The stilbene derivative *INN: Clomiphene* (*E-Z*)-2[*p*-(2-chloro-1,2-diphenylvinyl)phen-oxytriethylamine, acts as an antioestrogen; its use can lead to an increase in multiple pregnancies.

---

[1] See *P. Morand* and *J. Lyall*, The Steroidal Estrogens, Chem. Rev. *68*, 85 (1968).

6.12.2.2 Hormones of the Corpus Luteum (Gestagens)

The specifically female hormone from the *corpus luteum* or pregnancy hormone is INN: *Progesterone* (prolutone), $C_{21}H_{30}O_2$, which is found in the *corpus luteum* and in the placenta. It is derived from cholesterol by breakdown of the side-chain to an acetyl group, giving pregnenolone, $C_{21}H_{32}O_2$, which, by oxidation of the 3β-hydroxy-group and shifting the double bond from the 5,6- to the 4,5-position, gives progesterone, which is a common precursor of both male and female sex hormones. Pregnenolone, which is found in pig testes, has weak gestagenic activity. Both *gestagens* are derivatives of the hydrocarbon 5β-pregnane, $C_{21}H_{36}$, which has an ethyl substituent at C-17 and can be prepared from cholanic acid.

| 5β-Pregnane (m.p. 84°C [357 K]) | INN: *Progesterone* (m.p. 121°C [394 K]) (Pregna-4-ene-3,20-dione) | Pregnenolone (m.p. 190°C [463 K]) (3β-Hydroxypregna-5-ene-20-one) |

6.12.2.3 Contraceptive Steroids[1]

Modified gestagens, *e.g. 19-norethisterone acetate*, in combination with oestrogens, can prevent conception. Studies on this behaviour by *Pincus* (1955) provide the basis of most modern contraceptives, which have become of importance in controlling the fertility of the world's population.

| 19-Norethisterone acetate | 17α-Ethynyloestradiol | RU 486 (INN: *Mifepristone*) |

19-Norethisterone acetate (norpregnenin) has an ethynyl group at C-17; in many steroids this allows them to be used orally (*Inhoffen*). This compound is made from oestradiol, a starting material which does not have an angular 19-methyl group.

Oestrogenic components of the 'pill' include, for example, 17α-ethynyloestradiol (INN: *Ethynyloestradiol*, 19-nor-17α-pregna-1,3,5(10)-trien-20-yne-3,17-diol) which can be made by reaction of the *Grignard* derivative of acetylene with oestrone. The progesterone-antagonist INN: *Mifepristone* (RU 486, *mifegyne*, 11β-[4-(dimethylamino)phenyl]-17-

---

[1] See *R. Wiechert*, The Role of Birth Control in the Survival of the Human Race, Angew. Chem. Int. Ed. Engl. *16*, 506 (1977); C. Djerassi, The Bitter Pill, Science *245*, 356 (1989).

hydroxy-21-methyl-19-nor-17α-pregna-4,9-dien-20-yn-3-one) allows a safer ending to a pregnancy up to the end of seven weeks if it is used in conjunction with prostaglandin (*E. E. Baulieu*, 1981; for its mode of action see p. 873 *ff*.).

6.12.3 Corticoids

Closely related biogenetically to progesterone are the numerous hormones known as the *corticoids* (Latin *cortex* = rind, crust) which are obtained from extracts of the adrenal cortex. They are among the essential materials in the organ; removal of the gland from an animal leads to severe damage and, after a few days, death.

Corticoids have 21 carbon atoms, but only the three following 11β-hydroxy-compounds are regarded as true *adrenocortical hormones*, i.e. *cortisol*, $C_{21}H_{30}O_5$, *corticosterone*, $C_{21}H_{30}O_4$ and INN: **aldosterone**, $C_{21}H_{28}O_5$[1]:

Cortisol (INN: **Hydrocortisone**)
(m.p. 220°C [493 K])

Corticosterone (11β,21-Dihydroxypregna-4-ene-3,20-dione)
(m.p. 182°C [455 K])

INN: **Aldosterone**
(m.p. 155–158°C [428–431 K])

The corticoids especially influence the metal ion balance and the carbohydrate metabolism; they stimulate the formation of glycogen in the liver. Regulation of the corticoid products in the organism is controlled primarily by the *adrenocorticotropic hormone* (ACTH, Corticotropin) in the pituitary. In situations of stress the level of cortisol rises markedly; cortisol is a stress hormone.

Further corticoids or their chemically modified analogues include INN: **Cortisone, Prednisone, Prednisolone** (1-dehydrocortisol) and **Triamcinolone**, which have proved to be valuable *therapeutics* for a number of disorders (arthritis, rheumatism, bronchial asthma, allergies, diseases of the skin, inflammation), but the action of each individual corticoid is not sharply defined. The activity has been increased by the introduction of fluorine atoms or hydroxy groups, for example triamcinalone is a powerful inhibitor of inflammation, but it has relatively little effect on the balance of mineral ions.

---

[1] In addition to those shown, depending upon the solvent, other isomers and also dimers exist. See *K. Lichtwald* et al., Angew. Chem. Int. Ed. Engl. *24*, 130 (1985).

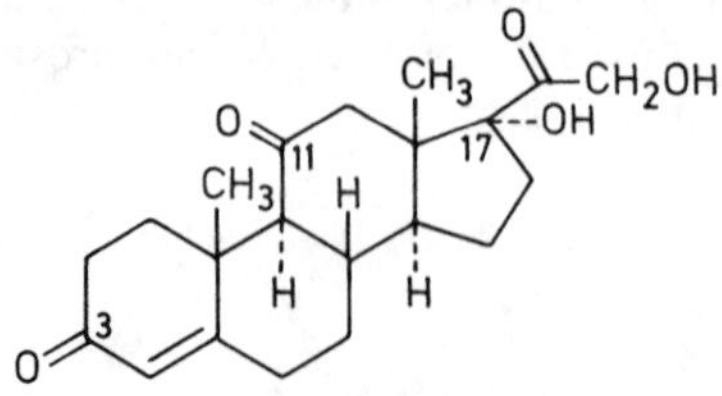

INN: **Cortisone**
(17α,21-Dihydroxy-
pregna-4-ene-
3,11,20-trione)

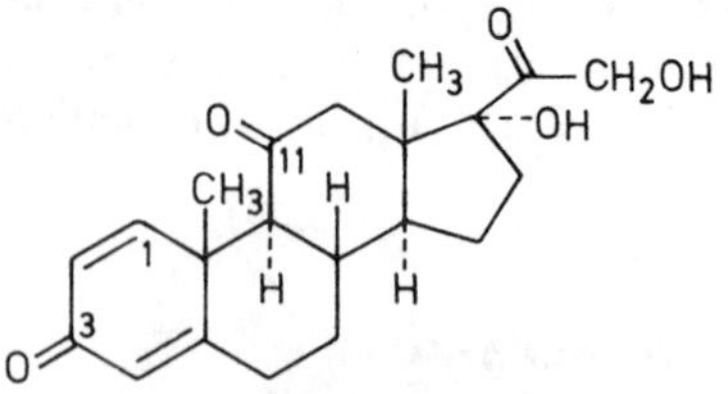

INN: **Prednisone**
(17α,21-Dihydroxy-
pregna-1,4-diene-
3,11,20-trione)

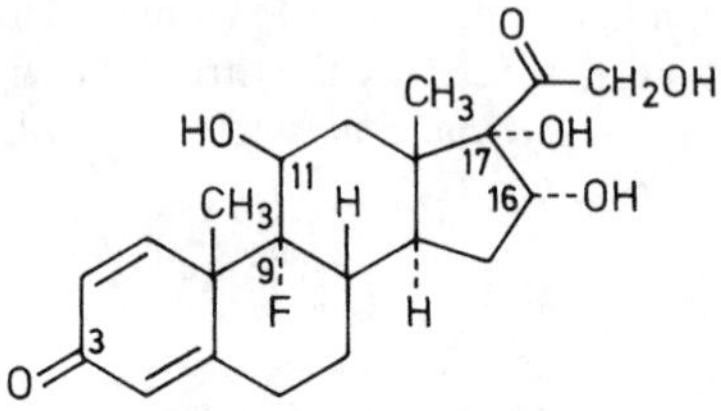

INN: **Triamcinolone**
(9-Fluoro-11β,16α,17α,21-tetrahydroxypregna-
1,4-diene-3,20-dione)

Commercially corticoids are for the most part obtained by biochemical methods. For example, cortisone and cortisol are obtained from progesterone by the action of certain micro-organisms which bring about selective hydroxylation. Likewise cortisone can be microbiologically dehydrogenated to prednisone (*Decortin*).

In addition, however, *total syntheses* of a number of steroids have been achieved, *e.g.* of oestrone (*Miescher*, 1948; *W. S. Johnson*, 1950), androsterone (*R. Robinson*, 1951; *W. S. Johnson*, 1953), cholesterol (*Woodward*, 1951), cortisone (*Woodward*, *Sarett*, 1951), Vitamin D_2 (*I. T. Harrison*, *Inhoffen*, 1957), etc.

It is of interest that steroids derived from lower organisms also function as hormones. Among these is ecdysone, $C_{27}H_{44}O_6$, which is known to be the *hormone associated with the shedding of their skin* by insects;[1] it has the following structure with an oxo- and five hydroxy groups:

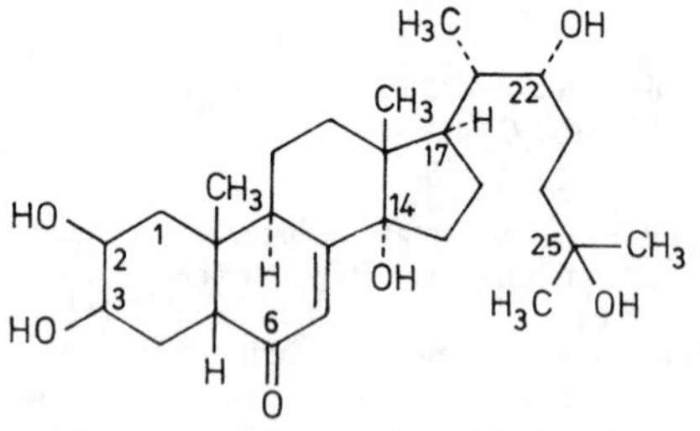

Ecdysone, (22R)-2β,3β,14α,22,25-Pentahydroxy-5β-cholest-7-en-6-one

---

[1] See *K. Slama*, Insect Hormones and Bioanalogues (Springer-Verlag, Vienna 1974).

6.13 Cardiotonic (Cardiac-Active) Steroids

This group of steroids includes the *cardenolides* and the *bufadienolides*. In contrast to what obtains in other steroids, the C- and D-rings in both of these classes of compounds are *cis*-fused.

6.13.1 Cardenolides

Particular contributions to the study of these cardiac-active compounds were made by *W. A. Jacobs, Tschesche, Reichstein* and *Stoll*.

Cardenolides exist as glycosides in the leaves of plants of the *digitalis* family (*Digitalis purpurea* = foxglove). The most important digitalis glycosides[1] are INN: *Digitoxin*, *Digoxin* and *Gitoxin*. Acid hydrolysis of these glycosides removes the sugar residue which is attached to the 3β-hydroxy-group, leaving the *genins* (aglycones), respectively *digitoxigenin*, $C_{23}H_{34}O_4$, *digoxigenin* and *gitoxigenin*, $C_{23}H_{34}O_5$.

These 5β-cardenolides, which have 23 carbon atoms, are related structurally to the bile acids. A particular feature is the α,β-unsaturated γ-lactone ring attached to C-17. The rings are joined to one another as follows: A/B *cis*, B/C *trans*, C/D *cis*.

Digitoxigenin   Strophanthidin

Unlike digitoxigenin, digoxigenin has a 12β-hydroxy group; gitoxigenin has a 16β-hydroxy group.

Strophanthidin, $C_{23}H_{32}O_6$, the aglycone from strophanthin, has a similar structure to that of digitoxigenin. A characteristic feature is the 19-formyl group. The glycoside strophanthin is obtained from the seeds of *Strophanthus kombé*. Like digitoxin it has cardiac activity, but it is only poorly absorbed orally. These glycosides are used therapeutically in very small doses to regulate the activity of the heart.

---

[1] See *T. Reichstein*, Chemie der herzactiven Glykoside (Chemistry of Cardio-active Glycosides), Angew. Chem. *63*, 412 (1951); Special Properties of the Sugars in Cardioactive Glycosides, Angew. Chem. Int. Ed. Engl. 1, 572 (1962).

6.13.2 Bufadienolides

Bufadienolides occur as poisons both in some plants and in secretory glands of toads. Structurally they are closely related to the cardenolides with the difference that the 17-substituent is a doubly unsaturated 6-membered lactone ring.

An example of a plant bufadienolide is *scillaren A* from white squill or sea-onion (*Scilla maritima*), which is a glycoside of scillarenin.

The toad poisons of the steroid series, which were studied especially by *H. Wieland*, do not occur naturally as glycosides. One of the known toad poisons is *bufotalin*, $C_{26}H_{36}O_6$, which also has a β-acetoxy group at C-16.

Scillarenin

Bufotalin

6.14 Steroid Sapogenins

The sapogenins, which are aglycones of saponins, belong structurally to two different polycyclic ring systems. One of these, which has a pentacyclic triterpene structure, has already been mentioned (see amyrin). The other group consists of the *steroid sapogenins*, which form the aglycones of the steroid saponins occurring in various plants of the *digitalis* family; they are not, however, cardiac active.

The most important of the steroid saponins are *digitonin*, *gitonin* and *tigonin*, which are split by acid hydrolysis into the related sugar and the aglycones *digitogenin*, $C_{27}H_{44}O_5$, *gitogenin*, $C_{27}H_{44}O_4$ and *tigogenin*, $C_{27}H_{44}O_3$, respectively. Tigogenin has the following structural formula:

Tigogenin

Gitogenin contains an extra 2α-hydroxy group and digitogenin also has a 15β-hydroxy group. The spiro-ketal grouping attached to C-16 and C-17 is typical of steroid sapogenins.

An important starting material for the partial synthesis of progesterone is *diosgenin*, $C_{27}H_{42}O_3$, from several Mexican plants of the *dioscorea* family, which differs from tigogenin by having a C=C double bond linking the 5,6-positions. 95% of all steroid hormones are prepared from diosgenin (*R. Marker*).

Steroid saponins (spirostanes) are widely distributed glycosidic plant substances which go into colloidal solution in water and form soap-like foams (Latin *sapo* = soap). Their aqueous solutions are poisonous since they cause *haemolysis* (destruction of the red blood cells). They form molecular complexes with sterols such as cholesterol which are rather insoluble and they can be detoxified in this way.

6.15 Steroid Alkaloids

These alkaloids are isolated from plants of the *solanum* family, where they exist as glycosides. Acid hydrolysis splits them into a sugar and an aglycone, which has a steroid residue as the basis of its structure.

The most important alkaloids of this sort are *tomatidin* and *solanidin*.

Tomatidin, $C_{27}H_{45}NO_2$, is the aglycone from *tomatin*, which occurs in the leaves of the wild tomato. On hydrolysis tomatin provides two mol of D-glucose and one mol of each of D-galactose and D-xylose. Structurally tomatidin resembles tigogenin, but the oxygen atom in the 6-membered ring of the latter has been replaced by an NH-group.

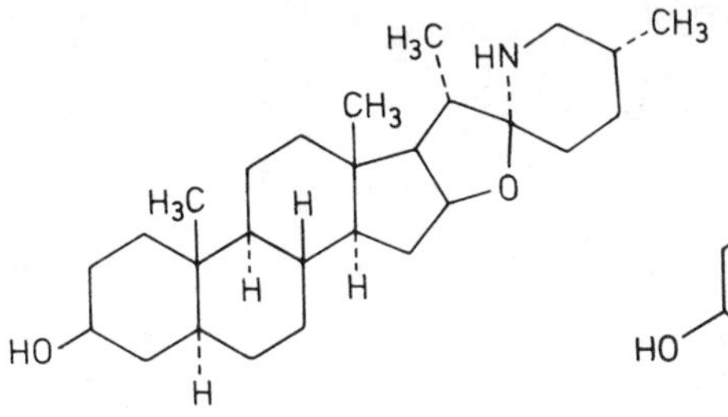

Tomatidin

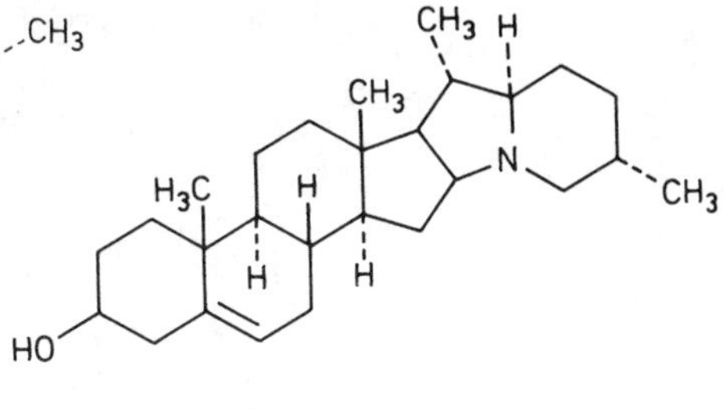

Solanidin

Solanidin, $C_{27}H_{43}NO$, occurs in the leaves, tubers and shoots of potatoes (*Solanum tuberosum*). It is formed by hydrolysing the glycoside α-*solanin* with dilute mineral acid, the other products being 1 mol each of D-glucose, D-galactose and D-rhamnose.

Conessin, $C_{24}H_{40}N_2$, is obtained from the bark of the Indian shrub *Holarrhena antidysenteria*, in which it occurs in appreciable amounts. This shrub is used in the treatment of amoebic dysentery. Its carbon skeleton resembles that of pregna-5-ene, but a 3β-dimethylamino-group takes the place of a 3β-hydroxy-group. Especially characteristic is the further fact that the angular 18-methyl group found in other steroids is replaced here by a nitrogen bridge linking C-18 to C-20.

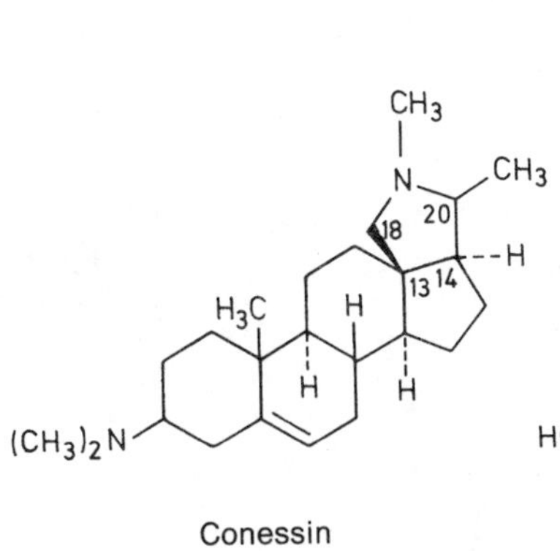

Conessin

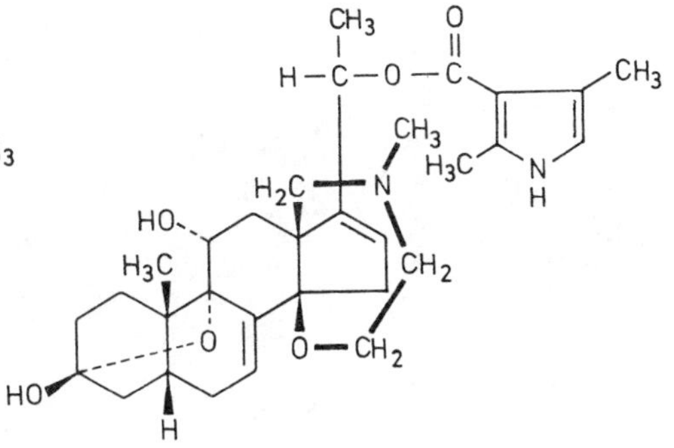

Batrachotoxin

Compounds of the conessin type can be used as starting materials in the preparation of steroid hormones.

Batrachotoxin, which occurs in the frog *Phyllobates aurotaenia*, is a steroid derivative with a remarkably high activity. It is one of the strongest known poisons, being 5000 times more toxic than sodium cyanide. This compound acts as a neurotoxin, affecting the transport of ions in nerves and muscles, and for this reason it has been an important tool for neurophysiologists.

7 Heterocyclic Compounds

This is a large family of cyclic compounds in which the ring atoms are not all carbon atoms, but include *heteroatoms* (i.e. atoms other than carbon), the commonest of these being nitrogen, oxygen and sulphur.[1] As in the case of carbocyclic compounds, the most stable rings are the five- and six-membered rings. There are also bi- and poly-cyclic ring systems in which benzene and heterocyclic rings are fused together.

The earliest classifications of heterocyclic compounds were based on either the overall numbers of atoms in the ring, or the number and types of heteroatom. *A. Albert* introduced a system whereby heterocycles were classed according to their structure and properties into three large groups, *heterocycloalkanes, heteroaromatics* and *heterocycloalkenes*.

(a) Heterocycloalkanes have saturated heterocyclic rings. Their properties differ very little, if at all, from those of their acyclic analogues. Thus 1,4-dioxan and tetrahydrofuran are typical ethers, lactones resemble esters and lactams resemble amides, and so on. Only small rings, especially three-membered rings, show some peculiarities due to ring strain.

Examples of heterocycloalkanes are:

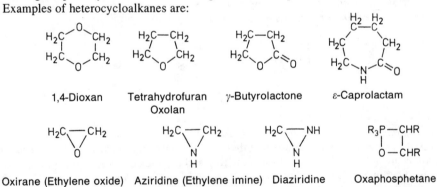

1,4-Dioxan	Tetrahydrofuran Oxolan	γ-Butyrolactone	ε-Caprolactam
Oxirane (Ethylene oxide)	Aziridine (Ethylene imine)	Diaziridine	Oxaphosphetane

The preparation and reactions of these compounds closely resemble those of their aliphatic analogues and have therefore already been treated.

---

[1] See *M. H. Palmer*, The Structure and Reactions of Heterocyclic Compounds (Arnold, London 1967); *A. Albert*, Heterocyclic Chemistry, 2nd edn (Athlone Press, London 1968); *L. A. Paquette*, Principles of Modern Heterocyclic Chemistry (W. A. Benjamin, New York 1968); *J. A. Joule, K. Mills* and *G. F. Smith*, Heterocyclic Chemistry, (Chapman & Hall, London 1994); *T. L. Gilchrist*, Heterocyclic Chemistry, 2nd edn. (Longmans, London 1992); *A. R. Katritzky* and *C. W. Rees* (Eds), Comprehensive Heterocyclic Chemistry, Vols. 1–8 (Pergamon Press, Oxford 1984).

The ground state of pyrrole can thus be symbolised by the formula overleaf. The *net atomic charges* on each of the ring atoms have been calculated to be as shown.[1] All the atoms in pyrrole, including the hydrogen atom attached to nitrogen, are coplanar; bond lengths indicate some, but not complete, delocalisation of the π-electrons around the ring.

(b) Heteroaromatics (heteroarenes) consist almost entirely of compounds with five- or six-membered rings, which, like benzene, have a *sextet* of *π-electrons* and therefore show some similarities in their reactions. They form the largest and most important group of heterocyclic compounds. They are subdivided into *π*-electron-rich and *π*-electron-deficient heteroaromatics.

The *π-electron-rich heteroaromatics* include the unsaturated five-membered ring heterocycles such as *pyrrole, furan* and *thiophene*.

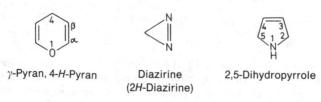

| Pyrrole | Furan | Thiophene | X = NH, O, S |

Numbering of these rings starts from the heteroatom which is atom-1, and then proceeds around the ring. If there is more than one sort of heteroatom in the ring, the order of precedence, *i.e.* the atom from which numbering commences, is oxygen first, then sulphur, then nitrogen.

In the above five-membered rings, each carbon atom contributes one electron whilst the heteroatom contributes two electrons to provide a *sextet of π-electrons*; cyclic interaction of all these electrons leads to all the ring-atoms becoming sp^2-hybridised. As in the case of benzene this can be represented by a circle inscribed within the ring. Since six π-electrons are spread over only five atoms the π-electron density on each carbon atom is greater than one, leading to their description as *π-electron rich* or *π-excessive heteroaromatics*.

The residual groups left if a hydrogen atom is removed from the rings are known, respectively, as *pyrrolyl, furyl* and *thienyl* groups.

In contrast, six-membered ring heterocycles such as **pyridine** or **pyrimidine** are examples of *π-electron-deficient heteroaromatics*.

| Pyridine | Pyrimidine |

In these six-membered rings, each of the ring atoms, whether carbon or nitrogen, contributes *one* electron to the sextet of π-electrons. The nitrogen atoms are already sp^2-hybridised in the skeleton and thus show their usual electronegativity; in consequence the π-electrons are not equally shared around the ring and the π-electron density on the carbon atoms is lower than it is on the heteroatom (*π-electron-deficient heteroaromatics*).

An excess or a deficiency of π-electrons each has its effect on the reactivity of heteroaromatics.

(c) Heterocycloalkenes are structurally intermediate between heterocycloalkanes and heteroaromatics. Examples are *γ-pyran* and *diazirine*. Some other heterocycloalkenes are named as partially reduced heteroaromatics, *e.g.* **2,5-dihydropyrrole**.

| γ-Pyran, 4-*H*-Pyran | Diazirine (2*H*-Diazirine) | 2,5-Dihydropyrrole |

The ring-size and degree of unsaturation is indicated in the names of the different types of heterocycles by the *endings* of the names. These may also change depending on the identity of the heteroatom. The endings are listed in Table 18; the first ending shown in each case refers to rings containing oxygen, sulphur, or selenium. If the ending is changed for some other elements, the second ending refers to this alternative and the element(s) concerned are shown in brackets (X = halogens).

Table 18

Ring-size	Unsaturated	Saturated
3-ring	-irene, -irine(N only)	-irane, -iridine(N only)
4-ring	-ete	-etane, -etidine(N only)
5-ring	-ole	-olane, -olidine (N only)
6-ring	-ine (N*,O*,S,Se,Te),	ane (O,S,Se,Te)†
	-inine (B,P,As,Sb,Bi)	inane (B,N,P,As,Sb,X)
7-ring	-epine	-epane

*The trivial names are normally used for pyridine and pyran.
†The final 'e' is frequently omitted.

The naming procedure may be illustrated by a few examples, *e.g. oxirane* represents a saturated three-membered ring containing one oxygen atom, *diazirine* is an unsaturated 3-membered ring containing two nitrogen atoms, dioxane is a saturated six-membered ring containing two oxygen atoms.

In recent years there has also been further development of the 'a'-*nomenclature*; this is applicable to both cyclic and acyclic compounds. In this the name of the hydrocarbon which would result if all the heteroatoms were replaced by carbon atoms forms the basis of the name and the presence of heteroatoms is shown by prefixes ending in 'a', *e.g.* oxa-, thia-, aza-, etc., in each case, if needed, preceded by a number signifying the location of the heteroatom (for examples, see pp.775,810).

The endings listed under the heading 'saturated' in Table 18 are used for *heterocycloalkanes*. Those in the column headed 'unsaturated' are used for both *heterocycloalkenes* and *heteroaromatics*. Isolated saturated sites in heterocycloalkenes, whether occupied by carbon or heteroatoms, are indicated by adding an '*H*' to the name, preceded by a number defining the position concerned, i.e. 1*H*-, 2*H*-, etc., as, for example, in 4*H*-pyran (p.702). Partially reduced heteroaromatic systems are also described by adding prefixes such as 'dihydro-' or 'tetrahydro-' to the name of the parent heteroaromatic compound, for example, *ε*-caprolactam = hexahydro-2*H*-azepin-2-one. Numbering always begins with a heteroatom and continues around the ring in such a direction that attached functional groups or further heteroatoms have the lowest possible numbers; this also applies to the numbering of 'extra' Hs. In constructing any name, the prefixes oxa-, thia-, aza- must always be used in this order.

The following ring systems provide some examples of this nomenclature as outlined in Table 18:

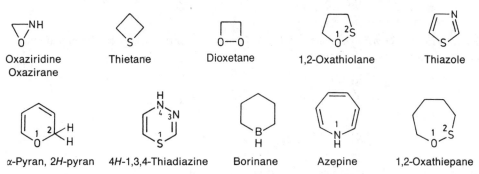

Oxaziridine   Thietane   Dioxetane   1,2-Oxathiolane   Thiazole
Oxazirane

α-Pyran, 2*H*-pyran   4*H*-1,3,4-Thiadiazine   Borinane   Azepine   1,2-Oxathiepane

In addition, established trivial names, such as pyrrole, thiophene and furan are very commonly used.

7.1 Five-Membered Rings with One Heteroatom

7.1.1 Pyrroles[1]

Among the five-membered ring heterocyclic compounds some derivatives of pyrrole are of special interest, since they form the building blocks from which blood pigments, bile pigments, leaf pigments and vitamin B_{12} are constructed, and are also present in various alkaloids, proteins and nucleic acids and enzymes.

Pyrrole (azole), C_4H_5N, was first discovered in 1834 by *Runge* in coal-tar. Later on, it was also discovered by *Anderson* in bone oil, the product obtained by distilling bone meal. If a pine-splint is moistened with hydrochloric acid it turns red when exposed to the vapour of pyrrole. This test reaction was responsible for giving pyrrole its name (Grk: *pyrros* = fiery red).

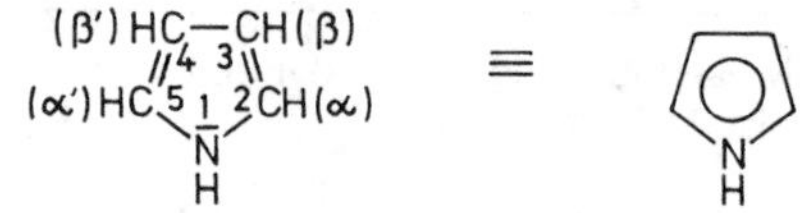

Pyrrole

In an older form of nomenclature, now only seldom used, positions 2 and 5 were called the α- and α'-positions and positions 3 and 4 were called the β- and β'-positions. When the hydrogen atom on the nitrogen atom is replaced by a substituent, this is named as an *N*-substituted product.

Preparation

7.1.1.1 Pyrrole can be obtained by *heating but-2-yn-1,4-diol with ammonia* under pressure:

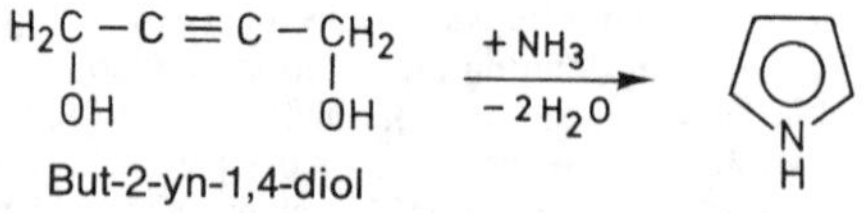

But-2-yn-1,4-diol

7.1.1.2 Industrially pyrrole is prepared from furan and ammonia:

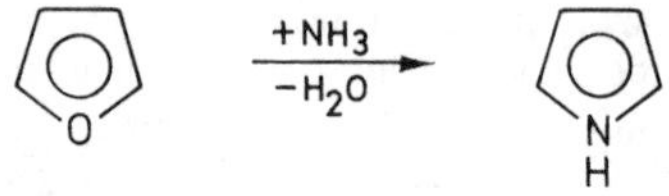

7.1.1.3 *Paal – Knorr* **Synthesis**. 2,5-Dialkyl- and 2,5-diaryl-pyrroles can be prepared by heating 1,4-diketones with ammonia:

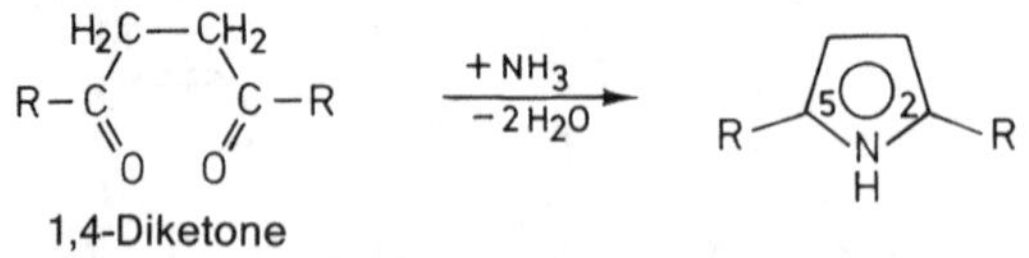

1,4-Diketone

---

[1]See *R. A. Jones* et al., The Chemistry of Pyrroles (Academic Press, New York 1977); *C. W. Bird* and *G. W. H. Cheeseman* (Eds), Comprehensive Heterocyclic Chemistry, Vol. *4* (Pergamon Press, Oxford 1984); *R. A. Jones* (Ed.), Pyrroles, Part 1 (Wiley, New York 1990).

If a primary amine is used instead of ammonia the product is the corresponding *N*-substituted pyrrole. Similar processes for making furan and thiophene will be mentioned.

7.1.1.4 *Knorr* **Synthesis**. This involves condensation of an α-aminoketone with a compound which has both a carbonyl group and a reactive methylene group. The formation of the aminoketone and the condensation reaction are usually carried out in one step. For example, if a hydroxyiminoacetoacetic ester (isonitrosoacetoacetic ester) is reduced with zinc dust and acetic acid in the presence of an acetoacetic ester, the latter undergoes a cyclo-condensation with the α-aminoacetic ester which is formed to give 3,5-dimethylpyrrole-2,4-bis(carboxylic acid ester). Alkaline hydrolysis and decarboxylation of this ester provides **2,4-dimethylpyrrole**.

Hydroxyimino-acetoacetic ester   Acetoacetic ester

Diethyl 3,5-dimethylpyrrole-2,4-dicarboxylate (R = C$_2$H$_5$)   2,4-Dimethylpyrrole

This method is particularly important for the preparation of substituted pyrroles for inclusion in porphyrin pigments.

Properties

Freshly distilled pyrrole is a colourless liquid, b.p. 131°C (404 K). It turns brown in the air and gradually resinifies. It is only slightly soluble in water but is totally miscible with ethanol or ether.

Pyrrole can be described in terms of the following limiting structures:

---

[1] See *R. D. Brown* et al., Theoret. Chim. Acta 7, 259 (1967). The use of an inscribed circle to represent the sextet of π-electrons provides no indication of the fact that most of the limiting structures suggest a polar structure for pyrrole. There is often advantage in using a formula with single and double bonds to represent pyrrole.

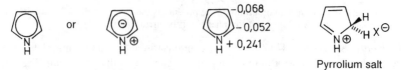

Pyrrolium salt

Because the lone pair of electrons of the nitrogen atom is involved in setting up a sextet of π-electrons pyrrole is only a weak base ($pK_a = -3.8$); it is very unstable to acid. When a mineral acid reacts with pyrrole protonation takes place predominantly at an α-carbon atom, giving rise to an unstable pyrrolium cation. Electrochemical polymerisation of pyrrole leads to the formation of a blue-black film of *polypyrrole*; if the counter-ion is $BF_4^{\ominus}$ or $ClO_4^{\ominus}$, this material conducts electricity and it finds use in metal-free batteries.[1]

Since there is a high π-electron density on the carbon atoms, pyrrole undergoes facile *electrophilic substitution*, especially on the 2,5-carbon atoms. *Halogenation*, using either sulphuryl chloride in ether or bromine in acetic acid, proceeds extremely rapidly to give *2,3,4,5-tetrahalogenopyrroles*. *Nitration* with nitric acid in acetic acid (acetyl nitrate) at 5°C (275 K) gives *2-nitropyrrole*. *Sulphonation* with the pyridine–sulphur trioxide complex at 90°C (360 K) provides *pyrrole-2-sulphonic acid*. Acetic anhydride *acetylates* pyrrole at the 2-position *without* any *Friedel–Crafts* catalyst present.

2,3,4,5-Tetrabromo
-pyrrole

$+2\,Br_2$

$H_3CCO_2^{\ominus}\ NO_2^{\oplus}$

2-Nitropyrrole

$(H_3CCO)_2O$

Pyrrole

$+SO_3/Pyridine$

2-Acetylpyrrole

Pyrrole-2-sulphonic acid

2,5-Disubstituted pyrroles undergo similar reactions at the 3- and 4-positions.

Pyrrole can also act as a weak acid; its acidity is about the same as that of acetylene. If pyrrole is heated with metallic potassium in n-heptane, stable potassium pyrrolide is formed. The corresponding *lithium* and *sodium* salts are made by warming pyrrole with either lithium or sodium hydrides in tetrahydrofuran. Likewise, treatment of pyrrole with an alkyl magnesium halide provides a salt-like *Grignard* reagent, *e.g.*

$+ H_3C\,MgBr$
$- CH_4$

$+K$
$-1/2\,H_2$

Magnesium bromide-pyrrolide   Pyrrole   Potassium pyrrolide

[1]See *J.-C. Moutet*, Functionalized Polypyrroles, Acc. Chem. Res. *22*, 249 (1989).

Whereas in the formation of a pyrrolium cation the sextet of π-electrons in the ring is lost, in the *anion* it remains. It resembles the cyclopentadienide anion.

Potassium pyrrolide has use in synthesis, for example treatment with alkyl halides at about 60°C (330 K) provides *N-alkylpyrroles*; when these are heated to about 200°C (475 K) they readily rearrange to *C-alkylpyrroles, e.g.*

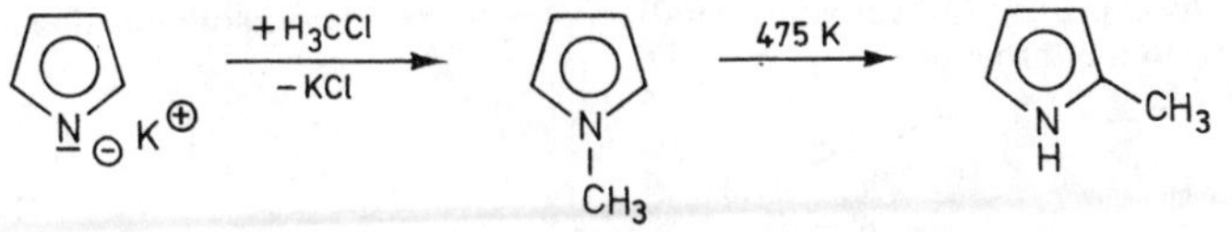

Potassium pyrrolide   1-Methylpyrrole   2-Methylpyrrole

In general, on pyrolysis *N*-alkyl, *N*-aryl and *N*-acyl groups migrate to the 2-position. *N*-Acylpyrroles are very easily hydrolysed.

Potassium pyrrolide reacts with carbon dioxide to give a salt of *pyrrole-2-carboxylic acid* (*cf. Kolbe–Schmitt* synthesis):

Potassium pyrrolide   Potassium salt of pyrrole-2-carboxylic acid

In contrast to furan, pyrrole has very little diene character; in general it does not take part in *Diels–Alder* reactions. However it does react with the very reactive *dienophile hexafluorobicyclo[2.2.0]hexa-2,5-diene* (hexafluoro-*Dewar* benzene) to form the following 1 : 1-adduct:

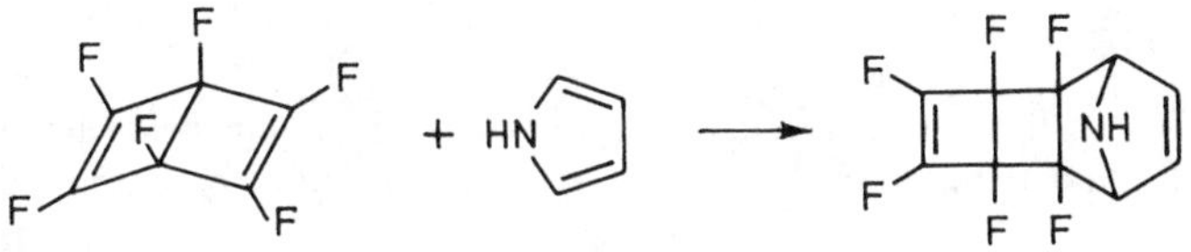

Hexafluoro-*Dewar* benzene   1:1-Adduct

Because of its aromatic character pyrrole is only reduced rather sluggishly. Reduction with zinc dust and acetic acid gives *2,5-dihydropyrrole*, while catalytic reduction gives either 2,5-dihydropyrrole or pyrrolidine:

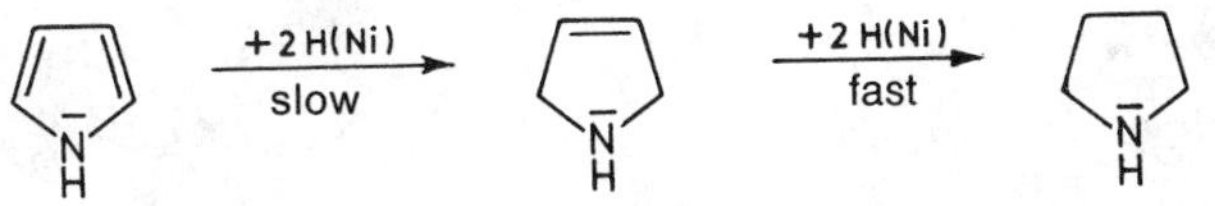

Pyrrole   2,5-Dihydropyrrole   Pyrrolidine
(b.p.129°C [402 K])   (b.p. 91°C [365 K])   (b.p. 88°C [361 K])

This reaction sequence thus proceeds from a heteroaromatic via a heterocycloalkene to a heterocycloalkane.

Both of these reduction products are liquids with amine-like smells. They are basic and have the properties typically associated with secondary amines. They show no tendency to polymerise.

Pyrrolidine is more simply made by letting tetrahydrofuran react with ammonia:

Tetrahydrofuran   Pyrrolidine

Two important cyclic amino acids are derivatives of pyrrolidine, **proline** (pyrrolidine-2-carboxylic acid) and **hydroxyproline** (4-hydroxypyrrolidine-2-carboxylic acid); both function as building-blocks in the construction of proteins.

Proline   Hydroxyproline   N-Methyl-2-pyrrolidinone   INN-*Captopril*

N-Methyl-2-pyrrolidinone (*N*-methylbutane-4-lactam, see p.281) is used as an *extraction solvent*, for removing butadiene from the C_4-fraction and aromatics from the catalytic reformation products (*Arosolvan process*). (2S)-1-(3-mercapto-2-methylpropionyl)-L-proline, INN: *Captopril*, inhibits the formation of angiotensin II (ACE-inhibitor, see p.839) and lowers the blood pressure.

7.1.2 Porphin Pigments[1]

This biochemically important group of naturally occurring pigments includes in particular the red pigments of blood, the green pigments of leaves, and vitamin B_{12}. These compounds have very complicated structures which were elucidated especially by the work of *Nencki, Piloty, Küster, Willstätter* and *Hans Fischer*.

7.1.2.1 Porphin

Porphin, $C_{20}H_{14}N_4$, has a *mesomerically stabilised* ring system made up of four pyrrole rings linked by four methine bridges (—CH=); this provides the chromophoric group of the system.

Porphin (IUPAC: Porphyrin)

This molecule contains a cyclic system of 18 π-electrons often denoted as shown by the grey background. Other 18-π-electron circuits are possible which can also contribute to the

[1] See *K. Zeile*, Neuere Entwicklungen in der Chemie der Porphin-Farbstoffe (Recent Developments in the Chemistry of Porphin Pigments), Angew. Chem. *68*, 193 (1956); *A. W. Johnson*, Macrocyclic Tetrapyrrolic Pigments, Chem. in Britain *3*, 253 (1967); *G. Buse*, The Present Position of Hemoglobin Research, Angew. Chem. Int. Ed. Engl. *10*, 663 (1971); *A. R. Battersby* and *F. J. Leaper*, Pyrrole Pigments, Biosynthesis and Total Synthesis, Chem. Rev. *90*, 1261 (1990).

mesomeric stabilisation. In addition, prototropic shift of the inner hydrogen atoms between the nitrogen atoms takes place. In other words, a number of circuits of 18 π-electrons may contribute to the overall electronic structure of the molecule.[1]

The numbering corresponds to an IUPAC rule, which also recommends that in IUPAC nomenclature the name *porphin* should be replaced by *porphyrin*.[2]

The *synthesis of porphin* from four molecules of pyrrole-2-carboxaldehyde, albeit in very low yield, was achieved by heating it with formic acid (*H. Fischer*, 1929). It forms dark red leaflets, which do not decompose below 360°C (630 K). The deep colour of porphin and of its derivatives is associated with the 18-π-electron circuits.

7.1.2.2 Haemoglobin

Haemoglobin (hemoglobin), the pigment of red blood corpuscles, consists of up to about 4% of a coloured component, *haem*, and about 96% of a protein component, *globin*; the two components interact with one another to form a complex. In protein chemistry haemoglobin is included among the *chromoproteins*. Haem is based on a porphin skeleton in which the hydrogens attached to carbon in the pyrrole rings have been replaced by organic groups while those on the nitrogen atoms are replaced by an iron (II) ion, thereby forming an inner complex. Oxidation of the haem portion of haemoglobin readily converts it into *haematin*, in which the iron atom has been oxidised to a +3 oxidation state.

Haemoglobin   Haemin

Based on its iron content, haemoglobin must have a relative molecular mass of about 17 000. However, by using an ultra-centrifuge a value of 68 000 has been obtained, which implies that a haemoglobin molecule contains four atoms of iron, *i.e.* four haem units.

In order to separate the pigment of haemoglobin from the protein material, blood is run into hot acetic acid to which saturated sodium chloride solution has been added. Oxidation of the haem ensues and brown crystals with blueish lustrous surfaces separate out. These

---

[1] See *D. Lloyd*, The Chemistry of Conjugated Cyclic Compounds. To Be or Not To Be Like Benzene, p. 78 (Wiley, Chichester 1989).
[2] See Nomenclature of Tetrapyrroles (Recommendations 1986), Pure Appl. Chem. *59*, 779 (1987).

are the so-called *Teichmann's* crystals, made up of the chloride salt of haematin, *haemin*, $C_{34}H_{32}FeN_4O_4]^\oplus Cl^\ominus$, which is an Fe(III) equivalent of haem.

In haemin the hydrogen atoms of the four pyrrole nuclei are all replaced by methyl, vinyl or 2-carboxy-ethyl groups. The two acidic hydrogens of the NH-groups are replaced by an *iron(III) ion*. The positive charge associated with this ion is counterbalanced by the chloride ion. With alkali, haemin is converted into the corresponding hydroxide haematin, mentioned above, $C_{34}H_{32}FeN_4O_4]^\oplus HO^\ominus$.

In neutral solution globin and haematin form *methaemoglobin*, which is brown-red and can also be obtained by oxidising haemoglobin. In methaemoglobin the iron is in a $+3$ oxidation state. It can be reduced back to haemoglobin, for example by formic acid.

7.1.2.3 The Physiological Function of Haemoglobin

In the body, haemoglobin carries out the important function of oxygen transport. In this process, for each atom of iron one molecule of oxygen is loosely bound to form *oxyhaemoglobin*, but the iron atoms do *not* change their oxidation state from $+2$. The red blood corpuscles take up atmospheric oxygen in the lungs and carry it in the blood stream into the tissues, where it is used up in the respiration of the cells. Since haemoglobin has four iron atoms the reversible exchange of oxygen can be represented by the following equilibrium:

$$Haemoglobin + 4O_2 \rightleftharpoons Oxyhaemoglobin$$

The state of this equilibrium is controlled by the partial pressure of oxygen, which in turn depends upon the amino acid sequence in the protein.[1] Oxyhaemoglobin is not itself directly involved in the actual oxidation process. Activation of the molecular oxygen involves first of all the catalytic action of *oxidoreductases* (see flavine enzymes and cytochrome). The human body contains about 4–5 g of iron, of which about 75% is contained in haemoglobin. The reserve iron in the body is attached to *ferritin*, a protein which is circulating in the blood stream.

Carbon monoxide binds to haemoglobin even more readily than oxygen does, forming *carbon monoxide–haemoglobin*, which is more stable than oxyhaemoglobin formed in normal respiration. Hence the normal role of haemoglobin as a transporter of oxygen is interfered with, respiration in the cells is stopped and the whole metabolism is disturbed. Prolonged inhalation of carbon monoxide brings about death (carbon monoxide poisoning).

7.1.2.4 Porphyrins[2]

Closely related to haemin is a class of metal-free substituted porphins, the porphyrins. These compounds, which are mostly red or brownish red, can be obtained from, for example, the haemin of blood by treating it with acid, which splits off the iron atom. If haemin is heated with formic acid in the presence of iron powder, the iron-free compound **protoporphyrin**, $C_{34}H_{34}N_4O_4$ is produced, whose molecules contain two protons in place of the $Fe^{3\oplus}$ and $Cl^\ominus$ of haemin. Thus the haem of haemoglobin is an iron(II) complex of the protoporphyrin. On the other hand, if haemin is treated with hydrogen bromide in acetic acid, the iron is removed and the product is **haematoporphyrin**, $C_{34}H_{38}N_4O_6$, which differs from the protoporphyrin by the addition of two molecules of water to the two vinyl groups. If a similar reaction is carried out using hydrogen iodide in acetic acid then reduction of the two vinyl groups of the haemin molecule takes place. The product is **mesoporphyrin**, $C_{34}H_{38}N_4O_4$; when heated with alkali this loses two molecules of carbon dioxide to give **aetioporphyrin III**, $C_{32}H_{38}N_4$, which contains no oxygen atoms.

---

[1] See *J. Hiebel, Hannibals* Alpenübergang aus Molekularbiologischer Sicht (*Hannibal's* Crossing of the Alps from a Molecular Biological Standpoint), Naturwiss. Rundschau *40*, 14 (1987).
[2] See *E. B. Fleischer*, The Structure of Porphyrins and Metalloporphyrins, Acc. Chem. Res. *3*, 105 (1970).

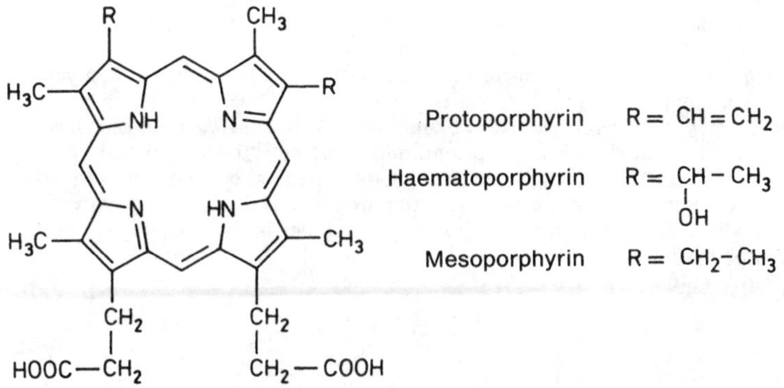

Protoporphyrin   R = CH=CH$_2$

Haematoporphyrin   R = $\overset{\displaystyle CH-CH_3}{\underset{\displaystyle OH}{|}}$

Mesoporphyrin   R = CH$_2$-CH$_3$

H. Fischer has proposed a simplified way of representing these compounds; for example, the four isomeric **aetioporphyrins I–IV** are drawn as follows:

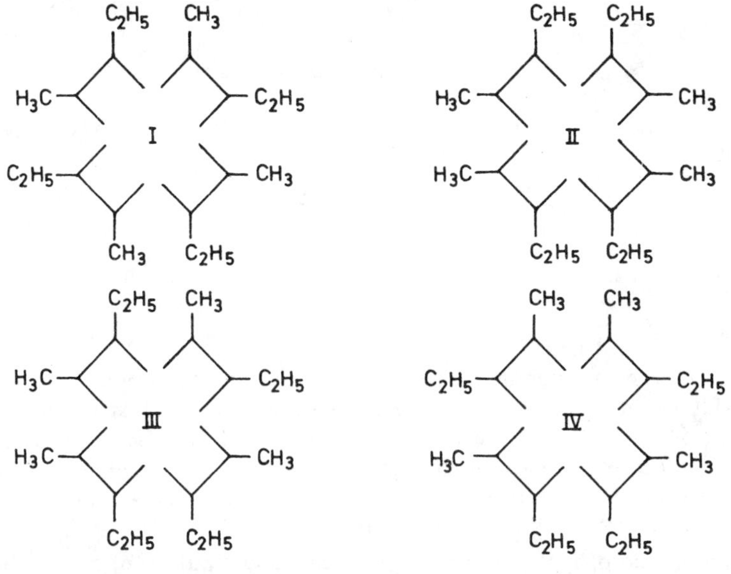

Comparison of their *absorption spectra* suggests that the synthetic *aetioporphyrin III* is identical with the aetioporphyrin isolated from haemin. This in turn provides evidence for the sequence in which the substituent groups in haemin are arranged.

It is important that the iron atom can be reintroduced into the porphin residue, for example by using iron(II) ions dissolved in acetic acid. The initially formed haems are oxidised by air to the stable haemins. Thus it is possible to regenerate haemin-like pigments from porphyrins. These researches reached their pinnacle in the *synthesis of haemin* (*H. Fischer* and *K. Zeile*, 1929).

Free Porphyrins. Some porphyrins also occur free in nature, for example in very old sedimentary rocks (2.6×10^9 years old) and in certain meteorites (carbonaceous chondrites). *Uroporphyrin* and *coproporphyrin* have been isolated in crystalline form from the urine and faeces of persons suffering from porphyrinuria, and, in acute stages, so has the partial building-block, *porphobilinogen* (*Westall*, 1952). These porphyrins also occur in small amounts as breakdown products from the normal metabolism. Since in higher concentrations they are carried by the blood into tissue, they cause a marked sensitivity to light (*sensitisers*).

Both porphyrins have great importance in the biochemical synthesis of the pigments of haemoglobin and of chlorophyll (see also vitamin B_{12}).

The *biosynthesis of porphyrins* in mammals takes place as follows. First of all, glycine and activated succinic acid (succinyl coenzyme A) condense, with loss of coenzyme A, to produce α-amino-β-oxoadipic acid, which is decarboxylated to δ-aminolevulinic acid (*Shemin*). Enzymatic condensation of two molecules of this acid leads to **porphobilinogen**, which is the biogenetic key compound for the preparation of this class of compounds. On standing *in vitro* this is converted into a mixture of different uroporphyrins. An intermediate in the process is porphyrinogen III which does not yet contain methine bridges and in which the pyrrole rings are linked by methylene bridges (given grey backgrounds in the formula of porphobilinogen).[1]

Glycine   Succinyl-CoA α-Amino-β-oxo--adipic acid δ-Aminolevulinic acid

Porphobilinogen

7.1.2.5 Chlorophyll[2]

Chlorophyll, the green colouring material of leaves, contains magnesium and occurs together with the red and yellow pigments, the carotenes and xanthophylls, in the chloroplasts of plant cells. Biochemically it is concerned with *photosynthesis* and the assimilation of carbon dioxide by the green parts of plants. Within the plant cells chlorophyll is bonded to a protein to form a chromoprotein which is stable to light and to oxygen and carbon dioxide (*W.N. Lubimenko*).

Separation of chlorophyll from the protein occurs under mild conditions, for example when fresh leaves dry out or on precipitation by salts. Ether extracts *chlorophyll*, the *carotenoids* and a lipid fraction which is a complex mixture. The protein part, plastin, remains in the aqueous phase or flocculates at the solvent boundary.

Naturally occurring chlorophyll consists of two components, *chlorophyll a*, which is blue-green, together with small amounts of *chlorophyll b*, which is yellowish-green (*Willstätter*). They can be separated by *adsorption chromatography*, best on elutriated saccharose as adsorbent (*Stoll*). Pure chlorophylls a and b have a waxy consistency which is connected with the presence of the phytyl group, derived from the unsaturated alcohol *phytol*, $C_{29}H_{39}OH$, which forms the alkyl portion of an ester group. Alkaline hydrolysis of

---

[1]See *B. Frydman* et al., Biosynthesis of Uroporphyrinogens from Porphobilinogen. Mechanism and Nature of the Process. *Acc. Chem. Res. 8*, 201 (1975).

[2]See *T. W. Goodwin* (Editor), Chemistry and Biochemistry of Plant Pigments, Vol. I (Academic Press, New York 1976).

Chlorophyll a (m.p. 117–120°C [390–393 K])

Vanadyl deoxophyllo-erythro-aetioporphyrin

$$C_{20}H_{39} = -CH_2-C=C-(CH_2)_3-C-(CH_2)_3-C-(CH_2)_3-CH-CH_3$$

Phytyl residue [(2E)-(7R,11R)-3,7,11,15-tetramethylhexadec-2-enyl]

either chlorophyll provides not only 1 mol of phytol but also 1 mol of methanol; the chlorophylls are diesters.

Whereas in the blood pigments the central atom of the complex is iron, in the chlorophylls there is an atom of magnesium which forms a complex and is responsible for the sensitivity of the chlorophylls to acid. The magnesium ion can be removed by careful hydrolysis using dilute acid, whereupon the colour of chlorophyll a changes from blue-green to olive-green and that of chlorophyll b from yellowish-green to claret-coloured. The difference between the two chlorophylls is that *one* of the *methyl groups* of chlorophyll a is replaced by a *formyl group* in chlorophyll b.

In 1940 the structural formula given above was suggested for chlorophyll a (*H. Fischer, Linstead*); it was shown to be correct by a total synthesis of chlorophyll a (*Woodward*).[1]

As porphin was defined as the parent compound of the haemins, so *chlorin* (dihydroporphin, IUPAC: dihydroporphyrin) was recognised as the parent compound of chlorophyll and its derivatives. This *chromophore* differs from porphyrin in that ring A has extra hydrogen atoms attached to the 2- and 3-positions, making these into saturated carbon atoms. However, despite this, the double bonds in chlorin are so arranged that *complete cyclic conjugation* is still a feature of the molecule. In the formula given, not only the trivial name chlorophyll a is retained but also the traditional numbering of the atoms.

Also characteristic of chlorophyll is the *fused five-membered ring*, which can be regarded as being derived from haemin by β-oxidation of the propionic acid residue attached to ring C, and its condensation with the 15-methine group to give a substituted cyclopentenone ring. Also, the vinyl group at position 8 in haemin (see p.709) is reduced to an ethyl group in chlorophyll. And finally, unlike haemin, the chlorophylls are *optically active* (strongly laevorotatory). There are three chiral centres each marked with an asterisk in the formula. Vanadyl deoxophyllo-erythro-aetioporphyrin has been found in mineral oil; it lacks all the oxygen atoms which are present in chlorophyll (*Treibs*).

---

[1]See Totalsynthese des Chlorophylls, Angew. Chem. 72 651 (1960).

7.1.2.6   Vitamin B$_{12}$[1]

Vitamin B$_{12}$, the *anti-perniciosa* factor, is closely related to the naturally occurring porphyrins. As long ago as 1926 *Minot* and *Murphy* observed that raw liver was efficacious in dietetic treatment of *pernicious anaemia*. In 1948, working quite independently of one another, *K. Folkers* and *E. L. Smith* isolated the active ingredient, vitamin B$_{12}$, as a crystalline substance. Later it was found in other biological materials such as milk powder, meat extracts, and in the culture solutions of various strains of fungi and bacteria, especially *Streptomyces griseus*.

Vitamin B$_{12}$, $C_{63}H_{88}CoN_{14}O_{14}P$ (CN—Cbl, INN: *Cyanocobalamin*), crystallises as dark red needles. The parent unsubstituted ring which forms the basis of vitamin B$_{12}$ is known as **corrin**. All the naturally occurring compounds derived from it, the *corrinoids*, contain *cobalt* in the oxidation state + 3 as the *central atom* as well as substituent acetamide and propionamide groups. In addition, vitamin B$_{12}$ contains a *cyanide ion*, which may also be replaced by other ions such as HO$^{\ominus}$ (HO—Cbl) and Cl$^{\ominus}$ and, bound like a *nucleotide*, the base *5,6-dimethylbenzimidazole*. Because it contains cobalt, vitamin B$_{12}$ is called *cobalamine*. The structure shown below was determined by *X-ray analysis* (*D. Crowfoot-Hodgkin*, 1955).

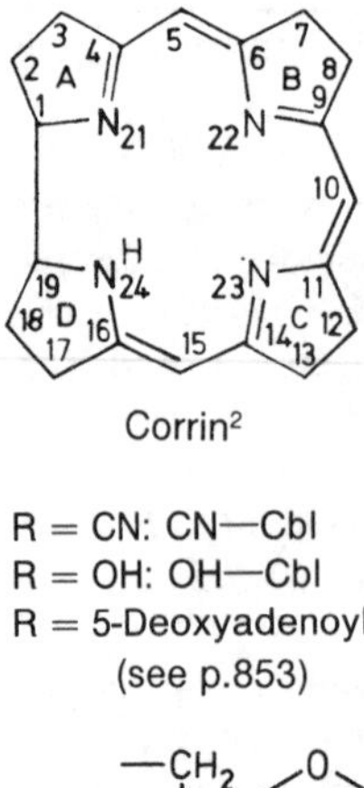

Corrin[2]

R = CN: CN—Cbl
R = OH: OH—Cbl
R = 5-Deoxyadenoyl:
   (see p.853)

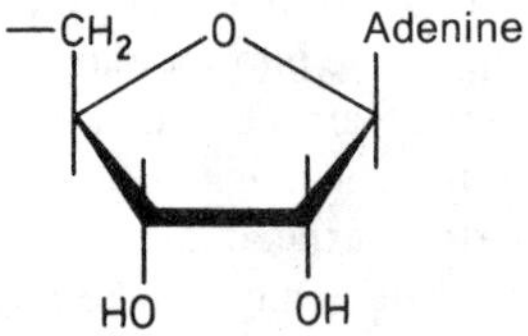

Ado–Cbl, oxidation
state of Co +2

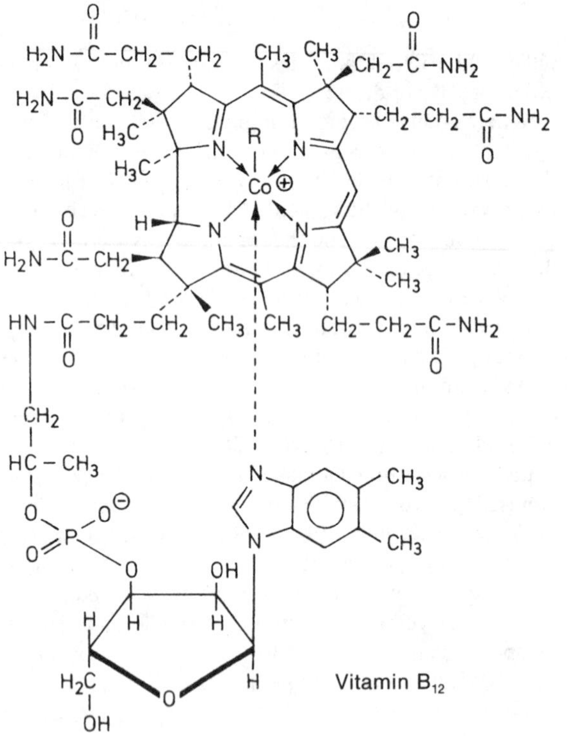

Vitamin B$_{12}$

---

[1] See *R. Bonnett*, The Chemistry of the Vitamin B$_{12}$ Group, Chem. Rev. 63, 573 (1963); *D. Dolphin* (Ed.), B$_{12}$, 2 Vols. (Wiley, New York 1982).
[2] In its numbering system corrin is treated as a contracted porphyrin from which the 20-methine group has been lost. In this way the *N*-atoms are numbered in the same order as they are in porphyrins. The same system is used for linear tetrapyrroles (see p.715 bilirubin).

This was confirmed by a total synthesis, by *Eschenmoser* and *Woodward*,[1] wherein the major challenge was the construction of the corrin nucleus of the molecule. This was built up from rings A–D which had previously had the appropriate substituents introduced into them and, in contrast to an earlier described *linear synthesis* (p.449 *ff.*), this is an example of a *convergent synthesis*.

In organisms the corrinoids are synthesized from δ-aminolevulinic acid.[2] They act as coenzymes, for example in methylation and in the conversion of ribonucleotides into deoxyribonucleotides. Instead of an anion they contain another ligand, for example in the case of vitamin B_{12} the 5'-deoxyadenosyl residue (Ado—Cbl). This results in a *metallo-organic compound* with a C—Co bond. Homolytic cleavage produces a free radical bonded to a protein which plays a decisive part in the reactions catalysed by vitamin B_{12}-coenzyme.[3]

7.1.2.7 Bile Pigments

This group of pyrrole derivatives includes yellow, green or red to brown iron-free pigments whose formation in mammals has been shown to derive from the oxidative breakdown of haemins. The most important example of this group is **bilirubin**, $C_{33}H_{36}N_4O_6$, which is orange-red and occurs in bile. Its constitution has been proved by synthesis (*H. Fischer*).

Bilirubin

As a result, bilirubin has the same arrangement of substituents in the pyrrole rings as is found in haemin.

It is formed from haemin or protoporphyrin as follows. The first step is oxidative removal of the 5-methine group, which is eliminated as *carbon monoxide* (hence the absence of the number 20 in the numbering of bilirubin). This opens the porphin ring to form *biliverdin* and is followed by hydrogenation of the C=C double bond at C-10, producing a central methylene bridge between rings C and D. The hydrogen atoms of the two remaining methine groups are each *trans* to the NH-groups of rings A and B.

Bilirubin has been shown to be an antioxidant which can interrupt radical chain reactions that lead to the formation of hydroperoxides.

7.1.2.8 Phthalocyanins

A useful series of pigments whose structures involve porphin rings are the synthetically produced *phthalocyanins* (*H. de Diesbach, Linstead*). They are made by heating phthalonitrile (1,2-dicyanobenzene) with a metal or a metal salt at 150–200°C (425–475 K), *e.g.* with copper(II) chloride blue needles of **copper phthalocyanin** (*Monastral Fast Blue B, Pigment Blue 15, Heliogen B*) are formed. This pigment is made industrially by heating *phthalic anhydride* with *urea* and copper(I) chloride in the presence of boric acid in a high-boiling solvent at about 200°C (475 K). The first step in this process is the formation of *3-amino-1-imino-1H-isoindole* as an intermediate:[4]

[1] See *R. B. Woodward*, Pure Appl. Chem. *33*, 145 (1973); *A. Eschenmoser*, Naturwiss. *61*, 513 (1974).
[2] See *A. R. Battersby* et al., Biosynthesis of Vitamin B_{12}, Acc. Chem. Res. *19*, 147 (1986); *26*, 15 (1993); Angew. Chem. Int. Ed. Engl. *34*, 385 (1995); *A. I. Scott*, Angew. Chem. Int. Ed. Engl. *32*, 1223 (1993).
[3] See *B. T. Golding*, The B_{12}-mystery, Chem. in Britain, *26*, 951 (1990).
[4] See *F. Baumann* et al., Isoindolenine als Zwischenprodukt der Phthalocyanin-Synthese. (Isoindolenines as Intermediates in the Synthesis of Phthalocyanins), Angew. Chem. *68*, 133 (1956).

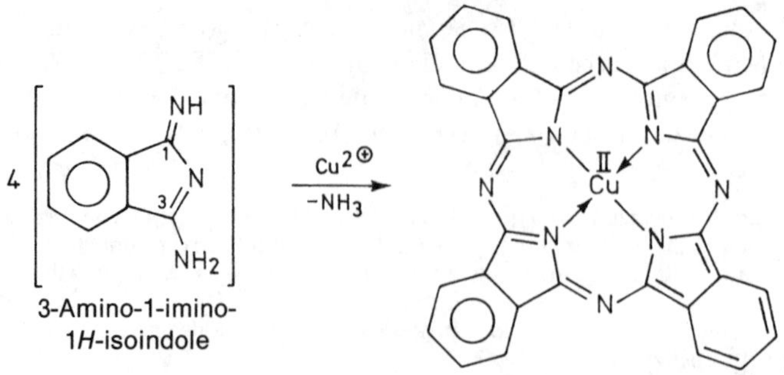

3-Amino-1-imino-
1*H*-isoindole

Copper phthalocyanin

These compounds are tetra-azaporphins in which the four methine groups of the porphins have been replaced by —N═bridges. In addition each pyrrole ring has a benzene ring fused to it. The central ring includes two NH-groups whose acidic hydrogen atoms are replaced by a metal atom of oxidation state +2. The intense colours of phthalocyanins are associated with *mesomerism* in the chemical bonding.

Since they have good colours, phthalocyanins are used as pigments, dyestuffs and printing inks; they represent >20% of the entire production of pigments. They are especially suitable for the colouring of paper, wallpaper, plastics and lacquer pigments. Some copper and nickel phthalocyanins, many with chloro-substituents in the benzene rings, can be generated on the fibre and are used for colouring textiles. Phthalocyanin molecules are remarkably stable to heat and to chemical reagents. Metal-free and metal-containing compounds sublime practically unchanged at 550–580°C (825–855 K). For a survey of the properties of these pigments see reference 1 below.

7.1.3   Furans[2]

Furan (oxole) is the analogue of pyrrole with an oxygen atom in a five-membered ring, which, like that of pyrrole, is planar. The bond lengths in the ring indicate that there is less delocalisation of the π-electrons than in pyrrole.

This ring system is found in *furanoses*.

Furan

Preparation

7.1.3.1   Industrially furan is made from *furfural* by loss of carbon monoxide at about 400°C (675 K) over a zinc oxide/chromium oxide catalyst:

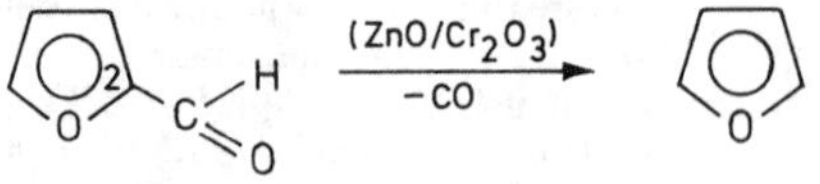

Furfural

Furan

[1] See *H. G. Volz*, Optical Properties of Pigments — Objective Methods for Testing and Evaluation, Angew. Chem. Int. Ed. Engl. *14*, 688 (1975).
[2] See *F. M. Dean*, Recent Advances in Furan Chemistry, Adv. Heterocyclic Chem. 30, 167; *31*, 237 (Academic Press, New York 1982); *C. W. Bird* and *G. W. H. Cheeseman* (Eds), Comprehensive Heterocyclic Chemistry, Vol. 4 (Pergamon Press, Oxford 1984).

7.1.3.2 By *decarboxylation of furoic acid* (furan-2-carboxylic acid) at about 200°C (475 K) (*Limpricht*, 1870):

Furoic acid   Furan

7.1.3.3 *Paal–Knorr* Synthesis. Furan derivatives are most commonly made by heating 1,4-diketo-compounds with dehydrating agents such as zinc chloride or phosphorus(V) oxide, *e.g.*

Hexane-2,5-dione (Acetonylacetone)   2,5-Dimethylfuran (b.p. 94°C [367 K])

Conversely the furan ring readily undergoes hydrolytic ring-opening, for example, addition of a small amount of sulphuric acid to a solution of 2,5-dimethylfuran in acetic acid converts it almost quantitatively into hexane-2,5-dione.

Properties

Furan has an odour resembling that of chloroform. It is a liquid, b.p. 32°C (305 K), which is stable to air and to alkali. A pine shaving which has been dipped into hydrochloric acid is turned green by its vapour. In the presence of acids it polymerises to a dark, insoluble resin.

Furan is a π-electron-rich heteroaromatic. In the absence of acid it readily undergoes *electrophilic substitution*; for example, it undergoes *Friedel–Crafts* acylation with acid anhydrides in the presence of *boron trifluoride* as a catalyst to give **2-acylfurans**. *Pyridine-sulphur trioxide complex* converts it into **furan-2-sulphonic acid**.

2-Acylfuran   Furan   Furan-2-sulphonic acid

Furan can behave as a diene and reacts with maleic anhydride to give both *exo* and *endo* *Diels–Alder* adducts:

(Diene)   (Dienophile)   Adduct   2,5-Dihydrofuran   Maleic anhydride

Maleic anhydride can also be regarded as a dioxo-derivative of 2,5-dihydrofuran.

Tetrahydrofuran (oxolan, THF) can be made either by catalytic hydrogenation of furan or by intramolecular removal of water from butane-1,4-diol. It is a poisonous liquid, b.p. 65°C (338 K). It is a cyclic ether and can be used instead of diethyl ether as a *solvent* in synthetic work. As such it is very useful in *Grignard* reactions and in reductions which make use of metal hydrides. Industrially tetrahydrofuran is made by cyclising butane-1,4-diol in the presence of phosphoric acid:

$$HO-CH_2-CH_2-CH_2-CH_2-OH \xrightarrow[-H_2O]{(H_3PO_4)}$$

Butane-1,4-diol   Tetrahydrofuran (b.p. 65°C [338 K])

It is a starting material for the manufacture of nylon and is also used as a solvent for rechlorinated poly(vinylchloride) and for poly(vinylcarbazole).

7.1.3.4   Important Derivatives of Furan

Furfural, furfuraldehyde, 2-furaldehyde (2-furylmethanal), is the most important industrial derivative of furan.

Preparation

Furfural is formed when pentoses are treated with dilute mineral acid. It was first obtained in 1831 by *Döbereiner* by distilling a mixture of bran (Latin: *furfur*) and dilute sulphuric acid:

Furfural

It is made industrially from oat-husks, corncobs, rice and groundnut husks, reeds, etc. by treating them with 5% sulphuric acid under pressure (4 bar, 0.4 MPa). The *pentosans* in these materials are hydrolysed to *pentoses* which are in turn converted into furfural.

Properties

Furfural is a colourless liquid, b.p. 162°C (435 K), which rapidly turns brown in air. With anilinium salts it gives a red-violet product (furfural test for pentoses).

In its chemical properties furfural resembles benzaldehyde; for example, with ethanolic potassium hydroxide it undergoes a *Cannizzaro* reaction to form **furfuryl alcohol** and the **potassium salt of 2-furoic acid** (furan-2-carboxylic acid) which had been obtained by *Scheele* in 1781 by heating mucic acid. The 2-furylmethyl group is known as the *furfuryl* group.

Furfural takes part in a reaction analogous to the benzoin condensation. In the presence of *cyanide ions* it is converted into *furoin*. The latter is very readily oxidised to *furil*. When heated with potassium hydroxide, furil undergoes a benzilic acid-type rearrangement to give the **potassium salt of furilic acid**.

Furfural reacts with acetic anhydride in a *Perkin* synthesis to give *furan-2-acrylic acid*.

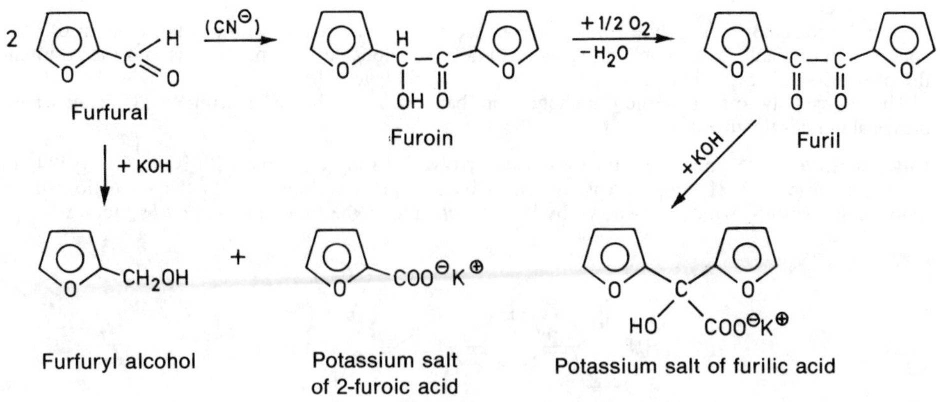

Furfural

Furoin

Furil

+ KOH

Furfuryl alcohol   +   Potassium salt of 2-furoic acid   Potassium salt of furilic acid

Uses

Furfural is used as a selective solvent in oil-refining and for the preparation of plastics of the phenol-aldehyde type. It is also the starting material for the preparation of tetrahydrofurfuryl alcohol (2-hydroxymethyltetrahydrofuran) which is an important solvent. *1-[(5-nitrofurfurylidene)amino]2,4-imidazoline-dione* (INN: **Nitrofurantion**) is used as a bactericide. Furfuryl mercaptan (2-furylmethanethiol) is a component of the aroma of roasted coffee; it is not present in fresh coffee.

When hexoses, especially ketohexoses (*cf.* levulinic acid), are treated with very dilute acid **5-hydroxymethylfurfural** is formed. It forms colourless needles, m.p. 35°C (308 K), which liquefy in air and give a red colour with resorcinol and hydrochloric acid (*Selivanov* reaction). It is oxidised to **furan-2,5-dicarboxylic acid** (dehydromucic acid):

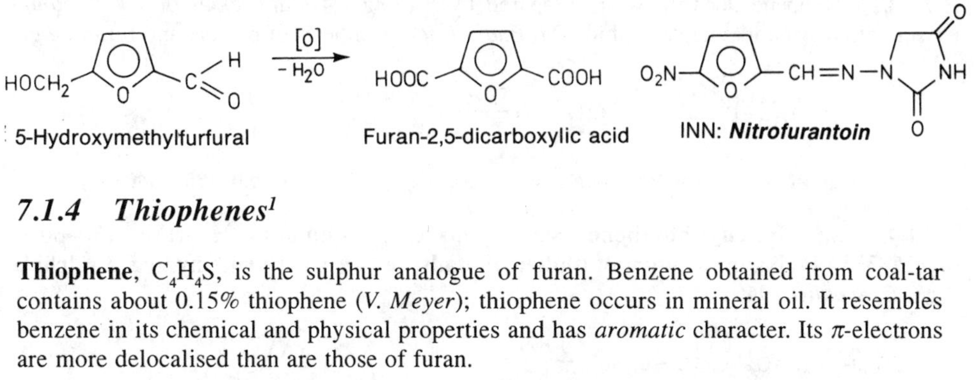

5-Hydroxymethylfurfural   Furan-2,5-dicarboxylic acid   INN: **Nitrofurantoin**

7.1.4   Thiophenes[1]

Thiophene, C_4H_4S, is the sulphur analogue of furan. Benzene obtained from coal-tar contains about 0.15% thiophene (*V. Meyer*); thiophene occurs in mineral oil. It resembles benzene in its chemical and physical properties and has *aromatic* character. Its π-electrons are more delocalised than are those of furan.

Thiophene

---

[1] See *S. Gronowitz*, Recent Advances in the Chemistry of Thiophenes, Adv. Heterocycl. Chem. *1*, 1 *ff.* (Academic Press, New York 1963); The Chemical Society Specialist Periodical Report *5*, 247 (1978); *C. W. Bird* and *G. W. H. Cheeseman* (Ed.), Comprehensive Heterocyclic Chemistry, Vol. *4* (Pergamon Press, Oxford 1984); *S. Gronowitz* (Ed.), Thiophene and its Derivatives, Parts 1–3 (Wiley, New York 1985, 1986).

In order to remove thiophene from benzene, the benzene is shaken with a small amount of cold conc. sulphuric acid. The thiophene is more rapidly sulphonated than benzene is and the resultant thiophene-2-sulphonic acid dissolves in the sulphuric acid layer.

The easiest way to remove the thiophene from benzene is by shaking it with *Raney* nickel which brings about desulphurisation.

Indophenin test. A colour test for thiophene involves heating a sample with isatin (see p.724) in conc. sulphuric acid. If thiophene is present a blue colour develops, due to the formation of an indophenin dyestuff, which was shown by *W. Schlenk* to have the following *indigoid* structure:

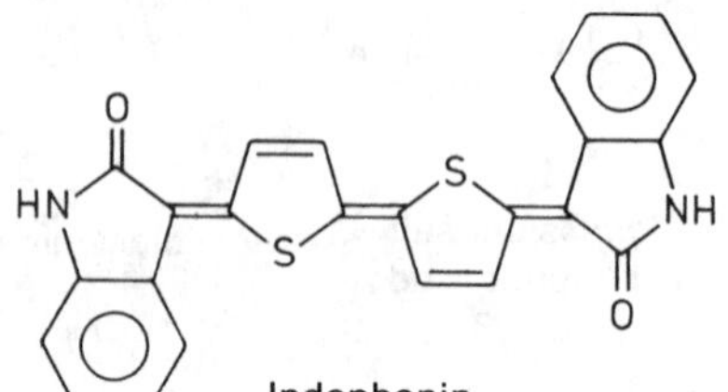

Indophenin

Interestingly, this was originally used as a test for benzene until it was shown that the colour was actually due to the presence of thiophene in benzene which had come from coal-tar.

Preparation

7.1.4.1 Industrially from *n-butane* and *sulphur* in the gas phase at 560°C (830 K).

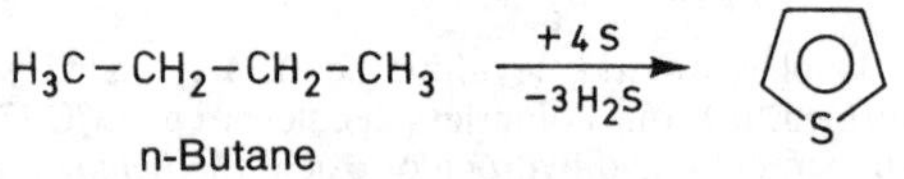

n-Butane

7.1.4.2 Thiopene derivatives are prepared by *heating 1,4-dialdehydes* or *1,4-diketones* with tetraphosphorus heptasulphide (*cf. Paal–Knorr* synthesis of pyrrole and furan), *e.g.*

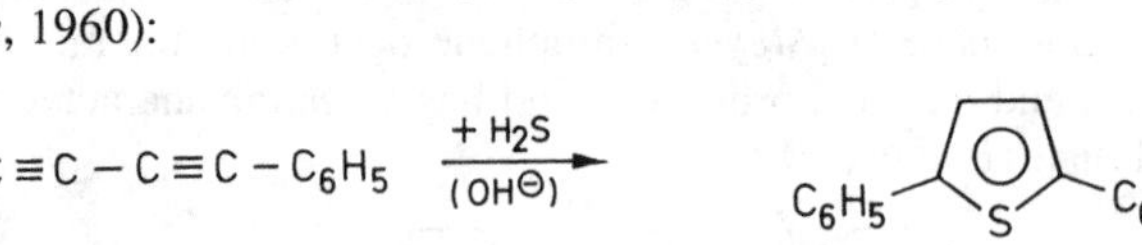

Hexane-2,5-dione (Acetonylacetone)   2,5-Dimethylthiophene (b.p. 136°C [409 K])

7.1.4.3 2,5-Diphenylthiophene can be made by saturating a warm (50–60°C [325–335 K]) alkaline solution of **diphenylbutadiyne** (see p.620) with hydrogen sulphide (*K. E. Schulte*, 1960):

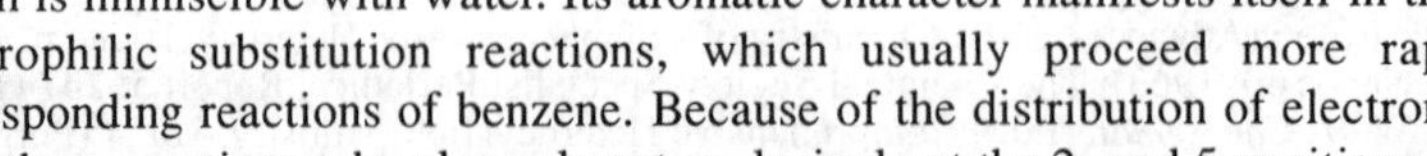

Diphenylbutadiyne   2,5-Diphenylthiophene (m.p. 152°C [425 K])

Properties

Thiophene is a colourless liquid, b.p. 84°C (357 K) (*cf.* benzene, b.p. 80.4°C [353.6 K]) which is immiscible with water. Its aromatic character manifests itself in that it undergoes electrophilic substitution reactions, which usually proceed more rapidly than the corresponding reactions of benzene. Because of the distribution of electron density in the ring, these reactions take place almost exclusively at the 2- and 5-positions and at the 3- or 4-positions only if the 2- and 5-positions already bear substituents.

It undergoes halogenation, nitration, sulphonation, mercuriation and *Friedel–Crafts* acylation with acid anhydrides or acid chlorides in the presence of phosphoric acid or tin(IV) chloride, *e.g.*

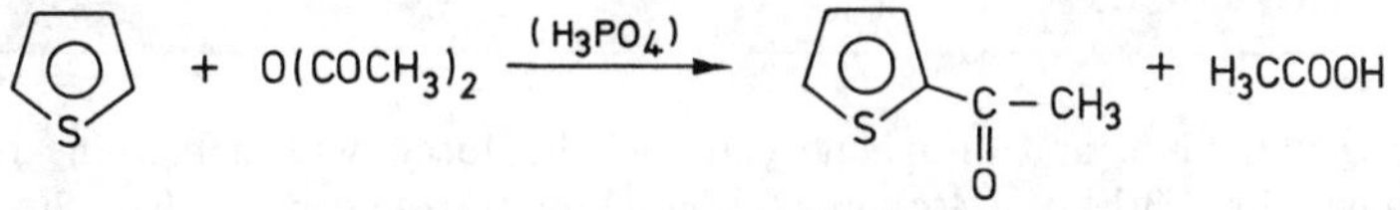

Thiophene

2-Acetylthiophene
(Methyl 2-thienyl ketone)

Birch **reduction** of thiophene with sodium and liquid ammonia in the presence of methanol provides **2,3-** and **2,5-dihydrothiophenes**, while catalytic hydrogenation gives **thiolane** (tetrahydrothiophene, thiophane). It is necessary to use a large excess of catalyst (Pd) to overcome the catalyst-poisoning effect of sulphur (*Mozingo*, 1945).

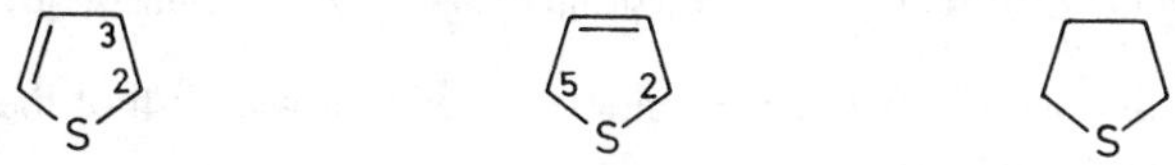

2,3-Dihydrothiophene   2,5-Dihydrothiophene   Thiolane

Under normal conditions thiophene does not take part in *Diels–Alder* reactions, but **thiophene dioxide** can act as a *diene* or as a *dienophile* and undergoes dimerisation with loss of sulphur dioxide, giving the following adduct:

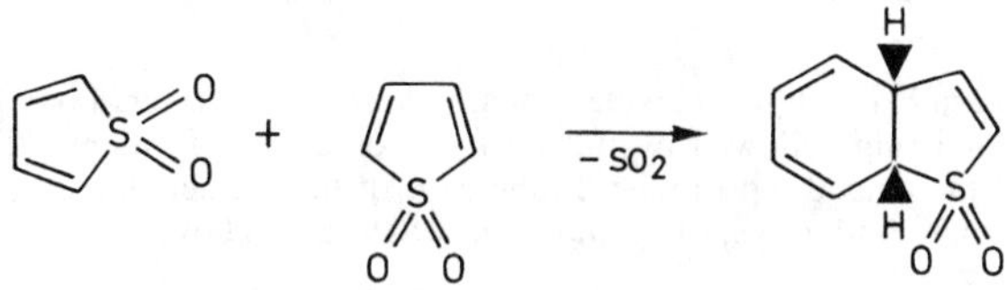

Thiophene dioxide   Adduct

Thiophene dioxide, unlike thiophene, does not possess a cyclic conjugated system of six π-electrons.

2,5-Dihydrothiophene dioxide can act as a source of an open-chain diene which undergoes a *Diels–Alder* reaction with benzyne. If 2,5-dihydrothiophene dioxide and benzenediazonium-2-carboxylate (see p.637) are heated together, butadiene and benzyne are generated and react together to give **1,4-dihydronaphthalene**:

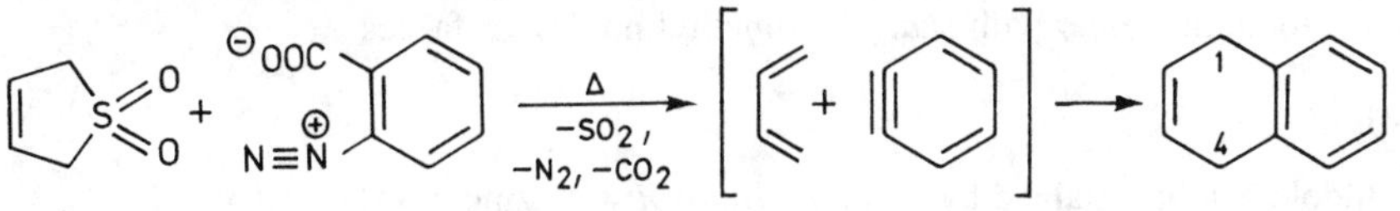

1,4-Dihydronaphthalene

Tetrahydrothiophene-1,1-dioxide (sulpholane, tetramethylenesulphone) is used in the purification of hydrocarbons, especially of benzene (*sulpholan process*), not only in extraction but also in extractive distillation processes.

The thiolane ring occurs naturally, condensed with an imidazolone ring, in *biotin* (vitamin H).

7.2 Benzopyrroles, Benzofurans and Benzothiophenes

In theory, pyrrole, furan and thiophene can each be fused with a benzene ring in two different ways, with 2,3(b) or 3,4(c) annelation. Only pyrrole can also be 1,2(a)-fused and the result is then a pyrrolopyridine. The following three *isomers* are possible in the case of pyrrole; **indole** is the most important of them:

| Indolizine | Indole | Isoindole |
| Pyrrolo[1,2-*a*]pyridine | Benzo[*b*]pyrrole | Benzo[*c*]pyrrole |

The furan and thiophene analogues of indole are as follows; both of them are sometimes described by their trivial names:

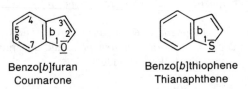

Benzo[*b*]furan   Benzo[*b*]thiophene
Coumarone   Thianaphthene

The numbering of these benzo[*b*]heterocycles usually starts at the heteroatom and then proceeds around the five-membered ring followed by the six-membered ring. In older literature the 2- and 3-positions were called the α- and β-positions. In the case of the isomers having the hetero-atom at the 2-position or common to both rings, the numbering is as shown above.

7.2.1 Indoles[1]

The development of indole chemistry stemmed from the β-glucoside *indican*, which is found in some indigo plants. Acid or enzymatic hydrolysis splits it into glucose and indoxyl.

Indole is found in small amounts in coal-tar and also in the oils of jasmine and orange flowers. It is a biochemical breakdown product of tryptophan which occurs in most proteins and is hence found, together with *skatole* (3-methylindole), in faeces.

Preparation

7.2.1.1   Indole can be obtained by *reducing **indoxyl*** with zinc powder and alkali:

---

[1]See *R. J. Sundberg*, The Chemistry of Indoles (Academic Press, New York 1976); *W. J. Houlihan* (Ed.), Indoles, Parts 1–3 (Wiley, New York 1972, 1979); *C. W. Bird* and *G. W. H. Cheeseman* (Eds), Comprehensive Heterocyclic Chemistry, Vol. 4 (Pergamon Press, Oxford 1984).

Indoxyl, 3(2H)-Indolone   Indole

7.2.1.2 Madelung Synthesis. If 2'-methylformanilide is heated to 350–360°C (625–635 K) in an atmosphere of nitrogen with potassium t-butoxide as a condensing agent, intramolecular ring-closure ensues and indole is formed in 79% yield (*Tyson*, 1950).

2'-Methylformanilide   Indole

7.2.1.3 Fischer Indole Synthesis. Indole derivatives are formed when phenylhydrazone derivatives of keto compounds are heated at about 180°C (450 K) with zinc chloride (or sulphuric acid or boron trifluoride) as a condensing agent. The first step involves a diaza-*Cope* rearrangement (see p.108) followed by loss of ammonia and the formation of a five-membered ring. Labelling experiments, using ^{15}N, show that the ammonia which is lost contains the β-nitrogen atom of the hydrazine group (*Clusius*).

Using *acetone phenylhydrazone* as an example, the *mechanism* of the reaction is as follows:

2-Methylindole

Indole is obtained similarly from acetaldehyde phenylhydrazone and *3-methylindole* (skatole) from propionaldehyde phenylhydrazone.

Properties

Indole crystallises in colourless leaflets, m.p. 53°C (326 K). It is weakly basic. On treatment with acids it does not form salts but instead, and especially if heated, *polymerises* to resinous products. On the other hand the hydrogen atom of the NH-group is weakly acidic and can be replaced by an alkali metal to form a salt. The *pine-shaving test* for pyrrole can also be used as a test for indole which produces a red colouration.

Indole is an *enamine* and electrophilic substitution reactions take place almost exclusively at the 3-position. The benzene ring is substantially less reactive than the pyrrole ring and is not activated by the NH-group. Substitution reactions only take place in the benzene ring if the reactive sites in the pyrrole ring are blocked by substituents.

7.2.1.4 Indoxyl [3(2H)-indolone], as its sulphate ester or linked to glucuronic acid as a glycoside, is a normal constituent of urine. It forms deep yellow prisms, m.p. 85°C (358 K). In alkaline media it is readily oxidised in the air to *indigo*. It is an intermediate in the industrial preparation of indigo.

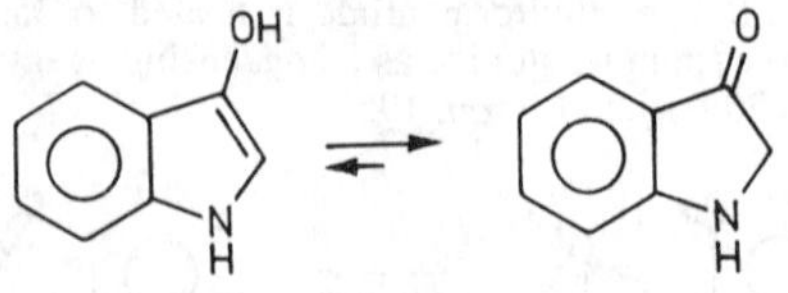

Indoxyl

Spectroscopic studies show that the above equilibrium lies very much on the side of the keto form. The carbonyl group is not adjacent to the heteroatom and it reacts with relatively weak *nucleophilic* reagents for keto groups, for example with hydroxylamine or phenylhydrazine to form an oxime or a phenylhydrazone. The 3-keto group of isatin behaves similarly.

7.2.1.5 Isatin crystallises in orange-red prisms, m.p. 203°C (476 K). It is best made from *aniline, chloral hydrate* and *hydroxylamine* in solution in conc. hydrochloric acid. In the first step **hydroxyiminoacetanilide** is formed which on treatment with conc. sulphuric acid is converted into **isatin-3-imide** (3-imino-2-indolinone) and finally isatin (yield = 75%).

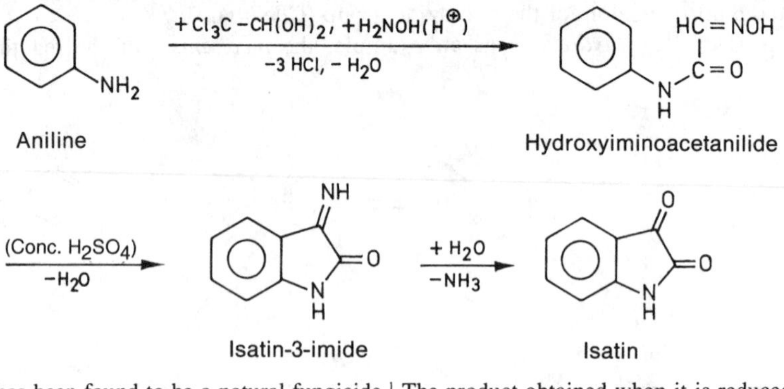

Aniline   Hydroxyiminoacetanilide

Isatin-3-imide   Isatin

Isatin has been found to be a natural fungicide.[1] The product obtained when it is reduced depends on the reducing agent used. *Reduction* proceeds via **dioxindole** and **oxindole** to **indole**:

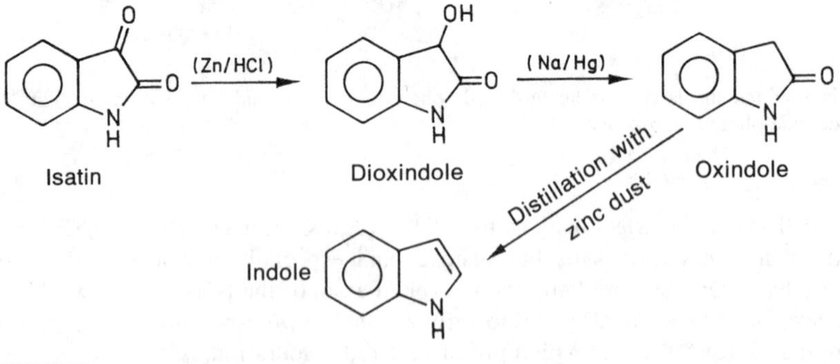

Isatin   Dioxindole   Oxindole

Indole

---

[1]See *M. S. Gil-Turnes* et al., Symbiotic Marine Bacteria Chemically Defend Crustacean Embryos from a Pathogenic Fungus, Science *246*, 116 (1989).

7.2.1.6 L(−)-**Tryptophan**, β-[3-indolyl]alanine, is found in small amounts as a constituent amino acid in almost all proteins. Since it is destroyed when proteins are hydrolysed with acid it was only discovered relatively late on when *trypsin* was broken down enzymatically (*Hopkins*, 1902). Its constitution has been confirmed by synthesis.

Preparation. It is best obtained starting from 3-dimethylaminomethylindole, which as an alkaloid is known as **gramine** and can be made readily from indole, formaldehyde and dimethylamine by means of the *Mannich* reaction. Gramine reacts with an acetamidomalonic ester in the presence of a catalytic amount of sodium hydroxide in an inert solvent to give a **skatylacetamidomalonic ester**, dimethylamine being evolved as a gas. Hydrolysis of this ester provides DL-**tryptophan** (*Snyder* et al.):

Indole   Gramine

Skatylacetamidomalonic ester   DL-Tryptophan

Properties. L-Tryptophan crystallises in silky leaflets; it is readily soluble in hot water. When heated rapidly it melts, with decomposition, at 293°C (566 K). It fluoresces in UV light.

In organisms tryptophan is broken down by enzymatic reactions (deamination and decarboxylation) into skatole or indole. The unpleasant smell of faeces is principally due to skatole.

Tryptophan undergoes a biologically interesting oxidation in the bodies of dogs, in which the indole ring undergoes ring-enlargement to a quinolone ring to give **kynurenic acid** (4-hydroxyquinoline-2-carboxylic acid, 1,4-dihydro-4-oxoquinoline-2-carboxylic acid); **kynurenine** is formed as an intermediate:

Kynurenine   Kynurenic acid

7.2.1.7 Tryptamine,   3-(2-aminoethyl)indole,   2-(3-indolyl)ethylamine,   arises biochemically from the decarboxylation of tryptophan. It is prepared synthetically by reducing 3-indolylacetonitrile.

Amines which are formed by loss of carbon dioxide from amino acids are described as *proteinogenic* or *biogenic* amines (see p.525, 741).

5-Hydroxytryptamine (serotonin, 5-HT) is found in the serum and in the intestinal mucous of mammals. It is a neurotransmitter and brings about a marked contraction of the blood vessels. *N*-Acetyl-5-methoxytryptamine is the pineal gland hormone *melatonin*, which is concerned with the diurnal rhythms of the 'internal clock', and is secreted overnight. Amides of serotonin with long-chain carboxylic acids (carboxylic acid-5-hydroxytryptamides = CHT) occur in the waxes on the surface of coffee beans, *e.g.* arachidic acid-5-hydroxytryptamide (eicosanoic acid-5-hydroxytryptamide).

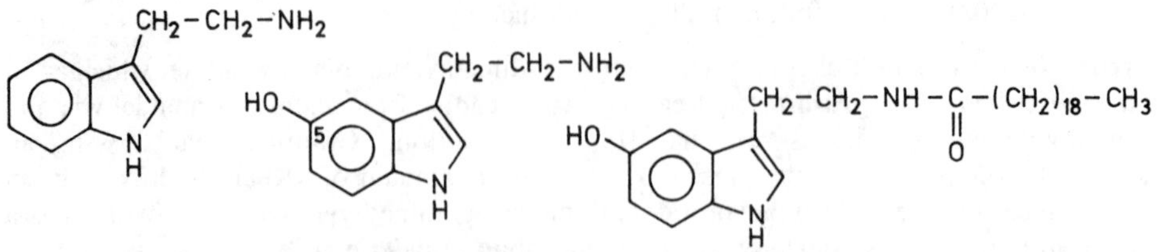

Tryptamine   5-Hydroxytryptamine   Arachidic acid-5-hydroxytryptamide
                                                 Eicosanoic acid-5-hydroxytryptamide

7.2.1.8 Indole-3-acetic acid (indolyl-3-acetic acid, IAA) was isolated from urine and from yeast (*Kögl*, 1934) and also from ripening maize grains (*Haagen-Smit*, 1946) and is a *growth factor* in plants. It can be synthesised by the reaction of **indolyl magnesium iodide** with chloroacetonitrile and hydrolysis of the resultant nitrile.

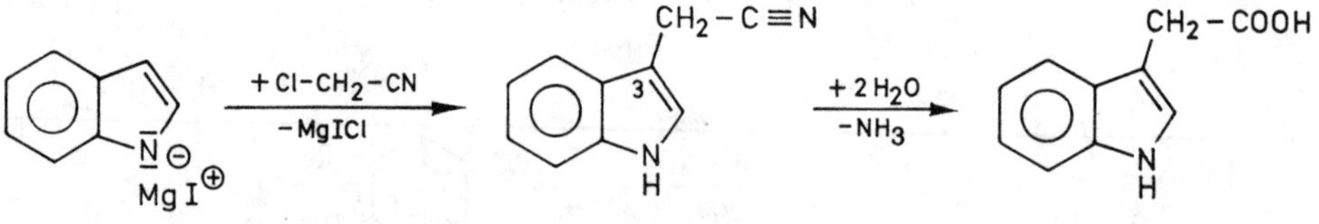

Indolyl magnesium iodide   Indole-3-acetonitrile   Indole-3-acetic acid

Indole-3-acetic acid and some related indole derivatives are plant growth factors (phytohormones, allelopaths), which are called **auxins.**[1] They have decisive effects on the apical growth and many of the development processes of higher plants. Indole-3-acetic acid is made in plants from tryptophan via indole-3-acetamide.

7.2.1.9 Indigo Dyestuffs[2]

The discovery of this class of dyestuffs is associated above all with the fundamental work of A. *Baeyer*, who also carried out a series of syntheses of indigo, which, however, found no industrial application.

Indigo was valued in ancient times as a very fast blue dyestuff. In those times it was obtained from the glucoside *indican* which occurs in the woad plant (*Indigofera tinctoria*). Nowadays synthetic indigo has completely replaced the natural product. Indigo has the following structural formula which includes two *intramolecular hydrogen bonds*:

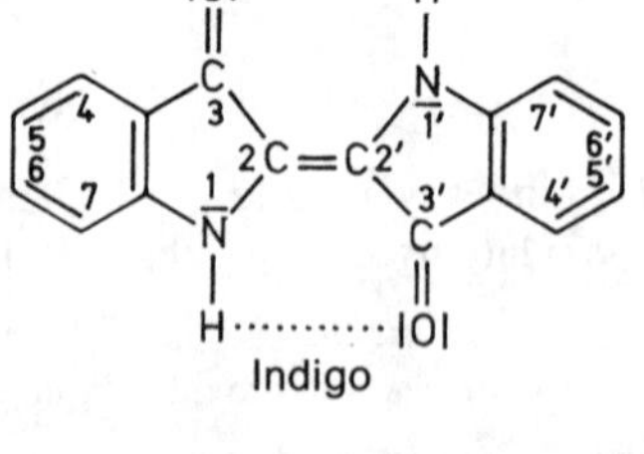

Indigo

---

[1] See *N. J. Leonard*, The Chemistry and Biochemistry of Plant Hormones, Vol. 7 (Academic Press, New York 1974).
[2] See *M. Seefelder*, Indigo, Culture, Science and Technology (Ecomed, Landsberg 1994).

In view of the central C=C double bond, indigo might be expected to exist as *E*- and *Z*-isomers, but only one form has ever been isolated from natural sources, and this has been shown by *X-ray analysis* to have the *E*-configuration, which is stabilised by the hydrogen bonding.

The deep colour of indigo is a result of the bonding of its atoms in the ground state (ground state chromophore).

Ground state chromophore
of indigo

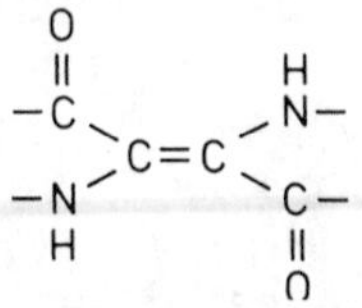

It consists of a two-fold cross-conjugated system, with both an electron-donating group (NH) and an electron-withdrawing group (C=O) at each end of the C=C double bond. These groups have been called, respectively, *dative* and *capto* groups, so that each end of the molecule is *captodative* substituted. The two benzene rings are of little importance for the properties of indigo (*Klessinger, Lüttke,* 1963).[1]

7.2.1.10 Industrial Syntheses of Indigo

(a) In the **1st *Heumann* synthesis** (1890), **aniline** was condensed with chloroacetic acid to give **phenylglycine**, which on being melted with alkali at 300°C (575 K) produced **indoxyl**; atmospheric oxidation of the latter gives indigo:

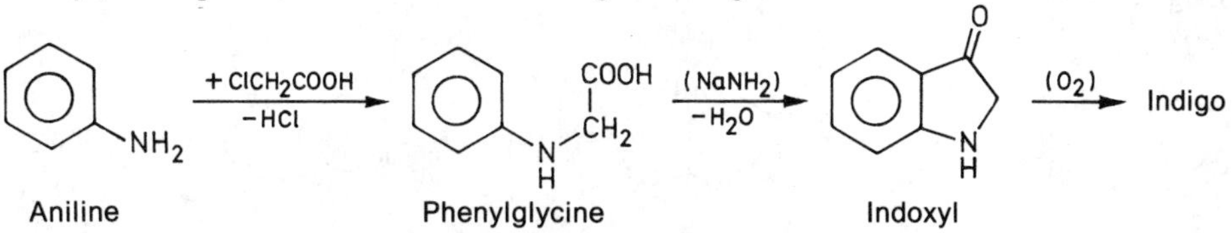

Aniline   Phenylglycine   Indoxyl

The profitability of this process was first achieved by using *sodium amide* as the condensing agent; this enabled ring-closure to be carried out at temperatures of 180–200°C (455–475 K) (*Pfleger,* 1901).

Phenylglycine can be made in appreciably better yield from aniline by treating it with the formaldehyde sodium bisulphite adduct at 50–70°C (325–345 K), followed by sodium cyanide solution, and hydrolysis of the resultant nitrile.

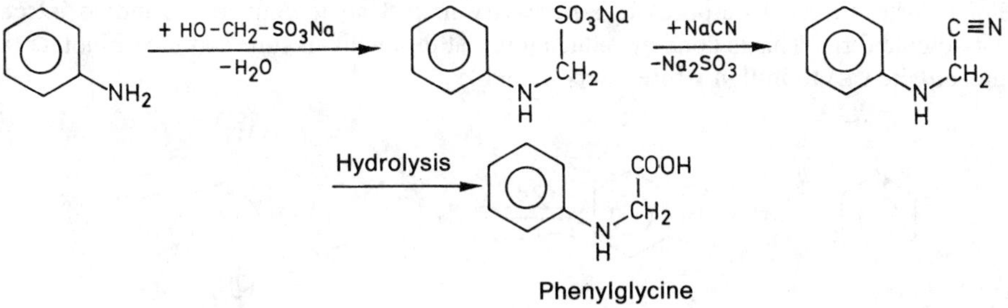

Phenylglycine

(b) In a further process, developed by Hoechst, *aniline* reacts with *ethylene oxide* to form β-hydroxyethylaniline, **N-phenylethanolamine**, which forms a disodium derivative

---

[1] See *M. Klessinger*, Captodative Substituent Effects and the Chromophoric System of Indigo, Angew. Chem. Int. Ed. Engl. *19*, 908 (1980).

when melted with alkali at 200°C (475 K). When heated rapidly to 300°C (575 K) and then cooled again to 240°C (510 K) this cyclises with loss of hydrogen to give **indoxyl**, which is oxidised by air to indigo:

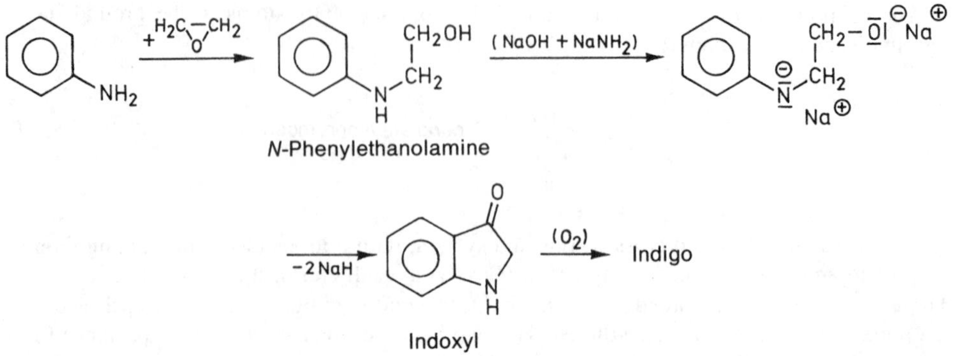

N-Phenylethanolamine

Indoxyl

(c) The **2nd *Heumann* synthesis** starts from *phenylglycine-o-carboxylic acid* which can be made from anthranilic acid and chloroacetic acid. When **phenylglycine-o-carboxylic acid** is melted with alkali, **indoxyl-2-carboxylic acid**, (3-oxoindolinone-2-carboxylic acid), is produced in quantitative yield. When this is warmed it loses carbon dioxide and is converted into indoxyl, which is again oxidised in alkaline solution by air to indigo (BASF process).

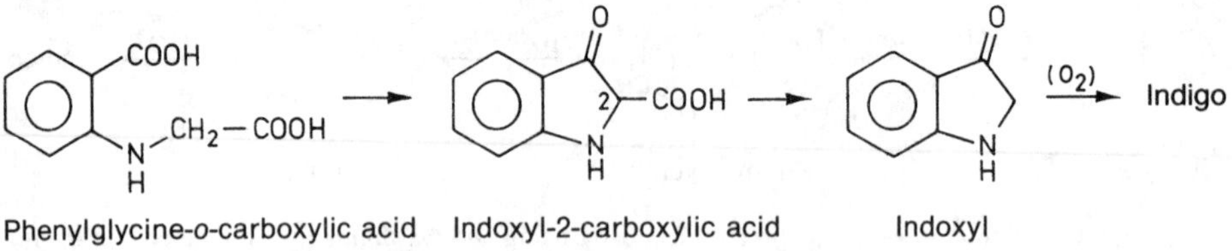

Phenylglycine-*o*-carboxylic acid   Indoxyl-2-carboxylic acid   Indoxyl

Properties. Indigo (**indigotin**, indigo blue, vat blue 1) forms blue crystals with a coppery lustre, m.p. 390–392°C (663–665 K). They dissolve well in boiling acetone or acetic acid as well as in warm aniline, and can be sublimed *in vacuo*.

For industrial dyeing purposes it is necessary first of all to convert the indigo into a water-soluble form. This is done by reducing it (vatting) with alkaline sodium dithionite (a hydrosulphite vat) to **indigo white** (leuco-indigo):

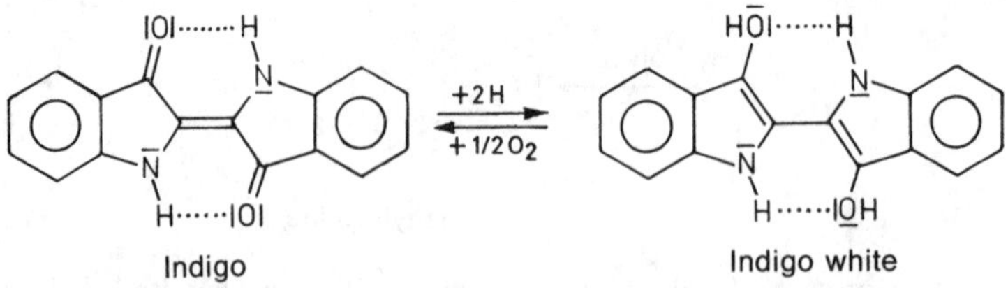

Indigo   Indigo white

This goes into solution and provides an *indigo vat* and can be absorbed directly onto cotton or viscose fibres (see vat dyeing, p.641). Indigo is characterised by its high fastness to washing and to light. The main applications of indigo dyestuffs are for *dyeing cotton* and for *calico printing*.

6,6′-Dibromoindigo. If two hydrogen atoms in the benzene rings of indigo are replaced by bromine atoms, the colour is changed, with a hypsochromic shift of the absorption bands from blue to red-violet. The resultant dyestuff is identical with the *Tyrian purple (royal purple, 'Purple of the Ancients')*, a dyestuff of great antiquity obtained from the mollusc *Murex trunculus L.* which occurs in the Mediterranean Sea.

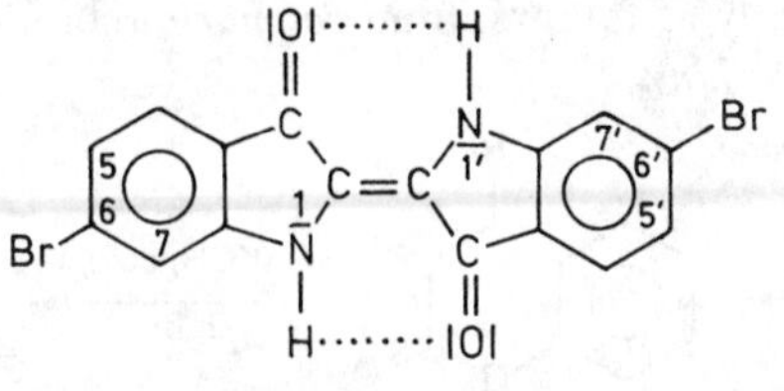

6,6′-Dibromoindigo

Indigo derivatives that have more halogen substituents are noted for their bright colours, *e.g.* 5,5′,7,7′-*tetrabromoindigo* (Ciba Blue G, Brilliant G, Brilliant Indigo 2B).

7.2.2 *Indolizines*

Whilst only *N*-alkyl- and *N*-aryl-derivatives of *isoindole* are known, *indolizine* and its derivatives are important as sensitisers and stabilisers in the production of film.

Indolizine (pyrrocoline) forms colourless crystals, m.p. 75°C (348 K). It is only weakly basic. Its stability and aromatic character are due to mesomerism involving the lone pair of electrons associated with the nitrogen atom:

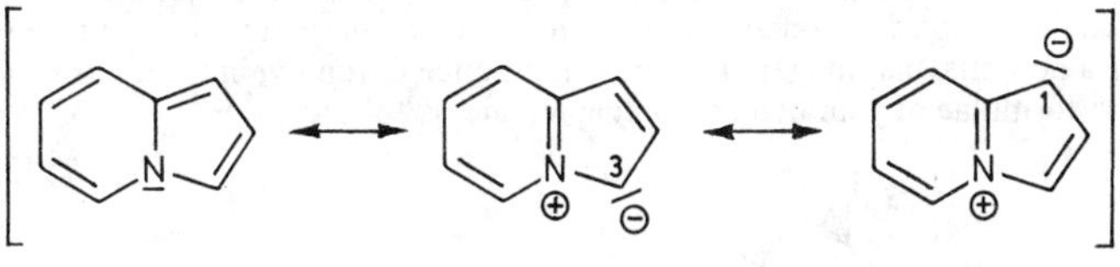

Indolizine

Together with the eight electrons from the formal C=C double bonds they make up a decet of electrons, *i.e.* $(4n+2)$ π-electrons ($n=2$).

Electrophilic substitution takes place at the 1- and 3-positions. Indolizine can be regarded as either a pyrrole or a pyridine derivative.

2-Methylindolizine acts as a sensitiser for the development of *polymethine dyestuffs* in photographic emulsions. It is made from 2-methylpyridine (α-picoline). This is converted into a quaternary salt by reaction with bromoacetone and this salt reacts with sodium hydrogen carbonate (sodium bicarbonate) to give **2-methylindolizine**.

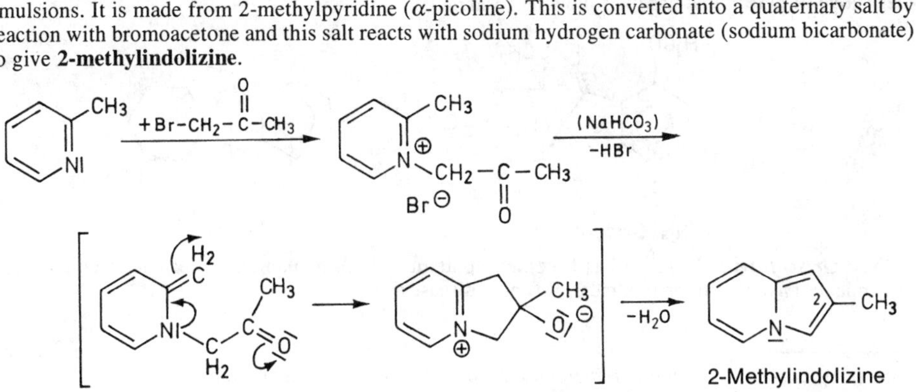

2-Methylindolizine

7.2.3 Coumarones

Coumarone (benzo[b]furan), like indole, is present in coal-tar. It is a colourless oil, b.p. 175–177°C (448–450 K) which tends to polymerise readily, especially in the presence of acids, to give industrially useful *coumarone resins*.

It can be made synthetically as follows, from *coumarin* (chromen-2-one) by means of a ring-contraction reaction (*Perkin*):

Coumarin   3,4-Dibromo-3,4-dihydrocoumarin   3-Bromocoumarin

o-Hydroxy-α-bromocinnamic acid   Coumarilic acid   Coumarone
(Benzofuran-2-carboxylic acid)

This reaction was responsible for the name 'coumarone'.

The coumarone ring is widely distributed in nature. Some compounds of interest found in plants are furocoumarin derivatives that have a coumarin–coumarone structure. Examples are **psoralen** (see p.556) and **bergapten**, from bergamot oil (*E. Späth*). Since the latter promotes the browning of the skin by sunlight, it has been used as a constituent in many cosmetic preparations. Because furocouramin is now regarded as a potential carcinogen, it is present practice to remove it from citrus oils and from oil of bergamot. The formulae of psoralen and bergapten are as follows:

Psoralen (R = H)
Bergapten (R = OCH$_3$)

Bergapten is a *plant fish-poison*.

Griseofulvin,[1] C$_{17}$H$_{17}$ClO$_6$, m.p. 222°C (495 K), $[\alpha]_D = +340°C$, is a colourless, optically active spiro-derivative of 2,3-dihydrocoumarone which was isolated from the mycelium of *penicillium griseofulvum* and has later been found to be a metabolic product from various *penicilliums*. It is a derivative of **grisan** and has the following formula:

INN: *Griseofulvin*   Grisan

INN: *Griseofulvin* (*Likuden*) has become important in both human and veterinary medicine as a systemic *antimycotic* in the treatment of fungal diseases.

---

[1]See *D. Woodcock*, in Systemic Fungicides, *R. W. Marsh* (Ed.) (Longmans, London 1977).

7.2.4   Thianaphthenes

Thianaphthene (benzo[b]thiophene, thionaphthene) also occurs in coal tar and resembles naphthalene in its chemical properties. It forms colourless leaflets, m.p. 32°C (305 K) and boils at 221°C (494 K).

Preparation

7.2.4.1   By *oxidising o-mercaptocinnamic acid* with potassium hexacyanoferrate(III):

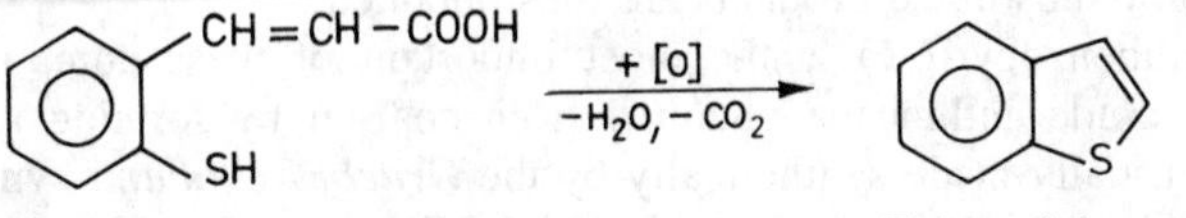

o-Mercaptocinnamic acid   Thianaphthene

7.2.4.2   A better method is from *styrene* and *hydrogen sulphide* at 600°C (875 K) over an iron sulphide – aluminium oxide catalyst (yield = 60%).

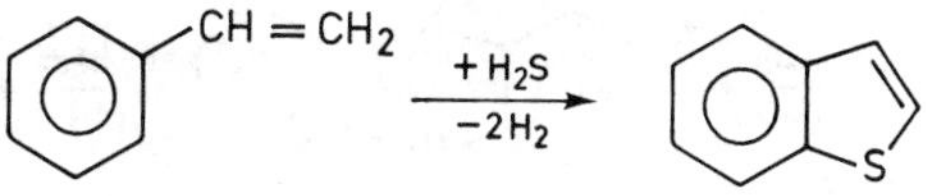

Properties

As in the case of indole, *electrophilic substitution* takes place at the 3-position of thianaphthene. Nitric acid oxidises it to a *sulphone* at room temperature. Thianaphthene is of great importance as a starting material in the making of **thioindigo dyestuffs**.

Thioindigo was first prepared from thiosalicylic acid (*Friedländer*, 1905). The method resembles the 2nd *Heumann* synthesis of indigo and involves the following intermediates:

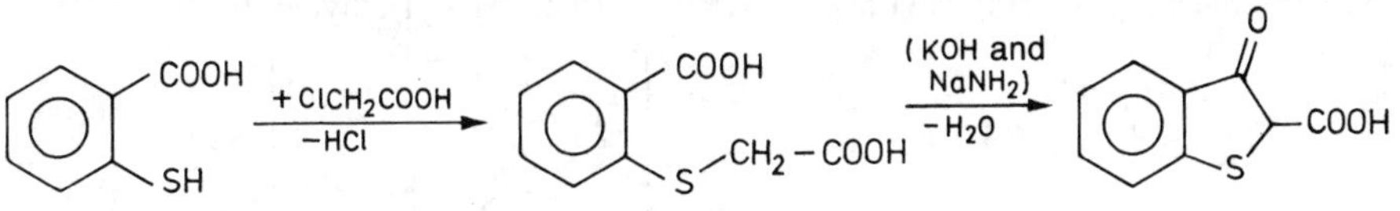

Thiosalicylic acid   *o*-Carboxyphenylthioglycolic acid   'Thioindoxyl-2-carboxylic acid'

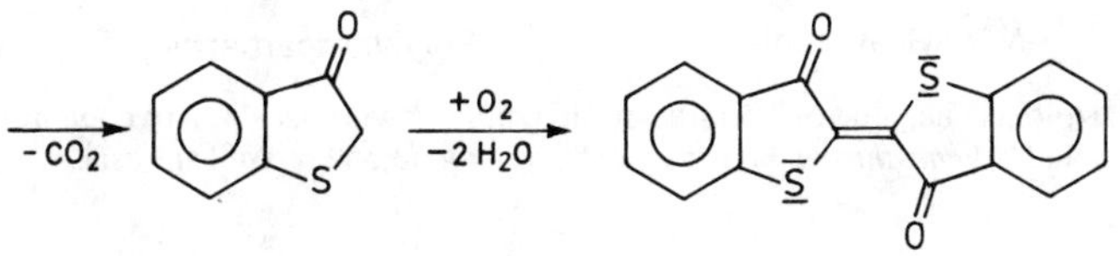

Thioindoxyl   Thioindigo

Thioindigo [$\Delta^{2,2'}$-bis(2,3-dihydrobenzo[b]-thiophene)-3,3'-dione] was the first pure red vat dyestuff; it is more fast to light and resistant to oxidation than indigo is.

7.2.5   Tricyclic Condensed Systems

There are derivatives of pyrrole, furan and thiophene in which two benzene rings are *ortho*-condensed to the heterocyclic ring:

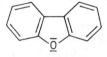

Carbazole (Dibenzopyrrole)
(m.p. 247°C [520 K])

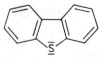

Dibenzofuran
(m.p. 83°C [356 K])

Dibenzothiophene
(m.p. 97°C [370 K])

In these compounds the heterocyclic rings are tetra-substituted and *electrophilic substitution* takes place in the benzene rings, usually at the positions *p* to the hetero-atoms. Either 3- or 3,6-di-substituted products are thus obtained.

Carbazole (dibenzopyrrole) is the most important of these compounds and can be separated from crude anthracene obtained from coal-tar by forming its potassium salt. Carbazoles are usually made synthetically by the **Graebe–Ullmann synthesis**. Carbazole itself is obtained in this way from **1-phenyl-1,2,3-benzotriazole**, which, when heated, loses nitrogen to give an almost quantitative yield of **carbazole**. 1-Phenyl-1,2,3-benzotriazole is made by diazotisation of *o*-aminodiphenylamine followed by intramolecular ring-closure:

o-Aminodiphenylamine   1-Phenyl-1,2,3-benzotriazole   Carbazole

Carbazole has industrial use in the manufacture of blue sulphur dyestuffs, *e.g. Hydron Blue R* (see p.791).

N-Vinylcarbazole, which is obtained from carbazole and acetylene in the presence of a mixture of potassium hydroxide and zinc oxide as catalyst, is used for the manufacture of the *thermoplastic* **poly(vinylcarbazole)**; KRE: [1-(9-carbazoyl)ethylene]:

N-Vinylcarbazole   Poly(vinylcarbazole)

N-Vinylcarbazole has become of importance in synthetic work as a reactive *enamine*. *Dibenzofuran* and *dibenzothiophene* have not as yet found any significant uses.

7.3 Five-Membered Rings containing Two Nitrogen Atoms

If a methine group of pyrrole is replaced by a nitrogen atom then, depending upon whether this is at the 2- or the 3-position, the compounds are known as **pyrazole** or **imidazole**.

| Pyrrole (Azole) | Pyrazole (1,2-Diazole) | Imidazole (1,3-Diazole) |

Both pyrazole and imidazole are π-electron-rich heteroaromatic compounds. They can also be formulated as follows:

Pyrazole   Imidazole

Each of these two isomeric compounds is more basic than pyrrole, imidazole being the stronger base of the two. This *basicity* is associated with there being a further nitrogen atom in the ring. It is only necessary for one of the nitrogen atoms to contribute its lone pair of electrons to make up the *sextet of π-electrons* in the ring; the other lone pair is available to take up a proton.

The salts formed from either pyrazole or imidazole have mesomeric cations:

Pyrazole hydrochloride (readily hydrolysed)   Imidazole hydrochloride

Pyrazole and imidazole also have acidic properties, for example, like pyrrole they both form potassium salts. Loss of a proton gives rise to *anions* which are also *mesomerically stabilised*:

Pyrazolyl potassium   Imidazolyl potassium

It is thus apparent that neither in the cation nor in the anion of either of these heterocycles can the positive or negative charge be associated with one particular nitrogen atom. In solution in either acid or alkali the two nitrogen atoms in the ring are *equivalent*; in the cation the proton can shift from one nitrogen atom to the other. As a result it is impossible to differentiate between isomeric 3- and 5-substituted pyrazoles. Likewise substituents at the 4- and 5-positions of imidazole cannot be distinguished from one another.

7.3.1 Pyrazoles

For a long time no pyrazole derivative had been found in nature, but in 1959 β-[1-pyrazolyl]alanine was isolated from the seeds of water melons (*Citrullus lanatus*) (*L. Fowden*).

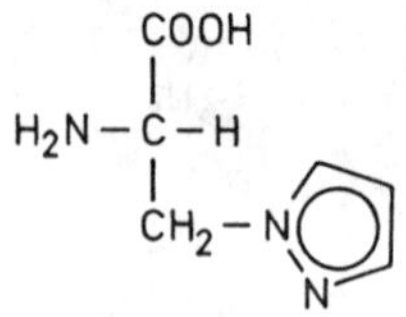

This amino acid is an isomer of histidine.

Preparation

7.3.1.1 Pyrazole derivatives are obtainable from 1,3-diketo-compounds and hydrazine, for example **3,5-dimethylpyrazole** from **acetylacetone**:

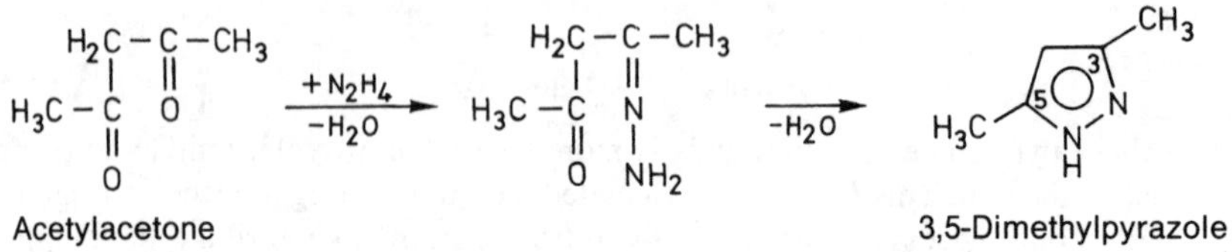

Acetylacetone   3,5-Dimethylpyrazole

Pyrazole itself can be obtained similarly, in 70% yield, from malonaldehyde diacetal and hydrazine (*Noyce*, 1955).

7.3.1.2 Substituted 5-aminopyrazoles can be obtained from *dichloromethyleneiminium salts* and *hydrazones* (*Viehe*, 1974), e.g.

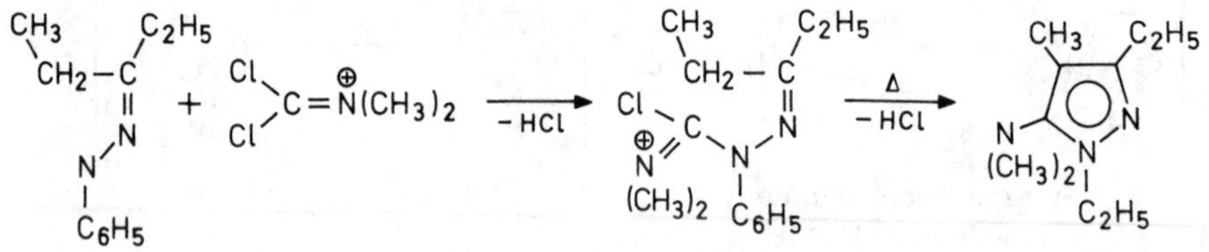

7.3.1.3 Esters of pyrazole-3,4,5-tricarboxylic acid result from the reaction of esters of acetylenedicarboxylic acid with diazoacetic acid esters:

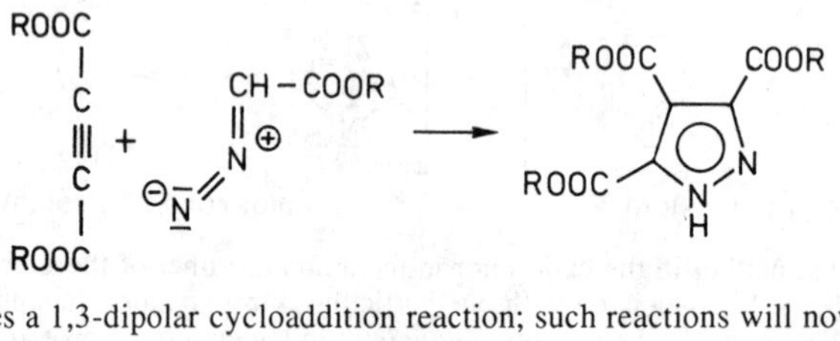

Reaction involves a 1,3-dipolar cycloaddition reaction; such reactions will now be discussed.

7.3.1.4   1,3-Dipolar Cycloaddition Reactions; (3 + 2)Cycloaddition Reactions[1]

Examples of this type of reaction are those of ozone with alkenes to form ozonides and of diazomethane with alkene and alkyne derivatives. They involve the reaction of a *mesomeric three-atom system* (a *1,3-dipole*) with an unsaturated *dipolarophile* to give a five-membered ring.

---

[1] See *R. Huisgen*, 1,3-Dipolar Cycloadditions, in *A. Padwa* (Ed.), 1.3-Dipolar Cyclisation Chemistry, Vol. 1, 1 (Wiley, New York 1984).

Mesomerically stabilised compounds such as *diazoalkanes, azides* and *ozone* can function as 1,3-dipoles; they can be represented by the following limiting structures:

$$\left[R-CH=\overset{\oplus}{N}=\bar{N}I^{\ominus} \longleftrightarrow R-\overset{\ominus}{\bar{C}}H-\overset{\oplus}{N}\equiv NI \longleftrightarrow R-\overset{\ominus}{\underset{|}{C}}H-N=\overset{\oplus}{\boxed{N}} \right]$$

$$\left[R-N=\overset{\oplus}{N}=\bar{N}I^{\ominus} \longleftrightarrow R-\overset{\ominus}{\bar{N}}-\overset{\oplus}{N}\equiv NI \longleftrightarrow R-\overset{\ominus}{\bar{N}}-\bar{N}=\overset{\oplus}{\boxed{N}} \right]$$

$$\left[I\bar{O}-\overset{\oplus}{\underset{}{O}}=\bar{O} \longleftrightarrow \bar{O}=\overset{\oplus}{\underset{}{O}}-\bar{O}I^{\ominus} \longleftrightarrow {}^{\ominus}I\bar{O}-\bar{O}-\underset{}{\boxed{\bar{O}}}^{\oplus} \right]$$

The heteroatoms which have been given a grey background have only a sextet of electrons; in each of their other limiting structures they are associated with an octet of electrons. These octet forms contribute more to the overall mesomeric structures than the sextet forms. The octet forms of the 1,3-dipoles may be compared to the limiting structures for allyl anions, but with the central carbon atom replaced by a positively charged heteroatom $X^{\oplus}$.

$$[\overset{\ominus}{\bar{C}}H_2-CH=CH_2 \longleftrightarrow CH_2=CH-\overset{\ominus}{\bar{C}}H_2] \quad [\overset{\ominus}{\bar{C}}H_2-\overset{\oplus}{X}=CH_2 \longleftrightarrow CH_2=\overset{\oplus}{X}-\overset{\ominus}{\bar{C}}H_2]$$

Allyl anion   1,3-Dipole

Dipolarophiles are compounds having multiple bonds, *e.g.*

$$\underset{}{>}C=C\underset{}{<}, \quad -C\equiv C-, \quad \underset{}{>}C=\bar{O}, \quad -C\equiv NI, \quad -\overset{\oplus}{N}\equiv NI.$$

In the course of forming a pyrazole ring this cycloaddition involves the loss of two π-bonds and the formation of two new σ-bonds. Kinetic studies show that in many cases formation of these bonds is a concerted one-step process.

Inasmuch as 1,3-dipolar cycloaddition reactions provide five-membered rings they can be regarded as falling between the dimerisation of alkenes to cyclobutane derivatives (see 1,2-cycloaddition, p.391) and *Diels–Alder* reactions (1,4-cycloaddition). For an account of the related cycloreversions see reference 1 below.

Properties

Pyrazole, m.p. 70°C (343 K), is a weak base with an odour akin to that of pyridine. It is stable to oxidising agents and to strong acids, and shows no tendency to polymerise.

In pyrazole the NH-group is electron-donating whilst the second, double-bonded nitrogen atom is electron-withdrawing, although more feebly. The overall result is that there is an excess of π-electrons on the carbon atoms, especially at the 4-position. In consequence, pyrazole undergoes *electrophilic substitution* at this position. For example, chlorination provides **4-chloropyrazole** and treatment with nitrating acid leads to the formation of **4-nitropyrazole**. This can be reduced to **4-aminopyrazole**, which can be diazotised to undergo coupling reactions.

4-Chloropyrazole   Pyrazole   4-Nitropyrazole   4-Aminopyrazole

---

[1]See *G. Bianchi*, 1,3-Dipolar Cycloreversions, Angew. Chem. Int. Ed. Engl. *18*, 721 (1979).

Pyrazole is fairly resistant to reducing agents. It is slowly reduced by sodium and an alcohol to **4,5-dihydropyrazole**. The latter compound is better obtained by the reaction of **acrolein** with hydrazine.

Acrolein   4,5-Dihydropyrazole

4,5-Dihydropyrazole is more strongly basic than pyrazole. It is readily attacked by oxidising agents, for example on treatment with bromine it is dehydrogenated back to pyrazole. Catalytic hydrogenation over palladium converts pyrazole into *pyrazolidine* (tetrahydropyrazole).

Both dihydropyrazoles and pyrazolidines are easily dehydrogenated and in consequence can be used as photographic developers. In particular *1-phenyl-3-pyrazolidone* (phenidone) is used for this purpose. It is made by letting acrylonitrile react with phenylhydrazine in alkaline conditions followed by acid hydrolysis:

3-Amino-1-phenyl-   1-Phenyl-3-pyrazolidone
4,5-dihydropyrazole   (Phenidone) (m.p. 121°C [394 K])

1-Phenyl-pyrazole-3(2*H*)-one

Dehydrogenation of 1-phenyl-3-pyrazolidone gives 1-phenylpyrazole-3(2*H*)-one.

7.3.1.5 Pyrazolone Derivatives

Pyrazolones are important derivatives of 4,5-dihydropyrazole. They are synthesised by intramolecular condensation reactions from hydrazine derivatives of β-ketoesters. For example, ethyl acetoacetate reacts with phenylhydrazine, with loss of water and ethanol, to give **3-methyl-1-phenyl-2-pyrazolin-5-one** (*L. Knorr*, 1883):

3-Methyl-1-phenyl-2-pyrazolin-5-one can exist in three *tautomeric* forms:

CH-form   OH-form   NH-form

In non-polar solvents such as chloroform it is principally in the CH-form, but in water it is an equilibrium mixture of 90% of the NH-form with 10% of the OH-form (*Katritzky*, 1964).

A 4,4-dimethyl derivative, which is a derivative of the CH-form, is obtained by treating the pyrazolinone with dimethyl sulphate in alkali. Nitrosation also takes place at the 4-position to give an isonitroso-compound (4-hydroxy-imino derivative). Oxidation with iron(III) chloride produces an *indigoid dyestuff*, **pyrazolone blue** (pyrazole blue, *Knorr*).

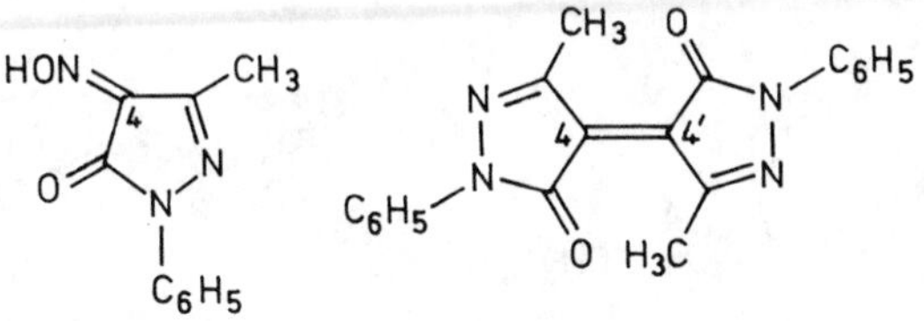

Isonitroso-compound (4-Hydroxyimino derivative)   Pyrazolone blue

Derivatives based on the OH-form include the *O*-methyl derivative, which is the main product obtained by methylation with diazomethane, and *O*-acyl derivatives. There are *N*-alkyl derivatives which arise from the NH-form.

The following derivatives are important *antipyretics* and *analgesics*.
INN: Phenazone, antipyrine, 2,3-dimethyl-1-phenyl-3-pyrazolin-5-one, is made by *N*-methylation of 3-methyl-1-phenyl-2-pyrazolin-5-one with methyl iodide in methanol under pressure at 100°C (375 K) (*Knorr*, 1884):

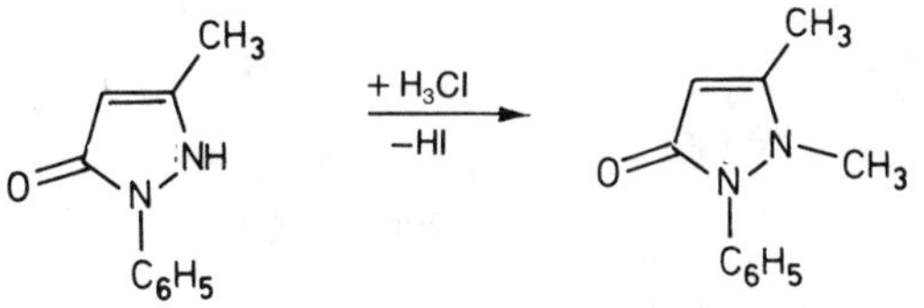

INN: Phenazone

Antipyrine

This compound melts at 114°C (387 K). It is readily soluble in water and in ethanol and has a bitter taste. With iron(III) chloride its aqueous solutions produce a brownish red colour.

Aminophenazone, 4-aminoantipyrine, pyramidone, 4-dimethylamino-2,3-dimethyl-1-phenyl-2-pyrazolin-5-one, is made from phenazone. Phenazone is first nitrosated to give a green 4-nitroso compound. The nitroso group is reduced and methylated with formaldehyde in the presence of formic acid (*Eschweiler – Clarke* reaction) to give aminophenazone:

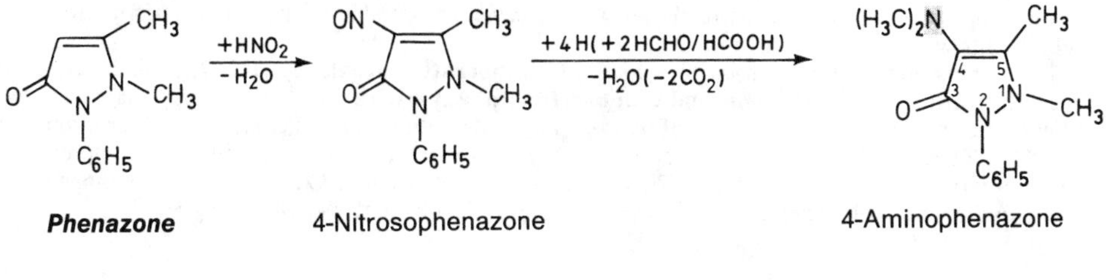

Phenazone   4-Nitrosophenazone   4-Aminophenazone

Because of the risk of some *N*-nitrosodimethylamine, which is carcinogenic, arising from the dimethylamino group of aminophenazone, the latter compound is no longer used in medicine. Its place has been taken very largely by 4-isopropylphenazone (*INN: **Propyphenazone***), wherein no such risk is involved. In replacing the nitrogen atom of a t-amino group by the corresponding substituted CH group (shaded grey in the formulae) it is often found that similar biological activity is retained (*Bioisosteres*)).

Condensation of hydrazobenzene with n-butylmalonic ester in the presence of a sodium alkoxide leads to the formation of the sodium salt of 4-n-butyl-5-hydroxy-1,2-diphenylpyrazolin-3-one, 4-n-butyl-1,2-diphenylpyrazolidine-3,5-dione, (*INN: **Phenylbutazone**, butazolidin*). This is an active antirheumatic agent but has dangerous long-term side-effects.

4-Isopropylphenazone   INN: **Phenylbutazone**

7.3.1.6 Pyrazole Azo-dyestuffs

This very light-fast set of dyestuffs arises by coupling pyrazolin-5-ones that have unsubstituted 4-positions with diazonium salts.

Tartrazine is an important yellow *dyestuff for wool* which is a member of this group. It is made by a cyclic condensation of an oxalacetate ester with *p*-sulphophenylhydrazine in an alkaline medium. This gives a 1-(*p*-sulphophenyl)-pyrazolin-5-one-3-carboxylic ester; coupling of this with diazotised sulphanilic acid provides tartrazine.

Tartrazine

A 4-unsubstituted *5-aminopyrazole* can be used instead of the pyrazolin-5-one (5-hydroxypyrazole) as the coupler in the preparation of these *azo-dyestuffs*. The products have characteristic greenish-yellow tints which are very fast to light.

Their preparation starts from *diacetonitrile* (**3-aminocrotononitrile**, see p.267). This reacts with phenylhydrazine to give ammonia and **cyanoacetone phenylhydrazone**. This intermediate can also be obtained in another way. Reaction of hydrogen cyanide with propargyl chloride (3-chloropropyne) at 30°C (300 K) in the presence of a copper catalyst provides propargyl cyanide (3-cyanopropyne) which rearranges to cyanoallene. This reacts with phenylhydrazine to give cyanoacetone phenylhydrazone. This latter compound cyclises to **5-amino-3-methyl-1-phenylpyrazole**:

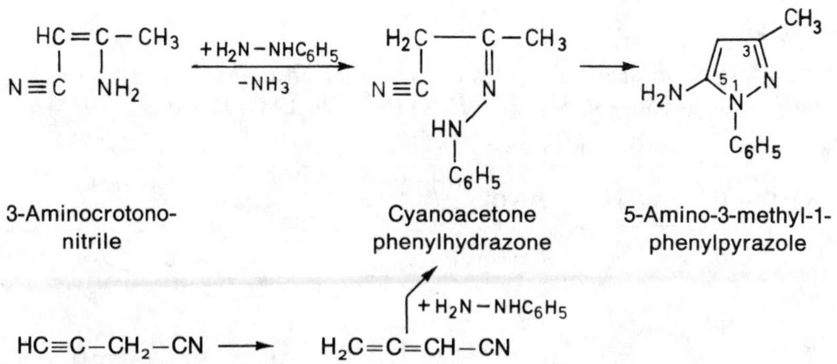

3-Aminocrotono-
nitrile

Cyanoacetone
phenylhydrazone

5-Amino-3-methyl-1-
phenylpyrazole

Propargyl cyanide (3-Cyanopropyne)   Cyanoallene

If the 5-amino-3-methyl-1-phenylpyrazole is allowed to react with an acetoacetic ester a substituted crotonic ester results and when this is heated in acetic acid it is cyclised to 4-*hydroxy-3,6-dimethyl-1-phenylpyrazolo[3,4-b]-pyridine*:

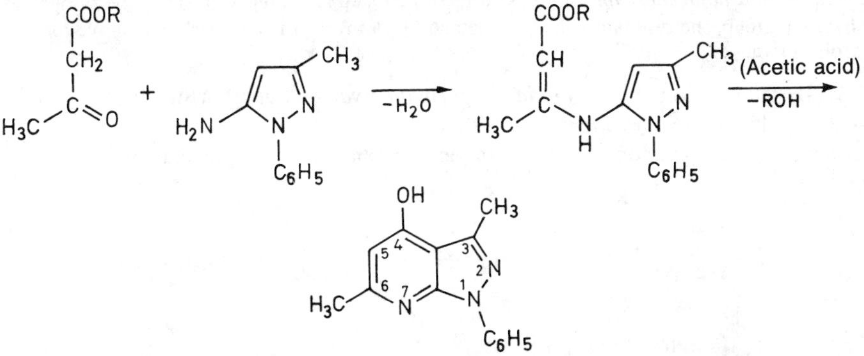

This bicyclic compound is used as the *coupling component* in the preparation of *disperse dyestuffs* that have high fastness to heat and are used for dyeing polyester fibres (see p.600).

7.3.1.7 Indazoles, in which a benzene ring and a pyrazole ring are fused together, are important in the synthesis of *dyestuffs* for use on poly(acrylonitrile) fibres.

Indazoles (benzopyrazoles) arise from the spontaneous cyclisation of *o*-acylphenylhydrazines:

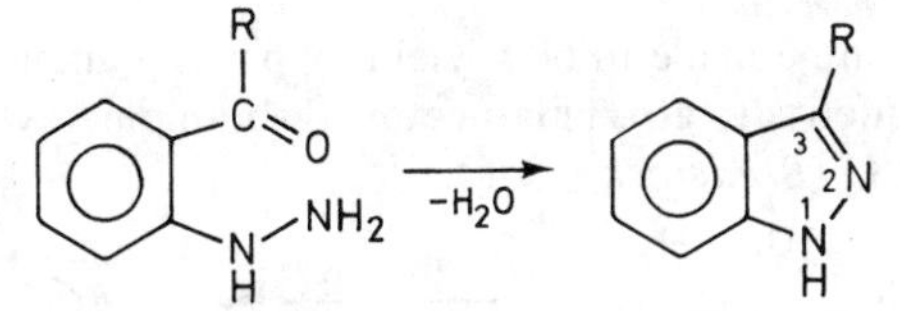

o-Acylphenylhydrazine   3-Alkylindazole

7.3.2   *Imidazoles[1]*

The imidazole ring system is present in a number of natural products, and especially in the purines.

[1] See *M. Grimmett*, Adv. Heterocycl. Chem. *27*, 241 (Academic Press, New York 1980); *K. T. Potts* (Ed.), Comprehensive Heterocyclic Chemistry, Vol. 5 (Pergamon Press, Oxford 1984).

Preparation

7.3.2.1 4,5-Disubstituted imidazoles can be prepared in good yield by treating **acyloins** with *formamide* at 150–180°C (420–450 K) (*Novelli*, 1939; *Bredereck*, 1953).

The reaction *mechanism* is as follows:

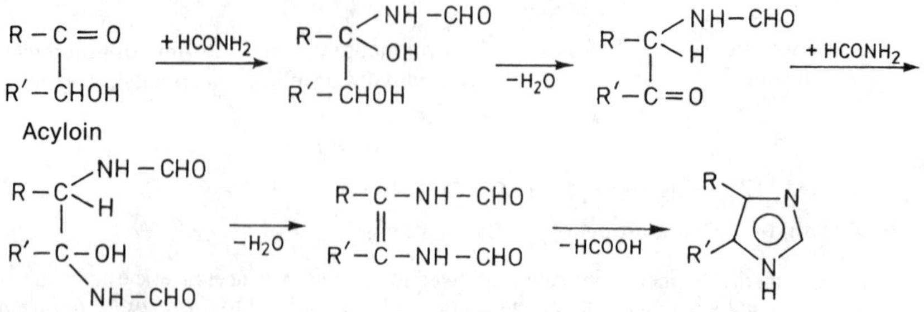

4,5-Disubstituted and 4(5)-monosubstituted imidazoles can be made similarly by the reaction of *formamide* with α-*halogenoketones*. In this case the first step is the replacement of the halogen atom by a hydroxy group; the resultant α-hydroxyketone (acyloin) then reacts as shown above to give an imidazole derivative.

7.3.2.2 By the reaction of *formamide* and *formaldehyde* on **benzil** or substituted benzils at 180–200°C (450–470 K) (*Bredereck*, 1959).

Formamide loses ammonia at this temperature and the reaction can be portrayed as follows:

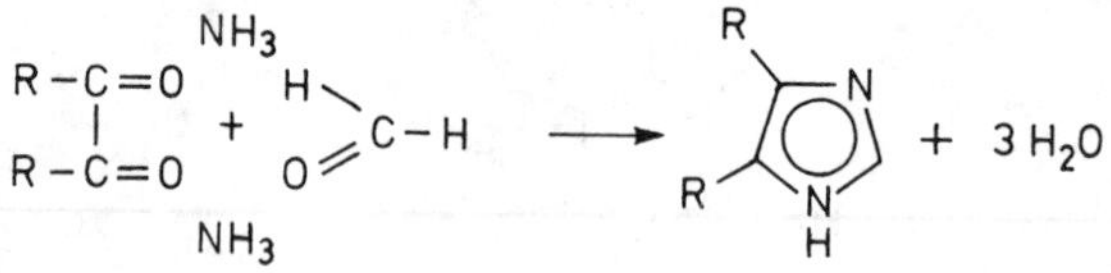

Benzil (R = C₆H₅)

7.3.2.3 By *condensation of α-halogenoketones with amidines* (*H. Beyer*, 1970):

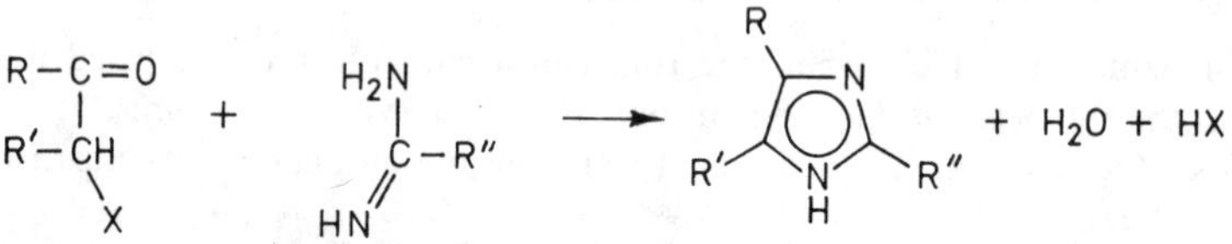

7.3.2.4 *Imidazole* itself can be made in 60% yield by passing ammonia through a heated mixture of **bromoacetaldehyde ethylene-acetal** (2-bromomethyl-1,3-dioxolane) and formamide at 180°C (450 K) (*Bredereck*, 1958).

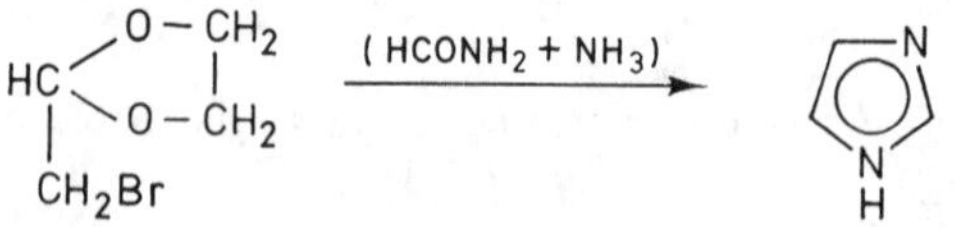

Bromoacetaldehyde ethylene-acetal   Imidazole

Properties

Imidazole (an older name was glyoxaline) is a colourless substance, m.p. 90°C (363 K) which is readily soluble in both water and ethanol. It forms stable salts with mineral acids. The markedly higher basicity of imidazole compared to pyrazole is attributable to its cyclic amidine structure.

As in pyrazole, the electron-density on the carbon atoms in imidazole is lower than it is in pyrrole. *Electrophilic substitution* (halogenation, nitration, sulphonation) takes place mainly at the 4- or 5-positions, but no reaction occurs with acid chlorides under *Friedel–Crafts* conditions. Diazonium cations couple with imidazole, above all at the 2-position; in alkaline solution orange, red or blue *azo-dyestuffs* are formed (*Pauly* reaction).

N-Acetylimidazoles are readily hydrolysed and can be used as *acetylating agents* (*Staab*, 1956).[1] It is not possible to hydrogenate imidazoles and *4,5-dihydroimidazoles* and *imidazolidines* have to be made synthetically.

7.3.2.5 The Most Important Imidazole Derivatives

The amino acid histidine is an imidazole derivative.

L(–)-**Histidine**, β-(4-imidazolyl)alanine is widely distributed as a protein building-block. The histidine content of the proteins present in blood amounts to 11%, so that L-histidine can be isolated from blood. Under the influence of putrefying bacteria histidine is readily broken down into **histamine**, 2-(4-imidazolyl)ethylamine:

L-Histidine (decomp. 287°C [560 K])   Histamine (m.p. 85°C [358 K])

Histamine is a biogenic amine, *i.e.* it is concerned with the processes of life; it has many-faceted activities. It expands the blood vessels, contracts the smooth muscular system, increases glandular secretion and behaves as a neurotransmitter. Its release is also responsible for allergic symptoms (hay fever).

Remedial materials, which neutralise the action of histamine by displacing it from the receptor, are known as *antihistamines*. Examples include ethylenediamine derivatives such as *INN: Antazoline* (antistine) or ethanolamine derivatives such as *INN: Diphenhydramine* (Benadryl). Because of its sedative and antiemetic action the latter is being used increasingly. The 2-aminoethylmercaptan derivative, *INN: Cimetidine* (Tagamet) has gained a lot of use in ulcer therapy since it stops the acid secretion of the stomach at its source.

INN: *Antazoline*   INN: *Diphenhydramine*   INN: *Cimetidine*

---

[1]See Synthesen mit heterocyclischen Amiden (Azoliden), Angew. Chem. *74*, 407 (1962); Neuere Methoden der präparativen Organischen Chemie, *5*, 53 (1967).

Other imidazole and imidazoline derivatives have become important as medicaments, *e.g.* 1-(2-hydroxyethyl)-2-methyl-5-nitroimidazole (*INN:* **Metronidazole**, *Clont*) against *Trichomonas* infections, especially *Trichomonas vaginalis*, and 1-(*o*-chloro-α,α-diphenylbenzyl)imidazole (*INN:* **Clotrimazole**, *Canesten*) as an antimycotic against fungal infections. 2[(2,6-dichlorophenyl)imino]imidazoline (*INN:* **Clonidine**, *Dixarit*, *Catapresan*) is used to treat high blood pressure and migraine.

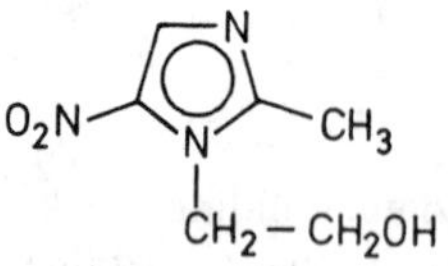

INN: **Metronidazole** 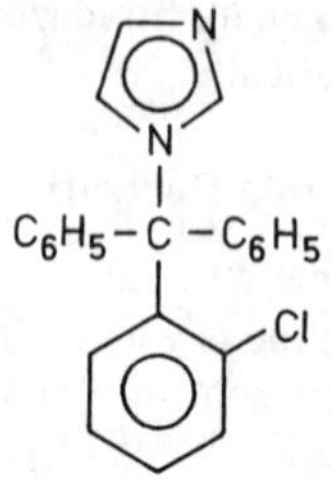   *INN:* **Clotrimazole** 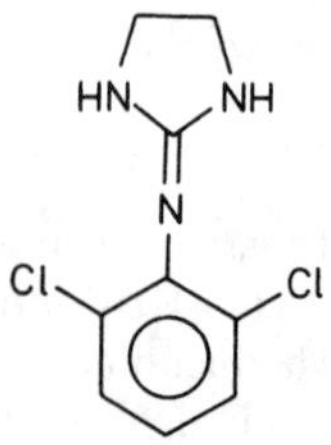   *INN:* **Clonidine**

7.3.2.6 Hydantoin (imidazolidine-2,4-dione) is made from the reaction of glycine with potassium cyanate (*cf.* Wöhler's synthesis of urea) and loss of water from the hydantoic acid which is formed:

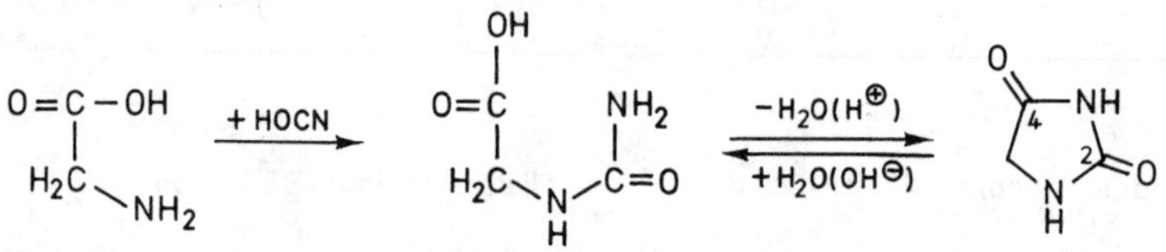

Glycine   Hydantoic acid   Hydantoin (m.p. 220°C [493 K])

Allantoin, a decomposition product from uric acid, and *creatinine* both contain hydantoin nuclei. The methylene group at position 5 of hydantoin is *reactive* and reacts with aldehydes or ketones to give condensation products.

Some hydantoin derivatives are pharmacologically active, *e.g.* **5,5-diphenylhydantoin** (*INN:* **Phenytoin**, *Zentropil*, *Epanutin*, *Phenhydan*) is used for severe epilepsy.

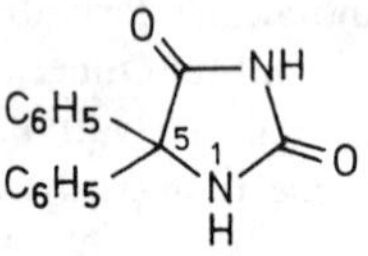

5,5-Diphenylhydantoin

7.3.2.7 Parabanic acid (imidazoline-2,4,5-trione, oxalylurea) is made from oxalyl chloride and urea:

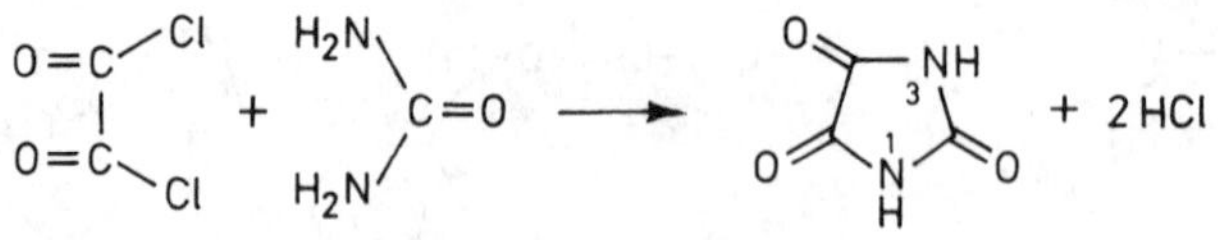

Parabanic acid

Another derivative of imidazolidine is biotin, in which an imidazolidone ring and a thiolane ring are fused together.

7.3.2.8 Biotin (vitamin H) belongs to the group of biological growth agents which regulate the growth of cells, and is an indispensable factor for the propagation of yeast and of other microorganisms. In both plants and animals it occurs only in very low concentrations. In humans it has been found to be essential for the normal functioning of the skin and it is called *vitamin H*. The basic protein *avidine*, which occurs in raw eggs, binds to biotin irreversibly and acts as an *antivitamin*.

Biotin was first isolated from egg yolk (*Kögl*, 1936) and later from liver (*du Vigneaud*). It contains three chiral centres (*) and can therefore in theory exist in eight stereoisomeric forms. Naturally occurring biotin forms colourless needles and is *dextrorotatory*. Its two rings are *cis*-fused. Its structure has been confirmed by several syntheses. The synthesis from D-mannose which was described on p.450 and another starting from L-cysteine can be represented by retrosyntheses:

| L-Cysteine | (+)-Biotin | α-D-Mannose |

Both syntheses start from materials from the 'chiral pool'. In that starting from mannose, the three chiral centres of biotin arise from the 'chiral template' it provides; they are represented in the formula above by the three C—H bonds drawn with heavy wedges, which are derived from the three HCOH-groups of mannose whose OH are marked with heavy type on p.450. In contrast, in the case of the synthesis from cysteine only the one chiral centre of biotin, given a grey background in the above formula, is derived from the chiral template provided by cysteine. The other two chiral centres are generated in the synthesis by means of stereoselective reactions.[1]

The chiral building blocks which are prepared in the course of the synthesis by means of derivatisation, activation and transformation of the chiral templates (see p.450, I–IV) are described as *chirons*.[2]

7.3.2.9 **Benzimidazoles** have a benzene ring fused to an imidazole ring. They can be made by condensing *o*-phenylenediamine with anhydrous carboxylic acids; *benzimidazole* itself is made from formic acid:

| *o*-Phenylenediamine | Benzimidazole | Carbendazim |

It is a colourless substance, m.p. 170°C (443 K). Its 5,6-dimethyl derivative forms part of the structure of vitamin B_{12}. Methyl benzimidazole-2-carbamate (BMC, Carbendazim) is approved for use in protecting plants destined to be foodstuffs.

7.4 Five-Membered Rings with Two Different Heteroatoms

Five-membered ring compounds of this sort are derived from furan or thiophene by replacing either the 2- or 3-methine group by a nitrogen atom. The parent compounds are called *oxazole, isoxazole, thiazole* and *isothiazole*.

---

[1]See *P. N. Confalone* et al., Stereospecific Total Synthesis of d-Biotin from L(+)-Cysteine, J. Am. Chem. Soc. 99, 7020 (1972).

[2]See *S. Hanessian*, Total Synthesis of Natural Products. The Chiron Approach (Pergamon Press, Oxford 1983).

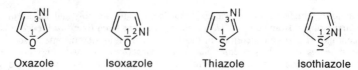

| Oxazole | Isoxazole | Thiazole | Isothiazole |

All of these ring systems are examples of π-electron-rich heteroaromatic compounds.

7.4.1 Oxazoles[1]

Oxazole was first obtained by decarboxylation of synthetically prepared oxazole-4-carboxylic acid (*J. W.* and *R. H. Cornforth*, 1947). A later synthesis started from oxazole-4,5-dicarboxylic acid.[2]

Preparation

7.4.1.1 *Treatment of acylated aminoketones with phosphorus(V) chloride* or thionyl chloride provides oxazole derivatives (*Gabriel*):

$$
\text{An acylaminoketone} \xrightarrow[-H_2O]{(PCl_5)}
$$

This reaction resembles the formation of furans from 1,4-diketones.

Acylated aminoketones are best made by the reduction of hydroxyiminoketones (isonitrosoketones) with zinc powder and acetic acid in the presence of acetic anhydride (*A. Treibs*), e.g.

$$
H_3C-CO-\overset{\overset{\displaystyle H}{|}}{C}=NOH + 2(H_3CCO)_2O + H_3CCOOH + 2\,Zn
$$

Hydroxyiminoacetone

$$
\rightarrow H_3C-CO-CH_2-NH-CO-CH_3 + 2(H_3CCOO)_2Zn
$$

Acetylaminoacetone

7.4.1.2 By heating *acylated acyloins with ammonium acetate* in acetic acid (*Theilig*, 1953):

$$
\begin{array}{c} R-C=O \quad O \\ | \qquad \quad \| \\ R'-CH \quad C-R'' \\ \diagdown_{\ \ \ O} \end{array} \xrightarrow{(H_3CCOONH_4)}
$$

7.4.1.3 Oxazole derivatives can also be made from α-halogenoketones and formamide or other acid amides by heating them together at 100°C (375 K) in the absence of a solvent (*Theilig*, 1953). First of all an imido-ester hydrochloride is formed which either cyclises to an oxazole or gives an α-hydroxyketone; the latter reacts with formamide giving an imidazole (see p.740). Addition of sulphuric acid promotes formation of the oxazole.

[1] See *J. W. Cornforth* in *R. C. Elderfield*, Heterocyclic Compounds, Vol. 5, 298 (Wiley, New York 1957); *I. J. Turchi* et al., Chem. Rev. 75, 389 (1975); *G. V. Boyd*, Comprehensive Heterocyclic Chemistry, Vol. 6, 177 (Pergamon Press, Oxford 1984); *I. J. Turchi* (Ed.), Oxazoles (Wiley, New York 1986).

[2] See *H. Bredereck* et al., Synthese des Oxazols, Chem Ber. 97, 1414 (1964).

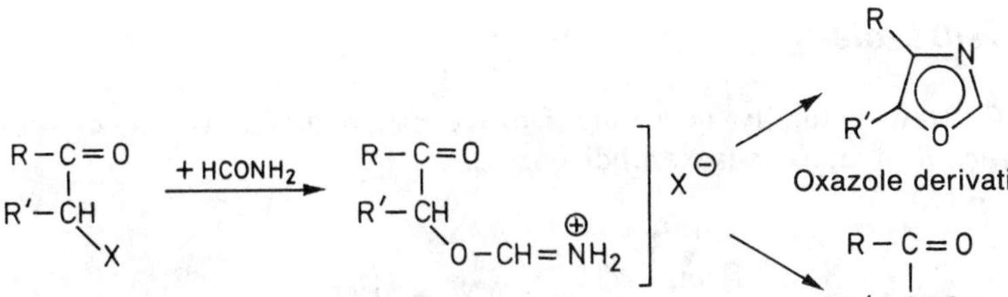

$$R-C=O \atop R'-CH-X \quad \xrightarrow{+ HCONH_2} \quad {R-C=O \atop R'-CH-O-CH=\overset{\oplus}{N}H_2}$$

α-Halogenoketone   Imido-ester
                                 hydrochloride

Oxazole derivative

$$R-C=O \atop R'-CHOH$$

α-Hydroxyketone

Properties

Oxazole is a liquid which boils at 70°C (343 K). Its derivatives are weak bases which smell rather like pyridine. They are stable to alkali. Their rings are relatively easily opened by heating them with strong acids; in this they are appreciably less stable than imidazoles. The reduction products of oxazole, **4,5-dihydro-oxazole** and **oxazolidine**, have been of use in unravelling the constitution of proteins. Oxazolidines are prepared by reducing oxazoles with sodium and an alcohol. **2-Oxazolidinone** (2-oxazolidone) is an oxo-derivative of oxazolidine.

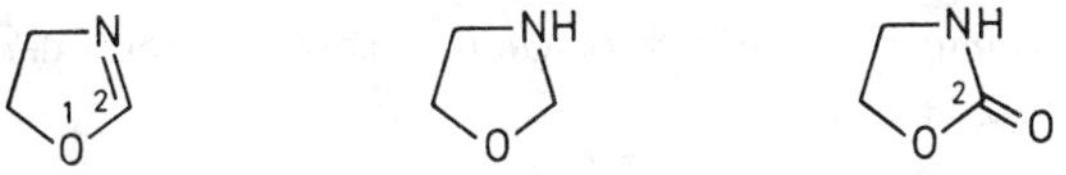

4,5-Dihydro-oxazole   Oxazolidine   2-Oxazolidinone

4,5-Dihydro-oxazoles can be made by the action of thionyl chloride on the appropriate 2-acylaminoalcohol, *e.g.*

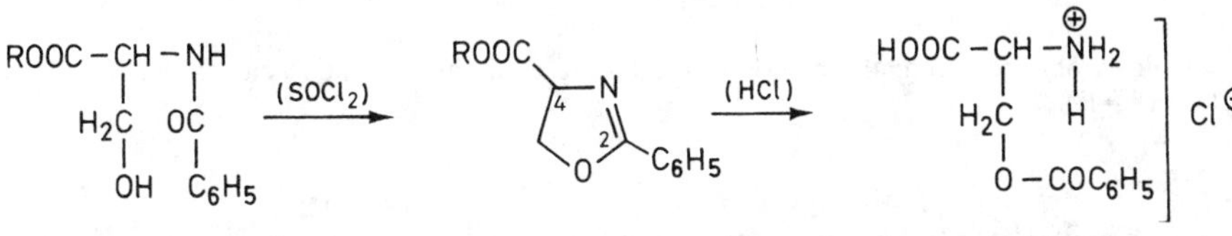

N-Benzoylserine   2-Phenyl-4,5-dihydro-oxazole-4-   o-Benzoylserine
methyl ester   carboxylic acid ester   hydrochloride

4,5-Dihydro-oxazoles are hydrolysed to open-chain O-acyl-derivatives by dilute mineral acid; they are stable to alkali.

Oxazolidines and 2-oxazolidinones are obtained by the reaction of 2-aminoalcohols with, respectively, aldehydes or phosgene, *e.g.*

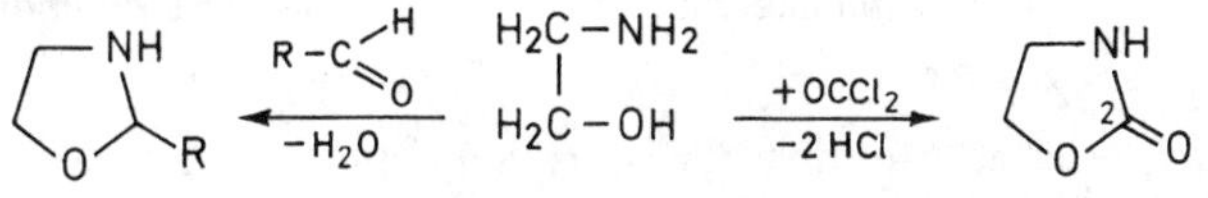

2-Alkyloxazolidine   2-Aminoethanol   2-Oxazolidinone

7.4.2   Isoxazoles[1]

Among the few naturally occurring isoxazole derivatives is the *antibiotic INN:* **Cycloserine**, (*R*)-4-amino-3-isoxazolidinone.

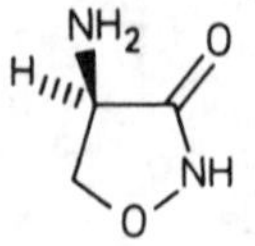

INN: Cycloserine

Preparation

7.4.2.1   The reaction of *1,3-dicarbonyl compounds* with *hydroxylamine* provides isoxazole derivatives:

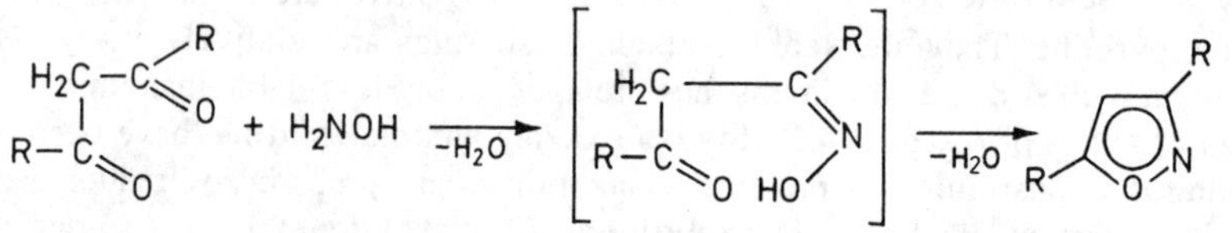

This standard method resembles that for the preparation of pyrazoles (see p.734). Isoxazole itself is obtained by using a malonaldehyde diacetal.

7.4.2.2   By a *1,3-dipolar cycloaddition reaction* of acetylene (or a derivative) with a nitrile oxide:

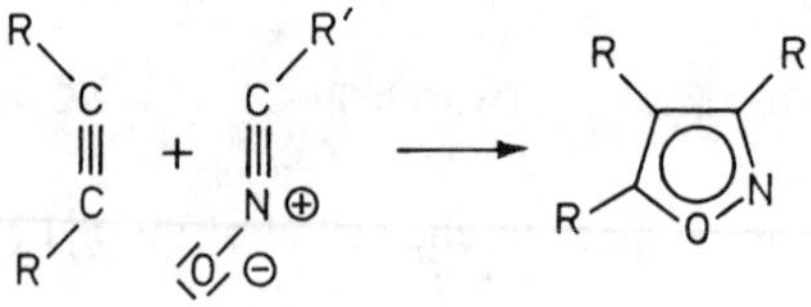

Nitrile oxide

Isoxazole is obtained from the reaction of fulminic acid (hydrocyanic acid oxide, R′=H) with acetylene (R=H).

Properties

Isoxazole is a liquid, b.p. 95°C (368 K) which has a strong smell reminiscent of pyridine. This ring system is stable to acid but is cleaved by bases. Reduction of isoxazoles usually leads to opening of the ring:

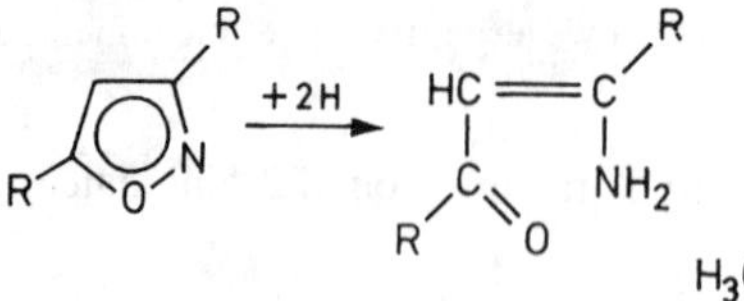

(*β*-Aminoketone) 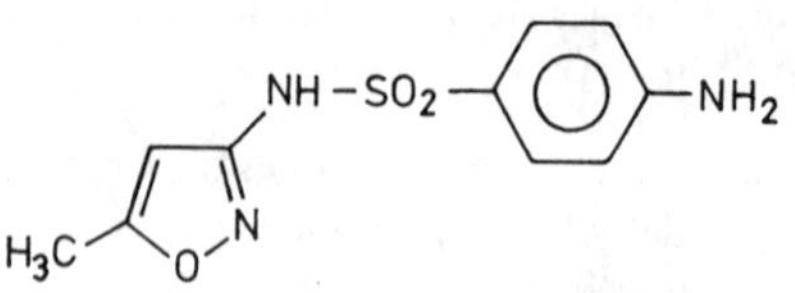   INN: **Sulphamethoxazole**

[1]See *B. J. Wakefield* et al., Isoxazole Chemistry Since 1963, Adv. Heterocycl. Chem. **25**, 147 (Academic Prs, New York 1979); *S. A. Lang* and *Yi Lin*, Comprehensive Heterocyclic Chemistry, Vol. 6, 1 (Pergamon Press, Oxford 1984).

N'-(5=Methylisoxazol-3-yl)sulphanilamide (*INN*: **Sulphamethoxazole**) has use as a *slow-acting sulphonamide.*

7.4.3   Thiazoles[1]

The thiazole ring is of biochemical importance, especially as a constituent of vitamin B$_1$ and of the coenzyme *carboxylase*. There is also a thiazolidine ring present in the *penicillins.*

Preparation

7.4.3.1   Hantzsch Thiazole Synthesis. Thiazole derivatives result from the condensation of α-halogenoketones with various sorts of thioamides. The mechanism is as follows and involves a nucleophilic substitution of the halogen atom by sulphur as the first step:

(X = halogen)

Thiazole itself is made in this way from chloroacetaldehyde and thioformamide. Rather than chloroacetaldehyde it is better to use α,β-dichloroethyl acetate, $ClCH_2$— CHCl— O —COCH$_3$, which is obtainable by the addition of chlorine to vinyl acetate, and which on hydrolysis yields chloroacetaldehyde. *Thioformamide* is obtained in 90% yield from ethyl thioformate and ammonia at 0°C (273 K).[2]

7.4.3.2 By treating acylated aminoaldehydes or aminoketones with tetraphosphorus decasulphide (*Gabriel*), e.g.

Acetylaminoacetone   2,5-Dimethylthiazole

This process is reminiscent of the preparation of thiophene derivatives from 1,4-dioxo compounds.

7.4.3.3   Preparation of 2,5-Dihydrothiazoles. These derivatives are obtained in good yield by the action of elemental sulphur and gaseous ammonia on the appropriate carbonyl compounds, such as isobutyraldehyde (2-methylpropanal). This reaction makes possible an industrial preparation of **D-penicillamine** (*Asinger* process):[3]

---

[1] See *J. V. Metzger*, Thiazole and its Derivatives, in *A. Weissberger* et al., The Chemistry of Heterocyclic Compounds, Vol. *34*, parts 1–3 (Interscience, New York 1979); *J. V. Metzger*, Comprehensive Heterocyclic Chemistry, Vol. 6, 235 (Pergamon Press, Oxford 1984).
[2] See *R. Mayer* et al., Einfache Synthese der Thioformamide (A Simple Synthesis of Thioformamide, Z. Chem. *4*, 457 (1964).
[3] See *W. M. Weigert* et al., D-Penicillamine — Production and Properties, Angew. Chem. Int. Ed. Eng. *14*, 330 (1975).

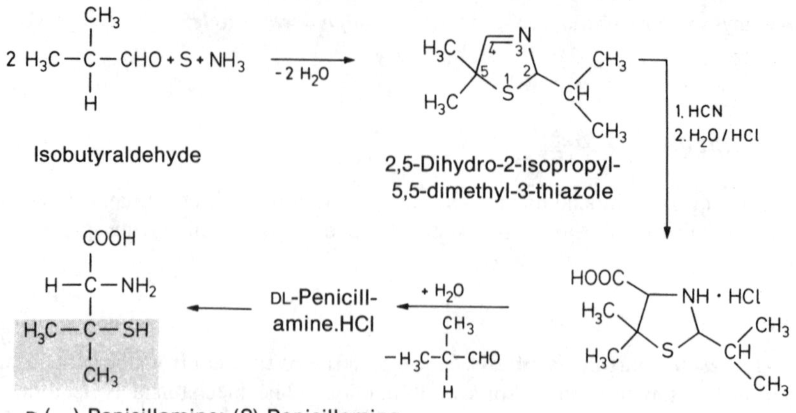

Isobutyraldehyde

2,5-Dihydro-2-isopropyl-
5,5-dimethyl-3-thiazole

1. HCN
2. H₂O / HCl

DL-Penicill-
amine.HCl

D-(−)-Penicillamine: (S)-Penicillamine

The ring-opening of the 2-substituted thiazolidine ring, which is very resistant to hydrolysis, is brought about by azeotropic removal of isobutyraldehyde with steam. Resolution of the racemic form is achieved by salt formation with a chiral base, *e.g.* L-lysine. It should be noted that in applying the *CIP* system the $(H_3C)_2CSH$ group (shown with a grey backing) takes precedence over the COOH group.

D-(−)-Penicillamine [D-(−)-2-amino-3-mercapto-3-methylbutyric acid, D-(−)-β,β-dimethylcysteine, D-(−)-β-mercaptovaline] was formerly obtained from penicillin, which contains this structural group (shadowed with grey in the formula of penicillin). The L-enantiomer is poisonous, which means that the racemic form cannot be used for therapeutic purposes. D-Penicillamine (*Metalcaptase, Trolovol*) is effective as a chelating agent for heavy metals; it also has use in the treatment of chronic arthritis.

7.4.3.4 Synthesis of Rhodanine.

When sodium chloroacetate reacts with ammonium dithiocarbamate and the intermediate is treated with hydrochloric acid, cyclisation takes place to give rhodanine:

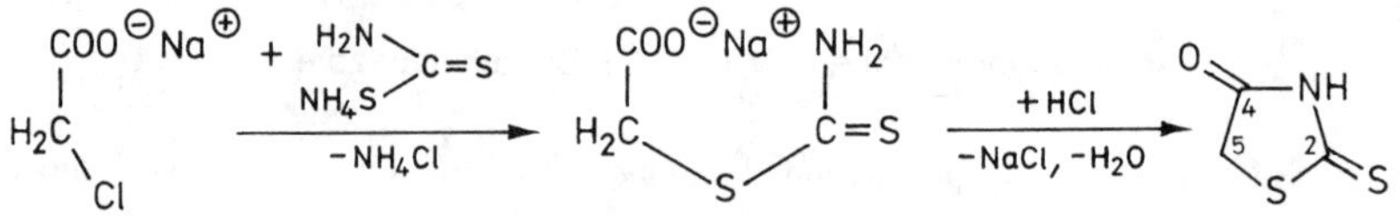

Rhodanine

Rhodanine, 2-thioxo-4-thiazolidinone, has a reactive 5-methylene group which makes it a *reagent* for aldehydes or nitroso compounds. It is also a useful synthetic agent.

Properties

Thiazole is a liquid, b.p. 117°C (390 K), which is very similar to pyridine (b.p. 115°C [388 K]) in its properties. When dissolved in water it gives a neutral solution but with mineral acids it forms stable salts, which are partially hydrolysed in solution. *Electrophilic substitution* (nitration, halogenation, sulphonation) is possible only at the 5-position and only if there is an electron-donating substituent such as NH₂ or OH, which shows a + M-effect and increases the electron density in the ring, at the 2-position. Thiazole is resistant to oxidising agents, even hot nitric acid has little effect. It is either unaffected by reducing agents or at most some ring-opening occurs. It is therefore not possible to make *dihydrothiazoles* or *thiazolidines* by catalytic hydrogenation but they must needs be prepared by means of condensation reactions.

7.4.3.5 2-Aminothiazole is best made, in 88% yield, by the reaction of one equivalent of vinyl acetate with thiourea in the presence of sulphuryl chloride at 80°C (350 K) (see p.747, method 7.4.3.1).

7.4.3.6 β-Lactam Antibiotics.[1] In 1928 *Fleming* discovered that certain mould fungi, especially *Penicillium notatum*, were active as *antibiotics* and cured bacterial infections (see p.848). Determination of the structure of natural **penicillin** showed that it had the following formula:

| Pencillin | Cephalosporin | *INN: Imipenem* |

The *penam system* (4-thia-1-azabicyclo[3.2.0]heptan-7-one) is based on a thiazolidine ring to which is fused a β-lactam ring. The penicillins known to date differ only in the identity of the substituent R. The dotted line shows that the ring system contains D-valine and L-cysteine moieties. The same building blocks are found in the ring system of **cephalosporin**, in which one of the methyl groups from valine has become part of the dihydrothiazine ring. In carbopenems, *e.g. INN: Imipenem*, the sulphur atom of the penem system has been replaced by a carbon atom, leading to a broadening of the antibacterial spectrum.

The penicillamine part of the penicillin formula has been given a grey background.

Of the naturally occurring penicillins, as yet only *penicillin G* (benzylpenicillin, $R = CH_2—C_6H_5$) has been used medicinally, in the form of a potassium salt or as the difficultly soluble *N,N'*-dibenzylethylenediamine salt (*Benzathin-Penicillin G, Tardocillin*).

Chemical or enzymatic removal of the phenylacetyl group (given a shaded background in the formula) from penicillin G produces *6-aminopenicillanic acid* (6-APA), which has a free amino group onto which other side-chains can be introduced by acylation. Of these half-synthetic antibiotics, which can be taken orally, the following are examples:

| *INN: Ampicillin* (D-form) n) | (S)-3-Aminobactamic acid | *INN: Aztreonam* |

α-Phenoxymethylpenicillin (*INN: Phenoxymethylpenicillin, Beromycin, Penicillin V*), $R = CH_2OC_6H_5$
α-Phenoxymethylpenicillin (*INN: Propicillin, Baycillin*), $R = CH(C_2H_5)–OC_6H_5$

---

[1] See *J. C. Sheehan*, The Enchanted Ring, The Untold Story of Penicillin (MIT, Cambridge, MA, USA 1983); *W. Dürckheimer* et al., Recent Developments in the Field of β-Lactam Antibiotics, Angew. Chem. Int. Ed. Engl. *24*, 180 (1985).

At present, the partially synthetically obtained *INN*: **Ampicillin** (*Amblosin, Binotal*) has considerable importance. It is stable to acid and can be used orally, and it is also stable to the bacterial enzyme penicillinase. It can be used against both gram-positive and gram-negative bacteria, *i.e.* it is a 'broad-band antibiotic'. Other penicillins which have been developed in recent years are for the most part chemically similar to ampicillin. (For the mode of action of penicillins, see p.847).

In the course of a large concerted investigation of microorganisms and their abilities to produce new active β-lactams, *bacteria* were found in 1981 which contained a *monocyclic* β-lactam ring. These are derivatives of the zwitterionic (*S*)-*3-aminobactamic acid*. By changing the basic framework and also the aminoacyl side-chain, INN: **Aztreonam** (*Azactam*) was discovered, which is very active against gram-negative aerobic bacteria.

7.4.3.7 **Benzothiazoles** contain a thiazole ring fused to a benzene ring, *e.g.*

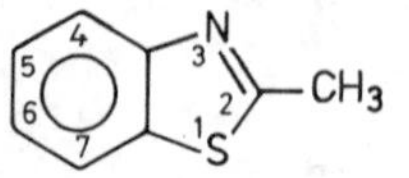

2-Methylbenzothiazole 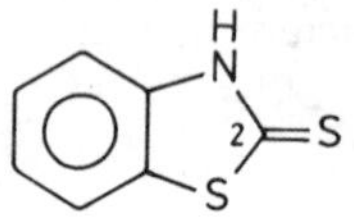   2-Mercaptobenzothiazole 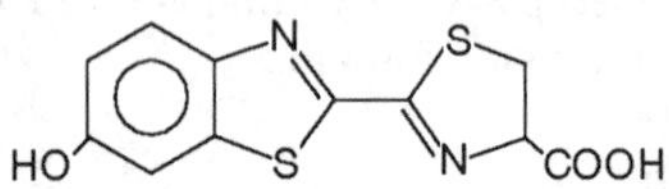   Luciferin

Benzothiazolium salts which have a reactive methyl group at the 2-position are of photochemical importance and are used in the preparation of polymethine dyestuffs. 2-Mercaptobenzothiazole, which actually exists as a 2(3*H*)-benzothiazolethione, is a vulcanisation accelerator (Captax) of industrial importance. The luminous substance of glow-worms is (*S*)-2-(6-hydroxy-2-benzothiazolyl)-2-thiazoline-4-carboxylic acid, luciferin.

7.4.3.8 **Polymethine Dyestuffs**

The photographic silver halide emulsion is sensitive only to ultraviolet, violet and blue light, and to extend its sensitivity to light of longer wavelength it must be spectroscopically sensitised. This can be done by adding certain light-sensitive dyestuffs, called *sensitisers*, which are able to transfer the light energy they receive to the crystal lattice of the silver halide.[1] Among such compounds the most important are the *polymethine dyestuffs*, which consist of a conjugated polymethine chain made up of an odd number of carbon atoms linking two nitrogen heterocycles which can function either as donor or acceptor groups in the conjugated π-electron system. These compounds are the **cyanine dyes** (cyanines):

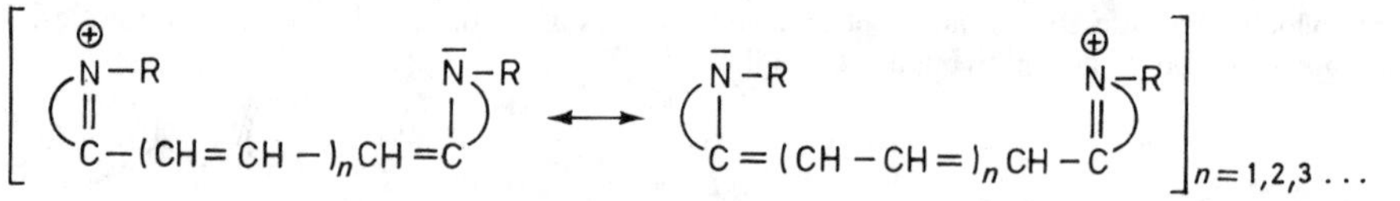

The absorption maxima of these cyanine dyes embrace the whole range of visible light and also extend into the near infrared (infrared photography) and ultraviolet (see Fig. 116, p.592).

Benzothiazolium salts are important heterocycles for use in sensitising dyestuffs. As an example, the following *cyanine dye* is made by condensing two molecules of **3-ethyl-2-**

---

[1] See *J. Bailey* and *B. A. J. Clark*, Uses of Photographic and Reprographic Techniques, Comprehensive Heterocyclic Chemistry, Vol. *1*, 361 (Pergamon Press, Oxford 1984); *H. Zollinger*, Color Chemistry: Syntheses, Properties and Application of Organic Dyes and Pigments (VCH Verlag, Weinheim and New York 1991).

methylbenzothiazolium iodide with one molecule of ethyl orthoformate in the presence of pyridine; the orthoformate ester provides the middle carbon atom of the polymethine chain:

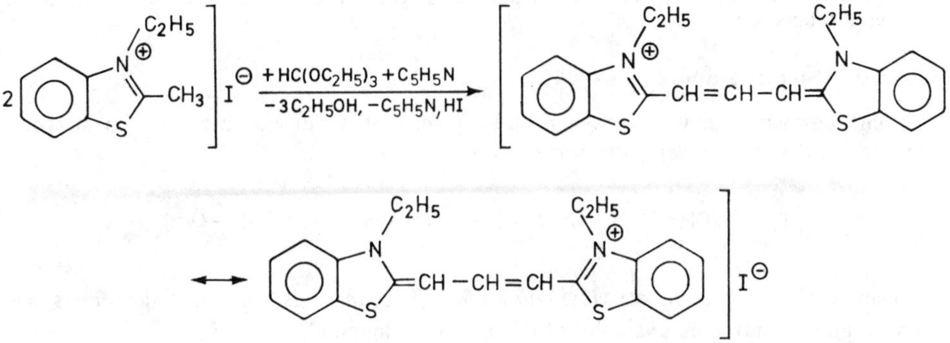

This particular benzothiazole dyestuff is a sensitiser for yellow and red light. If the hydrogen atom of the central methine group is replaced by a methyl group the resultant dyestuff sensitises the range of spectra from green to red (panchromatic film). By using appropriate starting materials it is possible to make polymethine dyes containing eleven or more methine groups.

Cyanine dyestuffs can be thought of as vinylogous amidines, the merocyanine depicted in Fig. 116 (p.592) as a vinylogous amide, and the anion of malonaldehyde as a vinylogous formate ion.

7.4.3.9 Azacyanines

When light-sensitive cyanine dyes are absorbed onto *poly(acrylonitrile) fibres* this results in a surprising fastness to light. This observation prompted a much more close investigation into heterocyclic dyestuffs of this sort, the *azacyanines*. Their preparation is made difficult by the fact that heterocyclic amines are rarely easy to diazotise, and that often this is not possible at all. It is also necessary for the ring nitrogen atom to be quaternised. The best method of obtaining azacyanines involves *oxidative coupling*[1] of heterocyclic hydrazones with arylamines or phenols (*Hünig*, 1957).

The hydrazones can be made in a variety of ways, for example by methylating 2-mercaptobenzothiazole with dimethyl sulphate and letting the resultant quaternary salt react with hydrazine, thus providing the hydrazone of N-methyl-2-benzothiazolinone. This can be oxidatively coupled, for example with N,N-dimethylaniline in solution in hydrochloric acid in the presence of potassium hexacyanoferrate(III) (or iron(III) chloride or hydrogen peroxide), and produces a violet **azo-dyestuff**:

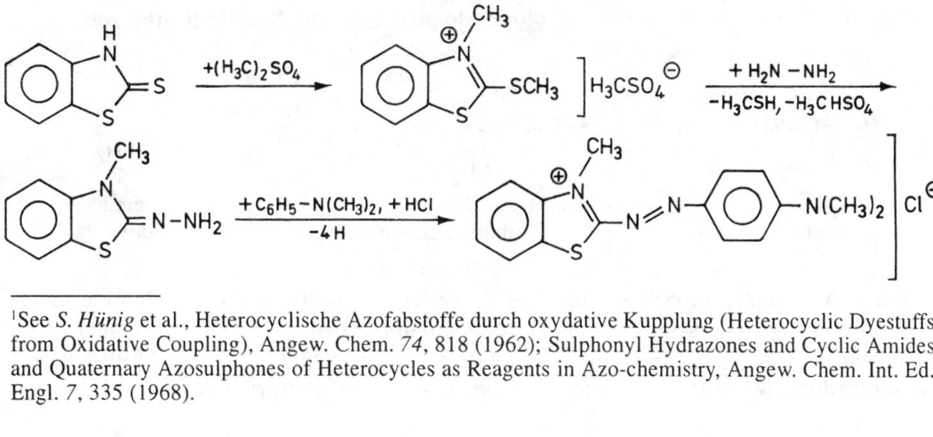

[1]See *S. Hünig* et al., Heterocyclische Azofabstoffe durch oxydative Kupplung (Heterocyclic Dyestuffs from Oxidative Coupling), Angew. Chem. *74*, 818 (1962); Sulphonyl Hydrazones and Cyclic Amides and Quaternary Azosulphones of Heterocycles as Reagents in Azo-chemistry, Angew. Chem. Int. Ed. Engl. *7*, 335 (1968).

As cationic dyestuffs, cyanines and azacyanines are useful for dyeing poly(acrylonitrile) fibres. They form salt-like links with anionic groups present (either end-groups or introduced as comonomers). The oxidative coupling reaction can serve as a *test* for the presence of *hydrazone groups* in heterocyclic systems.

7.4.3.10 Merocyanines

Chromogens which have an oxygen atom instead of a nitrogen atom in the electron-accepting group are known as *merocyanines*:

$$[>\bar{N}-(CH=CH)_n-CH\bar{\underset{.}{O}} \quad \longleftrightarrow \quad >\overset{\oplus}{N}=CH-(CH=CH)_n-\bar{\underset{.}{O}}|^{\ominus}]$$

They can be regarded as *vinylogues of carboxylic acid amides*. Indoanilines and indoaniline dyestuffs are examples of this type of compound.

7.4.4 Isothiazoles

Isothiazole and thiazole are related to pyridine in the same way that thiophene is related to benzene.

Preparation

7.4.4.1 Isothiazole is best obtained by a *gas-phase reaction between propene, ammonia and sulphur dioxide* over an aluminium oxide catalyst at 200°C (475 K):

$$H_3C-CH=CH_2 \quad \xrightarrow[(-H_2O,\,-H_2S)]{(NH_3+SO_2/Al_2O_3)} \quad \underset{\text{Isothiazole}}{\boxed{{}_{5}\overset{4\quad 3}{\bigcirc}{}_{2}N}}$$

Propene   Isothiazole

Alkenes with more than three carbon atoms can be used similarly.

7.4.4.2 The action of hydrogen sulphide on *diacetonitrile* (β-aminocrotononitrile, see p.267 f.) provides β-aminothiocrotonamide, which, on treatment with, for example, iodine in pyridine undergoes oxidative ring-closure to give **5-amino-3-methylisothiazole**:

$$\underset{\substack{| \\ NH_2}}{H_3C-C}=CH-C\equiv N \quad \xrightarrow{+H_2S} \quad \underset{\substack{| \qquad || \\ NH_2 \quad S}}{H_3C-C}=CH-\overset{}{C}-NH_2 \quad \xrightarrow[-2H]{(I_2)} \quad \underset{}{N\underset{S}{\bigcirc}NH_2}$$

β-Amino-   β-Amino-   5-Amino-3-methyl-
crotononitrile   thiocrotonamide   thiazole

7.4.4.3 A simple synthesis of the thiazole system is by the action of chlorine on the disodio-derivative of *1,1-dicyano-2,2-dimercaptoethylene*, itself obtained from malononitrile and carbon disulphide, in tetrachloromethane to give **3,5-dichloro-4-cyanoisothiazole**. Reaction proceeds via the intermediate as shown (*Hatchard*, 1964):

3,5-Dichloro-4-cyanoisothiazole

Treatment with ammonia converts this into the corresponding 5-aminoisothiazole derivative, which can be diazotised.

Properties

Isothiazole is a neutral liquid, b.p. 115°C (388 K), which smells like pyridine. It is miscible with organic solvents. *Electrophilic substitution* (halogenation, nitration, sulphonation) takes place mainly at the 4-position. 4-Nitroisothiazole can be reduced by iron and hydrochloric acid to 4-aminoisothiazole. Both the 4- and 5-aminoisothiazoles can be diazotised. The resultant diazonium salts couple to give *azo-dyestuffs*, which are useful for dyeing synthetic fibres.

7.5 Five-Membered Rings containing Three or more Heteroatoms

These compounds include the five-membered heterocyclic rings containing three, four, or five nitrogen atoms (triazoles, tetrazole, pentazole) as well as those with two nitrogen atoms and either an oxygen or a sulphur atom (oxadiazoles and thiadiazoles).

7.5.1 Triazoles

The compound with three adjacent nitrogen atoms in the ring is called **1,2,3-triazole**, that with two adjacent nitrogen atoms separated from the third by a CH-group is called **1,2,4-triazole**.[1] Like pyrazole and imidazole both of the triazoles can exist in two different *tautomeric* forms. Both are π-electron-rich heteroaromatic compounds:

| 1*H*-Form | 2*H*-Form | 1*H*-Form | 4*H*-Form |

1,2,3-Triazole   1,2,4-Triazole

---

[1]See *T. L. Gilchrist* et al., Adv. Heterocycl. Chem. *16*, 33 (1974); *J. B. Polya*, Comprehensive Heterocyclic Chemistry, Vol. *5*, 733 (Pergamon Press, Oxford 1984).

Physical studies indicate that in the case of 1,2,3-triazole the equilibrium favours the 2*H*-form, whereas in the case of 1,2,4-triazole the 1*H*-form is favoured. When a substituent is present on a nitrogen atom, positional isomers are possible depending upon the site of the substituent.

Preparation

7.5.1.1   1,2,3-Triazole can be made by a *1,3-dipolar cycloaddition* reaction between acetylenedicarboxylic acid and benzyl azide. The initially formed 1-benzyl-1,2,3-triazole-4,5-dicarboxylic acid (yield = 92%) can be quantitatively decarboxylated to 1-benzyl-1,2,3-triazole which, on hydrogenation at 150°C (425 K), is split into 1,2,3-triazole (77%) and toluene (*R. H. Wiley*):

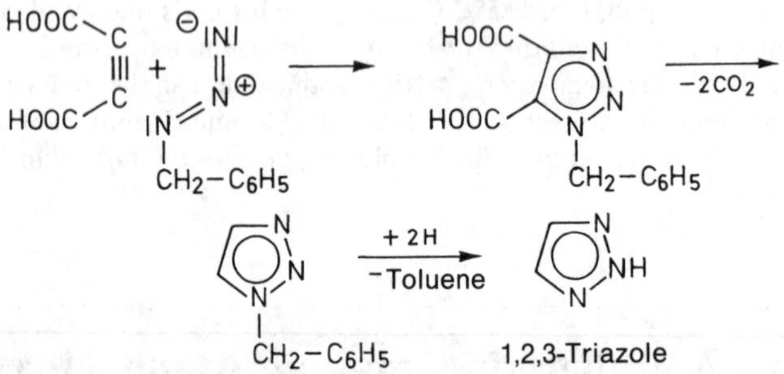

7.5.1.2   1,2,4-Triazole is most simply made by *deamination*, using nitrous acid, of 4-*amino-1,2,4-triazole*, which is obtained in 80% yield by heating formic acid with hydrazine.

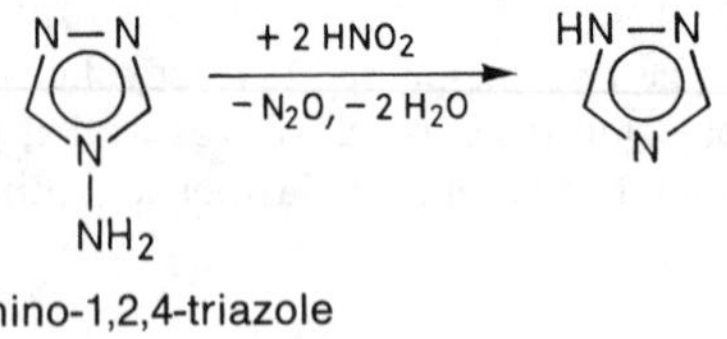

4-Amino-1,2,4-triazole

3-Amino-1,2,4-triazole (Amitrol), which is a useful herbicide, is made by heating aminoguanidine with formic acid, followed by alkaline cyclisation of the resultant formylaminoguanidine (*Thiele* and *Manchot*, 1898):

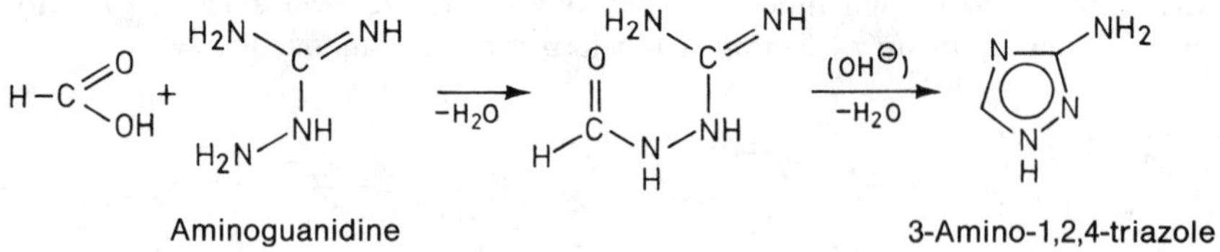

Aminoguanidine   3-Amino-1,2,4-triazole

This compound has been found to be carcinogenic.

4-Phenyl-4*H*-1,2,4-triazole-3,5-dione is one of the most reactive *dienophiles*. *Ethyl 3-phenylcarbamoylcarbazate*, which is readily obtained from *ethoxycarbonylhydrazine* and *phenyl isocyanate*, undergoes ring-closure in alkaline solution to give *4-phenylurazole* (4-phenyl-4*H*-1,2,4-traizolidine-3,5-dione); this can be oxidised to 4-phenyl-4*H*-1,2,4-triazole-3,5-dione by means of t-butyl hypochlorite.

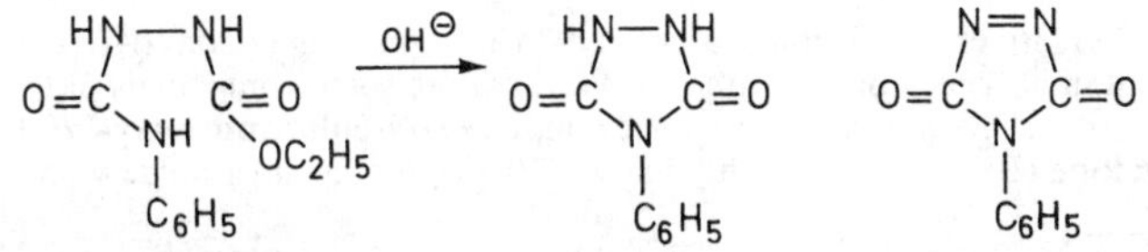

Ethyl 3-phenyl-
carbamoylcarbazate

4-Phenylurazole

4-Phenyl-4*H*-1,2,4-
triazole-3,5-dione

Nitron, a reagent which is used for estimating nitric acid, is a derivative of 1,2,4-triazole. It arises from the condensation of triphenylaminoguanidine with formic acid:

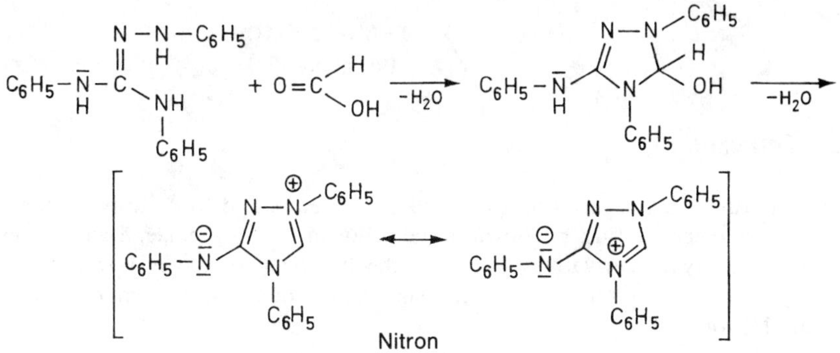

Nitron

Nitron (*N*-[1,4-diphenyl-3-(1,2,4-triazolino)]anilide) is a *mesoionic* compound (see p.759) which is readily soluble in water to give a strongly basic solution. It forms an insoluble nitrate, which can be used for the *quantitative estimation* of nitric acid. Its name is derived from the latter reaction.

Properties and Uses

1,2,3-Triazole is a colourless, hygroscopic substance, m.p. 23°C (296 K), b.p. 204°C/0.98 bar (477 K/980 hPa); 1,2,4-triazole has m.p. 121°C (394 K), b.p. 260°C (533 K). The two π-*electron-rich heteroaromatic compounds* are amphoteric. They are resistant to oxidising and reducing agents. The π-electron density on their carbon atoms is less than it is in pyrrole, pyrazole or imidazole. As yet only a few examples of *electrophilic substitution* reactions are known, *e.g.* hydroxymethylation with formaldehyde, and halogenation, the latter especially if an acid removing reagent is present, *e.g.*

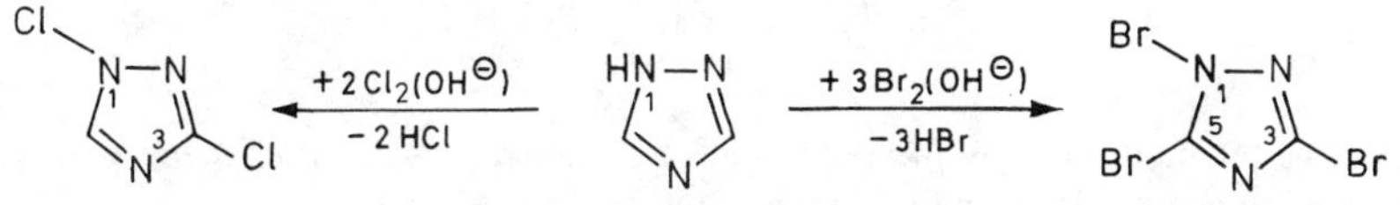

1,3-Dichloro
1,2,4-triazole

1,2,4-Triazole

1,3,5-Tribromo-
1,2,4-triazole

Among the halogenation products the *N*-chloro-, *N*-bromo and *N*-iodo-1,2,4-triazole derivatives are of particular interest. As in the case of *N*-bromosuccinimide, they readily lose the halogen atom attached to nitrogen and because of this they can be used as halogenating agents. In addition to the reactions with electrophiles some *nucleophilic substitution* reactions can take place with suitably substituted triazoles. When, for example, 3-halogeno-1,2,4-triazoles are treated with ammonia in a sealed tube they are converted into *3-amino-1,2,4-triazole*. In recent years this has had some use as a plant-growth regulator; for example, it is used to defoliate cotton plants before they are machine-harvested.

Arylsulphonyl derivatives of triazoles such as 1-(4-nitrophenylsulphonyl)-1H-triazole (4-NBST) have become useful in polynucleotide syntheses. Furthermore some compounds of this sort have been found to be active *plant-protection agents*, for example, (**1-[4-chlorophenoxy-1H-1,2,4-triazolyl])-methyl-t-butyl ketone** (Bayletone) is highly active (20–40 g/ha) against mildew and rust.

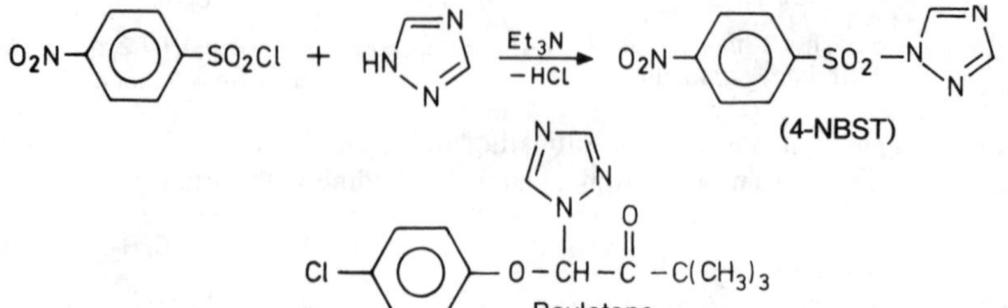

$$O_2N-\langle\bigcirc\rangle-SO_2Cl \ + \ HN\langle\text{triazole}\rangle \xrightarrow[-HCl]{Et_3N} \ O_2N-\langle\bigcirc\rangle-SO_2-N\langle\text{triazole}\rangle$$

(4-NBST)

$$Cl-\langle\bigcirc\rangle-O-CH-\overset{\displaystyle O}{\overset{\displaystyle \|}{C}}-C(CH_3)_3$$

Bayletone

7.5.2   Tetrazoles[1]

The tetrazole ring system, made up from one carbon atom and four nitrogen atoms, can exist in two *tautomeric* forms, as shown in the following two formulae. Such isomers are distinguishable only in derivatives in which the hydrogen of the NH-group has been replaced by a substituent group. Physical measurements indicate the tautomeric equilibrium favours the 1H-form:

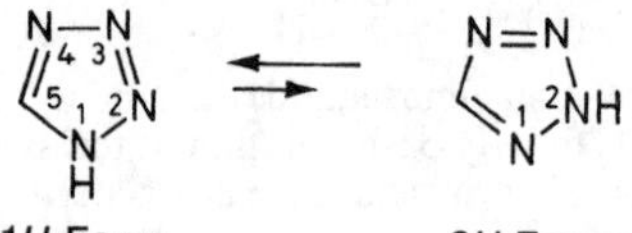

1H-Form   2H-Form

Tetrazole

Preparation

7.5.2.1   Tetrazole can be made analogously to 1,2,3-triazole, by a *1,3-dipolar cyclo-addition* reaction between anhydrous hydrogen cyanide and hydrazoic acid:

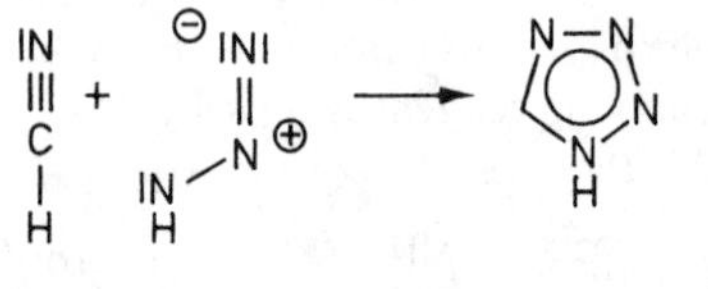

7.5.2.2   A better method is by the reduction of *diazotised 5-aminotetrazole* with ethanol. 5-Aminotetrazole is made from aminoguanidine and nitrous acid:

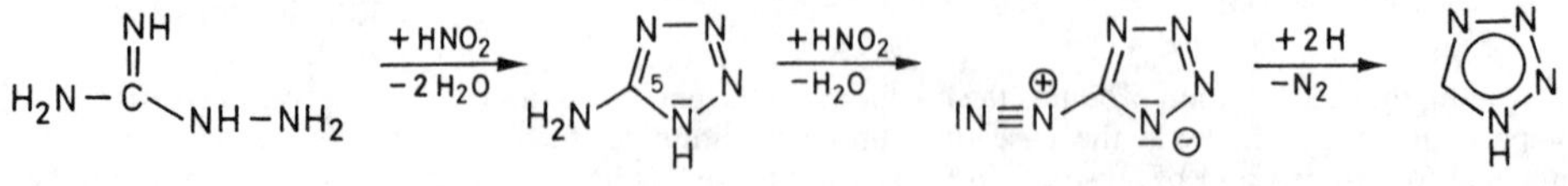

$$\underset{\text{Aminoguanidine}}{H_2N-\overset{\overset{\displaystyle NH}{\displaystyle \|}}{C}\diagdown_{NH-NH_2}} \xrightarrow[-2H_2O]{+HNO_2} \underset{\text{2-Aminotetrazole}}{H_2N-\langle\text{tetrazole}\rangle} \xrightarrow[-H_2O]{+HNO_2} IN\equiv N\langle\text{tetrazole}\rangle \xrightarrow[-N_2]{+2H} \underset{\text{Tetrazole}}{\langle\text{tetrazole}\rangle}$$

---

[1] See *R. N. Butler*, Comprehensive Heterocyclic Chemistry, Vol. 5, 791 (Pergamon Press, Oxford 1984).

Properties

Tetrazole forms colourless crystals, m.p. 156°C (429 K). Its aqueous solution is acidic and it forms stable metal salts. Only certain *N*-substituted tetrazoles can be brominated directly on the carbon atom, *e.g.*

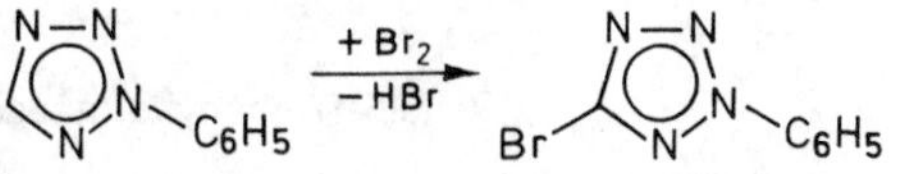

2-Phenyltetrazole   5-Bromo-2-phenyltetrazole

N-Acetyltetrazole can be used as a strong *acetylating agent*.

6,7,8,9-Tetrahydro-5*H*-tetrazolo-azepine, a bicyclic derivative with one of the nitrogen atoms common to both rings, is prepared by the reaction between one molecule of cyclohexanone and two molecules of hydrazoic acid:

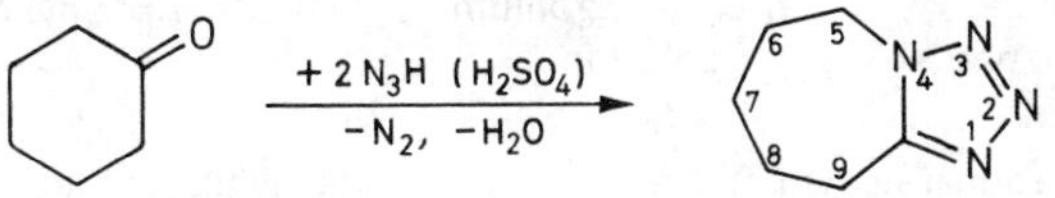

Cyclohexanone

6,7,8,9-Tetrahydro-
5*H*-tetrazolo-azepine

It is used as a blood-circulation stimulant (INN: **Pentetrazole**, Cardiazole) and stimulates the respiratory centres in the brain and the activity of the heart.

The first step in the reaction of nitrous acid with aminoguanidine, in process 7.5.2.2, involves formation of an azide which then cyclises. An azide can be in equilibrium with a tetrazole if the nitrogen atom which becomes common to both rings in the latter is part of an electron-deficient heterocyclic ring (**azido-tetrazolo isomerism**)[1] *e.g.*

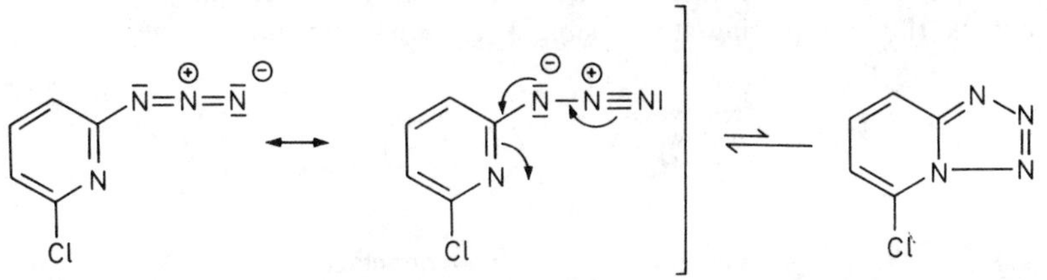

2-Azido-6-chloro-
pyridine

5-Chlorotetrazolo-
[1,5-a]pyridine

In dimethylsulphoxide as solvent at room temperature the equilibrium lies on the side of the azido form and is displaced even further in this direction as the temperature is raised.

Tetrazolium salts.[2] *Triphenyltetrazolium chloride* (TTC) has proved useful as a reduction indicator in biological work, since it undergoes phytochemical reduction to the deep red air-stable *1,3,5-triphenylformazan*:

---

[1] See *M. Tišler*, Some Aspects of Azido–Tetrazolo Isomerization, Synthesis *1973*, 123).

[2] See *W. Ried*, Formazane und Tetrazoliumsalze, ihre Synthesen und ihre Bedeutung als Reduktionsindikatoren und Vitalfarbstoffe (Formazans and Tetrazolium Salts, their Synthesis and Importance as Reduction Indicators and Biological Dyestuffs) Angew. Chem. *64*, 391 (1952); *A. W. Nineham*, The Chemistry of Formazans and Tetrazolium Salts, Chem. Rev. *55*, 355 (1955).

TTC (colourless)   1,3,5-Triphenylformazan (deep red)

Conversely, tetrazolium salts are made by dehydrogenating formazans.

Formazans are prepared by coupling diazonium salts with aldehyde-hydrazones. In this reaction the aldehydic hydrogen atom is replaced by a diazonium group (*v. Pechmann, Bamberger*), *e.g.*

Benzaldehyde   Benzenediazonium   1,3,5-Triphenylformazan
phenylhydrazone   chloride

Phenyl-substituted formazans are orange to deep red compounds and are usually crystalline. On oxidation with mercury(II) oxide, isoamyl nitrite (isopentyl nitrite), lead tetra-acetate, or, best of all, *N*-bromosuccinimide, formazans are dehydrogenated and converted into tetrazolium salts. The latter can be reduced again, for example by ammonium sulphide, to formazans (see above).

If cress seeds, for example, are germinated in a solution of a triphenyltetrazolium salt, then the cotyledons and the tips of the roots are deep red in colour (*F. Moewus*). As a result of this triphenyltetrazolium salts are used to test the ability of seeds to germinate (*Lakon*).

7.5.3   *Pentazoles*[1]

The 5-membered ring of **pentazole** is interesting because it is a homocyclic ring rather than a heterocyclic ring, all the ring atoms being of the same element, nitrogen:

Pentazole   Phenylpentazole

Evidence for the existence of a phenyl-substituted derivative of pentazole came from kinetic studies of the reaction of benzenediazonium chloride with lithium azide (*R. Huisgen*, 1956) and from labelling experiments using ^{15}N (*K. Clusius*, 1956). A little later it was found that if this reaction was carried out at a low temperature (*ca.* $-30°C$ [240 K]), substituted phenylpentazoles could be isolated in crystalline form (*I. Ugi*, 1958):

---

[1] See *R. Huisgen*, Thermische Stabilität und aromatischer Charakter: Ringöffnungen der Azole (Thermal Stability and Aromatic Character: Ring Opening of Azoles) Angew. Chem. *72*, 359 (1960); *I. Ugi*, Comprehensive Heterocyclic Chemistry, Vol. *5*, 839 (Pergamon Press, Oxford 1984).

It has been found that the presence of substituents such as *p*-dimethylamino- or *p*-ethoxy, which are strongly electron-donating (+ M-effect), has a stabilising effect on the molecule.

p-Dimethylaminophenylpentazole forms pale yellow leaflets which decompose at 50–54°C (323–327 K). Even gentle touching causes arylpentazoles to explode above − 10°C (263 K).

7.5.4   Sydnones, Mesoionic Compounds[1]

Sydnones (so-called because the first example was made in Sydney) contain one oxygen atom and two nitrogen atoms in the ring and are derivatives of 1,2,3-oxadiazole. The first one to be obtained was **N-phenylsydnone**, which was prepared by a process which has yet to be bettered (*Earl* and *Mackney*, 1935), by heating *N*-nitroso-*N*-phenylglycine with acetic anhydride. Loss of water is accompanied by a ring-closure reaction:

N-nitroso-   N-Phenylsydnone (R = C_6H_5)
N-phenylglycine

Sydnones are examples of **mesoionic compounds**, *i.e.* five-membered ring heterocyclic compounds whose structure can only be explained in terms of *mesomeric polar limiting structures* (*Baker* and *Ollis*, 1946).

These compounds have *high dipole moments* and readily undergo electrophilic substitution, *e.g.* nitration, halogenation and sulphonation, at the 4-position. These properties are perhaps best expressed by the following formulae, which summate the limiting structures given above:

Sydnoneimine   INN: *Molsidomine*

Substitution of the exocyclic oxygen atom by a nitrogen atom provides a *sydnoneimine*; these are stable if R′ is an acyl group. *N*-Ethoxycarbonyl-3-(4-morpholinyl)sydnoneimine, *INN: Molsidomine*, Corvaton, provides NO, which is a neurotransmitter, in the body; NO dilates the cardiac blood vessels and is effective against *angina pectoris*. **Molsidomine** differs from nitroglycerol as a prodrug in that no reducing agent is necessary to generate NO (*cf.* p.313).

Mesoionic compounds of this sort have been shown to be reactive in 1,3-dipolar cycloaddition reactions (*Huisgen*, 1967). An example of their preparative use is provided by the following pyrazole synthesis; the sydnone is represented by one of its dipolar limiting structures:

[1]See *W. D. Ollis*, Mesoionic Compounds, Adv. Heterocycl. Chem. *19*, 1 (1976).

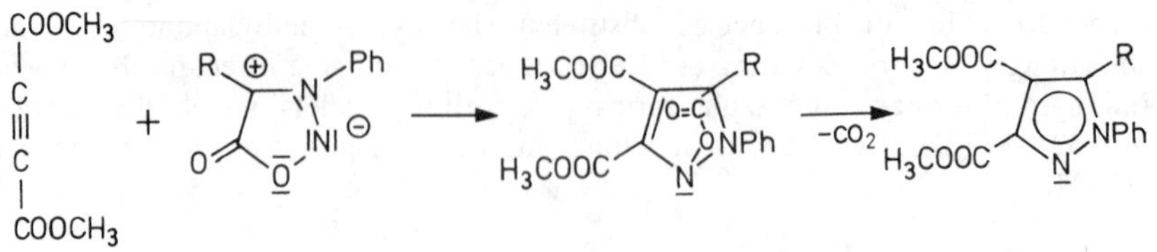

7.5.5 Thiadiazoles[1]

It should in theory be possible to have four different arrangements of the heteroatoms in thiadiazole rings; one example is **1,2,4-thiadiazole**:

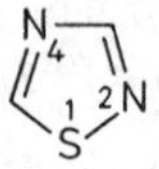

1,2,4-Thiadiazole

The synthesis of 1,2,4-thiadiazole starts from 5-amino-1,2,4-thiadiazole. This is first converted by means of a *Sandmeyer* reaction into the 5-bromo-derivative which is then catalytically hydrogenated to 1,2,4-thiadiazole (*Goerdeler*):

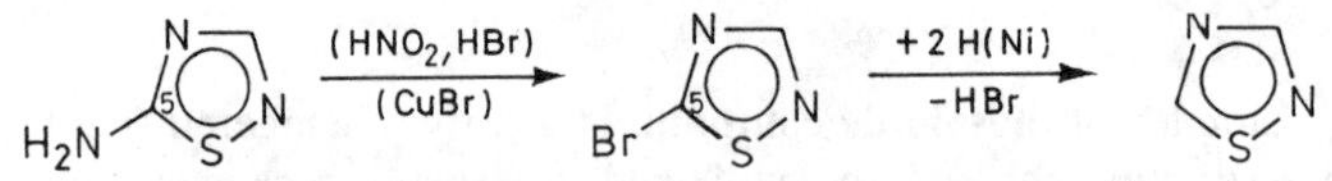

| 5-Amino-1,2,4- | 5-Bromo-1,2,4- | 1,2,4-Thiadiazole |
| thiadiazole | thiadiazole | |

1,2,4-Thiadazole is a colourless liquid which boils at 121°C (394 K). When treated with a reducing agent it readily undergoes ring-opening.

7.6 Six-Membered Rings Heterocycles with One Heteroatom

The parent compounds of this series of heterocycles are **pyridine**, the **pyrans** and **thiopyrans**.

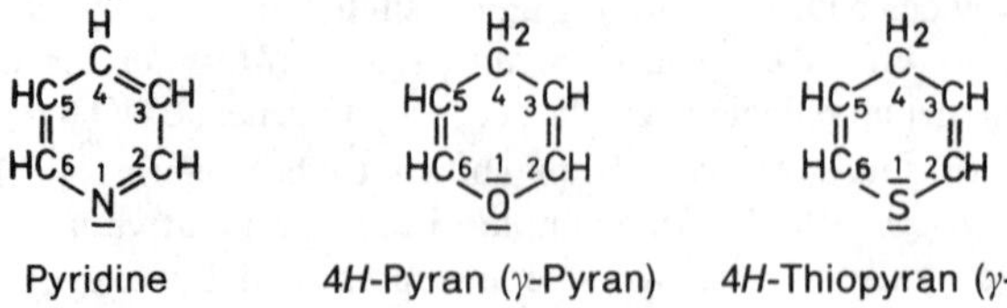

Pyridine   4H-Pyran (γ-Pyran)   4H-Thiopyran (γ-Thiopyran)

---

[1] See *J. E. Franz* and *D. P. Dhingra*, Comprehensive Heterocyclic Chemistry, Vol. *6*, 463 (Pergamon Press, Oxford 1984).

Positions 2 and 6 in the ring are sometimes described as the α-positions, positions 3 and 5 as the β-positions, and position 4 as the γ-position.

Many important natural products are pyridine and pyran derivatives. *4H-Pyran* (γ-pyran), b.p. 80°C (353 K) and *4H-thiopyran* (γ-thiopyran), b.p. 30°C (303 K)/16 mbar (16 HPa) were prepared from glutaraldehyde in 1962 by *Strating*.

7.6.1 Pyridines

Pyridine, C_5H_5N, occurs in coal tar ($\sim0.1\%$) and in the distillate from bones (bone oil) (*Anderson*, 1851) and has been produced industrially from these sources. Pyridine and methylpyridines are manufactured synthetically by a modification of the *Hantzsch* synthesis (see p.766) which involves passing a mixture of acetaldehyde, formaldehyde (as its methyl hemiacetal) and ammonia over a heated Al_2O_3/SiO_2 catalyst.

Like benzene, pyridine has a sextet of delocalised π-electrons. It can be described in terms of the following limiting structures.

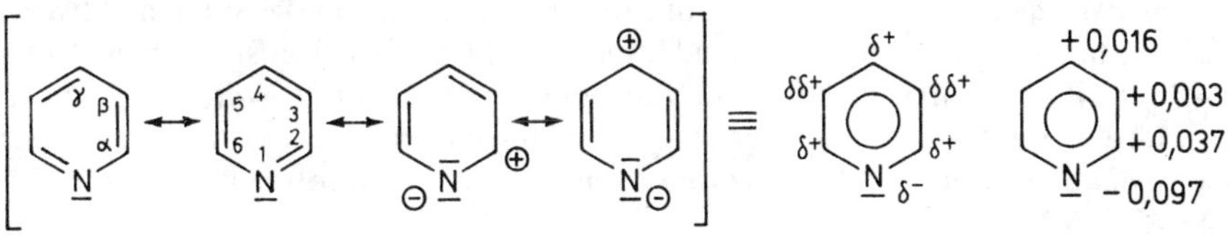

The first two formulae are analogous to the *Kekulé* formulae for benzene. However the electronegative nitrogen atom tends to draw the π-electrons towards itself, resulting in a lowering of the electron density on the carbon atoms. As a result the molecule as a whole is polarised (*dipole moment* $= 7.5 \times 10^{-30}$ Cm). This drift of electrons is expressed in the *dipolar* limiting structures and results in pyridine being an *electron-poor heteroaromatic compound*. The precise distribution of charge in the molecule has been calculated to be as shown above (cf. p.705). Its stabilisation energy amounts to 130 kJ/mol.

Properties

Pyridine is a weak tertiary base, b.p. 115°C (388 K), with a disagreeable odour. It is miscible in all proportions with water. It is much less basic (pK_a 5.2) than piperidine (pK_a 11.12, see p.768). This is a result of the sp^2-hybridisation of the lone pair of electrons in pyridine.

Electrophilic reagents interact with the nitrogen atom to form pyridinium salts. The resultant positive charge on the nitrogen atom increases the drift of electrons towards it, and affects in particular carbon atoms 2, 4 and 6. Hence *electrophilic substitution reactions* take place, if at all, only under extreme conditions, and then at the 3- and 5-positions. For example, *bromination* of pyridine in the gas phase at 300°C (570 K) provides *3-bromopyridine* together with a small amount of *3,5-dibromopyridine*. Sulphonation is achieved by using fuming sulphuric acid in the presence of mercury(II) sulphate at 230°C (500 K); a 70% yield of *pyridine-3-sulphonic acid* is obtained. Nitration under strong conditions, using potassium nitrate in fuming sulphuric acid at 370°C (640 K), gives *3-nitropyridine*. Pyridine does *not* participate in *Friedel–Crafts* reactions.

Pyridine-3-sulphonic acid   Pyridine   3-Nitropyridine

3-Bromopyridine   3,5-Dibromopyridine

In contrast to benzene, pyridine readily undergoes *nucleophilic substitution* at the 2- and 6-positions by *anions* such as $NH_2^\ominus$ or $OH^\ominus$.

7.6.1.1 *Chichibabin* reaction (1914). In this direct *amination* reaction[1] pyridine is heated with *sodium amide* in either dimethylaniline or liquid ammonia until evolution of hydrogen ceases. The first step takes place by an addition–elimination mechanism with nucleophilic attack by an $NH_2^\ominus$ anion on carbon atom 2 accompanied by loss of a *hydride ion*. The latter reacts with a proton from the amino group to form dihydrogen. This results in the formation of the *sodium salt of* **2-aminopyridine**, which can be hydrolysed to the free base (yield = 75–80%).

2-Aminopyridine

Further amination provides *2,6-diaminopyridine* (yield = 80–90%).

If there are substituents at the 2- and 6-positions of the pyridine ring reaction takes place instead at the 4-position to give the corresponding *4-aminopyridine derivative*.
3-Aminopyridine is best obtained by *Hofmann* degradation of nicotinamide, using sodium hypobromite.

An analogous nucleophilic substitution reaction occurs when pyridine is treated with *phenyl lithium* or *n-butyl lithium* in an atmosphere of nitrogen. This provides a way of introducing phenyl or n-butyl substituents at the 2-position (*K. Ziegler*), e.g.

2-Phenylpyridine   Pyridyne (a hetaryne)

[1]See *R. G. Shepherd* et al., Adv. Heterocycl. Chem. *4*, 145 (1965); *M. Wozniak* and *H. C. van der Plas*, Acta Chem. Scand. *47*, 95 (1993).

Nucleophilic substitution in electron-poor heterocycles can take place not only by an addition–elimination mechanism of this sort; other examples have been found to involve formation of a hetaryne or ring-opening.[1]

2- and 4-Hydroxypyridines [2(1*H*)- and 4(1*H*)-pyridones] can be made by treating the corresponding *pyrones* with ammonia, *e.g.*

2(2*H*)-Pyrone	2(1*H*)-Pyridone	2-Hydroxypyridine
(2-Pyrone) (α-Pyrone)	(2-Pyridone) (α-Pyridone)	Pyridin-2-ol

These compounds are tautomeric. In solution, equilibrium lies largely on the side of the *pyridones* whereas in the gas phase the hydroxy compounds are favoured.[2] The latter are examples of iminol forms of amides.

7.6.1.2   When pyridine reacts with alkyl halides **N-alkylpyridinium salts** are formed. In these salts, the positive charge on the nitrogen atom results in the ring being even more electron-poor than in pyridine itself. As a result *N-alkylpyridinium salts* undergo *nucleophilic attack* at the 2-position by hydroxide ions. The initial product is a 'pseudo-base' which can be converted into 1-methyl-2-pyridone by oxidation with potassium hexacyanoferrate(III), *e.g.*

N-Methylpyridinium salt	Pseudo-base	1-Methyl-2-pyridone

1,1′-Dimethyl-4,4′-bipyridinium dichloride (**Paraquat**, Methyl viologen) is a very powerful herbicide.

Paraquat	1,1′-*p*-Xylylene-bis(4,4′-bipyridinium) hexafluorophosphate

Paraquat is very poisonous. It has proved valuable for constructing supramolecular structures with macrocyclic electron-rich aromatic compounds.[3] 1,1′-*p*-Xylylene-bis(4,4′-bipyridinium) hexafluorophosphate is a participant in a one-stage catenane synthesis (see p.513).

---

[1] See *H. C. van der Plas*, The S$_N$ (ANRORC, Addition of the Nucleophile, Ring Opening and Ring Closure) Mechanism: A New Mechanism for Nucleophilic Substitution, Acc. Chem. Res. *11*, 462 (1978).

[2] See *P. Beak*, Acc. Chem. Res. *10*, 186 (1977).

[3] See *J. F. Stoddart* et al., A Polymolecular Donor–Acceptor Stack made of Paraquat, Angew. Chem. Int. Ed. Engl. *28*, 1402 (1989).

If there is a strongly electron-withdrawing group, such as —C≡N, attached to the nitrogen atom in pyridine, then attack by *nucleophiles* usually results in opening of the ring and elimination of the nitrogen atom. An example of this sort of reaction is provided by preparation of the so-called **Zincke aldehyde**. When *N*-cyanopyridinium bromide, which is obtained from pyridine and cyanogen bromide, is treated with *N*-methylaniline, cyanamide is lost and a deep red salt with a delocalised cation, sometimes called *König*'s salt, is formed. Alkali converts this salt into *N*-methylaniline and 5-(*N*-methylanilino)penta-2,4-dienal (*Zincke* aldehyde);

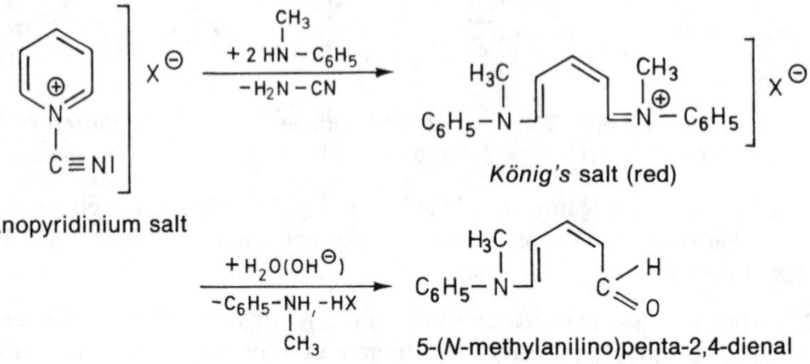

N-Cyanopyridinium salt

König's salt (red)

5-(*N*-methylanilino)penta-2,4-dienal
(*Zincke* aldehyde, yellow)

This aldehyde has played an important role in the synthesis of azulenes.

Uses

Pyridine is a good solvent for many organic compounds and for some inorganic salts. It acts as a catalyst for acylation reactions involving acid chlorides and also takes up the hydrogen halide which is liberated in the process. It is also used for denaturing ethanol.

Important Derivatives of Pyridine

7.6.1.3 Pyridine-*N*-oxides[1] are formed when pyridine or its derivatives are oxidised by peracids or, more simply, by a 30% solution of hydrogen peroxide in acetic acid (*Ochiai*, 1947; *Den Hertog*, 1951; see also amine oxides) *e.g.*

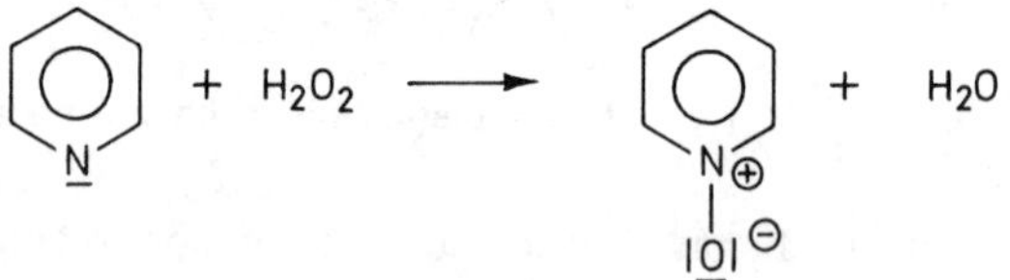

Pyridine-N-oxide also undergoes nucleophilic substitution at the 2- and 6-positions but, unlike pyridine, it can also undergo *electrophilic substitution* at the 4-position. This is because polarisation of the molecule permits the localisation of either positive or negative charge on the 2, 4- and 6-sites:

---

[1]See *A. R. Katritzky* et al., Chemistry of the Heterocyclic N-Oxides (Academic Press, New York 1971).

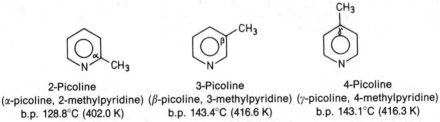

$$
\left[
\begin{array}{ccccc}
\end{array}
\right]
$$

Pyridine-*N*-oxide

The reaction of pyridine-*N*-oxide with nitrating acids provides *4-nitropyridine-N-oxide* in 85% yield. If this is treated with excess phosphorus(III) chloride in chloroform it is converted into *4-nitropyridine*. If removal of the oxygen is carried out reductively, using *Raney* nickel in acetic acid/acetic anhydride (*Jerchel*, 1958) then, in addition, the nitro group is reduced, giving **4-aminopyridine**:

4-Nitropyridine   4-Nitropyridine-*N*-oxide   4-Aminopyridine   4-Dimethylamino-pyridine

$$+ PCl_3 \quad -POCl_3 \qquad (Raney-Ni)$$

4-Nitropyridine-*N*-oxide reacts with acyl halides to give *4-halogenopyridine-N-oxides* and with sodium alkoxides or phenoxides to give *4-alkoxy-* or *4-phenoxy-pyridine-N-oxides*.

4-Dimethylaminopyridine (DMAP) is a useful catalyst for acylation reactions.[1]

7.6.1.4   There are three *isomeric* **picolines** (methylpyridines):

2-Picoline   3-Picoline   4-Picoline
(α-picoline, 2-methylpyridine) (β-picoline, 3-methylpyridine) (γ-picoline, 4-methylpyridine)
b.p. 128.8°C (402.0 K)   b.p. 143.4°C (416.6 K)   b.p. 143.1°C (416.3 K)

Methyl substituents at the 2-, 4- or 6-positions (α- or γ-positions) react readily with aldehydes or with nitroso compounds. Loss of water gives rise to condensation products. Thus **2-vinylpyridine** is made industrially from 2-picoline and formaldehyde; reaction proceeds via an aldol-like intermediate:

$$+ HC\overset{H}{\underset{O}{\lessgtr}} \qquad -H_2O$$

2-Picoline   2-(2-Pyridyl)ethanol   2-Vinylpyridine

---

[1]See *G. Höfle* et al., Angew. Chem. Int. Ed. Engl. *17*, 569 (1978).

Copolymerisation of 2-vinylpyridine and butadiene provides *elastomers* which adhere tightly to nylon fabrics and are used in the making of *tyre cords*. If vinylpyridine is copolymerised with acrylonitrile, introduction of the basic groups improves the take-up of dyestuffs by the poly(acrylonitrile) fibres.

In a similar way 4-picoline condenses with *p*-nitrosodimethylaniline to give an **azomethine**:

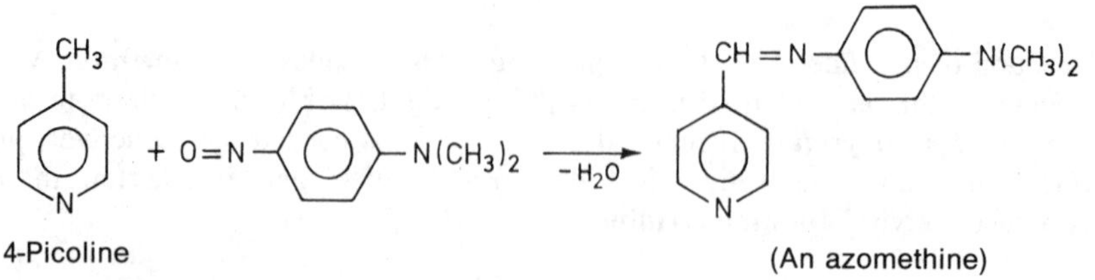

4-Picoline   (An azomethine)

3-Picoline does *not* undergo this type of condensation reaction.

The picolines occur together with pyridine and higher homologues, especially **2,4-lutidine** (2,4-dimethylpyridine) and **collidine** (2,4,6-trimethylpyridine), in the basic fractions of coal-tar and brown-coal-tar. Partial isolation of the individual bases by fractional distillation is possible.

7.6.1.5 Homologues of pyridine are usually *prepared* by making use of the *Hantzsch dihydropyridine synthesis*. This involves a ring-closure condensation reaction between a β-ketoester (2 mol), an aldehyde (1 mol) and ammonia (1 mol).

The course of the reaction may be illustrated by the condensation of an acetoacetic ester with acetaldehyde and ammonia. On the one hand, the acetoacetic ester undergoes a *Knoevenagel* reaction to give an ethylidene-acetoacetic ester; on the other, it reacts with ammonia to give an ester of β-aminocrotonic acid. The latter, as an *enamine*, takes part in a *Michael* addition reaction with the ethylidene-acetoacetic ester and cyclisation to 1,4-dihydrocollidine-3,5-dicarboxylic ester ensues. This can be oxidised by nitric acid to a collidinedicarboxylic ester.

Hydrolysis and decarboxylation of the latter provides *collidine*:

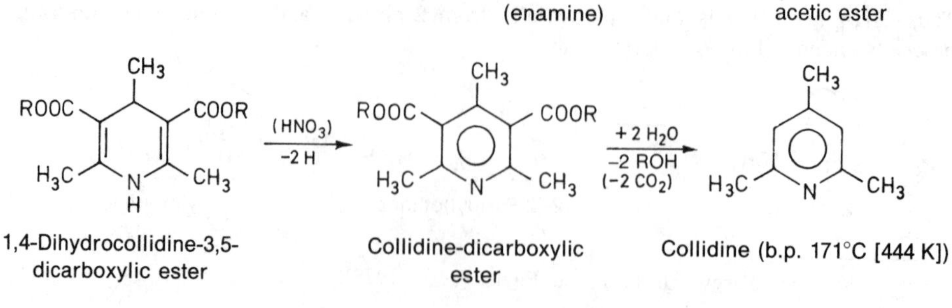

β-Aminocrotonic acid ester   Ethylidene-aceto-
(enamine)   acetic ester

1,4-Dihydrocollidine-3,5-   Collidine-dicarboxylic   Collidine (b.p. 171°C [444 K])
dicarboxylic ester   ester

7.6.1.6   Pyridine aldehydes can be made in about 40% yield by atmospheric oxidation of the corresponding picolines in the presence of vanadium(V) oxide as a catalyst (*Mathes* and *Sauermilch*), *e.g.*

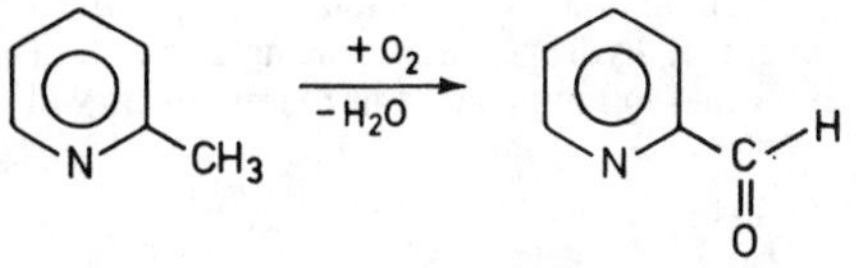

2-Picoline

Pyridine-2-carboxaldehyde
(b.p. 181°C [454 K])

Like benzaldehyde or furfural, in the presence of cyanide ions two molecules of pyridine-2-carboxaldehyde readily condense to form **pyridoin**, which does not exist as a hydroxyketone, but instead takes up a doubly chelated ene-diol structure that is stabilised by hydrogen bonding and has a deep orange colour. If air is passed through a methanolic solution of pyridoin it oxidises it to **pyridil**.

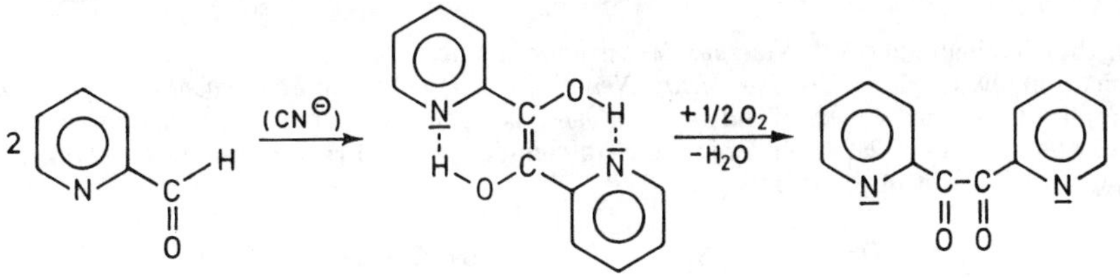

Pyridoin

2,2'-Pyridil

7.6.1.7   Pyridinecarboxylic acids.   Oxidation of 2-, 3- or 4-picoline with potassium permanganate produces the corresponding carboxylic acids, commonly known, respectively, as **picolinic acid**, INN: *Nicotinic acid* and **isonicotinic acid**. All three are valuable synthetic reagents.

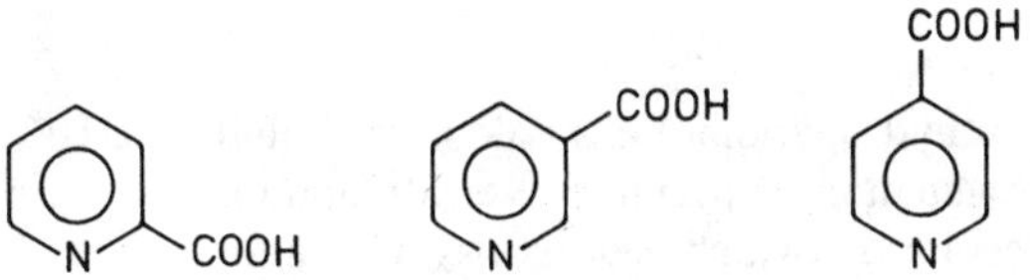

Picolinic acid

INN: *Nicotinic acid*
(Niacin)

Isonicotinic acid

Pyridinecarboxylic acids are more readily decarboxylated than is benzoic acid, the ease depending on the location of the carboxyl group, *i.e.* 2>4>3.

Vitamin B$_6$ (INN: *Pyridoxine*), 4,5-bis(hydroxymethyl)-2-methylpyridin-3-ol, was isolated, together with vitamins B$_1$ and B$_2$, from yeast by *R. Kuhn*.

Rats which are lacking in vitamin B$_6$ develop a special kind of dermatitis.

Microbiological investigations have shown that extracts of **vitamin B$_6$** contain also two other related pyridine derivatives, **pyridoxal** and **pyridoxamine** which likewise have vitamin character and for some microorganisms are more active *growth factors* than pyridoxine (*Snell*, 1945).

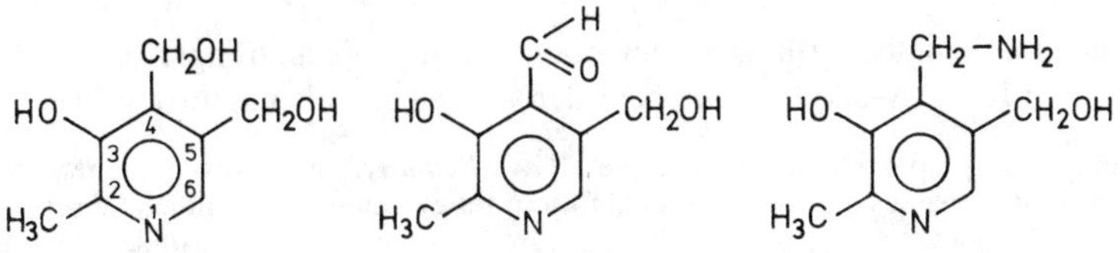

INN: **Pyridoxine**

Pyridoxal

Pyridoxamine

Pyridoxal phosphate and pyridoxamine phosphate are *transamination* catalysts. Pyridoxal phosphate also appears as a coenzyme in *decarboxylase*, an enzyme which brings about the decarboxylation of amino acids.

INN: *Nicotinamide*, pyridine-3-carboxamide, is the human *anti-pellagra* vitamin and is also of biological importance as a component of hydrogen-transporting enzymes (*Warburg, v. Euler*), see p.886. It occurs especially in the liver and in the muscles of mammals as well as in yeast, milk and in cereal germ.

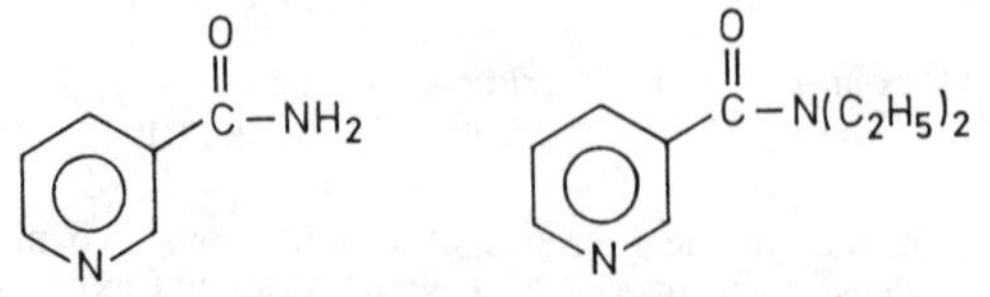

INN: *Nicotinamide*
Pyridine-3-carboxamide

INN: *Nicethamide*
N,N-diethylnicotinamide

N,N-Diethylnicotinamide (*INN: Nicethamide*) is an analeptic.

Isonicotonic hydrazide (*INN: Isoniazide, Neoteben, INH*) is an important *chemotherapeutic agent* against all forms and stages of *tuberculosis* and is used as such in combination with other tuberculostatic agents. 2-Propylisonicotinic thioamide (*INN:* **Protionamide**, *Ektebin*) is also sometimes used as a stronger agent.

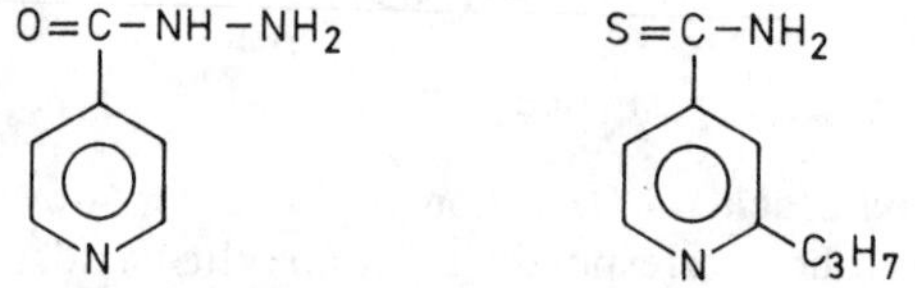

INN: *Isoniazide*

INN: *Protionamide*

7.6.1.8 Piperidine (hexahydropyridine) is a colourless liquid, b.p. 106°C (379 K). It can be made either by hydrogenation of pyridine over a platinum catalyst or by heating 1,5-diaminopentane dihydrochloride, which causes loss of ammonia as its hydrochloride and cyclisation:

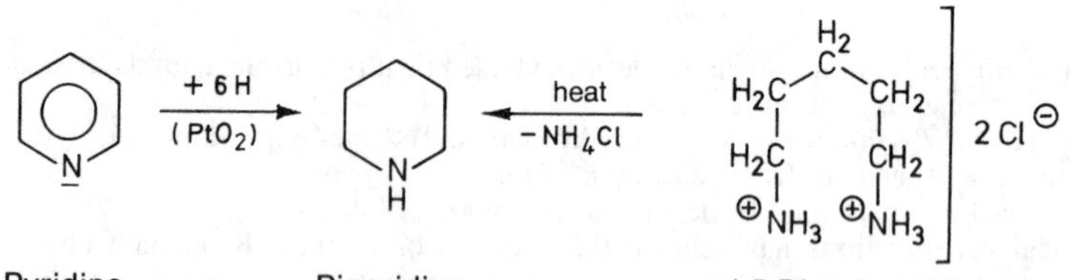

Pyridine

Piperidine
(or its hydrochloride)

1,5-Diaminopentane
dihydrochloride

Piperidine is a secondary aliphatic amine and, as such, is a stronger base ($pK_a = 11.12$) than pyridine. It forms *N*-alkyl- and *N*-acyl-derivatives and also a nitrosamine.

1-Phenylethyl-4-*N*-propionylanilinopiperidine (INN: *Fentanyl*) has proved to be of value as a strong analgesic; it surpasses *morphine* in its ability to relieve pain. The synthetic anaesthetic **phencyclidine**, 1-(1-phenylcyclohexyl)piperidine is misused as a narcotic drug under the name 'angel dust'.

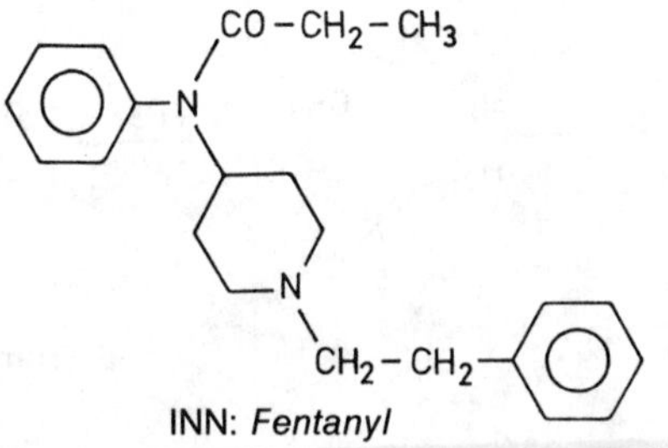

INN: *Fentanyl* 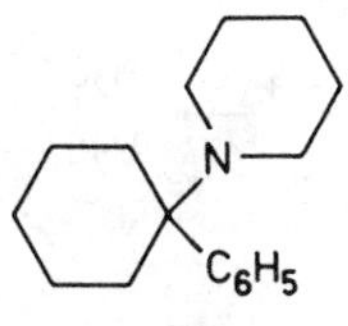   Phencyclidine

N,N-Dimethylpiperidinium chloride (mepiquat chloride, Pix) controls plant growth and is used in agriculture to prevent excessive growth in cotton.

Use has been made of the following *methods of ring-opening*, shown here using piperidines to exemplify them, in determining the structure of various alkaloids:

(a) Exhaustive methylation (*A.W. v. Hofmann*). Reaction of *N-methylpiperidine* with methyl iodide and silver oxide converts it into the quaternary salt *N,N-dimethylpiperidinium iodide*; treatment of this with silver oxide provides the corresponding hydroxide which, when dry distilled, loses water and undergoes ring-opening to give *N,N-dimethylpent-4-enylamine:*

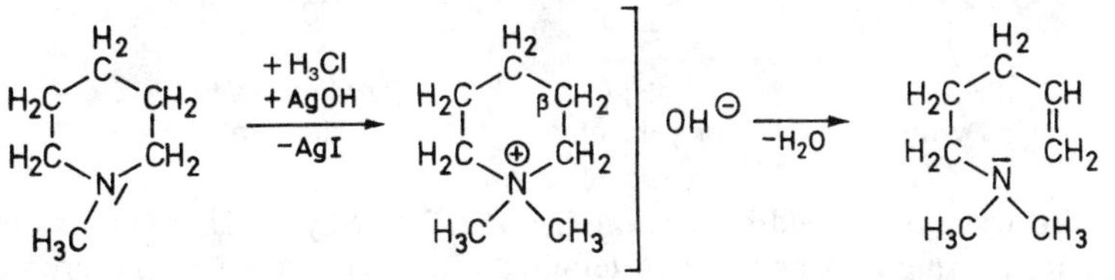

N-Methylpiperidine   *N,N*-dimethylpent-4-enylamine

In this reaction a proton is lost from the β-carbon atom.

If the product is treated again with methyl iodide and silver oxide another quaternary ammonium salt is formed which, when heated, splits into trimethylamine, water and penta-1,3-diene.

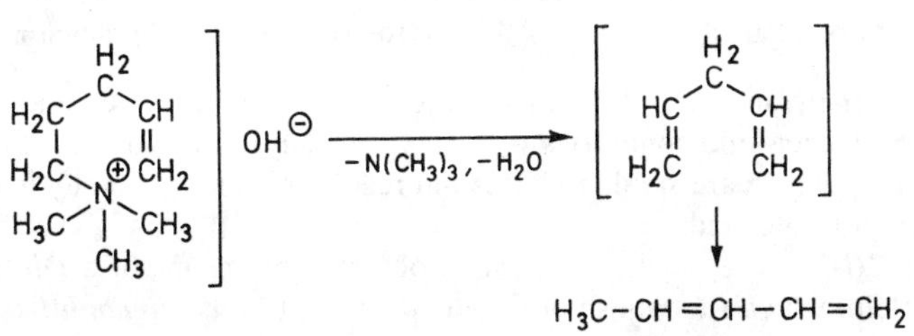

$$H_3C-CH=CH-CH=CH_2$$

Penta-1,3-diene

This method does not inevitably lead to ring-opening.

(b) Cyanogen bromide method (*v. Braun*).[1] Addition of cyanogen bromide to, for example, *N-phenylpiperidine* leads to the formation of a labile adduct which stabilises itself by concomitant breaking of a ring C—N bond and addition of the bromide ion to the terminal carbon atom to give a substituted cyanamide. If this is hydrolysed by heating it with boiling hydrobromic acid the cyanide group is eliminated in the form of cyanogen bromide leaving the secondary amine **5-bromopentyl-aniline**, which can then be further degraded:

---

[1] See *H. A. Hagemann*, The *v. Braun* Cyanogen Bromide Reaction, Org. React. *7*, 198 *ff.* (1953).

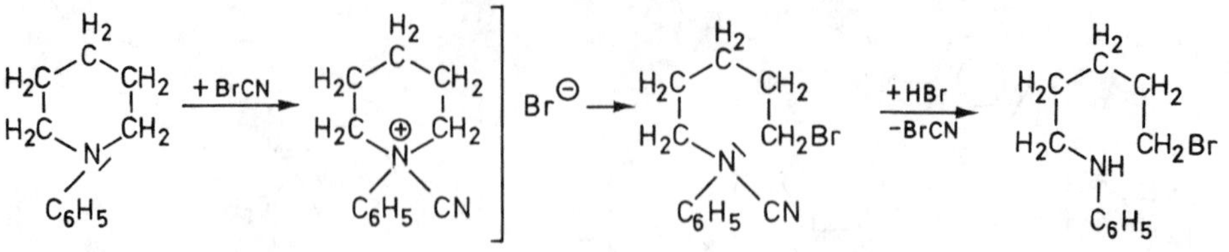

N-Phenylpiperidine

5-Bromopentylaniline

7.6.2 Pyrans[1]

Important derivatives of the six-membered ring heterocycles containing one oxygen atom and two C=C double bonds, the 2H- (or α-) and 4H- (or γ-) **pyrans**, include the α- and γ-pyrones and **pyrylium salts**.[2]

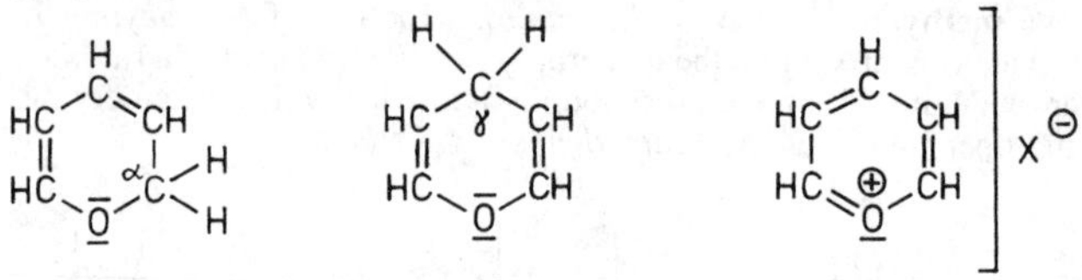

[2H]-
or α-Pyran

[4H]-
or γ-Pyran

Pyrylium salts
(oxonium salts)

7.6.2.1 **γ-Pyran** was synthesised by *Strating* in 1962. 3,4-Dihydro-2H-pyran is obtained by loss of water and ring expansion from tetrahydrofurfuryl alcohol in the presence of aluminium oxide as catalyst.

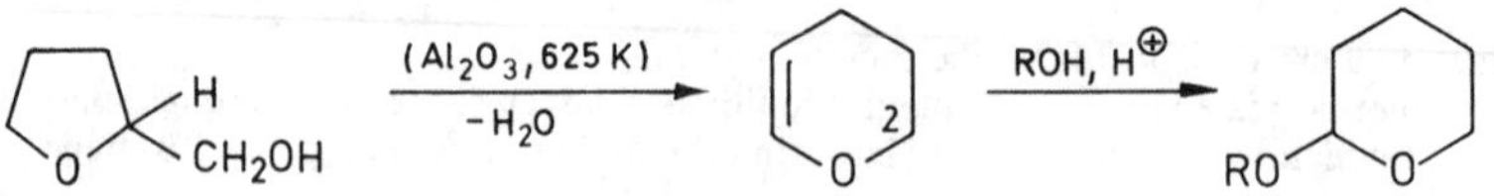

Tetrahydrofurfuryl alcohol   3,4-Dihydro-2H-pyran   Tetrahydropyranyl ether

3,4-Dihydro-2H-pyran is a good protecting group for primary, secondary and phenolic hydroxy groups. Under mild conditions with an acid catalyst they add to it to form *tetrahydropyranyl ethers*, which are stable to bases but readily regenerate the hydroxy compound in the presence of dilute acid.

3,4-Dihydro-2H-pyran-2-carboxaldehyde is obtained by means of a *Diels–Alder* reaction from acrolein, which functions both as the *diene* and as the *dienophile*:

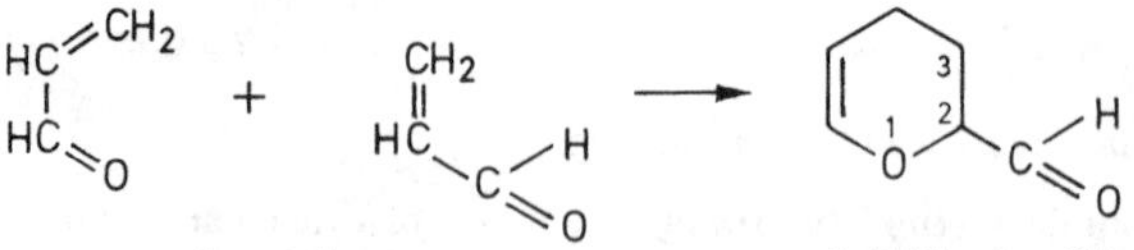

Acrolein

3,4-Dihydro-2H-
pyran-2-carboxaldehyde

[1] See *G. P. Ellis*, Comprehensive Heterocyclic Chemistry, Vol. 3, 647 (Pergamon Press, Oxford 1984).
[2] See *K. Dimroth*, Aromatische Verbindungen aus Pyryliumsalzen (Aromatic Compounds from Pyrylium Salts), Angew. Chem. 72, 331 (1960); *N. Schroth* et al., Pyrylium-Synthesen, 65 Jahre Pyryliumsalze, Z. Chem. 4, 281 (1964); *A. T. Balaban* et al., Pyrylium Salts, Adv. Heterocycl. Chem. 10, 241 (Academic Press, New York 1969).

Catalytic hydrogenation of dihydropyran provides **tetrahydropyran**; *pyranose* sugars are derivatives of this ring system.

7.6.2.2   γ-Pyrone (4-pyrone, 4H-pyran-4-one). Among naturally occurring derivatives of γ-pyrone are kojic acid (5-hydroxy-2-(hydroxymethyl)-4H-pyran), which is formed by the action of certain bacteria (*Aspergillus oryzae*) on solutions of carbohydrates such as glucose or starch, maltol, which was first isolated from larch-bark and later was detected as a constituent of the aroma of bread, and **meconic acid**, which forms part of an opium alkaloid:

| Kojic acid | Maltol | Meconic acid |

In a broad sense, the poison derived from *derris-root*, **rotenone**, is also a member of this class of compounds. It is an important naturally occurring *insecticide* and is obtained from the roots of a bean-like plant from East Asia called tuba (*Derris eliptica*). The molecule of rotenone contains five fused rings, two benzenoid and one each based on dihydropyran, dihydro-γ-pyrone and dihydrofuran (*Butenandt, La Forge*, 1932).

Rotenone

2,6-Dimethyl-γ-pyrone is the most readily accessible γ-pyrone derivative. It is obtained by heating dehydracetic acid, which is itself made by the self-condensation of two molecules of acetoacetic ester in the presence of sodium hydrogen carbonate, with sulphuric acid. The lactone ring is thereby opened to give a β-keto-acid which rearranges to 2,6-dimethyl-4-oxo-4H-pyran-3-carboxylic acid; this in turn is decarboxylated to form 2,6-dimethyl-γ-pyrone:

Dehydracetic acid (m.p. 109°C [382 K])   2,6-Dimethyl-γ-pyrone

Properties.   2,6-Dimethyl-γ-pyrone is a colourless compound, m.p. 132°C (405 K), which dissolves in water to give a neutral solution. It is of interest that the carbonyl group in γ-pyrones does not react with the normal reagents for carbonyl compounds; for example it does *not* form an oxime or a semicarbazone.

In fact 2,6-dimethyl-γ-pyrone (and other γ-pyrones) does not possess a normal carbonyl group, because of electronic interaction from the ring oxygen atom. It is best described in terms of the two *mesomeric* forms shown below. In the presence of mineral acids, a proton adds to the negatively charged oxygen atom, giving rise to an oxonium salt. This oxygen atom also reacts with methyl iodide to form 4-methoxy-2,6-dimethylpyrylium iodide.

2,6-Dimethyl-γ-pyrone   4-Methoxy-2,6-dimethyl-pyrylium iodide

7.7 Benzo-Derivatives of Pyridine and γ -Pyrone

A benzene ring can be fused to a pyridine in two ways, providing the two parent compounds **quinoline** and **isoquinoline**:

Quinoline   Isoquinoline

Both of these ring-systems are found in alkaloids. Isoquinoline (pK_a 5.42) is a stronger base than quinoline (pK_a 4.90). The bonding in these rings closely resembles that in naphthalene.

7.7.1 Quinolines

Quinoline was first found in coal-tar, in 1834, and shortly afterwards (1842) was obtained from the distillation of quinine with alkali.

Preparation

7.7.1.1 *Skraup*'s Synthesis.[1] Aniline is heated with dry glycerol, conc. sulphuric acid and iron(II) sulphate, with nitrobenzene also present as a dehydrogenating agent. First of all water is removed from glycerol to give acrolein. This then condenses with aniline to form 1,2-dihydroquinoline, which undergoes dehydrogenation to quinoline:

---

[1] See *R. H. F. Manske* and *M. Kulka*, The Skraup Synthesis of Quinolines, Org. React. *7*, 59 *ff.* (1953).

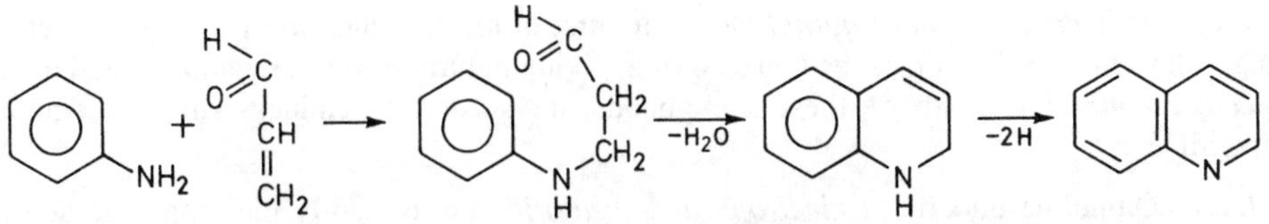

Aniline   Acrolein   1,2-Dihydroquinoline   Quinoline

Other mild dehydrogenating agents, *e.g.* arsenic acid, can be used instead of nitrobenzene. The presence of the iron(II) sulphate prevents the reaction from becoming violent.

7.7.1.2 *Doebner–von Miller* Quinaldine Synthesis. In a similar manner to the *Skraup* synthesis, aniline is condensed with crotonaldehyde in the presence of conc. sulphuric acid to give 2-methyl-1,2-dihydroquinoline. The crotonaldehyde is prepared *in situ* by the aldol condensation of two molecules of acetaldehyde. In the presence of a dehydrogenating agent the dihydroquinoline is oxidised to quinaldine (2-methylquinoline):

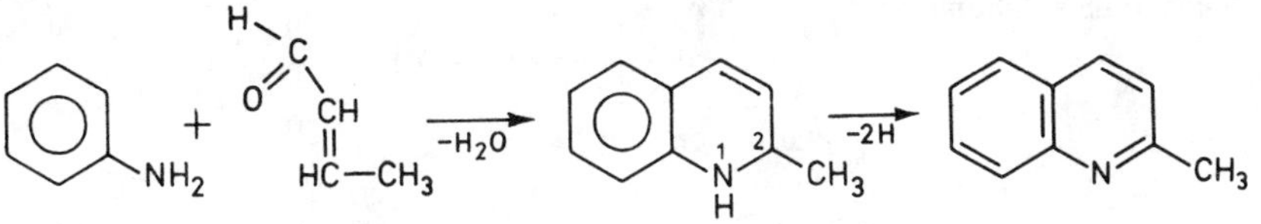

Aniline   Crotonaldehyde   2-Methyl-1,2-dihydroquinoline   Quinaldine

7.7.1.3 *Friedlander* Synthesis.[1] This is based on the alkaline condensation of *o-amino-benzaldehyde* with *aldehydes* or *ketones* which have a reactive CH_2-group next to the carbonyl group:

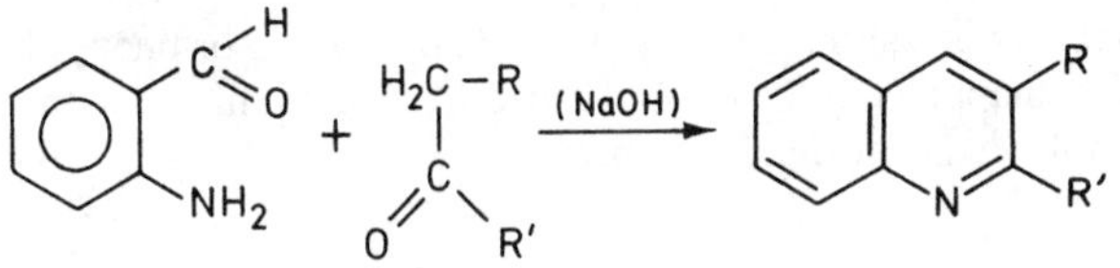

This reaction takes place particularly smoothly with β-keto acids in almost neutral solution under physiological-type conditions. Ring-closure is accompanied by decarboxylation (*C. Schöpf*). Anthranilic acid can also be used as a starting material.

Properties and Uses

Quinoline is a colourless liquid, b.p. 237°C (510 K), with a pungent smell. Its basicity is about the same as that of aniline. When it is oxidised with potassium permanganate the benzene ring is broken down and the product is pyridine-2,3-dicarboxylic acid (quinolinic acid):

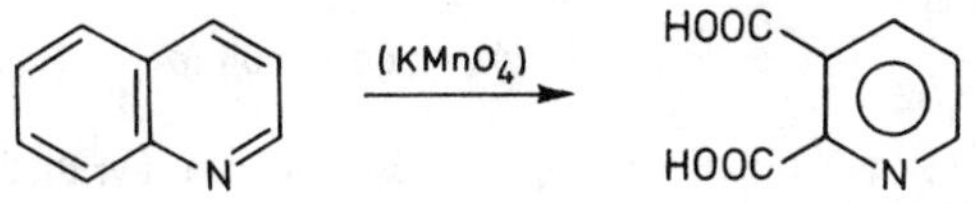

Quinolinic acid

---

[1] See *Chia-Chung Cheng* et al., The *Friedlander* Synthesis of Quinolines, Org. React. *28*, 37 (1982).

Catalytic hydrogenation of quinoline results first of all in reduction of the heterocyclic ring, followed by reduction of the benzene ring, giving in turn *tetrahydroquinoline* and then *decahydroquinoline*. Both of these compounds are secondary amines which resemble piperidine.

7.7.1.4 Quinoline undergoes *electrophilic substitution* more readily than pyridine does, attack taking place preferentially in the benzene ring. For example, nitration leads to the formation of *5-nitroquinoline* (53%) and *8-nitroquinoline* (47%), and sulphonation gives *quinoline-8-sulphonic acid* as the main product. Only in the case of halogenation is there substitution in the pyridine ring, *3-halogenoquinolines* being formed:

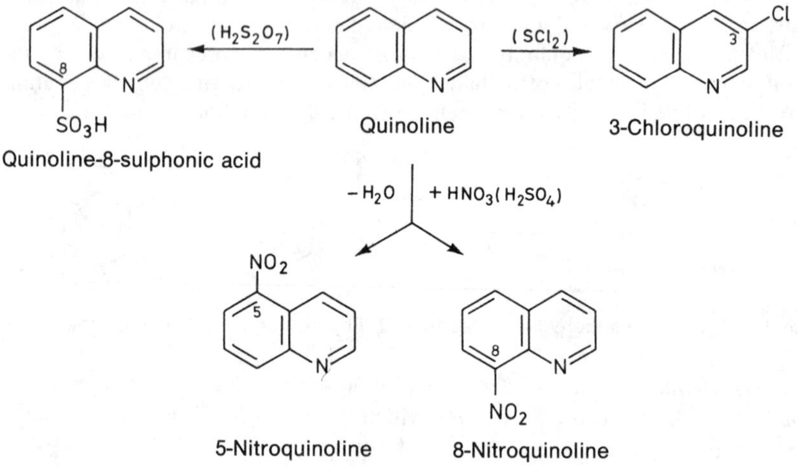

Quinoline-8-sulphonic acid   Quinoline   3-Chloroquinoline

5-Nitroquinoline   8-Nitroquinoline

Nucleophiles such as potassium amide, potassium hydroxide or phenyl lithium attack the 2- and 4-positions, as happens in the case of pyridine. For example, with potassium amide 2- and 4-aminoquinolines are formed (see p.762):

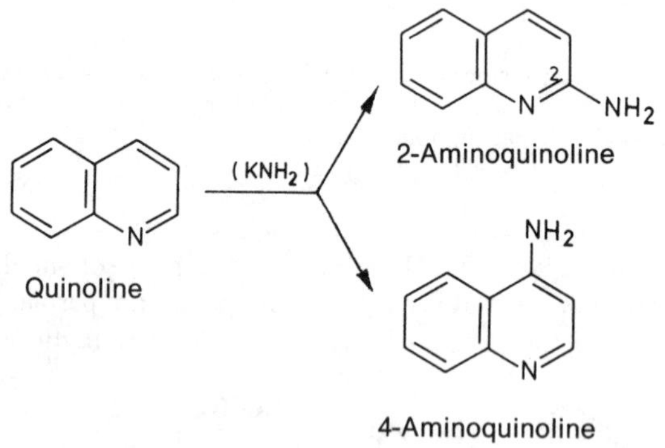

2-Aminoquinoline

Quinoline

4-Aminoquinoline

2- and 4-Methylquinolines (quinaldine and lepidine) resemble their pyridine analogues, the 2- and 4-picolines, in that their methyl groups undergo condensation reactions with aldehydes, ketones, nitroso compounds, etc.

7.7.1.5 8-Hydroxyquinoline (8-quinolinol) is readily prepared from *o*-aminophenol by means of the *Skraup* synthesis. With many metal ions it forms complexes, known as **oxinates**, which precipitate out quantitatively, and for this reason it has been used as a reagent in analytical chemistry.

8-Hydroxyquinoline   Oxinate   INN: **Norfloxacin**

8-Hydroxyquinoline is also a powerful *fungicide* and *antiseptic*.

Inhibition of *DNA-gyrase* provides vital interference with the development of bacteria (see p.858) and in consequence gyrase inhibitors such as 1-ethyl-6-fluoro-1,4-dihydro-4-oxo-7(1-piperazinyl)-quinoline-3-carboxylic acid (INN: **Norfloxacin**) are very active bactericides.

A number of active *antimalarials* are members of the quinoline series, *e.g.* INN: **Chloroquine** (*Resoquin*), 7-chloro-4-(4-diethylamino-1-methylbutylamino)quinoline and INN: **Primaquine**, 8-(4-amino-1-methylbutylamino)-6-methoxyquinoline. 4-Aminoquinoline derivatives of this series are, like quinoline, active against schizonts (in the asexual multiplication phase), whereas 8-aminoquinoline derivatives act on the gametes (in the sexual cycle). These stages have developed an increasing *resistance* to the effects of the antimalarials. Chloroquine has proved effective not only against malaria but especially in the treatment of rheumatic illnesses.

INN: *Chloroquine*   INN: *Primaquine*

Mention must also be made of N^1-*p*-chlorophenyl-N^5-isopropylbiguanide (Paludrine), which is a good antimalarial that is active against the asexual stage of development.

7.7.2 Benzoquinolines

Fusion of a benzene ring to side *b* of quinoline provides *acridine* (benzo[*b*]quinoline), and to side *c* *phenanthridine* (benzo[*c*]quinoline):

Acridine   Phenanthridine

In these cases the numbering of the atoms in the ring does not begin at the heteroatoms. Using the 'a'-nomenclature acridine can also be described as *10-aza-anthracene*.

7.7.2.1 Acridine occurs in small amounts in coal tar. It forms colourless needles, m.p. 111°C (384 K). Its pK_a is 5.58 and it forms salts with strong mineral acids. Solutions of acridine are yellow and show a strong blue fluorescence.

The best way of preparing acridine is by oxidising 9,10-dihydroacridine. The dihydro compound is obtained from diphenylamine-2-carboxylic acid; the first product is **acridone** and this is then reduced with sodium and amyl alcohol:

Diphenylamine-2-carboxylic acid   Acridone   9,10-Dihydro-acridine

The yellow mordant dyes derived from acridine are no longer used for dyeing. On the other hand the quinacridone pigments based on *acridone* are used for making red-violet pigments for car enamels. The parent compound is *trans*-quinacridone (5,12-dihydroquinolino[2,3-*b*]acridine-7,14-dione), and this can be modified by introduction of methyl groups or halogen atoms at positions 1–4 and 8–11.

trans-Quinacridone   INN: **Ethacridine**   Acridinium yellow

The acridine dyestuff 6,9-diamino-2-ethoxy-acridine (INN: **ethacridine**) is used as a disinfectant for wounds. Combined with *acridinium yellow* it forms the active ingredient of pastilles for relieving sore throats.

7.7.2.2 Phenanthridine can be prepared by heating *N*-(2-biphenylyl)formamide with phosphorus oxychloride (*cf. Bischler–Napieralksi* reaction). It can also be obtained by photochemical cyclisation and dehydrogenation of benzylidene-aniline.

N-(2-Biphenylyl)formamide   Phenanthridine   Benzylidene-aniline

It crystallises from aqueous ethanol as colourless needles, m.p. 108°C (381 K). In aqueous solution it shows a blue fluorescence. It is weakly basic (pK_a 5.58).

Amination of phenanthridine with *sodium amide* provides *6-aminophenanthridine*. When this is diazotised with nitrous acid and the solution is boiled **phenanthridone** is formed:

6-Aminophenanthridine   Phenanthridone

7.7.3 *Isoquinolines*

Isoquinoline also occurs in coal-tar. Its smell is reminiscent of that of benzaldehyde. It melts at 26.5°C (299.7 K), boils at 243°C (516 K) and resembles quinoline in its properties.

The following processes provide the best ways of *making* isoquinoline derivatives:

7.7.3.1 *Bischler–Napieralski* Reaction (1893).[1] If *N*-acylated β-phenylethylamines are heated with a dehydrating agent such as phosphorus(V) oxide or phosphorus oxychloride in benzene or tetralin an intramolecular cyclisation takes place and 3,4-dihydroisoquino-lines are formed. Either oxidation with potassium permanganate or dehydrogenation over palladium converts the dihydroisoquinolines into the corresponding isoquinoline deriva-tives. Thus, for example, *N*-phenylethylacetamide gives *1-methylisoquinoline*:

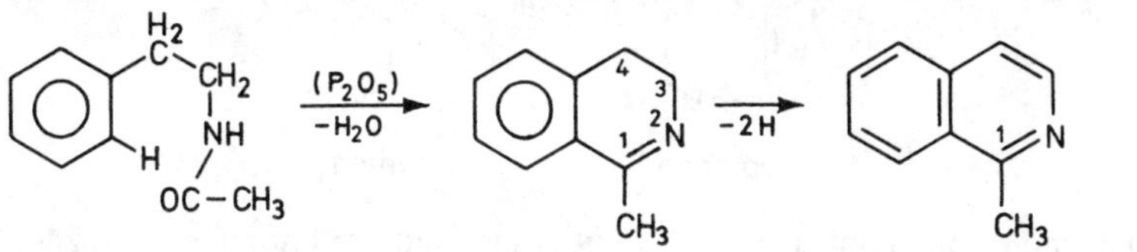

N-Phenylethylacetamide   1-Methylisoquinoline

1-Methylisoquinoline has a reactive methyl group which, for example, condenses with aldehydes.

7.7.3.2 1,4-Dipolar Cycloaddition Reactions.[2] Although *Diels–Alder* reactions play an important part in syntheses of carbocyclic compounds, they play a much smaller role in the synthesis of six-membered ring heterocycles. One possible way of carrying out such syntheses is by making use of *1,4-dipolar cycloaddition reactions*. For example, if iso-quinoline is treated with dimethyl acetylenedicarboxylate and phenyl isocyanate a yellow, crystalline adduct is formed, as follows:

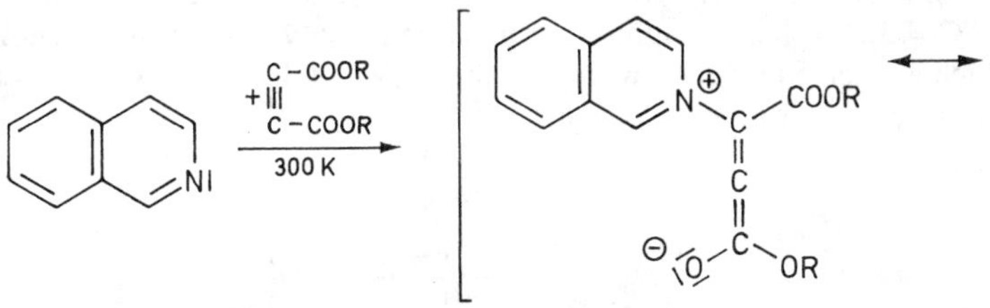

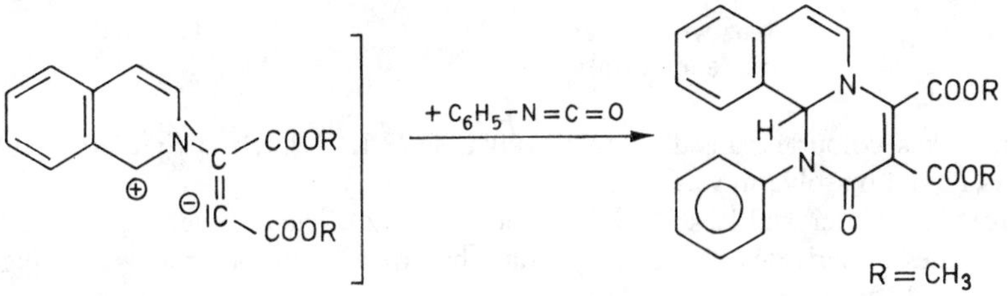

$R = CH_3$

[1] See *W. M. Whaley* et al., Org. React. **6**, 74 (1951).
[2] See *R. Huisgen*, Synthese von Heterocyclen mit 1,4-dipolaren Cycloadditionen, Z. Chem. **8**, 290 (1968).

The mechanism of the reaction can be explained as follows. First of all the acetylene-ester, as an electrophile, bonds to the nucleophilic nitrogen atom of isoquinoline to give a mesomeric *zwitterion*, which can behave as a 1,4-dipole. This reacts as such with a molecule of phenyl isocyanate to give a 6-membered ring adduct.

7.7.4 Chromans[1]

The compound in which a benzene ring is *o*-fused to a 4*H*-pyran ring is known as **4*H*-chromene**. Its dihydro-derivative is **chroman**, an oil, b.p. 215°C (488 K), which is volatile in steam and has a peppermint-like odour.

4*H*-Chromene   Chroman

7.7.4.1   Far more important than the parent compound are chromone (benzo-γ-pyrone), **flavone** (2-phenylchromone) and **3-hydroxyflavone** (flavanol).

Chromone   Flavone   3-Hydroxyflavone

Chromone is isomeric with *coumarin* (benzo-α-pyrone). It crystallises in colourless needles, m.p. 59°C (332 K). Its derivatives are most commonly made from phenols and β-keto esters, which are heated together with phosphorus(V) oxide (*Simonis*). For example phenol and benzoylacetic ester produce *flavone*:

Benzoylacetic ester   Flavone
(enol form)

Flavone forms colourless needles, m.p. 100°C (373 K), which are readily soluble in ethanol but do not dissolve in water.

3-Hydroxyflavone crystallises as yellow needles (Lat. *flavus* = yellow), m.p. 169°C (442 K). A series of derivatives containing further hydroxy groups occur as yellow pigments in flowers, wood and roots.

---

[1] See *J. B. Harborne* (Ed.), The Flavonoids: Advances in Research since 1980 (Chapman and Hall, London 1988).

Quercitin, 3,5,7,3′,4′-pentahydroxyflavone, is the most important plant pigment of this sort. It occurs in the bark of the American oak *Quercus velutina* as an L-rhamnoside, bonded at the 3-position. It also occurs either in the free state or bonded as a glycoside in many plants, for example in the flowers of wallflowers, pansies, snapdragons and roses, and also in tea and in hops.

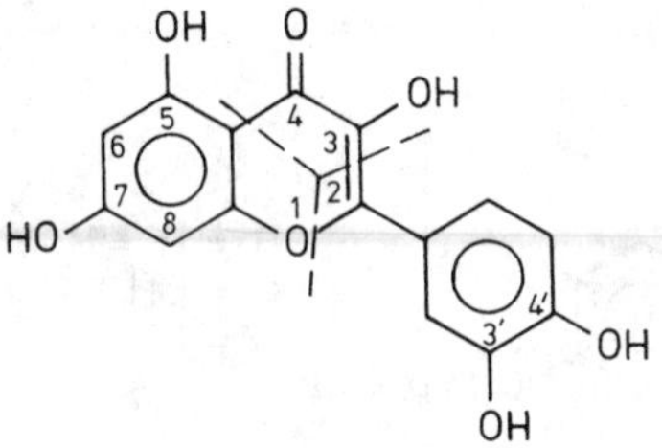

Quercitin

Elucidation of the structure of quercitin depended very much on the products obtained by melting it with alkali. This splits the molecule into *phloroglucinol* and *protocatechuic acid* (3,4-dihydroxybenzoic acid), thereby indicating the positions of the substituent hydroxy groups.

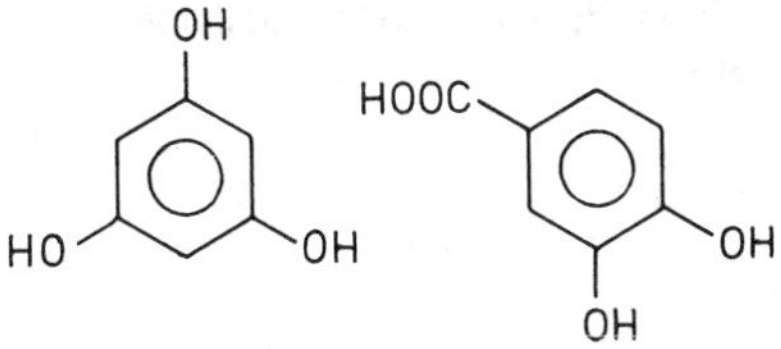

Phloroglucinol   Protocatechuic acid

Morin, 3,5,7,2′,4′-pentahydroxyflavone, is a sensitive reagent for the presence of aluminium ions. In ethanolic solution a characteristic green fluorescence develops.

7.7.4.2   Anthocyanins (Flavylium Salt Pigments)

Anthocyanins are responsible for the red, violet and blue colours of many flowers, fruits and other parts of plants. They are based on the ring skeleton of **flavene** (2-phenylchromene). Both chemically and plant-physiologically they are closely related to the hydroxyflavanols and they also exist as glycosides. Treatment of the glycosides with either acid or a glycoside-splitting enzyme breaks them down into the related sugar and the actual pigments (aglycones) which are **anthocyanidins.** In acid hydrolysis of an anthocyanin the anthocyanidin is converted into a flavylium salt whose cation is mesomerically stabilised.

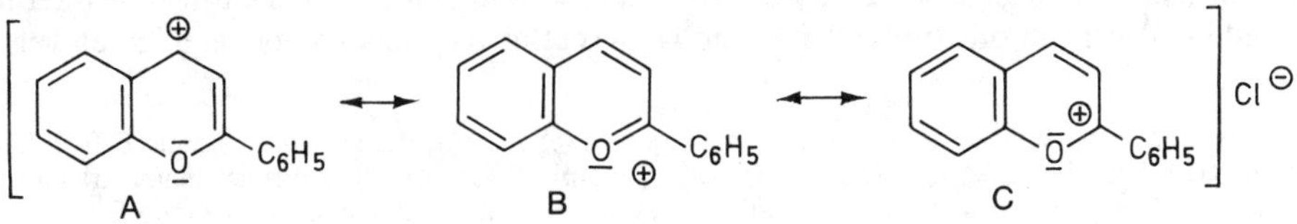

Flavylium chloride

In representing the salt the *oxonium formula B* is usually used, but condensation reactions which the anthocyanins undergo involve carbocationic character as evinced by the carbenium formulae A and C.

Although there is a wide range of shades of colour in flowers, the anthocyanidins are all related to the following three types of flavylium salts, which differ only in the number of hydroxy groups attached to the 2-phenyl group. They are: **pelargonidin chloride** (3,5,7,4'-tetrahydroxyflavylium chloride), **cyanidin chloride** (3,5,7,3',4'-pentahydroxyflavylium chloride) and **delphinidin chloride** (3,5,7,3',4',5'-hexahydroxyflavylium chloride).

$R^1 = R^2 = H$: Pelargonidin chloride
$R^1 = OH, R^2 = H$: Cyanidin chloride
$R^1 = R^2 = OH$: Delphinidin chloride

The sugar residues of anthocyanins (glucose, glucuronic acid) are usually attached to the HO-groups in the 3-position or in the 3- and 5-positions. *Succinylcyanins* and a *malonylflavone* have also been isolated from cornflower blooms. These are *half-esters* of, respectively, *succinic acid* and *malonic acid* with a glucose group (*H. Tamura*). Esters of *anthocyanidin-3-glucosides* with *acetic acid*, *coumaric acid* and *caffeic acid* (3,4-dihydroxycinnamic acid) are found in red wine.

Whilst the mineral acid salts of anthocyanidins are more or less red-coloured, *e.g.* pelargonidin chloride is yellowish red, cyanidin chloride red to violet, and delphinidin chloride blueish red, the free anthocyanidins obtained by neutralisation of the salts have violet to blue colours. Anthocyanidins contain a quinonoid group:

Pelargonidin chloride (red)   Pelargonidin (blue-violet)

The alkali salts (phenolates) of the neutral anthocyanidins are blue. Because of these changes of colour, it was for a long time assumed that the colour of a flower depended in the first place upon the particular hydrogen ion concentration in the cell fluid, especially since the red rose and the blue cornflower contained the same anthocyanin, cyanin. However it has now been shown that the variations in colour and the stabilisation of the anthocyanins is a result of self-association, co-pigmentation with flavones, and sandwich-like stacking of the molecules.[1]

7.7.4.3 Vitamin E (Tocopherol)

This substance belongs to the group of fat-soluble vitamins. It is abundant in wheat-germ oil and in other plant oils (up to 0.5%) and is present in very small amounts in green vegetables.

Its absence in rats renders them sterile. Female rats which have been fed with a vitamin E-free diet are unable to bear live offspring. The importance of vitamin E for fertility is further demonstrated by the fact that honey bees produce a queen only if they receive food which contains vitamin E.

Vitamin E deficiency in human beings causes neurological troubles. It is used in therapy for habitual abortions.

---

[1] See *T. Goto* and *T. Kondo*, Structure and Molecular Stacking of Anthocyanins — Flower Colour Variation, Angew. Chem. Int. Ed. Engl. *30*, 17 (1991).

So far four variants of the vitamin E type discovered by *Evans* have been isolated from natural products, known as α-, β-, γ- and δ-tocopherols. They have been shown to be derivatives of *chroman* (*Fernholz, W. John, Karrer, Todd*) and have an *isoprenoid* side chain at the 2-position (*cf.* also vitamin K$_1$).

α-**Tocopherol**, $C_{29}H_{50}O_2$, is an oil having the following constitution:

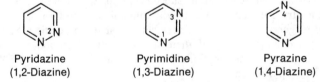

α-Tocopherol

A synthesis of ($\pm$)-α-tocopherol started from 2,5,6-trimethyl-1,4-hydroquinone and phytyl bromide, with zinc chloride present as a catalyst (*Karrer*). It is used as an antioxidant in food chemistry.

β- **and** γ-**Tocopherols**, $C_{28}H_{48}O_2$, each contain one fewer methyl group than α-tocopherol. In the case of β-tocopherol the methyl group at position 7 is missing; in γ-tocopherol that at position 5 is absent. In δ-**tocopherol**, $C_{27}H_{46}O_2$, both of these methyl groups, at the 5- and 7-positions, are missing.

7.8 Six-Membered Rings with Two Heteroatoms

Introduction of a further nitrogen atom into a pyridine ring provides three different isomers, depending upon the relative positions of the two nitrogen atoms. Of these aza-analogues of pyridine, the *diazines,* the most important is pyrimidine.

Pyridazine	Pyrimidine	Pyrazine
(1,2-Diazine)	(1,3-Diazine)	(1,4-Diazine)

Because of the higher electronegativity of the nitrogen atoms, these ring systems, like pyridine, are *π-electron deficient heteroaromatics.* As a result, electrophilic substitution reactions can be achieved only with great difficulty. They are only successful if electron-donating substituents such as H_2N-, HO-, HS-, RO- or RS-groups are present in the ring (+M-effect).

7.8.1 Pyridazines[1]

Pyridazine, b.p. 208°C (481 K), is a base with an odour similar to that of pyridine. It is readily soluble in water. It can be made from 3-hydroxy-6-pyridazinone by reaction with phosphorus oxychloride and replacement of the chlorine atoms by hydrogen:

---

[1] See *M. Tišler* and *B. Stanovnik*, Comprehensive Heterocyclic Chemistry, Vol. *3*, 1 (Pergamon Press, Oxford 1984).

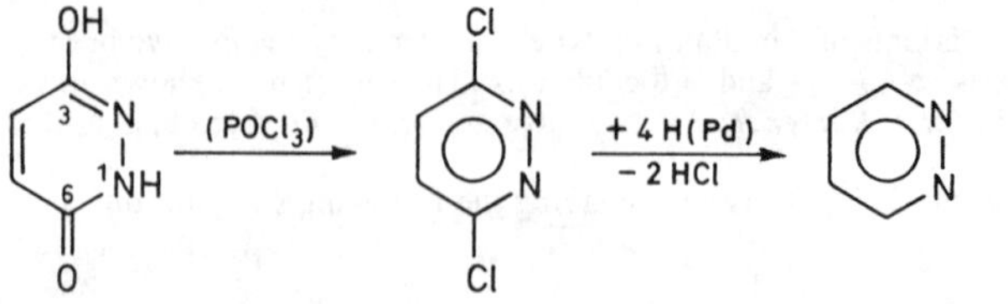

3-Hydroxypyridazin-6-one   Pyridazine

7.8.2 Pyrimidines[1]

The pyrimidine ring forms part of the structure of a number of important natural products, *e.g.* vitamin B$_1$, the purines, and products derived from the splitting of nucleic acids.

7.8.2.1 Preparation. 4,6-Disubstituted pyrimidines can be made by *heating 1,3-diketones with formamide* at 180–200°C (450–470 K) (*Bredereck* and *Gompper*, 1957). The reaction proceeds as follows:

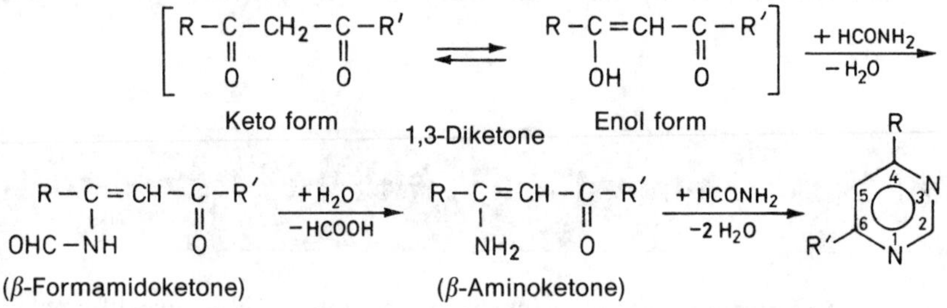

Keto form   Enol form

1,3-Diketone

(*β*-Formamidoketone)   (*β*-Aminoketone)

Pyrimidine itself can be made in 65% yield starting from malonaldehyde diacetal.

Properties. Pyrimidine is a crystalline substance, m.p. 22°C (295 K), which dissolves in water to give a neutral solution and reacts with mineral acids to form salts.

Electrophilic reagents attack pyrimidine, if at all, at the 5-position. Such substitution reactions take place more readily if there are strongly *electron-donating substituents* attached to the ring, thus 2-aminopyrimidine reacts with nitrating acid at 50°C (325 K) to form 2-amino-5-nitropyrimidine.

Important Derivatives of Pyrimidine

7.8.2.2 Barbituric acid (malonylurea) is a *cyclic ureide* and can be made by *condensing diethyl malonate with urea* in the presence of a sodium or magnesium alkoxide:

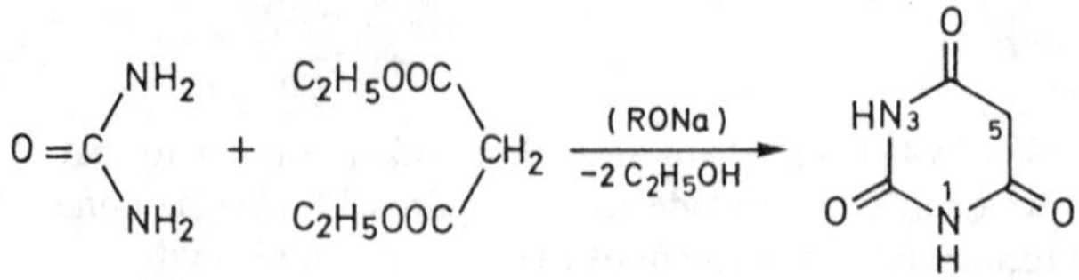

Barbituric acid

[1] See *D. J. Brown*, Comprehensive Heterocyclic Chemistry, Vol. *3*, 57 (Pergamon Press, Oxford 1984).

It forms colourless prisms, m.p. 245°C (518 K) which are soluble in hot water. Its pK_a^1 value is 4.01. Replacement of the hydrogen atoms of the *reactive* methylene group at the 5-position by alkyl groups lowers the acidity of the barbituric acid.

5,5-Dialkylbarbituric acids can be obtained using the above method by reaction of urea with diethyl dialkylmalonates. For example, diethyl diethylmalonate gives 5,5-diethylbarbituric acid, pK_a^1 7.89, m.p. 191°C (464 K), which was formerly widely used as a soporific under the name Veronal (INN: **Barbital**). It is still used as a component of medicaments which act as sedatives. (*E. Fischer* and *v. Mering*, 1903).

Other important **barbiturates** include 5-ethyl-5-phenylbarbituric acid (INN: **Phenobarbital**, *Luminal*), 5-(1-cyclohexenyl)-5-ethylbarbituric acid (*INN:* **Cyclobarbital**, *Phanodorm*) and, for use as a soporific and short-lived anaesthetic, 5-(1-cyclohexenyl)-3,5-dimethylbarbituric acid (*INN:* **Hexobarbital**, *Evipan*).

INN: *Barbital*   INN: *Phenobarbital*   INN: *Cyclobarbital*   INN: *Hexobarbital*

The pyrimidine ring is also present in some important sulphonamide *chemotheropeutics*. Among these are N^1-(4-methyl-2-pyrimidinyl)sulphanilamide (*INN:* **Sulphamerazine**) and N^1-(2,6-dimethyl-4-pyrimidinyl)sulphanilamide (*INN:* **Sulphisomidine**, *Aristamide*).

INN: *Sulphamerazine*   INN: *Sulphisomidine*

Other chemotherapeutics of this sort are N^1-(2,6-dimethoxy-4-pyrimidinyl)sulphanilamide (*INN:* **Sulphadimethoxine**, *Madribon*) and N^1-(5-methoxy-2-pyrimidinyl)sulphanilamide (*INN:* **Sulphamethoxydiazine**, *Durenat*), which is used as a long-term sulphanilamide.

2,4-Diamino-5-(3,4,5-trimethoxybenzyl)pyrimidine (*INN:* **Trimethoprim**) inhibits the reduction of folic acid to folinic acid in bacterial metabolism. It is used, together with sulphonamides such as *INN:* **Sulphamethoxazole**, which disturb the bacterial metabolism at another point, in a combined preparation, *e.g. Bactrim*.

INN: *Trimethoprim*

7.8.2.3   Pyrimidinium Bases from Nucleic Acids

Hydrolytic cleavage of nucleic acids gives rise to, among other products, the following pyrimidine bases:

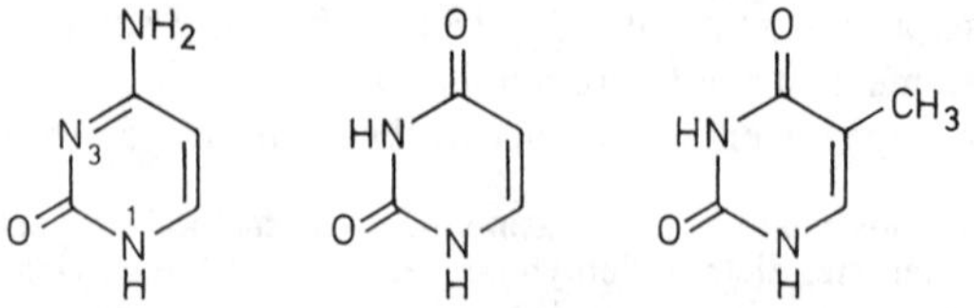

Cytosine   Uracil   Thymine

Nitrous acid converts cytosine into uracil, which can also be made by condensing urea with formylacetic acid. The latter compound is obtained by heating malic acid with conc. sulphuric acid:

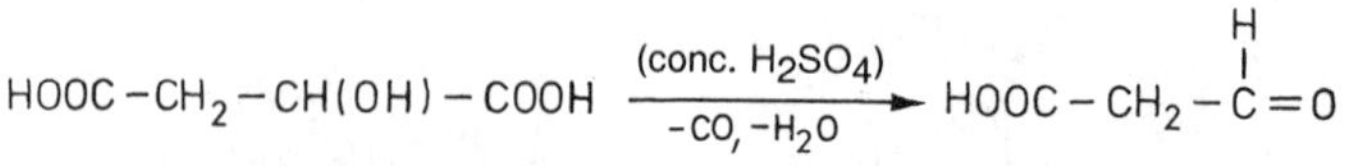

$$\text{HOOC}-\text{CH}_2-\text{CH(OH)}-\text{COOH} \xrightarrow[-\text{CO, }-\text{H}_2\text{O}]{\text{(conc. H}_2\text{SO}_4\text{)}} \text{HOOC}-\text{CH}_2-\overset{\text{H}}{\underset{}{\text{C}}}=\text{O}$$

Malic acid   Formylacetic acid (oxo form)

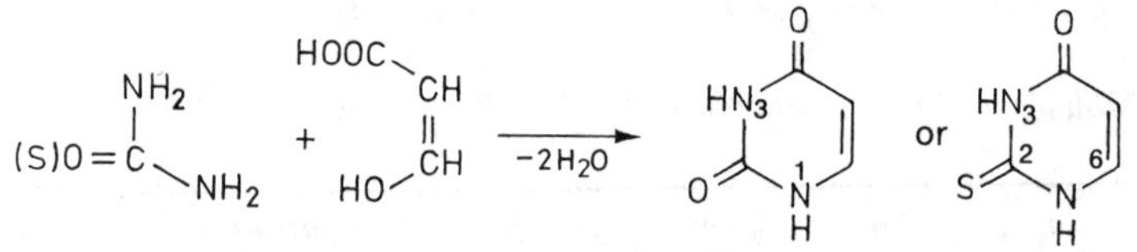

Formylacetic acid   Uracil   Thiouracil
(enol form)

Thiouracil, which has a sulphur atom at the 2-position, can be made similarly from formylacetic acid and thiourea (see formulae above).

6-Propylthiouracil (INN: *Propylthiouracil*, Thyreostat II), is an effective *thyrostatic*, which is used for treating over-activity of the thyroid gland, but its activity is exceeded 10- to 15-fold by that of 1-methylimidazole-2-thiol (INN: *Thiamazole, Methimazole*).

7.8.2.4   Thiamine (Vitamin B₁)

Thiamine (aneurine) contains a pyrimidine ring which is linked by a methylene bridge to a thiazole ring. It is found in almost all plant and animal tissues, and especially in grain husks, yeast and potatoes. It was first isolated by *Jansen* and *Donath* in 1926 from rice bran and later (1932) by *Windaus* from yeast.

Thiamine is broken down by sodium hydrogen sulphite into 5-(4-amino-2-methyl)pyrimidinylmethanesulphonic acid (I) and 5-(2-hydroxyethyl)-4-methylthiazole (II).

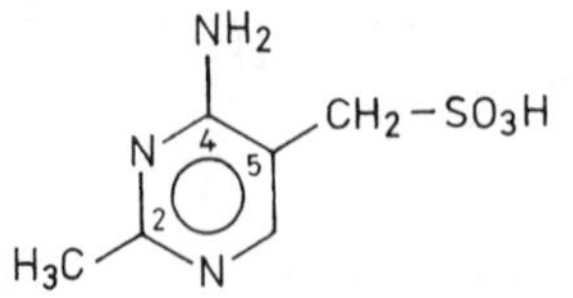

(I) 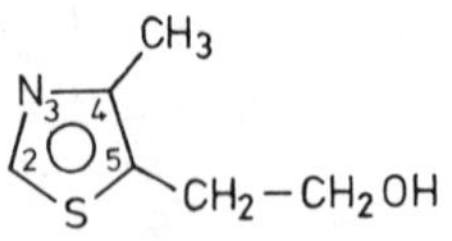   (II)

Both of the products could be prepared synthetically and this played an essential role in determining the structure of the parent molecule (*Williams*). Eventually thiamine was shown to be a quaternary *thiazolium chloride* with the following structure:

Thiamine

Thiamine was synthesised at more or less the same time by several groups of workers, all independently of each other (*Williams, Andersag* and *Westphal, Grewe*).

Thiamine forms a colourless chloride–hydrochloride which is readily soluble in water and decomposes at 250°C (525 K). It is used in the treatment of neuritis and neuralgia. It is also an important *growth material* for many microorganisms. *Thiamine diphosphate* is of great biochemical importance. It acts as a coenzyme of the enzyme *pyruvate decarboxylase* (INN: *Cocarboxylase*), which takes part in alcoholic fermentation and converts pyruvic acid into 'active acetaldehyde'.

7.8.3 Pyrazines[1]

Atmospheric oxidation of the condensation products obtained from α-aminoaldehydes or α-aminoketones leads to the formation of pyrazines:

α-Aminoketone   2,5-Dialkylpyrazine

Pyrazine itself can be made by dehydrogenating piperazine (hexahydropyrazine) over a copper chromite catalyst at high temperatures (*Dixon*, 1946).

7.8.3.1   Piperazine, m.p. 104°C (377 K), is a strong base (pK_a 9.83). It is obtained by heating ethylenediamine dihydrochloride, and is used as an *anthelmintic* (*i.e.* for the treatment of worms) against *oxyuriasis* and *ascaridiasis*.

The bicyclic compound *triethylenediamine* (1,4-diazabicyclo[2.2.2]octane, **DABCO**) can be made by heating *N*-hydroxyethylpiperazine. It forms hygroscopic crystals, m.p. 159.8°C (443 K), b.p. 174°C (447 K). DABCO is a *Lewis* base (pK_{a1} 8.60; pK_{a2} 2.95) which is much used as a catalyst.

7.8.3.2   2,5-Dioxopiperazines (2,5-diketopiperazines) are cyclic amides and are formed when α-amino acids are heated (see p.287).

Piperazine   N-Hydroxyethyl   DABCO   2,5-Dioxopiperazine
                    piperazine   Piperazine-2,5-dione

[1] See *A. E. A. Porter*, Comprehensive Heterocyclic Chemistry, Vol. *3*, 157 (Pergamon Press, Oxford 1984).

These readily obtainable heterocycles are well suited for transferring the chirality of one amino acid to another by asymmetric induction in the *Schöllkopf* reaction (1979).[1]

For this purpose a mixed dioxopiperazine obtained from, for example, a homochiral amino acid, L-valine, as the chiral auxiliary, and a DL-amino acid or glycine is used (see p.833). This is converted by trimethyloxonium tetrafluoroborate (see p.143) into a bislactim ether.

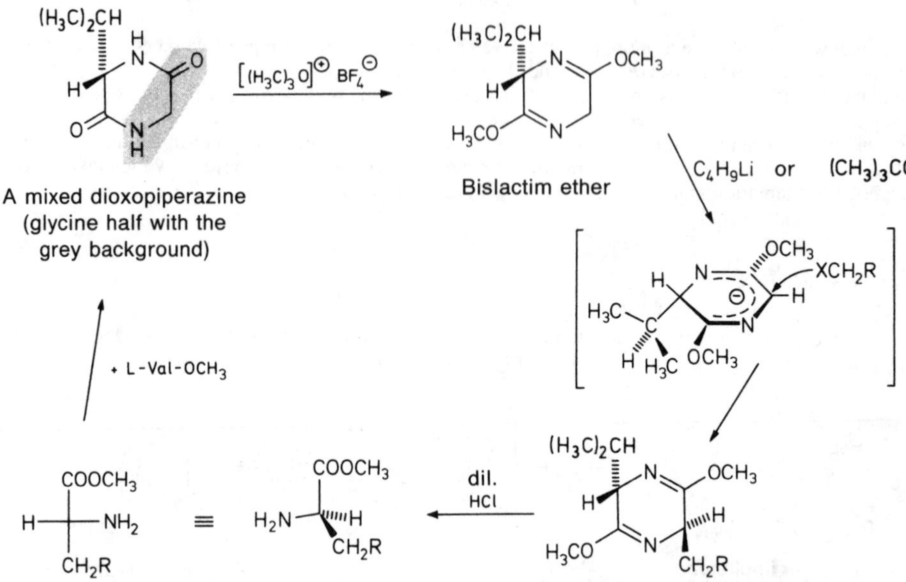

A mixed dioxopiperazine
(glycine half with the
grey background)

Bislactim ether

+ L-Val-OCH₃

dil.
HCl

This gives a delocalised carbanion that is planar, but whose two faces differ, in that one of them is shielded by the isopropyl group of the L-valine moiety. An electrophilic reagent attacks mainly from the more readily accessible side, namely that *trans* to the isopropyl group, and thus one diastereomer is produced in large excess. Careful hydrolysis gives an ester of a D-amino acid and an ester of L-valine. These can be separated by distillation, so that the chiral auxiliary is recovered. The optical yield in this enantioselective synthesis is high, *e.g.* for phenylalanine (R = C_6H_5) ee = 91–95%.

7.8.4 Benzodiazines

The more important members of this class of compounds are generally known by their trivial names and the atoms are numbered as follows:

Cinnoline
(m.p. 40°C [313 K])

Phthalazine
(m.p. 90°C [363 K])

Quinazoline
(m.p. 48°C [321 K])

Quinoxaline
(m.p. 30°C [303 K])

Only a few derivatives of these four ring systems, the *diazanaphthalenes*, which are of special interest, are mentioned here.

7.8.4.1   Phthalazine can be made by condensing phthalaldehyde with hydrazine.

---

[1] See *U. Schöllkopf*, Enantioselective Synthesis of Nonproteinogenic Amino Acids, Top. Curr. Chem. *109*, 65 (1983).

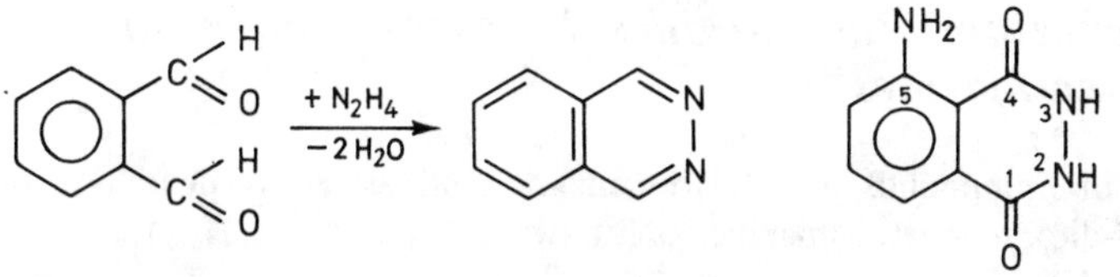

Phthalaldehyde   Phthalazine   Luminol

When an alkaline solution of the phthalazine derivative 5-amino-2*H*,3*H*-phthalazine-1,4-dione (3-aminophthalhydrazide,[1] luminol) is treated with hydrogen peroxide and a small amount of haemin, an intense blue-violet chemiluminescence results which lasts several seconds.

7.8.4.2   Quinazolines may be prepared by the reaction of *o*-acylanilines with acid amides:

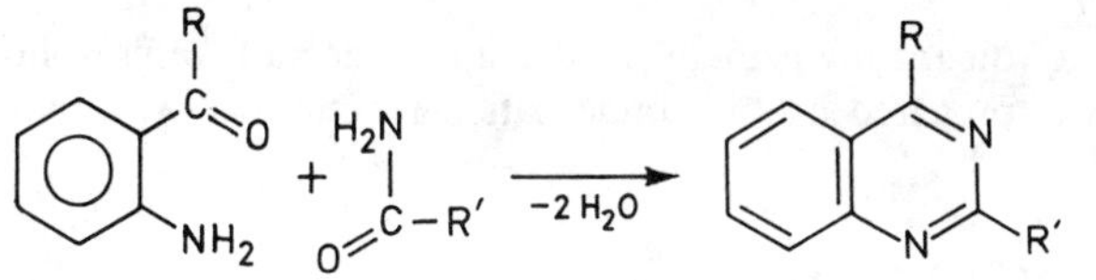

o-Acylaniline   Quinazoline (R = R' = H)

Similarly if anthranilic acid is heated with acid amides *quinazolin-4-ones* are produced.

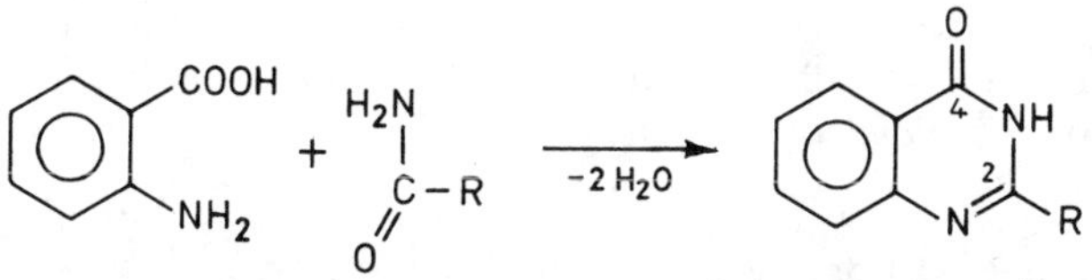

Anthranilic acid   Quinazolin-4-one (R = H)

2-Methyl-3-(*o*-tolyl)-4-(3*H*) quinazilinone (INN: ***Methaqualone***) acts as a *soporific*.

Rotation about the bond linking the rings in this compound is restricted and it can be resolved into enantiomers (atropisomerism).

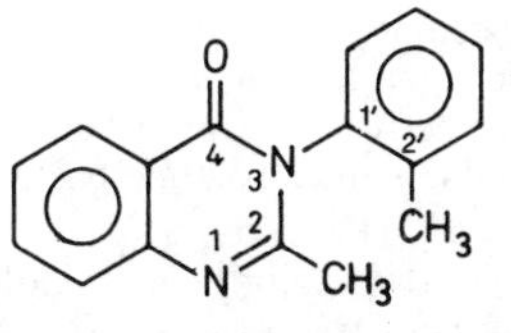

INN: *Methaqualone*

7.8.4.3   Quinoxalines arise from the condensation of *o*-phenylenediamine with 1,2-diketones, *e.g.*

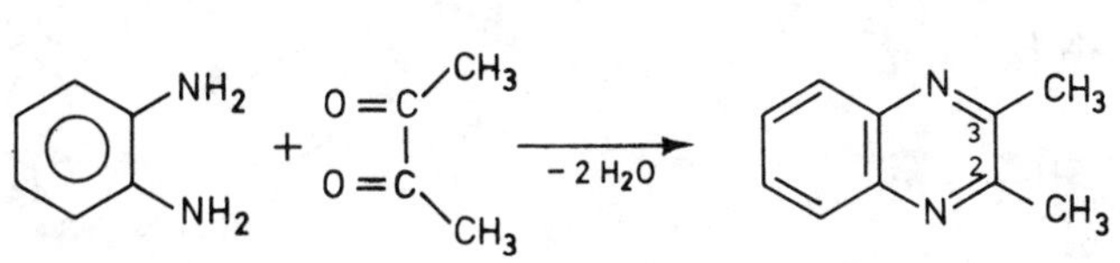

o-Phenylenediamine   Diacetyl   2,3-Dimethyl-quinoxaline

---

[1] See *K.-D. Gundermann* et al., Chemiluminescence in Organic Chemistry (Springer-Verlag, Berlin 1987).

7.8.5 Phenazines, Phenoxazines, Dibenzo-p-dioxins and Phenothiazines

These sets of compounds are the dibenzo-derivatives of, respectively, pyrazine, 1,4-oxazine, 1,4-dioxin and 1,4-thiazine, as follows:

Phenazine   Phenoxazine   Dibenzo-*p*-dioxin   Phenothiazine

7.8.5.1 Phenazine (dibenzopyrazine, 9,10-diaza-anthracene) forms yellow needles, m.p. 171°C (444 K). It is prepared by the condensation reaction of *o*-phenylenediamine with *o*-benzoquinone:

o-Benzoquinone   Phenazine

If *o*-phenylenediamine is oxidised by iron(III) chloride in acetic acid it undergoes self-condensation to give *2,3-diaminophenazine*:

2,3-Diaminophenazine   Pyocyanine

A more highly condensed ring system of this sort is found in the already mentioned compound indanthrene (see p.641), which can be regarded as a derivative of 5,10-dihydrophenazine. In addition, some *bacterial pigments* are phenazine derivatives, for instance *pyocyanine* (5-methyl-5*H*-phenazine-1-one, an inner salt), which is blue and obtained from the bacterium *Pseudomonas aeruginosa*.

7.8.5.2 Phenoxazine (dibenzo-5,10-oxazine) is a colourless substance, m.p. 156°C (429 K). It can be made by heating a mixture of *o*-aminophenol and its hydrochloride (*Gilman* and *Moore*, 1957):

Phenoxazine

1,4-Oxazine itself is only known in the form of derivatives. Tetrahydro-1,4-oxazine, *morpholine*, b.p. 128°C (401 K) is a secondary amine (pK_1 8.33) which is used as a base and as a solvent for certain chemical reactions. 4-Methylmorpholine *N*-oxide monohydrate, m.p. 71–73°C (344–346 K), is a solvent for cellulose and is a useful reagent. *1,4-Thiazine* is a colourless liquid, b.p. 76°C (349 K).

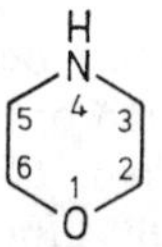

Morpholine

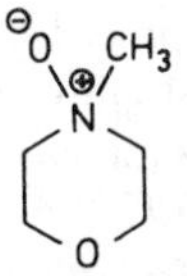

4-Methylmorpholine
N-oxide

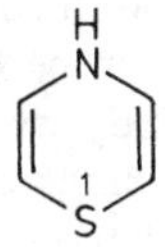

1,4-Thiazine

7.8.5.3 Phenoxazine Pigments

Of these pigments, which are derivatives of phenoxazonium chloride, only the *gallocyanines* have some importance. The simplest of these, **gallocyanine**, can be prepared by condensing gallic acid with *p*-nitrosodimethylaniline hydrochloride.

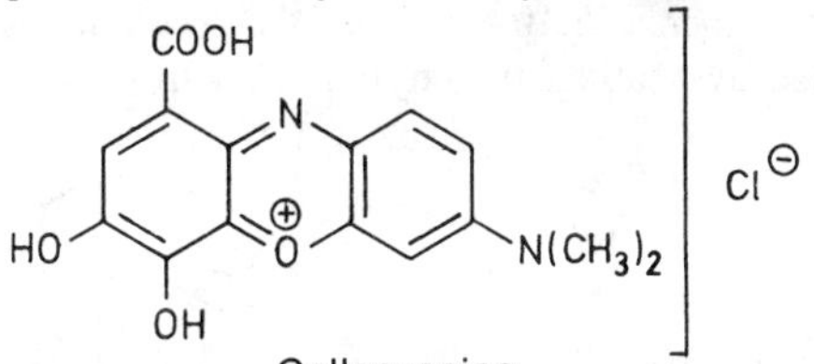

Gallocyanine

Cationic phenoxazine pigments are used for colouring poly(acrylonitrile) fibres.

Mention should also be made here of the **ommochrome pigments** which occur widely as pigments in the eyes, skin and wings of insects. They are regarded as the end-products of tryptophane metabolism and result from the oxidation of 3-hydroxykynurenine (*Butenandt*) (see p.725). *Ommatines* are the first examples of pigments of the phenoxazone type to be found in nature. An example is **xanthommatine** which has been isolated from the hatching secretion of the small tortoiseshell butterfly (*Vanessa urticae*).

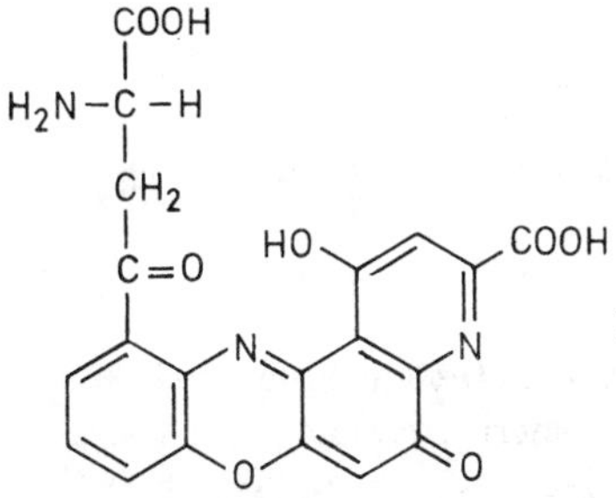

Xanthommatine

In addition, this type of compound has some significance as the chromophore of **actinomycin**, which is present as a metabolic product in some actinomycetes of the *Streptomyces* group; for example, *actinomycin D* (AMD) has a phenoxazone structure (*Brockmann*).[1] Actinomycins are powerful antibiotics, but they are also highly toxic and in consequence their only use is as *cytostatics*.

7.8.5.4 2,3,7,8-Tetrachlorodibenzo-*p*-dioxin (TCDD)[2]

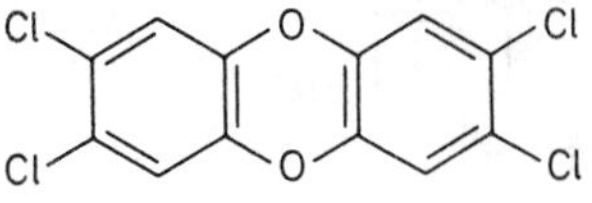

TCDD

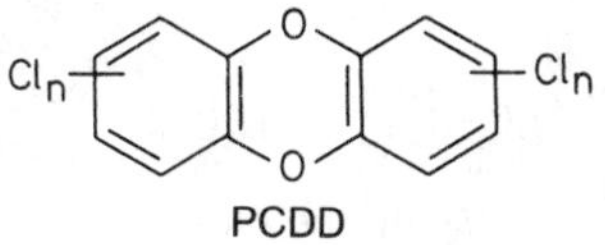

PCDD

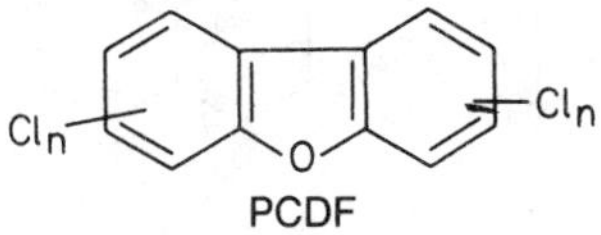

PCDF

---

[1] See *H. Lackner*, Three-dimensional Structure of the Actinomycins, Angew. Chem. Int. Ed. Engl. *14*, 375 (1975).
[2] See *R. A. Hiles*, Environmental Behaviour of Chlorinated Dioxins, Acc. Chem. Res. *23*, 194 (1993).

This compound is formed in industrial processes as a condensation product of 2,4,5-trichlorophenol. It can be formed in small amounts in combustion processes if chlorine-containing compounds are present, for example in the incineration of refuse. To burn it completely, temperatures exceeding 1200°C (1470 K) are required. TCDD ('dioxin') is very poisonous and produces skin blisters which only heal with difficulty. It has been shown to act as a carcinogen in some animals.[1]

Polychlorinated dibenzo-*p*-dioxins (PCDD) and dibenzofurans (PCDF) have also often been lumped together as 'dioxins'; this can lead to misunderstandings since on this reckoning some 200 different substances are involved.

It has been found that certain bacteria can break down polychlorinated aromatics, including 'dioxins', thereby rendering them inocuous.

7.8.5.5   Phenothiazine (dibenzo-5,10-thiazine)[2] is a colourless substance, m.p. 180°C (453 K) which is obtained by heating diphenylamine with sulphur:

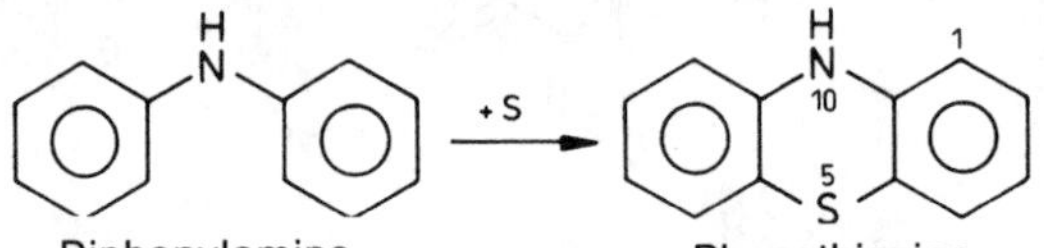

Diphenylamine   Phenothiazine

It is used for exterminating gnat larvae and as an intestinal antiseptic for animals.

Some phenothiazine derivatives have been tested for use in producing potentiation in narcosis, thus reducing the use of narcotics.

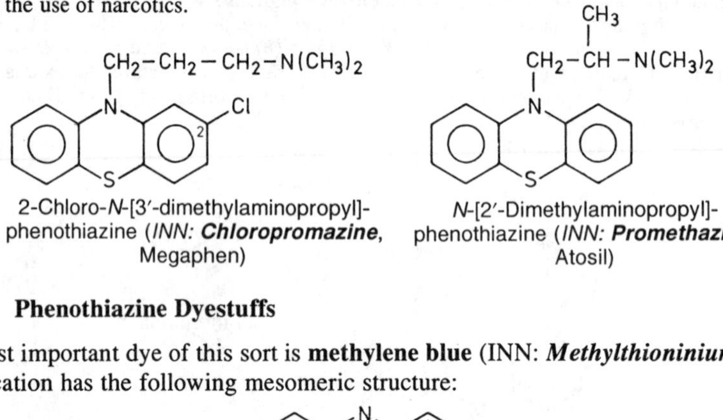

2-Chloro-*N*-[3'-dimethylaminopropyl]-phenothiazine (*INN:* **Chloropromazine**, Megaphen)

N-[2'-Dimethylaminopropyl]-phenothiazine (*INN:* **Promethazine**, Atosil)

7.8.5.6   Phenothiazine Dyestuffs

The most important dye of this sort is **methylene blue** (INN: *Methylthioninium chloride*), whose cation has the following mesomeric structure:

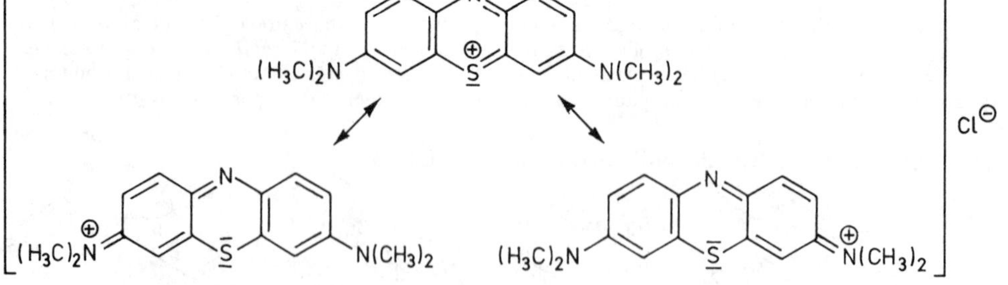

Methylene blue

---

[1] See *J. Sambeth*, Der Seveso-Unfall (The Seveso Accident), Chimia *36*, 128 (1982).
[2] See *R. R. Gupta*, Phenothiazines and 1,4-Benzothiazines: Chemical and Biomedical Aspects (Elsevier, Amsterdam 1988).

Industrially methylene blue is made by oxidising a mixture of *p*-aminodimethylaniline, dimethylaniline and sodium thiosulphate with chromic acid in the presence of zinc chloride and hydrochloric acid (*Bernthsen*, 1888). It is sold mostly as a double salt with zinc chloride and dyes silk a bright blue.

Methylene blue forms lustrous red crystals which dissolve readily in water. On reduction it gives a colourless leuco-base, *leucomethylene blue*:

Methylene blue   Leucomethylene blue

Because of the ease with which it can be reduced, methylene blue acts as a hydrogen acceptor in biochemical redox processes.

Methylene blue is used not only for dyeing textiles, but also has the ability to stain selectively the grey material in peripheral nerve systems (*Ehrlich*, 1885). This process is described as *vital staining*, since it can be carried out on living organisms. Methylene blue is thus rated as a 'vital stain'.

7.8.5.7 Sulphur Dyes

This important class of dyestuffs embraces a large number of valuable industrial compounds which are for the most part obtained by treating aminophenols or indoanilines with sulphur (*thionation*). The first dyestuff of this sort, *Vidal Black*, was discovered by *Vidal* (1893), who made it by melting *p*-aminophenol with sulphur and solid alkali at 200°C (470 K), and this opened the way to the development of sulphur dyestuffs. They are made either by heating the solid reactant with sulphur or sodium polysulphide or at lower temperatures by treating it with sodium polysulphide in aqueous or ethanolic solution.

Sulphur dyestuffs resemble vat dyes. In this case vatting depends upon the presence of either mercapto groups or sulphur bridges which are converted by sodium sulphide into SNa groups. In practice, sulphur dyes are used only for the dyeing of cotton and are characterised by lower working costs, ease of dyeing and fastness.

The most important dyestuff of the whole group, which exceeds any other synthetic dyestuff in the quantity in which it is made, is **Sulphur Black T**, which is made from sulphur and 2,4-dinitrophenol, and which is fast to both light and washing. *Green* sulphur dyestuffs are obtained by the reaction of sulphur with naphthol- or naphthylamine-sulphonic acids in the presence of copper(II) salts.

Especially fast to light and washing are various dyes of the hydron-blue series. Among them is, for example, **Hydron Blue R**, obtained by thionation of *carbazole-indoaniline* in butanol. This starting material can be made from *p*-nitrosophenol and carbazole in conc. sulphuric acid at −20°C (250 K).

Carbazole-indoaniline   Thianthrene

The coupling together of arene rings which takes place in thionation involves the formation of sulphur bridges. Hydron Blue R is formed in this way from the ring systems of phenothiazine and thianthrene. If more highly condensed arene ring systems take the place of carbazole then thionation provides genuine shades of *brown*.

Sulphur dyes are also known as 'immedial dyes'.

7.9 Six-Membered Rings Containing Three Heteroatoms

7.9.1 Triazines

Introduction of three nitrogen atoms into a benzene ring can be done in three ways, giving rise to three *isomeric triazines*:

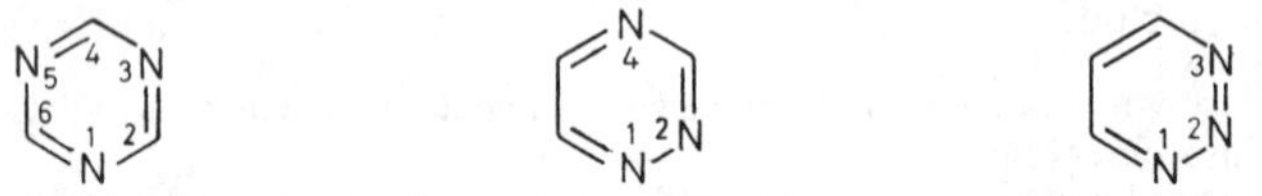

1,3,5-Triazine (*sym.*)   1,2,4-Triazine (*asym.*)   1,2,3-Triazine (*vic.*)

Of these three, only **1,3,5-triazine**, also known as *s*-triazine, is of significance.

Preparation

7.9.1.1 By *acid-catalysed trimerisation of hydrogen cyanide* (*Grundmann* and *Kreutzberger*, 1954).

7.9.1.2 Vacuum distillation of equimolecular amounts of *O-benzylimidoformate hydrochloride* and *N,N-diethylaniline* at 80°C (350 K) leads to breakdown of the free imido-ester and simultaneous trimerisation to 1,3,5-triazine (*F. Cramer*, 1956) (yield = 50%):

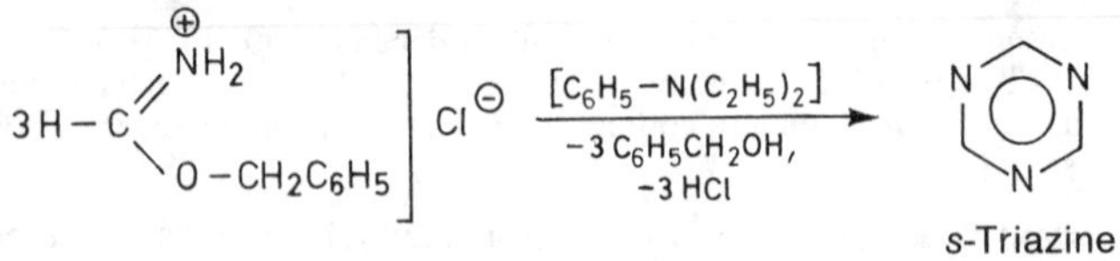

$$3\,H-C\underset{O-CH_2C_6H_5}{\overset{\oplus NH_2}{<}}\Bigg]\ Cl^{\ominus}\ \xrightarrow[\substack{-3\,C_6H_5CH_2OH,\\ -3\,HCl}]{[C_6H_5-N(C_2H_5)_2]}\ \text{s-Triazine}$$

7.9.1.3 *Thermal decomposition of formamidine hydrochloride* under reduced pressure in the presence of tri-n-butylamine provides 1,3,5-triazine in 70% yield (*F. C. Schaefer* et al.):

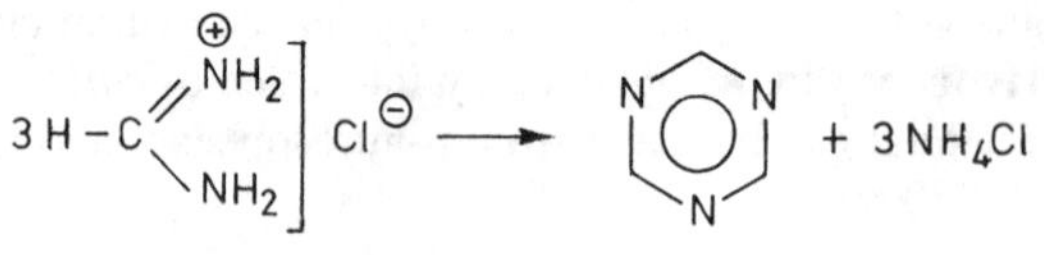

$$3\,H-C\underset{NH_2}{\overset{\oplus NH_2}{<}}\Bigg]\ Cl^{\ominus}\ \longrightarrow\ \text{(s-triazine)}\ +\ 3\,NH_4Cl$$

Processes 7.9.1.2 and 7.9.1.3 also lead indirectly to the trimerisation of hydrogen cyanide.

Properties

1,3,5-Triazine forms colourless, strongly refracting rhombic crystals, m.p. 86°C (359 K), which are readily soluble in organic solvents. Its high thermal stability is noteworthy; only at temperatures above 600°C (870 K) does it depolymerise to hydrogen cyanide. 1,3,5-Triazine is very sensitive to *nucleophiles, e.g.* water; the initial products decompose with splitting of the triazine ring. This nucleophilic ring-cleavage of the *s*-triazine ring can be

made use of for the syntheses of a variety of heterocycles.[1] For example, treatment of
o-aminothiophenol with s-triazine leads to *methinylation* and formation of benzothiazole:

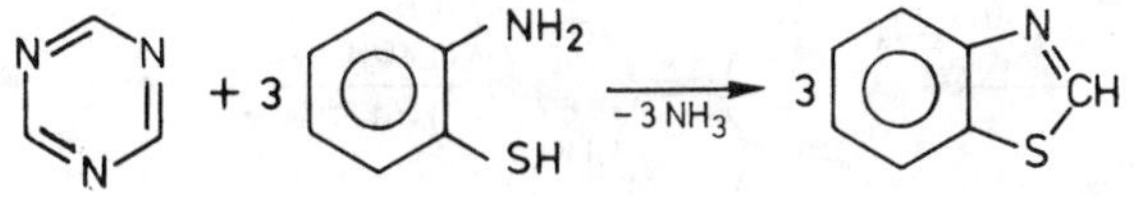

s-Triazine   o-Aminothiophenol   Benzothiazole

Electrophilic substitution reactions with s-triazine are extremely difficult. Nitration and
sulphonation are not possible, for the reagents which are usually employed hydrolyse it so
rapidly that it has no chance to undergo substitution reactions. With bromine at 0°C
(273 K) 1,3,5-triazine forms a stable perbromide, $C_3H_3N_3Br_3$; at 120°C (390 K) 2,4-dibro-
mo-1,3,5-triazine hydrobromide is formed in very good yield.

7.9.1.4 Derivatives of 1,3,5-Triazine

The following derivatives of cyanuric acid (see p.364) are all derivatives of s-triazine.

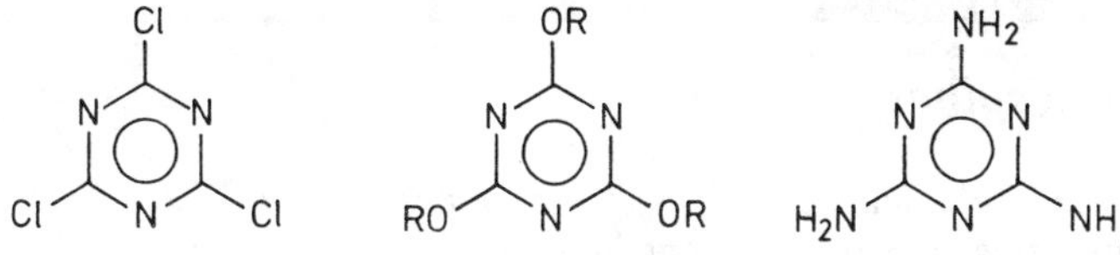

Cyanuric chloride   Cyanurate ester   Cyanuramide

Cyanuric chloride (trichloro-s-triazine, 2,4,6-trichloro-1,3,5-triazine) arises from the
trimerisation of cyanogen chloride and is an intermediate in the preparation of reactive
dyestuffs.

Cyanuric chloride reacts with sodium alkoxides to give **cyanurate esters** and with
ammonia to form **cyanuramide** (*melamine*).

Melamine is prepared industrially from urea:

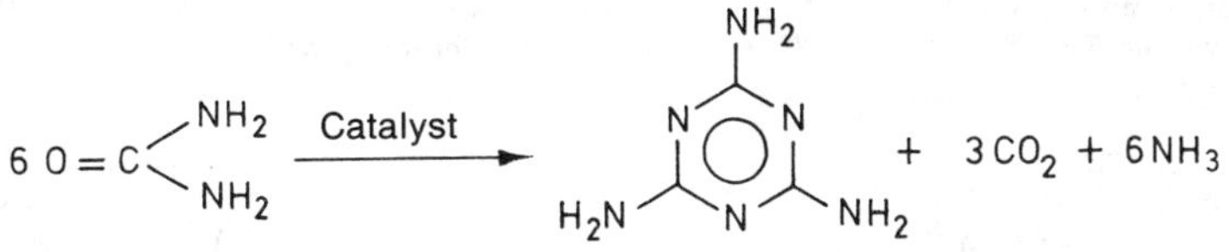

$$6\ O=C\begin{matrix} NH_2 \\ NH_2 \end{matrix} \xrightarrow{\text{Catalyst}} \text{Melamine} + 3\ CO_2 + 6\ NH_3$$

Melamine

Polycondensation of melamine with formaldehyde provides the important **melamine
resins** (Melaware, Meladur, Ultrapas) which find large usage as electrical insulators and for
tableware (plates, cups, *etc.*).

7.9.2 Oxathiazines

The sweetener *acesulpham K*, the potassium salt of oxathiazinedioxide [6-methyl-1,2,3-
oxathiazine-4(3H)-one-2,2-dioxide], is a derivative of oxathiazine. It is prepared from
t-butyl acetoacetate and fluorosulphonyl isocyanate (itself made from chlorosulphonyl
isocyanate and sodium fluoride):

---

[1] See *C. Grundmann*, Synthesen mit s-Triazine, Angew. Chem. *75*, 393 (1963); *A. Krentzberger*, Die
Chemie des s-Triazins, Fortschr. Chem. Forsch. *4*, 273 *ff.* (1963).

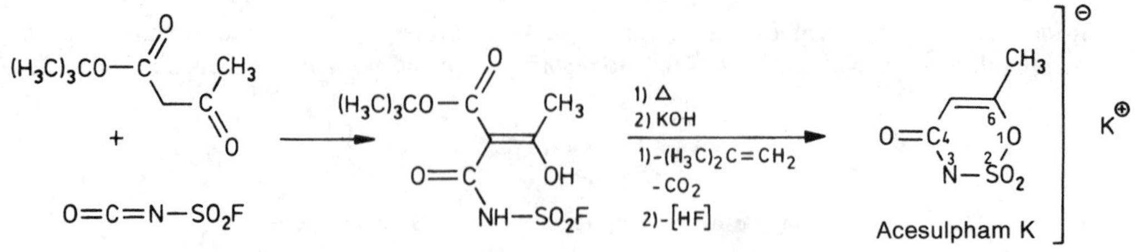

Acesulpham K

This compound forms colourless crystals which decompose at about 225°C (498 K). It is about 200 times as sweet as sugar and has no food value.

7.10 Benzo-Annelated Seven-Membered Rings Containing One or Two Heteroatoms

7.10.1 Benzazepines

In contrast to benzazoles and benzazines, *benzazepines*, the benzo-annelated derivatives of seven-membered ring heterocycles containing a nitrogen atom, have only been studied in detail in more recent years, although the first example was prepared in 1907. Benzo-derivatives of seven-membered ring heterocycles containing two heteroatoms, such as the *benzodiazepines*, *benzoxazepines* and *benzothiazepines*, have also only been investigated relatively recently. The recent rapid development of studies on these compounds is closely associated with unique psychopharmacological activity of some derivatives of these ring systems.

5-(3-Dimethylaminopropyl)-10,11-dihydro-5H-dibenz[b,f]azepine (*INN: **Imipramine**, Tofranil) may be taken as an example; it has an anti-depressant and calming effect. Another example is 5H-dibenz[b,f]azepine-5-carbonamide (*INN: **Carbamazepine**, Tegretal*) which is one of the most important antiepileptics.

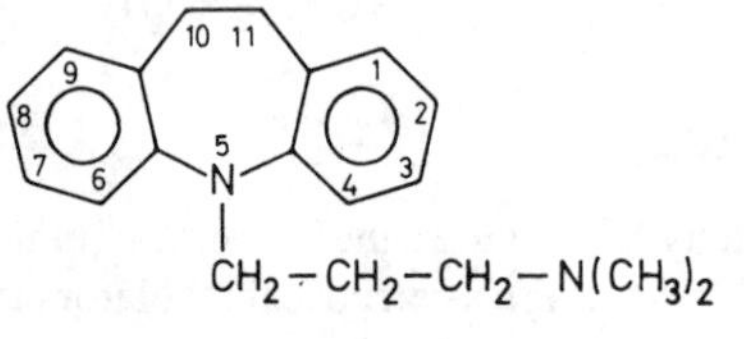

INN: *Imipramine*

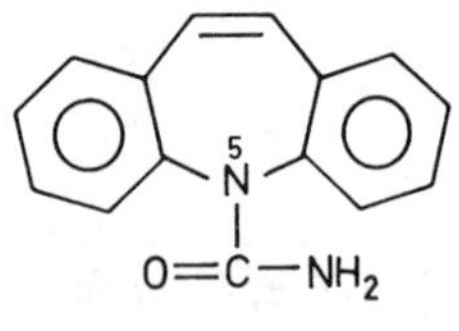

INN: *Carbamazepine*

7.10.2 Benzodiazepines[1]

1,4-Benzodiazepines are made in one step from 2-(halogenomethyl)quinazoline-3-oxides by treating them with primary aliphatic amines. The reaction involves an interesting ring-enlargement leading to the formation of 3H-1,4-benzodiazepine-4-oxides (*L. H. Sternbach* et al. 1961), *e.g.*

---

[1] See *V. Snieckus* et al., 1,2-Diazepines, A New Vista in Heterocyclic Chemistry, Acc. Chem. Res. *14*, 348 (1981).

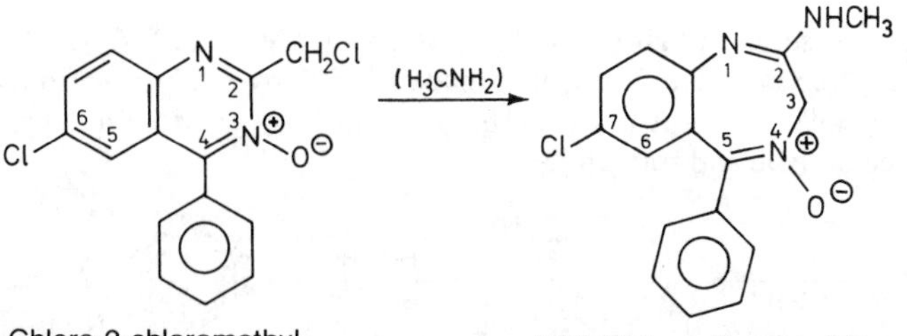

6-Chloro-2-chloromethyl-
4-phenylquinazoline-3-oxide

INN: *Chlorodiazepoxide*

7-Chloro-2-methylamino-5-phenyl-3H-1,4-benzodiazepine-4-oxide (INN: *Chlorodiazepoxide*, *Librium*), which is obtained in this way, is nowadays one of the most widely used *tranquilisers, i.e.* compounds which, without having a narcotic effect, depress only certain areas of the central nervous system and are used to combat states of over-excitement and anxiety.[1]

A further method for the preparation of 1,4-benzodiazepines is by the reaction of *o*-aminobenzophenones with glycine ester hydrochloride in boiling pyridine, leading to the formation of *1,3-dihydro-2H-5-phenyl-1,4-benzodiazepin-2-ones,* e.g.

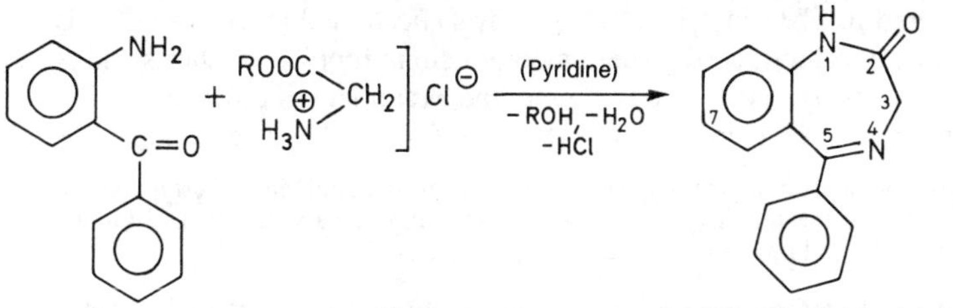

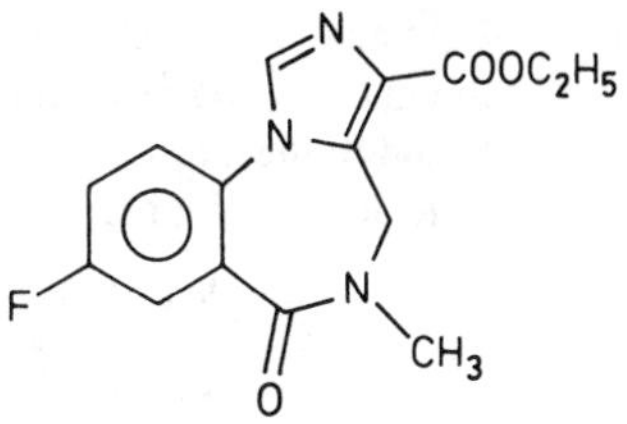

o-Aminobenzophenone

Glycine ester
hydrochloride

1,3-Dihydro-2*H*-5-phenyl-
1,4-benzodiazepin-2-one

INN: *Flumazenil*

The 7-chloro-1-methyl-derivative of this compound is the tranquiliser *INN: Diazepam* (*Valium*).
The imidazobenzodiazepine derivative *INN: Flumazenil* (Anexate) is a specific benzodiazepine antagonist which makes it possible to neutralise rapidly any general anaesthesia or sedation induced by benzodiazepines.

7.11 Bicyclic Heterocyclic Systems

This section deals with compounds in which two different heterocyclic rings are fused together. It includes some groups of natural products, in particular the purines, pterines and riboflavin.

7.11.1 Purines

This class of compounds, which includes important natural products from both plants and animals, has as its basic structure an imidazole ring fused to a pyrimidine ring. The parent

---

[1] Benzodiazepines occur in very small amounts in many foodstuffs, *e.g.* potatoes.

compound of the series, **purine**, a colourless substance, m.p. 217°C (490 K), does not, however, occur free in nature. It was first made in 1898 by *E. Fischer* from uric acid. More recently (*Bredereck*, 1962), it has been prepared in 35% yield by a simple route starting from aminoacetonitrile and formamide.

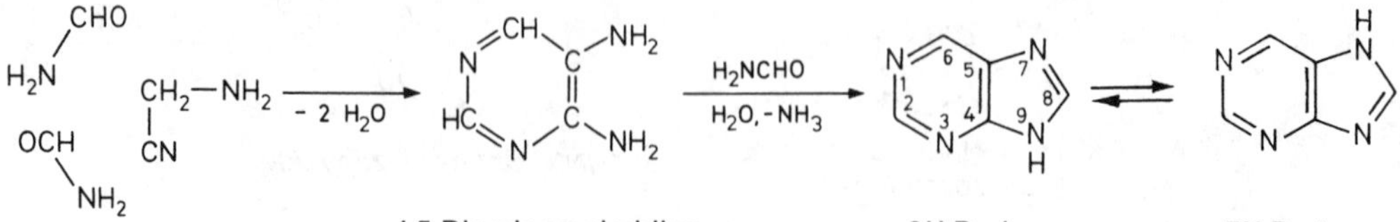

4,5-Diaminopyrimidine   9*H*-Purine   7*H*-Purine

Unless there is a substituent at the 7- or 9-position the substance exists as an *equilibrating mixture of tautomers* as shown above.

Purine, or imidazo[4,5-*d*]pyrimidine, has an aromatic 10-π-electron system. The following physiologically important compounds possess this ring system as their skeleton.

7.11.1.1   Uric Acid

Uric acid was first discovered in 1776 in urinary calculi by *Scheele* and by *Bergman*. It is the main nitrogen-containing end-product of protein metabolism in reptiles and birds and is found in the excrement of snakes (*ca.* 90%) and in *guano*, the excrement of birds (*ca.* 25%). Uric acid is obtained from these natural sources by extraction followed by addition of acid.

The urine of humans and of mammals contains only small amounts of uric acid, but in certain pathological conditions it is deposited in the joints (gout) as well as forming urinary calculi and kidney stones, which consist largely of the sodium or ammonium salts of uric acid.

The *Traube* synthesis is most commonly used for the preparation of uric acid and of other purine derivatives.

Preparation. *Traube* synthesis. When urea is condensed with ethyl cyanoacetate the product is cyanoacetylurea, which cyclises to 6-aminouracil in the presence of alkali. This can be converted, *via* its nitroso-derivative, into 5,6-diaminouracil, which reacts with ethyl chloroformate in the presence of sodium hydroxide to give the corresponding urethane. This loses ethanol when heated, forming uric acid:

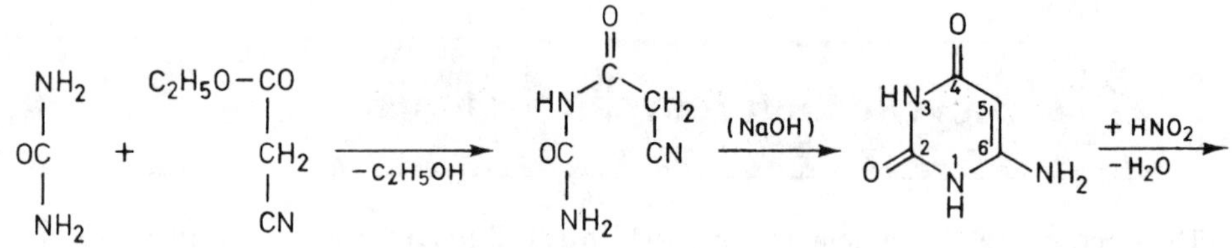

Cyanoacetylurea   6-Aminouracil

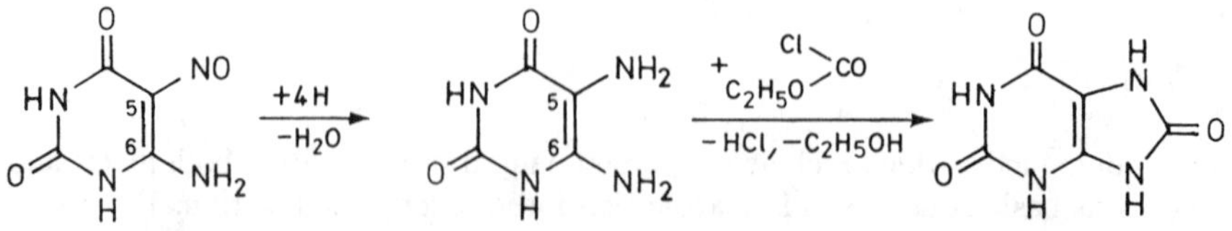

6-Amino-5-   5,6-Diaminouracil   Uric acid
nitrosouracil

The final step of the reaction can also be carried out by melting 5,6-diaminouracil with urea. Two molecules of ammonia are eliminated and uric acid is formed.

Properties. Uric acid is a colourless crystalline substance which, in contrast to barbituric acid, is only a weak acid (pK_a^1 5.75) and hardly soluble in water. It forms two series of salts, the acid and the neutral *urates*.

Its structure was shown by oxidative decomposition. Depending upon the conditions used (nitric acid or lead(IV) oxide) the product is either **alloxan** or **allantoin**, the first of these retaining the pyrimidine ring and the second the imidazole ring.

Alloxan   Allantoin

A *test* for uric acid is to add a small amount of dilute nitric acid to a sample and evaporate it to dryness. Ammonia is added to the cooled residue and, if uric acid is present, a purple colour is produced (*murexide test*); under the appropriate conditions this colour can arise without the necessity of adding the ammonia. Murexide is the *ammonium salt of purpuric acid*.

Murexide   Purpuric acid

7.11.1.2 Xanthine and Hypoxanthine

Xanthine and *hypoxanthine* are found in blood, in urine and in the liver.

Xanthine   Hypoxanthine   *INN: Allopurinol*

Xanthine is oxidised metabolically to uric acid. The action of the xanthinoxidase which brings this about is inhibited by *INN: Allopurinol*. This compound, 1*H*-pyrazolo[4,5-*d*]-4-pyrimidone, 4-hydroxy-pyrazolo[3,4-*d*]pyrimidine, differs from hypoxanthine in having a pyrazole ring instead of an imidazole ring. It is used in the treatment of gout.

7.11.1.3 Guanine and Adenine

Guanine is found in the free state in milk as well as in the excrement of mammals and of birds (guano). It has a silvery sheen and is a constituent of the scales of fishes and reptiles.

Adenine (6-aminopurine) is found in many natural products, for example in tea leaves, in beetroot juice, in yeast and in mushrooms (*Boletus edulis*).

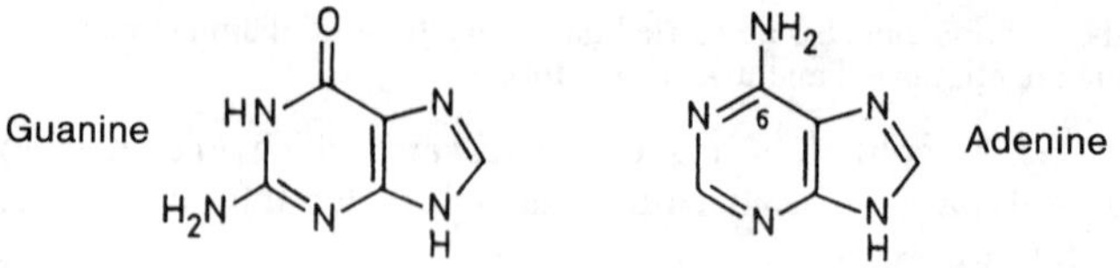

Guanine   Adenine

Both of these compounds form part of the structure of nucleic acids. In addition, adenine is a component of important coenzymes.

The easiest method for *preparing* purine starts from *4,5-diaminopyrimidine* or its derivatives. Acylation converts them, *via* the 5-acylamino-derivative, on heating, into purine derivatives, *e.g.*

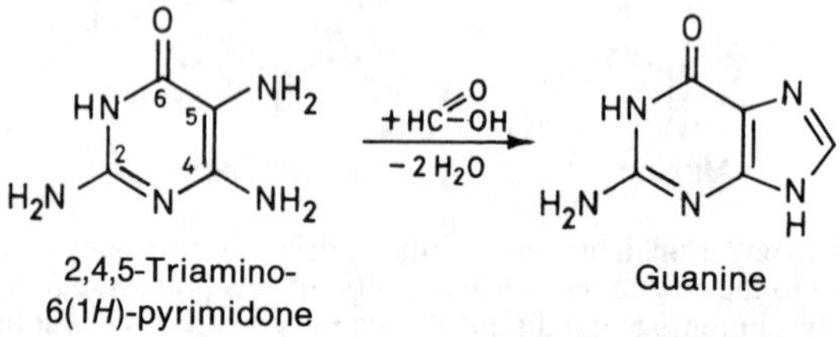

2,4,5-Triamino-
6(1*H*)-pyrimidone

Guanine

Formamide can be used in place of formic acid.

In 1955, kinetin (6-furfurylaminopurine) was isolated from aged deoxyribonucleic acid; originally thought to be of natural occurrence it was later found to be an artefact. This highly active physiological agent was the first example of a new group of *phytohormones*, the *cytokinins*,[1] which promote cell division and growth in plant tissue cultures. In addition, they promote germination and influence the transport of substances. More recently, another cytokinin, **zeatin**, was isolated from unripe maize seeds and identified as 6-(4-hydroxy-3-methyl-*trans*-2-butenyl)aminopurine. It plays a role in *cell division*.

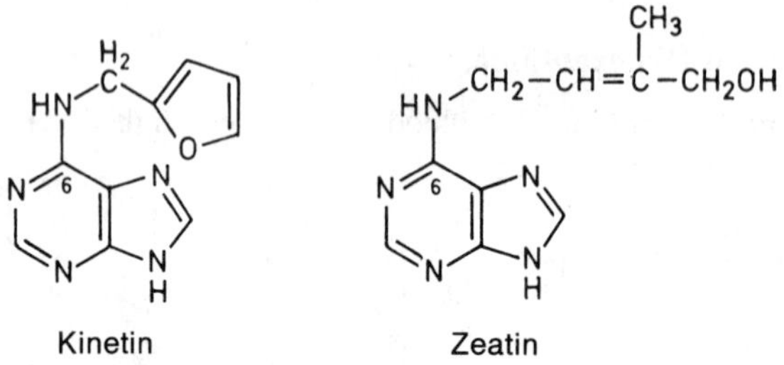

Kinetin   Zeatin

A synthetic 6-benzyl analogue, N^6-benzyladenine (*Verdan*), appears to have potential for storage of plants, and keeps green vegetables such as lettuces fresh for extended periods.

7.11.1.4   *N*-Methylated Xanthines

The three most important purine bases of this sort are *theophylline, theobromine* and *caffeine*, which differ from the previously mentioned purine derivatives in their activity as stimulants. They are derivatives of xanthine in which some or all of the hydrogen atoms attached to nitrogen have been replaced by methyl groups. Their formulae are as follows:

---

[1] See *S. Matsubara*, Phytochem. *19*, 2239 (1980).

Theophylline   Theobromine   Caffeine

Theophylline, 1,3-dimethylxanthine, is present in small amounts in tea leaves. It is active against bronchial asthma, improves the performance of the heart and the circulation in the coronary arteries and is one of the most important stimulants. It is also a diuretic. It forms colourless plates, m.p. 268°C (541 K), and dissolves readily in hot water.

The ethylenediamine salts of theophylline (*Euphylline*) and of 7-(β-hydroxyethyl)theophylline (*INN: Etofilline, Cordaline*) are used as *cardiotonics*.

Theobromine, 3,7-dimethylxanthine, is the principal alkaloid of cocoa beans; it is also found in tea and in cola nuts. It is a strong *diuretic*.

Caffeine, 1,3,7-trimethylxanthine, occurs in coffee beans (1–1.5%) and in tea (up to 5%). It is obtained from tea residues and as a by-product in the preparation of caffeine-free coffee. It is also made synthetically.

Caffeine exerts a stimulating and reviving effect on the central nervous system and on the functioning of the heart and has medical use for this purpose. It is a weaker diuretic than theophylline or theobromine. Because of their specific and in part similar physiological activity these three purine derivatives are also treated as alkaloids.

7.11.2 Pterins

Pteridine,[1] a yellow well-crystalline base (pK_a 4.05), m.p. 135°C (411 K) has a bicyclic ring system closely related to that of the purines. Pteridine derivatives were first isolated from the wings of butterflies and other insects (Grk. *pteron* = wing). In pteridine, a pyrimidine ring is fused to a pyrazine ring. The whole series of pteridine derivatives are known as *pterins*. The parent compound of the naturally occurring pterins is 2-amino-4-(3*H*)-pteridinone or **pterin.**

Pteridine   Pterin

The most important pterins are **xanthopterin** (7-desoxyleucopterin), $C_6H_5N_5O_2$, which is yellow, and **leucopterin**,[2] $C_6H_5N_5O_3$, which is colourless.

---

[1] See *W. Pfleiderer*, Neuere Entwicklungen in der Pteridinchemie (Recent Developments in the Chemistry of Pteridines), Angew. Chem. *75*, 993 (1963).
[2] See *F. Weygand* et al., Über die Biogenese des Leucopterins (On the Biogenesis of Leucopterin, Angew. Chem. *73*, 402 (1961).

Xanthopterin   Leucopterin

Leucopterin is the white wing pigment of cabbage white butterflies and xanthopterin is the yellow pigment in the wings of brimstone butterflies. Their structure has been confirmed by synthesis.

Xanthopterin can be oxidised to leucopterin by the enzyme *xanthinoxidase*.
Our knowledge of the pterins owes much to the work of *H. Wieland, Schöpf* and *Purrmann*. These deposition products of insect metabolism were of importance in the discovery of a new active member of the vitamin B group, namely folic acid.

INN: *Folic acid* was first isolated from a liver extract (*Stokstad*, 1938). Later it was obtained from spinach leaves (Lat. *folium* = leaf) and in a number of microorganisms (*Lactobacillus casei, Streptococcus lactis, etc.*) for which it is a *foodstuff*.

As a growth factor of leucocytes, folic acid affects the formation of blood. Combined with vitamin B_{12} it is used in the treatment of *pernicious anaemia*.

Its structure has been confirmed by synthesis. Folic acid is built up from pteroic acid (itself made up from 6-methylpterin and *p*-aminobenzoic acid) linked to L-glutamic acid by means of a peptide link. It is thus also known as *pteroylglutamic acid*[1] (*Angier, Hutchings, Tschesche, F. Weygand*).

*INN: **Folic acid** (Pteroylglutamic acid)*

Folic acid crystallises in thin yellow needles which are only very slightly soluble in either water or organic solvents. It is unstable to acids but with alkalies forms fairly stable salts.

The action of *sulphonamides* depends upon their interfering with the formation of *p*-aminobenzoic acid in bacterial metabolism.
In organisms folic acid actually acts in the form of *5,6,7,8-tetrahydrofolic acid* (folinic acid, *leucovorin*). Under the influence of ATP it is changed into *N*-10-formyltetrahydrofolic acid (rhizopterin), the so-called 'active formic acid' or 'active formaldehyde' (see p.894). Small changes to folic acid provide folic acid antagonists, e.g. *INN:* **Methotrexate** (this has=NH instead of =O attached to C-4, and CH_3 instead of H attached to N-10).
5,6,7,8-Tetrahydropterine (THP) acts as a catalyst in enzymatic oxidations.[2]

[1] See *R. Tschesche* and *F. Korte*, Zum biochemischen Syntheseweg der Pteroylglutaminsäure (Towards a Biochemical Synthesis of Pteroylglutamic acid), Z. Naturforsch. 8b, 87 (1953); *L. Jaenicke*, Die Folsäure im Stoffwechsel der Einkohlenstoff-Einheiten (Folic acid in the Metabolism of One Carbon-atom Units), Angew. Chem. 73, 449 (1961).
[2] See *M. Viscontini*, Tetrahydropterine, Katalysatoren der Phenylalanine-Hydroxylierung, Fortschr. chem. Forsch. 9, 605 (1967).

7.11.3 Flavins (Isoalloxazines)

The flavin or isoalloxazine ring system, which can also be looked on as a benzopteridine system, occurs in particular in INN: *Riboflavin* (vitamin B_2).

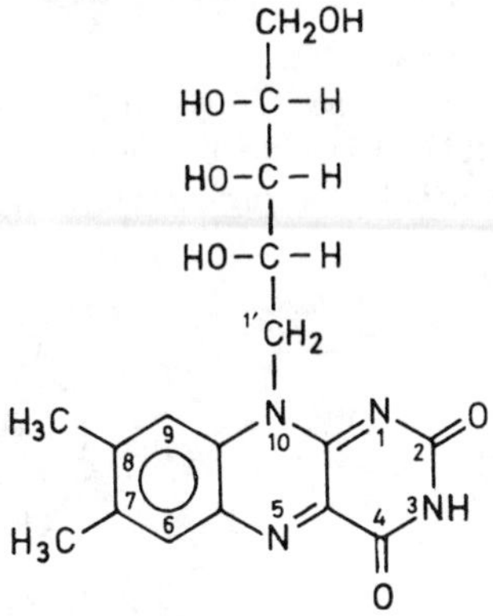

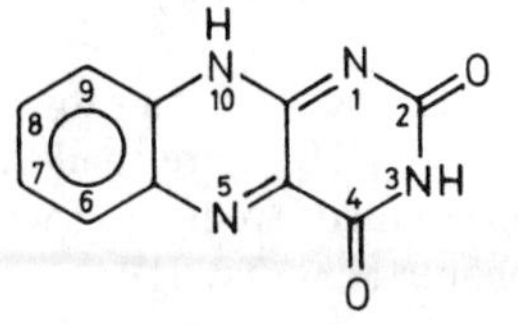

Isoalloxazine (Flavin)

INN: *Riboflavin*

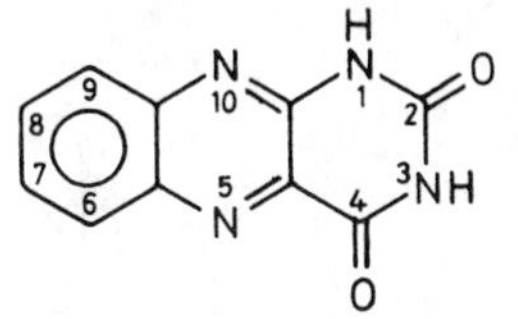

Alloxazine

Riboflavin (*lactoflavin*) is a yellow pigment which is widely distributed in nature. It dissolves in water to give solutions which show a strong yellow-green fluorescence. It was first isolated in 1933 from whey and from egg-white (*R. Kuhn, P. György* and *T. Wagner-Jauregg*). Only 70 mg of riboflavin were produced from 1000 litres of milk. Two years later its structure was proved by synthesis (*R. Kuhn, Karrer*). Riboflavin is *7,8-dimethyl-10-(D-1'- ribityl)flavin*; carbon atom 1' of the sugar alcohol D-ribitol is directly attached to the nitrogen atom at the 10-position of the flavin ring.

Riboflavin is a member of the *vitamin B complex* which also includes thiamine, pyridoxine, pantothenic acid, *myo*-inositol, p-aminobenzoic acid, nicotinamide, biotin, folic acid and cobalamine. It is an important growth factor in both animals and microorganisms. Riboflavin-5'-phosphate is a constituent of enzymes.

7.11.4 1,4-Diketopyrrolo[3,4-c]pyrroles (DPP)

The diaryl-1,4-diketopyrrolo[3,4-*c*]pyrroles were introduced in industry as pigments in 1983. This was the third time this century, following the copper phthalocyanins (1933) and the quinacridones (1955), that a completely novel class of pigments of high significance had been discovered.

DDP and its derivatives are made using a variant of the *Stobbe* condensation, starting from succinate esters and benzonitriles. The best yields are obtained when t-alkoxides are used as bases.

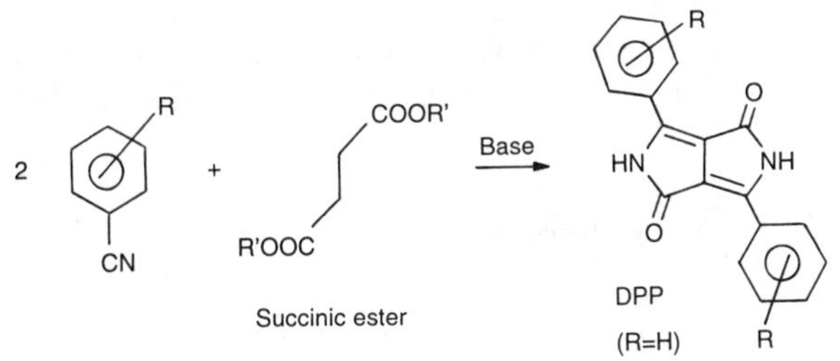

Succinic ester

DPP
(R=H)

This provides a set of pigments with colours ranging from yellowish-red to blue-violet.[1] The 4,4'-dichlorodiphenyl derivative of DPP (R = 4-Cl, Irgazin, DPP Red BO) has a physiologically pure red colour, which has neither a yellow nor a blue tinge.

7.11.5 Bicyclic Amidines

The bicyclic amidines 1,5-diazabicyclo[4.3.0]non-5-ene (DBN) and 1,8-diazabicyclo-[5.4.0]undec-7-ene [DBU] are made from the related lactams by base-catalysed condensation with acrylonitrile, reduction with *Raney* nickel, and ring-closure in the presence of *p*-toluenesulphonic acid (TsOH):

Both of these compounds have proved useful as dehydrohalogenating agents.[2]

7.12 Alkaloids[3]

The term 'alkaloids', introduced by *Meissner* in 1819, embraces a group of nitrogen-containing natural products which occur in many plants and in some species of animals. The name of this large family of compounds is derived from the 'alkali-like' properties of many of its members.

In plants, alkaloids almost always exist as salts of the so-called *plant acids*, such as oxalic, acetic, lactic, malic, tartaric, citric and aconitic acids. In some cases the alkaloids form salts with acids which are characteristically associated with the plant in question, *e.g.* quinic acid, meconic acid and others. In a specific family of plants it is common to find a series of related alkaloids (major and minor alkaloids). The alkaloid content of plants varies and is dependent upon the season of the year.

The isolation of individual alkaloids from plant material is usually a difficult and time-consuming operation. A common procedure starts by extracting dry plant material with boiling methanol. The

---

[1] See *A. Iqbal* et al., The Synthesis and Properties of 1,4-diketopyrrolo(3,4-*c*)pyrroles, Bull. Soc. Chim. Belg, *97*, 615 (1988).
[2] See *H. Oediger* et al., Synthesis *1972*, 591.
[3] See *I. W. Southon* and *J. Buckingham*, Dictionary of Alkaloids (Chapman and Hall, London 1989).

extract is concentrated and strongly acidified. The resultant acid solution is next extracted with ether in order to remove water-soluble non-basic organic material. The acid solution, which contains the alkaloids, is basified. At this stage some alkaloids crystallise out; those which are water-soluble are extracted out with chloroform. In order to separate out and purify the various different bases, use is made of steam distillation and of thin-layer and gas chromatography, carried out on both analytical and preparative scales.

The structure of most alkaloids is based on a more or less complicated heterocyclic ring system. Their structures are determined and are then often confirmed by synthesis. For the most part they are crystalline bases which are only slightly soluble in water; they form salts with acids and are nearly all optically active and almost exclusively laevo-rotatory.

As yet little information has been obtained about the *biochemical function* of alkaloids in plants. Many alkaloids have profound physiological effects, even in very small doses, on animal organisms. In consequence they have been in use since earliest times, principally for healing purposes, but also as condiments and intoxicants. Some are highly toxic.

In recent years there has been interest, not only in the determination of their structure and their synthesis, but especially in their *biosynthesis*.[1] Although in most cases the biosynthetic route in plants has not been fully uncovered in detail, some major questions have been clarified, making use of *isotopic labelling*, and earlier hypotheses have been in part experimentally confirmed.

The biosynthesis of alkaloids is closely tied up with *amino acid metabolism*. Radioactive amino acids have been successfully incorporated into alkaloids produced by plants. They have been classed as secondary metabolites as distinct from the primary metabolites such as carbohydrates, fats and essential oils.

By carrying out cleavage reactions on the labelled alkaloid molecules it has been possible to discover the positions of the labelled atoms and in this way knowledge of their biosynthetic formation in the plants has been obtained.

The *classification* of alkaloids is best done on the basis of the nature of their ring skeletons. In addition, the purine derivatives previously mentioned on p.798, together with some bases related to β-phenylethylamine, such as tyramine, hordenine and ephedrine, and some other compounds which have physiological and toxic properties that are typical of alkaloids, are also classified as alkaloids.

7.12.1 Alkaloids Based on Tetrahydropyrrole, Pyridine and Piperidine

The two most important alkaloids of this group are *nicotine* and *anabasine*, whose molecules contain a pyridine ring together with an *N*-methyltetrahydropyrrole ring or a piperidine ring as a substituent at the 3-position. They have the following structures:

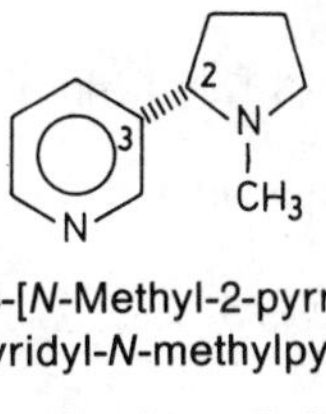

Nicotine(3-[*N*-Methyl-2-pyrrolidyl]pyridine;
3-Pyridyl-*N*-methylpyrrolidine)

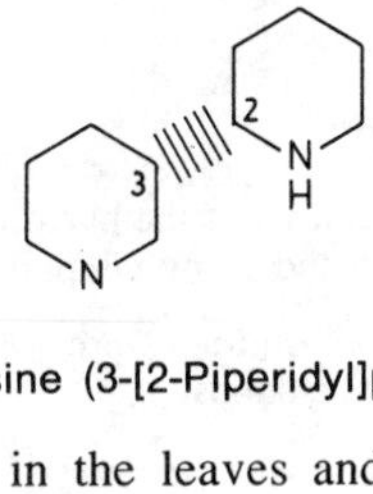

Anabasine (3-[2-Piperidyl]pyridine]

7.12.1.1 Nicotine, $C_{10}H_{14}N_2$, occurs as the major alkaloid in the leaves and roots of tobacco plants (*Nicotiana tabacum*). It is a colourless laevo-rotatory oil which is miscible in water and has a tobacco-like smell. In the air it quickly turns brown. Its salts are dextrorotatory.

---

[1] See K. *Mothes*, Zur Geschichte unserer Kenntnis über die Alkaloide (The History of our Knowledge about Alkaloids), Pharmazie 366, 199 (1981).

Nicotine is oxidised to *nicotinic acid* by the action of chromic acid. However, if the nitrogen atom of the pyridine ring in nicotine is methylated, oxidation with potassium hexacyanoferrate(III) in alkaline solution gives instead L-*hygrinic acid*, [(−)-*N*-methylpyrrolidine-2-carboxylic acid, L(−)-*N*-methylproline], thus proving the *absolute configuration* of nicotine.

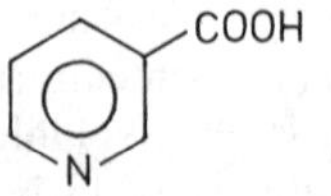

Nicotinic acid 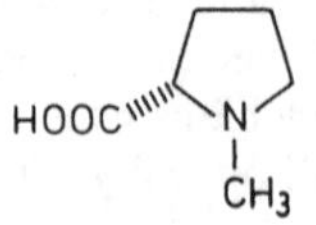   L-Hygrinic acid

The structure of nicotine was confirmed by synthesis (*Pictet*, 1905). The first-formed racemic form was resolved into *enantiomers* by means of the tartrate salts.

In small quantities nicotine acts on the nervous system as a stimulant, resulting in increased glandular secretion and a *raising of the blood pressure*. In large doses it produces paralysis of the breathing system and death. The fatal dose for humans is about 100 mg.

7.12.1.2   Anabasine, $C_{10}H_{14}N_2$, is an alkaloid which is isomeric with nicotine and also found in tobacco. It was later also found in the Asian goosefoot plant *Anabasis aphylla*. Like nicotine it is used as an *insecticide*.

The *biosynthesis* of (−)-anabasin proceeds as follows. The amino acid L-lysine is enzymatically converted into cadaverine and 5-aminopentanal, which cyclises to form 3,4,5,6-tetrahydropyridine. This reacts with the partially hydrogenated nicotinic acid (I), which is formed by the action of NADH, giving an addition product (II). Under the influence of ADP⊕ this is dehydrogenated and decarboxylated to give (−)-anabasin.

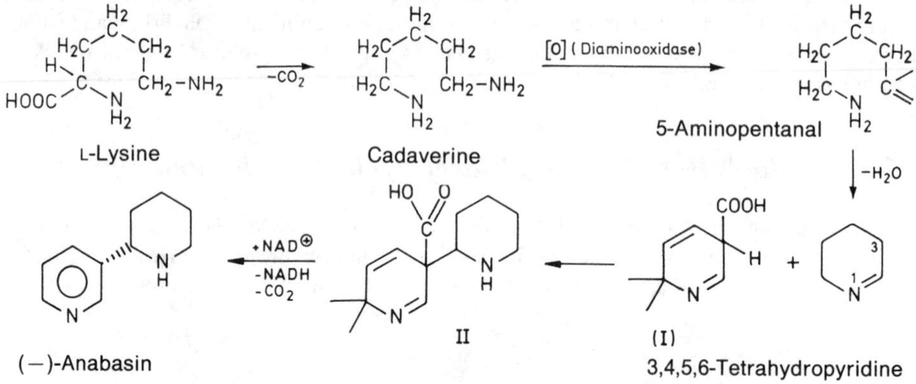

L-Lysine   Cadaverine   5-Aminopentanal

(−)-Anabasin   II   (I)   3,4,5,6-Tetrahydropyridine

The *biosynthesis* of the pyrrolidine ring in nicotine follows a similar path, starting from the amino acid ornithine and going *via* putrescine to 3*H*-4,5-dihydropyrrole.[1]

7.12.1.3   (+)-Coniine, 2-propylpiperidine, $C_8H_{17}N$, occurs in hemlock (*Conium maculatum*) and is exceedingly poisonous.

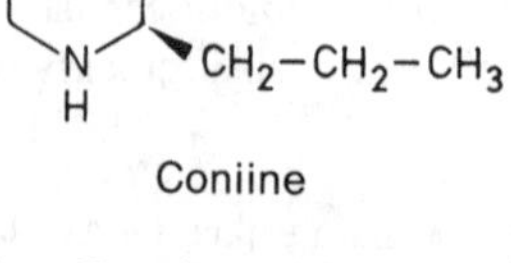

Coniine

---

[1] See *K. B. G. Torssell*, Natural Product Chemistry, p.272 *ff*. (Wiley, Chichester 1983).

Natural coniine is an oily liquid which is slightly soluble in water and readily soluble in ethanol.

The *preparation* of (±)-coniine by *Ladenburg* (1886) was the first synthesis of an alkaloid to be achieved. It was carried out by condensing α-picoline with acetaldehyde to give 2-(1-propenyl)pyridine which was then reduced to coniine.

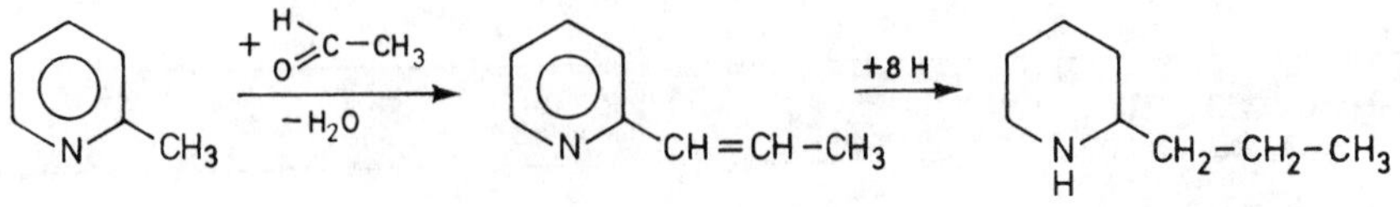

α-Picoline   2-(1-Propenyl)pyridine   (±)-Coniine

It was resolved by means of the tartrate salts.

7.12.1.4   INN: (–)-*Lobeline*, $C_{22}H_{27}N_2O$, is, like coniine, a relatively simple derivative of piperidine. It is obtained from *Lobelia inflata*, and was shown by *H. Wieland* to have the following structure; this has been confirmed by synthesis:

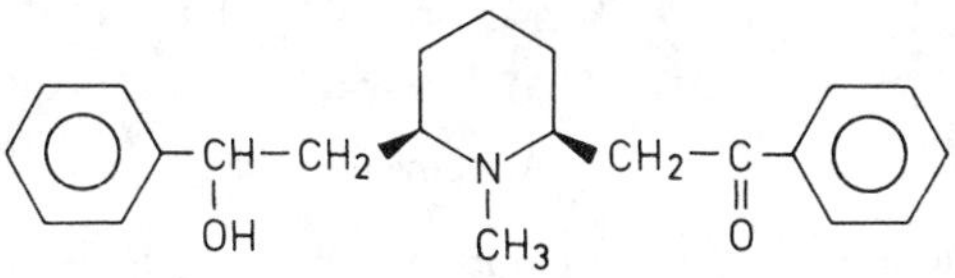

INN: *Lobelin*

Lobelin stimulates respiration and is used medicinally for this purpose. It has also been tried out for use in weaning people away from tobacco.

7.12.1.5   Piperine, $C_{17}H_{19}NO_3$, is the principal alkaloid in black peppers (*Piper nigrum*) and is also responsible for the hot taste of peppers. Hydrolysis splits it into piperidine and a doubly unsaturated carboxylic acid *piperinic acid*, $C_{10}H_{12}O_4$. Piperine is thus the amide formed from piperinic acid and piperidine.

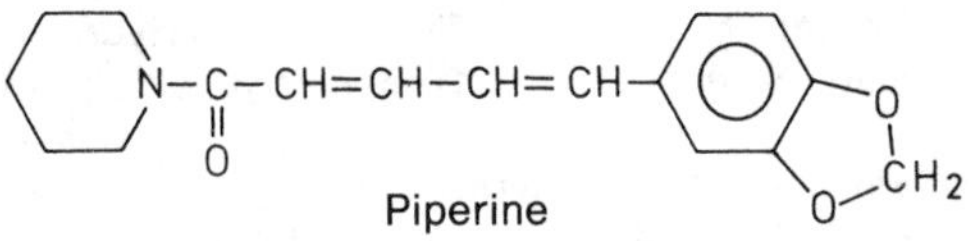

Piperine

Piperine is a crystalline substance, m.p. 129°C (402 K) which is only slightly soluble in water. When piperinic acid is oxidised with potassium permanganate, piperonal is formed.

7.12.2 Alkaloids Based on Tropane

Tropane, which provides the fundamental ring system of this group of alkaloids, contains both a pyrrolidine ring and a piperidine ring. Its structural and conformational formulae are as follows:

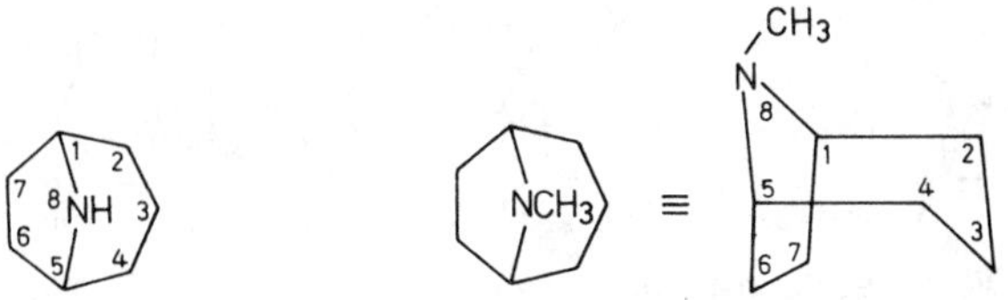

Nortropane   Tropane

In nortropane (8-azabicyclo[3.2.1]octane) there is no methyl group attached to the nitrogen atom.

Important alkaloids of the atropine and cocaine type are derivatives of tropane.

Atropine Group

7.12.2.1   Atropine [DL-tropyl tropate, (±)-hyoscyamine], $C_{17}H_{23}NO_3$, is the racemic form of (−)-hyocyamine and occurs in deadly nightshade (*Atropa belladonna*) and in the seeds of thorn apples (*Datura stramonium*). Its formula is as follows:

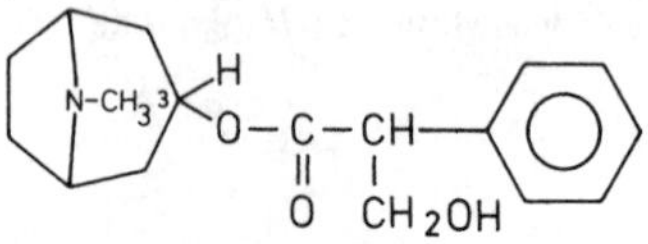

Atropine

Insight into the structure of atropine came from its acid hydrolysis products, the secondary alcohol **tropine** and (±)-**tropic acid**:

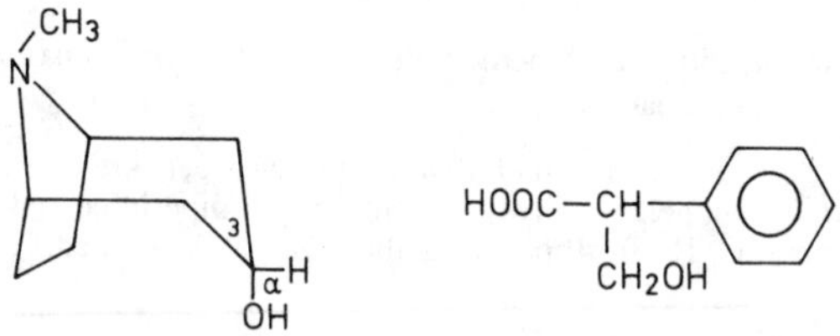

Tropine
(Tropan-3α-ol)

Tropic acid (3-Hydroxy-
2-phenylpropionic acid)

In order to define the conformation of tropane derivatives, the suggestion of *Cahn* that the N-methyl bridgehead should be taken as the reference point, and that a 3-hydroxy group *cis* to the N-methyl group should be denoted by β, and one *trans* to the N-methyl group by α, was adopted. Thus tropine is tropan-3α-ol; tropan-3β-ol is called pseudotropine.

Tropine, which has a plane of symmetry, is *optically inactive*. It is oxidised to the ketone *tropinone*.

The *structure* of tropine was elucidated by *Willstätter* and others, who showed that the N-methyl bridge could be removed by the exhaustive methylation procedure.

Later *R. Robinson* carried out an elegant simple *synthesis* of tropinone from succinaldehyde (butane-1,4-dial), diethyl acetone-1,3-dicarboxylate and methylamine:

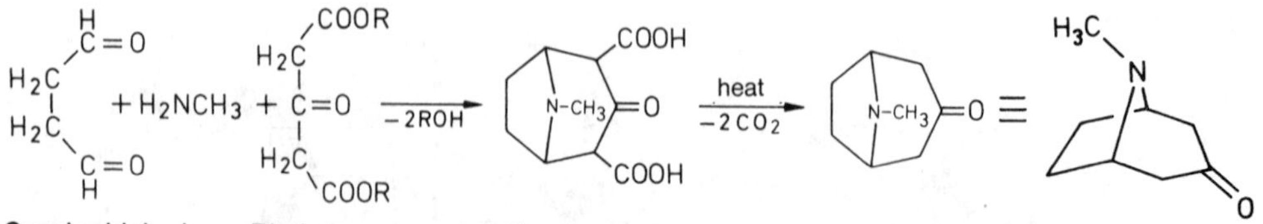

Succinaldehyde   Diethyl acetone-1,3-dicarboxylate
(R = C_2H_5)

Tropinone

Schöpf carried out the same reaction under physiological conditions, *i.e.* at room temperature and a controlled pH near to neutral.

Atropine crystallises in prisms, m.p. 115°C (388 K), and shows strong *pharmacological activity*. It was first prepared by *Willstätter* in 1901 by a rather complicated route. Nowadays it is made from tropine, which is available from the reduction of tropinone in acidic solution. Tropine is esterified with *O*-acetyltropoyl chloride, and the acetyl group is then hydrolysed off to leave atropine.

Atropine is an antidote against insecticides and nerve agents (nerve gases) that are derivatives of phosphoric acid, *i.e.* parathion or Tabun, Sarin and Soman. A dilute solution of atropine dilates the pupil of the eye and is therefore used in medical treatment of the eyes.

(–)-**Hyoscyamine**, $C_{17}H_{23}NO_3$, was isolated from henbane (*Hyoscyamus niger*) and has the same constitutional formula as atropine but is the tropyl ester of (*S*)-(–)-tropic acid. It is easily racemised to atropine.

7.12.2.2   Scopolamine, (–)-hyoscine, $C_{17}H_{21}NO_4$, like atropine and hyoscamine, is obtained from plants of the nightshade family (*Solanaceae*), especially in varieties of *Scopolia*. Structurally it is closely related to atropine but it contains two fewer hydrogen atoms and in their place an epoxide (oxiran) ring.

Scopolamine, (−)-Hyoscine

Scopolamine is used medicinally as a sedative and as an initial narcotic before operations. Some derivatives, such as INN: ***N-Butylscopolaminium bromide*** (*Buscopane*), are used as anticonvulsants (spasmolytics).

7.12.2.3   Cocaine Group

(–)-Cocaine, $C_{17}H_{21}NO_4$, is the major alkaloid in leaves of *Erythroxylum coca* shrubs, which are natives of South America and Java.

The *determination of the structural formula* of cocaine was closely associated with that of atropine. Acidic or alkaline hydrolysis of cocaine splits it into **ecgonine** (tropine-2-carboxylic acid), benzoic acid and methanol. Oxidation and decarboxylation of ecgonin converts it into tropinone. Hence cocaine must be the methyl ester of benzoylecgonine.

(–)-Ecgonine and (–)-cocaine have the following *conformational formulae* (*Fodor, Hardegger*):

(−)-Ecgonine   (2*R*,3*S*)-(−)-Cocaine

On the *Cahn–Ingold–Prelog* system carbon atoms 1 and 2 of (–)-cocaine have (*R*)-configurations, while carbon atoms 3 and 5 have (*S*)-configurations. (–)-Cocaine is thus methyl (2*R*,3*S*)-3-benzoyl-oxytropane-2-carboxylate.

The structure of cocaine was confirmed by synthesis (*Willstätter, R. Robinson*). *Robinson* carried out a simple synthesis of ecgonine, similar to his synthesis of tropinone, starting from succinaldehyde (butane-1,4-dial), monomethyl acetone-1,3-dicarboxylate and methylamine. Methyl tropinone-2-car-boxylate is formed as an intermediate and is converted into (±)-ecgonine by reduction and hydrolysis. Industrially ecgonine is obtained from the crude mixture of alkaloids from coca leaves, and is then esterified with methanol and benzoylated to provide cocaine.

(–)-Cocaine is a crystalline compound, m.p. 98°C (371 K), which had great importance as a local anaesthetic with a long-lasting effect. It paralyses the peripheral nerves and renders the skin and tissues insensitive.

In addition, however, cocaine has the ability to induce a feeling of well-being. It can thus readily be used as a narcotic drug and this leads to severe damage to the nervous system (cocainism). Eventually it brings about paralysis of the respiratory centres and death. Because of this, nowadays a series of active synthetic *anaesthetics*, which lack these bad effects, is used instead (see p.549).

Closely related in chemical structure to the two alkaloids just discussed are the plant bases of the pomegranate tree.

7.12.2.4   Alkaloids from the Pomegranate Tree

The most important alkaloid found in the bark of pomegranate trees (*Punica granatum*) is **pseudopelletierine** (*N*-methylgranatonine), $C_9H_{15}NO$, which can be regarded as a higher ring homologue of tropinone. It differs from the alkaloids discussed in the previous section in having *two* condensed piperidine rings. It can also be regarded as an aza-bridged cyclo-octane.

The basic framework of this class of alkaloids is that of *norgranatane* (9-azabicy-clo[3.3.1]nonane) or *granatane:*

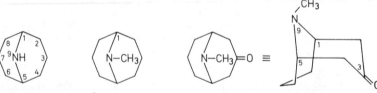

Norgranatane   Granatane   Pseudopelletierine

Pseudopelletierine is thus granatan-3-one. This was also synthesised by *R. Robinson*, using a similar way to that of making tropinone, and starting in this case from glutaral-dehyde (pentane-1,5-dial), diethyl acetone-1,3-dicarboxylate and methylamine. It is a strong base which melts at 48°C (321 K). It has a symmetric structure and is therefore *optically inactive*.

It is of interest historically in that it was from pseudopelletierine that cyclo-octatetraene was first prepared by *Willstätter*.

Isopelletierine, C_8H_5NO, an accompanying alkaloid, resembles coniine in its chemical structure. It can be prepared from the lithium derivative of α-picoline by letting it react with acetic anhydride followed by catalytic hydrogenation of the pyridine ring (*Wibaut*):

Isopelletierine

7.12.3 Alkaloids Based on Quinolizidine

The alkaloids of this group are derivatives of either **quinolizine** (4a-aza-4a,8a-dihydro-naphthalene) or **octahydroquinolizine**, *quinolizidine* (1-azabicyclo[4.4.0]decane).

9a*H*-Quinolizine
(9a*H*-Pyridol[1,2-*a*]pyridine)

Octahydroquinolizine
(Quinolizidine)

The numbering of the atoms in this ring system is as shown above. The bridgehead carbon atom is numbered 9a.

The best-known derivative of quinolizine is its *1,2,3,4-tetracarboxylic acid* which is the product of the reaction between dimethyl acetylenedicarboxylate and pyridine:

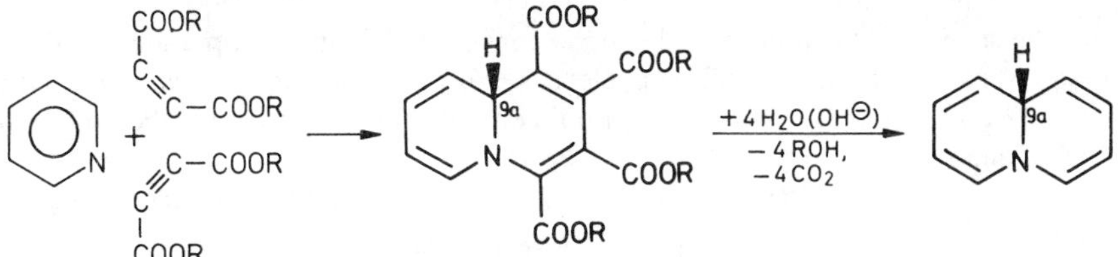

$R = CH_3$

Tetramethyl 9a*H*-quinolizine-1,2,3,4-tetracarboxylate

$+4H_2O(OH^{\ominus})$
$-4ROH,$
$-4CO_2$

9a*H*-Quinolizine

The initial reaction involves a *1,4-dipolar cycloaddition reaction*. When the tetra-ester is distilled in the presence of lime it undergoes saponification of the ester groups and decarboxylation to provide quinolizine. Perhydroquinolizine, quinolizidine, is the parent compound of the *lupinene alkaloids*.

(–)-**Lupinine**, $C_{10}H_{19}NO$, is the major alkaloid of the yellow lupin (*Lupinus luteus*). It is a crystalline substance, m.p. 70°C (343 K) which has the following structure:

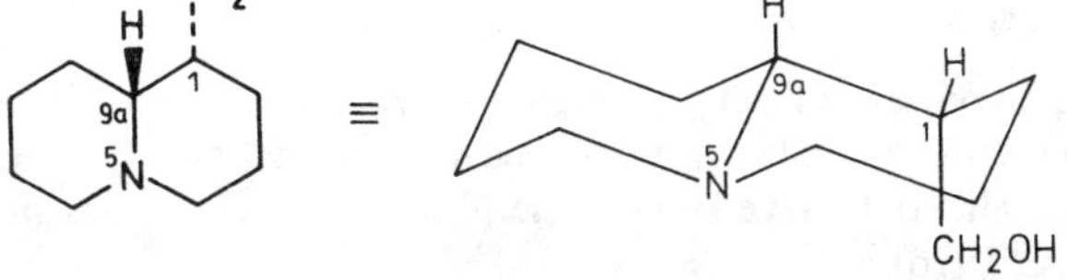

(−)-Lupinine (1-Hydroxymethylperhydroquinolizine)

Lupinine is oxidised to lupinic acid, proving the presence of a primary alcohol group in the molecule. *Schöpf* (1957) carried out a simple *synthesis* of lupinine starting from 3,4,5,6-tetrahydropyridine. It had a close relationship to the *biosynthesis* of lupinine from two molecules of L-lysine (*cf.* anabasine).

7.12.4 Alkaloids Based on Quinoline

The *cinchona alkaloids* comprise about thirty different, some stereoisomeric, plant bases which have been isolated for the most part from the bark of subtropical trees of the species

Cinchona and *Remijia*. The most important of this group of alkaloids is *quinine*, which has *chemotherapeutic use* in the treatment of malaria.

(–)-**Quinine**, $C_{20}H_{24}N_2O_2$, was first discovered by *Pelletier* and *Caventou* in 1820 in cinchona bark.

Koenigs, Skraup and *P. Rabe* were the principal contributors to the elucidation of the structures of the cinchona alkaloids. As a result of these researches the following structure was ascribed to quinine:

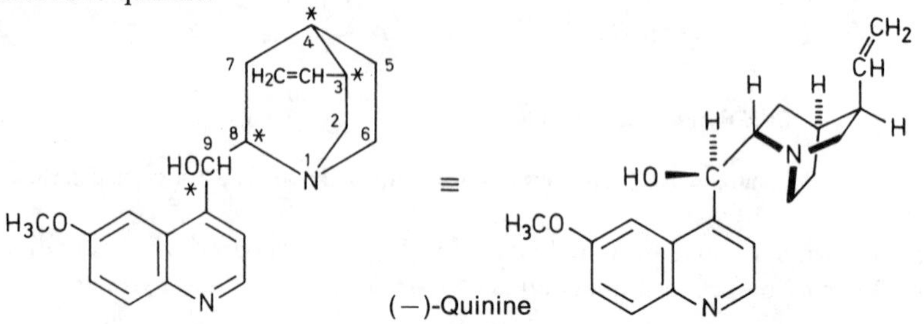

(–)-Quinine

It may be seen that the quinine molecule is constructed from two separate and different ring systems, namely from a quinoline nucleus (see p.774) and from a *quinuclidine ring* (1-azabicyclo[2.2.2]octane]. It possesses 4 chiral centres (*) which are described thus: (R)-3, (S)-4, (S)-8 and (R)-9.

Strong oxidation of quinine with chromic acid provides *quininic acid* and *meroquinene*.

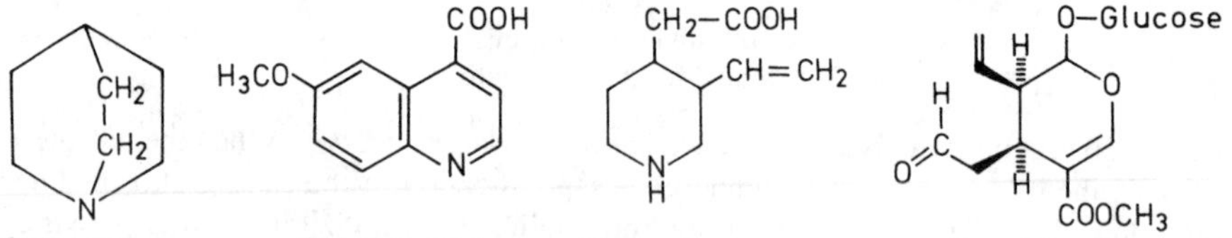

Quinuclidine   Quininic acid   Meroquinene   Secologanin ı

Final evidence for the structure of quinine was provided by its total synthesis (*Woodward, Doering,* 1944).

In plants, the quinuclidine ring is formed *via secologanin*, which is the key biogenetic intermediate, derived from mevalonic acid. The skeletal carbon atoms of secologanin less the vinyl group provide the seven carbon atoms of the quinuclidine ring.[1]

(–)-Quinine is a colourless substance which is only slightly soluble in water. In its anhydrous form it melts at 177°C (450 K). Its salts have an exceedingly bitter taste. (–)-Quinine and other cinchona alkaloids have proved useful as chiral auxiliaries in enantioselective syntheses (*Wynberg*, 1976).[2]

(+)-**Quinidine**, $C_{20}H_{24}N_2O_2$, is a stereoisomer of quinine. It differs in the stereochemistry at chiral centres 8 and 9, which in the present case have (R)-8 and (S)-9 configurations. It is used for treating disorders of the heart-rhythm.

(+)-**Cinchonine**, $C_{19}H_{22}N_2O$, an important minor alkaloid of the cinchona series, differs from quinine only in the absence of a methoxy group. It was discovered in cinchona bark by *Gomez* in 1811. Oxidative breakdown provides *cinchoninic acid* and *meroquinene*. Benzylation of the nitrogen atom in the quinuclidine ring gives rise to N-benzylcinchoninium salts, which are active as chiral phase-transfer catalysts in enantioselective alkylations.[3]

---

[1] See *A. R. Battersby* et al., Chem. Comm. *1970*, 194.
[2] See *W. Bartmann* and *B. M. Trost* (Ed.), Selectivity — a Goal for Synthetic Efficiency, p.365 (Verlag Chemie, Weinheim 1984).
[3] See *U.-H. Dolling* et al., J. Am. Chem. Soc. *106*, 446 (1984).

(–)-**Cinchonidine**, $C_{19}H_{22}N_2O$, has the same steric relationship to cinchonine as quinidine has to quinine.

(–)-**Cupreine**, $C_{19}H_{22}N_2O_2$, is demethylated quinine and can be converted into quinine by methylation.

7.12.5 Alkaloids Based on Isoquinoline

The main source of the about twenty-five alkaloids which make up the group of alkaloids based on isoquinoline is the milky sap from unripe poppies (*Papaver somniferum*), namely opium. Consequently these plant bases, which have very diverse structures are classified together with the overall name of *opium alkaloids*. This large class of substances is subdivided into various subgroups, depending upon their chemical structure.

Papaverine Group

The parent compound of this group of plant bases is *1-benzylisoquinoline*.

7.12.5.1   Papaverine, $C_{20}H_{21}NO_4$ was shown to be 1-(3′,4′-dimethoxybenzyl)-6,7-dimethoxy-isoquinoline. Its constitution was derived from the breakdown products which were obtained when it was melted with alkali, namely 6,7-dimethoxyisoquinoline and homoveratrole (3,4-dimethoxytoluene).

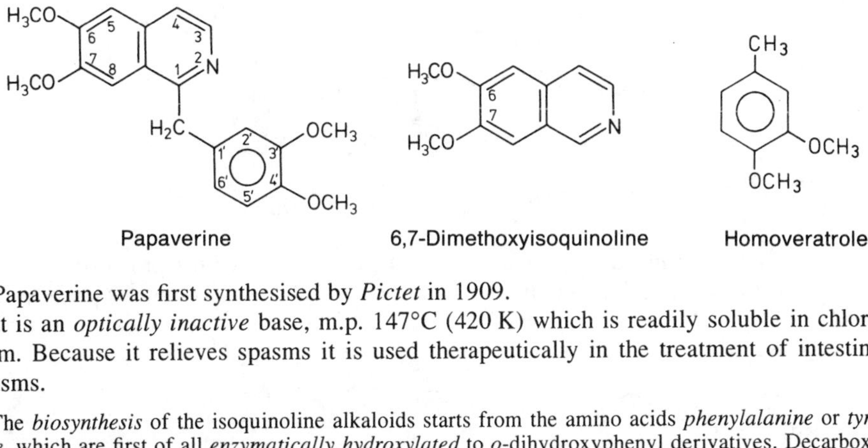

Papaverine   6,7-Dimethoxyisoquinoline   Homoveratrole

Papaverine was first synthesised by *Pictet* in 1909.

It is an *optically inactive* base, m.p. 147°C (420 K) which is readily soluble in chloroform. Because it relieves spasms it is used therapeutically in the treatment of intestinal spasms.

The *biosynthesis* of the isoquinoline alkaloids starts from the amino acids *phenylalanine* or *tyrosine*, which are first of all *enzymatically hydroxylated* to o-dihydroxyphenyl derivatives. Decarboxylation then leads to the formation of phenylethylamines. Also oxidative deamination can lead to aldehyde formation, *e.g.*

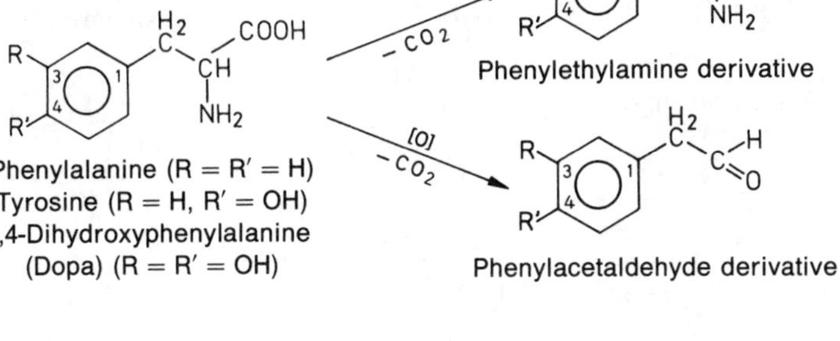

Phenylalanine (R = R′ = H)
Tyrosine (R = H, R′ = OH)
3,4-Dihydroxyphenylalanine
(Dopa) (R = R′ = OH)

Phenylethylamine derivative

Phenylacetaldehyde derivative

Condensation of the aldehyde with the phenylethylamine, coupled with an intramolecular *Mannich*-type reaction, provides the corresponding 1-benzyl-1,2,3,4-tetrahydroisoquinoline, *e.g.*

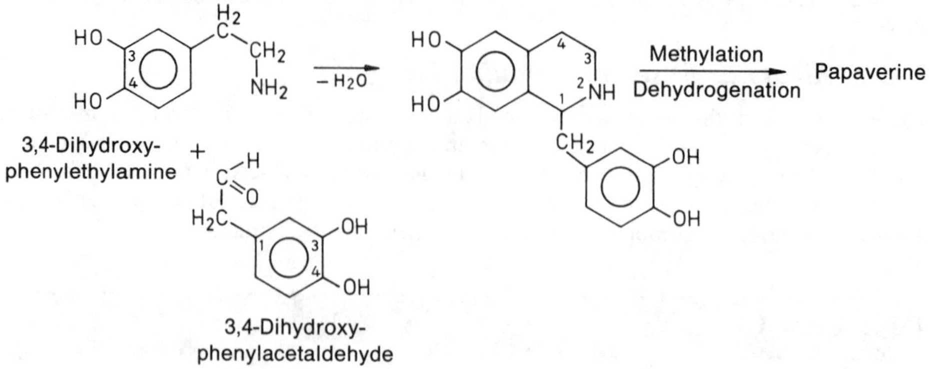

3,4-Dihydroxy-
phenylethylamine

3,4-Dihydroxy-
phenylacetaldehyde

Methylation
Dehydrogenation

Papaverine

Four-fold *enzymatic methylation* converts the resultant isoquinoline derivative into 1,2,3,4-tetra-hydropapaverine, which then undergoes dehydrogenation to papaverine.

7.12.5.2 **(+)-Laudanosine**, $C_{21}H_{27}NO_4$, m.p. 89°C (362 K) is the *N*-methyl-1,2,3,4-tetrahydro-derivative of papaverine. It occurs in small amounts as a minor alkaloid in opium. It has an asymmetric carbon atom (*) which is in an (*S*)-configuration.

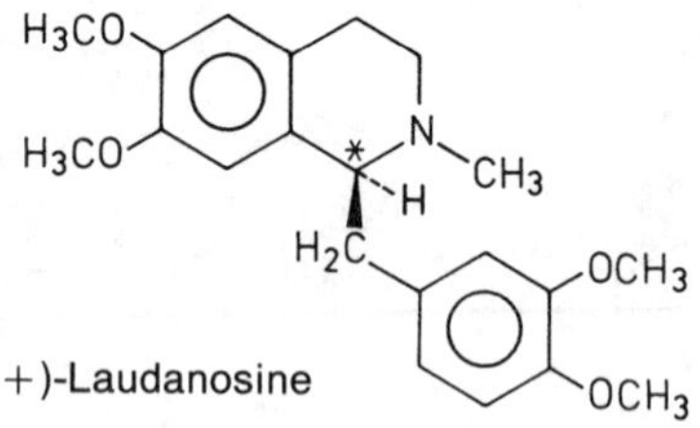

(*S*)-(+)-Laudanosine

7.12.5.3 **(−)-Narcotine**, INN: *Noscapine*, $C_{22}H_{23}NO_7$, is one of the principal alkaloids found in opium and was discovered by *Robiquet* in 1817. It forms colourless prisms, m.p. 176°C (449 K) and is a derivative of 1-benzyl-1,2,3,4-tetrahydroisoquinoline with the following structure:

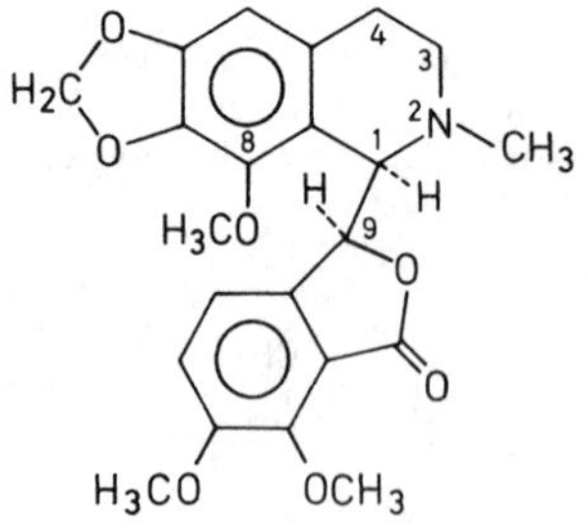

(1*R*, 9*S*)-Narcotine

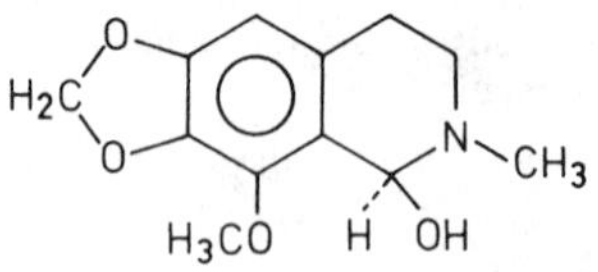

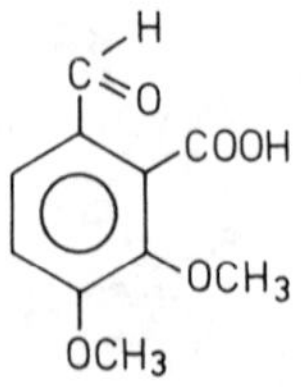

Opianic acid

Cotarnine

The principal products formed when narcotine is oxidised are **cotarnine** and **opianic acid**. The structure of cotarnine has been confirmed by synthesis. When opianic acid is heated it loses carbon dioxide to give veratraldehyde (3,4-dimethoxybenzaldehyde).

(−)-**Hydrastine**, $C_{21}H_{21}NO_6$, differs from narcotine in having no methoxy substituent at position 8 of the isoquinoline ring. It occurs in the roots of *Hydrastis canadensis*, but is not found in opium. Hydrastine acts as a styptic, *i.e.* it checks bleeding.

7.12.5.4   Berberine Group

The plant bases of this group have structures rather akin to those of narcotine and hydrastine. **Berberine**, $C_{20}H_{19}NO_5$, is present especially in barberry roots (*Berberis vulgaris*). In polar solvents it exists as a quaternary iminium base (I), but in the crystal state and in apolar solvents it exists as the pseudobase (B).

Berberine is *optically inactive*. It is a weak base, m.p. 145°C (418 K), which forms yellow salts. It is used in histology as a luminescent stain.

I   Berberine   B

Morphine Group

Among this group are the opium alkaloids *morphine* and *codeine* which have much more complicated structures than those of the alkaloids discussed hitherto. The fundamental ring system of morphine is *morphinane*,[1] which was synthesised by *Grewe* in 1946. In the formula below the hydrogenated isoquinoline system is shaded in in grey and the phenanthrene framework is outlined. When morphine is distilled in the presence of zinc powder it is broken down to give phenanthrene (*Vongerichten*).

7.12.5.5   (−)-Morphine, $C_{17}H_{19}NO_3$, was the first alkaloid to be isolated in crystalline form from plant sources (*Sertürner*, 1806). The correct molecular formula was found in 1848 (*Laurent*) but another 77 years had to pass before the constitution was finally proved.

Morphinane   Morphine   INN: *Naloxone*

---

[1] See *R. Grewe*, Synthetische Arzneimittel mit Morphin-Wirkung (Synthetic Medicines which have Activity like that of Morphine), Angew. Chem. *59*, 194 (1947).

In 1952, *Gates* and *Tschudi* achieved a *total synthesis* of morphine, thus bringing one chapter of the chemistry of morphine to a close after a period of nearly 150 years. Its biochemistry is now opening up another field, since its formation in different tissues in warm-blooded animals has now been detected.

(–)-Morphine crystallises in prisms, m.p. 254°C (527 K). It has a bitter taste and is only slightly soluble in water.

Among its salts, the *hydrochloride*, which crystallises in fine needles, in particular finds use in the relief of severe pain. Frequent use of this *narcotic* leads to the risk of addiction (morphinism), which causes a rapid physical deterioration of the body. *Diacetylmorphine* (heroin) is a particularly dangerous narcotic poison.

Colour reactions are used to test for the presence of morphine in solution. It gives a deep blue colour with iron(III) chloride, and if it is dissolved in conc. sulphuric acid and then a small amount of nitric acid is added, a red colour results.

7.12.5.6   Codeine, $C_{18}H_{21}NO_3$, the 3-monomethyl-ether of morphine, can be made by methylating morphine with diazomethane. It is less poisonous than morphine, but it has weak narcotic properties and is used in medicine as a pain-killer and to alleviate coughing.

At this point it is worth mentioning two *analgesics*, which show morphine-like activity in the body.

INN: Pethidine (Dolantin), ethyl 1-methyl-4-phenylpiperidinecarboxylate hydrochloride, a crystalline substance which melts at 188°C (461 K), combines the relief of muscular cramp shown by atropine with the alleviation of pain effected by morphine.

INN: Levomethadone (L-*Polamidone*), (–)-6-dimethylamino-4,4-diphenylheptan-3-one hydrochloride, is an even stronger analgesic than morphine. The withdrawal symptoms associated with methadone addiction are less strong than those associated with heroin withdrawal; because of this, methadone is used in some countries in the treatment of heroin addicts.

If the *N-methyl* group of morphine is replaced by an *allyl group* the resultant material, *Nalorphine*, is an *opium antagonist*; this is of some significance in understanding the mode of operation of the opium alkaloids. *INN: Naloxone*, L-17-allyl-4,5-epoxy-3,14-dihydroxymorphinan-6-one, (see p. 813) is used as an opium antagonist in therapeutic treatment of morphine poisoning.

INN: Pethidine (Dolantin)   *INN: Levomethadone* (Polamidone)

7.12.5.7   Curare Group

This group of alkaloids is found in *curare*, which is obtained from the South American *Strychnos* species and was used as an arrow poison; in particular it was investigated by

INN: (+)-Tubocurarine chloride

King, Späth and *Wintersteiner.* INN: **(+)-*Tubocurarine chloride*** is an important member of the group, in which two 1-benzyl-1,2,3,4-tetrahydroisoquinoline rings are joined to one another by means of oxygen bridges (see p. 814).

It resembles papaverine in its structure.

Intravenous injection leads to relaxation of the muscles and it is used for this purpose in surgical work.

7.12.6 Alkaloids Based on Indole

7.12.6.1 Pyrido-indole Alkaloids

Alkaloids based on the *9H-pyrido[3,4-b]indole* ring skeleton (formerly called β-carboline), which consists of a pyridine ring fused to an indole ring, are found in, for example, the bark of the species *Sikingia* (**harmane**) and *Peganum harmala* (**harmine** and **harmaline**). The structure of these three alkaloids was determined both by degradation and by synthesis (*Perkin* and *Robinson*).

9-*H*-Pyrido[3,4-*b*]indole (β-Carboline)   9*H*-1,2,3,4-Tetrahydropyridoindole

Harmane   Harmine   Harmaline (Dihydroharmine)

Tetrahydropyridoindoles can be made by intramolecular *Mannich* reactions, starting from tryptophane and acetaldehyde:

Tryptophane

Decarboxylation and dehydrogenation of the initial product provides harmane. It is probable that this route is also followed in the *biosyntheses* of pyridoindole alkaloids.

(+)-Yohimbane (yohimbine), $C_{21}H_{26}N_2O_3$, is an example of a pyridoindole alkaloid. It was discovered by *Spiegel* (1896) in *Corynanthe yohimbé* and proved to be identical with quebrachine which was obtained by *Hesse* (1880) from white *Quebracho* bark. It is a crystalline base, m.p. 241°C (514 K), which is soluble in ethanol and in chloroform. Its structure was confirmed by its *total synthesis (van Tamelen)* to be as follows:

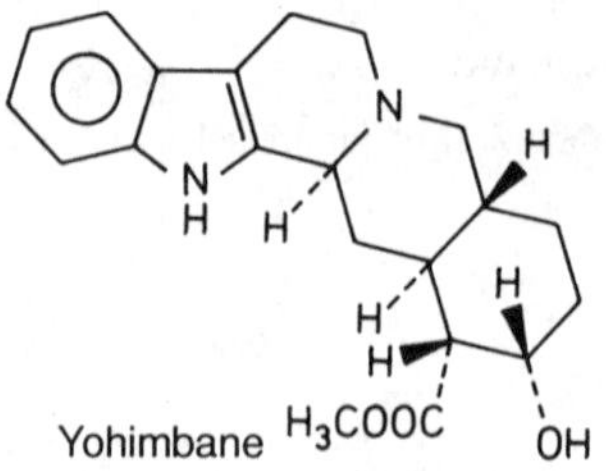

Yohimbane H_3COOC OH

Yohimbane is an α-blocker (see p.527); it acts by expanding the blood vessels and lowers the blood pressure.

7.12.6.2 Rauwolfia Alkaloids

Reserpine, $C_{33}H_{40}N_2O_9$, is the principal alkaloid of this group, and occurs in the Indian bush *Rauwolfia serpentina* and in other species of *Rauwolfia*. It was first isolated in a pure state in 1952, and its structure was elucidated in 1956 by *Schlittler*, and a total synthesis was carried out by *Woodward*.

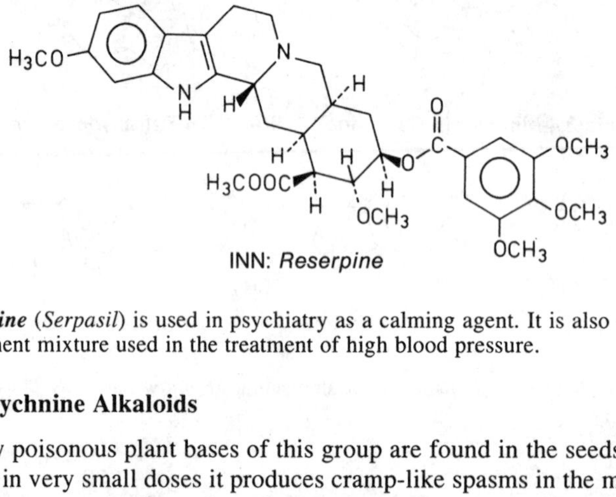

INN: *Reserpine*

INN: *Reserpine (Serpasil)* is used in psychiatry as a calming agent. It is also a constituent part of a multi-component mixture used in the treatment of high blood pressure.

7.12.6.3 Strychnine Alkaloids

The extremely poisonous plant bases of this group are found in the seeds of *Strychnos nux vomica*. Even in very small doses it produces cramp-like spasms in the muscles. The elucidation of its structure, which consists of seven fused rings, was accomplished first of all by the pioneering work of *H. Leuchs* and later by the work of *R. Robinson*.

(–)-**Strychnine,** $C_{21}H_{22}N_2O_2$, forms colourless prisms, m.p. 268°C (541 K). It has a very bitter taste. It is hardly soluble in water but dissolves in ethanol or chloroform. Strychnine gives a blue-violet colouration when treated with chromic acid and conc. sulphuric acid.

On the basis of an X-ray crystallographic analysis and of the total synthesis carried out by *Woodward*,[1] strychine has been assigned the following formula, originally suggested by *Robinson*:

---

[1] See *R. B. Woodward* et al., Die Totalsynthese des Strychnins, Angew. Chem. *75*, 456 (1963).

Strychnine (R = H)
Brucine (R = OCH₃)

(–)-**Brucine**, $C_{23}H_{26}N_2O_4$, is the 10,11-dimethoxy derivative of strychnine. It is a colourless substance, m.p. 178°C (451 K). With nitric acid it gives a red colouration, which is associated with the conversion of the two methoxy-substituted carbon atoms into an *o*-quinone structure.

Both strychnine and brucine have been used a lot for the resolution of racemic acids into their enantiomers.

7.12.6.4 Ergot Alkaloids[1]

A series of alkaloids which are α-blockers (see p.527), and which have powerful uterus-contracting action, has been found in ergot (*Secale cornutum*), in the sclerotium of the parasitic fungus *Claviceps purpurea* which grows on rye and other cereals. Their structures were worked out above all by *Jacobs* and *Craig* and by *A. Stoll*; they showed that they were amides of (+)-**lysergic acid**, $C_{16}H_{16}N_2O_2$, which is made up from indole and highly hydrogenated quinoline rings fused together. The structure was confirmed by synthesis (*Woodward*, 1954); it has a (5*R*, 8*R*)-configuration.

The most important of the ergot alkaloids is *INN: Ergotamine*, $C_{33}H_{35}N_5O_5$, which is a *tripeptide* built up from (+)-lysergic acid, L-(–)-phenylalanine, L-(–)-proline and L-(–)-α-hydroxyproline. It has the following structural formula:

INN: Ergotamine

The simplest alkaloid of this kind to be isolated is *ergometrine* (Ergonovine, *Ergobasine*), $C_{19}H_{23}N_3O_2$, in which lysergic acid forms an amide with L-2-aminopropanol.

[1] See A. *Hoffmann*, Die Mutterkornalkaloide (Ergot Alkaloids) (F. Enke Verlag, Stuttgart 1964).

(+)-*Lysergic acid diethylamide* (LSD), which can be prepared synthetically, has the remarkable property of inducing colour vision and hallucinations in healthy humans after absorbing only very small amounts (30 μg) of it.

Psilocin (4-hydroxy-*N,N*-dimethyltryptamine) and **psilocybin**, in which the 4-hydroxy group of psilocin has been converted into a phosphate ester, have been found to have similar psychotropic action. Both of these indole derivatives were isolated from the Mexican magic mushroom *Psilocybe mexicana* (*A. Hofmann*[1] and *R. Heim*, 1959). They are closely related to the neurotransmitter serotonin.

Psilocin   Psilocybin

---

[1] See Die psychotropen Wirkstoffe der mexikanischen Zauberpilze (The Psychotropic Action of Mexican Magic Mushrooms), Chimia *14*, 309 (1960).

8 Amino Acids, Peptides and Proteins

Proteins play an important role as natural building-blocks in the construction of the living cells of both animals and plants. Together with carbohydrates and fats they form the greater part of the food of human beings and of animals. Proteins serve not only as sources of energy for the living bodies, but also act as hormones and as enzymes. Because of this the chemistry of amino acids forms a vital part of biochemical research.[1]

Proteins (*Mulder*, 1838) can be defined as colloidal natural products of high molecular weight, which are made up from large numbers of different α-amino acids. These amino acids are linked together in long chains by means of *peptide links*, —CO—NH—, which have half-lives with respect to hydrolysis in pure water at room temperature of about seven years.[2]

Only plants are capable of building up proteins from simple inorganic compounds. The necessary elements, carbon, oxygen, hydrogen, nitrogen and sulphur are obtained by plants from carbon dioxide in the atmosphere, water and salts present in the soil such as nitrates, ammonium compounds and sulphates. Certain microorganisms and some plants, for example the *leguminosae*, can *directly* absorb nitrogen from the air with the help of bacteria in nodules on their roots.

In contrast, animal organisms are incapable themselves of synthesising all the various α-amino acids which are required to build up the proteins which are essential to them. In order to maintain normal bodily functions it is therefore essential for animal organisms to have a constant supply of proteins, especially since in the course of nitrogen metabolism nitrogen, which forms a vital component of the proteins, is lost in the form of urea, ureic acid and other by-products. On the other hand, animal organisms themselves are able to make some of the amino acids which are necessary for protein syntheses; this is achieved by means of the citric acid cycle, in which *transamination* plays an important part, for example resulting in the conversion of α-ketoglutaric acid into glutamic acid (see p.891).

The protein content of different foodstuffs is thus of great importance in human nutrition. The following products are especially rich in proteins: blood (~ 20%, blood serum ~ 7–8%) muscle (19%), fish (13%), eggs (12%) and milk (3%). Plant proteins are concentrated in the seeds, *e.g.* soya-beans 36%, cereal seeds 7–12%, and potatoes 2%. These numbers should only be regarded as average values.

The amino acids glycine, alanine, glutamic acid, valine, proline and aspartic acid have been found in meteorites. They are thought to have been formed by means of abiotic chemical synthesis in interstellar space.

---

[1] See *G. C. Barrett*, Chemistry and Biochemistry of the Aminoacids (Chapman and Hall, London 1985); *G. E. Schulz* et al., Principles of Protein Structure (Springer-Verlag, Berlin 1978).
[2] See *D. Kahne* et al., Hydrolysis of a Peptide Bond in Neutral Water, J. Am. Chem. Soc. *110*, 7529 (1988).

8.1 Amino Acids as Protein Building Blocks

The first insight into the complicated structures of proteins came from studies of their hydrolysis by either mineral acid or by alkali. In this way *Braconno*t, in 1820, isolated the amino acids glycine and leucine from glue. Decisive results were obtained by *Emil Fischer*, using the *ester method*. In this the protein is hydrolysed by heating it with conc. hydrochloric acid or 25% sulphuric acid and the resulting mixture of amino acids is converted by means of ethanolic hydrogen chloride into the corresponding esters. These were then separated by repeated fractional distillation. By this method it was shown in particular that the proteins were built up *solely from α-amino acids*.

With the exception of glycine, all the *α*-amino acids isolated from proteins contained at least one asymmetric carbon atom. As a result they are *optically active* and they belong exclusively to the L-series, *i.e.* they have the following projection formula:

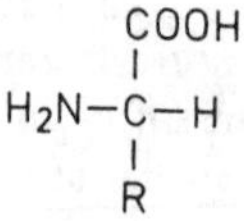

$$
\begin{array}{c}
\text{COOH} \\
| \\
\text{H}_2\text{N} - \text{C} - \text{H} \qquad \text{L-Configuration} \\
| \\
\text{R}
\end{array}
$$

This is an (S)-configuration according to the *Cahn–Ingold–Prelog* system, so long as the group R does not include a substituent which gives it higher priority than the carboxyl group. The latter situation obtains in the cases of L-cysteine and L-cystine.

The naturally occurring *α*-amino acids can be classified, depending upon their constitutions, as aliphatic, aryl or heterocyclic compounds. The more important ones are listed and described here.

Twenty of them are involved in the *genetic coding of information for the biosynthesis of proteins*. Such *proteinogenic amino acids* are numbered from **1** to **20** in the succeeding text. The pK_a values for the carboxyl and amino groups are shown alongside the relevant formulae (pK_1–pK_3).

8.1.1 Aliphatic Amino Acids

8.1.1.1 Monoamino-monocarboxylic Acids

The simple aliphatic amino acids have already been discussed (see p.283) so only their names are listed here. The shortened forms of the names of the different acids shown here enclosed in brackets (the three-letter abbreviations and one-letter abbreviations) are generally used in the representation of peptide chains.

1: Glycine (Gly, G), *INN: Aminoacetic acid*
glycocoll

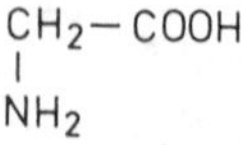

$$
\begin{array}{l}
\text{CH}_2 - \text{COOH} \\
| \\
\text{NH}_2
\end{array}
$$

pK_1:2.35
pK_2:9.78

2: L(+)-**Alanine** (Ala, A), *α*-aminopropionic acid, 2-aminopropanoic acid

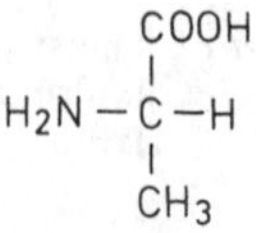

$$
\begin{array}{c}
\text{COOH} \\
| \\
\text{H}_2\text{N} - \text{C} - \text{H} \\
| \\
\text{CH}_3
\end{array}
$$

pK_1:2.35
pK_2:9.87

3: L(+)-**Valine** (Val, V), α-aminoisovaleric
acid, 2-amino-3-methylbutanoic acid

$$\begin{array}{c} COOH \\ | \\ H_2N-C-H \\ | \\ CH(CH_3)_2 \end{array}$$

pK_1:2.29
pK_2:9.72

4: L(−)-**Leucine** (Leu, L), α-aminoisocaproic
acid, 2-amino-4-methylpentanoic acid

$$\begin{array}{c} COOH \\ | \\ H_2N-C-H \\ | \\ CH_2-CH(CH_3)_2 \end{array}$$

pK_1:2.33
pK_2:9.74

5: (L(+)-**Isoleucine** (Ile, I), α-amino-β-
methylvaleric acid, 2-amino-3-methylpentanoic acid

äuı

$$\begin{array}{c} COOH \\ | \\ H_2N-C-H \\ | \\ H_3C-C-H \\ | \\ CH_2-CH_3 \end{array}$$

pK_1:2.32
pK_2:9.76

8.1.1.2 Hydroxy-amino-monocarboxylic Acids

6: L(−)-**Serine** (Ser, S), α-amino-β-hydroxypropionic acid, 2-amino-3-hydroxypropanoic
acid (β-hydroxyalanine), is a constituent of silk. DL-Serine can be made by a cyanohydrin
synthesis from hydroxyacetaldehyde (glycolaldehyde) and hydrogen cyanide in the presence of ammonia (*E. Fischer* and *H. Leuchs*).

7: L(−)-**Threonine** (Thr, T), α-amino-β-hydroxybutyric acid, 2-amino-3-hydroxybutanoic
acid, was first found in proteins only relatively late on. Its configuration corresponds to that
of D-threose:

pK_1:2.19
pK_2:9.44

$$\begin{array}{c} COOH \\ | \\ H_2N-C-H \\ | \\ CH_2-OH \end{array}$$

L-Serine

pK_1:2.09
pK_2:9.10

$$\begin{array}{c} COOH \\ | \\ H_2N-C-H \\ | \\ H-C-OH \\ | \\ CH_3 \end{array}$$

L$_s$-Threonine
(D$_g$-Threonine)

$$\begin{array}{c} C{\overset{H}{\underset{O}{<}}} \\ | \\ HO-C-H \\ | \\ H-C-OH \\ | \\ CH_2OH \end{array}$$

D$_g$-Threose

In amino acids which have more than one asymmetric carbon atom, their allocation to either the D-
or L-series is usually based on the configuration of the α-carbon atom which bears the amino group.
However, if L-threonine is related to D-threose then it must be considered as having a D-configuration.
In order to resolve this ambiguity, it is common practice to add to the letters D or L the subscript g or
s. This makes it clear whether the designation refers to *glyceraldehyde* or to *serine* for its basis. L(−)-
Threonine can thus be described as L$_s$-threonine or as D$_g$-threonine.

8.1.1.3 Mercapto-amino-monocarboxylic Acids

8: L(+)-**Cysteine** (Cys, C), α-amino-β-mercaptopropionic acid, 2-amino-3-mercapto-
propanoic acid (thioserine) is of biochemical interest, since it is very readily dehydrogenated, *e.g.* in air, to the disulphide L(−)-**cystine**. Together they form a reversible *redox
system*:

$$pK_1\!:\!1.86$$
$$pK_2\!:\!10.34$$

L-Cysteine   L-Cystine

Unlike most amino acids, cystine is only very slightly soluble in water. It was first discovered in urinary calculi (*Wollaston*, 1810) and, with methionine, is the main source of the sulphur content of proteins. In particular it is found in quantity in hair.

Selenocysteine, in which the HS-group of cysteine is replaced by an HSe-group, forms part of *glutathione peroxidase* and plays an important role in protecting the organism from the destructive activity of hydroperoxides.[1]

9: L(−)-**Methionine** (Met, M), α-amino-γ-methylmercaptobutyric acid, 2-amino-3-methylthiobutanoic acid, occurs in a variety of proteins, especially in casein.

$$pK_1\!:\!2.17$$
$$pK_2\!:\!9.27$$

N-Formylmethionine

It is the methylthio ether of homocysteine (shaded grey in the above formula) from which it is formed in the body by the action of the vitamin B_{12}-coenzyme. It serves an important function in *transferring methyl groups*; for this purpose it is activated by reaction with ATP (see *S*-adenosylmethionine). *N*-Formylmethionine participates in the biosynthesis of proteins in bacteria in which it always appears as the *N*-terminal amino-acid unit.

Sulphur-containing amino acids can be detected by thin-layer chromatography by means of the iodine azide reagent (*E. Chargaff*).[2]

8.1.1.4 Diamino-monocarboxylic Acids

These amino acids are *basic*, since they contain two amino groups but only one carboxyl group.

L(+)-**Ornithine** (Orn), 2,5-diaminopentanoic acid, does not occur in proteins, but is a decomposition product of arginine.

L-Ornithine

In birds, ornithine acts as a detoxifying agent for benzoic acid, by forming a dibenzoyl derivative. In the presence of putrefactive bacteria it is decarboxylated to putrescine, H_2N—$(CH_2)_4$—NH_2.

L(+)-**Citrulline** (Cit), 2-amino-5-ureidopentanoic acid, occurs in only a few proteins, one of which is casein.

---

[1] See *L. Flohé* et al., Selen in der enzymatischen Katalyse, Chemie in unserer Zeit *21*, 44 (1987).
[2] See *Houben-Weyl*, Methoden der Organischen Chemie 2, 578 (Thieme Verlag, Stuttgart 1953).

10: *INN:* L(+)-**Arginine** (Arg, R), 2-amino-5-guanidinopentanoic acid, differs from ornithine in having a guanidine residue in place of the 5-amino group. It is strongly basic and hence distinct from all other protein derivatives. It can be prepared by the addition of cyanamide to ornithine.

COOH   H$_2$N   COOH   H$_2$N

H$_2$N−C−H   \   H$_2$N−C−H   \   pK$_1$:1.82

| C=O | C=NH   pK$_2$:8.99

CH$_2$−CH$_2$−CH$_2$−NH   /   CH$_2$−CH$_2$−CH$_2$−NH   /   pK$_3$:13.20

L-Citrulline   *INN:* L-**Arginine**

The enzyme **NO-synthase** acts on L-arginine to give nitrogen monoxide (NO), which is involved in the body as a neurotransmitter in a variety of activities. NO is also active as a vasodilator (see pp.313,759) and in the immune system. Particularly important is its function in the brain as a retrograde messenger.

11: L(+)-**Lysine** (Lys, K), 2,6-diaminohexanoic acid, does not occur in all proteins.

COOH

| pK$_1$:2.16

H$_2$N−C−H   pK$_2$:9.20

| pK$_3$:10.80

CH$_2$−CH$_2$−CH$_2$−CH$_2$−NH$_2$

L-Lysine

Lysine, which is a homologue of ornithine, like the latter is converted into a diamine in the putrefaction of proteins, in this case into *cadaverine*, H$_2$N—(CH$_2$)$_5$—NH$_2$.

8.1.1.5   Monoamino-dicarboxylic Acids

These amino acids are acidic compounds since they have two carboxyl groups and only one amino group.

12: L(+)-**Aspartic acid** (Asp, D), aminosuccinic acid, 2-aminobutanedioic acid, was first obtained by hydrolysis of **13:** L(−)-**asparagine** (Asn, N), its mono-amide. Asparagine is found in the seeds of *leguminosae* and in asparagus (*Asparagus officinalis*). L-Asparagine also forms one of the units from which insulin is made up. L-Asparagine has a sweet taste whereas D-asparagine tastes bitter.

COOH   COOH

pK$_1$:1.99   H$_2$N−C−H   H$_2$N−C−H   pK$_1$:2.02

pK$_2$:3.90 | | pK$_2$:8.80

pK$_3$:10.00   CH$_2$ CH$_2$

| |

COOH   CONH$_2$

L-Aspartic acid   L-Asparagine

14: *INN:* L(+)-**Glutamic acid** (Glu, E), α-aminoglutaric acid, 2-aminopentanedioic acid, is obtained by hydrolysing **15:** L-**glutamine** (Gln, Q), which is a homologue of asparagine. L-Glutamine occurs in germinating seeds. It is an important constituent of muscle protein and of gluten from grain; L-glutamic acid is a neurotransmitter. L-Glutamine is also one of the units which make up insulin. Together with asparagine it forms a large part of the free amino acids present in human blood.

pK_1:2.13
pK_2:4.32
pK_3:9.95

$$\begin{array}{c} COOH \\ | \\ H_2N-C-H \\ | \\ (CH_2)_2 \\ | \\ COOH \end{array}$$

$$\begin{array}{c} COOH \\ | \\ H_2N-C-H \\ | \\ (CH_2)_2 \\ | \\ CONH_2 \end{array}$$

pK_1:2.17
pK_2:9.13

INN: L-*Glutamic acid*   L-Glutamine

Glutamic acid and its sodium salts are used as taste-enhancers in foodstuffs.

8.1.2 Aryl Amino Acids

16: L(–)-**Phenylalanine** (Phe, F), α-amino-β-phenylpropionic acid, 2-amino-3-phenyl-propanoic acid, was first isolated from the seeds of *leguminosae* and later from the hydrolysis products from various proteins.

17: L(–)-**Tyrosine** (Tyr, Y), 2-amino-3-(4-hydroxyphenyl)propanoic acid, is found in almost all proteins. It was first isolated from the products obtained by melting cheese with alkali (*J. v. Liebig*, 1846) (Grk. *tyros* = cheese). Today it is obtained by hydrolysing silk waste with conc. hydrochloric acid. It is only very slightly soluble in water. When heated it loses carbon dioxide to form *tyramine*.

pK_1:2.58
pK_2:9.24

$$\begin{array}{c} COOH \\ | \\ H_2N-C-H \\ | \\ CH_2 \end{array}$$

L-Phenylalanine

$$\begin{array}{c} COOH \\ | \\ H_2N-C-H \\ | \\ CH_2 \end{array}$$—OH

pK_1:2.20
pK_2:10.07

L-Tyrosine

In order to test for the presence of aryl amino acids, *e.g.* phenylalanine or tyrosine and also tryptophane, conc. nitric acid is added to the protein and, if they are present, a *yellow colour* develops. In this so-called *xanthoprotein reaction* nitration of the benzene rings takes place. Conc. nitric acid also turns the skin yellow, because of its action on the proteins of the epidermis.

Closely related to tyrosine is L-**3,5-di-iodotyrosine** (iodogorgic acid) which is found in the proteins of certain marine organisms such as corals and sponges. It can be prepared by iodinating tyrosine. It is thus, as it were, an intermediate on the way to *thyroxine*, the thyroid hormone.

L-**Thyroxine** (Thx), β-[(4-hydroxy-3,5-di-iodophenoxy)-3,5-di-iodophenyl]-α-alanine is a constituent of *thyroglobulin* in the thyroid gland, which acts as a *metabolic-regulating hormone*. Thyroxine was first isolated in a pure state by *Kendall* (1915) from the hydrolysis products derived from thyroid gland. It forms colourless needles which decompose at 231–233°C (504–506 K). Its structure has been confirmed by synthesis (*Barger and Harington*, 1927).

In the course of studies of the physiological function of the thyroid gland (*Glandula thyreoidea*) *Schiff* (1856) observed that its absence leads to severe damage to the nervous system of animals and to convulsions, not uncommonly resulting in death. In human beings the activity of the thyroid gland can be disturbed in two different ways. On the one hand underactivity of the gland (hypothyreosis) brings about diseases such as goitre, myxoedema and cretinism, whilst overactivity (hyperthyreosis) causes *Graves's* disease (exophthalmic goitre).

The *iodine-free* derivative of thyroxine, L-**thyronine**, shows no hormonal activity, so that the activity of thyroxine must be associated with the iodine atoms.

Other iodine-containing compounds as well as thyroxine are found in the thyroid gland, for example L-*3,5,3′-tri-iodothyronine*, which is in fact five times more active physiologically than is

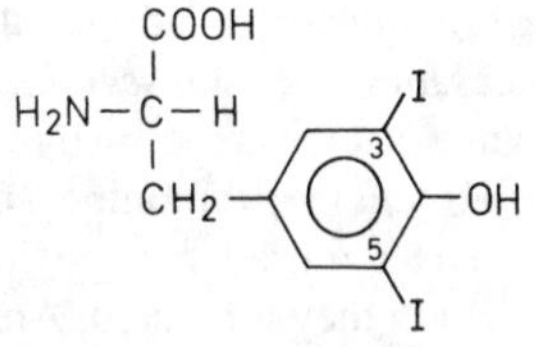

COOH
|
H₂N─C─H
|
CH₂─(ring, 3,5 positions)─OH, with I at positions
L-3,5-Di-iodotyrosine

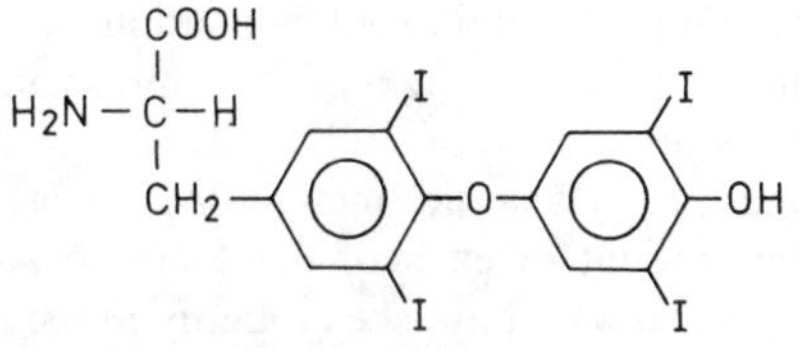

COOH
|
H₂N─C─H
|
CH₂─(ring)─O─(ring)─OH, with I substituents
L-Thyroxine

L-thyroxin (*INN: Sodium levothyroxin*) (*J. Roche*).
Over-production of iodothyronines can be restricted by certain compounds, for example thiourea, thiouracil or sulphaguanidine. In medical practice the main ones used are derivatives of thiouracil.

L-β-(3,4-Dihydroxyphenyl)-α-alanine (L-*Dopa*, INN: *Levodopa*) is formed enzymatically from tyrosine by the action of *tyrosinehydroxylase*. It is used in the treatment of *Parkinson's* disease (*Larodopa*).

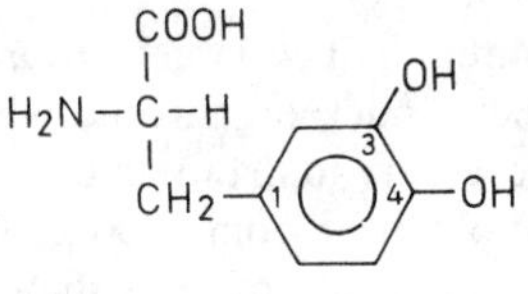

COOH
|
H₂N─C─H
|
CH₂─(ring 1,3,4 positions)─OH, ─OH

Dopa

8.1.3 Heterocyclic Amino Acids

18: L(−)-**Proline** (Pro, P), pyrrolidine-2-carboxylic acid (tetrahydropyrrole-2-carboxylic acid) and L(−)-**hydroxyproline** (Hyp, P̲), 4-hydroxypyrrolidine-2-carboxylic acid, are made largely by hydrolysing gelatine with hydrochloric acid.

pK_1:1.95
pK_2:10.64

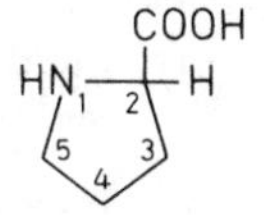

COOH
HN─(ring 1,2,5,4,3)─H
L(−)-Proline

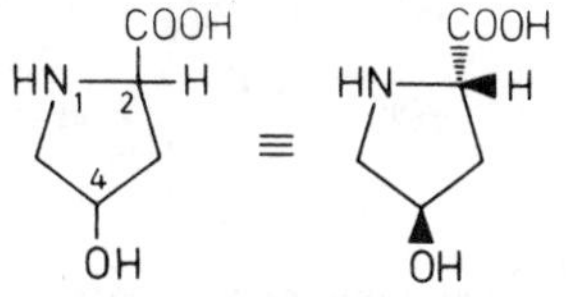

COOH
HN─(ring 1,2,4)─H
OH
≡
COOH
HN─H
OH
L(−)-*trans*-Hydroxyproline

Unlike all other L-amino acids, in this case the amino group takes the form of a cyclic *imino group*. L(−)-Hydroxyproline is formed from the proteinogenic L(−)-proline *after* transcription.

Two other important heterocyclic amino acids are **19:** L(−)-**tryptophane** (Trp, W), β-[3-indolyl]alanine, 2-amino-3-(1*H*-indol-3-yl)propanoic acid, and **20:** L(−)-**histidine** (His, H), β-[4-imidazolyl]alanine, 2-amino-3-(1*H*-imidazol-4-yl)propanoic acid:

pK_1:2.43
pK_2:9.44

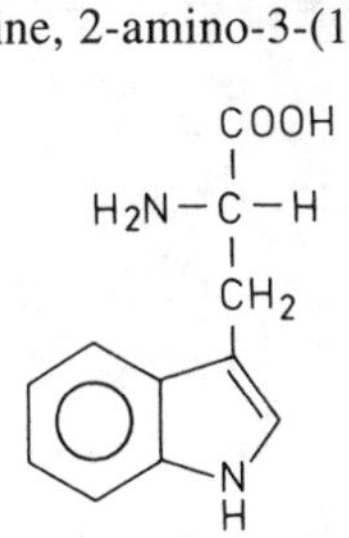

COOH
|
H₂N─C─H
|
CH₂
(indole ring)
N
H
L-Tryptophane

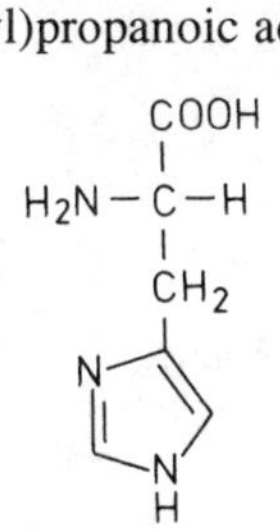

COOH
|
H₂N─C─H
|
CH₂
(imidazole ring)
N
N
H
pK_1:1.81
pK_2:6.05
pK_3:9.15

L-Histidine

Since not all of the amino acids mentioned, by a long way, occur in individual proteins, the question arises as to which of them are seen to be *necessary* or even *essential* to *life*, as nutritional requirements for adults. The following eight amino acids are known to be essential: leucine, phenylalanine, methionine, lysine, valine, isoleucine, threonine and tryptophane. The amount of each of these amino acids which is required for use in proteins (1.1–0.25 g per day), diminishes roughly in the order in which they are listed. Amino acids of this group are prepared industrially and added to foodstuffs. Aspartic acid and glutamic acid, and also γ-aminobutyric acid (GABA), are known to be neurotransmitters.

8.1.4 Industrial Sources of Amino Acids

The industrial production of amino acids has developed enormously.[1] Those produced in the greatest quantity are L-*glutamic acid*, DL-*methionine* and L-*lysine* (in that order), in amounts of 250,000 to 25,000 tonnes per year.

L-Lysine and L-glutamic acid are obtained largely from fermentation processes, following successful searches for suitable microorganisms, which produce the required amino acids in larger amounts than is necessary to maintain their own growth. For example, about 500 kg of L-*glutamic acid* can be obtained from 1000 kg of D-glucose. Since in the case of methionine, the racemic form is equally as active as the L-form for the purposes for which it is intended (for supplementation of animal foodstuffs), DL-methionine is the main *synthetic* product. This is made in three steps starting from acrolein:

$$H_2C=CH-CHO \xrightarrow{H_3CSH} H_3CS-CH_2-CH_2-CHO \xrightarrow[NH_4HCO_3]{NaCN}$$

Acrolein   3-Methylthiopropionaldehyde

DL-Methionine

Lysine is obtained by hydrolysing DL-α-*amino-ε-caprolactam* (3-aminohexahydroazepin-2-one), which is itself prepared from cyclohexene as follows:

DL-α-Amino-
ε-caprolactam   Lysine

---

[1] See Y. *Izumi*, Production and Utilisation of Amino Acids, Angew. Chem. Int. Ed. Engl. *17*, 176 (1978); B. *Hoppe*, Aminosäuren-Herstellung und Gewinnung (Amino Acids — Manufacture and Production), Chemie in unserer Zeit *18*, 73 (1984).

If the hydrolysis is carried out in the presence of L-*amino-ε-caprolactamhydrolase*, the product is exclusively L-lysine. The D-α-amino-ε-caprolactam which remains can be converted by a specific racemase into DL-α-amino-ε-caprolactam so that L-lysine is the sole product of the process. In some instances, seeding of a solution of the DL-amino acid with the L-enantiomer results in only the L-amino acid crystallising out. Racemisation of the mother liquors can in this case also lead to a recycling process and the sole final product is once again the L-amino acid. Lysine is present in only limited amounts in many human and animal foodstuffs; its supplementation greatly increases the nutritional value of the materials concerned.

8.2 Peptides

As *Emil Fischer* had deduced from the breakdown of proteins and from the synthesis of peptides, proteins are made up from large numbers of α-amino acids joined together by peptide links. In the same way that saccharides are classified as oligo- and poly-saccharides, so, according to the number of amino acid units they contain, peptides can be subdivided into *oligopeptides* (2–9 amino acid units), *polypeptides* (10–100 units) and *macropeptides* (proteins) (more than 100 units) (*Th. Wieland*).

The side-chains of the different amino acids (R_1–R_n) and also the terminal amino and carboxyl groups project alternately from each side of the linear peptide chain:

Peptide chain

Although only a limited number of amino acids take part in the building of these chains, by arranging them in different orders a very large number of variations is possible.

The methyl ester of α-L-*aspartyl*-L-*phenylalanine* (APM, aspartame) is used as a *sweetener*. This compound is about 200 times as sweet as cane sugar and has a clean taste. Because it is a dipeptide aspartame is susceptible to hydrolysis.

$$H_2N-CH-CO-NH-CH-CH_2C_6H_5$$
$$CH_2COOH COOCH_3$$

Aspartame

8.2.1 Synthesis of Peptides[1]

Despite its notable stability to pure water (see p.819), the peptide link is an energy-rich bonding. Because of this their formation requires the presence of at least one activated

---

[1] See *Th. Wieland* et al., Peptidsynthesen (V), Angew. Chem. *75*, 539 (1963); *G. R. Pettit*, Synthetic Peptides, 2 Vols. (van Nostrand, Reinhold, New York 1970–71); *E. Wünsch* (Ed.), *Houben-Weyl*, Methoden der Organischen Chemie *15/1,2* (Thieme Verlag, Stuttgart 1974); *M. Bodansky* et al., Peptide Chemistry. A Practical Textbook (Springer-Verlag, Berlin 1988).

amino acid derivative, usually one in which the carboxyl group has been converted into an acyl halide, an acyl azide or an ester group. The first method of peptide synthesis, carried out by *E. Fischer* in 1903, involved the reaction of an α-halogenoacyl chloride with the amino group of an unprotected amino ester, and then replacement of the halogen atom by an amino group by means of ammonia, *e.g.*

$$
\begin{array}{c}
\underset{\underset{Cl}{|}}{CH_2}-CO-Cl \; + \; H_2N-\underset{\underset{CH_3}{|}}{CH}-COOR \xrightarrow{-HCl} \underset{\underset{Cl}{|}}{CH_2}-CO-NH-\underset{\underset{CH_3}{|}}{CH}-COOR \\[2mm]
\xrightarrow[\text{(hydrolysis)}]{+NH_3} \quad \underset{\underset{NH_2}{|}}{CH_2}-CO-NH-\underset{\underset{CH_3}{|}}{CH}-COOH
\end{array}
$$

Glycylalanine

This method did not prove to be very satisfactory, however, since the resultant halogenoacyl derivative was not easy to obtain and furthermore was unstable.

Modern methods of peptide synthesis make use of amino acids with protected amino groups. In this method it is essential that the *protective group* which is used should be easily removed again once the peptide link is formed without that link being cleaved at the same time. In addition attempts are made to raise the yield of peptide by having a suitable catalyst present and also to prevent any extensive racemisation of the amino acids.

8.2.1.1   Introduction of the Protective Group

(a)   Benzyloxycarbonyl method (*M. Bergmann*, 1932). In this case the protective group attached to the nitrogen atom is the *benzyloxycarbonyl group*, formerly known as the carbobenzoxy (Cbo or Z) group. The benzyloxycarbonylamino acid is obtained by treating the amino acid in question with benzyl chloroformate, which is itself made from phosgene and benzyl alcohol:

$$
\text{C}_6\text{H}_5-CH_2-O-\underset{Cl}{\overset{O}{C}} \; + \; H_2N-\underset{\underset{R}{|}}{CH}-COOH \longrightarrow \text{C}_6\text{H}_5-CH_2-O-CO-NH-\underset{\underset{R}{|}}{CH}-COOH
$$

Benzyl chloroformate   Benzyloxycarbonylamino acid

(b)   t-Butyloxycarbonyl method (*G.W. Anderson*, 1957). The t-butyloxycarbonyl (BOC) group is finding increasing use as a protective group on the nitrogen atom. Acylation of the amino-group can best be carried out by using t-butyloxycarbonyl azide under mild conditions:

$$
\underset{\underset{CH_3}{|}}{\overset{\overset{CH_3}{|}}{H_3C-C}}-O-\underset{N_3}{\overset{O}{C}} \; + \; H_2N-\underset{\underset{R}{|}}{CH}-COOH \xrightarrow{-N_3H} \underset{\underset{CH_3}{|}}{\overset{\overset{CH_3}{|}}{H_3C-C}}-O-CO-NH-\underset{\underset{R}{|}}{CH}-COOH
$$

t-Butyloxycarbonyl azide   t-Butyloxycarbonylamino-acid

(c)   Fluoren-9-ylmethoxycarbonyl method (*L.A. Carpino*, 1970). The fluorenyl group absorbs UV radiation strongly and this makes it possible to follow very easily reactions involving the fluoren-9-ylmethoxycarbonyl (*Fmoc*) group. This group is introduced by means of either fluoren-9-ylmethyl chloroformate (X = Cl) or fluoren-9-ylmethoxycarbonyl azide (X = N_3):

$$+ \ H_2N-\underset{\underset{R}{|}}{CH}-COOH \ \xrightarrow{-HX}$$

$$CH_2-O-COX \qquad\qquad CH_2-O-CO-NH-\underset{\underset{R}{|}}{CH}-COOH$$

Fluoren-9-ylmethoxycarbonylamino acid   R

Instead of using the above three protective groups, which are all of the urethane type (shown with grey backgrounds), or the benzyl group, the tosyl (*du Vigneaud*), phthalyl (*Sheehan*) and the trifluoroacetyl (TFA) (*Weygand*) groups have also been used.

(d) Allyl ester method (*H. Kunz*, 1983). In the preparation of *glycopeptides* with *O*-glycosidic links between carbohydrates and the HO-groups of serine and threonine it is necessary, because of the labile nature of the glycosidic links, to use protective groups which can be removed under neutral conditions. Allyl esters meet this requirement. In the presence of *Wilkinson's* catalyst they rearrange to propenyl esters which, as enol esters, can be hydrolysed in aqueous ethanol:

$$R-\underset{\underset{O}{\|}}{C}-O-CH_2-CH=CH_2 \ \xrightarrow{\left[(C_6H_5)_3\,P\right]_3 RhCl} \ \left[R-\underset{\underset{O}{\|}}{C}-O-CH=CH-CH_3 \right] \xrightarrow{C_2H_5OH/H_2O} R-COOH$$

Allyl ester (R-C-OAll)   Propenyl ester
      ‖
      O

8.2.1.2 Formation of the Peptide Link

Only the most important of the many ways of forming peptide links are discussed here. They are all based on the use of high-energy derivatives of *N*-acylamino acids, especially acid chlorides, acyl azides and esters.

(a) Acid chloride (acyl chloride) method. An acylamino acid is first of all converted into its acid chloride which then reacts with an amino ester:

$$R-CO-NH-\underset{\underset{R_1}{|}}{CH}-C\!\!\overset{\overset{O}{\diagup}}{\diagdown}_{Cl} \ \xrightarrow[-HCl]{+\,H_2N-\underset{\underset{}{|}}{\overset{\overset{R_2}{|}}{CH}}-COOR'} \ R-CO-NH-\underset{\underset{R_1}{|}}{CH}-CO-NH-\underset{\underset{}{|}}{\overset{\overset{R_2}{|}}{CH}}-COOR'$$

Nowadays this process is often modified as follows: if the acylamino acid is heated with α,α′-dichlorodimethyl ether [bis(chloromethyl)ether] the acid chloride is formed as an intermediate and this then reacts with an amino ester to form a peptide.

(b) Azide method. This method was introduced by *Curtius* in 1902, and involves reaction of an amino acid azide having a protected amino group with an amino ester:

$$R-CO-NH-\underset{\underset{R_1}{|}}{CH}-C\!\!\overset{\overset{O}{\diagup}}{\diagdown}_{N_3} \ \xrightarrow[-N_3H]{+H_2N-\underset{\underset{R_2}{|}}{CH}-COOR'} \ R-CO-NH-\underset{\underset{R_1}{|}}{CH}-CO-NH-\underset{\underset{R_2}{|}}{CH}-COOR'$$

Use of the azide is especially suitable when possible racemisation of an amino acid must be avoided.

(c) Carbodiimide method (*Sheehan*, 1959). A very good method of forming peptide links involves the reaction of an acylamino acid with an amino ester in the presence of an equimolecular amount of *dicyclohexylcarbodiimide* (DCC),[1] which is converted into *N,N'*-dicyclohexylurea:

$$R-CO-NH-\underset{\underset{R_2}{|}}{\overset{\overset{R_1}{|}}{CH}}-COOH + H_2N-\underset{\underset{R_2}{|}}{CH}-COOR' + C_6H_{11}-N=C=N-C_6H_{11}$$

Dicyclohexylcarbodiimide

$$\longrightarrow R-CO-NH-\underset{\underset{R_2}{|}}{\overset{\overset{R_1}{|}}{CH}}-CO-NH-\underset{\underset{R_2}{|}}{CH}-COOR' + C_6H_{11}-NH-\overset{\overset{O}{\|}}{C}-NH-C_6H_{11}$$

N,N'-Dicyclohexylurea

An *O*-acylisourea derivative is formed as an intermediate. This contains an activated acyl group, which is transferred to the amino group of the amino ester:

$$R-CO-NH-\underset{\underset{R_1}{|}}{CH}-\underset{\underset{O}{\|}}{C}-O-C\underset{NH-C_6H_{11}}{\overset{N-C_6H_{11}}{<}}$$

Activated acyl group

(d)   Azolide method.[2] If an acylamino acid is treated with an equimolecular quantity of *N,N'-carbonyldiimidazole* in tetrahydrofuran at room temperature an energy-rich amino acid imidazolide is formed, which reacts readily with an amino ester to form a peptide:

$$R-CO-NH-\underset{\underset{R_1}{|}}{\overset{\overset{COOH}{|}}{CH}} + \underset{N,N'\text{-Carbonyldiimidazole}}{[imidazole]-C-[imidazole]} \xrightarrow[\text{- Imidazole}]{-CO_2,} R-CO-NH-\underset{\underset{R_1}{|}}{CH}-CO-N[imidazole]$$

N,N'-Carbonyldiimidazole   Amino acid imidazolide

$$\xrightarrow[\text{- Imidazole}]{+ H_2N-\underset{\underset{}{|}}{\overset{\overset{R_2}{|}}{CH}}-COOR'} R-CO-NH-\underset{\underset{R_1}{|}}{CH}-CO-NH-\underset{\underset{}{|}}{\overset{\overset{R_2}{|}}{CH}}-COOR'$$

(e)   *N*-Carboxylic acid anhydride method. When glycine is treated with either phosgene, diphosgene or triphosgene a glycine-*N*-carbonyl chloride is formed as an intermediate and this then cyclises to give *glycine-N-carboxylic acid anhydride* (Gly-NCA, oxazole-2,5(3*H*,4*H*)-dione, *Leuchs* anhydride). Small amounts of water act on this compound to form an insoluble high molecular weight polypeptide; carbon dioxide is evolved in the process:

$$\underset{\underset{OH}{|}}{\overset{\overset{H_2C-NH_2}{|}}{CO}} \xrightarrow{\underset{Toluene}{COCl_2,}} \underset{\overset{|}{OH}\ \overset{|}{Cl}}{\overset{H_2C-NH}{CO\ CO}} \xrightarrow[-HCl]{340\,K} \underset{Gly-NCA}{\overset{H_2C-NH}{OC-O-CO}} \xrightarrow{+H_2O} \underset{\underset{COOH}{|}}{\overset{\overset{H_2C-NH-COOH}{|}}{COOH}} \xrightarrow{-CO_2} \underset{\underset{COOH}{|}}{\overset{\overset{H_2C-NH_2}{|}}{COOH}}$$

$$+ \overset{H_2C-NH}{OC-O-CO} \longrightarrow \underset{\underset{COOH}{|}}{\overset{H_2C-NH-CO-CH_2}{\underset{HN-COOH}{}}} \xrightarrow{-CO_2} \underset{\underset{COOH}{|}}{H_2C-NH-CO-CH_2-NH_2}$$

---

[1] See *F. Kurzer* et al., Advances in the Chemistry of Carbodiimides, Chem. Rev. *67*, 107 (1967).
[2] See *G. W. Anderson* et al., J. Am. Chem. Soc. *80*, 4423 (1958), *82*, 4596 (1960); *H. A. Staab*, Angew. Chem. *74*, 407 (1962).

The dipeptide reacts with more Gly-NCA to form a tripeptide, and then a tetrapeptide, and so on.

If ^{13}C-labelled glycine-N-carboxylic acid anhydride is made to undergo thermal polycondensation it may be shown that the carbon dioxide which is evolved comes from the carboxyl group attached to the nitrogen atom (*Heyns*).

In nature polypeptides built up from only one particular amino acid are rare. Up to now only one such polypeptide is known, based on D-glutamic acid residues. This is found in the capsule of the *anthrax bacillus*.

N-Carboxylic acid anhydrides have use in the preparation of mixed dioxopiperazines (see p.786):

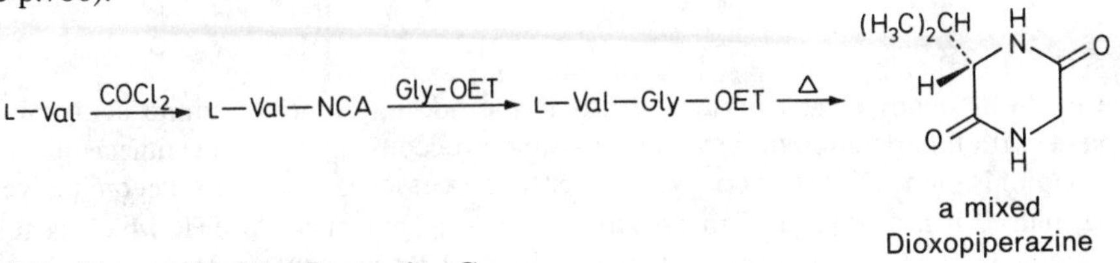

L—Val $\xrightarrow{\text{COCl}_2}$ L—Val—NCA $\xrightarrow{\text{Gly-OET}}$ L—Val—Gly—OET $\xrightarrow{\Delta}$

a mixed Dioxopiperazine

8.2.1.3 Removal of the Protective Group

After a peptide link has been formed the Cbo protective group can easily be removed by catalytic hydrogenolysis (*i.e.* hydrogenation accompanied by bond-fission) which produces toluene, carbon dioxide and the dipeptide. The benzyl group can also be removed by hydrogenolysis. This is the reason why these groups are valuable as protective groups (see p.828).

The t-butyloxycarbonyl group can readily be removed by hydrogen bromide in acetic acid or by moist trifluoroacetic acid. The BOC group, unlike the Cbo group, *cannot* be removed by catalytic hydrogenolysis, so that single amino acids can be attached to a polypeptide in this way.

The Fmoc group is distinctive in that it can be removed under mildly basic conditions, for example with piperidine or morpholine in DMF (in 5 min. at 20°C [293 K]) or with liquid ammonia.

8.2.1.4 *Merrifield* Synthesis[1]

If a growing peptide chain is anchored at one end to a polystyrene support, then successive additions of new units can be reduced to a process involving consecutive dosing with the appropriate reagents and washing off of the excess reagent; the three stages described above can thus be automated (solid-phase synthesis or polymer-supported synthesis, *Merrifield*, 1963).

The polystyrene support is made by copolymerising styrene with 0.2% divinylbenzene. Chloromethyl groups are introduced into some of the benzene rings, to provide *spacer groups*. The amino group of the initial amino acid is protected, and the carboxylate group reacts with the spacer to form an ester bond which links the amino acid to the support.

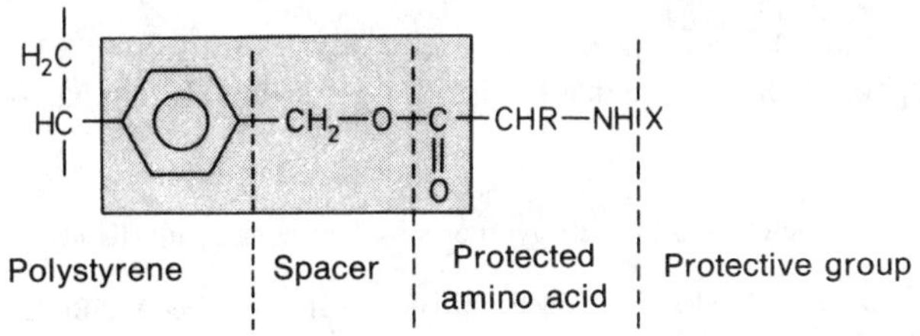

Polystyrene | Spacer | Protected amino acid | Protective group

---

[1] See *C. Birr*, Aspects of the Merrifield Peptide Synthesis (Springer-Verlag, Berlin 1978); *R. B. Merrifield*, Solid Phase Synthesis, Angew. Chem. Int. Ed. Engl. *24*, 799 (1985).

The anchorage must remain stable throughout the entire peptide synthesis, and must also, at its completion, be capable of being released under conditions which leave untouched the peptide links that have been formed. These conditions can be fulfilled if it involves a benzyl group (shown with a grey background in the above diagram). After removal of the protective group X (*e.g.* Cbo or BOC) the next *N*-protected amino group is joined on by the carbodiimide method, using dicyclohexylcarbodiimide (DCC):

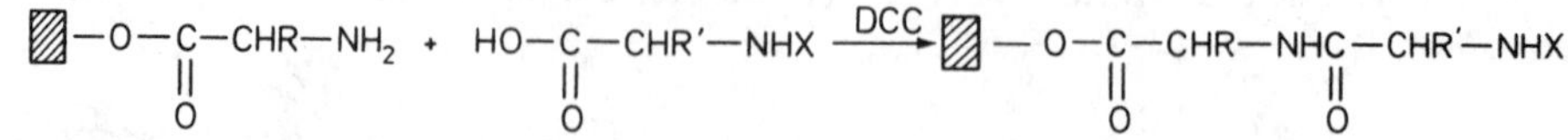

$$\boxed{/\!/}-O-\underset{\underset{O}{\|}}{C}-CHR-NH_2 \;+\; HO-\underset{\underset{O}{\|}}{C}-CHR'-NHX \xrightarrow{\;DCC\;} \boxed{/\!/}-O-\underset{\underset{O}{\|}}{C}-CHR-NHC-\underset{\underset{O}{\|}}{C}HR'-NHX$$

The cycle of removing the protective group and adding a further amino acid unit is continued, with interpolation of necessary washing procedures, until the synthesis is completed. Optimisation of each step in the overall process has led to its becoming very reliable, indeed it has been used to prepare a sample of human α-ACTH, which is fully active biologically. The reliability is increased by making segments containing about ten amino acid units automatically which are then linked together in the correct order.[1]

The synthesis of proteins by means of *genetic engineering* (see p.875) is becoming of increasing importance.

8.2.2 *Amino acid Analysis of Peptides*[2]

A peptide or protein is isolated and purified and is then hydrolysed into its constituent amino acids, which can be analysed both qualitatively and quantitatively by chromatographic methods. This operation is known as **amino acid analysis.** It has been developed from the fundamental work of *Moore* and *Stein* to the separation of amino acids on ion-exchange resins and complete automation of the process.

The first step in investigating the structure of a peptide chain is **end-group analysis,** *i.e.* discovering the identity of the groups at the end of the chain, the *N*-terminal amino acid which has a free amino group and the *C*-terminal amino acid which has a free α-carboxyl group. The most commonly used methods to achieve this are as follows:

8.2.2.1   DNP Method (*Sanger, 1954*), Dansyl Method (*Hartley, 1956*)

The terminal amino group must be substituted in such a way that the resultant derivative is resistant to hydrolysis. A suitable reagent for this purpose is *1-fluoro-2,4-dinitrobenzene.* When the protein is hydrolysed the dinitrophenyl group remains attached to the terminal amino acid. As a result this amino acid is readily separated out from the resultant mixture of amino acids and can then be identified chromatographically (Dinitrophenyl or DNP method).

In the Dansyl method the terminal amino acid is made to react with 5-dimethylamino-naphthalene-1-sulphonyl chloride (dansyl chloride). After the protein has been hydrolysed

---

[1] See *E. T. Kaiser* et al., Peptide and Protein Synthesis by Segment Synthesis-Condensation, Science *243*, 187 (1989).

[2] See *S. B. Needleman*, Protein Sequence Determination (Springer-Verlag, Berlin 1975); *H. Nau*, Sequence Analysis of Polypeptides and Proteins by Combined Gas Chromatography–Mass Spectrometry, Angew. Chem. Int. Ed. Engl. *15*, 75 (1976).

the resultant *dansylamino acid* fluoresces when irradiated with UV light. Its detection is 100 times more sensitive by this method than by the DNP method.

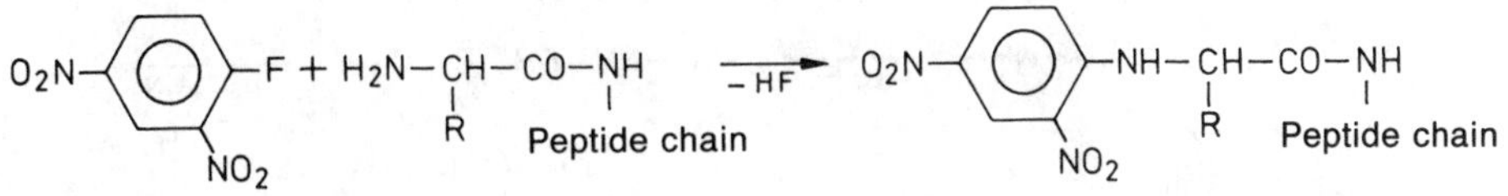

1-Fluoro-2,4-dinitrobenzene

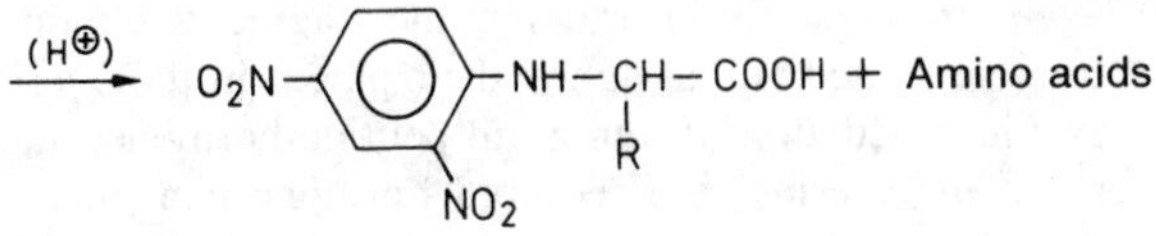

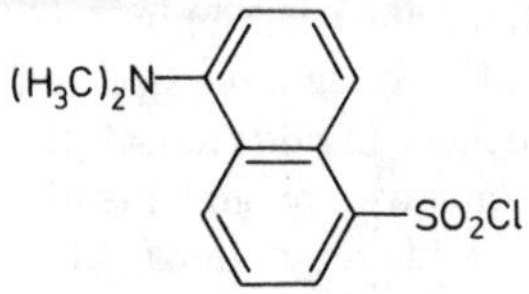

Dansyl chloride

8.2.2.2 *Edman* Degradation

This is the most important degradation process. It depends upon the selective removal of *N*-terminal amino acids from peptides or proteins. Reaction with *phenyl isothiocyanate* provides the corresponding *phenylisothiouronium cyanate*. With anhydrous hydrogen chloride this undergoes cleavage of the peptide link to form a *phenylthiazolinone*. Hydrolysis of the latter leads to ring opening and recyclisation to give a stable *phenylthiohydantoin* of the *N*-terminal amino acid; this is identified by HPLC.

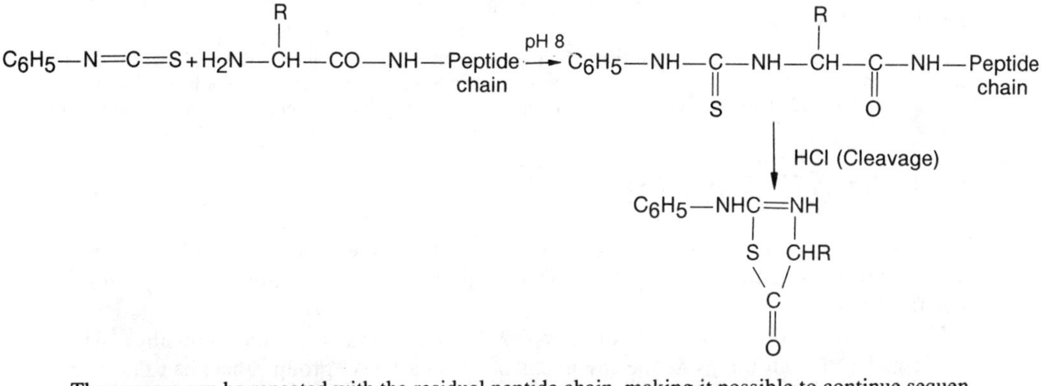

The process can be repeated with the residual peptide chain, making it possible to continue sequencing the amino acids in the chain. Its efficacy has been greatly increased by automating the process.

8.2.3 *Sequence Analysis of Peptides*

When the end-group has been identified, there remains the more difficult process of stepwise degradation of the peptide chain and of determining the order of the amino acids in the chain. The principle of sequence analysis depends upon acid or enzymatic hydrolysis of the macropeptide chain into smaller fragments (polypeptides).

The most important non-enzymatic method of breaking the chains involves the use of cyanogen bromide (*B. Witkop*, 1961). This reagent only breaks peptide links whose *acid groups are derived from methionine*, e.g.

$$
\begin{array}{ccc}
\underset{\displaystyle \begin{array}{c}CH_2-CH_2SCH_3\\ |\end{array}}{}\\
\sim CO-NH-CH-CO-NH\sim \quad\xrightarrow[-Br\ominus]{BrCN}\quad \sim CO-NH-CH-\underset{\|}{\overset{}{C}}-NH\sim \quad\xrightarrow{-H_3CSCN}\\
\end{array}
$$

$$
\sim CO-NH-CH-C\overset{\oplus}{\underset{NH\sim}{}} \quad\xrightarrow{2\,H_2O}\quad \sim CO-NH-CH-COOH \;+\; H_2N\sim
$$

Since the methionine unit only occurs infrequently in proteins, the fragments obtained are of a convenient size and are readily separated and assigned their places in the overall sequence. In addition, in the biosynthetic production of important peptide hormones such as somatostatin and insulin by genetic engineering, the required hormone can only be released in an undamaged form from the longer peptide chain which is formed initially, by making use of cyanogen bromide to split the chain.

Another method for splitting the peptide chain at specific sites involves the use of substrate-specific *enzymes*; for example, in general, *trypsin* only breaks the chain between arginine and lysine units. *Chymotrypsin* breaks the chain in such a way that the terminal *C*-amino acids of the resultant polypeptides are always phenylalanine or tyrosine. *Proteases* (C–N hydrolases) can bring about splitting to provide a variety of overlapping oligopeptide and polypeptide fragments which can be analysed further by *Edman* degradation. The separation of individual amino acids and their identification is usually achieved by means of HPLC. By linking together the oligopeptides whose structures have been analysed, it is possible to obtain the complete *sequence of amino acids* or the primary structure of the *whole peptide chain*.

Insulin was the first polypeptide to have its complete amino acid sequence determined, by *Sanger* in 1947. Since that time the structures of many other polypeptides and proteins have been elucidated;[1] the indirect method involving genetic-code peptide sequencing has become important in achieving this.

8.2.4 Naturally Occurring Peptides

Peptides which are involved as metabolic intermediates in the formation and breakdown of proteins are present in only very low concentrations, since they rapidly undergo further transformations.

Peptides isolated from animals or from plants often show structural anomalies, which may consist of such things as the involvement of a carboxyl group other than that at the α-position in forming peptide links. In such cases, the peptide avoids normal enzymatic attack in the organism and can hence undertake a specific physiological function. An example is provided by the *tripeptide* glutathione, which has an anomalous link involving the γ-carboxyl group of glutamic acid.

8.2.4.1   Glutathione

γ-Glutamyl-cysteinyl-glycine was discovered by *Hopkins* (1921) in yeast and in muscles; in addition it is found in blood and in most cells. Its structure was proved by synthesis (*Harington*):

---

[1] See M. O. *Dayhoff*, Atlas of Protein Sequence and Structure (National Biochemical Research Foundation, from 1967); A. *Haeberle*, Human Protein Data (VCH, Weinheim 1992).

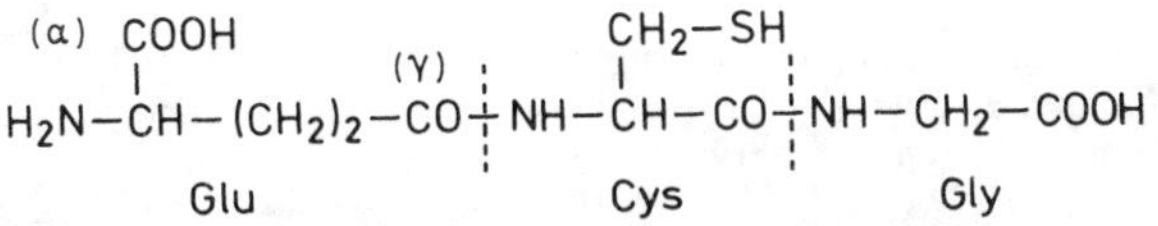

(α) COOH   CH₂—SH
     | (γ) |
H₂N—CH—(CH₂)₂—CO┼NH—CH—CO┼NH—CH₂—COOH   Glutathione (GSH)

     Glu   Cys   Gly

$$\text{(α)}\quad \begin{matrix}\text{COOH}\\|\\\text{H}_2\text{N—CH—(CH}_2)_2\text{—CO}\end{matrix}\overset{(\gamma)}{\vdots}\ \text{NH—}\begin{matrix}\text{CH}_2\text{—SH}\\|\\\text{CH—CO}\end{matrix}\ \text{NH—CH}_2\text{—COOH}$$

Glutathione readily takes part in a reversible conversion into a corresponding disulphide (GSSG), analogous to the interchange between cysteine and cystine. In consequence it acts as a transfer agent in *redox processes*. Its physiological action as an antioxidant stems from the fact that, as an easily oxidised compound it reduces other substances before they can attack and damage other groups which are sensitive to oxidation.

8.2.4.2 Neuropeptides (Endorphins)

The two pain-relieving (analgesic) polypeptides *methionine enkephalin* (I) and *leucine enkephalin* (II) were isolated by *Hughes* (1975) from pigs' brains, in the course of a search for the natural substrates of opiate receptors:

I: L-Tyr-Gly-Gly-L-Phe-L-Met   II: L-Tyr-Gly-Gly-L-Phe-L-Leu

Shortly afterwards peptides made up from 20–30 amino acids were obtained, which showed comparable activity to that of *morphine* and were therefore called endorphins (from 'endogenous morphines') (*Guillemin*). The opiate-antagonist INN: *Naloxone* also acts as an antagonist to endorphins. Unfortunately the hope that endorphins might provide harmless pain relievers has not been fulfilled.

8.2.4.3 Peptide Antibiotics

Within the last decade a number of *cyclic polypeptides* which are highly poisonous to almost all higher organisms and also to some bacteria have been isolated from micro-organisms and from fungi. As a result they have been considered as *antibiotics* (see p.455). Their resistance to the action of degradative enzymes is due above all to their having amino acids foreign to normal proteins, for example D-amino acids. Examples are *gramicidin* (*Dubos*, 1939) and *valinomycin*.

Gramicidin A is a linear pentadecapeptide with the following sequence of alternating L- and D-amino acids and a formyl group on the *N*-terminal amino acid and a β-ethanolamide group at the *C*-terminal acid (*Witkop*, 1965).

HC—L-Val—Gly—L-Ala—D-Leu—L-Ala—D-Val—L-Val—D-Val—L-Trp—
‖
O
                  D-Leu—L-Phe—D-Leu—L-Trp—D-Leu—L-Trp—NH—CH₂CH₂-OH
                                    Gramicidin A

Gramicidin has the ability to transport metal ions ($Na^\oplus$, $K^\oplus$) through biological membranes and as a result is of considerable biochemical importance.

The *depsipeptide* **valinomycin** is built up from *valine* and *hydroxy acids* and is a cyclic trimer of the following building blocks which contain ester groups as well as amide groups:

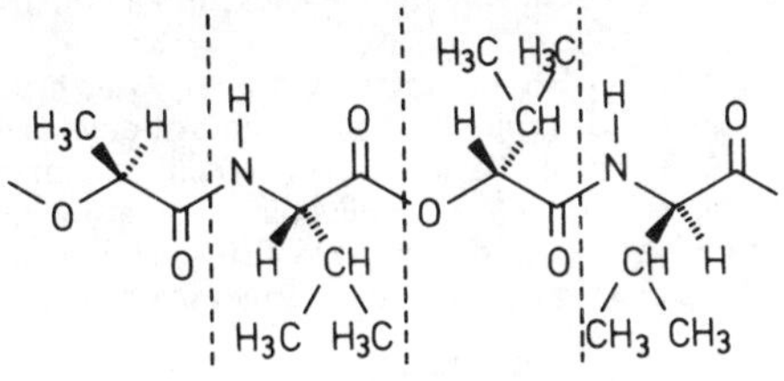

(L-Lactic acid) (L-Valine) (D-α-Hy- (D-Valine)
                           droxy-iso-
                           valeric
                           acid)

L-Lac   L-Val   D-Hiv   D-Val

L-Lac — L-Val — D-Hiv
---┼-- |
D-Val   D-Val
 | ---┼--
D-Hiv   L-Lac
 | |
L-Val   L-Val
 | |
L-Lac ┼ D-Val — D-Hiv

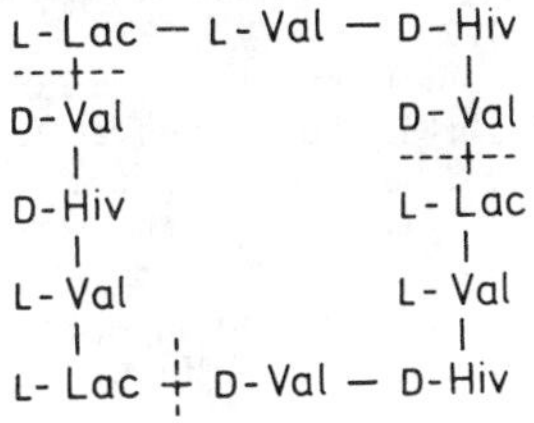

Valinomycin

Valinomycin transports $K^\oplus$ ions but not $Na^\oplus$ ions through cell membranes. The $K^\oplus$ ions form a lipophilic cryptate, similar to that obtained with [18]crown-6. In this instance valinomycin takes the part of the cryptand which acts here as a transport antibiotic.

8.2.4.4   Peptide Hormones

Especially important members of this group of compounds are the pancreatic hormones (insulin, glucagon) and the hormones of the pituitary gland (oxytocin, vasopressin, corticotropin).

Insulin *controls the amount of sugar in the blood* and is used medicinally in the treatment of diabetes (Diabetes mellitus). Its name is derived from its occurrence in the islets of *Langerhans* in the pancreas, from which it was first isolated in 1921 by *Banting, Best, Collip* and *Macleod*. Work by *Sanger*[1] showed that it was one of the smallest protein molecules and contained 51 L-amino acid units which form two polypeptide chains.

Sequence analysis using the DNP method showed that the A-chain consists of 21 amino acid units and the B-chain of 30 units.

The A-chain starts with the *N*-terminal amino acid glycine and is linked to the B-chain by two disulphide bridges which join cysteine units. The *N*-terminal amino acid of the B-chain is phenylalanine. There is also a third disulphide bridge which links two cysteine units within the A-chain. The relative molecular mass of insulin is about 6000. It very readily bonds zinc, leading to the formation of dimers and higher aggregates. The *structure* of insulin from the human pancreas can be shown schematically as follows (Fig. 122):

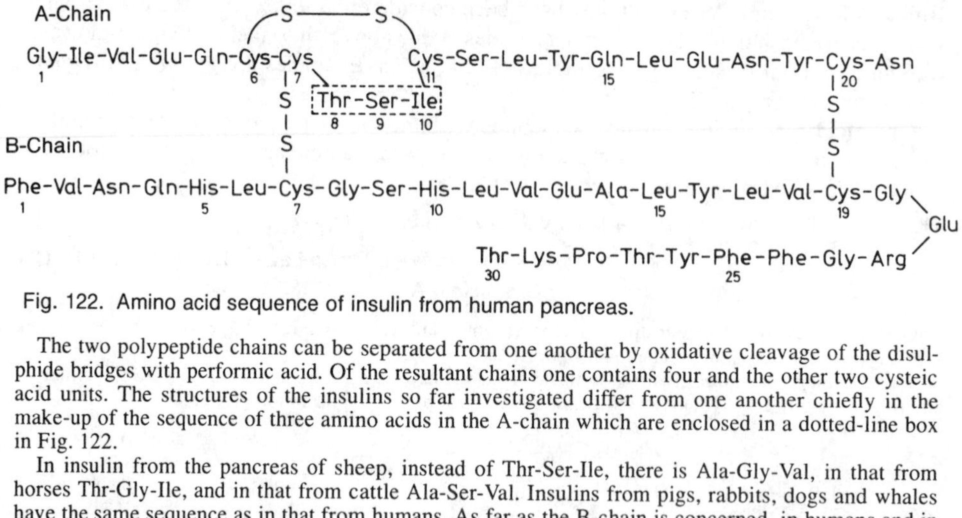

Fig. 122. Amino acid sequence of insulin from human pancreas.

The two polypeptide chains can be separated from one another by oxidative cleavage of the disulphide bridges with performic acid. Of the resultant chains one contains four and the other two cysteic acid units. The structures of the insulins so far investigated differ from one another chiefly in the make-up of the sequence of three amino acids in the A-chain which are enclosed in a dotted-line box in Fig. 122.

In insulin from the pancreas of sheep, instead of Thr-Ser-Ile, there is Ala-Gly-Val, in that from horses Thr-Gly-Ile, and in that from cattle Ala-Ser-Val. Insulins from pigs, rabbits, dogs and whales have the same sequence as in that from humans. As far as the B-chain is concerned, in humans and in elephants it is identical, whereas in cattle, pigs, sheep, horses and whales the Thr unit at position 30 is replaced by Ala. Using one-letter abbreviations this variant is described as T30A. Transpeptidisation has made it possible to replace the Ala in insulin from pigs by Thr, thus achieving a semi-synthetic production of human insulin.

8.2.4.5   INN: *Glucagon* is also produced in the islets of *Langerhans* and acts as a hyperglycemic antagonist to insulin, *i.e.* it raises the blood-sugar level. It contains 29 amino acid

---

[1] See *F. Sanger* et al., The Structure of Insulin, Endeavour *16*, 48 (1957); *H. Klostermeyer* et al., Chemistry and Biochemistry of Insulin, Angew. Chem. Int. Ed. Engl. *5*, 807 (1966).

units; their sequence is also known.

Since 1955 some *antidiabetics* that can be administered orally have been discovered; they are sulphonylureas, *e.g.* 1-butyl-3-tosylurea (INN: *Tolbutamide, Rastinon, Artosin*). An important step forward was provided with the introduction of INN: *Glibenclamide* (*Euglucon*) which requires the use of smaller doses. Glibenclamide and other similar neutral preparations have been described as the 'second generation' of antidiabetics. Other orally administered active antidiabetics, biguanide derivatives, which have been used, are now forbidden because of their dangerous side effects.

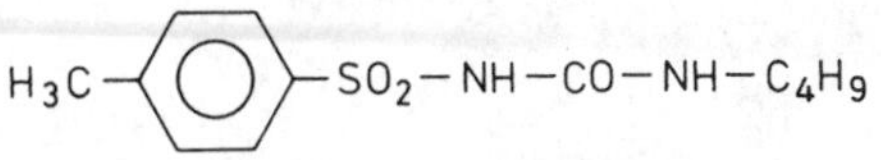

$$H_3C \text{---} \langle\bigcirc\rangle \text{---} SO_2-NH-CO-NH-C_4H_9$$

INN: Tolbutamide

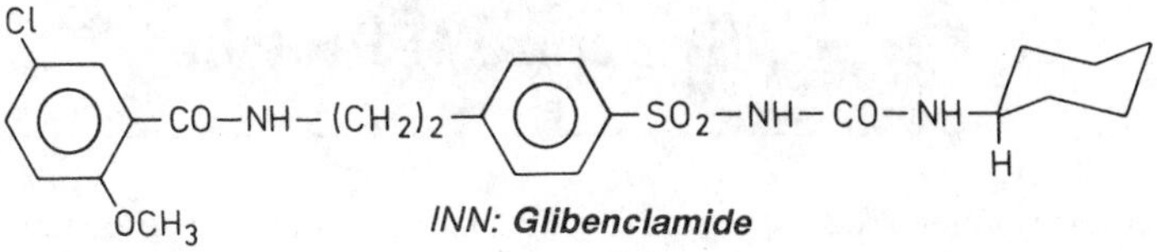

$$\langle\bigcirc\rangle\text{---}CO-NH-(CH_2)_2\text{---}\langle\bigcirc\rangle\text{---}SO_2-NH-CO-NH\text{---}$$

Cl (upper left), OCH₃ (lower)

INN: Glibenclamide

8.2.4.6 Among the most important hormones secreted by the posterior lobes of the pituitary gland are *INN: Oxytocin*, which stimulates contraction of the uterus and is also concerned with the sensation of satiation when food is absorbed, and **vasopressin**, which raises the blood pressure and acts as a neurotransmitter.

The two hormones were obtained pure in 1952 and shortly afterwards they were identified as *nonapeptides* (*Du Vigneaud, Tuppy*, 1953). Oxytocin is built up from amino acids as follows:

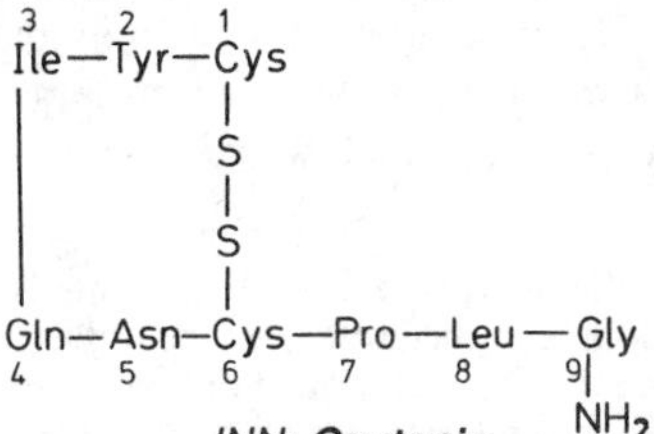

```
3    2    1
Ile—Tyr—Cys
              |
              S
              |
              S
              |
Gln—Asn—Cys—Pro—Leu—Gly
 4    5    6    7    8    9|
                         NH2
   INN: Oxytocin
```

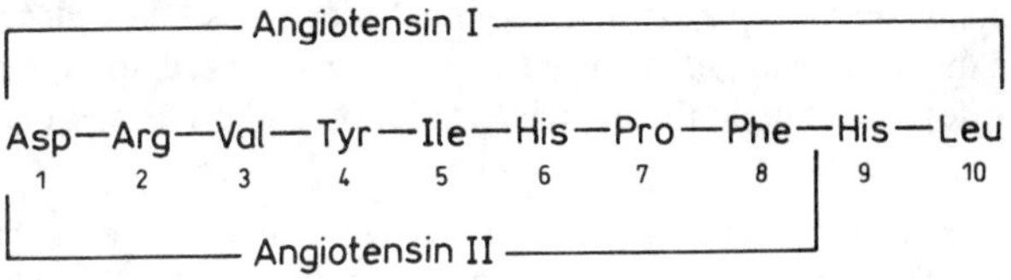

```
┌─────────── Angiotensin I ───────────┐
Asp—Arg—Val—Tyr—Ile—His—Pro—Phe—His—Leu
 1    2    3    4    5    6    7    8    9   10
└─────── Angiotensin II ───────┘
```

The ring structure of this protein is a result of a disulphide bridge between two cysteine units, one of which bears a terminal amino group. This results in the formation of the 20-membered ring *pressic acid*, which has a side chain consisting of three amino acid units attached to Cys^6. The structure of oxytocin was confirmed by synthesis (*Du Vigneaud*, 1954).

The hormone angiotensin II, which occurs in blood, brings about marked raising of blood pressure. It is a linear *octapeptide* which is formed in a stepwise enzymatic breakdown of a serum globulin. The last step of this breakdown involves the splitting of the dipeptide His-Leu from a decapeptide by the 'angiotensin converting enzyme' (ACE). The sequence of amino acids shown above is found in horses, pigs and human beings; in cattle Ile^5 is replaced by Val^5.

8.2.4.7 INN: *Corticotrophin* (α-corticotropin, a constituent of the adrenocorticotropic hormone, **ACTH**) is one of the hormones of the anterior lobes of the pituitary gland; it stimulates the formation of corticoids in the adrenal cortex. It is made up from 39 amino acid residues in the sequence shown below. The first part of this sequence, up to unit 24, shows the same activity and is produced synthetically (Synacthen).

Ser-Tyr-Ser-Met-Glu-His-Phe-Arg-Trp-Gly-Lys-Pro-Val
1 |
Gly-Asn-Pro-Tyr-Val-Lys-Val-Pro-Arg-Arg-Lys-Lys-Gly
| 24 23
Ala-Glu-Asp-Glu-Ser-Ala-Glu-Ala-Phe-Pro-Leu-Glu-Phe
                                                    39
Human α-Corticotropin (α-ACTH)

As in the case of insulin there are small differences in the above sequence of amino acids in various animals as compared to that in human ACTH.

8.3 Properties and Structure of Proteins[1]

Chemically uniform proteins occur but rarely. Crystalline proteins also contain colloidal impurities which are trapped in the crystals. The relative molecular masses of these macromolecules have been determined with the help of techniques such as osmotic pressure, ultracentrifuge sedimentation (*Svedberg*) or light scattering, and provide values between 10 000 and many millions. The relative molecular masses are often quoted in *Svedberg units* (*S*), the sedimentation coefficients in the ultracentrifuge.

Most proteins are amphoteric electrolytes (ampholytes). In so far as they are soluble, they form *colloidal* solutions, from which they can be 'salted out' again by addition of neutral salts such as sodium chloride or magnesium sulphate. In doing this no significant change in their physical or chemical properties takes place. The precipitation is reversible; if the flocculant *gel* is stirred with water it reverts to being a *sol*. The solubility of proteins is dependent upon the *p*H of the solution and on its salt content. It is at its lowest at the isoelectric point.

Lower or higher $H^\oplus$ concentrations can bring about irreversible **denaturation** of proteins. This is associated with a change in the structure and the biochemical functioning of the protein. In the process the peptide chain unwinds and some disulphide bridges may be ruptured. Denaturation of proteins also ensues at higher temperatures, by powerful irradiation or by certain coagulating agents such as urea or guanidinium salts; the latter have been described as chaotropic reagents. Irreversible coagulation occurs in the curdling of milk and in the hardening of eggs when they are boiled or fried.

There are two generally applicable *colour tests* for the *qualitative recognition* of proteins in solution:

(a)   Ninhydrin reaction. When amino acids, peptides or proteins are heated with a dilute aqueous solution of ninhydrin a *blue-violet* colour appears (see p.624).

(b)   Biuret reaction. If a drop of very dilute copper(II) sulphate solution is added to an alkaline solution of a protein a red- or blue-violet colour is produced. All peptides and proteins give positive results in this test.

The following diagram of portions of a peptide chain shows the structure of the copper(II) complex which is formed:

---

[1] See *H. Neurath* et al., The Proteins, 3rd edn, 4 vols. (Academic Press, New York 1975–79); Brookhaven Protein Bank, Brookhaven National Laboratory; Chemistry Department, *M. F. Perutz*, Protein Structure: New Approaches to Disease and Therapy (Freeman, New York 1992).

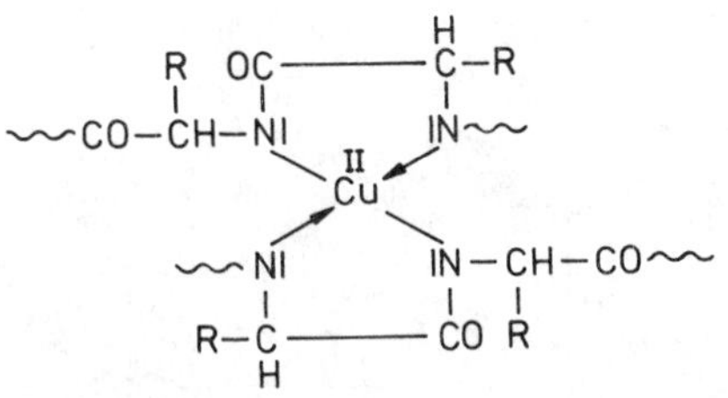

```
                        H
        R   OC--------C--R
        |   |         |
  ~~CO--CH--NI        IN~~
                 II
                 Cu
        ~~~NI         IN--CH--CO~~
           |          |   |
         R-C--------CO  R
           H
```

The term 'biuret reaction' is misleading in that the grouping of atoms found in biuret, H_2N—CO—NH—CO—NH_2, is not present in proteins.

Structurally, proteins consist of several peptide chains, some of which may be identical, and each of which contains up to several hundred amino acid units. Determination of the sequence of these units involves the same principles that were described for peptides. Within the protein molecule the various chains are held together by means of *disulphide bridges*, *hydrogen bonding* and/or *ionic bonding* (between acidic and basic chains of amino acids).

The biological activity of proteins depends not only upon the chemical structure but especially upon the *steric fine structure*,[1] which can be determined with the help of physical methods such as X-ray crystallography, electron microscopy, etc. It is possible to distinguish between different stages of their structural organisation (*Linderström-Lang*, 1959):

(a) The *primary structure* or amino acid sequence (see p.833 *f*).

(b) *Secondary structure*, which is associated with regularities in the *conformation* of the polypeptide chains (pleated sheet structure: Fig. 123a, α-helix: Fig. 123b).

(c) The *tertiary structure* of the globular proteins, which goes beyond the secondary structure to consider steric *folding* which is often held in place by disulphide bridges (see insulin: Fig. 122, p.836) and other intramolecular interactions.

Depending upon their shape, protein molecules are classified into *fibrous* proteins (scleroproteins), in which the peptide chain, as in polypeptides, is arranged in either a folded or a spiral fashion, and *globular* proteins (spheroproteins), whose peptide chains are curled around to form a shape resembling a sphere or an ellipsoid.

8.3.1 Fibrous Proteins

Characteristic features of *fibrous proteins* are their mechanical strength and their insolubility in aqueous media. They form skeletal and connective tissues in animals and, like cellulose in plants, have a *fibrous structure*.

X-ray crystallographic studies have shown that the peptide chains in fibrous proteins can, depending upon their repeat units, be arranged sterically in various ways.

---

[1] See *R. E. Dickerson* et al., Struktur und Funktion der Proteine (Verlag Chemie, Weinheim, 1971; *H. D. Jakubke* et al., Aminosaüren, Peptide, Proteine (Verlag Chemie, Weinheim, 1982).

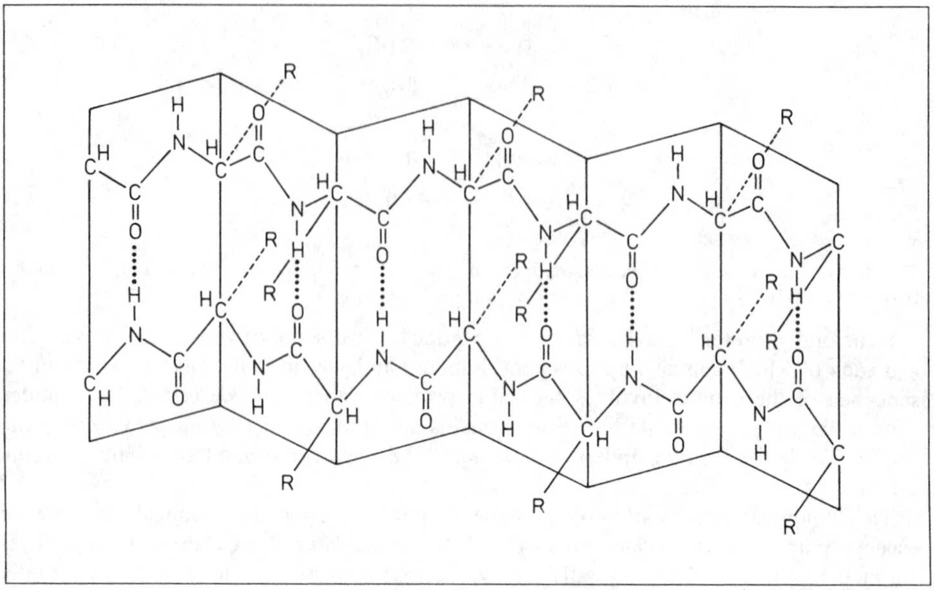

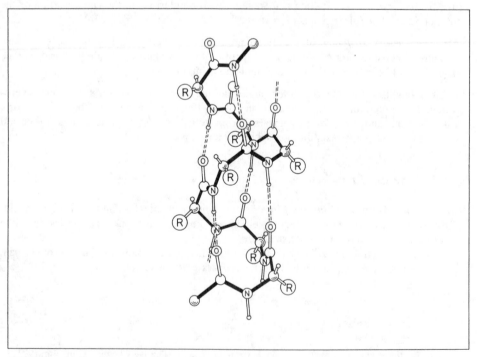

Fig. 123a. Pleated sheet structure of silk fibroin: β-keratin type.

Fig. 123b. α-Helix of the α-keratine type.

8.3.1.1   Silk Fibroin. β-Keratin Type

Silk fibroin is the fibre from the cocoons of the silk moth (*Bombyx mori*), which are sealed by means of a more readily soluble protein, *sericin*. Other β-keratins include the hard constituents of horny tissue, as in feathers, nails, hooves, horns and scales.

Adjacent peptide chains in fibrous proteins take up a zigzag arrangement and are held together by hydrogen bonding between their CO- and NH-groups. Whilst a normal polypeptide chain that is planar, *e.g.* that of polyglycine, has a repeat unit of about 727 pm, in silk fibroin it amounts to only 708 pm. To explain this shortening, and also from an examination of models, it is considered that silk fibroin takes up a *pleated sheet structure* in which the planes of alternate peptide groups are at an angle to each other, and that the α-carbon atom which bears the side chain lies in both of these planes (*Pauling* and *Corey*). The side chains, all represented here by R, project in almost vertical directions, alternately on either side of the chain (Fig. 123a).

8.3.1.2   α-Keratin Type

In the case of α-keratins, which occur in protein fibres, hair and wool, hydrogen bonding takes place not between two different peptide chains but *within one chain*. In 1951 *Pauling* and *Corey* suggested a model in which the basic shape of the chain is an α-helix (screw). This comes about if the peptide chain is twisted into a helical form resembling a cylinder. In doing so, CO- and NH-groups on two adjacent turns are brought close to one another widely so that formation of hydrogen bonds becomes possible. The two types of keratin are examples of *secondary structure* in proteins. One turn of the helix involves 3.6 amino acid units, resulting in a repeat unit (pitch) of 540 pm and a radius of 230 pm (see Fig. 123b). The side chains R are directed outwards. Hydrogen bonding is represented by double broken lines. X-ray studies have shown that only the right-handed screw, as shown in Fig. 123b, is found; a left-handed helix is disfavoured because it involves the unsatisfactory placing of the side chains close to one another. Glycine and proline units destabilise the α-helix.

Since the repeat unit of α-keratin is only 510 pm, it is thought that a number of helices are twisted together like the strands of a rope or cable to form a *superhelix*.

When wet hair is stretched to about double its length, it goes from an α-keratin structure (α-helix) into a β-keratin structure (pleated sheet structure). In both of these structures, all the peptide bonds exist in (Z)-configurations.

Among proteins of the α-keratin type there is also **fibrin**, which is formed when blood coagulates. When this happens, **fibrinogen**, which is soluble in blood plasma, is converted under the influence of the enzyme *thrombin*, in a complicated series of reactions, into insoluble fibrin, which has a *fibrous structure*. For this to take place it is essential to have vitamin K_1 present. Coagulation of blood is arrested by *heparin*, *coumarin* and some of their derivatives (see p.556 *f*), as well as by some poisons from certain animals, for example the leech poison *hirudin*; the action of the latter depends upon the formation of a complex with thrombin, whose catalytic activity is thereby prevented.[1]

8.3.1.3   Collagen Type

Collagen is the main constituent of the structural and connective tissues of tendons, ligaments, cartilage and the organic material of bones. The steric structure of the peptide chain resembles that of α-keratin, but there is a left-handed helix with spirals that are noticeably

---

[1] See *T. J. Rydel* et al., The Structure of a Complex of Recombinant Hirudin and Human α-Thrombin, Science *249*, 277 (1990).

further apart than in the α-helix; the repeat unit is about 860 pm. In this case also three strands are twisted together as in a rope (triple helix).

Whilst most fibrous proteins are insoluble in water, collagen, when heated with water, especially in the presence of acid, is converted into a colloidal *glue*. When this glue is very pure and dried it provides colourless **gelatin.** In aqueous media gelatin swells and goes into solution. If this solution is cooled it sets to form a jelly. Collagen in hides is converted into leather by tanning.

Closely related to collagen is **elastin,** the main constituent of the elastic tissue in tendons. It differs from collagen in that it does not swell up in water; the difference is associated with cross-linking of the polypeptide chains.

8.3.2 Globular Proteins (Spheroproteins)

Globular proteins usually exist in the form of spheroids or ellipsoids. They include a number of domains in which amino acid sequences occur in the form of single strands and with α-helices alternating with β-strands. A classification of globular proteins on the basis of their structures is not yet possible and they are best classified on the basis of their sources, *e.g.* as *plasma-, milk-* or *egg-proteins* or as *plant-seed proteins*. Albumins, globulins, histones and protamines all form part of this class of proteins, as do prolamines and glutelins which are found in grain.

8.3.2.1   Albumins are soluble in pure water but can be precipitated out again by adding a solution of ammonium sulphate. Among animal albumins are **serumalbumin** in blood plasma, **ovalbumin,** the main constituent of egg-white, and **lactalbumin** in milk. Human serumalbumin (HSA) has a relative molecular mass of 65 000; it contains a sequence of 585 amino acid units and has seven disulphide cross-linkages (see p.866).

Plant albumins are found in smaller amounts, for the most part in seeds, *e.g.* **leucosin** in wheat grains and **ricin** in castor beans. The latter is highly toxic because it agglutinates blood corpuscles.

8.3.2.2   Globulins differ from albumins in being insoluble in pure water, but they are soluble in dilute aqueous solutions of neutral salts, *e.g.* in 5% sodium chloride solution, and in acids and bases. They are salted out by semisaturated ammonium sulphate solution. Sometimes globulins exist as metal proteins, containing iron, copper or zinc, but also as *glycoproteins* or as *lipoproteins* (see p.844 *f.*). They are ellipsoidal in shape and have relative molecular masses which vary between 130 000 and 200 000.

Animal globulins are found in the body fluids as humoral globulins; these include the *immunoglobulins*, for example **serum globulin** in blood plasma. Cell globulins are found in the tissues, and milk contains **lactoglobulin.**

Important examples of the cell globulins are **myosin,** the main constituent of transversely striated muscles, and **thyroglobulin** from the thyroid gland.

Almost all the plant globulins are acidic; this is due to their high content of monoaminodicarboxylic acids. They coagulate less readily when they are heated than do animal globulins, and show less tendency to crystallise. They include the reserve proteins in the plant seeds; an example is **edestin** from hemp seed.

8.3.2.3   Histones are nucleoproteins which can be isolated from the cell nuclei of red and white blood corpuscles and from the spermatozoa of higher animals. Together with nucleic acids they are concerned in the synthesis of chromatin in the cell nuclei. They are basic on account of their containing a high proportion of arginine units. The five different histones H1, H2A, H2, H3 and H4, whose amino acid sequences are known, provide the basis of the structure of the *compact units of chromosomes*, known as *nucleosomes*, in cell nuclei (eukaryotic chromosomes).[1]

---

[1] See A. *Klug*, From Macromolecules to Biological Assemblies, Angew. Chem. Int. Ed. Engl. *22*, 565 (1983).

8.3.2.4 **Protamines** are the main protein components, coupled with nucleic acids, (nucleoprotamines) in the sperm of certain kinds of fish; examples are **clupeine** (from herring) and **salmine** (from salmon). They have a high content of arginine and as a result are strongly basic and can be extracted out with dilute mineral acid (INN: **Protamine sulphate**, protamine chloride). Compared to most proteins they have relatively low molecular masses (*ca.* 5000) and simple constitutions. In their properties they are more like polypeptides.

Prolamines and **glutelins** form the major protein constituents of grainstuffs and of flour. Prolamines, unlike all other proteins, can be extracted from flour using 80% aqueous ethanol, but they are insoluble in either absolute ethanol or water. Glutelins are only soluble in dilute alkali.

Prolamines are rich in glutamic acid (~ 35%) and proline (up to 13%); they contain very little arginine or histidine and no lysine. Among the most important prolamines are **gliadin**, from wheat and rye, **zein**, from maize, and **hordenine**, from barley. The simultaneous presence of gliadin and glutelin in wheatflour and ryeflour is responsible for their being able to be used in baking. Together the two proteins form *gluten protein*. Glutelins contain lysine and tryptophane and in this way supplement the prolamines.

A property of **lectins**, which is specifically associated with their structure, is that they can link to carbohydrates. They occur in very many plants and some, for example concanavalin A, from the Jack bean, *Canavalia ensiformis*, have the ability to agglutinate erythrocytes. Lectins are therefore agglutinating agents which have bonding sites for sugar, and recognise structures whose surfaces contain carbohydrates.[1]

8.3.3 Conjugated Proteins

As well as simple proteins made up only from amino acids there are also *conjugated proteins* (also called proteids) which have in addition to the protein component a non-protein moiety, known as a *prosthetic group*. These groups include phosphoric acid, pigments, carbohydrates or lipids. They are bonded to the protein material through a carboxyl or other acid grouping.

Conjugated proteins are classified into the following groups, depending upon the specific nature of the prosthetic group:

8.3.3.1 Phosphoproteins

The most important phosphoprotein is **casein** which is present in milk in the form of its soluble calcium salt. In these proteins the phosphoric acid group is joined by an ester linkage to a hydroxy group. For example, **phosvitin**, which was isolated from egg yolk, has as its prosthetic group serinephosphoric acid (*O*-phosphoserine):

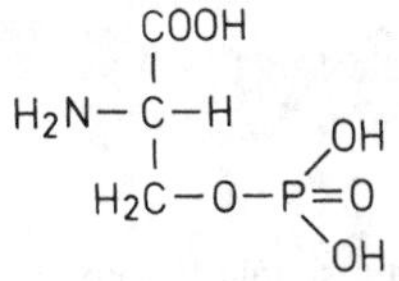

$$\begin{array}{c} COOH \\ | \\ H_2N-C-H \\ | \quad\quad OH \\ H_2C-O-P=O \quad \text{Serinephosphoric acid (O-Phosphoserine)} \\ \quad\quad OH \end{array}$$

The phosphoric acid group can be readily removed, for example by aqueous sodium hydroxide. This can also be achieved by hydrolysis with ester-splitting enzymes, and in this case it is not also accompanied by proteolysis (hydrolysis of the protein).

---

[1] See *N. Sharon* et al., Lectins as Cell Recognition Molecules, Science *246*, 227 (1989); *C. Mandal* et al., Sialic Acid Binding Lectins, Experientia *46*, 433 (1990).

8.3.3.2   Chromoproteins

In this important class of compounds, the protein is linked to a *pigment*. Among the latter are the animal *respiratory pigments* such as haemoglobin, myoglobin, and the cytochromes, as well as *enzymes* with haemin-like functional groups, *e.g.* cytochrome oxidases, catalases and peroxidases. The pigment component of phytochromes in green plants is a compound of the biliverdin type.

Haemoglobin, the carrier protein for oxygen, is made up from four haem groups and four polypeptide chains (see p.709), namely from two identical A-chains and two identical B-chains, which are linked to one another by means of hydrogen bonding and ionic inter-actions. The A-chain consists of 141 amino acid units and the B-chain of 146; their sequences are known. It has been found that sickle-cell anaemia, which is a hereditary dis-ease of black people, is due to *valine* taking the place of glutamine in position 6 of the B-chain.

Myoglobin, the red pigment of muscle cells, is closely related to haemoglobin, but it contains only one peptide chain and one haem group. It has a greater affinity for oxygen than the blood pigments do. The physiological function of myoglobin is not as an oxygen-carrier but rather it acts as an oxy-gen store for the muscle.

X-ray studies[1] have provided more precise information about the structure of haemoglobin and of myoglobin.

Reaction of nitrite with myoglobin produces nitrosomyoglobin which is red and more stable to atmospheric oxygen than myoglobin itself. This explains the reddening of meat which comes about when it is treated with so-called pickling or preserving salts.

8.3.3.3   Glycoproteins and Mucopolysaccharides

Numerous proteins, and especially the animal albumins and globulins, contain a carbohy-drate component which is firmly bonded chemically to the protein as an integral part of the structure. These **glycoproteins**[2] have side-chains attached to the protein residue, which are joined either by *O-glycosidic links* to the hydroxy group of either *serine* or *threonine* (*O*-glycoproteins) or by *N-glycosidic* links to the amide group of *asparagine* (*N-glyco-proteins*).

The saccharide chains of *O*-glycoproteins have the following basic structure: β-D-Gal(1→3)α-D-GalNAc(1→0)Ser(or Thr).

N-Glycoproteins have the following branched basic structures:

```
Lactosamine      -(1 → 2)α-D-Man(1 → 6)
or Mannose                              \
                                          β-D-Man(1 → 4)β-D-GlcNAc(1 → 4)-β-D-GlcNAc-Asn
Lactosamine      -(1 → 2)α-D-Man(1 → 3) /
or Mannose
```

These basic structures show variations in their terminal units, for example *N*-acetylneuraminic acids (AcNeu, sialic acids). If the sialic acids are removed by means of neuraminidases, the characteristics of the glycoprotein are changed, with the result that it is decomposed in the liver and this interferes with the circulation of the blood.

---

[1] See *M. F. Perutz*, Röntgenanalyse des Hämoglobins (X-ray Structure Determination of Haemoglobin, Angew. Chem. *75*, 589 (1963); *G. Fermi* et al., The Crystal Structure of Human Deoxyhaemoglobin, J. Mol. Biol. *175*, 159 (1984); *G. Buse*, The Present Position of Haemoglobin Research, Angew. Chem. Int. Ed. Engl. *10*, 663 (1971).
[2] See *H. Paulsen*, Syntheses, Conformations and X-ray Structure Analyses of the Saccharide Chains from the Core Regions of Glycoproteins, Angew. Chem. Int. Ed. Engl. *29*, 823 (1990).

Murein is a peptidoglycan whose structure is that of a disaccharide made up from *N-acetylglu-cosamine* and N-acetylmuraminic acid (the 3-*O*-ether from glucosamine and lactic acid) which is bound by a peptide-like link to branched peptides made up from L-*alanine*, D-*alanine*, D-*glutamic acid* and L-*lysine*. It is the main constituent of bacterial cell walls. *Penicillin* stops the biosynthesis of *murein* in bacteria; this is the reason for its effectiveness.

(a) The **mucoids** occur principally in supporting and connective tissues, and also in mucins in animal organs and secretions. They have a small protein content in an essentially *carbohydrate* structure made up from glucosamine or galactosamine (= chondrosamine), mannose or galactose, and also D-glucuronic acid or L-iduronic acid. The compounds of this group are also known as **mucopolysaccharides.** They are classified according to their structures as *acidic* or *neutral* mucopolysaccharides.

The simplest mucopolysaccharide is **hyaluronic acid**, which is present especially in the vitreous humour of the eye (Grk. *hyalin* = vitreous, glassy), as well as in the umbilical cord. It is made up from *N*-acetylglucosamine and D-glucuronic acid. Its physiological importance depends upon its ability to bring about changes in the permeability of cell membranes and to prevent invasion by infectious germs. These activities are neutralised by the enzyme *INN: Hyaluronidase* which brings about cleavage of the polysaccharide into smaller fragments, at the same time lowering the viscosity. Hyaluronidase, which is abundant in sperm, also plays a role in the fertilisation process.

Heparin, which was first obtained from liver (Grk. = *hepar*) is an *acidic* mucopolysaccharide. It is built up from glucosamine which has been polyesterified with sulphuric acid, together with sulphate esters of D-glucuronic acid and L-iduronic acid. It is an *anticoagulant*, preventing the coagulation of blood, and is used in the treatment and prevention of thrombosis.

Chondroitin sulphate is a *sulphate ester* of a mucopolysaccharide which has been isolated from cartilage and from tendons. It is constituted from *N*-acetylchondrosamine, D-glucuronic acid or L-iduronic acid and sulphuric acid. In chondroitin sulphate C the sulphuric acid group is attached to the primary alcohol group of chondrosamine.

Neutral mucopolysaccharides contain glucosamine units and, instead of uronic acids or sulphuric acid, neutral sugars, *e.g.* mannose or galactose.

(b) The **blood group substances** (agglutinogens)[1] are glycolipids, which are found on the surfaces of *erythrocytes*. They cause flocculation (agglutination) in the transfusion of incompatible blood groups. The *antigenic specificity* of the blood group substances against the antibodies (agglutins) which are present in the blood serum is associated with the terminal *oligosaccharides*, the *determinants*, which are joined through a polysaccharide chain (core) with ceramide. This is intercalated but not covalently bound in the lipid membrane of the erythrocytes.

The antigens of the blood groups A, B and O have the following determinants, which all bear L-*fucose* as an end-group.

A-Substance

α-D-GalNAc(1 → 3)
β-D-Gal(1 → 4)D-GlcNAc ------- Core
α-L-Fuc(1 → 2)

B-Substance

α-D-Gal(1 → 3)
β-D-Gal(1 → 4)D-GlcNAc ------- Core
α-L-Fuc(1 → 2)

O = H-Substance   α-L-Fuc(1 → 2)-β-D-Gal(1 → 4)D-GlcNac -------- Core

---

[1] See *S. Hakomori* and *A. Kobata* in 'The Antigens' (*M. Sela*), Vol. II, p.79 (Academic Press, New York 1974); *R. U. Lemieux*, Chem. Soc. Rev. *1978*, 423.

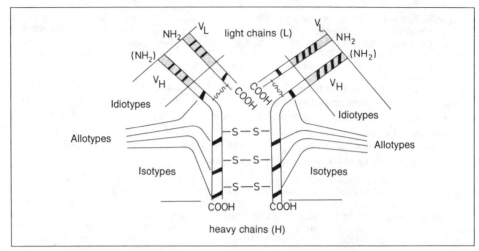

Fig. 124. Schematic representation of an antibody molecule. Portions given a grey background are variable regions, those in black are hypervariable regions.

The action of blood group substances may be explained using blood group A. This blood contains in its serum *anti-B-agglutinin* which is a protein that can react with B-erythrocytes to bring about agglutination, thus causing incompatability. The serum of blood of group B contains an anti-A-agglutinin which is similarly incompatible with A-erythrocytes.

Antigen-antibody reactions which produce *allergies* follow a similar course. Lower molecular weight *allergens* become bonded to proteins as *haptens*, and in doing so can also act as antigens. This can happen, for example, in the case of *penicillin*, if it becomes attached to a particular protein in the body, thereby becoming a hapten that in this way becomes a foreign body and behaves as an antigen.

(c)   **Antibody molecules**[1] of the IgG type are obtained in crystalline form from aqueous solutions of monoclonal antibodies which contain poly(ethyleneglycol) and hence have been made accessible for X-ray structure determination (MAB, derived from 'Monoclonal Antibody'). They consist of two pairs of polypeptide chains which are held together by disulphide bridges to form a Y-shaped structure. This is shown diagrammatically in Fig. 124. In solution they exist as globular structures also having disulphide bridges *within* the individual chains. Each of the two light chains (L) contains 220 amino acid units whilst each of the heavy chains (H) has 440 or 550 such units. The common *structural unit* of both chains is a *domain* of 110 amino acid units of which the light chains each contain two and the heavy chains each contain four or five. The two chains differ in the domains at the amino end (V_H and V_L) in having more variability in their amino acid sequences. Within the V_H and V_L domains there are regions of especially great variability, known as the *hypervariable regions*. V_H and V_L (given grey backgrounds in Fig. 124) are the sectors where the antigen is detected because of the characteristic *determinants* or *epitopes* and in which it is involved in an *antigen–antibody reaction*.

In addition to the carbohydrates mentioned above, characteristic sequences of amino acids or lipid constituents of the antigen can also function as determinants or epitopes. The

[1] See *R. Huber*, Structural Basis for Antigen–Antibody Recognition, Science *233*, 702 (1986); *R. C. Kennedy* et al., Anti-idiotypes and Immunity, Scientific American *255*, No. 1, 40 (1986).

high specificity of the antibody depends upon the extraordinary variety in the amino acid sequence in domains V_H and V_L, which constitute the *idiotype* or the paratope of the antibody molecule and in which a specific amino acid sequence brings about the antigen–antibody reaction for any particular determinant. A role is played not only by the amino acid sequence but also by the conformational properties of the molecule, for example the formation of cavities or pockets.

In the other domains, in which there is appreciably less variability in the amino acid sequences, are the isotypes of the antibody. They are the same for antibodies of one class, but the antibodies of different individuals differ from one to another in this region. The resultant characteristic variable regions are known as the *allotypes* of the antibody.

Antibody molecules are formed in the white blood corpuscles, the B-lymphocytes, and are passed on to blood plasma. They form the main constituent of serum globulin and are known in general as immunoglobulins.

The specificity of the antigen–antibody reaction is combined with the possibility of detecting radioactive labelled compounds in **radioimmunoassay** (RIA).[1] Using this technique, compounds such as oestradiol, prostaglandins, and also proteins which are of themselves antibodies, can be estimated quantitatively in *nanomolar amounts*. For this purpose a radioactively labelled compound is needed which has the same structure as the immunologically active molecule which is to be investigated. The labelled molecule forms an antigen (AG*) which competes with the antigen AG at the bonding site of the antibody (AK) for the material under investigation.

$$AG^* + AG + AK \rightleftharpoons AG^*–AK + AG–AK + AG + AG^*$$

The ratio of the labelled to the unlabelled molecules in the antigen–antibody complex (AG–AK) corresponds to the ratio of their concentrations in the analytical sample and obeys the law of mass-action. It follows therefore that the fewer radioactive antigen molecules there are bonded to the antibody the larger is the concentration of unlabelled antigen molecules in solution.

A further test which has been developed is **solid-phase immunoassay**, which involves the bonding of a particular antigen-specific antibody onto a solid matrix. The sample for testing is brought into contact with the solid support material which has the specific antibody anchored to its surface. The compounds which have not been involved in an antigen–antibody reaction are washed off again, and a second antibody is added, which detects another region on the fixed antigen and is, for example, radioactively labelled. The radioactivity which remains attached to the surface after it has been washed again is proportional to the amount of antigen in the test sample. If the second antibody is labelled with an enzyme that can change a colourless or non-fluorescent substrate into a coloured or fluorescent compound, then a test which is relatively simple to operate is available (EIA = enzyme immunoassay). This has found general use under the name of the ELISA-test (enzyme-linked immuno sorbent assay), for example for the detection of pregnancy or of infection by the HIV virus.[2]

(d)   The term **antibody enzymes, abzymes** (from 'antibody enzymes') is used for situations which involve enzymatic reactions in the cavities of the hypervariable regions of antibodies. It is in the nature of these regions that not only is the specificity of that sort of enzyme action very variable but also that it can be influenced by appropriate antigens in

---

[1] See *H. G. Eckert*, Radioimmunoassay, Angew. Chem. Int. Ed. Engl. *15*, 525 (1976).
[2] See *D. M. Kemeny* et al. (Ed.), ELISA and Other Solid Phase Immunoassays (Wiley, Chichester 1988).

predictable ways. If, for example, the *transition state* of a reaction is known, then it can be imitated in the antigen by a stable atomic grouping having a similar structure. Thus a tetrahedral transition state such as is involved in the hydrolysis of an ester (see p.258), can be imitated by a phosphonic ester group which has at its centre a tetrahedral phosphorus atom in place of a carbon atom.

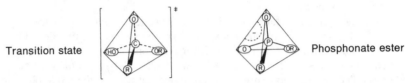

If a haptene containing a phosphonic ester group is attached to a protein an antigen is formed which causes the formation of antibodies which can enzymatically cleave many esters.[1] The fascination of the method lies in the fact that antibody enzymes can be induced which bring about reactions which do not take place normally in the organism; an example is a *Diels–Alderase*.[2]

8.3.3.4 Lipoproteins

This class of materials includes proteins which occur in blood plasma, in mitochondria and in egg yolk. They are complexes of proteins with lipids and have varied constitutions whose densities increase as the protein content rises. The following fractions have been separated by means of the ultracentrifuge: chylomicrons (density 0.94; protein content 2%), 'very low density lipoprotein' VLDL (0.94–1.006; 9%), 'low density lipoprotein' LDL (1.006–1.063; 21%), 'high density lipoprotein' HDL (1.063–1.21, 33%). The amount of phospholipids in the lipid protein increases through the series given above while the amount of neutral fats decreases. The proportion of cholesterol esters is at its highest in LDL, which is the most important transport for cholesterol in the organism. Increased concentration of LDL in blood increases the risk of arteriosclerotic changes to the arteries. By bonding to protein in blood and other aqueous body fluids the lipids become more readily transportable, in the form of oil droplets which are enclosed in a monomolecular layer of phospholipids and unesterified cholesterol, and are stabilised by a protein, apolipoprotein B-100 (apo B-100). Fig. 125 illustrates this for LDL.[3]

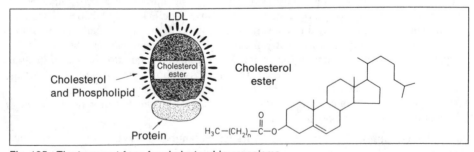

Fig. 125. The transport form for cholesterol in organisms.

[1] See *R. A. Lerner* et al., Observations in the Interface between Immunology and Chemistry, Chemtracts-Organic Chemistry *3*, 1 (1990).
[2] See *D. Hilvert*, Antibody Catalysis of a *Diels–Alder* Reaction, J. Am. Chem. Soc. *111*, 9261 (1989).
[3] See *M. S. Brown* and *J. L. Goldstein*, A Receptor Mediated Pathway for Cholesterol Homeostasis, Angew. Chem. Int. Ed. Engl. *25*, 583 (1986).

9 The Chemistry and Function of Nucleic Acids

Nucleic acids are present in all cells of both animals and plants. They control the preparation of proteins which are vital for the life and functioning of each cell. The instruction (blueprint) for the synthesis of the proteins is stored in specific parts of the *chromosomes* called *genes* which are made up from long chains of **deoxyribonucleic acids** (**DNA**). Every human cell contains in its nucleus 46 chromosomes, whose DNA would, if stretched out in a straight line, be about 4 cm long. The compact structure of chromosomes is due to mutual interactions between the DNA and histones and other proteins which result in the formation of *nucleoproteins* (see p.842).

Ribonucleic acids (RNA) are involved in transmission of the information controlling protein synthesis, which is stored in DNA.

Of the four classes of compounds, carbohydrates, lipids, proteins and nucleic acids, that form the main constituents of living cells, it is only in the case of nucleic acids that it is possible to give a date for their initial discovery. This was in 1869, when *F. Miescher*, who had separated the cell nuclei from the contents of pus cells, isolated from them a phosphorus-containing substance which he called a *nuclein*. Seventy-five years later the *chemistry of inheritance* was established, based on the discovery that the characteristics of a strain of *pneumococcus* could be transformed by DNA, and that these transformations were hereditary (*O.T. Avery*, 1944).

In 1950 *Chargaff* recognised that there were two pairs of DNA-building blocks whose members were always present in identical amounts, namely that there was always a 1:1 ratio of adenine and thymine and a 1:1 ratio of guanine and cytosine. This observation, the X-ray structural investigations of *R. Franklin* and *M. Wilkins*, and the conclusion by *Pauling* that proteins had α-helical structure, led *Watson* and *Crick* to propose the double helix as the secondary structure of DNA.[1]

9.1 Structural Units of Nucleic Acids

The products obtained when nucleic acids are broken down are heterocyclic bases which are derivatives of pyrimidine and of purine, a pentose and phosphoric acid, in the molecular ratio 1:1:1. The pentose is either D-ribose or 2-deoxy-D-ribose. These two sugars are found in, respectively, *ribonucleic acids* (RNA) and *deoxyribonucleic acids* (DNA).

---

[1] See *E. Chargaff*, Vorwort zu einer Grammatik der Biologie (A Foreword to a Grammar of Biology), Experientia *26*, 810 (1970); *J. D. Watson*, The Double Helix (Weidenfeld and Nicolson, London 1968).

9.1.1 Nucleosides

Bases isolated from ribo- and deoxyribo-nucleic acids are the pyrimidine derivatives *cytosine, uracil, thymine* and *5-methylcytosine* and the purine derivatives *adenine* and *guanine*:

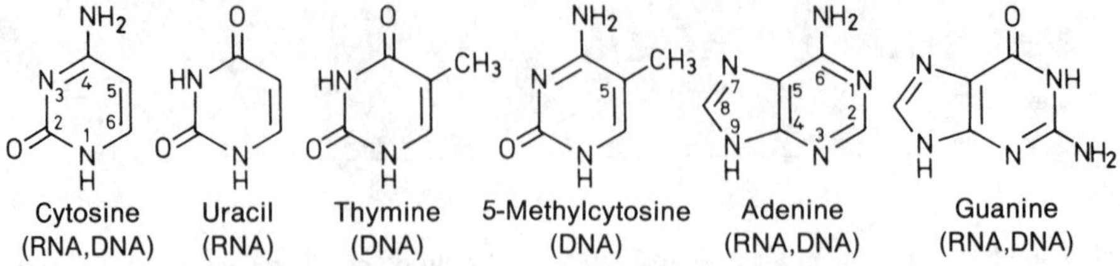

Cytosine	Uracil	Thymine	5-Methylcytosine	Adenine	Guanine
(RNA,DNA)	(RNA)	(DNA)	(DNA)	(RNA,DNA)	(RNA,DNA)

These bases are attached to the sugar (D-ribose or 2-deoxy-D-ribose) by an *N*-glycosidic link, thereby forming **nucleosides,** for example, in the presence of polyphosphate esters adenine and D-ribose form *adenosine*, 9-(β-D-ribofuranosyl)adenine (*G. Schramm*, 1962).

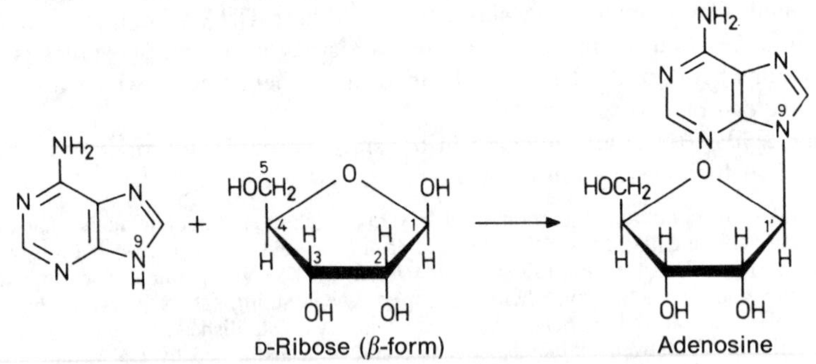

D-Ribose (β-form)   Adenosine

They are prepared synthetically by reaction of the 1-halogeno derivative of the protected sugar with reactive derivatives of the appropriate bases.[1]

The names of the nucleosides derived from pyrimidine bases end in *-idine, e.g.* cytidine, uridine, thymidine, and of those derived from purine in *-osine, e.g.* adenosine, guanosine, inosine. In order to distinguish the numbering of atoms in the carbohydrate ring from those in the *N*-heterocycles, the carbon atoms of the sugar moiety are numbered 1' to 5'.

Shown below are some of the nucleotides which are the principal constituents to be isolated from the mixture obtained on partial hydrolysis of nucleic acids:

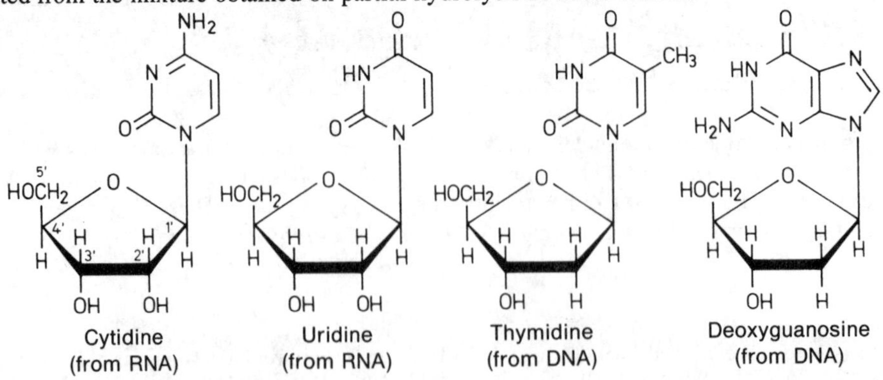

Cytidine	Uridine	Thymidine	Deoxyguanosine
(from RNA)	(from RNA)	(from DNA)	(from DNA)

[1] See *G. M. Blackburn* and *M. J. Gait* (Eds.), Nucleic Acids in Chemistry and Biology (IRL Press, Oxford 1990).

In addition it is found that in plant DNA part of the cytosine fragment is replaced by 5-methylcytosine and that in bacterial RNA 6-methyladenine takes the place of about 1% of all the adenine fragments. Modification of the sugar residue of the nucleoside and in the mode of linkage between the sugar portion and the base have also been observed in t-RNA (see p.868). Examples are, respectively, 2'-O-methylribose and pseudouridine (ψ); in the latter, uracil is linked *via* carbon atom 5 to the ribose.

3'-Azido-3'-deoxythymidine (*Azidothymidine, AZT, Retrovir, Zidovudin*) protects cell cultures against the AIDS virus, but its mode of clinical action is unclear. The same applies in the case of 2',3'-dideoxycytidine (DDC).

2'-O-Methylribose   Pseudouridine (ψ)   Azidothymidine

9.1.2 Nucleotides

If the 5'—OH group of the sugar portion of a nucleoside is esterified with phosphoric acid the product is the corresponding *nucleoside-5'-phosphate* or **nucleotide**, which should be regarded as the real building blocks from which nucleic acids (polynucleotides) are constructed.

Table 19. The most important types of building blocks in nucleic acids

Base*	Nucleoside		Nucleotide:Nucleoside 5'-phosphate†	
	Ribo-	Deoxyribo-	Ribo-	Deoxyribo-
Adenine (A)	Adenosine	Deoxyadenosine	Adenylic acid Adenosine monophosphate (AMP)	Deoxyadenylic acid Deoxyadenosine monophosphate (d-AMP)
Guanine (G)	Guanosine	Deoxyguanosine	Guanosine monophosphate (GMP)	Deoxyguanosine monophosphate (d-GMP)
Uracil (U)	Uridine	—	Uridine monophosphate (UMP)	—
Cytosine (C)	Cytidine	Deoxycytidine	Cytidine monophosphate (CMP)	Deoxycytidine monophosphate (d-CMP)
Thymine (T)	—	Deoxythymidine (Thymodine)	—	Deoxythymidine monophosphate d-TMP

*The letters A, G, U, C, T are used to denote the polynucleotide building blocks which contain that particular base.
†Nucleoside-5'-phosphate, NMP; nucleoside-5'-diphosphate = nucleoside-5'-pyrophosphate, NDP; nucleoside-5'-triphosphate, NPT; d = deoxy, *e.g.* d-NMP, d-ATP = 2'-deoxyadenosine-5'-triphosphate.

Quite apart from their function as building blocks, some nucleotides, for example the adenylic acids which are made up from D-ribose, adenine and phosphoric acid (or di- or tri-phosphoric acid) play an important role in metabolism. **Muscle-adenylic acid (AMP)** arises from anaerobic degradation of carbohydrates in muscle. The precursors of adenosine-5'-monophosphate (AMP) are adenosine diphosphate (ADP) or adenosine triphosphate (ATP), which are converted into muscle-adenylic acid when contraction of the muscles takes place (*Lohmann*). Adenosine triphosphate plays an important role as a *source of energy* in enzymatic metabolic processes.

In living organisms the purine base adenine that forms part of adenosine or of AMP is converted by the enzyme adenosyldeaminase (ADA) into *hypoxanthin*. This results in **inosin** being the nucleoside and **inosinic acid** (inosinic acid monophosphate, IMP) the nucleotide; this was the first nucleotide to be isolated from meat extract (*Liebig*, 1847).

These nucleotides have the following structures:

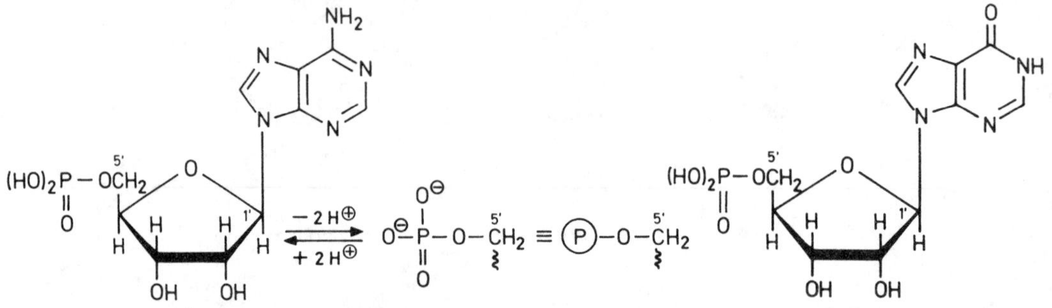

Muscle-adenylic acid or Adenosine-5'-monophosphate (AMP)   Inosinic acid (IMP)

The 5'-monophosphate esters are strong acids with pK_a values of about 1 and 6 for the two possible dissociation steps, *e.g.* for inosinic acid $pK_a = 1.54, 6.04$. Under normal physiological conditions with pH ~ 7, they thus exist almost entirely in the dianionic form. This should always be remembered since the formulae usually display these molecules as undissociated phosphate esters.

The names of the important building block types of nucleic acid are gathered together in Table 19, which also includes the one-letter symbols for the bases that are important because they are widely used for depicting the nucleotide sequence in nucleic acids.

Deoxyribonucleic acids are formed in living organisms by enzymatic reduction of ribonucleotides.[1]

9.2 The Structure of Nucleic Acids[2]

Nucleic acids are macrocyclic molecules built up from nucleotides. The basic structure is that of a polyester, in which a sugar residue (ribose or deoxyribose) alternates with a phos-

---

[1] See *H. Follmann*, Enzymatic Reduction of Ribonucleotides: Biosynthesis Pathway of Deoxy-ribonucleotides, Angew. Chem. Int. Ed. Engl. *13*, 569 (1974).
[2] See *W. Saenger*, Principles of Nucleic Acid Structures (Springer-Verlag, Berlin 1984).

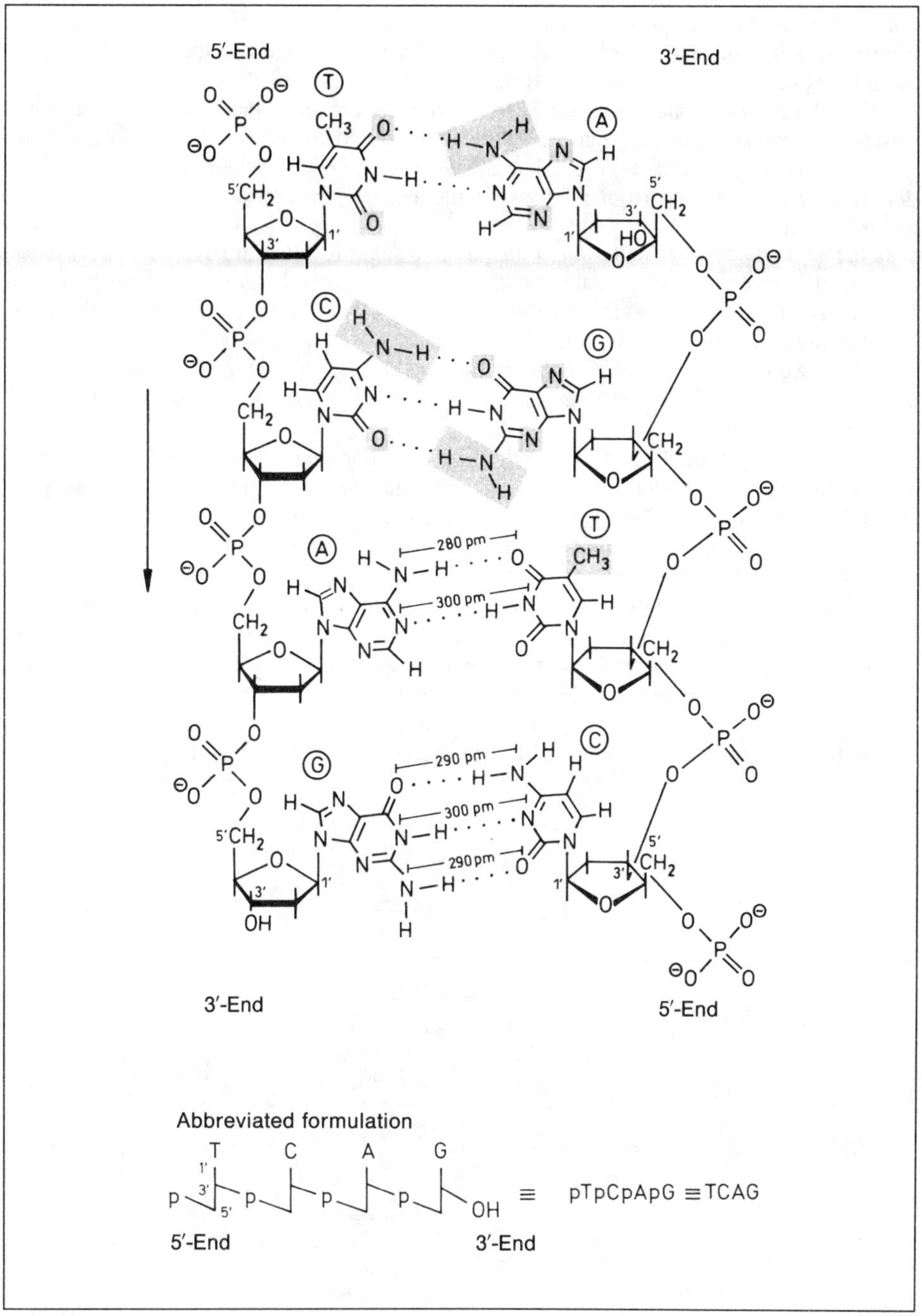

Fig. 126. Schematic representation of a portion of a DNA double strand and a simplified presentation of the nucleotide sequence in the two chains. In contrast to the conventional usage in other skeletal formulae, the angles in the abbreviated formula above do *not* represent CH_2 groups.

phate ester residue. The phosphoric acid residue forms an ester link with the 3'-OH group of one sugar residue and with the 5'-OH group of the other sugar residue of the neighbouring nucleotides. It may also be seen from Fig. 126 that each of the two nucleotide chains which make up the primary structure has a certain direction as indicated by the arrow. A distinction can be made between the two ends, one of which has a 5'-end unlinked to a neighbouring unit, whilst the other has a free 3'-end. In the secondary structure of the double strand, the directions of the two complementary single strands are opposite to one another.

In the abbreviated representation of a strand the sugar residues are symbolised by means of vertical lines and the phosphate residues by a small p flanked by two sloping lines. In further stepwise simplifications the abbreviation is finally reduced to just the symbols of the bases, expressed in the order in which the nucleotides are arranged in a nucleic acid strand. The sequence d (TCAG) shown at the bottom right-hand corner of Fig. 126 denotes the sequence in DNA unequivocally; in this representation the sequence begins with the 5'-end of the nucleic acid strand.

The bases are directed inwards and the hydrogen bonding which occurs between them ensures that there are linkages between the complementary single strands. Although the strength of each of these linkages is only small there are so many of them that the total effect is to make the overall linkage very stable. This in turn leads to the stability with which information is stored in DNA.

From Fig. 126 it may be seen that a purine ring system is always paired with a pyrimidine ring system. In the resulting base pairs there are two hydrogen bonds between adenine (A) and thymine (T) and three between guanine (G) and cytosine (C). The complementary pairs of bases are arranged like the rungs of a ladder, whose stringers (uprights) are the polyester chains made up from phosphate and deoxyribose. This can be seen in the schematic representation of a double helix portrayed in Fig. 127. Fig. 128, which is based

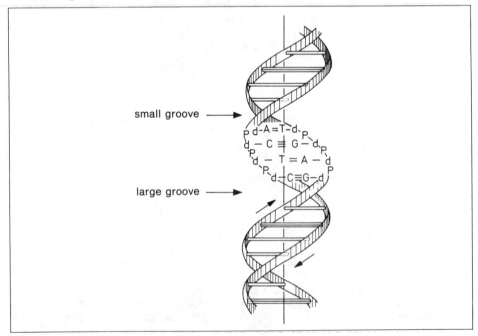

Fig. 127. Model of a double helix, A-helix (d = deoxyribose; p = phosphate)

on space-filling models, demonstrates that the complementary base-pairs lie on top of one another like books in a pile which is not lined up and is arranged like a spiral staircase.

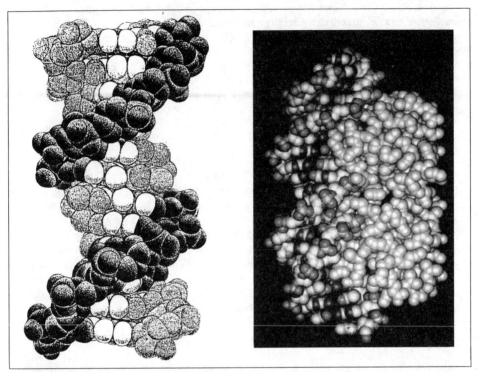

Fig. 128a. Space-filling model of a section of B-DNA as an example of a dodecamer of the sequence CGCGAATTCGCG. Ten of the edges of the 12 base-pairs are partially visible (unshaded domes)[1]

Fig. 128b. Space-filling model of the complex B-DNA of a bacteriophage with a protein which is located in the large groove (repressor-operator complex, see p.872)[2]

To date, X-ray studies have indicated three types of DNA double helix, of which the A-helix and the B-helix have right-hand spirals whilst the Z-helix has a left-hand spiral.[1] The B-DNA shown in Fig. 128a is stable at high humidities and has a 'step-height' of 340 pm; one turn is completed by a sequence of ten base-pairs (bp). It may be seen from Fig. 127 that on the surface of the double helix there are a large groove and a small groove which wind themselves, like the grooves of a screw, around the helical axis; they can be observed using a scanning tunnelling microscope.[3]

In the middle of Fig. 128a it is possible to see into the large groove and pick out base-pairs, represented by unshaded domes, which are stacked on top of another, and which cover the surface of the large groove. Here there can be attractive forces between atoms on the edges of the bases, shown on a grey background in Fig. 126, and the side chains of proteins, which have infiltrated the large groove (see Fig. 128b). In this way hydrogen bonding can take place to protein residues such as glutamic acid, glutamine, serine and threonine, and also hydrophobic interaction with, for example, phenylalanine or leucine. Structures of higher order, including catenanes and knots, derived from the double helix, can be detected.[3]

[1] See R. E. Dickerson, The DNA Helix and How It Is Read, Scientific American 249, No. 6, 86 (1983).
[2] See A. K. Aggarwal et al., Recognition of a DNA Operator by the Repressor of Phage 434: A View at High Resolution, Science 242, 899 (1988).
[3] See T. B. Beebe et al., Direct Observation of Native DNA Structures with the Scanning Tunnelling Microscope, Science 243, 370 (1989).

Because of the great chain length of poly(deoxyribonucleotide) chains in chromosomes, segments of 10^3 base-pairs (kilobase-pairs, kbp) or 10^6 base-pairs (megabase-pairs, Mbp) are used as dimensions. Since the average relative molecular mass of a nucleotide unit is about 310, a double strand has a relative molecular mass of about 6.2×10^5 per kb. The chromosome of the bacterium *Escherichia coli* (*E.coli*) contains 3.8 Mbp; multiplication by the 'step-height' of the *B*-double helix provides a length for the DNA-chain of about 1.2 mm and a relative molecular mass of 2.3×10^9.

DNA of bacteria exist as double-stranded loops which are coiled into superspirals (super-helices). These compact topologies, which are vital to the bacteria, are produced by the enzyme *DNA-gyrase*.

9.2.1 Sequence Analysis of DNA

A determination of the base sequence of DNA is of particular importance in order to under-stand how it functions. Since 1975 rapid chemical methods have become available to carry out sequence analysis, *e.g.* the *Maxam–Gilbert* process, which bears a certain resemblance to the methods used for determining the primary structure of proteins.

First of all, the enormous molecule of DNA is broken down enzymatically into smaller pieces, which can be separated out on the basis of their relative molecular masses by means of gel-electrophoresis on agarose and poly(acrylamide). The partial sequences which are to be analysed are labelled at the 5'-end with ^{32}P and then selectively modified by chemical reactions so that the resultant products can be cleaved at the modified sites in either neutral or weakly alkaline solution. This can be achieved, for example, by methylation of adenine and guanidine or by splitting cytosine or thymine with hydrazine to give deoxyribosylurea. This must be done in such a way that the sequence is only changed at a small number of places.

Separation of the fragments according to their relative molecular masses by means of gel-chromatography indicates which was the piece that contained the labelled 5'-end of the sequence and the site adjacent to that of the base which has been broken down. The same process can also be carried out labelling the 3'-end with ^{32}P.

The process has been automated and it is now possible to sequence DNA strands con-taining more than a hundred nucleotide units in this way.[1] The utilisation of electronic data processing is very promising in this field and the first *data banks* already hold millions of sequence data.[2] A huge project has been set up to elucidate and map the 3×10^9 base-pairs which make up human genomes (see p.878).

9.3 Synthesis of Nucleic Acid Sequences

Chemical syntheses of oligonucleotides provide important information that is basic for deciphering genetic codes. It is of great significance in the further analysis of DNA (see

[1] See *G. M. Church* et al., Multiplex DNA Sequencing, Science *240*, 185 (1988); *C. Gautier* et al., Nucleic Acid Sequences Handbook (Praeger, New York 1981).
[2] See *T. R. Gringeras* et al., Steps Towards Computer Analysis of Nucleotide Sequences, Science *209*, 1322 (1980); *M. Sprinzl*, Nucleinsäure-Datenbanken (Nucleic Acid Data Banks), Nachr. Chem. Tech. Lab. *32*, 212 (1984).

Fig. 128) and for supplying DNA-fragments which can be used in genetic engineering. A vital step involves the preparation of phosphate diesters at the 5'- and 3'-OH-groups of the respective ribose or deoxyribose nucleosides or nucleotides; these need protective groups for similar purposes to those described in the discussion of peptide synthesis. The same applies to the activation of phosphoric acid, which is not esterified under the conditions normally used for the preparation of carboxylic acid esters; as in the case of peptides, dicyclohexylcarbodiimide (DCC), for example, has proved useful as a coupling agent, but activated derivatives of phosphoric acid are also available which form phosphate esters directly. This may be exemplified by the synthesis of AMP from adenosine which has been suitably protected at positions 2' and 3':

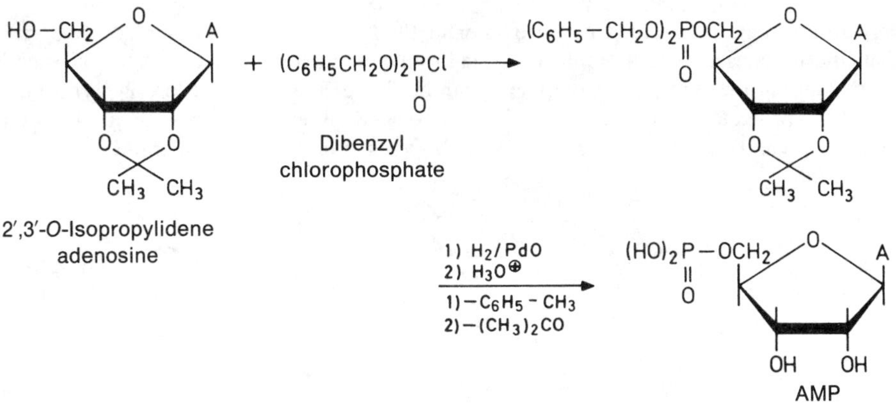

2',3'-O-Isopropylidene
adenosine

Dibenzyl
chlorophosphate

1) H₂/PdO
2) H₃O⊕

1) −C₆H₅ − CH₃
2) −(CH₃)₂CO

AMP

The benzyl groups of the dibenzyl 2',3'-O-isopropylideneadenosine-5'-phosphate which is formed can be removed as toluene by hydrogenolysis and the isopropylidene group can then be hydrolysed by acid.

Specific protection of the 2'-OH-group of ribose in the *synthesis* of *oligoribonucleotides*, which is difficult, is unnecessary in the preparation of *oligodeoxyribonucleotides* for which, among others, the diester and triester methods have been used. The first dinucleotide d(TPT) ≡ d(pT)₂ was prepared in this way in 1955 (*Michelson* and *Todd*).

9.3.1 Diester Method

In this method, two appropriately protected deoxynucleosides, one that may be called the *phosphate component* and the other the *HO-component*, are joined together using a suitable coupling reagent. Benzoyl or acetyl groups act as the protecting groups for the amino-

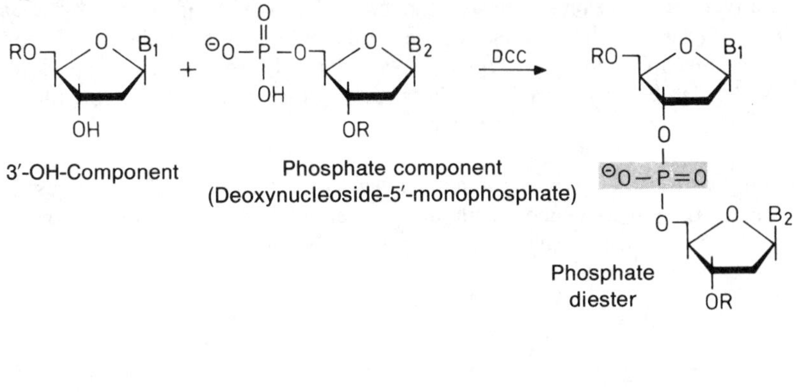

3'-OH-Component

Phosphate component
(Deoxynucleoside-5'-monophosphate)

Phosphate
diester

groups of the bases B_1 and B_2 (A, C or G), an acetyl group is used to protect the 3'-OH-group, and a 4-monomethoxytrityl group is used to protect the 5'-OH-group; in the accompanying formulae they are not differentiated but are all represented by the letter R.

This reaction leads directly to the formation of the 3',5'-phosphate ester group as found in DNA. A drawback is the reactivity of the $PO^\ominus$ function (given a grey background above) which competes with the 3'-OH-group, giving rise to pyrophosphate esters as troublesome byproducts.

9.3.2 Triester Method

This disadvantage is avoided in the triester method in which a deoxynucleoside-3'-phosphate diester is used, leading to the reaction product being a phosphate triester. The second ester group on the phosphate component can be, for example, a 4-chlorophenyl group (a phosphate protective group). Especially reactive coupling reagents such as 1-(4-nitrophenylsulphonyl)-1H-1,2,4-triazole (4-NBST) are used in the triester method.

Preparatively the phosphate triester can be replaced with advantage by a *phosphite component* (*Letsinger*). The latter can be made by reaction of the 3'-OH-component with a dichlorophosphite ($ROPCl_2$) or with chloro-diisopropylaminophosphite ($RO\text{—}P\text{—}N[CH(CH_3)_2]_2$). The phosphite triester which results from coupling with the

$$RO\text{—}P\text{—}N[CH(CH_3)_2]_2$$
$$\mid$$
$$Cl$$

5'-OH-component can be oxidised to a phosphate triester under mild conditions.

The main problem associated with the triester method arises from the necessity of removing the phosphate protective group selectively and completely at the end of each internucleotide bonding step.

Automated processes on insoluble supports, similar to those used in *Merrifield* syntheses, are in use for both of the above methods.[1]

As the length of the synthetic oligonucleotide increases, so do the problems of purification which arise from the drawbacks in the diester and triester methods.

None the less it is possible to synthesise high molecular weight DNA carrying defined biological information. This is only so because of the capacity for *self-organisation* which always appears when oligonucleotides with complementary sequences and sufficient length are present. In such circumstances a double helix forms spontaneously, as illustrated in Fig. 128.

The model shown there is based on X-ray analysis of a double helix which has been built up *in vitro* from the *self-complementary* oligomer CGCGAATTCGCG. If the sequence of bases is written down in a line from left to right they may be seen to be complementary base-sequences which interact with one another all along the line to form a double helix

5′ CGCGAATTCGCG 3′
3′ GCGCTTAAGCGC 5′

in which there are twelve complementary base pairs. A self-complementary nucleic acid sequence is called a *palindrome*.

9.3.3 Sticky-end Method *(Khorana, 1972)*

The possibilities just mentioned may also apply to shorter sequences of oligonucleotides which have regions with complementary bases. Such regions are given grey backgrounds in the following three oligonucleotides:

I: AATTCATGGGT GGTTGTAAGAACTTCT   II:

III: GTACCGACCAACATT

If such partial sequences occur at the ends of the oligonucleotides which already exist as double helices, they are described as 'sticky ends' (cohesive ends).

The spontaneous formation of 15 base pairs, on account of their complementarity, provides the partially double-stranded sequence IV.

IV: 5′ AATTCATGGCT↓GGTTGTAAGAACTTCT
GTACCGA CCAACATT 5′

This has another sticky end (shown with a grey background) and reacts in the same way with a further oligonucleotide which has a suitable complementary sequence of bases available. Further repetition of this process can generate larger molecular complexes, which, however, have gaps in single strands; such a gap is identified by the arrow in diagram IV. These gaps can be filled by enzymatic forging of diester links with DNA-ligase, thus providing a dual stranded DNA segment with strands of equivalent length.

The partial sequence IV is a synthetic initial part of a *gene* which contains information for the biosynthesis of the peptide hormone and neurotransmitter *somatostatin*, which inhibits the release of somatotropin (a growth hormone), insulin and glucagon.

---

[1] See *H. G. Gassen* et al., Chemical and Enzymatic Syntheses of Genefragments (Verlag Chemie, Weinheim, 1982).

Biosynthesis itself will be considered a little later. At this point it is necessary to make clear the way in which the complementary ends of the next oligonucleotide must be constructed so that spontaneous aggregation with the end shaded grey in IV can take place.

The double helix can be dissociated into single strands by 'melting' the hydrogen bonds. This *denaturation* can be reversed (*renaturation*) by slow cooling and spontaneous aggregation of the complementary base pairs.

9.4 Viruses

Viruses are submicroscopic pathogens which pass more or less readily through bacterial filters (filterable viruses). The diameters of these small particles are usually less than 300 nm. (The diameters of bacteria are commonly 500–2000 nm). A peculiarity of viruses is that they only multiply in living animal or plant organisms and not in inanimate media or in media from which cells are absent.

Among viruses there are a series of pathogens which are responsible for, for example, smallpox, measles, herpes, influenza and yellow fever in human beings, in animals for foot and mouth disease, cowpox and rabies, and in plants for various sorts of mosaic diseases which attack tobacco, beet, tomatoes, potatoes, etc. The infective action is associated with their nucleic acid content.

Simple viruses, which can often be obtained in crystalline form, consist solely of nucleic acids and proteins; they are *nucleoproteins* and contain either DNA or RNA.

The most thoroughly investigated example of the smallest phytopathogenic viruses is the **tobacco mosaic virus** which has a relative molecular mass of about 40 million. It consists of about 95% pro-

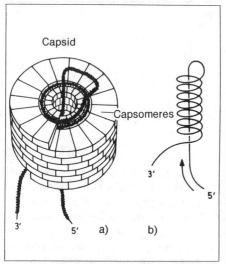
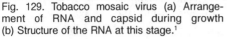
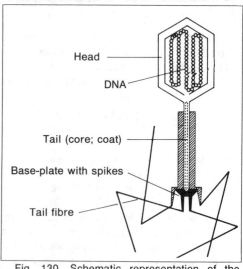

Fig. 129. Tobacco mosaic virus (a) Arrangement of RNA and capsid during growth (b) Structure of the RNA at this stage.[1]

Fig. 130. Schematic representation of the bacteriophages T_2 and T_4.

[1] See *A. Klug*, From Macromolecules to Biological Assemblies, Angew. Chem. Int. Ed. Engl. *22*, 565 (1983).

tein and 5% RNA, and forms rods which are about 280 nm long and 15 nm thick (*Stanley*, 1947). The product from virus propagation, which consists of the nucleic acid and the coat, is also known as a *virion*. In each particle long fibres of nucleic acid are stretched in helical form through the encapsulating protein, which is called the *capsid* (see Fig. 129). Careful treatment with phenol allows separation of the ribonucleic acid. The entire virus molecule is made up from about 2100 identical polypeptides (*capsomeres*) with relative molecular masses of about 17500. The capsomeres show a strong tendency to aggregate. Fig. 129 shows a model of the arrangement of RNA and capsomeres during development of the virus and the structure of the RNA at this stage.

If a healthy tobacco plant is infected with even the smallest amount (10^{-9} g) of the tobacco mosaic virus, then rapid synthesis of the viral RNA and protein ensues, to the cost of the plant. After only a few days the invading virus has multiplied itself a millionfold.

Research has provided valuable insights into the infective mechanisms of bacterial viruses or **bacteriophages** (*Delbrück, Luria, Hershey*). Some bacteriophages are quite different from the tobacco mosaic virus (see Fig. 130). The hexagonal head of phages T_2 and T_4 consists of a protein coat which contains a strand of DNA and has a tail which at its far end has a base-plate to which are attached spikes bearing long bent filaments. If these touch a bacterium then the end-plate of the phage is brought into contact with its surface and holds fast there until an enzyme exuded from the tail section has dissolved the contiguous portion of the bacterial membrane. The DNA from the head of the phage can then be injected through the tail into the bacterial cells (*infection*). The protein coat of the phage stays behind on the surface of the bacterium. The DNA from the phage now interacts by means of the information it carries with the metabolism of the bacterium and produces more phage-DNA and coatproteins for new phages. In this way a new generation of phages arises, which ultimately dissolve the wall of the host cell and come out of the dead bacterium into the surrounding medium.

DNA from bacteriophage T_4 has a length of about 61 μm or 180 kb; more than a hundred of the genes present in this double strand are known. Smaller viruses are better suited for investigating the mechanism by which the process outlined above takes place, since their stocks of genes are more convenient to survey and can be more readily changed experimentally. An example is provided by the *Simian virus 40* (SV 40), which is found in the kidneys of apes. The DNA from this virus contains only 5243 nucleotide pairs and their sequence is known. It belongs to the group of *tumour viruses*, which can release various forms of tumours. *Temin and Baltimore*, quite independently of each other, showed (1970) that some RNA viruses, which contain the enzyme *reverse transcriptase*, also belong to this group, and so, with the RNA acting as a template, are able to synthesise DNA which is described as *proviral DNA* or as the *provirus*. Because of these *retroviruses* the normal direction of the *transcription* (see p.866) of DNA into RNA that takes place in the cell is reversed, with the result that an infected cell contains within it DNA which is derived from the retrovirus, and therefore cannot be completely removed from the body by chemotherapy. AIDS (acquired immuno deficiency syndrome) is transferred by a virus which is a member of the group of slow-acting lentiviruses, discovered in 1983 by *Montagnier* and named LAV (lymphadenopathy AIDS virus). It proved to be identical to HTLV III (human T-cell lymphotropic virus),[1] discovered by *Gallo* in 1984. It seemed appropriate to describe this virus henceforth as HIV (human immunodeficiency virus).

In the meantime a second 'HIV2' virus has been identified in West Africa; its genome (see p.875) is clearly different from that of the HIV virus.

Retroviruses are characterised by at least three genes:

— the *gag* gene for the so-called core or inner capsid proteins (group-specific antigen)
— the *pol* gene for reverse transcriptase (polymerase)
— the *env* gene for the outer envelope glycoproteins (envelope), which is known to be the antigen of the organism.

---

[1] See *R. C. Gallo*, The Aids Virus, Scientific American *256, No. 1*, 38 (1987); *H. Varmus*, Retroviruses, Science *240*, 1427 (1988).

Compared to the other genes of the virus, the *env* gene is very variable and therefore the outer envelope glycoproteins are for ever changing. This makes it difficult for the organism to form neutralising antibodies.

In the making of the specific sequence of cell DNA, reverse-transcriptase causes thymidine to be replaced by azidothymidine, and further preparation of DNA is interrupted at this site. This is due to the effect of the azidothymidine.

9.5 Function of Nucleic Acids[1]

As has already been mentioned, nucleic acids are of crucial importance biologically, in heredity and in protein synthesis.

A **gene** is a DNA fragment of a chromosome which carries the genetic information required for the synthesis of a polypeptide. Chromosomes are stored in cell nuclei and are passed on from generation to generation. They have the capacity to provide identical reproduction (propagation) or mutation (gene-change).

The genetic material of chromosomes consists of *deoxyribonuleic acids* having a double helix structure. By breaking the hydrogen bonds which link pairs of bases in the two polynucleotide chains DNA is able to couple either to a new *complementary* DNA single strand or to the old one, as shown in the following simplified scheme (Fig. 131). Thus by the action of the enzyme *DNA-polymerase* (*Kornberg*, 1955) two new double strands are formed, one strand of which comes from the original parent double helix (shown in black in Fig. 131) and the other is a daughter strand (shown in grey in Fig. 131). This process is known as semiconservative regeneration; it has been confirmed by labelling experiments using ^{15}N.

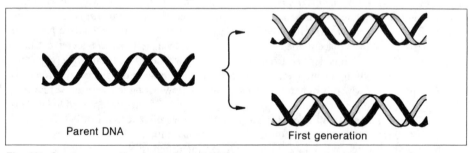

Parent DNA   First generation

Fig. 131. Semiconservative replication of DNA.

The principle which applies to the identical *replication* (duplication) of DNA, also applies to the transfer of genetic information from a DNA chain to an RNA chain. This *transcription* leads to the formation of **messenger-ribonucleic acid**, m-RNA, which carries the information from the cell nucleus into the cytoplasm. In m-RNA ribose takes the place of 2-deoxyribose, and, in place of the thymine residues of DNA, the complementary RNA

[1] See *R. Knippers et al.*, Molekulare Genetik, 5th edn (Thieme Verlag, Stuttgart 1990); *B. Lewin*, Genes IV (Oxford University Press, Oxford 1990).

has uracil. For example, a partial chain GGACGGAT gives rise to a chain CCUGCCUA in m-RNA. In contrast to DNA, in RNA not only G and C but also G and U can pair, but the hydrogen bonding is weaker.

9.5.1 The Genetic Code

In the process of protein biosynthesis (translation) RNA provides the working plan for the order in which amino acids combine to form the primary structure. This plan, which is stored in the DNA, and is transferred as described above to the RNA, is the *genetic code*. It was shown quite conclusively in 1966 (*Khorana, Nirenberg*)[1] that the *diversity of heredi-tary characteristics* is caused by *variations in the arrangement of just a few links* and *not by a corresponding diversity of different links*.[2]

Nature makes use of the operation of a code which is based on the disposition and arrangement of the four bases adenine, guanine, cytosine and uracil. These bases, A, G, C and U, may be regarded as the letters of this code, and 'words' which are built up from them provide the relationship of the 20 proteinogenic amino acids. In order to make up 20 words from only four letters, these words must each contain at least three letters. Table 20 lists the

Table 20. The codons found in messenger-ribonucleic acids (m-RNA) and their significance

First letter of the codons		Second letter of the codons								Third letter of the codons
		U		C		A		G		3'-end
5'-end										
	U	UUU	Phe	UCU	Ser	UAU	Tyr	UGU	Cys	U
		UUC	Phe	UCC	Ser	UAC	Tyr	UGC	Cys	C
		UUA	Leu	UCA	Ser	UAA	Ende	UGA	Ende	A
		UUG	Leu	UCG	Ser	UAG	Ende	UGG	Trp	G
	C	CUU	Leu	CCU	Pro	CAU	His	CGU	Arg	U
		CUC	Leu	CCC	Pro	CAC	His	CGC	Arg	C
		CUA	Leu	CCA	Pro	CAA	Gln	CGA	Arg	A
		CUG	Leu	CCG	Pro	CAG	Gln	CGG	Arg	G
	A	AUU	Ile	ACU	Thr	AAU	Asn	AGU	Ser	U
		AUC	Ile	ACC	Thr	AAC	Asn	AGC	Ser	C
		AUA	Ile	ACA	Thr	AAA	Lys	AGA	Arg	A
		AUG	Met*	ACG	Thr	AAG	Lys	AGG	Arg	G
	G	GUU	Val	GCU	Ala	GAU	Asp	GGU	Gly	U
		GUC	Val	GCC	Ala	GAC	Asp	GGC	Gly	C
		GUA	Val	GCA	Ala	GAA	Glu	GGA	Gly	A
		GUG	Val*	GCG	Ala	GAG	Glu	GGG	Gly	G

*Part of the start signal. Grey background: homocodon

[1] See *M. Nirenberg*, Der genetische Code, Angew. Chem. *81*, 107 (1969); *H. G. Khorana*, Nucleinsäure Synthesis als Werkzeug für das Studium des genetischen Codes (Nucleic acid synthesis as the tool for studying the genetic code), *ibid 81*, 1027 (1969).

[2] See *H. Blumenberg*, Die Lesbarkeit der Welt (The Legibility of the World) (Suhrkamp, Frankfurt-am-Main 1981, p. 381). This book presents and combines intellectual and scientific concepts out of which the genetic code can be seen to have developed.

specific 24 triplets (base triplets) which go to make up the glossary of the genetic code; they are called codons. To prepare a protein from 50 amino acids it is necessary to have 50 triplets or 150 bases. The *homocodons* which are shown on a grey background in Table 20 are in all probability the oldest constituents of the genetic code.

An important feature of the code is its degeneracy, as a result of which it shows amino acids which have more than one codon. This is indeed the case for the majority of them; only methionine and tryptophane each have only one specific codon. There is, however, no codon which is characteristic of more than one amino acid. Several codons which apply to one and the same amino acid differ from one another for the most part in the third letter, *i.e.* that representing the 3'-terminal base. The specificity of the first two letters of a codon is thus greater than that of the third letter (see in Table 20, for example, Val, Ala, Gly). This is the result of stabilisation of the code during evolution, thereby minimising potential sources of error. The 3 codons UAG, UAA and UGA signify the discontinuation of protein synthesis. AUG is characteristic for the introduction of methionine into a chain; it can also function as a starting signal, when it specifies that in bacteria or viruses *N*-formylmethionine, which is an amino acid with a protected amino-group (see p.828 f.), should be introduced as the first building block of a bacterial protein or of a viral protein. In this way the direction of growth of the protein sequence is ensured. GUG, which in the normal way codes for valine, can also act as a starter for the introduction of *N*-formylmethionine, but this is much rarer.

The sequence of bases must be followed from the start, triplet by triplet. If at the start one or two bases are out of place then all the succeeding codons will lead to the introduction of other amino acids and this usually results in the formation of a protein that has no biological activity (frame shift mutation).

The genetic code is the same for all living organisms, in other words, a given nucleotide sequence provides the same protein in all organisms if it is interpreted correctly. Researches on the statistical geometry of the building-block sequences of t-RNA gives an estimated age of the genetic code of 3.8 ($\pm$0.6) thousand million years.[1]

9.5.2 *Mutations*

Changes in the nucleotide sequence can lead to mutations. From the chemical point of view, disorders which may be brought about in the genetic code by *intercalation,* or by *chemical* or *photochemical reactions,* are important.

Intercalation involves the insertion of a planar molecule between the base pairs of the double helix. For example the phenoxazone system of *actinomycin D* intercalates between two consecutive nucleotide pairs, *viz* $\frac{G{:}{:}{:}C}{G{:}{:}{:}C}$, which causes deformation of the DNA at this point. In subsequent replication this can lead to the insertion of an additional nucleotide, causing a frame shift mutation. Intercalation of actinomycin can result in transcription of DNA into RNA being interrupted at this point, because the RNA-polymerase which is dependent upon the DNA can migrate no further over the deformed region.

Cytosine (C)   Uracil (U)

[1] See *M. Eigen* et al., Science *244,* 673 (1989).

The most important chemical changes in DNA are a result of the action of deaminating agents, and alkylating agents such as diazomethane or methanesulphonate esters.

In this way cytosine can be converted into uracil (see p.864).

As a result, in DNA replication a base pair $C\vdots G$ is changed into a daughter strand $C\vdots G$ opposite to $A\vdots U$:

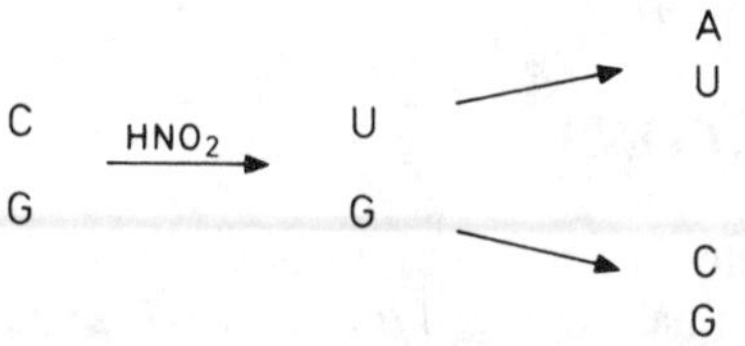

Decisive changes in the bases are brought about by alkylation of the carbonyl groups of cytosine or of guanine, *e.g.*:

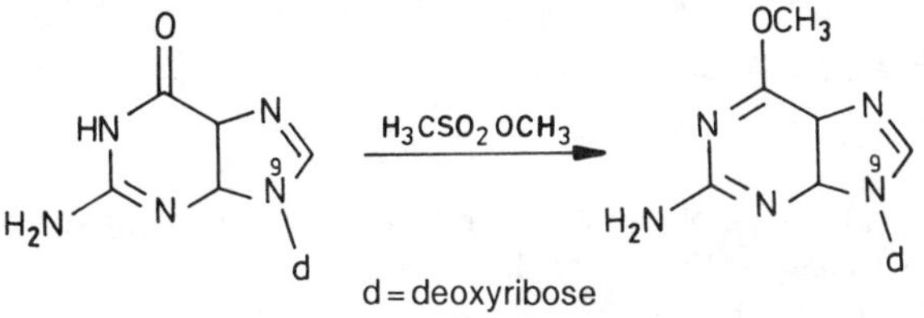

d = deoxyribose

Methyl groups can also be introduced onto the nitrogen atoms except N-9.

Among the changes which can arise due to irradiation of DNA with UV-light, one which is of the greatest importance is the formation of cyclobutanoid dimers between neighbouring pyrimidine rings of one strand:

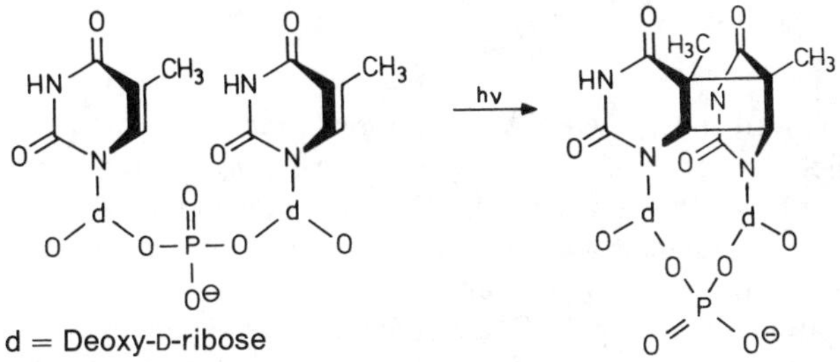

d = Deoxy-D-ribose

If furocoumarins such as psolaren (R = H) or bergapten (R = OCH_3) are intercalated and then irradiated with UV-light they form covalent bonds with pyrimidine bases and bring about cross-linking of two DNA strands:

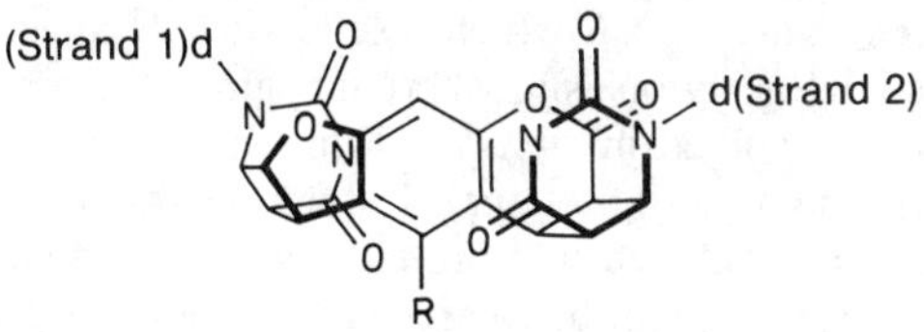

In this case the furocoumarin forms a bis-adduct with two thymine rings, one from each strand of the DNA.

Frame shift flaws arise because of these changes in the bases, with results as described above. The organism has at its disposal a repair mechanism, which can remove such deformations of the information content of nucleic acids by exchanging the modified bases.

DNA-polymerase plays an important role in the repair of DNA. If such repairs are unsuccessful, mutation ensues. In this way several bases may be deleted from a DNA strand. If denatured DNA strands (see p.860) are subjected to renaturation in the presence of deleted single strands of DNA, a double helix is formed in which longer portions of DNA without complementary sequences remain as single strands. The result is a *heteroduplex*.

9.5.3 *Transcription of DNA*

In transcription of DNA, only *one* strand of the double helix, the *matrix strand*, is copied; it is represented in the right half of Fig. 126. The DNA sequence shown in I behaves as follows in transcription:

```
                                    RNA-              _ - TTA
 Coded strand  5'---TTAGTAA---3'  polymerase   5' -´        GTAA---3"
 Matrix strand 3'---AATCATT---5'  ───────►     3'  `        CATT---5'
                                                    `- AAT
                  I                                II ──────►
```

The DNA-dependent RNA-polymerase loosens the hydrogen bonding between the complementary bases in a small section of the double helix to give II and moves along the chain in the direction indicated by the arrow. In the region which has been opened up the making of RNA commences and continues towards the 5'-end.

```
     5'--TTAG--3'                       --TTA GTA A--
     3'--AAT C--5'      ─────────►      --AAT CAT T--
     5'  UU    3'                        UUAGU
             ↘                                ↘
              A                                A
```

The formulae show that the new RNA strand forms a temporary base pair with the matrix DNA strand, which can be thought of as a short piece of a *hybrid* DNA–RNA double helix. As soon as the last base of the transcribing RNA is attached at the 3'-end it removes itself from the DNA. The beginning and end of the duplication are signalled by nucleotide sequences whose function is not as yet certain.

The resultant primary transcript of DNA from nucleus-containing cells (eukaryotes) has long sequences which contain no information referring to amino acids (intervening sequences = *introns*). Messenger RNA usually contains only sequences coded for amino acids, which occur in various regions of the primary duplicate (*exons*).[1] Therefore the introns must be removed from the primary duplicate and the exons must be spliced in in the correct manner. The processed m-RNA which results after treatment (processing) is conveyed from the nucleus into the cytoplasm, where it controls the preparation of proteins at the ribosomes and also is involved in *translation* in which the 'language of the nucleic acids' is replaced by the 'language of proteins'. The transposition of the information stored in the DNA in genes into m-RNA and into proteins as described above is known as *gene expression*. This is the crossover point between genotype and phenotype.

It is possible, with the help of *reverse transcriptase*, to produce DNA copies of an m-RNA. These are described as c-DNA and are used in genetic engineering.

---

[1] See *P. P. Minghetti* et al., Molecular Structure of the Human Albumin Gene is Revealed by Nucleotide Sequence within q11-22 Chromosome 4, J. Biol. Chem. *261*, 6747 (1986).

9.5.4 Translation

Protein synthesis takes place at the ribosomes, which also form the 'protein factories' in cells in which they are present in large numbers (10^4–10^6). They consist of a large and a small sub-unit which both contain *ribosomal* RNA (r-RNA). Fig. 132 shows the disposition of the sub-units of the m-RNA, how the *m-RNA* arranges itself between the sub-units, and

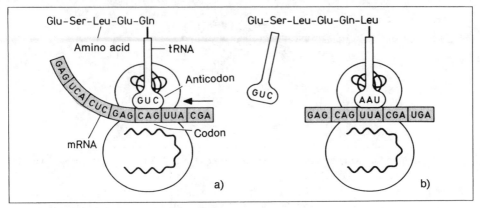

Fig. 132. Model of the process of translation

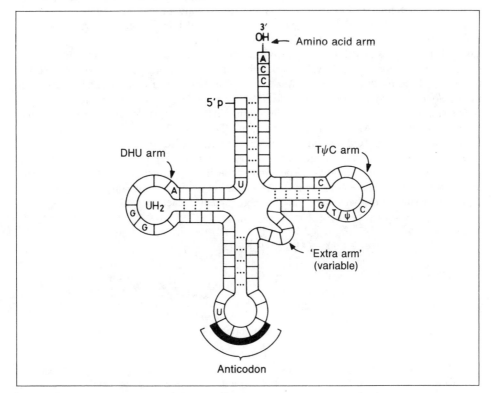

Fig. 133a. Generalised formulation of a t-RNA molecule.

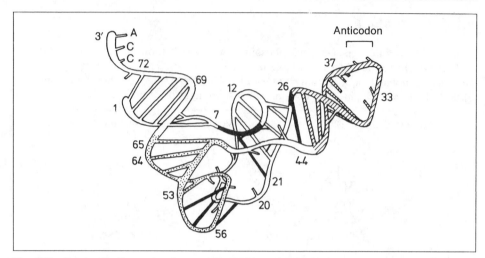

Fig. 133b. Schematic X-ray structure representation of the t-RNA for phenylalanine in yeast.

how the third type of RNA, *transfer*-RNA (t-RNA), which is involved in protein synthesis, takes part in the process. This provides the next amino acid to be incorporated into the protein chain and has at the appropriate place a base triplet which is complementary to the codon for a particular amino acid in the m-RNA; it is known as an *anticodon*.

There is at least one t-RNA for each amino acid and this is made up from about 80 nucleotides and has a cloverleaf tertiary structure which results from there being maximal base pairing. A generalised picture is provided by Fig. 133a and may be compared with the schematic representation of the X-ray structure of a t-RNA in Fig. 133b. The tertiary structure is stabilised by the presence of $Mg^{2\oplus}$ ions. The presence of unusual nucleotides such as *pseudouridine* (ψ) and *dihydrouridine* (DHU or U_h or UH_2) is characteristic.

The site of the anticodons is clearly discernible in the X-ray structure diagram, as is the 'amino acid arm' CCA, whereas the other arms (DHU 14–20, variable 45, 46 and TψC 56–61) can only be picked out with more difficulty in the L-form of the molecule.

As in the case of peptide formation *in vitro*, the biosynthesis of proteins likewise requires *energy-rich* amino acid derivatives. The amino acids in an organism are activated by means of adenosine triphosphate (ATP); an *activated* amino acid-adenosine monophosphate and diphosphate (pyrophosphate) are formed in an enzymatic reaction:

$$
\begin{array}{ccc}
\text{COOH} & & \text{O} \qquad \text{O} \\
| & & \diagdown \quad \parallel \qquad \qquad 5' \\
H_2N-\overset{|}{C}-H + ATP \longrightarrow & H_2N-\overset{|}{C}-H \quad & C-O-P-O-CH_2 \\
| & | & OH \\
R & R & \\
& & \text{OH} \quad \text{OH}
\end{array}
$$

+Diphosphate
(A = Adenine)

Aminoacyl-AMP

The α-aminoacyl group is then transferred, with loss of the adenosine monophosphate (AMP), onto the amino acid arm of the t-RNA, and is bonded onto the 3'-OH-group of the D-ribose of the terminal adenosine residue:

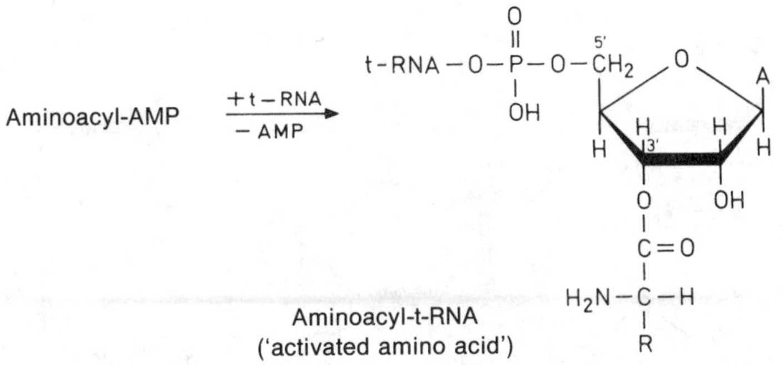

$$\text{Aminoacyl-AMP} \xrightarrow[- \text{AMP}]{+ \text{t-RNA}}$$

Aminoacyl-t-RNA
('activated amino acid')

A sufficient supply of amino-acyl-t-RNA is always available in the cell plasma of an adequately fed organism to migrate to the ribosomes and there to seek out its appropriate complementary codon on the m-RNA. In Fig. 132 this is illustrated schematically for glutamine on the left and for leucine on the right; the bonding with the preceding amino acid in the sequence is also shown. Linking proceeds enzymatically with guanosine triphosphate also present as co-factor. The growing chain remains bonded to the already coupled t-RNA until the succeeding t-RNA takes over by linking to its amino acid. Only then is the t-RNA of the preceding amino acid removed. In this case the synthesis is interrupted by the triplet UGA (see Table 20) and the completed protein detaches itself.

In the course of this procedure about ten amino acids are added to the growing peptide chain every second. Only the twenty *proteinogenic amino-acids* (see p.820) can take part in the process; they are consequently known as *ribosomal amino acids*.

The replacement of glutamic acid by valine which is associated with sickle-cell anaemia is brought about by the fact that in the 6. codon of the DNA in the gene responsible for the β-chain of haemoglobin an A has been replaced by a T (point mutation).

9.5.5 RNA Enzymes (Ribozymes)

The idea that all enzymes are proteins was overturned in 1982, quite independently of one another by *Altmann* in studies on the t-RNA from *Escherichia coli* and by *Cech* on r-DNA from the protozoan *Tetrahymena*. The processed RNA in both organisms, which has been prepared as previously described in the case of m-RNA (see p.866) and which has come from the *primary transcript* by removing the *introns* and splicing in the exons, shows the functional activity of t-RNA or of r-RNA. These facts are consistent with *biocatalysis*, that is catalysis *by the RNA* itself.

Fig. 133c shows a section of an unprocessed ribosomal RNA from the eukaryotes of *Tetrahymena*. For this catalytic activity it is important that steric folding takes place in such a way that parts of the single-stranded RNA so arrange themselves, with respect to complementary sequences opposite to them, that short double-stranded regions may arise. In addition, because of other interactions indicated by the double-headed arrow, the molecule takes up a complicated steric form (see Fig. 133b). This allows the unprocessed r-RNA to behave like an enzyme which cuts out the part shown in Fig. 133c with a grey background, namely the intron, and to splice together the broken ends. This takes place in the region enclosed in a frame in Fig. 133c, wherein only magnesium ions and guanosine monophosphate (GMP) must be available. Chemically this results in transesterification involving the phosphate diester:

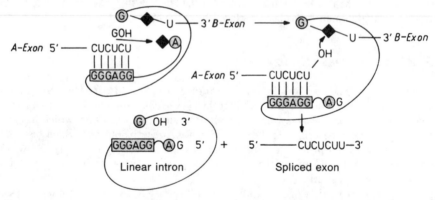

$\textcircled{P} = PO_3^{2\ominus}$

The letters A on a grey background refer to the base at the 5′-end of the intron, which is indicated by an arrow in Fig. 133c. In the following formulae phosphate diester groups which participate in transesterification (enclosed in a frame above and given a grey background) are indicated by ♦ and the intron strand is given a grey background.

The GMP (shown above as GOH), which is essential as a co-factor, attacks the 5′-end of the intron and binds itself to it. In this way a 3′-OH-group is formed in the *A-exon*, which bonds to the *B-exon* in a second transesterification reaction, and provides the *spliced, processed exon*. In this way an intron made up of 414 nucleotides is released and it undergoes cyclisation, losing a fragment containing 15 nucleotides in the process; after ring-opening and further ring-closure it is again shortened, this time by 4 nucleotides. The final result is an extended molecule containing 19 fewer nucleotides than the original intron. It is known as L-19IVS (Linear intervening sequence minus 19 nucleotides).

The manner in which activation is brought about by transesterification of the phosphate esters involved is not yet certain. It appears that the tertiary structure of the RNA strand

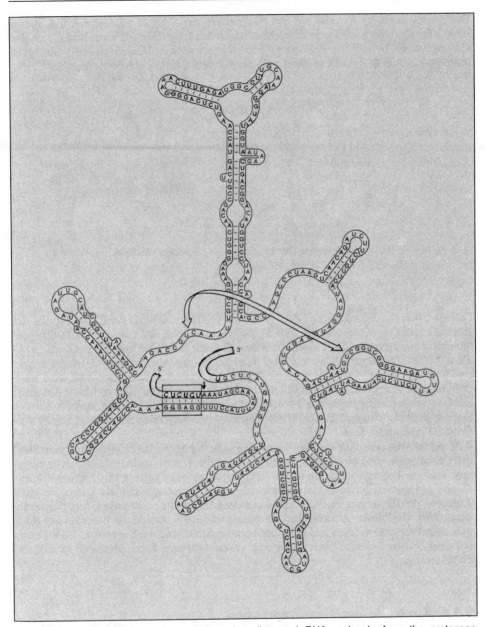

Fig. 133c. Two-dimensional representation of a ribosomal RNA molecule from the protozoan *Tetrahymena*.[1] The intron strand has a grey background.

---

[1] See *T. R. Cech*, RNA as an Enzyme, Scientific American *255*, *No. 5*, 76 (1986).

plays a role analogous to that of the tertiary structure of proteins in enzyme activity. Strictly speaking the process which is described here does not accord with the definition of a catalyst, since an intron which splits itself off from a molecule does not remain unchanged. In order to get over this difference the term *ribozyme* has been introduced. Meanwhile it has turned out that the sequence L-19IVS can cleave short ribonucleotide chains and reassemble them again, without itself undergoing any change

9.5.6 *Control of Gene Expression*

The entire genetic information for the whole organism is contained in the genes of every constituent cell. Therefore a control system of *promoters* and *repressors* is required so that only the proteins which are required for a particular situation of the cells or of the organism may be prepared. Such controls can be effected either in transcription or translation. Most studied is transcription control. *Jacob* and *Monod* suggested the *operon model* for this, which consists of a *regulator*, a *promoter*, an *operator*, and *structure genes*. The promoter contains the previously mentioned nucleotide sequence which provides the signal for the start of the transcription of the structure gene. So long as the structure gene is not expressed, the promoter is blocked by bonding of the *repressors* to the operator (repressor–operator complex, see p.855). Oligonucleotides can also intercalate into the large grooves, as a result of which a triple helix with repressor action is formed. The repressor can consist of a protein which intercalates in the large groove in this region of the double helix, and results in hydrogen bonding and in hydrophobic interaction with the groups on the bases shown on grey backgrounds in Fig. 126. The repressor is coded in a different region of the DNA from the regulator gene and remains on the operator until the protein belonging to the structure gene is needed. Then a promoter enters the cell and forms a complex with the repressor. In the course of a change of structure it detaches itself from the operator with the result that the structure gene is no longer blocked, and transcription of the following structure gene can commence from the place where the DNA-dependent RNA-polymerase is linked to the *promoter*. The process is as follows, using the case of the lactose operon (lac.-operon) as an example:

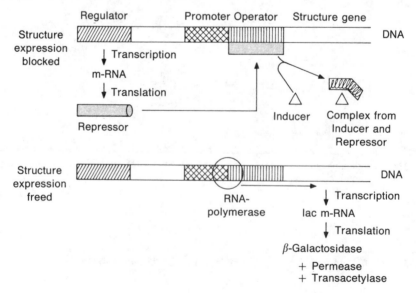

Expression of the three enzymes β-galactosidase, permease and transacetylase, which are necessary for the breaking down of lactose, is blocked so long as a cell containing sufficient glucose is provided. A deficiency of glucose and a sufficient supply of lactose can thus, depending on the utilisation of the lactose, result in a modification of the metabolism of the cell so that, for example, allolactose functions as the inducer and deactivates the repressor by forming a complex. The synthesis of the three enzymes necessary for metabolising lactose then proceeds unhindered.

The progesterone antagonist RU486 engages the receptor for progesterone and changes it so that it interrupts the normal transcription of the gene. In this way the synthesis of proteins which are necessary for maintaining pregnancy is hindered. The embryo dies, and, under the influence of prostaglandin, which facilitates contraction of the muscles of the uterus, is ejected together with its mucous membrane.

The development of RU486 was only possible as a result of an insight into the results of gene expression.

The insight which has been gained into processes like those described in this chapter, for example the fact that the same molecules and the same processes are involved whether they take place in humans or in amoeba, lends strong support to the understanding of *Darwin's* theory of evolution.[1]

9.5.7 Gene Analysis (DNA Diagnosis)

Hybridisation, which was developed from *renaturation* (see p.860), provides the methodological core of gene analysis. It turns out that the recombination of base pairs which is accomplished in this way also takes place with partial sequences of DNA, and that it also can be carried out on *supporting membranes*.

The term *hybridisation* is used, since by this procedure DNA and RNA strands, for example, can be renatured with one another (see p.866); in genetics 'hybrid' signifies 'from two different sources'.[2]

9.5.7.1   Hybridisation on nitrocellulose membranes, as described by *Southern* in 1975, is called *blotting*; this method is of great significance in the chemistry of proteins and nucleic acids. *Southern* describes the transfer of a separated collection of nucleic acid fragments obtained by gel-electrophoresis on agarose onto a nitrocellulose membrane which acts as a matrix. The fragments are pressed onto the gel and then elution from the gel onto the matrix can be speeded up by, for example, renewed electrophoresis (electroblotting).

In *Southern* **blotting** (Southern blot) a *denaturation* of the dissociated DNA fragments takes place *before* transfer onto the matrix, so that only a single strand is transferred. A very complex mixture of fragments results and the DNA sequences which are sought are located by hybridisation with a labelled single strand of DNA or RNA (gene probe) which contains the required complementary sequence. This is done *after* washing off the unhybridised gene probes, for example by autoradiography, if radioactive ^{32}P has been used as the marker. In this way the few DNA fragments from the analysis sample which contain a DNA sequence complementary to that of the gene probe are located. *Biotin*, which can be coupled to the DNA probe, also serves as a marker; in this case the strong bonding of *avidin* to biotin is used for location purposes. If the method is applied to the location and detection of RNA

---

[1] See *Wen-Hsiung Li* et al., Fundamentals of Molecular Evolution (Simauer, Sunderland, MA 1991).
[2] On this basis the word is also used for combinations of *s*- and *p*-orbitals (p.23) and therewith for the description of molecular properties by using *orbital theory*. In contrast the hybrids described here are *substances*, as in the case of hybridomes described on p.879.

fragments it is described as **Northern blotting. Western blotting** is the transfer of *proteins* which have been separated by electrophoresis onto the matrix, as is often done in electroblotting. Locating is done in this case by *antibodies* specifically labelled with ^{125}I.[1]

9.5.7.2   Applications. *HIV antibodies* can be detected especially reliably by Western blotting. This test is used for confirming a positive result from the ELISA test.

Detection of a *genetic disease* is possible if the site of the gene and the kind of mutation are known, for example in the case of sickle-cell anaemia. The corresponding marked gene can be used as a probe in this case.

Genetic fingerprinting depends upon the presence of frequent repeating short sequences which are in uncoded segments (introns) of DNA (mini-satellite regions). The analysis makes use of *Southern* blotting in which DNA fragments of the mini-satellite regions are used as gene probes.[2]

9.5.7.3   The **p**olymerase **c**hain **r**eaction (**PCR**) is a gene-copying process, whose use for gene diagnosis can hardly be overrated. Starting from a single DNA molecule it is possible, with little expense and in a short time, to prepare copies in any required amount (*Mullis*, 1985).[3]

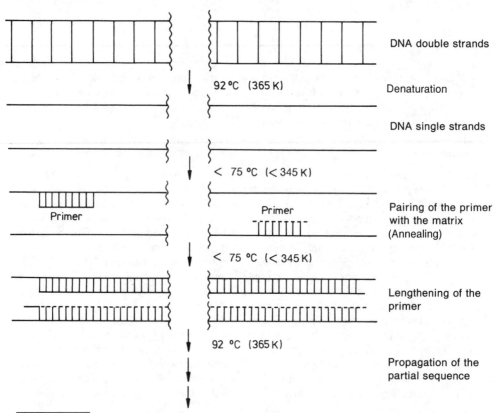

[1] See *U. Beisiegel*, Protein Blotting 7, 1 (1986).
[2] See *P. Gill* et al., Forensic Application of DNA 'Fingerprints', Nature *318*, 577 (1985).
[3] See *K. B. Mullis*, Spektrum der Wissenschaft *1990*, Vol. 6, p.60; Angew. Chem. Int. Ed. Engl. *33*, 1209 (1994); *A. Graham*, A Haystack of Needles: Applying the Polymerase Chain Reaction, Chem. Ind. (London), 718 (1994).

In the process, the previously mentioned *DNA-polymerases* play a key role in the *replication* and *repair* of DNA. It is possible to lengthen a short nucleotide sequence (primer) if it is paired with an already prepared strand (matrix strand). Pairing takes place spontaneously on the part of the matrix strand which has a base sequence complementary to that of the primer. (See sticky-end method, p.859). The medium must also contain the four required nucleotide 5'-phosphates, dATP, etc. Polymerase builds on to the primer those bases which are complementary at each time; diphosphate ($\textcircled{P}$—O—$\textcircled{P}$, see p.889) is displaced in the process. If the next nucleotide of the matrix is A, then T is built onto the primer, likewise a G onto a C. In replicating or repairing the DNA one strand serves as the matrix for the other.

Before using PCR a primer for the strand with 3'- and 5'-end groups is always synthesised; this always consists of an oligonucleotide made up of about twenty building blocks, whose sequence is complementary to the corresponding sequence at the end of the nucleic acid that is to be copied.

The scheme on p.874 shows the reaction stages accomplished by the PCR. They take place without interruption if the temperature is regularly raised to 92°C (365 K) and then lowered again to 75°C (345 K). At 92°C (365 K) the DNA-copying sequence (target sequence) is denatured and only the single strand is present. When the temperature is lowered the primer pairs itself with the particular matrix strand of the target sequence (annealing); lengthening of the primer sequence along the two matrices then takes place. The resultant double helices are denatured by another rise in temperature. At each repetition of the process shown in the Scheme, the number of resultant DNA molecules is doubled. After carrying out n cycles there are 2^n copies of the target molecule, *i.e.* after 30 cycles, 2^{30} or more than 10^9 copies. Since one cycle takes about ten minutes this operation would take about five hours.

The decisive factor in the continuous operation of the polymerase chain reaction is the thermal stability of DNA-polymerase. Bacteria of the species *Thermus aquaticus* (Taq), which are found in the hot springs in the Yellowstone Park, contain a DNA-polymerase which is fully functional at 92°C (365 K) (Taq-polymerase).

9.6 Genetic Engineering and Biosynthesis[1]

Genetic engineering or 'synthetic biology' are terms used to describe processes in which attempts are made to manipulate the genetic material of an organism. For chemists this presents the fascinating possibility of using *bacteria* to synthesise *proteins* which can only be made otherwise in *animal or human cells*. The basis of such processes is the transformation of bacteria which lack nuclei (prokaryotes), by transferring small double-stranded DNA molecules between the bacteria and building them into the genome[2] of the recipient. For example, the resistance of bacteria to antibiotics has been transmitted in this way

---

[1] See *J. E. Davies* and *H. G. Gassen*, Synthetic Gene Fragments in Genetic Engineering — The Renaissance of Chemistry in Molecular Biology, Angew. Chem. Int. Ed. Engl. *22*, 13 (1983), *H. G. Gassen* et al., Der Stoff aus dem Gene sind (Materials Produced from Genes) (J. Schweitzer Verlag, Munich 1986); *H. J. Rehn, G. Reed, A. Pühler* and *P. Stadler* (Eds.), Biotechnology, 2nd edn., Vols. 1 & 2 (VCH, Weinheim 1993).

[2] The term genome denotes the entire gene content of an organism.

(*Lederberg* and *Tatum*, 1946). In the same way bacterial cells can be induced to synthesise a foreign protein. This is done with the help of a cyclic DNA molecule or *plasmid* that can be isolated from bacteria. The plasmid acts as a *vector* to introduce a gene that contains the blue-print for the required protein.

9.6.1 Modification of Plasmids

In order to build the DNA sequence that contains the information required to induce the synthesis of a protein which is foreign to a bacterium, reactions are used in the plasmid to cut and splice or ligate the DNA which are like those used standardly in the working up of the primary copy to processed m-RNA in each eukaryotic cell. Enzymes known as restriction-endonucleases can recognise DNA sequences of plasmids and sever them in characteristic ways (*Smith*, 1975).[1] Thus the endonuclease 'Eco RI' from *Escherichia coli* RY 13 cuts the DNA double strand at the places where the sequence GAATTC, which is palindromic, appears, as indicated by the portion of the double strand shown with a grey background:

$$-- \text{CGGAATTCCC} -- \quad \xrightarrow{\text{Eco RI}} \quad -- \text{CGG} \qquad \text{AATTCCC} --$$

$$-- \text{GCCTTAAGGG} -- \qquad\qquad -- \text{GCCTTAA} \qquad \text{GGG} --$$

It may be seen that this staggered cutting of the complementary single strands generates --TTAA and AATT-, which are overlapping ends that can be used in the 'sticky-end' method for the synthesis of polynucleotides (see p.859 *f.*). In contrast the restriction endonuclease 'Hae III' brings about a cut wherever the palindromic sequence GGCC appears, but in this case the cuts are opposite to one another, providing 'blunt ends' at the scission point.

$$--- \text{GAGGCCCG} --- \quad \xrightarrow{\text{Hae III}} \quad -- \text{GAGG} \qquad \text{CCCG}$$

$$--- \text{CTCCGGGC} --- \qquad\qquad \text{CTCC} \qquad \text{GGGC}$$

The first type of cleavage creates short sequences of linear DNA at the ends of the ring-opened plasmid. One of these linear strands can now recombine with a DNA sequence, for example c-DNA, which contains the required information for the synthesis of the desired protein; it is described as foreign DNA. Recyclisation by means of a ligase provides a plasmid which has been modified by recombination, and in this way the newly introduced foreign-DNA segments are incorpor-ated into the bacterial cell and thus can play an active part.

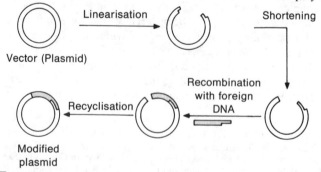

Vector (Plasmid)   Linearisation   Shortening

Recombination with foreign DNA

Recyclisation

Modified plasmid

---

[1] See *C. Kessler*, Restriktionsenzyme, Chemie in unserer Zeit *22*, 37 (1988).

A precise knowledge of the nucleotide sequence of the plasmid which is to be used, and of the sites at which the restriction endonucleases attack, is essential for the planned utilisation of this method. The diagram in Fig. 134 shows the large number of sites at which cleavage may be carried out, including those cleavages brought about by the two enzymes just discussed, which in this case differ not only in the nature of the cuts they produce but also in the numbers of them. The previously mentioned virus SV 40 can also act as a vector.

In contrast to the *transduction* of DNA by bacteriophages, which occurs by injection, the transfer of protein-free DNA in bacterial cells is known as *transfection*, and the infiltration of plasmids into bacterial cells as *transformation*. In these there is no specific mechanism for penetrating the cell walls and their permeability must be increased by chemical means if an efficient transformation is to be achieved; dilute calcium chloride solution, for example, is effective for this purpose. Cells which are enabled to undergo transformation are described as *competent cells*.

Prescribed sequences can be cut out of any DNA with the help of restriction endonucleases, and after inclusion in a plasmid their genetic properties can be examined. It is also now possible to find out the primary structure of polypeptides and proteins indirectly in this way. A suitable sequence of bases is synthesised and translated into the amino acid sequence by translation; this is a procedure which is of increasing use in synthesis.[1] Unnatural amino acids can also be built into proteins in this way by suitable modification of the m-RNA.[2]

9.6.2 *Cloning*

By the use of competent cells, the transformation yield is in the ppm region, since only a portion of the plasmids are involved. The next problem for gene technology is therefore to detect the transformed cells and to propagate them; this is possible by means of *cloning*. A clone is a population of cells derived from a single ancestor cell. All the cells belonging to a clone must have the ability to produce protein which is foreign to bacteria and whose blue prints have been inserted by transformation into bacterial cells. The efficiency of the process depends upon the fact that under favourable conditions bacteria divide themselves every 20 minutes, so that in 24 hours it is possible to obtain about 10^{20} descendants of one cell. Numerous methods are available for recognising the clones which have arisen from the successfully transformed cells but they are only mentioned briefly here.

For example, use can be made of the resistance genes found in the plasmid pBR 322, whose form is shown in Fig. 134. If the foreign DNA is inserted in the region of the tetracycline-resistant gene, then, after transformation, the bacterium is only resistant to ampicillin; if it is cultivated in the presence of ampicillin then the only cells to survive are those which contain the plasmid. If this culture is then treated with tetracycline, those cells which contain the modified plasmid no longer grow since they are no longer resistant to this bacteriostatic antibiotic. However the cells which contain unmodified plasmid are resistant to tetracycline and continues to grow; they are destroyed by adding *cycloserine*, so that only cells which contain foreign DNA now survive. If these transgenic cells are isolated and cul-

---

[1] See *M. J. Bishop* et al. (Ed.), Nucleic Acid and Protein Sequence Analysis (I.R.L. Press, Oxford 1987).
[2] See *P. G. Schultz* et al. A General Method for Site-Specific Incorporation of Unnatural Amino Acids into Proteins, Science *244*, 182 (1989).

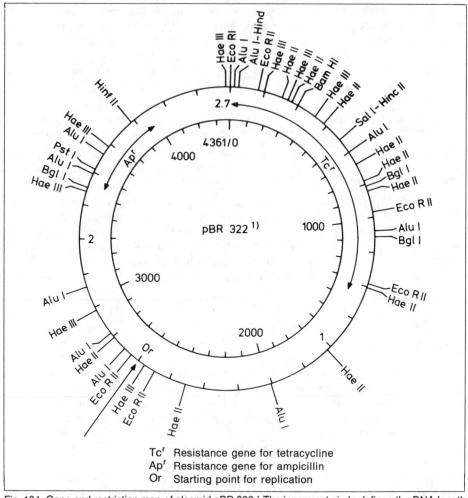

Fig. 134. Gene and restriction map of plasmid pBR 322.[1] The innermost circle defines the DNA length in terms of base pairs. The numbers on the outer ring signify the relative molecular masses in units of 10^6. The cleavage sites for a number of restriction endonucleases are also shown.

tivated further, clones are obtained from which the bacterial foreign protein can be obtained. *Radioimmunoassay* (see p.847) can be used as a check at this stage of the process.

9.6.2.1 Mapping of human genes has begun, making use of morphological methods, but methods based on gene technology are of ever-growing importance. They depend upon restriction mapping (Fig. 134) and on the building of DNA fragments into plasmids. It is possible to accommodate DNA fragments up to 45 kbp into *cosmids*, which, in addition to the plasmid structure, also contain the packing information of phages, the so-called cos-sites; in yeast, artificial chromosomes (YAC) fragments up to 100 kbp can be accommodated. Such plasmids have been cloned and stored in gene banks. In order to combine the necessarily fragmented information-carriers it is necessary to have reliable and generally accepted markers; the following methods have been proposed for getting them.[2] Short

---

[1] Plasmids are indicated by a small p followed by two capital letters (usually the initial letters of the constructor) and a code number.
[2] See *M. Olson* et al., A Common Language for Physical Mapping of the Human Genome, Science **245**, 1434 (1989).

DNA sequences, which only occur once in the genome and which can be reproduced at any time by means of a *polymerase chain reaction* (PCR, see p.874), have been suggested as the markers. Such 200–500 bp long labelled sequences (STS = sequence tagged site) can be recognised as follows. Two partial sequences of about 20 nucleotides (primer) from the beginning and from the end of the STS were synthesised and tested with a DNA probe to see whether they could act as templates for a synthesis of the required STS by means of a polymerase chain reaction in the presence of the two characteristic sequences. The investigation of the sequence between the markers would then put further development of the mapping onto a secure foundation.

9.6.2.2 Monoclonal Antibodies.[1] Fusion of (hybridising) a cancer cell with an antibody-producing cell (*e.g.* a spleen cell) can produce non-toxic cells (hybridomes) which multiply in a nutrient medium and can be cloned (hybridome technique, *Köhler, Milstein*). The *monoclonal antibodies* produced from selected clones are of value not only for investigating the structure of antibody molecules (see p.846 *f.*), but also in radioimmuno assays and for the purification of natural products on a large scale. They are also gaining increasing importance as catalysts in organic synthesis.[2]

9.6.3 Biosynthesis of Hormones

Somatostatin, which is made up from fourteen amino-acids, was the first hormone to be prepared by biosynthesis, from the following synthetic gene fragment.[3] This made use of the synthetic plan[4] worked out for the synthesis of the structure gene for the hormone angiotensin II.

Somatostatin:

```
a    Start   Ala Gly Cys Lys Asn  Phe Phe Trp Lys  Thr Phe Thr Ser Cys  Stop Stop
Eco RI                                                                              Bam HI
    ─(A)─    ─ ─    ─(B)─    ─ ─   ─(C)─    ─ ─   ─(D)─
5'AATTCATGGCTGGTTGTAAGAACTTCTTTTGGAAGACTTCACTTCGTGTTGATAG
   GTACCGACCAACATTCTTGAAGAAACCTTCTGAAAGTGAAGCACAACTATCCTAG 5'
   ───(E)───    ─ ─   ───(F)───    ─ ─   ───(G)───    ─ ─   ───(H)───
```

The oligonucleotides (A), (B) and (E) have been joined by means of the 'sticky-end' method; it can now be seen how this process can be applied to the ends. Of the sixteen underlined base triplets of the coding strand (see p.863 *f.*), fourteen are correlated with the component amino acids of somatostatin. They can be allotted to the codons specified in Table 20 if it is borne in mind that, in the DNA, T occurs at the position which is occupied by U in RNA. Thus, in the coding strand of somatostatin, the triplet GCT corresponds to the codon CGU in Table 20, and ATG to the codon AUG for methionine. Triplet ATG is a constituent of the necessary starting signal for commencing protein synthesis with methionine, which is located at the 5'-end of the coding strand. Since somatostatin contains no other methionine building block, it can be selectively removed by the action of cyanogen bromide (see p.833 *f.*) after the biosynthesis has been completed.

---

[1] See *C. Milstein*, From the Structure of Antibodies to the Diversification of the Immune Response, Angew. Chem. Int. Ed. Engl. *24*, 816 (1985).
[2] See *A. D. Napper* et al., A Stereospecific Cyclisation Catalyzed by an Antibody, Science *237*, 1041 (1987).
[3] See *K. Itakura* et al., Expression in *Escherichia Coli* of a Chemically Synthesised Gene for the Hormone Somatostatin, Science *198*, 1056 (1977).
[4] See *H. Köster* et al., Hoppe-Seyler's Zeitschrift für Physiologische Chemie *356*, 1585 (1975).

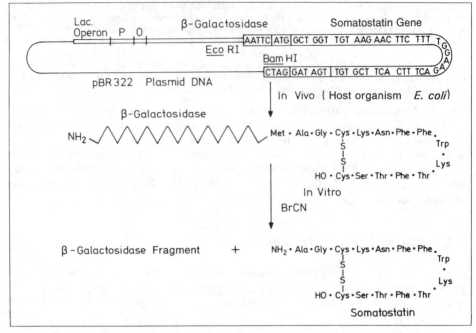

Fig. 135. Scheme for the biosynthesis of somatostatin. P = promoter, O = operator.

The specific sequences of the restriction endonucleases are indicated at the ends of the gene fragments (see Fig. 134). With their help the somatostatin gene can be built into a plasmid as foreign DNA, so that the expression of the genetic information takes place under the control of the lactose operon from *E. coli*. This itself is attached directly to the 5'-end of the somatostatin gene in the plasmid. The expression of somatostatin is shown schematically in Fig. 135. The important role of methionine, which in this instance has no formyl group attached to it, should be noted, since it is included directly after the sequence of β-galactosidase in the fusion peptide from *β-galactosidase* and *somatostatin*. The cloning and selecting of the successfully transformed cells proceeds as described above (Section 9.6.2). The yield of somatostatin amounts to between 0.001 and 0.03% of the total protein content of the bacterial protein, *i.e.* on average about 10 000 somatostatin molecules are produced in each cell.

A similar type of scheme for the biosynthesis of *insulin* makes use of *Escherichia coli* or yeast cells as the host organism. Since 1982 this process has been carried out on a technical scale to provide human insulin (humilin) for use in medicine. In addition, **Tumour Necrosis Factor** (TNF), a protein made up from 157 amino acids, is made industrially in a similar way with the help of the TNF gene isolated from human genome.

A second method for the biosynthesis of appropriate genes avoids chemical synthesis and sets out from m-RNA. In this, the nucleotide sequence in the m-RNA is transferred to DNA by transcription with the help of the enzyme reverse transcriptase mentioned on p.861. Consequently, inversion of the transcription provides first the matrix strand of the DNA, which becomes a complete double strand and then can be expressed in the same way as has been described for synthetic gene fragments.

The two processes can be combined and a gene fragment obtained in the way just described can be attached to a synthetically prepared sequence to form a completed gene. This method is used, for example, in the biosynthesis of human growth hormone. By this means it is possible to exchange almost any amino acid residue in an enzyme by a different one; the choice is limited to proteinogenic amino acids. In this way it is possible to gain important insights into the functioning of enzymes.[1]

The application of gene technology opens up new possibilities for the preparation of valuable organic compounds based on renewable raw materials. It makes great demands on the sense of responsibility of scientists, especially in connection with the risks associated with the cloning of genes. Fortunately, original worries about the safety of cloning bacteria have proved to be groundless.

---

[1] See *J. R. Knowles*, Tinkering with Enzymes: What are we learning? Science *236*, 1252 (1987).

10 Enzymes

In the natural processes of life, as well as vitamins and hormones there is a third class of catalytically active substances known as *enzymes*.[1] In older literature they were sometimes described as *ferments*.

Enzymes are colloidal protein or nucleic acid catalysts of high molecular weight (see p.115), which, as described in Chapter 9, are produced in living organisms. Even when present in minute amounts they affect the rates and the directions of biochemical reactions; it is not uncommon for rates to be increased by factors of 10^{10} to 10^{14} compared to the rates of corresponding uncatalysed reactions. They are responsible for large-scale metabolic processes concerning, for example, proteins, fats and other metabolic products, and exhibit high reaction specificity, substrate specificity and stereoselectivity. Their action is strongly dependent on the temperature and on pH.

Enzyme systems are mixtures or complexes of enzymes which catalyse a reaction sequence. **Isoenzymes** differ in the primary structure of their protein moiety, and catalyse only one reaction.

Any substance which is changed by enzymatic action and forms part of the balanced equation concerned, is described as a *substrate*.

Investigations into the structure of enzymes led to the hypothesis that an enzyme molecule is made up from a colloidal 'carrier' and a *prosthetic group*, which is a non-protein group closely combined with the protein. This concept was put forward especially by *Willstätter* and *v. Euler*. Later on it proved possible to identify a series of *active prosthetic groups* in certain enzymes. These lower molecular weight compounds do not themselves show any enzymatic activity; this appears only if they are bonded to *specific proteins*.

Because the active group is reversibly separable by chemical means from the overall enzyme molecule, the protein part of the enzyme is known as the **apoenzyme** and the prosthetic group as the **coenzyme** or better, as discussed on p.885, as the cosubstrate.

According to older ideas, the coenzyme *alone* should be responsible for the actual reactions (reaction specificity) whilst the apoenzyme served only as a mediator with the substrate in question (substrate specificity). However, more recent work shows that the protein also takes an important part in the process, especially where the stereochemistry of reaction processes is concerned.

The fact that certain *derivatives of vitamins*, especially those of the B-series such as thiamine diphosphate, appear to be constituents of coenzymes is of great biological importance.

---

[1] See The Enzymes, *P. D. Boyer* et al. (Eds), Vols. 1–13, 3rd edn. (Academic Press, New York 1971–1976); Enzymes, *M. Dixon* et al. (Eds), 3rd Edn. (Academic Press, New York 1980); Enzyme Handbook, *T. E. Barman*, (Ed.), Vols. 1–2, Suppl. 1 (Springer-Verlag, Berlin 1969–1974).

Sometimes the action of an enzyme also depends upon the presence of inorganic salts or of metal ions, which catalyse the progress of the enzymatic processes and can be regarded as *inorganic cofactors*. It has turned out that enzymes are also active not only in non-aqueous solvents such as ethyl acetate, but also in supercritical liquid carbon dioxide.

Nomenclature and Classification of Enzymes

Some enzymes, whose activity has been known for a long time, are still known by trivial names, *e.g. pepsin* or *trypsin*, but it has become necessary to set up a form of *systematic nomenclature for enzymes*.[1] This divides enzymes into the following *six main divisions*:

1. Oxidoreductases
2. Transferases
3. Hydrolases

4. Lyases
5. Isomerases
6. Ligases (Synthetases)

A group of enzymes showing a particular type of action is also named after the reaction which they catalyse, by attaching the suffix 'ase', *e.g.* those enzymes which transfer a group from one compound to another are known as transferases.

10.1 Oxidoreductases

Oxidoreductases are involved in the transfer of hydrogen, or of electrons, from one substrate to another. If transfer takes place with the involvement of an inorganic acceptor they are called *dehydrogenases*; if oxygen is directly involved they are described as *oxidases*.

In German text the term dehydrases may be used in place of dehydrogenases.

The action of dehydrogenases in tissues may be observed by means of *redox dyestuffs* such as methylene blue; the dyestuff is converted into a colourless leuco-compound (*Thunberg*). It was in this way, for example, that the enzymatic dehydrogenation of succinic acid to fumaric acid in the absence of air was established:

$$
\begin{array}{c}
H_2C-COOH \\
| \\
H_2C-COOH
\end{array}
+ \text{Methylene blue}
\xrightarrow[\text{dehydrogenase}]{\text{Succinate-}}
\begin{array}{c}
H-C-COOH \\
|| \\
HOOC-C-H
\end{array}
+ \text{Leucomethylene blue}
$$

Above all, pyridine nucleotides and flavin nucleotides and also cell haemins are among the **coenzymes of oxidoreductases** whose constitutions have been established.

10.1.1 Pyridine Nucleotides

Investigations of fermentation dehydrogenases, by *v. Euler* and *Warburg*, led to the elucidation of the constitutions of two coenzymes which were called **codehydrase I** and **codehydrase II.** The former is made up from one molecule of nicotinamide, one molecule of

---

[1] See Enzyme Nomenclature. Recommendations (1964, 1972, 1978) of the IUPAC and the IUB (Elsevier, Amsterdam 1965, 1973; Academic Press, New York 1979).

adenine, two molecules of D-ribose and two molecules of phosphoric acid. It is a *dinucleotide* consisting of adenosine monophosphate (AMP) and a nucleotide of nicotinamide. The bond to the diphosphate is at the 5'-position in both ribose molecules. In codehydrase II the hydroxy group at the 2'-position of the D-ribose attached to adenine is esterified by phosphoric acid.

The former designations of these two coenzymes, diphosphopyridine nucleotide (DPN) and triphosphopyridine nucleotide (TPN), have been replaced in modern nomenclature by the names **nicotinamide adenine dinucleotide (NAD$^{\oplus}$)** and **nicotinamide adenine dinucleotide phosphate (NADP$^{\oplus}$).** They have the following structures:

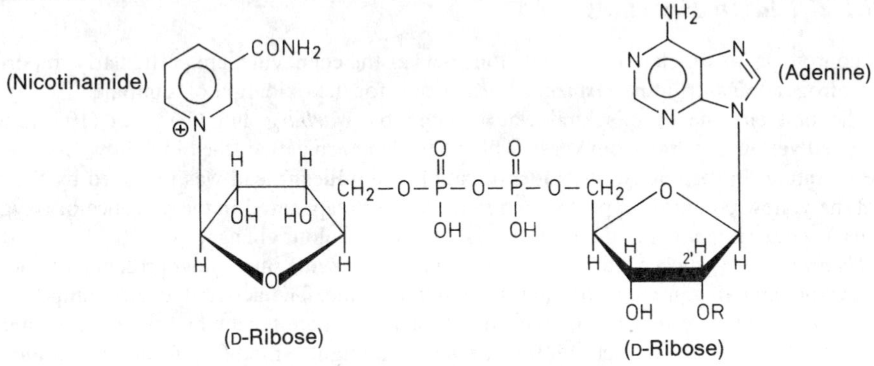

Nicotinamide adenine dinucleotide (NAD $^{\oplus}$) R = H

Nicotinamide adenine dinucleotide phosphate
(NADP $^{\oplus}$) R = P(OH)$_2$
    ‖
    O

The main function of this coenzyme consists of *reversible transfer of hydrogen* in the form of a hydride ion, as a result of which the pyridine ring is reduced and the nitrogen atom loses its positive charge:

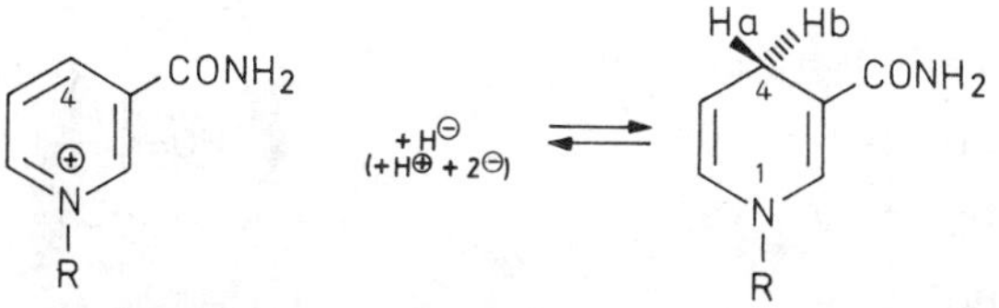

R stands for the remainder of the molecule of the adenine nucleotide. The *hydrogenated coenzymes* are called NADH and NADPH respectively. These *coenzymes* undergo change in the course of their action and are therefore, in strict terms, *not catalysts*. They are thus better described as *cosubstrates*.

The *1,4-dihydropyridine system* shows a broad absorption in its UV spectrum at 340 nm, whilst the pyridine system itself does not absorb in this region. As a result the enzymatic conversion of NAD$^{\oplus}$ into NADH can be followed spectroscopically, the increase in absorption per unit time being proportional to the rate of reaction.

NAD$^{\oplus}$ and NADP$^{\oplus}$ can be combined with various specific proteins to form relatively loosely bound adducts, in which they exist in the linear forms shown above. They are readily regenerated in an equilibrium reaction. These active enzymes bring about a series of *redox processes*; for example, they catalyse the action of *alcohol dehydrogenase* in the following equilibrium:

$$\text{Alcohol} + \text{NAD}^{\oplus} \rightleftharpoons \text{Aldehyde} + \text{NADH} + \text{H}^{\oplus}$$

This reaction occupies a pivotal position in the breakdown of alcohol in organisms.[1]

Conversely, NADH reduces acetaldehyde to ethanol in the fermentation process. Hydrogen transfer from the prochiral centre 4 proceeds *stereoselectively*. Whether H_a or H_b is transferred from NADH depends upon the enzyme involved. It appears from kinetic isotope effects that a tunnelling effect is involved in the reaction.

10.1.2 Flavin Enzymes

Flavoproteins, in which a flavin nucleotide acts as the coenzyme, are particularly important in hydrogen transfer in the *respiratory chain* and for the oxidation of xanthine.

The first enzyme of this kind was isolated by *Warburg* and *Christian* (1932), who obtained **yellow enzyme** from yeast, which has also been called 'the old yellow enzyme' in the literature. In the course of its identification a significant lead was provided by the fact that the yellow colour disappeared on reduction and reappeared in the presence of oxygen. Thus a *redox system* lay at the root of this reversible colour change.

Theorell (1934) succeeded in splitting the yellow enzyme into a flavoprotein part and the coenzyme, and also in reconstituting the active enzyme. The active dyestuff component is the monophosphate ester of riboflavin. It was observed here for the first time that a vitamin (riboflavin or vitamin B_2, see p.801) can act, in a slightly modified form, as a *coenzyme* (*György* and *R. Kuhn*). The phosphate residue is joined to riboflavin at the 5'-position in the D-ribityl moiety. This coenzyme is hence called **riboflavin-5'-phosphate** or **flavin mononucleotide (FMN)**.

This name is not strictly correct since it is *not* a nucleotide, *i.e.* an *N*-glycoside of the ribose phosphate, but rather a derivative of the pentahydroxy sugar alcohol ribitol. It has however passed into use on account of its resemblance to the genuine nucleotides. Its chemical name is *7,8-dimethyl-10-ribitylisoalloxazine-5'-dihydrogenphosphate*.

In these particular oxidoreductases the isoalloxazine ring in riboflavin acts as the reversible *flavin redox system*; the two hydrogen atoms add to nitrogen atoms 1 and 5:

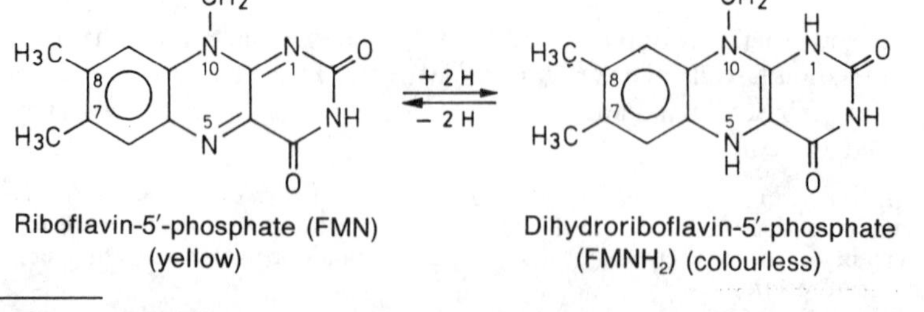

Riboflavin-5'-phosphate (FMN)   Dihydroriboflavin-5'-phosphate
(yellow)   (FMNH₂) (colourless)

---

[1] The ability of the body to break down alcohol is greatly lowered by medicaments such as aspirin, barbiturates or tranquilisers. The action of the alcohol is therefore markedly increased by the influence of such substances.

Riboflavin-5′-phosphate (FMN) can be prepared enzymatically from riboflavin by means of adenosine triphosphate (ATP) (*Kearney* and *England*, 1954):

$$\text{ATP} + \text{Riboflavin} \xrightarrow{\text{Flavokinase}} \text{ADP} + \text{FMN}$$

The yellow enzyme does not itself bring about dehydrogenation, but on the other hand it enables hydrogen from dihydronicotinamide adenine nucleotide phosphate (NADPH) to transfer to molecular oxygen or some other hydrogen acceptor, for example to the cytochrome system in the respiratory chain.

Most flavin enzymes contain **flavin adenine dinucleotide (FAD)** instead of flavin mononucleotide (FMN) as the prosthetic group. In the former, riboflavin-5′-phosphate and adenosine monophosphate are linked through a diphosphate group. FAD has the following structure:

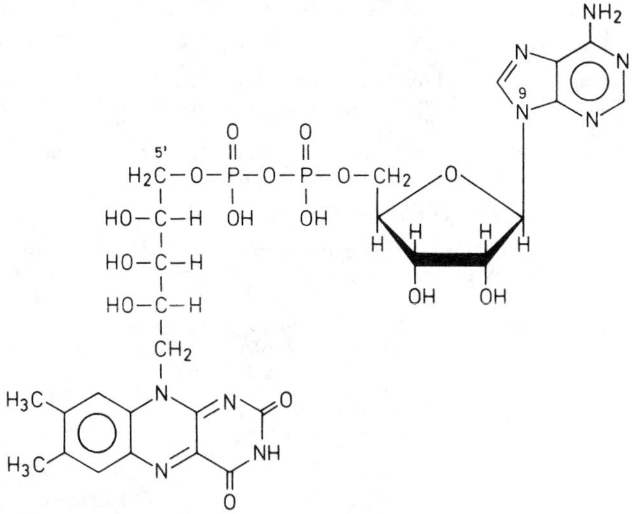

Flavin adenosine dinucleotide (FAD)

In this case also, on reduction there is a colour change from yellow to colourless.

Kornberg (1950) achieved the *biosynthesis* of flavin adenine dinucleotide in the following way, with the aid of an enzyme from yeast:

$$\text{ATP} + \text{FMN} \rightleftharpoons \text{FAD} + \text{Diphosphate}$$

In the presence of certain specific proteins, FAD dehydrogenases bring about very diverse dehydrogenations. Thus *Schardinger* (1902) showed that methylene blue is decolourised by formaldehyde in the presence of fresh milk. In this case, the so-called *Schardinger* enzyme in milk brings about dehydrogenation of the aldehyde hydrate to a carboxylic acid, at the same time reducing the dyestuff:

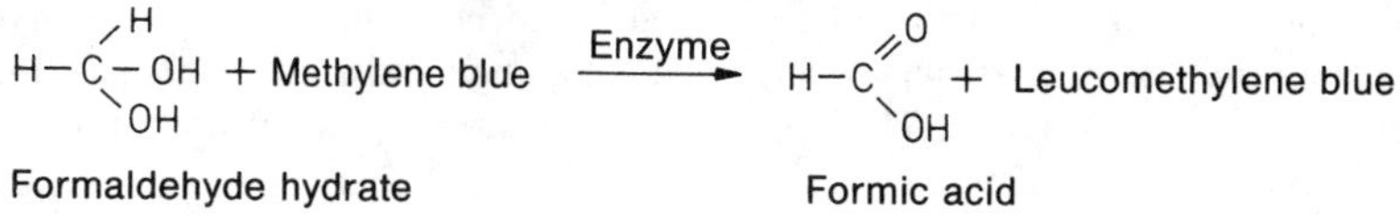

$$\underset{\text{Formaldehyde hydrate}}{\text{H}-\overset{\text{H}}{\underset{\text{OH}}{\text{C}}}-\text{OH}} + \text{Methylene blue} \xrightarrow{\text{Enzyme}} \underset{\text{Formic acid}}{\text{H}-\text{C}\overset{\text{O}}{\underset{\text{OH}}{}}} + \text{Leucomethylene blue}$$

Another FAD enzyme is **xanthineoxidase** which catalyses the oxidation of xanthine to uric acid.

The following enzyme systems concerned in the respiratory chain are closely related to the flavin enzymes.

10.1.3 Cell Haemins

The most important examples of this group of enzymes are the **cytochromes**, which are found in all animal cells. They have *iron-containing porphyrins* as coenzymes. In 1925 *Keilin* established that, in the redox chain concerned with respiration in cells, at least three enzymes are involved which act consecutively; they were called cytochrome a, b and c, and could be distinguished from one another by means of their absorption bands. Later on, a further enzyme, cytochrome d, was discovered, but so far only **cytochrome c** has been isolated in a pure state. In the step concerned with the oxidation of iron and involving haemin (Fe^{3+}) or haem (Fe^{2+}), its prosthetic group is behaving identically to the blood pigments.

The *biochemical function* of the cytochromes is concerned with the transfer of *electrons* (reduction equivalents) from suitable substrates *via* various intermediates to molecular oxygen. The *respiratory chain* commences with the transfer of hydrogen from the substrate to NAD^{+}, forming NADH, which can give up the hydrogen to FAD. The reduced flavin system ($FADH_2$) is oxidised again by the cytochrome c-Fe^{3+}, and this results in the liberation of two protons:

$$FADH_2 + 2 \text{ Cytochrome c-Fe}^{3+} \rightarrow FAD + 2 \text{ Cytochrome c-Fe}^{2+} + 2H^{+}$$

The enzyme chain ends with the reoxidation of cytochrome c-Fe^{2+} by *cytochrome oxidase*, which is identical to '*Warburg's* respiratory enzyme'. The electrons thereby released then reduce oxygen absorbed from the cells to O^{2-}, which combines with two H^{+} to give water. The individual reaction steps of the cell respiration can be illustrated in simple form by the following *redox chain* (U = ubiquinone, U^{2-} = ubihydroquinone dianion) (Fig. 136).

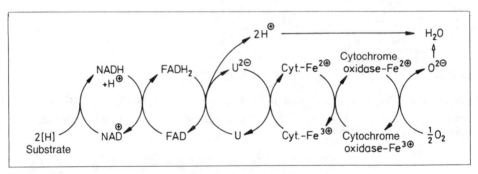

Fig. 136. Redox system sequence in the respiratory chain.

The energy which is released in the successive redox steps is stored in the adenosine triphosphate. This process is known as oxidative or *respiratory chain* phosphorylation.

Among other haemoproteins are the **catalases** which catalytically decompose hydrogen peroxide, which acts as poison to cells, into water and molecular oxygen. **Peroxidases**, which act similarly, like catalases occur in almost all animal and plant cell tissue; they also have haem as a prosthetic group. They catalyse the oxidation of a substrate (Don. H_2) by hydrogen peroxide (see p.822).

$$H_2O_2 + H_2O_2 \xrightarrow{\text{Catalases}} O_2 + 2\,H_2O \qquad \text{Don. } H_2 + H_2O_2 \xrightarrow{\text{Peroxidases}} \text{Don.} + 2\,H_2O$$

10.2 Transferases

Transferases transfer a molecular group from one substrate to another. The names of these enzymes are derived from those of the groups which are transferred together with the suffix 'transferase'. For example, an enzyme which transfers a methyl group is known as a methyltransferase. Transferases are classified into the following groups, depending upon the nature of the group which is transferred.

10.2.1 Phosphotransferases

The largest group of this type of enzyme is that of the phosphotransferases, which transport a phosphate or a diphosphate from one molecule to another. Adenosine triphosphate (ATP), adenosine diphosphate (ADP) and adenosine monophosphate (AMP; INN: *Adenosine phosphate*) function as **coenzymes.**

Adenosine triphosphate provides a source of energy for most metabolic processes which take place in living organisms. It has a high group transfer potential and is the phosphate donor in almost all phosphorylations.

Adenosine triphosphate (ATP)   Cyclic AMP (cAMP)

Transphosphorylation plays an important role in, for example, alcoholic fermentation and in the respiratory chain, and ATP can transfer phosphoric acid to alcoholic hydroxy groups, to carboxyl groups and to amide groups. Such transfer reactions are catalysed by specific enzymes, which are known in general as *kinases*. The action of *adenylate cyclase* on ATP brings about loss of diphosphate (P)—O— (P)) and the formation of cyclic adenosine monophosphate (cAMP). The latter is responsible for the activity of many hormones and also of some medications which pass through impermeable cell walls, and hence has been described as a 'second messenger'.[1]

---

[1] See *H. P. Baer* et al. (Eds), Physiological and Regulatory Functions of Adenosine and Adenine Nucleotides (Raven Press, New York 1979).

The chemical shifts of ^{31}P-NMR signals are dependent upon the chemical environment and the α-, β- and γ-positions in ATP can be assigned unequivocally, and can be distinguished from inorganic phosphate. With the introduction of horizontal superconducting magnets it has become possible to trace metabolic processes involving phosphorylation in human beings directly, for example its efficiency in muscles.[1]

10.2.2 Acyltransferases

Acyltransferases are involved in transferring acyl groups from a donor to an acceptor molecule (see p.306). The most important of these enzymes is *coenzyme A*, which is effective for all biological acylations.

Lipmann and *Snell* have shown that **coenzyme A** is made up from the following building blocks: adenine, D-ribose, pantothenic acid, 2-aminoethanethiol (cysteamine) and three molecules of phosphoric acid. It is interesting that once again in this coenzyme there is a vitamin of the B-group, pantothenic acid. Coenzyme A has the structure shown below.

The HS-group of cysteamine, which has been shown to be the active group of coenzyme A, can react with carboxylic acids to give thiolesters (R—CO—S—CoA); this gives rise to a very *energy-rich S*-acyl bond. When it is hydrolysed about 34 kJ/mol is liberated, which is about the same amount of energy that is evolved in the decomposition of adenosine triphosphate.

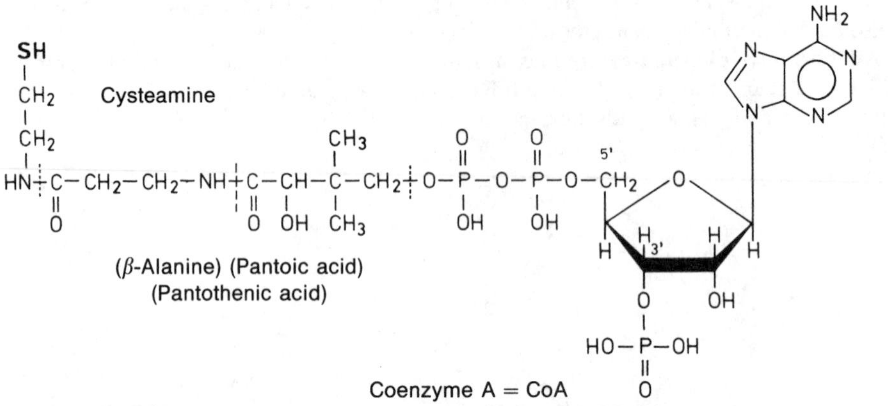

Coenzyme A = CoA

Coenzyme A also plays a key role in carbohydrate metabolism (glycolysis) and the metabolism of fats, in the conversion of carbohydrates into fatty acids, as well as in the biosynthesis of terpenes and steroids and of natural products which proceed *via* polyketides.[2] Acetylcoenzyme A ('activated acetic acid', H_3C—CO—SCoA) acts here as the carrier of the H_3CCO-group.

10.2.3 Aminotransferases

Aminotransferases are of great importance as intermediates in amino acid metabolism. They occur in almost all animal organs. They bring about reciprocal amination and deami-

---

[1] See *G. K. Radda*, The Use of NMR Spectroscopy for the Understanding of Disease, Science *233*, 640 (1986).

[2] See *K. B. G. Torssell*, The Polyketide Pathway in Natural Product Chemistry, p.107 (Wiley, Chichester 1983).

nation between keto acids (especially pyruvic acid) and glutamic acid, without any evolution of free ammonia. This type of reversible reaction is known as **transamination**[1] (*Braunstein*, 1937), *e.g.*

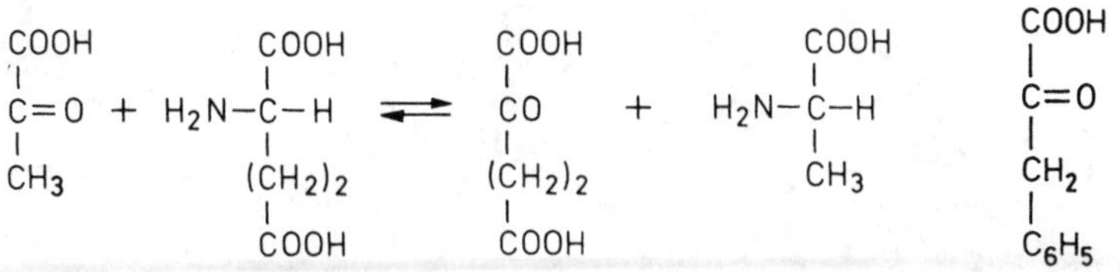

$$
\underset{\substack{\text{Pyruvic}\\\text{acid}}}{\overset{\substack{\text{COOH}\\|\\\text{C}=\text{O}\\|\\\text{CH}_3}}{}} + \underset{\substack{\text{L}(+)\text{-Glutamic}\\\text{acid}}}{\overset{\substack{\text{COOH}\\|\\\text{H}_2\text{N}-\text{C}-\text{H}\\|\\(\text{CH}_2)_2\\|\\\text{COOH}}}{}} \rightleftharpoons \underset{\substack{\alpha\text{-Ketoglutaric}\\\text{acid}}}{\overset{\substack{\text{COOH}\\|\\\text{CO}\\|\\(\text{CH}_2)_2\\|\\\text{COOH}}}{}} + \underset{\substack{\text{L}(+)\text{-Alanine}}}{\overset{\substack{\text{COOH}\\|\\\text{H}_2\text{N}-\text{C}-\text{H}\\|\\\text{CH}_3}}{}} \qquad \underset{\substack{\text{Phenylpyruvic}\\\text{acid}}}{\overset{\substack{\text{COOH}\\|\\\text{C}=\text{O}\\|\\\text{CH}_2\\|\\\text{C}_6\text{H}_5}}{}}
$$

α-Ketoglutaric acid is an intermediate in the citric acid cycle.

Pyridoxal phosphate (see vitamin B$_6$) acts as coenzyme for aminotransferases:

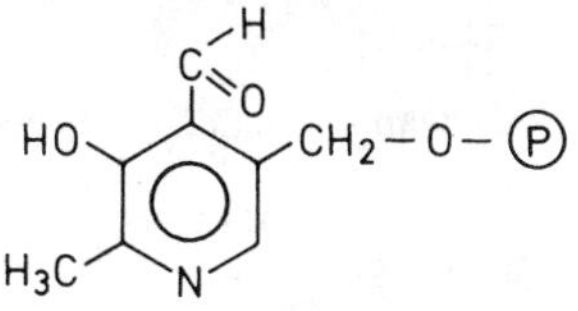

Pyridoxal phosphate P = PO$_3^{2\ominus}$

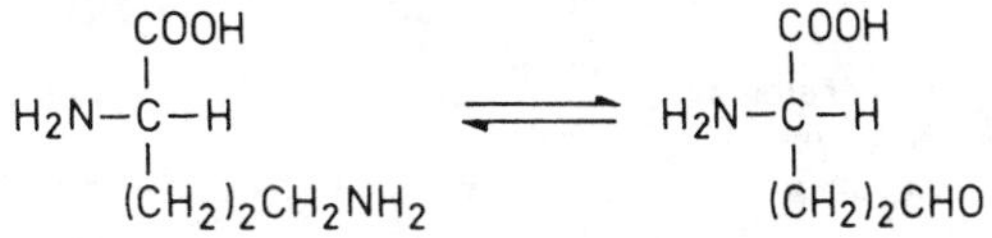

$$
\underset{\text{L-Ornithine}}{\overset{\substack{\text{COOH}\\|\\\text{H}_2\text{N}-\text{C}-\text{H}\\|\\(\text{CH}_2)_2\text{CH}_2\text{NH}_2}}{}} \rightleftharpoons \underset{\substack{\text{L-Glutamic acid}\\\text{semialdehyde}}}{\overset{\substack{\text{COOH}\\|\\\text{H}_2\text{N}-\text{C}-\text{H}\\|\\(\text{CH}_2)_2\text{CHO}}}{}}
$$

Transamination plays a particular role in the biochemical build-up and breakdown of proteins. L-Glutamic acid acts as the provider of nitrogen in the synthesis of amino acids from keto acids in animal organisms. L-Aspartic acid performs a similar function in the preparation of amino acids in plants. Transamination reactions of *ornithine* mainly involve the δ-amino group and in such δ-*transamination* the ornithine is converted into L-*glutamic acid semialdehyde* (L-2-amino-5-oxopentanoic acid, glutaminal). Transamination proceeds very rapidly; for example, the turnover number for the conversion of pyruvic acid into L(+)-alanine amounts to 66 000 min^{-1}, *i.e.* 66 000 molecules are converted every minute. Transamination of phenylalanine provides *phenylpyruvic acid*, which is decarboxylated to *phenylacetaldehyde*. This is a constituent of the *aroma of honey* and arises in the metabolic processes of honey bees by the successive action of phenylalanine aminotransferase and phenylpyruvate decarboxylase.

10.2.4 Methyl- and Formyl-transferases

Of the **methyltransferases** the most important donor is **S-adenosylmethionine** ('active methyl') which is formed from the sulphur-containing amino acid methionine and adenosine triphosphate (ATP) (*Cantoni*). It has the following structure:

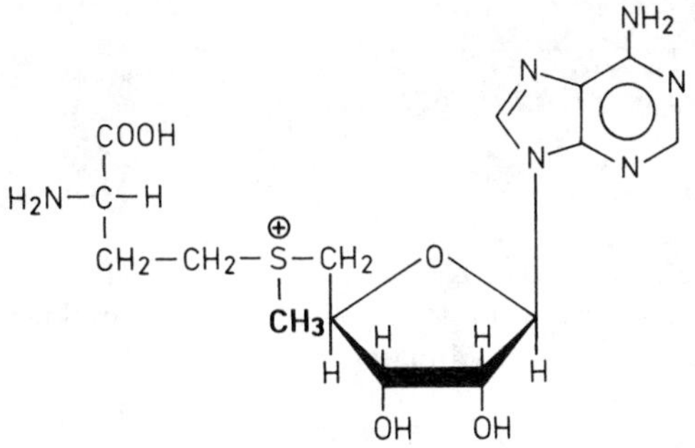

S-Adenosylmethionine

[1]See *K. Heyns*, Die Bedeutung der Transaminierung in Stoffwechsel (The Importance of Transamination in Metabolism), Angew. Chem. *61*, 474 (1949).

Since the methyl group in activated *S*-adenosylmethionine is attached to a positively charged sulphur atom it can easily be detached as a *carbenium ion equivalent* and transferred to an atom having a lone pair of electrons, *e.g.* nitrogen. An example of *enzymatic transmethylation* is provided by the conversion of *guanidinoacetic acid* into *creatine*:

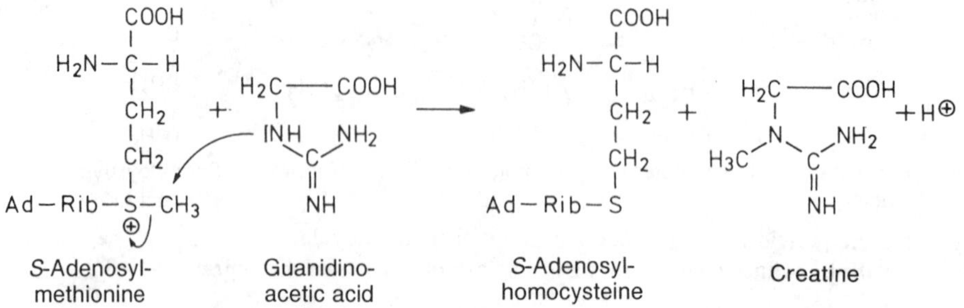

| *S*-Adenosyl-methionine | Guanidino-acetic acid | *S*-Adenosyl-homocysteine | Creatine |

In an analogous reaction which takes place in living organisms, *2-aminoethanol* (colamine) is converted by threefold methylation into *choline*.

A cyclopropane ring occurs in numerous natural products and these arise from the transfer of a *methylene group* from the *methyl group* of *S*-adenosylmethionine to a carbon–carbon double bond.[1] This is the way in which dihydrosterculic acid is formed from oleic acid.

Formyltransferases, which transfer the formyl group as a C_1-unit, have as their prosthetic group N-10-formyl-5,6,7,8-tetrahydrofolic acid (**coenzyme F**). In living organisms it arises from folic acid by the action of ATP. It has the following structure:

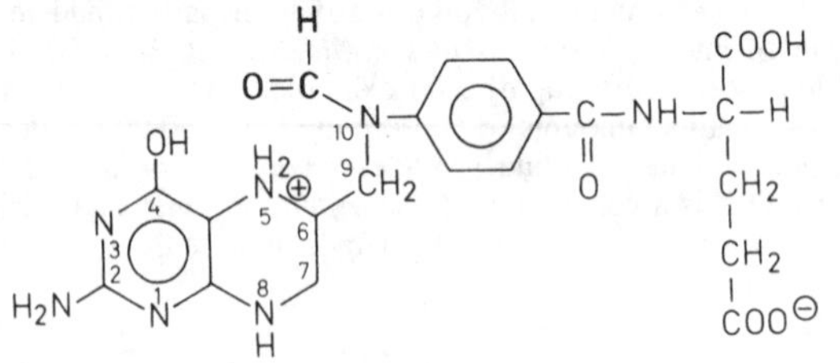

These enzymes are especially active in the *biosynthesis of purines* where formyl groups contribute to the formation of both rings. The formyl group, which can act as either an 'active formic acid' or as an 'active formaldehyde' is provided by, for example, the hydroxymethyl group of L-serine.

10.3 Hydrolases

The hydrolases are subdivided into C–O hydrolases and C–N hydrolases, according to which of these types of bonding is involved. In present literature they are known as ester hydrolases, glycoside hydrolases and peptide hydrolases.

---

[1] See *J. H. Law*, Biosynthesis of Cyclopropane Rings, *Acc. Chem. Res.*, *4*, 199 (1971).

10.3.1 Ester Hydrolases

Depending upon the type of ester which is hydrolysed these enzymes, which were formerly called esterases, are described as, for example, carboxylic ester hydrolases (lipases), phosphoric ester hydrolases (phosphatases) or sulphohydrolases (sulphatases).

Lipases are fat-splitting enzymes which are found in warm-blooded animals especially in the pancreas, the intestinal wall and the liver. They hydrolyse fats to free fatty acids. This reaction is also catalysed by other enzymes such as *porcine liver esterase*. Lipases are widely distributed in microorganisms and are obtained technically from this source. Such enzymes make it possible to carry out enantioselective partial hydrolysis of prochiral dicarboxylic esters to provide chiral synthetic building blocks, *e.g.*

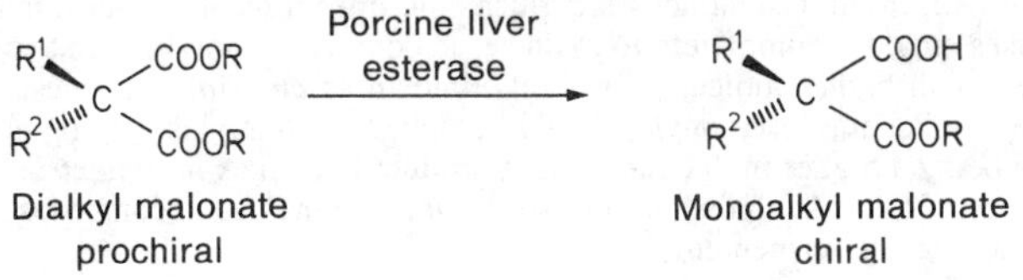

Dialkyl malonate
prochiral

Monoalkyl malonate
chiral

The enantiomeric excess (ee) exceeds 86% if R^1 and R^2 are of greatly different sizes.[1]

Phosphatases are present in almost all living cells and cleave ester-linked phosphoric acid from nucleotides, phosphatides and sugar phosphates. Their activity is strongly dependent on pH.

Sulphatases are found in all animal tissues, especially in the kidneys. They are able to hydrolyse naturally occurring sulphate esters such as phenol sulphates or indoxyl sulphates, chondroitin sulphate and sinigrin.

10.3.2 Glycoside Hydrolases (Glycosidases)

Depending upon their particular substrate specificity, these enzymes bring about the hydrolysis of glycosidic bonds in mono-, oligo- and poly-saccharides as well as N- and S-glycosidic links. The chemical *stereospecificity* of an enzyme was observed for the first time with glycoside hydrolases in that they broke either only α- or only β-glycosidic bonds. The following enzymes are among the most important glycoside hydrolases (glycosidases):

α-Glucosidase (maltase) is found in the pancreas and the intestinal juices and also in barley malt. In particular it breaks the α-glucosidic link in maltose.

β-Glucosidase occurs in, for example, *emulsin*, a mixture of enzymes obtained from bitter almonds. It hydrolyses β-glucosides such as β-methylglucoside, amygdalin and cellobiose.

β-Galactosidase (lactase) is obtained from the intestinal juices of young animals or from certain yeasts. Especially it attacks the β-galactosidic linkage in lactose, forming glucose and galactose.

β-Fructosidase (saccharase, invertase) above all hydrolyses sucrose to glucose and fructose, but also hydrolyses other *furanoid* β-fructosides, for example raffinose. This enzyme also occurs in yeast and in the intestinal juices.

---

[1] See *M. Schneider* et al., Angew. Chem. Int. Ed. Engl. *23*, 66 (1984).

β-Glucuronidases are widely distributed in both animals and plants and cleave glucuronides, including mucoids.

Among the glycoside hydrolases which hydrolyse polysaccharides, ones of particular importance are amylases, cellulase and inulinase.

Amylases break down starch and glycogen. A distinction must be made between α-*amylase*, which is widely distributed, but is found especially in the salivary and pancreatic juices, and β-*amylase*, which occurs in germinating seeds (barley malt). The mixture of α- and β-amylase which is found in barley malt was formerly known as *diastase*. Both amylases have been isolated in crystalline form (*K. H. Meyer*) and have very different actions on starch molecules.

Whilst β-amylase decomposes starch starting from the ends of the molecule and attacking only the second link from the end, α-amylase attacks the glucosidic links almost indiscriminately. As a result β-amylase decomposes amylose completely to maltose; in contrast, α-amylase first splits it irregularly into larger fragments with higher molecular weights, known as *dextrins*, and these are then further degraded to maltose. Unlike amylose, *amylopectin* is attacked by β-amylase giving only up to 60% of maltose, the residue being a higher molecular weight product known as *limit dextrin*. The α(1,6)-glycosidic branching sites of the side chains are not split. Only α-amylase decomposes the limit dextrin further to sugars which can be fermented.

10.3.3 Peptide Hydrolases

The breakdown of proteins is brought about by proteolytic enzymes (proteases, C—N hydrolases) in the stomach and intestinal tract. Depending upon the substrate specificity of the enzymes involved, this begins in the interior of the polypeptide chain and results in the formation of smaller peptide fragments, which can then be broken down further, starting from the ends of the peptide chains. The first group of enzymes were originally known as *proteinases* and the second as *peptidases*. Nowadays the enzymes which cleave high molecular weight proteins at specific places, depending upon the amino acid sequence, in the interior of the peptide chain are called *peptide–peptide hydrolases*. Among these are the *digestive enzymes*, namely **pepsin** in the mucous membrane of the stomach, **trypsin** and INN: *Chymotrypsin* from the pancreas, the enzyme complex *cathepsin*, which has intercellular action in the organs and tissues of animals, the rennet enzyme *rennin* from calves' stomachs, and *papain* from the sap of papaya trees (*Carica papaya*).

Of the enzymes which also occur in the stomach and intestinal tract and which act to break down oligo- and poly-peptides, a distinction is made between α-amino-peptide-amino-acid hydrolases and α-carboxy-peptide-amino-acid hydrolases, which, respectively, attack the peptide chain at either the *N*- or the *C*-terminal. There are also dipeptide hydrolases which hydrolyse only dipeptides. Of all these enzymes the mode of action of chymotrypsin has been the most investigated.[1]

The proteinase *renin*, which is obtained from the kidneys, is concerned with hormonal regulation. It splits off *angiotensin I* from the plasma protein *angiotensinogen*. This is the first step of the angiotensin cascade; if this is stopped at various stages it leads to lowering of the blood pressure (see p.708, 837).

---

[1] See *D. M. Blow*, Structure and Mechanism of Chymotrypsin, Acc. Chem. Res. **9**, 145 (1976).

10.4 Lyases

Lyases catalyse the breaking of various bonds. They are classified according to the type of bond whose cleavage they cause, and are divided into C—C, C—O, C—N, C—S and C—halogen lyases. Only the first two of these are considered here.

10.4.1 C—C Lyases

One of the most important of this type of enzyme is *pyruvate decarboxylase*. This splits pyruvic acid, which arises as an intermediate in alcoholic fermentation, into acetaldehyde and carbon dioxide (*Neuberg*):

$$H_3C-\underset{\underset{O}{\|}}{C}-COOH \xrightarrow{\text{Pyruvate decarboxylase}} H_3C-C\overset{H}{\underset{O}{\diagdown}} + CO_2$$

10.4.1.1 Thiamine diphosphate (TDP), which was isolated from yeast by *Lohmann* in 1937, was identified as the coenzyme of pyruvate decarboxylase. It acts as follows: first of all, an 'active pyruvate' is formed from the thiamine diphosphate and this undergoes decarboxylation to an 'active acetaldehyde'.[1]

2-(α-Hydroxyethyl)-TDP ('Active acetaldehyde') R = H
2-(α-Hydroxy-α-carboxyethyl)-TDP ('Active pyruvate') R = COOH

In a living organism the carbohydrate breakdown *via* pyruvic acid does not stop at the stage of 'active acetaldehyde' but continues to give 'active acetic acid'. In this process α-lipoic acid and nicotinamide adenine dinucleotide (NAD®) are necessary as dehydrogenating coenzymes.

10.4.1.2 α-Lipoic acid[2] (6,8-thioctic acid, 1,2-dithiolane-3-pentanoic acid) has been found to be a nutrient for certain microorganisms. It contains a cyclic disulphide group, which can be very readily split reductively.

α-Lipoic acid (Lip S₂)   Dihydro-α-lipoic acid [Lip(SH)₂]

[1] See H. *Holzer*, Wirkungsmechanism von Thiaminpyrophosphat (Mode of Action of Thiamine Pyrophosphate), Angew. Chem. *73*, 721 (1961).
[2] See A. L. *Fluharty*, Biochemistry of the Thiol Group in S. *Patai* (Ed.), The Chemistry of the Thiol Group, pp. 590 *ff*. (Wiley, New York 1974).

The free carboxyl group is usually bound to a protein by an amide link involving L-lysine.

α-Lipoic acid can take up hydrogen enzymatically from 'active acetaldehyde', forming acetyl-TDP and dihydro-α-lipoic acid:

$$
\underset{\text{'Active acetaldehyde'}}{\overset{\displaystyle H}{\underset{\displaystyle OH}{H_3C-C-TDP}}} + LipS_2 \longrightarrow \underset{\text{Acetyl-TDP}}{\overset{\displaystyle O}{H_3C-C-TDP}} + Lip(SH)_2
$$

Acetyl-TDP then transfers its acetyl group to dihydro-α-lipoic acid:

$$
\underset{\displaystyle O}{H_3C-C-TDP} + Lip(SH)_2 \longrightarrow \underset{\displaystyle O}{H_3C-C-S-LipSH} + TDP
$$

The energy-rich 5-acetyl-dihydro-α-lipoic acid then takes part in another acetyl-transfer reaction with coenzyme A to form acetylcoenzyme A and dihydro-α-lipoic acid:

$$
\underset{\displaystyle O}{H_3C-C-S-LipSH} + CoA-SH \longrightarrow \underset{\text{Acetylcoenzyme A}}{\underset{\displaystyle O}{H_3C-C-S-CoA}} + Lip(SH)_2
$$

Oxidation of dihydro-α-lipoic acid to α-lipoic acid is catalysed by $NAD^\oplus$:

$$
Lip(SH)_2 + NAD^\oplus \longrightarrow LipS_2 + NADH + H^\oplus
$$

10.4.1.3   Another C—C lyase is **amino acid decarboxylase**, which occurs in both animal and plant tissues as well as in microorganisms. It brings about *anaerobic decarboxylation* of various amino acids:

$$
\underset{\displaystyle R}{\overset{\displaystyle COOH}{H_2N-C-H}} \longrightarrow R-CH_2-NH_2 + CO_2
$$

Similar amines also arise from the decay of proteins.

Pyridoxal phosphate (*cf.* vitamin B_6), which also plays a role in transamination reactions, serves as the prosthetic group for this enzyme. It takes part in the reaction as shown on p.897.

The exchange of CO_2 for $H^\oplus$ on the α-carbon atom of the amino acid involves an *electrophilic reaction* on an atom which is normally a centre for *nucleophilic attack*. This biochemical process is associated with an *umpolung* of an *acceptor* (*electrophilic*) site into a *donor* (*nucleophilic*) site (note the dipolar limiting structure of the intermediate).[1]

---

[1] See p.376; *D. Seebach* and *D. Enders*, Umpolung of Amine Reactivity, Angew. Chem. Int. Ed. Engl. *14*, 15 (1975).

Pyridoxal phosphate

$\textcircled{P} = PO_3{}^{2\ominus}$

10.4.2 C—O Lyases

The most important enzymes belonging to this group are the hydrolyases which break C—O bonds with elimination of a molecule of water. Such enzymes participate in, for example, alcoholic fermentation, glycolysis and the citric acid cycle.

An example is provided by the *enzymatic dehydration of L-malic acid to fumaric acid* which is brought about by L-*malate hydrolyase (fumarate hydratase)*:

L-Malic acid   Fumaric acid

10.5 Isomerases

Isomerases catalyse intramolecular rearrangements and a distinction can be made between **racemases**, **epimerases** and *cis–trans*-**isomerases**.

The catalytic action of racemases is limited to the conversion of *optically active* compounds, especially amino acids and hydroxy acids, *e.g.* lactic acid, into their racemic forms (see p. 827).

An example of *epimerisation* (see p. 342 *f.*) is the equilibrium reaction which occurs in the photosynthetic cycle between D-xylulose-5-phosphate and D-ribulose-5-phosphate (see p. 901).

Z–E-Rearrangement of maleic acid into fumaric acid can be brought about *enzymatically* with the help of *maleate isomerase* which has been isolated from bacteria.

10.6 Ligases (Synthetases)

In contrast to lyases which break certain types of bonds, ligases are able to make new chemical bonds. An example is provided by *enzymatic carboxylation* which occurs with *biotin* and ATP. **Carboxybiotin** (CO_2-biotin) acts as the prosthetic group for the carboxylase; it has a carboxyl group attached to a nitrogen atom of the imidazolidone ring and the carboxyl group in the side chain is bound by a peptide link to the enzyme involved.[1]

Formation of the carboxybiotin occurs as follows:

Biotin enzyme   Carboxybiotin enzyme

The active CO_2-group of this enzyme can, for example, carboxylate acetyl-CoA to form malonyl-CoA:

Acetyl-CoA (CoASAc)   Malonyl-CoA

In the biosynthesis of fatty acids the active methylene group of malonyl-CoA reacts readily with acyl-CoA and is decarboxylated to a β-keto-acid.

---

[1] See *F. Lynen* et al., Die biochemische Funktion des Biotins, Angew. Chem. *71*, 491 (1959); *F. Lynen*, Der Weg von der 'aktiven Essigsäure' zu den Terpenen und Fettsäuren (The 'Active Acetic Acid' Route to Terpenes and Fatty Acids), Angew. Chem. *77*, 929 (1965).

11 Metabolic Processes

Because carbohydrate metabolism[1] plays such an important role in all organisms, it will be used to exemplify metabolic processes. *Carbon dioxide* and *water* are the principal end-products of the combined reaction paths of metabolic processes. Carbon dioxide results largely from decarboxylation reactions in the citric acid cycle, water from the biochemical oxidation in the respiratory chain. Conversely carbohydrates are built up in plants from carbon dioxide which they have taken up in the presence of water. The amount of carbon dioxide which is fixed annually by photosynthesis in plants has been conservatively estimated to be 7.5×10^{11}t (replaceable raw material).

The energy which is liberated in the individual reactions is either converted into heat, *e.g.* in muscular activity, or stored as chemical energy in energy-rich compounds, for example in the diphosphate bonding in adenosine triphosphate (ATP).

11.1 Photosynthetic Cycle[2]

In the assimilation of carbon dioxide, or *photosynthesis*, the radiant energy of light is converted, in the presence of chlorophyll, which is bonded to proteins in the plants, into chemical energy. According to the overall equation for photosynthesis:

$$6\,CO_2 + 6\,H_2O \xrightarrow[\text{(Chlorophyll)}]{h\nu} (CH_2O)_6 + 6\,O_2 \quad \Delta H = 2830\,kJ$$

glucose and molecular *oxygen* are the products obtained from carbon dioxide and water. Unlike plants, animals are not able to manufacture carbohydrates from these simple inor-

---

[1] Within *metabolism* it is possible to differentiate between *catabolism*, involving degradative processes, and *anabolism*, which involves synthetic processes of, for example, muscle proteins.

[2] See *M. Calvin*, Der Weg des Kohlenstoffs in der Photosynthese (The Carbon Pathway in Photosynthesis), Angew. Chem. *74*, 165 (1962); *N. I. Bishop*, Photosyntheses, The Electron-transport System of Green Plants, Ann. Rev. Biochem. *40*, 197 (1971).

ganic compounds. Hence photosynthesis occupies a pivotal role in energy retention in nature.[1]

Blackman (1906) suggested that there were two phases which could be distinguished from one another, one involving a light reaction and the other a dark reaction.

In the first, photochemical, phase (the light reaction) the light energy brings about a photolysis of water, being assisted in this by the assimilation pigment which functions as an energy-transferring agent. In this way, through the redox system of the plastoquinone, two electrons and oxygen are formed. These are not derived from carbon dioxide but from water. This has been confirmed by research using isolated chloroplasts (*Hill*, 1937) and by the use of isotopic oxygen ^{18}O (*Ruben*, 1941).

$$H_2O \xrightarrow[\text{(Chlorophyll)}]{h\nu} 2H^{\oplus} + \tfrac{1}{2}O_2 + 2e^{\ominus}$$

Reduction of carbon dioxide then follows with the help of an enzyme having NADPH as its active grouping. NADPH is itself formed by reaction of the electrons resulting from the action of light and NADP$^{\oplus}$. ATP is formed in the latter reaction (*Vishniac* and *Ochoa*, 1951):

$$2H^{\oplus} + 2e^{\ominus} \xrightarrow{\quad ADP + \textcircled{P} \quad ATP \quad NADP^{\oplus} \quad} NADPH + H^{\oplus}$$

This provides the overall reaction:

$$6CO_2 + 12 NADPH + 12 H^{\oplus} \longrightarrow C_6H_{12}O_6 + 12 NADP^{\oplus} + 6H_2O$$

More detailed studies of the second reaction phase (the dark reaction) have resulted from the work on assimilation by *M. Calvin*, using ^{14}C-labelled carbon dioxide and the unicellular green algae *Chlorella pyrenoidosa*. Even after only short irradiation (a few seconds or minutes) a number of ^{14}C-labelled compounds could be separated by paper chromatography and detected by autoradiography, *e.g. 3-phosphoglyceric acid*,[2] D-*ribulose-1,5-bisphosphate* and D-*sedoheptulose-7-monophosphate*. Among other ^{14}C-labelled compounds detected were phospho-*enol*-pyruvic acid, dihydroxyacetone phosphate, glucose-, mannose- and fructose-phosphates and also malic acid, citric acid, glycine, alanine, aspargic acid, etc. (*Calvin* and *Benson*, 1955).

Tracer methods have shown that it is not the carbon dioxide itself which is reduced, but rather 3-phosphoglyceric acid. This arises from ribulose-1,5-bisphosphate, which is a substrate for fixation of the carbon dioxide, by addition of water. 3-Phosphoglyceric acid is reduced to 3-phosphoglyceraldehyde by the NADPH which is formed in the light reaction; the aldehyde isomerises to dihydroxyacetone phosphate. The action of *aldolase* on these two triose phosphates provides fructose-1,6-bisphosphate and the action of *phosphatase* on the latter converts it into fructose-6-phosphate, which is used

---

[1] There are bacteria which can satisfy their requirements for carbon compounds, starting from carbon dioxide, without recourse to the action of light. For this *chemosynthesis* (*Pfeffer* 1897) or *chemolithotrophy*, hydrogen sulphide, for example, can act as a source of energy:

$$CO_2 + H_2S + O_2 + H_2O \rightarrow [CH_2O] + H_2SO_4$$

One place where this process is involved is in the depths of the ocean.

[2] The reactions described in this section take place in the cells at a pH of about 7. Under these conditions the *carboxylic acids* are present almost entirely in the form of their *anions*; they are formulated here as carboxylic acids purely for the sake of simplicity.

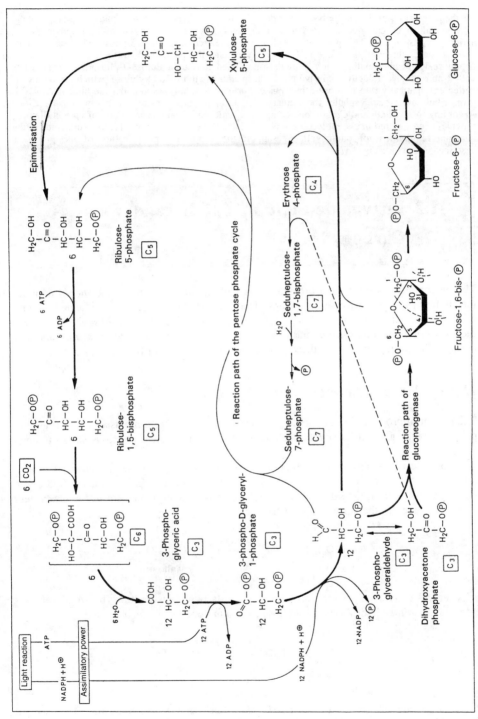

Fig. 137. Photosynthetic cycle

for the preparation of di- and poly-saccharides. On the other hand fructose-6-phosphate and 3-phosphoglyceraldehyde can be converted into erythrose-4-phosphate and xylulose-5-phosphate by the action of *transketolase*. This pentose phosphate is epimerised to ribulose-5-phosphate. The action of *aldolase* on erythrose-4-phosphate and dihydroxyacetone phosphate leads to the formation of sedoheptulose-1,7-bisphosphate, which is dephosphorylated to sedoheptulose-7-phosphate. This reacts with 3-phosphoglyceraldehyde to give the 5-phosphates of ribose and xylulose, which are both converted enzymatically into ribulose-5-phosphate. Further phosphorylation under the influence of ATP gives ribulose-1,5-bisphosphate, the primary CO_2-acceptor, thus closing the photosynthetic cycle. According to *Bassham* and *Calvin*, one complete circuit, based on one molecule of carbon dioxide, provides energy stored in two molecules of NADPH and three molecules of ATP. Measurements of the quantum requirement of photosynthesis support the simplified interpretation of the processes described above (see Fig. 137).[1]

11.2 Enzymatic Build-up and Breakdown of Carbohydrates

The biochemical build-up and breakdown of carbohydrates is one of the most important metabolic processes in living organisms. The first insight into these complicated enzymatic processes came from the *decomposition of carbohydrates in the absence of air, i.e.* their *anaerobic decomposition*, and especially from their *alcoholic fermentation* and *glycolysis*. The starting points for these processes are glucose and glycogen.

11.2.1 Alcoholic Fermentation

The industrial process does not start from pure glucose, but from molasses or from natural products which are rich in starch, such as potatoes or the seeds of various sorts of cereal crops.

Before the actual fermentation process can be carried out, the *starch* must be broken down enzymatically into lower sugars. The enzyme *diastase* brings about the hydrolytic cleavage of starch into the disaccharide *maltose*; this enzyme is present in the *green malt* which is formed in the germination of barley:

$$(C_6H_{10}O_5)_n + \frac{n}{2} H_2O \xrightarrow{\text{(Diastase)}} \frac{n}{2} C_{12}H_{22}O_{11}$$

$$\text{Starch} \qquad\qquad\qquad\qquad\qquad \text{Maltose}$$

After this conversion of the starch into a sugar, a pure yeast culture (brewers' yeast) is added and *maltase* hydrolyses the maltose to two molecules of *glucose*:

$$C_{12}H_{22}O_{11} + H_2O \xrightarrow{\text{(Maltase)}} 2\,C_6H_{12}O_6$$

$$\text{Maltose} \qquad\qquad\qquad\quad \text{Glucose}$$

---

[1] For the pentose phosphate cycle see *D. Matthies*, Biochemische Formelsammlung (Biochemical Formula Collection) (Akademische Verlagsgesellschaft, Wiesbaden 1979).

Fermentation of the hexose to *ethanol* and *carbon dioxide*, which proceeds according to the following overall equation:

$$C_6H_{12}O_6 \rightarrow 2C_2H_5OH + 2CO_2$$

involves a number of individual reactions, each of which results from the action of a specific enzyme. So far a dozen or more enzymes have been isolated from the enzyme complex of yeast, *zymase*, all of which participate in the fermentation process. Ordinary yeast can ferment D-glucose, D-mannose and D-fructose.

Mechanism of Alcoholic Fermentation

Alcoholic fermentation begins with a *prefermentation* stage involving the phosphorylation (esterification) of glucose. In this, glucose first of all reacts with adenosine triphosphate (ATP) to give, together with adenosine diphosphate (ADP), *glucose-1-phosphate* (*Cori*-ester) which rearranges to *glucose-6-phosphate* (*Robison* ester, Glc-6-*p*) (see Fig. 138). This is isomerised by the action of *glucose-6-phosphate isomerase* to *fructose-6-phosphate* (*Neuberg* ester), which, catalysed by phosphofructokinase, reacts with ATP to form *fructose-1,6-bisphosphate* (*Harden-Young* ester).

After this prefermentation stage *fermentation* proper starts, in which the initial fructose-1,6-bisphosphate (a hexose) is split into two triose phosphates, dihydroxyacetone phosphate and glyceraldehyde-3-phosphate, by the action of *aldolase*. The two products are in reversible *equilibrium* with one another. In the next step, glyceraldehyde-3-phosphate reacts with NAD$^\oplus$, in the presence of inorganic phosphate, to form 1,3-diphosphoglyceric acid. This mixed anhydride of phosphoric acid and glyceric acid is dephosphorylated by ADP to 3-phosphoglyceric acid which rearranges to 2-phosphoglyceric acid. Then, in the presence of Mg$^{2\oplus}$ ions, loss of water ensues, providing phospho-enol-pyruvic acid, which is converted by dephosphorylation into pyruvic acid (keto form). *Pyruvate decarboxylase* decarboxylates this to acetaldehyde; at the same time *carbon dioxide*, one of the end-products of alcoholic fermentation, is evolved. This reaction is irreversible and determines the rate of reaction of the whole fermentation process. The final step of fermentation, the hydrogenation of acetaldehyde to *ethanol*, is brought about by NADH, as mentioned above.

The decisive step in this series of reactions is the *reversible splitting and forming* of the bond between carbon atoms 3 and 4 in *fructose-1,6-bisphosphate* by *aldolase*. This involves an *aldol addition* in which the *methylene component* is generated by the interaction of the ε-amino group of a *lysine residue* (H$_2$N—R) which is present at the active site of the enzyme:

Dihydroxyacetone   Schiff's   Enamine
phosphate   base

A *Schiff's* base is formed first, and this rearranges to an enamine (see p.267 *f.*), whose dipolar limiting structure (in the neutral molecule) provides the *methylene component* in a reaction with glyceraldehyde-3-phosphate.

Glyceraldehyde-   Fructose-1,6-bisphosphate
-3-phosphate

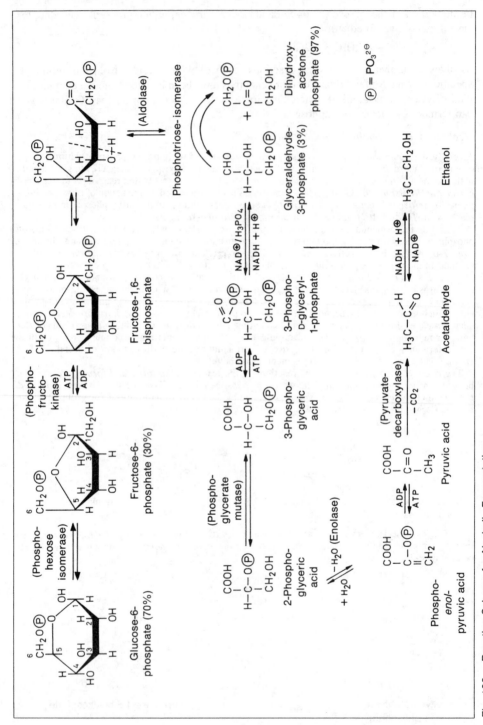

Fig. 138.   Reaction Scheme for Alcoholic Fermentation.

In the aldol reaction two chiral centres (shown with a grey background) are involved, whose correct configuration is brought about by the combined action of the substrate and of the inner surface of the cavity at the active centre of the enzyme. Since this is the surface of a protein it is chiral, and this ensures that, with respect both to its own *chirality* and to that of the *cavity*, the substrate molecule undergoes an unequivocal and unique adsorption and reaction.[1]

This reaction scheme was advanced in principle in 1931 by *Ohle* and later on was experimentally confirmed and supplemented by *Warburg, Meyerhof* and *Lohmann*.

Glycerol (~ 3%) arises as a by-product in alcoholic fermentation under the usual conditions. It results from the reduction of dihydroxyacetone phosphate or glyceraldehyde-3-phosphate and subsequent hydrolysis. It is likely that the reagent involved in this is NADH. In support of this hypothesis it has been established by *Neuberg* that, if the acetaldehyde formed is trapped with sodium hydrogen sulphite, then the percentage of glycerol rises. In these circumstances, more active hydrogen (from NADH) is available for the reduction of the triose phosphate.

11.2.2 Glycolysis (Degradation of Sugars in Living Organisms)

The anaerobic breakdown of glycogen, which provides a carbohydrate reserve in muscles and in superfluous animal tissue, leads to the formation of lactic acid according to the following overall equation:

$$C_6H_{12}O_6 \rightarrow 2CH_3{-}CHOH{-}COOH$$

It commences with the stepwise cleavage of polysaccharide chains accompanied by the take-up of phosphoric acid, brought about by *phosphorylase*. The first step, which results in the formation of glucose-1-phosphate (*Cori*-ester) is also a phosphorolysis reaction and is termed *glycogenolysis*:

$$\text{Glycogen} + H_3PO_4 \underset{H^{\oplus}}{\overset{\text{Phosphorylase}}{\rightleftharpoons}} \text{Glucose-1-phosphate}$$

Further breakdown of glucose-1-phosphate, or *glycolysis*, to pyruvic acid proceeds analogously to alcoholic fermentation (see p. 902 f.). Pyruvic acid is broken down into acetaldehyde and carbon dioxide by the *decarboxylase* in yeast, but muscle contains no comparable enzyme system. Instead the pyruvic acid is hydrogenated to L(+)-*lactic acid* (sarcolactic acid) by NADH which is formed in an intermediate stage of the glycolysis:

$$\begin{array}{ccc} \text{COOH} & & \text{COOH} \\ | & & | \\ \text{O=C} \quad + \text{ NADH} + \text{H}^{\oplus} \rightleftharpoons & \text{HO-C-H} + \text{NAD}^{\oplus} \\ | & & | \\ \text{CH}_3 & & \text{CH}_3 \end{array}$$

Pyruvic acid   L(+)-Lactic acid

Lactic acid fermentation takes place anaerobically in active muscle and provides energy. Conversely, glycogen is reconstituted in the liver from L(+)-lactic acid (*glycogenesis*) by the respiratory process. In the cells of embryos, lactic acid fermentation plays an essential role until birth; cancer cells also meet their energy requirements in this way.

Aerobic carbohydrate degradation takes place mainly as part of the citric acid cycle, which is discussed in the next section.

---

[1] See *J. Rebek*, Model Studies in Molecular Recognition, Science *235*, 1478 (1987).

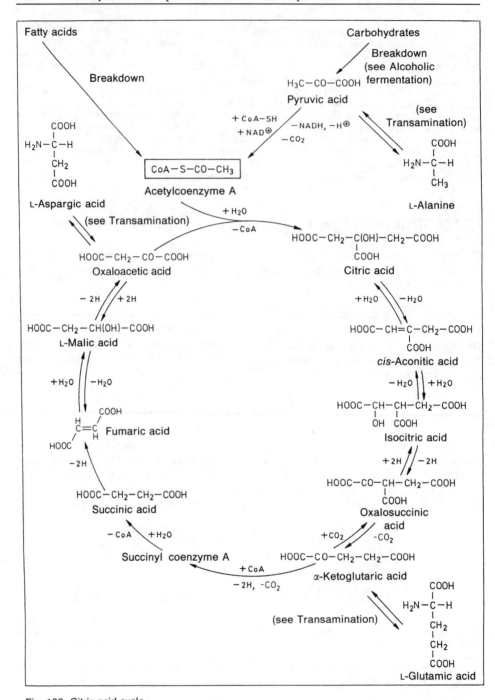

Fig. 139. Citric acid cycle.

11.2.3 Citric Acid Cycle

Pyruvic acid, which arises as the main intermediate in glycolysis, is oxidised completely to *carbon dioxide* and *water* in the *mitochondria* of animals in the presence of atmospheric oxygen (*i.e.* in respiration). Researches by *Knoop, Martius, Krebs* and *D. E. Green* showed that citric acid has an extremely important function in aerobic carbohydrate metabolism. Above all, *pyruvic acid* is not hydrogenated to L-lactic acid, as happens in glycolysis, but is split by *dehydrogenative decarboxylation*, and the remaining acetyl group bonds to coenzyme A to form acetylcoenzyme A (*Lipmann*).

This reaction requires the presence of coenzyme A and the co-factors cocarboxylase, α-lipoic acid and nicotinamide adenine dinucleotide ($NAD^\oplus$) as a hydrogen acceptor. The hydrogen is ultimately transferred by the agency of other enzyme systems to molecular oxygen, a process which is described as the *respiratory chain*.

Acetylcoenzyme A undergoes an aldol-type reaction with *oxalacetic acid* (oxosuccinic acid) to give citric acid. Now, under the influence of a number of enzyme systems, the real *citric acid cycle* commences; the individual steps involved are shown in Fig. 139.

First of all, citric acid loses water and is converted into *cis-aconitic acid*, which adds water to give *isocitric acid*. This is dehydrogenated to *oxalosuccinic acid* (1-oxopropane-1,2,3-tricarboxylic acid) which in turn is decarboxylated to α-*ketoglutaric acid*. Dehydrogenative decarboxylation converts this, *via succinylcoenzyme A*, into *succinic acid*. The latter is dehydrogenated to *fumaric acid* which adds a molecule of water to give L-*malic acid*. L-Malic acid is then dehydrogenated to *oxaloacetic acid*, thereby completing the cycle.

The overall result is that the acetyl group which is derived from pyruvic acid is indirectly burned to *carbon dioxide* and *water*. If one starts again from the first step, where acetylcoenzyme A is used up, in the complete cycle two carbon atoms are eliminated as *carbon dioxide* and eight hydrogens are burnt up by means of the respiratory system to form four molecules of water, two of which re-enter the reaction scheme. In the course of the process, adenosine triphosphate (ATP) is at the same time resynthesised. The whole process takes place with a small gain in energy; in particular, much energy is liberated in burning hydrogen to water.

Originally the citric acid cycle was worked out as a scheme for the oxidation of carbohydrates, but later it transpired that the oxidation of proteins follows, in part, the same process. Thus, for example, L-alanine, L-aspargic acid and L-glutamic acid are very rapidly oxidised in muscle, during the course of which *transamination* results in their conversion into the corresponding keto acids which then enter as intermediates into the citric acid cycle. Conversely, in the synthesis of proteins, keto acids which are present in a living organism are withdrawn from the citric acid cycle, so that their presence is only maintained by a continuing supply of oxaloacetic acid.

Furthermore there is a close relationship between the fatty acid cycle and the citric acid cycle, since all the carbon atoms of the fatty acids are introduced into the cycle in the form of acetylcoenzyme A. Finally it should be mentioned that in addition the terpenes and steroids are synthesised biochemically from coenzyme A.

Thus the citric acid cycle is of vital importance for the metabolism of carbohydrates, fats and proteins in both animals and plants and also in microorganisms.

Appendix: Hazardous Substances and Carcinogenic Materials. Named Reactions and Concepts

Foreword

A glance at the list of dangerous compounds in Hazardous Substances Regulations[1] should provide the reader with some idea of the way in which chemical and legalistic matters are interconnected. One result of this is that the characteristic properties of substances which seem to be unequivocal in chemical terms, are only relative in legal terms.

For example, croton oil, mentioned on p.677, is a hazardous material which is denoted by a skull when it is provided for use for lubricating clocks, but is not when it is used as a purgative for horses. In the latter case it is not subject to the directives concerning hazardous materials, but to medical regulations, and is not marked with the skull symbol. The same chemical properties are regarded in different lights when seen from differing legal viewpoints.

This has the result that legal specifications and regulations may be made which differ from country to country and are themselves always in the course of change and development. For example, amendments to the list from which the following table has been made up have already been introduced and this should always be borne in mind. Further future changes must also be expected and further amendments to the regulations may follow, and it is necessary to keep one's eyes open for them[1].

In the following list of carcinogenic materials the first column gives the relevant page number in the main text. As a rule the statement in the text is derived from the German DFG-MAK list of 1988; reference to its supplementation is made in footnote 4 to the list. The carcinogenic materials mentioned in the list, insofar as they apply to organic compounds, are chemicals which, for the most part, are made synthetically. It should not be concluded from this that most carcinogenic compounds are man-made ones resulting from synthetic processes. In fact the number of carcinogenic natural products is far greater but they are not subject to hazard regulations. The article by *B.N. Ames* cited on p.556 of the main text provides a good insight into this field.

---

[1]See, for example, Authorised and Approved List: Information approved for the classification, packaging and labelling of dangerous substances for supply and conveyance by road, 2nd edtn. (HMSO, London 1988)

A large amount of practical information is provided by *L. Roth*, Krebserzeugende Stoffe (Carcinogenic Substances), 2nd edtn. (Wissenschaftliche Verlagsgesellschaft, Stuttgart 1988).

The criterion for inclusion in the list of named reactions and concepts is the extent to which they are used in everyday speech. Thus *Adams* catalyst and *Raney* nickel are included, but other hydrogenation catalysts such as those of *Paal* and *Skita* or of *Sabatier* and *Sanderens* are not. Further details may be found on the pages whose numbers are given alongside.

These page numbers represent only the most important entries in the text. The subject index lists all the pages in the text on which 'named' concepts are mentioned.

This supplement and the formula index can certainly be added to and developed, and the author would be grateful to all readers for their criticisms and suggestions.

Hazardous Substances: Extract from Addendum VI to Regulations concerning Hazardous Substances

Extracted entries referring only to those substances mentioned in the main text of the book.

In accordance with the rules of the EEC, hazardous materials are designated by the following symbols, black on a yellow background, which must be used in all packaging of recognised hazardous materials.

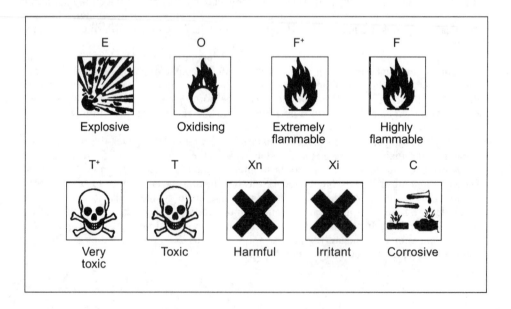

The special hazards associated with different substances are given the following R numbers (R = risk) and combinations of numbers. The risk phrases are taken from Annex III of EEC Directive 67/548/EEC.

1.3 List of Special Hazards (R numbers)

Indication of particular risks

R 1	Explosive when dry
R 2	Risk of explosion by shock, friction, fire or other sources of ignition
R 3	Extreme risk of explosion by shock, friction, fire or other sources of ignition
R 4	Forms very sensitive explosive metallic compounds
R 5	Heating may cause an explosion
R 6	Explosive with or without contact with air
R 7	May cause fire
R 8	Contact with combustible material may cause fire
R 9	Explosive when mixed with combustible material
R10	Flammable
R11	Highly flammable
R12	Extremely flammable
R13	Extremely flammable liquified gas
R14	Reacts violently with water
R15	Contact with water liberates highly flammable gas
R16	Explosive when mixed with oxidising substances
R17	Spontaneously flammable in air
R18	In use, may form flammable/explosive vapour-air mixture
R19	May form explosive peroxides
R20	Harmful by inhalation
R21	Harmful in contact with skin
R22	Harmful if swallowed
R23	Toxic by inhalation
R24	Toxic in contact with skin
R25	Toxic if swallowed
R26	Very toxic by inhalation
R27	Very toxic in contact with skin
R28	Very toxic if swallowed
R29	Contact with water liberates toxic gas
R30	Can become highly flammable in use
R31	Contact with acids liberates toxic gas
R32	Contact with acids liberates very toxic gas
R33	Danger of cumulative effects
R34	Causes burns
R35	Causes severe burns
R36	Irritating to eyes
R37	Irritating to respiratory system
R38	Irritating to skin
R39	Danger of very serious irreversible effects
R40	Possible risk of irreversible effects
R41	Risk of serious damage to eyes
R42	May cause sensitisation by inhalation
R43	May cause sensitisation by skin contact
R44	Risk of explosion if heated under confinement
R45	May cause cancer
R46	May cause heritable genetic damage
R47	May cause birth defects
R48	Danger of serious damage to health by prolonged exposure

Combination of particular risks

R14/15	Reacts violently with water, liberating highly flammable gases
R15/29	Contact with water liberates toxic, highly flammable gas
R20/21	Harmful by inhalation, and in contact with skin
R20/21/22	Harmful by inhalation, in contact with skin and if swallowed
R20/22	Harmful by inhalation and if swallowed
R21/22	Harmful in contact with skin and if swallowed

R23/24	Toxic by inhalation and if swallowed	R26/28	Very toxic by inhalation and if swallowed
R23/24/25	Toxic by inhalation, in contact with skin and if swallowed	R27/28	Very toxic in contact with skin and if swallowed
R23/25	Toxic by inhalation and if swallowed	R36/37	Irritating to eyes and respiratory system
R24/25	Toxic in contact with skin and if swallowed	R36/37/38	Irritating to eyes, respiratory system and skin
R26/27	Very toxic by inhalation and in contact with skin	R36/38	Irritating to eyes and skin
		R37/38	Irritating to respiratory system and skin
R26/27/28	Very toxic by inhalation, in contact with skin and if swallowed	R42/43	May cause sensitisation by inhalation and skin contact

The following S-numbers (S = safety) and their combinations denote the appropriate safety precautions for dealing with hazardous substances. The security phrases are taken from Annex IV of EEC Directive 67/548/EEC.

1.4 Safety Precautions (S-numbers)

Indication of safety precautions required

S 1	Keep locked up	S18	Handle and open container with care
S 2	Keep out of reach of children	S20	When using do not eat or drink
S 3	Keep in a cool place	S21	When using do not smoke
S 4	Keep away from living quarters	S22	Do not breathe dust
S 5	Keep contents under ... (appropriate liquid to be specified by the manufacturer)	S23	Do not breathe gas/fumes/ vapour/spray (appropriate wording to be specified by manufacturer)
S 6	Keep under ... (inert gas to be specified by the manufacturer)	S24	Avoid contact with skin
		S25	Avoid contact with eyes
S 7	Keep container tightly closed	S26	In case of contact with eyes, rinse immediately with plenty of water and seek medical advice
S 8	Keep container dry		
S 9	Keep container in a well ventilated place	S27	Take off immediately all contaminated clothing
S12	Do not keep the container sealed	S28	After contact with skin, wash immediately with plenty of ... (to be specified by the manufacturer)
S13	Keep away from food, drink and animal feeding stuffs		
S14	Keep away from ... (incompatible materials to be indicated by manufacturer)	S29	Do not empty into drains
		S30	Never add water to this product
S15	Keep away from heat	S33	Take precautionary measures against static discharges
S16	Keep away from sources of ignition - No Smoking		
S17	Keep away from combustible material	S34	Avoid shock and friction

		Combination of safety precautions required	
S35	This material and its container must be disposed of in a safe way	S1/2	Keep locked up and out of reach of children
S36	Wear suitable protective clothing	S3/7/9	Keep container tightly closed, in a cool well ventilated place
S37	Wear suitable gloves	S3/9	Keep in a cool, well ventilated place
S38	In case of insufficient ventilation, wear suitable respiratory equipment	S3/9/14	Keep in a cool, well ventilated place away from ... (incompatible materials to be indicated by the manu-
S39	Wear eye/face protection		facturer)
S40	To clean the floor and all objects contaminated by this material use ... (to be specified by the manufacturer)	S3/9/14/49	Keep only in the original container in a cool, well ventilated place away from ... (incompatible materials to be indicated by the manufacturer)
S41	In case of fire and/or explosion do not breathe fumes	S3/9/49	Keep only in the original container in a cool, well ventilated place
S42	During fumigation/spraying wear suitable respiratory equipment (appropriate wording to be specified)	S3/14	Keep in a cool place away from ... (incompatible materials to be indicated by the manufacturer)
S43	In case of fire, use ... (indicate in the space the precise type of fire-fighting equipment. If water increases the risk, add - Never use water)	S7/8	Keep container tightly closed and dry
S44	If you feel unwell, seek medical advice (show the label where possible)	S7/9	Keep container tightly closed and in a well ventilated place
S45	In case of accident or if you feel unwell, seek medical advice immediately (show the label where possible)	S20/21	When using, do not eat, drink or smoke
		S24/25	Avoid contact with skin and eyes
S46	If swallowed seek medical advice immediately and show this container or label	S36/37	Wear suitable protective clothing and gloves
S47	Keep at temperature not exceeding ... °C (to be specified by the manufacturer)	S36/37/39	Wear suitable protective clothing, gloves and eye/face protection
S48	Keep wetted with ... (appropriate material to be specified by the manufacturer)	S36/39	Wear suitable protective clothing and eye/face protection
S49	Keep only in the original container	S37/39	Wear suitable gloves and eye/face protection
S50	Do not mix with ... (to be specified by the manufacturer)	S47/49	Keep only in the original container at temperature not exceeding ....°C (to be specified by the manufacturer)
S51	Use only in well ventilated areas		
S52	Not recommended for interior use on large surface areas		
S53	Avoid exposure - obtain special instructions before use		

List of Hazardous Substances

The following information can be obtained from the list of graded and hazardous substances:

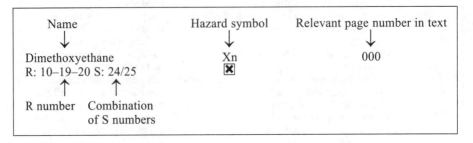

The list of entries is extracted from the Approved List of Dangerous Substances and includes only the substances which are mentioned in the text of the book.

Acetaldehyde R: 12–36/37 S: 9–16–29–33	F, Xi	203	**Acetylene** R: 5–6–12 S: 9–16–33	F	90 f.
Acetic acid, >90% R: 10–35 S: 2–23–26	C	235	**Acrolein** R: 11–23–3/37/38 S: 29–33–44	F, T	205
Acetic acid, 25-90% R: 34 S: 2–23–26	C	235	**Acrylonitrile** R: 45–11–23/24/25–38 S: 53–16–27–44	F, T	240
Acetic anhydride R: 10–34 S: 26	C	253	**Acrylic acid** R: 10–34 S: 26–36	C	238
Acetone R: 11 S: 9–16–23–33	F	217	**Adipic acid** R: 36	Xi	332
Acetone cyanohydrin R: 26/27/28 S: 7/9–27–45	T	210	**Allyl alcohol** R: 11–26–36/37/38 S: 16–39/45	F, T	122
Acetonitrile R: 11–23/24/25 S: 16–27–44	F, T	266	**Allyl chloride** R: 11–26 S: 16–29–33–45	F, T	68
Acetyl chloride R: 11–14–34 S: 9–16–26	F, C	252	**Allyl isothiocyanate** R: 23/25–33 S: 1–13	T	370

Allyl isothiocyanate in ether R: 23/25–33 S: 1–13	T ☠	371
Aluminium alkyls R: 14–17–34 S: 16–43	F, C 🔥 ☣	189
Aluminium chloride R: 34 S: 7/8–28	C ☣	475
Aluminium isopropoxide R: 11 S: 8–16	F 🔥	194
3-Aminobenzene-sulphonic acid R: 20/21/22 S: 25–28	Xn ✖	575
4-Aminobenzene-sulphonic acid R: 20/21/22 S: 25–28	Xn ✖	574
2-Aminoethanol R: 20–36/37/38	Xi ✖	305
Aminophenols R: 20/21/22 S: 28	Xn ✖	564, 788
Ammonia, anhydrous R: 10–23 S: 7/9–16–38	T ☠	158
Ammonia, >35% solution R: 34–36/37/38 S: 7–26	C ☣	158
Ammonia, 10–35% solution R: 36/37/38 S: 2–26	Xi ✖	158
Ammonium chloride R: 22–36 S: 22	Xn ✖	563
Amyl alcohol R: 10–20 S: 24/25	Xn ✖	117
Aniline R: 23/24/25–33 S: 28–36/37–44	T ☠	569

Arsenic acid R: 23/25–45 S: 1/2–20/21–28–44	T ☠	176
Arsenic compounds R: 23/25 S: 1/2–20/21–28–44	T ☠	176, 576
Atropine R: 26/28 S: 1–25–45	T ☠	806
Aziridine R: 11–26/27/28–45 S: 9–29–36–45	F, T 🔥 ☠	307 f.
Azobenzene R: 20/22 S: 28	Xn ✖	566
Azoxybenzene R: 20/22 S: 28	Xn ✖	565
Benzal chloride, Benzylidene chloride R: 36/37/38 S: 39	Xi ✖	486
Benzaldehyde R: 22 S: 24	Xn ✖	529
Benzaldehyde cyanohydrin R: 26/27/28 S: 2–13–27–46	T ☠	529
Benzene R: 45–11– 23/24/25–48 S: 53–16–29–44	F, T 🔥 ☠	473
Benzidine R: 45–22 S: 53–44	T ☠	567
Benzonitrile R: 21/22 S: 23	Xn ✖	584
***p*-Benzoquinone** R: 23/25–36/37/38 S: 26–28–44	T ☠	515
Benzoyl chloride R: 34 S: 26	C ☣	546
Benzyl alcohol R: 20/22 S: 26	Xn ✖	522

Benzylamine R: 34 S: 26	C	524
Benzyl benzoate R: 22 S: 25	Xn	533
Benzyl chloride R: 36/37/38 S: 39	Xi	486
Benzyl chloroformate R: 34–37 S: 26	C	828
BHC R: 21–25–40 S: 22–36/37–44	T	478
Bis(chloromethyl) ether R: 45–10–22–24–26 S: 53–45	T⁺	829
Bisphenol-A epichlorohydrin R: 36/38–43 S: 28–37/39	Xi	312
Bisphenol-A epichlorohydrin (reaction product in the form of an epoxide resin with an average molecular weight <700) R: 36/38–43 S: 28–37/39	Xi	311
Bis(4-isocyanato-phenyl)-methane R: 20–36/37/38–42 S: 26–28–38–45	Xn	355
Bis(tri-n-butyltin) oxide R: 23/24/25 S: 2–13–44	T	190
Bitter almond oil (contains hydrogen cyanide) R: 26/27/28 S: 7/9–13–45	T	529
Boron trifluoride R: 14–26–35 S: 9–26–28–36–45	T	143

Bromine R: 26–35 S: 7/9–26	C	68
Bromobenzene R: 10–38	Xi	480
Bromoethane R: 20/21/22 S: 28	Xn	124
Bromoform R: 23–36/38 S: 28–44	T	135
Bromomethane R: 26 S: 1/2–7/9–24/25–27–45	T	125
1-Bromopropane R: 11–26/27/28 S: 7/9–29–45	F, T	70
Brucine R: 26/28 S: 1–13–45	T	816
Buta-1,3-diene R: 13–45 S: 9–16–33	F, T	100
Butanal R: 11 S: 9–29–33	F	191
Butane R: 13 S: 9–16–33	F	55
Butanols (except t-butanol) R: 10–20 S: 16	Xn	116
t-Butanol R: 11–20 S: 9–16	F, Xn	116
Butanone R: 11 S: 9–16–23–33	F	219
But-2-enal R: 11–23–36/37/38 S: 29–33–44	F, T	206
Butene R: 13 S: 9–16–33	F	61
Butyl alcohols (except t-butanol) R: 10–20 S: 16	Xn	116

Butylene	F	61
R: 13 S: 9–16–33		
But-2-yne-1,4-diol	T	122
R: 25–34		
S: 22–36–44		
Butyraldehyde	F	191
R: 11 S: 9–29–33		
Butyric acid	C	236
R: 34 S: 26–36		
n-Butyronitrile	T	266
R: 10–23/24/25		
S: 44		
Butyryl chloride	F, C	250
R: 11–34		
S: 16–23–26–36		

Calcium carbide	F	90
R: 15 S: 8–43		
Calcium hypochlorite	O, C	134
R: 8–31–34		
S: 2–26–43		
Carbaryl	Xn	354
R: 20/22–37		
S: 2–13		
Carbon disulphide	F, T	360
R: 12–26		
S: 27–29–33–43–45		
Carbon monoxide	F, T	373
R:12–23		
S: 7–16		
Carbonyl chloride	T	352
R: 26		
S: 7/9–24/25–45		
Catechol, 1,2-dihydroxybenzene	Xn	507
R: 21/22–36/38		
S: 22–26–37		
Chloral hydrate	T	205
R: 25–36/38		
S: 25–44		

Chloramine-T	Xi	494
R: 36/37/38		
S: 2–7–15		
Chlorine	T	125
R: 23–36/37/38		
S: 7/9–44		
Chlormequat salts	Xn	306
R: 20/21/22 S: 2–13		
Chloroacetic acid	T	270
R: 23/24/25–35		
S: 22–36/37/39		
Chloroacetonitrile	T	726
R: 23/24/25 S: 44		
Chloroacetyl chloride	C	527
R: 34–37 S: 9–26		
Chlorobenzene	Xn	479
R: 10–20 S: 24/25		
2-Chlorobuta-1,3-diene	F, Xn	102
R: 12–20		
S: 9–16–29–33		
Chlorodinitro-benzene	T	503
R: 23/24/25–33		
S: 28–37–44		
1-Chloro-2,3-epoxypropane	T	309
R: 45–10–23/24/25–34–43		
S: 53–9–44		
Chloroethane	F	125
R: 13 S: 9–16–33		
2-Chloroethanol	T	239
R: 26/27/28		
S: 7/9–28–45		
(2-Chloroethyl)-trimethyl-ammonium salts	Xn	306
R: 20/21/22 S: 2–13		
Chloroform	Xn	134
R: 20 S: 2–24/25		

Chloromethane R: 13–20 S: 9–16–33	F, Xn	125
1-Chloro-4-nitro-benzene R: 23/24/25–33 S: 28–37–44	T	482
Chlorophenol R: 20/21/22 S: 2–28	Xn	484, 507
Chloropicrin R: 26/27/28–36/37/38 S: 26–36–45	T	135
Chloroprene R: 12–20 S: 9–16–29–33	F, Xn	102
Chloropropane R: 11–20/21/22 S: 9–29	F, Xn	125
3-Chloropropene R: 11–26 S: 16–29–33–45	F, T	68
N-Chloro-p-toluene-sulphonamide sodium salt R: 36/37/38 S: 2–7–15	Xi	494
Chlorosulphonic acid R: 14–35–37 S: 26	C	492
2-Chlorotoluene R: 36/37/38 S: 39	Xi	486
Chromium trioxide R: 8–35–43 S: 28	O, C	191
Colchicine R: 26/28 S: 1–13–45	T	653
Coniine R: 23/24/25 S: 1–13–45	T	804
Copper(I) chloride R: 22 S: 22	Xn	584

Coumarin derivatives R: 23–24–25 S: 1–13–45	T	556
Cresol R: 24/25–34 S: 2–28–44	T	505
Crotonaldehyde R: 11–23–36/37/38 S: 29–33–44	F, T	206
Croton oil R: 27/28 S: 1–13–45	T	677
Cumene R: 10–37	Xi	477
Cumene hydroperoxide R: 11–35 S: 3/7/9–14–27–37/39	O, C	496
Curare R: 23/24/25 S: 1–13–44	T	814
Cyanamide R: 25–36/38–43 S: 3–22–36–44	T	367
Cyanide salts excluding complex cyanides such as cyanoferrate(II) and (III) R: 26/27/28–32 S: 1/2–7–28–29–45	T	265
2-Cyanopropan-2-ol R: 26/27/28 S: 7/9–27–45	T	210
Cyanuric chloride R: 36/37/38 S: 28	Xi	793
Cyclobutane-1,3-dione R: 11 S: 9–16–33	F	391
Cyclohexane R:11 S: 9–16–33	F	398

Cyclohexanol R: 20/22–37/38 S: 24/25	Xn ☒	401
Cyclohexanone R: 10–20 S: 25	Xn ☒	402
Cyclohexylamine R: 10–21/22–34 S: 36/37/39	C ▧	571
Cyclopentane R: 11 S: 9–16–29–33	F 🔥	394
Cyclopentanone R: 10–36/38 S: 23	Xi ☒	395
Cyclopropane R: 13 S: 9–16–33	F 🔥	387

2,4-D R: 20/21/22 S: 2–13	Xn ☒	500
DDT R: 25–40–48 S: 22–36/37–44	T ☠	604
Diacetonealcohol R: 36 S: 24/25	Xi ☒	218
4,4′-Diaminobiphenyl R: 45–22 S: 53–44	T ☠	567
1,2-Diaminoethane R: 10–21/22–34–43 S: 9–26–36/37/39	C ▧	166
o-**Dianisidine** R: 26/27/28–33 S: 28–36/37–45	T ☠	603
1,2-Dibromoethane R: 45–23/24/25– 36/37/38 S: 53–44	T ☠	60
Dibromomethane R: 20 S: 24	Xn ☒	135
Dichloroacetic acid R: 35 S: 26	C ▧	270
1,2-Dichlorobenzene R: 20 S: 24/25	Xn ☒	479

1,4-Dichloro- benzene R: 22 S: 2–24/25	Xn ☒	479
α,α′-Dichloro- dimethyl ether R: 45–10–22–24–26 S: 53–45	T⁺ ☠	829
1,2-Dichloroethane R: 11–20 S: 7–16–29–33	F, Xn 🔥☒	80
1,2-Dichloro- ethylene, **1,2-Dichloroethene** R: 11–20 S: 7–16–29	F, Xn 🔥☒ F, Xn	65 95
Dichloromethane R: 20 S: 24	Xn ☒	135
2,4-Dichloro- phenoxyacetic acid R: 20/21/22 S: 2–13	Xn ☒	500
Dichloropropane R: 11–20 S: 9–16–29–33	F, Xn 🔥☒	89
2,3-Dichloro- propene R: 11–22 S: 9–16–29–33	F, Xn 🔥☒	99
α,α-Dichloro- toluene R: 36/37/38 S: 39	Xi ☒	486
Dicyanogen, Cyanogen R: 11–23 S: 23–44	F, T 🔥☠	371
Diethanolamine R: 36/38 S: 26	Xi ☒	305
N,N-**Diethylaniline** R: 23/24/25–33 S: 28–37–44	T ☠	577
Diethyl ether R: 12–19 S: 9–16–29–33	F 🔥	142
Diethyl ketone R: 11 S: 9–16–33	F 🔥	211

***O,O*-Diethyl-*o*- (4-nitrophenyl) thiophosphate** R: 26/27/28 S: 1–13–28–45	T ☠	140
Diethyl oxalate R: 22–36 S: 23	Xn ✖	326
Diethyl sulphate R: 45–46–20/21/22–34 S: 53–26–44	T⁺ ☠	138
Digitoxin R: 23/25–33 S: 1–44	T ☠	444, 697
1,2-Dihydroxy- benzene R: 21/22–36/38 S: 22–26–37	Xn ✖	507
1,3-Dihydroxy- benzene R: 22–36/38 S: 26	Xn ✖	508
1,4-Dihydroxy- benzene R: 20/22 S: 2–24/25–39	Xn ✖	509
Di-isopropyl ether R: 11–19 S: 9–16–33	F 🔥	116
Di-isopropyl ketone R: 11 S: 16–23	F 🔥	216
Diketene R: 10–20 S: 3	Xn ✖	255
3,3′-Dimethoxy- benzidine R: 26/27/28–33 S: 28–36/37–45	T ☠	603
1-(3′,4′-Dimethoxy- benzyl)-6,7-dimeth- oxy-isoquinoline R: 22 S: 22	Xn ✖	811
1,2-Dimethoxyethane R: 10–19–20 S: 24/25	Xn ✖	302

2,3-Dimethoxy- strychnine R: 26/28 S: 1–13–45	T ☠	817
***N,N*-Dimethyl- aniline** R: 23/24/25–33 S: 28–37–44	T ☠	577
α,α-Dimethyl- benzyl hydroperoxide R: 11–35 S: 3/7/9–14–27–37/39	O, C 🔥 ⟁	496
1,1′-Dimethyl-4,4′- bipyridinium salts R: 26/27/28 S: 1–13–45	T ☠	763
Dimethyl carbonate R: 11–20/21/22 S: 9–29	F, Xn 🔥 ✖	352
Dimethyl ether R: 13 S: 9–16–33	F 🔥	142
***N,N*-Dimethyl- hydrazine** R: 45–11–23/25–34 S: 53–16–33–44	F, T 🔥 ☠	170
Dimethyl- nitrosamine R: 45–25–26–48 S: 53–45	T⁺ ☠	162
2,4-Dimethyl- pentan-3-one R: 11 S: 16–23	F 🔥	216
Dimethylpropane R: 13 S: 9–16–33	F 🔥	56
Dimethyl sulphate R: 45–25–26–34 S: 53–26–27–45	T⁺ ☠	138
2,4-Dinitroaniline R: 26/27/28–33 S: 28–36/37–45	T ☠	590

Dinitrobenzene R: 26/27/28–33 S: 28–36/37–45	T ☠	485
Dinitrophenol R: 23/24/25–33 S: 28–37–44	T ☠	502
Dinitrotoluene R: 23/24/25–33 S: 28–37–44	T ☠	490
1,4-Dioxan R: 11–19–20 S: 9–16–33	F, Xn 🔥✖	301
Dipentene R: 10–38 S: 28	Xi ✖	665
Diphenylamine R: 23/24/25–33 S: 28–36/37–44	T ☠	578
Diphosphorus pentasulphide R: 11–20/22–29	F, Xn 🔥✖	263
Ephedrine R: 22 S: 22–25	Xn ✖	526
Epichlorohydrin R: 45–10–23/24/25– 34–43 S: 53–9–44	T ☠	309
1,2-Epoxypropane R: 12–20/21/22–45 S: 9–16–26–29	F, T 🔥☠	312
L-6,7-Epoxytropyl tropate R: 26/27/28 S: 1–25–45	T ☠	806
Ergot alkaloids and their salts and derivatives R: 23/25 S: 2–13–45	T ☠	817
L-Erythro-2-methyl-amino-1-phenyl-propan-1-ol R: 22 S: 22–25	Xn ✖	526

Ethane R: 12 S: 9–16–33	F 🔥	54
Ethanediol R: 22 S: 2	Xn ✖	300
Ethanethiol R: 11–20 S: 16–25	F, Xn 🔥✖	145
Ethanol R: 11 S: 7–16	F 🔥	114
Ethanolamine R: 20–36/37/38	Xi ✖	305
Ethene R: 13 S: 9–16–33	F 🔥	60
Ether R: 12–19 S: 9–16–29–33	F 🔥	141
Ethoxides, sodium/ potassium Methoxides, sodium/potassium R: 11–14–34 S: 8–16–26–43	F, C 🔥⌧	115
Ethyl acetate R: 11 S: 16–23–29–33	F 🔥	259
Ethyl acrylate R: 11–20/22–36/37/ 38–43 S: 9–16–33	F, Xi 🔥✖	337
Ethyl alcohol R: 11 S: 7–16	F 🔥	114
Ethylamine R: 13–36/37 S: 16–26–29	F, Xi 🔥✖	157, 161
N-Ethylaniline R: 23/24/25–33 S: 28–37–44	T ☠	577
Ethylbenzene R: 11–20 S: 16–24/25–29	F, Xn 🔥✖	477
Ethyl bromide R: 20/21/22 S: 28	Xn ✖	125

Ethyl carbamate	T	354
R: 23/24/25–45		
S: 2–13		
Ethyl chloride	F	125
R: 13 S: 9–16–33		
Ethyl chloroformate	F, T	353
R: 11–23–36/37/38		
S: 9–16–33–44		
Ethylene	F	60
R: 13 S: 9–16–33		
Ethylene chloride,	F, Xn	80
Dichloroethane		
R: 11–20		
S: 7–16–29–33		
Ethylene	T	239
chlorohydrin		
R: 26/27/28		
S: 7/9–28–45		
Ethylenediamine	C	166
R: 10–21/22–34–43		
S: 9–26–36/37/39		
Ethyleneimine	F, T	307
R: 11–26/27/28–45		
S: 9–29–36–45		
Ethylene oxide	F⁺, T	302
R: 45–46–13–23–		
36/37/38		
S: 52–3/7/9–16–33–34		
Ethyl formate	F	530
R: 11 S: 9–16–33		
Ethyl mercaptan	F, Xn	145
R: 11–20		
S: 16–25		
Ethyl methyl ether	F	141
R: 13 S: 9–16–33		
Ethyl methyl ketone	F	219
R: 11		
S: 9–16–23–33		
Ethyne	F	90
R: 5–6–12		
S: 9–16–33		

Fluorine	T	136
R: 7–26–35		
S: 7/9–36–45		
Formaldehyde,	T	201
>30%, Formalin		
R: 23/24/25–40–43		
S: 2–26–28–51		
Formic acid,	C	233
>90%		
R: 35 S: 2–23–26		
Formic acid,	C	233
25-90%		
R: 34 S: 2–23–26		
Fumaric acid	Xi	334
R: 36 S: 26		
Furfural	T	718
R: 23/25		
S: 24/25–44		
Furfuryl alcohol	Xn	718
R: 20/21/22		
2-Furylmethanal	T	718
R: 23/25		
S: 24/25–44		

Glycol	Xn	300
R: 22 S: 2		

HCH	T	478
R: 21–25–40		
S: 22–36/37–44		
Heptane	F	52
R: 11		
S: 9–16–23–29–33		
γ-1,2,3,4,5,6-Hexa-	T	478
chlorocyclohexane		
R: 23/24//25		
S: 2–13–44		
Hexachlorophene	T	604
R: 24/25		
S: 20–37–44		

Hexamethylenedi-isocyanate	T	354
R: 23–36/37/38–42/43		
S: 26–28–38–45		
Hexane	F, Xn	52
R: 11–20/21–40		
S: 9–16–23		
Hydrazine	T	160
R: 10–26/27/28–		
34–45		
S: 36/37/39–45		
Hydrochloric acid	C	820
>25%		
R: 34–37 S: 2–26		
Hydrogen	F	54
R: 12 S: 7/9		
Hydrogen bromide,	C	69
anhydrous		
R: 35–37		
S: 7/9–26–44		
Hydrogen bromide,	C	69
>40%		
R: 34–37 S: 7/9–26		
Hydrogen chloride,	C	68
anhydrous		
R: 35–37		
S: 7/9–26–44		
Hydrogen cyanide	F, T	265
R: 12–26/27/28		
S: 7/9–13–16–45		
Hydrogen iodide,	C	124
>25%		
R: 34 S: 26		
Hydrogen peroxide	O, C	71
solution, >60%		
R: 8–34		
S: 3–28–36/39		
Hydrogen sulphide	F, T	146
R: 13–26		
S: 7/9–25–45		
Hydroquinone	Xn	509
R: 20/22		
S: 2–24/25–39		

Hydroxy-ammonium chloride, hydroxylamine hydrochloride	Xn	197
R: 20/22–36/38		
S: 2–13		
4-Hydroxy-4-methyl-pentan-2-one	Xi	218
R: 36 S: 24/25		
8-Hydroxy-quinoline sulphate	Xn	774
R: 20/21/22 S: 2–13		
Hyoscyamine	T	807
R: 26/28		
S: 1–24–45		
Iodine	Xn	135
R: 20/21 S: 23–25		
Iodomethane	T	141
R: 23/24/25–34–45		
S: 26–44		
Isobutyric acid	Xn	235
R: 21/22		
Isodrin	T	421
R: 26/27/28		
S: 1–13–28–45		
Isopentane	F	56
R: 11		
S: 9–16–29–33		
Isoprene	F	100
R: 12		
S: 9–16–29–33		
Isopropanol	F	116
R: 11 S: 7–16		
Isopropylbenzene	Xi	477
R: 10–37		
Lead alkyls	T	190
R: 26/27/28–33		
S: 13–26–36/37–45		

Lindane	T	478
R: 23/24/25		
S: 2–13–44		
Lithium	F, C	183
R: 14/15–34		
S: 8–43		
Lithium aluminium hydride	F	194
R: 15		
S: 7/8–24/25–43		
Magnesium alkyls	F, C	184
R: 14–17–34		
S: 16–43		
Maleic acid	Xi	334
R: 22–36/37/38		
S: 26–28–37		
Maleic anhydride	Xi	334
R: 22–36/37/38–42		
S: 22–28–39		
Malononitrile	T	752
R: 23/24/25		
S: 23–27		
Manganese dioxide	Xn	515
R: 20/22 S: 25		
p-**Mentha-1,8(9)-diene**	Xi	664
R: 10–38 S: 28		
Mercury alkyls	T	188
R; 26/27/28–33		
S: 2–13–28–36–45		
Mercury compounds, inorganic	T	188
R: 26/27/28–33		
S: 1/2–13–28–45		
Mesitylene	Xi	477
R: 10–37		
Mesityl oxide	Xn	219
R: 10–20/21/22		
S: 25		
Metanilic acid	Xn	575
R: 20/21/22		
S: 25–28		
Methacrylates	Xi	219
R: 36/37/38		
S: 26–28		
Methacrylic acid	C	241
R: 34 S: 15–26		
Methacrylonitrile	F, T	240
R: 11–23/24/25–43		
S: 9–16–18–29–45		
Methane	F	53
R: 12 S: 9–16–33		
Methanesulphonic acid	C	149
R: 34 S: 26–36		
Methanethiol	F, Xn	145
R: 13–20 S: 16–25		
Methanol	F, T	113
R: 11–23/25		
S: 2–7–16–24		
Methyl acetate	F	113
R: 11		
S: 16–23–29–33		
Methyl alcohol	F, T	113
R: 11–23/25		
S: 2–7–16–24		
Methylamine	F, Xi	159
R: 13–36/37		
S: 16–26–29		
N-**Methylaniline**	T	577
R: 23/24/25–33		
S: 28–37–44		
Methyl bromide	T	125
R: 26		
S: 1/2–7/9–24/25–27–45		
2-Methylbuta-1,3-diene	F	100
R: 12		
S: 9–16–29–33		
Methylbutane	F	56
R: 11		
S: 9–16–29–33		

Methyl chloride   F, Xn   125
R: 13–20
S: 9–16–33

Methyl chloroformate   F, T   352
R: 11–23–36/37/38
S: 9–16–33–44

Methylcyclohexane   F   476
R: 11 S: 9–16–33

2-Methylcyclo-hexanone   Xn   213
R: 10–20 S:25

2,2′-Methylene-bis (3,4,6-trichloro-phenol)   T   604
R: 24/25
S: 20–37–44

Methylene bromide   Xn   135
R: 20 S: 24

Methylene chloride   Xn   135
R: 20 S: 24

4-Methylene-2-oxetanone   Xn   256
R: 10–20 S: 3

Methyl formate   F   234
R: 12 S: 9–16–33

Methyl iodide   T   141
R: 23/24/25–34–45
S: 26–44

Methyl isocyanate   F, T   366
R: 12–23/24/25–
36/37/38
S: 9–30–43–44

Methyl mercaptan   F, Xn   145
R: 13–20 S: 16–25

Methyl methacrylate   F, Xi   241
R: 11–36/37/38–43
S: 9–16–29–33

N-Methyl-1-naphthyl carbamate   Xn   354
R: 20/22–37
S: 2–13

4-Methylpent-3-en-2-one   Xn   220
R: 10–20/21/22
S: 25

2-Methylpropan-2-ol   F, Xn   116
R: 11–20
S: 9–16

2-Methylprop-2-ene nitrile   F, T   240
R: 11–23/24/25–43
S: 9–16–18–29

2-Methylpyridine   Xn   765
R: 10–20/21/22–
36/37
S: 26/36

4-Methylpyridine   T   765
R: 10–20/22–24–
36/37/38
S: 26–36–44

N-Methyl-2-pyrrolidinone   Xi   708
R: 36/38 S: 41

Methyltrichloro-silane   F, Xi   178
R: 11–14–36/37/38
S: 26/39

Methyl vinyl ether   F   320
R: 13
S: 9–16–33

Mineral oil and coal tar distillates with flash points below 21°C   F   84
R: 11
S: 9–16–29–33
with flash points between 21 and 55°C
R: 10

Morpholine   C   788
R: 10–20/21/22–34
S: 23–26

2-Naphthol, *β*-**Naphthol** R: 20/22 S: 24/25	Xn ✖	630
2-Naphthylamine R: 45–22 S: 53–44	T ☠	631
Neopentane R: 13 S: 9–16–33	F 🔥	56
Nickel tetracarbonyl R: 11–26–45 S: 9–23–45	F, T 🔥 ☠	236
Nicotine R: 25–26/27 S: 28–36/37/39–42–45	T ☠	803
Nitrating acid R: 8–35 S: 23–26–30–36	O, C 🔥 ☢	487
Nitric acid 20–70% R: 35 S: 2–23–26–27	C ☢	487
Nitric acid-sulphuric **acid mixture with** **>30% nitric acid** R: 8–35 S: 23–26–30–36	O, C 🔥 ☢	487
Nitroaniline R: 23/24/25–33 S: 28–36/37–44	T ☠	572
Nitrobenzene R: 26/27/28–33 S: 28–36/37–45	T ☠	487
Nitrocellulose **containing at most** **12.6% N** R: 11 S: 16–33–37/39	F 🔥	464
Nitroethane R: 10–20/22 S: 9–25–41	Xn ✖	152
Nitromethane R: 5–10–22 S: 41	Xn ✖	151
2-Nitronaphthalene R: 45 S: 53–44	T ☠	628

4-Nitrophenol, *p*-**Nitrophenol** R: 20/21/22–33 S: 28	Xn ✖	501
1-Nitropropane R: 10–20/21/22 S: 9	Xn ✖	152
2-Nitropropane R: 45–10–20/22 S: 53–9–44	T ☠	152
2-Nitrotoluene, **4-Nitrotoluene** R: 23/24/25–33 S: 28–37–44	T ☠	489
Octane R: 11 S: 9–16–29–33	F 🔥	56
Oleum, 20–65% **SO₃** R: 14–35–37 S: 26–30	C ☢	490
Osmium tetroxide R: 26/27/28–34 S: 7/9–26–45	T ☠	71
Oxalic acid R: 21/22 S: 2–24/25	Xn ✖	325
Oxalate salts R: 21/22 S: 2–24/25	Xn ✖	325
Oxalonitrile R: 11–23 S: 23–44	F, T 🔥 ☠	371
Oxiran R: 45–46–13–23– 36/37/38 S: 52–3/7/9–16– 33–44	F⁺, T 🔥 ☠	302
Papaverine R: 22 S: 22	Xn ✖	811

Paraldehyde R: 11 S: 9–16–29–33	F 🔥	201
Paraquat and its salts R: 26/27/28 S: 1–13–45	T ☠	763
Parathion R: 26/27/28 S: 1–13–28–45	T ☠	140
Pentane R: 11 S: 9–16–29–33	F 🔥	56
Pentane-2,4-dione, Acetylacetone R: 10–22 S: 21–23–24/25	Xn ✖	322
Pentan-3-one R: 11 S: 9–16–33	F 🔥	211
Peracetic acid, >10% R: 5–22–34 S: 3–27–36	O, C 🔥	71
Perchloric acid, 10–50% R: 34 S: 23–28–36	C	581
Perchloroethylene R: 20/22 S: 2–25	Xn ✖	95
Phenol R: 24/25–34 S: 2–28–44	T ☠	495
Phenol salts R: 24/25–34 S: 2–28–44	T ☠	550
Phenylenediamine R: 23/24/25–43 S: 28–44	T ☠	579
1,3-Phenylenediamine dihydrochloride **1,4-Phenylenediamine dihydrochloride** R: 23/24/25 S: 28–44	T ☠	580

Phenylhydrazine R: 23/24/25–36 S: 28–44	T ☠	586
1-Phenyl-3-pyrazolidinone R: 22	Xn ✖	736
Phosgene R: 26 S: 7/9–24/25–45	T ☠	352
Phosphoric acid >25% R: 34 S:26	C	718
Phosphorus oxychloride R: 34–37 S: 7/8–26	C	251
Phosphorus pentachloride R: 34–37 S: 7/8–26	C	251
Phosphorus, red R: 11–16 S: 7–43	F 🔥	271
Phosphorus tribromide R: 14–34–37 S: 26	C	271
Phosphorus trichloride R: 34–37 S: 7/8–26	C	172
Phosphoryl chloride R: 34–37 S: 7/8–26	C	251
Phthalic anhydride R: 36/37/38	Xi ✖	527
2-Picoline R: 10–20/21/22–36/37 S: 26–36	Xn ✖	765
4-Picoline R: 10–20/22–24–36/37/38 S: 26–36–44	T ☠	765

Polychlorinated biphenyls Xn ⊠ R: 33 S: 35	603	**β-Propiolactone** T⁺ ☒ R: 45–26–36/38 S: 53–45	280

Polychlorinated biphenyls Xn 603
R: 33 S: 35

Potassium F, C 183
R: 14/15–34
S: 5–8–43

Potassium dichromate Xi 191, 203
R: 36/37/38–43
S: 22–28

Potassium fluoride T 136
R: 23/24/25
S: 1/2–26–44

Potassium hydroxide C 233
R: 35
S: 2–26–37/39

Potassium permanganate O, Xn 71
R: 8–22 S: 2

Potassium sulphide C 146
R: 31–34 S: 26

Propanal F, Xi 204
R: 11–36/37/38
S: 9–16–29

Propane F 55
R: 13 S: 9–16–33

Propan-1-ol F 116
R: 11 S: 7–16

Propan-2-ol F 116
R: 11 S: 7–16

Propan-3-olide T⁺ 280
R: 45–26–36/38
S: 53–45

Propargyl alcohol, prop-2-yn-1-ol T 122
R: 10–23/24/25–34
S: 26–28–36–44

Propenal F, T 205
R: 11–23–36/37/38
S: 29–33–44

Propene F 60
R: 13 S: 9–16–33

β-Propiolactone T⁺ 280
R: 45–26–36/38
S: 53–45

Propionaldehyde F, Xi 204
R: 11–36/37/38
S: 9–16–29

Propionic acid C 236
R: 34 S: 2–23–26

Propionic anhydride C 253
R: 34 S: 26

Propionyl chloride F, C 250
R: 11–14–34
S: 9–16–26

Propyl alcohol F 116
R: 11 S: 7–16

Propyl bromide F, T 70
R: 11–26/27/28
S: 7/9–29–45

Propylene, Propene F 59
R: 13 S: 9–16–33

Propylene-1,2-oxide F, T 312
R: 12–20/21/22–45
S: 9–16–26–29

Pyrethrin I Xn 396
R: 20/21/22
S: 2–13

Pyridine F, Xn 761
R: 11–20/21/22
S: 26–28

3-Pyridyl-N-methyl-pyrrolidine T 803
R: 25–26/27
S: 28–36/37/39–42–45

Pyrogallol Xn 513
R: 20/21/22

Quinone T 515
R: 23/25–36/37/38
S: 26–28–44

Resorcinol	Xn	508
R: 2–36/38 S: 26	☒	
Rotenone	T	771
R: 26/27/28	☠	
S: 36/37/39–42–45		

Scillarenin glycosides	T	698
R: 23/24/25	☠	
S: 2–13–44		
Scopolamine	T	807
R: 26/27/28	☠	
S: 1–25–45		
Selenium	T	643
R: 23/25–33	☠	
S: 20/21–28–44		
Selenium compounds	T	822
R: 23/25–33	☠	
S: 20/21–28–44		
Silver nitrate	C	94
R: 34 S: 2–26	☒	
Sodium	F, C	142
R: 14/15–34	☀☒	
S: 5–8–43		
Sodium azide	T	171
R: 28–32 S: 28	☠	
Sodium carbonate	Xi	247
R: 36 S: 22–26	☒	
Sodium dithionite	Xn	203
R: 7–22–31	☒	
S: 7/8–26–28–43		
Sodium hydride	F	297
R: 15	☀	
S: 7/8–24/25–43		
Strophantin	T	697
R: 23/25–33 S: 44	☠	
Strychnine	T	816
R: 26/28	☠	
S: 1–13–45		
Styrene	Xi	617
R: 10–36/37	☒	

Succinic anhydride	Xi	325
R: 36/37 S: 25	☒	
Sulphanilic acid	Xn	574
R: 20/21/22	☒	
S: 25–28		
Sulpholane	Xn	721
R: 22 S: 25	☒	
Sulphur dioxide	T	150
R: 23–36/37	☠	
S: 7/9–44		
Sulphuric acid >15%	C	490
R: 35 S: 2–26–30	☒	
Sulphuryl chloride	C	251
R: 14–34–37	☒	
S: 26		

2,4,5-T	Xn	501
R: 20/21/22	☒	
S: 2–13		
1,1,2,2-Tetrachloroethane	T	95
R: 26/27	☠	
S: 2–38–45		
Tetrachloro-ethylene	Xn	95
R: 20/22 S: 2–25	☒	
Tetrachloro-methane, Carbon tetrachloride	T	135
R: 26/27	☠	
S: 2–38–45		
Tetrahydrofuran	F, Xi	717
R: 11–19–36/37	☀☒	
S: 16–29–33		
Tetrahydrofurfuryl alcohol	Xi	719
R: 36 S: 39	☒	
Tetrahydro-thiophene-1,1-dioxide	Xn	721
R: 22 S: 25	☒	

Thiocarbamide, Thiourea ⊠ R: 22–40 S: 22–24	Xn	361
Thiocyanates ⊠ R: 20/21/22–32 S: 2–13	Xn	369
Thiocyanic acid ⊠ R: 20/21/22–32 S: 2–13	Xn	369
Thionyl chloride ⊡ R: 14–34–37 S: 26	C	251
Tin tetrachloride ⊡ R: 34–37 S: 7/8–26	C	190
Titanium tetrachloride ⊡ R: 14–34–36/37 S: 7/8–26	C	187
Toluene 🔥⊠ R: 11–20 S: 16–29–33	F, Xn	476
p-Toluenesulphonic acid ⊠ R: 36/37/38 S: 26–37	Xi	490, 550
Toluidine ☠ R: 23/24/25–33 S: 28–36/37–44	T	569
Trialkylboranes 🔥⊡ R: 17–34 S: 7–23–36–43	F, C	179
Tribromomethane ☠ R: 23–36/38 S: 28–44	T	135
Tributyltin compounds ☠ R: 23/24/25 S: 26–27–28–44	T	190
Trichloroacetaldehyde monohydrate ☠ R: 25–36/38 S: 25–44	T	134, 205
Trichloroacetic acid ⊡ R: 35 S: 24/25–26	C	271
1,1,1-Trichloro-2,2-bis(4-chlorophenyl) ethane ☠ R: 25–40–48 S: 22–36/37–44	T	604
1,1,2-Trichloroethane ⊠ R: 20/21/22 S: 9	Xn	80
Trichloroethene, Trichloroethylene ⊠ R: 20/22 S: 2–25	Xn	95
Trichloromethane ⊠ R: 20 S: 2–24/25	Xn	135
Trichloronitromethane ☠ R: 26/27/28–36/37/38 S: 26–36–45	T	135
2,4,5-Trichlorophenoxyacetic acid ⊠ R: 20/21/22 S: 2–13	Xn	501
2,4,6-Trichloro-1,3,5-triazine ⊠ R: 36/37/38 S: 28	Xi	793
Tricresyl phosphate ☠ R: 23/24/25–39 S: 20/21–28–44	T	140
Triethylamine 🔥⊠ R: 11–36/37 S: 16–26–29	F, Xi	192
Trifluoroacetic acid, >10% ⊡ R: 20–35 S: 9–26–27–28	C	831
1,2,3-Trihydroxybenzene ⊠ R: 20/21/22	Xn	513

1,3,5-Trimethyl-benzene ☒ Xi R: 10–37		477
Trimethyl borate ☒ Xn R: 10–21 S: 23–25		140
2,4,4-Trimethylpent-1-ene 🔥 F R: 11 S: 9–16–29–33		86
2,4,6-Trimethyl-1,3,5-trioxan 🔥 F R: 11 S: 9–16–29–33		201
1,3,5-Trioxan, Trioxymethylene ☒ Xn R: 22 S: 24/25		201
Triphenyltin acetate ☠ T R: 23/24/25 S: 2–13–44		190
L-Tropyl tropate ☠ T R: 26/28 S: 1–24–25		806
DL-Tropyl tropate ☠ T R: 26/28 S: 1–25–45		805
Turpentine oil ☒ Xn R: 10–20/21/22 S: 2		665
Urethane ☠ T R: 23/24/25–45 S: 2–13		354

Valeric acid, Pentanoic acid C R: 34 S: 26–36		237
Vinyl acetate 🔥 F R: 11 S: 16–23–29–33		81
Vinyl chloride 🔥☠ F, T R: 13–45 S: 9–16–33		79
Xylenol ☠ T R: 24/25–34 S: 2–28–44		505
Xylidine ☠ T R: 23/24/25–33 S: 28–36/37–44		569
Xylene ☒ Xn R: 10–20 S: 24/25		477
Zinc alkyls 🔥 F, C R: 14–17–34 S: 16–43		186
Zinc chloride C R: 34 S: 7/8–28		598
Zinc powder - Zinc dust (not stabilised) 🔥 F R: 15–17 S: 7/8–43		563
Zinc powder - Zinc dust (stabilised) R: 10–15 S: 7/8–43		563

Carcinogenic Materials

In the following list of carcinogenic materials the first column shows the page on which they are mentioned in the main text. In general, the mention which is made in the main text is derived from the (German) DFG-MAK list of 1988.

Page in Main Text	Carcinogenic Material	Groups		
		I (very great risk)	II (great risk)	III (risk)
		Amount of material in parts per 100		
240 f.	Acrylonitrile		≥1	<1 −0,1
	o-Aminoazotoluene 4-Aminobiphenyl Salts of 4-Aminobiphenyl Antimony trioxide[2]	≥1 ≥1	≥0,1 <1 −0,1 <1 −0,1 ≥1	<0,1 −0,01 <0,1 −0,01 <0,1 −0,01 <1 −0,1
176	Arsenic pentoxide, Arsenious acid, Arsenic acids and their salts (Arsenites, Arsenates)[2] Asbestos[2]		≥3 ≥1	<3 −0,3 <1 −0,1
567	Benzidine (4,4′-Diaminobiphenyl) Salts of Benzidine	≥1 ≥1	<1 −0,1 <1 −0,1	<0,1 −0,01 <0,1 −0,01
473	Benzene		≥1	
646	Benzo(a)pyrene[4]		≥0,1	<0,1 −0,005
	Beryllium[2] Beryllium compounds[2] Bis(chloromethyl) ether	≥0,05	≥1 ≥1 <0,05−0,005	<1 −0,1 <1 −0,1 <0,005−0,0005
100	Buta-1,3-diene			≥1
	Cadmium chloride[2] Calcium chromate[2]	≥1	<1 −0,1 ≥1	<0,1 −0,01 <1 −0,1

Page in Main Text	Carcinogenic Material	Groups		
		I (very great risk)	II (great risk)	III (risk)
		Amount of material in parts per 100		
309	1-Chloro-2,3-epoxypropene (Epichlorohydrin)			≥1
	N-Chloroformylmorpholine Chloromethyl methyl ether (Chlorodimethyl ether)[1] Chromium(III) chromate[2] Cobalt[2,3] (as cobalt metal, cobalt oxide and cobalt sulphide) Diarsenic trioxide (Arsenic trioxide)	≥1	≥0,005 <1 −0,1 ≥1 ≥1 ≥3	<0,005−0,0005 <0,1 −0,01 <1 −0,1 <1 −0,1 <3 −0,3
166	Diazomethane		≥1	<1 −0,1
	1,2-Dibromo-3-chloropropane 1,2-Dibromoethane (Ethylene dibromide) Dichloroacetylene 3,3′-Dichlorobenzidine Salts of 3,3′-Dichlorobenzidine		≥1 ≥1 ≥1 ≥1 ≥1	<1 −0,1 <1 −0,1 <1 −0,1 <1 −0,1 <1 −0,1
102	1,4-Dichlorobut-2-ene		≥0,1	<1 −0,01
	2,2′-Dichloro-4,4′-methylenedianiline [4,4′-Methylene-bis(2-chloroaniline)] Salts of 2,2′-Dichloro-4,4′-methylenedianiline [Salts of 4,4′-Methylene-bis(2-(chloroaniline)]		≥1 ≥1	<1 −0,1 <1 −0,1
138	Diethyl sulphate		≥1	<1 −0,1
603	3,3′-Dimethoxybenzidine (o-Dianisidine) Salts of 3,3′-Dimethoxybenzidine (Salts of o-Dianisidine)		≥0,5 ≥0,5	<0,5 −0,05 <0,5 −0,05
603	3,3′-Dimethylbenzidine (o-Tolidine) Salts of 3,3′-Dimethylbenzidine (Salts of o-Tolidine)		≥0,5 ≥0,5	<0,5 −0,05 <0,5 −0,05
	Dimethylcarbamoyl chloride 3,3′-Dimethyl-4,4′-diamino-diphenylmethane	≥0,05	<0,05−0,005 ≥1	<0,005−0,0005 <1 −0,1

Page in Main Text	Carcinogenic Material	Groups		
		I (very great risk)	II (great risk)	III (risk)
		Amount of material in parts per 100		
170	N,N-Dimethylhydrazine			≥5
170	1,2-Dimethylhydrazine		≥0,1	<0,1 −0,01
162	Dimethylnitrosamine (N-Nitrosodimethylamine)	≥0,01	<0,01−0,001	<0,001−0,0001
	Dimethylsulphamoyl chloride			≥1
138	Dimethyl sulphate		≥1	<1 −0,1
312	1,2-Epoxypropane (Propene-1,2-oxide)			≥1
354	Ethyl carbamate		≥1	<1 −0,1
307	Ethyleneimine		≥1	<1 −0,1
301	Ethylene oxide			≥0,1
450	Hexamethylphosphoramide (hexa methylphosphoric triamide, HMPT)	≥0,05	<0,05−0,005	<0,005−0,0005
160	Hydrazine			≥5
	2-Methylaziridine (Propeneimine)		≥1	<1 −0,1
631	2-Naphthylamine	≥1	<1 −0,1	<0,1 −0,01
	Salts of 2-Naphthylamine	≥1	<1 −0,1	<0,1 −0,01
	Nickel[2,3] (as nickel metal, nickel sulphide and sulphide ores, nickel oxide and nickel carbonate), as well as nickel compounds in the form of respirable droplets		≥5	<5 −0,5
236	Nickel tetracarbonyl		≥1	<1 −0,1
	5-nitroacenaphthene		≥1	<1 −0,1
	4-Nitrodiphenyl	≥1	<1 −0,1	<0,1 −0,01
628	2-Nitronaphthalene		≥1	<1 −0,1
152	2-Nitropropane		≥1	<1 −0,1
	1,3-Propanesultone	≥1	<1 −0,1	<0,1 −0,01

Page in Main Text	Carcinogenic Material	Groups		
		I (very great risk)	II (great risk)	III (risk)
		Amount of material in parts per 100		
280 f.	Propan-3-olide (1,3-Propiolactone)		≥1	<1 −0,1
	Strontium chromate[2] 2,3,4-Trichlorobut-1-ene		≥1 ≥0,1	<1 −0,1 <0,1 −0,01
79	Vinyl chloride		≥1	<1 −0,1
	Zinc chromate (including zinc potassium chromate[2])		≥1	<1 −0,1

[1] The grade referred to is technical monochlorodimethyl ether, which in practice may contain up to 7 parts per 100 of bis(chloromethyl) ether as impurity.

[2] When present in an inhalable form (in the case of asbestos as fine dust).

[3] Not including alloys.

[4] As a standard reference for carcinogenic polycyclic aromatic hydrocarbons (PAH) obtained as pyrolysis products from organic material.

In footnote[1] the name "monochlorodimethyl ether" has been replaced by "chloromethyl methyl ether".

In footnote[4] the following statement has been appended to the list of carcinogenic materials: "Since the meeting of 19 December 1985 of the Senate Committee for the testing of industrial materials which are a danger to health, the newly established reliable scientific information concerning carcinogenic materials remains unaltered".

Compendium of Named Reactions and Concepts

The numbers refer to the appropriate pages in the text.

A

Adams catalyst
Hydrogenation catalyst. Platinum which is formed from platinum(IV) oxide (PtO_2) in the course of the hydrogenation.   73

Amadori rearrangement
The rearrangement of an aldose-N-glycoside (N-glycosylamine) to a 1-amino-1-deoxy-2-ketose.   431

Andrussov process
Catalytic oxidation of methane and
ammonia to hydrogen cyanide.   265

Armstrong formula
A variant of the diagonal formula for
benzene.   469

Arndt-Eistert synthesis
Conversion of a carboxylic acid into
its next higher homologue by the
reaction of its acid chloride with
diazomethane.   252

Asinger process
Preparation of D-penicillamine by the
action of elemental sulphur and
gaseous ammonia on 2-methyl-
propanal (isobutyraldehyde).   747

Avogadro number N_A
The number of particles contained in
a mol, earlier described also as the
Loschmidt number;
$N_A = 6.022 \times 10^{23}$.   14

B

Baeyer strain
The ring-strain which results from the
Baeyer strain theory.   385

Baeyer strain theory
The deviation of the ring-angles from
the tetrahedral angle (109° 28') in cyclic
hydrocarbons in their planar forms,
which leads to strain in the ring.   385

Baeyer test
Qualitative test for $C = C$ double
bonds by their decolourising
potassium permanganate.   72

Baeyer-Villiger reaction
Oxidation of ketones to esters by
peracids.   531

Balz-Schiemann reaction
Preparation of fluorobenzene by heating dry benzenediazonium
tetrafluoroborate.   584

Bart reaction
Preparation of benzenearsonic acid
from benzene diazonium chloride and
sodium arsenate(III).   583

Barton reaction
Formation of a carbonyl group on
an unactivated carbon atom at the
4-position to a nitrite group by
photolysing the nitrite.   139

Barton rule
Concerns the stereochemistry of the
reduction of cyclohexanones to cyclohexanols by $LiAlH_4$ or $NaBH_4$.   402

Beckmann method
Determination of the relative
molecular mass from the lowering of
the freezing point.   14

Beckmann rearrangement
Conversion of oximes into substituted
amides.   544

Beilstein test
Qualitative test for halogens in organic
compounds.   11

Birch reduction
Selective 1,4-reduction of arenes, e.g.
by sodium in liquid ammonia.   627
Application to heteroaromatic compounds.   721

Bischler-Napieralski reaction
Intramolecular cyclisation of *N*-acyl-
β-phenylethylamines to form 3,4-
dihydroisoquinolines.   777

Blanc reaction
Introduction of a chloromethyl group
into a benzene ring using formaldehyde and hydrogen chloride in the
presence of, for example, zinc
chloride.   486

***Blanc's* rule**
Rule concerning the nature of the products formed (anhydrides or ketones) in ring-closure reactions of dicarboxylic acids.   325

***Bohn-Schmidt* reaction**
Production of polyhydroxyan-thraquinones from simple hydroxyan-thraquinones by oxidation, e.g. with fuming sulphuric acid.   640

***Boltzmann* constant k_B**
The general gas constant (R) divided by the *Avogadro* number N_A
($k_B = R/N_A$).   560

***Boudouard* equilibrium**
Equilibrium between, on one hand, solid carbon and gaseous carbon dioxide, and, on the other, carbon monoxide.   373

***Bouveault-Blanc* reduction**
Reduction of carboxylate esters to primary alcohols using sodium and an alcohol (usually ethanol).   259,523

***von Braun* reaction**
Treatment of tertiary amines with cyanogen bromide to give disubstituted cyanamides.   365

***Bredt's* rule**
Limitations on the formation of double bonds linked to bridgehead atoms in bicyclic compounds.   414

***Bredt* compounds**
Bicyclic compounds that follow *Bredt's* rule   414

***anti-Bredt* compounds**
Compounds which do not follow *Bredt's* rule and have double bonds at the bridgehead positions in bicyclic compounds.   415

***Bucherer* reaction**
Formation of β-naphthylamine from β-naphthol by reaction with ammonium hydrogen sulphite and concentrated ammonia.   631

C

***Cahn-Ingold-Prelog* system (*CIP* system)**
Rules concerning the unambiguous designation of the absolute configur-ation of a chiral compound.   274, 346
(Application of the *CIP* system to possible diastereomers of C = C derivatives).   66

***Cannizzaro* reaction**
Base-catalysed disproportionation of aldehydes with no hydrogen atom on the carbon atom next to the carbonyl group, leading to the formation of the corresponding alcohols and carboxy-late anions.   199, 226
Cannizzaro reaction of formalde-hyde.   202
Crossed *Cannizzaro* reactions.   313

***Caro's* acid**
Peroxosulphuric acid.   280, 481

***Chichibabin* hydrocarbon**
A biphenyl derivative with two unpaired electrons which can interact with one another through the benzene rings, setting up a quinonoid structure, which contributes 90% of the overall structure.   613

***Chichibabin* reaction**
Direct amination of pyridine to 2-aminopyridine.   762

***Claisen* condensation**
Base-catalysed condensation of a car-boxylate ester containing a vicinal activated methylene group with another ester molecule to provide a β-ketoester.   295
Claisen condensation involving a ketone with a reactive methylene group.   322
Reverse *Claisen* condensation and acid cleavage.   298

Claisen rearrangement
Thermal rearrangement of allyl ethers of phenols and enols to give *o*-allylphenols or γ,δ-unsaturated carbonyl compounds. It consists of an *O*-allyl $\rightarrow$ *C*-allyl rearrangement to distinguish it from the   498, 508

Thio-*Claisen* rearrangement in which a sulphur atom takes the place of an oxygen atom.   499

Claus diagonal formula
A centrosymmetric formula proposed for benzene.   469

Clemmensen reduction
Reduction of the carbonyl group of ketones (and of aldehydes) to a methylene group by amalgamated zinc and hydrochloric acid.   214, 226

Collins reagent
A reagent prepared from chromium trioxide in pyridine which is used for the oxidation or dehydrogenation of primary alcohols to aldehydes.   191

Cope elimination
Thermal decomposition of amine oxides into an alkene and a hydroxy-lamine derivative.   163

Cope rearrangement
Thermal isomerisation of 1,5-dienes in a sigmatropic [3,3] reaction.   108

Diaza-*Cope* rearrangement
Rearrangement step of the *Fischer* indole synthesis.   723

Corey-House synthesis
Formation of hydrocarbons from alkyl halides and lithium dialkyl cuprates. 55

Corey-Seebach method
Conversion of aldehydes into ketones which involves umpolung of the carbonyl group in the reaction.   209

Cori ester
Glucose-1-phosphate; an intermediate in alcoholic fermentation and in carbohydrate metabolism.   903, 905

Craig's principle of counter-current separation
Fractional partition of substances between two liquid phases which are only partially miscible with one another, in which a mobile phase is led over a stationary phase.   7

Cram's rule
Prediction of the preferred diastereomer which is formed by asymmetric induction when, for example, a chiral ketone is reduced by $LiAlH_4$, and a new chiral centre results.   344

Criegee mechanism
The three-step mechanism involved in the formation of ozonides by the action of ozone on alkenes.   72

Criegee oxidation
Oxidative cleavage of 1,2-diols with lead tetra-acetate in acetic acid.
          236, 302

Curtius reaction
When acid azides are heated in alcoholic solution and the initial products are hydrolysed, the products are an amine, carbon dioxide and nitrogen. Intermediates are an isocyanate and a urethane.   159

D

Demjanov reaction
Ring contraction and ring expansion reactions resulting from the treatment of cyclopropyl and cyclobutyl amines with nitrous acid.   392

Dewar benzene
A formula which was proposed for benzene. The compound with this for-mula is a non-planar valence isomer of benzene.   417, 469

Dieckmann **condensation**
Intramolecular cyclisation of dicar-
boxylate esters under the conditions
used for the *Claisen* condensation.
                                    395

Diels-Alder **reaction (Diene**
synthesis)
Cycloaddition reactions of conjugated
dienes leading to the formation of
cyclohexene derivatives.   336

  Diels-Alder **adduct**
  The product from a *Diels-Alder*
  reaction.   717

Doebner-von Miller **synthesis of**
quinaldine
Preparation of quinoline derivatives,
e.g. of quinaldine from aniline and
crotonaldehyde.   773

Dumas **method**
Method for the quantitative determination of the amount of nitrogen in
organic compounds.   12

E

Edman **degradation**
Stepwise degradation by phenyl isocyanate of a protein chain, starting
from the *N*-terminal amino-acid.   833

Einhorn **procedure**
Introduction of a benzoyl group using
benzoyl chloride and pyridine.   547

Erlenmeyer **rule**
According to this rule it is usually not
possible to isolate compounds with
two hydroxy groups attached to the
same carbon atom; a molecule of
water is lost to leave a carbonyl
group.   112, 260
Exceptions to the *Erlenmeyer* rule.
                               205, 350

Eschweiler-Clarke **methylation**
Preparation of tertiary amines by
reductive methylation of primary or
secondary amines with formaldehyde
and formic acid (see *Leuckart-
Wallach* reaction).   227, 737

Evans **principle**
This states that thermal pericyclic
reactions take place preferentially *via*
aromatic transition states.   659

F

Fehling **reaction**
Test reaction for aldehydes, and also
for reducing sugars such as glucose,
by means of the reagent: 432

  Fehling's **solution** which is made
  up from copper sulphate solution
  and an alkaline solution of
  potassium sodium tartrate.   193

Fenton's **reagent**
Hydrogen peroxide in the presence of
iron(II) salts; this oxidises, for example,
1,2-diols to hydroxyaldehydes.   316

Fischer **indole synthesis**
Preparation of indole derivatives by
ring closure of phenylhydrazones. 723

Fischer **projection**   119
A convention for the projection of the
substituents at the asymmetric centre
of open-chain enantiomeric compounds
by *Fischer* projection formulae.   120
It is of particular importance in connection with the stereochemistry of
sugars.   314

> *Fischer* **configuration**
> **formulae**   425

Fischer-Tropsch **synthesis**
Preparation of fuel-oils from synthesis
gas enriched with hydrogen.   87

Fourier **transformation method**
in NMR spectroscopy.   44
in IR spectroscopy.   34

Frank-Caro **process**
Preparation of calcium cyanide from
calcium carbide and nitrogen.   368

Friedel-Crafts **acylation**
Conversion of arenes into ketones by
the use of acid chlorides or acid anhydrides in the presence of aluminium
chloride.   539

with naphthalene   629, 643
with heteroaromatics:
  furan   717
  thiophene   721

Friedel-Crafts **alkylation**
Alkylation of arenes by reaction with
alkyl halides or alkenes in the
presence of aluminium chloride.   475

> **Reductive** *Freidel-Crafts*
> **reaction**   649

Freidel-Crafts **catalysts**
Aluminium chloride and other *Lewis*
acids such as conc. sulphuric acid,
phosphoric acid.   475

Friedländer **synthesis**
Preparation of quinoline derivatives
by base catalysed condensation of
o-aminobenzaldehyde with aldehydes
or ketones.   773

Fries **rearrangement**
Rearrangement of phenyl esters of
arenecarboxylic acids into acylphenols in the presence of *Lewis* acids
such as aluminium chloride.   498

G

Gabriel **synthesis**
Preparation of primary amines from
alkyl halides and potassium phthalimide.   160

Gattermann **aldehyde synthesis**
Preparation of phenol aldehydes or
phenol ether aldehydes by the action
of hydrogen cyanide and hydrogen
chloride on phenols or phenol ethers.
                                    536

Gattermann-Koch **synthesis**
Formylation of arenes by reaction
with carbon monoxide and hydrogen
chloride.   535

Gilman's **reagent**
Reagent used to test for *Grignard*
derivatives.   184

Girard's **reagent**
With carbonyl compounds these form
water-soluble hydrazones that can be
separated off from accompanying
insoluble materials.   256

Gomberg-Bachmann **reaction**
Arylation of diazonium salts in the
presence of alkali.   585

Graebe-Ullmann **synthesis**
Formation of carbazole by diazotisation of *o*-aminodiphenylamine
and associated intramolecular
ring-closure.   732

Gram **colouring**
Charge-transfer complex formed from
crystal violet and iodine which makes
possible the classification of bacteria
into gram-positive and gram-negative
bacteria.   607

Graves's **disease**
Overactivity of the thyroid gland
leading to upset of metabolism.   824

Grignard **compounds**
Organomagnesium compounds which
exist in solution predominantly as
bimolecular complexes.   184
Test for them using *Gilman's*
reagent.   184

Grignard **reaction**
Addition of organomagnesium compounds to polar multiple bonds,
especially carbonyl groups.   185
Polar or radical intermediates may be
involved in the reaction.   186, 616

Grignard **reagents**
Organomagnesium compounds, usually
made up freshly from magnesium and
an alkyl halide and used in ethereal
solution.   184

H

Hammett **equation**
Expression for a correlation system in
which the positions of equilibria and
the rates of chemical reactions are
related to the effects of substituents,
particularly in benzene derivatives.
                                  560

Hammett **substituent constants**
A parameter for characterising the
effects of substituents in the *Hammett*
equation.   561

Hammond **postulate**
The transition state in a chemical
reaction more closely resembles the
reactants, the less stable the latter
are.   127

Hantzsch's **dihydropyridine synthesis**
Cyclising condensation reactions
involving a β-ketoester, an aldehyde
and ammonia in the ratio 2:1:1, which
are used for the preparation of homologues of pyridine.   766

Hantzsch's **thiazole synthesis**
Cyclising condensation reactions of
α-halogenoketones with thioamides to
give thiazoles.   747

Harden-Young **ester**
Fructose-1,6-bisphosphate.
Intermediate in alcoholic fermentation
and in carbohydrate metabolism. 903

Haworth **ring formulae**
Ring formulae for sugars based on
Fischer projection formulae.   436

Haworth **synthesis**
Preparation of phenanthrene from
naphthalene.   643

Heisenberg's **uncertainty principle**
A quantum theory postulate concerning the exact location of an electron
at any particular time.   18

Hell-Volhard-Zelinsky **reaction**
Method for the selective preparation
of α-bromocarboxylic acids.   271

Henderson **equation**
Relationship between the pH and the
pK_a values.   232

Heumann's **indigo syntheses**
Synthesis of indigo from
1. Phenylglycine in an alkaline melt
   (sodamide).   727
2. Phenylglycine-*o*-carboxylic acid in
   an alkaline melt.   728

Hinsberg reaction
Separation of primary and secondary
amines by the use of benzenesulphonyl chloride.   158

Hoesch ketone synthesis
Preparation of aryl ketones from polyhydric phenols and nitriles in the
presence of hydrogen chloride.   541

Hoffmann's drops
A 3:1 mixture of ethanol and ether. 144

Hofmann elimination reaction
Breakdown of tetra-alkylammonium
hydroxides involving a β-elimination
reaction and formation of a tertiary
amine, an alkene and water.   157

Hofmann reaction
Conversion of primary amides into
primary amines containing one less
carbon atom, using sodium hypobromite.   159

Hofmann rule
Statement about the preferred product
formed in *Hofmann* elimination reactions in which the formation of more
than one isomeric alkene is possible.
                                    157

Horner-Emmons reaction
Formation of alkenes from carbonyl
compounds, using phosphonate esters
or phosphine oxides (PO-activated
alkene formation).   175

Houdry process
Catalytic cracking of high-boiling
mineral oil fractions to give petrol. 85

Huang-Minlon method
A preparatively valuable variant of the
Wolff-Kishner reduction reaction. 215

Hückel model
Description of C=C double bonding
in terms of sp^2-hybridisation and
overlap of the $2p_z$-orbitals.   63

Hückel's rule
Statement about the stability of different
fully conjugated cyclic polyenes. 655

Hünig's base
The non-alkylating base ethyldiisopropylamine.   158

Hund's rule
Concerns the occupancy of degenerate orbitals by electrons to give the
state of highest multiplicity.   19, 21

Hunsdiecker reaction (Borodin reaction)
Preparation of alkyl halides from the
silver salts of carboxylic acids.
                                124, 390

I

Ivanov reaction
Treatment of phenylacetic acid with
excess *Grignard* reagent to give an

intermediate which reacts with carbonyl compounds to form substituted
3-hydroxycarboxylic acids.   553

J

Jablonski diagrams
Schematic representation of photochemical processes.   223, 224

Jones reagent
Chromium(VI) oxide in sulphuric
acid, used to obtain ketones from
secondary alcohols.   208

K

Kekulé formula
The most important way of formulating benzene derivatives.   467
Use of Kekulé formulae to represent naphthalene.   626
Connection with the stabilisation energy of benzene.   469

Kjeldahl method
Method for the quantitative estimation of nitrogen, used in food and agricultural chemistry.   12

Knoevenagel reaction
Condensation reactions of the active methylene group of malonic ester with aldehydes and ketones.   328

Knorr pyrrole synthesis
Preparation of substituted pyrroles by condensation of an α-aminoketo-compound with a β-ketoester.   705

Koch-Haaf reaction
Preparation of tertiary carboxylic acids, e.g. trimethylacetic acid, by carbonylation of alkenes with carbon monoxide in acid solution.   230

Kolbe electrolysis
Anodic decarboxylation by electrolysis of a concentrated solution of sodium acetate, leading to the formation of ethane by dimerisation of the resultant methyl radicals.   54

Kolbe nitrile synthesis
Preparation of nitriles from alkyl halides and potassium cyanide.   266

Kolbe-Schmitt synthesis
Preparation of salicyclic acid from sodium phenoxide and carbon dioxide.   550

König's salt
Product obtained by treating N-cyanopyridinium bromide with N-methylaniline.   764

Koopman's theorem
Makes possible the correlation of results obtained from photoelectron spectroscopy with orbital energies.   38

L

Lambert-Beer law
Basis for quantitative UV and visible spectroscopy.   38

Langerhans' islets
Cells of the pancreas in which insulin is formed.   836

Lassaigne test
Qualitative test for nitrogen in organic chemistry.   11

Lawesson reagent
A very active reagent for replacing the oxygen atoms in amides by sulphur. 263

Leblanc soda process
Basis for the general use of soaps.   247

Leuchs anhydride
Glycine-N-carboxylic acid anhydride, prepared from glycine and phosgene.   830

Leuckart-Wallach reaction
Reductive amination of ketones, e.g. with ammonium formate, in which formic acid acts as the reducing agent.   525

Lewis acids
Compounds which are short of electrons and can behave as electron acceptors.   380

Lewis bases
Compounds which can act as donors of electron pairs.   381

Liebermann reaction
Colour reaction used as a test for
nitrite and nitroso-groups.   505

Linde process
Low temperature separation of synthesis gas.   373

**Lobry de Bruyn-van Ekenstein
rearrangement**
Base-catalysed equilibration between
glyceraldehyde and dihydroxyacetone.
                                       317
Importance in carbohydrate chemistry.
                                       433

Lossen reaction
Thermal rearrangements of hydroxamic acids to isocyanates, which are
hydrolysed, with loss of carbon dioxide,
to give primary amines.   160

M

Madelung synthesis
Preparation of indole derivatives by
intramolecular cyclisation of N-acyl-
toluidines, e.g. 2'-methylformanilide,
with strong bases at higher temperatures.   723

Maillard reaction
Non-enzymatic browning of reducing
sugars by reaction with the amino-
group of amino-acids, peptides or
proteins.   431

Mannich base
A β-aminoketone arising from a
Mannich reaction.   543

Mannich reaction
Condensation of a compound having
an acidic C-H with formaldehyde and
primary or secondary amines; this
results in α-aminomethylation at the
acidic CH.   544, 725, 812
Use of acetaldehyde in Mannich reactions.   815

Markovnikov's rule
Rule dealing with regioselectivity in
the addition of hydrogen halides to
C=C double bonds in which the two
carbon atoms carry different numbers
of hydrogen atoms.   69, 95, 189

Anti-Markovnikov product
Product of an addition reaction to a
C=C double bond in which the regioselectivity is the opposite of that predicted by the Markovnikov rule, e.g.
the peroxide effect;   69
hydroboration with subsequent oxidation.   180

Maxam-Gilbert process
A process for sequence analysis of
DNA.   856

McConnell equation
Expression of the relationship
between the coupling constant in an
ESR spectrum and the spin density or
spin population.   47

McLafferty rearrangement
A rearrangement that frequently occurs
when unsaturated compounds which
have a CH-group at the γ-position are
subjected to mass spectrometry. 32, 223

McMurry reaction
Titanium catalysed reductive coupling
of carbonyl compounds leading to the
formation of an alkene.   216, 228

Meerwein-Ponndorf-Varley reaction
Reduction of a carbonyl group to an
alcohol group in the presence of an
aluminium alkoxide, e.g. aluminium
isopropoxide.   194, 214, 226
The reverse of this reaction is
Oppenauer oxidation.   194, 208

Meerwein-Schuster reaction
Arylation of alkenes (especially α,β-
unsaturated carbonyl compounds)
with arenediazonium salts in the presence of copper salts.   585

Merrifield synthesis
Automated solid-phase synthesis of
medium-sized proteins.   831, 859

Michael addition or Michael reaction
Nucleophilic addition of compounds
having reactive methylene groups
(acetoacetic esters, malonic esters,
etc.) to activated C=C double bonds
(e.g. α,β-unsaturated carbonyl compounds).   328, 333, 687, 766
Industrial use in *group-transfer polymerisation*.   329

  Michael acceptor
  α,β-Unsaturated carbonyl compounds
  in Michael addition reactions.   328

Michael donor
Compounds with reactive methylene groups taking part in *Michael*
reactions.   328, 333

Michaelis-Arbusov reaction
A method for forming carbon-phosphorus bonds by making esters of
alkanephosphonic acids from trialkyl
phosphites and alkyl halides.   173

Michler's ketone
4,4'-Bis(dimethylamino)benzo-
phenone, prepared from N,N-dimethyl-
aniline and phosgene in the presence
of zinc chloride.   578, 608
Use of *Michler's* ketone in preparing
Gilman's reagent.   184

Mössbauer effect
Characteristic frequency shifts in the
γ-ray resonance spectroscopy of solid
bodies.   11

Müller-Rochow synthesis
Preparation of dichlorodimethysilane
from dichloromethane and silicon
with copper as a catalyst.   178

N

Natta catalyst
Catalyst for the stereoregular polymerisation of unsymmetric monoalkenes
such as propene in the
  Natta process.   77

Nef reaction
Preparation of aldehydes and ketones
from the reaction of salts of nitroalkanes and sulphuric acid (*aci-*
nitroalkane cleavage).   154
Use in carbohydrate chemistry.   434

Nesmeyanov reaction
Preparation of aryl mercury chlorides
from diazonium salts and mercury(II)
chloride.   584

Neuberg ester
Fructose-6-phosphate; an intermediate
in alcoholic fermentation and in carbohydrate metabolism.   903

Newman projection
Projection formulae of two neighbouring atoms looking along the line
of the bond linking the atoms.   57

Newman's reagent
Makes possible the separation of
enantiomeric forms of *Lewis* bases
which contain no functional
group.   348

Norrish type I cleavage
Homolytic cleavage of a photochemically excited carbonyl compound to
give a radical pair consisting of an acyl
radical and an alkyl radical.   222

Norrish type II cleavage
Intramolecular migration of a hydrogen atom in photochemically excited
aldehydes or ketones leading to the
formation of a diradical that decomposes to an alkene and an enol.   223

Nylander reaction
Test reaction for reducing sugars
using   432

Nylander's reagent
an alkaline solution of bismuth(III)
hydroxide with tartaric acid present
as a complexing agent.   193

O

Oppenauer oxidation
Oxidation of a secondary alcohol to a
ketone catalysed by an aluminium
alkoxide.   208

Reverse of the *Meerwein-Ponndorf-
Varley* reaction.   194

P

Paal-Knorr synthesis
Formation of pyrroles by heating 1,4-
diketones with ammonia.   704
Formation of furans by heating 1,4-
diketones alone with dehydrating
agents.   717

Parkinson's disease   825

Pascal's triangle
Guide to the relative intensities of the
lines which are seen in the hyperfine
structure in an ESR spectrum.   46

Paterno-Büchi reaction.
Formation of oxetanes by the
cycloaddition of photochemically
excited ketones to alkenes.   223, 224

Pauli principle
Each orbital can contain no more than
two electrons.   19
Significance in ESR spectroscopy.   45
Significance in the formula of
benzene.   470

Pauly reaction
Formation of orange, red or blue
dyestuffs by coupling diazonium ions
with imidazole.   741

Perkin synthesis
Formation of cinnamic acid by the
reaction of benzaldehyde with acetic
anhydride under basic
conditions.   555

Peterson reaction
Conversion of a carbonyl compound
into an alkene by activating an acidic
CH-group; it is an alternative to the
Wittig and *Horner-Emmons* reactions.
                                     178

Pitzer strain
Conformational strain arising from
steric interaction between hydrogen
atoms on neighbouring carbon atoms
which are unable to take up preferred
staggered conformations (see *Baeyer*
and *Prelog* strain).   394

Platonic bodies
Regular polyhedra. Three of the five platonic bodies have been made as hydrocarbons $(CH)_n$, namely tetrahedrane (a tetrahedron), cubane (a cube) and dodecahedrane (a dodecahedron).
421

Prelog strain
Conformational interaction between non-vicinal hydrogen atoms across a ring (transannular interaction) (see *Baeyer* and *Pitzer* strain).   406

Prilezhaev reaction
Formation of epoxides by the reaction of peracids with alkenes.   71

Prins reaction
Acid-catalysed addition of formaldehyde to an alkene, giving a 1,3-diol.
101

Pschorr synthesis
Preparation of phenanthrene derivatives from diazotised α-phenyl-*o*-aminocinnamic acids in the presence of copper powder.   642

Pummerer rearrangement
Rearrangement of sulphoxides to α-acyloxysulphides in the presence of acid anhydrides.   148

R

Raman spectroscopy   35
Supplement to IR spectroscopy since IR-inactive vibrations can be detected in
Raman spectra.   373, 626

Raney nickel
A reduction catalyst made by alkaline decomposition of a nickel-aluminium alloy.   73

Raoult's law
Basis of the cryoscopic and ebullioscopic methods for the determination of relative molecular masses.   15

Raschig process
Process for making chlorobenzene from benzene by oxychlorination, and hydrolysing chlorobenzene to phenol.
479, 495

Rast method
Determination of relative molecular masses by the freezing-point method, using camphor as solvent.   15

Reeves conformation formulae
These formulae present the chair form of the tetrahydropyran ring in sugars.
438

Reformatsky reaction
Preparation of β-hydroxycarboxylate esters from the reaction of organozinc compounds obtained from halogenocarboxylic acid esters with carbonyl compounds.   278

Reimer-Tiemann synthesis
Preparation of phenol aldehydes from phenol, chloroform and alkali.   537

Reppe syntheses
Four syntheses involving acetylene: vinylation, ethynilation, cyclisation and carbonylation.   96

Robinson annelation
Preparation of bicyclic ketones by a combination of the *Michael* reaction and the aldol reaction.   687

Robison ester
Glucose-6-phosphate; an intermediate in alcoholic fermentation and in carbohydrate metabolism.   903

Rochelle salt or *Seignette* salt
Potassium sodium tartrate; constituent of *Fehling's* solution.   193, 347

Rosenheim-Wieland structural formula
Correct proposal for the formula of cholesterol.   688

Rosenmund reduction
Catalytic reduction of acid chlorides to aldehydes.   192

Ruff process
Oxidative shortening of the carbon chain of an aldose by one carbon atom.   434

Ruggli-Ziegler dilution principle
Method for the preparation of large rings from difunctional compounds (e.g. dinitriles) in which dilution favours intramolecular cyclisation as compared to intermolecular polymerisation.   409 *f.*

Rutherford-Bohr-Sommerfeld model of the atom
A clear, but not completely true, model of the atom.   17

S

Sachsse-Bartholomé process
Industrial process for making acetylene.   93

Sandmeyer reaction
Replacement of the diazonium group in arenediazonium salts by a halogen atom or a cyano group using the appropriate copper(I) salt.   583

Saytzeff rule
Prediction of the preferred product obtained in β-elimination of hydrogen halide from alkyl halides when the formation of a number of isomeric alkenes is possible.   157

Schardinger enzyme
Occurs in milk which has not been heated and catalyses the decolourisation of methylene blue by formaldehyde.   887

Schiff's bases
Condensation products of aldehydes and ketones with primary amines.
                                           197, 211, 532
Participant in enamine-imine tautomerism.   213
Role as intermediates in alcoholic fermentation.   903

Schiff's reagent
An aqueous solution of fuchsin which has been decolourised by sulphur dioxide.   607

Schlenk equilibrium
The equilibrium between different organometallic compounds which exists in solutions of *Grignard* reagents.   184

Schlenk tube
Glass apparatus for carrying out reactions in the absence of air, e.g. in an atmosphere of nitrogen.   636

Schmidt reaction
Action of hydrazoic acid in the presence of conc. sulphuric acid on carboxylic acids, leading to the formation of primary amines.   160

Schöllkopf reaction
Enantiomeric syntheses of amino-acids via bislactim ethers of dioxopiperazine; their anions have two diastereotopic sides, one of which is efficiently shielded.   786

Schöniger method
A method for the quantitative estimation of sulphur, chlorine or iodine in organic compounds.   13

Schotten-Baumann method
A way of introducing benzoyl groups, using benzoyl chloride in aqueous alkali.   547

Schrödinger equation
The fundamental equation of wave mechanics.   17

Schweizer's **reagent**
An ammoniacal solution of copper(II) hydroxide; it is a solvent for cellulose.
462, 465

Selivanov **reaction**
Colour test for the detection of hydroxymethylfurfural.   719

Shapiro **reaction**
Preparation of alkenes by the action of strong bases on tosylhydrazones. 211

Sharpless **reagent**
Reagent for the stereoselective epoxidation of substituted allyl alcohols.
344, 348

Simmons-Smith **reaction**
Synthesis of cyclopropanes from alkenes by means of an organozinc compound prepared from diiodomethane.   186

Skraup **synthesis**
Preparation of quinoline from aniline and glycerol in the presence of nitrobenzene as a dehydrogenating agent.   772

Sommelet **reaction**
Preparation of benzaldehyde from benzyl chloride and hexamethylenetetramine.   529

Stephen **reaction**
Reduction of arylnitriles with tin(II) chloride to give aryl aldehydes, e.g. benzaldehyde.   530

Stobbe **condensation**
Condensation of aldehydes or ketones with succinate esters in the presence of a base.   331

Stork **reaction**
Synthesis of α-alkyl- or α-acyl-carbonyl compounds from enamines. 213

Strecker **degradation**
Degradation of amino-acids by carbonyl compounds (e.g. ninhydrin), resulting in decarboxylation and oxidative deamination, and giving an aldehyde containing one fewer carbon atoms.   625

Strecker **synthesis**
Preparation of amino-acids by the action of hydrogen cyanide and ammonia on aldehydes, and hydrolysis of the initially formed aminonitrile.   285
Preparation of ethylenediaminetetraacetic acid from hydrogen cyanide and ethylenediamine.   289
Preparation of glucosamine by means of the *Strecker* synthesis.   445

Süs **reaction**
Photochemical rearrangement of *o*-benzoquinone diazide derivatives to the corresponding cyclopentadiene ketenes; this forms the basis for the diazotype process.   601

Svedberg **unit**
Sedimentation coefficients of proteins in the ultracentrifuge.   838

Swarts **reaction**
Conversion of alkyl halides into alkyl fluorides, using metal fluorides as fluorinating agents.   136

Swern **oxidation**
Oxidation of primary alcohols to aldehydes using dimethyl sulphoxide activated by oxalyl chloride.   191

T

Tebbe **reagent**
Organometallic reagent for the preparation of enol ethers from carboxylic acid esters.   187

Teichmann's **crystals**
Crystals of haemin obtained from haemoglobin.   710

W

Wacker-Hoechst process
Oxidation of alkenes by air in an
aqueous palladium chloride/copper(II)
chloride solution.   203
In this way:
ethylene → acetaldehyde   203
propene → acetone   218
but-1-ene → butanone   219
Reaction mechanism involving participation of a transition metal complex.383

Wacker process
Industrial manufacture of acetic
anhydride from ketene and acetic
acid.   254

Wagner-Meerwein rearrangement
Nucleophilic 1,2-rearrangement in a
carbon skeleton, of particular significance in terpene chemistry.   217, 670

Walden inversion
Inversion of the configuration at a
chiral centre in the course of substitution reactions;   342
characteristic for S_N2 reactions.
                              129, 450

Walsh model
A model for the electronic structure in
cyclopropane involving sp^2-hybridised
orbitals.   387

Wilkinson catalyst
Chloro-tris(triphenylphosphine)-
rhodium(I); a homogeneous catalyst
for hydrogenations using hydrogen at
normal pressure.   73, 381

Willgerodt-Kindler reaction
Conversion of alkyl aryl ketones into
thioamides with the same number of
carbon atoms by heating them with
sulphur in morpholine.   544

Williamson synthesis
Synthesis of ethers by the reaction of
alkali metal alkoxides with alkyl
halides.   141, 497

Wittig reaction
Conversion of a carbonyl group into
an alkene by treating a ketone with
methylenetriphenylphosphorane; in
general reactions of carbonyl compounds with phosphoranes.   174

Wohl process
Shortening of the carbon chain of an
aldose by one carbon atom by forming its oxime, converting the oxime
into a nitrile, and eliminating hydrogen cyanide.   434

Wohl-Ziegler reaction
Selective bromination of alkenes at
the allyl position by means of N-bromosuccinimide.   331

Wöhler's synthesis of urea
Preparation of urea from ammonium
cyanate.   1, 355

Wolff rearrangement
Rearrangement of diazoketones to
ketenes.   253

Wolff-Kishner reduction
Reduction of carbonyl groups to
methylene groups by heating a ketone
or aldehyde hydrazone.   215

Wood-Fildes theory
Theory of the role of sulphonamides
and p-aminobenzoic acid in inhibition
of bacterial growth.   575

Woodward-Hoffmann rules
Principle of the retention of orbital
symmetry.   104
Application to electrocyclic reactions,
                                   106
to sigmatropic migrations,   106
to the Cope rearrangement,   108
to the Diels-Alder reaction,   338
to ene reactions,   339
to the cycloaddition of ketenes to
cyclopentadiene.   405

Wurster's red
A radical cation formed by oxidation
of N,N-dimethyl-p-phenylenediamine.
                                    616

Wurtz-Fittig reaction
Reaction of a halogenobenzene with
an alkyl halide in the presence of
sodium to give the corresponding
alkylbenzene.   474

Wurtz synthesis
Formation of hydrocarbons from alkyl
halides and sodium.   55
Intramolecular
applications.   388, 420

Wurzschmitt method
Used for the quantitative determination of halogens in organic
compounds.   13

Y

Young process
Industrial process for obtaining

absolute alcohol.   115

Z

Zeisel method
A method for the quantitative estimation of alkoxy groups.   144

Zemplén method
An improvement of the last step of
the Wohl process.   435

Zerevitinov method
Test reaction, using methyl magnesium iodide, for reactive hydrogen
atoms.   185

Ziegler catalyst
Organometallic mixed catalyst for the
polymerisation of ethylene at normal
pressure by the

Ziegler process.   76

Ziegler-Natta catalyst
Organometallic mixed catalyst for co-
ordinative chain polymerisation,
which can lead to the formation of a
stereoregular product.   75
Application in copolymerisation,   78
to polymerisation of acetylene,   95
to stereoregular polymerisation of
butadiene.   683

Zincke aldehyde
5-N-Methylaminopenta-2,4-dienal;
reacts with cyclopentadiene in the
synthesis of azulene.   654, 764

Index of Names

Subject Index

Bonding (*continued*)
  ionic 22
  semipolar 163
σ-Bonding 23
π-Bonding 64
Bonds
  angles
    compression 386
    deformation 386
  banana 64
  bent 64
  distances 93
  lengths 93
σ-Bonds, migration 106
Bone oil 704, 761
Bones 841
9-Borabicyclo-[3.3.1]nonane (9-BBN) 180
Borane 180
Boric acid, esters 140
Borinane 703
Bornane 667
Bornane-4-carboxylic acid lithium salt 415
Bornane group 669
(1R,4R)-Bornan-2-one 669
Borneol 671
Bornyl chloride 670
4-Bornyl lithium 415
Borodin reaction 124, 390
Boron compounds, organo 179
Boron trialkyls 179
Boron triaryls 179
Boron trifluoride 78
Bottled gas 55
Boudouard equilibrium 373
Bouveault-Blanc reduction 259, 523
Brain 306
  tissue 307
Brake fluid 309
Bran 441, 463, 718
Brandy 115
Brassica 448
Brassica nigra 371
Bread aroma 771
Bredt's rule 414, 415, 416
Brestan 190
Brewer's yeast 902
Brickettes 84
Bridged-ring compounds 413
  nomenclature 414
Bridgehead carbon atoms 414, 415
Bridgehead double bonds' non-formation 415
Brilliant G 729
Brilliant Indigo 2B 729
Bristles 81
British-Anti-Lewisite (BAL) 313
Bromal 205
Bromal hydrate 205
Bromine, analysis 13
Bromoacetaldehyde ethylene-acetal 740
α-Bromoacetophenone 541
α-Bromoacids 271
2-Bromo-alkanes, alkaline hydrolysis 129
N-Bromoamide 159
p-Bromoazobenzone 565
Bromobenzene 474, 479, 480, 584
p-Bromobenzoic acid, pK_a 546
γ-Bromobutyric acid 272

4-Bromobutyric acid 272
3-Bromocamphor 671
(+)-3-Bromocamphor-10-sulphonic acid 348, 671, 672
1-Bromo-2-chloro-1-iodoethylene 66
Bromo-chloro-iodo-methane 274
1-Bromo-3-chloropropane 388
2-Bromo-2-chloro-1,1,1-trifluoroethane 137
Bromocoumarin 730
Bromocyclobutane 391
Bromocycloheptatriene 651
2-Bromocyclopentadienone 420
Bromocyclopentane 396
1-Bromo-1,2-dichloroethylene 66
Bromoethane 124
2-Bromoethanol 301
o-Bromofluorobenzene 481
Bromoform 134, 135
Bromomalonic acid, hydrolysis 340
Bromomethane 125
(Bromomethyl)allyl cation 101, 102
2-Bromomethyl-1,3-dioxolane 740
Bromonium ion 67
5-Bromopentylaniline 769, 770
p-Bromophenacyl bromide 542
5-Bromo-2-phenyltetrazole 757
1-Bromopropane 70, 388
D(+)-α-Bromo-propionic acid 284
3-Bromopropionic acid 239, 271
3-Bromopyridine 761, 762
β-Bromostyrene 620
N-Bromosuccinimide (NBS) 331
  brominating agent 331
  dehydrogenating agent 331
5-Bromo-1,2,4-thiadiazole 760
o-Bromotoluene 480
p-Bromotoluene 480
Bromotropylium bromide 651
Bronchial asthma 528, 695, 799
Brown coal 82, 330
  conversion into petrol and diesel oil 87
Brucine 817
Brufen 236
Bucherer reaction 631
Buckminsterfullerene 646
Bufadienolides 698
Bufotalin 698
Bullvalene 421
Buna 100, 682
Buna CB 683
Buna S 617
Buscopane 807
Butadiene 61, 79, 89, 98, **100**f, 101, **103**, 332, 336, 397, 638
  1,2-addition 101
  1,4-addition 101
  addition of bromine 101, 102
  addition of hydrogen halides 103
  bond lengths 103
  conformations 104
  copolymerisation with acrylonitrile 682
  copolymerisation with styrene 682, 683
  manufacture 88
  metal-π-complexes 409
  polymerisation 682
  stereoregular polymerisation 683
  structure 103